Organic Geochemistry

Methods and Results

Edited by
G. Eglinton and M. T. J. Murphy

Springer-Verlag Berlin Heidelberg GmbH 1969

With 246 Figures

ISBN 978-3-642-87736-0 ISBN 978-3-642-87734-6 (eBook)
DOI 10.1007/978-3-642-87734-6

Originally published by Springer-Verlag Berlin · Heidelberg 1969
Softcover reprint of the hardcover 1st edition 1969

Foreword

For many years, the subject matter encompassed by the title of this book was largely limited to those who were interested in the two most economically important organic materials found buried in the Earth, namely, coal and petroleum. The point of view of any discussions which might occur, either in scientific meetings or in books that have been written, was, therefore, dominated largely by these interests. A great change has occurred in the last decade. This change had as its prime mover our growing knowledge of the molecular architecture of biological systems which, in turn, gave rise to a more legitimate asking of the question: "How did life come to be on the surface of the Earth?" A second motivation arose when the possibilities for the exploration of planets other than the Earth—the moon, Mars, and other parts of the solar system—became a reality. Thus the question of the possible existence of life elsewhere than on Earth conceivably could be answered.

In the search for the answers to these questions the role of the organic compounds found amongst the various types of geological structures buried in the Earth became a more central source of knowledge. It became necessary for the geologist who could recognize and define the nature of the rock to consult with the organic chemist who could isolate and determine the ultimate architecture of the organic molecules in the rock; and the two together to listen to the biologist as to the possible significance of their findings with regard to the basic questions of evolution. Thus, what had once been the primarily economic motive for trying to determine the origin of petroleum or of coal, that is, to facilitate further discovery and use is giving way to a much broader and deeper question which will require the best efforts of all three kinds of scientists.

This book constitutes really the first textbook to attempt to bring together all three of the aspects—chemistry, geology and biology—which have a bearing on the subject, following this new motivation. The last book bearing the same title, that of Breger, was written in 1962 and was a sort of transition volume between the older interest and the new. While the present volume is a multiple-authored one, its prime editors are to be congratulated on having produced a selection of essays which form a comprehensive whole in a way that no other such collection that I know of has done. Each of the essayists have written a thoroughly authoritative piece of work in an area in which he is competently working, and Drs. Eglinton and Murphy have organized them so as to make them a unitary whole. In fact, the first chapter, which is called "Introduction", is an extended table of contents containing a brief statement about the nature of each chapter as well as an extensive table from which one can immediately comprehend the place of each essay in the system.

There is no doubt in my mind that this field will continue to grow and expand, both in depth and in breadth, as well as in interest, and that this book may very well provide the first textbook for an organized academic study of the subject.

October, 1969.

MELVIN CALVIN
Laboratory of Chemical Biodynamics
University of California
Berkeley, California, USA

Preface

This book has proved to be a much bigger and more difficult enterprise than we had foreseen and we thank the authors and publishers for their patience in seeing it through to completion. The overall scheme was drawn up in Glasgow in 1965 but various events, notably trans-Atlantic moves of both editors, delayed the gathering and processing of the manuscripts. Most of the editing of the manuscripts was handled at Glasgow University. Since then we have worked on the book in Hartford (Connecticut), Berkeley (California), Austin and Houston (Texas) and Bristol (England). During this time, several authors have moved their laboratories and we are sorry to report that Dr. R. I. MORRISON, the author of the chapter on "Soil Lipids", has died.

Initially we set ourselves the task of gathering in the chapters within the space of about six months. This, not unexpectedly, proved to be over-optimistic and it was not until the middle of 1968 that most of the chapters were in our hands and in their final form. In fact, the majority of the chapters cover the literature up to about the end of 1967 and some to the end of 1968. In several cases the authors have provided recent addenda.

In inviting contributors and in our editorial procedures, we have attempted to bring about some uniformity in format and approach throughout the book. The authors, though mainly from Great Britain and the United States of America, belong to eight countries. Consequently, we have not been able to introduce an entirely systematic presentation or bring about a consistent style. In any case, differences would be expected in view of the range of disciplines represented in the volume.

We thank Miss DEE WITNEY for her assistance in copy-editing, Miss ISABEL MCGEACHIE, Mrs. WENDY HARRISON and Mrs. SUE SNEDDON for coping with the seemingly interminable correspondence and Dr. K. DOURAGHI-ZADEH, Dr. A. G. DOUGLAS, Mr. W. HENDERSON, B. Sc., Mr. D. H. HUNNEMAN, B. Sc., Dr. A. MCCORMICK, Dr. I. MACLEAN, and Mr. B. J. URQUHART, B. Sc., for their assistance in preparing the manuscripts for publication.

We are especially grateful to BRENDA KIMBLE, B. Sc., for the extensive and detailed copy-editing of the galley and page proofs at Bristol during 1969. We also thank Dr. J. R. MAXWELL for the Subject Index.

Finally, we are most grateful to Professor CALVIN for the foreword, which puts so neatly into words what we have tried to do.

November 1969

GEOFFREY EGLINTON,
University of Bristol,
Bristol, England.

Sister MARY T. J. MURPHY,
St. Joseph College,
Hartford, Connecticut, USA.

Contents

CHAPTER 1

G. EGLINTON and M. T. J. MURPHY

Introduction

CHAPTER 2

G. EGLINTON

Organic Geochemistry. The Organic Chemist's Approach

CHAPTER 3

M. T. J. MURPHY

Analytical Methods

CHAPTER 4

A. L. BURLINGAME and H. K. SCHNOES

Mass Spectrometry in Organic Geochemistry

CHAPTER 5

A. G. DOUGLAS

Gas Chromatography

CHAPTER 6

W. E. ROBINSON

Isolation Procedures for Kerogens and Associated Soluble Organic Materials

CHAPTER 7

M. O. DAYHOFF and R. V. ECK

Paleobiochemistry

CHAPTER 8

W. D. I. ROLFE and D. W. BRETT

Fossilization Processes

CHAPTER 9

B. J. BLUCK

Introduction to Sedimentology

CHAPTER 10

D. H. WELTE

Organic Matter in Sediments

CHAPTER 11

L. R. MOORE

Geomicrobiology and Geomicrobiological Attack on Sedimented Organic Matter

CHAPTER 12

E. T. DEGENS

Biogeochemistry of Stable Carbon Isotopes

CHAPTER 13

W. G. MEINSCHEIN

Hydrocarbons — Saturated, Unsaturated and Aromatic

CHAPTER 14

P. L. PARKER

Fatty Acids and Alcohols

CHAPTER 15

F. M. SWAIN

Fossil Carbohydrates

CHAPTER 16

M. STREIBL and V. HEROUT

Terpenoids — Especially Oxygenated Mono-, Sesqui-, Di- and Triterpenes

CHAPTER 17

R. B. SCHWENDINGER

Carotenoids

CHAPTER 18

P. E. HARE

Geochemistry of Proteins, Peptides, and Amino Acids

CHAPTER 19

E. W. BAKER

Porphyrins

CHAPTER 22

F. J. STEVENSON and J. H. A. BUTLER

Chemistry of Humic Acids and Related Pigments

CHAPTER 23

R. I. MORRISON

Soil Lipids

CHAPTER 24

V. WOLLRAB and M. STREIBL

Earth Waxes, Peat, Montan Wax and Other Organic Brown Coal Constituents

CHAPTER 25

B. R. THOMAS

Kauri Resins — Modern and Fossil

CHAPTER 26

W. E. ROBINSON

Kerogen of the Green River Formation

CHAPTER 27

G. C. SPEERS and E. V. WHITEHEAD

Crude Petroleum

CHAPTER 28

E. EISMA and J. W. JURG

Fundamental Aspects of the Generation of Petroleum

CHAPTER 29

B. S. COOPER and D. G. MURCHISON

Organic Geochemistry of Coal

CHAPTER 30

P. E. CLOUD JR.

Pre-Paleozoic Sediments and Their Significance for Organic Geochemistry

CHAPTER 31

A. WEISS

Organic Derivatives of Clay Minerals, Zeolites, and Related Minerals

List of Contributors

BAKER, E. W.	Mellon Institute, Carnegie-Mellon University, Pittsburgh, Pennsylvania, USA
BLUCK, B. J.	University of Glasgow, Glasgow, Scotland
BRETT, D. W.	Botany Department, Bedford College, London University, London, England
BURLINGAME, A. L.	Space Sciences Laboratory, University of California, Berkeley, California, USA
BUTLER, J. H. A.	Division of Soils, C.S.I.R.O., Adelaide, Australia
CLOUD, P. E., JR.	Department of Geology, University of California, Santa Barbara, California, USA
COOPER, B. S.	Department of Geology, University of Newcastle upon Tyne, England
DAYHOFF, M. O.	National Biochemical Foundation, Silver Spring, Maryland, USA
DEGENS, E. T.	Department of Chemistry, The Woods Hole Oceanographic Institution, Woods Hole, Mass., USA
DOUGLAS, A. G.	Organic Geochemistry Unit, University of Newcastle upon Tyne, England
ECK, R. V.	National Biomedical Research Foundation, Silver Spring, Maryland, USA
EISMA, E.	Koninklijke/Shell Exploratie en Produktie Laboratorium, Rijswijk, The Netherlands. (Present address: Amsterdam 2, R. Vinkeleskade 4)
EGLINTON, G.	Organic Geochemistry Unit, School of Chemistry, University of Bristol, Bristol, England
FLORKIN, M.	University of Liège, Liège, Belgium
HARE, P. E.	Geophysical Laboratory, Carnegie Institute, Washington, D. C., USA
HEROUT, V.	Czechoslovak Academy of Sciences, Institute of Organic Chemistry and Biochemistry, Prague, Czechoslovakia
JURG, J. W.	Koninklijke/Shell Exploratie en Produktie Laboratorium, Rijswijk, The Netherlands
KRANZ, R.	Kernforschungsanlage Jülich, Institut für Reaktorwerkstoffe, Jülich, Germany
MEINSCHEIN, W. G.	Geology Department, Indiana University, Bloomington, Indiana, USA

MOORE, L. R.	Geology Department, University of Sheffield, Sheffield, England
MORRISON, R. I.	The Macaulay Institute for Soil Research, Aberdeen, Scotland. (Deceased on November 23rd, 1967)
MURCHISON, D. G.	Department of Geology, University of Newcastle upon Tyne, England
MURPHY, M. T. J.	Department of Chemistry, St. Joseph's College, West Hartford, Connecticut, USA
PARKER, P. L.	University of Texas, Marine Science Institute at Port Aransas, Texas, USA
ROBINSON, W. E.	The Laramie Petroleum Research Center, U.S. Bureau of Mines, Laramie, Wyoming, USA
ROLFE, W. D. I.	Hunterian Museum, Glasgow University, Glasgow, Scotland
SCHNOES, H. K.	The Biochemistry Department, College of Agriculture, University of Wisconsin, Madison, Wisconsin, USA
SCHWENDINGER, R. B.	Esso Agricultural Products Laboratory, Linden, New Jersey, USA
SPEERS, G. C.	The British Petroleum Co., Research Centre, Sunbury-on-Thames, Middlesex, England
STEVENSON, F. J.	Department of Agronomy, University of Illinois, Urbana, Illinois, USA
STREIBL, M.	Czechoslovak Academy of Sciences, Institute of Organic Chemistry and Biochemistry, Prague, Czechoslovakia
SWAIN, F. M.	University of Minnesota, Minneapolis, Minnesota, USA
THOMAS, B. R.	Chemistry, Division, DSIR, Petone, New Zealand. (Present address: Organic Chemistry Dept., TH, Lund, Sweden)
WEISS, A.	Institut für Anorganische Chemie der Universität, München, Germany
WELTE, D. H.	Institut für Geologie, Universität Würzburg, Germany
WHITEHEAD, E. V.	The British Petroleum Co., Research Centre, Sunbury-on-Thames, Middlesex, England
WOLLRAB, V.	Czechoslovak Academy of Sciences, Institute of Organic Chemistry and Biochemistry, Prague, Czechoslovakia. (Present address: 625 Limburg, Parkstr. 22, Germany)

CHAPTER 1

Introduction

G. EGLINTON and M. T. J. MURPHY

Organic Geochemistry Unit, School of Chemistry
The University of Bristol, Bristol 8, England
and
Department of Chemistry
Saint Joseph College, West Hartford, Connecticut

Organic chemistry made its first major impact on geology and the earth sciences in 1934 when the German chemist, Treibs, isolated and identified biologically important pigments, the metal porphyrins, from many crude oils and shales. The nature of these porphyrins led him to claim two major findings – first, that the oils were certainly, in part at least, biological in origin and second, that the conditions of formation of the oil could not have involved high temperatures.

Organic geochemistry, the study of the fate and distribution of carbon compounds in contemporary environments and in Recent and ancient sediments, is now progressing rapidly, spurred on by such interests as the search for deposits of fossil fuels, the control of pollution, the problem of the origins of life and the prospective analysis of the surfaces of other planets and of returned lunar samples. Advances in instrumentation for separative and analytical methods have made it possible to handle the extremely complex mixtures of organic compounds present in most sediments. The previous rough characterizations of the organic matter in terms of percentages of carbon, hydrogen, oxygen, etc., are being replaced by detailed information relating to the distribution of individual compounds.

This book is intended to provide an overall view of the field of organic geochemistry, giving current methods and techniques, research areas and results. The chapters are written by scientists of several nations and of a wide range of backgrounds, but in general, the approach is that of the organic chemist rather than the geologist. So far, those who call themselves "geochemists" are few in number and the chapters have been written with two types of reader in mind: geologists who have some chemical knowledge and organic chemists who have an interest in, but little knowledge of geology. The multi-authorship presents some difficulties in that uniform style, presentation and depth of coverage for the chapters have not been possible, but this is the first book covering this particular area of science in which the specific subjects for review have been carefully chosen and interrelated. Further, many of the chapters give practical details for the isolation, separation, characterization and identification of organic matter and of individual organic compounds, thereby bringing together within one book a good deal of the current expertise in organic geochemistry. In some cases, short glossaries are provided. We have attempted to apportion space roughly in relation to the extent of present knowledge in the individual areas of research.

The authors are all active researchers in the subject matter of their chapters. In writing about their interests, their aim has been to produce surveys which are critical and selective rather than exhaustive. Most chapters cover the literature up to early 1967. Very recent important papers are cited in the 1968 addenda which appear at the end of the appropriate chapters, and goes some way, we hope, to make up for the delays seemingly inevitable in the compiliation and publication of a multi-author book of this size.

Layout

The layout of the book, emphasizing the relationships between the chapters, is summarized in the Table. The majority of topics have not been dealt with before within the covers of one book. Some interesting points of contrast emerge; for example, to the geomicrobiologist, the insoluble brown debris, known as kerogen, present in many sediments, is often a mass of fungal filaments and other microbiological debris, while to the chemist, it is an insoluble amorphous and polyfunctional polymeric material. Clearly some account must be taken of both views. There is overlap between chapters, but this is not excessive and generally serves to bring out useful comparisons in approach, methods and interpretation.

The chapters divide very roughly into four groups, arranged sequentially:

Analytical Methods

Chapters 2–6 provide the organic chemist's approach and explain the analytical tools needed to isolate and characterize organic compounds: general (Ch. 3), mass spectrometry (Ch. 4), gas chromatography (Ch. 5) and, in some depth, their application to the extraction of organic matter from sediments and fossils (Ch. 6).

General Geological Processes and Principles

Chapters 6–12 emphasize the biological and geological aspects: the evolutionary process (Ch. 7), fossilization (Ch. 8), sedimentology (Ch. 9), sedimented organic matter (Ch. 10), microbiological attack on organic matter (Ch. 11), and isotopic factors (Ch. 12).

Geological Abundance of Specific Classes of Organic Compounds

Chapters 13–19 provide a fairly straightforward coverage of the distribution of discrete classes of organic compounds, arranged according to biogenetic theory – i. e., those compounds are grouped together which are produced by biochemical pathways which are related and common to many types of organisms, an example being terpene biogenesis. Similar discussions continue for Chapters 20–25, but in part they are very much concerned with specific geological situations; thus, shells (Ch. 20), gas-inclusions in minerals (Ch. 21), soils (Chs. 22 and 23), and lignites (Chs. 24 and 25).

Specific Geological Situations

Specific geological situations are dealt with in Chapters 20–25, as indicated above, but receive special attention in the remaining chapters of the book (Chs. 26–30). Chapter 31, dealing with organic complexes of clays, is somewhat different in concept and is placed at the end of the book for editorial convenience.

Individual Chapters

The contents of the chapters are summarized very briefly in the following paragraphs:

Chapter 2
Organic Geochemistry – The Organic Chemist's Approach

(G. EGLINTON)

This chapter summarizes the basic assumptions and techniques of the organic chemist where they impinge on the subject of organic geochemistry. Organic matter should as far as is possible be characterized at the molecular level, i.e., in terms of single substances of established chemical structure including relative and absolute molecular geometry (stereochemistry). With this information the molecular composition of geological materials can be interpreted in terms of the known reactivity of the functional groups present (which affect the processes of diagenesis and maturation) and probable biological origin of the original organic compounds.

Chapter 3. Analytical Methods

(M. T. J. MURPHY)

Analytical methods are at the heart of organic geochemistry, for the basic problem is the separation and identification of the complex mixtures encountered. The separation methods, largely comprised of the various types of chromatography, are reviewed in relation to the geochemical needs of high sensitivity, small-scale operation, and high resolving power. Some examples are given here but later chapters carry illustrations specific to their topics which are also referred to in this chapter.

Chapter 4. Mass Spectrometry in Organic Geochemistry

(A. L. BURLINGAME and H. SCHNOES)

Mass spectrometry has revolutionized identification of small amounts of organic compounds and accordingly now plays a major role in organic geochemistry. The chapter surveys the technique and application of the method, important practical aspects and the type of chemical information most readily derived. Examples are drawn from recent geochemical studies. High resolution mass spectrometry is dealt with in detail, providing one of the few authoritative treatments of this topic. The powerful combination of a gas chromatograph directly coupled to a mass spectrometer is also discussed: complex mixtures are then readily analyzed without prior isolation of individual components.

Chapter 5. Gas Chromatography

(A. G. DOUGLAS)

Gas chromatography is an extremely powerful analytical tool and is given more detailed treatment than are the other separatory techniques. As in Chapter 3, the technical terms and apparatus in use are explained and, following a section on columns and column packings, examples of applications to geochemistry are provided from recent papers. Much of the work in this field has been concerned with hydrocarbon and fatty acid analyses; however, the technique could be applied to most other organic geochemical problems.

Chapter 6. Isolation Procedures for Kerogens and Associated Soluble Organic Materials

(W. E. ROBINSON)

The isolation of organic matter from sediments and fossils is not an easy task, for the mineral and organic matter are often intimately associated. In this chapter, methods are discussed for the isolation of the insoluble organic "kerogen" and of the soluble organic material, sometimes referred to as "bitumen" or "geolipids". For the former, procedures include physical techniques such as flotation and chemical treatments such as acid demineralization. The bitumens on the other hand are generally removed by extraction. The chapter gives methods for each of the major classes of lipids, such as hydrocarbons, fatty acids, etc.

Chapter 7. Paleobiochemistry

(M. DAYHOFF and R. V. ECK)

This chapter is concerned almost entirely with conceptual material and is rather different from the others in this book. Paleobiochemistry relates to the possibility of revealing something of the origins of life and the subsequent phylogeny by first examining the existing biochemistry found for a wide range of organisms and then inferring from the analytical data (for example, the amino acid sequences of key enzymes) the phylogenetic position of the particular present-day organisms. The expectation is that it will be possible to work back from the present situation to elucidate a great many details of that pertaining at earlier times. Our intention in including this subject in this book is to provide a stimulus for chemical and geological research in the Precambrian. It should also stimulate attempts to correlate geochemical results for fossils, for example as discussed by FLORKIN in Chapter 20.

Chapter 8. Fossilization Processes

(W.D.I. ROLFE and D.W. BRETT)

ROLFE and BRETT present a brief survey of the factors – physical, chemical, biological, environmental – affecting the preservation of organic matter and organisms in sediments. The information should serve as a guide to the geochemist as to what he may expect to encounter in the way of preserved material, and which types of chemical are likely to have survived the conditions of interment and subsequent diagenesis.

Chapter 9. Introduction to Sedimentology

(B.J. BLUCK)

Here the emphasis shifts more to the physical factors and the geological processes. BLUCK discusses the various factors affecting the form and content of sedimentary rocks, the ways in which different types of sediment are laid down and the processes of diagenesis when they occur at water interfaces and in newly-buried and older sediments. The main emphasis in this discussion is on carbonate rocks and the part played by certain organisms, such as algae, in their formation.

Chapter 10. Organic Matter in Sediments

(D.H. WELTE)

Although this chapter is in outline form only, it is a useful summary of the situation relating to the incorporation, nature, distribution and role of organic matter in sediments. It emphasizes the point that organic matter plays an important part in the formation and diagenesis of most sediments.

Chapter 11. Geomicrobiology and Geomicrobiological Attack on Sedimented Organic Matter

(L.R. MOORE)

This chapter is mainly concerned with the microbiological attack which takes place in organic matter during and after deposition. Extensive and repeated attack is revealed by microscopic examination of Recent and ancient sediments. Thus, the insoluble and supposedly amorphous organic matter can sometimes be seen to include masses of fungal hyphae and bacteria-like organisms which presumably invested the debris during diagenesis. Many other types of microfossils, such as spores and pollen, are observed. The great significance for organic geochemistry, of course, lies in the guide-lines it provides for chemical study of the organic matter. There should surely be a correlation between the composition of the resistant plant or animal debris, the metabolites of the microorganisms and the chemical composition of the organic matter of which they represent the last stage in its biological alteration. The nature of the attack and the state of preservation of the organic debris are indications of the environmental conditions – such as marine, brackish, or fresh-water and terrestrial – pertaining at the time of deposition.

Chapter 12. Biogeochemistry of Stable Carbon Isotopes

(E. T. DEGENS)

As DEGENS remarks in this chapter, "information on the distribution of carbon isotopes in biological and geological material is essential to understand the complex biogeochemical cycle of carbon through time and space." Thus, it is necessary to study the isotopic fractionation by living matter – as introduced during photosynthesis – in terms of environmental and metabolic effects in order to understand the isotopic distribution found in the organic matter of sediments, both Recent and ancient. The chapter also discusses the problem of estimating the $\delta\,^{13}C$ values for the original magmatic carbon source.

Chapter 13. Hydrocarbons – Saturated, Unsaturated, and Aromatic

(W. G. MEINSCHEIN)

In recent years hydrocarbons have been frequent "shuttlecocks" in the controversy over a biological versus a non-biological origin for organic matter in sediments and meteorites. In this chapter, MEINSCHEIN discusses the distribution of hydrocarbons in living things and sediments in terms of possible lipid precursors, metabolic processes and natural environments. Hydrocarbons are treated as biological, environmental, and chemical indicators, for they are ubiquitous but minor products of plants and animals. One salient query which MEINSCHEIN asks is "what proportion of the alkanes in ancient rocks was initially derived as alkanes from biological lipids?" His opinion is that organisms have played a dominant role, either in the anabolic production of alkanes or in the catabolic control of specific lipid precursors that are possible sources of $>C_{15}$ alkanes in ancient rocks. His discussion in support of this opinion hinges on the chemical equivalences of the individual members of specific homologous series and the distributional differences between the structurally related alkanes, acids, alcohols, and olefins that are present either in biological lipids or the organic extracts of sediments. This chapter deals predominantly with hydrocarbons that contain more than 15 carbon atoms per molecule.

Chapter 14. Fatty Acids and Alcohols

(P. L. PARKER)

This chapter concentrates on the attempts which have been made to correlate the biological and geological occurrences of long-chain carboxylic acids, such as the n-fatty acids, $CH_3(CH_2)_nCO_2H$. Fatty acids, particularly unsaturated fatty acids, are prominent constituents of the lipid fractions of most plants and animals. Sediments, especially ancient sediments, contain fatty acids which are almost always saturated, but the chain-lengths are still in the same carbon number range and generally show some alternation of carbon number (even/odd). Acids having branched carbon chains, notably those of the isoprenoid type,

$$CH_3{-}\overset{\overset{\displaystyle CH_3}{|}}{C}H{-}CH_2{-}CH_2{-}(CH_2{-}\overset{\overset{\displaystyle CH_3}{|}}{C}H{-}CH_2{-}CH_2)_n{-}CH_2{-}\overset{\overset{\displaystyle CH_3}{|}}{C}H{-}CH_2{-}CO_2H$$

are important too and are possibly derived from chlorophyll.

Chapter 15. Fossil Carbohydrates

(F. M. SWAIN)

It is quite remarkable that individual sugars have been isolated from ancient sediments and fully characterized by the classical methods of organic chemistry, crystallization and mixed melting point, etc. More recent work relies on chromatographic and enzymatic methods, and in this chapter these methods and the resulting data are summarized for carbohydrates in sediments and discrete fossils. The results are compared with those to be expected for the particular fossil source materials, assuming a parallelism with present-day biochemistry. This chapter makes a point of the use of carbohydrates as potential geochemical markers, in view of their thermal instability, and as paleoenvironmental indicators.

Chapter 16. Terpenoids – Especially Oxygenated Mono-, Sesqui-, Di-, and Triterpenoids

(M. STREIBL and V. HEROUT)

Terpenoids have already become of some considerable importance in organic geochemistry. This chapter reviews the geological distribution so far revealed of terpenoids (excluding carotenoids), i.e., the $(C_5)_n$ derivatives. These compounds are especially valuable as biological markers, for their distribution in living things has been extensively studied and they are important chemotaxonomically. The terpenoids in earth waxes, peat, lignites, etc., show evidence of hydrogen transfer processes; compounds bearing similar or identical carbon skeletons appear in fully reduced form together with the corresponding partially and fully-aromatized compounds.

Chapter 17. Carotenoids

(R. B. SCHWENDINGER)

Carotenoids present another biogenetically homogeneous class of compounds, though their functionality may be very diverse. They are based on a $(C_5)_8 = C_{40}$ carbon skeleton. The biological production of some carotenoids such as fucoxanthin, the major xanthophyll present in algae, is enormous. Recent sediments from all environments so far studied contain carotenoids. Analytical methods are again important, but only UV and visible spectroscopy have been widely applied. Carotenoids are unstable compounds and have not been recorded for sediments older than a few thousand years. Aromatic hydrocarbons such as toluene are easily formed from them by heating in the presence of water and clay. Reactions of this type may in part explain the presence of aromatic compounds in petroleum and other sediments.

Chapter 18. Geochemistry of Proteins, Peptides, and Amino Acids

(P. E. HARE)

HARE's chapter reveals important advances in the understanding of the fate of amino acids, polypeptides and proteins. Such compounds form part of the nitrogen cycle involving living things and their relationship with the geosphere.

It seems that many of the literature reports for amino acids in fossils, sediments, and meteorites may have been occasioned by Recent contamination, either *in situ* in the field or in the laboratory. Certain biologically-produced amino acids decompose faster than others and thereby act as markers for Recent contamination or Recent origin. Laboratory experiments with shells provide supporting evidence for these "geogenetic processes". Optical activity changes are similarly important and informative. Thus the β-carbon atom of L-isoleucine is not readily racemised while the α- is. Hence, the biological amino acid, L-isoleucine, eventually gives rise to an equilibrium mixture of L-isoleucine and D-alloisoleucine which may then be found in geological specimens.

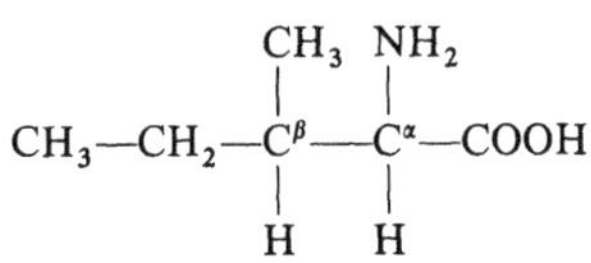

L-isoleucine

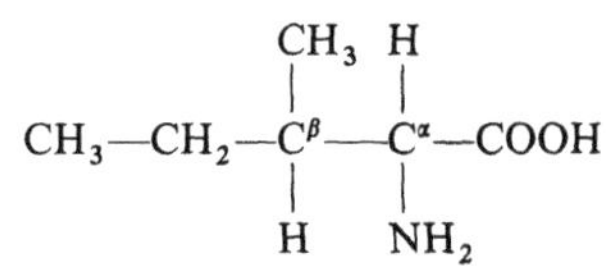

D-alloisoleucine

These results, which are quite recent, present important possibilities for geochronology and for a more critical evaluation of the nitrogen content of ancient rocks.

Chapter 19. Porphyrins

(E. W. BAKER)

The latest developments in the story of the porphyrins, first initiated so masterfully by TREIBS, are reviewed by BAKER. Extensive treatment is given to the sections giving experimental methods for the concentration, purification and identification of petroporphyrins. A geochemical section extends TREIB'S correlations of plant origin, diagenetic changes and porphyrin content of sediments. The petroporphyrin fractions are now known to be complex mixtures of alkyl-substituted porphyrins rather than single compounds. This field is developing rapidly.

Chapter 20
Fossil and Shell "Conchiolin" and other Preserved Biopolymers

(M. FLORKIN)

The physical state – morphology – of the apparently well-preserved biopolymers found intimately associated with the mineral matrix of fossil shells has been extensively studied by FLORKIN, GREGOIRE and their collaborators. They have used light and electron microscopy and have compared their results with those for contemporary materials. The good state of preservation of the paleoproteins (e. g., conchiolin in mother-of-pearl) preserved in ancient shells is not so apparent when chemical studies of the material are made. The amino acid composition reveals changes. There is still hope of being able to compare the amino acid sequences of fossil proteins and their modern and Recent counterparts. Such comparisons would lead to a study of the composition of the biopolymers along phylogenetic lines and therefore provide a parallel for the studies outlined in Chapter 7. Other preserved biopolymers considered in this chapter include those of bones, graptolites and dinosaur eggs.

Chapter 21
Organic Compounds in the Gas-Inclusions of Fluorspars and Feldspars

(R. KRANZ)

KRANZ's chapter is concerned with the gasses released from gas bubbles and liquid inclusions of certain minerals when these are broken up. They prove to contain mixtures of low molecular weight compounds such as CO_2 and H_2, and include hydrocarbons such as CH_4, $CH_2=CH_2$ and even cyclic compounds and amines. Fluorspars contain fluorohydrocarbons, such as CH_3F, and thionyl-sulfurylfluoride. The chapter discusses the formation of minerals and the laboratory observation of the increased solubility of inorganic matter in dilute solutions of organic amino compounds.

Chapter 22. Chemistry of Humic Acids and Related Pigments

(F. J. STEVENSON and J. H. A. BUTLER)

Humic acids and related pigments represent a class of polymeric material which does not have regular repeating monomeric units but rather contains many different building blocks linked in a random fashion through $-O-$, $-CH_2-$, $-NH-$, $-S-$, and other groups and having a wide range of functional groups. They are probably formed in soils and sediments by secondary synthesis reactions and are not therefore a direct metabolic product. Humic acids play an important role in soil in relation to metal ion transport.

Chapter 23. Soil Lipids

(R. I. MORRISON)

The lipids present in soils still await detailed study of the non-humic fraction but long chain esters, $CH_3(CH_2)_nCOO(CH_2)_nCH_3$, seem to be important constituents. They are probably derived from plant waxes. Little attempt has been made so far to correlate the constituents of soils with the particular type of soil and plant contribution.

Chapter 24. Earth Waxes, Peat, Montan Wax, and Other Organic Brown Coal Constituents

(V. WOLLRAB and M. STREIBL)

The methods used in this chapter for the isolation and characterization of lipids from earth waxes, peat, etc., are similar to those already discussed in other chapters (Soil Lipids, Ch. 23; Terpenoids, Ch. 16; Fatty Acids and Alcohols, Ch. 14).

There is now a good deal of information relating the composition of modern plant waxes and the fossil compounds extractable from peat and brown coal. The application of powerful methods such as gas chromatography and mass spectrometry has revealed that there is still much to be learned about the origin of these materials.

Chapter 25. Kauri Resins – Modern and Fossil

(B. R. THOMAS)

A single specific geological material, kauri resin, is discussed in detail in this chapter. The chemistry of the modern resin fresh from the tree is compared with that of the fossil material, both Recent (a few hundred years) and ancient (as old as Cretaceous). This study is noteworthy for bringing together the botanical, geological and geographical contexts and dealing with them alongside the composition and biogenesis of the chemical components and provides a model for further researches on other important resins and ambers.

Chapter 26. Kerogen of the Green River Formation

(W. E. ROBINSON)

This chapter concerns one particular kerogen deposit. The methods used in this study provide a cross-section of those presently available in organic geochemistry. The particular deposit was formed in Eocene times in a stratified lake under highly reducing conditions. Application of a range of techniques reveals that this kerogen is mainly macromolecular with long aliphatic chains and alicyclic structures, cross-linked by oxygen and other heteroatoms. Even at this stage much remains to be done concerning the precise nature of this material.

Chapter 27. Crude Petroleum

(G. C. SPEERS and E. V. WHITEHEAD)

This chapter presents an extensive survey of the chemical work on the constituents of petroleum, summarizing the compounds present and the procedures available for their analysis. Special attention is given to triterpanes and their isolation and identification. The biochemical correlation, which may permit inferences to be drawn concerning the ultimate origin of the oils, are also outlined.

Chapter 28. Fundamental Aspects of the Generation of Petroleum

(E. EISMA and J. W. JURG)

The problem of the genesis of petroleum is here tackled mainly from the laboratory stand-point. Chemical evidence for the thermal/catalytic conversion of fatty acids into hydrocarbons is given. This type of experiment should serve as a model for other geogenetic studies.

Chapter 29. Organic Geochemistry of Coals

(B. S. COOPER and D. G. MURCHISON)

Transmitted and reflected light microscopy can be used to establish the presence of different types of petrographic constituents in coals of all levels of rank. The varying morphologies and properties of these constituents indicate differing progenitors and early genesis. Coalification is discussed in terms of a biochemical stage that normally lasts from deposition until a mature peat forms, but which may be prolonged into the time of brown-coal formation. The biochemical stage is followed by a geochemical stage during which temperature and pressure influences predominate. This chapter will be of particular value to geochemists because of the discussion of the flora and the environment of deposition of the original material.

Chapter 30. Pre-Paleozoic Sediments and their Significance for Organic Geochemistry

(P. E. CLOUD)

The Precambrian or pre-Paleozoic makes up the first seven-eighths of the earth's history. CLOUD summarizes the opportunities for research in organic geochemistry afforded by the sedimentary rocks that record these ancient times – the oldest of which still contain microstructures inferred to be of vital origin. He points out that "even though we may be denied a record of life itself, we are on the verge of a new era of investigation of pre-Paleozoic rocks in which organic geochemistry and paleo-microbiology will continue to play interacting and significant parts." Extensive areas on all continents display vast thicknesses of relatively unaltered sedimentary rocks of pre-Paleozoic age that await detailed study.

Chapter 31. Organic Derivatives of Clay Minerals, Zeolites and Related Minerals

(A. WEISS)

Much is now known about the way carbon compounds can be adsorbed into the lattices of clays and other minerals. Clay complexes must be of great importance in determining which organics accumulate in sediments. It is fairly certain that catalysis by clays play some part in the diagenesis and maturation of the organic matter. This close relationship between certain carbon compounds and clay minerals is suspected by WEISS to have been "concerned with the abiotic events leading to the origin of life." These silicate minerals differ in their capabilities for accommodating molecules. The weathering of rocks and soils must be very dependant upon such effects. Organic compounds are selectively taken up from dilute solution into the interlayer spaces where they may be protected against microbial degradation. Organic derivatives of clay minerals, etc., are undoubtedly of considerable academic and industrial interest.

TABLE

No.	Chapter title	Methods for study of organic matter		
		isolation	separation/ fractionation	analysis/ identification
2.	Organic Geochemistry – The Organic Chemist's Approach. G. EGLINTON	*	*	–
3.	Analytical Methods. Sister M. T. J. MURPHY	* Derivatives	*Chromatography, clathration, reactions*	* *Derivatives*
4.	Mass Spectrometry in Organic Geochemistry. A. L. BURLINGAME and H. SCHNOES	–	GC – MS	*Mass spectrometry*
5.	Gas Chromatography. A. G. DOUGLAS	*	*Gas chromatography*	*Gas chromatography*
6.	Isolation Procedures for Kerogens and Associated Soluble Organic Materials. W. E. ROBINSON	*Kerogen, soluble organics*	*	*
7.	Paleobiochemistry. M. DAYHOFF and R. V. ECK	–	–	–
8.	Fossilization Processes. W. D. I. ROLFE and D. W. BRETT	–	–	–
9.	Introduction to Sedimentology. B. J. BLUCK	–	–	–
10.	Organic Matter in Sediments. D. H. WELTE	–	–	–

Discussion of organic matter			Discussion of geological data	
classes of compound	insoluble polymer/ kerogen	biological origin	processes	situations/ materials
*	*	*	*	*
–	–	–	–	–
Hydrocarbons, carboxylic acids, nitrogen compounds, porphyrins, sulphur compounds	–	–	–	Extraterrestrial samples
Hydrocarbons, carboxylic acids, carbohydrates, alcohols	–	–	–	–
Hydrocarbons, carboxylic acids, porphyrins, amino acids carbohydrates	*	–	–	–
Proteins, nucleic acids	–	Phylogeny, evolutionary biochemistry	–	Pre-biological era
–	–	Phylogeny	*Fossilisation, biostratinomy, diagenesis, weathering, metamorphism*	Depositional environments
–	–	*	Sedimentology, *deposition, diagenesis, weathering, metamorphism*	Depositional environments; siliceous, iron-bearing and phosphatic sediments; evaporites, and carbonate rocks
–	–	*	Formation and diagenesis of sediments	*

Table

No.	Chapter title	Methods for study of organic matter		
		isolation	separation/ fractionation	analysis/ identification
11.	Geomicrobiology and Geomicrobiological Attack on Sedimented Organic Matter. L. R. MOORE	Kerogen, *particulate debris*	–	–
12.	Biogeochemistry of Stable Carbon Isotopes. E. T. DEGENS	–	–	Mass spectrometry
13.	Hydrocarbons – Saturated, Unsaturated, and Aromatic. W. G. MEINSCHEIN	–	–	*Gas chromatography*
14.	Fatty Acids and Alcohols. P. L. PARKER	Carboxylic acids	Gas chromatography	IR, MS
15.	Fossil Carbohydrates. F. M. SWAIN	Derivatives	*Chromatography*	IR; colorimetric and enzymatic assays
16.	Terpenoids – Especially Oxygenated Mono-, Sesqui-, Di-, and Triterpenoids. M. STREIBL and V. HEROUT	*Terpenoids*, C_{10} upwards	*Chromatography*	IR, NMR, MS
17.	Carotenoids. R. B. SCHWENDINGER	*Carotenoids*	*Chromatography*	UV, vis.
18.	Geochemistry of Proteins, Peptides, and Amino Acids. P. E. HARE	*Amino acids*	*Ion-exchange, chromatography*	Colorimetric and enzymatic assays
19.	Porphyrins. E. W. BAKER	Acid extraction	TLC, GC, gel permeation	Spectrometric, UV, vis., *MS*
20.	Fossil and Shell Conchiolin and Other Preserved Biopolymers. M. FLORKIN	*Protein membranes*	Ion-exchange	Colorimetry, *electron-microscopy*

(Continued)

Discussion of organic matter			Discussion of geological data	
classes of compound	insoluble polymer/ kerogen	biological origin	processes	situations/ materials
–	Cutin, chitin, lignin, cellulose, kerogen	*Bacteria, actino-mycetes, fungi, algae* etc.	Environments; *geo-microbiological attack;* formation and diagenesis of sediments	Depositional environments
Sugars, amino acids, proteins	kerogen, lignin	Plankton	Deposition, dia-genesis maturation and δ ^{13}C *values*	Depositional environments; coal, petroleum, gases, kerogen, igneous rock, meteorites
Lipids, *alkanes* (normal, branched and cyclic), *alkenes, aromatics*	–	Micro-organisms, plants, biosynthesis	Maturation of geolipids	Petroleum
Carboxylic acids, alcohols	*	Plants, bacteria, plankton, algae	Deposition and diagenesis	*Algal mats, marine bays, ocean water*
Mono-saccharides, oligo-saccharides, polysaccharides	Chitin, cellulose	Plants, algae	*	Fossil plants
Hydrocarbons, carboxylic acids, alcohols, ketones	–	Plants	*	Peat, brown coal, earth waxes and resins
Hydrocarbons, oxygenated compounds	–	Algae, phytoplankton, zooplankton	Diagenesis	Recent sediments
Amino acids, peptides, proteins, nitrogen compounds	Chitin, collagen, etc.	Microbiology, shells, nitrogen cycle	Diagenesis, amino acid alterations	Fossils (shells, bones, teeth), sediments and soils, meteorites, water, atmosphere
Porphyrins	–	Plants, bacteria, animals	Diagenesis	Recent and ancient sediments
Amino acids, peptides, proteins	Chitin, nacroin, conchiolin	*Shells*, bones, teeth	Diagenesis, "paleisation" of proteins	*Fossils* (graptolites, shells, bones, teeth)

Table

No.	Chapter title	Methods for study of organic matter		
		isolation	separation/ fractionation	analysis/ identification
21.	Organic Compounds in the Gas-Inclusions of Fluorspars and Feldspars. R. KRANZ	*Low molecular weight organics*	Vac-line, gas chromatography	*Mass spectrometry*
22.	Chemistry of Humic Acids and Related Pigments. F. J. STEVENSON and J. H. A. BUTLER	Precipitation, solubility	Acid/base	Derivs., *functional groups* $-OH$, $-O-$, $-CO_2H$, $-CO-$, UV, IR
23.	Soil Lipids. R. I. MORRISON	Solvent extraction	Chromatography	Spectrometry, IR
24.	Earth Waxes, Peat, Montan Wax, and Other Organic Brown Coal Constituents. V. WOLLRAB and M. STREIBL	*Solvent extraction*	*Chromatography*, TLC, ion-exchange	Spectrometry
25.	Kauri Resins – Modern and Fossil. B. R. THOMAS	Solvent extraction	*Chromatography*, TLC, GLC	*Spectrometry*, *NMR*, *IR*, *MS*, *UV*
26.	Kerogen of the Green River Formation. W. E. ROBINSON	*	*Reactions*, oxidation/reduction, thermal degradation, reduction, solvent extraction, functional groups, hydrolysis	IR, UV, X-ray, carbon isotope
27.	Crude Petroleum. G. C. SPEERS and E. V. WHITEHEAD	–	Chromatography, thermal diffusion, urea/thiourea clathration	Spectrometry, MS, NMR, IR, X-ray, optical activity
28.	Fundamental Aspects of the Generation of Petroleum. E. EISMA and J. W. JURG	–	Chromatography	Mass spectrometry

(Continued)

Discussion of organic matter			Discussion of geological data	
classes of compound	insoluble polymer/ kerogen	biological origin	processes	situations/ materials
Inert gases, hydrocarbons, fluorocarbons, amines	–	*	Silicification; hydrothermal and radiation geochemistry	Feldspars, fluorides, gas and fluid inclusions, mineral genesis, silicified wood
Humic acids, *fulvic* acids etc.	–	*	–	Soils and peats
Glycerides, waxes, phospholipids, hydrocarbons, steroids, triterpenes, carotenoids	*	Plants, micro-organisms	*	*Soils*
Hydrocarbons, alcohols	*	Plants, micro-organisms	Coalification	Earth waxes, peat, montan wax, brown coal
Di-terpenoids, hydrocarbons, alcohols, carboxylic acids	Polymeric resin	*Kauri tree*	*	*Fossil resins*
Many classes of compound	*Kerogen*	Phytoplankton, micro-organisms	–	*Kerogen of Eocene lacustrine sediment*
Hydrocarbons, sulphur, oxygen and nitrogen compounds, porphyrins	Asphalt etc.	*	Genesis of petroleum	*Petroleum*
Hydrocarbons, *fatty acids*, porphyrins	–	*	*Clay catalysis*	–

Table

No.	Chapter title	Methods for study of organic matter		
		isolation	separation/ fractionation	analysis/ identification
29.	Organic Geochemistry of Coal. B. S. COOPER and D. G. MURCHISON	–	Density, float/sink	Petrographic
30.	Pre-Paleozoic Sediments and Their Significance for Organic Geochemistry. P. E. CLOUD	–	–	–
31.	Organic Derivatives of Clay Minerals, Zeolites and Related Minerals. A. WEISS	–	–	–

Topics in *italics* are given special emphasis. *Abbreviations:* * = Brief, general discussion; IR = metry; GC = gas chromatography; TLC = thin layer chromatography; GC–MS = combined gas

(Continued)

Discussion of organic matter			Discussion of geological data	
classes of compound	insoluble polymer/ kerogen	biological origin	processes	situations/ materials
Humic acids, aromatic compounds, functional groups	Cutinite	Plants, algae, micro-organisms	*Depositional environments*, *coalification* biochemical, geochemical	*Coal, peat*
–	–	Algae, micro-organisms	Diagenesis, metamorphism	*Cherts, carbonates, sulphates, phosphates*
Hydrocarbons, oxygenated, and nitrogen-containing compounds	–	–	Weathering, origin of petroleum, origin of life	*Clay minerals, zeolites* and organic complexes thereof

infrared spectrometry; UV = ultraviolet spectrometry; vis. = visible spectrometry; MS = mass spectro-chromatography – mass spectrometry.

CHAPTER 2

Organic Geochemistry
The Organic Chemist's Approach

G. EGLINTON

Organic Geochemistry Unit, School of Chemistry, The University of Bristol
Bristol 8, England

Contents

I. Introduction

My purpose in writing this chapter has been to outline the premises and methods of "molecular organic geochemistry". This term encompasses the characterization of organic matter at the molecular level as it occurs in sediments and other natural environments. The likelihood of establishing firm correlations between a compound and its past history increases the more completely the structure is known. Many of the later chapters exemplify this approach and provide a detailed key to the literature, whereas this chapter is concerned more with presenting the overall scheme.

Geologists and inorganic geochemists unfamiliar with organic chemistry should find that the treatment of the subject given here progresses in a fairly regular fashion after an initial short section describing the basic structural concepts relating to carbon compounds. The rapid pace of development of the field owes much to the availability of improved techniques for the separation and proper characterization of individual compounds, particularly of the lipid (fat-soluble) type.

This chapter deals with lipids originating from geological sources – "geolipids" – and presents structural and presumed causal relationships between these lipids and the biologically-formed "biolipids".

With organic geochemistry, "natural product" organic chemistry is extended to the whole complex path of carbon in Nature. Analytical techniques, such as gas chromatography and mass spectrometry, enable the organic chemist to go beyond the examination of the constituents of single plants or animals and to deal with the complex mixtures of compounds present in sediments and rocks. Radiolabelling techniques, now a standard tool in following biogenetic pathways, should be directly applicable to "geogenetic" studies, involving the tracking-down of the reaction sequences in the geological materials.

The strong carbon-carbon covalent bonds often persist through the accumulation and diagenesis of a sediment, thereby leading to the idea of the preservation of all or part of the carbon skeleton of a lipid molecule. Oxidation, reduction, and other reactions evidently occur, but the altered molecule may still be correlated with that of the original biological precursor(s) by comparing their carbon skeletons. Such compounds have been termed "biological markers".

All organic compounds present in the geosphere are not necessarily of biological origin; much current interest has centered on the search for truly abiogenic material, which would have direct relevance to problems such as the origins of life and the composition of extraterrestrial samples. Organic geochemistry encompasses all such studies, including the examination of natural environments. The term "biogeochemistry" which is in current use, is more limited in scope and, in the author's view, refers to studies of the part played by living organisms which are directly concerned with the deposition of sediments and the formation of minerals.

Organic geochemistry has its contribution to make to the understanding of the processes of natural evolution. One aspect concerns comparative biochemistry, studied not only for its relevance to the present day but also to past eras.

Two main approaches can be taken to the biochemistry of organisms in past times. The first is to examine fossil specimens in the hope that some of the information is still there in the form of protein structure, secondary metabolites, etc. (Chapters 7, 18, and 20). The second is to examine the biochemistry of present-day living organisms and make phylogenetic comparisons. From these, one can attempt to infer evolutionary sequences in the development of biochemical processes (Chapter 7); this approach is known variously as paleobiochemistry, evolutionary biochemistry and chemical paleogenetics.

The first approach, the direct examination of fossil material, is the main topic of this book, but the emphasis is on secondary metabolites, which, unfortunately, have less significance for evolutionary studies than do, for example, the nucleic acids. The second approach, evolutionary biochemistry, on the other hand, has had considerable success in comparing the amino acid sequences of certain key proteins isolated from different species. While some of these molecules, such as hemoglobin, vary considerably in amino acid composition and sequencing from one species to another, others, such as cytochrome c, show few differences.

Organic geochemistry has immediate economic and social significance. The understanding of the origin of fossil fuels (peat, coal, oil, etc.), the search for them, and their proper chemical utilization, are major fields for the application of knowledge gleaned in this field. Much information evidently resides in the archives of petroleum companies and is presently inaccessible. Environmental studies are possibly of much greater value in the long term, since careful planning and chemical control of our use of natural and artificial environments will become more and more essential as civilization increases in complexity. The degree of interdependence, part naturally enforced by air and water movements and part man-contrived, between countries, continents and oceans will soon necessitate planetary management of our resources, production and waste. The understanding and control of the movement, alteration and fate of organic compounds in the atmosphere, hydrosphere and geosphere will be an essential part of such management.

> "What would the world be, once
> bereft
> Of wet and of wildness? Let
> them be left,
> O, let them be left, wildness and
> Wet;
> Long live the weeds and the
> Wilderness yet."
>
> GERARD MANLEY HOPKINS

II. Organic Chemistry: Significance of Molecular Structure and the Representation of Formulae

Some knowledge of organic chemistry is required for an appreciation of the aims, methods and results of organic geochemistry. A detailed introduction is not within the scope of this book, and the reader is referred to texts such as those authored by LLOYD [1], GRUNDON and HENBEST [2], TEDDER and NECHVATAL [3], FIESER and FIESER [4], MORRISON and BOYD [5], ROBERTS and CASSERIO [6], and CRAM and HAMMOND [7]. A brief account is given here for ready consultation during study of the present book.

A. Carbon Compounds

Organic chemistry is the chemistry of carbon compounds. It derives its name, historical background, and much of its interest and impetus from the fact that living organisms rely on the element carbon for the make-up of their building materials and their reactive constituents. The four tetrahedrally disposed valencies (I) mean that an infinite variety of three-dimensional structures is possible by utilizing bonds formed with other carbon atoms and with atoms of other elements such as hydrogen, nitrogen, oxygen, sulphur, phosphorus, and so on. Elemental carbon as diamond (II) is an extended three-dimensional array of atoms interlinked rigidly through the four covalent bonds (sp^3 orbitals; σ bond); as graphite (III) on the other hand, carbons are interlinked by only three covalent bonds

(I) (II) (III)

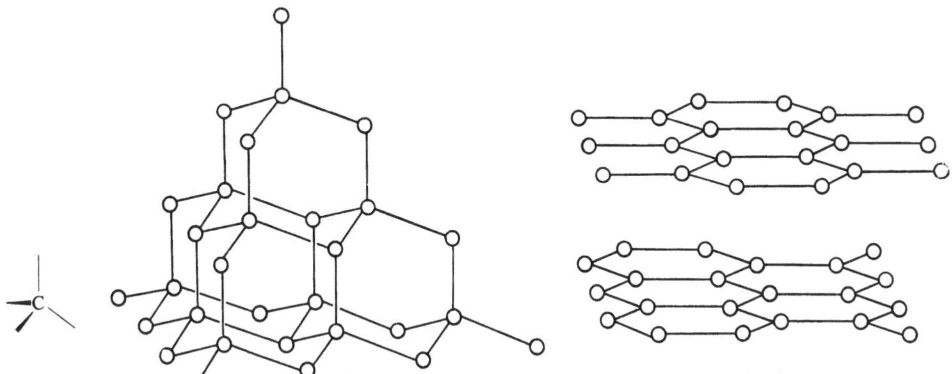

Fig. 1. Representations of molecular formulae. (I) Perspective representation of the four valencies of carbon; (II) A portion of the lattice structure of diamond; (III) A portion of the layered structure of graphite

(IV) (V) (VI) (VII) (VIII)

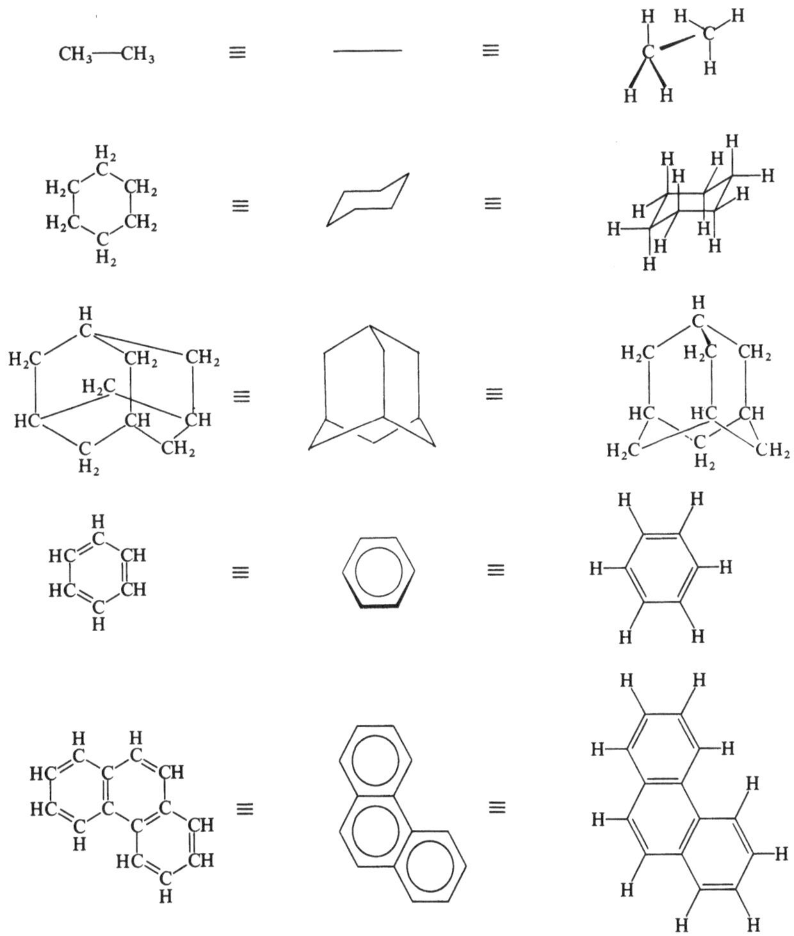

Fig. 2. Representations of several simple hydrocarbons. (From left to right: conventional formula, shorthand skeletal formulation and full perspective formula). (IV) Ethane, C_2H_6; (V) Cyclohexane, C_6H_{12}; (VI) Adamantane, $C_{10}H_{16}$; (VII) Benzene, C_6H_6; (VIII) Phenanthrene, $C_{14}H_{10}$. (The structures have not been drawn out completely in every case. Thickened lines are closer to observer.)

(sp^2 type; σ bonds and distribution of π electrons) per atom, and the two-dimensional result is a flat extended sheet of tightly bonded carbon atoms. These sheets stack loosely together and slip easily over one another. Carbon-carbon single (σ) bonds are generally about 1.5 Angströms (1 Å $=10^{-8}$ cms) long. Compounds of carbon and hydrogen only, hydrocarbons, possess interlinked carbon atoms which carry a number of hydrogen atoms appropriate to satisfy the remaining valencies. Organic compounds can be described and depicted in a number of ways – common name and empirical formula, perspective formula, shorthand skeletal formulation, and conventional formula.

Some examples of hydrocarbons are given in Fig. 2. In ethane (IV) there is more or less unrestricted rotation about the carbon-carbon single bond. This rotational freedom in cyclohexane (V) leads to a ring structure which is flexible and can be "flipped" from one preferred shape (conformer) to another, with only small inputs of energy. But in adamantane (VI) the carbon framework is rigid. The similarity between the diamond lattice and the adamantane structure is obvious, also that between the graphite, benzene (VII) and phenanthrene (VIII) structures. Representation of compounds involving other elements is achieved in a similar fashion (Fig. 3). The amino acids, glycine (IX) and alanine (X), have the structures shown. A portion of a polypeptide molecule having glycine and alanine as units in it is also indicated (XI).

(IX) (X) (XI)

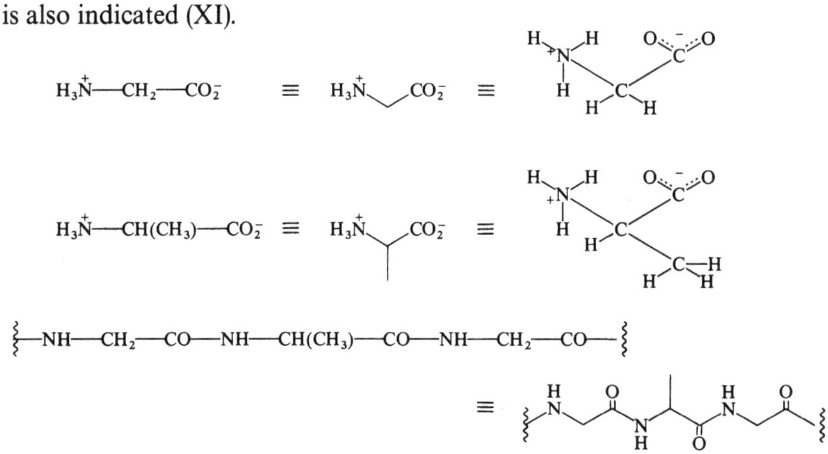

Fig. 3. Representations of α-amino acids and polypeptides. (IX) Glycine, $C_2H_5O_2N$; (X) Alanine, $C_3H_7O_2N$ shown as zwitterion structures; (XI) Portion of chain of polypeptide, showing two units of glycine and one of alanine (*– Gly – Ala – Gly –*)

Isomeric compounds (same empirical formulae) differing in respect to which atoms are attached to which other atoms, are *structural* or *positional isomers.* [e.g., n- and isobutane, see Section II.C]. Where isomeric compounds have the same gross structure but differ only in the way the atoms are oriented in space, then they are termed *stereoisomers* [See Section II.C]. Subdividing this term, where the constituent molecules are asymmetric, and are mirror images of one another, then they are *enantiomers.* Where they are not mirror images, they are called diastereoisomers, or *diastereomers. Geometric* isomerism, around a double bond, is a particular case of diastereoisomerism (e.g., *cis*- and *trans*-but-2-ene are a pair of geometric isomers).

B. Stereochemistry

The importance of stereochemistry, consequent upon the rigid directional nature of the carbon valencies, cannot be over-emphasized, for it explains many important features and properties of organic compounds [8]. The well-known *optical isomerism* of biologically-formed organic compounds derives in almost every instance from the presence of *asymmetric* carbon atoms. This concept applies where four different atoms or groups are linked tetrahedrally to a carbon atom. Molecules containing these atoms are asymmetric (no plane, center or axis of symmetry) and exist as two different kinds or *configurations*, which are related, one to the other, as mirror images but are not superposable. These two different molecules are said to be enantiomeric; each alone would rotate plane-polarized light [by the same amount, but in opposing directions, to the right (+) or the left (−)] and is therefore said to be *optically active*. An equimolar mixture of the pair, (+) and (−), of the enantiomers is termed a *racemate* (±), and has no effect on polarized light, and is said to be *optically inactive*. Laboratory syntheses using optically inactive compounds and reagents always produce asymmetric products which are racemic, for the enantiomers (mirror image isomers) do not differ in energy content or other physical and chemical properties (other than asymmetry). Syntheses effected by biologically-produced enzymes or by whole cells or organisms almost invariably use and produce single enantiomers, where the molecules possess asymmetric carbon atoms. Thus, only L-amino acids are utilized biologically, a consequence of the stereochemical control effected by the asymmetric molecular structure of the enzymes (which are themselves constructed solely from L-amino acids). In the L-series of amino acids, the variable substitution is always on the side of the molecule shown (Fig. 4). The NH_3^+ and CO_2^- groups are the reactive polar portions of the L-alanine molecule and presumably interact directly with the asymmetric surface of the enzyme during reaction, such as the introduction into a growing polypeptide chain. Optical activity observed in organic compounds is taken as prime evidence for their prior formation by living organisms.

Where there is more than one (say, n) asymmetric centers (e.g., carbon atoms) in a molecule, then several (2^n) configurational *stereoisomers* are possible. They can be grouped into enantiomeric pairs (*racemic modifications*) of diastereomers. In some compounds, however, the full number (2^n) of stereoisomers is not achieved owing to the existence of *meso* compounds (where one half of the molecule mirrors the other half). The orientation of the centers with respect to each other in the same molecule can be defined as the *relative configuration*. Defining the *absolute configuration* requires that the true orientation of the bonds at the centers be known [e.g., in the compound known as L(+) alanine (Fig. 4), the NH_3^+ group is on the left when the molecule is viewed so that the methyl and carboxyl groups

L(+)Alanine $[\alpha]_D + 14.7$ D(−)Alanine $[\alpha]_D - 14.7$

Fig. 4. Absolute configurations of L- and D-alanine. The L-enantiomers of the α-amino acids are predominant in living organisms

are pointing away from the observer]. Two conventions exist for the assignment of absolute configuration to the asymmetric centers in a molecule, the D and L convention and the newer R (rectus) and S (sinister) convention.

C. Structural Isomerism

For each of the empirical formulae: CH_4, C_2H_6, C_3H_8, there is only one way of arranging the atoms so that the valencies of carbon (4) and hydrogen (1) are satisfied. There is a unique molecular structure for each empirical formula. The three compounds are the first three fully saturated hydrocarbons, the alkanes. Ethane and propane are also formal members of a *homologous* series – the normal (n-)alkanes, $CH_3(CH_2)_nCH_3$, where n = 0, 1, 2 ... – one methylene (CH_2) group being added for each additional carbon number (Fig. 5). The fourth member of the series, n-butane, is *not* the only molecule having the formula C_4H_{10}: a *structurally* (positionally) *isomeric* compound, isobutane, is possible and is known. This *isomer* does not belong to the homologous series of n-alkanes since its carbon skeleton is branched. The number of isomeric alkanes, C_nH_{2n+2}, possible for any given empirical formula, increases rapidly with increasing carbon number (value of n), e.g. (the number of structural isomers is given in parentheses) for n = 5 (3), 8 (18), and 10 (75). By the time 19 carbon atoms are available for arrangement, the alkanes possible (structural isomers) number in the region of 100,000.

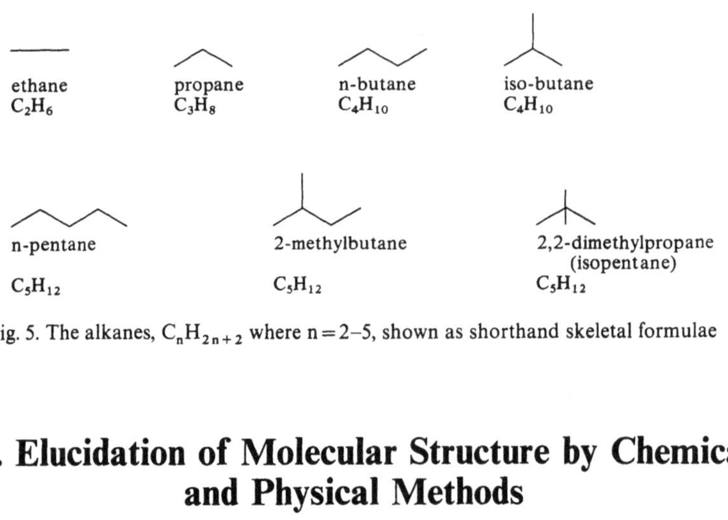

Fig. 5. The alkanes, C_nH_{2n+2} where n = 2–5, shown as shorthand skeletal formulae

III. Elucidation of Molecular Structure by Chemical and Physical Methods

A. Isolation and Separation of Lipids

Only methods capable of application at the micro level (say, 10^{-6} gm, μg) need be considered, even in this general discussion, for the quantities of geolipids to be characterized are often very small. The particular problems associated with the extraction of geolipids from their source materials are discussed in later sections of this chapter.

Procedures for the isolation and separation of lipids rely partly on the predominantly "lipid" nature of the molecules and partly on the quantity and kind of any polar groupings present. The hydrocarbon portions present insure the solubility of lipid molecules in relatively non-polar organic solvents such as isooctane, benzene and toluene ("like dissolves like"). The more numerous the polar substituents carried by the lipid molecule, the lower the solubility in such hydrocarbon solvents. More polar solvents such as ether, chloroform, and methanol are then effective.

Specific interactions between molecular groupings result in enhanced solubility or, if we are dealing with a liquid-solid situation, adsorption. Examples of such effects are hydrogen-bonding and complex formation (Fig. 6): a lipid bearing a hydroxyl group will show enhanced solubility in an ether solvent, due to hydrogen bonding; similarly, chloroform is a good solvent for ethers; aromatic hydrocarbons are selectively adsorbed as a result of weak π-complex formation during thin layer chromatography over silica impregnated with silver ions ($AgNO_3$-SiO_2).

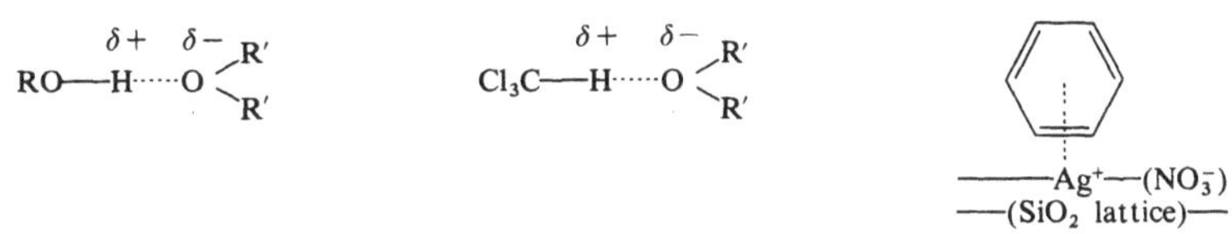

Fig. 6. Hydrogen bonding and π-bonding associations

Lipids bearing acidic or basic groups are conveniently isolated and then separated from the neutral lipids by converting them into the corresponding salts, the ionic character of which normally confers water solubility. Thus, carboxylic acids can be converted to their sodium or potassium salts, and these taken up into a separate water layer prior to acidification and re-isolation free from neutral lipids. The hydrocarbon portions of the salts still confer some degree of non-polarity which may lead to difficulties in separation – soap foams, etc.

Reactive groupings like the carboxyl group are readily converted into derivatives leading to a further change in polarity, solubility, and molecular weight.

$$RCO_2H + CH_2N_2 \rightarrow RCO_2CH_3 + N_2\uparrow$$

Molecular weight is in itself an importance feature, for on the microscale it permits some separation by adsorption chromatography (in TLC, the higher molecular weight compounds have smaller R_f values – cover a shorter distance in the same time) and gas-liquid chromatography (here the lower vapor pressure of the heavier molecules leads to longer retention time – cover a shorter distance down the column in the same time). Molecular weights, and more specifically, molecular shape, are the main features controlling separation involving inclusion into the cavities of synthetic zeolites (molecular sieving), growing crystals such as urea and thiourea (clathration) and polymer networks (gel permeation chromatography).

Actual procedures and examples of the application of these principles are dealt with in Chapter 3. Some topics (e.g., gas chromatography, Chapter 5) are given detailed treatment because of their great practical importance. To summarize, micro amounts of lipids are handled by carrying them as dilute solutions in a

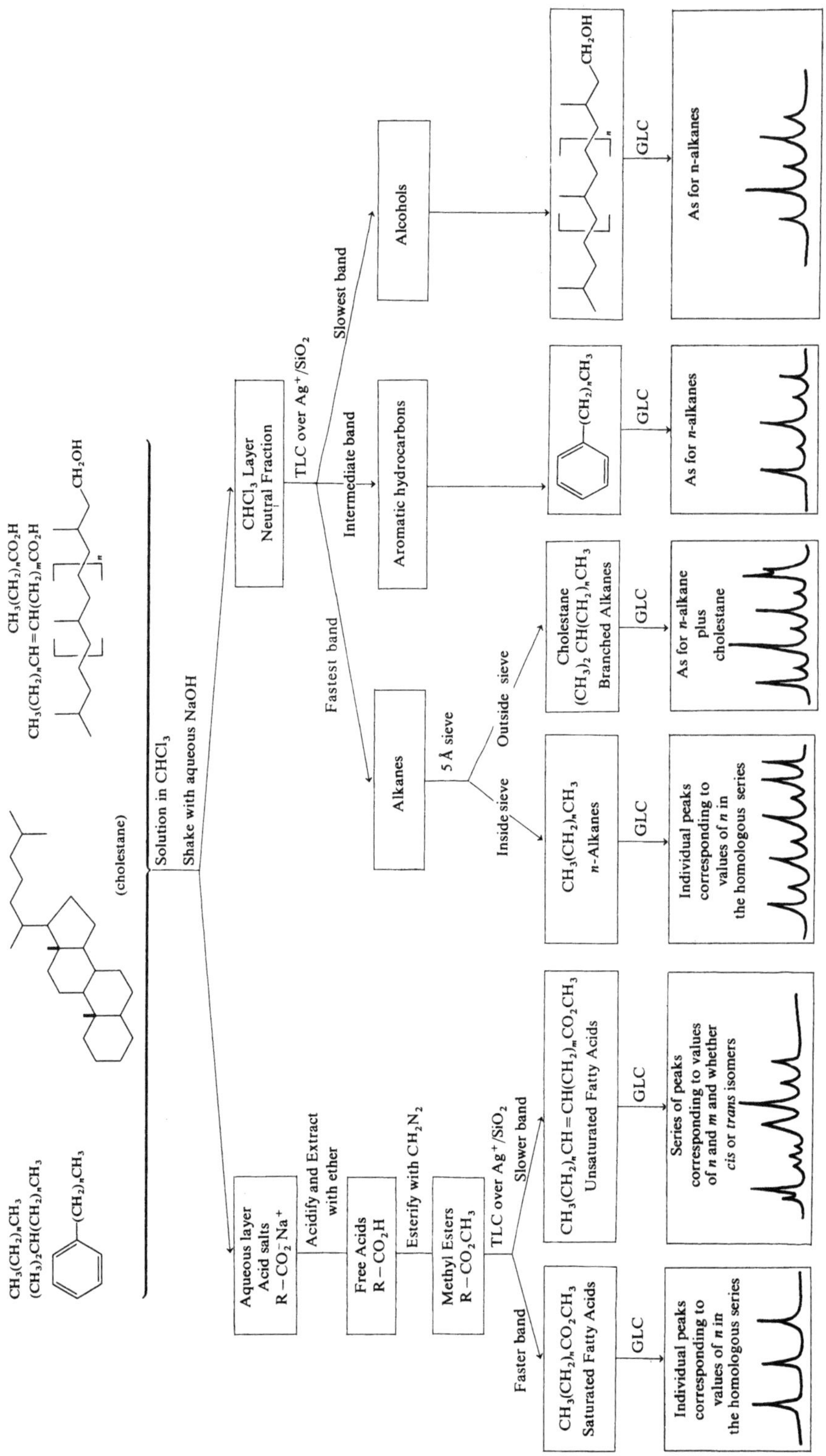

Fig. 7. Example of a separation scheme for a trial mixture of lipids

variety of organic solvents; they are separated into fractions based on the nature and number of polar or hetero atom groupings, and certain considerations of molecular size and shape; chromatographic procedures, especially gas-liquid chromatography, then further subdivide these fractions – frequently as classes of compounds or their derivatives – according to a number of molecular characteristics, but largely by molecular weight.

Fig. 7 illustrates one example of a separation procedure suitable for a particular mixture of lipids. At the final stage, gas-liquid chromatography produces a series of individual peaks corresponding to the individual homologues of that particular fraction or class of compounds.

One alternative to the above kind of separation scheme, which is effectively based on separation according to classes of compounds, is to bring the whole mixture into a condition suitable for gas chromatography. Thus, the mixture examined in the figure could be treated with diazomethane (acids → methyl esters) and then with silylating agent [alcohols → trimethylsilyl derivatives (TMSi)]. The resulting mixture of hydrocarbons, esters and trimethylsilyl ethers might then be analyzed by gas-liquid chromatography. The problem here is that, first, many peaks (corresponding to the very large number of compounds) would overlap, and second, no information would be available from other procedures (such as Ag^+/TLC) to indicate the class of compounds involved. As ORO has suggested, high resolution capillary GC–MS provides a partial answer [9]. Given very efficient columns and adequate information from the mass spectra, this procedure has the major advantages of simplicity and relative freedom from contamination.

B. Characterization and Identification of Lipids

The restrictions of applicability on the micro-scale that we have placed on our discussion so far, apply with particular emphasis to problems of lipid characterization and identification. (Characterization of an organic compound implies the revelation by experiment of some of its molecular features, such as reactive groups. Identification calls for recognition of the unknown as being identical in all respects to a known compound. It would also apply to the identification of a compound in terms of a particular structure.) Information on the constitution of a compound can be gathered in many ways, quite often during the isolation and separation procedures. Thus, much can be inferred from gas-liquid chromatographic retention times on different stationary phases. Thin layer chromatography can be similarly informative, especially where standard compounds are run for direct comparison on the same plate. Subjection of the unknown to reaction conditions known to bring about certain functional group changes (e.g., $-CO_2H \rightarrow CO_2Me$) is very effective when carried out in conjunction with these chromatographic procedures.

Spectrometric methods are undoubtedly the key to structure determination. Table 1 summarizes the situation as it is presently. The composite technique of sequential GC–MS is both powerful and extremely sensitive, and its use is discussed in several chapters (e.g., Chapters 4 and 5).

Table 1. *Summary of criteria for compound identification, and quantities needed for measurement*

Criteria	Measurements	Molecular parameter measured	State during examination	Minimum quantities required (gm)[a]
Chromatographic				
Thin layer chromatography (TLC)	Distance travelled on plate (R_f value)	Solubility and adsorption behavior, molecular size	In solution	10^{-9}
Gas-liquid chromatography (GLC or GC)	Time taken to emerge (RT)	Molecular size, solubility, and adsorption effects	Gas phase/solution in liquid phase	10^{-9}
Spectrometric				
Ultraviolet and visible (UV and Vis)	Absorption maxima (λ max) and Intensity (ε)	Excitation of non-bonded-electrons and conjugated systems	In solution	10^{-6}[b]
Infrared (IR)	Complex spectrum	Absorption of energy of vibration of interatomic bonds	In solution, liquid state or solid	10^{-4}
Nuclear magnetic resonance (NMR)	Complex spectrum	Interaction of the nuclear spins of protons	In solution or liquid state	10^{-3}
Optical rotation (ORD, CD)	Single value or curve	Degree of asymmetry of molecule and consequent effect on polarized light	In solution or liquid state	10^{-3}
Mass spectrometry (MS)	Complex spectrum	Ease of fragmentation of molecule and stability of fragments	Gas phase	10^{-9}
Other				
Melting Point (M. pt.)	Single value	Sum of all interactions in crystal lattice	Solid	10^{-4}
X-ray structure (X-ray)	Complex intensity pattern	Electron density in unit cell	Solid	10^{-4}[c]

[a] Without specialized micro equipment.
[b] Strongly absorbing molecules.
[c] Suitable crystal.

As we have already seen in this chapter, full characterization of an organic compound calls for definition of the stereochemistry in addition to the gross structure. Ideally, one would hope to completely separate a single pure lipid component by gas chromatography, collect it from the exit port of the chromatograph, and then subject it to the full set of spectrometric procedures. Invariably, if there is enough for GC, then there is enough for GC–MS. Unfortunately, it is often difficult to obtain sufficient lipid for IR, NMR, etc., and this means that potentially very valuable positional and stereochemical information is unavailable. Judicious use of derivative formation and of GLC phases of differing character can make up for this difficulty somewhat. Later in this chapter, this approach is applied to the isoprenoid acids. Very occasionally, enough of a compound having crystals of the right symmetry, etc., is available, and full X-ray structural analysis becomes possible, and the complete structure is revealed unambiguously (see Chapter 27).

IV. Special Significance of Molecular Structure in Relating Geolipids to Biolipids

A. Carbon Skeleton Philosophy. Chemical Fossils [11] and Biological Markers

We have seen that the tetrahedral carbon atom confers on organic compounds the possibility of great diversity in structure, and consequently, in the physical and chemical properties of the individual compounds. A large number of individual compounds may be structurally isomeric, possessing the same empirical formula; furthermore, where one (or more) carbon atoms is asymmetrically substituted, then the substance may exist as an enantiomeric pair of optical isomers.

Advances in technique now permit the study of complex mixtures of organic compounds at the molecular level. The type of organic geochemistry we are discussing is concerned with the complete characterization of individual organic compounds, including relative stereochemistry and, where possible, absolute stereochemistry. The more complete the characterization of the compound, then the more likely is a useful correlation with its past history obtained.

All living things use carbon as a building unit. The C–C single bond is strong, and therefore, portions of the original molecules – the carbon skeletons – often survive passage through several organisms or incorporation into sediments. Emphasis is therefore on the carbon skeleton of organic compounds (in the sense that "organic" means compounds containing carbon atoms – it does not imply a biological origin). Approximate calculations give a lifetime of about 10^9 years for alkanes subjected to 150° C in the absence of catalysts. Hence, biological information represented by a particular carbon skeleton should be preserved over very long periods of time [12].

The biosynthetic systems, involving enzymes, coenzymes, etc., of living organisms produce a great variety of carbon compounds, possessing a range of functional groups. However, the carbon skeletons of these compounds are assembled according to a very small number of basic processes – acetate biosynthesis, isoprenoid biosynthesis, etc. [Section V.A] – and the types of carbon skeletons formed are correspondingly few; thus, most biological fatty acids are straight chain, most biological polyene pigments have a highly branched (methyl substituted) carbon chain made up of n-repeating isoprenoid (C_5) units, and so on. Hence, if we could perform a perhydrogenation experiment on a single living organism (such as an alga), we would find that the lipid constituents would now be represented in our analysis by an alkane fraction comprised very largely of n-alkanes, branched alkanes bearing two or three methyl substituents, families of acyclic isoprenoid alkanes and cyclic isoprenoid alkanes (terpanes, sesquiterpanes, diterpanes, steranes, and triterpanes and tetraterpanes). Such an experiment could provide a useful chemotaxonomic tool for the classification of organisms, since it is a "data reduction" step, and would therefore permit a more rapid and extensive survey. Most organisms would give the above range of compounds, but qualitative and quantitative differences would reflect their own biolipid idiosyncrasies.

To some extent, it would appear from the limited studies available that the anaerobic reducing type of sedimentary environment affects such a carbon skeleton experiment. Certainly, biolipids already possessing reduced structures, such as hydrocarbons, survive well. Most Recent (the last 10,000 years) and some ancient sediments contain hydrocarbon fractions to some extent resembling those of higher plants. More significantly, the same extracts contain other alkanes and cycloalkanes so far not detected in any plant or animal but which have carbon skeletons identical with or closely resembling those of known unsaturated, oxygen- and nitrogen-containing constituents of living organisms. It is in this sense that the concept of preservation of organic compounds as "chemical fossils" has arisen. In fact, unreduced compounds, such as amino acids, sugars, fatty acids, etc., also appear as chemical fossils and the term encompasses any organic compounds ultimately of biological origin. The terms "chemical fossil" and "biological marker" are practically synonymous in that the presence of such a compound can be taken as indicating a biological origin for at least part of the material. The characteristics required of a biological marker include stability and a sufficient uniqueness of structure such that it is difficult to envisage other than a biological origin for the compound when encountered in quantity in Nature. Specific correlations involving particular compounds occuring in organisms of particular phyla, classes, orders, families, genera or species may be possible, as in chemotaxonomy (Section V.C).

Studies aimed at direct and firm correlations between biolipids and geolipids are in their infancy, but we can perceive some important principles. Thus, we can treat the conversion of the hypothetical biolipid precursor to the geolipid product in terms of biological information destroyed by alteration or partial destruction of the carbon skeleton and by removal of hetero atoms or unsaturation. Again, the *absence* of many other homologous, isomeric, or related compounds can be very significant and affect the status of the compound under study as a biological marker. The biolipid-geolipid relationship is subject to experimental study in several ways. For example, we might (1) survey present-day ecosystems for the turnover of organic compounds and then predict or establish which organisms make significant contributions to a forming sediment; (2) we could conduct experiments in suitable microenvironments with radio-labelled compounds and whole organisms and then follow the path of the products of diagenesis; (3) study the geolipid fractions, identifying specific compounds and inferring from their structures appropriate, but hypothetical, biolipid precursors. Only this latter approach is presently represented in the literature to any extent.

Some specific examples of the application of the proposed biolipid-geolipid relationship are given in the sections which follow.

B. Possible Relationships between Biolipids and Geolipids. Alkanes of the Formula $C_{31}H_{64}(C_nH_{2n+2}; n=31)$

So far, extensive studies have failed to reveal more than three biologically-produced C_{31} alkanes – the normal alkane and the 2- and 3-methyl isomers (Fig. 8). Millions of structural isomers are possible, but only these three are known to be produced by living organisms*. This high degree of specificity in biosynthesis contrasts with the very low specificity of abiogenic processes for alkane synthesis, which in the laboratory generally produce very complex mixtures. For most plants, anyway, the normal alkane, n-C_{31} greatly exceeds in abundance the iso-C_{31}, and this in turn exceeds the abundance of the anteiso-C_{31}. This sequence is characteristic for odd carbon number alkanes, but even carbon number alkanes show a reversal in the relative abundance of the iso and anteiso hydrocarbons.

(XII) 

(XIII) 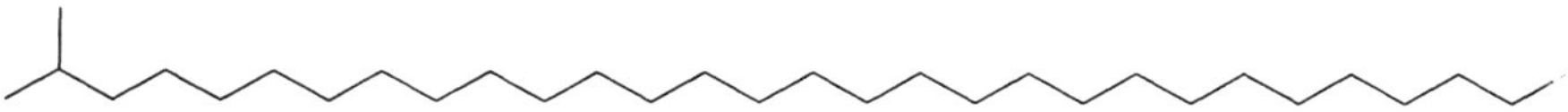

(XIV) 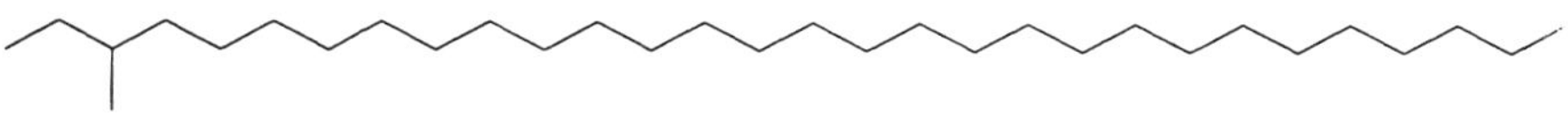

Fig. 8. Skeletal formulations for the three positionally isomeric alkanes, $C_{31}H_{64}$, n-hentriacontane (XII), 2-methytriacontane (iso-; XIII) and 3-methyltriacontane (anteiso-; XIV). The zig-zag arrangement is largely for convenience of representation: the molecules are flexible and free to take up an almost unlimited number of shapes (conformations)

Much more obvious is the regular alternation of abundance of the n-alkanes with respect to carbon number. For example, for higher plants, major amounts of n-C_{27}, C_{29}, C_{31}, and C_{33} and minor amounts of n-C_{26}, C_{28}, C_{30}, C_{32}, and C_{34} are encountered. The maximum of the distribution centers around C_{31} in most cases. The biosynthetic rationalization of these observations will be discussed in Section V.A. Our present concern is to compare biological (higher plant) alkane patterns with the alkane patterns found for Recent and ancient sediments. We might expect that alkanes with their stable and unreactive structures should be good biological markers, and the large number of carbons (*circa* 30 in these cases) in each molecule means that the number of isomers possible is extremely large. The retention of biological information is then recognizable as (1) the presence of the particular normal and branched alkanes of the appropriate carbon number and structure; (2) the abundance of these alkanes in the appropriate proportions of odd over even carbon number; and (3) the absence of the vast number of isomeric compounds expected for most abiogenic processes.

Most Recent sediments and some ancient sediments from both freshwater and marine environments show alkane patterns similar to that described above, thereby justifying the role of biological marker for the alkanes. However, some ancient sediments, especially those which have been subjected to some alteration or transport, such as crude oils, coals, and the older paleozoic sediments, show different patterns. The higher alkanes are much less prominent; the odd over

* There are a few exceptions [13].

even carbon number dominance of the n-alkanes is weak or non-existent and a much wider variety of isomeric alkanes is present. There is a lack of general agreement at this point, but we hold to the view that the principal cause of these differences is the thermal alteration of the original organic content of the sediments.

C. Possible Relationships between Biolipids and Geolipids. Cycloalkanes of the Formula $C_{27}H_{48}(C_nH_{2n-6}; n=27)$

This formula requires four rings of carbon atoms and again a vast number of structural isomers are theoretically possible. No $C_{27}H_{48}$ hydrocarbon is yet known as a plant or animal product, but oxygenated and more unsaturated compounds with the same number of carbons are biosynthesized in quantity.

(XV)

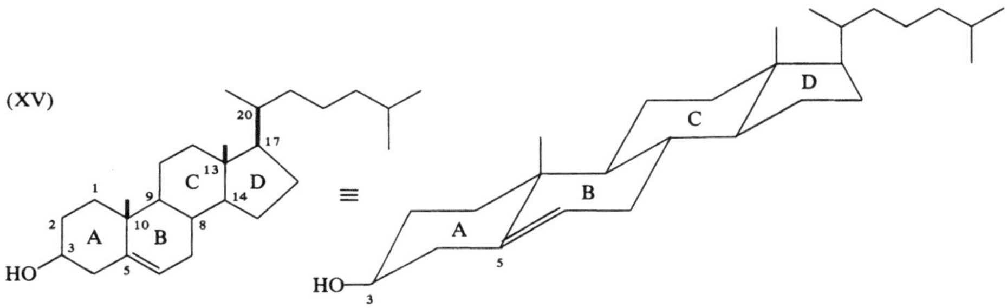

(XVI)

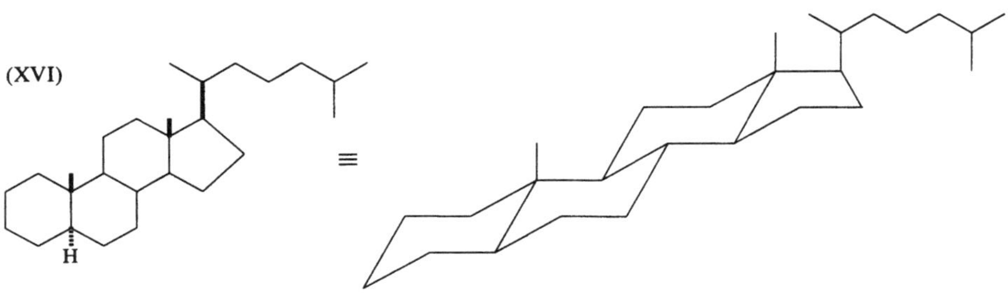

(XVII)

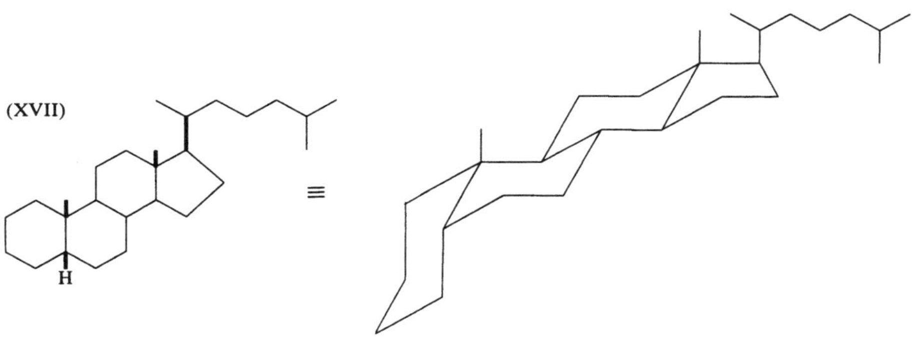

Fig. 9. Skeletal formulae and perspective representations of cholesterol (3β-hydroxycholest-5-ene; XV) and of the diastereomeric $C_{27}H_{48}$ cycloalkanes, 5α-cholestane (XVI) and coprostane (5β-cholestane; XVIII). (Heavy lines in the skeletal formulae indicate bonds directed *above* the plane of the ring and dashed lines bonds directed *below* the ring: Part of the steroidal ring labelling and numbering scheme is also indicated). The system of fused rings is fairly rigid but the side chain is free to rotate, fold up, etc.

One example is the steroid cholesterol, infamous as a constituent of the fatty deposits investing diseased human arteries. The *absolute stereochemistry* of the molecule is indicated for the single, naturally-occurring, optically active enantiomer (Fig. 9). There are eight asymmetric centers in the molecule (positions 3, 8, 9, 10, 13, 14, 17, 20) and hence 2^7 (=128) pairs of enantiomers, stereochemically different but with the same overall skeleton of four fused alicyclic rings, substituent hydroxyl and methyl groups and alkyl side chain. The enzymes producing cholesterol direct the biosynthesis so specifically that not only is the single gross skeletal structure selected out of the vast number possible, but also, the stereochemical choice is limited to one out of the 256 possibilities. Part of this control is inherent in the structure of the already biologically pre-formed molecule squalene epoxide, which undergoes cyclization followed by degradation to lead to cholesterol [14]. It seems probable that the flexible squalene epoxide molecule, $C_{30}H_{50}O$, is energetically most stable in folded conformations suited to cyclization processes which would produce the right ring fusions and stereochemistry.

Steroids have been detected in Recent sediments, but ancient sediments contain cycloalkanes which presumably derive by reduction of the original biologically-produced steroids. Two cycloalkanes, $C_{27}H_{48}$, have been identified, cholestane and coprostane (Fig. 9). The absolute stereochemistries of these two geolipids are not known, e.g., we do not know which is present, (+)-cholestane or (−)-cholestane. Cholestane (XVI) and coprostane (XVII) differ only in the stereochemistry of the A/B ring fusion and might well have arisen by natural (geological) reduction of cholesterol, since the same carbon atom (5) is there involved in a double bond (5,6). In the laboratory, simple chemical reduction of cholesterol furnishes a mixture of the two isomers.

Cholestane and coprostane can be regarded as biological markers since their skeletons can be directly related to that of a compound, cholesterol, of known biological origin. This type of rigid, complex, yet highly stable skeleton with its many centers of asymmetry is well suited to being a biological marker. Once such a structure has been assembled, drastic treatment is needed to rearrange it or break it down. If one could make the presently unjustifiable assumption that the cholestane and coprostane derive solely from cholesterol by reduction in the sediments, then certain generalizations concerning the retention and loss of biological information follow: (1) positional substitution (i.e., the gross carbon skeleton) is retained; (2) seven asymmetric centers have retained their relative stereochemistry (and, presumably, their absolute stereochemistry); (3) the information represented by the 3β-hydroxyl of cholesterol has been lost completely; (4) the presence of the 5,6-double bond in the biolipid precursor, cholesterol, can be tentatively inferred from the presence of the pair of diastereomers, cholestane and coprostane, differing in their stereochemistry at C_5.

D. Possible Relationships between Biolipids and Geolipids. Branched Chain Carboxylic Acids of the Isoprenoid Type

The long alkyl side chain of the chlorophyll molecule (XVIII) is provided by the C_{20} unsaturated alcohol, phytol (XIX) (Fig. 10). This compound has two asymmetric carbons and natural phytol is an optically active substance, a single enantiomer, with the D-configuration at carbons 7 and 11, i.e., it is 7D, 11D-. Catalytic reduction will provide dihydrophytol as a mixture of two diastereomers, 3D, 7D, 11D (XX) and 3L, 7D, 11D- (XXI). Any stereospecificity in the reduction would, however, give unequal mixtures of the diastereomers or even, in the extreme case, only one.

Oxidative degradation of the diastereomeric pair of dihydrophytols would be expected to give rise to a regular series of methyl-branched carboxylic acids of

(XVIII)

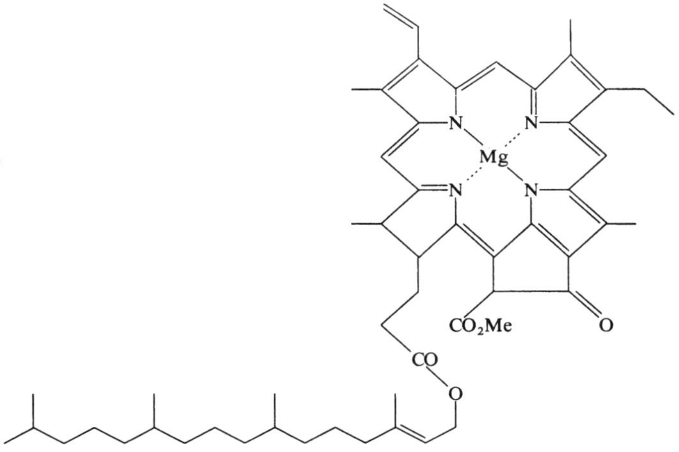

(XIX)

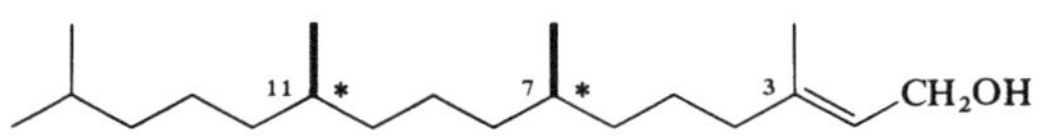

(XX)

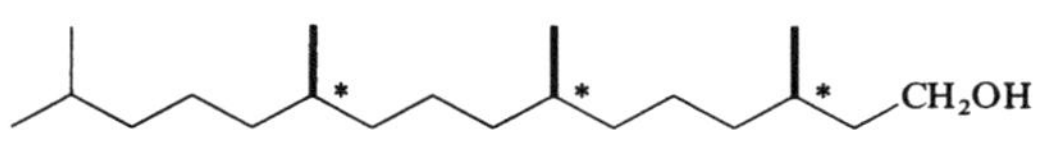

(XXI)

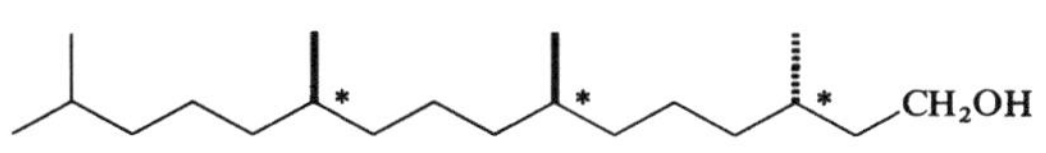

Fig. 10. Skeletal formulae for chlorophyll *a* (XVIII), natural phytol ($C_{20}H_{40}O$; 7D, 11D-; XIX), and two diastereomers ($C_{20}H_{42}O$) of reduced natural phytol, dihydrophytol (3D, 7D, 11D-, XX and 3L, 7D, 11D-, XXI). The asterisks indicate asymmetric carbon atoms, thickened bonds being directed above the plane of the paper and hatched ones below. The zig-zag arrangement is largely for convenience of representations: the molecules are flexible and free to take up an almost unlimited number of shapes (conformations)

the isoprenoid type. The higher members (C_{20} and C_{19} and C_{17} to C_{14}) are shown in Fig. 11; (XXII–XXVII). The series is not, strictly speaking, a homologous series for the homology (i.e., regular repetition of units) is not complete. However, the righthand end of the molecule can be regarded as being systematically oxidized away. If we assume that the asymmetric centers (other than the first one, i.e., carbon 3 of dihydrophytol) are not affected by the oxidation process, then the number of stereoisomers present, and theoretically capable of identification, would be as follows: (XXII) and (XXIII) – 2 each, DDD and LDD; (XXIV) to (XXVII) – 1 each, DD. However, in compounds (XXIII) and (XXVII) one of the asymmetric carbon atoms is alpha to the carboxylic acid group and, therefore, is subject to ready epimerization; if this process should occur, the numbers would remain the same, except for XXVII, when two diastereoisomers (2D, 6D and 2L, 6D) would again be generated. Actually, if we assume that the initial dihydrophytol molecules were a mixture of all possible stereoisomers (for example, when synthesized by non-

(XXII)

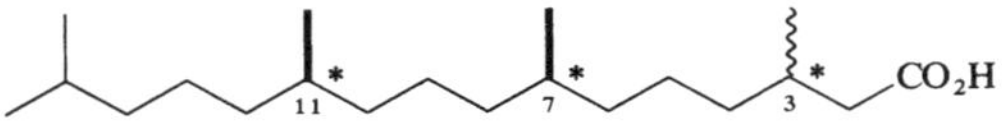

(XXIII)

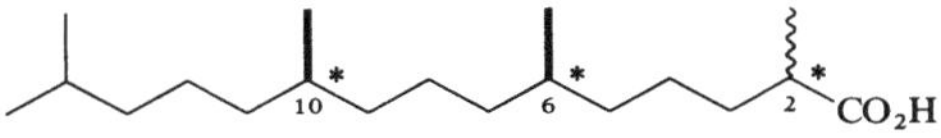

(XXIV)

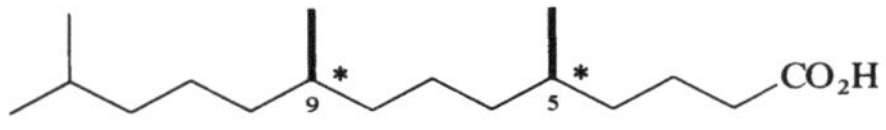

(XXV)

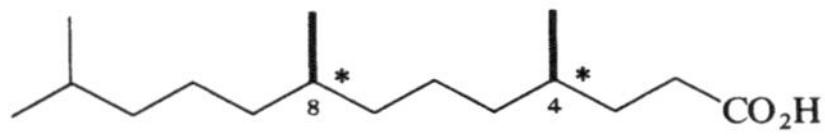

(XXVI)

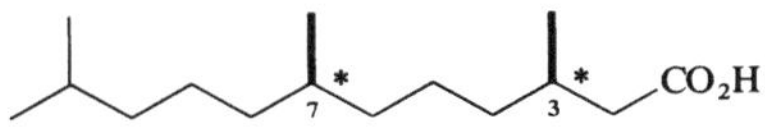

(XXVII)

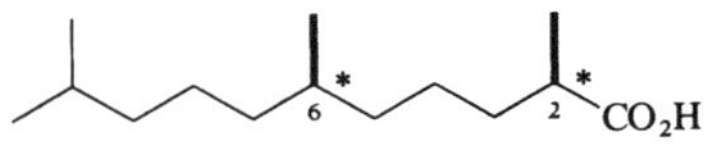

Fig. 11. Series of carboxylic acids of isoprenoid type, theoretically derivable by progressive oxidation of dihydrophytol, (XX) and (XXI). The skeletal formulae correspond to the following acids: (XXII) 3,7,11,15-tetramethylhexadecanoic acid (phytanic), $C_{20}H_{40}O_2$; (XXIII) 2,6,10,14-tetramethylpentadecanoic acid (pristanic or norphytanic), $C_{19}H_{38}O_2$; (XXIV) 5,9,13-trimethyltetradecanoic acid, $C_{17}H_{34}O_2$; (XXV) 4,8,12-trimethyltridecanoic acid, $C_{16}H_{32}O_2$; (XXVI) 3,7,11-trimethyldodecanoic acid (farnesanic), $C_{15}H_{30}O_2$; (XXVII) 2,6,10-trimethylundecanoic acid (norfarnesanic), $C_{15}H_{30}O_2$. (For presentation see legend for Fig. 10. The irregular bond in XXII and XXIII indicates that the methyl group can have either stereochemistry.)

stereospecific processes in the laboratory) then the number of stereoisomers (2^n, where n = the number of asymmetric centers) would be as follows: (XXII) and (XXIII) 8 (i.e., 4 enantiomeric (±)-pairs); (XXIV) to (XXVII) 4 (i.e., 2 enantiomeric (±) pairs).

Full definition of the molecular architecture of these acids requires identification of the gross carbon skeleton and the stereochemical characterization of each asymmetric carbon atom. For instance, the terminology is such that molecules of (XXII) with the centers being 3D, 7D, and 11D are said to have the same *relative stereochemistry* as do the enantiomeric molecules bearing centers 3L, 7L and 11L; the two enantiomers have opposite *absolute configurations.*

To return to the geolipids, we know little of the present-day fate of the vast quantity of chlorophyll biosynthesized by plants. A tiny proportion of the total plant debris is preserved in sediments, the rest undergoing degradation by micro-organisms or being consumed by animals. BLUMER, and others [15, 16], have obtained some information for the marine environment. For example, the important biological marker, the C_{19} hydrocarbon, pristane (Fig. 12, XXVIII), is found in certain copepods which feed on the phytoplankton. SEVER and PARKER [17] have very recently identified the reduced alcohol, dihydrophytol, in algal mats from a marine bay, other Recent sediments and in ancient sediments. This alcohol

(XXVIII)

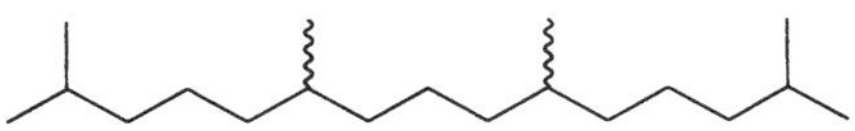

(XXIX)

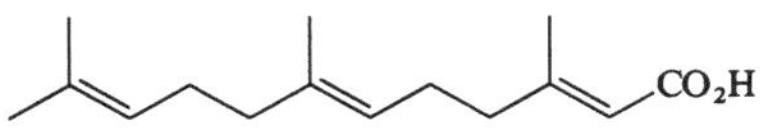

(XXX)

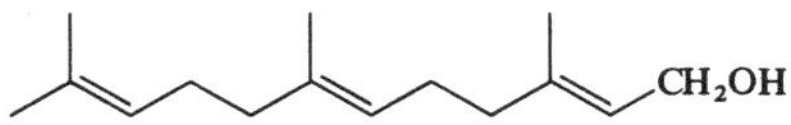

Fig. 12. Skeletal formulae for the norisoprenoid hydrocarbon pristane ($C_{19}H_{40}$, XXVIII), the isoprenoid acid, farnesoic acid ($C_{15}H_{24}O_2$, XXIX) and the isoprenoid alcohol, farnesol ($C_{15}H_{26}O$, XXX)

presumably represents an early stage in the diagenesis of phytol and its stereochemistry awaits study.

The isoprenoid carboxylic acids detailed above have been detected in some organisms and in several Recent and ancient sediments. We do not know if they are incorporated directly into sediments or formed later. Some information is available on the relative stereochemistries of the fossil acids isolated from the 50 million year old Green River shale. These acids bear two centers which display the same relative stereochemistry as do the centers removed from the double bond in phytol, where they are D,D. We do not yet know the absolute stereochemistry (D,D or L,L). Relative to phytol, the biological information has been retained in so far as:

(1) gross structure (placement of methyls on the carbon chain);

(2) relative stereochemistry at C-7 and C-11 (phytol numbering);

(3) the presence of two diastereoisomers of the C_{20} acid (XXII), possibly indicating non-specific reduction of the 2, 3 double bond. [The situation for the acids (XXIII) and (XXVII) is rendered uncertain because the asymmetric α-carbon atom bears an easily epimerizable hydrogen];

(4) the C_{15} acid (XXVI) is an interesting case. It can be regarded as the reduced form of farnesoic acid (XXIX), which can be derived from the biologically-abundant alcohol farnesol (XXX). If the acid (XXVI), isolated from a sediment, had been derived from farnesol by reduction and oxidation, then one could expect

a number of diastereomers. Only one diastereomer is found in the Green River shale extract, thus arguing in favor of a phytol derivation. The isolation and characterization of these acids is discussed further in Section VII.C.

For want of a better starting point, if we continue to make the assumption that the series of isoprenoid acids discussed above are derived from phytol, then the results again illustrate the principles outlined in preceding sections [Sections IV.A–C]. Thus, gross positional structure is retained, and in addition, the stereochemical data clearly indicate when centers have been retained unchanged, where new ones have been created, and others epimerized. Significantly, when portions of the presumed biological precursor molecule have been removed, probably by oxidation processes, the remaining portions are still capable of carrying the original information in the form of positional and stereochemical isomerism. Of course, only the first tentative correlations relating to the fate of isoprenoids in natural environments and sediments are being made. The true relationships are certain to be highly complex. Isoprenoids are ubiquitous and a fair number of classes, such as carotenoids and prenols may contribute degraded and reduced isoprenoids to the environment or sediment.

E. Possible Relationships between Biolipids and Geolipids. Biopolymers. Proteins and Polypeptides

The macromolecules carrying the genetic information, (i.e., the nucleic acids) undoubtedly have the largest information content of all the different types of organic compounds in living organisms. The information is encoded in the sequence of the four DNA bases, connected in a series of many millions of nucleotide units. These molecules are labile and there seems to be little chance of ever recovering more than a few segments of sequentially connected units from geological materials. There is no report of any such isolation.

The next most important source of information is the precise sequence of amino acids – some twenty-odd different ones – in proteins. Some proteins are structural materials but more significant for our purposes are the enzymes, the essential catalysts used by all cells. The synthesis of the enzymes is controlled by the DNA message, *via* the messenger RNA and the other equipment for information transfer. The enzymes in turn control the stereospecific syntheses of the secondary metabolites, such as the hydrocarbons and fatty acids, and the steroids and other isoprenoids discussed in the preceding sections. Hence, in identifying the secondary metabolites in a biological system, we are indirectly reading part of the information content of the enzyme and, in turn, a very small part of the genetic information encoded in the DNA (though the redundancy in the code would not permit unambiguous conclusions, should we ever attempt to go that far).

We have seen how the whole or a part of a molecule can be treated as a source of information. In the relatively small molecules of the secondary metabolites, the information resides in the gross structure and in the stereochemical configuration. Both are aspects of the carbon skeleton. In the following paragraphs, the identification of unaltered polypeptides and proteins from biological sources will be discussed and the information content contrasted with that to be derived from the degraded materials encountered in sediments and fossils.

In general, the primary structure of a polypeptide or protein (the particular sequence of amino acids) determines the formation of any covalent cross links, hydrogen bonds, hydrophobic (non-polar) group associations, and dipolar charge interactions between amino acid units in different portions of the chain. The correct amino acid sequence, therefore, automatically determines the shape of the eventual, biologically-active entity, which may involve secondary, tertiary and higher, structural orders of spirals, coils, etc. Turning to the geochemical possibilities, we have first to recognize that the protein content of most organisms is rapidly destroyed by microorganisms soon after death. The peptide links are fairly readily hydrolized, and the liberated amino acids are degraded further by deamination and decarboxylation. However, careful examination of some fossil shells, teeth and bones under the electron microscope has revealed organic material closely resembling the original protein layers (Chapter 20) in its relationship with the mineral matrix. Fossil shells have particularly good preservation of microstructure where the "protein" interleaves the calcium carbonate layers, as does mortar in a brick wall. Even here, results from several laboratories indicate that the state of molecular preservation may be relatively poor (Chapter 18), substantial hydrolysis, together with loss and structural alteration of individual units, having occurred even after a few thousand years.

The type of experiment which would seem to be desirable, though it may not be feasible, is outlined in Fig. 13. The individual fractions of the polypeptide material might then be studied in several different ways.

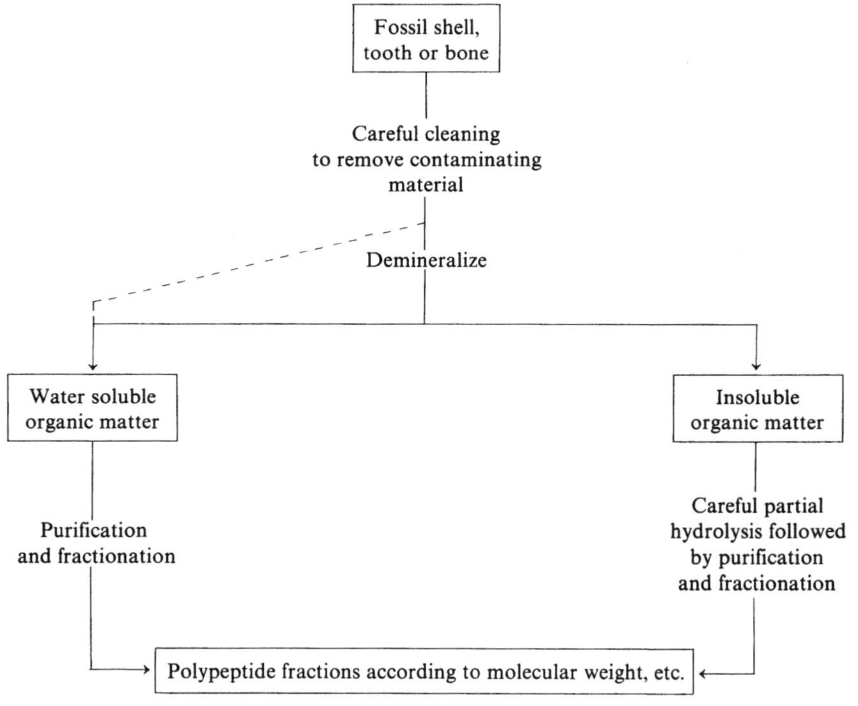

Fig. 13. Projected isolation scheme for polypeptide fractions from fossil shell, tooth or bone

Let us assume that we are attempting to establish the degree of preservation in such material of a single nonapeptide, oxytocin, which is an animal hormone of known structure (Fig. 14), fully confirmed by laboratory synthesis. The identity, and hence for our purposes, the full "information content" of this biopolymer is contained in (a) the absolute configuration of each amino acid residue, and (b) the gross structure, which includes the precise sequence of these residues. Assuming we have one fraction comprising the pure nonapeptide, then we can attempt:

(1) Full identification of the complete, biologically-active nonapeptide. We might proceed by first establishing the presence of a single molecular species and then measuring its molecular weight. We could go on to determine the precise

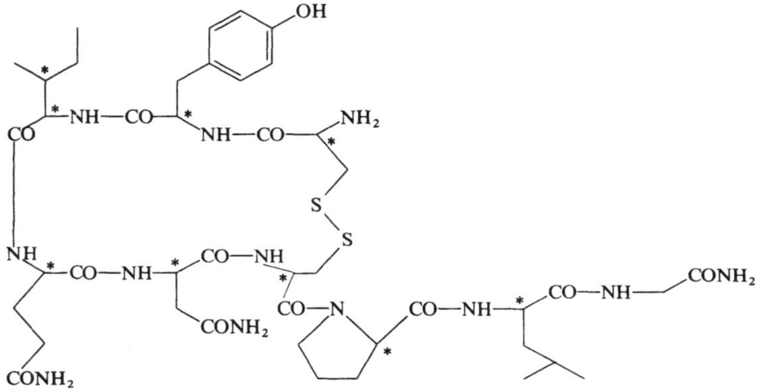

Oxytocin

Fig. 14. Skeletal formula for the nonapeptide, oxytocin. Amino acids are all of the L configuration. Asymmetric carbons are indicated by a star (*). The conformation shown is purely schematic and is arranged in this way for convenience only

arrangement of the amino acids by undertaking a partial hydrolysis (enzymic or acidic conditions to avoid racemization), followed by a sequence determination involving of the resulting mixture of smaller peptides by terminal residue (end group) analysis. The absolute configuration (L) of each amino acid unit would have to be checked in each of the smaller peptides in the resulting mixture. [Alternatively, though probably more laboriously, a crystalline, heavy metal derivative, such as the zinc complex, might be subjected to an X-ray structural analysis].

(2) Incomplete identification as a result of ignoring some of the above criteria. Thus, partial hydrolysis of the fraction without first establishing its molecular weight and purity, followed by subsequent identification of the various oligopeptides released, would provide only limited information. Such data would not permit a reconstruction on paper of the original polypeptide and the yield of "information" is greatly diminished. Again, complete hydrolysis down to the individual amino acids, with or without characterization of configuration, would provide even less information.

The yield of information is much less when it is not possible to first isolate a biopolymer fraction having the appropriate chemical and physical properties (especially molecular weight). Present-day techniques commonly involve subjecting the whole sediment to hydrolysis prior to the amino acid determination.

Consequently, one cannot assign a given amino acid to any particular polypeptide or protein. In fact, it may even have been present as the free amino acid or as a condensation product with molecules of another type. The situation is alleviated somewhat where a single fossil is examined. Thus, shell proteins, such as conchiolins, are protected to some extent by the calcium carbonate matrix, and hydrolytic treatment of thoroughly cleaned fragments of fossil shells can be expected to yield amino acid compositions which bear some relationship to those of modern shells (Chapters 18, 20).

What are the mechanisms by which the information content of a biopolymer, such as oxytocin, may be degraded? Firstly, there is hydrolytic cleavage whereby the sequence of amino acids in the oligopeptides is destroyed and, if there is preferential loss of individual amino acids to the environment, the overall proportions of the amino acids will be changed. Secondly, there is a wide variety of reactions which can change the nature of functional groups (e.g., $-SH$ may be oxidized to $-SO_3H$). Amino acids are thereby altered and the original composition is difficult to infer. Another example would be the phenyl group undergoing oxidation to a phenolic group. Thirdly, asymmetric centers can be racemized. The stereochemistry at some carbon atoms is readily changed; thus, the asymmetric alpha-carbon of an alpha amino acid, when it bears a hydrogen atom, is readily epimerized. The equilibrium is made possible by the weakened C–H link, loss of H^+ resulting in a loss of the absolute stereochemistry (racemization). The same situation applies wherever a C–H link is alpha to a carbonyl group.

Equilibration can take place through the enolate anion or the enol form (Fig. 15a). The hydrogen on the beta position *is not activated* and the center is difficult to racemize. Hence, we can predict that of the two types of optically active acids shown (Fig. 15b), the alpha-substituted acid would suffer much more

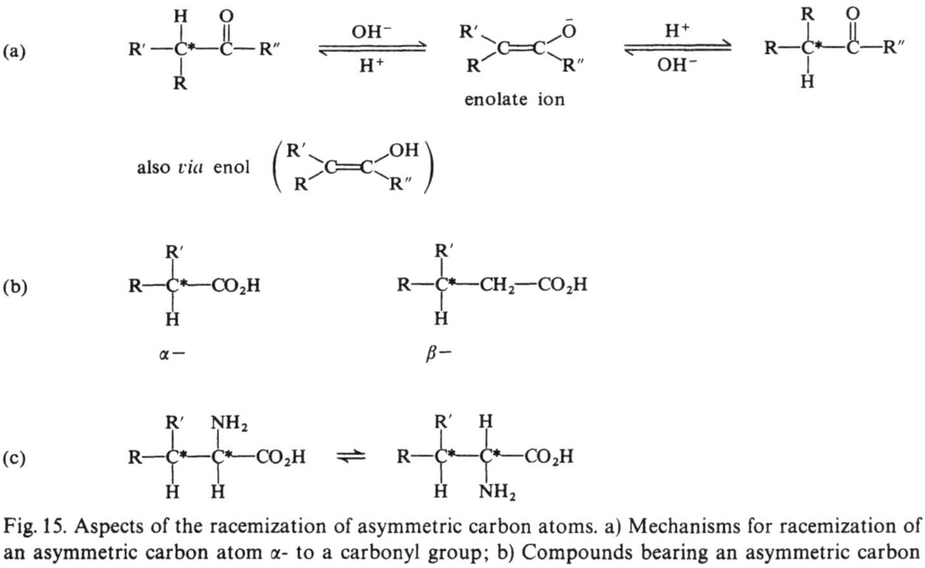

Fig. 15. Aspects of the racemization of asymmetric carbon atoms. a) Mechanisms for racemization of an asymmetric carbon atom α- to a carbonyl group; b) Compounds bearing an asymmetric carbon atom α- and β-, respectively, to a carboxylic acid group; c) Equilibration of two diastereomers, each bearing twin asymmetric centers. (The planar formulae shown are often used when depicting asymmetric centers)

rapid racemization. Again, where there are *two* asymmetric centers in the *same* molecule (Fig. 15c), then one may be much more rapidly racemized: a mixture of two stereoisomers results, the original substance and another diastereomer bearing the inverted labile center. The two isomers are not mirror images of each other and have slightly different physical and chemical properties and are formally capable of separation and independent identification. In such a case, the relative proportions of these two compounds in a fossil or sediment could provide information on the degree of Recent contamination and/or the age and past history of the sample. This point is well brought out by HARE in Chapter 18.

V. Biosynthetic Rationalization of Biolipid Patterns

A. Main Biosynthetic Pathways and Resulting Biological Patterns

Lipid fractions from biological sources generally contain a range of long chain hydrocarbons, fatty acids, alcohols, and esters. Though complex, these fractions are not random mixtures, but exhibit a high degree of order in that the molecules present reflect in their structures the chemical reaction paths systematically followed in the originating organisms. Two types of natural products [14], wherein long chains of carbon atoms are linked together, are the straight-chain lipids, $CH_3(CH_2)_nX$ (where $X = -CH_3$, $-CHO$, $-CH_2OH$, for example) and the acyclic isoprenoid lipids $H[CH_2CH(CH_3)CH_2CH_2]_nX$.

The straight-chain lipids originate biochemically in the enzymatically controlled addition of acetate units (C_2) to a C_2 starter, an acetyl derivative. The result is an homologous series of long-chain, fatty acids (n-alkanoic acids; XXXI) in which n usually varies from 8 to 30, depending on how many C_2 units are added (Fig. 16). Thus, if this biological, "polyacetate" or "polyketide", pathway is being followed we see a series of even-numbered fatty acids with the odd-numbered members missing. Long-chain alcohols (n-alkanols; XXXII) are also found and are believed to be formed by reduction of the fatty acids, as they are of even carbon number. However, the hydrocarbons (n-alkanes; XXXIII), which are of odd carbon number, are probably formed by decarboxylation of the acids or of an intermediate. There is, therefore, a definite carbon number pattern for the long-chain acids, alcohols and hydrocarbons formed by the polyacetate route. However, it should be mentioned that if a "3-carbon starter" were to be incorporated, then a similar pattern would appear, but the dominant carbon numbers would be odd, when previously even, and *vice-versa*; there is evidence that this incorporation often occurs to a small extent. Again, branched starter acids produce the branched 2- and 3-methyl compounds described earlier (Section IV,B). Branching at points along the chain is also observed. It is brought about either by methylation (methionine) of double bonds or by incorporation of C_3 units (derived from methylmalonyl coenzyme A) during chain lengthening. One example of each of these processes is provided (Fig. 16b and c, respectively). Acids and hydrocarbons of type (XXXIV–XXXVI) have been reported for bacteria and algae [18] and acids of type (XXXVII) from the preen glands of birds.

The isoprenoids have branched chains comprised of 5-carbon units, assembled in a regular sequence. In the living system the reactive C_5 derivative, isopentenyl

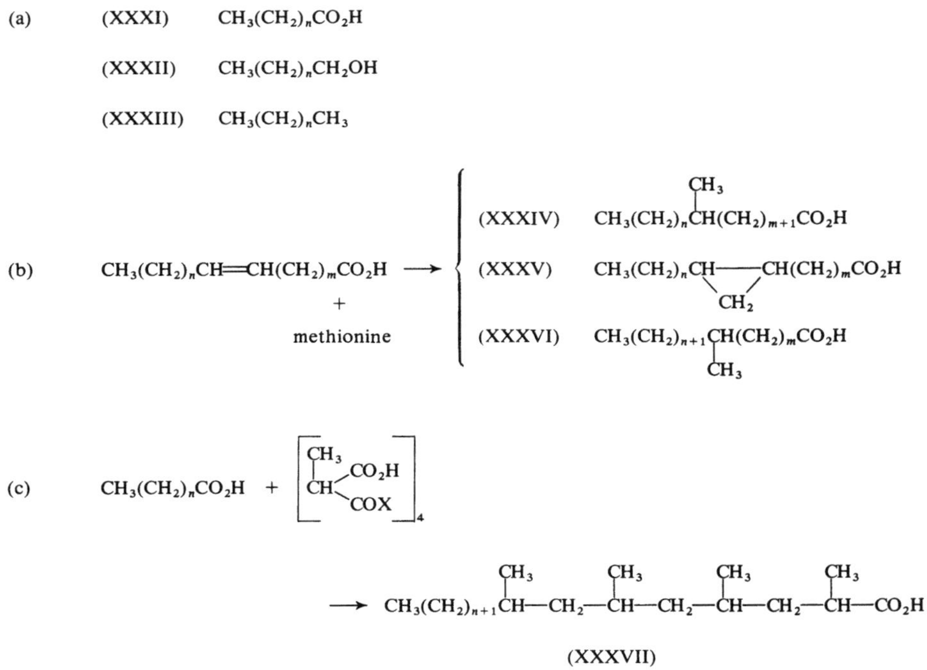

Fig. 16. Straight and branched-chain biolipids originated by the polyacetate type of biogenesis

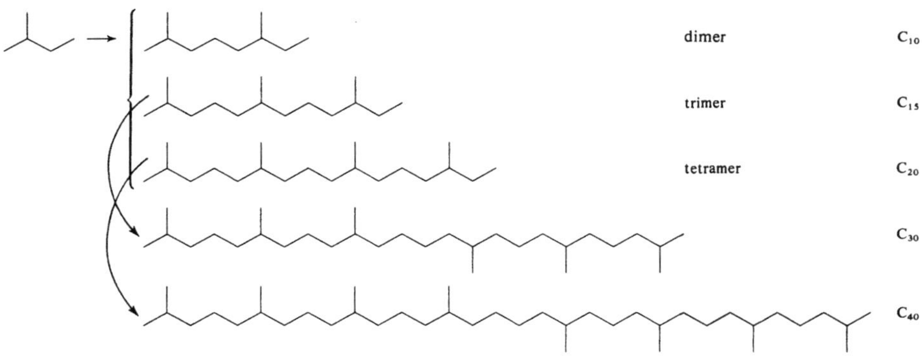

Fig. 17. Carbon skeletons for isoprenoid biolipids originated by the terpenoid type of biogenesis

pyrophosphate, which is derived from acetate *via* a C_6 compound (mevalonic acid), polymerizes under the influence of the enzyme and cofactors to form the C_{10} dimer (terpene), the C_{15} trimer (sesquiterpene) and the C_{20} tetramer (diterpene). The "head to tail" fashion of addition results in three methylene units between each pair of carbons bearing a methyl group. (Fig. 17). When this "terpenoid" skeleton is found in a naturally-occurring molecule, it is a very reasonable assumption that the compound has been formed by this particular biological pathway. Larger isoprenoid skeletons are mainly built up in living organisms by "tail to tail" coupling of two units: C_{15} units give C_{30} triterpenoids, and C_{20} give C_{40} tetraterpenoids. Thus, the very important group of compounds, the steroids, C_{27} to C_{29}, are degraded and alkylated triterpenoids. Rearrangements, generally involving movement of methyl groups and changes in ring size and fusion, further extend the range of carbon skeleton ultimately originating in the isoprenoid route. However, careful inspection of the structure and application of the so-called "biogenetic isoprenoid rules" often enables the origin to be discussed. Higher "head to tail" polymers are known, including the prenols reaching C_{110} and extending to the high polymer hydrocarbon, rubber.

B. "Natural Product" Chemistry

To the organic chemist, "natural product chemistry" connotes the study of the secondary metabolites of plants and animals, their structures, and the routes by which they are biosynthesized. Thus, most of our knowledge relating to the long-chain lipids and the multitude of terpenoids, briefly mentioned in the preceding section, is the result of the activities of natural product chemists. Their approach, and the methods characteristic of it, are now being extended to situations involving chemical exchange *between* organisms, for example, as with sex attractants and growth inhibitors. The range of activity so controlled extends to defense mechanisms, instinctive behavior patterns, communication and the attraction and repulsion of other organisms. The term "chemical ecology" has been coined for such chemical studies of the interrelation between organisms and of their interaction with their environments [19].

Organic geochemistry is also an extension of natural product studies, but encompasses more extensive areas of science, together with the whole span of geological time. The task is, in part, to trace the fate of biosynthesized organic compounds in natural environments and sediments. In the widest sense, we may try to express in molecular terms the path of carbon in Nature and to reassert thereby the original definition of Nature – the whole natural world around us, biosphere and geosphere. We thereby widen the view of natural product chemistry so that it includes the study of organic compounds wherever found in the natural world, whether they are inside or outside organisms, incorporated in igneous rocks as bitumen veins, present in fossils and sediments, or issuing from volcanoes as gasses. This definition seems both logical and useful, and renders the terms "biolipid" and "geolipid" of use as subdivisions of all natural lipids. The sphere of interest then covers the path of carbon – within single organisms, through communities of organisms, entombment in sediments and recycling as a result of weathering and metamorphic processes (Section VI, A).

By way of an example of this philosophy, the C_{19} isoprenoid hydrocarbon, pristane (Fig. 12, XXVIII) is one biological marker compound, the path of which is presently being tracked from its origin in the phytol (Fig. 10, XIX) of phytoplankton, through various organisms in the marine food web to interment in marine sediments [20, 21]. Baldwin and Braven [22] have recently proposed that an unusual lipid and protein constituent of marine phytoplankton might be another "biochemical integrator". This compound bears a carbon-phosphorus bond; it is 2-aminoethylphosphonic acid, $NH_2CH_2CH_2P{=}O(OH)_2$.

C. Chemotaxonomy, Paleochemotaxonomy, Chemosystematics, and Evolutionary Processes

The uniformity of present-day biochemistry, whereby the same biochemical pathways operate in all living organisms, means that there are many carbon compounds which are potential "biological markers". Organisms do differ, of course, in the relative amounts of these compounds they produce and, further, there are also many compounds which are produced only by members of particular groups of organisms—at the family, genus, species, variety, or even race, level. Such compounds permit chemotaxonomic (comparative biochemical) studies which form a valuable background for organic geochemistry. Thus, for Recent sediments, we can confidently extrapolate from the chemotaxonomic characteristics of contemporary plants and animals and draw inferences about the presence or absence of particular organisms in their original environments of deposition—using species and family markers, for example. To make one rather general point from the data presently available: it would appear that the pentacyclic triterpenoids are largely restricted to higher plants. Their presence in a sediment may be tentatively construed as indicating a contribution from higher plants [23, 24].

Conversely, analysis of ancient sediments, when properly interpreted, may provide us with paleochemotaxonomic data on extinct organisms. Instances of direct correlations between the organic constituents of present-day and fossil species have been provided by the work of KNOCHE and OURISSON [25, 26].

Chemotaxonomic studies provide one view of the results of evolution. The chemical constitution and products of an organism must surely reflect the evolutionary process. Chemosystematics, involving the careful interrelation of chemical data and biological data (morphology, genetic information, etc.) is presently making great progress. For example, the cross breeding of plants and the genetic interchange of individuals in a population can now be followed chemically [29–32]. It is too early to say yet whether organic geochemistry can do much to throw light on the problems of evolutionary biochemistry (Chapter 7). The fossil record provides a glimpse of the phylogenetic development of the present-day flora and fauna. Clearly, we should thus expect to find in ancient sediments and fossils those geolipids originating from biolipids appropriate to the type of organisms known to have evolved at that time. Thus, in the earliest Precambrian sediments, we should see geolipids corresponding to the metabolites of unicellular organisms [18, 33, 34]. The blue-green algae examined so far have hydrocarbon fractions dominated by the C_{17} n-alkane, thus providing a possible marker.

VI. Organic Chemistry of Natural Environments

A. Path of Carbon Compounds in Nature

One key to the understanding of the geolipid content of ancient sediments certainly lies in the lipids of present-day environments and Recent sediments. Most, but certainly not all, paleoenvironments have a contemporary equivalent—at least in gross features. "Recent" in the geological sense encompasses the last

10,000 years, or thereabouts, and "ancient", the whole span of time prior to that. Our planet has on, and in, its surface today a great range of situations which can be classified and recognized as distinct "environments" of different kinds [35, 36]. The passage of carbon compounds from individual organisms, through ecosystems and finally into sediments, is subject to direct observation and experiment. The resulting data may then be used to interpret the organic content of ancient sediments. Such an approach would be normal geological practice – the application of the principle of uniformitarianism, but as with other geological matters, it is clear that we must now think in planetary terms, with interrelated events spread over millions of years.

The time is ripe for such an appraisal of the path of carbon compounds in Nature. The separation and identification procedures now to hand provide us with the means to examine the complex mixtures of organic compounds encountered in natural environments and in sediments. The products of a single species of plant or animal no longer represent the feasible limit for effective study. In addition, radiochemical labelling techniques are available which will permit experiments designed to follow the path of a given compound, and its derived products, through an environment. Undoubtedly, all of the numerous organic compounds produced by living organisms are not preserved in the sediments being deposited from most present-day environments. If they were, the analytical situation would be an impossible one, with our present tools. Most water-soluble compounds seem to undergo selective destruction, presumably brought about by bacterial attack and chemical reactions in the system. More than half of the present-day biomass of this planet is said to be bacterial [31]. The need for rapid bacterial degradation of lipids during sewage treatment and water purification became particularly obvious some years ago with the widespread use of synthetic detergents. The highly branched alkylated benzene derivatives used initially were resistant to attack (hard detergents) and accumulated in rivers, reservoirs, etc. Bio-degradable (soft) detergents, having straight, unsubstituted side chains had to be introduced in their place to mollify a general public justifiably alarmed at the sight of foaming drinking water.

Mason [38] has reviewed the estimates made for the amounts of carbon in circulation in the various parts of the carbon cycle – a concept developed by Goldschmidt and amplified by Wickman [39]. Fig. 18 represents in diagrammatic form a very much over-simplified idea of the way carbon is cycled on this planet, driven largely by radiation from the sun and heat from the earth's interior. Similar diagrams have been constructed for other biologically-important elements. Inter-relationships between these cycles (C, H, O, N, P, etc.) await detailed study. For example, Russian workers have given considerable attention to the inorganic but not the organic aspects of the biogeochemical transformation of sulfur in Nature [40–42].

Taking atmospheric carbon dioxide as the starting point, photosynthetic organisms grow and multiply by building the carbon atoms from this source into their lipids, proteins, carbohydrates, etc. An elaborate food web of animals is dependent on the ingestion of the primary photosynthetic organisms. Death of the organisms and immediate burial provides carbon compounds directly to a forming sediment. Microbial attack and chemical processes in the young sediment

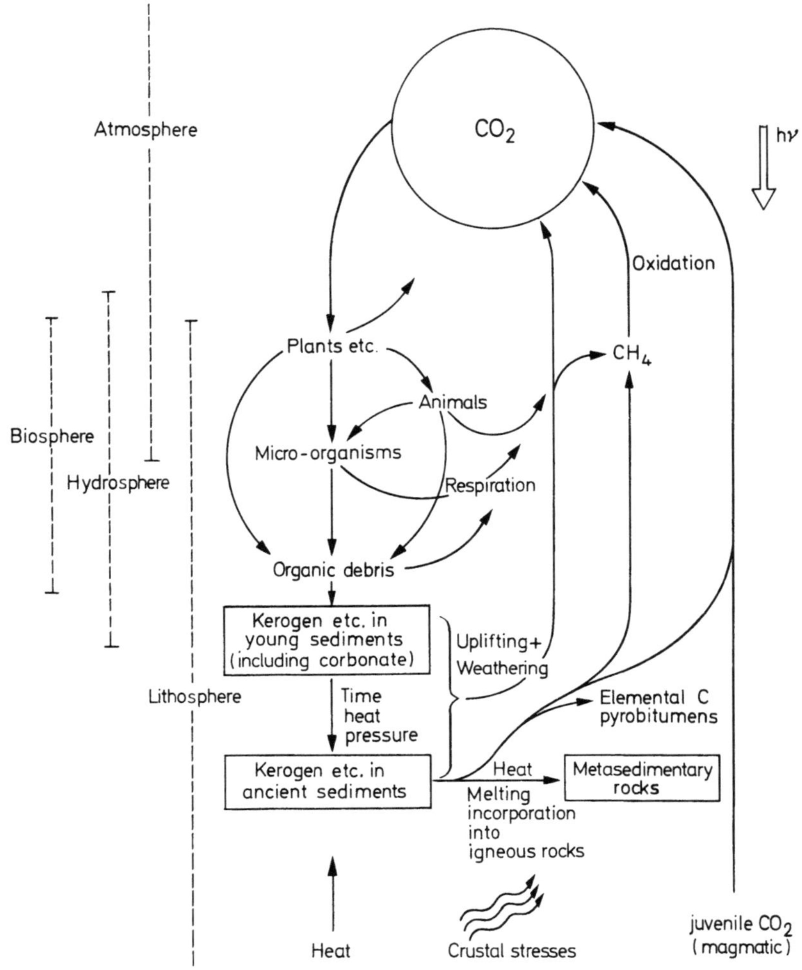

Fig. 18. Path of carbon compounds in Nature

will normally degrade these compounds to smaller molecules such as carbon dioxide and methane. Some of the carbon dioxide is returned to the atmosphere and some is incorporated into the sediment as carbonates. Insoluble high molecular weight organic matter – kerogen – may also be formed. However, some compounds are relatively resistant to degradation and they may persist as related compounds, the biological markers, bearing unaltered or partially altered carbon skeletons. The biological markers may be recognized within the living and nonliving components of an ecosystem and the sediment formed from it [15]. Reactions continue in the sediment during diagenesis and the processes of consolidation, burial and metamorphism, which may take place with the passage of time.

The erosion of ancient sediments, which have been uplifted and thereby exposed to the action of wind, rain, heat, frost, and microbial attack, constantly recycles carbon compounds into the biosphere. This process must have taken place ever since sediments were first deposited. Microorganisms in soils and other environments use these carbon sources. Kerogen, crude petroleum, and bitumens contain a vast range of compounds, mostly complex hydrocarbons whose structures now bear only a partial resemblance to those of biolipids, the presumed source material, but certain species of microorganisms have enzyme systems which enable them to catabolize these compounds. A contemporary occurrence illustrating this process is provided by the Torrey Canyon disaster [43, 44]. This giant oil tanker (about 100,000 tons) foundered on rocks near Cornwall, England and the cargo of crude petroleum spread towards England and France. The oil slicks were dispersed by application of detergent sprays (incidentally, the detergent proved to be very toxic to marine life) along the English coasts, but the French used powdered rock and other materials to absorb the oil and carry it to the bottom. The sites of some of these mass sinkings of crude petroleum were marked and then revisited about nine months later. Bottom samples revealed no trace of oil – microbial action and bottom currents had been fast and complete.

B. Pollution of Environments

Pollution on the scale of the Torrey Canyon incident serves to emphasize the drastic effect modern industrial society is having on the contemporary environment. More gradual changes, such as the spread of intensive agriculture and urban living, are less dramatic but more insidious. The northern hemisphere is suffering particularly rapid change as the pace of industrial development quickens and it will become increasingly difficult to find natural environments relatively free from pollution and other man-made effects. We need urgently to set aside areas of land and water representative of different environments so that we can study the operation of ecosystems with and without the influx of the debris of civilization – detergents, insecticides, herbicides, fungicides, etc. Studies must not stop short at the stage of the first disappearance of a particular compound, such as one of the chlorine-containing insecticides, but should also attempt to follow the fate of its degradation products [45]. Such "path of carbon" researches would seem to be particularly essential where it is likely that the precise compound is unlikely to have been previously encountered by living organisms.

"Pollution"

by Tom Lehrer

(... Time was when an American about to go abroad would be warned by his friends or the guide books not to drink the water, but times have changed, and now a foreigner coming to this country might be offered the following advice ...)

If you visit American city, you will find it very pretty –
Just two things of which you must beware –
Don't drink the water and don't breathe the air!

Pollution, pollution –

They got smog and sewage and mud,
Turn on your tap and get hot and cold running crud.

See the halibuts and the sturgeons being wiped out by detergents.
Fish got to swim and birds got to fly,
But they don't last long if they try.

Pollution, pollution –

You can use the latest toothpaste,
And then rinse your mouth with industrial waste.

Just go out for a breath of air, and you'll be ready for Medicare
The city streets are really quite a thrill –
If the hoods don't get you, the monoxide will!

Pollution, pollution –

Wear a gas mask and a veil,
Then you can breathe – long as you don't inhale.

Lots of things there that you can drink – but stay away from the
kitchen sink.
The breakfast garbage that you throw into the Bay*,
They drink at lunch at San José

So, go to the city – see the crazy people there
Like lambs to the slaughter, they're drinking the water
And breathing the air!

VII. Organic Chemistry of Sediments

A. The Time-Scale

Organic chemists are accustomed to reactions which reach equilibrium in seconds, minutes, and occasionally, days. The vast spans of time against which the geologist measures his phenomena seem quite alien and unimaginable. Little is known about the organic reactions proceding in sediments and we cannot ignore the possibility that ultra-slow reactions make significant contributions. Correlation with the geological time scale is essential for the characterization of rock samples. A shortened version of the geological succession is given in Table 2. In essence, this scale is based on major evolutionary changes, which resulted in characteristic faunal differences in the sediments deposited all over the world. Geological formations are classified according to their age by being assigned to a particular period.

* San Francisco Bay.

Table 2. *Geological society Phanerozoic time scale, 1964*

	Approximate oldest age in millions of years before present
	Cainozoic[a]
Quaternary	
Pleistocene	2
Tertiary	
Pliocene	7
Miocene	26
Oligocene	38
Eocene	54
Palaeocene	65
	Mesozoic[a]
Cretaceous	
Upper	100
Lower	162
Jurassic	
Upper	162
Middle	172
Lower	195
Triassic	
Upper	205
Middle	215
Lower	225
	Palaeozoic[a]
Permian	
Upper	240
Lower	280
Carboniferous	
Upper (Silesian)	325
Lower (Dinantian)	345
Devonian	
Upper	359
Middle	370
Lower	395
Silurian	440
Ordovician	
Upper	445
Lower	500
Cambrian	
Upper	515
Middle	540
Lower	570

[a] "Era"; the sub-headings, e.g. Quaternary, correspond to "Periods" and the remainder to "Epochs". The table is based on English usage and is drawn from the Geological Society literature [16]. American versions are readily available in many textbooks of geology.

B. State of Organic Matter in Sediments

Organic matter makes up about 0.7% of many sediments, largely in the form of insoluble polymeric material. The organic chemist is conditioned to think in terms of discrete molecules, each of which is free to move more or less independently of its neighbors. Loose association with adjacent molecules occurs largely through hydrogen bonds and dipolar interactions. The limits of the molecule are, by convention, defined as the extent of the covalent linkages. The lipids extracted from sediments with organic solvents are just as acceptable for study as those extracted from living organisms. But, the highly insoluble, heterogeneous and presumably cross-linked material – kerogen – making up the bulk of sedimentary organic material is outside the experience of the majority of organic chemists. Kerogen is somewhat unattractive as a problem largely because at first sight there appears to be little in its structure which is capable of definition and explanation. This dislike is understandable, but there is one problem within the normal purview of the organic chemist – the actual state, within the sediment, of the extractable compounds or those easily removed by simple hydrolytic procedures, etc. Ultimately, we should not really consider the analysis of a sediment complete until the actual nature of the organic material has been established – as it exists *in situ* in the sediment.

For example, carboxylic acids, RCO_2H, might be present as follows [41–50]:

1. RCO_2-matrix. Covalently bound – esterified – to alcoholic or phenolic groups of the main kerogen matrix. Extraction with most solvents would be ineffective, but extraction with an alcohol might result in ester interchange and consequent extraction.

$$\mathrm{RCOO-matrix + R'OH \rightleftharpoons RCO_2R'' + HO-matrix}$$

Such a process might be catalyzed by acidic or basic catalysts present in the sediment.

2. RCO_2R'. The acids are present as the esters of simple alcohols, the sterols, glycerol, or long chain wax alcohols. Such esters are soluble in most of the solvents used in the extraction of sediments.

3. $RCO_2^-M^+$. Ionically bound, as in salts and complexes with clay minerals. Here, aqueous or alcoholic extraction, or treatment with acid and subsequent solvent extraction would be necessary.

4. RCO_2H. Present as the free acid. Organic solvents are effective in extracting free acids.

From the above, it is clear that the success and degree of discrimination of extraction procedures will be very dependent upon the precise state of involvement of the carboxylic acid.

Further difficulties arise when one considers the actual physical site of entombment of the organic compounds. Occlusion in mineral grains and crystals is a well established phenomenon and results in this fraction being inaccessible to solvents, unless there is prior dissolution of the host crystals. Clay lattices are known to expand and accomodate organic molecules (Chapter 31), but there has been comparatively little study of the nature of the compounds trapped in clays obtained directly from natural environments and sediments. Some selective incorporation seems inevitable. The specific location of organic compounds where they occur within a fossil and the surrounding matrix may prove to be both feasible and important. Special micro-sampling techniques, including the use of the micro-probe mass spectrometer, are likely to be required.

C. Collection of Geological Samples and Extraction and Separation of Geolipids

Contamination is the ever-present and dominating hazard of organic geochemistry. Whenever possible, the chemist should take part in obtaining the geological sample so that the possibility of natural and introduced contaminants can be estimated at the site. These are numerous and include direct and major biological contaminants, as with one meteorite specimen, which was found to be contaminated with the feces of a dog-fox! Microbiological growth (bacteria, fungi, algae, lichens) and water and atmospherically-borne contaminants (water-solubles, diesel and auto fumes, etc.) are harder to detect but, one suspects, quite serious in most cases. Many rock exposures are badly weathered and fresh rock may require extensive quarrying. Little is known of the penetration of organic compounds into rocks; slow diffusion is certainly possible, especially into layered lattices and amorphous minerals, but a likely source of trouble must surely be along crystal boundaries and microscopic cracks and fissures in otherwise solid rock. Rather barren samples, such as most of the Precambrian rocks, would be especially suspect for these reasons. Even bacteria can penetrate some rocks.

These and other problems continue after the sample has been collected, and contact with all sorts of lipids has to be studiously minimized – fingerprints, dandruff, grease pencils, marking inks, newsprint (printing oils) and the range of the additives present in plastic tubing, bags, and other equipment (plasticizers, antioxidants, UV absorbers, pigments and fillers, to name but a few). Diffusion of volatile compounds and adsorption of relatively non-volatile substances by direct contact could both be serious over long periods of storage. Hamilton's classic paper [51] on the amino acids of a single wet thumbprint is an indication of some of the difficulties inherent in microanalytical work. Aluminum foil (prewashed with distilled solvents) is a suitable wrapping material for bulk samples, prior to temporary bagging in a Teflon bag or, less suitable, a polythene bag. Glass jars are satisfactory storage receptacles.

Sound, unfractured and unweathered rock samples require careful cleaning prior to extraction (Fig. 19). When the content of organic compounds is believed to be low, as in most Precambrian samples, washing with organic solvents or powerful oxidizing solutions (such as chromic acid) is normally performed, with or without prior etching (HF) or cutting away of the outer surface. The cleaned fragments may then be pulverized (with or without further processing) in one or two stages and the fine powder exhaustively extracted with organic solvents to furnish the geolipid extract. Alternatively or subsequently, the mineral matrix can be dissolved with hydrofluoric acid (silicate minerals) and hydrochloric acid (carbonates) and the then-released insoluble organic debris further extracted with

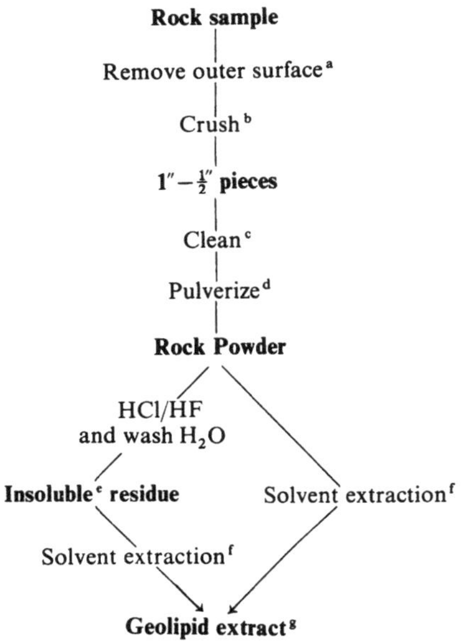

[a] e.g., with water-cooled, diamond-tipped, circular rock saw.
[b] e.g., with pre-cleaned jaw crusher.
[c] e.g., with chromic acid and/or organic solvents.
[d] e.g., with capsule mill (Tema mill).
[e] Mainly organic debris.
[f] Toluene/methanol.
[g] Other geolipid fractions can be obtained by hydrolytic extraction (alkali, acid, etc.), oxidative and reductive attack, and sublimation and pyrolysis.

Fig. 19. Generalized procedure for extraction of geolipids from a rock

organic solvents. Direct sublimation at temperatures below 300° C has been used on the powdered rock and on the demineralized material. Provided the temperature is kept low, the sublimate should contain many compounds of unaltered structure and contamination problems are much reduced. Higher temperatures, such as 500° C bring about pyrolytic decomposition of the organic material, including the main kerogen fraction, but the original organic content may still be inferred to some extent from the structures of the pyrolysis products.

I have selected one specific case to illustrate the isolation and identification of individual compounds in a geolipid extract, namely, the acyclic isoprenoid acids in the molecular size range C_{14} to C_{20} (Section IV.D). These branched acids have been isolated from sediments and their stereochemical configurations correlated with that of phytol. Fig. 20 outlines the procedure used [52]. Each step is discussed and explained in the ensuing paragraphs.

Step 1 ensures that ester linkages, $-CO_2-$, in the lipid material are hydrolyzed and all potential carboxylic acids converted to the potassium salts. Step 2 eliminates most compounds other than carboxylic acids, using the particular acidity of the acids as the basis for separation (salt formation; more acidic than phenols and alcohols; less acidic than mineral acids). The acid salts (soaps) are adsorbed on the silica while other organic compounds pass into the solvent eluate. Step 3

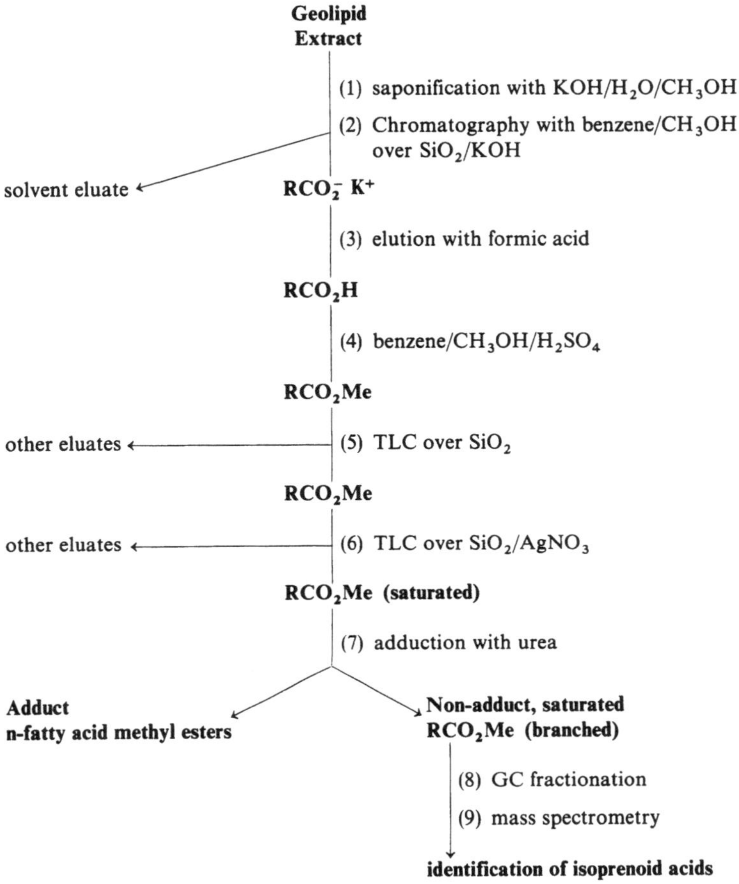

Fig. 20. Isolation and identification procedure for isoprenoid acids

liberates the acids from the column, but the fraction is still complex and contains compounds other than simple carboxylic acids. Step 4 converts the functional group (CO_2H) to the ester group (CO_2Me), thereby ensuring successful chromatography in step 5, wherein the band corresponding in R_f value to model carboxylic esters (e.g., $CH_3(CH_2)_{16}CO_2Me$) is collected and extracted. Step 6 selects only saturated esters, by preferentially holding back unsaturated and aromatic esters through silver ion coordination. The saturated methyl ester fraction is now subjected to a fractionation step based on molecular geometry.

This step (7) selects highly branched (methyl) or cyclic esters, since these molecules are too bulky to pack inside the spiral structure of the unit cell in the special urea crystal as it grows. The straight chain $[CH_3(CH_2)_nCO_2Me]$ and slightly branched $[(CH_3)_2CH(CH_2)_nCO_2Me]$ methyl esters are adducted by the urea crystals, which form round them. The crystals of adduct are then removed, leaving the isoprenoids in the mother liquors. (Actually, the isoprenoids can be further purified by adduction with thiourea; this time other types of compound are rejected in the mother liquors.)

Step 8 finally separates the individual branched saturated fatty acid methyl esters. Gas chromatography does this largely on the basis of molecular weight (Section III.A and Chapter 5). Chromatographic retention times on different stationary phases here afford quite good evidence for gross structure, bearing in mind the class separations already carried through to reach this step. However, standard samples of the esters are necessary to provide reference values for the retention times.

Step 9, mass spectrometry, preferably conducted directly on the effluent from the gas chromatographic columns (coupled GC—MS), but otherwise on small trapped fractions, provided a great deal of structural evidence (Chapter 4). Molecular weight, functional group, position of substituents, etc., are all determined by GC—MS. With molecules (XXII) to (XXVII) (Section IV.D) the position of the methyls can be deduced fairly unambiguously, but the stereoisomeric complexities cannot be settled by mass spectrometry alone, for the fragmentations of acyclic systems show little reponse to relative stereochemistry and none to absolute stereochemistry. The quantities of such compounds available for study in biological and geological materials are often too small for effective study by measurement of optical rotations. Further, such measurements can be relatively uninformative on three counts: (i) the rotational contributions of fairly symmetrically substituted carbon atoms, such as carbons 7 and 11 in (XXII), are normally extremely small; (ii) the overall asymmetry of the molecule is minimized by the fairly free rotation of the acyclic framework of carbon-carbon single bonds; (iii) the polar, light-absorbing carboxyl group is close (alpha position) to an asymmetric carbon atom (and hence effective in producing a significant rotation) only in compounds (XXIII) and (XXVII). The effect falls away rapidly with each carbon atom inserted between the asymmetric carbon and the polar grouping.

One technique we have found useful for establishing the *relative stereochemistry* of some of these acids is high resolution capillary gas chromatography [52]. For example, very efficient columns readily discriminate between the methyl esters of the two diastereomers of the C_{15} acid (XXVI), producing two peaks on the gas chromatographic trace, corresponding to the 3D, 7D (and/or 3L, 7L) and 3L, 7D (and/or 3D, 7L) acids.

Only very small amounts of sample are necessary; the isolation procedures, approximate GC retention times and mass spectra, establish the gross structure while the precise coincidence of retention time with those of stereochemically pure samples of established structure settles the relative stereochemistry. With the GC—MS technique, no isolation step is required.

The *absolute stereochemistry* can be determined by similar procedures in favorable cases. In theory, two techniques can be used:

(i) High resolution gas chromatography using a phase containing asymmetric molecules or groupings (Fig. 21, *a*). The solubilities (reflecting molecular interactions) of the two enantiomers of the solute sample in, say, an all L-polyester stationary phase may differ sufficiently for separation to occur.

(ii) High resolution gas chromatography over an optically inactive, or, rather, a non-asymmetric phase, of a derivative of the acid where the attached group bears an additional stereochemically pure center. The example shown in Fig. 21 *b*

(a)

(b)

(c)

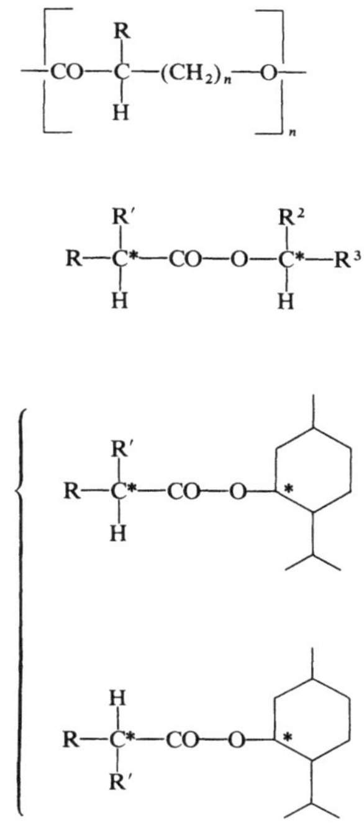

Fig. 21. Gas chromatographic determination of relative and absolute stereochemistry of branched acids. a) Example of an ester polymer consisting of asymmetric monomer units; b) Diastereomer formed by esterification of an acid (bearing an asymmetric center) with an optically active alcohol; c) Pair of diastereomeric esters formed by esterification of the racemic modification of the acid with L-menthol

is an ester prepared from an optically active alcohol. We have found the L-menthyl ester to be effective (Fig. 21, *c*).

The same techniques have been applied to the recognition of the D or L nature of amino acids occurring in sediments, including those of Precambrian age. The method is precise and requires only a few nanograms of each amino acid. Attempts to effect similar separations of the stereoisomers of acyclic isoprenoid hydrocarbons have so far been unsuccessful.

D. Recent Sediments

Recent sediments provide the key to an understanding of ancient sediments. They reflect the immediate environment of deposition. Organic matter is best preserved in sediments deposited in aquatic environments and is generally found to be intimately associated with very fine-grained sediments. Cores withdrawn from sediments at the bottom of lakes, etc., vary widely in appearance and composition. The upper layers are frequently semi-fluid, the middle more jelly-like in consistency and the deepest layers like a stiff mud. Slow compaction gradually expels the water.

Our first concern must be to study the environment giving rise to a particular sediment, for we need to know which organic compounds are reaching the sediment and how they are presented – as part of decaying organisms, in solution, adsorbed on organic debris or on clay particles. Some workers believe that the contribution of organic matter from the terrestrial plants and animals may be dominant even at some distance from the shore. The aquatic system provides a means for fairly rapid and selective destruction of organic compounds by irradiation, oxidation, etc., and the organic matter eventually contributed to the sediment forming at the base of the water column may be quite unrepresentative of the biomass above it and in the catchment area. Most organically-rich Recent sediments are reducing, but there is generally an oxidizing zone in the upper layers of the sediment and the water column above it. Not much is known about the decay processes of plants and invertebrates, but careful studies by ZANGERL [53] and others of the fossil death assemblages (taphonomy, Chapter 8) formed in an aquatic sediment indicate that vertebrate carcasses can decompose in the space of a few days. A concretion is left, formed partly of iron carbonate and sulfide produced by interaction of the decomposition gasses with iron-bearing interstitial waters. Such processes should permit chemical studies in the laboratory. We must also allow for other types of individual occurrences such as the rapid addition of fresh sediments, perhaps bringing an end to a steady state system of microbial attack – storm waters, floods, mud slumps and clouds of volcanic pumice are observed to produce such effects. Microorganisms and burrowing animals are very active in the upper meter or so of Recent sediments, and undoubtedly play a major part in degrading some compounds and synthesizing others. This biological activity presents an analytical problem in that it is difficult, if not presently impossible, to distinguish between the organic compounds extracted from the living organisms inhabiting the sediment and those extracted from the non-living content of the sediment. ^{14}C measurements are of no use, for the carbon turned over will still remain substantially of the same "time label" whether in a living organism or not. Comparative biochemical studies may offer the best hope – for example, anaerobes lack photosynthetic pigments. We need to know a great deal more about the turnover of organic compounds in Recent sediments. Thus, most important biolipids, including very sensitive compounds such as the carotenes and the chlorophyll pigments, have been detected, but they may have been extracted from living organisms present in the sediment. Biological activity within the sediment is held to be effectively zero at greater depths as the sediment becomes compressed. At this point, Recent unconsolidated fine-grained muds show overall contents of organic carbon approximating to those of ancient shales. As in the shales, most of the organic matter is already insoluble – the "heteropolycondensate" described by DEGENS (Chapter 12).

Fig. 22 illustrates one type of data acquired from a Recent sediment; it is a gas chromatographic trace for the total alkanes extracted from a section of a core removed from Mud Lake in Florida. This lake is entirely surrounded by semi-tropical vegetation. The undisturbed sediment is many feet deep and consists of very dark, finely-divided, somewhat gelatinous and largely organic material. Most of the organic matter is, of course, already insoluble. Different sections cut from the core do differ in their chemical composition, but for the extract shown, we see that the dominant hydrocarbons are in the C_{30} region and are odd-numbered, in accordance with the findings for plant waxes [54, 55]. It is possible, therefore, that plant waxes made a prominent contribution at the time that particular section was laid down – a few thousand years ago. In addition, there are fair quantities of even-numbered alkanes, especially in the region just above C_{20}. This prevalence may be due in part to contribution from organisms which are known to provide even-numbered alkanes in addition to odd-numbered (for example, certain bacteria [56]), and in part to geochemical reactions occurring in the sediment [15]. As an example of the latter, I refer to the direct reduction of even-numbered acids and alcohols to the corresponding even-numbered hydrocarbons. Laboratory data are available, but the process has yet to be demonstrated

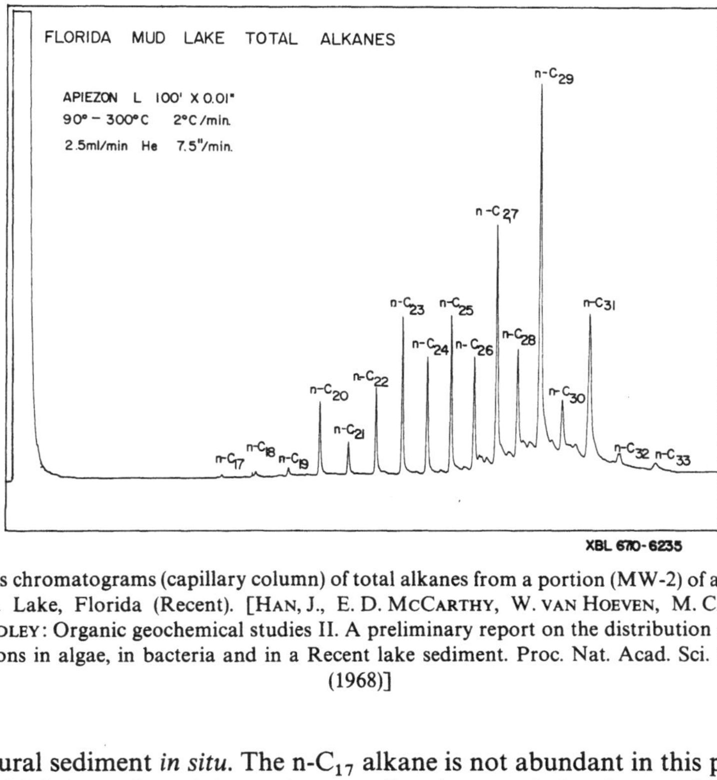

Fig. 22. Gas chromatograms (capillary column) of total alkanes from a portion (MW-2) of a core taken from Mud Lake, Florida (Recent). [HAN, J., E. D. MCCARTHY, W. VAN HOEVEN, M. CALVIN, and W. H. BRADLEY: Organic geochemical studies II. A preliminary report on the distribution of aliphatic hydrocarbons in algae, in bacteria and in a Recent lake sediment. Proc. Nat. Acad. Sci. U.S. **59**, 29 (1968)]

for a natural sediment *in situ*. The n-C_{17} alkane is not abundant in this particular section which is, at first sight, rather puzzling because it has been said that this sediment consists largely of algal debris. However, other layers do show a more prominent n-C_{17} contribution and also an abundance of the isoprenoid hydrocarbons.

E. Ancient Sediments

By far the greatest part of fossil organic matter is trapped in ancient shales, largely as finely-divided insoluble debris: coal, petroleum and natural gas represent minor quantities. A fair number of ancient sediments has now been examined for organic geochemical content. The general picture which emerges, for the shales anyway, is that the amount of soluble organic compounds, the geolipid fraction, diminishes with increasing age and that the overall composition tends increasingly towards a preponderance of hydrocarbons, with a corresponding progress towards graphitization of the insoluble organic debris. These progressive changes can be ascribed largely to slow thermal and catalytic processes taking place in the sediment after burial, compaction, and lithification [57–61]. Absolute age appears to be of much less significance than the thermal history of the rocks

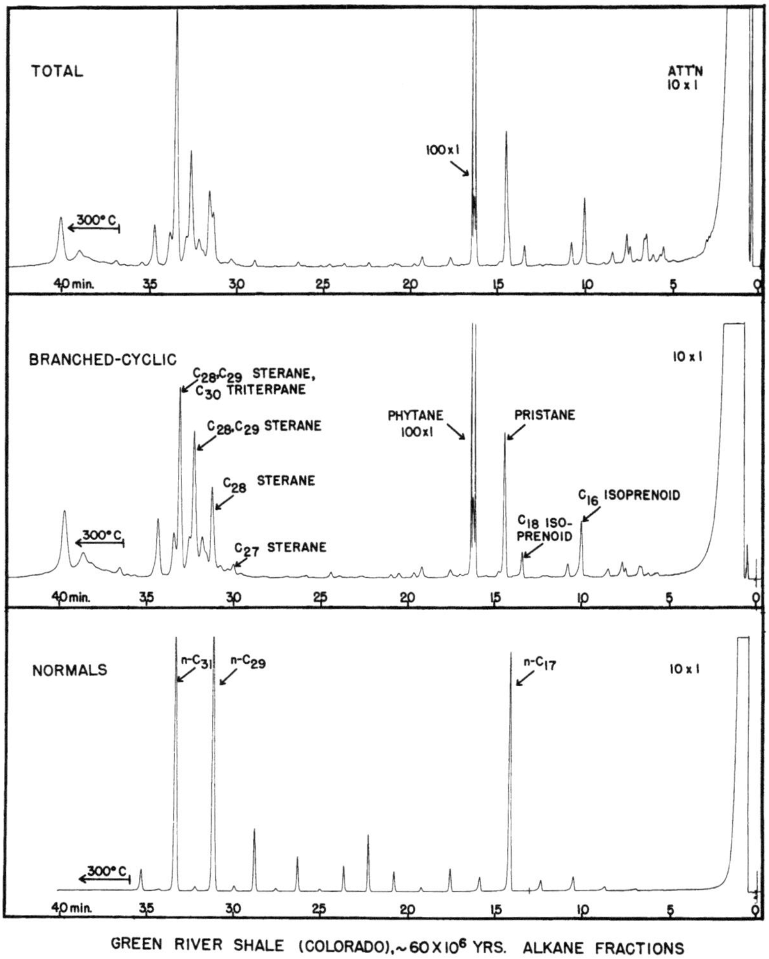

Fig. 23. Gas chromatograms of alkane fractions from the Green River shale (Eocene). [JOHNS, R. B., T. BELSKY, E. D. MCCARTHY, A. L. BURLINGAME, P. HAUG, H. K. SCHNOES, W. RICHTER, and M. CALVIN: The organic geochemistry of ancient sediments. – Part II. Geochim. Cosmochim. Acta **30**, 1191 (1966)]

as brought about by deep burial or contact with igneous intrusions. Two examples of ancient sediments will be discussed here briefly by way of illustration of the approach used.

The Green River shale, of Eocene age (50×10^6 years), is held to have been laid down in shallow lakes. It is not known what the actual contributions of the land plants and aquatic plants were. The gas chromatograms of the alkane fractions from this shale are shown in Fig. 23. The *n*-alkanes show the marked dominance of the odd carbon numbers in the C_{30} region, corresponding to the hydrocarbons of higher plants, while the n-C_{17} alkane, which is so prominent, may correspond to the hydrocarbon production of algae. Of course, these correlations are only tentative, but they should serve to stimulate further research. The branched and cyclic alkanes, which are largely isoprenoid, parallel the known pattern of abundance of isoprenoids in plants and animals; thus, abundant C_{15}, C_{20}, C_{30}, and C_{40}. The peak in the branched cyclic fraction around 40 minutes elution time corresponds to perhydrocarotene (C_{40}) [62]. The peaks in the C_{30} region are now being studied by combined high resolution capillary gas chromatography – mass spectrometry [63]. We hope to identify each of the steranes and triterpanes and then

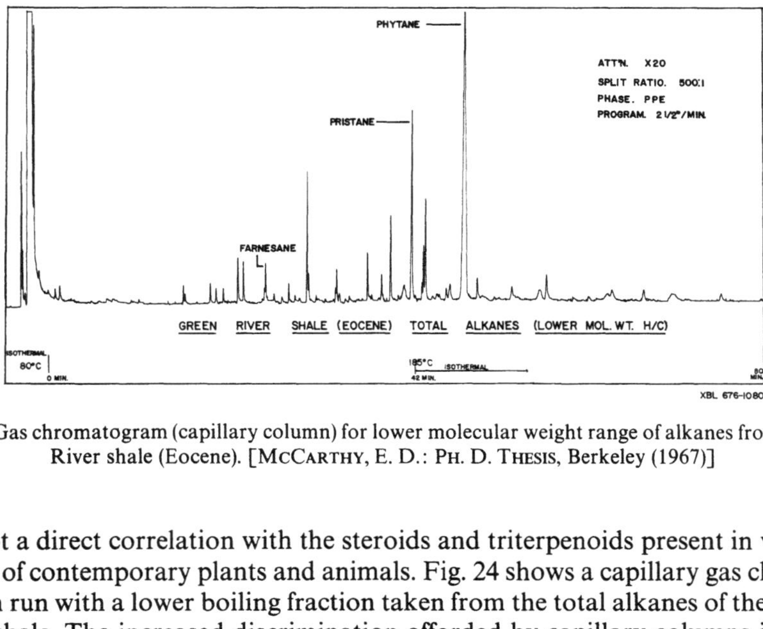

Fig. 24. Gas chromatogram (capillary column) for lower molecular weight range of alkanes from Green River shale (Eocene). [McCarthy, E. D.: Ph. D. Thesis, Berkeley (1967)]

attempt a direct correlation with the steroids and triterpenoids present in various genera of contemporary plants and animals. Fig. 24 shows a capillary gas chromatogram run with a lower boiling fraction taken from the total alkanes of the Green River shale. The increased discrimination afforded by capillary columns is noteworthy. The peaks due to pristane and phytane are quite sharp and it is evident that these alkanes are not mixtures of several positional isomers. Their relative and absolute stereochemistries remain to be settled, but even the present results give strong support to the organic geochemist in his claim that these hydrocarbons are adequate biological markers.

The Soudan shale (Precambrian, $>2 \times 10^9$ years) provides the second example of an extract from an ancient sediment [64, 65]. Fig. 25 illustrates gas chromatograms obtained (capillary column) for the total alkane fraction. The pattern is remarkably simple, although, greater detail is revealed when the chromatogram is run under conditions of higher resolution. The normal alkanes are centered around C_{18} and are of a fairly narrow spread. The isoprenoids, pristane and phytane, are very prominent. The pattern is different from that of the Green River shale, and there is the further complication that the rock itself may have been heated by igneous contact at some time or other. The history of Precambrian rocks is often little understood. These very old sediments yield only small quantities of extract, increasing thereby the everpresent possibility of contamination, either during past geologic eras, during Recent times, or even during the processes of

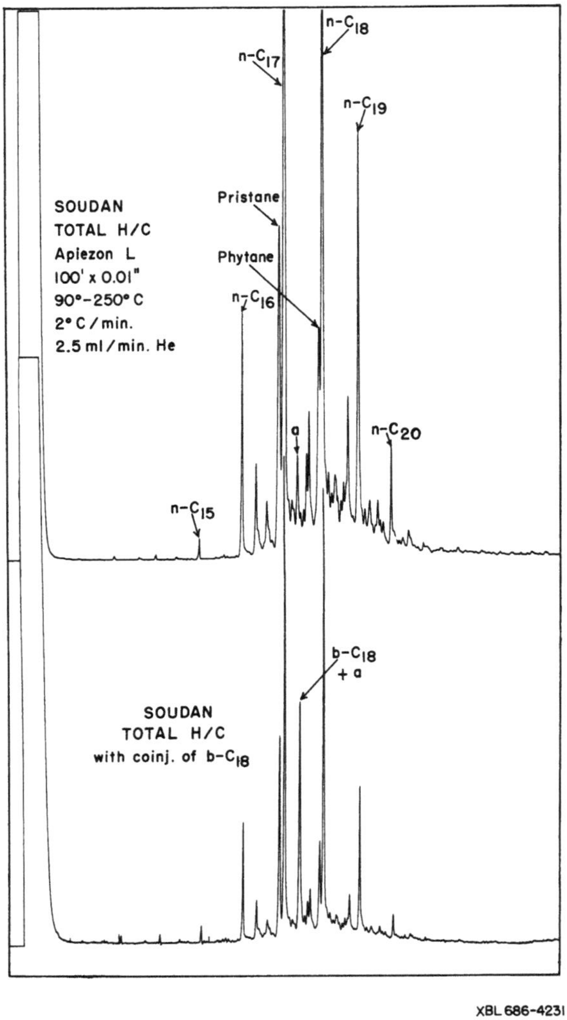

Fig. 25. Gas chromatogram (capillary column) of total alkanes from Soudan shale (Precambrian, $> 2 \times 10^9$ years). [HAN, J.: Unpublished data from Space Sciences Laboratory, University of California, Berkeley; also CALVIN, M.: Chemical evolution. London: Oxford University Press 1969.]

extraction and separation of the geolipids. However, the importance of the information to be derived, especially in relation to the appearance and early development of life on this planet, will continue to serve as an incentive for further work.

VIII. Geolipids – Biological or Non-Biological Origin?

A. The Problem

One of the major questions asked now and in the past is – can one distinguish lipids of biological (biogenic) and non-biological (abiogenic) origin? This question is clearly of great interest on two counts: first, in relation to the discovery of that point in time when life began, and second, in characterizing extraterrestrial samples as being life-originated or not. As might be anticipated, such issues have proved highly controversial, and there is no unified view. One difficulty is that very little is known of the products of non-biological processes which might conceivably have taken place on and within the earth's crust, on meteorite bodies, and on planetary surfaces exposed to solar and cosmic radiation and a variety of atmospheres and temperatures. At present, we have rather limited experimental data, drawn from reactions such as the Fischer-Tropsch. The structural and stereochemical uniqueness, including optical activity, which we believe characterizes the operation of biochemical systems (in the form of the "biological marker substances"), is not apparent in these laboratory products but we need much more direct experimentation under simulated primitive earth and extraterrestrial conditions. However, the problem is further complicated by the non-biological reactions which take place during diagenesis, maturation, and metamorphism of organic matter in sediments and thereby alter and sometimes destroy the original record, biogenic or abiogenic, contained therein.

B. Pre-Biological Era and the Origin of Life

We cannot at present ascertain where the borderline exists between pre-biological and biological times in the succession of Precambrian sediments [66, 67]. However, if we assume that pre-biological sediments did exist and are still to be found in the crustal rocks, then their content of organic material should reflect (a) the original pre-biological organic matter incorporated into the sediment at the time of deposition, and (b) the diagenetic and maturational changes induced since then by the thermal and other stresses experienced during the long span of time. The latter changes should presumably resemble those experienced by biogenic sediments. The original pre-biological matter might have been produced by a great variety of abiogenic processes. On the primitive earth the accumulating products would not be subject to the rapid attack by living microorganisms which would be the case today in any open environment. Again, the atmospheric and terrestrial conditions must have been rather different from now. Both the backward search through time (organic geochemical studies) and the forward search to the origin of life (chemical evolution studies) have a part to play in any examination of this difficult problem. Present day biosynthetic processes are no more than elegantly controlled syntheses and degradations – highly specific in relation to the particular atoms united or disunited and in the direction (stereochemistry) of bond formation. It seems highly likely that at least some prebiotic reactions were similar in general nature and products formed, but lacked the high speed and precise control of the later enzyme-catalyzed reactions of living systems. Hence, residual organic geochemicals in Precambrian rock may not reveal any significant discontinuity at what may turn out to be a rather diffuse boundary between abiogenic and biogenic times. May we reasonably expect to distinguish abiogenic and biogenic pristane? It is too early to say, but as indicated later, it may be that overall patterns of abundance of different hydrocarbons and their isomers may be more significant than the presence of a single compound.

To take a specific class of biological compounds: it is, of course possible for the chemist to synthesize, with some difficulty, oligopeptides which are identical with the natural, enzymatically-produced biopolymers. WALD [68] has pointed out that such laboratory feats still count in one sense as biologically-controlled syntheses since the careful and deliberate invention, selection and attention of the human organism is required throughout. Close approximation to truly abiogenic conditions is afforded by the now classic "prebiotic" soup type of experiment in which a mixture of simple compounds, such as methane, ammonia, and water, is subjected to energy input (hv, electrical discharge, bombardment with high energy particles, heat, sound, etc.) and the products analyzed. Amino acids are produced in appreciable yields and the wide variety formed include all the biologically important ones. Those bearing asymmetric carbon atoms are racemic mixtures. Suitable conditions have been derived (e. g., clay catalysts, presence of dehydrating agents) for self-condensation of these, and of somewhat simpler mixtures. There is evidence for some non-random sequencing, i.e., amino acid units show some preference in regard to the choice of the next unit to be attached [69]. Presumably, the course of pre-biological chemical evolution was to some extent controlled by such preferences. Anyway, the main point to emerge from the above is that we can presently distinguish amino acid biopolymers from the products of current abiogenic type of experiments by the following criteria: (1) most of the biopolymers contain a very limited number of different α-amino acids in contrast to the very large number of all types in the complex abiogenic mixture; (2) nearly all the biopolymers contain only L-α-amino acids in contrast to the completely racemic abiogenic product; (3) the sequencing of the biopolymers is precise and almost infinitely improbable in abiogenic terms.

These studies and their counterparts with other biological products have great significance for the study of Precambrian sediments and extraterrestrial materials (Chapters 7, 13, 18, 30).

C. Biogenic – Abiogenic: The Present Situation

The sedimentary, erosional and igneous processes (Fig. 26) afford a more or less constant turnover of the outer few miles' thickness of the earth's crust (Section VI, A). Many geologists and geophysicists now believe that the continents have fractured, drifted and rejoined several times in the past [10]. This turnover, and the efficiency of the biological carbon cycle, mean that almost all of the carbon accessible to us has been utilized at one stage or another by living organisms. In that sense, this carbon (and its compounds) are biogenic. Exceptions could arise with (1) any carbon emerging from deeper layers of the crust—juvenile, presumably magmatic, carbon; and (2) carbon and carbon compounds still present in Precambrian formations predating the origin of life. There is no obvious major

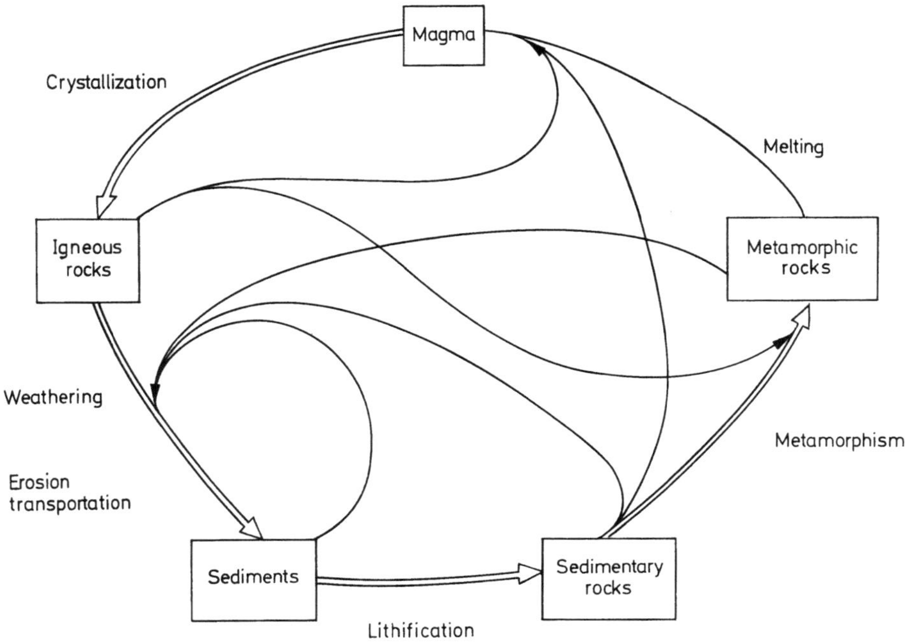

Fig. 26. The rock cycle. (Adapted with permission from a display in the Department of Geology, The University of Texas, Austin, Texas.)

present-day contribution of magmatic carbon, though WICKMAN's estimates [39] of the amounts of carbon in the different sections of the carbon cycle are held to indicate that large amounts of carbon dioxide have been released from magmatic sources throughout geological time [38].

There are extensive deposits of carbonaceous shales of Precambrian age, some of great age, such as the Fig Tree and Onverwacht series of South Africa. Here the rocks are dated radiometrically in excess of 3000×10^6 years and the biogenic origin of the carbon is still controversial. The present situation is such that we cannot seek to use these rocks as a guide to even older ones – rather, we need to improve our definitions and methods, probably by examining younger rocks containing organic remains of acceptable biological origin. These very ancient sedimentary rocks (up to 3400×10^6 years) [71] certainly contain organic matter, but it is mostly graphitic, insoluble material, rather close in its properties to high grade (high percentage carbon) coal. Extremely small amounts of soluble lipids can be extracted, but as HOERING and others have pointed out [72], the extent of carbonization – presumably effected mainly by temperature and time – is discouraging. However, one can argue that the dehydrogenation and graphitization processes need not completely erase the biological record, so that even small amounts of compounds which retain most of their carbon skeletons, would still be adequate records. Thus, in the laboratory, the prolonged thermal treatment of a pure *n*-alkane (*n*-$C_{28}H_{58}$ at 375° for hundreds of hours) will convert most of the hydrocarbon into a graphitic powder and free hydrogen. But even after a week, a small percentage survives, and more importantly, is accompanied by a smooth series of *n*-alkanes of lower carbon number and series of alkanes and aromatic hydrocarbons [73]. Similar reactions appear to take place in sediments containing *n*-alkanes when they are heated.

In the same way, the long history at much lower temperatures would lead to graphitization of almost all of the organic material in the Precambrian rocks, but a small portion of the compounds of possible biogenic origin could survive. One test for this would require the analysis of sediments known to have been exposed to high temperatures over established periods of time. Igneous dykes and sills cut many sediment formations; some correlation between thermal stress and loss of biological information should be possible. It is difficult to make reliable estimates of the rate at which carbon skeletons are broken down in a rock, but simple calculations have been made for hydrocarbons. It is possible, but not certain, that the small amounts of normal and isoprenoid alkanes detected in the older Precambrian rocks represent the residuum of carbon skeletons (not necessarily alkanes) synthesized at that time. They could have migrated in at some later date, but if we assume they were indigenous, then the second question is: were they biosynthesized or were they synthesized by some abiogenic process?

As already mentioned, all – or most – of the carbon atoms at the earth's surface have been processed at one time or another by living organisms. We could say that the compounds built from them are therefore "biogenic" but this description is not particularly helpful, for biological history is reflected much more closely at the molecular rather than the atomic level. Once again, we return to the fundamental problem: which of the compounds present in a given situation are "biogenic"? How can we discriminate between biologically-formed compounds and those formed abiologically, and the various intermediary classifications – altered biological, altered abiological – which presently require definition? The problem is obviously complex and, to some extent, an exercise in semantics. However, we must attempt to define this problem, for it represents the principle hurdle for the biological marker concept.

Consider a metasedimentary rock (a thermally altered sediment) which originally contained biologically-derived organic debris. For simplicity, let us concern ourselves only with the hydrocarbon extract, which can be assumed to contain complex mixtures of normal, branched, and cyclic alkanes and a range of aromatic hydrocarbons. Let us assume further that some of these alkanes will have retained in whole or part the biologically-derived carbon skeleton (biogenic), but some of this will have suffered drastic rearrangement or breakdown (degraded biogenic), while some alkanes will have been synthesized from carbon-containing molecules, such as carbon monoxide, or from graphitic sheets, during the thermal processes (abiogenic).

The basis for definitions used here is whether or not most of the carbon-carbon bonds in the given compounds were formed biologically or abiologically. Of course, although we may now agree on a suitable nomenclature for compounds where we know their origins, we are not able to apply the reverse procedure and deduce the origins from the structure of compounds, for we as yet have insufficient information about the products of abiogenic synthesis in rock matrices. In other words, we may discuss the system of nomenclature we should like to apply, but we are not in a position to make routine *use* of this nomenclature. Mixtures of hydrocarbons similar to those encountered in metasedimentary rocks can be synthesized by processes like the Fischer-Tropsch, starting with carbon monoxide as the carbon source:

$$CO + H_2 \xrightarrow{\text{catalyst}} CH_3(CH_2)_nCH_3 + \text{other hydrocarbons} + H_2O$$

Presumably geological conditions exist which would generate mixtures of hydrocarbons. The experiments of ANDERS and his colleagues [74] with iron meteorite catalyst are relevant here; also, the hydrogenation of coal, some of which possesses a predominantly graphitic structure. Some of the hydrocarbons generated in these and other processes [75] are similar in structure to those formed biologically or by degradation of biological material.

The search for true, abiogenic lipids in geological materials does not look promising. Most supposedly abiogenic oils and carbonaceous residues in igneous and hydrothermal deposits are likely to be degraded biogenic or mixtures of biogenic and degraded biogenic materials. Table 3 is an attempt to provide some sort of classification for naturally occurring organic compounds.

D. Biogenic – Abiogenic: Possible Criteria

It has been pointed out [37] that no single criterion can be used to characterize life and the distribution of organic compounds is unlikely to provide a water-tight exception. Of course, a whole set of biological marker compounds would provide many criteria, though admittedly of a similar type. Their detection would serve to indicate, not life, but rather its functioning, past or present. Throughout this chapter, we have been attempting to correlate molecular structure and biological origin; the correlation must become progressively more difficult and unreliable the more altered the sample.

In summary, the criteria which may be used are:

(1) Precise molecular structure of individual compounds, including relative and absolute stereochemistry, and the consequential optical activity or isotopic composition;

(2) The relative abundance of the various positional and steric isomers of an individual compound;

(3) The relative abundance of individual members of homologous series of compounds;

(4) The relative abundance of different types and classes of compounds.

Another way of putting the above is as follows: If a particular structure is identified, then the biogenically significant feature would not really be its presence, but rather the absence or low abundance of related or isomeric structures, which would be expected to be present if abiogenic processes had been operating. Absence of certain compounds may be highly significant for biogenicity.

We can illustrate the above points by reference to the generalized distribution of isoprenoid compounds in living organisms. The following points emerge (Cf. V,A): (i) the biological monomer unit is the C_5 compound, the isopentenyl pyrophosphate; in terms of utilization of these units by telomerization and polymerization, most organisms contain C_{10}, C_{15}, C_{20}, C_{30}, and C_{40} isoprenoids

Table 3. *Tentative classification of naturally occurring organic*[a] *compounds*

Compounds	Examples	Geologic occurrences	Suggested classification
Original, slightly rearranged, or degraded, biosynthesized carbon skeletons with or without functional groups, but with largely correct stereochemistry	Pristane, phytane, isoprenoid acids, sterane and triterpane hydrocarbons	As in the Green River shale (Eocene Age)	Biogenic (biological markers)
Original biosynthesized carbon skeletons heavily degraded; small fragments recognizable as smaller molecules with correct positional substitution and stereochemistry	Range of *n*-alkanes formed by thermal alteration of single *n*-alkanes?	For example, hydrothermally altered sediments and extracts – Abbott Mercury Mine and oil from deep wells at 200° C	Degraded biogenic
Original biosynthesized carbon skeletons heavily degraded and small fragments reacted further to form range of complex products and polymeric material	Organic debris in metasedimentary rocks	For example, Precambrian Thucolites	Degraded biogenic
Original biological material completely degraded to volatile compounds and graphitic residues	CH_4, CO_2, C, etc.	Volcanic gases, diamonds?	Degraded biogenic but structurally indistinguishable from abiogenic [unless isotopic data informative]
Original magmatic carbon	Carbon, CO_2, etc.	Volcanic gases, igneous rocks?	Magmatic
Small molecules, methane, etc., carbon sources subject to nonbiological reactions or catalysis	?	?	Abiogenic. The carbon source material would likely be degraded biogenic[b]
Organic constituents of meteorites	Hydrocarbons	Infall of carbonaceous meteorites	Meteoritic

[a] Propose that the terms "organic" and "inorganic" be restricted to compounds containing carbon and those not containing carbon, respectively. On this definition, CO_2 and calcium carbonate would be termed "organic", although the latter is certainly borderline. The value of this terminology (Section 4, i) is that confusion with biogenic/abiogenic is avoided.

[b] One can envisage further variations on these items in the table. Thus, thermal alteration of an abiogenic mixture could well produce organic debris almost indistinguishable from degraded biogenic. Again, partial biological reworking of abiogenic material could produce a similarly confusing situation where both types of molecule are present.

in reasonable abundance, and in some cases, a further series, $C_{45}, C_{50} \rightarrow (C_5)_n$ as rubber, with values of n in the tens of hundreds of thousands. Compounds based on the C_{25} and C_{35} skeletons are very rare. (ii) Most isoprenoid compounds display a characteristic pattern of carbon-carbon linkage for the C_5 units of head-to-tail or tail-to-tail, depending on the total carbon number. Thus, the C_{10}, C_{15}, C_{20} and $(C_5)_n$, where $n>8$, compounds are mostly linked head-to-tail, while the C_{30} and C_{40} compounds have two head-to-tail residues joined tail-to-tail. (Head is the branched carbon end.) McCarthy and Calvin [12] have discussed this point very thoroughly. (iii) Where stereochemical variations are possible, certain features generally prevail, and the positions of oxygenation tend to be consistently related in the same series of compounds. Steroids and triterpenoids illustrate this point well, almost invariably having an oxygen at the 3 position, which is residual from the enzymic opening and cyclization of squalene epoxide.

E. Meteorites, the Lunar Sample, and Extraterrestrial Planetary Analysis

From time to time over the last hundred years, organic chemists have applied their techniques to the study of the carbon compounds of meteorites, which at present provide us with our only samples of extraterrestrial matter. Those meteorites which contain appreciable carbon, the carbonaceous chondrites, make up only a very small percentage of the total number of falls. Porous and friable, these chondrites are very readily contaminated with terrestrial debris, vapors, and small particles, such as pollen grains and microorganisms. There have been many bitter exchanges over the validity and meaning of analyses reported for meteorites, including the very well known Orgueil, which fell near the village of Orgueil, in France in 1864. The present consensus of opinion seems to be that Orgueil and certain other carbonaceous chondrites do contain an appreciable fraction (say, 2 percent) of genuinely extraterrestrial, and presumably abiogenic, organic matter, partly solvent soluble, but that the biological markers, which can be detected only in very small amounts, are terrestrial contaminants, introduced largely during storage and handling prior to analysis. Interest remains high in this area of meteorite analysis and gives some idea of what we may encounter with the lunar sample, but the contamination problem seems to be presently an insuperable obstacle to generally acceptable and definitive studies.

Even the first samples of lunar rock will be subject to some uncertainty, for the conditions for collection will not absolute preclude contamination by terrestrial organic matter and microorganisms introduced by the lunar vehicle and the astronauts and their equipment. However, the level of contamination should be low and to some extent, measurable by controlled experiments made in vacuum chambers here on earth. Debris from meteoritic impact will almost certainly be distributed about the lunar surface, and this will present a further problem in discrimination. Few scientists expect any life form or even evidence from extinct life forms, though the one-time presence of substantial water bodies has some supporters. The organic analysis is expected to take the form of a straight survey of the composition of the organic matter without expectation of the presence of biological markers. The composition will be of interest whatever is revealed, for it should represent the result of intense solar bombardment of carbon compounds under ultra-high vacuum, extremes of temperature and prolonged contact with the lunar surface. If the amounts of organic material are very low, then great care will be necessary to detect lunar compounds to the exclusion of terrestrial laboratory contaminants.

The study of the organic constituents of the atmospheres and surfaces of the other planets in our solar system will remain a challenge for many years to come. The results, which will refer, in effect, to "parallel, but separate experiments", should tell us much about the development of our own planet and of the life on it. Organic geochemists can take some part in the "fly-by" sampling and earth-based spectroscopic studies which give limited information on planetary atmospheres and surfaces; the principle scope comes in the automated landers designed to analyze actual surface material *in situ*. The Mars lander proposed by NASA for 1973 is such a case. It will carry a number of experiments, including a television camera, biological experiments designed to detect the growth of any micro-organisms in nutrient broths and bio-organic experiments. Bio-organic studies have as their main aim a survey of any Martian organic compounds, principally in the hope of arriving at a decision as to whether their composition indicates a life origin or not. The problem here is vastly more difficult than the "abiogenic/biogenic" question posed for terrestrial materials, for we have no right to assume that a Martian biochemistry would appreciably resemble our own. Chemical evolution experiments encourage the belief that pre-biological chemical developments wherever and whenever they occur, might well take a similar path, employing similar basic units – amino acids, purines, etc. But we should be prepared to view dispassionately the analytical data telemetered back from Mars; indications of unexpected "order" in the relative abundances of individual compounds or series of compounds [76], lack of thermodynamic equilibrium, and the presence of only a few positional isomers and stereoisomers of an empirical formula capable of providing a large number of isomers, would all be suggestive of the operation of processes comparable in their selectivity to terrestrial biochemistry – even though the classes and individual compounds might be very different from our own. Organic geochemists working in this area propose a variety of experiments for Mars landers; there is fairly general agreement that a miniaturized GC–MS system provides the most versatile analytical facility. Pulverized Martian rock could be subjected to various pretreatments, such as pyrolysis [77, 78], which would liberate organic compounds or their degradation products in the vapor state. Immediate sequential analysis by the GC–MS equipment and on board data reduction would follow. The telemetered data received on the earth could then be compared with data obtained from similar terrestrial experiments using desert soil, for example. It seems too much to expect that the first lander will provide an unequivocal "yes" or "no" to the age-old question of extraterrestrial life, whether it be in existence or long since dead, but any data obtained on Martian carbon compounds will be of great intrinsic interest.

Acknowledgements

Some of the information given in this chapter has been kindly provided by colleagues at the University of Bristol and by Professor MELVIN CALVIN and his colleagues at the University of California, Berkeley. I am grateful to them all, and to the sponsors of our work at Bristol, namely, the Natural Environment

Research Council, the Science Research Council, Shell International, the National Aeronautics and Space Administration (NsG 101-61, to the University of California, Berkeley), the Petroleum Research Fund and Quaker Chemicals Limited.

References

1. LLOYD, D.: Structure and reactions of simple organic compounds. London: University Press 1967.
2. GRUNDON, M. F., and H. B. HENBEST: Organic chemistry – an introduction, second edit. London: Oldbourne 1968.
3. TEDDER, J. M., and A. NECHVATEL: Basic organic chemistry. Parts I and II. London: Wiley 1966.
4. FIESER, L. F., and M. FIESER: Introduction to organic chemistry. Heath 1957.
5. MORRISON, R. T., and R. N. BOYD: Organic chemistry, second edit. Boston: Allyn and Bacon, Inc. 1966.
6. ROBERTS, J. D., and M. C. CASERIO: Basic principles of organic chemistry. New York: Benjamin 1964.
7. CRAM, D. J., and G. S. HAMMOND: Organic chemistry, second edit. New York: McGraw-Hill 1964.
8. MISLOW, K.: Introduction to stereochemistry. New York: Benjamin 1966.
9. ORÓ, J.: Personal Communication.
10. EGLINTON, G.: Applications of gas chromatography – mass spectrometry in organic geochemistry. Proc. Soc. Analyt. Chem. **4**, 111 (1967).
11. –, and M. CALVIN: Chemical fossils. Sci. Am. **216**, 32 (1967).
12. MCCARTHY, E. D., and M. CALVIN: Organic geochemical studies. I. Molecular criteria for hydrocarbon genesis. Nature **216**, 642 (1967).
13. MOLD, J. D., R. E. MEANS, R. K. STEVENS, and J. M. RUTH: The paraffin hydrocarbons of wool wax. Homologous series of methyl alkanes. Biochemistry **5**, 455 (1966).
14. BU'LOCK, J. D.: The biosynthesis of natural products. London: McGraw-Hill 1965.
15. BLUMER, M.: Organic pigments: Their long-term fate. Science **149**, 722 (1965).
16. –, and W. D. SNYDER: Isoprenoid hydrocarbons in recent sediments: Presence of pristane and probable absence of phytane. Science **150**, 1588 (1965).
17. SEVER, J., and P. L. PARKER: Normal and isoprenoid fatty alcohols in sediments. Science (in press).
18. HAN, J., E. D. MCCARTHY, and M. CALVIN: The hydrocarbon constituents of the blue-green Algae, *Nostoc muscorum*, *Anacystis nidulans*, *Phormidium loridum* and *Chlorogloea fritschii*. J. Chem. Soc. (in press).
19. Syracuse University 1968 Lecture series. Perspectives in chemical ecology. (To be published by Academic Press.)
20. AVIGAN, J., G. W. A. MILNE, and R. J. HIGHET: The occurrence of pristane and phytane in man and animals. Biochim. Biophys. Acta **144**, 1127 (1967).
21. GÖHRING, K. E. H., P. A. SCHENK, and E. D. ENGELHARDT: A new series of isoprenoid isoalkanes in crude oils and cretaceous bituminous shales. Nature **215**, 503 (1967).
22. BALDWIN, M. W., and J. BRAVEN: 2-aminoethylphosphonic acid in *Monochrysis*. J. Marine Biol. Assoc. U.K. **48**, 603 (1968).
23. HILLS, I. R., G. W. SMITH, and E. V. WHITEHEAD: Optically active spiroterpane in petroleum distillates. Nature **219**, 243 (1968).
24. HILLS, I. R., and E. V. WHITEHEAD: Pentacyclic triterpanes from petroleum and their significance. Advances in Organic Geochemistry (HOBSON and LOUIS, editors). London: Pergamon Press 1966.
25. KNOCHE, H., P. ALBRECHT, and G. OURISSON: Organic compounds in fossil plants (*Voltzia brongniarti*, Coniferales). Angew. Chem. Intern. Ed. **7**, 631 (1968).
26. –, and G. OURISSON: Hydrocarbons in modern and fossil *Equisitum*. Angew. Chem. Intern. Ed. **6**, 1085 (1967).
27. BLOCH, K.: Lipid patterns in the evolution of organisms. From Evolving genes and proteins (BRYSON and VOGEL, editors). New York: Academic Press 1965.
28. STRANSKY, K., M. STREIBL, and V. HEROUT: On natural waxes. VI. Distribution of wax hydrocarbons in plants at different evolutionary levels. Coll. Czech. Chem. Commun. **32**, 3213 (1967).
29. ALSTON, R. E., T. J. MABRY, and B. L. TURNER: Perspectives in chemotaxonomy. Science **142**, 545 (1963).

30. – Biochemical systematics. Chap. 7 in: Evolutionary biology, vol. 1 (TH. DOBZHANSKY, M. K. HECHT, and W. C. STEERE, editors). New York: Appleton-Century-Crofts 1967.
31. TURNER, B. L.: Pure Appl. Chem. **14**, 189 (1967).
32. MABRY, T. J., and J. A. MEARS: Alkaloids and plant systematics. Chap. in: The chemistry of the alkaloids. New York: Reinhold Book Corporation 1969.
33. ORÓ, J., T. G. TORNABENE, D. W. NOONER, and E. GELPI: Aliphatic hydrocarbons and fatty acids of some marine and freshwater microorganisms. J. Bacteriol. **93**, 1811 (1967).
34. GELPI, E., J. ORÓ, H. J. SCHNEIDER, and E. O. BENNETT: Olefins of high molecular weight in two microscopic algae. Science **161**, 700 (1968).
35. ODUM, E. P.: Ecology. New York: Holt, Rinehart, and Winston 1963.
36. PHILLIPSON, J.: Ecological energetics. London: Arnold 1966.
37. MARQUAND, J.: Life: Its nature, origins, and distribution. Edinburgh and London: Oliver and Boyd 1968.
38. MASON, B.: Principles of geochemistry, third edit. New York: Wiley & Sons 1966.
39. WICKMAN, F. E.: The cycle of carbon and the stable carbon isotopes. Geochim. Cosmochim. Acta **9**, 136 (1956).
40. HOLSER, W. T., and I. R. KAPLAN: Isotope geochemistry of sedimentary sulfates. Chem. Geol. **1**, 93 (1966).
41. IVANOV, M. V.: Microbiological processes in the formation of sulphur deposits. Israel programs for scientific translations. Jerusalem (1968). Available from U.S. Department of Commerce, CFSTI, Springfield, USA.
42. SOKOLOVA, G. A., and G. I. KARAVAIKO: Physiology and geochemical activity of Thiobacilli. Israel programs for scientific translations. Jerusalem 1968. Available from U.S. Department of Commerce, CFSTI, Springfield, Va. 22151, USA.
43. PETROW, R.: The black tide. London: Hodder and Stoughton 1968.
44. SMITH, J. E., editor: Torrey Canyon pollution and marine life. Marine Biological Association of the United Kingdom (1968).
45. ROBINSON, J., A. RICHARDSON, A. N. CRABTREE, J. C. COULSON, and G. R. POTTS: Organochlorine residues in marine organisms. Nature **214**, 1307 (1967).
46. Geological Society Phanerozoic Time-Scale 1964: Quart. J. Geol. Soc. Lond. **120** S, 260 (1964).
47. EGLINTON, G., A. G. DOUGLAS, J. R. MAXWELL, J. H. RAMSAY, and S. STALLBERG-STENHAGEN: Occurrence of isoprenoid fatty acids in the Green River shale. Science **153**, 1133 (1966).
48. HAUG, P., H. K. SCHNOES, and A. L. BURLINGAME: Isoprenoid and dicarboxylic acids from the Colorado Green River shale (Eocene). Science **158**, 772 (1967).
49. KVENVOLDEN, K. A.: Normal fatty acids in sediments. J. Amer. Oil Chem. Soc. **44**, 628 (1967).
50. BURLINGAME, A. L., and G. R. SIMONEIT: Isoprenoid fatty acids isolated from the kerogen matrix of the Green River formation (Eocene). Science **160**, 531 (1968).
51. HAMILTON, P. B.: Amino acids on hands. Nature **205**, 284 (1965).
52. MACLEAN, I., G. EGLINTON, K. DOURAGHI-ZADEH, and R. G. ACKMAN: Correlation of stereo-isomerism in present-day and geologically ancient isoprenoid fatty acids. Nature **218**, 1019 (1968).
53. ZANGERL, R., and E. S. RICHARDSON: Paleoecological history of two Pennsylvanian Black shales. Fieldiana Geol. Mem. **4** (1963).
54. EGLINTON, G., and R. J. HAMILTON: Leaf epicuticular waxes. Science **156**, 1322 (1967).
55. KOLATTUKUDY, P. E.: Biosynthesis of surface lipids. Science **159**, 498 (1968).
56. TORNABENE, T. G., E. GELPI, and J. ORÓ: Identification of fatty acids and aliphatic hydrocarbons in *Sarcinia lutea* by gas chromatography and combined gas chromatography – mass spectrometry. J. Bacteriol. **94**, 333 (1967).
57. DOUGLAS, A. G., and B. J. MAIR: Sulfur: Role in genesis of petroleum. Science **147**, 499 (1965).
58. PHILIPPI, G. T.: On the depth, time and mechanism of petroleum generation. Geochim. Cosmochim. Acta **29**, 1021 (1965).
59. WELTE, D. H.: Relation between petroleum and source rock. Bull. Am. Assoc. Petrol. Geologists **49**, 2246 (1965).
60. BROOKS, J. D., and J. W. SMITH: The diagenesis of plant lipids during the formation of coal, petroleum and natural gas – I. Changes in the n-paraffin hydrocarbons. Geochim. Cosmochim. Acta **31**, 2389 (1967).
61. DEGENS, E. T.: Diagenesis of organic matter. Chap. 7 in: Diagenesis in sediments (G. LARSEN and G. V. CHILINGAR, editors). Amsterdam: Elsevier Publ. Co. 1967.

62. Murphy, Sister M. T. J., McCormick, and G. Eglinton: Perhydro-carotene in the Green River shale. Science **157**, 1040 (1967).
63. Henderson, W., V. Wollrab, and G. Eglinton: Identification of steroids and triterpenes from a geological source by capillary gas-liquid chromatography and mass spectrometry. Chem. Comm. **13**, 710 (1968).
64. Burlingame, A. L., P. Haug, T. Belsky, and M. Calvin: Occurrence of biogenic steranes and pentacyclic triterpanes in an Eocene shale (52 million years) and in an Early Precambrian shale (2.7 billion years): A preliminary report. Proc. Nat. Acad. Sci. U. S. **54**, 1406 (1965).
65. Cloud, P. E., Jr.: Carbonaceous rocks of the Soudan Iron formation (Early Precambrian). Science **148**, 1713 (1965).
66. – Atmospheric and hydrospheric evolution on the primitive earth. Science **160**, 724 (1968).
67. Calvin, M.: Chemical evolution. London: Oxford University Press 1969.
68. Wald, G.: The origin of life. Sci. Am., August issue (1954).
69. Steinman, G., and M. N. Cole: Synthesis of biologically pertinent peptides under possible primordial conditions. Proc. Nat. Acad. Sci. U.S. **58**, 735 (1967).
70. Wilson, J. T.: A revolution in earth science. Can. Mining Met. Bull. February, 1 (1968).
71. Engel, A. E. J., B. Nagy, L. A. Nagy, C. G. Engel, G. O. W. Kremp, and C. M. Drew: Alga-like forms in Onverwacht series, South Africa: Oldest recognized lifelike forms on earth. Science **161**, 1005 (1968).
72. Hoering, T. C.: Chap. in: Researches in geochemistry, vol. 2, editor P. H. Abelson. New York: Wiley 1967.
73. Henderson, W., G. Eglinton, P. Simmonds, and J. E. Lovelock: Thermal alteration as a contributory process to the genesis of petroleum. Nature **219**, 1012 (1968).
74. Studier, M. M., R. Hayatsu, and E. Anders: Organic matter in early solar system. – I. Hydrocarbons. Geochim. Cosmochim. Acta **32**, 151 (1968).
75. Munday, C., K. Pering, and C. Ponnamperuma: Synthesis of acyclic isoprenoids by the γ-irradiation of isoprene. Science (in press).
76. Lovelock, J. E.: A physical basis for life detation experiences. Nature **207**, 568 (1965).
77. Simmonds, P. G., G. P. Shulman, and C. H. Stembridge: Organic analysis of soil by pyrolysis-gas chromatography-mass spectrometry – A candidate experiment for a Mars Lander. Science (in press).
78. Bayliss, G. S.: The formation of pristane, phytane and related isoprenoid hydrocarbons by the thermal degradation of chlorophyll. Am. Chem. Soc. Meeting, San Francisco, Spring (1968).

CHAPTER 3

Analytical Methods

M. T. J. MURPHY

Department of Chemistry
Saint Joseph College
West Hartford, Connecticut

Contents

I. Introduction

Organic geochemists from varying backgrounds focus several specialities on the analysis of geological samples in order to reach conclusions on the nature of the components present in a rock or sediment, their fate since burial, and the original form of the material in a living organism. The complexity of compounds extracted from rock and sediment specimens requires the discipline of analytical chemistry with the assistance of modern instrumentation for good results and meaningful conclusions. The kinds and relative weights of various component classes in a geological sample provide the specific data on which the geochemist infers the history of a sample. Any valid conclusion depends upon the accuracy and reliability of the methods employed in the analysis. The geochemist attempts to discover *how much* is present before stating *how come* it is present in the sample. Often it is important to know the quantity of a component before any interpretation is made on how it is possible for the compound to be present in the specimen.

Organic geochemistry, in spite of its recent development, commonly employs several analytical techniques not as yet recognized as standard methods. Many of these procedures have been adapted from the earlier methods developed by petroleum and lipid chemists [1, 2]. Complexity of components, small amounts of contamination-free extracts, and the extended implications of any results demand a level of sophistication beyond most analytical work.

Briefly the analytical method proceeds as follows: the organic matter in any natural sample is first extracted from the rock, then separated into component classes and identified by accepted procedures.

II. Extraction

Either the nature of the sample itself or the type components may determine the best extraction method for a given specimen. Solvent extraction may consist of removing the organic matter by simply shaking the sample with particular solvents, placing the sample in a solvent in an ultrasonic energy source, or by Soxhlet apparatus. Such an extract represents soluble or free organic matter as distinguished from the insoluble kerogen portion of the rock. Choice and purity of solvent is important since many geological samples contain little organic matter. Hexane itself is often used to remove hydrocarbons, including triterpenes, from rocks. Perhaps more commonly employed is the mixture benzene-methanol (4:1; v/v) which dissolves many more classes than does hexane [3]. In general the polarity of the extraction solvent will determine to a large extent the compounds removed from the specimen. Extraction at elevated temperatures usually removes more organic matter from the rock matrix than does a similar extraction carried out at room temperature, provided this is the first extraction attempted on the sample.

The author's experience with Soxhlet extraction of aromatic hydrocarbons from East Berlin Shales, Connecticut, shows that a prior extraction of the pulverized rock (200 μ) with a solvent and using ultrasonic energy will remove most of the soluble organic matter from matrix, leaving little still bound to the clay particles for removal in the Soxhlet procedure. The methods using ultrasonic energy are inherently faster than the other procedures. For example, 1 hour using ultrasonic energy, if properly done, gives results equal to 72 hours employing elevated temperatures in a Soxhlet extraction. For a specimen rich in organic matter, such as an oil shale, second and third extractions with fresh solvent effectively remove most solvent extractable matter. Experimental procedures for the removal of the bound or kerogen matter are more varied and less generally applicable to all samples [4, 5].

VAN HOEVEN *et al.* [4], modifying an older procedure of KWESTROO and VISSER, has developed a simple method for removing the bound organic matter from the rock matrix. Hydrofluoric acid, prepared in concentrated solutions, free of contamination, dissolves the rock matrix, thus freeing the solvent-insoluble or bound organic matter.

ROBINSON in Chapter 26 describes methods for removal of kerogen in specimens such as the Green River Oil Shale. In addition ozone is commonly used to remove some kerogen material from older specimens, especially Precambrian rocks which usually contain very little soluble organic matter. BITZ and NAGY have modified the standard ozonolysis procedure used by chemists for geological samples [6]. In this method the soluble organic matter is removed in a Soxhlet apparatus prior to ozonolysis. ABLESON and HOERING have also employed ozone to degrade organic matter from several geological samples [7]. Solvent extraction supposedly leaves the organic matter from the rock unaltered, even if the extraction has been carried out at elevated temperatures, whereas the ozonolysis drastically alters the structure of kerogen by reacting with double bonds, thus freeing a smaller molecule from the original material, perhaps now modified because of the ozone treatment.

Pyrolysis attempts a similar approach in that the products of a reaction are fragments or rearrangements of the original starting material. This latter technique drastically changes the original organic material, even though the technique is much simpler in practice than ozonolysis. Several mass spectrometers have pyrolysis units preceding the inlet system.

Extraction methods provide the relative percentage of organic matter in the rock. Many analyses in the literature fail to indicate this relationship, and as a result make interpretation and the significance of a particular class of compounds, such as free fatty acids, difficult when the weight of the total extract or rock is unknown. Oil Shales, such as the Green River Shale, rich in organic matter, contain 33 percent total organic matter, but the solvent-soluble portion, however, is less than 10 percent of the shale [8].

III. Separation Methods

Identification of components in a geological extract and any resulting conclusions concerning their origin are more definite when individual compound classes are separated prior to spectroscopic analysis. Chromatography, clathration and specific chemical reactions comprise the most important geochemical techniques for separating complex mixtures of organic compounds. Petroleum chemists developed many of these separation procedures, which are widely accepted analytical methods for geochemistry today.

An analytical thin-layer chromatogram of the crude extract provides much information on the nature or complexity of the mixture, particularly which chemical classes are present as well as which are absent. A chromatogram of 100–200 μg of extract on Silica Gel – G (0.25 mm), developed in a solvent mixture of hexane-ether-acetic acid (95:5:1) will separate hydrocarbons, esters, acids, alcohols from one another, and from nitrogen compounds which remain at the origin.

The hydrocarbon distribution of the mixture is evident from a similar thin-layer chromatogram developed in hexane. Such initial chromatograms provide a relative distribution of compound classes in the sample extract. This perspective is helpful in any discussion on the significance of any single component present in a natural sample. For example, whether the isoprenoid hydrocarbons are a major or minor constituent of the extract is important in any interpretation of the analytical results [8, 9].

Chromatographic techniques, separation methods based on adsorption and partition principles, are extensively employed to fractionate complex geological sample extracts. Gas [10] or solution chromatography [11], including column [12], paper, and thin layer [13–15], may be carried out on an analytical or preparative scale depending on the size of sample [16, 17]. Special types of chromatography also include ion-exchange and gel permeation chromatography. In ion-exchange chromatography, the stationary phase is a polymer chain containing acidic or basic functional groups which attract to varying degrees ionic species such as acids, bases, and salts in the moving phase, thus causing differences in migration characteristics of the components of the mixture [18].

Gel filtration chromatography separates molecules according to size by providing pore sites on the stationary phase, such as cross-linked polydextran (Sephadex) [19]. These pores include certain size molecules and exclude others, thus effecting separation of components [20].

The nature of the mixture determines the type of column chromatography necessary to separate individual compound classes [21, 22]. The initial thin-layer chromatogram provides information on the relative distribution of classes, thus pointing to a particular adsorbent and solvent series for effective separation of components [23, 24]. Silica Gel – G, 60–80 mesh, usually activated ca. 120° C is effective in separating unsaturated from saturated substances, whereas alumina is recommended for mixtures of aliphatic and aromatic hydrocarbons. A sample to adsorbent ratio of 1:250 usually separates mixtures without tailing [25]. A common eluotropic series of solvents for column chromatography is hexane > carbon tetrachloride > benezene > ether > methanol > water.

Because of trace impurities remaining on the adsorbent from the manufacturing process, it is important to pre-wash the column with hexane or a more polar solvent to remove them from the gel before the sample is applied to the column. The bed volume of a column is dependent upon mesh size and packing characteristics. A rule of thumb useful for determining the bed volume is 1 ml per gram of adsorbent. A procedure blank examined before a sample is analyzed will determine the extent of contamination, or the lack thereof, for any column. Any thin-layer chromatogram requires the similar prewashing procedure before samples are applied to the adsorbent.

Most solvents used in geochemistry require distillation. A recent study by KVENVOLDEN and HAYES indicates that pesticide grade solvents are contamination-free and suggest that they may be used directly from the bottle [26].

For certain separations of aromatic hydrocarbons, for example, alumina Grade I may need a certain amount of deactivation. Alumina Grades II, III, IV, and V (Brockman Index) can be prepared by adding 3, 6, 10, 15 percent water to the grade I adsorbent. A simple technique, recommended by WOELM involves adding the weight percent of water to the adsorbent in a stoppered round-bottom flask, rotating for a short time, and allow it to equilibrate for two hours.

Some laboratories find alumina activated at 170° C for 2 hours as effective as deactivated alumina for aromatic hydrocarbons. Monitoring the effluent with ultraviolet light helps signal the presence of aromatic compounds.

Column chromatography is ordinarily followed by an analytical thin layer chromatogram to check on effectiveness of a column, and if necessary by preparative thin-layer [14]. Often helpful at this point in the scheme of the analysis is a silver-ion impregnated thin-layer chromatogram to detect unsaturated compounds within a given fraction. A 2 percent $AgNO_3$ aqueous solution is used to prepare the slurry [29]. Unsaturated compounds may be removed by preparative Ag^+ thin-layers, or by Ag^+ impregnated columns. The former technique is much more practical to many geochemists. Following this separation a gas chromatogram may give evidence of several members of a single column or thin-layer fraction.

Details on gas chromatography applications to geochemistry are presented by DOUGLAS in Chapter 5 (see Figs. 1–3).

The best record of a thin-layer is the chromatogram itself and prepared chromaplates made with plastic backing are easily filed for reference. A photograph of a glass-backed chromatogram also provides an invaluable record for future reference. Any Polaroid camera with manual settings and close-up lenses is easily mounted for black and white or color photographs. If the chromatogram is photographed under ultraviolet light, a Wratten 2E filter effectively absorbs the excessive blue light. Pola Pan type 200, a fine-grain film, gives good even definition of trace compounds on chromatograms, and a pencilled label, sprayed with the same fluorescent dye serves as a good identification for the chromatogram.

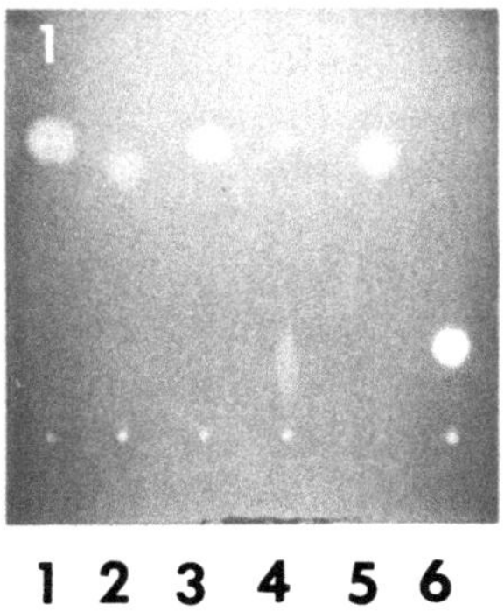

Fig. 1

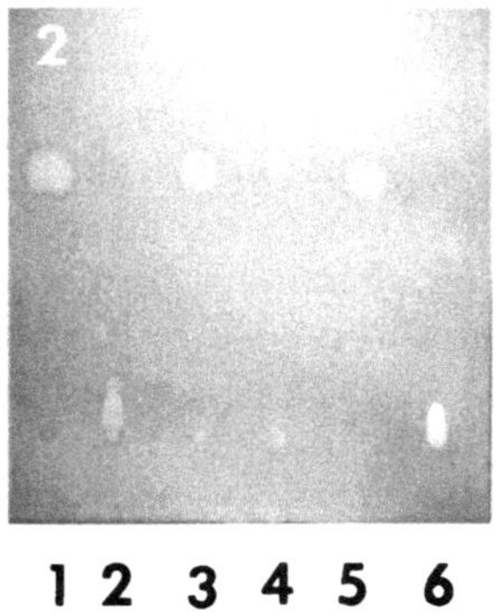

Fig. 2

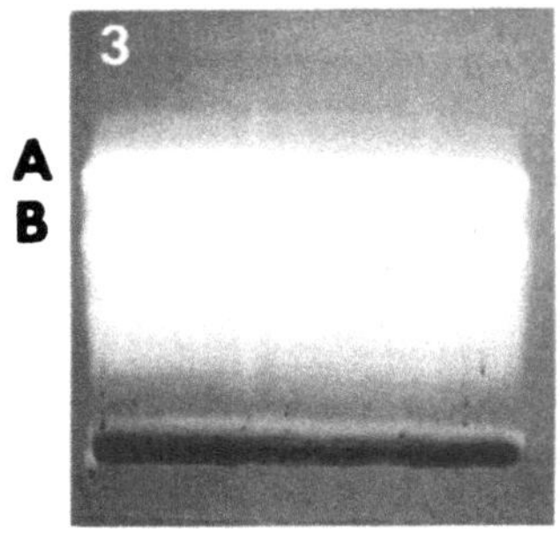

Fig. 3

Fig. 1. Thin layer chromatogram of hydrocarbons: *1* octadecane, *2* 1-octadecene, *3* squalane, *4* squalene, *5* cholestane, *6* phenanthrene. Conditions: Silica Gel – G; *n*-hexane developer; 0.0005 percent Rhodamine 6 – G visualizer; observed under ultraviolet light.

Fig. 2. Thin layer chromatogram of hydrocarbons on layer impregnated with 2 percent silver nitrate: Same samples and conditions as Fig. 1. Note differences in migration values for unsaturated compounds.

Fig. 3. Preparative thin layer chromatogram for East Berlin gray shale. Conditions: Silica Gel – G; *n*-hexane-benzene (9:1; v/v); viewed under ultraviolet light. Components A and B eluted from adsorbent for gas chromatography-mass spectrometry. (See Fig. 9.)

For color photography, 35 mm film, which is more economical than Polaroid, lends itself to a permanent set-up for color film (see Figs. 4–6).

On thin layer chromatograms, the detection level of trace components depends upon the sensitivity of the particular spray reagent, its concentration, and quality of the spray application. For example, for hydrocarbons, Rhodamine 6G is three times as sensitive as dichlorofluorescein, and a 0.0005 percent solution of the former fluorescent spray is more sensitive than the usually recommended 0.001 percent solution. The latter spray gives a green background instead of a faint purple color which provides high contrast for the yellow lipid components. As little as 0.25 μ hydrocarbon or fatty acid is visible with Rhodamine 6G under ultraviolet light. The darker background of the Ag^+ chromatograms lessens the sensitivity by two and requires a 0.001 percent spray solution. Alumina adsorbent usually shows less sensitivity than Silica Gel – G.

Sulfur, often present in natural samples, particularly Recent sediments, appears intensely purple on Silica Gel–G when sprayed with Rhodamine 6G [30, 31]. It R_f value is somewhat less than saturated hydrocarbons, but greater than organic sulfides and thiols which resemble the yellow lipid color with this reagent.

A. Clathration

For the separation of straight chain hydrocarbons, and other similar lipids, techniques such as clathration effectively separate certain classes of compounds according to molecular radius. Methods such as urea and thiourea adduction as well as molecular sieves further separate a chromatographic preparative

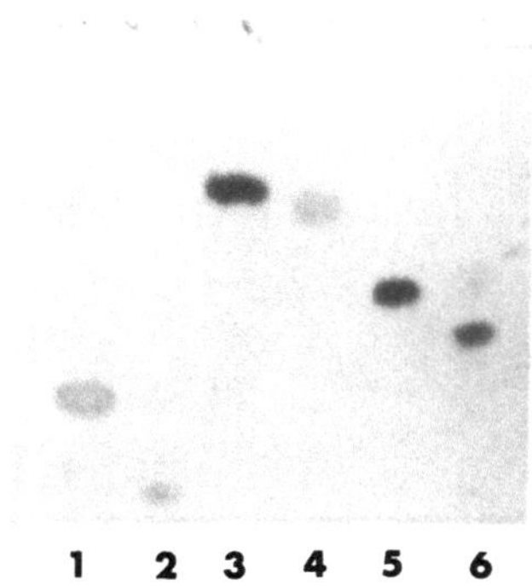

Fig. 4

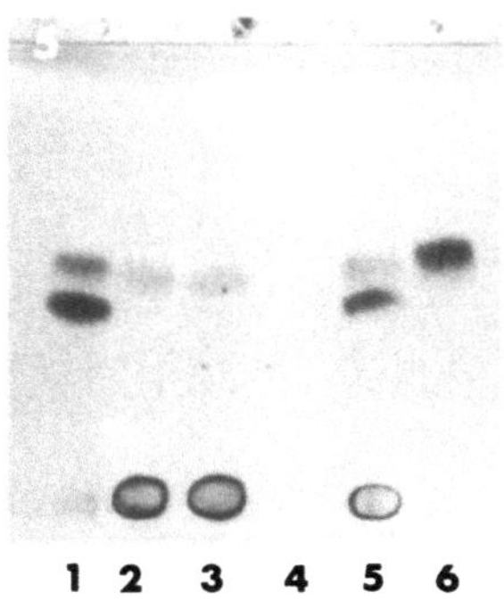

Fig. 5

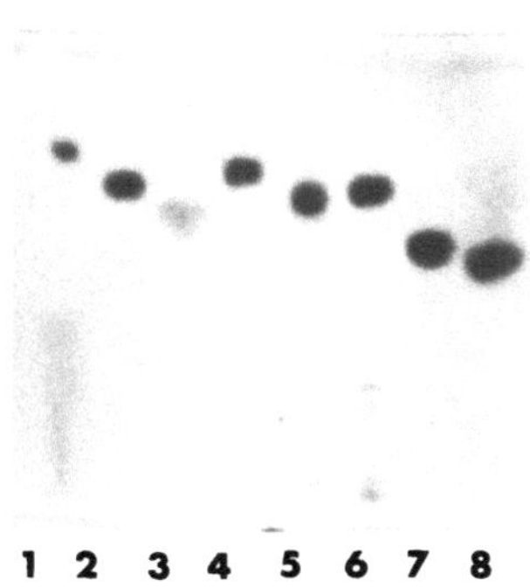

Fig. 6

Fig. 4. Thin layer chromatogram of hydrocarbon and sulfur compounds: *1* squalene, *2* β-carotene, *3* dotriacontane, *4* octadecane, *5* di-*n*-octadecyl trisulfide, *6* octadecyl thiol. Conditions: Silicic acid-magnesium silicate; *n*-hexane developer; sulfuric acid-dichromate visualizer. Compare R_f values of organic sulfur compounds to saturated and aromatic hydrocarbons [30].

Fig. 5. Thin layer chromatogram of derivatives of N-ethyl maleimide (NEM). *1* octadecyl thiol, *2* and *3* NEM derivative of octadecyl thiol, *4* NEM, no spot, *5* octadecyl thiol derivative with NEM, reaction on thin layer, *6* mixture of elemental sulfur and octadecyl thiol. Conditions: Silicic acid-magnesium silicate; *n*-hexane developer; sulfuric acid-dichromate visualizer [30].

Fig. 6. Thin layer chromatogram of hydrocarbons and sulfur compounds: *1* dotriacontane, *2* octadecane, *3* elemental sulfur, *4* di-*n*-octadecyl trisulfide, *5* octadecyl thiol, *6* squalene, *7* p,p′-bitolyl, *8* retene. Conditions: Silicic acid-magnesium silicate; *n*-hexane-diethyl ether (94:6; v/v) developer; phosphomolybdic acid visualizer. Note the R_f for elemental sulfur in this solvent system compared to hexane [30].

fractions [2, 32, 33]. Perhaps the only limitation is the scarcity of adduction agents or molecular sieves for molecules of increasing diameter. Urea will adduct normal and iso-compounds and alkyl benzenes compounds; thiourea accepts isoprenoid, small ring compounds and steranes, such as cholestane, sitostane, ergostane. The larger steranes will not fit into the hexagonal thiourea crystal lattice [34, 35]. Molecular sieves (5 Å) effectively separate normal hydrocarbons from other type compounds [36]. Larger diameter synthetic zeolithes are available in 10 X and 13 X sizes, but available solvents for reflux systems become a problem with these materials [37–39]. A 7 Å sieve removes acyclic isoprenoid hydrocarbons from a branched-cyclic mixture; if necessary, further fractionation of the cyclic fraction may be achieved using thiourea (Figs. 7 and 8).

A recommended procedure for the separation of a hexane eluate containing hydrocarbons consists of a urea adduction to remove the straight chain, iso- and alkyl benzene components, followed by a clathration with 5 Å molecular sieve to separate the normal hydrocarbons from the branched and iso-hydrocarbons and alkyl benzenes.

The non-adduct from the urea preparation is then subjected to thiourea for adduction of the isoprenoid and sterane hydrocarbons. The thiourea non-adduct

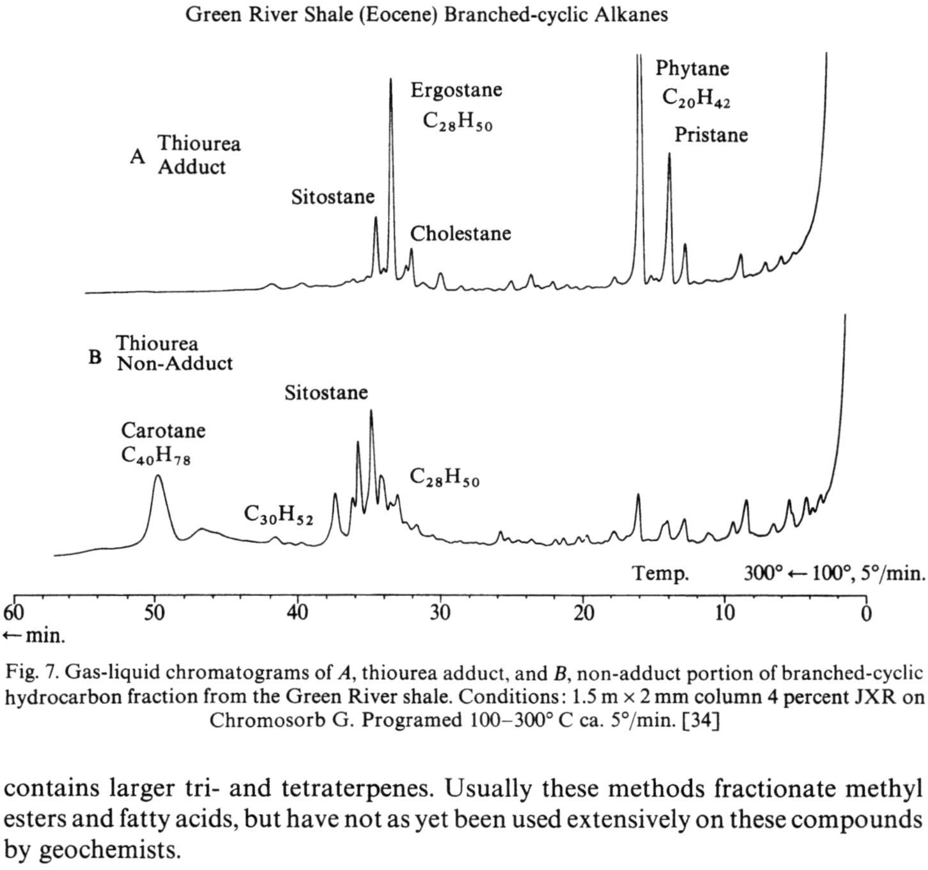

Fig. 7. Gas-liquid chromatograms of *A*, thiourea adduct, and *B*, non-adduct portion of branched-cyclic hydrocarbon fraction from the Green River shale. Conditions: 1.5 m × 2 mm column 4 percent JXR on Chromosorb G. Programed 100–300° C ca. 5°/min. [34]

contains larger tri- and tetraterpenes. Usually these methods fractionate methyl esters and fatty acids, but have not as yet been used extensively on these compounds by geochemists.

B. Derivatives

Derivative formation modified to a micro scale facilitates separations and identifications otherwise impossible or difficult. Often these methods precede instrumental methods in the scheme of analysis. For geochemical derivative applications, the mercury or silver ion complex with olefins and methylation of fatty acids find wide use on both macro and micro scales. As mentioned above, silver ion impregnated on thin layers effectively separates unsaturated compounds from saturated substances. Often the final step in a procedure preceeding gas chromatography is the isolation of unsaturated components by preparative chromatography on silver impregnated layers [29]. Also in the literature are several methods for incorporating silver ion on silica gel columns to effect a separation of unsaturated components from a mixture [40, 41].

C. Saponification

Glycerides of higher fatty acids occurring in geological samples are converted to potassium salts by refluxing in methanolic potassium hydroxide. The saponification proceeds smoothly in a one-step extraction-saponification tech-

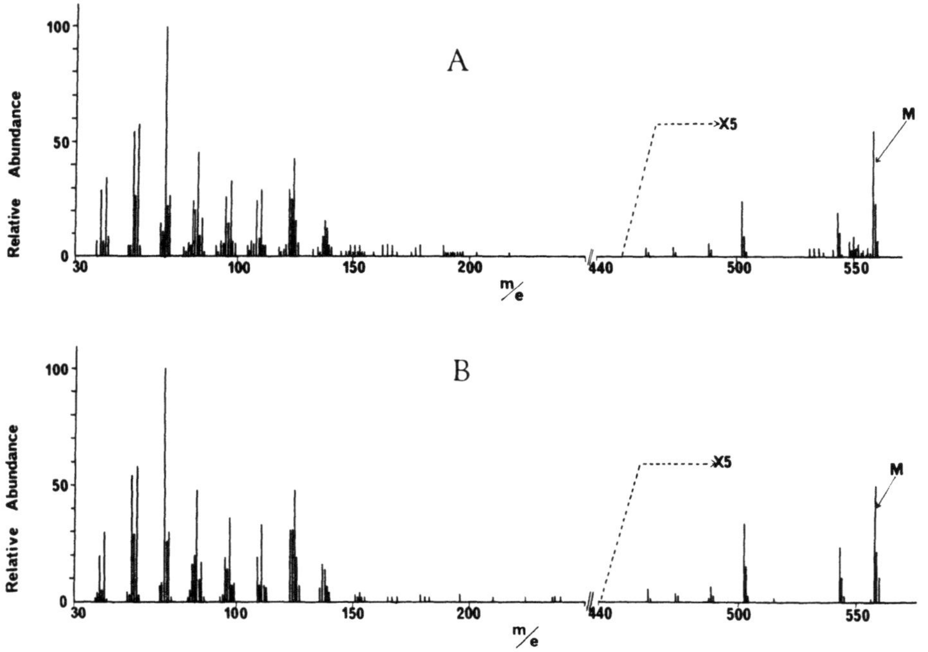

Fig. 8. Mass spectra of *A*, $C_{40}H_{78}$ from Green River shale and *B*, authentic perhydro-β-carotene. Conditions: LKB-9000 mass spectrometer; 70 ev; direct insertion probe; evaporation temperature 125°; ion-source temperature 300° C [34].

nique, using a Soxhlet apparatus with a benzene-methanolic KOH (10 percent) reflux solution [42]. Separation of the aqueous and benzene phases isolates the fatty acids or their K^+ salts in the aqueous phase, while the hydrocarbons and less polar compounds remain in the organic phase. Conversion of the acid salts to the free fatty acid with HCl and subsequent extraction with ether concentrates the fatty acids for further chromatographic or spectroscopic analysis.

If the organic matter is first extracted from the rock matrix by other suitable methods, a subsequent saponification in methanolic KOH (10 percent) will enable separation of the acid salts from other extractables.

D. Esterification

Because of the polarity and boiling points of fatty acids it is convenient to convert them to their respective methyl esters for gas chromatography. A simple reflux of the fatty acids in methanol containing a trace of sulfuric acid, followed by extraction in benzene, comprises a simple classical method to obtain fatty acid methyl esters for subsequent chromatographic analysis [43].

Some investigators prefer boron trifluoride in methanol as an esterification reagent [44]. The reaction takes place in minutes compared to the conventional two hour reflux described above and is readily adapted to micro quantities of fatty acids. Because the reaction is complete in minutes, gas chromatography may follow immediately after conversion to the esters [45–47].

E. Special Derivatives

Silylation reactions of a wide variety of polar compounds containing an active hydrogren permit analysis of the trimethylsilyl ether derivatives by gas-liquid chromatography. Such compounds as fatty acids, alcohols, sterols, and steroids, plus complex molecules having poly-functional groups, such as sugars, amino acids, phenols, etc., form trimethylsilyl ether derivatives with silylating agents such as hexamethyldisilazane, trimethyl chlorosilane, and bis-dimethylsilylacetamide. The resulting silyl ether derivatives are more volatile, and lend themselves to gas chromatographic analysis. These latter compounds have lower boiling points and thus lend themselves to gas chromatography. The reaction is rapid, works well on micro amounts, and permits injection onto a column almost immediately after conversion to the silyl ether.

BOYLAN introduced an interesting extension of silyl derivatives by exchanging the metal ion in porphyrins with silicon. A similar lowering of boiling point occurs, enabling analysis by gas chromatography. Before this technique, most research on porphyrins occurring in geological samples has been limited to ultraviolet and visible spectroscopy.

2,4,7-Trinitrofluorenone (I) complexes with various aromatic compounds, permitting isolation from more complex mistures [52, 53]. These adducts possess characteristic colors which makes the reagent useful as a spray visualizer for aromatic hydrocarbons developed on thin layers [54, 55].

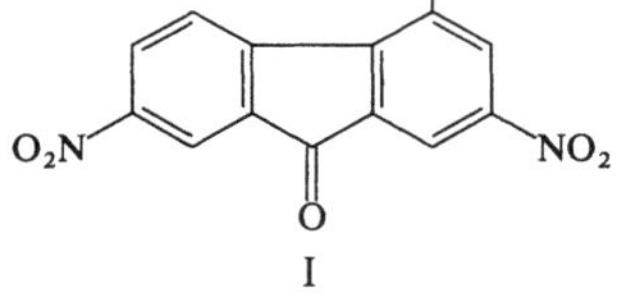

I

Some geological samples, particularly Recent sediments, contain sulfur which is extracted by some organic solvents. The origin of the sulfur may be organic sulfur compounds such as sulfides, thiols or disulfides, etc. N-ethylmaleimide selectively reacts with thiols; such a derivatization is easily carried out on micro samples prior to application to thin layer chromatograms [30, 31].

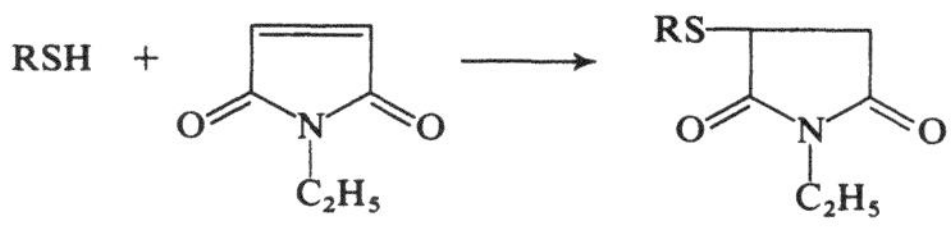

Sulfur compounds in geological extracts are often undetected because of their similarity to hydrocarbons with respect to polarity and reaction to some color visualizers. Elemental sulfur, purple under ultraviolet light, is distinguishable, but its *Rf* value is almost that of saturated hydrocarbons with thiols and sulfides migrating closely behind sulfur. After reaction with diethylmaleiate, the thiol adduct remains at the origin. The reaction is easily carried out on a preparative scale.

F. Ozonolysis

The amount of extractable organic matter from a rock is often only a small percentage of the total carbon content of the rock, excluding any graphite which may be present. The solvent-insoluble material, kerogen, described in Chapter 26 by ROBINSON, may be partially removed from the rock matrix by subjecting solvent-extracted rock to ozone which attacks carbon–carbon double bonds to form ozonides, later isolated as esters or acids. Thus the smaller fragment molecules are identified by conventional methods and attributed to the original kerogen material [6, 56].

Older rocks, particularly those from Precambrian times, contain only trace amounts of extractable organic matter, but ozone reacting with kerogen provides information on the indigenous carbon material. This technique will become more useful as interest increases in the very early terrestrial rocks. The compounds thus isolated are varied but present in trace quantities. Any analysis of ozonized extract probably requires gas chromatography followed by mass spectrometry. Even with mass spectra, interpretation of results is difficult.

IV. Identification Methods

Organic geochemistry utilizes the major analytical techniques available for identifying individual components in mixtures isolated from geological speciments. Ultraviolet, visible, infrared, and nuclear magnetic resonance spectroscopy, mass spectrometry, and optical rotation comprise the more important methods routinely employed by geochemists at one time or another. For all of these methods, perhaps with the exception of high resolution mass spectrometry, a pure sample in varying amounts must be available for analysis. Some samples can only be analyzed by gas chromatography in combination with mass spectrometry because of the one or so milligrams available. Even when recognized identification methods are applied to geochemical problems, the use of standards cannot be recommended too highly, as well as complete procedure blanks. Isomers, for example, present difficult analytical problems; several methods must be employed before definitive identification can be made.

Infrared analysis readily identifies functional groups, but is less useful for hydrocarbons without functional groups [57]. Esters, fatty acids, unsaturated and hetero compounds absorb characteristically in the infrared region of the electromagnetic spectrum. The fingerprint region is also helpful for identifying hydrocarbons, such as steranes and triterpanes. Infrared analysis requires relatively large amounts of sample, for example, 1 mg of pure material, unless microtechniques are used.

Compounds containing chromophores absorb in the ultraviolet region of the spectrum. Many unsaturated conjugated systems readily absorb, resulting in high sensitivity [58, 59]. Olefins possess end absorption and fatty acids absorb at wavelengths less than 200 mμ, which is below the operational level of most spectrophotometers. Hetero and conjugated systems readily absorb ultraviolet radiation

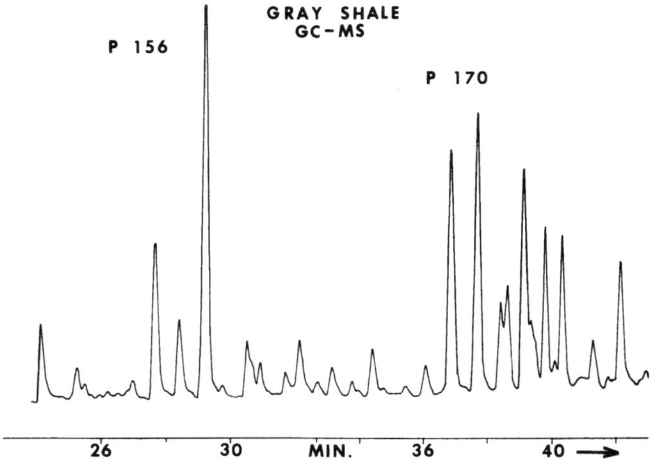

Fig. 9. Gas-liquid chromatogram of hexane fraction from East Berlin Gray shale separated on thin layer chromatograms in Fig. 3, *A* and *B*. Note several isomers for parent peak 156 and 170. Conditions: Perkin-Elmer 226 chromatograph-Hitachi RMU 6 mass spectrometer; 150′ × 0.020″ SCOT column, Apiezon L, programed 100–300° C ca. 1°/min.

with high extinction coefficients in many cases [60, 61]. Aromatic hydrocarbons have characteristic absorption in the ultraviolet region as well as fluorescence, which often help identify the particular compound. Identification from fluorescence is somewhat more difficult as this emission is a smooth distribution, but maxima often provide some supporting evidence for identification. A trace of a more strongly fluorescent compound may mask a weaker component, and thus requires independent confirmation by another technique.

Nuclear magnetic resonance determines proton environment, enabling the analyst to identify many organic compounds isolated from geological specimens. Some of the problems with isomers are solved with nuclear magnetic resonance spectroscopy. Several dimethyl naphthalene isomers in the East Berlin Shale, Connecticut, have been identified from chromatography retention times, mass and nuclear magnetic resonance spectra. To date the method requires at least 1–3 mg of pure compound, requiring usually a gas chromatography preparation preceeding the analysis. For this reason geochemistry has not as yet taken full advantage of the capabilities of nuclear magnetic resonance spectroscopy.

Mass spectrometry has been most helpful for significant research in geochemistry [62, 63]. It is a powerful key to structure as well as molecular weight. Purity of the compound is important, but micro amounts are easily analyzed by this method [64, 65]. Gas chromatography followed by mass spectrometry is ideal for complex mixtures available in micro amounts. Eighty or ninety components have been so separated in an aromatic fraction of the same East Berlin Shale, on a capillary column with mass spectra obtained on each individual component. Such components are often trace amounts, and difficult to collect for further analysis. Nevertheless, many identifications are possible with mass spectrometry which would otherwise be impossible [66, 67]. Isomers are occasionally separated on a good capillary column, but the mass spectra are too similar

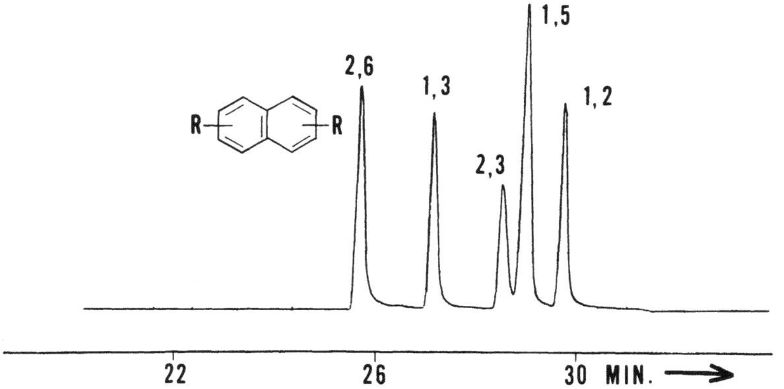

Fig. 10. Gas-liquid chromatogram of several dimethyl naphthalene isomers for comparison with East Berlin Gray shale in Fig. 9. Conditions: Same as for Fig. 9.

to rely on mass spectrometry alone for identification. Such data requires careful interpretation as well as reference compounds for identification (see Fig. 9, 10).

Optical rotary dispersion is a technique seldom used in geochemistry, although it has been applied to biogenic vs. abiogenic problems in the past. Several milligrams of pure material are required for meaningful data. The newer instruments capable of measurement at various wavelengths make the method attractive for identifications of certain compounds possessing an asymmetric center [68–70].

In summary, modern instrumentation facilitates identification of components isolated in trace amounts from geological samples. It is important to remember that more than one method may be necessary for identification, and any conclusion is further supported by other analytical data, such as derivative formation and physical constants, such as refractive index, melting and boiling points.

V. Contamination

All too often geological samples are unavailable in large amounts, and usually contain relatively small amounts of organic matter. Any laboratory procedure must be contamination free if the results are valid. Usually contamination is introduced from solvent impurities, silicone grease or plasticizers, and every now and then, fingerprints. A check on these sources cannot be made too often, as most geochemists know from experience. A better procedure is to limit any sample contact with grease or plastic material. Silicones and plasticizers, especially phthalate esters, are often contaminants. Even when this latter procedure is followed, a procedure blank is strongly recommended for all geochemical work. Such a blank follows the same experimental conditions as the sample, and any gross contamination is recognized and eliminated before the sample is analyzed. When presenting results, the data becomes more reliable when a procedure blank has been used.

In conclusion, even the best method becomes that much more reliable when reference compounds have been used to confirm the presence of a particular component. More studies could easily use standard reference materials, but often the specific compound is unavailable or never has been available. As time passes, however, the geochemist himself is coming to realize that procedure blanks and reference compounds are routine measures for good analyses. It is hoped that more reference compounds will become available for this type research; the synthetic organic chemist has an important role in organic geochemistry for the future.

References

1. MORRIS, L., and B. NICOLS: Chromatography of lipids. In: Chromatography, 2 ed. (E. HEFTMANN, ed.), p. 466–509. New York: Reinhold 1967.
2. MEINSCHEIN, W., and G. KENNY: Analyses of a chromatographic fractions of organic extracts of soils. Anal. Chem. **29**, 1153–1161 (1957).
3. EVANS, E., G. KENNY, W. MEINSCHEIN, and E. BRAY: Distribution of n-paraffins and separation of saturated hydrocarbons from Recent marine sediments. Anal. Chem. **29**, 1858–1861 (1957).
4. HOEVEN, W. VAN, J. R. MAXWELL, and MELVIN CALVIN: Fatty acids and hydrocarbons as evidence of life processes in ancient sediments and crude oils. Geochim. Cosmochim. Acta **33**, 877 (1969).
5. KWESTROO, W., and J. VISSER: Ultrapurification of hydrofluoric acid. Analyst **90**, 297–298 (1965).
6. BITZ, SR. M. C., and B. NAGY: Ozonolysis of "polymer-type" material in coal, kerogen, and in the Orgueil meteorite: A preliminary report. Proc. Nat. Acad. Sci. U.S. **56**, 1383–1390 (1966).
7. ABELSON, P.: Researches in geochemistry, vol. 2. New York: Wiley 1967.
8. CUMMINS, J., and W. ROBINSON: Normal and isoprenoid hydrocarbons isolated from oil-shale bitumen. J. Chem. Eng. Data **9**, 304–307 (1964).
9. LIDA, T., E. YOSHII, and E. KITATSIYI: Identification of normal paraffins, olefins, ketones, and nitriles from the Colorado shale oil. Anal. Chem. **38**, 1224–1227 (1966).
10. ETTRE, L., and A. ZLATKIS: The practice of gas chromatograph. New York: Wiley 1967.
11. HEFTMANN, E.: Chromatography, 2nd ed. New York: Reinhold 1967.
12. SNYDER, L.: Principles of adsorption chromatography. New York: Marcel Dekker 1968.
13. STAHL, E.: Thin layer chromatography. New York: Springer 1962.
14. BOBITT, J.: Thin layer chromatography. New York: Reinhold 1963.
15. RANDERATH, K.: Thin layer chromatography (D. D. LIBMAN, trans.). New York: Academic Press 1964.
16. MANGOLD, H.: Thin layer chromatography of lipids. J. Am. Oil Chem. Soc. **38**, 708–727 (1961).
17. – Thin layer chromatography of lipids. J. Am. Oil Chem. Soc. **41**, 762–777 (1964).
18. WALTON, H.: Techniques and applications of ion-exchange. In: Chromatography, 2nd ed. (E. HEFTMANN, ed.), p. 325–342. New York: Reinhold 1967.
19. GELOTTE, B., and J. PORATH: Gel filtration. In: Chromatography, 2nd ed. (E. HEFTMANN, ed.), p. 343–372. New York: Reinhold 1967.
20. BLUMER, M., and W. SNYDER: Porphyrins of high molecular weight in a Triassic oil shale: Evidence by gel permeation chromatography. Chem. Geol. **2**, 35–45 (1967).

21. SNYDER, L.: Applications of linear elution adsorption chromatography to the separation and analysis of petroleum. Anal. Chem. **33**, 1527–1534 (1961).
22. – Applications of linear elution adsorption chromatography to the separation and analysis of petroleum. Anal. Chem. **33**, 1535–1538 (1961).
23. DOUGLAS, A., and G. EGLINTON: The distribution of alkanes. In: Comparative phytochemistry (T. SWAIN, ed.), p. 57–77. New York: Academic Press 1966.
24. EGLINTON, G., P. SCOTT, T. BELSKY, A. BURLINGAME, W. RICHTER, and M. CALVIN: Occurence of isoprenoid alkanes in a Precambrian sediment. In: Advances in organic geochemistry (M. LOUIS, ed.), p. 41–74. New York: Pergamnon Press 1966.
25. HAMWAY, P., M. CEFOLA, and B. NAGY: Factors affecting the chromatographic analysis of asphaltic petroleum and Recent marine sediment organic matter. Anal. Chem. **34**, 43–48 (1962).
26. KVENVOLDEN, K., and J. HAYES: The solvent battle-solvent purification procedures used in organic geochemistry laboratories, Group for the analyses of carbon compounds in carbonaceous chondrites and the returned lunar sample meeting, Tempe, Arizona, Dec. 1968.
27. COMMINS, B.: A modified method for the determination of polycyclic hydrocarbons. Analyst **83**, 386–389 (1958).
28. – Interim report on the study of techniques for the determination of polycyclic aromatic hydrocarbons in air. Nat. Cancer Inst. Monograph **9** (1961).
29. MORRIS, L.: Impregnated adsorbent chromatography. In: New Biochemical separations (A. T. JAMES and L. J. MORRIS, eds.), p. 295–320. New York: Van Nostrand 1964.
30. MURPHY, Sr. M., B. NAGY, G. ROUSER, and G. KRITCHEVSKY: Identification of elementary sulfur and sulfur compounds in lipid extracts by thin layer chromatography. J. Am. Oil Chem. Soc. **42**, 475–480 (1965).
31. – – Analysis for sulfur compounds in lipid extracts from the Orguiel meteorite. J. Am. Oil Chem. Soc. **43**, 189–196 (1966).
32. BARON, M.: Analytical applications of inclusion compounds. In: Physical methods in analytical chemistry, vol. 4 (W. BERL, ed.), p. 226–232. New York: Academic Press 1961.
33. NICOLAIDES, N., and F. LAVES: Structural information of long-chain fatty material obtained by x-ray diffraction studies of single crystals. J. Am. Oil Chem. Soc. **40**, 400 (1963).
34. MURPHY, Sr. M., A. MCCORMICK, and G. EGLINTON: Perhydro-β-carotene in the Green River shale. Science **157**, 1040–1042 (1967).
35. MONTGOMERY, D.: Thiourea adduction of alkylated polynuclear aromatic hydrocarbons and heterocyclic molecules. J. Chem. Eng. Data **8**, 432–436 (1963).
36. THOMAS, T., and R. MAYS: Separations with molecular sieves. In: Physical methods in chemical analysis, vol. 4 (W. BERL, ed.), p. 45–98. New York: Academic Press 1961.
37. MAIR, B., and M. SHAMAIENGAR: Fractionation of certain aromatic hydrocarbons with molecular sieve adsorbents. Anal. Chem. **30**, 276–279 (1958).
38. FENSELAU, C., and M. CALVIN: Selectivity in zeolite occlusion of olefins. Nature **212**, 889–891 (1966).
39. WIEL, A. VAN DER: Molekularsiebe und Gas-Flüssigkeits-Chromatographie als Hilfsmittel zur Bestimmung der Konstitution von Paraffin. Erdoel, Kohle, Erdgas, Petrochem. **8**, 632–636 (1965).
40. DEVRIES, B.: Quantitative separations of higher fatty acid methyl esters by adsorption chromatography with silver nitrate, J. Am. Oil Chem. Soc. **40**, 184 (1963).
41. PRIVETT, O., and E. NICKELL: Preparation of highly purified fatty acids via liquid-liquid partition chromatography. J. Am. Oil Chem. Soc. **40**, 189 (1963).
42. NAGY, B., and Sr. M. C. BITZ: Long-chain fatty acids from the Orguiel meteorite. Arch. Biochem. Biophys. **101**, 240–248 (1963).
43. GEHRKE, C., and D. GOERLITZ: Quantitative preparation of methyl esters of fatty acids for gas chromatography. Anal. Chem. **35**, 76–80 (1963).
44. DURON, O., and A. NOWONTNY: Microdetermination of long-chain carboxylic acids by transesterification with boron trifluoride. Anal. Chem. **39**, 370–372 (1963).
45. VORBECK, M., L. MATTICK, F. LEE, and C. PEDERSON: Preparation of methyl esters of fatty acids for gas-liquid chromatography. Anal. Chem. **33**, 1512–1514 (1961).
46. LEO, R., and P. PARKER: Branched-chain fatty acids in sediments. Science **152**, 649–650 (1966).
47. EGLINTON, G., A. DOUGLAS, J. MAXWELL, and J. RAMSAY: Occurence of isoprenoid fatty acids in the Green River shale. Science **153**, 1133–1135 (1966).

48. COOPER, J.: Fatty acids in Recent and ancient sediments and petroleum reservoir waters. Nature **193**, 744–746 (1962).
49. SWEELEY, C., R. BENTLEY, M. MAKITA, and W. WELLS: Gas-liquid chromatography of trimethylsilyl derivatives of sugars and related substances. J. Am. Chem. Soc. **85**, 2497–2507 (1963).
50. SUPINA, W., R. KRUPPA, and R. HENLY: Dimethylsilyl derivatives in gas chromatography. J. Am. Oil Chem. Soc. **44**, 74–76 (1967).
51. HORNING, M., E. BOUCHER, and A. MOSS: The study of urinary acids and related compounds by gas phase analytical methods. J. Gas Chromatog. **5**, 297–302 (1967).
52. NEUMANN, M., and P. JOSSANG: Emploi des complexes π en chromatographie sur couche mince derives polynitres aromatiques et hydrocarbures aromatiques à noyaux condenses. J. Chromatog. **14**, 280–283 (1964).
53. KUCHARCZYK, N., J. FOHL, and J. VYMETAL: Dunstschichtchromatographie von aromatischen Kohlenwasserstoffen und einigen heterocyclischen Verbindungen. J. Chromatog. **11**, 55–61 (1963).
54. BERG, A., and J. LAM: Separation of polycyclic aromatic hydrocarbons by thin layer chromatography on impregnated layers. J. Chromatog. **16**, 157–166 (1964).
55. INSCOE, M.: Photochemical changes in thin layer chromatographs of polycyclic aromatic hydrocarbons. Anal. Chem. **36**, 250–255 (1964).
56. NAGY, B., and Sr. M. C. BITZ: Analysis of bituminous coal by a combined method of ozonolysis, gas chromatography, and mass spectrometry. Anal. Chem. **39**, 1310–1313 (1967).
57. COLTHUP, N., L. DALY, and S. WIBERLEY: Infrared and Raman spectroscopy. New York: Academic Press 1964.
58. JAFFE, H., and M. ORCHIN: Theory and applications of ultraviolet spectroscopy. New York: Wiley 1962.
59. SCHOPF, J., K. KVENVOLDEN, and E. BARGHOORN: Amino acids in Precambrian sediments: An assay. Proc. Nat. Acad. Sci. U.S. **59**, 639–646 (1968).
60. MAIR, B., R. ZALMAN, E. EISENBRAUN, and A. HORDOYSKY: Terpenoid precursors of hydrocarbons from the gasoline range of petroleum. Science **154**, 1339–1341 (1966).
61. COMMINS, B.: Polycyclic aromatic hydrocarbons in carbonaceous metoerites. Nature **212**, 273–274 (1966).
62. BIEMANN, K.: Mass spectrometry organic chemical applications. New York: McGraw-Hill 1962.
63. BEYNON, J., R. SAUNDERS, and A. WILLIAMS: The mass spectra of organic molecules. New York: American Elsevier 1968.
64. MACLEOD, W., and B. NAGY: Deactivation of polar chemisorption in a fritted-glass molecular separator interfacing a gas chromatograph with a mass spectrometer. Anal. Chem. **40**, 841–842 (1968).
65. BURLINGAME, A., P. HAUG, T. BELSKY, and M. CALVIN: Occurrence of biogenic steranes and pentacyclic triterpanes in an Eocene shale (52 million years) and in an early Precambrian shale (2.7 billion): A preliminary report. Proc. Nat. Acad. Sci. U.S. **54**, 1406–1412 (1965).
66. MODZELESKI, V., W. MACLEOD, and B. NAGY: A combined gas chromatograph-mass spectrometric method for identifying *n*- and branched chain alkanes in sedimentary rocks. Anal. Chem. **40**, 987–989 (1968).
67. EGLINTON, G., P. SCOTT, T. BELSKY, A. BURLINGAME, and M. CALVIN: Hydrocarbons of biological origin from a one-billion year old sediment. Science **145**, 263–264 (1964).
68. HILLS, I., E. WHITEHEAD, D. ANDERS, J. CUMMINS, and W. ROBINSON: An optically active triterpane, gammacerane in Green River, Colorado oil shale bitumen. Chem. Commun. 752–753 (1966).
69. HILLS, I., and E. WHITEHEAD: Triterpanes in optically active petroleum distillates. Nature **209**, 977–979 (1966).
70. –, G. SMITH, and E. WHITEHEAD: Optically active spirotriterpane in petroleum distillates. Nature **219**, 243–246 (1968).

CHAPTER 4

Mass Spectrometry in Organic Geochemistry *

A. L. BURLINGAME and H. K. SCHNOES **

Department of Chemistry and Space Sciences Laboratory, University of California, Berkeley, California

Contents

I. Introduction

Mass spectrometry does not represent a novel physical tool in geochemical research. Petroleum chemists have made use of mass spectrometry for many years in the analysis of the components of crude oil. Such applications were concerned primarily with quantitative analytical problems, with emphasis on the determination of compound types which constitute a given petroleum sample, and the number of papers on this subject is legion. In recent years – and the developments of recent years are primarily the subject of this review – the tremendous potential of mass spectrometry for the solution of structural problems has been recognized by the organic chemical community in general, and since then rather extensive and sophisticated use has been made of the technique for structural studies on organic compounds. The widespread application to problems of organic chemistry has in turn sparked intense interest in mechanistic studies on the fragmentation

* The authors acknowledge generous financial support from the National Aeronautics and Space Administration, Grants NsG 101 and NGR 05-003-134.

The authors wish to thank *Analytical Chemistry, Nature, Proceedings of the National Academy of Sciences, Science*, and *Geochimica Cosmochimica Acta* for permission to reproduce selected figures.

The authors are indebted to Drs. K. BIEMANN and J. HAYES for figures concerning the Murray meteorite; Dr. J. ORÓ for Fig Tree Formation; Drs. M. CALVIN and E. D. MCCARTHY for the mass spectra of the C_{19} isomers; Dr. G. EGLINTON and Mr. D. H. HUNNEMAN for the mass spectrum of the dihydroxy acid, trimethyl silyl ether; and Dr. D. W. THOMAS for discussions regarding the phytadienes.

** Department of Biochemistry, University of Wisconsin, Madison, Wisconsin.

processes of ionized molecules in the vapor state. Practical applications and fundamental studies combined have thus within a few years produced a substantial body of information on the nature of mass spectra of organic compounds, information which makes feasible a fruitful exploitation of mass spectrometric data for the solution of structural problems. The considerable number of volumes [1–10] addressing themselves wholly or in part to the application of mass spectrometry to problems of organic and natural products chemistry reflects the pace of progress and the scope of this field; it is recommended that these works be consulted for a thorough initiation into all phases of organic mass spectrometric research and its ramifications.

An earlier review by HOOD [11] discussed the contribution of mass spectrometric data to research on the molecular composition of petroleum, and the results of compound type analyses are presented in some detail there. In contrast, this chapter is intended to concentrate more heavily on modern mass spectrometric techniques developed during the past few years for the solution of molecular structures of isolated pure compounds, and to bring into focus the salient information from correlations of mass spectra with molecular structure and stereochemistry from an increasingly vast body of mass spectrometric literature.

Topics of interest include modern instrumentation currently employed in geochemical research, the exploitation of high resolution mass spectrometry and computer techniques, and the coupling of gas chromatographic and mass spectrometric instruments. Structural problems in which mass spectrometry has played an important role will receive particular attention and the information available on the mass spectra of compounds of geochemical interest will be presented.

It is hoped that the discussion will serve to underscore the fact that mass spectrometry is a major and basic physical tool available to the organic geochemist. It provides information complementary to that obtained from other techniques, but does not necessarily replace them. We shall stress the type of structural information most readily gleaned from a mass spectral fragmentation pattern, and also the ambiguities and limitations inherent in such data.

II. Instrumentation and Techniques

A. Conventional Instrumentation

Mass spectrometers most commonly utilized for geochemical work have been of the single focussing ("low resolution") type. In recent years double focussing ("high resolution") instruments have become commercially available and several reports describe their application to various aspects of organic geochemistry (cf. section II, D and E). Technical discussions on the designs of mass spectrometers and the historical developments in this field can be found in the monograph by BEYNON [1] and ion optical considerations in the volume by EWALD and HINTENBERGER [12] and that edited by MCDOWELL [13]; less technical summaries of the principles of mass spectrometer operations and designs are included in several other monographs (for example, KISER [14]; BRUNNÉE and VOSHAGE [15]).

In general, quantitatively reproducible fragmentation patterns are best obtained employing rapid magnetic scanning (0.1–10 sec per decade in mass) while operating the ion source at high static accelerating voltage (preferably 8–10 Kv for highest sensitivity) and the ion multiplier at high gain ($\sim$150–200 volts per stage). Such operation demands utilization of fast response recording systems ($\geqq$5 Kc band width), but obviates the problem caused by the dependence of the gain of the multiplier on kinetic energy (changing accelerating voltage) of the impinging ions [16].

For quantitatively reproducible mass spectra, the sample pressure must be maintained constant for the duration of the scan of the entire spectrum; in addition, all instrument operating parameters must remain constant. In a conventional inlet system this is assured by maintaining a reservoir sample size which is large compared to the leak rate into the mass spectrometer (generally in the order of 1 percent sample depletion per hour). Using direct introduction to the ion source of either collected or g.l.c. effluent samples, fluctuations in sample pressure in the source chamber are usually rapid (in order of seconds) and, therefore, direct inlet temperature control and very rapid scanning are necessary to obtain a representative spectrum. Mention should be made of the variations in fragmentation pattern produced by changes in the ion source temperature [17, 18].

We may restrict the discussion here, then, to a brief summary of certain instrument characteristics of importance for geochemical applications:

1. High sensitivity. This is an important aspect, since only submicrogram or even nanogram amounts of pure substance may be available for study. Many currently available mass spectrometers are quite capable of obtaining high quality mass spectra on submicrogram samples and interpretable patterns can be produced with samples in the nanogram range, when ion multipliers are used for detection.

2. Resolution, $M/\Delta M$ in excess of 1:2,000 (10 percent valley definition) and a mass range of approximately 2,000 mass units. Many "conventional" or low resolution mass spectrometers demonstrate such performance routinely.

3. Fast scanning capability and correspondingly fast response recording systems. Submicrogram samples introduced directly into the ion source (see below) will often be volatilized and pumped off very quickly, and a single scan of the entire mass range of interest should thus be accomplished within seconds (0.1–10 seconds). This is of particular importance for g.l.c.-mass spectrometer coupling arrangements, where it is desirable to obtain several consecutive scans while a component is emerging from a g.l.c. column. Current mass spectrometers may be scanned both electro-dynamically (by varying the accelerating voltage) and magnetically (by varying magnet current); the latter mode has certain advantages discussed above.

The recording of a decade in nominal mass in tenths to a few seconds is normally accomplished via a recording oscillograph utilizing 2–5 Kc galvanometers, which play u.v. light on sensitive linograph paper. Since the dynamic range of ion intensities is in the order of 1 to 10^4 and since it is desirable to measure all of these intensities to the same order of precision (0.1–1 percent), the use of 5 Kc galvanometers which display a full deflection of 2.0 inches may be a disadvantage in fast scanning, compared with the $\leqq 2.5$ Kc galvanometers which display an 8–10 inch deflection. In the future, this problem will best be solved by digital data acquisition techniques [19]. A step in this direction has been reported for low resolution mass spectrometers where digitized data are recorded on gapless magnetic tape for subsequent batch processing by computer [20].

4. A direct ion source introduction system, discussed in more detail below.

5. The capability of direct coupling of a gas chromatograph to the ion source of a mass spectrometer. One of the ensuing sections describes this technique in greater detail. In principle, any mass spectrometer could be used for such a tandem arrangement, but relatively accessible source construction is of distinct technical advantage. Sector instruments, in which the ion source is outside the magnet gap, are thus more conveniently coupled to a gas chromatograph than the corresponding 180° Dempster-type mass spectrometer. Cycloidal instrument geometry, which places the collector slit inside the magnetic field, precludes the use of standard ion multipliers (due to effect of varying magnetic field on ion multiplier response) and is consequently not capable of scanning the entire mass range in seconds.

6. Ionization. Most analytical applications involve ionization of the sample by electron bombardment. For a number of compound types (i.e., normal or branched alkanes, amines, alcohols, etc.) electron bombardment ionization results in molecular ions of relatively low abundance. While this is not a serious shortcoming, for certain applications (i.e., mixture analysis) alternative ionization methods, such as field ionization [21] and chemical ionization [22], which repress the fragmentation pattern itself and produce relatively intense molecular ion peaks and $M-1$ peaks, may be of distinct advantage. BECKEY and COMES [23] give a thorough discussion of ionization methods.

In an electron bombardment ion source, the approximate energy of the bombarding electrons can be controlled (or set) from below the ionization potential of a molecule (~ 4 e.v.) to approximately 70 e.v. As the energy bombarding energy is increased, a value is reached at which ionization takes place which is referred to as the appearance potential of the ion in question [24, 25]. Thereafter, upon further increase in electron energy, some fraction of the energy above the appearance potential is imparted to the molecule from the impact, resulting in a molecular ion with excess internal energy of both electronic and vibrational nature which is responsible for the concomitant fragmentation.

Electron energies from the value at the appearance potential to approximately 25 electron volts sometimes alter the fragmentation patterns drastically, and the pattern reproducibility is correspondingly poor. Of course, at energies near the appearance potential, it is often possible to suppress the entire fragmentation pattern leaving only molecular ions. It should be cautioned that the sensitivity does not increase with decreasing electron energy (suppression of the fragmentation), but does in fact decrease. Using low electron bombarding energies, the analysis of complex mixtures is facilitated [18] particularly at high resolution and will be discussed in that section further. Applications of low voltage techniques at low resolution will be discussed under appropriate sections.

B. Sample Handling and Introduction Systems

The general techniques for sample handling and introduction for the mass spectrometric analysis of organic compounds described in some detail by BEYNON [1] and BIEMANN [2] are also applicable to geochemical research with perhaps minor modifications depending on the size and nature of sample.

Geochemical samples of pure compounds destined for mass spectrometric analysis are usually obtained directly from the effluent of a gas chromatographic column. It is, then, most convenient to collect such samples in an ordinary melting point capillary, which can be introduced directly into the ion source inlet system of the mass spectrometer. This straightforward collection technique avoids the necessity of further transferral of sample into a special introduction tube fitted for the mass spectrometer. Samples collected in devices from which they cannot be introduced into the mass spectrometer (for instance, some of the g.l.c. traps commercially available) can be transferred in solution (e.g., dichloromethane) to a melting point capillary sealed at one end, with subsequent removal of solvent under a stream of pure nitrogen. It is often most convenient in such cases to reinject the sample and recollect it in a capillary. Compounds of high volatility must be trapped in U-tube capillaries with appropriate cooling; the section of tubing containing the condensed sample can then be cut and introduced into the instrument. For geochemical samples which are obtained in microgram quantities or less, the common methods of sample introduction into the mass spectrometer by means of a reservoir (0.5–3 liter) and valve system, are unsatisfactory. The sample reaches the mass spectrometer ion source as a stream of gas through a pin-hole leak designed such that at any given time very little of it will be in the ionizing chamber. For such samples, direct introduction is mandatory. Most commercial mass spectrometers are now fitted with "direct introduction probes" as standard instrument accessories. In Fig. 1 the probe arrangement utilized in the authors' laboratory for geochemical analyses is illustrated. A melting point capillary containing the sample fits into the probe tip, which is inserted directly (through the valve arrangement shown and without breaking the instrument vacuum) into the ion source region. Upon volatilization (accomplished by the ion source and/or auxillary probe heater), all of the sample passes through the region of the ionizing electron beam. Intense spectra can thus be obtained on very small quantities of material. A further advantage of the direct inlet is the possibility of partial fractionation of multiple component samples upon carefully controlled heating*.

Indeed, if a quantitative mixture analysis is desired, the direct inlet system cannot be used, since such analyses assume a constant partial pressure of each component [18]. The direct probe inlet system is an absolute necessity for very non-volatile and heat sensitive samples [26], since the pressure in the ion source region (order of 10^{-6} to 10^{-7} mm Hg) is considerably lower than in a reservoir system such that lower temperatures are sufficient for the sublimination of such compounds. The lower temperature and the fact that collisions with reservoir

* For example, preparation of volatile derivatives, e.g., silyl ethers of hydroxyl functions, acetylation or methylation of amino functions, etc., often yields mixtures of the starting material and product which differ greatly in volatility and many times can be fractionated in the direct inlet as indicated.

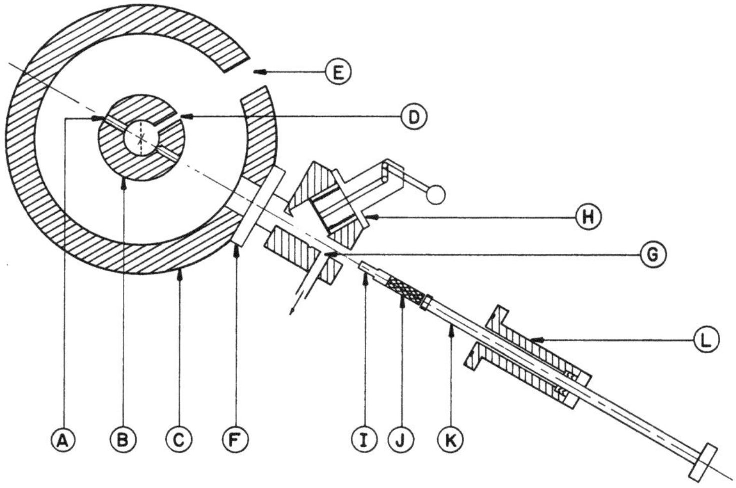

Fig. 1. Direct inlet system (C.E.C. 21–110B mass spectrometer). a Pump out port. b Ion source block. c Source housing. d Gas inlet port. e Heated cover plate port. f Direct inlet attached to source housing. g Pump out line to rough pump and diffusion pump. h Inline valve. i Silver sample tip. j Ceramic insulator. k Probe shaft. l Quick disconnect shaft holder

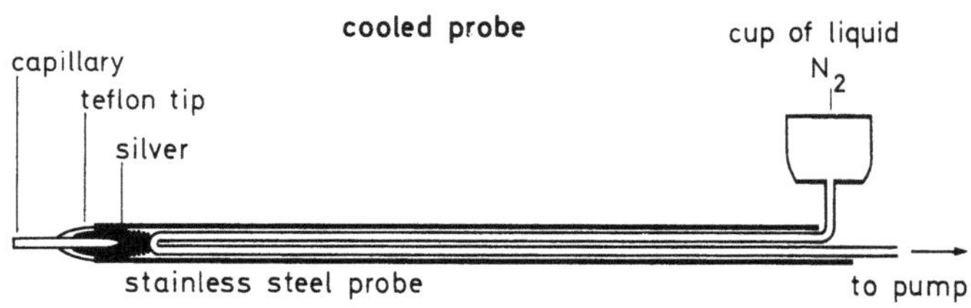

Fig. 2. Liquid nitrogen cooled direct inlet probe

walls after volatilization are avoided minimizes the decomposition of labile molecules.

For the analysis of volatile samples (e.g., hydrocarbons, esters, etc.) the direct probe poses some difficulties, since volatilization may occur very rapidly even at the lowest operating temperatures of the ion source (approximately 70–100°). To overcome this problem, a liquid nitrogen cooled probe may be used (Fig. 2 illustrates such a probe constructed in the authors' laboratory). Spectra of hydrocarbons in the range from C_{10} to C_{16} have been obtained successfully using this technique. The operation and use of such a probe is, however, somewhat more cumbersome and time-consuming. An alternative method, in which volatile samples are adsorbed on molecular sieves, has been described [27]. The desorption process from the sieve in the source chamber is sufficiently slow to permit the determination of mass spectra. Since it has been suggested [28] that liquid phase coated solid support (g.l.c. packing) may be used for efficient g.l.c.-fraction collection, it may also be useful for direct introduction of samples into the ion source of a mass spectrometer. The most convenient, although instrumentally most elaborate, system for introducing volatile samples directly into the ion source is the g.l.c.-mass spectrometer coupling arrangement. This technique (described more fully below) is utilized ordinarily for the analysis of complex mixtures but would be of equal advantage for the introduction of individual samples. Wherever multiple port construction of the ion source of the mass spectrometer permits the simultaneous connection of several inlet systems (a direct probe, all-glass inlet, stainless steel inlet*, and a g.l.c. inlet), this latter method becomes routinely feasible and is to be recommended. Of course, this requires that the compound being investigated is thermally stable under the operating conditions and temperatures of the gas chromatograph and also that it is not catalytically decomposed by the hot metal components of either the chromatograph or the mass spectrometer, e.g., dehydration of β-amino alcohols, etc.

C. Principles of High Resolution Mass Spectrometry and Techniques of Data Acquisition and Processing

The utilization of conventional, high sensitivity, fast scanning instrumentation with nominal mass resolutions in excess of 2,000 is not only an essential basic approach to the identification of relatively pure components isolated from geologic sources, but is also an extremely powerful physico-chemical technique of ubiquitous application to the myriad of organic structural problems.

Turning to a discussion of high resolution mass spectrometry and the techniques involved in data acquisition and processing, we will see in the next section that such data adds new dimensions and scope to the fragmentation pattern recognition approach of conventional mass spectrometry.

However, in this section we must now consider a brief discussion of the defect of nuclidic masses from integral mass numbers. The accurate atomic weights [29] of selected elements and isotopes of interest are given in Table 1, with reference to the major natural isotope of carbon defined to be $^{12}C = 12.000000$. One quickly notes that each element and isotope has a small but unique deviation from its integral mass such that if one had the capability of sufficiently accurate measurement of a mass in question, this would permit determination of the atomic composition. If all possible nuclidic combinations for the elements commonly occurring in organic molecules are considered, it may be noted that various nominal (integral) mass numbers may be composed of more than one elemental and/or isotopic combination. In Table 2 such common nominal mass multiplets are given and it can be seen that they differ sufficiently in mass such that accurate mass measurement would permit a distinction among, and actual determination** of, the

* A stainless steel inlet system usually consists of a 3-liter reservoir, valving system and gold leak to the ion source chamber and a micromanometer (see BEYNON, 1960, pp. 147–160) for the accurate measurement of the sample pressure. A micromanometer is necessary for quantitative analysis of mixtures of compounds whose fragmentation patterns and sensitivities are known [18].

** This demands certain qualification with reference to the so-called virtual doublets in Table 2. Those doublets which differ by more than 10 m.m.u. may be separated easily by current instrument resolutions up to at least mass 300, and their mass measured with sufficient accuracy to distinguish those possibilities unambiguously. However, there are quite a few compositional doublets which differ in mass by less than 5 m.m.u. which cannot be resolved currently at higher masses and require the most sophisticated techniques and high accuracy of mass measurement ($\sim$1 ppm) to assign elemental compositions unambiguously. Such is not routinely the case.

Table 1. *Atomic weights*[a] *and approximate natural abundance of selected isotopes*[b]

Isotope	Atomic weight ($^{12}C = 12.000000$)	Natural abundance (percent)
^{1}H	1.007825	99.985
^{2}H	2.014102	0.015
^{12}C	12.000000	98.9
^{13}C	13.003354	1.1
^{14}N	14.003074	99.64
^{15}N	15.000108	0.36
^{16}O	15.994915	99.8
^{17}O	16.999133	0.04
^{18}O	17.999160	0.2
^{19}F	18.998405	100
^{24}Mg	23.985045	78.80
^{25}Mg	24.985840	10.15
^{26}Mg	25.982591	11.06
^{28}Si	27.976927	92.2
^{29}Si	28.976491	4.70
^{30}Si	29.973761	3.1
^{31}P	30.973763	100
^{32}S	31.972074	95.0
^{33}S	32.971461	0.76
^{34}S	33.967865	4.2
^{35}Cl	34.968855	75.8
^{37}Cl	36.965896	24.2
^{50}V	49.947165	0.24
^{51}V	50.943978	99.76
^{54}Fe	53.939621	5.84
^{56}Fe	55.934932	91.68
^{57}Fe	56.935394	2.17
^{58}Fe	57.933272	0.13
^{63}Cu	62.929594	69.1
^{65}Cu	64.927786	30.9
^{79}Br	78.918348	50.5
^{81}Br	80.916344	49.5
^{127}I	126.904352	100

[a] KÖNIG, L. A., J. H. E. MATTAUCH, and A. H. WAPSTRA: Nucl. Phys., **31**, 18 (1962).
[b] BEYNON, J. H.: Mass Spectrometry and Its Application to Organic Chemistry. Appendix 3, p. 554. Amsterdam: Elsevier 1960.

respective nuclidic composition. We will return to this point again in the next section.

At this point, mention should be made of the necessity of incorporating ion energy focussing properties in addition to the direction focussing provided by the magnetic sector of a "single focussing" mass spectrometer to obtain high resolution. Such a combination of energy and direction focussing ion optical paths is termed "double focussing" and permits high resolution and concomitant accurate mass measurement.

Advantage may be taken of the ion optical properties of a radial electric field to provide energy (velocity, kinetic energy) focussing of an ion beam having an initial energy distribution. In double focussing mass spectrometer configurations, an electrostatic sector (radial electric field) is combined in tandem with a magnetic

Table 2

Mass	Doublet	Mass difference ΔM	Approximate max. ident. mass
2	H_2-D	-0.001548[a]	77
12	$H_{12}-C$	-0.09390	4,600
13	$CH-{}^{13}C$	-0.00446[a]	223
14	$N-CH_2$	-0.01258	730
16	$O-NH_2$	-0.02381	1,190
16	$O-CH_4$	-0.03639	1,820
19	$H_{19}-F$	-0.15027	7,500
32	$S-O_2$	-0.01776	890
32	$S-C_2H_8$	-0.09055	4,524
28	$CO-N_2$	-0.01123	560
28	$CO-{}^{28}Si$	-0.01798	900
30	$NO-C_2H_6$	-0.04897	2,450
36	$C_3-{}^{32}SH_4$	0.0033[a]	165
36	$HFO-C_3$	0.0011[a]	55
42	$C_2H_2O-N_3$	-0.00134[a]	77
48	$H_4{}^{28}SiO-C_4$	$+0.003142$[a]	157
50	$C_3H-H_2O_3$	-0.00268[a]	130
60	$C_5-N_2O_2$	0.00402[a]	200
60	${}^{13}CHNO-C_5$	0.0040[a]	200

[a] Origin of some virtual doublets in computer sort of possible empirical compositions, given current instrument performance.

sector [30]. Thus, aberration of the isobaric ion beam due to an initial kinetic energy spread (from kT, electron bombardment and fragmentation dynamics) may be eliminated to first order and, therefore, resolutions ($M/\Delta M$) of 10,000 to 60,000 may be attained routinely with concomitant measurement of isobaric ion beams to an accuracy approaching 1 part per million.

Such capability of accurate mass measurement provides the potentiality for "unique" determination of the corresponding elemental composition of the species in question, *assuming that the resolving power is sufficient to separate beams with the least inherent mass defect into homogeneous (isobaric) components.*

The most common elemental compositions encountered in organic structural studies in which the same nominal mass can have multiple components are listed in Table 2. Also listed are their relative mass differences and the highest mass at which each respective multiplet could be identified, assuming a resolution ($M/\Delta M$) of 50,000.

From ion optical and geometric considerations, two commercial double focussing instrument designs have employed tandem arrangements of electrostatic (energy focussing) and magnetic (velocity focussing) fields. These differ in the locus of points which bring the isobaric ion beams into coincident double focus, thereby achieving high resolution. The Mattauch-Herzog geometry consists of opposed radii of curvature for electrostatic and magnetic fields and provides not only a point, but a plane of double focus [31]. Thus, instruments employing such ion optical geometry display the versatility of operation in the static mode as a spectrograph (photographic plate in plane of focus, allowing simultaneous spectrum registration over an approximate 30:1 mass range), and in the dynamic mode (scan of ion accelerating voltage or, more commonly, magnetic field strength) as a spectrometer employing ion multiplier ion detection (assume sufficient amplifier band width for rapid scanning, few seconds per decade in mass with resolution, $M/\Delta M$, in excess of 10,000). The Nier-Johnson geometry results in a point of double focus [32], requiring instrument operation as a spectrometer for determination of the entire mass spectrum. Accurate mass measurement with either ion-optical geometry with a precision of a few parts per million in mass is easily accomplished for peaks of interest by superposition of the unknown mass upon that of a known mass standard peak, e.g., that of perfluorocarbon ion C_nF_{2n+1}, by applying a periodic perturbation to the electric or magnetic sector, i.e., "peak matching" [33]. This is obviously a time-consuming, tedious process which may be best accomplished for a few relatively intense peaks of the probable several hundred in any given mass spectrum. True operation of such double focussing instrumentation as a spectrometer requires very high band width amplification (excess 10 Kc) for ion multiplier detection and high data transfer rates in data acquisition either on frequency modulated analogue magnetic tape [34] or into intermediate storage after analogue to digital conversion for transfer onto digital magnetic tape [35, 36].

Automated techniques for the reading and measurement of accurate masses from mass spectrograms are well understood and now quite routinely accomplished [26, 37–39]. Accuracy of mass measurement in the ten parts per million range results from comparable linear distance measurement, but relative ion abundances are poor due to the non-linearity of photographic emulsions over wide dynamic ranges, e.g., $1-10^4$ [38]. However, simultaneous recording of a mass range (30:1) via the integrating nature of photographic emulsions may be employed where the wide dynamic range is important even though the relative abundance measurements are poor. Current mass dispersions ($\sim 10^3$ microns/a.m.u.), preserving the 30:1 mass range, limit the accuracy of mass measurement when the photographic emulsion grain size (1–5 microns) becomes comparable to the beam width (a problem in practice with $M/\Delta M$ in excess of 30,000).

In the preceding discussion attention was directed to acquisition of high resolution mass spectra via magnetic tape recording of either the analogue signal (electron multiplier voltage profile) or corresponding converted digital data (timing increments of digitized multiplier voltage profile) on analogue or digital magnetic tape, respectively. Although the analogue mode is feasible [34], the dynamic range and accuracy of mass measurement is mediocre, compared with current, more extensive data on direct digitization and transfer to core utilizing an on-line real-time high speed digital computer [36, 40, 41]. Real-time digital data acquisition shows great promise of increased accuracy of mass measurement (probably ten- to a thousand-fold over photographic and analogue magnetic tape) and excellent relative peak intensity measurements with a dynamic range in excess of 1 part in 10^4 even for instrument resolutions in excess of 20,000 [41]. Such advantages are extremely desirable from the point of view of obtaining reproducible fragmentation patterns at high resolution for pattern comparison and structural identifications (also location of stable isotopes, e.g., ^{13}C, D, ^{18}O, ^{15}N, by mass measurement), and particularly for the analyses of mixtures of compounds where the fragmentation pattern of the components is indeed additive and could be deconvoluted by matrix algebraic techniques (analogous to petroleum type analyses [11, 42] mentioned in the sections on hydrocarbons).

D. High Resolution Mass Spectra: Their Nature, Presentation and Interpretation

We have seen the tasks involved in obtaining high quality data for all peaks in a high resolution mass spectral fragmentation pattern. Such a capability coupled with the usual advantages of minute sample size (microgram range) yields a knowledge of primary nuclidic composition, relative ionic abundance and structural information in precision and detail previously unprecedented in organic chemical research. It hardly needs to be emphasized that molecular components (and ionic fragments), which differ in elemental composition, do not interfere (are resolved) and, therefore, the usual constraints on sample purity and mixture analyses applying to the use of other spectroscopic tools are alleviated significantly. Only compounds isomeric in elemental composition become severe analytical problems. Furthermore, the coupling of a gas-liquid chromatographic inlet to a high resolution mass spectrometer, while then demanding computer techniques of data acquisition, processing and presentation (either from spectrograms or real-time instrument operation) for its full exploitation, brings the most potent physicochemical techniques together for utilization on any volatile organic matter regardless of its elemental or structural complexity.

Studies in natural products and organic chemistry, as well as geochemistry, which attempt to exploit the vast amount of compositional and relative ionic abundance data revealed in a high resolution mass spectral fragmentation pattern, are only now being realized primarily due to the developments in sophisticated computer data acquisition and processing facilities [26, 36, 37, 40, 41].

Since the high resolution mass spectrum of a complex organic molecule (not to mention mixture) may consist of several hundred ions of differing elemental composition and relative abundance, there arises the task of their mental assimilation and subsequent interpretation in terms of salient features of molecular structure and stereochemistry.

Due to the digital nature of mass spectral data, calculation of accurate mass and intensity from the raw data is easily accomplished by high speed digital computer; also subsequent selection of all possible elemental compositions, sorting of data and output in any desired format is best accomplished routinely by computer techniques [41]. Of course, having the elemental compositions of all ions in the fragmentation pattern obviates any interpretative dependence upon the actual numerical value of the accurate (or nominal) mass, and the emphasis shifts to the information sought initially, e.g., the atomic compositions themselves.

At this point, the mention of certain phenomenological differences between the interpretation of conventional and high resolution mass spectral fragmentation patterns may be appropriate. Interpretation of conventional spectra consists chiefly of an attempt to deduce the atomic compositions of fragments by taking arithmetic differences between relatively intense peaks of interest. Generally, only the relatively intense peaks are of concern, since interpretation of minor peaks would be both unwarranted and impossible. Furthermore, minor peaks could be due to impurities. For example, consider the molecular ion and the sequences of peaks below it, representing neutral species lost in fragmentation, e.g., M^+, $(M\text{-}15)^+$, $(M\text{-}18)^+$, $(M\text{-}43)^+$, $(M\text{-}59)^+$, etc. One is then faced with the necessity of making the "best guesses" as to whether 28 a.m.u. is CH_2CH_2, CO, N_2, CH_2N and whether M-43 is C_3H_7, C_2H_3O, etc. Other spectral information is of some help in making certain decisions, e.g., consideration of the natural ^{13}C, ^{15}N, ^{18}O isotopic abundances [43, 44] and metastable transitions [45] *, odd versus even molecular weight representing odd numbers of nitrogen atoms present, etc. *Suffice it to assert at this point that the quantitatively reproducible feature of conventional spectra provide a pattern of fragments which are, in many cases, uniquely characteristic of the molecular structures in question and should be compared in every detail with authentic samples for postulation of gross structural identity.*

Utilizing high resolution techniques obviates the former aspect of the "interpretative" task by rendering directly all fragment compositional data and, in addition, yields (under on-line, real-time conditions) reproducible patterns of fragments generally sorted according to their heteroatomic content, e.g., C/H, C/HO...C/HN...C/HNO, etc. The interpretation of such data must be considered another realm, since not only are the compositional ambiguities ** minimized, but the salient criteria for the determination of unknown structures still include the initial pattern recognition approach and subsequent verification of the structural identity by comparison of the fragmentation pattern in question with that of an authentic compound or a structurally closely related compound ***.

An important point concerns the fact that the larger the fragment being considered, the more significant it may be in terms of yielding information regarding the molecular structure. For example, peaks at low masses ($<m/e$ 100) tend to have their genesis in multiple processes, whereas at higher masses the formation of a fragment from a particular part of the molecule is often unique.

With knowledge of the elemental compositions there is no longer justification to phenomenologically disregard structurally very significant peaks of very low relative intensity, especially when they are at higher masses in the spectrum.

It is imperative, therefore, that these data be manipulated so that they can be presented for study in a format which will permit ready visual grasp of the salient features, not only those contained within the pattern of relative abundances of various carbon and hydrogen combinations, $C_xH_yX_z$, within heteroatomic groups,

* Metastable ions, m^*, occur as relatively low intensity, broad peaks (often at non-integral masses) in a mass spectrum and are the result of an ion, m_1^+, decomposing after acceleration into m_2^+ and a neutral species m_0, e.g.,

$$m_1^+ \rightarrow m_2^+ + m_0 .$$

These are related by the semi-empirical relation:

$$m^* \cong \frac{(m_2^+)^2}{m_1^+} .$$

They are of significance in demonstrating that the ions m_1^+ and m_2^+ are related by the fragmentation process resulting in the loss of the neutral species, m_0.

** Refer to footnote **, p. 95.

*** For example, methyl or methoxyl substitution in the aromatic nucleus of an indole alkaloid "shifts" the aromatic fragments by either 14 or 30 a.m.u. respectively. See K. BIEMANN, *op. cit.*, pp. 305–312.

X_z, but those revealed *among* any of the given heteroatomic groups. This format should not only be designed to reveal the kinds of ions present (odd versus even electron ions, degree of unsaturation, hydrogen rearrangement ions, elimination of neutral species, e.g., CO, H_2O, molecular ions, etc.), but should preserve the reference frame (pattern) for ease of intercomparison of total high resolution mass spectral data on related compounds.

Two computer approaches have been utilized considerably in aiding mental comprehension and interpretation of high resolution mass spectra, one originating in Biemann's group termed "element mapping" [46], the other in the authors' laboratory termed "heteroatomic plotting" [19, 39, 47]. Another approach, which combines the disadvantages of both these methods, termed "topographical element mapping" has been suggested [48]. A more detailed discussion of techniques for the presentation of high resolution mass spectra, including digital cathode ray tube (C.R.T.) display in real-time, may be found elsewhere [41].

To illustrate the presentation of high resolution mass spectral data, the real-time high resolution mass spectrum of the methyl ester of a keto acid studied in connection with the Colorado Green River Formation [49, 50] has been chosen and the element map of authentic methyl 14-oxopentadecanoate is depicted in Fig. 3*. In such a two dimensional array the ordinate represents increasing nominal mass and the abscissa increasing heteroatom content in discreet heteroatomic groupings, such that the most carbon rich species would be the first term ("a_{11}") on the "main diagonal" of a matrix and the molecular ion the last term ("a_{nm}"). Associated with each C/H XY... entry is a relative abundance term, which in Biemann's presentation is of a semilogarithmic nature due to the inherent nonlinearity of the photographic emulsion (see above).

KETC ESTER   REAL TIME CATA   CORRECTEC PEAK AREA INTENSITIES

	C/H	C/HC	C/HC2	C/HC3
26	2/ 2 5			
27	2/ 3 55			
28	2/ 4 38			
29	2/ 5 107	1/ 1 7		
31		1/ 3 21		
39	3/ 3 61			
40	3/ 4 16			
41	3/ 5 355			
42	3/ 6 90	2/ 2 21		
43	3/ 7 186	2/ 3 999		
44	3/ 8 3	2/ 4 17		
45		2/ 5 36		
52	4/ 4 5			
53	4/ 5 42			
54	4/ 6 50			
55	4/ 7 462	3/ 3 88		
56	4/ 8 77	3/ 4 2		
57	4/ 9 111	3/ 5 41		
58	4/1C 4	3/ 6 589		
59		3/ 7 132	2/ 3 12C	
60		3/ 8 3		
66	5/ 6 6			
67	5/ 7 120			
68	5/ 8 67			
69	5/ 9 313	4/ 5 16		
70	5/1C 35	4/ 6 18		
71	5/11 43	4/ 7 288		
72		4/ 8 19		
73		4/ 9 35		
74			3/ 6 384	
75			3/ 7 23	
77	6/ 5 3			

Fig. 3. Element map: Real time high resolution mass spectrum of methyl 14-oxopentadecanoate

* The authors wish to thank Dr. D. H. SMITH and Mr. B. R. SIMONEIT for these unpublished real-time data taken using a modified C.E.C. 21–110 B mass spectrometer coupled in a real-time manner to an S.D.S. Sigma 7 high speed digital computer, resolution 20,000, magnetic scan 10 seconds. Attention is directed to the excellent accuracy (< 1 percent) of the relative abundances as determined by area measurements which were obtained via direct ion multiplier digitization [36].

KETC ESTER REAL TIME DATA CORRECTEC PEAK AREA INTENSITIES

	C/H	C/HC	C/HC2	C/HC3
79	6/ 7 28			
80	6/ 8 11			
81	6/ 9 122			
82	6/1C 54			
83	6/11 279	5/ 7 38		
84	6/12 25	5/ 8 168		
85	6/13 21	5/ 9 78		
86	6/14 2	5/10 1		
87		5/11 12	4/ 7 284	
88			4/ 8 21	
91	7/ 7 7			
93	7/ 9 16			
94	7/1C 21			
95	7/11 1C6	6/ 7 4		
96	7/12 47	6/ 8 36		
97	7/13 212	6/ 9 114		
98	7/14 14	6/10 38C		
99		6/11 50		
100		6/12 1C	5/ 8 3	
101			5/ 9 38	
102			5/1C 3	
107	8/11 25			
108	8/12 17			
109	8/13 52	7/ 9 2		
110	8/14 14	7/10 10		
111	8/15 31	7/11 82		
112	8/16 2	7/12 1C4		
113		7/13 15	6/ 9 5	
114		7/14 8	6/1C 4	
115		7/15 2	6/11 35	
116			6/12 6	
119	9/11 2			
120	9/12 2			
121	9/13 30			
122	9/14 2			
123	9/15 26	8/11 8		
124	9/16 3	8/12 6		
125		8/13 47		
126		8/14 52		
127		8/15 6	7/11 4	
128		8/16 4	7/12 9	
129			7/13 26	
130			7/14 15	
132				6/12 6
133	1C/13 1			
135	1C/15 8			
136	1C/16 14			
137	1C/17 7	9/13 7		
138	1C/18 34	9/14 6		
139	1C/19 37	9/15 18		
140		9/16 68		
141		9/17 3	8/13 4	
143		9/19 1	8/15 21	
144			8/16 4	
149	11/17 10			
150	11/18 3			
151		1C/15 2		
152	11/2C 6			
153	11/21 3	1C/17 6		
154		1C/18 9	9/14 1	
155		1C/19 2		
157			9/17 5	
161	12/17 1			
162	12/18 19			
163	12/19 82			
164	12/2C 15			
167		11/19 1		
169			1C/17 2	
171			1C/19 3	
177	13/21 4			
178	13/22 2	12/18 2		
179	13/23 7			
180		12/20 136		
181		12/21 161		
185			11/21 5	
194		13/22 3		
195		13/23 1C		
196		13/24 5	12/2C 8	
197		13/25 3C		
198		13/26 5		
199			12/23 2	
200			12/24 11	
201			12/25 1	
210		14/26 3	13/22 1	
212			13/24 96	
213			13/25 461	
215			13/27 1	
227			14/27 1	
228			14/28 1	
237			15/25 4	
238			15/26 18	
239			15/27 176	
240			15/28 3C	
252			16/28 1	
254			16/3C 1	
255				15/27 30
256				15/28 4
270				16/3C 93
271				16/31 13
272				16/32 1

Fig. 3.

In Fig. 3 the nominal mass ranges from m/e 26 to the molecular ion region at m/e 270, and the grouping of ions according to heteroatomic content yields four columns, e.g., C/H, C/HO, C/HO_2 and C/HO_3 containing entries for composition and corresponding linear relative abundances.

For comparison, the heteroatomic plots in Fig. 4 depict the high resolution fragmentation pattern for methyl 14-oxopentadecanoate, I. As in element mapping, the ions are grouped according to heteroatomic content and then plotted as bar

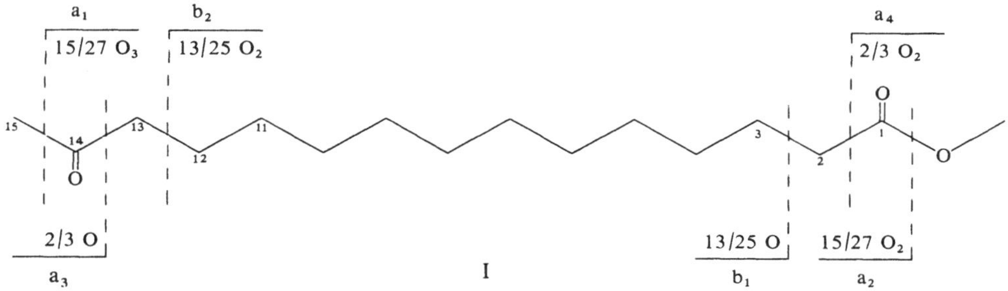

I

graphs. For compound I four graphs are obtained: the hydrocarbon ions, C/H; the mono-oxygenated ions, C/HO; the dioxygenated ions, C/HO_2; and the trioxygenated ions, C/HO_3. The ordinates display both relative ion intensity normalized to the most intense peak (base peak) in the spectrum, and also the percent of total ionization [51], percent Σ_{26}. The carbon-hydrogen ratios are then plotted on the abscissa in such a way that the completely saturated alkyl fragments occur as major divisions, i.e., C_nH_{2n+1}; $C_{n+1}H_{2n+3}$, etc., providing 14 mass units between major divisions which are then divided into seven degrees of unsaturation (double bonds and/or rings). For example, a saturated alkane would have all major ions except the molecular ion falling at major C_nH_{2n+1} divisions. One may readily ascertain certain information regarding the types of ions [52] of which the fragmentation pattern is composed.

The molecular ion of Compound I is found in the C/HO_3 plot and is, therefore, of composition $C_{16}H_{30}O_3$. Loss of methyl radical gives ion a_1, $C_{15}H_{27}O_3$ (see structure I and Fig. 4), which has two degrees of unsaturation, i.e., containing both oxo groups.

Other simple bond cleavages alpha to the carbonyl functions yield the acyl fragments: a_2, $C_{15}H_{27}O_2$; a_3, C_2H_3O; and a_4, $C_2H_3O_2$; see Structure I and Fig. 4.

Another group of fragments arises formally by fission of the 2,3-bond (b_1) and 12,13-bond (b_2) which is beta to each carbonyl function but with charge retention on the long chain moiety. Such processes are in fact characteristic of a long-chain aliphatic di-oxo-functionality and represent quite complex fragmentation processes [53].

A third group of fragments, representatives of a common hydrogen rearrangement process (McLafferty rearrangement [54]) occurring whenever a carbonyl function has available a gamma-hydrogen, are present at the small tic marks* and are labelled c_1, $C_3H_6O_2$, and c_2, C_3H_6O, and may be depicted as shown:

* Fragments occurring at small tic marks are odd-electron ions (ion radicals) which include molecular ions, fragments that arise by loss of neutral molecules, e.g., CO, H_2O, NH_3, HCN, etc., and single hydrogen rearrangement ions.

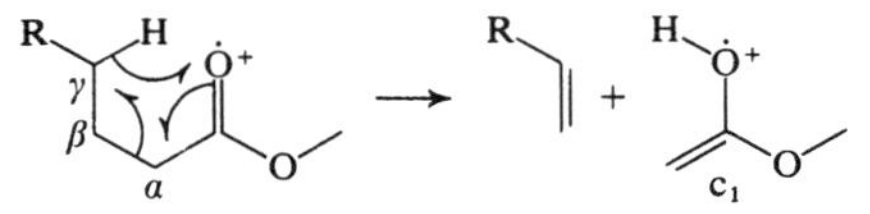

This rearrangement is useful in determining whether there is substitution on the carbon alpha to a carbonyl group, since this fragment would then shift by the mass of the substituent (see discussion of alpha-methyl dicarboxylic acids, section III B).

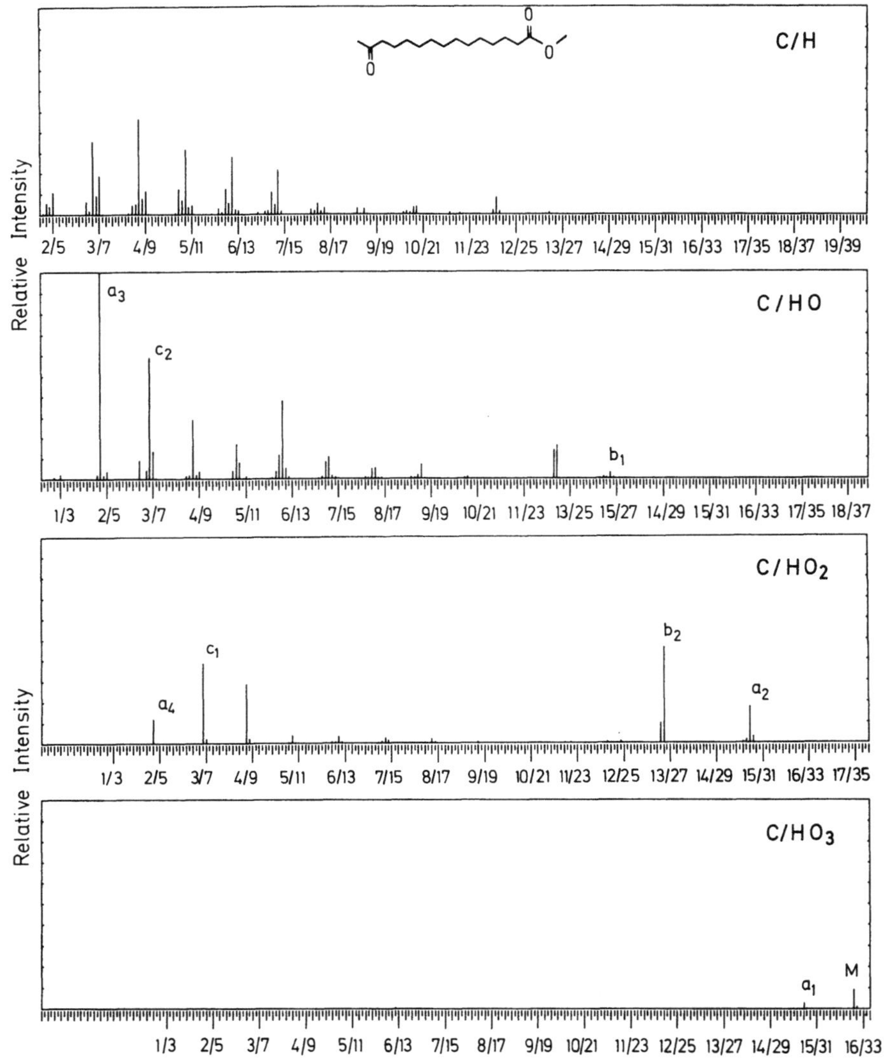

Fig. 4. Heteroatomic plots: real time high resolution mass spectrum of methyl 14-oxopentadecanoate

Turning from the high resolution mass spectrum of a pure compound, let us consider the information which may be ascertained from the high resolution mass spectra of complex mixtures isolated from carbonaceous sediments.

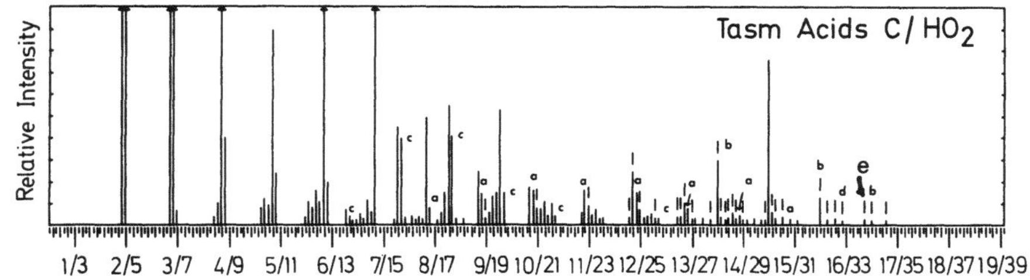

Fig. 5a. Partial high resolution mass spectrum of the mixture of acids isolated after first ultrasonic solvent extraction of Alaskan Tasmanite

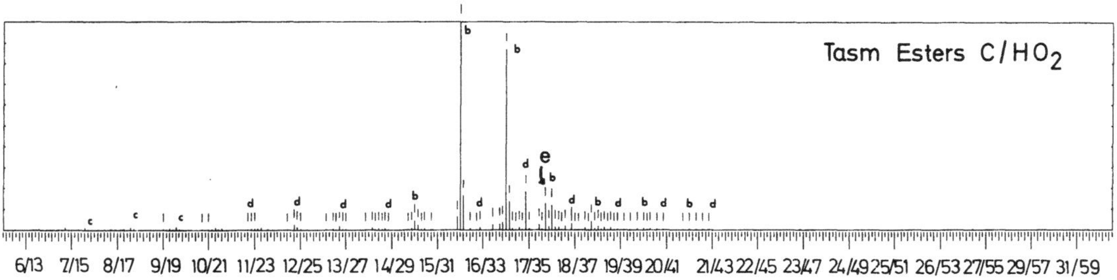

Fig. 5b. Partial high resolution mass spectrum of the mixture of esters derived from esterification of the acids (Fig. 5a) and subsequent clathration, i.e., branched-cyclic esters

The spectra of three mixtures have been chosen to illustrate the vast quantity of specific information derivable from a total acid fraction, its branched-cyclic esters and a total basic fraction.

The first concerns the C/HO_2 ions (Fig. 5a) from the total free acid fraction isolated from ultrasonic extraction (benzene/methanol, 3:1) of pulverized Alaskan Tasmanite [55]; the second, the C/HO_2 ions (Fig. 5b) from the branched and cyclic esters remaining after clathration of the normal esters from the previous fraction.

In Fig. 5a several series of molecular ions may be noted: *a*, $C_nH_{2n+2}O_2$ ($n=8$–15); *b**, $C_nH_{2n-20}O_2$ ($n=15$–18); *c*, $C_nH_{2n-8}O_2$ ($n=7$–13) and a series not in this plot (would be coincident with series *a*); *d*, $C_nH_{2n-14}O_2$ ($n=14$–17). Series *a* are saturated carboxylic acids; *b* are tricyclic aromatic acids (phenanthrene or anthracene type); *c* are phenyl alkanoic acids and *d* are naphthyl alkanoic acids. The intense peaks in Fig. 5a occurring at $C_nH_{2n-1}O_2$ are fragments of the series *a* compounds.

Finally, one should note ion *e*, whose composition is $C_{18}H_{14}O_2$ (Fig. 5a) and $C_{19}H_{16}O_2$ (Fig. 5b). This composition would correspond to a cycloaromatic acid, such as:

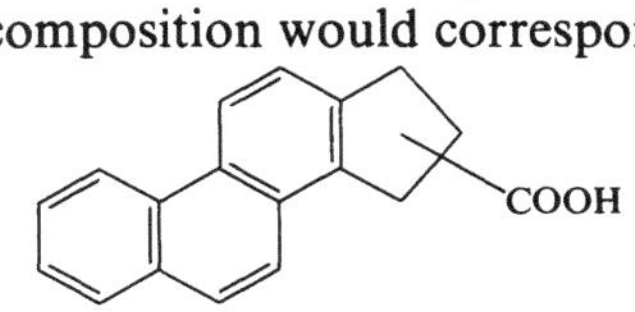

* In Figs. 5a and b it should be noted that many of the peaks have a short line above them and this indicates more than seven degrees of unsaturation. For example, fragments whose C/H ratio display more than seven degrees of unsaturation would fall below the next lower major saturated division, i.e., $C_{n-1}H_{2n-1}$; hence, such C/H ratios are indicated by the vertical line notation. Thus, the "*b*" series are plotted 21 divisions (mass units) below the major division with the correct carbon number, i.e., $C_nH_{2n-20}O_x$; and the "*c*" series 9 divisions (mass units) below, i.e., $C_nH_{2n-8}O_x$.

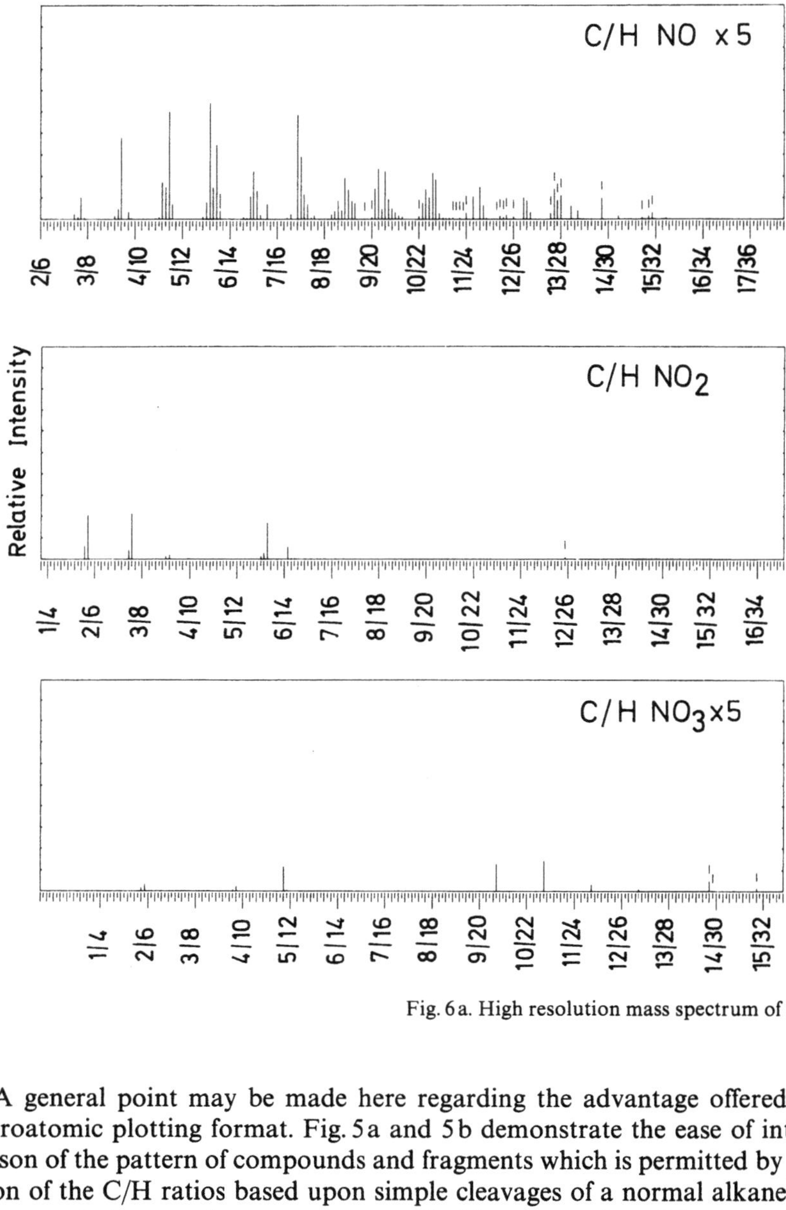

Fig. 6a. High resolution mass spectrum of Colorado

A general point may be made here regarding the advantage offered by the heteroatomic plotting format. Fig. 5a and 5b demonstrate the ease of intercomparison of the pattern of compounds and fragments which is permitted by presentation of the C/H ratios based upon simple cleavages of a normal alkane model. For pure substances, of course, one can with this preservation of the fragmentation pattern make quantitative abundance comparisons to establish gross structural identity.

As a result of urea clathration of the esters of the total acid fraction, the normal fatty acid esters (series *a*) should be absent. Of course, the molecular ions of these branched-cyclic esters will occur one methylene number higher (CH_2) and Fig. 5b is lined up such that the molecular ions fall at the same place as those of Fig. 5a. Also series *d* now occurs in this C/HO_2 plot since series *a* was removed.

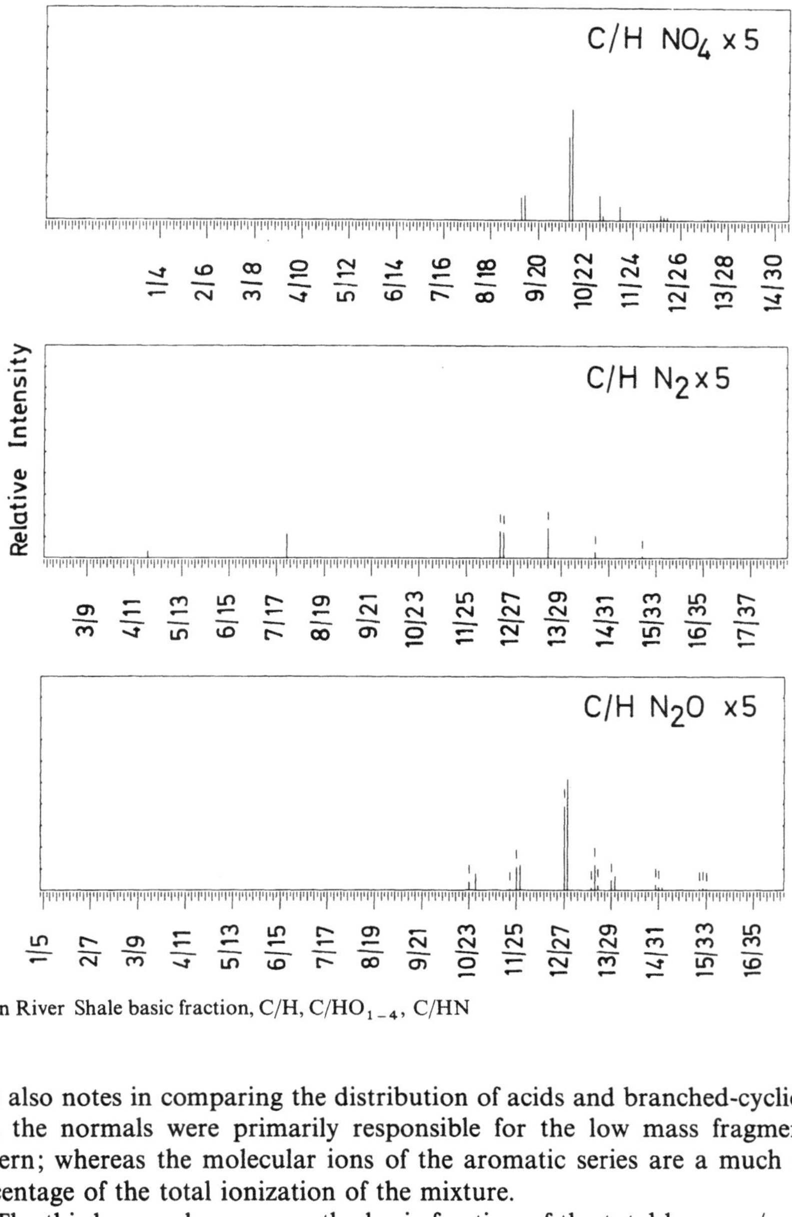

Green River Shale basic fraction, C/H, C/HO_{1-4}, C/HN

One also notes in comparing the distribution of acids and branched-cyclic esters that the normals were primarily responsible for the low mass fragmentation pattern; whereas the molecular ions of the aromatic series are a much greater percentage of the total ionization of the mixture.

The third example concerns the basic fraction of the total benzene/methanol extract of the oil shale of the Green River Formation [56]. It shows particularly intense peaks in the C/HN plot of the high resolution mass spectrum (Figs. 6a and b). [It should be noted that the C/HO_3 and C/HN plots are very similar; this is due to the fact that distinction between combinations of C/HO_3 and C/HN is difficult *a priori*, since such peaks fall within the error limit of the mass measurement discussed previously (Table 2); the computer, therefore, plots both combinations.] Fragments corresponding to C_7H_9N, $C_8H_{11}N$, and $C_9H_{13}N$ in the C/HN plot

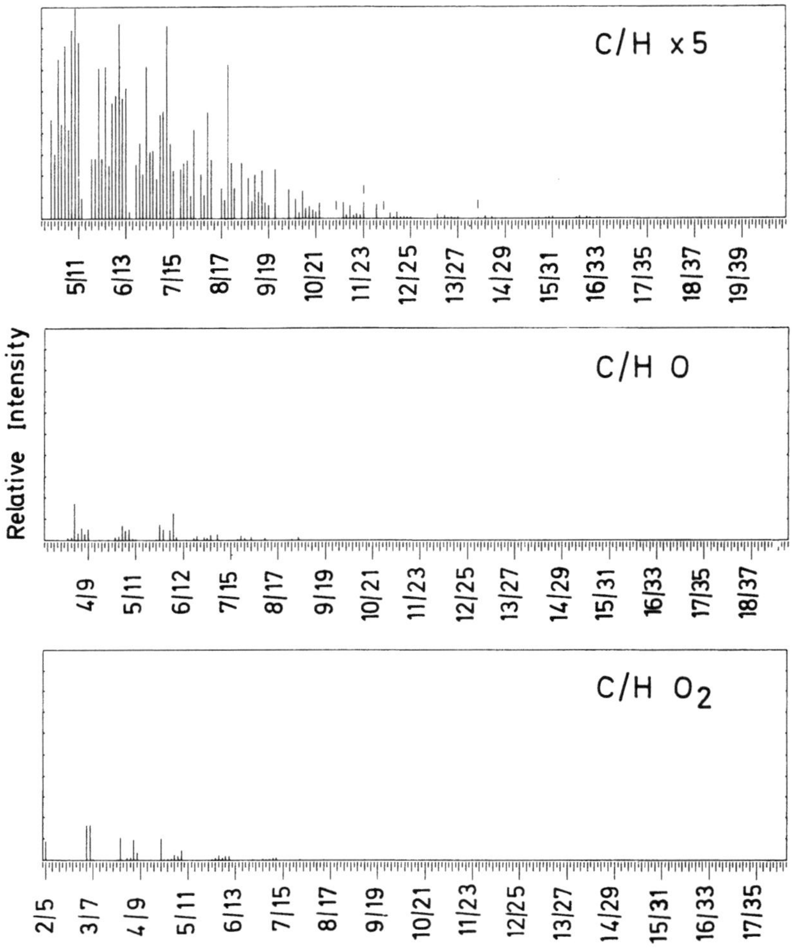

Fig. 6b. High resolution mass spectrum of Colorado Green

could correspond to molecular ions of simple alkyl pyridines, and fragment ions at C_7H_8N, $C_9H_{12}N$, $C_{10}H_{14}N$ would be a further indication of this group of compounds.

A series of alkyl indoles seems to be indicated by the fragment peaks of C_9H_8N, $C_{10}H_{10}N$, $C_{11}H_{12}N$, $C_{12}H_{14}N$, $C_{13}H_{16}N$, $C_{14}H_{18}N$, and $C_{15}H_{20}N$, and the series at $C_{10}H_9N$, $C_{11}H_{11}N$, $C_{12}H_{13}N$, $C_{13}H_{15}N$, $C_{14}H_{17}N$ and $C_{15}H_{19}N$ would represent the molecular ions of C_1-C_6 substituted alkyl quinolines or iso-quinolines. Fragment ions at even mass corresponding to this series can be noted also. The C_3, C_4, and C_6 alkyl quinolines ($C_{12}H_{13}N$, $C_{13}H_{15}N$, $C_{15}H_{19}N$) appear to be present in greatest abundance.

A fourth series is apparent from the spectrum, commencing with the peak of m/e $C_9H_{11}N$ and containing $C_{10}H_{13}N$, $C_{11}H_{15}N$, $C_{12}H_{17}N$, $C_{13}H_{19}N$, $C_{14}H_{21}N$, $C_{15}H_{23}N$, to which the structure of tetrahydroquinolines (ranging from the unsubstituted to the C_6 substituted) could be ascribed but, of course, other

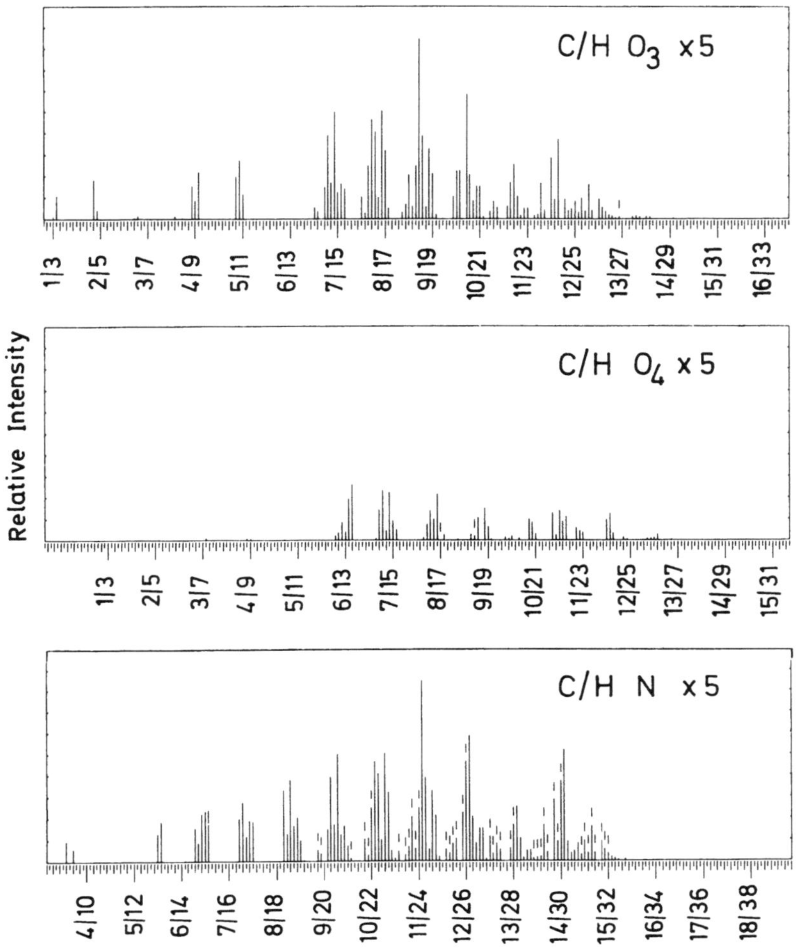

River Shale basic fraction, C/HNO_{1-4}, C/HN_2, C/HN_2O

compound types, such as cycloalkylpyridines, would also give rise to this series. Even mass peaks, such as the strong ions at $C_{10}H_{12}N$, $C_{11}H_{14}N$, $C_{12}H_{16}N$ could represent fragments. Alkyl pyrroles may be present, since the peaks at $C_{15}H_{13}N$ and $C_{14}H_{11}N$ would correspond to the C_{10} and C_9 alkyl pyrrole molecular ions. The interpretation of other plots of the base fraction is difficult since many possibilities must be considered, and no firm conclusions can be drawn from a single high resolution mass spectrum.

It should be clear from the foregoing examples and discussion that high resolution mass spectral data lend themselves very favorably (compared with other physico-chemical techniques) to programmed logical manipulations, e.g., computer-aided interpretation – particularly in the sense of making correlations with ion types and possible fragmentation sequences, such that the predicted pattern and fragmentation sequence possibilities can be checked much more quickly than by hand (visual and mental inspection). Such attempts find example in the mass spectrometric literature already, the most successful being work on computer programmed amino acid sequencing in small peptides initiated in three laboratories [57, 58a, 59–62]. Other studies hold promise for checking (or determining) possible molecular ions from the fragmentation pattern alone [63].

E. Application of High Resolution Mass Spectrometry

Since BEYNON [64] first demonstrated the usefulness of accurate mass measurement in determining the elemental composition of an ion, several groups have led in the development and further application to organic, bio- and geochemistry. To cite a few typical studies from leading laboratories, LEDERER and BARBER *et al.* [58] have investigated amino acid sequencing; BIEMANN and co-workers have elucidated structures of peptides, indole alkaloids and antibiotics [61, 62, 65, 66]. BURLINGAME and co-workers have studied the sesquiterpenols, gaillardin [67] and widdrol [68], mechanistic correlations with photochemical processes [69] and Amaryllis alkaloids [26, 70, 71]; DJERASSI and co-workers have studied complex fragmentation mechanisms [72]; SAUNDERS and WILLIAMS [73] have utilized the technique in the chemical industry. BOMMER and VANE [74] have used high resolution spectra for identification of drug metabolites in the presence of impurities of differing elemental composition.

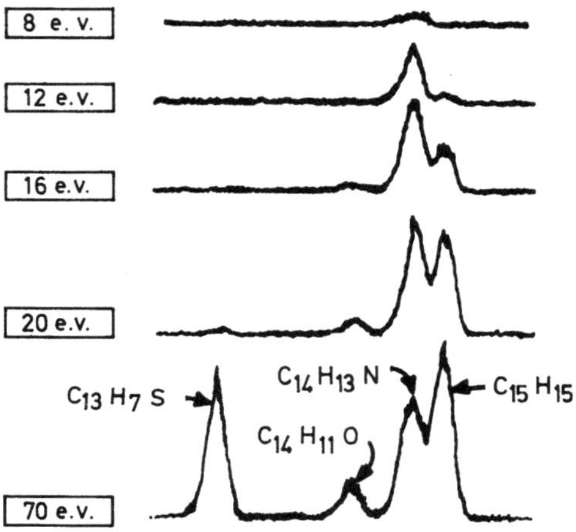

Fig. 7. Low voltage — high resolution mass spectrum of m/e 195 [after LUMPKIN, Anal. Chem. **36**, 2399 (1964)]

The earliest application to organic geochemistry was by CARLSON *et al.* [75] who demonstrated the resolution of normal hydrocarbons from naphthyl aromatic hydrocarbons of the same nominal mass (^{12}C *vs.* H_{12}; see Table 2), and alkyl benzenes from alkyl benzothiophenes of the same nominal mass (^{32}S *vs.* $^{12}C_2H_8$; see Table 2). CARLSON *et al.* [75] also reported the identification of alkyl thiophenes in virgin gas oil (multiplet with monocycloalkanes). LUMPKIN [76] first demonstrated the combined use of high resolution and low ionizing voltage for the analysis of a complex mixture. The potential of low voltage techniques combined with high resolution can be seen from Fig. 7 taken from LUMPKIN'S paper. At

8.0 e.v. only the molecular ion of a substituted carbazole appears at m/e 195. As the bombarding energy is increased to 70 e.v., four peaks are present due to the sulfur-, oxygen- and nitrogen-containing, as well as the hydrocarbon, species. Further studies by MEAD, MEAD and BOWEN [77] and JOHNSON and ACZEL [78] have extended the high resolution at low voltage techniques to the more detailed analyses of complex petroleum fractions.

GALLEGOS, GREEN, LINDEMANN, LE TOURNEAU and TEETER [42] have demonstrated a group-type quantitative analysis of 19 components: seven saturated hydrocarbons with zero to six rings, nine aromatic hydrocarbons with one to four rings, and three aromatic sulfur compound types, without requiring silica gel chromatographic fractionation of high boiling petroleum stocks. (See section III E, and Fig. 44.) DRUSHEL and SOMMERS [79] have described the analysis of aromatic sulfur compound concentrates from high boiling petroleum fractions. BAKER *et al.* [80] have investigated petroporphyrins and determined the elemental compositions of selected peaks in their spectra.

MEAD and REID [81] have reported the analysis of petroleum waxes using field ionization at high resolution. This technique, combined with low ionizing voltage at high resolution, holds promise for the quantitative analysis of complex mixtures.

F. Gas Chromatograph-Mass Spectrometer Coupling

A valuable method of analyzing geochemical samples by mass spectrometry consists of the direct coupling of a gas chromatographic apparatus to the mass spectrometer.

Independently, gas chromatographic and mass spectrometric instrumentation have reached a high degree of sophistication, and their combination in one instrumental package furnishes an analytical technique of great versatility and power. The outstanding advantages of the combined operation of gas chromatographic and mass spectrometric equipment are its speed, sensitivity, avoidance of an isolation step, considerable elimination of contamination and its adaptability to routine analyses and data handling.

A complete mass spectrometric analysis can be obtained in essentially the time required for an ordinary g.l.c. run. Every component can be analyzed, and the homogeneity of a single g.l.c. peak can be checked by obtaining successive mass spectral scans as the compound emerges from the column.

All components enter the mass spectrometer directly and isolation of individual peaks is avoided – a very time consuming step in ordinary g.l.c.-mass spectrometric analysis. Furthermore, possibilities of contamination, always a danger in the isolation of microgram samples, are very considerably reduced. The sensitivity of the mass spectrometer makes possible the analysis of trace components (i.e., nanogram range), which would be difficult or impossible to isolate in the conventional way. This also permits the use of very efficient, but low capacity columns (capillary columns) for the separation of a mixture. The g.l.c.-mass spectrometer finally provides an ideal means for the routine analysis of mixture samples, and computers could be used effectively for both the qualitative and quantitative interpretation of data.

Gas chromatographs can be connected directly to the mass spectrometer, with all of the column effluent passing into the ion source region. Such an arrangement is practical for capillary columns with carrier gas flow rates of 1–3 ml/min. The routine use of packed columns requires the interface of an enrichment device between the gas chromatograph and the mass spectrometer. Four such devices — the "molecular separator" of RYHAGE [82], the fritted glass tube of WATSON and BIEMANN [83], the semipermeable membrane of LLEWELLYN [84] and the Teflon "membrane" of LIPSKY [85] — have been introduced recently and some are now in common use. These separators preferentially remove the light carrier gas molecules and a gas stream enriched in compound enters the mass spectrometer ion source, without prohibitive loss of resolution or sample. The separator functions also as a pressure reduction unit such that ion source pressures of 10^{-5} mm Hg or less can be maintained conveniently.

If the gas chromatograph is equipped with an effluent splitter, part of the effluent can be diverted to an ionization detector to produce a conventional gas chromatogram. Alternatively, or in addition, the mass spectrometer total ion current monitor can be used as a detector by connecting its output to a pen and ink recorder. In the authors' laboratory a dual-pen recorder produces simultaneously gas chromatograms both from the output of the chromatograph ionization detector and the total ion monitor, providing a convenient and immediate check on the time-lag between gas chromatograph and mass spectrometer and possible loss of resolution*.

Several complete mass spectra (from three to ten) are usually obtained from a single g.l.c. peak emerging from the column. Extremely rapid mass scanning (in the order of 0.1 to several seconds) is thus a necessity. Accurate identification of mass number (by simple counting) may be quite difficult under these circumstances and the simultaneous recording of the spectrum of a marker compound — bled in from a reservoir, or by the technique of LEEMANS and MCCLOSKEY [86] — is usually necessary. Consecutive runs both with and without marker can be used to eliminate any uncertainties or ambiguities in the interpretation of the spectrum [20].

The first attempts to couple a gas chromatograph to a mass spectrometer have been described as early as 1957 [87], and since then the technique has found considerable application in a variety of fields. Several excellent review papers [86, 88, 89] should be consulted for detailed accounts of the technical aspects and applications of the method.

Some important applications to geochemical problems are illustrated by the separation on a capillary column of saturated hydrocarbons up to C_{11} and their mass spectrometric identification [90]. The same group analyzed olefinic compounds [91] and hydrocarbons in the C_8 to C_9 range. Sixteen C_9 hydrocarbons from petroleum were identified by the combination of a cycloidal focussing mass spectrometer (equipped with Wien filter and ion multiplier) and capillary gas chromatograph [92]. The combination of a gas chromatograph with a Time-of-Flight mass spectrometer was used in the experiments of ANDERS and co-workers [93].

* F. C. WALLS and D. B. BOYLAN, unpublished results, this laboratory.

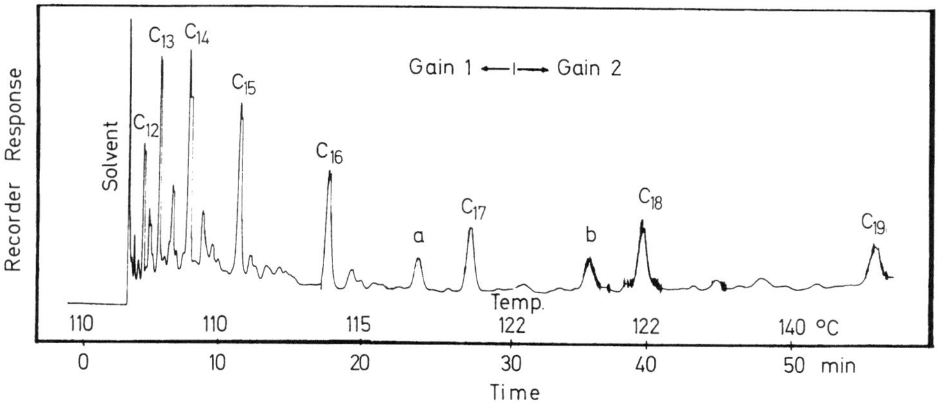

Fig. 8. Capillary gas-liquid chromatogram of Fig Tree Shale

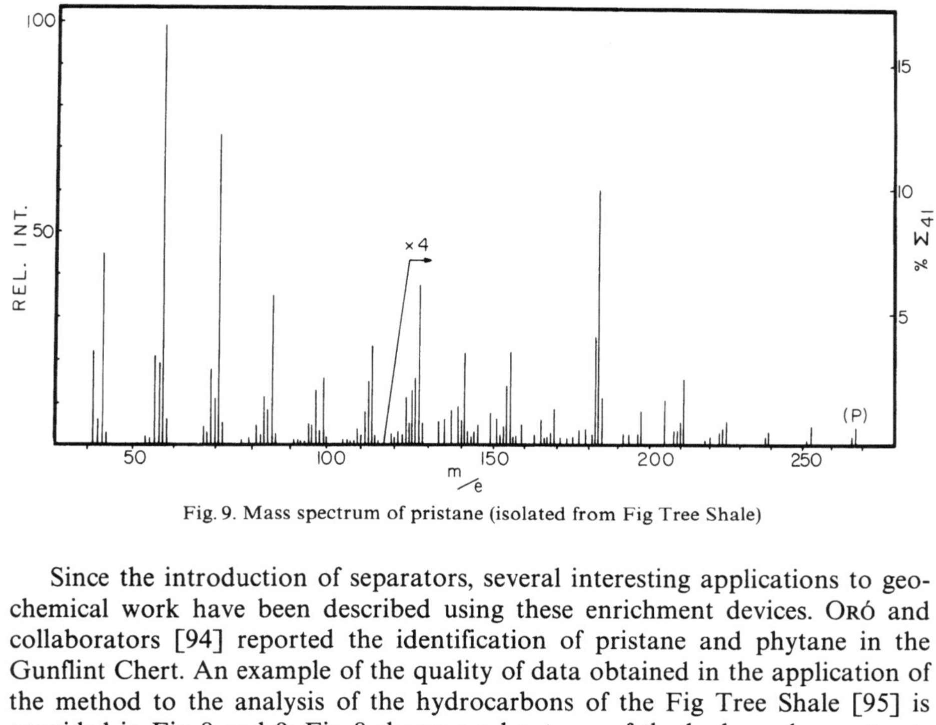

Fig. 9. Mass spectrum of pristane (isolated from Fig Tree Shale)

Since the introduction of separators, several interesting applications to geochemical work have been described using these enrichment devices. ORÓ and collaborators [94] reported the identification of pristane and phytane in the Gunflint Chert. An example of the quality of data obtained in the application of the method to the analysis of the hydrocarbons of the Fig Tree Shale [95] is provided in Fig. 8 and 9. Fig. 8 shows a g.l.c. trace of the hydrocarbon extract; Fig. 9 is the corresponding mass spectrum of component a, scanned as compound a emerged from the gas chromatograph. The spectrum can be identified clearly as that of the C_{19} isoprenoid, pristane. EGLINTON and his co-workers [96, 97] have used the technique for the analysis of isoprenoid fatty acids from extracts of the oil shale from the Green River Formation.

A very important extension of this technique is the combination of gas chromatography and high resolution mass spectrometry [83, 98–101]. An example of the results taken from BIEMANN'S laboratory is illustrated in Figs. 10 and 11,

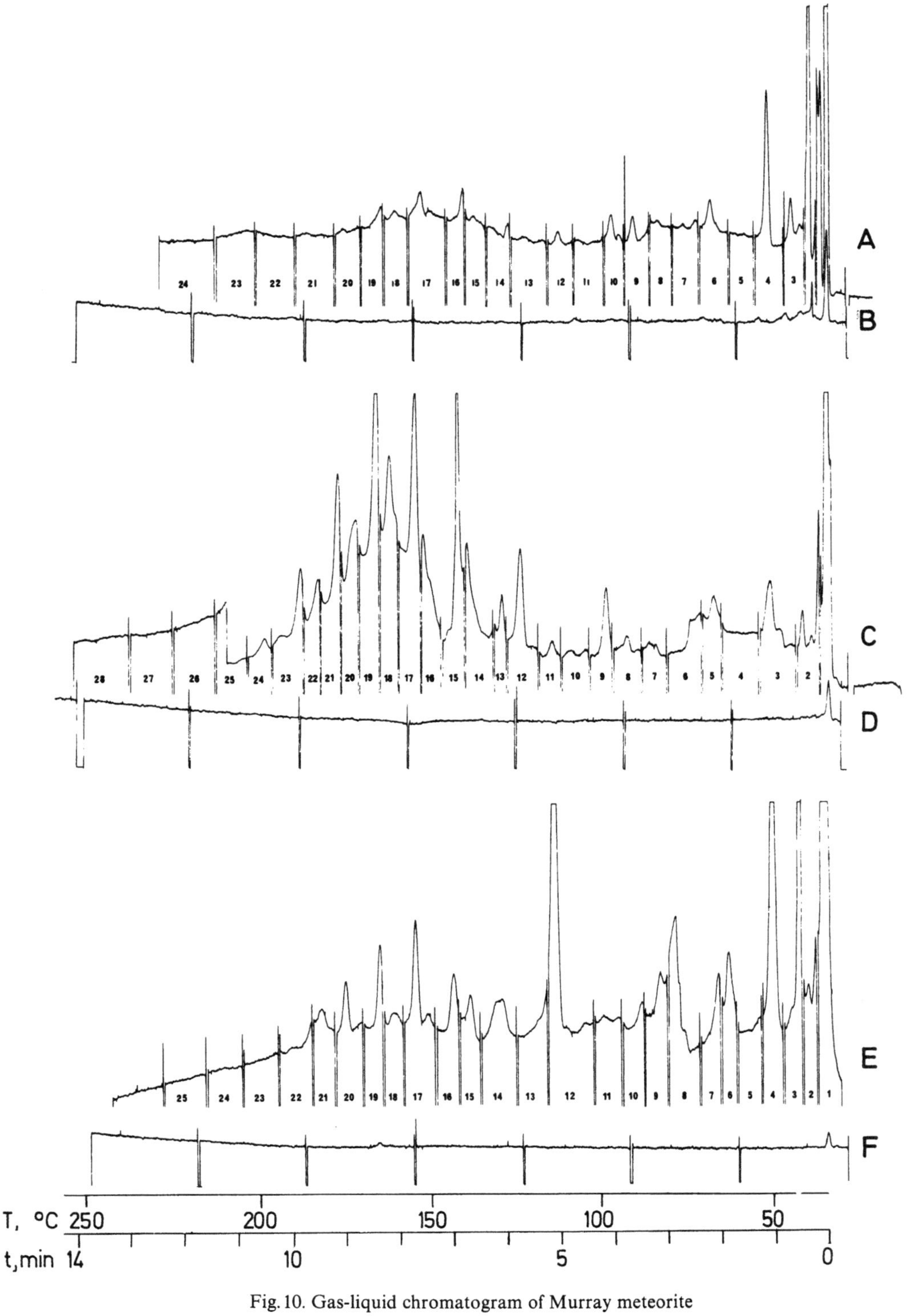

Fig. 10. Gas-liquid chromatogram of Murray meteorite

```
                        ELEMENT MAP532-40-1     MURRAY GLC-III-254-15     J.H.

      CH(AL)            CH(AR)          CHS              CHO            CHO2
117                     9/ 9 0****
118                     9/10 0***
119                     9/11 0*****
120                     9/12-1****
121 9/13 0****
122 9/14 0***
123 9/15 0******
124 9/16 0*****
125 9/17 0******
126 9/18 0******
127 9/19 0*******
128                     10/ 8 0****
129                     10/ 9-1***
131                     10/11 0***
133                     10/13 0*****
134                     10/14 0***
13510/15 0***
13610/16 0***
13710/17 0*****
13810/18 0****
13910/19 0*****         11/ 7 0****
14010/20 0*****
14110/21 0******        11/ 9 0******
142                     11/10 1*******
145                     11/13 0***
147                     11/15 0*****    9/ 7 0***
148                                     9/ 8 0***
14911/17 0**
15011/18 0***
15111/19 0****
15211/20-2***
15311/21-1****
15411/22-1*****
15511/23 0******
159                     12/15 0***
160                     12/16 0***
161                     12/17 0***
16512/21 0***
16912/25 0****
176                     13/20 0****
17913/23 0****
18013/24-1***
18113/25 0***
18213/26 0****
18413/28 0******
19614/28 0****
207

      CH(AL)            CH(AR)          CHS              CHO            CHO2

      CH(AROMATIC) IONS ARE THOSE WITH ION TYPE X LESS THAN OR EQUAL TO -6
```

Fig. 11. Element map of peak 15 in Fig. 10

showing the g.l.c.-high resolution mass spectrometer analysis of the organic constituents of the Murray meteorite. The analytical technique involves the pyrolysis of a sample of the meteorite, the condensation of the evolved organic matter in a cold trap, and subsequent flash evaporation of the condensate into a gas chromatograph connected to a mass spectrometer. The gas chromatograms illustrated in Fig. 10 are the traces obtained when a sample of Murray chondrite was heated over a temperature range of 25°–146° C (A), 146°–254° C (C) and 255°–409° C (after admixture of zinc dust – E). Traces B, D and F represent blanks. High resolution spectra recorded on a photoplate were obtained for each peak; the vertical lines in each chromatogram indicate successive exposures. The photoplate data are subsequently reduced by computer techniques.

The end result of these manipulations is illustrated in Fig. 11, the element map (see section on high resolution) of peak 15 in chromatogram C. It will be noted that compositions of ions are plotted in vertical columns, labelled CH-aliphatic, CH-aromatic, CHS, CHO, CHO_2. In the case at hand, the element map permits the conclusion that the major component of g.l.c. peak 15 is tridecane (*m/e* 184, $C_{13}H_{28}$; the asterisks indicate abundance on a logarithmic scale [46]*) and that minor components are a methylnaphthalene ($C_{11}H_{10}$), a C_7-alkyl benzene ($C_{13}H_{20}$) and a monounsaturated or monocyclic aliphatic compound of composition $C_{14}H_{28}$ (m/e 196). Furthermore, trace amounts of methylbenzothiophene (C_9H_8S) are seen to be present, but no oxygen-containing molecules or ions.

This one example may suffice to illustrate the wealth of information that can be obtained from high resolution mass spectra of mixtures of g.l.c. fractions. However, it is also clear that the routine use of high resolution instruments and the necessary analysis of the data requires the extensive utilization of computer methods (see section II C, D). The work of HAYES [101] should be consulted for more details on the analysis of carbonaceous chondrites by g.l.c.-high resolution mass spectrometer and computer techniques.

III. Characterization of Organic Matter from Geologic Sources

A. Hydrocarbons

The application of mass spectrometry to the analysis of hydrocarbon fractions obtained from petroleum represents the earliest exploitation of the technique for the solution of organic chemical problems. This chapter will not review the extensive literature in this area. Many papers are concerned with the quantitative determination of compounds in mixtures; others with qualitative type analysis (i.e., the detection and estimation of certain classes of compounds — saturated hydrocarbons, olefins, aromatics — in a given petroleum fraction). The technique of low ionizing voltage-electron bombardment mass spectrometry often utilized in this connection will be discussed later (see section II E). For many of these applications mass spectrometry has now been replaced by gas chromatographic analysis, a method which lends itself equally to rapid and sensitive quantitative analysis of mixtures containing predominantly known components but requiring much less elaborate instrumentation. The review will thus cover mainly those investigations in which mass spectrometry played a major role in the structural characterization of individual compounds.

A very extensive collection of hydrocarbon mass spectra is available [102], and fairly detailed discussion of aliphatic, cyclic and aromatic hydrocarbons and their type analysis in petroleum fractions can be found in the older literature [103–106]. For a review of the mechanistic aspects of hydrocarbon mass spectrometry the volume of BUDZIKIEWICZ *et al.* [3] should be consulted.

* The numbers immediately preceeding them indicate error in measurement to the nearest millimass unit, m. m. u.

1. Normal Hydrocarbons

Mass spectrometric identification of individual straight-chain saturated hydrocarbons is usually quite straightforward, since the smooth envelope of major peaks at intervals of 14 mass units (i.e., 43, 57, 71, 85, etc.) corresponding to ions of the type $CH_3CH_2CH_2^+$, $CH_3CH_2CH_2CH_2^+$, etc., whose intensity maximizes around C-4 and decreases with increasing mass, represents a characteristic pattern for all compounds of this type*. A typical example is illustrated in Fig. 12. Identification of a specific normal alkane essentially depends on the correct recognition of the molecular ion. Unambiguous interpretation is trivial, unless the compound is impure; in particular, the identification may present a problem, if isomers are present which may contribute intense peaks to the fragmentation pattern. Since the molecular ion peaks of normal alkanes resulting from electron bombardment

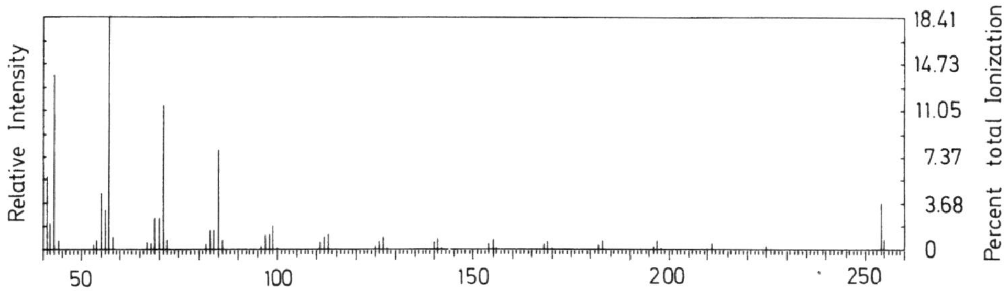

Fig. 12. Mass spectrum of *n*-octadecane

ionization are of quite low intensity, the correct recognition of the molecular ion may occasionally pose a problem for very impure samples. These general remarks are, of course, equally valid for other types of compounds, but it should be stressed here, since the simple pattern of normal alkanes is very easily obscured by branched impurities.

Normal hydrocarbons [108] appear to occur in all carbonaceous geological environments thus far investigated; it appears pointless to enumerate the great number of studies which have utilized mass spectrometry for their identification.

2. Branched Alkanes: Iso, Anteiso, Isoprenoidal Alkanes

Mass spectrometry yields much more significant structural information for more complex hydrocarbons. In these cases, indeed, mass spectrometry is essential to furnish an initial clue to the structure of the isolated compound which can then be checked by comparison with synthetic material. Gas chromatographic analysis alone would provide rapid information only in more routine cases where the composition of a mixture can be partially surmised from previous work.

Since fragmentation of branched hydrocarbons occurs preferentially at the site of branching, peaks corresponding in mass to the fragment including the branching point appear as abundant ions in the mass spectrum relative to their homologues

* This pattern is only typical for spectra resulting from electron impact at relatively high temperatures: 180–250° C [17, 18] (see section II A). BECKEY has reported the comparison of the chemical ionization and field ionization spectra of several decane isomers [107].

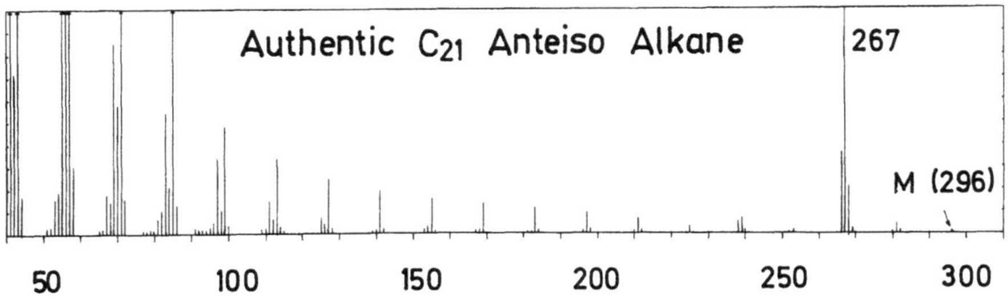

Fig. 13. Mass spectrum of authentic 18-methyl eicosane (C_{21} *anteiso-*)

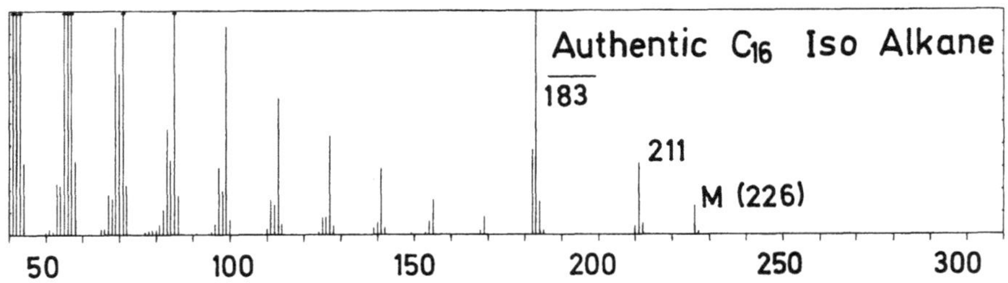

Fig. 14. Mass spectrum of authentic 14-methylpentadecane (C_{16}-*iso*)

14 mass units above or below. This fact permits, therefore, the determination of the location of a branch in the carbon chain—information which cannot be obtained readily by other means. The principle is well-illustrated by the spectrum of an *anteiso* alkane (II, Fig. 13), where cleavage at the branching point, resulting in the loss of 29 mass units ($CH_3CH_2^{\cdot}$) produces a very intense ion; the remainder of the spectrum follows the pattern expected from a simple *n*-alkane. A compound of this type is thus identified unambiguously by its fragmentation pattern. For the *iso*-alkane structure, III, by analogy, an intense peak at M-15 would be expected and is observed (Fig. 14). The more pronounced loss of 43 mass units would not be predicted necessarily since the fragment corresponding to M-43 should be a primary (and, therefore, not favored) carbonium ion; its great intensity probably results from the stability of the secondary propyl radical ($CH_3\dot{C}H-CH_3$) formed in this process.

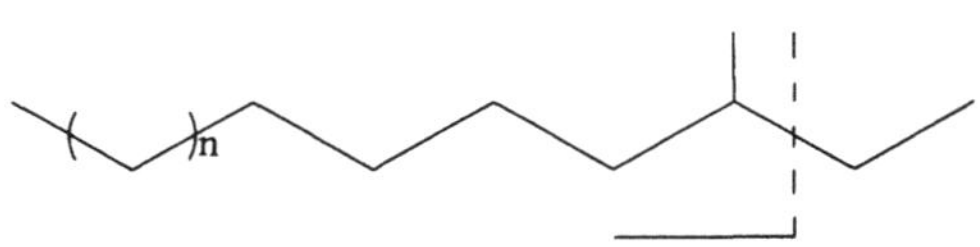

II M-29

M-43

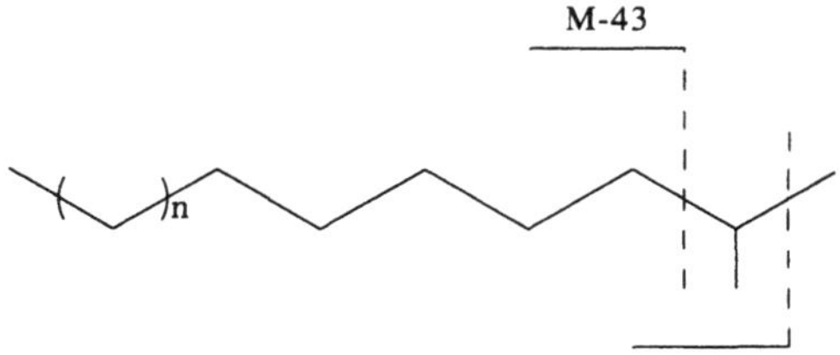

III M-15

The mass spectra of other branched alkanes are equally characteristic; 4-methyl-alkanes exhibit intense M-43 peaks, for example, and the structure of 5-methyl-alkanes would be evident from the pronounced loss of 57 mass units.

Mass spectrometry has permitted the identification of a number of *anteiso* and *iso*-alkanes from geological sources. JOHNS *et al.* [109] reported the identification of several *anteiso*- (C_{16} to C_{18}) and *iso*-alkanes (C_{16} to C_{18}) from the Nonesuch Seep Oil and VAN HOEVEN *et al.* [110] identified the C_{15}-*iso*-alkane and the C_{18}-*anteiso* compound in the Australian Moonie Oil by these methods. LEVY *et al.* [111] have shown the presence not only *iso* and *anteiso* alkanes, but also the 4- and 5-methyl-alkanes in paraffin wax by g.l.c. and mass spectrometric techniques. These papers, as well as the reports by MOLD *et al.* [112] and EGLINTON *et al.* [113] present mass spectrometric data on compounds isolated from various natural sources.

Multibranched hydrocarbons present, of course, much more complex fragmentation patterns, whereby peaks corresponding to cleavages of the carbon chain at the site of branching will again furnish the more abundant ions of the spectrum. Isoprenoid alkanes are compounds of particular geochemical interest belonging to this class, since they are usually taken as fairly conclusive evidence of the biological origin of the hydrocarbon material extracted from an ancient sediment [109, 113, 114]. An unambiguous identification is thus of great importance, especially in the case of very old rocks. For the case of two typical isoprenoids, the C_{16}- (IV) and the C_{20}-compound (V, phytane), the intense peaks expected in their mass spectra are illustrated graphically in structures IV and V:

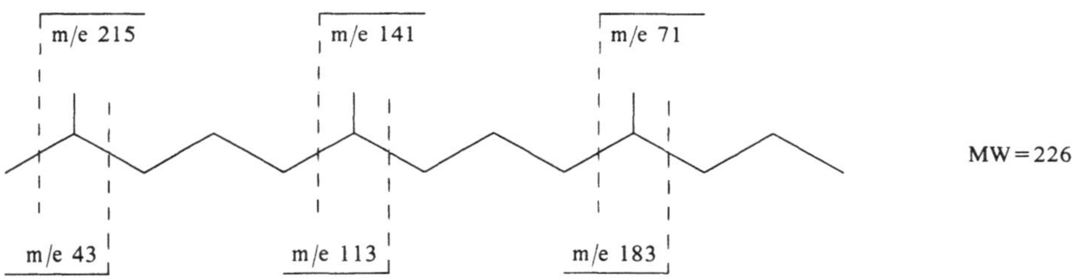

IV

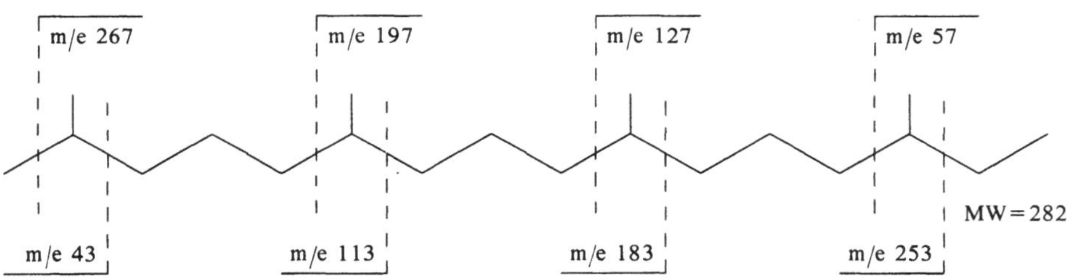

V

Figs. 15 and 16, the mass spectra of IV and V isolated from different geological sources, show that the actual pattern follows prediction. Photographs of original mass spectrometric records of three isoprenoids isolated from the Moonie Oil [110]

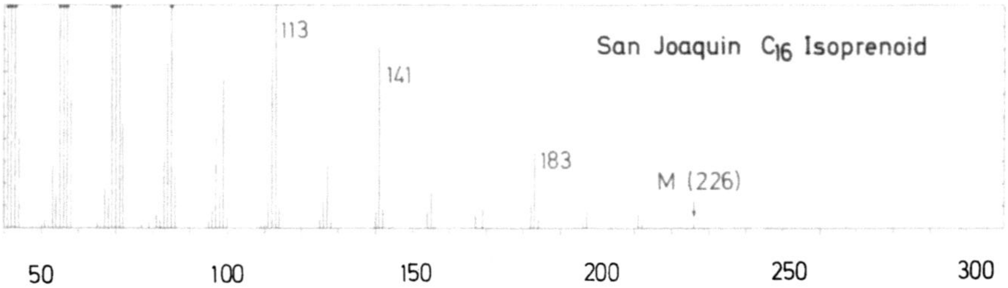

Fig. 15. Mass spectrum of C_{16} isoprenoid alkane (San Joaquin Oil)

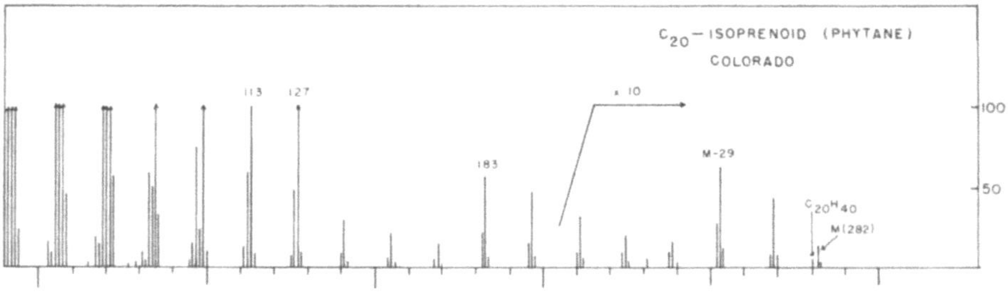

Fig. 16. Mass spectrum of C_{19} isoprenoid alkane, phytane (Colorado Green River Formation)

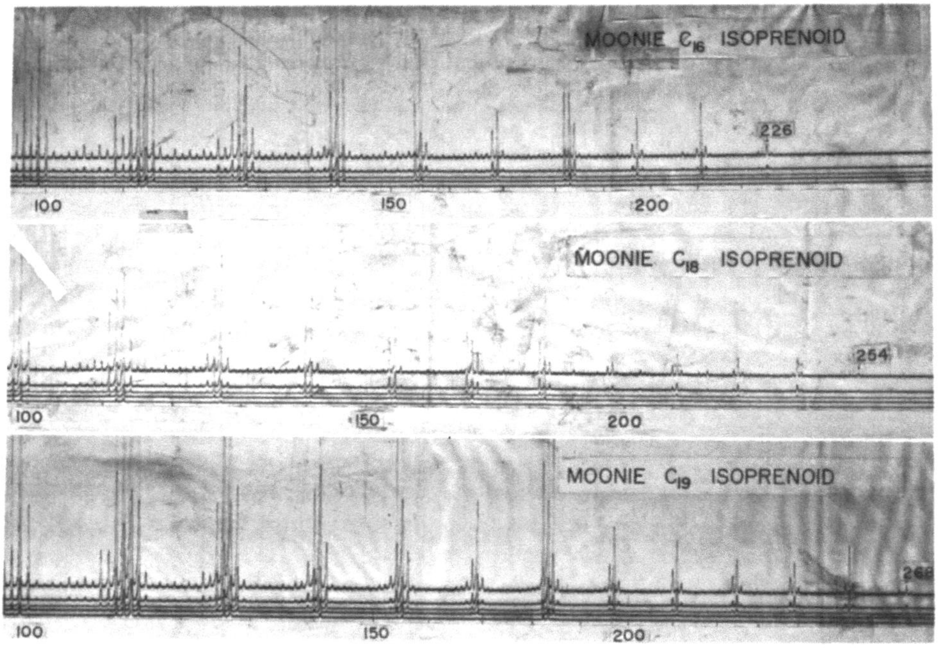

Fig. 17. Oscillograph original mass spectra of Moonie C_{16}, C_{18}, C_{19} isoprenoid alkanes (VAN HOEVEN *et al.*)

are presented in Fig. 17 to illustrate some further typical patterns and the appearance of such isoprenoid mass spectra as they present themselves for interpretation.

The mass spectral fragmentation pattern is quite characteristic and sensitive to the position of substituents on the carbon chain. For example, the C_{16} isoprenoid (IV) could be distinguished easily from its modified isomer VI, since for the latter a mass spectrum exhibiting intense peaks at m/e 197, 127 and 57 would be expected.

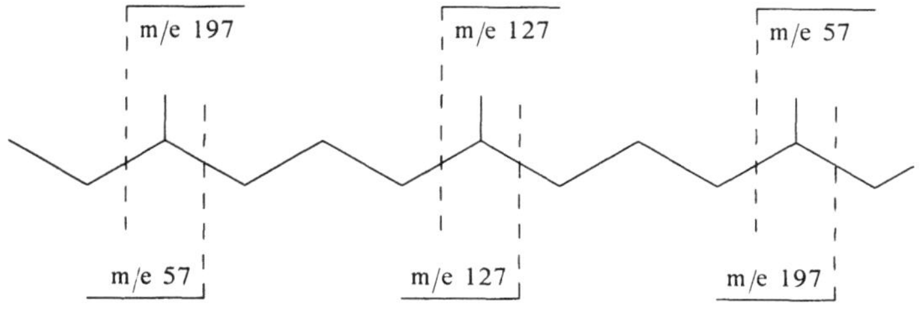

MW=226

VI

In practical cases several factors combine to make interpretation somewhat less straightforward. One common problem is contamination by other hydrocarbon material, which, if it does not obscure correct recognition of the molecular ion, may lead to ambiguities in the interpretation of the fragmentation pattern should the contaminant give rise to abundant fragments. For instance, contamination of an isoprenoid alkane with its *iso*-alkane isomer would produce a mass spectral pattern showing inordinately intense M-15 and M-43 peaks, such that the correct identification of the isoprenoid might be difficult or even impossible. Most often, nevertheless, even when small amounts of other hydrocarbons are present (and it is rare that geochemical g.l.c. fractions are ever really "pure"), the typical fragmentation pattern of the isoprenoid chain is characteristic enough to permit certain identification [109, 113]. Furthermore, problems of this nature can often be overcome by obtaining mass spectra of several "cuts" of a g.l.c. peak and comparing their fragmentation patterns, or else by obtaining a series of mass spectrometer scans: as one fraction enters the ion source (by direct introduction or g.l.c. inlet) comparison of successive scans provides a check on sample homogeneity and may lead to unambiguous identification of components. A much more serious difficulty may be encountered in distinguishing mass spectrometrically between isoprenoidal alkanes and other branched or multibranched hydrocarbons of non-isoprenoidal skeleton. The close similarity of the mass spectra of 2-methyl-octane and 2,6-dimethyl-heptane is an informative example [102]. The mass spectra of Figs. 21–23 illustrate another case. In connection with work on the C_{17} isoprenoid from the Antrim Shale, McCarthy and Calvin [115] synthesized several branched hydrocarbons to compare their g.l.c. retention times and mass spectra. The g.l.c. trace (capillary column) of three C_{19}-branched hydrocarbons is shown in Fig. 18. The naturally occurring isomer, pristane (VII), is eluted first, followed by synthetic 2,6,10,13-tetramethyl pentadecane (VIII) and 2,6,10-trimethyl hexadecane (IX). The respective mass spectra (Figs. 19–21) exhibit strikingly similar fragmentation patterns: compound VIII differs from its isomers in the intensity of the peak at m/e 183 and the abundance of the M-29 ion, the latter being readily explained by the structure of the compound; however, a distinction between compounds VII (pristane) and IX on the basis of the mass spectrum would be difficult indeed and, in a practical case, probably impossible.

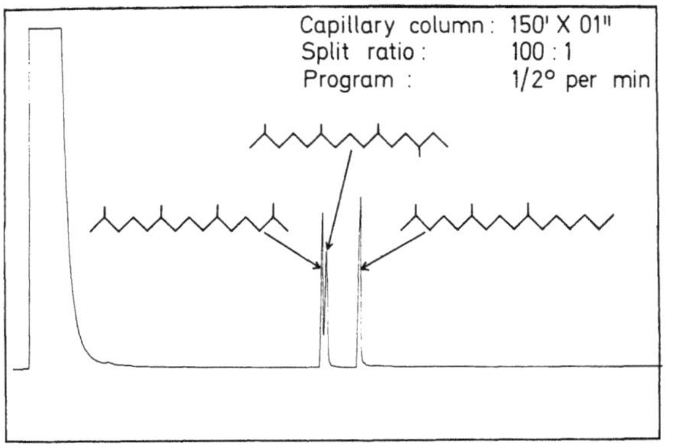

Fig. 18. Capillary gas-liquid chromatogram of C_{19} isoprenoid isomeric alkanes

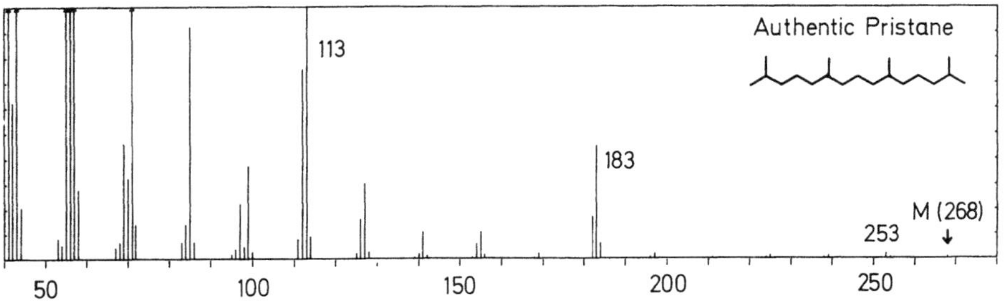

Fig. 19. Mass spectrum of authentic pristane *vs.* m/e 113 base peak

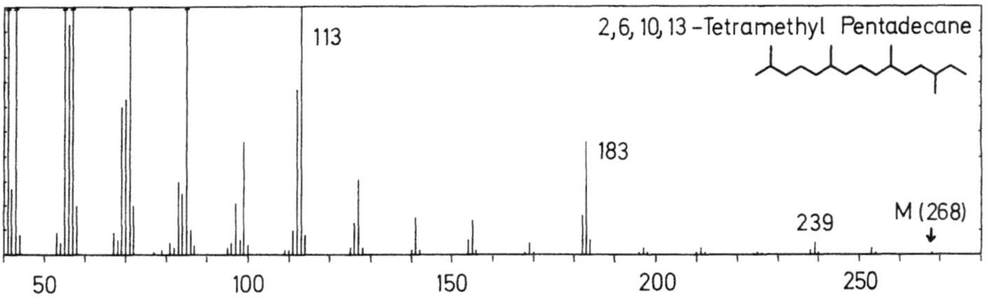

Fig. 20. Mass spectrum of 2,6,10,13-tetramethylpentadecane *vs.* m/e 113 base peak

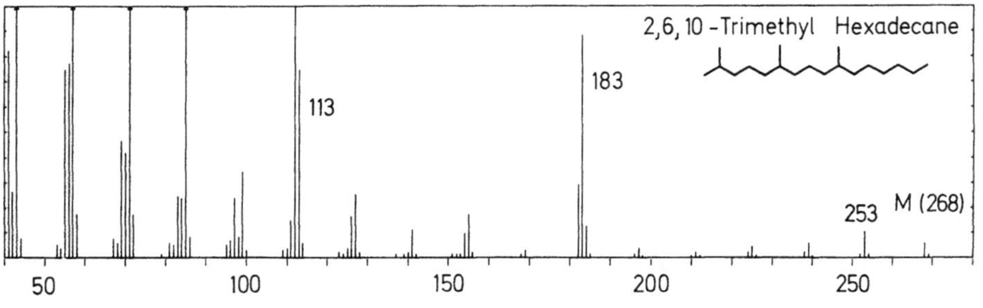

Fig. 21. Mass spectrum of authentic 2,6,10-trimethylhexadecane *vs.* m/e 113 base peak

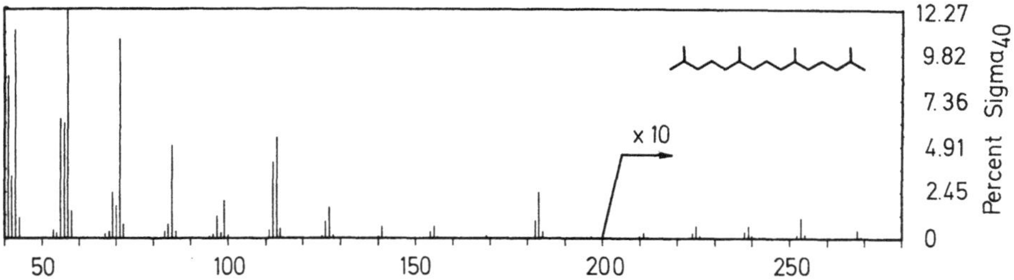

Fig. 22. Mass spectrum of pristane *vs.* percent $\sum_{40}$

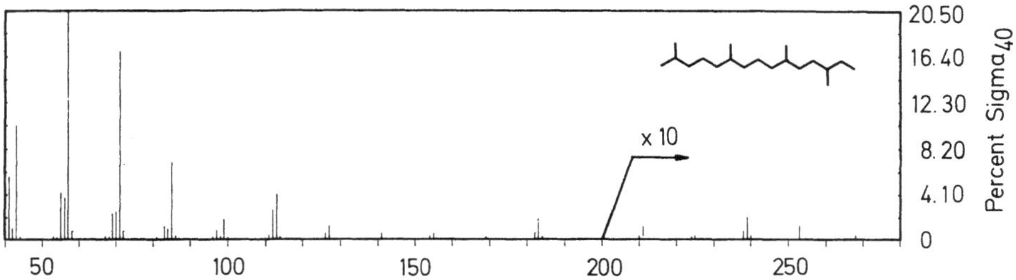

Fig. 23. Mass spectrum of 2,6,10,13-tetramethylpentadecane *vs.* percent $\sum_{40}$

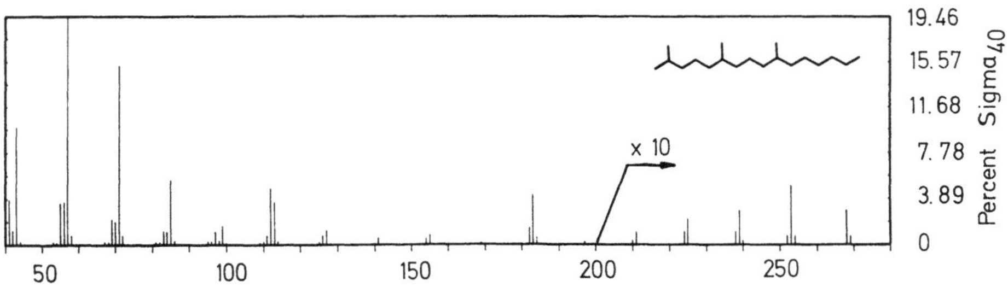

Fig. 24. Mass spectrum of 2,6,10-trimethylhexadecane *vs.* percent $\sum_{40}$

Spectra of the same compounds plotted with m/e 59 as the base peak and showing the percent of the total ionization (percent Σ_{40}) are shown in Figs. 22–24 for comparison. The differences in relative intensity and percent of the total ionization are somewhat more apparent here; in particular, it can be seen that certain peaks differ considerably in their contribution to the total ionization, e.g., m/e 59 is approximately 12 percent in Fig. 22, but approximately 20 percent in Figs. 23 and 24. There are similar differences in m/e 113 and 183, and particular differences appear above m/e 183 reflecting the change in structure at one end of the chain. However, it would be rather dangerous to postulate structural details without authentic comparison spectra. The limitations of the mass spectrometric method are clearly evident here, and the need for corroborative data (at least in some cases), such as comparison of g.l.c. retention times, needs no further emphasis. Unfortunately, only for the more common branched hydrocarbons are synthetic samples available and some caution should, therefore, be exercised in the assignment of definite structural terms, such as "pristane", to hydrocarbons identified only by mass spectrometry.

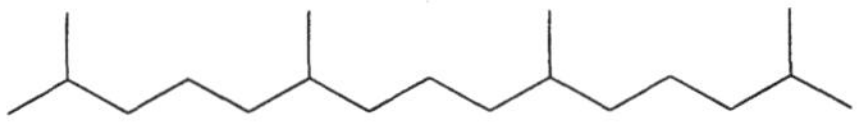

VII

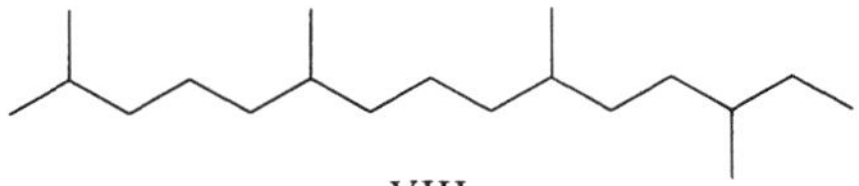

VIII

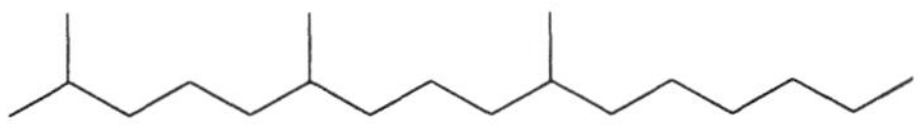

IX

In the last few years mass spectrometry has been applied extensively to the characterization of isoprenoidal hydrocarbons. Pristane (C_{19}, VII) has been reported in Texas gas oil and Mid-Continent Oil [117]. In a subsequent paper by these authors [118], mass spectrometry was used in the identification of the C_{14}, C_{15}, C_{16}, C_{18}, C_{19}, C_{20} and C_{21} regular isoprenoids from East Texas Oil. DEAN and WHITEHEAD [119] first reported phytane, and MAIR *et al.* [120] identified the C_{14} and C_{15} isoprenoids in a light gas oil fraction. From the Colorado Green River Formation the C_{15}, C_{16}, C_{18}, C_{19} and C_{20} isoprenoids have been reported [113, 121, 122] and a series of isoprenoids has been isolated from the Precambrian Nonesuch Shale [113, 122]. Extensive mass spectral data are given also in the papers by JOHNS *et al.* [109] and VAN HOEVEN *et al.* [110] reporting on the isolation of a range of isoprenoids from a great variety of shales and oils, including the Precambrian Soudan formation [123]. Using g.l.c. and mass spectrometer-g.l.c. coupling, ORÓ and collaborators have identified pristane and phytane in the Gunflint Chert [94] and in the Fig Tree Shale [95], both of Precambrian age. MEINSCHEIN and co-workers [124, 125] suggested the presence of normal and isoprenoid alkanes in the Soudan Shale and the Nonesuch formation on the basis of mass spectra of extracts. Other applications of mass spectrometry to the identification (or partial identification) of isoprenoid alkanes include the papers of MAIR [126], PONNAMPERUMA and PERING [127], BLUMER *et al.* [128] and GÖHRING *et al.* [129]. Reports of higher isoprenoids from geological sources include identification of carotane (C_{40}) in the Green River Formation [130] and of carotene and other pigments in Florida Mud extracts [131].

Isoprenoids thus far isolated have been of "regular" structure, i.e., possessing methyl substitution at C-2,6,10,14. It is quite conceivable, however, that compounds of the isoprenoid type will be found which possess different substitution patterns, for example, structures derived from such precursors as squalene [115]. Indeed, compounds which do not bear a definite structural relationship to the isoprenoid class at all, but which might be derivable from more complex isoprenoidal precursors may be present in geological formations, as the recent finding [132] of 2-methyl-3-ethyl-heptane in petroleum, which could be derived from the cyclic monoterpene limonene, demonstrates. Correct identification of such hydrocarbons is essential, however, if any valid conclusions as to biological precursors or diagenetic history are to be drawn. Mass spectrometry will play a major role in these efforts, and it is for this reason that the ambiguities and potential uncertainties of the method need to be clearly understood.

3. Cyclic Hydrocarbons

The mass spectra of cyclic hydrocarbons are more complex than those of acyclic compounds and, while certain structural features may often be apparent, a clear-cut prediction of the fragmentation pattern in terms of structure (as in the case of branched alkanes) is usually not possible. Thus, the information which a mass spectrum furnishes on an unknown compound may be limited to the exact molecular weight (giving an indication of number of carbon atoms and rings or double bonds) and the nature of alkyl substituents (from the loss of groupings, i.e., M-15, M-29, etc.); more detailed deductions concerning the gross structure of the unknown would be dangerous, at best, except in simple cases. Mechanistic aspects of the fragmentation of cyclic hydrocarbons, including monoterpenoid hydrocarbons, are discussed by BUDZIKIEWICZ *et al.* [3] and an up-to-date bibliography on the mass spectra of these compounds is available therein. A fairly detailed summary of the application of mass spectrometry to structural studies on monoterpenoids has been presented by MCFADDEN and BUTTERY [133].

Examples of the use of mass spectrometry for the structural elucidation of simple cycloalkyl compounds are provided by the work of JOHNS *et al.* [109], reporting on the occurrence of cyclohexylalkanes in the Nonesuch Seep Oil (C_{16} to C_{19} compounds), and by VAN HOEVEN *et al.* [110], who identified these substances in the Moonie Oil. The mass spectrum of a representative of this series, Nonesuch C_{16}-cyclohexyl alkane (X) is shown in Fig. 25.

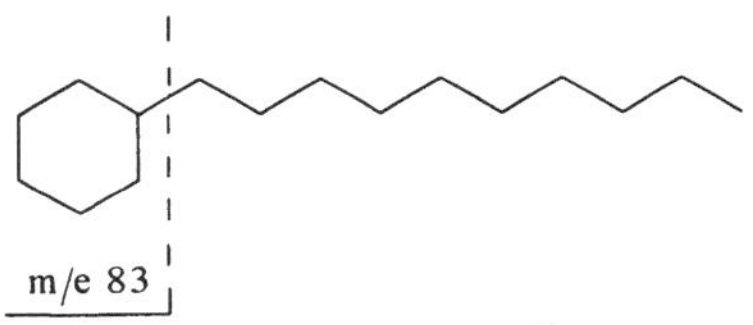

MW = 224

X

The molecular ion m/e 224 identifies the compound as a C_{16} alkane possessing one ring, and the nature of the ring is evident from the intense peak at m/e 83, which corresponds to a cyclohexyl carbonium ion. A molecular ion of m/e 224 would be observed also, of course, for an acyclic compound possessing a double bond; an olefin, however, would show a distinctly different fragmentation pattern (see below). The alkyl chain of X must necessarily comprise ten carbons and the pattern of peaks from m/e 97 to m/e 224 suggests the absence of branching in the chain. The peak at m/e 83 could also be due to a methylcyclopentyl ion; a distinction between these possibilities would be difficult *a priori*, and the identification would remain tentative unless confirmed by appropriate standards. The example illustrates, however, the power of mass spectrometry to limit possibilities for an

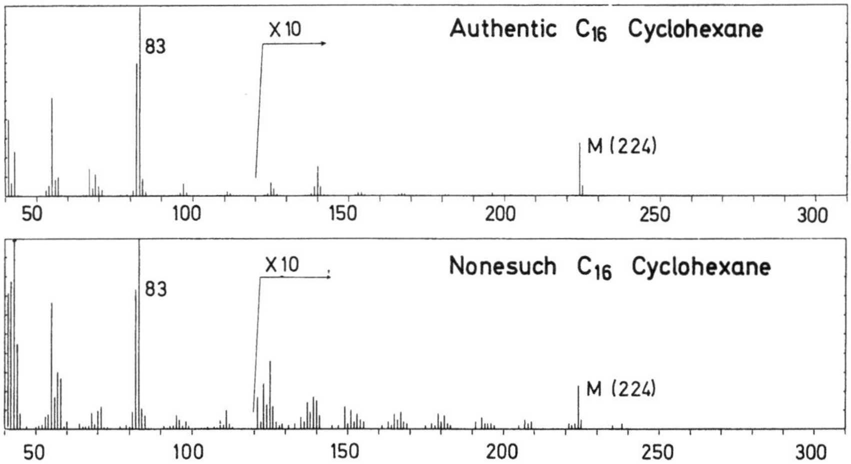

Fig. 25. Mass spectrum of *n*-hexadecylcyclohexane (Nonesuch)

unknown to a few structural variations among which a choice can then be made on the basis of other data or comparison with standards. Cyclohexyl alkanes have also been reported on the basis of mass spectra of petroleum fractions [134] and in the Athabasca petroleum deposit [135]. An analysis of paraffin wax [111] indicated cyclopentyl- and cyclohexyl alkanes; this work, as well as that of MEYERSON *et al.* [136], also reports mass spectrometric results obtained for several authentic compounds of this class.

More complex cyclic and bicyclic compounds have been isolated from petroleum. LINDEMANN and LE TOURNEAU [137] report the mass spectra of fourteen novel compounds, including various methyl bicycloheptanes, bicyclooctanes and bicyclononanes, as well as methyl-substituted cycloheptane and cyclohexane derivatives obtained from oil fractions. In this work a gas chromatograph-mass spectrometer combination was utilized.

4. Steranes and Triterpanes

Recent identifications of a number of steroid and triterpenoid hydrocarbons in geologic materials provide a very interesting illustration of the application of mass spectrometry to geochemical structure problems. First indications of the presence of such molecules were obtained on the basis of mass spectra of complex hydrocarbon mixtures. O'NEAL and HOOD [138], SCHISSLER *et al.* [139] and MEINSCHEIN [140, 141] reported the presence of sterane-type compounds in oil and sediments on the basis of characteristic peaks (see below) observed in mixture spectra; MEINSCHEIN's data also suggested the presence of hexacyclic and C_{40} polycyclic structures. More recently, CARLSON *et al.* [142] presented both high and low resolution mass spectral data to support the conclusion of sterane-type structures in petroleum. The first report on the isolation and mass spectral identification of individual compounds of this class was made by BURLINGAME and collaborators [143], who isolated the C_{27}, C_{28} and C_{29} steranes and a C_{30}-triterpane (lupane) from extracts of the Colorado Green River Formation. The

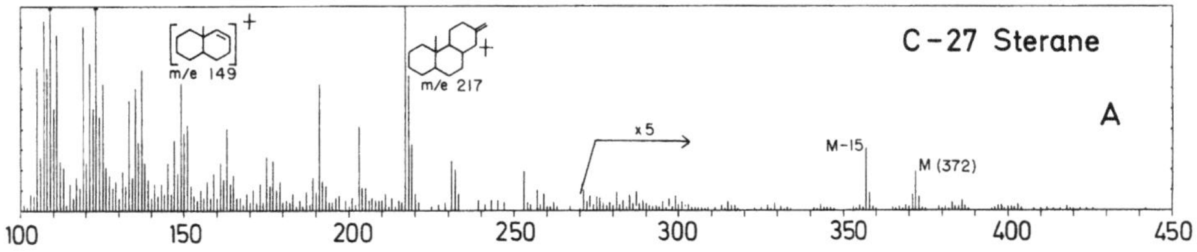

Fig. 26a. Mass spectrum of C_{27}-sterane Green River Formation

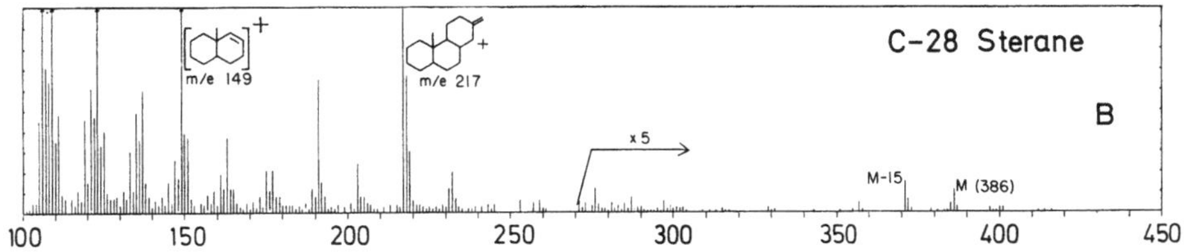

Fig. 26b. Mass spectrum of C_{28}-sterane Green River Formation

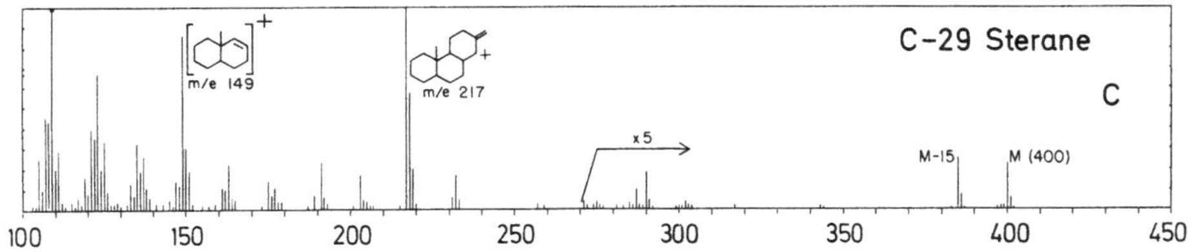

Fig. 26c. Mass spectrum of C_{29}-sterane Green River Formation

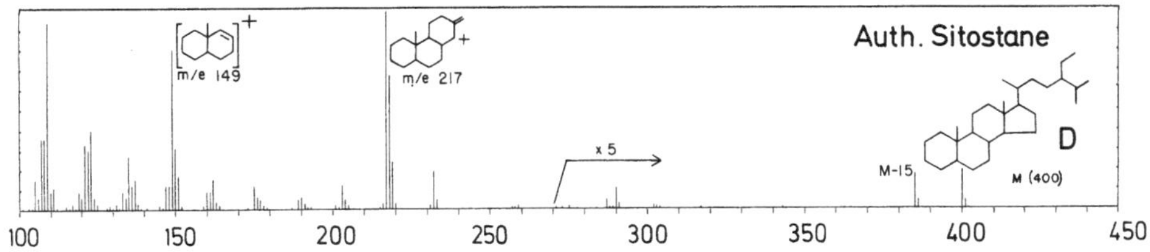

Fig. 26d. Mass spectrum of authentic sitostane

mass spectra of these compounds, together with that of an authentic C_{29}-sterane, sitostane (XI), are presented in Fig. 26. The fragmentation pattern (peaks at M-15, 218, 217, 149) is typical for saturated sterane structures. A partial interpretation based on the work of DJERASSI and co-workers [144] is presented below, with sitostane (XI) as an example:

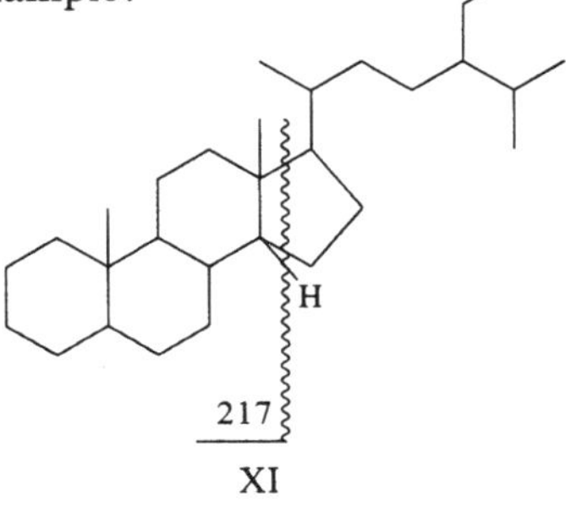

XI

The base peak at m/e 217 arises by a complex fragmentation process involving the loss of the side chain, C-15, -16, -17 from ring D and the hydrogen attached to C-14, as depicted schematically in structure XI. The ion of m/e 149 includes rings A and B and the 19-methyl group.

The mass spectra of the three tetracyclic hydrocarbons shown in Fig. 26 permitted the structural assignment of cholestane (XII) to the compound of Fig. 26a, ergostane (XIII) to Fig. 26b and sitostane (XI) to Fig. 26c.

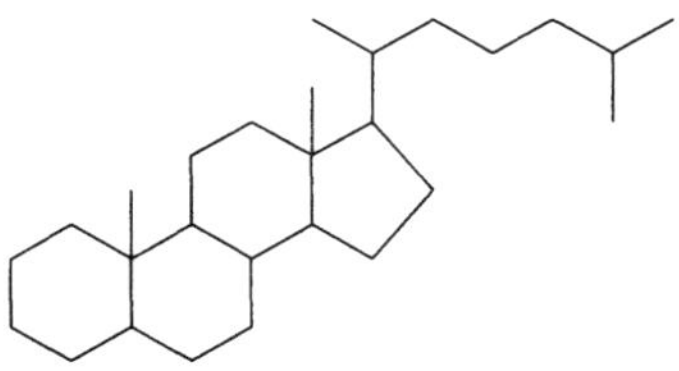

XII

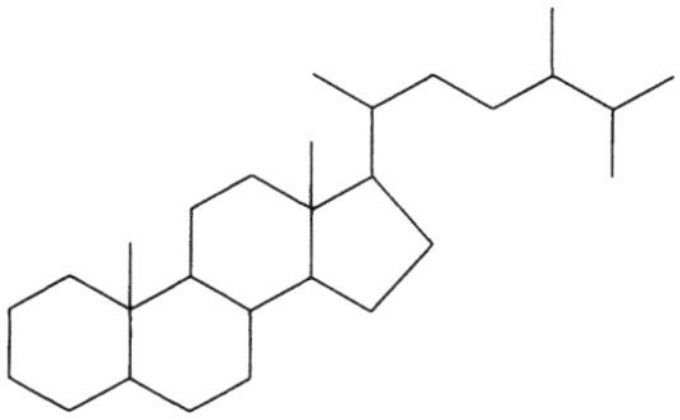

XIII

By a combination of i.r., g.l.c. retention times and mass spectrometry, EGLINTON *et al.* [130] have recently confirmed these assignments. The triterpane (Fig. 27a) was not identified, but the mass spectrum leaves no doubt about the basic skeleton of this compound. Peaks at m/e 191 and m/e 149, 137 are very characteristic for pentacyclic terpenoid hydrocarbons; indeed, as comparison with the spectrum of authentic lupane (XIV), Fig. 27b, shows, a structure of the lupane-type would seem to be indicated.

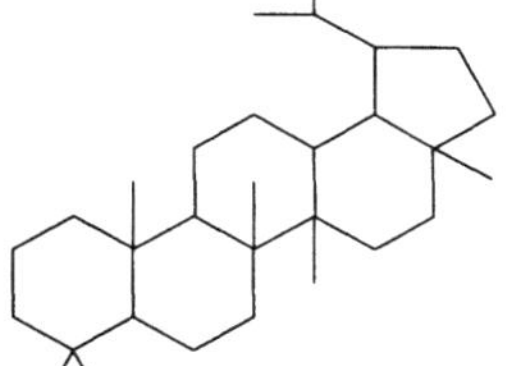

XIV

For more detailed discussion of the mass spectra of triterpenoid hydrocarbons, the papers by DJERASSI and co-workers [145, 146] should be consulted.

Mass spectrometric evidence for the presence of tetracyclic and pentacyclic hydrocarbons in a Precambrian sediment [143] is illustrated in Fig. 27c. The spectrum is that of a mixture of compounds, but it can be interpreted as indicating the presence of C_{27} (MW = 372), C_{28} (MW = 400) steranes (note peaks at m/e 218, 217, 149), as well as of some C_{30} triterpane (MW = 412).

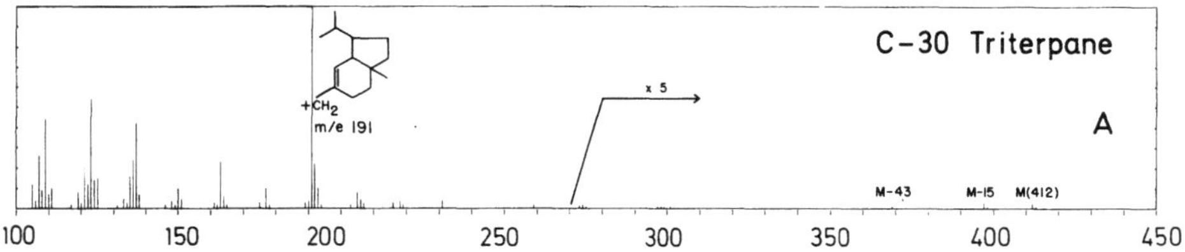

Fig. 27a. Mass spectrum of C_{30} triterpane

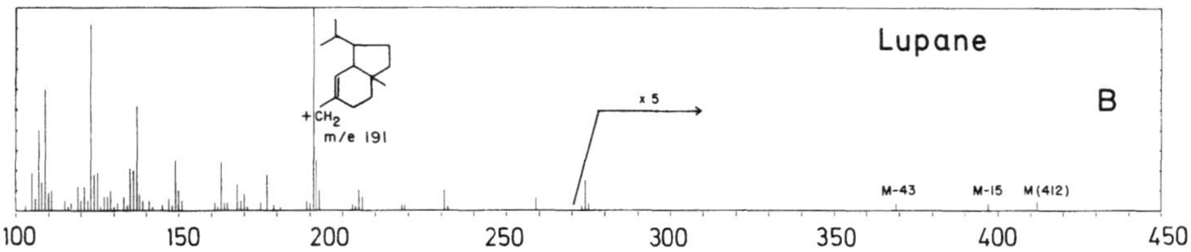

Fig. 27b. Mass spectrum of authentic lupane

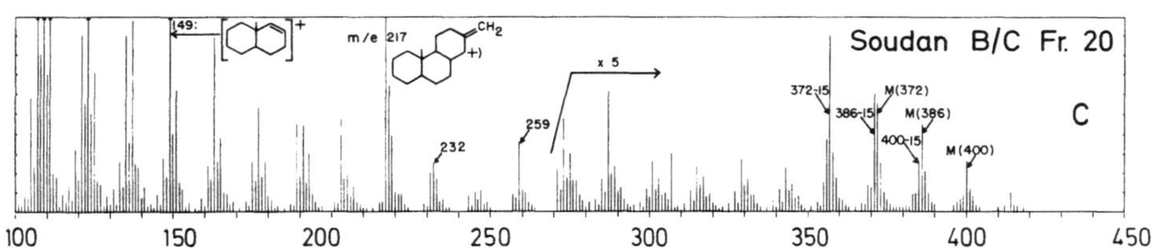

Fig. 27c. Mass spectrum of Soudan B/C Fr. 20

Other examples of the use of mass spectrometry for the investigation of this type of compound include the identification of the triterpane gammacerane [147]

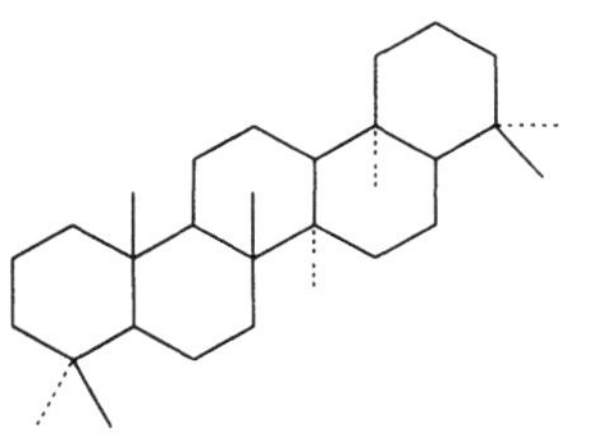

XV

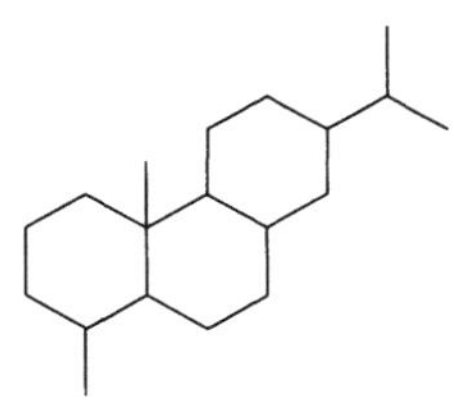

XVI

(XV) and the isolation of four new triterpanes from Nigerian Crude Oil [148]. Mass spectral data presented for the latter compounds clearly identify them as triterpenoid hydrocarbons but do not permit definite structural assignments. More recently, MAXWELL [97b] reported the identification of the diterpenoid hydrocarbon, fichtelite (XVI) by mass spectral and i.r. methods; extensive mass spectral data on other unidentified triterpanes are reported here as well as in the work of HAUG [56]. Mention should be made here also of the use of mass spectrometry by CARRUTHERS and WATKIN [149] and MAIR and MARTINEZ-PICO [150] in the identification of 1,2,3,4-tetrahydro-2,2,9-trimethylpicene, and cyclopentano-

and 3-methylcyclopentano-phenanthrene, respectively, which, although not sterane hydrocarbons, may be considered degradation products of triterpenoid and steroid compounds.

It should be stressed that mass spectrometry by itself is insufficient for the unambiguous structural determination of compounds of this complexity, unless direct comparison with known substances is possible. Even in such cases, however, stereochemical differences between natural product and standard may not be reflected in the fragmentation pattern. But it should be equally evident that mass spectrometric data are essential for structural work in that they provide at least an initial lead to the type of compound at hand.

5. Olefins

Reports on the isolation and characterization of individual olefins from geological sources are not numerous. Compound type analyses have been performed extensively and the work of FIELD and HASTINGS [151], FRISQUE *et al.* [152], and MIKKELSEN *et al.* [153] is illustrative of the methods and problems involved in the quantitative analysis of olefins in petroleum fractions.

An interesting technique for the analysis and structure determination of individual olefins has been developed by LINDEMANN [91]. It involves a gas chromatograph-mass spectrometer combination with the addition of a "micro reactor" (either before or after the gas chromatograph), which has the function of reducing olefin material to the respective saturated hydrocarbon. Structures of the saturated hydrocarbons were determined by mass spectrometry, and correlation of results for different runs (with the "micro reactor" both before and after the gas chromatograph), as well as g.l.c. retention times, permitted deduction of structure and stereochemistry of the original olefin. This method was successfully applied to the analysis of a cracked gasoline olefin-paraffin mixture and a considerable number of paraffins, cyclo-paraffins, aliphatic mono-olefins and cyclic olefins could be identified.

Olefins from sources other than petroleum do not appear to have received much attention thus far (see, for example, the reports of olefins in shale oil [154, 155]). From zooplankton – not a strictly geochemical source but pertinent here since it illustrates the use of mass spectrometry for structural studies on olefins – BLUMER and THOMAS [156] isolated four isomeric phytadienes (XVII–XX). The carbon skeleton of these compounds was established by mass spectrometry after hydrogenation to the saturated derivative, and the position of the double bonds could be ascertained by i.r. spectroscopic methods and ozonolysis, which yielded a C_{17} acid from XVII, the identical C_{16}-aldehyde from XVIII and XIX, and a C_{15}-aldehyde from XX.

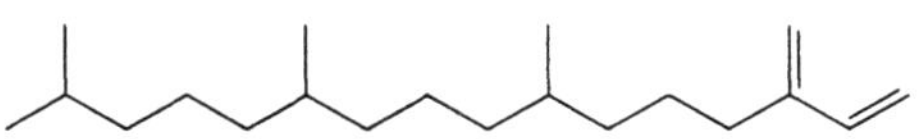

XVII

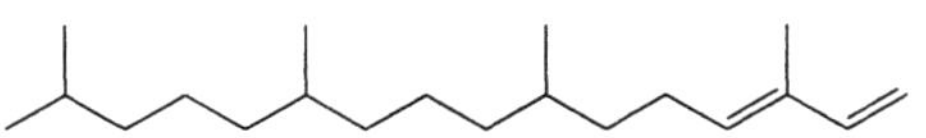

XVIII, XIX

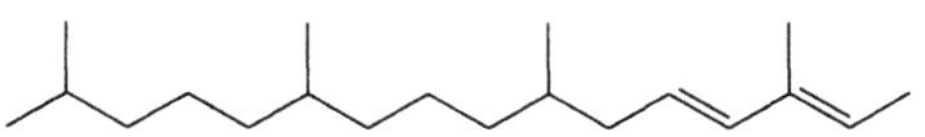

XX

These experiments permitted a definite structural assignment and indicated also that compounds XVIII, XIX differed only in double bond geometry. A similar approach revealed the structure of three mono-olefins with the pristane skeleton [157].

Olefins, even those of simple structure, cannot be characterized completely by mass spectrometry, since the position and geometry of the double bond is usually not apparent from the fragmentation pattern; spectra of isomeric olefins are very similar, if not identical [1–3].

The problem of locating unsaturation in aliphatic olefins has attracted some attention and for mono-olefinic compounds several practical mass spectrometric solutions have been reported. In one of these, the unsaturated compound is transformed to a mixture of two ketones via the epoxide [158]. From the characteristic pattern of the ketone mass spectra the original double bond can be localized.

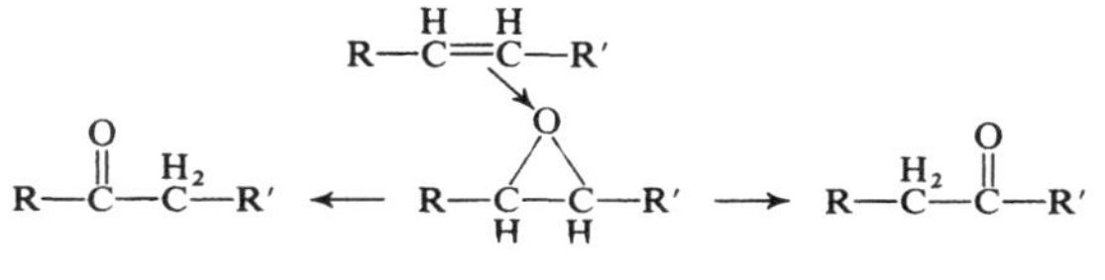

Another method [159] makes use of the fragmentation pattern of amino alcohols derived from the olefin, and in a third [160] the isopropylidine derivative (obtained by osmium tetroxide hydroxylation of the unsaturated compound followed by reaction with acetone) serves the same purpose. For example, prominent ions at m/e 185 and 213 in the spectrum of the isopropylidine derivative XXI of *cis*-hexadecene define the position of the double bond originally present. The method is intriguing especially for its ability to distinguish between *cis* and *trans* olefins [160].

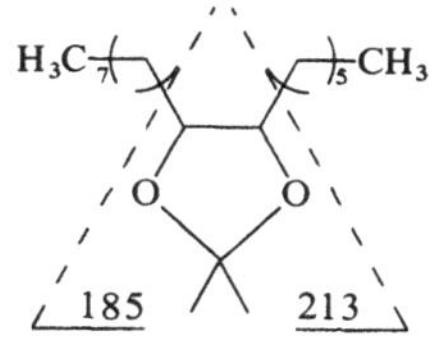

XXI

The method recently proposed by APLIN and COLES [162a] should be mentioned here also. Two recent methods of potential usefulness involve hydroxylation of the double bond followed by the determination of the mass spectra of the *vic*-methyl ethers or the *vic*-trimethyl silyl ethers [162b, 191].

The ozonolysis experiments of BLUMER and THOMAS [156, 157] illustrate another practical approach which, indeed, in combination with g.l.c. mass spectrometry might constitute the method of choice for geochemical studies, since techniques have been described [161] for the routine application of this method to microgram quantities of olefin.

6. Aromatic Compounds

Very extensive use has been made of mass spectrometry for the analysis of mixtures of aromatics from petroleum fractions. The pertinent data and their implications regarding the molecular composition of petroleum have been summarized previously [11, 163]. Mass spectra of individual aromatics have been discussed by O'NEAL and WEIR [103], and a very comprehensive review on the mass spectra of alkylbenzenes is available [164]. Mechanistic aspects of the fragmentation of aromatic hydrocarbons are summarized by BUDZIKIEWICZ *et al.* [3].

In general, aromatic hydrocarbons exhibit well-defined mass spectra with relatively intense molecular ion peaks and few, but abundant, fragment ions. The most characteristic fragments of alkyl substituted aromatic hydrocarbons correspond to cleavages *beta* to the aromatic ring. For mono-substituted alkylbenzenes, for example, the ion of m/e 91 is the most prominent feature of the spectrum; further substitution on the ring would be evident from the shift of this

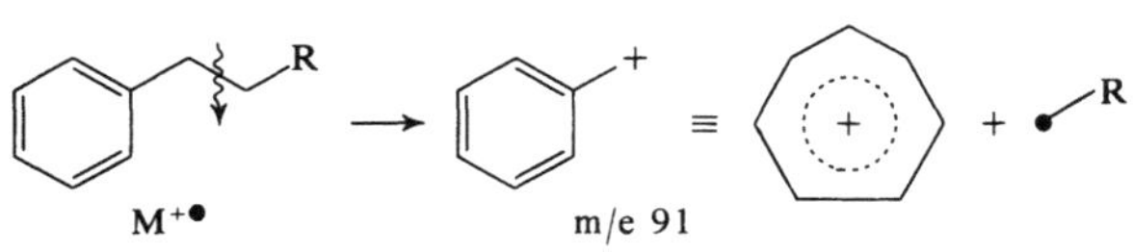

peak by the appropriate mass increment. Thus, for alkylbenzenes, peaks of the series m/e 105, 119, 133, etc. are characteristic, whereas for alkylnaphthalenes, peaks at m/e 141, 155, 169, etc., would be expected. Another characteristic feature of aromatics with alkyl chains of more than three carbon atoms is a rearrangement ion at even mass (i.e., 92, 106, 120, etc., for alkylbenzenes).

Recent applications of modern physical methods, in particular nuclear magnetic resonance and mass spectrometry, are well-illustrated by the extensive investigations of MAIR and co-workers into the mononuclear [165, 166], dinuclear [167, 168] and trinuclear aromatic [150] fractions of petroleum, studies which have led to the identification of a considerable number of individual aromatic hydrocarbons.

Since aromatic compounds exhibit relatively intense molecular ion peaks, low voltage techniques can be used to advantage in the analysis of mixtures. Aromatic hydrocarbons have ionization potentials below 10 electron volts, such that by the appropriate choice of ionization conditions, spectra can be obtained which consist almost exclusively of molecular ion peaks. The technique makes possible also the analysis of aromatics and olefins in mixtures with saturated compounds, since the latter possess appreciably higher ionization potentials (10–13 e.v.). First developed by FIELD and HASTINGS [151], the method has found wide application to the analysis of aromatic hydrocarbon mixtures [169–172]. Considering the developments in field ionization [21, 23] and the recent introduction of chemical ionization methods [22], it is to be expected that these techniques will find application to mixture analyses.

Only a limited number of compound classes can be distinguished when single focussing (low resolution) mass spectrometers are employed for low voltage studies. Ambiguities arise, in particular, when hetero-aromatic compounds are present in the mixture. High resolution mass spectrometers overcome this problem (see section II E). A number of geochemical studies have exploited the advantages high resolution mass spectrometry offers. For example, LUMPKIN [76] identified a series of acenaphthenothiophenes, dibenzofurans, acenaphthenes and carbazoles in the trinuclear aromatic petroleum fraction by obtaining spectra both at high and low resolution and high and low voltages (see Fig. 7).

A number of analyses of aromatics by high resolution mass spectrometry have already been discussed in section II E.

B. Carboxylic Acids

A considerable number of fatty acids have been isolated from petroleum. Only recently, however, have physical methods and mass spectrometry in particular played important roles in the elucidation of their structures. LOCHTE and LITTMAN [174], BALL *et al.* [175] and DINNEEN *et al.* [155] provide summaries of work on petroleum and shale oil up to 1959, when the classical methods of distillation and column chromatography for the isolation of pure samples, coupled with synthesis of the presumed acid and comparison of physical constants, were major routes of attack. In more recent publications mass spectrometry has played a major role in the structural elucidation of acids. Indeed, it can be said that a mass spectrum provides perhaps the most unambiguous information on the structure of normal and branched aliphatic acids, and oxy-aliphatic acids, of any single physical method. Routine application of mass spectrometry techniques has been made possible through the extensive studies by RYHAGE, STENHAGEN and collaborators

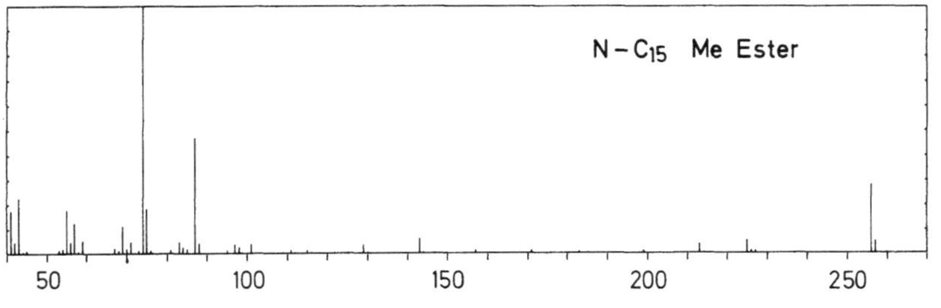

Fig. 28. Mass spectrum of methyl *n*-pentadecanoate

(for a review and leading references, see RYHAGE and STENHAGEN [176]) on the mass spectral fragmentation of normal, branched, dicarboxylic, unsaturated and variously oxygenated fatty acids.

Most commonly, acids are analyzed as their methyl esters, and the mass spectrum of methyl pentadecanoate (Fig. 28) illustrates the typical pattern of a normal methyl ester. Two features are characteristic: the intense even-mass rearrangement peak at m/e 74 and the series of odd-mass peaks of the general composition $C_nH_{2n-1}O_2$ at m/e 73, 87, 101, 115, 129, etc. As shown in structure XXII, the m/e 74 ion arises by a rearrangement process (the "McLafferty rearrangement") involving the γ-hydrogen atom, the latter series arises by simple cleavage of the alkyl chain. It should be noted (Fig. 28) that the fragments corresponding to cleavage after

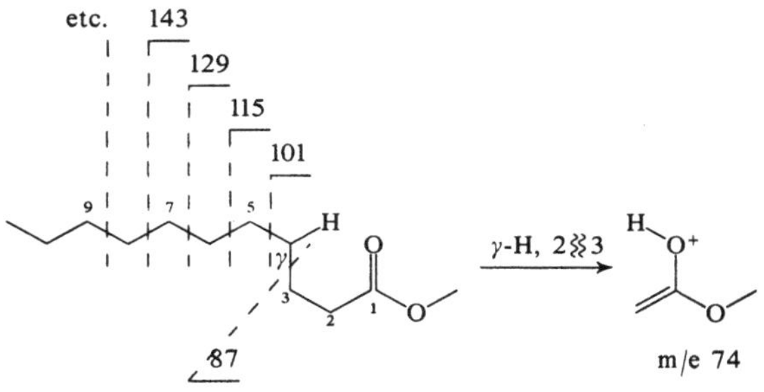

XXII

C-6 (m/e 143), C-10 (m/e 199) are formed in somewhat greater abundance than their immediate neighbors belonging to the same series. As in the case of hydrocarbons, a branch in the alkyl chain is immediately evident by the enhancement of the intensity of those peaks formed by fragmentation at the site of branching. The mass spectra of phytanic and norphytanic acid methyl esters (XXIII and XXIV) recently isolated from kerogen oxidation products [177] are illustrative.

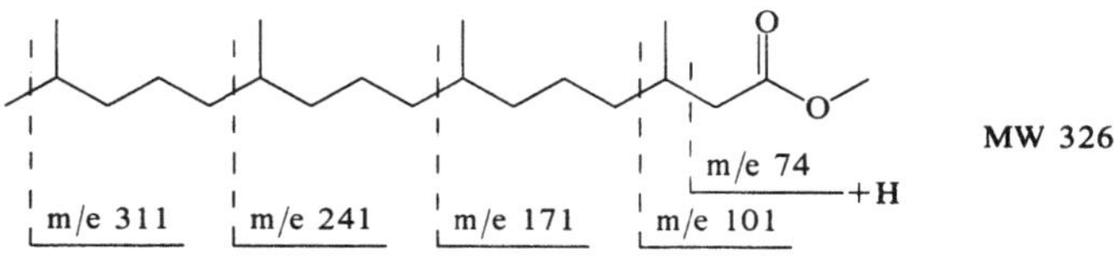

XXIII

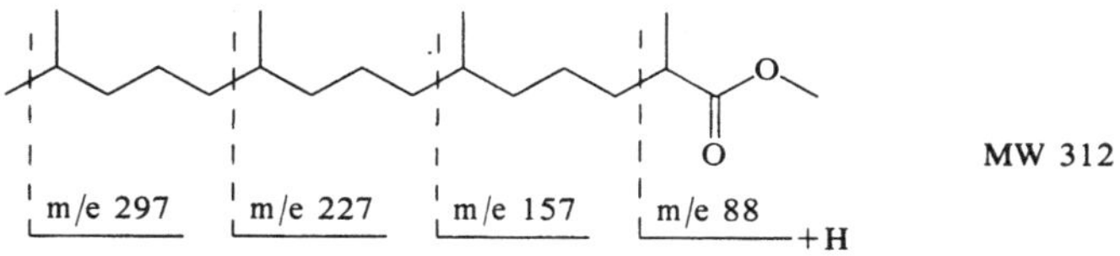

XXIV

The spectra (Figs. 29 a and b, respectively) also underscore two further important points: substitution of the alpha-carbon can be deduced from the mass of the rearrangement ion (i.e., m/e 74 shifts to m/e 88 in the spectrum of XXIV, Fig. 29 b) and fragmentation at the site of branching may in certain cases produce the most intense peak of the spectrum and suppress the usually dominant rearrangement ion of m/e 74 (i.e., the peak at m/e 101 in the spectrum of XXIII, Fig. 29 a).

The carbon number and, therefore, the structure of normal fatty acids can, of course, also be determined from the retention time on a gas chromatographic column by coinjection techniques. Definite assignments are, however, possible only for simple mixtures of acids.

In recent investigations on long-chain fatty acids, therefore, mass spectrometry has been utilized quite extensively. COOPER and BRAY [178] investigated the normal acids in marine sediments. The esters were analyzed by gas chromatography and their identity confirmed by mass spectrometry. COOPER [179] used

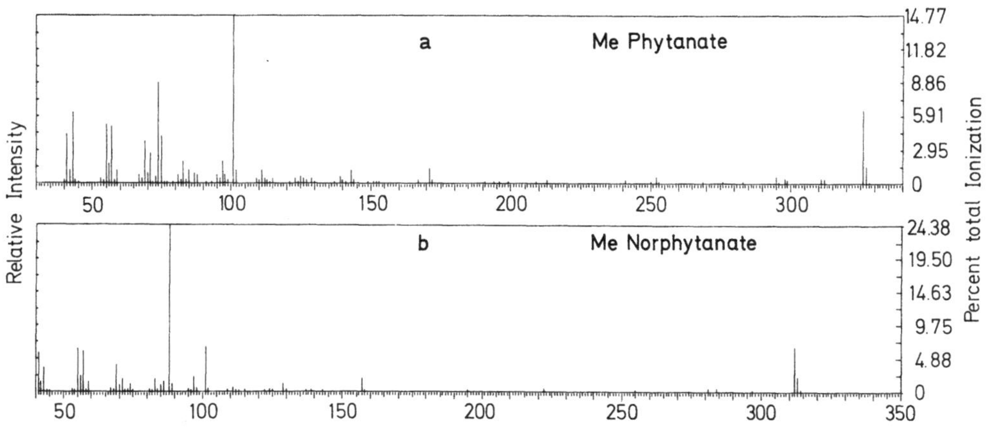

Fig. 29a and b. Mass spectra of phytanic acid methyl ester and norphytanic acid methyl ester

the same technique for the analysis of normal acids in recent and ancient sediments, and detected all members of the series in the carbon range 14–32. A number of investigations have been undertaken to establish the relationship between the hydrocarbon and carboxylic acid content in oils and sediments. KVENVOLDEN [180] determined the distribution of normal acids and hydrocarbons in Cretaceous shales, and compared it to that found for recent sediments. Carboxylic acids in a variety of other sediments [181–183] and those obtained from shale kerogen [177, 183] have been investigated similarly.

The studies on isoprenoid acids from petroleum and shales represent an interesting example of the potential of the mass spectrometric method. CASON and GRAHAM [184], using g.l.c. and mass spectrometry, isolated and identified the C_{14}, C_{15}, C_{19} and C_{20} isoprenoid acids from California petroleum. The identity of these substances was conclusively established by comparison with synthetic samples. EGLINTON *et al.* [96] have reported on a fairly extensive investigation of isoprenoid acids in shales. In the Green River Formation phytanic (C_{20}) and pristanic (norphytanic C_{19}) acids were isolated and identified by mass spectrometry (see Figs. 29a and b). The gas chromatograph-mass spectrometer combination system was used. Later work resulted in the identification of the C_{14}, C_{15}, C_{16}, C_{17} and C_{21} isoprenoid acids [97, 185]. These reports should be consulted for detailed mass spectral data on isoprenoid type acids. In a recent mass spectrometric investigation of the same shale, HAUG *et al.* [186] characterized two further isoprenoidal acids (C_9 and C_{10}), and CASON and KHODAIR [187, 188] have reported on the C_{11}-isoprenoid acid in petroleum. Oxidation experiments on kerogen material from the Green River Formation have yielded the C_{19} and C_{20} isoprenoid acids [177, 189], the mass spectra of which are illustrated in Figs. 29a and b. Mass spectrometry was used in the partial identification of mono-unsaturated acids obtained from a Scottish shale oil [97, 185]. Unsaturated esters are readily distinguishable from cyclic isomers by their characteristic peaks at M-32 and M-74, but as in the case of olefins, the position of the double bond cannot be specified from the fragmentation pattern itself. The technique described by MCCLOSKEY [190] involving formation of the isopropylidine derivative would be one method of characterizing unsaturated esters by mass spectrometry. Other methods which can be used to locate double bonds in an alkyl chain are described in the section discussing olefinic hydrocarbons. Other aliphatic acids which were identified by mass spectrometry include α,ω-dicarboxylic acids and keto acids. Members of the first class have been isolated from the Green River Formation extracts [97, 186, 191] and from Coorongite [185]. Their mass spectra have been studied and reviewed in some detail by RYHAGE and STENHAGEN [176]. They are characterized by molecular ions of low intensity and prominent M-31 and M-37 peaks, as well as ions at m/e 74 and (very characteristic for this class) at 98. Figs. 30

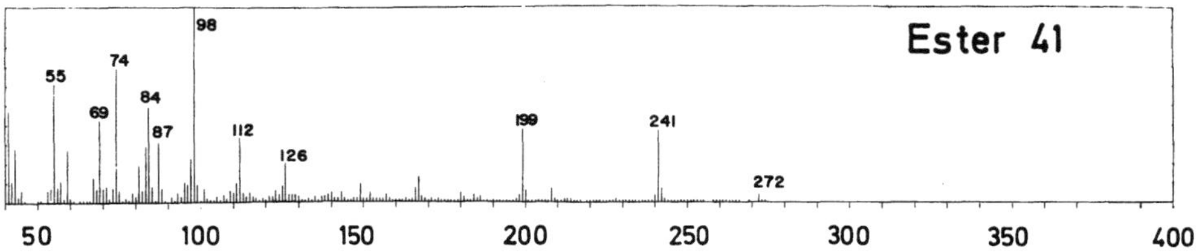

Fig. 30. Mass spectrum of dimethyl tridecane-1,13-dioate

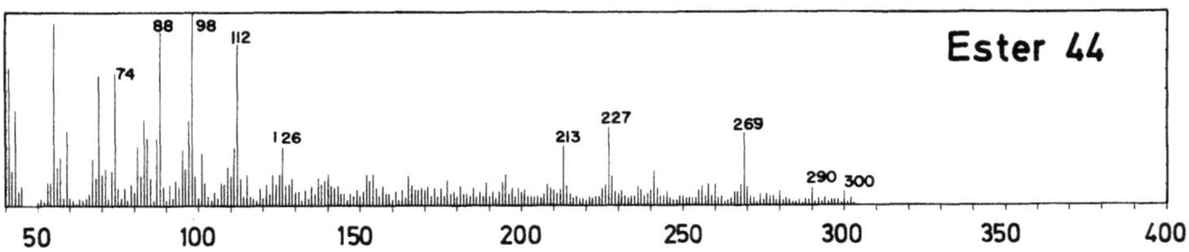

Fig. 31. Mass spectrum of dimethyl 2-methyltetradecane-1,14-dioate

and 31 illustrate this pattern, where Fig. 30 shows the spectrum of dimethyl tridecane-1,13-dioate, and the spectrum of Fig. 31 was identified as that of dimethyl 2-methylpentadecane-1,15-dioate (XXV, MW-300), both obtained from the Green River Formation oil shale acid fraction [186]. An interpretation of the spectrum is given below and illustrates well the information which can be derived from the mass spectra of these compounds.

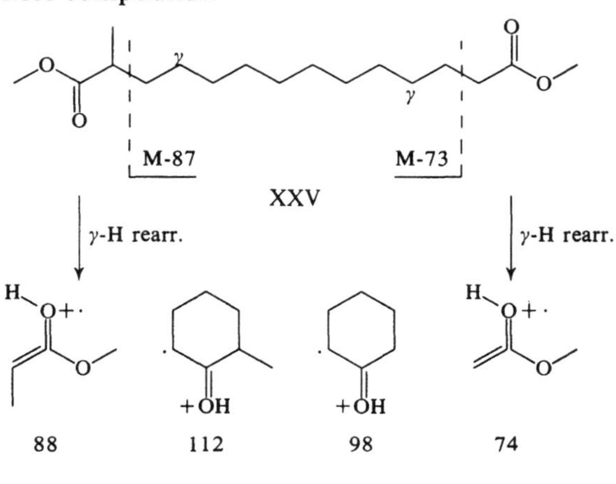

Interesting mass spectral results of possible future geochemical application have been obtained for some hydroxy fatty acids isolated from apple cutin [191]. For their mass spectrometric identification, these acids were converted to their trimethylsilylethers, a step which served two principal functions: it markedly facilitates their g.l.c. separation, and it permitted the position of hydroxyl functions to be determined. Trimethylsilyl ethers undergo the expected alpha-cleavage and the resulting fragments give rise to intense peaks.

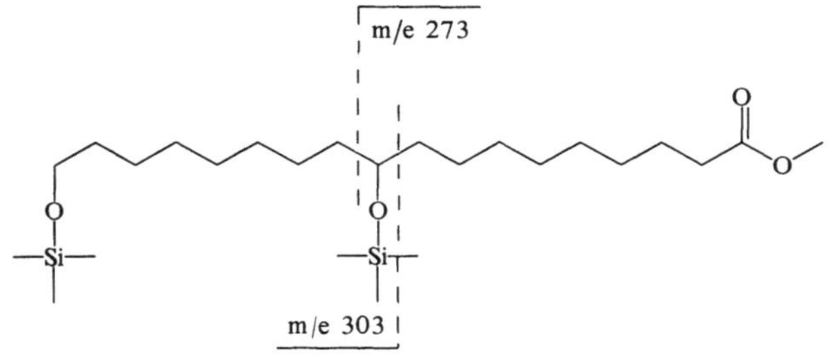

XXVI

The spectrum of TMSE (trimethylsilyl ether) of methyl 10,18-dihydroxy-octadecanoate, XXVI, Fig. 32, provides an illustration of a typical fragmentation pattern. The spectrum shows a relatively intense peak at m/e 459 (M-15), and the position of the 10-hydroxy function is evident from the peaks at m/e 273 and 303

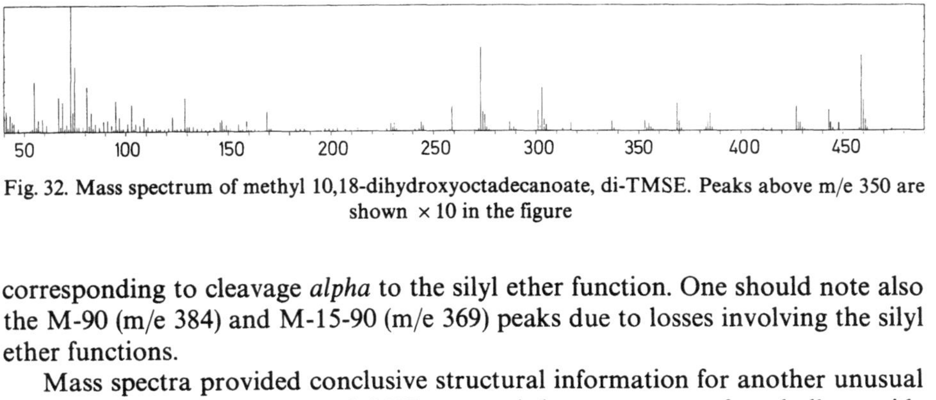

Fig. 32. Mass spectrum of methyl 10,18-dihydroxyoctadecanoate, di-TMSE. Peaks above m/e 350 are shown × 10 in the figure

corresponding to cleavage *alpha* to the silyl ether function. One should note also the M-90 (m/e 384) and M-15-90 (m/e 369) peaks due to losses involving the silyl ether functions.

Mass spectra provided conclusive structural information for another unusual class of fatty acids. Haug *et al.* [49] reported the occurrence of methylketoacids in Green River Formation oil shale extracts, and Burlingame and Simoneit [50] have reported a series isolated from its mineral matrix. Mass spectra of keto acids were available through the work of Ryhage and Stenhagen [176] and their identification proved to be a fairly straightforward matter. The spectrum of a C_{15} keto acid methyl ester is interpreted in section II D.

An extensive mass spectrometric study of aromatic esters isolated from geological sources has been undertaken by only one laboratory thus far [50, 192]. The mass spectra of only a small number of simple aromatic acids and esters have been studied [193–195] and the application to geochemical samples is thus not necessarily straightforward, although some general aspects of the fragmentation pattern of phenylalkanoic acids are quite predictable. An intense aromatic ion corresponding to β-cleavage of the alkyl chain and a rearrangement ion (if possible) are expected features.

The aromatic ion (at m/e 91, 105, 119, etc., for a hydrocarbon aromatic nucleus) immediately provides the mass of this moiety and, therefore, permits the deduction of possible substituents; the rearrangement ion yields information about the nature of *alpha*-substituents and also the nature of the chain (if such a rearrangement should be impossible, i.e., no hydrogen *gamma* to the ring). The spectra of selected aromatic acid methyl esters shown in Figs. 33–35 are instructive examples taken from our laboratory [192]. In Fig. 33 the mass spectrum of a methyl methyl benzoate, XXVII, is shown. The pattern excludes the *ortho*-isomer, but cannot

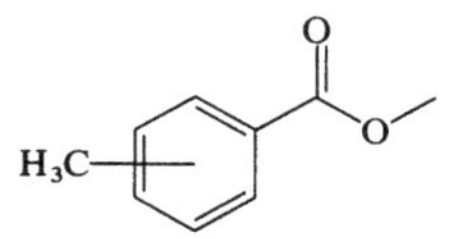

XXVII

distinguish between the *para*- and *meta*-isomers [193]. The intense rearrangement ion at m/e 88 in Fig. 34 indicates an alpha-methyl substituent (since the compound was analyzed as the methyl ester). The fact that the rearrangement occurs with

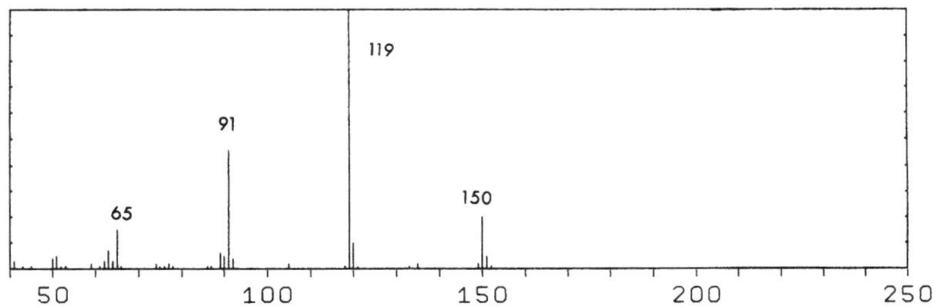

Fig. 33. Mass spectrum of methyl methyl benzoate

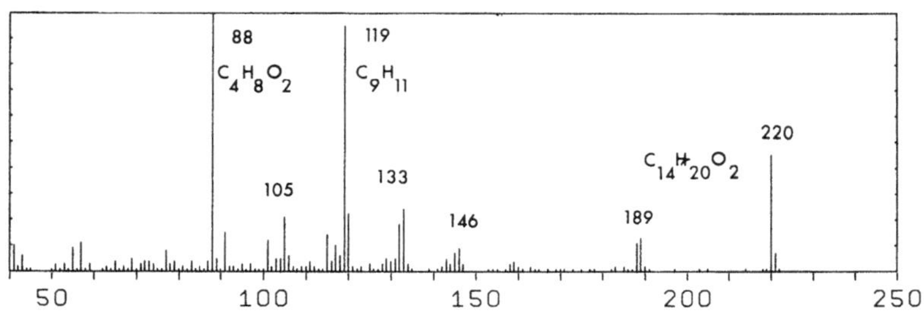

Fig. 34. Mass spectrum of methyl 2-methyl-4-(dimethylphenyl)-batanoate

the loss of 101 mass units to yield the peak at m/e 119 then demands a chain of four carbon atoms, whereby the carbon alpha to the benzene ring cannot be dissubstituted (since m/e 88 would then be impossible). The peak at m/e 119 requires an alkyl substituted (2 carbons) benzene ring, and from the absence of an appreciable M-15 peak, a dimethylbenzene moiety was indicated. StructureXXVIII combines these features, whereby it is, of course, quite evident that the position of these methyl substituents on the ring cannot be determined from the mass spectrum alone, since the spectra of all possible isomers were not available for comparison.

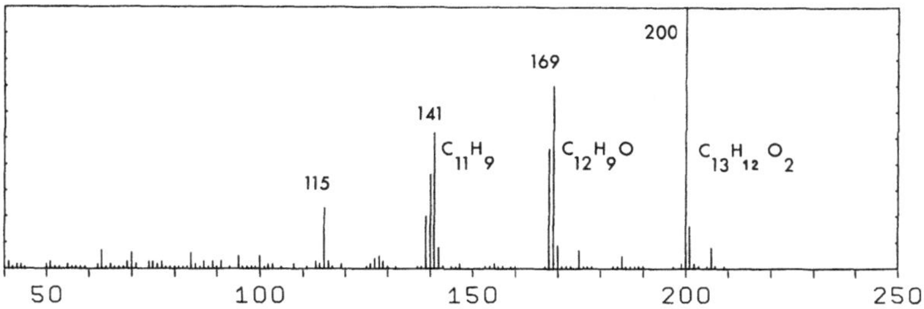

Fig. 35. Mass spectrum of methyl methylnaphthoate

The third spectrum shown, Fig. 35, is that of a naphthyl carboxylic acid methyl ester. The pattern demands the partial structure as shown in XXIX.

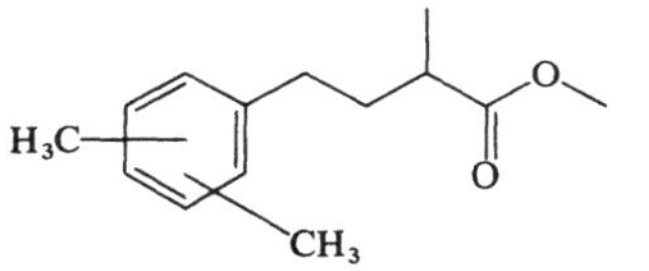

XXVIII

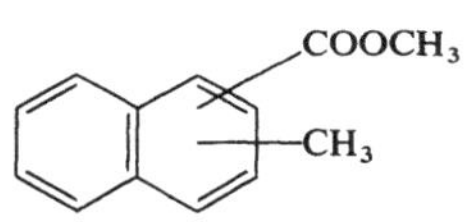

XXIX

In the work of CASON and collaborators [196, 197] on alicyclic acids from petroleum, mass spectrometry has been utilized extensively, although for this class of compounds the correct interpretation of the data is quite difficult. The pattern of a cyclic ester can be distinguished readily from its mono-unsaturated isomers, but such important structural features as ring size or nature and number of substituents on the ring are difficult to deduce from the mass spectrum alone. Thus, for a given unknown sample, unambiguous distinction between a cyclopentyl or cyclohexyl acid is often not possible on the basis of presently available mass spectral data of known acids. A further complication arises from the observation that stereochemical detail of substitution on the ring seems to have a pronounced effect on the intensity of certain peaks [188].

CASON and KHODAIR [196] have published a discussion of the mass spectra of various cyclopentylacetic acids and cyclopentenylacetic acids. The utilization of mass spectrometry in this area is illustrated by the identification of a C_{11} cyclic acid (XXX) in petroleum [197]. The mass spectrum of the ester and the corresponding amide indicated a cyclic C_{11} carbon skeleton with methyl substituents, but structure XXX could be established only after direct comparison with synthetic material.

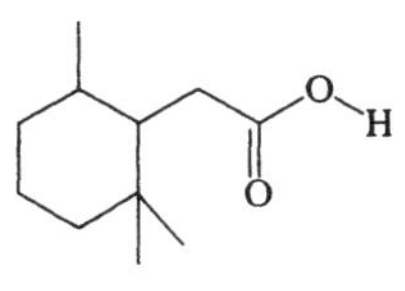

XXX

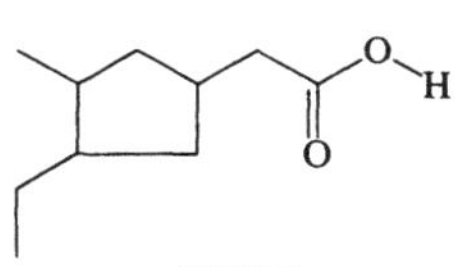

XXXI

Further work in CASON's laboratory [188, 198] has resulted in the characterization of XXXI, and an extensive mass spectral study of extract from the Green River Formation in our laboratory [56] has revealed a series of saturated cyclic acids ranging in molecular weight (of their methyl esters) from 156 (cyclopentylacetic) to 212 (C_{11}-monocyclic acid).

High resolution mass spectrometry holds great promise for future studies of acidic materials and, of course, heteroatomic compounds in general. Such applications of the technique have been carried on in the authors' laboratory for some time. Acidic material of Green River Formation Oil Shale extracts [56, 191] and kerogen material [189] has been investigated in some detail. Dicarboxylic acids, ketoacids, aromatic and cyclic acids were shown to be present, and particularly intriguing, the spectra also indicated the presence of pentacyclic acids (carbon range $C_{28}-C_{34}$) which could be considered degradation products of perhaps terpenoidal precursors. Similar studies on acids isolated from Alaskan Tasmanite have been undertaken [55]. A specific example from this work was discussed in section II D. It should be stressed that the routine exploitation of this technique is of major assistance in geochemical research since it rapidly provides excellent qualitative data on compound distribution and types.

C. Nitrogen Compounds

The literature on the isolation and characterization of nitrogen compounds is restricted to reports of analyses of petroleum or shale oil. Early studies depended on the separation of basic or non-basic nitrogen fractions, and the mixture analysis of such fractions by means of ultraviolet spectroscopy or chemical methods. The structures of individual compounds were established by synthesis of the substances presumed to be present and by direct comparison of physical data with unknowns. A considerable number of bases have been detected in petroleum by this approach. The reviews of LOCHTE [199], LOCHTE and LITTMANN [174], DEAL *et al.* [200], BALL *et al.* [175] and COSTANTITINIDES and ARICH [201] summarize the data and list individual compounds isolated. LOCHTE and LITTMANN [174] in 1955, for instance, list 25 pyridines and 32 quinolines isolated and identified from petroleum, shale oil and coal tar.

Mass spectrometry has been used in this effort primarily for type analysis of mixtures of nitrogen compounds. Such analyses are based on the fact that a particular nitrogen heterocyclic ring system of composition $C_nH_{2n+Z}N$ can be recognized by the value of Z. Alkylpyridines, for example, have a "Z-number" (i.e., hydrogen deficiency) of -5, indoles have $Z=-9$, quinolines have $Z=-11$. These Z-numbers are, of course, not always sufficient to establish a particular skeleton, since the given value is not necessarily unique for a structural type. Thus, dihydropyrindines, tetrahydroquinolines or cycloalkylpyridines would all have a Z-number of -7, and, again, bicycloalkylpyridines or cycloalkyldihydropyridines could not be distinguished from alkylindoles on the basis of their Z-numbers, which is -9 for both.

A paper by LALAU [202] discusses the application of such methods to the quantitative analysis of pyridine and quinoline distribution in petroleum distillates. Using the parent and M-1 peaks as primary data in a matrix calculation, the presence of mono and bicycloalkanopyridines and quinolines, as well as cycloalkanocarbazoles and sulfur-containing basic nitrogen compounds could be demonstrated. DINNEEN, COOK and JENSEN [203] performed a similar study on the gas oil fraction derived from shale oil. Alkylpyridines, cycloalkanopyridines, the corresponding quinolines and indoles were shown to be present, as well as more complex heterocyclic ring systems, although specific assignments for the latter could be made only tentatively. Infrared and ultraviolet spectra (the former for estimation of alkyl substituents, the latter for determination of carbon skeletons), as well as chemical data (e.g., basic *vs.* non-basic nitrogen) provided supporting evidence for type assignments. Pyrroles constituted only a very minor portion of the total nitrogen compounds identified. SNYDER and BUELL [204] combined adsorption chromatography, ultraviolet spectroscopy and low voltage mass spectrometry of fractions for an analysis of cracked gas oils. They could determine quantitatively the contribution of indoles, carbazoles and benzocarbazoles to the total nitrogen present in various distillate fractions of the oil. In an early paper by SAUER and collaborators [205], a combination of ultraviolet and mass spectroscopy permitted the identification of pyridines, pyrroles, quinolines, indoles and carbazoles in cracked and virgin Kuwait oils. Quantitative abundance data were obtained by matrix calculations and it could be shown that among the non-basic compounds, carbazoles were present in greatest proportion. More recently, JEWELL and HARTUNG [206] have presented a mass spectrometric study of nitrogen compounds in heavy gas oils. Low voltage mass spectra of fractions and ultraviolet absorption showed the presence of quinolines, and probably naphthenoquinolines. Furthermore, benzoquinolines, 1,10-phenanthroline and alkyl-1,2,3,4-tetrahydrocarbazolenines could be identified in mixtures, while the presence of indole, carbazole, α-hydroxypyrrole and α-hydroxybenzoquinolines was strongly suggested from the available ultraviolet and mass spectral evidence. The same authors [207], using gas chromatography for final fractionation of nitrogen compounds from a cracked oil, could demonstrate the presence of a homologous series of carbazoles from the appearance of peaks at m/e 167, 181, 195, 209. Similarly, phenazines were indicated by the mass spectral and ultraviolet data (e.g., m/e 180 = MW of phenazine; 194 = methylphenazine).

The application of mass spectrometry to the structural elucidation of individual bases has been handicapped somewhat by the lack of extensive mass spectral investigations of pure compounds. Aliphatic amino compounds have been studied in some detail and reviewed recently [3], but only a few papers deal with the mass spectra of aromatic nitrogen compounds which are the predominant types isolated from petroleum or shales. BIEMANN [208] discussed the fragmentation of simple alkylpyridines and pyrazines and noted that, whereas positional isomers exhibit similar fragmentation, the intensity of fragment peaks often differs considerably with substitution pattern. For example, the mass spectra of isomeric ethylpyridines [208] differ quite noticeably in the intensity of the M-1 and M-15 fragment peaks. For 2-ethylpyridine, the M-1 ion is the dominant species of the spectrum, whereas the M-15 ion is a very minor contributor; for 3-ethylpyridine, however, M-15 is the base peak and M-1 is much reduced in intensity; 4-ethylpyridine represents a spectrum intermediate between these two extremes. A further important mode of fragmentation for substituted pyridines possessing an alkyl chain of at least three carbon atoms is a McLafferty type rearrangement leading to an odd-mass ion. Quinolines and isoquinolines exhibit fragmentation patterns which can be rationalized by similar mechanisms: *beta*-cleavage and the McLafferty rearrangement are dominant pathways whenever possible, and γ-cleavages are observed for alkyl substituents at position C-4 and C-8 of the quinoline ring. The papers by CLUGSTON and MACLEAN [209] and SAMPLE *et al.* [210] should be consulted for detailed interpretation of the fragmentation of oxygenated quinolines [211] and of alkylquinolines and isoquinolines.

Mass spectra of C-alkyl- and N-alkylpyrroles and pyrrole carboxylic esters have been reported and analyzed by DJERASSI and his co-workers [3]. Spectra of indoles are available through the work of BEYNON [1]. Excellent summaries of mass spectral data on N-heterocyclic compounds have recently been published [3, 212] and should be consulted for the mechanistic aspects of the interpretation of these mass spectra and leading references to the literature.

The data available and the nature of these compounds ordinarily do not permit the unambiguous structural definition of aromatic bases by mass spectrometry. Successful structural elucidations of nitrogen compounds by physical methods have resulted, therefore, from a combination of techniques, particularly n.m.r., u.v. and mass spectrometry, whereby the last two provided the molecular weight (indicative of structural type) and fragmentation pattern (indicative of size of substituents and, in some cases, position). VAN METER *et al.* [213] isolated a number of pyridines and pyrroles from shale oil naphtha. A summary of results by the Bureau of Mines, Laramie, is available [155]. Mass spectra of the pure compounds gave the molecular weight and some indication of structural features, and final identification was made by comparison of other physical data (m.p., i.r.) with known samples. HELM *et al.* [214] reported the first definite identification of carbazole in virgin Wilmington petroleum; after purification of the compound by g.l.c., the structural assignment was made on the basis of its mass spectrum (MW = 167) and comparison of other spectral data. Alkyl substituted homologs of carbazole (C-1 – C-10) were found to be present also.

Individual nitriles have been identified by JEWELL and HARTUNG [215] in hydrogenated furnace oils. On the basis of ultraviolet and mass spectra, dicyanobenzene, cyano-2,3-dihydroindene, cyanonaphthalene, dicyano-2,3-dihydroindene and dicyanonaphthalene have been identified. The presence of other aromatic nitriles was suggested by the mass spectra of various fractions. The assignment of methylcyanonaphthalene to the substance obtained as the benzene eluate from an alumina column may be incorrect, however, since the spectrum published (peaks at m/e 167, 149, 113, 112, 104, etc.) shows striking resemblance to that obtained from "octoil", a common contaminant. Mention should be made here also of the identification of aliphatic nitriles (C_{12} to C_{15}) in Colorado Shale Oil [154], although mass spectrometry was not utilized in this study. The combination of fluorescence spectrometry, infrared and mass spectrometry was used for the identification of alkyl-pyridines, -pyrindines and -indoles and pyrroloquinolines [216] in petroleum. The now available modern methods of fractionation and analysis are well illustrated by the results of BRANDENBURG and LATHAM [217]. Individual bases belonging to the cyclohexyl pyridine, pyrindine and quinoline structural types were isolated by successive chromatography of a

Wilmington petroleum extract on several different g.l.c. columns, and structural assignments were made on the basis of n.m.r., u.v., i.r. and mass spectral data. Thus, 2-(2′,2′,6′-trimethylcyclohexyl)-4,6-dimethyl pyridine (XXXII) and 2,4,7a, 8,8-pentamethyl-*trans*-cyclopenta-(f)-pyrindane (XXXIII) have been reported, although, unfortunately, no detailed account of the data is available as yet.

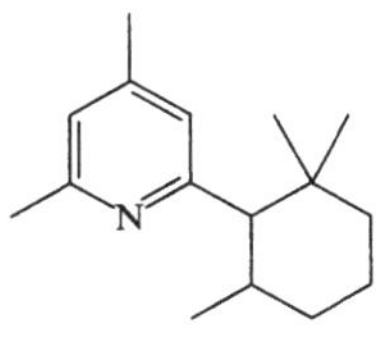

XXXII

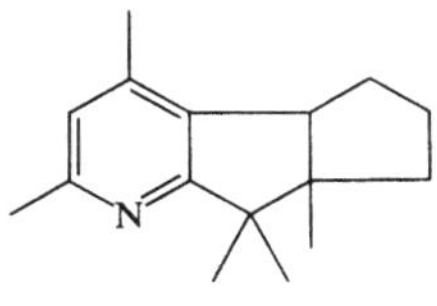

XXXIII

The former base (XXXII) had been isolated previously from petroleum [174, 199]. Both are rather interesting structurally, since their carbon skeletons appear to incorporate an isoprenoid arrangement. No studies have been reported as yet of bases isolated from shales themselves, although a preliminary mass spectrometric investigation utilizing both high and low resolution mass spectrometry has been undertaken on the bases from the Green River Formation [56]. An application of high resolution mass spectrometry to the analysis of a base mixture is illustrated in Fig. 6 and discussed in section II D. Similar studies on total base fractions of extracts have been performed on other sediments such as the Pierre Shale, Wyoming [56].

D. Porphyrins and Pigments

Not surprisingly, porphyrins, although of great geochemical significance, have not been subjected to intensive mass spectrometric investigations until very recently [218]. The non-volatility of these compounds precluded their study until direct introduction systems became available. Even then, the vaporization of porphyrins into the ion source requires very high temperatures (250–350° C). The literature contains a number of papers on the mass spectra of porphyrins. The spectra of nickel etioporphyrin [219], vanadyl etioporphyrin [220] and mesoporphyrin and ferric mesoporphyrin chloride dimethyl ether [221] have been reported and more recently quite extensive studies on porphyrins and chlorins have become available [222–227].

From these studies it appears that future geochemical work on these substances would benefit greatly from the utilization of this technique. The cyclic tetrapyrrole ring system is quite stable under electron impact and in simple derivatives the molecular ion is usually the most abundant species. In general, fragmentation proceeds with loss of substituents on the ring system, thus providing an excellent means of identifying the size and nature (i.e., ethyl, ester, etc.) of the substituent grouping.

In geochemical applications the complex mixtures, which all isolated porphyrin samples represent, pose a problem for the interpretation of the resulting spectra. However, even in cases where the structure and substitution of a single component may be impossible to determine, the method at least provides information on the number of components in a mixture and their molecular weights (also elemental compositions by accurate mass measurement), which, if a particular ring system can be recognized from the ultraviolet spectrum, already enables one to draw certain deductions about its substituents.

The literature on the application of mass spectrometry to porphyrin geochemistry is not extensive. The first use of the technique seems to have been made by THOMAS and BLUMER [226]. They isolated a pigment mixture from the Serpiano Oil Shale (Triassic), which was separated into groups by extraction techniques and further fractionated by thin layer chromatography. The resulting fraction was analyzed by ultraviolet and mass spectrometry. The mass spectra of one group (the 2N HCl extract) indicated the presence of homologous series belonging to the spectral "etio" and "deoxyphylloerythrin" types. Another fraction containing chlorins was similarly analyzed, but they were found to degrade partially to the porphyrin ring systems. Vanadium porphyrin complexes were found by this method to consist also of the deoxyphylloerythrin and etio types with molecular weights ranging from 438 to 504 for the former and 436 to 492 for the latter. A fraction containing unknown pigments absorbing at 630 mμ was analyzed and found to consist of two series of homologous compounds (molecular weight ranging from 430–500 and 456–498, respectively). The mass peaks observed suggested series of trivinyletioporphyrins, but from chromatographic and ultraviolet spectral evidence, the authors suggest that the vinyl grouping might have been generated in the mass spectrometer by loss of the elements of water. MORANDI and JENSEN [227] have analyzed the porphyrin constituents of Green River Formation Oil Shale, the shale oil produced from it and of Wilmington crude oil. The porphyrins from the oil shale were characterized from their ultraviolet spectra as being of the phyllo-type. Mass spectrometry (using an all-glass introduction system) indicated two series of compounds, one corresponding to homologs of etioporphyrins, the other corresponding to a series showing one further degree of unsaturation. When introduced through a direct inlet, the mass spectrum was quite different due to the (assumed) formation of indium complexes. The spectra were interpreted to indicate carboxy and alkyl substituents for one porphyrin-type in the mixture and carboxy, alkyl and exocyclic ring substituents for the other.

BLUMER and SNYDER [229] have reported further studies using Sephadex LH-20 gel permeation chromatography on fractions extracted from the Triassic oil shale of Serpiano (Switzerland). The technique has indicated the existence of homologous series of pigments extending into the high molecular weight region (above 20,000 mol wt.). It also proved effective with separation of porphyrin fractions of up to molecular weight 700. Low resolution mass spectra of these fractions showed the presence of porphyrin nuclei containing from 7 to 23 methylene groups attached to the periphery. However, complete resolution of fossil porphyrins into individual species has not been effected.

BAKER [228] established the nonhomogeneity of petroporphyrins extracted from petroleum sources by low voltage mass spectrometry. More recently, BAKER *et al.* [80] have utilized accurate mass measurement of selected major peaks in the spectra of asphaltenes of Agha Jari, Baxterville, Belridge, Boscan, Burgan,

Mara, Melones, Rozel Point, Santiago and Wilmington petroleums, in addition to a gilsonite, Athabasca tar sand and the Colorado Green River Formation oil shale. These results confirm the presence of two major and one minor homologous series of petroporphyrins, the major being a mono-cycloalkano series (visible spectra similar to deoxophylloerythroetioporphyrin) and alkyletioporphyrins (visible spectra indicative of either incomplete β-substitution or of bridge substitution). On the basis of mass and electronic spectral evidence an alkylbenzoporphyrin structure (XXXIV) is suggested for the minor series. Partial low voltage mass spectra depict the molecular weight distributions and selected accurate mass measurements determine the elemental compositions (see section II D). The reader is referred to the original paper for careful study of this survey of petroporphyrin data.

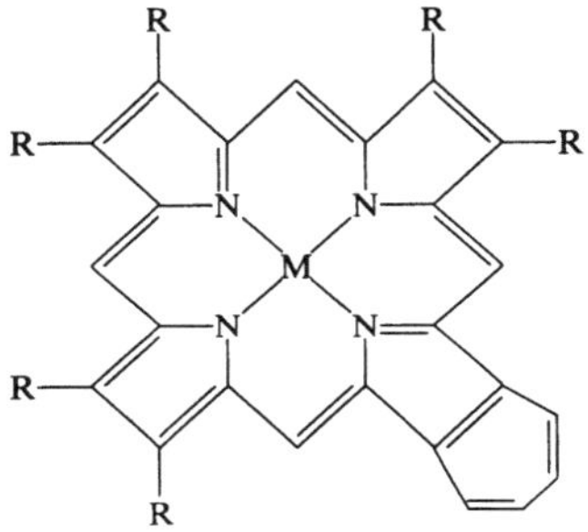

XXXIV

At the outset of this section, the lack of volatility of this class of compounds was noted to be a difficulty in studying their mass spectra. Recently, BOYLAN and CALVIN [230] have reported the preparation of volatile silicon complexes of etioporphyrin I, XXXV.

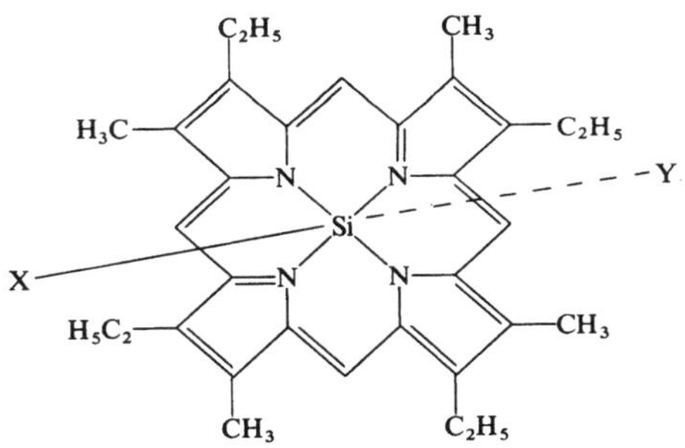

XXXV

The gas chromatographic retention times and predominant mass spectral ions of the X,Y-Si^{IV}-etioporphyrins are reported. Such an approach to the volatility problem should stimulate further research into the nature and distribution of geoporphyrins making use of gas chromatography and mass spectrometry for investigation of microquantities.

E. Sulfur Compounds

Sulfur-containing compounds are important minor constituents in petroleum fractions [201]. Thiols and sulfides occurring in petroleum have been summarized by BALL *et al.* [175], and the sulfur-containing constituents (mostly thiophenes) of shale oil are reviewed by DINNEEN *et al.* [155]. Several mass spectrometric studies on sulfur compounds of geochemical interest can be found in the current literature. Mass spectra of aliphatic [231] and aromatic thiols [232, 233] and aliphatic thioethers [231, 234, 235] have been investigated in some detail, as have various aromatic thioether systems [236, 237]. The mass spectra of thiophenes, another important class of geochemical compounds, have been discussed in a number of reports [238–244]. For a summary of the mass spectrometry of various sulfur compounds — thiols, sulfides, sulfones, thiophenes — and leading references, the volume by BUDZIKIEWICZ *et al.* [245] is recommended. An extensive report on the mass spectra of a great number of sulfur compounds is also available [246].

The number of individual sulfur compounds known in petroleum is considerable; for example, research under Project 48 of the American Petroleum Institute has resulted in the identification of about 195 individual sulfur compounds from four crude oils representing 13 different structural classes [247]. A review of the sulfur constituents of petroleum has appeared [201]. Mass spectrometry has found use especially for the type analysis of mixtures of sulfur compounds. Thus, THOMPSON *et al.* [247, 248] estimated the abundances of thiophenes in crude oil. HASTINGS, JOHNSON and LUMPKIN [249] developed a matrix scheme for group-type analysis of aromatics and sulfur compounds. THOMPSON and coworkers [248] used low voltage mass spectrometry of sulfur compound concentrate fractions and reported the presence of benzothiophenes, dibenzothiophenes, naphthanothiophenes, naphthenobenzothiophenes and homologues of these systems in high boiling petroleum distillates.

Application of mass spectrometry to other studies of sulfur compounds can be found in the papers of NAGY and GAGNON [135], BIRCH *et al.* [250], JEWELL and HARTUNG [206], AMBERG [251], and DRUSHEL and SOMMERS [79]. DINNEEN *et al.* made a detailed mass spectrometric investigation of the sulfur compounds of shale oil [155]. The utilization of mass spectrometers of relatively low resolving power (i.e., $M/\Delta M$: 1,000–2,000) can lead to ambiguities in the identification of sulfur compound types, however, since certain aromatic molecules give rise to similar peaks at the same nominal masses. For example, compounds of the general formulae C_nH_{2n+2}, $C_{n+1}H_{2n-10}$ and $C_{n-2}H_{2n-6}S$ should give the same parent peaks. The use of matrix calculations overcomes this problem to some extent but high resolution mass spectrometry constitutes by far the most elegant method for the recognition of individual compound types. This technique has been used repeatedly during the last few years for the analysis of aromatics and also sulfur compounds. The work of CLERC and O'NEAL [252] and CARLSON *et al.* [143] is illustrative of early attempts where resolution in the order of 3,000 was utilized. More recent efforts in that direction describing group type analysis of aromatic hydrocarbons and aromatic sulfur compounds are those of REID [173] and GALLEGOS *et al.* [42].

Low ionizing voltages at high resolutions can be used to advantage if appropriate calibration data are available [77].

A particularly impressive example of the application of high resolution mass spectrometry to the group-type analysis of petrochemical fractions has recently been described by GALLEGOS *et al.* [42] and is reproduced in Fig. 36. In Fig. 36 is shown a segment from a high resolution mass spectrum of the 800°–950° F fraction of an Arabian crude oil. Among the compound types identified in Fig. 36

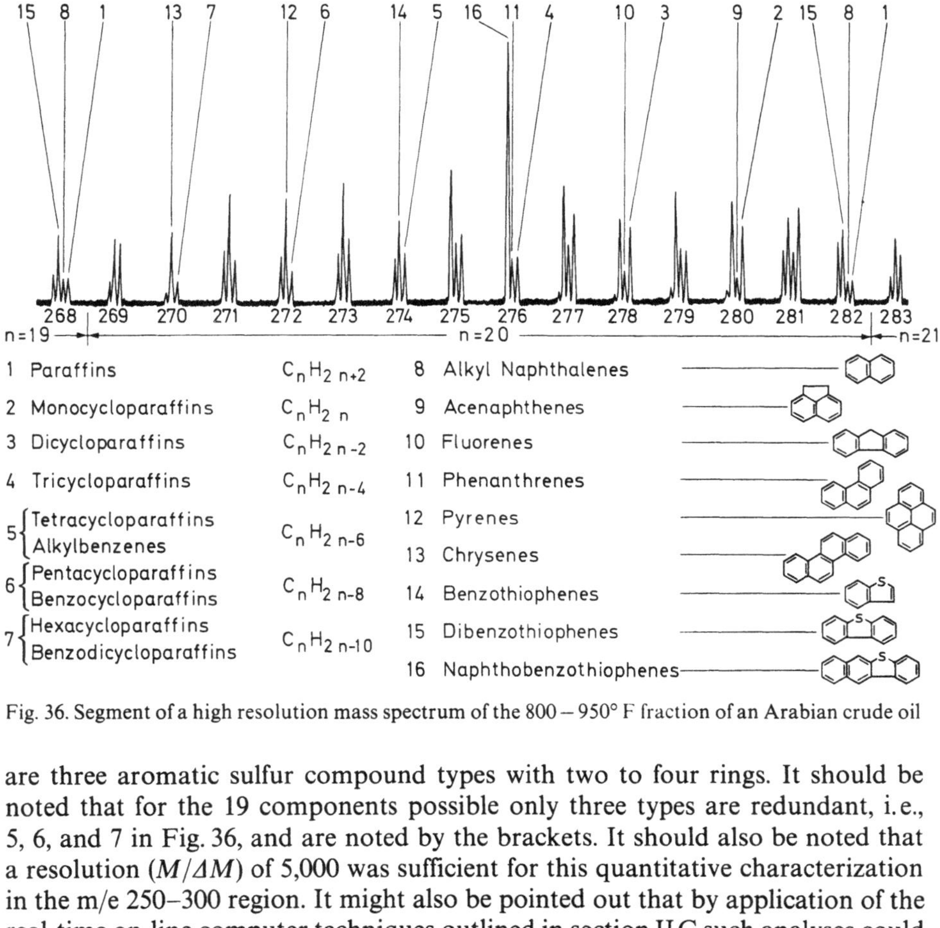

Fig. 36. Segment of a high resolution mass spectrum of the 800 – 950° F fraction of an Arabian crude oil

are three aromatic sulfur compound types with two to four rings. It should be noted that for the 19 components possible only three types are redundant, i.e., 5, 6, and 7 in Fig. 36, and are noted by the brackets. It should also be noted that a resolution ($M/\Delta M$) of 5,000 was sufficient for this quantitative characterization in the m/e 250–300 region. It might also be pointed out that by application of the real-time on-line computer techniques outlined in section II C such analyses could be completely automated with an analysis cycle time of seconds.

For other examples and discussion of the application of high resolution techniques to the analysis of sulfur compounds, see section II E.

F. Other Classes of Compounds

Among the non-hydrocarbon compound types not discussed thus far, the phenols represent a group of apparently fairly general occurrence in petroleum fractions [201]. Mass spectrometry, however, does not seem to have been utilized in structural studies of this class. Simple ketones have been reported in petroleum [174, 201] and two reports describe the utilization of mass spectrometric data for partial determination of some ketone structures. Thus, alkyl-substituted fluorenones (C_1-C_4) could be identified in Wilmington petroleum [253] by ultraviolet spectroscopy and a mixture mass spectrum which showed peaks at m/e 194, 208, 222 and 236. An alicyclic ketone could be identified as an acetyl-isopropyl-methylcyclopentane derivative from n.m.r. and mass spectral data [254].

Other oxygen-containing compounds, such as benzofurans and dibenzofurans, have been detected in petroleum fractions, in particular through studies utilizing high resolution mass spectrometry (see, for example, references [75, 76, 207, 213]). It can be expected that in the future mass spectrometry will be applied extensively to structural elucidation of oxygen heterocycles, ketones, alcohols and perhaps amino acid and sugar derivatives isolated from geologic sources.

IV. Application to the Analyses of Extraterrestrial Materials

In the preceding sections the results of mass spectrometric investigations into various classes of geochemically important compounds have been presented. The discussion is not exhaustive, but it is sufficiently complete to yield a good portrait of the spectrum of research activities. Some other research areas in which mass spectrometry plays a central role should be mentioned. The broad field of isotope ratio determinations, in particular $^{12}C/^{13}C$ and $^{16}O/^{18}O$ ratio measurements of both the inorganic and organic matter of sediments, is one of them. A separate chapter (12) is devoted to a full discussion of the utilization of mass spectrometry in these studies.

Another interesting and exciting application of the technique relates to the study of organic matter from extraterrestrial sources. The analysis of meteorite organics by various groups has been referred to in connection with other studies above. The techniques employed in meteorite studies fall roughly into three categories: the pyrolysis of meteoritic material and subsequent analysis of the total volatile organic matter thus liberated by conventional or high resolution mass spectrometry; the trapping of the pyrolysis products and subsequent separation and analysis via the g.l.c.-mass spectrometric coupling arrangement; the extraction of meteorites and analysis of either the total extract mixture or the mixture fractionated via g.l.c. into the mass spectrometer. NAGY, MEINSCHEIN and HENNESSEY [255, 256] first reported a mass spectrometric analysis of a total meteorite extract and found a distribution of hydrocarbons reminiscent of biogenic sources. ORÓ and collaborators [257, 258] have analyzed a great number of meteorites, primarily by g.l.c. methods and by the g.l.c.-mass spectrometer coupling technique; their results show a hydrocarbon distribution quite similar to that found for ancient sediments. The work by HAYES and BIEMANN [259] on the organic matter of the Murray and Holbrook meteorites has already been referred to. These authors utilized high resolution mass spectrometry both for the analyses of total pyrolysis fractions and of g.l.c. separated pyrolysis mixtures. Their work is of particular importance first because it gives an indication of the distribution and abundance of heteroatomic species in meteorites and, secondly, because it illustrates the utilization of computer pre-processing and presentation for the analysis of geochemical samples. The employment of a Time-of-Flight mass spectrometer, both with and without additional g.l.c. separation, for the study of organic pyrolysis products obtained from several chondrites (Orgueil, Murray and Felix) has recently been described by ANDERS and collaborators [93]. The detection and identification of aromatic compounds in meteorites, paralleling the results of HAYES and BIEMANN [259] to some extent, is a major finding of this work. Since HAYES [101, 260] presents a critical review of all organic chemical work which has been performed on meteorites, a detailed account of all findings in this field is unnecessary here.

In the near future, it is expected that lunar samples will be available for the analysis of possible organic matter. The extremely small quantity of organic material which these samples might contain makes mass spectrometry one of the critical tools for such studies. Several laboratories are presently engaged in the perfection of mass spectrometric techniques and analytical schemes in preparation for the eventual analysis of lunar samples. High resolution mass spectrometry, in particular, will play a crucial role in these studies since it provides a maximum of structural information on such extremely small samples.

Aside from applications to strictly geochemical work, mass spectrometry has found extensive use in related analytical investigations, the results of which have some bearing on geochemical speculations. The studies of products formed upon irradiation of methane, or the reaction of carbon monoxide and hydrogen, Fischer-Tropsch reaction products [261, 262], or the investigation of the constituents of living organisms of geochemical significance, e.g., zooplankton [156, 157], algae [131], bacteria [218], etc., are areas where the utilization of mass spectrometric techniques has been found of advantage.

It is felt that the present chapter serves its purpose if, besides providing a brief introduction to mass spectrometry and perhaps incomplete summary of currently available data and results, it gives an insight into the type of information most readily available by mass spectrometry and the techniques most fruitfully exploited to obtain such information.

References

1. BEYNON, J. H.: Mass spectrometry and its application to organic chemistry. Amsterdam: Elsevier 1960.
2. BIEMANN, K.: Mass spectrometry: Organic chemical applications. New York: McGraw-Hill 1962.
3. BUDZIKIEWICZ, H., C. DJERASSI, and D. H. WILLIAMS: Mass spectrometry of organic compounds. San Francisco: Holden-Day 1967.
4. — — — Structural elucidation of natural products by mass spectrometry, vol. 1, Alkaloids. San Francisco: Holden-Day 1964.
5. — — — Structural elucidation of natural products by mass spectrometry, vol. 2, Steroids, terpenoids and sugars. San Francisco: Holden-Day 1964.
6. SPITELLER, G.: Massenspektrometrische Strukturanalyse organischer Verbindungen. Weinheim: Verlag Chemie 1966.
7. MCLAFFERTY, F. W., ed.: Mass spectrometry of organic ions. New York: Academic Press 1963.
8. BURLINGAME, A. L., ed.: Topics in organic mass spectrometry. New York: Interscience (in press).

9. See also the series, Advances in mass spectrometry. (a) WALDRON, J. D., ed., vol. 1. London: Pergamon Press 1959: (b) ELLIOTT, R. M., ed., vol. 2. New York: The Macmillan Company 1963. (c) MEAD, W. L., ed., vol. 3. London: The Institute of Petroleum 1966. (d) KENDRICK, E., vol. 4. New York: The Institute of Petroleum 1968.
10. MCLAFFERTY, F. W., and J. PINZELIK: Index and bibliography of mass spectrometry, 1963 – 1965. New York: John Wiley & Sons 1967.
11. HOOD, A.: The molecular structure of petroleum. In: Mass spectrometry of organic ions (F. W. MCLAFFERTY, ed.). New York: Academic Press 1963.
12. EWALD, H., u. H. HINTENBERGER: Methoden und Anwendungen der Massenspektroskopie. Weinheim: Verlag Chemie 1953.
13. MCDOWELL, C. A., ed.: Mass spectrometry. New York: McGraw Hill 1963.
14. KISER, R. W.: Introduction to mass spectrometry and its applications. Englewood Cliffs, N. J.: Prentice-Hall 1965.
15. BRUNNÉE, C., u. H. VOSHAGE: Massenspektrometrie. München: Karl Thiemig KG 1964.
16. LALAU, C.: Mass discrimination caused by electron multiplier detectors. In: Topics in organic mass spectrometry (A. L. BURLINGAME, ed.), ch. 2. New York: Interscience (in press).
17. CASSUTO, A.: Temperature factors in mass spectrometry. In: Mass spectrometry (R. I. REED, ed.), p. 283. New York: Academic Press 1965.
18. BEYNON, J. H.: Mass spectrometry and its applications to organic chemistry, p. 428 – 431. Amsterdam: Elsevier Publishing Co. 1960.
19. BURLINGAME, A. L.: Current developments and applications of on-line digital computer techniques in reduction and presentation of complete high resolution mass spectral data. EUCHEM Conference on Mass Spectrometry, Sarlât, France, Sept. 7 – 12, 1965.
20. HITES, R., and K. BIEMANN: A computer-compatible digital data acquisition system for fast scanning, single-focusing mass spectrometers. Anal. Chem. **39**, 965 (1967).
21. (a) BECKEY, H. D.: Field ion mass spectra of organic molecules. In: Mass spectrometry (R. I. REED, ed.), p. 93. London: Academic Press 1965. (b) BECKEY, H. D., H. KNÖPPEL, G. METZINGER, and P. SCHULZE: Advances in experimental techniques, applications, and theory of field ion mass spectrometry. In: Advances in mass spectrometry, vol. 3, (W. L. MEAD, ed.), p. 35. London: The Institute of Petroleum 1966. (c) ROBERTSON, A. J. B., and B. W. VINEY: Production of field ionization mass spectra with a sharp edge. In: Advances in mass spectrometry, vol. 3 (W. L. MEAD, ed.), p. 23. London: The Institute of Petroleum 1966.
22. (a) MUNSON, M. S. B., and F. H. FIELD: Chemical ionization mass spectrometry I: General introduction. J. Am. Chem. Soc. **88**, 2621 (1966) and subsequent papers. (b) FIELD, F. H.: Chemical ionization mass spectrometry. In: Advances in mass spectrometry, vol. 4 (E. KENDRICK, ed.), p. 645. London: Institute of Petroleum, 1968. (c) GELPI, E., and J. ORÓ: Chemical ionization mass spectrometry of pristane. Anal. Chem. **39**, 388 (1967).
23. BECKEY, H., and F. J. COMES: Techniques of molecular ionization. In: Topic in organic mass spectrometry, chap. 1. (A. L. BURLINGAME, ed.). New York: Interscience (in press).
24. FIELD, F. H., and J. L. FRANKLIN: Electron impact phenomena and the properties of gaseous ions. New York: Academic Press 1957.
25. VEDENEYEV, V. I., L. V. GURVICH, V. N. KONDRAT'YEV, V. A. MEDVEDEV, and YE. L. FRANKEVICH: Bond energies, ionization potentials and electron affinities. New York: St. Martin's Press 1966.
26. BURLINGAME. A. L.: Application of high resolution mass spectrometry in molecular structure studies. In: Advances in mass spectrometry, vol. 3 (W. L. MEAD, ed.), p. 701. Amsterdam: Elsevier 1966.
27. SCHUMACHER, E., u. R. TAUBENEST: Massenspektrometrische Untersuchungen an einigen Hexahydrotriazinen. Helv. Chim. Acta **49**, 1439 (1966).
28. AMY, J. W., E. M. CHAIT, W. E. BAITENGEN, and F. W. MCLAFFERTY: A general technique for collecting gas chromatographic fractions for introduction into the mass spectrometer. Anal. Chem. **37**, 1265 (1965).
29. MATTAUCH, J.: Precision measurement of atomic masses and some of their implications on nuclear structure and synthesis. In: Advances in mass spectrometry, vol. 3 (W. L. MEAD, ed.), p. 1 – 19. London: The Institute of Petroleum 1966.
30. DUCKWORTH, H. G., and S. N. GOSHEL: High resolution mass spectroscopy, In: Mass spectrometry (C. A. MCDOWELL, ed.), p. 201. New York: McGraw-Hill 1963.

31. MATTAUCH, J., u. R. HERZOG: Über einen neuen Massenspektrographen. Z. Physik **89**, 786 (1934).
32. JOHNSON, E. G., and A. O. NIER: Angular aberrations in sector shaped electromagnetic lenses for focussing beams of charged particles. Phys. Rev. **91**, 10 (1953).
33. QUISENBERRY, K. S., T. T. SCOLMAN, and A. O. NIER: Atomic masses of H^1, D^2, C^{12} and S^{32}. Phys. Rev. **102**, 1071 (1956).
34. MCMURRAY, W. J., B. N. GREENE, and S. R. LIPSKY: Fast scan high resolution mass spectrometry. Anal. Chem. **38**, 1194 (1966).
35. OLSEN, R. W., and A. L. BURLINGAME: Entire high resolution mass spectra by direct digitization of multiplier output on a CEC 21 – 110 mass spectrometer. Thirteenth Annual Conference on Mass Spectrometry and Allied Topics, St. Louis, Mo., May 16 – 21, 1965, p. 192.
36. BURLINGAME, A. L., D. H. SMITH, and R. W. OLSEN: Real-time data acquisition, subsequent processing and display in high resolution mass spectrometry. Anal. Chem. **40**, 13 – 19 (1968).
37. BIEMANN, K., P. BOMMER, D. M. DESIDERIO, and W. J. MCMURRAY: New techniques for the interpretation of high resolution mass spectra of organic molecules. In: Advances in mass spectrometry, vol. 3 (W. L. MEAD, ed.), p. 639. Amsterdam: Elsevier 1966.
38. VENKATARAGHAVAN, R., F. W. MCLAFFERTY, and J. W. AMY: Automatic reduction of high resolution mass spectral data. Anal. Chem. **39**, 178 (1967).
39. SMITH, D. H.: High resolution mass spectrometry. Techniques and applications to molecular structure problems. Ph. D. Thesis, University of California, Berkeley, 1967.
40. BURLINGAME, A. L.: Data acquisition, processing, and interpretation via coupled high speed realtime digital computer and high resolution mass spectrometer systems. Internat. Mass Spectrometry Conference, Sept. 25 – 29, Berlin, 1967. In: Advances in mass spectrometry, vol. 4, p. 15 (E. KENDRICK, ed.). London: The Institute of Petroleum 1968.
41. – D. H. SMITH, T. O. MERREN, and R. W. OLSEN: Realtime high resolution mass spectrometry. In: Computers in analytical chemistry (D. H. ORR and J. NORRIS, eds.). New York: Plenum Press 1969.
42. GALLEGOS, E. J., J. W. GREEN, L. P. LINDEMAN, R. L. LETOURNEAU, and R. M. TEETER: Petroleum group-type analysis by high resolution mass spectrometry. Anal. Chem. **39**, 1833 (1967).
43. BIEMANN, K.: Mass spectrometry: organic chemical applications, p. 59 – 69. New York: McGraw-Hill 1962.
44. SILVERSTEIN, R. M., and G. L. BASSLER: Spectrometric identification of organic compounds. New York: John Wiley & Sons 1964.
45. BEYNON, J. H.: Mass spectrometry and its application to organic chemistry, p. 251 – 262. Amsterdam: Elsevier 1960.
46. BIEMANN, K., P. BOMMER, and D. M. DESIDERIO: Element-mapping, a new approach to the interpretation of high resolution mass spectra. Tetrahedron Letters, No. 26, 1725 (1964).
47. BURLINGAME, A. L., and D. H. SMITH: Automated heteroatomic plotting as an aid to the presentation and interpretation of high resolution mass spectral data. Tetrahedron **24**, 5749 (1968).
48. VENKATARAGHAVAN, R., and F. W. MCLAFFERTY: Topographical element map as a display for high resolution mass spectra. Anal. Chem. **39**, 278 (1967).
49. HAUG, P., H. K. SCHNOES, and A. L. BURLINGAME: Keto-carboxylic acids isolated from the Colorado Green River shale (Eocene). Chem. Comm., No. 21, 1130 (1967).
entrapped fatty acids isolated from the Green River Formation (Eocene). Nature **218**, 252 (1968).
50. BURLINGAME, A. L., and B. R. SIMONEIT: High resolution mass spectral analysis of the mineral
51. BIEMANN, K.: Mass spectrometry: Organic chemical applications, p. 42 – 45. New York: McGraw-Hill 1962.
52. BIEMANN, K., W. J. MCMURRAY, and P. V. FENNESSEY: Ion-types, a useful concept in the interpretation of high resolution mass spectra. Tetrahedron Letters, No. 33, 3997 (1966).
53. RICHTER, W. J., D. H. SMITH, and A. L. BURLINGAME: Intramolecular reaction of remote functionalities in long-chain aliphatic diketones upon electron impact. Proc. 16th Ann. Conf. on Mass Spectrometry and Allied Topics, Pittsburgh, Pa., May 13 – 17, 1968, p. 186.
54. For discussion of McLafferty rearrangements, see H. BUDZIKIEWICZ, C. DJERASSI, and D. H. WILLIAMS, Mass spectrometry of organic compounds. San Francisco: Holden-Day 1967.
55. BURLINGAME, A. L., P. C. WSZOLEK, and B. R. SIMONEIT: The fatty acid content of Tasmanites. In: Advances in organic geochemistry 1968 (I. HAVENAAR and P. A. SCHENCK, eds.). Braunschweig: Pergamon-Vieweg 1969.

56. HAUG, P. A.: Applications of mass spectrometry to organic geochemistry. Ph. D. Thesis, University of California, Berkeley, 1967.
57. BARBER, M., P. POWERS, and W. A. WOLSTENHOLME: Interpretation of high resolution mass spectra. Fourteenth Annual Conference on Mass Spectrometry and Allied Topics, Dallas, 1966. Abstracts, p. 618.
58. (a) BARBER, M., P. POWERS, M. J. WALLINGTON, and W. A. WOLSTENHOLME: Computer interpretation of high resolution mass spectra. Nature **212**, 784 (1966). (b) For introductory work, see M. BARBER, P. JOLLES, E. VILKAS, and E. LEDERER: Determination of amino acid sequences in oligopeptides by mass spectroscopy. I. The structure of fortuitine, an acylnonapeptide methyl ester. Biochem. Biophys. Res. Commun. **18**, 466 (1965).
59. SENN, M., and F. W. MCLAFFERTY: Automatic amino-acid-sequence determination in peptides. Biochem. Biophys. Res. Commun. **23**, 381 (1966).
60. — R. VENKATARAGHAVAN, and F. W. MCLAFFERTY: Mass spectrometric studies of peptides. III. Automated determination of amino acid sequences. J. Am. Chem. Soc. **88**, 5593 (1966).
61. BIEMANN, K., C. CONE, and B. R. WEBSTER: Computer-aided interpretation of mass spectra. II. Amino acid sequence of peptides. J. Am. Chem. Soc. **88**, 2597 (1966).
62. — — —, and G. P. ARSENAULT: Determination of the amino acid sequence in oligopeptides by computer interpretation of their high resolution mass spectra. J. Am. Chem. Soc. **88**, 5598 (1966).
63. —, and W. J. MCMURRAY: Computer-aided interpretation of high resolution mass spectra. Tetrahedron Letters, No. 11, 647, 653 (1965).
64. BEYNON, J. H.: High resolution mass spectrometry of organic materials. In: Advances in mass spectrometry, vol. 1 (J. D. WALDRON, ed.), p. 328. Oxford: Pergamon Press 1959.
65. BIEMANN, K., P. BOMMER, A. L. BURLINGAME, and W. J. MCMURRAY: The high resolution mass spectra of ajmalidine and related substances. Tetrahedron Letters, No. 28, 1969 (1963); and High resolution mass spectra of ajmaline and related alkaloids. J. Am. Chem. Soc. **86**, 4624 (1964).
66. RICHTER, W., u. K. BIEMANN: Hochauflösungsmassenspektren von Penicillinderivaten. Monatsh. Chem. **95**, 766 (1964).
67. KUPCHAN, S. M., J. M. CASSADY, J. E. KELSEY, H. K. SCHNOES, D. H. SMITH, and A. L. BURLINGAME: Structural elucidation and high resolution mass spectrometry of gaillardin, a new cytotoxic sesquiterpene lactone. J. Am. Chem. Soc. **88**, 5292 (1966).
68. BURLINGAME, A. L., C. C. FENSELAU, and W. J. RICHTER: High resolution mass spectrometry in molecular structure studies. IV. Mechanistic aspects of the fragmentation of widdrol. J. Am. Chem. Soc. **89**, 3232 (1967).
69. — — — W. G. DAUBEN, G. W. SHAFFER, and N. D. VIETMEYER: On the mechanism of electron impact induced elimination of ketene in conjugated cyclohexenones and correlations with photochemistry. J. Am. Chem. Soc. **89**, 3346 (1967).
70. SCHNOES, H. K., D. H. SMITH, A. L. BURLINGAME, P. W. JEFFS, and W. DÖPKE: Mass spectra of amaryllidaceae alkaloids: the lycorenine series. Tetrahedron **24**, 2825 (1968).
71. (a), (b) BURLINGAME, A. L., P. LONGEVIALLE, R. W. OLSEN, K. L. PERING, D. H. SMITH, H. M. FALES, and R. J. HIGHET: High resolution mass spectrometry in molecular structure studies. V, VI. The fragmentation of amaryllis alkaloids in the crinine series (unpublished results). (c) BURLINGAME, A. L., P. LONGEVIALLE, H. M. FALES, and R. J. HIGHET: High resolution mass spectrometry in molecular structure studies. VII. The fragmentation of amaryllis alkaloids in the crinine series (unpublished results).
72. BROWN, P., and C. DJERASSI: Electron-impact induced rearrangement reactions of organic molecules. Angew. Chem., Intern. Ed. **6**, 477 (1967).
73. SAUNDERS, R. A., and A. E. WILLIAMS: High resolution mass spectrometry. In: Mass spectrometry of organic ions (F. W. MCLAFFERTY, ed.), ch. 8. New York: Academic press 1963.
74. BOMMER, P., and F. M. VANE: The use of fragmentation patterns of 1,4-benzodiazepines for the structure determination of their metabolites by high resolution mass spectrometry. Proc. 14th Ann. Conf. on Mass Spectrometry and Allied Topics, Dallas, Tex., May 22–27, 1966, p. 606–609.
75. CARLSON, E. G., G. T. PAULISSEN, R. H. HUNT, and M. J. O'NEAL JR.: High resolution mass spectrometry. Anal. Chem. **32**, 1489 (1960).

76. LUMPKIN, H. E.: Analysis of a trinuclear aromatic petroleum fraction by high resolution mass spectrometry. Anal. Chem. **36**, 2399 (1964).

77. MEAD, W. K., W. L. MEAD, and K. M. BOWEN: Combination of high resolution and low ionizing voltages in the determination of hydrocarbon and sulfur compound types in petroleum fractions using mass spectrometry. In: Advances in mass spectrometry, vol. 3 (W. L. MEAD, ed.), p. 731. London: Institute of Petroleum 1966.

78. JOHNSON, B. H., and T. ACZEL: Analysis of complex mixtures of aromatic compounds by high resolution mass spectrometry at low-ionizing voltages. Anal. Chem. **39**, 682 (1967).

79. DRUSHEL, H. V., and A. L. SOMMERS: Isolation and characterization of sulfur compounds in high-boiling petroleum fractions. Anal. Chem. **39**, 1819 (1967).

80. BAKER, E. W., T. F. YEN, J. P. DICKIE, R. E. RHODES, and L. F. CLARK: Mass spectrometry of porphyrins II. Characterization of petro-porphyrins. J. Am. Chem. Soc. **89**, 3631 (1967).

81. MEAD, W. L., and W. K. REID: Field ionization mass spectrometry of heavy petroleum fractions at high resolution. Internat. Mass Spectrometry Conference, Berlin, Sept. 25 – 29, 1967.

82. RYHAGE, R.: Use of a mass spectrometer as a detector and analyzer for effluents emerging from high temperature gas-liquid chromatography columns. Anal. Chem. **36**, 759 (1964).

83. WATSON, J. T., and K. BIEMANN: High resolution mass spectra of compounds emerging from a gas chromatograph. Anal. Chem. **36**, 1135 (1964).

84. LLEWELLYN, P., and D. LITTLEJOHN: The separation of organic vapors from carrier gases. Pittsburgh Conference on Analytical Chemistry and Applied Spectroscopy, February 1966.

85. LIPSKY, S. R., C. G. HARWORTH, and W. J. MCMURRAY: Utilization of system emphasizing the selective permeation of helium through a unique membrane of teflon as an interface for gas chromatograph and mass spectrometer. Anal. Chem. **38**, 1585 (1966).

86. LEEMANS, F. A. J. M., and J. A. MCCLOSKEY: Combination gas chromatography-mass spectrometry. J. Am. Oil Chem. Soc. **44**, 11 (1967).

87. HOLMES, J. C., and F. A. MORRELL: Oscillographic mass spectrometric monitoring of gas chromatography. Appl. Spectry. **11**, 86 (1957).

88. STÄLLBERG-STENHAGEN, S., and E. STENHAGEN: Gas-liquid chromatography-mass spectrometry combination. In: Topics in organic mass spectrometry (A. L. BURLINGAME, ed.), ch. 4. New York: Interscience (in press).

89. MCFADDEN, W. H.: Introduction of gas chromatographic samples to a mass spectrometer. Separation Science **1**, 723 (1966).

90. LINDEMANN, L. P., and J. L. ANNIS: Use of a conventional mass spectrometer as a detector for gas chromatography. Anal. Chem. **32**, 1742 (1960).

91. – Combination of mass spectrometer, gas chromatograph, and catalytic hydrogenation for the analysis of olefins. Abstracts, Amer. Chem. Soc., Div. Petrol. Chem., Atlantic City, Sept. 9 – 14, 1962, Vol. 7, No. 3, p. 15.

92. DORSEY, J. A., R. N. HUNT, and M. J. O'NEAL: Rapid-scanning mass spectrometry. Continuous analysis of fractions from capillary gas chromatography. Anal. Chem. **35**, 511 (1963).

93. STUDIER, M. H., R. HAYATSU, and E. ANDERS: Organic compounds in carbonaceous chondrites. Science **149**, 1455 (1965).

94. ORÓ, J., D. W. NOONER, A. ZLATKIS, S. A. WIKSTROM, and E. S. BARGHOORN: Hydrocarbons of biological origin in sediments about two billion years old. Science **148**, 77 (1965).

95. – – Aliphatic hydrocarbons from the Precambrian of South Africa. Nature **213**, 1082 (1967).

96. EGLINTON, G., A. G. DOUGLAS, J. R. MAXWELL, J. N. RAMSAY, and S. STÄLLBERG-STENHAGEN: Occurrence of isoprenoid fatty acids in the Green River shale. Science **153**, 1133 (1966).

97. (a) RAMSAY, J. N.: The organic geochemistry of fatty acids. M. S. Thesis, University of Glasgow, 1966. (b) MAXWELL, J. R.: Studies in organic geochemistry. Ph. D. Thesis, University of Glasgow, 1967.

98. HENNEBERG, D.: High resolution mass spectra of compounds separated by capillary columns. Anal. Chem. **38**, 495 (1966).

99. WATSON, J. T., and K. BIEMANN: Direct recording of high resolution mass spectra of gas chromatographic effluents. Anal. Chem. **37**, 844 (1965).

100. – Direct recording of high resolution mass spectra of gas chromatographic effluents. Ph. D. Thesis, Massachusetts Institute of Technology, Cambridge, Mass., June 1965.

101. HAYES, J. M.: Techniques for high resolution mass spectrometric analysis of organic constituents of terrestrial and extraterrestrial samples. Ph. D. Thesis, Massachusetts Institute of Technology, 1966.
102. Catalog of Mass Spectral Data: American Petroleum Institute, Research Project 44, Carnegie Institute of Technology, Pittsburgh, Pennsylvania.
103. O'NEAL, M. J., JR., and T. P. WIER JR.: Mass spectrometry of heavy hydrocarbons. Anal. Chem. **23**, 830 (1951).
104. CLERC, R. J., A. HOOD, and M. J. O'NEAL JR.: Mass spectrometric analysis of high molecular weight saturated hydrocarbons. Anal. Chem. **27**, 868 (1955).
105. LUMPKIN, H. E., B. W. THOMAS, and A. ELLIOT: Modified method for hydrocarbon type analysis by mass spectroscopy. Anal. Chem. **24**, 1389 (1952).
106. BROWN, R. A., R. C. TAYLOR, F. W. MELPOLDER, and W. S. YOUNG: Mass spectrometer analysis of some liquid hydrocarbon mixtures. Anal. Chem. **20**, 5 (1948).
107. BECKEY, H. D.: Comparison of field ionization and chemical ionization mass spectra of decane isomers. J. Am. Chem. Soc. **88**, 5333 (1966).
108. CLARK, R. C., JR.: Occurrence of normal paraffin hydrocarbons in nature (tables and bibliography). Woods Hole Oceanographic Institution, Woods Hole, Massachusetts, Ref. No. 66–34, July, 1966.
109. JOHNS, R. B., T. BELSKY, E. D. MCCARTHY, A. L. BURLINGAME, P. HAUG, H. K. SCHNOES, W. J. RICHTER, and M. CALVIN: The organic geochemistry of Ancient sediments. Part II. Geochim. Cosmochim. Acta **30**, 1191 (1966).
110. HOEVEN, W. VAN, P. HAUG, A. L. BURLINGAME, and M. CALVIN: Hydrocarbons from an Australian oil, 200 million years old. Nature **211**, 1361 (1966).
111. LEVY, E. J., E. J. GALBRAITH, and F. W. MELPOLDER: Interpretive techniques for the determination of paraffin wax composition by mass spectometry and gas chromatography. In: Advances in mass spectrometry, vol. 2 (R. M. ELLIOTT, ed.), p. 395. New York: Macmillan 1963.
112. MOLD, J. D., R. K. STEVENS, R. E. MEANS, and J. M. RUTH: The hydrocarbons of tobacco: normal iso- and anteiso-homologues. Biochemistry **2**, 605 (1963).
113. EGLINTON, G., P. M. SCOTT, T. BELSKY, A. L. BURLINGAME, W. J. RICHTER, and M. CALVIN: Occurrence of isoprenoid alkanes in a Precambrian sediment. In: Advances in organic geochemistry 1964, Internat. Series of Monographs in Earth Sciences, vol. 24 (G. D. HOBSON and M. C. LOUIS, eds.), p. 41–74. Oxford: Pergamon Press 1966.
114. MCCARTHY, E. D., and M. CALVIN: Organic geochemical studies I. Molecular criteria for hydrocarbon genesis. Nature **216**, 642 (1967).
115. – – The occurrence of the C_{17}-isoprenoid in the Antrim Shale. Squalene as a possible precursor. Tetrahedron **23**, 2609 (1967).
116. EGLINTON, G., and M. CALVIN: Chemical fossils. Sc. Am. **216**, 32 (1967).
117. BENDORAITIS, J. G., B. L. BROWN, and L. S. HEPNER: Isoprenoid hydrocarbons in petroleum. Isolation of 2, 6, 10, 14-tetramethylpentadecane by high temperature g.l.c. Anal. Chem. **34**, 49 (1962).
118. – – – Isolation and identification of paraffins in petroleum. Sixth World Petroleum Congr., Sect. V-15, New York, 1963.
119. DEAN, R. A., and E. V. WHITEHEAD: The occurrence of phytane in petroleum. Tetrahedron Letters **21**, 768 (1961).
120. MAIR, B. J., N. C. DROUSKOP, and T. J. MAYER: Composition of the branched paraffin-cycloparaffin fraction of the light gas oil fraction. J. Chem. Eng. Data **7**, 420 (1962).
121. CUMMINS, J. J., and W. E. ROBINSON: Normal and isoprenoid hydrocarbons isolated from oil shale bitumen. J. Chem. Eng. Data **9**, 304 (1964).
122. EGLINTON, G., P. M. SCOTT, T. BELSKY, A. L. BURLINGAME, W. J. RICHTER, and M. CALVIN: Hydrocarbons of biological origin from a one-billion-year-old sediment. Science **145**, 263 (1964).
123. BELSKY, T., R. B. JOHNS, E. D. MCCARTHY, A. L. BURLINGAME, W. J. RICHTER, and M. CALVIN: Evidence for life processes in a sediment two and a half billion years old. Nature **206**, 446 (1965).
124. MEINSCHEIN, W. G.: Soudan formation: organic extracts of early Precambrian rocks. Science **150**, 601 (1965).
125. – E. S. BARGHOORN, and J. W. SCHOPF: Biological remnants in a Precambrian sediment. Science **145**, 262 (1964).

126. MAIR, B. J.: Terpenoids, fatty acids, and alcohols as source materials for petroleum hydrocarbons. Geochim. Cosmochim. Acta **28**, 1303 (1964).
127. PONNAMPERUMA, C., and K. L. PERING: Possible abiogenic origin of some naturally occurring hydrocarbons. Nature **209**, 979 (1966).
128. BLUMER, M., M. M. MULLIN, and D. W. THOMAS: Pristane in zooplankton. Science **140**, 974 (1963).
129. GÖHRING, K. L. H., P. A. SCHENCK, E. D. ENGLEHARDT: A new series of isoprenoid iso-alkanes in crude oils and Cretaceous bituminous shales. Nature **215**, 503 (1967).
130. MURPHY, M. T. J., A. MCCORMICK, and G. EGLINTON: Perhydro-β-carotene in the Green River shale. Science **157**, 1040 (1967).
131. HAN, J., E. D. MCCARTHY, W. VAN HOEVEN, M. CALVIN, and W. H. BRADLEY: Organic geochemical studies II. The distribution of aliphatic hydrocarbons in algae, bacteria and in a Recent lake sediment: A preliminary report. Proc. Nat. Acad. Sci. U.S. **57**, 29 (1968).
132. MAIR, B. J., Z. ROUEN, E. J. EISENBROWN, and A. G. HARODYSKY: Terpenoid precursors of hydrocarbons from the gasoline range of petroleum. Science **154**, 1339 (1966).
133. MCFADDEN, W. H., and R. G. BUTTERY: Application of mass spectrometry in flavor and aroma chemistry. In: Topics in organic mass spectrometry (A. L. BURLINGAME, ed.). New York: Interscience 1969.
134. LEVY, E. J., R. R. DOYLE, R. A. BROWN, and F. W. MELPOLDER: Identification of hydrocarbons by gas liquid chromatography and mass spectrometry. Anal. Chem. **33**, 698 (1961).
135. NAGY, B., and G. C. GAGNON: The geochemistry of the athabasca petroleum deposit. I. Elution and spectroscopic analysis of the petroleum from the vicinity of McMurray, Alberta. Geochim. Cosmochim. Acta **23**, 155 (1961).
136. MEYERSON, S., T. D. NEVITT, and P. N. RYLANDER: Ionization-dissociation of some cycloalkanes under electron impact. In: Advances in mass spectrometry, vol. 2 (R. M. ELLIOTT, ed.), p. 313. New York: Macmillan 1963.
137. LINDEMANN, L. P., and R. L. TOURNEAU: New information on the composition of petroleum. Proc. 6th World Petroleum Congr., Frankfurt, Germany, June 19–26, 1963.
138. O'NEAL, M. J., and A. HOOD: Mass spectrometric analysis of polycyclic hydrocarbons. Symposium on Polycyclic Hydrocarbons, Div. Petroleum Chemistry, Amer. Chem. Soc., 127 (1956).
139. SCHISSLER, D. O., D. P. STEVENSON, R. J. MOORE, G. J. O'DONNELL, and R. E. THORPE: Steranes in petroleum. 5th Ann. Conf. on Mass Spectrometry and Allied Topics, New York, 1957.
140. MEINSCHEIN, W. G.: Origin of petroleum. Bull. Am. Assoc. Petrol. Geologists **43**, 925 (1959).
141. – Significance of hydrocarbons in sediments and petroleum. Geochim. Cosmochim. Acta **22**, 58 (1961).
142. CARLSON, E. G., M. L. ANDRE, and M. J. O'NEAL: Characterization of the heavier molecules in a crude oil by mass spectrometry. In: Advances in mass spectrometry, vol. 2 (R. M. ELLIOTT, ed.), p. 377. New York: Macmillan 1963.
143. BURLINGAME, A. L., P. HAUG, T. BELSKY, and M. CALVIN: Occurrence of biogenic steranes and pentacyclic triterpanes in an Eocene shale (52 Million Years) and in an early Precambrian shale (2.7 Billion Years): A preliminary report. Proc. Nat. Acad. Sci. U.S. **54**, 1406 (1965).
144. DJERASSI, C.: Rearrangement reactions in organic mass spectrometry. In: Advances in mass spectrometry, vol. 4 (E. KENDRICK, ed.), p. 199. London: Institute of Petroleum 1968.
145. BUDZIKIEWICZ, H., J. M. WILSON, and C. DJERASSI: Mass spectrometry in structural and stereochemical problems. XXXII. Pentacyclic triterpenes. J. Am. Chem. Soc. **85**, 3688 (1963).
146. KARLINER, J., and C. DJERASSI: Terpenoids LVII. Mass spectral and nuclear magnetic resonance studies of pentacyclic triterpene hydrocarbons. J. Org. Chem. **31**, 1945 (1966).
147. HILLS, J. R., E. V. WHITEHEAD, D. E. ANDERS, J. J. CUMMINS, and W. E. ROBINSON: An optically active triterpane, gammacerane, in Green River, Colorado, oil shale bitumen. Chem. Comm. 752 (1966).
148. – – Triterpanes from Nigerian crude oil. Nature **209**, 977 (1966).
149. CARRUTHERS, W., and D. A. M. WATKIN: The constituents of high-boiling petroleum distillates, VIII. Identification of 1, 2, 3, 4-tetrahydro-2, 2, 9-trimethylpicene in an American crude oil. J. Chem. Soc. 724 (1964). – Identification of 1, 2, 3, 4-tetrahydro-2, 2, 9-trimethylpicene in an American crude oil. Chem. & Ind. (London) 1433 (1963).
150. MAIR, B. J., and J. L. MARTINEZ-PICO: The composition of the trinuclear aromatic portion of the heavy gas oil and light lubricating distillate. Proc. Am. Petrol. Inst. **42**, 173 (1962).

151. FIELD, E. H., and S. H. HASTINGS: Determination of unsaturated hydrocarbons by low voltage mass spectrometry. Anal. Chem. **28**, 1248 (1956).
152. FRISQUE, A. J., H. M. GRUBB, C. H. EHRHARDT, and R. W. VANDERHAAR: Total analysis of olefinic naphthas by mass spectrometry. Anal. Chem. **33**, 389 (1961).
153. MIKKELSEN, L., R. L. HOPKINS, and D. Y. YEE: Mass spectrometer-type analysis for olefins in gasoline. Anal. Chem. **30**, 317 (1958).
154. IIDA, T., E. YOSHII, and E. KITATSUJI: Identification of normal paraffins, olefins, ketones, and nitriles from Colorado shale oil. Anal. Chem. **38**, 1224 (1966).
155. DINNEEN, G. U., R. A. VAN METER, J. R. SMITH, C. W. BAILEY, G. L. COOK, C. S. ALLBRIGHT, and J. S. BALL: Composition of shale-oil naphtha. Bureau of Mines Bull. 593, U.S. Govt. Printing Office, 1961.
156. BLUMER, M., and D. W. THOMAS: Phytadienes in zooplankton. Science **147**, 1148 (1965).
157. — — Zamene, isomeric C_{19} mono-olefins from marine plankton, fishes, and mammals. Science **148**, 370 (1965).
158. KENNER, G. W., and E. STENHAGEN: Localization of double bonds by mass spectrometry. Acta Chem. Scand. **18**, 1551 (1964).
159. AUDIER, H., S. BORY, M. FETIZON, P. LONGEVIALLE et R. TOUBIANA: Orientation de la fragmentation en spectrometrie de masse par introduction des groupes fonctionnels. (6^{e} partie) Localisation des liaisons ethyleniques. Bull. Soc. Chim. France 3034 (1964).
160. WOLFF, R. E., G. WOLFF, and J. A. MCCLOSKEY: Characterization of unsaturated hydrocarbons by mass spectrometry. Tetrahedron **22**, 3093 (1966).
161. BEROZA, M., and B. A. BIERL: Rapid determination of olefin position in organic compounds in microgram range by ozonolysis and gas chromatography; alkylidine analysis. Anal. Chem. **39**, 1131 (1967).
162. (a) APLIN, R. T., and L. COLES: Simple procedure for localization of ethylenic bonds by mass spectrometry. Chem. Comm. No. 17, 858 (1967). (b) NIEHAUS, W. G., and R. RYHAGE: Determination of double bond positions in polyunsaturated fatty acids using combination gas chromatography-mass spectrometry. Tetrahedron Letters **49**, 5021 – 5026 (1967).
163. DEAN, R. A., and E. V. WHITEHEAD: The composition of high boiling petroleum distillates and residues. 6th World Petroleum Congr. Frankfurt, Germany, 1963, Sect. V, Paper 9.
164. GRUBB, H, M., and S. MEYERSON: Mass spectra of alkylbenzenes. In: Mass spectrometry of organic ions (F. W. MCLAFFERTY, ed.), p. 453. New York: Academic Press 1963.
165. MAIR, B. J., and J. M. BARNEWALL: Composition of the mononuclear aromatic material in the light gas oil range, low refractive index portion, 230° C to 305° C. J. Chem. Eng. Data **9**, 282 (1964).
166. — Structures of some mononuclear, aromatic hydrocarbons from a heavy gas oil and light lubricating distillate. J. Chem. Eng. Data **12**, 126 (1967).
167. —, and T. J. MAYER: Composition of the dinuclear aromatics, C_{12} to C_{14}, in the light gas-oil fraction of petroleum. Anal. Chem. **36**, 351 (1964).
168. YEW, F. F., and B. J. MAIR: Isolation and identification of C_{13} to C_{17} alkylnaphthalenes, alkylbiphenyls, and alkyldibenzofurans from the 275° C to 305° C dinuclear aromatic fraction of petroleum. Anal. Chem. **38**, 231 (1966).
169. LUMPKIN, H. E.: Low voltage techniques in high molecular weight mass spectrometry. Anal. Chem. **30**, 321 (1958).
170. —, and T. ACZEL: Low voltage sensitivities of aromatic hydrocarbons. Anal. Chem. **36**, 181 (1964).
171. GORDON, R. J., R. J. MOORE, and C. E. MULLER: Aromatic types in heavily cracked gas oil fractions. Anal. Chem. **30**, 1221 (1958).
172. CRABLE, G. F., G. L. KEARNS, and M. S. NORRIS: Low voltage mass spectrometric sensitivities of aromatics. Anal. Chem. **32**, 13 (1960).
173. REID, W. K.: Use of high resolution mass spectrometry in the study of petroleum waxes, microcrystalline waxes, and ozokerite. Anal. Chem. **38**, 445 (1966).
174. LOCHTE, H. L., and E. R. LITTMANN: The petroleum acids and bases. New York: Chemical Publishing Company 1955.
175. BALL, J. S., W. F. HAINES, and R. V. HELM: Minor constituents of a California petroleum. 5th World Petroleum Congr., New York, 1959, Sect. V, Paper 14.

176. RYHAGE, R., and E. STENHAGEN: Mass spectrometry of long chain esters. In: Mass spectrometry of organic ions (F. W. MCLAFFERTY, ed.), ch. 9. New York: Academic Press 1963.

177. BURLINGAME, A. L., and B. R. SIMONEIT: Isoprenoid fatty acids isolated from the kerogen matrix of the Green River Formation (Eocene). Science **160**, 531 (1968).

178. COOPER, J. E., and E. E. BRAY: A postulated fate of fatty acids in petroleum formation. Geochim. Cosmochim. Acta **27**, 1113 (1963).

179. – Fatty acids in recent and ancient sediments and petroleum reservoir waters. Nature **193**, 744 (1962).

180. KVENVOLDEN, K. A.: Molecular distributions of normal fatty acids and paraffins in some Lower Cretaceous sediments. Nature **209**, 573 (1966).

181. ABELSON, P. H., and P. L. PARKER: Fatty acids in sedimentary rocks. Carnegie Inst. Wash. Yearbook **61**, 181 (1961).

182. LAWLOR, D. L., and W. E. ROBINSON: Fatty acids in Green River formation oil shale. Paper presented at the Detroid Meeting, Amer. Chem. Soc., Div. Petrol. Chem., May 9, 1965.

183. HOERING, T. C., and P. H. ABELSON: Fatty acids from the oxidation of kerogen. Carnegie Inst. Wash. Yearbook **64**, 218 (1965).

184. CASON, J., and D. W. GRAHAM: Isolation of isoprenoid acids from a California petroleum. Tetrahedron **21**, 471 (1965).

185. DOUGLAS, A. G. K. DOURAGHI-ZADEH, G. EGLINTON, J. R. MAXWELL, and J. N. RAMSAY: Fatty acids in sediments including the Green River shale (Eocene) and Scottish torbanite (Carboniferous). In: Advances in organic geochemistry (1966) (G. D. HOBSON and G. C. SPEERS, eds.). London: Pergamon Press (1969).

186. HAUG, P., H. K. SCHNOES, and A. L. BURLINGAME: Isoprenoid and dicarboxylic acids from the Colorado Green River shale (Eocene). Science **158**, 772 (1967).

187. CASON, J., and A. I. A. KHODAIR: Isolation of the eleven-carbon acyclic isoprenoid acid from petroleum. Mass spectroscopy of its *p*-phthalimidophenacyl ester. J. Org. Chem. **32**, 3430 (1967).

188. KHODAIR, A. I. A.: Isolation and structure determination of certain acidic components from California petroleum. Ph. D. Thesis, University of California, Berkeley, 1965.

189. BURLINGAME, A. L., and B. R. SIMONEIT: High resolution mass spectrometry of Green River formation kerogen oxidations. Nature **222**, 741 (1969).

190. MCCLOSKEY, J. A., and M. J. MCCLELLAND: Mass spectra of O-isopropylidene derivatives of unsaturated fatty esters. J. Am. Chem. Soc. **87**, 5090 (1965).

191. EGLINTON, G., and D. H. HUNNEMAN: Gas chromatographic-mass spectrometric studies of long chain hydroxy acids. I. The constituent cutin acids of apple cuticle. Phytochemistry **7**, 313 – 322 (1968).

192. HAUG, P., H. K. SCHNOES, and A. L. BURLINGAME: Aromatic carboxylic acids isolated from the Colorado Green River formation (Eocene). Geochim. Cosmochim. Acta **32**, 358 (1968).

193. MCLAFFERTY, F W., and R. S. GOHLKE: Mass spectrometric analysis. Aromatic acids and esters. Anal. Chem. **31**, 2076 (1959).

194. ACZEL, T., and H. E. LUMPKIN: Correlation of mass spectra with structure in aromatic oxygenated compounds. Benzoate-type esters. Anal. Chem. **34**, 33 (1962).

195. GRIGSBY, R. D., M. C. HAMMING, E. J. EISENBRAUN, D. V. HERTZLER, and N. BRADLEY: Mass spectrometric fragmentation of substituted propionic and butyric acids. Abstracts, 14th Ann. Conf. on Mass Spectrometry and Allied Topics, Dallas, Tex., May 22 – 27, 1966, p. 574.

196. CASON, J., and A. I. A. KHODAIR: Mass spectra of certain cycloalkylacetates and of related unsaturated esters. J. Org. Chem. **32**, 575 (1967).

197. –, and K. L. LIAUW: Characterization and synthesis of a monocyclic eleven-carbon acid isolated from a California petroleum. J. Org. Chem. **30**, 1763 (1965).

198. –, and A. I. A. KHODAIR: Separation from a California petroleum and characterization of geometric isomers of 3-ethyl-4-methylcyclopentylacetic acid. J. Org. Chem. **31**, 3618 (1966).

199. LOCHTE, H. L.: Petroleum acids and bases. Ind. Eng. Chem. **44**, 2597 (1952).

200. DEAL, V. Z., F. T. WEISS, and T. T. WHITE: Determination of basic nitrogen in oils. Anal. Chem. **25**, 426 (1953).

201. COSTANTINIDES, G., and G. ARICH: Non-hydrocarbon constituents in petroleum. In: Fundamental aspects of some petroleum geochemistry (B. NAGY and U. COLOMBO, eds.), p. 109. Amsterdam: Elsevier Publishing Co. 1967.

202. LALAU, C.: Mass-spectrometric study of nitrogen compounds from petroleum distillates. Anal. Chim. Acta **22**, 239 (1960).
203. DINNEEN, G. U., G. L. COOK, and H. B. JENSEN: Estimation of types of nitrogen compounds in shale-oil gas oil. Anal. Chem. **30**, 2026 (1958).
204. SNYDER, L. R., and B. E. BUELL: Characterization and routine determination of nonbasic nitrogen types in cracked gas oils by linear elution adsorption chromatography. Anal. Chem. **36**, 767 (1964).
205. SAUER, R. W., F. W. MELPOLDER, and R. A. BROWN: Nitrogen compounds in domestic heating oil distillates. Ind. Eng. Chem. **44**, 2606 (1952).
206. JEWELL, D. M., and G. K. HARTUNG: Identification of nitrogen bases in heavy gas oil; chromatographic methods of separation. Chem. Eng. Data **9**, 297 (1964).
207. HARTUNG, G. K., and D. M. JEWELL: Carbazoles, phenazines and dibenzofuran in petroleum products; methods of isolation, separation and determination. Anal. Chim. Acta **26**, 514 (1962).
208. BIEMANN, K.: Mass spectrometry: Organic chemical applications, p. 130–136. New York: McGraw-Hill 1962.
209. CLUGSTON, D. M., and D. B. MACLEAN: Mass spectra of oxygenated quinolines. Canad. J. Chem. **44**, 781 (1966).
210. SAMPLE, S. D., D. A. LIGHTNER, O. BUCHARDT, and C. DJERASSI: Mass spectrometry in structural and stereochemical problems. CXXIV. Mass spectral fragmentation of alkylquinolines and isoquinolines. J. Org. Chem. **32**, 997 (1967).
211. STEVENSON, R. L., and M. E. WACKS: Mass spectrometric analysis of some 8-hydroxyquinolines. Abstracts, 14th Ann. Conf. on Mass Spectrometry and Allied Topics, Dallas, Tex., May 22–27, 1967, p. 581.
212. SPITELLER, G.: Mass spectrometry of heterocyclic compounds. In: Advances in heterocyclic chemistry, vol. 7 (A. R. KATRITZKY, ed.). New York: Academic Press, 1966.
213. METER, R. A. VAN, C. W. BAILEY, J. R. SMITH, R. T. MOORE, C. S. ALLBRIGHT, I. A. JACOBSON JR., V. M. HYLTON, and J. S. BALL: Oxygen and nitrogen compounds in shale-oil naphtha. Anal. Chem. **24**, 1758 (1952).
214. HELM, R. V., D. R. LATHAM, C. R. FERRIN, and J. S. BALL: Identification of carbazole in Wilmington petroleum through use of gas-liquid chromatography and spectroscopy. Anal. Chem. **32**, 1765 (1960).
215. HARTUNG, G. K., and D. M. JEWELL: Identification of nitriles in petroleum products. Complex formation as a method of isolation. Anal. Chim. Acta **27**, 219 (1962).
216. DRUSHEL, H. V., and A. L. SOMMERS: Isolation and identification of nitrogen compounds in petroleum. Anal. Chem. **38**, 19 (1966).
217. BRANDENBURG, C. F., and D. R. LATHAM: Separation and identification of basic nitrogen compounds in Wilmington petroleum. 4th National Meeting, Society for Applied Spectroscopy, Denver, Colorado, 1965.
218. DAS, B. C., and E. LEDERER: Mass spectrometry of complex natural products. In: Topics in organic mass spectrometry (A. L. BURLINGAME, ed.). New York: Interscience (in press).
219. HOOD, A., E. G. CARLSON, and M. J. O'NEAL: Petroleum oil analysis. In: Encyclopedia of spectroscopy (G. L. CLARK, ed.), p. 613. New York: Reinhold Publ. Co. 1960.
220. MEAD, W. L., and A. J. WILDE: Mass spectrum of vanadyl etioporphyrin I. Chem. & Ind. (London) 1315 (1961).
221. WHITTEN, D. G., K. E. BENTLEY, and D. KUWADA: Pyrolysis studies. Controlled thermal degradation of mesoporphyrin. J. Org. Chem. **31**, 322 (1966).
222. HOFFMAN, D. R.: Mass spectra of porphyrins and chlorins. J. Org. Chem. **30**, 3512 (1965).
223. JACKSON, A. H., G. W. KENNER, K. M. SMITH, R. T. APLIN, H. BUDZIKIEWICZ, and C. DJERASSI: Pyrroles and related compounds – VIII. Mass spectroscopy in structural and stereochemical problems. LXXVI. The mass spectra of porphyrins. Tetrahedron **21**, 2913 (1965).
224. SEIBL, J.: Mass spectral analysis of metal complexes of some organic compounds. In: Advances in mass spectrometry, vol. 4 (E. KENDRICK, ed.), p. 317. London: Institute of Petroleum 1968.
225. BUDZIKIEWICZ, H.: Mass spectrometric investigation of porphyrin derivatives. In: Advances in mass spectrometry, vol. 4 (E. KENDRICK, ed.), p. 313. London: Institute of Petroleum 1968.
226. THOMAS, D. W., and M. BLUMER: Porphyrin pigments of a triassic sediment. Geochim. Cosmochim. Acta **28**, 1147 (1964).

227. MORANDI, J. R., and H. B. JENSEN: Comparison of porphyrins from shale oil, oil shale, and petroleum by absorption and mass spectroscopy. J. Chem. Eng. Data **11**, 81 (1966).

228. BAKER, E. W.: Mass spectrometric characterization of petroporphyrins. J. Am. Chem. Soc. **88**, 2311 (1966).

229. BLUMER, M., and W. D. SNYDER: Porphyrins of high molecular in a Triassic oil shale: Evidence by gel permeation chromatography. Chem. Geol. **2**, 35 (1967).

230. BOYLAN, D. B., and M. CALVIN: Volatile silicon complexes of etioporphyrin I. J. Am. Chem. Soc. **89**, 5472 (1967).

231. LEVY, E. J., and W. A. STAHL: Mass spectra of aliphatic thiols and sulfides. Anal. Chem. **33**, 707 (1961).

232. MOMIGNY, J.: The mass spectra of monosubstituted benzene derivatives. Phenol, monodeuteriophenol, thiophenol, and aniline. Bull. Soc. Roy. Sci. Liege **22**, 541 (1953).

233. LAWESSON, S. O., J. O. MADSEN, G. SCHROLL, J. H. BOWIE, and D. H. WILLIAMS: Studies in mass spectrometry. Part XVI. Mass spectra of thiophenols. Acta Chem. Scand. **20**, 2325 (1966).

234. SAMPLE, S. D., and C. DJERASSI: Mass spectrometry in structural and stereochemical problems. CIV. The nature of the cyclic transition state in hydrogen rearrangements of aliphatic sulfides. J. Am. Chem. Soc. **88**, 1937 (1966).

235. DUFFIELD, A. M., H. BUDZIKIEWICZ, and C. DJERASSI: Mass spectrometry in structural and stereochemical problems. LXXI. A study of the influence of different heteroatoms on the mass spectrometric fragmentation of five-membered heterocycles. J. Am. Chem. Soc. **87**, 2920 (1965).

236. BOWIE, J. H., S. O. LAWESSON, J. O. MADSEN, G. SCHROLL, and D. H. WILLIAMS: Studies in mass spectrometry. XIV. Mass spectra of aromatic thio ethers. The effect of structural variations on the relative abundance of skeletal rearrangement ions. J. Chem. Soc. B, 951 (1966).

237. FISCHER, M., and C. DJERASSI: Massenspektrometrie und ihre Anwendung auf strukturelle und stereochemische Probleme, LXXXVII. Möglichkeit von Alkylwanderungen bei aromatischen Verbindungen. Chem. Ber. **99**, 750 (1966).

238. KINNEY, I. W., JR., and G. L. COOK: Identification of thiophene and benzene homologs. Anal. Chem. **24**, 1391 (1952).

239. FOSTER, N. G., D. E. HIRSCH, R. E. KENDALL, and B. H. ECCLESTON: The mass spectra and correlations with structures for 23 alkylthiophenes. Bureau of Mines Report No. 6433 (1964).

240. – – – – The mass spectra and correlations with structure for 14 alkylthiophenes. Bureau of Mines Report No. 6671 (1965).

241. – The mass spectra and correlations with structure for 2-t-butyl-, 3-t-butyl-, 2,5-di-t-butyl-, and 2,4-di-t-butylthiophenes. Bureau of Mines Report No. 6741 (1966).

242. COOK, G. L., and N. G. FOSTER: Relation of molecular structure to fragmentation of some sulfur compounds in the mass spectrometer. Proc. Am. Petrol. Inst. **41**, III, 199 (1961).

243. HANUS, V., and V. CERMAK: Note on the mass spectra of alkylthiophenes, and the structure of the ion $C_5H_5S^+$. Coll. Czech. Chem. Commun. **24**, 1602 (1959).

244. FOSTER, N. G., and R. W. HIGGINS: Mass spectrometric studies of selected ions in deuterated alkylthiophenes. Abstracts, 14th Ann. Conf. on Mass Spectrometry and Allied Topics, Dallas, Tex., May 22–27, 1966, p. 463.

245. BUDZIKIEWICZ, H., C. DJERASSI, and D. H. WILLIAMS: Mass spectrometry of organic compounds. San Francisco: Holden-Day 1967.

246. COOK, G. L., and G. U. DINNEEN: Mass spectra of organic sulfur compounds. Bureau of Mines Report No. 6698, 1965.

247. THOMPSON, C. J., N. G. FOSTER, H. J. COLEMAN, and H. T. RALL: Sulfur compound characterization studies on high boiling petroleum fractions. Bureau of Mines Report No. 6879 (1966).

248. – H. J. COLEMAN, H. T. RALL, and H. M. SMITH: Separation of sulfur compounds from petroleum. Anal. Chem. **27**, 175 (1955).

249. HASTINGS, S. H., B. H. JOHNSON, and H. E. LUMPKIN: Analysis of the aromatic fraction of virgin gas oils by mass spectrometer. Anal. Chem. **28**, 1243 (1956).

250. BIRCH, S. F., T. V. CULLUM, R. A. DEAN, and R. L. DENYER: Sulfur compounds in kerosine boiling range of Middle East crudes. Ind. Eng. Chem. **47**, 240 (1955).

251. AMBERG, C. H.: Organic sulfur compounds in Alberta cracked distillates. J. Inst. Petrol. **45**, 1 (1959).

252. CLERC, R. J., and M. J. O'NEAL JR.: The mass spectrometric analysis of asphalt. A preliminary investigation. Anal. Chem. **33**, 380 (1961).

253. LATHAM, D. R., C. R. FERRIN, and J. S. BALL: Identification of fluorenones in Wilmington petroleum by gas-liquid chromatography and spectrometry. Anal. Chem. **34**, 311 (1962).

254. BRANDENBURG, C. F., D. R. LATHAM, G. L. COOK, and W. E. HAINES: Identification of a cycloalkylketone in Wilmington petroleum through use of chromatography and spectroscopy. Chem. Eng. Data **9**, 463 (1964).

255. NAGY, B., W. G. MEINSCHEIN, and D. J. HENNESSEY: Mass spectrometric analysis of the Orgueil meteorite: Evidence for biogenic hydrocarbons. Ann. N. Y. Acad. Sci. **93**, 27 (1961).

256. MEINSCHEIN, W. G., B. NAGY, and D. J. HENNESSEY: Evidence in meteorites of former life: The organic compounds in carbonaceous chondrites are similar to those found in Recent sediments. Ann. N. Y. Acad. Sci. **108**, 553 (1963).

257. ORÓ, J., D. W. NOONER, A. ZLATKIS, and S. A. WIKSTROM: Paraffinic hydrocarbons in the Orgueil, Murray, Mokoia and other meteorites. Life Sciences and Space Research **4**, 63 (1966).

258. NOONER, D. W., and J. ORÓ: Organic compounds in meteorites I. Aliphatic hydrocarbons. Geochim. Cosmochim. Acta **31**, 1359 (1967).

259. HAYES, J. M., and K. BIEMANN: High resolution mass spectrometric investigations of the organic constituents of Murray and Holbrook chondrites. Geochim. Cosmochim. Acta **32**, 239 (1968).

260. – Organic constituents of meteorites – a review. Geochim. Cosmochim. Acta. **31**, 1395 (1967).

261. SHARKEY, A. G., J. L. SCHULZE, and R. A. FRIEDEL: Mass spectrometric determination of the ratio of branched to normal hydrocarbons up to C_{18} in Fischer-Tropsch products. Anal. Chem. **34**, 826 (1962).

262. FRIEDEL, R. A., and A. G. SHARKEY: Alkanes in natural and synthetic petroleums: Comparison of calculated and actual compositions. Science **139**, 1203 (1963).

CHAPTER 5

Gas Chromatography

A. G. DOUGLAS

Organic Geochemistry Unit, Geology Department
University of Newcastle upon Tyne, England

Contents

I. Introduction

The possibility that components of a mixture might be separated by partitioning them between a gas and liquid phase was proposed by MARTIN and SYNGE in 1941 [1], and demonstrated, by the separation of a mixture of fatty acids, by JAMES and MARTIN in 1952 [2]. The potentialities of this basically simple process were soon realised by biochemists and chemists working in many fields and in the past 15 years many thousands of papers devoted to the technique have appeared.

This devotion by the chemist and biochemist augurs well for the organic geochemist who, entering the field today, has at his disposal in gas chromatography the most powerful separatory tool available, one which is perhaps the most significant contribution ever to be made to analytical methods.

The technique is applicable to gases, and all liquids and solids (or derivatives of such) that can be distilled, and also, in pyrolysis gas chromatography, to compounds which can be decomposed in a reproducible manner. Moreover the speed, sensitivity, simplicity of apparatus and the resolution of extremely complex mixtures that can be obtained enhance its value to the organic geochemist.

In this technique the mixture to be analyzed is injected into a moving stream of carrier gas which flows through a column. The column normally contains a finely divided powder, the surface of which is coated with a thin film of liquid phase, and separation is effected by differences in the partition coefficients of the constituents of the mixture between the gas and liquid phases. The various solutes comprising the mixture thus appear at the distal end of the column at different times, depending on how long they have been dissolved in the liquid phase. A detector, which normally measures instantaneous changes in the composition of the effluent, is placed at the column exit and provides a signal for a strip chart recorder on which the chromatogram is obtained as a series of symmetrical peaks. Ideally, each peak represents a single solute and the time from injection to its maximum concentration in the detector is its retention time; for constant gas flow rates this is a characteristic of the compound. It follows that two compounds with identical boiling points, but with different partition coefficients, may easily be separated; on the other hand two compounds with the same partition coefficient will emerge as one peak. A single peak on a chromatogram should therefore be taken only as prima facie evidence that a single component is present; if only one peak is obtained on a high resolution column the evidence is better, and if only one peak is obtained on a variety of stationary phases the evidence for homogeneity is considered by many satisfactory. Similarly, superposition of peaks on coinjection of unknown and authentic samples requires the use of more than one stationary phase. Thus a mixture of n-heptadecane and pristane emerge as one peak from a $1/8'' \times 5'$ column containing either silicone rubber (SE-30) or Apiezon L, whereas they are easily separated on a column containing tetracyanoethylated pentaerythritol (TCEPE) [3]. However, they can be separated on SE-30, using a column of higher efficiency (cf. Fig. 3). The superposition of some branched hydrocarbons from honeybee wax has also been noted on columns containing either Apiezon or the fluorosilicone polymer QF-1 [4]. In practice, eluting solutes are often collected and submitted to some other spectroscopic or analytical examination.

The possibility of producing artifacts during the gas chromatographic process both in the injector [5, 6] and on the solid supports should be noted [7, 8], and special precautions should be taken to use appropriate temperatures and to remove catalytic sites on the solid phase.

The chromatographic process in which the stationary phase is an active solid rather than a liquid is called gas solid chromatography, solute separation being dependent on differential adsorption rather than partition. This technique, in which the solid adsorbent is commonly alumina, silica gel, molecular sieve, charcoal, etc., is used mainly for the separation of gases and low molecular weight hydrocarbons; details of these applications should be sought in appropriate articles [9].

II. Definition of Terms

Recommendations on nomenclature and presentation of data in gas chromatography were made [10] and subsequently modified [11]; space limitations allow only brief mention of a few important expressions.

Relative retention data are obtained from chromatograms containing internal standards which should preferably be n-alkanes, although current practice in steroid analysis is to use cholestane and in the fatty acid field a saturated n-fatty acid. Adjustments should be made for the dead volume or gas hold-up of the apparatus. Thus the relative retention (also called the separation factor) of solute 2, to an internal standard, e.g., solute 3 in Fig. 1 is given by AC/AD. This ratio may be expressed in terms of time (retention time) or volume (retention volume), the latter being the product of retention time and carrier gas flow rate. A plot of the logarithm of retention volume vs. carbon number is a straight line for a homologous series; such linear plots have been found *inter alia* for n-paraffins, alcohols, ketones and esters [12]. In fatty acid analysis, a modification of this plot is the concept of Equivalent Chain Length (E.C.L.) [13]; saturated normal fatty acid esters have integral E.C.L. values equal to the number of carbon atoms in the

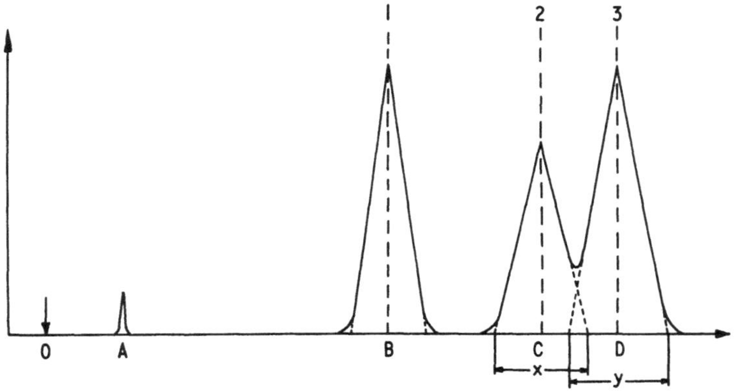

Fig. 1. Idealized isothermal gas chromatogram showing the injection point 0, and the position of the air peak *A*; points *B*, *C* and *D* represent the mid-point of the curves of solutes 1, 2 and 3 respectively

molecule. The E.C.L. values for branched and unsaturated esters are generally non integral, for the latter the values on polar stationary phases are greater and on non-polar phases less than their actual carbon numbers.

Preferably, relative retention data should be expressed as the Kovats Retention Index (I) [14] which relates the logarithm of the retention time of a solute to those of the n-alkanes (whose Retention Index is defined as 100 times the number of carbon atoms in the molecule). The Retention Index is given by the expression

$$I = 100\,N + 100\,n\,\frac{(\log R_x - \log R_N)}{(\log R_{N+n} - \log R_N)}$$

where R_x is the adjusted retention time of the unknown compound and R_N and R_{N+n} are the retentions of the n-alkanes of carbon number N and $N+n$. In Fig. 1, if solutes 1 and 3 are pentane and hexane, the Retention Index of solute 2 is given by

$$I_2 = 500 + 100\,\frac{(\log AC - \log AB)}{(\log AD - \log AB)}.$$

The separation of two solutes by g.l.c. depends mainly on their relative partition coefficients (equivalent to relative retention) and the column efficiency, the former as seen above is a measure of the position of the two peaks on the chromatogram, the latter is a measure of the peak width. If the bases of the solute peaks 1 and 3 in Fig. 1 were overlapping, their separation factor would be the same but the column efficiency would be much less. The resolution of solute 2 and 3 is given by the expression,

$$\text{Resolution} = \frac{2(AD - AC)}{(x+y)}.$$

Column efficiency, which relates peak width to retention time is usually expressed as the theoretical plate number,

$$n = 16\left(\frac{\text{Retention time}}{\text{Peak width}}\right)^2$$

and since it varies with the nature of the solute the latter should be specified; for solute 3 in Fig. 1,

$$n = 16\left(\frac{AD}{y}\right)^2.$$

The height of the column which is equal to one theoretical plate (H.E.T.P.) may be calculated from fundamental column parameters by the Van Deemter equation [15]. For high resolution packed columns, in which a narrow mesh range of solid support containing a small amount of liquid phase is used, it is possible to obtain efficiencies of 500 – 1,000 plates per foot. Efficiencies in excess of 1,000,000 theoretical plates have been claimed for a capillary column [16] but PURNELL has cautioned on the comparison of efficiencies of packed and capillary columns. He has shown that the separation of hydrocarbons obtained on a 500,000 plate capillary column were little better than on a 5,000 plate, 7 foot, packed column [17]; this anomaly is due to the large gas hold up (OA in Fig. 1) in capillary columns. Nevertheless, capillary columns of high resolving power may be prepared, and Fig. 3c represents a chromatogram obtained on such a column.

III. Apparatus

The basic design of a gas chromatograph incorporating a flame ionization detector is shown in Fig. 2. Molecular sieve traps are used in the gas lines to remove contaminants (commonly water vapor and hydrocarbon oils). Carrier gas reaching the injection port should preferably be heated; this is conveniently accomplished at the injection block which is normally maintained at a temperature higher than that of the column. Gases may be injected with a gas-tight syringe or with a gas sampling loop [18, 19]. Liquids and solids, dissolved in solvents of low boiling point, are most commonly injected with a leak-proof micro syringe by piercing a silicone rubber septum which seals the injection port. "Bleed" of plasticizers, etc., from the septum may give rise to ghost peaks under conditions of high injection port temperatures; they are often encountered during temperature programming and when operating at high sensitivity. Conditioning of septa [20, 21] to minimize this bleed, or water cooling of the septum holder is commonly employed. Liquids and solids may also be introduced into the chromatograph in special capillary crushing devices [22, 23], and syringes for solid sampling are available [24]. In on-column injection the sample is delivered directly to the top of the column packing; otherwise the liquid is "flash evaporated" in a

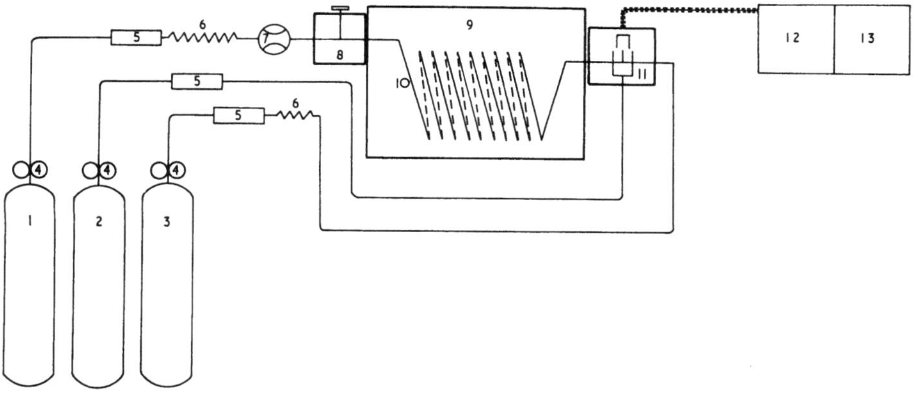

Fig. 2. Schematic diagram of a gas chromatograph incorporating a flame ionization detector. 1 – carrier gas; 2 – air; 3 – hydrogen; 4 – pressure regulators; 5 – molecular sieve traps; 6 – capillary restrictors; 7 – gas sampling loop; 8 – injection port and heater; 9 – column oven; 10 – chromatographic column; 11 – flame ionization detector; 12 – amplifier; 13 – recorder

hot zone and then delivered to the column. The column and its packing, which may be regarded as the heart of the apparatus (since it is here that the separation takes place and this ultimately limits the performance of the gas chromatograph), is described separately below. Effluent from the column passes to a detector, preferably thermostatted, the function of which is to respond to the appearance of a solute in the carrier gas effluent; this response should be convertible to an electrical signal which is displayed differentially by a potentiometric recorder (e.g., Fig. 3). Although many types of detectors have been described those most commonly used are undoubtedly: (a) thermal conductivity, (b) flame ionization, (c) argon ionization, and (d) electron capture detectors. Their requirements and achievements in terms of linearity, effective volume, time constant, sensitivity, etc., have been discussed many times [19, 25–27].

Thermal conductivity detectors (catharometers) employ either a resistance wire or thermistor bead in a cell (or cells) which forms part of a Wheatstone bridge network, and require hydrogen or helium as carrier gas [26]. In general their sensitivity is about 3 to 6 orders of magnitude less than ionization detectors, although thermistor detectors occupying a very small volume have been used with capillary columns [28]. Currently, perhaps the most popular detector is the flame ionization detector (F.I.D.) which depends for its operation on the production of ions when an organic substance burns in a hydrogen flame [29–31]. This detector has a linear response over a wide range, it is insensitive to a number of inorganic vapours and it is simple to construct. Argon ionisation detectors are somewhat more sensitive than the F.I.D. and the electron capture detector [32] can detect one halogen containing molecule in 10^{13} argon atoms [33]. This detector has the particular advantage that it will selectively respond to compounds which are able to capture free electrons such as aromatic hydrocarbons and compounds containing oxygen, nitrogen or halogens. Ionisation detectors have been reviewed and compared [34, 35]. Another detector developed for monitoring radioactive effluents has been described [36].

The need to characterize components separated by gas chromatography, other than by retention data, has been mentioned above. Two possibilities are (a) to trap the individual components for subsequent chemical investigation or (b) to couple a device with the gas chromatograph which will give some specific information regarding the emerging solute. The attachment of infrared spectrophotometers [37–39] (GC-IR), nuclear magnetic resonance spectrometers [40] and mass spectrometers [41, 42] (GC-MS) has been achieved. This last technique (see Chapter 4) has been applied to some geochemical problems, to which it is particularly suited [43, 44]; combined GC-MS and GC-IR instruments are now commercially available. Methods used for trapping eluting solutes vary with the size and type of samples being collected. Large samples of several grams are conveniently collected in variously designed, cooled, traps [18, 19, 45]; losses due to the formation of aerosols with high boiling solutes may be reduced by using electrostatic precipitators [46]. A systematic investigation for recovery improvement has been made [47]. Small samples may be very simply and qualitatively collected by directing the column effluent on to a cold surface such as a salt plate or millipore filter for direct infrared examination. We have conveniently used an empty glass melting point capillary tube for collecting solutes from $\frac{1}{8}''$ high resolution columns for subsequent infrared and mass spectrometric analyses; capillaries containing potassium bromide powder or column packing may also be used. Methods for recovering gas chromatography and thin layer chromatography fractions for infrared examination is the subject of a recent comprehensive review [49] and procedures for the identification of trapped components have been reviewed [50, 51].

For many years gas chromatograms were obtained with the column oven maintained at a constant temperature (isothermal operation); for mixtures boiling over a wide range this results in early peaks being sharp and closely spaced whereas late peaks are broad and widely spaced. Although such mixtures may be divided into a number of fractions by distillation or preparative gas chromatography, and subsequently analysed isothermally, an alternative procedure is to increase the temperature of the column oven progressively, usually in a linear fashion. Thus early peaks in the chromatogram are more spread out and later peaks appear sharper and closer, resulting in a chromatogram of more uniform appearance and reduced time of analysis. Since the viscosity of the carrier gas increases with increasing temperature, this technique requires a differential flow controller in the carrier gas line for the best results. The extension of programmed temperature techniques to include extremely low temperatures, so called cryogenic gas chromatography [52] offers a potential advantage to the geochemist examining wide range boiling mixtures such as may be found in the pyrolysis products of coals, kerogens, etc. A monograph [53] and other reviews of programmed temperature gas chromatography are available [54, 55].

A more recent technique than temperature programming is flow programming [56, 57], in which the carrier gas inlet pressure is increased continuously during the analysis. Consequences of flow programming are that wide boiling range mixtures may be chromatographed at lower isothermal temperatures with the attendant advantage of analysing mixtures containing thermally labile substances. This lower operational temperature increases the number of liquid phases that may be employed and decreases baseline drift as compared with temperature programming. Finally, SCOTT has shown that preparative scale separations may be achieved using analytical columns and flow programming [58].

Various pre-column techniques that have been used, include hydrogenation, dehydrogenation, and hydrogenolysis (so-called carbon skeleton chromatography) [59, 60] in which the hydrogen carrier gas, in the presence of a catalyst, saturates multiple bonds and removes sulfur, halogens, oxygen, nitrogen, phosphorus, silicon and metals, thus producing saturated hydrocarbons; by raising the catalyst temperature dehydrogenation can be made to occur [61]. Subtractive precolumns, e.g., molecular sieve for n-paraffins [62], packings containing sulfuric acid, silver or mercury for olefins [63], boric acid for alcohols [64], etc., have been described. One of the most important of the pre-column techniques is pyrolysis-gas chromatography in which the organic material is pyrolysed under carefully controlled conditions and the volatile products are swept on to the top of the column and subsequently separated; this method has recently been comprehensively reviewed [65] and a bibliography is available [66]. Typical substrates for this technique are polymers [67], proteins [68], micro-organisms [69], petroleum residues [70], etc.; the applicability of this method to large organic matrices of interest to the geochemist such as kerogens and coal macerals is evident. Pyrolysis of gas chromatographic eluates followed by further gas chromatography of the pyrolysates to give unequivocal fingerprints has been demonstrated for saturated and unsaturated hydrocarbons, fatty acids, etc. [71], comparison of fragmentation patterns obtained by electron impact and pyrolysis g.l.c. have been made and it is suggested that the latter method may give information not easily obtained by mass spectrometry [72].

IV. Columns and Column Packings

As indicated above, the essential part of the gas chromatograph is the column and the ultimate performance of the system will depend on it. Materials from which columns may be constructed include stainless steel, copper, aluminium, glass and plastic, their lengths may vary from a few inches to one mile and their diameters from 0.010 inches to several inches. By appropriate combinations of these dimensions we can provide preparative, analytical and high resolution capillary columns; it should be realised however that the first two terms are in some measure equivocal. Thus, although large preparative columns (diameter several inches) are used to separate large (e.g., 50 g) samples, columns of about $20' \times 0.08''$ i.d. which allow collection of enough material for mass spectrometry are used in this laboratory; for infrared spectroscopy collection from several analyses may be required. For the organic geochemist with sample sizes of one gram or very much less, preparative columns of $(5'–20') \times (\frac{1}{4}''–\frac{3}{8}''$ o.d.) should suffice for many separations; however each problem will determine specific requirements. For a recent discussion on preparative scale gas chromatography see VERZELE [73] (and references therein). Dimensions of high resolution packed columns are normally $(10'–50') \times (\frac{1}{16}''–\frac{1}{8}''$ o.d.), they require not more than about 1 µl of solution for analysis. Capillary columns (open tubular columns) are normally made of stainless steel or glass [74], ranging in length from about 50′–500′ with an internal diameter of 0.010″–0.030″. The load capacity of such columns is small and decreases with decreasing diameter; at the same time the column performance increases. Loads in the range 10^{-6}–10^{-7} g are usually placed on the capillary column by using a stream splitting device for injection; in this the homogeneous mixture is split, the major portion being vented to the atmosphere. Such small sample sizes demand sensitive detectors with a rapid response, and capillary chromatographs normally use flame ionisation or micro-argon detectors. Two monographs [19, 75] and an excellent review by DESTY [76], one of the pioneers in the field of capillary chromatography, are available. Chromatograms of a complex hydrocarbon mixture analysed on preparative, high resolution packed and capillary columns are shown in Fig. 3.

The solid support in packed columns should provide a large, inert surface on which the stationary phase is spread in a thin, uniform film; most commonly used are various proprietory brands of diatomaceous earths (Chromosorb, Celite, C-22 firebrick, etc.). Preparation of these supports before coating with liquid phase variously includes crushing, sieving within narrow mesh ranges, removal of fines, treating with acid and alkali, washing and treating with some substance which will combine with residual active sites (e.g., dichlorodimethylsilane, hexamethyldisilazane, etc.). Other support materials include glass and metal beads, polytetrafluorethylene powder, inorganic salts, etc.; porous polymer beads which may be used without a liquid phase have been developed. Preparative columns commonly use 30–60 mesh supports although recently, for large scale preparative work, 10–20 mesh supports and very long columns have been advocated [73, 78]. High resolution packed columns use packings ranging from 60–120 mesh (SCOTT [79] recommends 120–150 mesh); the preparation of a 100–120 mesh packing for a $20' \times \frac{1}{16}''$ o.d. column is described below. Selection of the appropriate liquid phase is most important and its choice will often determine whether or not a separation can be performed. Some general requirements of liquid substrates have been recommended [80] – they include thermal stability, low vapor pressure, ability to dissolve at least one of the components to be separated and chemical inertness; some studies on stationary phase temperature limitations are available [81, 82]. Generally, mixtures containing compounds of similar polarity but differing in boiling points, can be resolved on non-polar liquid phases and will usually elute in order of their boiling points. For the resolution of mixed polar and non-polar solutes, a polar liquid phase will retain the polar solute preferentially, thus Carbowax 400 with good aromatic selectivity has been shown to elute n-decane (b.p. 170° C) before benzene (b.p. 80° C) [83]. Similarly, the selective retardation of amines by the heavy metal salts of fatty acids has been demonstrated and a separation requiring 250,000 theoretical plates on a silicone column required only four theoretical plates on a zinc stearate column [84]. The terms polar and non-polar are relative only and generally ill-defined, although a number of workers have proposed quantitative relationships [85–87]. Examples of commonly used non-polar phases are squalane, methyl silicones (e.g., SE-30, SF-96) and the Apiezon greases, polyphenyl ethers are slightly polar and polar phases include the polyesters (e.g., polyethyleneglycol succinate) and polyglycols (e.g., Carbowaxes). Tables of liquid phases recommended for specific sample type

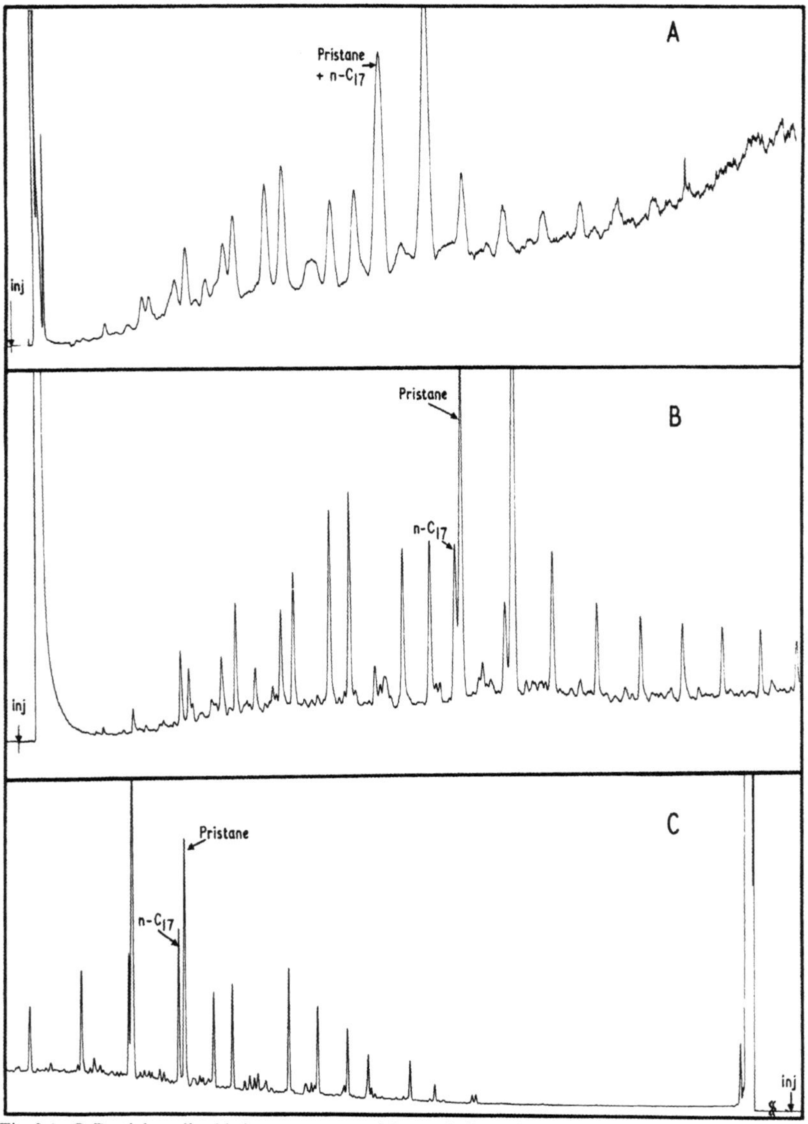

Fig. 3 A–C. Partial gas liquid chromatograms of the total alkane fraction of the Marl slate. A. Column $4' \times \frac{1}{4}''$ containing 10 percent SE-30 on 60–80 mesh Chromosorb-P. Nitrogen 30 ml./min. Temp. programmed from 120–270° C at 4° C/min. B. Column $20' \times \frac{1}{16}''$ containing 3 percent OV-1 on 100–120 mesh Gas Chrom Q. Nitrogen 8 ml./min. Temp. programmed from 80–320° C at 2° C/min. C. Column $200' \times 0.01''$ coated with Apiezon L. Nitrogen 1 ml./min. Temp. programmed from 70–250° C at 2° C/min.

The chromatogram illustrated in Fig. 3C was provided by Mr. W. Henderson, B. Sc., of the Geochemistry Unit. The University, Bristol 8. Note that this chromatogram runs from right to left, unlike the other two

analyses are given by Kaiser [19]; such tables are also available from most of the g.l.c. manufacturers and supply houses.

Methods of coating the solid support vary, but the following is generally satisfactory. For packings containing 10 percent or more stationary phase the support is mixed with a solution containing the correct amount of phase and the whole is evaporated under reduced pressure in a rotary evaporator. For packings containing less than 3 percent stationary phase, the following procedure may be followed. It is virtually that of HORNING *et al.* [88], and is given in some detail for the preparation of a $20' \times \frac{1}{16}''$ o.d. (0.040" i.d.) column containing 2 percent of a silicone gum (OV-1).

Gas Chrom P (100–120 mesh) is re-sieved to 100–120 mesh size and washed several times with concentrated hydrochloric acid and water, it is then washed with methanol and dried; at each step the slurry is evacuated (water pump) and air is allowed to enter the flask, thereby forcing the liquids into the pores of the support. The dry, hot support is poured into a 5–10 percent solution of dichlorodimethylsilane (or hexamethyldisilazane) in toluene, reduced pressure being applied as above. The support is ready for coating after thorough washing with toluene and methanol and drying. The dry, hot support is poured into a 2 percent solution of OV-1 in toluene and reduced pressure is applied as described above. After standing, the slurry is vacuum filtered on a Buchner funnel and then air and oven dried. (Recently, the use of fluidised drying has been claimed to increase column performance [89]). A $10' \times 1/16''$ o.d. length of stainless steel tubing, having one end plugged with quartz wool and a retainer of stainless steel mesh, is filled with the packing in small batches with attendant tapping and suction (oil pump). Two such lengths are connected with a copper sleeve, leaving no gap at the butt joint, the sleeve is swaged in place using an appropriate die and a KBr disc press. Primary ageing of the column, during which solvent and liquid phase impurities are removed, is carried out at about 340° C in a slow stream of inert carrier gas.

Various methods of coating adsorbents and packing columns are available [19, 88] and the influence of the mode of packing on the performance of preparative columns has been described [90]. Methods are available for coating capillary columns [19, 75, 76, 91, 92], by which the inner wall is coated with a thin unbroken film of liquid phase; preparation and use of the more recently developed support-coated open tubular columns has been reported [93, 94]. In these columns a thin layer of porous support is deposited on the inner wall of a capillary and the liquid phase is coated, as a thin film, on this supporting layer. Since these columns have about the same separating power as open tubular columns but have a larger capacity, their usefulness in coupled gas-chromatographic techniques [95] is evident. Capillary columns packed with support and coated with liquid phase have been described and evaluated [96–98].

V. Application of Gas-Liquid Chromatography to Organic Geochemistry

As indicated above, the literature on gas chromatography is now voluminous and reflects the wide variety of sample types that have been analysed. This new dimension in analytical methodology has not, as yet, been fully exploited by the organic geochemist, and this serves as a justification for reference to potential applications in the following discussion. Gas chromatography was invented by a biochemist and was quickly accepted and developed by the petroleum industry; it is not surprising therefore that the analysis of fatty acids and hydrocarbons by this method has received much attention. An example is DESTY's classic paper [99] on the use of capillary columns coated with squalane or Apiezon L for the separation of petroleum gases and hydrocarbons. Capillary columns with polar and non-polar coatings have been used for the separation of petroleum hydrocarbons [100], a mixed substrate containing squalane and the π complexing phase dipropyl tetrachlorophthalate has been used [101] for separating aromatic hydrocarbons in the boiling range 80°–180° C. Middle fractions of crude oil were separated on packed columns containing Apiezon or silicone oil [102] and waxes containing hydrocarbons ranging from C_{25}–C_{68} were analysed on a 3 percent SE-52 packed column [103]. High molecular weight aromatic hydrocarbons have been analysed including many alkyl phenanthrenes on both SE-30 and QF-1 [104]; glass beads with very low loadings of silicone oil separated polycyclic aromatic hydrocarbons up to and including chrysene [105] while coronene (b.p. 577° C) was separated on silicone grease [106]. Normal and isoprenoid hydrocarbons obtained from an oil and shales ranging in age from 60×10^6–1×10^9 years, were chromatographed on a variety of liquid phases [3] and *iso*, *anteiso*, cyclohexyl and isoprenoid alkanes were reported in oils and shales ranging in age from 3×10^6–2.7×10^9 years [107]; the authors collected fractions from SE-30 columns and obtained further resolution of these fractions on columns containing 7 ring meta-polyphenylether (Polysev). The resolving power of this latter phase for normal and isoprenoid hydrocarbons is illustrated in their separation from a variety of meteorites [108], thus resolution of phytane and eicosane is not complete on a 200′ capillary, coated with the virtually non-polar Apiezon L, but they are well separated on a 240′ column coated with Polysev. Complete separation was also obtained on a 100′ column coated with Carbowax 20M (terminated with terephthalic acid). Separation on SE-30 coated packed columns and capillaries of some triterpanes (and nortriterpanes) has been reported [109–111]. WOLLRAB *et al.* [112], have used a packing containing thermally stripped Apiezon L with a thermal stability up to 350° C for the analysis of paraffins in brown coal. Analyses by g.l.c. of mixtures containing normal and non-normal paraffins using a molecular sieve column in series with a partition column has been reported; a subtractive technique compares chromatograms obtained with and without the molecular sieve column [113], whereas elution and g.l.c. of the normal paraffins by a back flushing procedure has been used by others [114].

It is often convenient to convert compounds with functional groups into one or other of their derivatives [50]; these preferably should have greater vapor pressure and less polarity than the compounds themselves. Although free fatty acids may be analysed by g.l.c. [115], they are commonly converted to their methyl esters [18], a conversion which may be accomplished on as little as 30 µg of lipid [116]. Reviews on the analysis of fatty acids by g.l.c. are available [117, 118] and the use of high resolution capillary columns is noted [119, 120]. Normal fatty acids have been analysed as their methyl esters in recent and ancient sediments [121, 122], and in meteorites [123], using SE-30 columns, and in ocean water using polyester columns [124, 125]. Branched chain fatty acids in sediments [126] and marine bacteria [127] were reported, analysis being on columns containing diethyleneglycol succinate or Apiezon L. Isoprenoid fatty acids were isolated from petroleum [128] and an Eocene shale [129]; both teams of authors trapped fractions from polar columns [neopentylglycol succinate

(NGS) and Versamid] respectively and re-analysed the fractions on silicone columns.

The position of the double bond in unsaturated fatty acid esters has been determined by decomposition of the ozonides in the column injector followed by g.l.c. of their pyrolysis products [130, 131]. Gas chromatographic analyses of other acidic components have been well documented; reviews on methods of resin acid [18] and bile acid [132, 133] analysis are available. Acids from oleoresins have been chromatographed on SE-30 and ethyleneglycol succinate (EGS) [134] and from bled resins on NGS [135]. Other analyses of resin acids using Versamid (polyamide) [136], and QF-1 (fluoroalkylsilicone polymer) [137] have been made. Isomerisation and disproportionation of resin acid esters during g.l.c. has been noted [138]. Analysis of sediments for amino acids has until now been done by the conventional ion exchange methods; rapid g.l.c. techniques will no doubt be investigated by the geochemist. Conversion of the amino acids to the corresponding N-trifluoroacetyl methyl esters followed by g.l.c. on NGS has been reported [139, 140] and a number of reviews are available [141–143], their analysis on capillary columns is noted [144]. Aromatic acids obtained by ozonolysis of polymer-like materials in coal, kerogen and the Orgeuil meteorite were analysed, as their methyl esters, on SE-30 columns [145]; ozonization products of lignite tars were analysed, as esters, on Apiezon and polyester columns [146]. Benzene polycarboxylic acids, obtained by oxidation of alluvial soil, were analysed on a column containing SE-30 [147].

The analysis of sugars by g.l.c. does not appear to have been exploited by the organic geochemist. Monosaccharides are conveniently chromatographed as their trimethylsilylethers on succinate polyester phases [148, 149]; retention data for 22 carbohydrates chromatographed as their trifluoroacetates on silicone oil have been published [150]. Other chemical types of high molecular weight which have been analysed by g.l.c. on silicone gum columns include the mono-, di- and triglycerides [151, 152], and methods for their analyses have been reviewed [153]. Polyisoprenoid alcohols ranging up to C_{80} [154] and long chain α,ω-diols [155], in the range C_{10}–C_{26} were analysed on SE-30 as their acetates and as the derived saturated hydrocarbons; the free diols were also analysed on SE-30. A mixed phase containing SE-52 and Carbowax 20M was used to analyse methyl ketones, ranging from C_{17}–C_{37}, found in soil and peat waxes [156].

One of a number of arguments left to the organic geochemist in establishing the biological origin of "chemical fossils" is their retention of optical activity, and such optical activity has been confirmed in petroleum- and coal-derived hydrocarbons. The value of a gas chromatographic technique which would allow separation of optical isomers is evident, and this has now been accomplished in a number of laboratories. Enantiomeric alcohols and amino acids have been converted to diastereoisomeric derivatives and then analyzed on capillary columns coated with conventional polar and non-polar phases [157]. On the other hand enantiomers may be resolved using an optically active stationary phase coated on capillary columns [158, 159].

The interests of the geochemist, be they the gaseous constituents of volcanoes, rocks, soils, or the atmosphere of Mars, the high molecular weight compounds trapped within sediments or the polymeric diagenetic products of such entrapments, all come within the compass of this technique, and it is certain that its application will be extended beyond that of the geochemistry of hydrocarbons and fatty acids to some of the compound types which have been discussed above.

Bibliography

Of the many papers so far published on gas chromatography which are of potential application to organic geochemistry only a few have been mentioned in this short chapter. The reader is therefore encouraged to make use of the valuable reviews, abstracts and bibliographies which appear from time to time.

One journal devoted solely to gas chromatography is the *Journal of Gas Chromatography*. Two others which normally contain a number of articles wholly on the subject are the *Journal of Chromatography* and *Analytical Chemistry*.

One of the most valuable sources of information on gas chromatography is to be found in *Gas Chromatography Abstracts*, issued free, quarterly, to members of the Gas Chromatography Discussion Group of the Institute of Petroleum; these abstracts are also published yearly as a bound volume. The abstract indices are extremely well done and the first cumulative index (in press) should be of great value to the gas chromatographer. *Preston Technical Abstracts* are published as punched cards with two associated retrieval systems available, one uses cards based on an optical coincidence system, the other uses a computer tape; since this system adds about forty abstracts per week to its index these systems are continually updated. *Analytical Chemistry Reviews*, published in April of even years, contains a section on gas chromatography with a valuable bibliography covering the preceding two years publications, and *Chromatographic Reviews*, edited by M. Lederer, also contains review articles from time to time. Bibliographies of the years work are to be found in the December issues of the *Journal of Gas Chromatography* (bibliography discontinued after December 1966), and manufacturers newsletters, research notes, etc., occasionally have useful bibliographies.

There are many textbooks available now on gas chromatography which cannot conveniently be mentioned here; those included in the following reference section have been found very useful. Proceedings of the various symposia on gas chromatography are normally available as bound volumes and a number of these too have been mentioned in the reference section.

References

1. Martin, A. J. P., and R. L. M. Synge: A new form of chromatogram employing two liquid phases. 1. A theory of chromatography. 2. Application to the micro-determination of the higher mono-amino acids in proteins. Biochem. J. **35**, 1358 – 1368 (1941).
2. James, A. T., and A. J. P. Martin: Gas-liquid partition chromatography: the separation and micro-estimation of volatile fatty acids from formic acid to dodecanoic acid. Biochem. J. **50**, 679 – 690 (1952).
3. Eglinton, G., P. M. Scott, T. Belsky, A. L. Burlingame, W. Richter, and M. Calvin: Occurrence of isoprenoid alkanes in a Precambrian sediment. In: Advances in organic geochemistry, 1964 (G. D. Hobson and M. C. Louis, eds.), p. 41 – 74. Oxford: Pergamon Press, Inc. 1966.
4. Stránský, K., M. Streibl u. F. Šorm: Über Naturwachse. IV. Über einen neuen Typ verzweigter Paraffine aus dem Wachs der Honigbiene. Coll. Czech. Chem. Comm. **31**, 4694 – 4702 (1966).

5. Brochmann-Hanssen, E., and A. B. Svendsen: Gas chromatography of alkaloids, alkaloidal salts and derivatives. J. Pharm. Sci. **51**, 1095–1098 (1962).
6. Hoffman, N. E., and I. R. White: Gas chromatographic analysis of malonic acids. Anal. Chem. **37**, 1541–1542 (1965).
7. Tai, W. T., and E. W. Warnhoff: β-Keto esters from reaction of ethyl diazoacetate with ketones. Can. J. Chem. **42**, 1333–1340 (1964).
8. Gottfried, H.: Studies of the thermal decomposition of corticosteroids during gas chromatography. Steroids **5**, 385–397 (1965).
9. Scott, C. G., and C. S. G. Phillips: New potentialities in gas solid chromatography. In: Gas chromatography 1964 (A. Goldup, ed.), p. 266–284. London: The Institute of Petroleum 1965.
10. Ambrose, D., A. T. James, A. I. M. Keulemans, E. Kováts, R. Rock, C. Rouit, and F. H. Stross: Preliminary recommendations on nomenclature and presentation of data in gas chromatography. Pure Appl. Chem. **1**, 177–186 (1960).
11. Adlard, E. R., M. B. Evans, A. G. Butlin, R. S. Evans, R. Hill, J. F. K. Huber, A. B. Littlewood, W. G. McCambley, J. F. Smith, W. T. Swanton, and P. A. T. Swoboda: Recommendations of the data sub-committee for the publication of retention data. J. Gas Chromatog. **3**, 298–302 (1965).
12. Ray, N. H.: Gas chromatography I. The separation and estimation of volatile organic compounds by gas liquid partition chromatography. J. Appl. Chem. **4**, 21–25 (1954).
13. Miwa, T. K., K. L. Mikolajczak, F. R. Earle, and I. A. Wolff: Gas chromatographic characterisation of fatty acids. Identification constants for mono- and dicarboxylic methyl esters. Anal. Chem. **32**, 1739–1742 (1960).
14. Kováts, E.: Gas-chromatographische Charakterisierung organischer Verbindung. Teil 1. Retentionsindices aliphatischer Halogenide, Alkohole, Aldehyde und Ketone. Helv. Chim. Acta. **41**, 1915–1932 (1958).
15. Deemter, J. J. van, F. J. Zuiderwag, and A. Klinkenberg: Longitudinal diffusion and resistance to mass transfer as causes of nonideality in chromatography. Chem. Eng. Sci. **5**, 271–289 (1956).
16. Zlatkis, A., and H. R. Kaufman: Use of coated tubing as columns for gas chromatography. Nature **184**, 2010 (1959).
17. Purnell, J. H.: Gas Chromatography Discussions. Am. Chem. Soc. Meeting, Atlantic City, N. J. (1959).
18. Burchfield, H. P., and E. E. Storrs: Biochemical applications of gas chromatography. New York: Academic Press 1962.
19. Kaiser, R.: Gas phase chromatography. London: Butterworths 1963.
20. Tamema, A., F. E. Kurtz, N. Rainey, and M. J. Pallansch: Preconditioning of septums to reduce eluate contamination and anomalous peaks during temperature programming. J. Gas Chromatog. **5**, 271–272 (1967).
21. Kolloff, R. H.: Septum bleeding in flame ionization programmed temperature gas liquid chromatography. Anal. Chem. **34**, 1840–1841 (1962).
22. Nerheim, A. G.: Indium encapsulation technique for introducing weighed samples in gas chromatography. Anal. Chem. **36**, 1686–1688 (1964).
23. Wendenburg, J., and K. Jurischka: A simple device for breaking sealed glass ampoules in gas chromatography. J. Chromatog. **15**, 538–540 (1964).
24. Knight, J. A., and C. T. Lewis: A sampling device for viscous and solid materials for gas chromatography. J. Chromatog. **18**, 158–159 (1965).
25. Lovelock, J. E.: Ionisation methods for the analysis of gases and vapours. Anal. Chem. **33**, 162–178 (1961).
26. Keulemans, A. I. M.: Gas chromatography, 2nd edit. New York: Reinhold Publishing Co. 1959.
27. Karmen, A.: Ionization detectors for gas chromatography. In: Advances in chromatography, vol. 2 (J. C. Giddings and R. A. Keller, eds.), p. 293–336. London: Edward Arnold Ltd. 1966.
28. Camin, D. L., R. W. King, and S. D. Shawhan: Capillary gas chromatography using microvolume thermal conductivity detectors. Anal. Chem. **36**, 1175–1178 (1964).
29. McWilliam, I. G., and R. A. Dewar: Flame ionization detector for gas chromatography. In: Gas chromatography, 1958 (D. H. Desty, ed.), p. 142–147. London: Butterworths 1958.
30. Ongkiehong, L.: Investigation of the hydrogen flame ionization detector. In: Gas chromatography, 1960 (R. W. P. Scott, ed.), p. 7–15. London: Butterworths 1960.

31. DESTY, D. H., C. J. GEACH, and A. GOLDUP: An examination of the flame ionization detector using a diffusion, dilution apparatus. In: Gas chromatography, 1960 (R. W. P. SCOTT, ed.), p. 46–64. London: Butterworths 1960.
32. LOVELOCK, J. E., and S. R. LIPSKY: Electron affinity spectroscopy – A new method for the identification of functional groups in chemical compounds separated by gas chromatography. J. Am. Chem. Soc. **82**, 431–433 (1960).
33. SCOTT, R. P. W.: Gas-liquid chromatography: recent developments in apparatus and technique. Brit. Med. Bull. **22**, 131–136 (1966).
34. FOWLIS, I. A., R. J. MAGGS, and R. P. W. SCOTT: The critical examination of commercially available detectors for use in gas chromatography. I. The macro argon and flame ionisation detectors. J. Chromatog. **15**, 471–481 (1964).
35. CONDON, R. D., P. R. SCHOLLY, and W. AVERILL: Comparative data on two ionization detectors. In: Gas chromatography, 1960 (R. P. W. SCOTT, ed.), p. 30–45. London: Butterworths 1960.
36. JAMES, A. T.: Methods for the detection and estimation of radio-active compounds separated by gas-liquid chromatography. In: New biochemical separations (A. T. JAMES and L. J. MORRIS, eds.), p. 1–24. London: D. Van Nostrand Co. Ltd. 1964.
37. BARTZ, A. M., and H. D. RUHL: Rapid scanning infrared-gas chromatography instrument. Anal. Chem. **36**, 1892–1896 (1964).
38. SCOTT, R. P. W., I. A. FOWLIS, D. WELTI, and T. WILKINS: Interrupted-elution gas chromatography. In: Gas chromatography, 1966 (A. B. LITTLEWOOD, ed.), p. 318–336. New York: Academic Press 1967.
39. SCHMELTZ, I., C. D. STILLS, W. J. CHAMBERLAIN, and R. L. STEDMAN: Analysis of cigarette smoke fractions by combined gas chromatography and infrared spectrophotometry. Anal. Chem. **37**, 1614–1616 (1965).
40. BRAME, E. G.: Combining gas chromatography with nuclear magnetic resonance spectrometry. Anal. Chem. **37**, 1183–1184 (1965).
41. RYHAGE, R.: Fast recording of mass spectra of organic compounds of high molecular weight, using amounts of material in the microgram range. Arkiv Kemi. **20**, 185–191 (1962).
42. – Use of a mass spectrometer as a detector and analyser for effluents emerging from high temperature gas liquid chromatography columns. Anal. Chem. **36**, 759–764 (1964).
43. DOUGLAS, A. G., K. DOURAGHI-ZADEH, G. EGLINTON, J. R. MAXWELL, and J. N. RAMSAY: Fatty acids in sediments including the Green River Shale (Eocene) and Scottish Torbanite (Carboniferous). In: Advances in organic geochemistry, 1966 (G. D. HOBSON and G. C. SPEERS, eds.). Oxford: Pergamon Press (in press).
44. ORÓ, J., D. W. NOONER, A. ZLATKIS, S. A. WILKSTRÖM, and E. S. BARGHOORN: Hydrocarbons of biological origin in sediments about two billion years old. Science **148**, 77–79 (1965).
45. HUNT, R. J.: Efficient recovery in preparative chromatography. Column (W. G. PYE, Gas Chromatography Bulletin), **2**, 6–9 (1967).
46. DAL NOGARE, S., and R. S. JUVET: Gas liquid chromatography. New York: Interscience Publishers 1962.
47. VERZELE, M.: Substance recovery in preparative gas chromatography. J. Chromatog. **13**, 377–381 (1964).
48. THOMAS, P. J., and J. L. DWYER: Collection of gas chromatographic effluents for infrared spectral analysis. J. Chromatog. **13**, 366–371 (1964).
49. SNAVELY, M. K., and J. G. GRASSELLI: Methods for recovering gas chromatography and thin-layer chromatography fractions for infra red spectroscopy. In: Progress in infra red spectroscopy, vol. 3. New York: Plenum Press (in press).
50. CRIPPEN, R. C., and C. E. SMITH: Procedures for the systematic identification of peaks in gas-liquid chromatographic analysis. J. Gas Chromatog. **3**, 37–42 (1965).
51. PERRY, S. G.: Peak identification in gas chromatography. Chromatog. Rev. **9**, 1–22 (1967).
52. MERRITT, C., J. T. WALSH, D. A. FORSS, P. ANGELINI, and S. M. SWIFT: Wide range programmed temperature gas chromatography in the separation of very complex mixtures. Anal. Chem. **36**, 1502–1508 (1964).
53. HARRIS, W. E., and H. W. HABGOOD: Programmed temperature gas chromatography. London: John Wiley & Sons 1966.

54. Martin, A. J., C. E. Bennett, and F. W. Martinez: Linear programmed temperature gas chromatography. In: Gas chromatography, Second Internat. Symposium of the Instrument Society of America (H. J. Noebels, R. F. Wall, and N. Brenner, eds.), p. 363 – 373. New York: Academic Press 1959.
55. Mikkelsen, L.: Advances in programmed temperature gas chromatography. In: Advances in chromatography, vol. 2 (J. C. Giddings and R. A. Keller, eds.), p. 337 – 357. London: Edward Arnold Ltd. 1966.
56. Scott, R. P. W.: New horizons in column performance. In: Gas chromatography, 1964 (A. Goldup, ed.), p. 25 – 37. London: The Institute of Petroleum 1965.
57. Zlatkis, A., D. C. Fenimore, L. S. Ettre, and J. E. Purcell: Flow programming. A new technique in gas chromatography. J. Gas Chromatog. **3**, 75 – 81 (1965).
58. Scott, R. P. W.: Preparative scale chromatography with analytical columns. Nature **198**, 782 – 783 (1963).
59. Beroza, M., and R. A. Coad: Reaction gas chromatography. J. Gas Chromatog. **4**, 199 – 216 (1966).
60. Thompson, C. J., H. J. Coleman, R. C. Hopkins, and H. T. Rall: Hydrogenolysis – An identification tool. J. Gas Chromatog. **5**, 1 – 10 (1967).
61. Rowan, R.: Identification of hydrocarbon peaks in gas chromatography by sequential application of class reactions. Anal. Chem. **33**, 658 – 665 (1961).
62. Adlard, E. R., and B. T. Whitham: Behaviour of 5 A molecular sieve in subtractive gas chromatography. Nature **192**, 966 (1961).
63. Innes, W. B., W. E. Bambrick, and A. J. Andreatch: Hydrocarbon gas analysis using differential chemical absorption and flame ionisation detectors. Anal. Chem. **35**, 1198 – 1203 (1963).
64. Ikeda, R. M., D. E. Simmons, and J. D. Grossman: Removal of alcohols from complex mixtures during gas chromatography. Anal. Chem. **36**, 2188 – 2189 (1964).
65. Sixth Symposium on Pyrolysis and Reaction Gas Chromatography. Paris, 1966. Evanston, Ill.: Preston Technical Abstracts Company (in press).
66. McKinney, R. W.: Pyrolysis gas chromatography. A bibliography (1960 – 1963). J. Gas Chromatog. **2**, 432 – 436 (1964).
67. Perry, S. G.: Techniques and potentialities of pyrolysis-gas chromatography. J. Gas Chromatog. **2**, 54 – 59 (1964).
68. Merritt, C., and D. H. Robertson: The analysis of proteins, peptides and amino acids by pyrolysis-gas chromatography and mass spectrometry. J. Gas Chromatog. **5**, 96 – 102 (1967).
69. Oyama, V. A., and G. C. Carle: Pyrolysis gas chromatography application to life detection and chemotaxonomy. J. Gas Chromatog. **5**, 151 – 154 (1967).
70. Leplat, P.: Application of pyrolysis-gas chromatography to the study of the non-volatile petroleum fractions. J. Gas Chromatog. **5**, 128 – 135 (1967).
71. Levy, E. J., and D. G. Paul: The application of controlled partial gas phase thermolytic dissociation to the identification of gas chromatographic effluents. J. Gas Chromatog. **5**, 136 – 145 (1967).
72. Simon, W., P. Kriemler, J. A. Voellmin, and H. Steiner: Elucidation of the structure of organic compounds by thermal fragmentation. J. Gas Chromatog. **5**, 53 – 57 (1967).
73. Verzele, M.: Preparative scale gas chromatography. VII. Support material and column shape. J. Gas Chromatog. **4**, 180 – 189 (1966).
74. Desty, D. H., J. N. Haresnape, and B. H. F. Whyman: Construction of long lengths of coiled glass capillary. Anal. Chem. **32**, 302 – 304 (1960).
75. Ettre, L. S.: Open tubular columns in gas chromatography. New York: Plenum Press 1965.
76. Desty, D. H.: Capillary columns; trials, tribulations and triumphs. In: Advances in chromatography, vol. 1 (J. C. Giddings and R. A. Keller, eds.), p. 199 – 228. New York: Marcel Dekker, Inc. 1966.
77. Hollis. O. L.: Separation of gaseous mixtures using porous polyaromatic polymer beads. Anal. Chem. **38**, 309 – 316 (1966).
78. Verzele, M., and M. Verstappe: Preparative scale chromatography. V. Experiments with very long columns. J. Chromatog. **19**, 504 – 511 (1965).
79. Scott, R. P. W.: The construction of high-efficiency columns for the separation of hydrocarbons. In: Gas chromatography 1958 (D. H. Desty, ed.), p. 189 – 199. London: Butterworths 1958.

80. ADLARD, E. R.: An evaluation of some polyglycols used as stationary phases for gas liquid partition chromatography. In: Vapour phase chromatography, 1956 (D. H. DESTY, ed.), p. 98–113. New York: Academic Press 1957.
81. TAKÁCS, J., J. BALLA, and L. MÁZOR: The estimation of the highest operating temperature of stationary phases by means of a derivatograph. J. Chromatog. **16**, 218–220 (1964).
82. HAWKES, S. J., and E. F. MOONEY: Temperature limitations of stationary phases in gas chromatography. Anal. Chem. **36**, 1473–1477 (1964).
83. DURRETT, L. R.: Applications of Carbowax 400 in gas chromatography for extreme aromatic selectivity. Anal. Chem. **32**, 1393–1396 (1960).
84. PHILLIPS, C. S. G.: Gas chromatography instrumentation for the laboratory. In: Gas chromatography. First international symposium of the Instrument Society of America (V. J. COATES, H. J. NOEBELS, and I. S. FAGERSON, eds.), p. 51–63. New York: Academic Press 1958.
85. LITTLEWOOD, A. B.: The classification of stationary liquids used in gas chromatography. J. Gas Chromatog. **1**, 16–29 (1963).
86. LAZARRE, F., and S. ROUMAZEILLES: The polarity of liquid phases and the selectivity of chromatographic columns. Bull. Soc. Chim. France 3371–3373 (1965).
87. BROWN, I.: Identification of organic compounds by gas chromatography. Nature **188**, 1021–1022 (1961).
88. HORNING, E. C., W. J. A. VANDEN HEUVEL, and B. G. CREECH: Separation and determination of steroids by gas chromatography. In: Methods of biochemical analysis, vol. XI (D. GLICK, ed.), p. 69–147. New York: Interscience Publishers 1963.
89. KRUPPA, R. F., R. S. HENLY, and D. L. SMEAD: Improved gas chromatography packings with fluidized drying. Anal. Chem. **39**, 851–853 (1967).
90. HUYTEN, F. H., W. VAN BEERSUM, and G. W. A. RIJNDERS: Improvements in the efficiency of large diameter gas-liquid chromatography columns. In: Gas chromatography, 1960 (R. P. W. SCOTT, ed.), p. 224–241. London: Butterworths 1960.
91. MORGANTINI, M., and L. GUIDUCCI: Preparation of glass capillary columns and study of the conditions for the separation of some isomers of oleic acid. Riv. Ital. Sostanze Grasse **43**, 155–161 (1966).
92. HALÁSZ, I., and C. HORVATH: Columns for gas chromatography. German patent 1,183,716 (1964). Chem. Abstr. **62**, 9800 (1965).
93. — — Open tube columns with impregnated thin layer support for gas chromatography. Anal. Chem. **35**, 499–505 (1963).
94. ETTRE, L. S., J. E. PURCELL, and S. D. NOREM: Support-coated open tubular columns. J. Gas Chromatog. **3**, 181–185 (1965).
95. LIPSKY, S. R., W. J. MCMURRAY, and C. HORVATH: The analysis of complex organic compounds by fast electrical scanning high resolution mass spectrometry and gas chromatography. In: Gas chromatography, 1966 (A. B. LITTLEWOOD, ed.), p. 299–317. New York: Academic Press 1967.
96. LANDAULT, C., and G. GUIOCHON: Packed capillary columns in gas liquid chromatography. In: Gas chromatography, 1964 (A. GOLDUP, ed.), p. 121–139. London: The Institute of Petroleum 1965.
97. HALÁSZ, I., and E. HEINE: Packed capillary columns in gas chromatography. Anal. Chem. **37**, 495–500 (1965).
98. GRANT, D. W.: Packed capillary columns. Presented at 4th Wilkins Gas Chromatography Symposium, Manchester, England (1966).
99. DESTY, D. H., A. GOLDUP, and B. H. F. WHYMAN: The potentialities of coated capillary columns for gas chromatography in the petroleum industry. J. Inst. Petrol. **45**, 287–298 (1959).
100. MCTAGGART, N. G., and J. V. MORTIMER: The application of coated capillary columns to quantitative analysis in the petroleum industry. J. Inst. Petrol. **50**, 255–267 (1964).
101. SCHWARTZ, R. D., R. G. MATHEWS, and D. J. BRASSEUX: Resolution of complex hydrocarbon mixtures by capillary column gas chromatography – Composition of the 80°–180° C aromatic portion of petroleum. J. Gas Chromatog. **5**, 251–253 (1967).
102. ANTHEAUME, J., and G. GUIOCHON: Application of gas chromatography to a study of the middle fractions of a crude oil. Bull. Soc. Chim., France 298–307 (1965).
103. LUDWIG, F. J.: Analysis of microcrystalline waxes by gas-liquid chromatography. Anal. Chem. **37**, 1732–1737 (1965).

104. SOLO, A. J., and S. W. PELLETIER: Gas liquid chromatography of phenanthrenes. Anal. Chem. **35**, 1584–1587 (1963).

105. HISHTA, C., J. P. MESSERLY, and R. F. RESCHKE: Gas chromatography of high boiling compounds on low temperature columns. Anal. Chem. **32**, 1730–1733 (1960).

106. GUDZINOWICZ, B. J., and W. R. SMITH: High temperature gas liquid chromatography. Exploratory studies using an ionization detector chromatograph. Anal. Chem. **32**, 1767–1771 (1960).

107. JOHNS, R. B., T. BELSKY, E. D. MCCARTHY, A. L. BURLINGAME, P. HAUG, H. K. SCHNOES, W. RICHTER, and M. CALVIN: The organic geochemistry of ancient sediments – Part II. Geochim. Cosmochim. Acta **30**, 1191–1222 (1966).

108. ORÓ, J., D. W. NOONER, A. ZLATKIS, and S. A. WILKSTRÖM: Paraffinic hydrocarbons in the Orgueil, Murray, Mokoia, and other meteorites. In: Life sciences and space research, vol. IV, p. 63–100. Washington: Spartan Books 1966.

109. HILLS, I. R., and E. V. WHITEHEAD: Triterpanes in optically active petroleum distillates. Nature **209**, 977–979 (1966).

110. – –, D. E. ANDERS, J. J. CUMMINS, and W. E. ROBINSON: An optically active triterpane, gammacerane, in Green River, Colorado, oil shale bitumen. Chem. Comm. 752 (1966).

111. BURLINGAME, A. L., P. HAUG, T. BELSKY, and M. CALVIN: Occurrence of biogenic steranes and pentacyclic triterpanes in an Eocene shale (52 million years) and in an early Precambrian shale (2.7 billion years): A preliminary report. Proc. Nat. Acad. Sci. U.S. **54**, 1406–1412 (1965).

112. WOLLRAB, V., M. STREIBL, and F. ŠORM: Über die Zusammensetzung der Braunkohle. VI. Analyse von Wachskomponenten des Montanwachses mittels Höchsttemperatur-Gas-Verteilungschromatographie. Coll. Czech. Chem. Comm. **28**, 1904–1913 (1963).

113. WHITHAM, B. T.: Use of molecular sieves in gas chromatography for the determination of the normal paraffins in petroleum fractions. Nature **182**, 391–392 (1958).

114. EGGERSTEN, F. T., and S. GROENNINGS: Determinations of small amounts of n-paraffins by molecular sieve-gas chromatography. Anal. Chem. **33**, 1147–1150 (1961).

115. HRIVNAK, J., and V. PALO: Separation of C_2—C_{18} free fatty acids in dairy products by dual column programmed temperature gas chromatography. J. Gas Chromatog. **5**, 325–326 (1967).

116. ARCHIBALD, F. M., and V. P. SKIPSKI: Determinations of fatty acid content and composition in ultramicro lipid samples by gas liquid chromatography. J. Lipid Res. **7**, 442–445 (1966).

117. JAMES, A. T.: Methods of separation of long chain unsaturated fatty acids. Analyst **88**, 572–582 (1963).

118. HORNING, E. C., A. KARMEN, and C. C. SWEELEY: Gas chromatography of lipids. In: Progress in the chemistry of fats and other lipids, vol. VII (R. T. HOLMAN, ed.), p. 167–246. Oxford: Pergamon Press 1964.

119. MASSINGILL, J. L., and J. E. HODGKINS: Methyl esters of fatty acids on a QF-1 capillary column. J. Gas Chromatog. **3**, 110 (1965).

120. PALLOTTA, U., G. LASI, and C. ZORZUT: On the determination of elaidinic acid in crucifer plant oils. Riv. Ital. Sostanze Grasse **42**, 142–148 (1965).

121. COOPER, J. E.: Fatty acids in Recent and ancient sediments and petroleum reservoir waters. Nature **193**, 744–746 (1962).

122. KVENVOLDEN, K. A.: Molecular distributions of normal fatty acids and paraffins in some lower Cretaceous sediments. Nature **209**, 573–577 (1966).

123. NAGY, B., and Sister M. C. BITZ: Long chain fatty acids from the Orgueil meteorite. Arch. Biochem. Biophys. **101**, 240–248 (1963).

124. WILLIAMS, P. M.: Organic acids in Pacific Ocean waters. Nature, **189**, 219–220 (1961).

125. SLOWEY, J. F., L. M. JEFFREY, and D. W. HOOD: The fatty acid content of ocean water. Geochim. Cosmochim. Acta **26**, 607–616 (1962).

126. LEO, R. F., and P. L. PARKER: Branched-chain fatty acids in sediments. Science **152**, 649–650 (1966).

127. PARKER, P. L., C. VAN BALEN, and L. MAURER: Fatty acids in eleven species of blue-green algae: Geochemical significance. Science **155**, 707–708 (1967).

128. CASON, J., and D. W. GRAHAM: Isolation of isoprenoid acids from a California petroleum. Tetrahedron **21**, 471–483 (1965).

129. Eglinton, G., A. G. Douglas, J. R. Maxwell, J. N. Ramsay, and S. Stallberg-Stenhagen: Occurrence of isoprenoid fatty acids in the Green River Shale. Science **153**, 1133–1135 (1966).
130. Davison, V., and H. J. Dutton: Microreactor chromatography. Quantitative determinations of double bond positions by ozonization-pyrolysis. Anal. Chem. **38**, 1302–1305 (1966).
131. Blumer, M.: Personal communication.
132. Kuksis, A.: Newer developments in determination of bile acids and steroids by gas chromatography. In: Methods of biochemical analysis, vol. 14 (D. Glick, ed.), p. 325–454. London: John Wiley & Sons 1966.
133. Sjövall, J.: Gas liquid chromatography of bile acids. In: New biochemical separations (A. T. James and L. J. Morris, eds.), p. 65–79. London: D. van Nostrand Co., Ltd. 1964.
134. Carman, R. M., and D. E. Cowley: Diterpenoids. XII. Dundatholic acid. Australian J. Chem. **20**, 193–196 (1967).
135. Thomas, B. R.: The chemistry of the order Araucariales. Part 4. The bled resins of *Agathis australis*. Acta Chem. Scand. **20**, 1074–1081 (1966).
136. Brooks, T. W., G. S. Fisher, and N. M. Joye: Gas liquid chromatographic separation of resin acid methyl esters with a polyamide liquid phase. Anal. Chem. **37**, 1063–1064 (1965).
137. Chang, C. W. J., and S. W. Pelletier: Gas chromatographic study of the separation of resin acid methyl esters on a QF-1 column. Anal. Chem. **38**, 1247–1248 (1966).
138. Hudy, J. A.: Resin acids. Gas chromatography of their methyl esters. Anal. Chem. **31**, 1754–1756 (1959).
139. Cruikshank, P. A., and J. C. Sheehan: Gas chromatographic analysis of amino acids as N-trifluoroacetylamino acid methylesters. Anal. Chem. **36**, 1191–1197 (1964).
140. Makisumi, S., and H. A. Saroff: Preparation, properties and gas chromatography of the N-trifluoroacetyl esters of the amino acids. J. Gas Chromatog. **3**, 21–27 (1965).
141. Karmen, A., and H. A. Saroff: Gas-liquid chromatography of the amino acids. In: New biochemical separations (A. T. James and L. J. Morris, eds.), p. 81–92. London: D. van Nostrand Co., Ltd. 1964.
142. Potteau, B.: The analysis of amino acids and lower peptides after conversion to volatile derivatives. Recent applications of gas chromatography. Bull. Soc. Chim. France 3747–3756 (1965).
143. Weinstein, B.: Separation and determination of amino acids and peptides by gas-liquid chromatography. In: Methods of biochemical analysis, vol. 14 (D. Glick, ed.), p. 203–323. London: John Wiley & Sons 1966.
144. Halász, I., and K. Bünnig: Darstellung von N-Trifluoracetylaminosäure-Alkylestern für die gaschromatographische Trennung in Capillarkolonnen. Z. Anal. Chem. **211**, 1–5 (1965).
145. Bitz, Sister M. C., and B. Nagy: Ozonolysis of "polymer type" material in coal, kerogen, and in the Orgueil meteorite: A preliminary report. Proc. Nat. Acad. Sci. **56**, 1383–1390 (1966).
146. Landa, S., and L. Vodicka: Ozonisation des naturalen Anteiles des Mittelöls aus Braunkohlenschwelteer. Brennstoff-Chem. **43**, 366–371 (1962).
147. Hansen, E. H., and M. Schnitzer: Nitric acid oxidation of Danish illuvial organic matter. Soil Sci. Soc. Am. Proc. **31**, 79–85 (1967).
148. Richey, J. M., H. G. Richey, and R. Schraer: Quantitative analysis of carbohydrates by gas-liquid chromatography. Anal. Biochem. **9**, 272–280 (1964).
149. Oates, M. D. G., and J. Schrager: The use of gas-liquid chromatography in the analysis of neutral monosaccharides in hydrolysates of gastric mucopolysaccharides. Biochem. J. **97**, 697–700 (1965).
150. Vilkas, M., Jan-I-Hui, and G. Boussac: Chromatographic analysis of sugars as the trifluoroacetates. Tetrahedron Letters 1441–1446 (1966).
151. Jurriens, G., and A. C. J. Kroesen: Determination of glyceride composition of several solid and liquid fats. J. Am. Oil Chem. Soc. **42**, 9–14 (1965).
152. Lichfield, C., R. D. Harlow, and R. Reiser: Quantitative gas-liquid chromatography of triglycerides. J. Am. Oil Chem. Soc. **42**, 849–857 (1965).
153. Kuksis, A.: Gas-liquid chromatography of glycerides. J. Am. Oil Chem. Soc. **42**, 269–275 (1965).
154. Wellburn, A. R., and F. W. Hemming: Gas liquid chromatography of derivatives of naturally occurring mixtures of long chain polyisoprenoid alcohols. J. Chromatog. **23**, 51–60 (1966).

155. DANIELS, D. G. H.: Gas chromatography of long chain α,ω-diols and related compounds. J. Chromatog. **21**, 305–306 (1966).
156. MORRISON, R. I., and W. BICK: Long chain methyl ketones in soils. Chem. Ind. (London) 596–597 (1966).
157. GIL-AV, E., R. CHARLES-SIGLER, G. FISCHER, and D. NUROK: Resolution of optical isomers by gas liquid partition chromatography. J. Gas Chromatog. **4**, 51–58 (1966).
158. –, B. FEIBUSH, and R. CHARLES-SIGLER: Separation of enantiomers by gas liquid chromatography with an optically active stationary phase. In: Gas chromatography, 1966 (A. B. LITTLEWOOD, ed.), p. 227–239. New York: Academic Press 1967.
159. FEIBUSH, B., and E. GIL-AV: Gas chromatography with optically active stationary phases. Resolution of primary amines. J. Gas Chromatog. **5**, 257–260 (1967).

Addendum

The Journal of Gas Chromatography has been renamed the Journal of Chromatographic Science (since January 1969) and now publishes contributions on all aspects of chromatography. Preston Technical Abstract's punched card data are now available as bound volumes (from January 1968). A new journal named Chromatographia, publishing on all aspects of chromatography has appeared. A recent book entitled, "The Practice of Gas Chromatography" edited by L. E. ETTRE and A. ZLATKIS contains a useful ten page section in which text books and symposia publications available in 1967 are listed together with a list of relevant review articles, journals, organisations etc.

CHAPTER 6

Isolation Procedures for Kerogens and Associated Soluble Organic Materials

W. E. ROBINSON

Laramie Petroleum Research Center,
U.S. Bureau of Mines.
Laramie, Wyoming

Contents

I. Introduction

Kerogens are of geochemical interest because they represent one of the most abundant forms of carbonaceous materials. The term "kerogen", or "oil-former", was originally given to the organic matter in oil shales, torbanite, kukersite, and others. Recently, broader usage has been made of the term in describing the insoluble organic material present in nonreservoir sedimentary rocks and other rocks. The soluble organic materials associated with oil-shale kerogens are of particular interest because they are usually indigenous to the formation and have had little opportunity for migration because of the low porosity and low permeability of the formations.

DOWN and HIMUS [1] published a classification of shales and coals based upon a number of properties including the yield of oil on distillation. By this classification, the term "kerogen rocks" describes sedimentary deposits containing insoluble organic kerogen which on distillation yields an oil equivalent to more than

50 percent of the organic content of the rock. Similarly, the term "kerogen coals" describes sedimentary materials containing organic matter which upon distillation yields an oil equivalent to more than 10 percent and less than 50 percent of the organic content.

In the present discussion, the term "kerogen" will be used to describe the insoluble organic material present in kerogen rocks. The term "bitumen" will be used to describe the soluble organic material present in the kerogen rock based upon solubility in a hydrocarbon solvent. Other nomenclature will be self-explanatory. For example, the material solubilized by methanol will be referred to as "methanol-soluble material".

The discussion covers the isolation of kerogen, bitumen, fatty acids, porphyrins, amino acids, and carbohydrates from kerogen rocks. The procedures as described are intended to be informative without any implication of recommended usage. Other methods are applicable and in some circumstances may be preferred. Likewise, the articles reviewed were chosen for the subject matter discussed and no attempt has been made to completely review the subject.

II. Isolation of Kerogen

One important problem in kerogen constitutional studies is the isolation of unaltered kerogen from its associated mineral. The problem of the isolation of kerogen is complicated by extreme differences in mineral content and mineral composition of the kerogen rock. Some kerogen rocks contain as little as 5 percent mineral while others may contain as much as 95 percent mineral. Some sediments are highly argillaceous while others highly calcareous. Some kerogen rocks contain large amounts of pyrite but others contain little pyrite. Pyrite is extremely difficult to separate from the organic material because of an apparent attraction for the kerogen. Thus, it is understandable that the best available separation techniques do not find universal application.

Separation techniques should ideally remove all mineral matter from the organic kerogen without fractionating the kerogen into dissimilar components or altering the kerogen chemically. At the present time this has not been accomplished. The methods to be described consist of the modified Quass, sink-float, acid digestion, and pyrite removal. In studies of oil shale at the Bureau of Mines laboratory, techniques other than those to be described such as electrostatic, magnetic, ultrasonic, and flotation separations have been tried with only limited success.

Hand-picked samples of oil shale from the Bureau of Mines demonstration mine near Rifle, Colorado, are crushed to 1- to 2-inch pieces by using a large-size jaw crusher. The 1- to 2-inch pieces of sample are crushed to 8 mesh or smaller in a small-size jaw crusher. The 8-mesh oil-shale sample is then crushed to 100 mesh or smaller in a hammer mill or more recently in a disc mill, consisting of crushing discs contained in a barrel that vibrates in an eccentric manner. This method quickly crushes the oil shale to very small particle sizes. The possibility exists that minor degradation or oxidation of the organic material in the kerogen rock may occur during the crushing procedure.

A. Modified Quass Method

The method as used by QUASS [2] on South African kerogen rock is an adaptation of the amalgamation process used for coals. The method efficiently reduced the ash content of the South African kerogen rock from 40 percent to about 5 percent. HIMUS and BASAK [3] used the method on kerogen rocks and obtained significant mineral removal from only one of four different samples. The modified Quass method as used successfully on Green River Formation kerogen rock is described by SMITH and HIGBY [4].

This method is based upon the principles of differential wetting of the organic kerogen and the inorganic mineral by two immiscible liquids such as oil and water. The organic kerogen is wet by the organic liquid phase and the mineral is wet by the water phase resulting in the kerogen being retained in the oil phase and the mineral being released into the water phase. Periodic changing of the water phase results in mineral reduction. New surfaces on the oil-kerogen phase are created by some form of kneading, mixing, or grinding action.

Prior to this concentration procedure, the kerogen rock is crushed to pass a 100-mesh screen, then extracted with benzene to remove most of the soluble organic material in the kerogen rock. After air drying, the extracted and crushed rock is leached with dilute acid to remove mineral carbonates.

The next step in the procedure is the preparation of a pastelike ball of the kerogen rock and an oily phase. Considerable practice is required in preparing a paste ball that will retain its adhesive nature in the water phase. This is done by mixing the extracted and acid-leached kerogen rock with an oily phase in the container of an attrition grinder. Sufficient oily phase is stirred with the crushed rock to form a paste with the consistency of a light putty. Mineral oil or n-hexadecane have proven satisfactory for the oily phase; however, numerous other organic materials may be useful. The n-hexadecane has the advantage over mineral oil of contributing only one impurity of known composition in case the oily phase cannot be completely removed from the kerogen.

Water is added to the container along with ceramic balls and a grinding and mixing action is started. The mineral is preferentially wet by the water and tends to accumulate in the water phase. The organic kerogen is preferentially wet by the oily phase and tends to be retained in the oily phase. Periodic removal of and replenishing the water phase reduces the mineral content of the kerogen rock. The mineral "fallout" is quite rapid during the early part of the grinding; consequently, the water phase should be changed frequently during the first few hours of grinding. This process is continued until a minimum ash value is obtained for the sample. With some kerogen rocks, the disappearance of the X-ray diffraction peak for soda feldspar can be followed to show the reduction in mineral content.

At the end of the grinding, the water phase is removed by suction and sufficient acetone is added to destroy the adhesive nature of the oily phase and the kerogen. The kerogen concentrate, oily phase, and the acetone are removed from the grinding balls by suction after which the acetone and some of the oil is removed from the kerogen concentrate by filtration. The sample is air dried and extracted with benzene to remove the remaining oily phase from the kerogen concentrate. The sample is then dried at 60° C under reduced pressure. After drying, the sample is washed with hot water first by decanting and finally by extraction in a soxhlet extractor for 24 hours. The sample is then dried at 60° C under reduced pressure, flushing occasionally with nitrogen.

A comparison of the decrease in mineral content and other data obtained for 11 kerogen rocks is shown in Table 1. Reduction in ash content amounted to 95 percent or more for two of the kerogen rocks. In only one kerogen rock was the reduction in ash content less than 50 percent. There was considerable variation in the amount of the total organic material represented in the concentrate, ranging from 24 to 99 percent for the Brazilian and South African kerogen rocks, respectively. The assay oil/carbon ratios varied from 0.43 to 0.95 for the 11 kerogens which ranged in age from 12 to 300 million years. The oil/carbon ratios were not

Table 1. *Approximate age, assay oil to organic carbon ratios, and the decrease in ash content by a concentration procedure for 11 oil shales*

Name and location	Approximate age (million years)	Wt percent assay oil to wt percent organic carbon ratio*	Ash before concentration (percent)	Decrease in ash content by a concentration procedure (percent)**
Alaska (Howard Pass)	120–215	0.95	34.1	96
Argentina (San Juan)	190–215	0.61	82.6	58
Brazil (Sao Paulo)	12	0.52	75.0	83
Canada (New Glasgow)	300	0.44	84.0	60
Colorado (Piceance Ck)	60	0.85	65.7	80
France (St. Hilaire)	215	0.43	66.3	73
New Zealand (Orepuki)	60	0.54	32.7	70
Oregon (Shale City)	40	0.70	48.3	25
Scotland (Dunnet)	250	0.67	77.8	76
South Africa (Ermelo)	215–300	0.72	33.6	96
Spain (Puertollano)	215–300	0.68	62.8	59

* Reported by ROBINSON and DINNEEN [5].
** Concentrated by the Modified Quass Method [4].

related to age but undoubtedly are related to the molecular structure of the kerogens. The kerogens with low oil/carbon ratios have high contents of condensed aromatic structures that produce low yields of oil and high yields of carbon residue upon pyrolysis. On the other hand, kerogens with high oil/carbon ratios have high contents of oil-producing materials such as aliphatic and alicyclic structure.

The main advantage of this concentration procedure is the absence of drastic chemical treatment, producing essentially a chemically unaltered kerogen concentrate. However, the possibility of fractionating the kerogen of some kerogen rocks does exist. A disadvantage of the concentration procedure is the time required to prepare the concentrate, which is usually several days. Also, the mineral reduction for some kerogen rocks is very low and with most kerogen rocks the reduction in mineral content is not complete. However, the method is ideal for some kerogen rocks, for example Alaskan and South African kerogen rocks (Table 1).

B. Sink-Float Method

The sink-float method of concentrating kerogen takes advantage of differences in the specific gravity of the mineral components and the organic kerogen. For example, Green River Formation kerogen has a specific gravity of 1.07 [6], and the mineral components have a range of specific gravity from about 2 to 5. Separation is accomplished by centrifugation in a dense liquid medium where the lightweight kerogen tends to float and the minerals tend to sink.

The sink-float technique was used by LUTS [7] in lowering the ash content of an "oil shale" kerogen rock sample to about 5 percent. In this procedure, finely powdered rock is suspended in a solution of $CaCl_2$ having a specific gravity of 1.06 to 1.15. The mixture is centrifuged at 3,000 rpm until most of the organic material rises to the surface of the liquid and the mineral sinks to the bottom of the container. The organic material is filtered and washed free of $CaCl_2$.

HUBBARD *et al.* [8] centrifuged kerogen rock in mixtures of carbon tetrachloride and benzene having specific gravities from 1.15 to 1.40. In this method the kerogen rock, ground to pass a 100-mesh screen, is extracted with benzene to remove benzene-soluble material and leached with dilute acid to remove mineral carbonates. The kerogen rock, extracted, leached, and dried, is suspended by stirring in a 1.40-density carbon tetrachloride-benzene medium, then centrifuged. The floating material is removed from the centrifuge tube by aid of suction, then filtered, washed with acetone, and dried. The material that sinks is treated similarly. Portions of the "float" sample and the "sink" sample are retained for analyses. The remainder of the "float" sample from the first centrifugation is suspended in a 1.20-density carbon tetrachloride-benzene medium and centrifuged in a similar manner. Likewise, a portion of the material that floated on the 1.20-density medium is suspended in a 1.15-density carbon tetrachloride-benzene medium and centrifuged. By this procedure, a total of three "float" samples and three "sink" samples of different kerogen concentration are obtained.

The first float concentrate, representing 39 percent of the kerogen, contained 27 percent ash. The second float concentrate, representing 6 percent of the kerogen, contained 14 percent ash. The final float concentrate with 9 percent ash represented only 1 percent of the total kerogen. A total of 12 percent of the kerogen was obtained in concentrates having 15 percent or less ash.

The H/C ratios and the assay oil/organic carbon ratios of the sink and float concentrates suggest that little or no fractionation of the kerogen occurred. For example, the H/C ratios of the concentrates had maximum deviation of only 1.2 percent and the assay oil/C ratios of the concentrates had maximum deviation of only 3.4 percent. If the kerogen had been composed of materials of widely different compositions and these materials had been selectively fractionated by the concentration procedure, greater differences would have been obtained for the two properties.

The advantages of this concentration method are (1) the kerogen is not altered chemically, and (2) a concentrate of less than 10 percent mineral can be obtained. Elemental analyses can be determined easily at various stages of concentration; therefore, it is easy to extrapolate to zero ash content and obtain elemental compositions of the mineral-free kerogen. The disadvantages of the concentration method are (1) low yields of good concentrate are obtained, and (2) some kerogens may be fractionated.

C. Acid Treatment

One commonly used method of preparing kerogen concentrates is acid digestion using hydrochloric acid, hydrofluoric acid or combinations of these acids. The acid digestion method as described by SMITH [6] consists of treating the finely ground and benzene-extracted rock in a polyethylene or other resistant container with sufficient methyl alcohol to form a thick paste. The purpose of the alcohol treatment is to wet the sample so that good contact between acid and rock is achieved. Mineral carbonates are removed by treatment of the moist rock with a dilute solution of hydrochloric acid (1 part concentrated HCl to 9 parts water) at room temperature until evolution of CO_2 ceases. The suspension is filtered and the residue is washed with water. The residual rock is treated then with 1:1 concentrated HCl (37 percent) and water at room temperature for 1 hour followed by filtering and washing with water to remove mineral carbonates. The washed residue is treated with a 1:1 mixture of concentrated HCl (37 percent) and HF (48 percent) and evaporated to dampness on low heat of about 100° C to volatilize the reaction products of the silicates. With most rocks it is necessary to repeat the acid treatment and evaporation. The damp residue is suspended in saturated boric acid solution to prevent the formation of insoluble fluorides and then filtered. This procedure is repeated and the residue is washed finally with water. The residue is extracted then with 1:1 HCl and H_2O, filtered, and washed with water. The residue is suspended in hot water and washed successively by decanting and filtering until the filtrate is free of chloride ions. The kerogen concentrate is dried at

80° C under reduced pressure. With Green River oil shale, the concentrate from this procedure contains 3 to 5 percent mineral of which about 80 percent of the mineral is pyrite.

The advantages of this method are (1) most of the minerals (except pyrite) are removed from the kerogen rock, (2) there is little opportunity for fractionation of the organic material, and (3) the method is rapid. The main disadvantage of the method is the possible alteration of the organic kerogen by the action of strong mineral acids.

D. Pyrite Removal

Pyrite is one constituent of the rock mineral that is difficult to remove from the kerogen by physical methods because of an apparent attraction of the pyrite for the kerogen. Pyrite can be removed by nitric acid digestion; however, the nitric acid oxidizes and nitrates the kerogen. Reduction with zinc and hydrochloric acid removes the pyrite slowly but the extent of reduction of functional groups present in the kerogen is not easily predicted. LAWLOR *et al.* [9] removed pyrite quantitatively by reduction using lithium aluminium hydride, resulting in specific alteration of kerogen functional groups.

The latter method consists of treating finely crushed, benzene-extracted, and carbonate-free kerogen rock with a tetrahydrofuran solution of $LiAlH_4$. Five parts of sample to one part of $LiAlH_4$ is placed in a flask equipped with a reflux condenser. A volume of tetrahydrofuran equivalent to 7.5 ml per gram of sample is slowly introduced. The reaction mixture is refluxed for 30 minutes, cooled to room temperature, and vacuum filtered, stopping the filtration while the residue is still moist. The moist residue is transferred as quickly as possible in small portions to a beaker containing water. The water reacts with and destroys the excess $LiAlH_4$. The resulting mixture is acidified with 1 N HCl, heated to boiling, and filtered. As a result of the acid treatment the pyritic sulfur is evolved as H_2S and the pyritic iron remains in solution as ferrous ions. The lithium and aluminium of the complex are dissolved by treating the residue sufficiently with acid solution to assure complete removal of aluminium ions, as determined by testing the washings with ammonium hydroxide. The product is washed free of chloride ions by decanting and filter washing with hot water and is finally dried under vacuum at 60° C.

Lithium aluminium hydride treatment successfully removed pyrite from a carbonate-free kerogen rock and a kerogen concentrate containing originally 3.1 and 5.3 percent pyrite, respectively. Quantitative removal of pyrite from the samples was shown by chemical analysis for pyrite by a modification of the Mott method [10]. The disappearance of the major X-ray diffraction peak for pyrite in each of the samples substantiated the conclusion that the removal of pyrite was complete.

III. Isolation of Bitumen

An important and interesting phase of a study of a kerogen rock is the investigation of the composition of the soluble material associated with the kerogen. Currently, the geochemical aspects of this type study are of considerable interest because of the emphasis placed on finding new petroleum source beds by the aid of geochemical information and techniques. Most "oil shale" kerogen rocks are unique and are of particular interest because the organic material is indigenous to the formation with little opportunity to migrate to or from other locations.

Many different techniques for the extraction of soluble organic material from kerogen rocks have been devised and are being used [11–14]. For example, some investigators use a range of solvents and extract the rock successively with solvents of increasing polarity. This tends to separate the soluble material into fractions of different degrees of polarity. Other investigators use mixed solvents, usually a hydrocarbon solvent plus a polar solvent. These extractions provide quantities of polar materials not usually removed with hydrocarbon solvents.

FERGUSON [15] studied variables in the extraction of bitumen from sediments. Yields of bitumen roughly doubled when particle size of the sample was reduced from 30 to 10 microns by hammer-mill crushing. An extraction time of 24 hours was considered sufficient to remove more than 99 percent of the bitumen. Benzene was found to be more satisfactory for removing hydrocarbons than mixed solvents. A specially designed extraction apparatus where the sample was continuously stirred in boiling solvent was more efficient that a soxhlet extractor. BAKER [16] used the techniques of FERGUSON in studies of extracts from sediments of the Cherokee Group. CUMMINS and ROBINSON [17] extracted bitumen from Mahogany zone kerogen rock of the Green River Formation. ROBINSON *et al.* [18] studied the bitumen obtained from sections of a 900-foot core of the Green River Formation. BURLINGAME *et al.* [19] isolated steranes and triterpanes from Green River Formation kerogen rock and a Precambrian shale. TOURTELOT and FROST [20] studied the extractable organic material in nonmarine and marine shales of Cretaceous age. HILLS *et al.* [21] identified gammacerane, a pentacyclic triterpane, in bitumen of the Green River Formation kerogen rock. EGLINTON *et al.* [22] extracted hydrocarbons of biological origin from a 1-billion-year-old sediment. ROBINSON and DINNEEN [5] extracted bitumen from 12 different kerogen rocks of different age, different environment and of different organic content. ORÓ *et al.* [23] isolated and identified n-alkanes ranging from C_{14} to C_{37} compounds in fungal spores. Odd-carbon-numbered compounds predominated in the C_{25} to C_{29} range. CALVIN and MCCARTHY [24] isolated the C_{17} isoprenoid hydrocarbon (2,6,10-trimethyltetradecane) from an ancient sediment.

A. Extraction Technique

The method to be described is a simple benzene extraction of the kerogen rock. Variations of the extraction technique, used by other investigators, are satisfactory and for certain applications may be more desirable. Several investigators have made use of ultrasonic vibration in the extraction process.

The kerogen rock, crushed to pass a 100-mesh screen, is placed in an extractor of appropriate size. For samples up to 500 grams, a soxhlet extractor of size suitable for the sample is used. Considerable channeling occurs in the large extraction thimbles and it is desirable to periodically stop the extraction and stir the sample. Because of this channeling, it is best to use small samples. Equipment similar to that described by FERGUSON [15] would be desirable. Filter aids such as sand or other suitable materials can be used where analysis of the residual kerogen rock is not to be determined. For samples of several kilograms, it has been found convenient to extract in 20 liter carboys at room temperature with constant stirring. Three or four charges of fresh solvent removes most of the soluble material.

After placing the sample in a suitable extractor fresh purified benzene is added to the extractor. The extraction process is started and continued until very little colored material is removed in the extract. This usually is nearly complete in 8 hours, but, in most cases it is convenient to let the extraction continue for 24 hours. The solvent is distilled from the extract at reduced pressures.

B. Fractionation Technique

A modification of the fractionation technique used by CUMMINS and ROBINSON [17] follows. The crude extract is dissolved in a 40 to 1 volume ratio of n-pentane to sample and allowed to stand overnight at 0° C and then filtered. The pentane-insoluble material removed by filtration is washed with a small quantity of cold pentane (0° C) and is then dried.

One part of pentane-soluble material is placed on a prewetted column of 25 parts of alumina (Alcoa, F 20, 80–200 mesh, extracted with pentane, activated 400° C for 2 hours) and eluted with purified solvents. The column is exhaustively eluted first with pentane, followed by benzene, then a mixture of 10 percent methanol and 90 percent benzene. Each of the eluted fractions is stripped free of solvents and dried. The pentane-eluted material is referred to as hydrocarbon concentrate, the benzene-eluted materials as resins I and the benzene-methanol-eluted material as resins II.

One part of hydrocarbon concentrate is placed on a prewetted column of 25 parts of silica gel (Davidson, 200 mesh, extracted with pentane, dried at 100° C for 2 hours) and eluted with purified solvents. The column is exhaustively eluted first with isooctane, followed by benzene and finally 2-propanol. These fractions contain mainly alkane, aromatic, and polar compounds.

The aromatic oils were further fractionated on a column of alumina eluting successively with isooctene, benzene, and isopropanol. The alkanes are fractioned into normal alkanes and isoalkanes plus cycloalkanes on molecular sieves (Linde, 5 A, extracted with pentane, activated at 240° C under vacuum overnight [25]. The method consists of placing the alkanes on the sieves with hot isooctane and subsequently removing the trapped n-alkanes with n-pentane at room temperature.

When large quantities of branched-cyclic alkanes are prepared, the material is fractionated into 10° C cuts in a molecular still at 10^{-3} Torr pressure starting at room temperature. The first few cuts, being the lowest molecular weight material, are redistilled in a spinning-band distillation column at 1 Torr pressure. Fractions from the molecular distillation, having boiling points too high for the spinning-band distillation, are placed on a column of alumina (200 parts alumina to 1 part sample) and eluted with n-pentane, collecting 25 ml fractions of the eluting material. The solvent is removed from each fraction by evaporation on a hot-water bath, using dry nitrogen to speed evaporation. The resulting fractions from the vacuum distillation and the fractions from the alumina columns are suitable for trapping from GLC columns. The trapped fractions, from repeated trappings, are identified or classified by utilizing mass, infrared, and NMR spectral analysis.

Thermal diffusion [17] has been used in lieu of molecular distillation for fractionation of the isoalkanes and cycloalkanes. Although thermal diffusion offers some advantage in type separation, it is much more time-consuming than molecular distillation.

IV. Isolation of Acids

The presence of fatty acids in recent and ancient sediments is of geochemical interest, contributing evidence for the organic origin of petroleum and other fossil fuels. It is of special interest to study the relationship of fatty acids and the corresponding n-alkanes in the sediment. The presence of other alkane acids, such as the iso- and anteiso-acids, isoprenoid acids, and resin acids, are also of interest from geochemical considerations.

Fatty acids up to C_{35} were identified by MEINSCHEIN and KENNEY [13] in an aerobic soil sample. COOPER [26] identified fatty acids ranging from C_{14} to C_{30} from recent and ancient sediments. A mechanism explaining the conversion of even acids to odd acids and odd alkanes, was proposed by COOPER and BRAY [27]. ABLESON *et al.* [28] extracted fatty acids from several sources including the Green River Formation where fatty acids up to C_{18} were identified. LAWLOR and ROBINSON [29] identified C_{10} to C_{35} fatty acids from Mahogany zone oil shale of the Green River Formation. The distribution of Fatty acids in some Lower Cretaceous sections of the Powder River Basin was studied by KVENVOLDEN [30]. CASON and GRAHAM [31] isolated C_{14}, C_{15}, C_{19}, and C_{20} isoprenoid acids from a petroleum crude. EGLINTON *et al.* [32] isolated C_{19} and C_{20} chain isoprenoid acids from Green River Formation kerogen rock. Branched-chain fatty acids were isolated from marine sediments and from Green River Formation oil shale by LEO and PARKER [33].

A. Extraction Techniques

In a technique described by LAWLOR and ROBINSON [29] a sample of 100-mesh kerogen rock is treated with dilute acid to remove mineral carbonates and to convert salts of acids to free acids. The residual rock is treated then with 8 percent methanolic KOH at reflux temperature for 18 hours. The aqueous solution of the potassium salts of the free acids, acids formerly present as salts, and acids formerly present as esters is separated from the insoluble residue by filtration. Water and carbon tetrachloride are added to the filtrate and the mixture is shaken vigorously. By this treatment, the potassium salts of the acids are concentrated in the water phase and the nonacidic material is concentrated in the carbon tetrachloride phase. After separation of the water phase from the carbon tetrachloride phase, the water phase is acidified with HCl, and the free acids are removed by extraction with carbon tetrachloride. The solvent is removed leaving the acid product as a residue. The crude acids were converted to methyl esters using methanol and boron trifluoride.

In another technique, PARKER [34] extracted the rock sample with methanol while the sample was stirred and subjected to ultrasonic vibrations for 15 minutes. The methanol extract is recovered by filtration. The residue is stirred with chloroform and subjected to ultrasonic vibrations. The extracts removed by methanol and by chloroform are combined and saponified. The recovered acids are converted to methyl esters using methanol and boron trifluoride. This method does not saponify esters associated with any insoluble kerogen that may be present.

B. Fractionation Techniques

The methyl esters of fatty acids may be concentrated by adduction with urea. The straight-chain esters, the branched-chain esters or the total methyl ester fraction may be analyzed by gas-liquid chromatography (GLC). Comparison of retention times of unknown constituents with known standards are utilized for component identification. Trapped peaks from the GLC are analyzed by mass, infrared, and NMR spectra to provide positive identification of the constituents.

MCCARTHY and DUTHIE [35] described a method for the separation of free fatty acids (FFA) from other lipids using a column of silicic acid. Coarse particles of silicic acid are selected by suspending 100 g of the silicic acid in 400 ml of methanol and decanting and discarding the silicic acid not settling in 5 minutes. This is repeated once with methanol and once with 400 ml of acetone. The silicic acid is rinsed with diethyl ether and permitted to air dry.

Five grams of prepared silicic acid are mixed with 10 ml of isopropanol-KOH (50 mg KOH per ml) and 30 ml of diethyl ether. After standing 5 minutes, the silicic acid is slurried into a glass column and washed with 100 ml of diethyl ether.

The sample, dissolved in a small quantity of diethyl ether, is placed on the silicic acid and thoroughly washed into the packing by several small portions of diethyl ether. Sterols, sterol esters, and glycerides are eluted as one fraction with 100 to 150 ml of diethyl ether at a flow rate of 5 ml per minute. The FFA are removed from the column with 50 ml of 2 percent formic acid in diethyl ether (1:1) followed by 75 to 100 ml of diethyl ether. If phospholipids are present, they are removed from the column by methanol.

Attempts to isolate alcohols in a hydrolysate of kerogen by utilizing thin-layer chromatography with silica gel were made by LAWLOR and ROBINSON [36]. They also used thin-layer chromatography to purify methyl esters of kerogen acids prior to mass spectral analyses.

V. Isolation of Porphyrins

The study of fossil pigments as well as the study of other biological fossils provide information on the modification of organic debris under geological conditions. TREIBS [37] undertook a systematic search for well-defined organic pigments in rocks and found that extracts of "oil shale" kerogen rocks, as old as Triassic, exhibited spectral characteristics similar to those of complex metal salts of porphyrins. This was a particularly important discovery because porphyrins are regarded as remnants of biological precursors, such as chlorophyll and hemin.

More recently porphyrins have been studied in various other sediments and in petroleums. GROENNINGS [38] developed a technique for the quantitative determination of porphyrin aggregates in petroleum. The extraction reagent was a saturated solution of hydrogen bromide in glacial acetic acid. The porphyrin content of extracts from Green River "oil shale" kerogen rock was determined by MOORE and DUNNING [39]. SUGIHARA and MCGEE [40] isolated a crystalline porphyrin from gilsonite of the Green River Formation. BLUMER and OMENN [41] isolated porphyrins and chlorins from a Triassic "oil shale" kerogen rock of Serpiano, Switzerland. A comparison of porphyrins in "oil shale" kerogen rock, in shale oil, and in petroleum was made by MORANDI and JENSEN [42]. BAKER [43] extracted porphyrins from gilsonite and petroleum asphaltenes using methanesulfonic acid. HODGSON *et al.* [44] reviewed the role of porphyrins in petroleum, oil sands, oil shales, coal, and sedimentary rocks.

The method of extracting porphyrins from kerogen rocks consists of either (1) extracting the soluble material from the rock prior to treatment of the extract with a porphyrin extraction reagent, or (2) extraction of the rock directly with the porphyrin extraction reagent.

MOORE and DUNNING [39] used the first technique. The kerogen rock is crushed and is extracted successively with different boiling solvents. The first solvent is benzene followed by benzene-methanol azeotrope (60 percent benzene-40 percent methanol), benzene-chloroform azeotrope (87 percent chloroform-13 percent methanol), methanol and finally pyridine. The extraction is continued until the extract is colorless and shows no fluorescence under ultraviolet light. Each extraction requires 72 to 96 hours. The extracts are isolated by distilling off most of the solvent at atmospheric pressure followed by heating at 50° C under reduced pressure. The porphyrin aggregate from each extract is isolated by method of GROENNINGS [38].

MORANDI and JENSEN [42] utilized the second technique and extracted the porphyrins directly from the kerogen rock with the Groennings extraction reagent. Before the procedure could be used, it was necessary to treat the rock with 10 percent HCl to remove mineral carbonates.

BLUMER and OMENN [41] extracted the kerogen rock with 1:1 benzene-methanol mixture and isolated the porphyrins and chlorins by solvent partitioning using different concentrations of HCl. By this method, about 1.6 kg of the kerogen rock is exhaustively extracted with a benzene-methanol mixture followed by evaporation of the solvent. The dried extract is redissolved in a minimum amount of chloroform and the asphaltenes are precipitated by adding hot isooctane. The precipitate is filtered and washed with a mixture of chloroform and isooctane. The filtrate is evaporated to dryness and redissolved in warm diethyl ether. The ether solution is extracted successively with 2.4 N HCl, 4.9 N

HCl, and finally with 6.1 N HCl. The 2.4 N HCl extract is neutralized with sodium acetate and the porphyrin pigment is removed by ether extraction. The 4.9 and 6.1 N HCl extracts are diluted to 1.0 N acidity and the chlorins are recovered by ether extraction.

VI. Isolation of Amino Acids

Proteins and nucleoproteins are universal components of all living organisms. The isolation and identification of proteinaceous materials in sedimentary rocks is therefore of geochemical interest.

The amino acid content of soils has been studied by various investigators. STEVENSON [45] examined soils and obtained quantitative analyses of hydrolysates from the samples. ERDMAN *et al.* [46] made a comparative study of amino acids present in a recent marine deposit and those present in an Oligocene marine mud. VALLENTYNE [47] reviewed the occurrence of amino acids, purines, and pyrimidine in lakes, oceans, sewage, and terrestrial soils. SWAIN [48] analyzed Paleozoic rocks for amino acids.

A. Extraction Technique

STEVENSON [45] described a method for extracting amino acids from soil samples. A 50 g sample of finely ground, air-dry soil, from which the undecomposed plant residues had previously been removed by flotation on water, is refluxed with 150 ml of 6 N HCl for a period of 24 hours. The mixture is filtered through a Buchner funnel with suction and the residue washed thoroughly with distilled water. The filtrate and washings, containing the amino acids, are combined and the excess HCl is removed by repeated evaporation at 50° C under reduced pressure. Any humin present is separated by centrifugation.

B. Fractionation Techniques

An aliquot of amino acid solution containing from 20 to 25 mg of α-amino acid nitrogen is passed through a column of a cation exchange resin (Amberlite IR-120) in the hydrogen form. The material retained on the column is washed with distilled water until the eluate is neutral to litmus paper, then eluted with 2 liters of 5 percent aqueous ammonia followed by a washing of 500 ml of distilled water. The eluate from the second water washing is discarded until a positive ninhydrin test is obtained then the remaining eluate is retained. The eluate from the ammonia and second water washings are combined and the ammonia is removed by gentle evaporation. The solution is made slightly alkaline with $Ba(OH)_2$ and evaporated to a paste to remove all the combined ammonia, as indicated by a negative Nessler's test. The residue is taken up in distilled water and the barium is removed as the sulfate. The final solution is combined with the original water eluate.

STEIN and MOORE [49] used a column of Dowex-50 resin to separate amino acids. An aliquot of amino acids solution, containing approximately 6 mg of α-amino acid nitrogen, is added to the top of the column and is washed into the resin with three 2 ml portions of distilled water. The column is eluted with 1.5 N HCl until glutamic acid appeared in the eluate, with 2 N HCl until valine appears, and finally with 4 N HCl.

VII. Isolation of Carbohydrates

Because of the high content of carbohydrates in plants relative to lipids and proteins, the study of carbohydrates in sediments is particularly important to geochemical considerations. The geochemistry of carbohydrates has been studied extensively; however, the exact role played by carbohydrates in the formation of fossil fuels remains to be learned.

Free sugars ranging from 10 to 3,000 mg/kg of sedimentary organic material were found in fresh water sediments by VALLENTYNE and BIDWELL [50] and WHITTAKER and VALLENTYNE [51]. DUFF [52] obtained water-soluble polysaccharides from peat and soils. Trace amounts (less than 1 mg/kg) of glucose were found in hydrolysates of Green River kerogen rock by PALACAS [53]. SHABAROVA [54] studied the carbohydrates in sediments of marine and inland seas. Sugars released by acid hydrolysis of marine sediments from the Santa Barbara Basin were reported by PRASHNOWSKY *et al.* [55]. PALACAS *et al.* [56] reported the presence of carbohydrates in ancient sedimentary rocks. SWAIN and ROGERS [57] studied the distribution of carbohydrate residues in Middle Devonian Onondaga beds of Pennsylvania and western New York.

A. Extraction Techniques

SWAIN and ROGERS [57] described a method of extractive degradation of carbohydrates in sediments. Cleaned and crushed samples are treated with 50 percent sulfuric acid to degrade carbohydrate materials to furfurals. The resulting solution is tested in triplicate for total carbohydrates by adding a phenol solution [58]. The density of the colored solutions is measured in a colorimeter and the total carbohydrates, expressed as glucose equivalent, are calculated by comparison with known samples of D-glucose. Monosaccharides are extracted and separated from the rock by acid hydrolysis. The quantity of sulfuric acid necessary to neutralize the alkaline constituents of the rock is determined by preliminary treatment with a known amount of acid and subsequent back-titration with NaOH in the presence of phenolphthalein. An amount of 0.5 N H_2SO_4 acid is added to neutralize the alkaline constituents plus an amount required to hydrolyze the rock sample. The sample-acid mixture is boiled under reflux for 8–10 hours. Then the mixture is centrifuged and neutralized with $BaCO_3$. Desalting is done first by ethanolic precipitation then by passage through ion-exchange resins. The desalted hydrolysates are concentrated in a flash evaporator to standard volume and examined by paper chromatography.

WHITTAKER and VALLENTYNE [51] describe a semiquantitative method for the determination of free sugars in lake sediments. The sample, suspended in 70 percent ethanol, is boiled for 5 minutes, then filtered while the extract is still hot. After the filtrate cools it is filtered again to remove any material precipitated on cooling. The extracts are deionized on ion-exchange resins to remove inorganic materials. The deionized extract is concentrated by evaporation to dryness under reduced pressure at 45° C.

B. Fractionation Technique

WHITTAKER and VALLENTYNE [51] used unidimensional paper chromatography to separate sugars. A wad of folded filter paper is stapled to the bottom of a sheet of Whatman No. 1 filter paper in order to increase the flow and maintain an even flow during the development of the chromatograms. The equilibrated mixture of butanol, ethanol, and water (45:5:50) is used as the solvent. Ten spots (5 to 50 microliters) of standard sugar solution with three spots of unknowns interposed between the standard spots are placed on each of duplicate chromatograms. Chromatograms are developed for 72 hours. Prolonged development of 7 days improves separation of nonpentose sugars; however, the long development removes pentoses. After development, one chromatogram is sprayed with a benzidine spray which reveals the position of all the sugars in the standard mixture and their counterparts in the unknown. Other chromatograms are sprayed with indicators for specific sugars.

VIII. Concluding Remarks

Much progress has been made in isolating kerogens from their associated minerals; however, it is apparent that additional research in this area is needed. This would properly include fundamental research of the properties of the mineral and its relationship to the organic kerogen. It would be of interest to ascertain the extent and nature of any attractive forces between minerals and kerogens. This would require study of individual formations since it is unlikely that the properties of each kerogen rock are the same.

Much development work has been expended in modifying and improving techniques for isolating individual compounds and pigments associated with the insoluble kerogen in kerogen rocks. This research has resulted in the identification of a significant number of compounds which provide information concerning the origin and transformation of carbonaceous materials. Much more data need to be collected and organized before logical conclusions can be drawn. In relating a given organic deposit with another, some improvement in organization of data would result if isolation procedures were standardized.

References

1. DOWN, A. L., and G. W. HIMUS: Classification of oil shales and cannel coals. J. Inst. Petrol. **26**, 329–348 (1940).
2. QUASS, F. W.: The analysis of the kerogen of oil shales. J. Inst. Petrol. **25**, 813–819 (1939).
3. HIMUS, G. W., and G. C. BASAK: Analysis of coals and carbonaceous materials containing high percentages of inherent mineral matter. Fuel **28**, 57–64 (1949).
4. SMITH, J. W., and L. W. HIGBY: Preparation of organic concentrate from Green River oil shale. Anal. Chem. **32**, 17–18 (1960).
5. ROBINSON, W. E., and G. U. DINNEEN: Constitutional aspects of oil-shale kerogen. Proc. 7th World Petroleum Congress, Mexico City, Mexico No. 3, 669–680 (April 1967).
6. SMITH, J. W.: Ultimate composition of organic material in Green River oil shale. U.S. Bur. Mines Rept. Invest. **5725** (1961).
7. LUTS, K.: The use of the float and sink method for isolation of organic matter in oil shales. Brennstoff-Chem. **9**, 217–218 (1928).
8. HUBBARD, A. B., H. N. SMITH, H. H. HEADY, and W. E. ROBINSON: Method of concentrating kerogen in Colorado oil shale by treatment with acetic acid and gravity separation. U.S. Bur. Mines Rept. Invest. **4872** (1952).
9. LAWLOR, D. L., J. I. FESTER, and W. E. ROBINSON: Pyrite removal from oil-shale concentrates using lithium aluminium hydride. Fuel **42**, 239–244 (1963).
10. HESS, W. VAN, and E. EARLY: A modification of Mott's method for the determination of pyritic sulfur in coal. Fuel **38**, 425–428 (1959).
11. SMITH, P. V.: Studies on origin of petroleum: Occurrence of hydrocarbons in recent sediments. Bull. Am. Assoc. Petrol. Geologists **38**, 377–404 (1954).
12. HUNT, J. M., F. STEWART, and P. A. DICKEY: Origin of hydrocarbons of Uinta Basin, Utah. Bull. Am. Assoc. Petrol. Geologists **38**, 1671–1698 (1954).
13. MEINSCHEIN, W. G., and G. S. KENNEY: Analysis of a chromatographic fraction of organic extracts of soils. Anal. Chem. **29**, 1153–1161 (1957).
14. BRAY, E. E., and E. D. EVANS: Distribution of *n*-paraffins as a clue to recognition of source beds. Geochim. Cosmochim. Acta **22**, 2–15 (1961).
15. FERGUSON, W. S.: Determining hydrocarbons in sediments. Bull. Am. Assoc. Petrol. Geologists **46**, 1613–1620 (1962).
16. BAKER, D. R.: Organic geochemistry of Cherokee Group in Southeastern Kansas and Northeastern Oklahoma. Bull. Am. Assoc. Petrol. Geologists **46**, 1621–1642 (1962).
17. CUMMINS, J. J., and W. E. ROBINSON: Normal and isoprenoid hydrocarbons isolated from oil-shale bitumen. J. Chem. Eng. Data **9**, 304–307 (1964).
18. ROBINSON, W. E., J. J. CUMMINS, and G. U. DINNEEN: Changes in Green River oil-shale paraffins with depth. Geochim. Cosmochim. Acta **29**, 249–258 (1965).
19. BURLINGAME, A. L., P. HAUG, T. BELSKY, and M. CALVIN: Occurrence of biogenetic steranes and penta-cyclic triterpanes in an Eocene shale and in an early Precambrian shale. Proc. Nat. Acad. Sci. U.S. **54**, 5 (1965).

20. TOURTELOT, H. A., and I. C. FROST: Extractable organic material in non-marine and marine shales of Cretaceous age. U.S. Geol. Surv. Profess. Papers **525-D**, 73–81 (1965).
21. HILLS, I. R., E. V. WHITEHEAD, D. E. ANDERS, J. J. CUMMINS, and W. E. ROBINSON: An optically active triterpane, gammacerane, in Green River Colorado oil-shale bitumen. Chem. Commun. **20**, 752–754 (1966).
22. EGLINTON, G., P. M. SCOTT, T. BELSKY, A. L. BURLINGAME, and M. CALVIN: Hydrocarbons of biological origin from a one-billion-year-old sediment. Science **145**, 263–264 (1964).
23. ORÓ, J., J. L. LASETER, and D. WEBER: Alkanes in fungal spores. Science **154**, 399–400 (1966).
24. CALVIN, M., and E. D. MCCARTHY: The isolation and identification of the C_{17} Isoprenoid saturated hydrocarbon, 2, 6, 10-trimethyltetradecane: squalene as a possible precursor. Div. of Pet. Chem., Am. Chem. Soc., New York, N.Y., 87–91 (Sept. 11–16, 1966).
25. O'CONNOR, J. G., F. H. BUROW, and M. S. NORRIS: Determination of normal paraffins in C_{20} to C_{32} paraffin waxes by molecular sieve adsorption. Anal. Chem. **34**, 82–85 (1962).
26. COOPER, J. E.: Fatty acids in recent and ancient sediments and petroleum reservoir waters. Nature **193**, 744–746 (1962).
27. –, and E. E. BRAY: A postulated role of fatty acids in petroleum formation. Geochim. Cosmochim. Acta **27**, 1113–1127 (1963).
28. ABELSON, P. H., T. C. HOERING, and P. L. PARKER: Fatty acids in sedimentary rocks. Intern. Ser. Monographs Earth Sci. **15**, 169–174 (1964).
29. LAWLOR, D. L., and W. E. ROBINSON: Fatty acids in Green River formation oil shale. Div. of Pet. Chem., Am. Chem. Soc., Detroit, Mich., General Papers No. **1**, 5–9 (April 1965).
30. KVENVOLDEN, K. A.: Molecular distribution of normal fatty acids and paraffins in some Lower Cretaceous sediments. Nature **209**, 573–577 (1966).
31. CASON, J., and D. W. GRAHAM: Isolation of isoprenoid acids from a California petroleum. Tetrahedron **21**, 471–483 (1965).
32. EGLINTON, G., A. G. DOUGLAS, J. R. MAXWELL, J. N. RAMSAY, and S. STALLBERG-STENHAGEN: Occurrence of isoprenoid fatty acids in the Green River shale. Science **153**, 1133–1135 (1966).
33. LEO, R. F., and P. L. PARKER: Branched-chain fatty acids in sediments. Science **152**, 649–650 (1966).
34. PARKER, P. L., and R. F. LEO: Fatty acids in blue-green algal mat communities. Science **148**, 373–374 (1965).
35. MCCARTHY, R. D., and A. H. DUTHIE: A rapid quantitative method for the separation of free fatty acids from other lipids. J. Lipid Res. **3**, 117–119 (1962).
36. LAWLOR, D. L., and W. E. ROBINSON: Alkanes and fatty acids in Green River formation oil shale. Symposium "Interdisciplinary Aspects of Lipid Research". Am. Oil Chemists' Society, Los Angeles, Calif. (April 1966).
37. TREIBS, A.: Chlorophyll and hemin derivatives in organic mineral substances. Angew. Chem. **49**, 682 (1936).
38. GROENNINGS, S.: Quantitative determination of the porphyrin aggregate in petroleum. Anal. Chem. **25**, 938–941 (1953).
39. MOORE, J. W., and H. N. DUNNING: Interfacial activities and porphyrin contents of oil-shale extracts. Ind. Eng. Chem. **47**, 1440–1444 (1955).
40. SUGIHARA, J. M., and L. R. MCGEE: Porphyrins in gilsonite. J. Org. Chem. **22**, 795–798 (1957).
41. BLUMER, M., and G. S. OMENN: Fossil porphyrins: Uncomplexed chlorins in a Triassic sediment. Geochim. Cosmochim. Acta **25**, 81–90 (1961).
42. MORANDI, J. R., and H. B. JENSEN: Comparison of porphyrins from shale oil, oil shale, and petroleum by absorption and mass spectroscopy. Chem. Eng. Data **11**, 81–88 (1966).
43. BAKER, E. W.: Mass spectrometric characterization of petroporphyrins. J. Am. Chem. Soc. **88**, 2311–2315 (1966).
44. HODGSON, G. W., B. L. BAKER, and E. PEAKE: The role of porphyrins in the geochemistry of petroleum. 7th World Petroleum Congr. Mexico City, Mexico (April 1967).
45. STEVENSON, F. J.: Ion exchange chromatography of amino acids in soil hydrolysates. Soil Sci. Soc. Am. Proc. **18**, 373–377 (1954).
46. ERDMAN, J. G., E. M. MARLETT, and W. E. HANSON: Survival of amino acids in marine sediments. Science **124**, 1026 (1956).
47. VALLENTYNE, J. R.: The molecular nature of organic matter in lakes and oceans with lesser reference to sewage and terrestrial soils. J. Fisheries Res. Board Can. **14**, 33–82 (1957).

48. SWAIN, F. M.: Distribution of some organic substances in paleozoic rocks of Central Pennsylvania, in: Coal Science. Washington, D.C.: American Chemical Society 1966.
49. STEIN, W, H., and S. MOORE: Chromatographic determination of the amino acid composition of proteins. Cold Spring Harbor Symp. Quant. Biol. **14**, 179 (1949).
50. VALLENTYNE, J. R., and R. G. S. BIDWELL: The relation between free sugars and sedimentary chlorophyll in lake muds. Ecology **37**, 495–500 (1956).
51. WHITTAKER, J. R., and J. R. VALLENTYNE: On the occurrence of free sugars in lake sediment extracts. Limnol. Oceanog. **2**, 98–110 (1957).
52. DUFF, R. B.: Occurrence of methylated carbohydrates and rhamnose as components of soil polysaccharides. J. Sci. Food Agr. **3**, 140–144 (1952).
53. PALACAS, J. G.: Geochemistry of carbohydrates. Ph. D. Thesis, University of Minnesota (1959).
54. SHABAROVA, N. T.: The biochemical composition of deep-water marine mud deposits. Biokhimiya **20**, 146–151 (1955).
55. PRASHNOWSKY, A., E. T. DEGENS, K. O. EMERY, and J. PIMENTA: Organic materials in Recent and ancient sediments. Part I. Sugars in marine sediments of Santa Barbara Basin, California. Neues Jahrb. Geol. Palaeontol. Monatsh. **8**, 400–413 (1961).
56. PALACAS, J. G., F. M. SWAIN, and F. SMITH: Presence of carbohydrates and other organic compounds in ancient sedimentary rocks. Nature **185**, 234 (1961).
57. SWAIN, F. M., and M. A. ROGERS: Stratigraphic distribution of carbohydrate residues in Middle Devonian Onondaga beds of Pennsylvania and Western New York. Geochim. Cosmochim. Acta **30**, 497–509 (1966).
58. DUBOIS, M., K. A. GILLES, J. K. HAMILTON, P. A. REBERS, and F. SMITH: Colorimetric method for determination of sugars and related substances. Anal. Chem. **28**, 350–356 (1956).

CHAPTER 7

Paleobiochemistry

M. O. DAYHOFF and R.V. ECK

National Biomedical Research Foundation,
Silver Spring, Maryland

Contents

I. Introduction

"Relics" of ancient organisms can be found in the biochemical systems of their living descendants. The exceedingly conservative nature of the evolutionary process has preserved such relics in all living species. Many basic reaction pathways and even many features of complicated polymer structures are derived from extremely remote ancestors, far beyond the ordinary fossil record. This dynamic preservation of the biochemical components of living cells is often quite as rigorous as the preservation of sedimentary fossils. So infrequently have changes occurred which were acceptable to natural selection that some details of the metabolism of organisms of the Precambrian may be inferred with confidence from their living descendants. Unlike fossil evidence, all of the biochemical information pertains to *direct* ancestors. Special features of extinct collateral lines, or metabolic pathways that have been completely abandoned, would never be inferred from the living biochemical evidence.

Chemical characteristics of organisms, particularly sequences of polymers, can be readily quantified and correlated using logical and statistical methods. The sequences already known contain as much information as thousands of morphological traits. More and more data of this kind are being found by the very active programs in many biochemical laboratories [1]. From this wealth of molecular structural knowledge, it will certainly be possible to infer in detail the order of formation of the main phyla and classes as well as some of the biochemical details of common ancestor organisms.

In this chapter, some of the biochemical evidence for the relatedness of all forms of life will be sketched out. The only reasonable explanation for the observed detailed biochemical similarities seems to be an evolution of all organisms from a single common ancestor. The last such ancestor of all living things, the "proto-organism", could already manufacture precise protein enzymes to control its metabolic reactions. It ultimately gave rise to the simplest bacteria as well as to man. Table 1 shows the relationships among various biochemical components. Table 2 is an abbreviated glossary.

Table 1. *Cell reactions of proteins and nucleic acids*

Catalysts		Products	Subcategories	Starting materials
Chromosomes	→	Chromosomes	Highly coiled DNA double strands associated with structural and regulatory proteins	Proteins Deoxyribonucleotides
DNA	→	DNA		Deoxyribonucleotides
DNA	→	RNA	Messenger RNA, m-RNA Transfer RNA, t-RNA Ribosomal RNA, r-RNA	Ribonucleotides
m-RNA t-RNA Ribosomes Specific enzymes	→	Protein	Structural proteins Enzymes (catalytic proteins) Regulatory proteins	Amino acids
Enzymes	→	Metabolites	Nucleotides Sugars Amino acids Other small molecules	Minerals, CO_2 and water or foods

An investigation of the observed mechanisms of mutation and evolution shows that, in geological time, organisms generally increase the amount of their genetic material through chromosomal aberrations, thereby developing the ability to manufacture many different though related polymers. Since the proto-organism contained many such closely related polymers, it should be possible to infer ancestral polymer structures and evolutionary development. By a consideration of these polymers, some of the metabolic capabilities of the precursors of the proto-organism can be discovered.

Before complex specific proteins could be made, organisms must have been able to make precise nucleic acid polymers. This follows from the observed conservation of the nucleic acids of the mechanism which is required to produce a coded protein, as is explained later. Many details of this preprotein era will be deciphered when the voluminous information in the nucleic acid sequences is worked out. Even earlier, there must have been an era of lesser ability to produce precise and varied nucleic acid polymers, and before that an era of chemical evolution where reaction rates depended upon the unique chemical properties of organized small molecules rather than on the properties of systematically produced

Table 2. *Glossary*

Chromosome	A small structure observed in dividing cells under the optical microscope. It is an extremely long (perhaps several feet) multiply coiled double strand of DNA polymer. There are 46 chromosomes in normal human body cells.
Gene	A term first used by geneticists for an entity located on a chromosome which determines a heritable trait. It has been shown to be a portion of the double strand of the DNA polymer which precisely determines the structure of a particular protein. If it were unfolded it would be about 0.3 microns long.
Giant Chromosomes	These are found in the salivary glands and other organs of certain insects. They have characteristic striated patterns. They are favorite objects for genetic study and have been used to work out fine details of chromosomal aberrations. Portions of the pattern are sometimes observed to repeat, demonstrating an increase in the total amount of genetic material in the chromosome.
Gene Doubling	From observations on giant chromosomes and from genetic inheritance studies, it is found that a doubling may be represented as follows: A B C D E → A B C B C D E → A′B′C′B″C″D′E′ (chromosome aberration) (point mutations) The letters represent regions of a chromosome which may be of various lengths, from a few nucleotide bases to many genes. These regions are altered by subsequent point mutations so that B′ and B″ are slightly different from B.
DNA	A long polymer of 4 different monomeric units, the nucleotide bases. These bases form two complementary pairs. DNA typically occurs as double strands in which complementary pairs are weakly bonded at each position in the polymer. Each of these chains is the template for the reproduction of its complement during chromosome division. Enzymes to catalyze the polymerization, coupled to an energy supply are also necessary for the synthesis of these complementary strands.
RNA	A long polymer of four different monomeric units, the nucleotide bases. Three of these units are the same as those found in DNA. The fourth base and the sugar in the backbone connecting the bases are slightly different. These changes do not disturb the ability of the bases to form complementary pairs. Corresponding RNA strands called messenger RNA are produced by the chromosomal DNA. Each messenger RNA molecule leads to the production of a small number of unique protein chains by a complex coding process.
Proteins	Very large, complex polymeric molecules, which have many important structural, catalytic or regulatory capabilities. They are produced by an intricate, precise mechanism, which builds up hundreds of individual amino acid monomeric links into a precisely determined linear sequence, using 20 different kinds of amino acids. The chemical specificities of the various amino acids then interact to fold this long chain into a more or less compact three-dimensional structure, which is the actual functional entity. In the case of enzymes, this folded molecule acts as a kind of specialized machine for the manipulation of specific molecules, catalyzing metabolic reactions. The information determining the precise linear order of the amino acid links resides in the nucleotide sequences of the chromosomes, the genes. The genes produce a type of nucleic acid, messenger RNA, which carries this information to the cytoplasm, and participates, along with transfer RNA, ribosomal RNA, and certain enzymes in the protein-manufacturing complex. Some proteins function as regulators of the rate of production by the genes of each individual messenger RNA, in a highly-integrated feedback system.

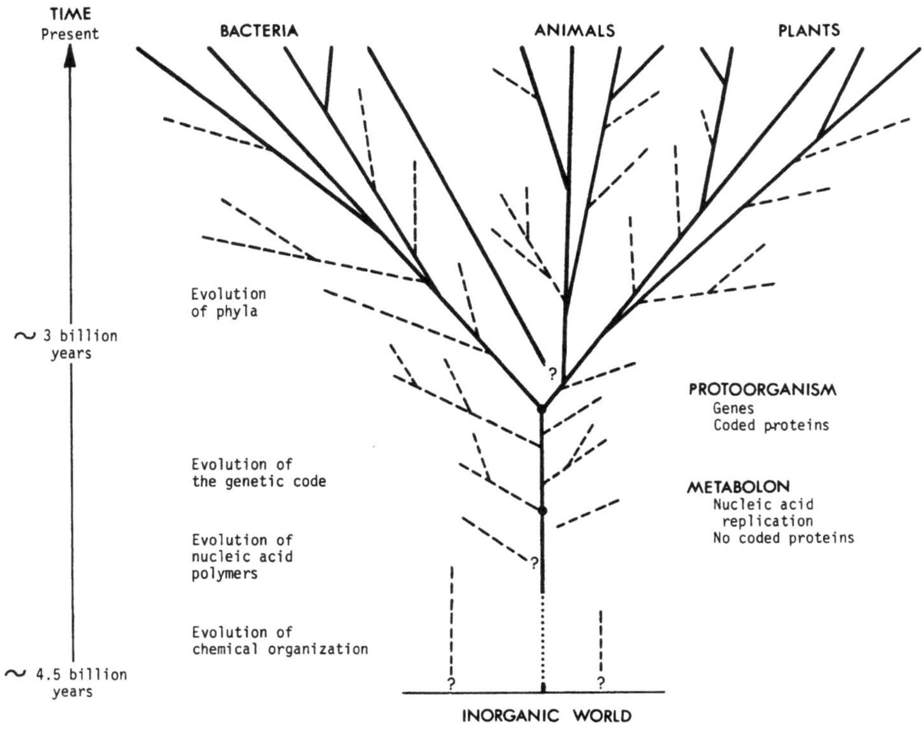

Fig. 1. General aspects of the phylogenetic tree of life inferred from the details of biochemical structures. All living phyla seem to be derived from a proto-organism with many metabolic capacities, nucleic acid genes, and protein enzymes produced using the genetic code. This genetic code is virtually unchanged in all organisms today. The proto-organism in turn is descended from a metabolon which could replicate its nucleic acids but could not produce coded proteins. Preceding this, the chemistry of nucleic acids and of small molecules evolved. Extinct organisms, whose nature cannot be inferred from biochemical evidence from living species are suggested by dashed lines

polymer catalysts (Fig. 1). The nature of the reactions of basic metabolism probably still reflects the potentialities of these small molecules.

For billions of years, evolutionary change has been a repetitive process producing many billions of generations of organisms, each organism differing but little from its parent(s) and from its offspring. Changes have generally occurred, and been selected for, one step at a time. There seems no one logical place in this sequence of stages to mark the division between living and nonliving. We will give the name "metabolons" to the organized chemical entities capable of growth and metabolism which were on the direct ancestral line to the proto-organism.

Since the metabolons of the era of chemical evolution were simple and contained few traits, it will be difficult to collect convincing evidence from biochemistry alone for any particular course of their development. However, it is possible that even the secrets of this remote era may be pried out using a combination of disciplines. A number of workers are attempting to describe this era with other approaches. Starting with the available elemental constituents and energy sources of a primitive planet containing no life, attempts are being made to work out theoretically [2] and experimentally [3] the nature of the organic compounds which might have been available for the first living things. Model systems are being investigated to determine the catalytic and organizing properties of simple biochemicals. Attempts are also being made to describe the stages by which chemical reaction, on the primitive earth, in the presence of sunlight, could have developed a series of discrete, autocatalytically growing metabolons. From these, complex life would develop by a continuation of the processes of growth, change, and selection. Many years ago such a development was proposed by HALDANE and by OPARIN [4] in broad, general terms. What is now being attempted is a step-by-step, detailed biochemical study, each step of which is simple enough to be examined in sufficient detail to achieve final scientific acceptance.

Simultaneously, a search is proceeding for fossil evidence of simple organisms, such as algae and bacteria, and for very ancient chemical residues [5]. One day the biochemical, chemical, and geological approaches may well support each other extensively. The evolution of life may then be traced convincingly through many understandable steps all the way back to chemical simplicity.

II. Biochemical Unity of Life

It has become clear that the basic metabolic processes of all living cells are very similar. A number of identical compounds, mechanisms, structures, and reaction pathways are found in all living things so far observed including such diverse cells as those of human liver or skin, or of paramecia, or bacteria, or onion root tip. Many experiments demonstrating these common features have been performed [6].

1. All cells utilize polyphosphates, particularly adenosine triphosphate, for energy transfer. These polyphosphates are manufactured in photosynthesis or in the oxidation of stored food. Their decomposition is coupled to the organic syntheses of thermodynamically unstable products needed by the cell. The experimental observation of these detailed synthetic reactions of scores of compounds, each unlikely to occur by chance, supports the theory of a common origin.

2. Cells synthesize and store similar compounds – sugars, fats, starches, and proteins – using similar reaction pathways. These compounds are degraded with release of energy in a similar way in most cells.

3. The metabolic reactions are catalyzed largely by proteins, which are linear polymers of definite structure composed of various sequences of twenty amino acid building blocks. A number of the proteins have identifiable counterparts known as homologues in most living things. The amino acid sequence of homologous proteins are often nearly identical and they have similar three-dimensional structures and functions. Experimental systems can be set up in which biochemical reactions involve cooperation between catalytic proteins from widely different biological species.

4. Proteins are manufactured in the cell by a complex coding process. The sequence of amino acids in each protein is determined by a sequence of nucleotides in the chromosomal DNA, the gene. Three consecutive nucleotides specify one amino acid. There is a complex mechanism, involving both protein and nucleic

acid enzymes, which catalyzes this translation process. The code, that is, the correspondence of amino acids to nucleotide triplets, is almost identical in all organisms which have been studied [7].

5. All organisms have a mechanism for duplicating their nucleic acid polymers (DNA) and for producing extra transcripts (RNA) called messengers which serve as templates in the process of protein synthesis. Such complex mechanisms and processes as these must have been devised by evolution from simpler yet similar components, traces of which necessarily remain in their modern descendants.

6. There is a relatively small number of ubiquitous small compounds which take part in metabolic processes: for example, nicotinamide, pyridoxal, glutathione, the flavenoids, the carotenes, the isoprenoid compounds, and iron sulfide. More complex components are lipid membranes, coenzyme A, the heme groups, and chlorophyll. Since there are millions of possible compounds of comparable size and energy, it seems most unlikely that these particular ones would have been chosen independently by organisms. These small compounds often act in conjunction with proteins and are the catalytic centers of enzymes. The protein portion increases the efficiency of the reaction and determines the specific compound, or small number of compounds, that will react, but does not alter the basic nature of the reaction.

That so many similar things could have originated independently in different organisms by chance is incredible. Consider the probability of duplicating the sequence of amino acids in just one polymer 100 units long. Since any one of 20 amino acids could occur in each position, there are 20^{100} different polymers that could be constructed! It is improbable in the extreme that two totally unrelated organisms would by chance have manufactured and selected two structures with a degree of similarity as great as that observed. The phenomenon of organic evolution is illustrated even more impressively by these aspects of biochemistry than by biological, morphological, embryological, or fossil evidence.

The biopolymers, the synthetic pathways, and the small molecules that are common to all living things have been preserved almost unchanged since the time, perhaps more than three billion years ago, when all the organisms now living, including bacteria, spinach, oysters, and people, had a single ancestor. This latest common ancestor of all life, the proto-organism, may have represented a kind of "common divisor" of the metabolic chemistry shared by present living things [1, 6]. It would have had a degree of complexity sufficient to include all these common components. Sooner or later it must have happened that one organism gave rise to two persisting lines of descent whereas all of its contemporaries have died out. At the time that it lived there may have been nothing outstanding about this organism. There is no reason to believe that this organism could have come together by chance; it was far too complex. The inferred biochemical structure of this primitive cell shows evidence that it was itself the product of many evolutionary steps, very similar in nature to steps which have occurred more recently. Implicit in the inferred structure of the proto-organism and the principles of evolution is a wealth of information about the far more primitive organisms which preceded it.

CYTOCHROME C

```
                                          1                   2                   3                   4                   5
                        1 2 3 4 5 6 7 8 9 0 1 2 3 4 5 6 7 8 9 0 1 2 3 4 5 6 7 8 9 0 1 2 3 4 5 6 7 8 9 0 1 2 3 4 5 6 7 8 9 0 1 2 3 4

Human                   - - - - - - G D V E K G K K I F I M K C S Q C H T V E K G G K H K T G P N L H G L F G R K T G Q A P G O S O
Rhesus Monkey           - - - - - -(G.D.V.E.K.G.K.K.I.F=I.M.K=C.S.Q.C.H.T.V.E.K=G.G.K.H=K.T.G.P.N.L.H.G.L)F.G R K T G Q A P G O.S O.
Pig, Bovine, Sheep      - - - - - - G D V E K G K K I F V Q K C A Q C H T V E K G G K H K T G P N L H G L F G R K T G Q A P G F S O
Horse                   - - - - - - G D V E K G K K I F V Q K C A Q C H T V E K G G K H K T G P N L H G L F G R K T G Q A P G F T O
Dog                     - - - - - - G D V E K G K K I F V Q K(C.A.Q.C.H.T.V.E)K G G K H K T G P N L H G L F G R K T G Q A P G F S O
Rabbit                  - - - - - - G D V E K G K K I F V Q K C A Q C H T V E K G G K H K T G P N L H G L F G R K T G Q A V G F S O
Kangaroo                - - - - - - G D V E K G K K I F V Q K C A Q C H T V E K G G K H K T G P N L H G L F G R K T G Q A P G F T O
Chicken, Turkey         - - - - - - G D I E K G K K I F V Q K C S Q C H T V E K G G K H K T G P N L H G L F G R K T G Q A E G F S O
Duck                    - - - - - - G D V E K G K K I F V Q K C S Q C H T V E K G G K H K T G P N L H G L F G R K T G Q A E G F S O
Turtle                  - - - - - - G D V E K.G K K I F.V Q K C A Q C H T V E K G G K H.K T G P N L N G L.I G R K T G Q A E G F.S O
Tuna Fish               - - - - - - G D V A K.G K K.T F V Q K.C A Q(C.H)T V E N G G K.H K(V.G.P.N)L W.G L F.G R.K T(G.Q)A E G O.S O
Samia Moth              - - G V P A G N A E N G K K I F V Q R C A Q C H T V E A G G K H K V G P N L H G F O G R K T G Q A P G F S O
Neurospora (Fungus)     - - G F S A G D S K K G A N L F K T R C A E C H G E G G N L T Q K I G P A L H G L F G R K T G S V D G O A O
Candida (Fungus)        P A P F E Q G S A K K G A T L F K T R C A E C H T I E A G G P H K V G P N L H G I F S R H S G Q A Q G O S O
Yeast                   - T E F K A G S A K K G A T L F K T R C E L C H T V E K G G P H K V G P N L H G I F G R H S G Q A Q G O S O

      COMMON                        G         G       F       C     C H                 K   G P   L   G       R     G       G     O

      ALLELES           - - - - - -   D V E K   K K I   V Q K   A Q     T V E K G G K H   T     N   H   L F G   K T   Q A P   F S
                        P A G F P A   S A K N   A T L   I T R   S E     G E G A N L P Q   V     A   N   I I S   H S   S V E   O T
                          T P V S Q   N I A       N T   K M     E L       I   N     T     I         W   F O               V     A
                            E   E       S                                     G                                           Q
                                K                                                                                         D

      SEQUENCES OF COMMON ANCESTORS

        NODE  1         - - - - - - G D V E K G K K I F I M K C S Q C H T V E K G G K H K T G P N L H G L F G R K T G Q A P G O S O
        NODE  2         - - - - - - G D V E K G K K I F V Q K C A Q C H T V E K G G K H K T G P N L H G L F G R K T G Q A P G F S O
        NODE  3         - - - - - - G D V E K G K K I F V Q K C A Q C H T V E K G G K H K T G P N L H G L F G R K T G Q A P G F S O
        NODE  4         - - - - - - G D V E K G K K I F V Q K C A Q C H T V E K G G K H K T G P N L H G L F G R K T G Q A P G F S O
        NODE  5         - - - - - - G D V E K G K K I F V Q K C A Q C H T V E K G G K H K T G P N L H G L F G R K T G Q A P G F S O
        NODE  6         - - - - - - G D V E K G K K I F V Q K C A Q C H T V E K G G K H K T G P N L H G L F G R K T G Q A   G F S O
        NODE  7         - - - - - - G D V E K G K K I F V Q K C A Q C H T V E K G G K H K T G P N L H G L F G R K T G Q A   G F S O
        NODE  8         - - - - - - G D V E K G K K I F V Q K C A Q C H T V E K G G K H K T G P N L H G L F G R K T G Q A E G F S O
        NODE  9         - - - - - - G D V E K G K K I F V Q K C S Q C H T V E K G G K H K T G P N L H G L F G R K T G Q A E G F S O
        NODE 10         - - - - - - G D V E K G K K I F V Q K C A Q C H T V E   G G K H K V G P N L H G L F G R K T G Q A   G O S O
        NODE 11         - - G       A G D A E K G K K I F V Q R C A Q C H T V E   G G K H K V G P N L H G   F G R K T G Q A   G O S O
        NODE 12         - - G F     A G D A K K G A   L F K T R C A E C H T V E   G G   H K V G P N L H G   F G R K T G Q A   G O S O
        NODE 13         -     F     A G S A K K G A T L F K T R C A E C H T V E   G G P H K V G P N L H G I F G R H S G Q A Q G O S O
```

Fig. 2

III. Biochemical Evolution

In overall aspect, a general description of the evolutionary process is applicable to all levels of complexity, including chemical evolution. Suppose that a functioning organism (whether of modern complexity or of primordial simplicity) already exists and is competing successfully with other organisms; then an evolutionary step requires the following succession of events:

1. A new characteristic, *X*, becomes available to the organism by some rare chance event – for example, the addition of an extra piece of genetic material to a chromosome, the synthesis of a critical high-energy small molecule, or the absorption of a single rare molecule from the environment.

2. The presence of *X* in the organism must encourage processes which lead to the production or absorption of more *X* by the organism; *X* is then available continuously, instead of rarely.

3. The new organism, incorporating *X*, is more successful than others in some niche available to it. This new organism therefore becomes established.

If successful, each such characteristic *X* involves reciprocity. Both *X* and the entire organism prosper. If the new and old organisms inhabit the same environment, in direct competition, the old one will become extinct. If the new organism is better fit to survive in an accessible slightly different environment but less fit in the original one, the two populations will both survive. By the accumulation of further evolutionary steps, the two populations will then become increasingly different.

The evolutionary process is extremely conservative because older cell components come to be relied upon by many later additions. For example, acetate is fundamentally involved in energy transfer systems and is also an essential starting material in the synthesis of such diverse metabolites as sugars, fats, and amino acids; it is extremely unlikely that any other compound could advantageously replace it in all four capacities simultaneously. Any change in such a very old component, even though the change might have some particular advantage, would coincidentally disturb so many other things that it would almost always be extremely disadvantageous to the organism. This conservation is well illustrated in the amino acid sequences of proteins. Fig. 2 shows a comparison of the first halves of the cytochrome *c* sequences from a number of diverse species. At many positions the amino acids are common to all. At other positions, replacements with chemically related amino acids are observed. Occasionally one or a few amino acids have been inserted or deleted (in this example, at the amino end). Changes in sequence, although relatively frequent, are accepted very rarely by

Fig. 2. Sequences of the first half of the cytochrome c proteins from 18 species. References to the original reports are given in Reference 1. The amino acids common to all sequences or else the alternative amino acids or "allele groups" are shown at each position. The inferred sequences of the common ancestors at each divergence point, or node, in the phylogenetic tree of Fig. 3, are also displayed. Sites are left blank where no single amino acid was most likely. The punctuation, described in detail in Reference 1, indicates degrees of uncertainty in some of the details of the sequences. One letter abbreviations for the amino acids are as follows: *A* – alanine; *C* – cysteine; *D* – aspartic acid; *E* – glutamic acid; *F* – phenylalanine; *G* – glycine; *H* – histidine; *I* – isoleucine; *K* – lysine; *L* – leucine; *M* – methionine; *N* – asparagine; *O* – tyrosine; *P* – proline; *Q* – glutamine; *R* – arginine; *S* – serine; *T* – threonine; *V* – valine; *W* – tryptophan

natural selection. On the average a single such change may be incorporated in a particular protein perhaps once in a few million years. An identical change observed in two sequences would usually be produced by only one mutational event, occurring in the common ancestor of both. From such inferences at all the 109 amino acid positions in these cytochromes, we have constructed a phylogenetic

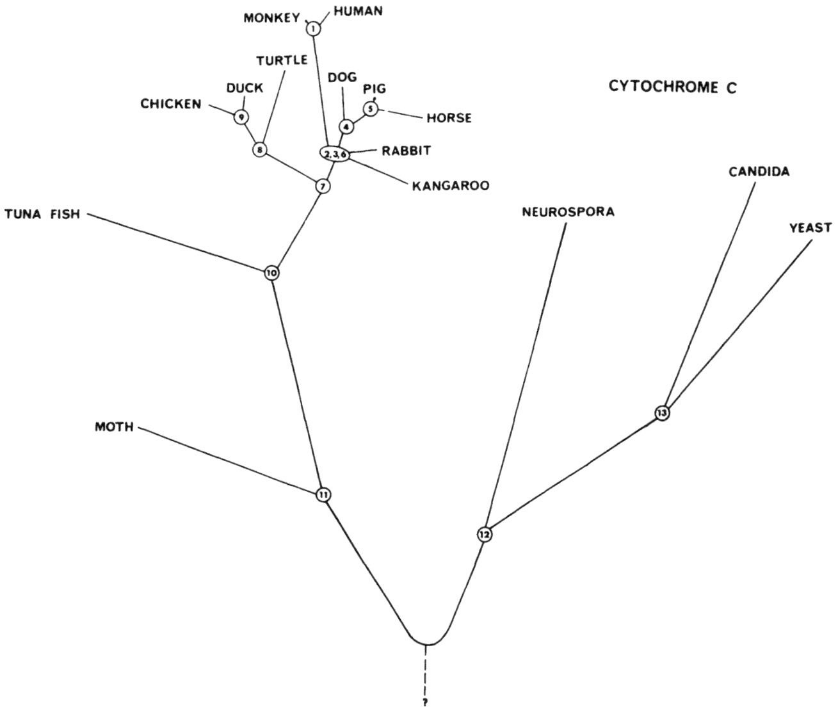

Fig. 3. The phylogenetic tree of species derived from the (complete) protein sequences of Fig. 2. The branch lengths are proportional to the number of amino acid changes between the observed sequences and the inferred ancestors. The point of earliest time cannot be inferred directly from the sequences. We have placed it from other considerations. Similar trees derived from other proteins such as hemoglobins, insulins and immunoglobulins may illuminate the finer details of the history of selected organisms

tree of these species. (By a somewhat different mathematical procedure, a similar tree was derived by FITCH and MARGOLIASH [8].) Our tree, similar to that generally accepted on classical grounds, is shown in Fig. 3. The topology of this tree is optimal in that it requires the occurrence of a minimum number of mutational events. The amino acid sequences of the ancestors at all branch points, the nodal sequences (see Fig. 2), were inferred with good confidence from the observed sequences on the three branches from each node. For each position, the amino acid observed in the majority of the three branches was chosen [1]. Most of the amino acids are unchanged from the ancestors. In some few places, indicated by blanks, the occurrence of two or more amino acids seemed equally likely.

The branch lengths shown are proportional to the number of mutations inferred. (The longer lengths are foreshortened by a statistical effect; the correction for this has not been applied to this figure.) The amount of change in human cytochrome *c* seems to be no greater than that in any of the far more primitive-looking organisms, including yeast, that have been studied. The branch lengths then bear a rough relationship to the geologic time which has elapsed. Each family of proteins has its own distinct rate of evolution, however, so that such branch lengths give only relative times.

The genetic complement of mammals contains perhaps 10^5 to 10^6 genes. The proteins from each of these genes, when compared with similar ones from other species, contain information about the homologous proteins of our ancestors. Those which have been added recently, such as the fibrinopeptides, are changing so rapidly that the ancestry of closely related species, such as sheep, cattle, and several kinds of deer, may be traced [1, 9]. By contrast, those which were already present in the proto-organism are changing very slowly.

The proto-organism may have contained about 200 protein catalysts or enzymes for its metabolic activities. This number is based on an estimate of the number of metabolic reactions held in common by many diverse phyla. The structures of most of these 200 primordial proteins would have been highly conserved in present day species, since so many later structures and functions would have come to depend on them. The great amount of information contained in these sequences, when they are worked out, should permit the reconstruction of the ancestral relationships of all of the main phyla. Coincidentally, many aspects of some extremely ancient protein and nucleic acid structures should also be obtained.

IV. Increase in Genetic Material

The evolutionary mechanisms by which an organism with 200 genes has developed into one with 10^5 or more genes have been intensively investigated by geneticists and biochemists. The study of the globins gives us some insight. In man, there are at least four different hemoglobin proteins which occur in the blood at some time during life. A fifth protein, myoglobin, which has functions similar to hemoglobin is found in muscle. All five of these proteins are recognizably related in their amino acid sequences, particularly in the region where the heme group which carries oxygen is attached. Their functions and their three-dimensional shapes as determined by X-ray crystallography are very similar. These sequences are so similar that they could not have originated independently by chance. The five genes that produce them must have had a common origin. The globin genes must have doubled on several different occasions. A known process which could account for this is chromosomal aberration which is of relatively frequent occurrence and which presumably is the basis for the great increase in the total amount of genetic material in modern organisms, compared to their primitive ancestors. At each doubling, the cell acquired the capacity to manufacture two identical proteins by means of two distinct genes. Subsequently, each of these genes was subject to independent "point" mutations, each producing a change of one amino acid. The protein structures gradually varied independently and came to perform slightly different, though related, functions.

Several other families of related proteins have already been investigated – digestive enzymes, immunoglobulins, and peptide hormones, to name a few. It appears from these investigations that chromosome or gene doubling is an important mechanism for the increase in the amount and variety of genetic material.

It is also possible to increase the amount of genetic material by duplicating portions of genes. A single protein of increased length is then produced. This latter process has clearly taken place in ferredoxin, an enzyme deeply involved in the photochemical processes. The reduction of its prosthetic group, FeS, is a key photochemical event in photosynthesis. The enzyme further catalyses the formation of ATP by radiation and the synthesis of pyruvate from CO_2. These functions suggest the incorporation of its prototype into metabolism very early in biochemical evolution, even before complex proteins and the complete modern genetic code existed. An examination of the amino acid sequence of ferredoxin shows clearly that it has evolved by doubling a shorter protein which may have contained only eight of the simplest amino acids. There is some evidence still visible that this half molecule in turn was derived from a short repeating sequence of four amino acids [10].

Through protein studies, the existence of the following previously known mechanisms for increasing the genetic material of organisms has been confirmed (see Table 2). In order of the frequency of their occurrence, there are three main types:

1. Doubling of a portion of a gene, or an insertion or deletion of a few nucleotides in a sequence.
2. The doubling of a gene, leading to the production of two separate loci, initially producing identical proteins.
3. The addition of extra, whole chromosomes.

The natural rate of each of these types of error in chromosome replication and the low probability that the resulting structures will be beneficial limit the rate of evolution. The products of each of these processes, when they occur, are subjected to intense natural selection and are seldom preserved. In general, it is the rare survivors that we observe.

The process of genetic mutation results from the chemical modification of a nucleotide in the gene sequence or from copying errors. When one nucleotide in the sequence of a gene is changed, it usually leads to the production of a protein with one changed amino acid. Compared to the processes of increasing the genetic material, this kind of change is relatively frequent.

The development of new metabolic capabilities is greatly facilitated by a duplication of already existing genetic material. Otherwise, the development of new metabolic capacities would require the synthesis and adaptation of random nucleotide sequences or the adaptation by point mutations alone of existing genes which already have other essential functions. Examples from protein sequence studies demonstrate that there are two favored processes for the addition of new metabolic capabilities.

First, when the gene for a protein is doubled, as in (2) or (3) above, only one of these new genes is required for the continued production of molecules with the previous function. Since the other gene product does not need to perform this exact same function, previously forbidden structural changes in it may now be accepted by natural selection. It can change fairly rapidly by mutations of the individual nucleotides and can come to fill a new and useful function, related to the old one but different.

Secondly, when a portion of a gene is doubled, as in (1) above, one end of the enlarged protein continues to perform the old function while the new section is free to adjust its structure by nucleotide change. The new section of the protein may increase the specificity or efficiency, or it may acquire some new function.

Of the almost infinite number of possible combinations of chromosomal aberrations which might have occurred, only a very insignificant fraction have actually occurred in living organisms. It seems very likely that many which could have occurred but in fact did not might have proved beneficial and been accepted. The evolutionary path that was actually followed was based on rare random events, the acceptable mistakes. At any given time the future of biochemical evolution is largely indeterminate because of these many possible alternatives.

On the other hand, the evolutionary past of an organism is unique. It can be inferred from living evidence, using evolutionary principles. Each organism or structure or function had ancestors which were very similar to itself, but simpler. (This is generally true even if it had more complex immediate ancestors.) In a particular case, generally only a few of the possible slightly simpler ancestors were plausible. In tracing a structure back through time, it must be remembered that all of the other structures and functions of the cell would also have been simpler, not only the one we happen to be considering. At a great distance in time, the information becomes noisy, but this limitation on inferences is partially compensated by the greater simplicity of the organism.

V. Before the Proto-Organism

Many of the principles of evolution determined by genetic studies are also applicable to biochemical details. Furthermore they can be applied to the inferred biochemical components of the proto-organism. Many of these components resemble one another so closely that it is evident that they must have had common ancestors. One example for which we have the best quantitative evidence is found in the nucleotide sequences and three-dimensional structures of the transfer RNA's. These are specific catalysts involved in the manufacture of proteins, acting to translate the information contained in the nucleic acid "messengers". Each member of the transfer-RNA family possesses two specific active sites. One site determines which amino acid it carries; the other determines which triplet of the nucleotide messenger sequence it will recognize. In the translation process, these enzymes line up consecutively along the messenger. As each achieves its proper position, its amino acid is attached to the growing protein chain (see Fig. 4).

It is known that all t-RNA's within one individual have a number of structural features in common, including their general size and shape. Their close similarity must be due to a common origin. The probability is extremely small that a cell could synthesize independently even two structures which would function smoothly in the coding process. However, it is relatively easy for nature to create such similar structures by the production of redundant genetic material (by doubling or other chromosomal aberrations) followed by the accumulation of independent

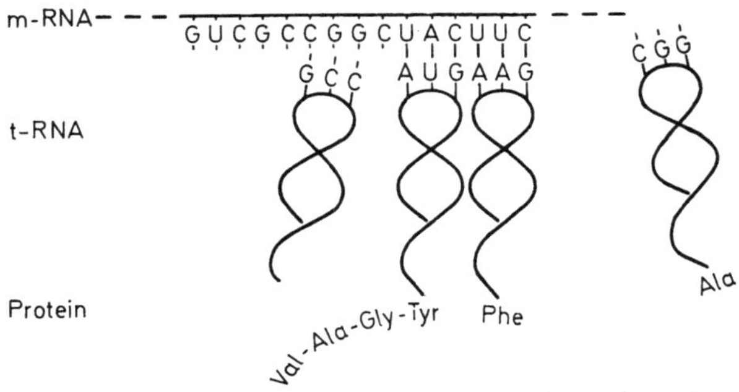

Fig. 4. Details of protein manufacture. The process occurs on the surface of a ribosome (r-RNA) using t-RNA which has previously been attached to its specific amino acid. In the translation process, the t-RNA enzymes line up consecutively along the messenger. As each achieves its proper position, its amino acid is attached to the growing protein chain. The letters *A*, *C*, *G* and *U* represent the four types of nucleotide bases. *A–U* forms one pair; *C–G* forms the other. The known chemical details are somewhat more complex than is indicated here, but follow the principles illustrated

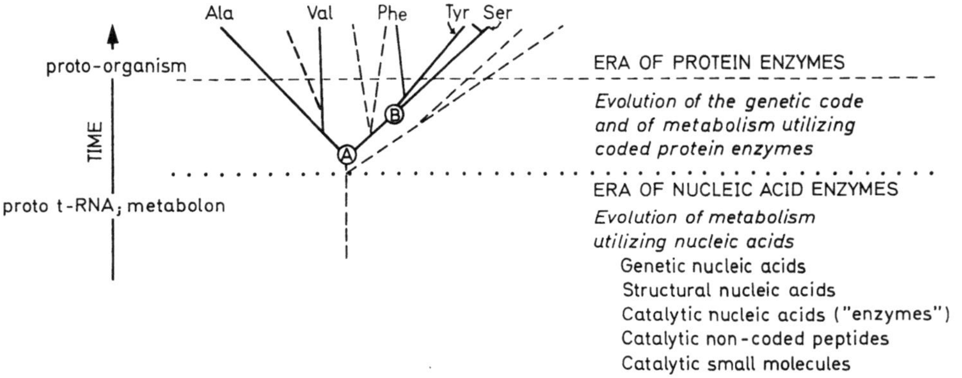

Fig. 5. The phylogenetic tree of t-RNA. These catalysts function in the translation of the messages of the genes into specific proteins. It is conjectured that the t-RNA structures now specifically recognizing the 20 different amino acids are derived from a single proto-t-RNA by successive doublings and mutations of the one gene which originally coded for it. The specific ability to incorporate the various amino acids into proteins developed gradually. The divergence in time of the t-RNA polymers can be deduced from their known structures. The lengths of the solid lines represent relative numbers of changes actually observed in the t-RNA's whose sequences have been determined. The ability to discriminate alanine from serine could not have developed before the branch point, *A*. The ability to discriminate between serine and tyrosine could not have come before the later branch point, *B*. The following amino acid abbreviations are used: Ala – alanine; Val – valine; Phe – phenylalanine; Tyr – tyrosine; Ser – serine. Dashed lines indicate possible locations of other t-RNA sequences yet to be worked out.

mutational changes in the separate genes. The development of the specificity of the genetic code seems to represent such a succession of doublings and minor changes in the genes which produce the t-RNA molecules.

The sequences of five t-RNA's, those for alanine, tyrosine, serine, phenylalanine, and valine, are completely known for yeast [1]. From these, ancestral sequences can be inferred and a tentative phylogenetic tree can be drawn (see Fig. 5). The lengths of the branches are proportional to the number of changes from the ancestral sequences. The trunk of the tree is arbitrarily placed in the middle. The other t-RNA's will fill out the tree in a manner somewhat like that indicated by the other dashed lines. This tree represents a complex "phylogeny" of macromolecules all within one ancestral lineage, before the proto-organism. The ancestral single molecule is proto-t-RNA occurring in the far more primitive metabolon.

From the observed similarity of the structures of all the t-RNA's and from the nature of the evolutionary process, we can infer several further properties of the metabolon which lived before the era of coded proteins. This metabolon had the ability to conserve the structure of its nucleotide enzymes; otherwise, we could not today still observe so much similarity among the different t-RNA's. It must therefore have had an accurate nucleotide-transcription mechanism. It may already have had a separate "memory", such as a proto-DNA chromosome.

Proto-t-RNA must have imparted some competitive advantage to the metabolon so that natural selection preserved it. Proto-t-RNA most probably acted as a nonspecific catalyst polymerizing amino acids. The mechanism would have been similar to the one still used today. (This polymerizing mechanism could hardly have arisen at a later time. Once the divergence of the specific t-RNA's had occurred, it would be almost impossible for nature to introduce a new feature; the simultaneous adaptation to this new feature by all the divergent t-RNA's would be extremely unlikely.) This primeval organism was capable of producing proto-ribosomes and at least one nucleic acid messenger, since both of these are essential to the polymerizing mechanism. The messenger corresponded to the first gene for a protein; the t-RNA's and the genes for protein enzymes then evolved together. Presumably, the protein produced by this messenger was advantageous, even though the translation process may well have been ambiguous. The particular amino acids in the protein would depend on mass balance and other chemical factors, as well as on the specificity of the amino acid-t-RNA reaction at that time. A number of other cell functions had probably already come to depend on protein before the code evolved. At this level of complexity, the metabolon had the following characteristics: nucleotide enzymes, including proto-t-RNA; structural nucleic acids, including proto-ribosomes; proto-messenger-RNA; and the genetic nucleic acids necessary for the reliable reproduction of all these. It also had amino acids, the above-mentioned protein, and such amino acid polymers as it could synthesize by other more primitive mechanisms. Some of these primitive polymers still survive, for example, glutathione and polyglutamic acid.

The organization, structures, and functions of the metabolon will be considerably clearer when many more of the polymer structures of the proto-organism have been inferred. The mathematics of inferring the common ancestry of polymers is straightforward, and the practical limit depends on our ability to collect a sufficient amount of suitable polymer-sequence data. To understand better the earliest stages of evolution, it will be advantageous to combine evidence from many sources.

VI. Evolution of Chemical Organization

A. Biochemical Inferences

Before the era of large, systematically coded polymers, the metabolon must already have achieved considerable complexity. The succession of reactions leading to the synthesis of the ubiquitous small molecules used in metabolism may have been strongly conserved by evolution just as the structures of protein and nucleic acid polymers were. If so, the succession of reactions producing each compound today may represent a recapitulation of the evolutionary events which originally built up the chain of reactions. It may be possible, then, to infer self-consistent functional ancestors with greatly simplified primitive metabolic pathways [11]. By a logical process of stripping away the peripheral components, those on which other components do not critically depend, we may be able to simplify the large body of known metabolic processes and arrange them in a rough chronological order of their addition to the metabolons. This would sketch out many details of the presently unknown era represented by the dotted lower trunk of the phylogenetic tree of Fig. 1. The very oldest components would become evident. These in turn could suggest possibilities for studies approaching this era from the opposite direction, that of the natural development of a more complex organization from a primordial, chemically simple world.

B. Astronomical and Geological Considerations

The compounds available to the first metabolons must obviously have been those which were produced by nonliving processes. Inferences about the nature of prebiological conditions may be drawn from thermodynamic equilibria, from special processes implicit in the evolutionary history of the earth, and from geological observations. Components evidently available from the beginning include phosphates, polyphosphates, CO_2, H_2O, N_2, H_2, CO, H_2S, NH_3, Mg^{++} and other metal ions, and FeS and other minerals. Various reactions have been proposed for the development of more complex compounds through the absorption of light or other energy [12]. The nonliving production of thermodynamically unstable substances by light would be the archetype of the energy utilization process in the metabolon, which evolved into the efficient, complex, highly regulated process of photosynthesis in modern plants.

C. Fossil Evidence

Recent studies [5] have been directed at the detection of preserved remnants of actual chemical components of ancient organisms. It is too soon to tell what the practical limit of these studies may be in tracing the actual chemistry of the earliest living things; but if fossils can be found in which degradative processes have not been too severe, they will be of much interest.

D. Model Laboratory Experiments

Complex structures can develop naturally from less structured ingredients because of the inherent chemical properties of systems. Cocrystallization and coacervate formation are well known examples. Intricate living processes have developed by the repeated operation of this potentiality of nature. Laboratory experiments have been directed at the self-organization of components into complexes such as coacervates, membranes, enzyme aggregates and polymers [4, 11]. Other experiments study the cooperative catalytic properties of these and other important primitive biological components.

The stages of self-organization must have been added to the metabolon one at a time. Once the self-organization and possible function of any such stage have been understood, then the plausibility of the previous existence of the necessary components must be shown. To study the first evolutionary stages, laboratory experiments with simulated primordial conditions, or theoretical considerations, such as chemical thermodynamics, are used to establish the nature of the essential first complex compounds, e.g., amino acids, sugars, fats, and nucleotides.

One day these four approaches may well support each other so thoroughly that the evolution of life can then be traced back through many understandable steps all the way to chemical simplicity. When this has been done, it will be seen that the "origin of life" was not a single unique event, but rather, a continuous development of the potentialities inherent in the building blocks and the energy flow of the universe.

References

1. DAYHOFF, M. O., and R. V. ECK: The atlas of protein sequence and structure 1967–68. National Biomedical Research Foundation, Silver Spring, Maryland 1968.
2. DAYHOFF, M. O., E. R. LIPPINCOTT, and R. V. ECK: Thermodynamic equilibria in prebiological atmospheres. Science **146**, 1461–1464 (1964).
   ECK, R. V., E. R. LIPPINCOTT, M. O. DAYHOFF, and Y. T. PRATT: Thermodynamic equilibrium and the inorganic origin of organic compounds. Science **153**, 628–633 (1966).
   SUESS, H. E.: Thermodynamic data on the formation of solid carbon and organic compounds in primitive planetary atmospheres. J. Geophys. Res. **67**, 2029–2034 (1962).
3. STUDIER, M. H., R. HAYATSU, and E. ANDERS: Organic compounds in carbonaceous chondrites. Science **149**, 1455–1459 (1965).
   FOX, S. W.: How did life begin? Science **132**, 200–208 (1960).
   MILLER, S. L., and H. C. UREY: Organic compound synthesis on the primitive earth. Science **130**, 245–251 (1959).
   PONNAMPERUMA, C., R. M. LEMMON, R. MARINER, and M. CALVIN: Formation of adenine by electron irradiation of methane, ammonia and water. Proc. Nat. Acad. Sci. U.S. **49**, 737–740 (1963).
   ORO, J.: Synthesis of organic compounds by high-energy electrons. Nature **197**, 971–974 (1963).
   PALM, C., and M. CALVIN: Primordial organic chemistry. 1. Compounds resulting from electron irradiation of $C^{14}H_4$. J. Am. Chem. Soc. **84**, 2115–2121 (1962).
   ABELSON, P. H.: Amino acids formed in "primitive atmospheres". Science **124**, 935 (1956). Presented at the NAS Autumn Meeting, Nov. 9–10, 1956.
4. HALDANE, J. B. S.: The origin of life. Rationalist Annual, 142–154 (1928).
   OPARIN, A. I.: The chemical origin of life. Springfield (Ill.): Ch. Thomas 1964.
5. BARGHOORN, E. S., and J. W. SCHOPF: Microorganisms three billion years old from the Precambrian of South Africa. Science **152**, 758–763 (1966).
   BARGHOORN, E. S., and S. A. TYLER: Microorganisms from the Gunflint Chert. Science **147**, 563–577 (1965).
   SCHOPF, J. W., and E. S. BARGHOORN: Alga-like fossils from the early Precambrian of South Africa. Science **156**, 508–512 (1967).
6. SALLACH, H. J., and R. W. MCGILVERY: Intermediary metabolism chart. Middleton, Wisconsin: Gilson Medical Electronics 1963.
7. MARSHALL, R. E., C. T. CASKEY, and M. NIRENBERG: Fine structure of RNA codewords recognized by bacterial, amphibian, and mammalian transfer RNA. Science **155**, 820–826 (1967).

8. MARGOLIASH, E., and W. M. FITCH: Construction of phylogenetic trees. Science **155**, 279–284 (1967).
9. DOOLITTLE, R. F., and B. BLOMBACK: Amino-acid sequence investigations of fibrinopeptides from various mammals: Evolutionary implications. Nature **202**, 147–152 (1964).
10. ECK, R. V., and M. O. DAYHOFF: Evolution of the structure of ferredoxin based on living relics of primitive amino acid sequences. Science **152**, 363–366 (1966).
11. OPARIN, A., ed.: Evolutionary biochemistry. Proc. 5th Intern. Congr. New York: MacMillan & Co. 1963.
12. EGLINTON, G., and M. CALVIN: Chemical fossils. Sci. Am. **216**, 32–43 (1967).

This work was kindly supported by a contract from the National Aeronautics and Space Administration U.S.A., NSR-21-003-002, and a grant from the General Medical Sciences Institute of the U.S. National Institutes of Health.

CHAPTER 8

Fossilization Processes

W. D. I. ROLFE and D. W. BRETT

Hunterian Museum, Glasgow University,
and Botany Department, Bedford College, London University

Contents

I. Introduction

Fossils have been simply defined as traces of ancient life. Paleontology, the study of these remnants, can thus well be thought of as four-dimensional biology [1]. To adapt one description of biology, "the aim of paleontology is to understand the structure, functioning and history of ancient organisms and of populations of such organisms" [2]. Because of the nature of their material, however, paleontologists have often needed information on aspects of modern life that have scarcely interested neontologists. Examples that might be cited are the whole field of actuopaleontology (with close links to forensic medicine – see below), the study of population structures not only of living communities but also of dead assemblages [3], and the detailed morphological study of preservable tissues.

Attempts to base a more precise definition of fossils upon their age or condition [4] are clearly artificial and largely undesirable. GEIKIE long ago pointed out that the term "fossil" is as applicable to the bones of a sheep buried under the gravelly flood deposit of a modern river as to extremely ancient organic remains [5], and this view is endorsed by workers on Recent death assemblages studying organisms which have been dead for only a matter of months or at most a few years [6]. The problem of whether to regard "chemical fossils" [7] as true fossils is largely a semantic one. BROUWER [8] follows STIRTON in redefining fossils as "recognizable remains of living creatures from the geological past", to exclude, for example, coal seam components that are no longer "recognizable" as the remains of organisms. Since the passage from the biosphere to the lithosphere is transitional, however, varying degrees of recognizability are inevitable.

The marked differences observed between fossils collected from strata that, from their superposition alone, were known to be of different ages, early led to the recognition that organisms have changed progressively with time. The demonstration of these changes is one of the aims of paleontology which in this way supplies knowledge of the course of evolution [9]. A recent work is available summarizing this distribution of organisms in geological time [10]. The recognition of these changes led to the deduction that the relative geological age of strata could be determined from their fossil content, and this method forms the basis for establishing the chronology needed to order earth history or stratigraphy [11, 12].

The study of fossils as originally living organisms in relationship to their ancient environment is known as paleoecology [13–15] a subject which is closely linked with paleobiogeography [16, 17], the study of the former distribution of plants and animals. In turn, stratigraphy, paleoecology and paleobiogeography form the basis for the reconstruction of paleogeography. Paleobiogeography may be assisted by geochemical means. Modern biogeochemical provinces are those regions of the earth's surface that differ from adjacent regions in their content of certain elements or compounds, and which thereby impress specific biological characteristics on the local fauna and flora. Major changes in paleogeography occurred in the past, and it should be possible to reconstruct paleobiogeochemical provinces from the petrology and geochemistry of the rocks as well as to seek independent evidence for them by geochemical analysis of their contained fossils [18].

The basic data for these studies are fossils systematically classified and accurately identified by name [19]. The reliability of identifications of fossils as well as the information that can be gained from their study may be increased by the use of a variety of analytical and observational methods [20] which have recently been summarized [21]. Successive attempts to lay proper emphasis on other than purely systematic approaches to fossils [22] were signified by the introduction of such terms as paleobiology and paleoethology, neither name now used extensively. With the fuller recognition that true taxonomy involves explanation as well as description [2] the need for such division ceased to exist. Within a unified paleontology, however, it is useful to recognize disciplines by adding the prefix paleo- to the relevant biological disciplines, as illustrated in the above paragraphs and by other more specialized examples, such as paleoneurology [23], paleopsychology [24], paleopathology [25], paleobiochemistry [26] and paleogenetics [27]. The subject of this volume, organic geochemistry, is a recent addition to these disciplines, and exemplifies reductionism in paleontology: the study of successively lower levels of organization. Explanation by reduction implies its opposite, which SIMPSON [2] has termed composition, "explanation goes up as well as down the scale of levels of organization". The relating of geochemical discoveries to geology is thus a compositional process, requiring knowledge of paleontology acquired from all its disciplines, as well as from the rest of geology.

When considering these and all other aspects of paleontology the effects of fossilization must be evaluated, even though their study is not a prime objective.

II. Taphonomy: The Study of Fossilization

The study of the processes involved in the conversion of a formerly living organism to a fossil is known as taphonomy [28–30]. The fossilization processes have ultimately to be reconstructed from a study of the available end products, the fossils and their enclosing sediments. Since these end products are the results of many processes, often conflicting or reversible, attempts have been made to investigate the several component processes separately and also by experiment. Almost all fossilization processes can be considered to have a selective filtering action, allowing certain remains to survive whilst destroying most others. For an organism to be preserved as a fossil is the exception rather than the rule and the fossil record is inevitably an incomplete one. SIMPSON [31] has estimated "that on the order of 1 to 10 percent of all the species that ever lived are recoverable as fossils and that, of those, on the order of 1 to 10 percent have so far been found and described". That these figures are of the right order of magnitude is suggested by the following examples. In two different modern environments, tropical river alluvium and reef limestone, NEWELL [32] has estimated that only about 0.1 percent and 2 percent of the numbers of species inhabiting the respective areas are preservable. In a well documented Recent example [33], the number of fish scales recovered from surface sediments of a shallow lake was only 4 percent of the number expected from fish that died in the pond. There are several sources of bias which have to be taken into account in estimates of survival in the fossil record. SIMPSON [31] has listed and discussed collection and other sources of bias in detail and some of them will be discussed in the present chapter.

Many accounts are available describing and classifying the states of preservation of fossils [5, 8, 34–36] and the reader is referred to them for a systematic

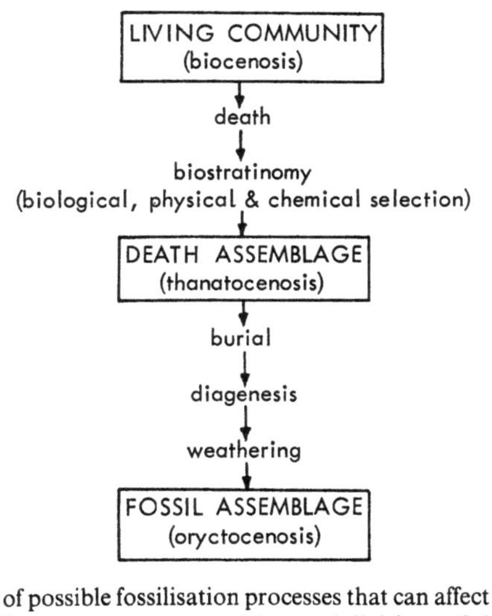

Fig. 1. Simplified flow chart of possible fossilisation processes that can affect a community of organisms; feedback can occur from all post mortem stages. Many subdivisions of the death assemblage can be recognised based on mode of formation [91, 95], and these are necessary for paleoecology. Additional assemblages that have been named include the remnant assemblage (liptocenosis) and burial assemblage (taphocenosis) but these appear to have little value

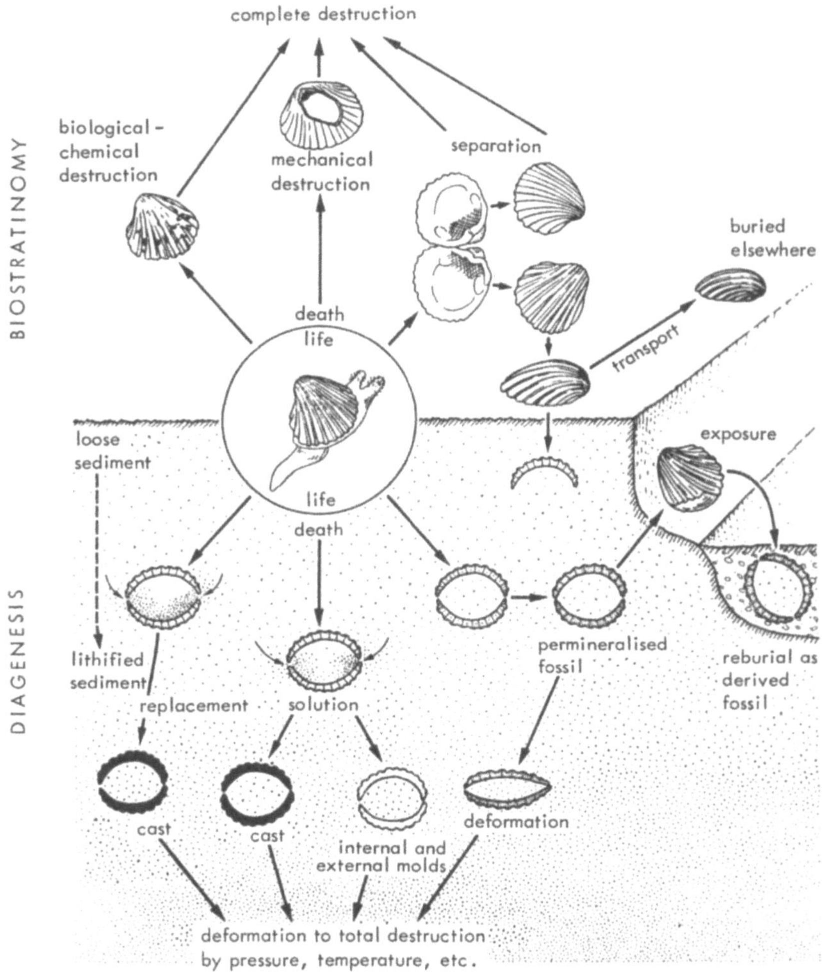

Fig. 2. Diagram to show some of the processes that may affect an organism after death, to suggest most of the field of taphonomy. (After THENIUS [201], Fig. 2, modified)

treatment since this topic will receive only incidental mention here. The processes of fossilization responsible for these various states of preservation, however, have only rarely been discussed outside the German literature, and a consideration of these processes is most important to organic geochemistry. An attempt will therefore be made to set out a framework for discussion of these processes, emphasizing factors which should be borne in mind by the geochemist. A flow chart and schematic illustration of some of the fossilization processes to suggest the field embraced by taphonomy are given as Figs. 1 and 2; Fig. 3 portrays the possible fates of components of living communities and death assemblages.

There are many incidental references to individual aspects of fossilization scattered throughout the literature of paleontology, and many other relevant facets may be found in literature on soil science, fuel technology, limnology,

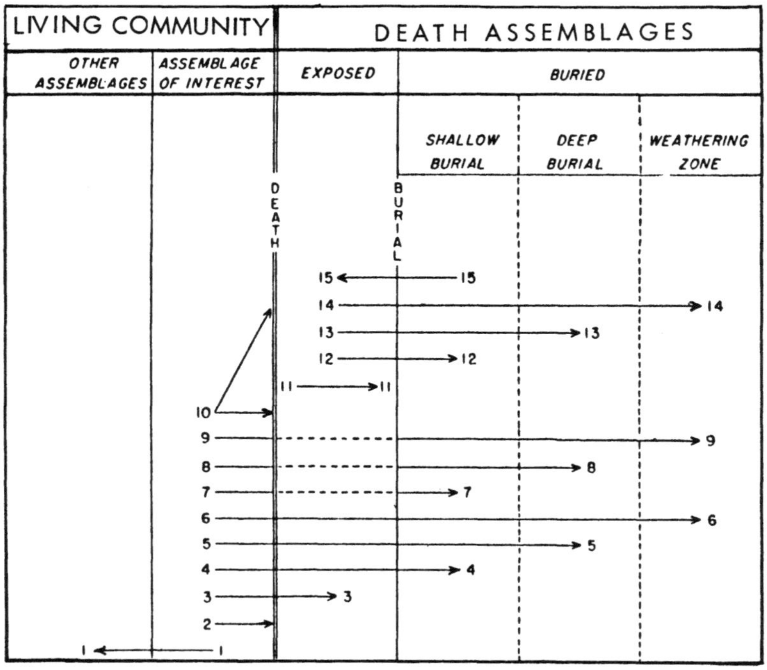

Fig. 3. The possible fates of components of life and death assemblages. Case 1 – emigration; 2 – death and immediate removal; 3 – remains enter exposed death assemblage but are destroyed or removed; 4, 5, 6 – remains enter exposed death assemblage and are buried; 7, 8, 9 – remains of infaunal organisms *in situ*; 10 – nonliving parts discarded during lifetime; 11 – remains from other death assemblages are introduced but destroyed or removed subsequently; 12, 13, 14 – remains from other death assemblages are introduced and persist to be buried; 15 – buried remains are reintroduced to exposed death assemblage by erosion. Cases 4, 7, 12 – remains are destroyed or removed while buried at a shallow depth; 5, 8, 13 – remains are destroyed or removed by diagenetic processes; 6, 9, 14 – remains persist and eventually are exposed to weathering. (After JOHNSON [202], Fig. 1)

microbiology and general biochemistry among others. It should be emphasized that this article is in no sense meant to be exhaustive and doubtless much will necessarily have been overlooked. It is likely that the organic geochemical approach will be a stimulus to research in taphonomy.

III. Biostratinomy*

Biostratinomy is the investigation of changes and processes which have affected the carcass of an animal or remains of a plant from the time of their death until burial in a sediment [37, 38]. The term may also be used to denote those changes and processes. The restriction of the term biostratinomy simply to a study of the relative dispositions of fossils to each other and to the enclosing sediment appears unjustifiable. The study of present day phenomena as a means of elucidating the biostratinomy of fossils has been called, paradoxically, actuo-paleontology [39].

* In the literature, terms dealing with fossilization have been used in different senses by different authors. The usage here follows that of the original authors wherever possible.

A. Biological Factors in Biostratinomy

All organisms occupy some place in a complex food web and large numbers of them will be destroyed by predation, scavenging or microbial decay in early stages of what may be regarded as the biostratinomic pathway. An example is afforded by the copepod crustaceans. Free swimming copepods are today among the most numerous of marine animals, being produced at an estimated rate of nine hundred per cc per year, but they form the base of the pyramid of marine life and are probably totally consumed by predators. Although their small size and lack of strongly mineralized skeleton are also important factors, this fact of predation helps to explain the great rarity of copepods as fossils in the geological record, and the fact that free-swimming forms are unknown until the Tertiary (and then only in freshwater deposits). Parasitic representatives are known much earlier in the form of cysts in Jurassic echinoid skeletons, a highly specialized adaptation but one favoring preservation (and incidentally implying a much earlier origin for the free swimming forms from which the parasitic forms evolved).

The nature of the living organism's adaptation to its environment contributes largely to its post-mortem fate: sessile benthos or infauna (animals living in sediments forming the sea floor) for example, are more likely to be preserved than other aquatic creatures, and aquatic creatures more likely to be preserved than terrestrial. One striking example of this is the concentration of amphibians in a Permian siltstone described by DALQUIST and MAMAY [40]. Of four hundred skeletons, 90 percent represent *Diplocaulis*, 8 percent *Trimerorhachis* and only 2 percent others. The deposit is thought to have formed by the drying out of a watercourse, allowing time for the truly terrestrial forms to walk away from the area, and thus accounting for their absence from the deposit, but killing the amphibians *en masse*. Such adaptations favouring fossilization may, however, be anulled by other factors. CRAIG and JONES [41] have suggested that although the chances of transportation of the shells of epifaunal benthos (those animals living on the sea floor) in the Irish Sea are greater than those of infaunal species, the probability of their destruction is not necessarily greater since infaunal skeletons are weaker. Furthermore, only one third of the infauna have preservable hard parts compared with over half the epifauna; overall, epifaunal species are about four times more abundant than infaunal.

1. Structure and Composition

Organisms with hard parts are more likely to be represented as fossils than those without. Approximately 44 percent of all known phyla lack such preservable hard parts [31] and the rarity of such phyla in the fossil record is one of the chief causes for the long recognized imperfection of that record. It is partly for this reason that plants are generally rarer than animals as fossils. A thorough review of the inorganic chemistry [42] and mineralogy of organic tissues has recently been provided by LOWENSTAM [43] and much data is available in the biological literature [44–46]. The structure of the hard parts themselves can influence their preservability and the fate of such hard parts cannot be considered simply in bulk physical or chemical terms; account has to be taken of microstructure and the relative disposition of hard parts and readily decomposable soft parts. The lamellibranch *Nucula nucleus* is common in the bottom fauna of the Deep Channel near Heligoland whereas the similar sized *Venus ovata* is rare. Yet quantitative determinations of shells occurring in the sediment show the reverse relationship. A recent change in fauna could explain this, but KESSEL [47] states that it is due to the relatively greater amount of organic matrix between the prisms of *Nucula* shells which leads to a more rapid breakdown of the aragonite shell than occurs in *Venus*. On the other hand the preservation of organic material may be favored by a microstructure which effectively seals off the compounds from the action of degradative agents.

With the exception of plant cuticles and the exines of spores (composed of the very stable cutin and sporopollenin), structures which appear tough and resistant in plants may not always be those least likely to be broken down. Many instances are known of the preservation of the delicate growing points and other non-woody parts of plants. The several layers of the plant cell wall differ not only in chemical composition but also in ultratexture due to the characteristic arrangement of the cellulose microfibrils [48–50]. This is of considerable importance in connexion with the degradation of plant material both before and after burial. The course of structural degradation of the thick-walled lignified elements of wood has been demonstrated by BARGHOORN [51–54], and by BARGHOORN and SCOTT [55]. It has been most frequently observed that the central layer of the secondary wall (the S_2 layer) is the first to decay, while the primary wall and adjacent outermost layer of the secondary wall (S_1) are only affected in badly degraded specimens.

Size and shape are also important factors affecting preservation: small, compact corpses tend to be preserved intact more readily than those of large organisms or organisms in which the body is diffuse or much branched.

In the unusual conditions of formation of the Eocene Geiseltal lignite, the tissues of small vertebrates tend to be better preserved than those of larger animals, presumably because the tanning fluids could permeate the bodies of these small animals more rapidly than larger corpses. Bulky plant tissues probably stand a better chance of good preservation because their permeable cell wall material (absent in animal tissues) permits rapid infiltration.

2. Autolysis

Many organisms, or their deciduous organs or tissues, will have undergone natural processes of senescence [56] before becoming involved in the biostratinomic process. This fact is particularly relevant to the geochemistry of fossil leaves. Before the leaves of the broad-leaved trees fall, there is generally a decline of proteins, nucleic acid and chlorophyll, accompanied by mobilization of carbohydrates and organic acids and an increase of yellow and red pigments. Many esoteric metabolites doubtless remain in fallen leaves in addition to the surface lipids, pigments and lignin skeleton that are common to most. Terpenoids, although commonly deposited in greatest quantity in the perennial parts of trees, are also present in leaves. This applies particularly to the conifers since their leaves, whether seasonally deciduous or evergreen, are not readily decayed and may build up into considerable deposits on the forest floor. The formation of oil-bearing soils through this process has been suggested.

Up to 90 percent of the dry weight of an arthropod cuticle may be resorbed during the rapid enzymatic autolysis that takes place before moulting [57]. The nature of the compounds resorbed from the old cuticle varies with the arthropod, but both inorganic and organic constituents may be removed. Some moults are distinguishable from corpses on morphological grounds. Decomposed moults of crabs differ from corpses in that the fourth cephalic segment remains attached to the thorax, rather than detaching with the carapace [58]. Chemical data from Recent moults and corpses could usefully be compared with data from their fossil counterparts. Many cells in a large organism will continue life processes for a considerable time after somatic death. Tissue death and complete disorganization eventually occurs when reserves have been metabolized. Living tissues deprived of oxygen will continue to operate enzyme systems resulting, for instance, in deaminations, decarboxylations and hydrolytic processes. The tissues of green plants may still respire sugars through lactic or alcoholic fermentation paths under anaerobic conditions, thus providing energy for other catabolic processes. Autolysis will result in a considerable alteration in the chemistry of the tissues and may also alter the subsequent course of microbial degradation where this follows. Little is known of the possibilities of inhibition of autolytic processes in biostratinomy, but it may be conjectured that in many cases the infiltration of substances inhibitory to enzyme activity has occurred early in the fossilization process. This is of course open to experiment.

3. Viability

The tissues of seeds and the spores of lower organisms and microbes may remain viable for lengthy periods [59, 60], and those of some species will be continuing many of the life processes in the state of dormancy when other parts of the same organism or species are well advanced towards fossilization. Seeds of the Indian Lotus (a water-lily, *Nelumbo nucifera* Gaert.) may still be viable, that is capable of germination if returned to favorable conditions, after burial in peat for over one thousand years [61], possibly as long as 3,000 years [62]. Clearly these seeds have resisted not only the degradative attack of microorganisms but also the chemical action of the peaty water which is often toxic to all but select groups of microorganisms. It must be remembered that these seeds are utilizing stored material in order to keep their life processes going and that however slowly these are proceeding, they will ultimately lead to the breakdown of structure. The longer a tissue remains alive without an external energy supply the greater will be the internal degradation. From the point of view of fossilization, the case of the Indian Lotus is of little consequence, for unless it is killed and chemically fixed early, it cannot be expected to produce a structurally well preserved fossil, despite its apparent resistance. The fact that quiescent tissues may retain their *chemical* integrity for so much longer than the material of the matrix in which they are buried (e.g., peat) is of some importance to the geochemist. The possibility of the presence of living material, particularly in young sediments and soils, should be constantly borne in mind since their inclusion in samples for analysis will substantially affect the conclusions regarding the persistance and degradation of certain organic compounds in the sediment. A simple reliable test for seed viability using tetrazolium salt has been described and may be of wide application [60, 63].

4. Attack by Other Organisms

Microbial decay (Chapter 11, this volume) of organisms begins immediately after death, unless inhibited by rapid burial or other less usual factors such as desiccation, the presence of antibiotic substances secreted by other organisms [64], or large populations of bacteriophages or predatory organisms which considerably reduce the population of the bacteria and fungi responsible for decay. Early suspension or exclusion of microbial decay may be partly or solely responsible for the retention of decomposable organic material in those fossils from which it has been isolated.

The rate and nature of decay, whether aerobic or anaerobic, are influenced by a number of factors. From forensic science it is known that moisture speeds the decomposition of human carcasses by the bacteria normally present in the body and that the optimum temperature range is 21°–38° C (70°–100° F). Diseased subjects decompose more quickly than healthy ones, fat subjects more quickly than lean, and old individuals rather slowly. Eight progressive stages have been recognized in aerobic decomposition of exposed human bodies and these stages doubtless apply to other terrestrial vertebrates [65]. Associated with each period of decomposition are different characteristic scavenging insects. From a study of the insects found associated with mammalian skeletons in the Californian Pleistocene asphalt pits, Pierce [65] has found that the first five stages are represented, indicating that the submergence of the carcasses was slow and took up to five months, whereas at present day slower-submerging seeps, seven stages can be observed. The eight stages may be summarized as follows (an asterisk indicates those genera found fossil at the La Brea pit).

Table 1

Period I	Fresh carcass; autolytic decomposition of proteins to amino acids by enzymes. Fermentation by larvae of flies: *Calliphora*, *Muscina*, *Musca* (*puparia of which are found fossil).
Period II	Bacterial decomposition for first three months, breaking down amino acids to ptomaines. Flies: *Sarcophaga*, *Lucilia* (*puparia); Bacteria: *Proteus*, *Escherichia*.
Period III	Formation of fatty acids and start of caseous product formation, third to sixth month. Beetle: **Dermestes*; Moth: *Aglossa*.
Period IV	Formation of caseous products such as adipocere, third to sixth month. Beetles: *Korynetes*, *Necrobia*, *Piophila*, *Anthomyia*.
Period V	Ammoniacal fermentation, black liquefaction in fourth to eighth months. Flies: *Phora*, *Lonchaea*; Beetles: *Ophyra*, **Silpha*, *Necrodes*, **Nicrophorus*, **Hister*, *Saprinus*; Bacteria: *Proteus*, *Bacillus*.
Period VI	Desiccation, sixth to twelfth months. Scavengers principally mites.
Period VII	Extreme desiccation, one to three years.
Period VIII	Debris, over three years. Beetles: *Ptinidae*, **Tenebrionidae* (significance unclear).

From such detailed palaeoecologial associations inferences can be made as to the state to which the organic material is likely to have decomposed. Conversely, the organic geochemical facies could independently suggest what processes of decomposition a fossil had undergone, and thus enable the reconstruction of a more detailed paleoecology.

Little is known of the detailed chemical changes which occur during decay in the geologically more important subaquatic environments, partly because of the difficulty of study under such conditions. KRAUSE [66] has studied post mortem *in vitro* breakdown of plankton under fresh water, and notes that one quarter of the mean weight is lost immediately after death, probably by the loss of soluble body substances due to an increase in permeability of the cell membrane, a normal accompaniment of cell senescence. Almost one third of the body substance is lost within a few hours, one half within one day, and only one quarter remains after one month. The breakdown of such plankton is only slightly faster under aerobic than anaerobic conditions. In natural waters, this dead plankton would approach the lake floor, at some time after the first day. SCHÄFER [39] has recently published detailed accounts of the grosser aspects of decay for one region, the North Sea. Whale corpses on a beach, for example, will suffer a rapid aerobic decay of organic compounds forming gases in the visceral cavities which lead to inflation of the body. Mummification of some soft parts may ensue due to the evaporation of water, but when the body cavity is eventually breached, viscous blubber escapes, forming a dark brown aureole around the carcass, impregnating and cementing the sand into concretions. Similarly, adipocere can accumulate by saponification of the fats, which are more resistant to decomposition than proteins. This can occur during decomposition under water or under moist conditions on land and has been responsible for the preservation as "pseudomorphs" within historic times of carcasses of terrestrial and marine animals [67–71]. The oldest fossil preserved by this means appears to be from the Geiseltal Eocene, where the brain and spinal chord of a frog were preserved. Later occurrences include the Pleistocene giant deer *Megaceros* and mammoth, with well-preserved subcutaneous fat; the literature has been summarised by BERGMANN [72].

Several observations have been made on the decay of fish [39, 67, 73] which suggest that articulated skeletons can only result from the slow processes of anaerobic decay. Under aerobic conditions decay is complete within a week or two, although increased salinity will retard decomposition. Carcasses with large visceral cavities float to the surface due to gases developing in the body cavity and the hard parts are strewn as fragments over a wide area. When anaerobic conditions prevail, however, as in stagnant basins, scavenging benthos is precluded, skeletons are articulated and well-preserved and putrefaction proceeds slowly, resulting in the preservation of organic material.

Some plant material dug out of lake sediments of Recent origin still appears green and little decayed. Chlorophyll has been identified in the Geiseltal Eocene lignite. Observations on the fine structure of such material as well as experimentally treated plant material are certainly required. In connection with their investigations on the origin of coal balls, STOPES and WATSON [74] carried out an experiment to test the suspected preservative powers of sea water [75]. Fresh plant parts were placed between layers of peat in a glass vessel and covered with sea water to which were added various living and dead mussels and cockles along with fragments of shells of these creatures. The jars were completely filled with sea water and covered. After ten months almost all traces of the animal soft parts had disappeared, leaving only the shells, and the water was clear and without smell. The plants were found to be green and fresh-looking, and when sectioned the cell contents were seen to be more or less perfectly preserved with chloroplasts and nuclei still intact. Some of the plant material was then sealed in a glass tube with some of the clear liquid and left exposed to light. A footnote in their paper shows that no change had taken place over a further two years.

Some idea of the length of time that has elapsed since death and before burial may be obtainable from the attitude of the preserved hard parts. During decay maceration of the hard parts commonly follows a definite sequence which can be determined for specific conditions, as MÜLLER has described. Vertebrates tend to have their jaws detached at a very early stage in this sequence, and this has been so frequently noted in fossil skeletons that WEIGELT [76] referred to it as his "lower jaw law", although there are certainly exceptions [77]. Skull and limbs then become disconnected, ribs loosen; limb disarticulation is followed by vertebral column disarticulation, but before this is completed, disintegration of the bones starts. Selective scavenging of large well-muscled limb bones has been invoked for their rarity in the Alaskan Pleistocene, and for the comparative abundance of the less-muscled, denser lower limb bones [78]. A Recent clupeid fish carcass arches dorsally, reaching a maximum recurvature three days after death, before the ventral side splits open. Subsequently the head recurves more or less strongly until finally it becomes detached from the body. All stages in this sequence are preserved in fossil clupeids from the Oligocene of the Mainz basin [30]. By applying such observations to fossil assemblages the least decayed specimens should be recognizable which should in turn yield the most organic chemical information. One such detailed study has been made of the Liassic ichthyosaurs of the Holzmaden Jurassic Posidonienschiefer, which has enabled the probable conditions of death and burial to be reconstructed [79]. Some organic geochemical data are also available for this deposit [80].

Saprophytic breakdown of tissues by fungi, algae and bacteria may form new organic compounds *in situ*. The paths of attack followed by these organisms may be dictated by the microstructure of the host organism. MOORE [81] (see Chapter 11, this volume) has described how Carboniferous Coal Measure plant structures may be completely pseudomorphed by filaments of fungi, presumably the organism responsible for much of the decay.

Parasites and symbionts may hasten degradation. Thus mycorhizal associates typically occur in situations rich in organic material, such as forests and heaths, and lead to rapid aerobic degradation of the underground parts when the host plant dies.

Boring organisms of at least eight invertebrate phyla and three plant groups may hasten the breakdown of hard tissues [46, 82, 83] partly by direct destruction of organic and inorganic tissue, but also by enabling bacteria to gain access [84] and by reducing their resistance to mechanical breakdown. Some of these borers feed off the organic matrix as they ramify through dead or living skeletons, but the majority do not [48]. Boring occurs by mechanical means, or through the agency of acids, CO_2, enzymes, or sequestering agents. Borings have been described from many fossils [85–87]. Boring marine algae, possibly also fungi and bacteria, are instrumental in causing the centripetal replacement of aragonite skeletons by micrite – a fine grained mosaic of aragonite crystals [88]. Skeletons are first bored or pitted, the penetrating filament then dies and decays, and the vacated tubes are filled, by some unknown process, with micrite [89]. By repetition of this process, observed by BATHURST in sediments from a Bahamas lagoon, much of the shell comes to be surrounded by a micrite envelope. For some reason, this envelope resists diagenetic solution and remains after the skeletal aragonite has been dissolved during diagenesis, forming a mould within which a growth of coarse calcite crystals can take place to form casts of the original skeleton. At least some of these algae secrete a chemically stable mucilage that enables the micritization process and thus plays an important part in the preservation of hard parts. SHEARMAN and SKIPWITH [90] have claimed that these mucilages may be recognizable as acid-insoluble organic residues in rocks of all ages back to the Lower Paleozoic, and that they account for a large part of the "kerogen" of limestones.

B. Physico-Chemical Factors in Biostratinomy

These factors vary according to the environment in which the organism dies, and their parameters are being defined as the result of recent work by geochemists, sedimentologists and paleontologists [11, 13, 91–93]. The effects of these factors on the dead organism will often be the same as on inorganic particles. Sometimes, however, the properties peculiar to organic remains will lead to their selective destruction or preservation under certain physico-chemical conditions.

1. Dynamic Selection and Mechanical Destruction

As during sedimentation of mineral particles, dynamic selection of organisms may take place during transport [29]. The dead remains of various forms may be transported together in associations that do not reflect any single life environment. Under these conditions there is much winnowing and sorting of organic remains by size, shape, and effective specific gravity. The remains of young and small organisms may thus be separated from those of older and larger forms, although HALLAM [94] has recently shown that for marine molluscan death assemblages, water-current sorting is less important than the selective destruction of small size grades. Less commonly, the fossils of an older rock formation are weathered free of rock matrix and become incorporated in a younger formation in association with fossil forms that lived much later and under quite different conditions [32]. All such assemblages have been termed death assemblages or thanatocenoses (Fig. 1) to distinguish them from the original community living in one area at one time, the biocenosis [91, 95]. Studies of the orientation of skeletons within the thanatocenosis are important in paleoecological reconstruction [30].

Mechanical destruction of invertebrate skeletons by abrasion under water has been investigated by several workers [94, 96–99]. CHAVE's results show that the durability of skeletal materials is more variable than that of normal sedimentary mineral grains. The skeletal mineralogy does not exert a clear influence on this durability, and although size has some effect on durability it is not a controlling factor. The major factor as noted in the last section is the microstructure and the correlated disposition of organic matrix in the skeleton. Thus dense, fine-grained skeletons are more durable than skeletons with much openwork. The destruction of mollusc shells has also been studied in Recent shallow seas, and the results of this applied to a number of fossil occurrences [100]. The shape of a skeleton will affect its dynamic properties; gastropods, for example, usually withstand fragmentation better than lamellibranchs [94].

2. Oxidation–Reduction and Hydrogen Ion Potentials

Oxidation–reduction (redox–Eh) and hydrogen ion potentials (pH) are important factors for the delimitation of stability fields of sedimentary and skeletal particles. A classification of non-clastic, non-evaporite sediments based on pH and Eh is shown as Fig. 4, and a similarly based diagram showing the field occupied by the natural environments as Fig. 5. The actual range and distribution

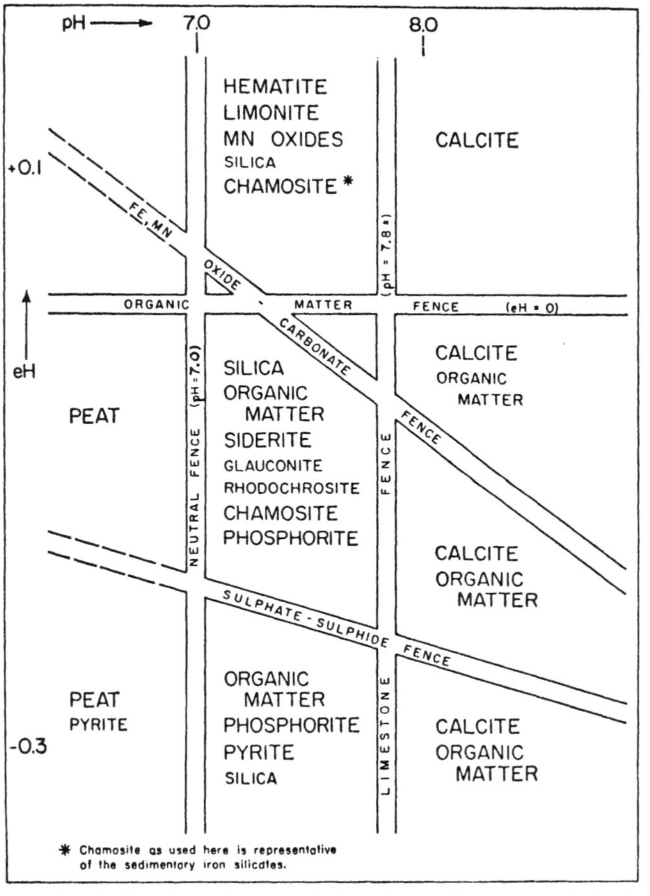

Fig. 4. Sedimentary chemical end-member associations (evaporites excepted) in their relations to environmental limitations imposed by selected Eh and pH values. (From PETTIJOHN [93], Fig. 155, after KRUMBEIN and GARRELS [203], Fig. 8)

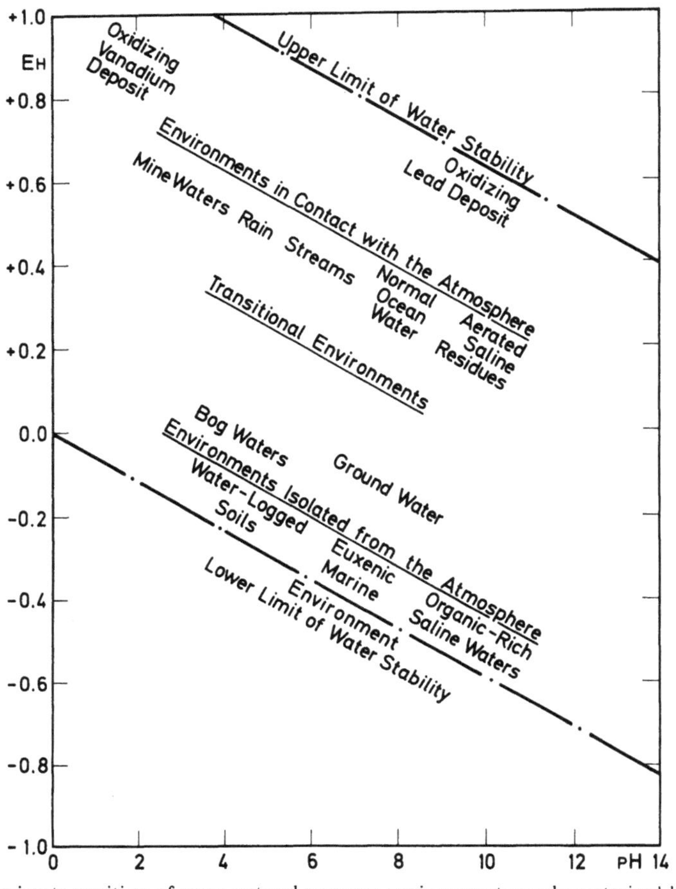

Fig. 5. Approximate position of some natural aqueous environments as characterised by Eh and pH. (After Garrels and Christ [174], Fig. 11.2)

of Eh–pH measurements made in natural aqueous environments is shown in Fig. 6. It is important to recall that Eh and pH are dependent variables [101], being symptomatic of biological and chemical reactions that are occurring. Since living systems play a major role in controlling the aqueous milieu, measurements of Eh–pH may often indicate ecology [102]. Conversely, if Eh–pH conditions could be reconstructed from the mineral species, and ultimately perhaps from the organic molecules present in a fossil and its surrounding sediment, detailed paleoecological reconstruction would be possible.

In the presence of oxygen (aerobic environment with positive Eh) rapid decomposition takes place and the organic content is largely destroyed, even the most resistant parts being metabolized [103]. In the absence of oxygen (anaerobic environment, negative Eh) putrefaction occurs with extensive resynthesis by micro-organisms commonly yielding a preservable body, often rich in organic material. Such anaerobic conditions often occur in poorly ventilated marine basins, the classic example being the Black Sea which is rich in H_2S below 150 m and consequently accumulates up to 35 percent organic material in the bottom deposits.

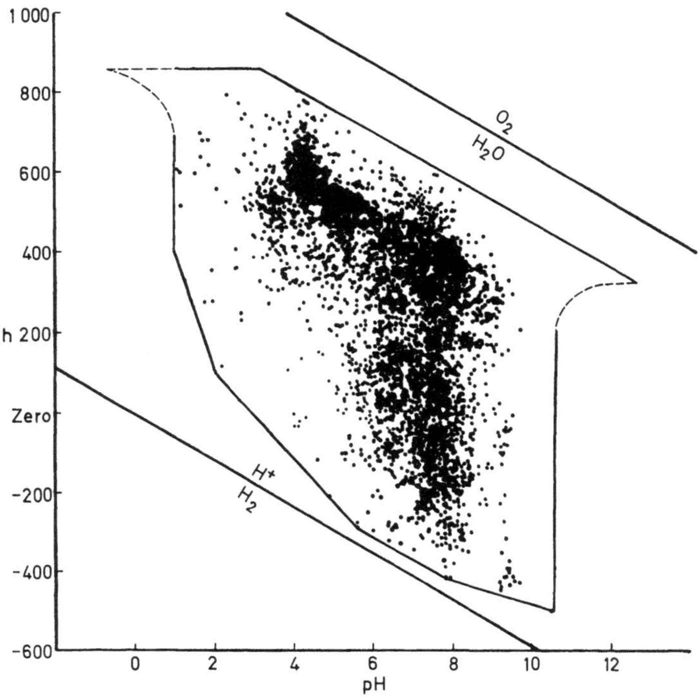

Fig. 6. Distribution of Eh–pH measurements of the natural aqueous environments. (After BAAS-BECKING *et al.* [102], Fig. 31)

The position of the zero oxidation–reduction surface relative to the sediment–water interface, whether above, coincident with, or below it, will determine whether chemical changes take place before, during or after burial (i.e., during diagenesis) in aquatic environments. It may be impossible to distinguish in a fossil when certain of these changes took place and hence their effects will be considered together here. It may also be difficult to determine whether particular features observed in fossils are the result of biological or physico-chemical activity. An example of this is the difficulty in assessing whether the removal of cellulose from the cell walls that occurs during structural degradation of plant tissues is due to microbial activity or to chemical hydrolysis. The anatomical characteristics and chemical composition of ancient wood from presumed sterile environments are very similar to those of wood that has been degraded by fungi. The course of structural degradation in the cell walls of the woody tissue has been mentioned briefly in the previous section in relation to the microstructure and chemical composition of the cell wall material. The sequence of degradation has been studied in most detail in ancient and buried wood which shows no clear evidence of extensive attack by microorganisms [51–54, 104–107]. The investigations of VAROSSIEAU and BREGER [108] are of special interest since the samples of degraded wood studied by them were of known age and had been buried under more or less sterile conditions and in similar circumstances. The wood, mostly piles under buildings ranging in age from 30 to 600 years, was exposed when Rotterdam was bombed in 1940. Their results clearly confirmed earlier reports that lignin is the most resistant of the cell wall constituents under the conditions of burial, and also demonstrated that in the most decayed parts of the older samples the methoxyl content of the lignin was reduced indicating some degradation of the lignin. (In this connection reference may be made to a recent Symposium on the degradation of lignin in geological environments [109].) Cellulose was found to disappear progressively with age resulting in the characteristic "lignin enrichment" of the decayed wood. This was greatest in the outermost layers of the wood (Fig. 7). Humic materials which are not present in new wood were found in samples 450 years old.

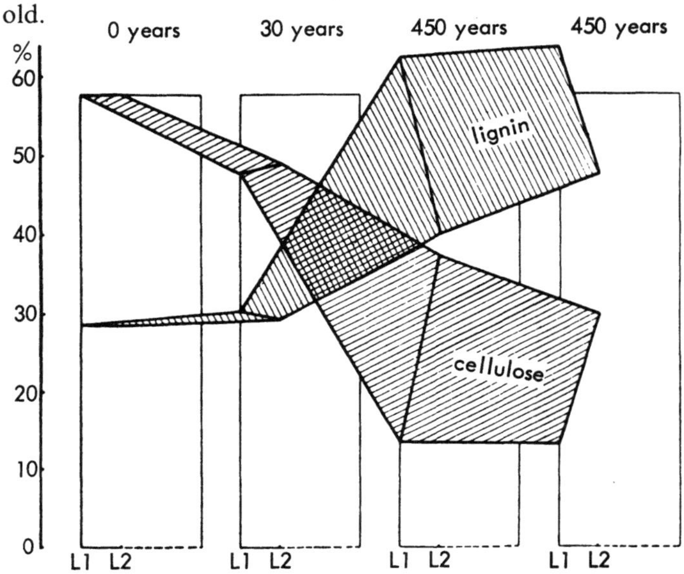

Fig. 7. Cellulose and lignin content of two successive, 1-cm outer layers (L 1, L 2) in four spruce samples which were buried for an increasing period of time. The cellulose content in the outermost 1-cm layers decreases rapidly with increasing age. The cellulose content of all second layers shows a similar although not as pronounced trend. The lignin content of two successive layers of each pile increases with increasing duration of burial and is greatest in the outermost layer.
(After VAROSSIEAU and BREGER [108], Fig. 6)

Cellulose has not been identified in fossils older than the Tertiary (up to 50 million years) and this has a much lower DP* than normal cellulose from living plants. This reduction in size of molecule however probably occurs fairly rapidly, geologically speaking, for a similar reduction in DP is found in cellulose from the wrappings of Egyptian mummies, *ca.* 5,000 years old. Lignin has been demonstrated in Paleozoic plants as dioxane-soluble fractions giving the characteristic color reactions for lignin, and as extracts with a typical (non-angiosperm) lignin UV absorption spectrum [50].

3. Microenvironment of the Decaying Organism

The decaying organism may create its own microenvironment exhibiting different Eh–pH conditions from that of the macroenvironment. EMERY [110] has noted the presence of pyrite internal molds of foraminifera, diatoms and radiolaria

* DP = degree of polymerization.

in the tops of cores from the sea off southern California. Such early pyrite is formed in neutral or alkaline environments by the reaction between the sulphide resulting from breakdown of organic matter and bacterial reduction of sulphates, and iron brought in as detrital minerals [111, 112]. Since the tops of these cores are above the level of zero Eh, EMERY has suggested that local spots of negative Eh were produced by the decomposing protoplasm of the organisms more or less isolated from the positive Eh of the matrix. Pyrite infilling of plant and animal cells or skeletons may occur very soon after death [113–118], and favor their preservation as fossils. One of the few experiments on fossilization that have been performed showed that humified wood could be replaced by marcasite by the action of colonies of anaerobic sulphur bacteria living in the wood cells in the presence of iron and calcium sulphate solution [119, 120].

The difficulty of interpreting when such mineralization of a dead organism took place is well illustrated by pyrite spherules. These are known from rocks of all geological ages and were recently suggested to be formed by some unknown organism, now represented by the organic material enclosing and/or enclosed by such spherules, rather than being deposited within resistant parts of other organisms after death. This suggestion was only disproved by the failure to demonstrate any such organism directly precipitating pyrite in Recent sediments rich in spherules [112]. The presence of organic material within the spherules (LOVE's "matrix body") is due either to the incorporation of original cell contents or to the immigration, deposition, and possibly polymerization of soluble organic material interstitially after formation of the pyrite [121].

Glauconite is another mineral that may form under marine conditions before burial (i.e., during halmyrolysis) and may be associated with decaying organic matter. It readily fills in small shells such as foraminifera and thus leads to their preservation as casts. Similarly early replacement of hard parts may take place by phosphatization [122], dolomitization, silicification, and ferruginization.

Analysis of alewife fish adipocere "concretions" shows them to contain more than thirty times the total amount of fatty acids present in carcasses, as well as showing many differences in elemental composition [70]. This relative increase of lipid material during adipocere formation seems to be due to concentration from the surrounding medium, although others believe it to derive autochthonously from proteins [72]. Such concentration (perhaps analogous to some permineralization) is believed to occur rapidly and illustrates the chemical changes that can take place at an early stage of fossilization. It is known that skeletal debris can "scavenge" elements preferentially from sea water [118, 123], from sediments of the sea floor [124] (and also during diagenesis [125]) with an efficiency partly related to the internal surface area of the debris, which in turn depends on the size of the crystallites comprising the skeleton. Little is known about the possibility for similar scavenging of free organic molecules.

4. Salinity

Increasing salinity retards bacterial activity and, in extreme cases, preservation can occur or be enhanced by "pickling" in salt. Specimens of woolly rhinoceros from the Pleistocene asphalt deposits of Galicia have the skin preserved with traces of rock salt in it. Salt impregnation during deposition is thought to have prevented bacterial decomposition of the flesh and skin whilst in the asphalt [30]. Since organisms with hard parts are normally rare in hypersaline environments, however, fossils will also be rare.

5. Temperature Effects

Temperature plays a direct role in the rare cases of preservation by freezing, exemplified by Pleistocene woolly mammoth and rhinoceros of the Siberian permafrost. Desiccation, too, demands abnormal temperatures, and is thus likely to be confined to tropical and subtropical regions. Specimens of the Cretaceous dinosaur *Anatosaurus* are known with the skin preserved, probably by desiccation, and show how unusually good preservation may result from the suspension of normal decomposition by extreme conditions before normal diagenetic processes lead to the mineralization of the fossil.

Under more normal conditions temperature may control some aspects of fossilization. Although the proportions of mineral species precipitated by marine organisms may vary with temperature [43, 126], and thus provide a basis for paleothermometry [127], there is commonly some "vital effect" which will mean that after death the hard parts will be metastable. In some parts of the Cretaceous Chalk, aragonitic fossils are almost totally lacking, their original presence being demonstrable only from impressions (external molds) left on the attachment areas of calcite-shelled oysters, which prove that aragonitic shells were destroyed by dissolution in the bottom waters of the Chalk sea. Lowering of the temperature of sea water increases the solubility of aragonite and calcite and of atmospheric CO_2, thus lowering the pH. JEFFERIES [128] calculates that sea water will dissolve aragonite but not calcite at temperatures below 10° C. Thus temperature can be responsible for the selective dissolution of hardparts under marine conditions.

6. Selective Dissolution

Skeletal minerals are unstable under the conditions prevailing at the sediment-water interface in the ocean, with the order of decreasing solubility: high magnesium calcite, aragonite, low magnesium calcite, opal, apatite [123]. This leads to a selective dissolution of skeletons, and accounts for the relative enrichment of phosphates in ocean floor sediments [123] and the production of chemical lag deposits [129]. Such selective dissolution is well documented from both Recent and fossil examples [130–135], and is confirmed by phase equilibrium studies [97, 98, 136].

Ammonite shells, for example, are composed of aragonite whereas the aptychi of these shells are composed of calcite. In bituminous shales such as the Jurassic Posidonienschiefer (*v.i.*) therefore, ammonites are found largely dissolved out whereas their aptychi are well preserved [130]. Similar differential dissolution may occur even in limestones: the Jurassic Solnhofen limestone has aptychi preserved in the apertures of ammonites only seen in "ghost" outline [137].

Dissolution by sea water may be active at the very depths and latitudes where it should be most highly supersaturated with calcium carbonate. CLOUD [138] has explained the corrosion of limestones in lime-precipitating regions by diurnal variation in pH (7.7–8.3), probably caused by the liberation of CO_2 by plants at night [139], implying solution and precipitation according to the direction of change. The normal "compensation depth" below which skeletal carbonates go into solution is about 4,000 m, and results from the high concentration of carbon dioxide caused by increased hydrostatic pressure and reduced temperature. Solubility of calcium carbonate in sea water increases with salinity and decreased temperature; the apparent solubility product increases about threefold on exposure to 1,000 atm. pressure [140]. In the colder waters of Antarctica the compensation depth shallows to *ca.* 500 m [140]. No such compensation depth appears to occur for biogenous silica [141]. In the marine environment calcareous shells may be dissolved before burial (according to JARKE [142] even during life), in which case no trace will be preserved, or during and after burial (subsolution of HEIM [143], HOLLMAN [144]) whilst the sediment is still plastic; this results in all stages from "ghost" preservation to total destruction [145, 146]. Since the speed of solution is proportional to the effective solution surface of the shell, it implies that if large, thick-shelled organisms found in a deposit show signs of corrosion then destruction of thinner shelled and smaller organisms must have occurred [135, 142]. Although CO_2 is responsible for solution in free sea water, during burial under anaerobic conditions H_2S has a solution effect at least equal to that of CO_2. MOSEBACH [145] has demonstrated that it is not necessary for the H_2S to be oxidised to sulphuric acid to cause such solution, and this also explains the lack of gypsum crystals that should otherwise result from dissolution of shells. HECHT's [67] experiments have shown that the richer a sediment is in decomposable organic substances, the more likely it is that calcareous shells will be dissolved. HELLER [80] found a similar inverse relationship between organic matter and calcium carbonate in the Jurassic Posidonienschiefer and explained it by: increased solution by H_2S resulting from breakdown of SH-containing amino acids, increased acidity due to presence of short-chain organic acids resulting from various degradative paths, and by formation of water-soluble chelates with calcium by amino acids.

The absence or brittleness of bones in and around adipocere-preserved fossils has been explained by decalcification brought about by the fatty acids [72]. Phosphatic and "horny" skeletal parts (such as some brachiopods, arthropod cuticles, nautiloid anaptychi, hooks of belemnite arms, surfaces of fish scales, graptolites and snail opercula) will survive in acid environments where calcareous parts will be dissolved. It is also in these environments that organic compounds are likely to be preserved and hence such hardparts should repay the attention of the geochemist.

Lamellibranchs with an outer conchiolin periostracum will soon lose their inner aragonitic nacreous layer, as SORBY first observed. Eventually the whole shell will become flexible, and it is at this stage that impressions may be made on the shell, either by foreign objects or from other regions of the same organism ("Palimpsestformation" of RICHTER [135]). Finally the periostracum detaches in wrinkles from the dissolving shell [67, 135, 142]. Such corrugated lamellibranchs are known from many geological horizons and fossils are known with the periostracum and ligament preserved when the rest of the shell is represented only by molds or casts [147]. Similarly wrinkled fossil arthropod cuticles and deformed bones have been shown by experiment to result from decalcification [67, 86, 148], although doubts have been expressed as to the possibility of leaching out calcium carbonate from the bone mineral [130].

Selective dissolution may result during early diagenesis from the advance of a "solubility front" through a recently deposited sediment. This front may exist at the transition zone between oxidation and reduction layers and be characterised by acid pH values [142].

IV. Diagenesis

A. Diagenesis and Metamorphism

Diagenesis is the term used to comprise all those changes that take place after burial within a sediment and thus to its contained fossils, without crustal movement being directly involved and excluding the effects of weathering [149, 150]. With increasing temperature and pressure, diagenesis passes into metamorphism, with extensive changes in mineralogy and fabric of the rock. There is no natural boundary between diagenesis and metamorphism [151], but Fig. 8 illustrates the approximate temperature and pressure conditions responsible for diagenesis and metamorphism [152]. Whereas metamorphism tends to produce more uniform rock series by obliterating previous differences in composition and structure,

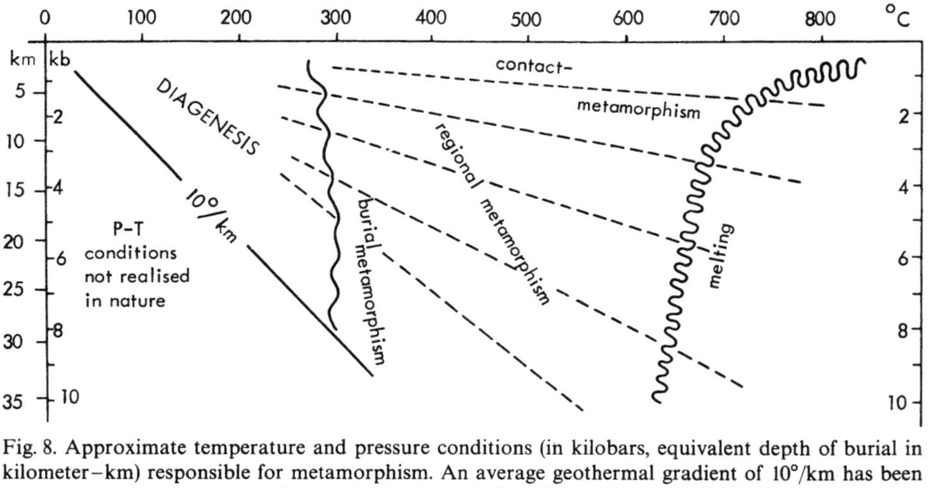

Fig. 8. Approximate temperature and pressure conditions (in kilobars, equivalent depth of burial in kilometer–km) responsible for metamorphism. An average geothermal gradient of 10°/km has been assumed in excluding the field below this line. (After WINKLER [152], Fig. 1)

diagenesis tends to differentiate sediments by unmixing components [153]. Although fossils are known from both contact and regionally metamorphosed rocks, they are rare. Their rarity in regional metamorphic rocks is largely explained by the fact that such metamorphism begins and centers in precisely that part of a geosynclinal belt which is characterized by largely unfossiliferous sediments [154]. Pressure-temperature fields for the formation of particular metamorphic rocks are known, and suggest that only the most stable families of organic compounds could survive for any length of time under such circumstances [155]. Experimental work has shown that after heating, mineral prisms in the lamellibranch shell will develop small gas bubbles released from the included conchiolin. Similar bubbles have been found in Middle Jurassic lamellibranch shells which are recrystallized by contact metamorphism [156]. Distorted fossils from metamorphic rocks have been widely used to determine the stress-field in which the rock formed.

B. Geochemical Phases of Diagenesis

FAIRBRIDGE [157] has suggested that diagenesis may be subdivided into three distinctive geochemical phases, each of which tends towards equilibrium, only to be upset by the introduction of a new set of environmental parameters. The first phase, syndiagenesis, is the early burial phase of bacterial activity in which the organic matter provides the nutrient for bacterial metabolism and also for infauna. The dissolution discussed in the previous section may be continued or initiated, and pyrite-preserving dead organisms may continue to grow below the zero redox potential level to form nodules. The second phase, anadiagenesis, is the deep burial phase of compaction and cementation, which may pass into metamorphism. Inorganic chemical reactions predominate, and authigenic (growth *in situ*) minerals form. The third phase is epidiagenesis, the meteoric phase which passes into weathering, and is initiated by tectonic emergence of the basin. Calcareous fossils that escaped syndiagenetic dissolution are commonly destroyed at this stage and are recognizable in the rock as hollow molds. Oxidation of pyrite takes place. Recurrences of anadiagenesis and epidiagenesis are to be expected in thick rock sequences containing unconformities.

C. Syndiagenesis

Syndiagenesis occurs within the top few meters of a newly formed sediment, and leads to great changes in the organic content of the sediment; these changes have been summarized by DEGENS [158]. Sediment inhabitants may rework the sediment and consume organic constituents, resynthesizing new compounds within the sediment, which may thus be thought of as "biometasomatism" [158]. The rate and completeness of these changes will be correlated with the position of the zero Eh level relative to the sediment-water interface, and be dependent on the sedimentation rate. The bacterial population falls off rapidly with depth, while Eh becomes increasingly negative and pH increases slightly, until more or less constant conditions are reached at a depth of a few meters. Porosity and water content continue to decrease downwards.

As has already been mentioned, many of the factors discussed under biostratinomy will operate also during diagenesis, particularly in syndiagenesis, and these need not be reiterated here. Attention will be confined to processes more characteristic of diagenesis.

Burial by some means is essential for fossilization, but the chances and rate of burial are obviously not the same in all environments. The rarity of upland animals and plants in the geological record is due to the fact that they inhabit regions of erosion rather than deposition [29]. On the other hand, aquatic organisms and those that live near water are in close association with the deposition of sediments and are favorably situated for quick burial. Consequently, the greater part of the fossil record consists of organisms that lived in water, especially in the sea or near the margins of streams, lakes or swamps. Thus most of the fossil record is the record of lowland or marine basins of aqueous sedimentation [32]. In such environments the decomposing organism may hasten its own burial since the ammonia released will create centres of alkalinity, around which silica will go into solution and carbonate precipitate [159, 160]. Further carbonate diffusing in will be precipitated, causing the early local growth of calcite-cemented concretions in which fossils will be protected from compaction, and in which residual organic compounds may be sealed [119]. The "burial" of organisms by lavas, although strictly non-diagenetic, has preserved a few fossils [161]. These occur largely as external molds although cell structure has been recorded from subsequently permineralized wood from such situations.

Unusual conditions of burial, although by definition rare, may be of exceptional importance in view of their preservation of organic compounds. Trapping within amber, known from several Mesozoic and Tertiary localities, or submergence in natural asphalt seeps are well documented examples. The Eocene lignite of Geiseltal is renowned for the remarkable preservation of skin, hair, connective tissue, muscle, cartilage, glands and vascular tissues, adipose cells, pigmented chromatophores, red blood corpuscles, and even epithelial cells with intracellular bridges and nuclei [38, 162–164]. This preservation probably results from the rapid tanning action of humic acids released during coalification, a process also responsible for the preservation of human carcasses in more recent peats [165]. Bones, although soft, are found in the Geiseltal lignite but not in nearby lignites. It is thought that calcareous groundwaters, derived from the Muschelkalk region to the south, permeated the brown coal swamp neutralizing the humic acids and enabling the bones to survive.

Such tanning is one means of histological fixation which can be accomplished by a variety of reagents and processes, which simultaneously inhibit enzyme autolysis and microbial action. Cell organelles may retain their integrity if the cytoplasm becomes fixed and subsequent mineralization would not need to take place to preserve them. The spiral chloroplast of a fossil *Spirogyra* from the Eocene [166], cells within the prothalli of Carboniferous ovules, and the cluster of pigment granules around the nucleus of the alga *Sphacelaria* from late Glacial clays, are examples of plant cell organelles preserved probably by fixation. It is possible that carbonized fossils include more material from the cytoplasm than permineralized fossils, although in both cases the cell contents will have been altered to humic or coaly material. Reports of recognizable cell inclusions, especially nuclei, occur in the older paleobotanical literature. Although some of these inclusions may be genuine [167], others may be masses of altered organic material derived from the original cell contents and inner parts of the cell walls. This is commonly seen in petrified tissues in which organic material may be forced to the centre by diagenetic crystallization within the cell lumen.

D. Later Diagenetic Processes

The syndiagenetic processes described above may be accompanied by processes mainly operative during later diagenesis which have been reviewed by BATHURST [168]. They can be summarized as cementation, recrystallization, authigenesis

[169], differentiation, replacement, solution (*v.s.*) and compaction. In reconstructing the course of these processes, the paragenesis (sequence of development of a mineral suite) can be deduced from microscopic study of the textural relations (fabric) which the minerals have to one another [170]. Diagenetic alteration may not always be detectable optically or mineralogically however. Recent work suggests that such cryptic alteration may be revealed: by using trace element data in a model to derive estimates of the degree of alteration [171]; by the occurrence of systematically "foreign" isotope abundances [172]; and by population analysis of variance in chemical composition indicating departure from the disequilibrium compositions characteristic of living populations, and indicating diagenetic approach to equilibrium [173]. Additional thermodynamic studies of mineral equilibria during diagenesis may enable the prediction of particular reactions [101, 174, 175].

1. Cementation

Cementation, the precipitation of minerals in voids of a sediment, reinforces skeletal hardparts by filling in spaces vacated by the decay of organic material (usually termed permineralization or impregnation), and forms casts (of drusy mosaic) where previous solution has left molds of organisms [168]. The increased permeability resulting from an abundance of skeletons at a particular horizon in a rock may provide a route for rapid permineralization. Also by differential cementation of the matrix, trace fossils may be enhanced [176]. Such cement is supplied from the solution and redeposition of minerals (especially skeletal minerals) at different levels of burial. The commonest cementing minerals are silica, calcite, dolomite, and siderite. The cement may grow syntaxially (in crystallographic continuity) on its skeletal foundation forming, for example, solid calcite crystals from the originally reticular crystals of echinoderm skeletons.

The early formation of concretions has already been referred to, but concretions and nodules may also form around fossils by local cementation and differentiation (redistribution by solution and diffusion) at later diagenetic stages.

Amorphous silicification, commonly resulting from impregnation by hot spring waters, is an extreme example of rapid cementation which may lead to exceptional preservation as in the Pre-Cambrian Gunflint chert [177], the Devonian Rhynie chert [178], and much silicified wood [159]. A striking feature of permineralized fossil plants when compared with animals is the frequency with which their "soft" tissues are preserved. This is undoubtedly due to the existence in the plant of fairly rigid cell walls which are not only responsible for the fossil retaining much of the original shape of the plant, but also serve as a permeable network through which the mineral solutions may penetrate all parts of even the most delicate tissues of the higher plants. The chemical nature of the plant cell wall probably has some influence on such silicification. Silica and sodium silicates are readily and firmly adsorbed onto cellulose as well as forming complexes with other organic substances. In undegraded cell walls silica may be deposited in the intercapillary spaces, whereas in partially degraded lignified cell walls it is deposited in the framework originally occupied by cellulose and therefore reflects the micellar structure of the wall [159, 179].

After dissolution of hardparts, remaining organic tissues may form nuclei around which specific cementing minerals are concentrated, perhaps creating a reactive environment for adsorption processes. Organic linings from foraminifera which have been revealed in Cretaceous flints by fluorescence microscopy may have been preserved in this way. Similar shell membranes described from silicified Paleozoic radiolaria, however, have been suggested to result from natural chromatographic separation [180] of degradation products by the silica gel, rather than being of primary origin [181, 182]. Many examples are known of such diagenetic formation of structures which can be interpreted as primary morphological features [177, 183]. It becomes increasingly difficult to be certain of the origin of such structures when dealing with organisms that have no close extant relatives. Part of the paleontologist's task is to recognize and disentangle these diagenetically superimposed features from original structures.

2. Recrystallization

Recrystallization (neomorphism of Folk [170]) is the transformation of crystal size, form or orientation within one mineral species and its polymorphs [170]. This normally leads to an increase in grain size, although under certain conditions grain diminution can occur. Fossils may ultimately be destroyed by such crystal enlargement or only distinguishable by a "dust line" from their matrix. At earlier stages small and fragile skeletons and original microstructure may be destroyed, and organic compounds hydrolysed. The relevance of recrystallization to the preservation of organic material is illustrated by the inversion of originally fibrous aragonite of mollusc shells to its more stable, coarsely crystalline polymorph, calcite. Fabric details show that this recrystallization may be completed before cementation has lithified the matrix [184], and that it takes place with no intermediate production of cavities. The process has been carried out experimentally [185]. Organic laminae and original microstructure of the original shell may then be retained in the replacing mosaic [156, 170, 184, 186–188]. This is in striking contrast to most originally aragonitic shells, which have suffered solution followed by cavity-filling by calcite (casts), in which case microstructure will be destroyed and organic material lost. Detailed analysis and codification of the processes of recrystallization have been given by Folk [170].

3. Replacement (Metasomatism)

Replacement (or metasomatism) is the replacement of one mineral by another through gradual substitution [189]. In its broadest sense this term thus includes some recrystallization [170], but it is customary to restrict it to changes in gross composition, such as occurs in the replacement of calcite by quartz. This process is responsible for the preservation of many fossils. The finest details of microstructure may still be recognizable, so that molecule-for-molecule replacement has been invoked, even though this is unlikely in view of the dissimilarity of the molecules often involved [159]. The replacing mineral may preserve the shape of the original skeletal mineral (pseudomorphism), but volume changes may take place [190] and be signified by changes in porosity, by bulging, compression, folding and tension- or contraction-cracking [189], resulting in features (such as beekite rings) which might be taken for primary skeletal morphology.

The processes of transport and reaction involved in replacement will be largely controlled by the available size of opening, which may range from super-capillary to intralattice. The most faithful replacement takes place in openings of smallest size, and is accomplished by diffusion [191] through the solid phase. Grosser replacement takes place by mechanical transportation in fluids carrying compounds either in solution or a dispersed state [192]. Any diagenetically stable mineral can be found replacing fossils, and numerous examples have been listed by LADD [35]. A well known example is that of dolomitization [193, 194]. This, like many replacement phenomena, is selective: originally aragonite skeletons are more readily replaced than the matrix, which in turn is more readily replaced than calcite skeletons. Dolomitization commonly destroys all microstructure in fossils, although exceptions have been recorded [195]. Ore minerals introduced from an igneous source, although beyond the sphere of diagenesis, may preserve fossils [196] but doubt exists as to whether the ore mineral primarily replaced the organism, or a pre-existing diagenetic mineral [197].

4. Compaction

Compaction may directly affect fossils by breaking, crushing and distorting them [198, 199], and such fossils form useful measures of the amount of compaction of the enclosing sediment [200]. Clean shell-breaks have been thought to develop under high burial pressure and to suggest rapid deposition, whereas plastic deformation of empty shells may suggest slow sedimentation with slow decay of organic substances [130]. Wood that must have been plasticized, perhaps by bitumens, has been found filling cracks in the Jurassic Posidonienschiefer [130], and compaction plays an important part in directing such mobilization. As local discontinuities within the sediment, fossils may dissolve very locally under increasing overburden pressure and cement the sediment to form an enveloping concretion [200]. Smaller skeletons may be destroyed by such pressure solution, and stylolites may develop within larger skeletons [135].

V. Weathering

When rocks are exposed to the atmosphere by erosion they undergo weathering, which upsets any equilibrium achieved during diagenesis. The chemical effects of weathering include hydration, hydrolysis, oxidation, reduction, carbonation, and attack by acid and alkaline solutions. These same processes have already been described as operating during biostratinomy and diagenesis. Fossils interrupted by the supervention of other factors before such processes could be completed may therefore resume earlier courses of chemical change during weathering, or have them reversed. Some fossils may have selectively withstood dolomitization during diagenesis, for example, and retained their original calcite skeletons, but groundwater solutions descending to the water table will leach out such skeletons and leave a hollow mold in the dolomite matrix [195]. For some distance below the water table these minerals may be redeposited and thus permineralize or replace other fossils.

References

1. OLSON, E. C., and R. L. MILLER: Morphological integration. Chicago: Chicago University Press 1958.
2. SIMPSON, G. G.: The status of the study of organisms. Am. Scientist **50**, 36–45 (1962).
3. CRAIG, G. Y., and G. OERTEL: Deterministic models of living and fossil populations of animals. Quart. J. Geol. Soc. Lond. **122**, 315–355 (1966).
4. COOK, S. F., S. T. BROOKS, and E. EZRA-COHN: The process of fossilization. Southw. J. Anthrop. **17**, 355–364 (1961).
5. THOMAS, G.: Processes of fossilization. New Biology **8**, 75–97 (1950).
6. CRAIG, G. Y., and A. HALLAM: Size-frequency and growth-ring analyses of *Mytilus edulis* and *Cardium edule*, and their palaeoecological significance. Palaeontology **6**, 731–750 (1963).
7. EGLINTON, G., and M. CALVIN: Chemical fossils. Sci. Am. **216**, 32–43 (1967).
8. BROUWER, A.: General palaeontology. Edinburgh: Oliver & Boyd 1967.
9. RHODES, F. H. T.: The course of evolution. Proc. Geologists Ass. (Engl.) **77**, 1–53 (1966).
10. HARLAND, W. B., *et al.* (eds.): The fossil record. Geological Society of London 1967.
11. KRUMBEIN, W. C., and L. L. SLOSS: Stratigraphy and sedimentation, 2nd ed. San Francisco: W. H. Freeman & Co. 1963.
12. SHAW, A. B.: Time in stratigraphy. New York: McGraw-Hill Book Co., Inc. 1964.
13. AGER, D. V.: Principles of paleoecology. New York: McGraw-Hill Book Co., Inc. 1963.
14. HECKER, R. F.: Introduction to paleoecology. New York: American Elsevier Publ. Co., Inc. 1965.
15. IMBRIE, J., and N. NEWELL (eds.): Approaches to paleoecology. New York: John Wiley & Sons, Inc. 1964.
16. CLOUD, P. E.: Paleobiogeography of the marine realm. In: Oceanography (M. SEARS, ed.). Publ. Am. Ass. Advan. Sci. **67**, Washington.
17. DURHAM, J. W.: The biogeographic basis of paleoecology. In: Approaches to paleoecology (J. IMBRIE and N. NEWELL, eds.). New York: John Wiley & Sons, Inc. 1964.
18. VINOGRADOV, A. P.: Biogeochemical provinces and their role in organic evolution. Geochemistry (USSR) **1963**, 214–228 (1963).
19. WHITTINGTON, H. B.: Taxonomic basis of paleoecology. In: Approaches to paleoecology (J. IMBRIE and N. NEWELL, eds.). New York: John Wiley & Sons, Inc. 1964.
20. BARGHOORN, E. S., W. G. MEINSCHEIN, and J. W. SCHOPF: Paleobiology of a Precambrian shale. Science **148**, 461–472 (1965).
21. KUMMEL, B., and D. M. RAUP (eds.): Handbook of paleontological techniques. San Francisco: W. H. Freeman & Co. 1965.
22. WELLER, J. M.: The status of paleontology. J. Paleontol. **39**, 741–749 (1965).
23. EDINGER, T.: Fossil brains reflect specialized behaviour. World Neurol. **2**, 934–941 (1961).
24. BERINGER, C. C.: Gedanken über eine Psychologie fossiler Tiere (Paläopsychologie). Neues Jahrb. Geol. Palaeont. Abhandl. **97**, 1–19 (1953).
25. TASNÁDI KUBACSKA, A.: Paläopathologie. 1965.
26. DEGENS, E. T., u. H. SCHMIDT: Die Paläobiochemie, ein neues Arbeitsgebiet der Evolutionsforschung. Pal. Z. **40**, 218–229 (1966).
27. PAULING, L., and E. ZUCKERKANDL: Chemical paleogenetics. Molecular "restoration studies" of extinct forms of life. Acta Chem. Scand. **17** (Suppl. 1), 9–16 (1963).
28. EFREMOV, J. A.: Taphonomy; a new branch of geology. Pan-Am. Geol. **74**, 81–93 (1940).
29. – Taphonomie et annales géologiques (première partie). Ann. Centre Étud. Doc. Paléont. **4** (1953).
30. MÜLLER, A. H.: Lehrbuch der Paläozoologie, 2. Aufl., Bd. 1. Jena: G. Fischer 1963.
31. SIMPSON, G. G.: The history of life. In: The evolution of life (S. TAX, ed.). Chicago: Chicago University Press 1960.
32. NEWELL, N. D.: The nature of the fossil record. Proc. Am. Phil. Soc. **103**, 264–285 (1959).
33. VALLENTYNE, J. R.: On fish remains in lacustrine sediments. Am. J. Sci. **258-A**, 344–349 (1960).
34. HARTZELL, J. C.: Conditions of fossilization. J. Geol. **14**, 269–289 (1906).
35. LADD, H. S.: Introduction. In: Treatise on marine ecology and paleoecology, vol. 2, Paleoecology. Mem. Geol. Soc. Am. **67** (1957).
36. WILLARD, B., and P. L. KILLEEN: Fossils and fossilization. Proc. Penn. Acad. Sci. **5**, 62–66 (1931).

37. WEIGELT, J.: Die Biostratonomie der 1932 auf der Grube Cecilie im mittleren Geiseltal ausgegrabenen Leichenfelder. Nova Acta Leopoldina, N.F. **1**, 157–175 (1933).

38. – Some remarks on the excavations in the Geisel Valley. Res. Progr. **1**, 155–159 (1935).

39. SCHÄFER, W.: Aktuo-Paläontologie nach Studien in der Nordsee. Frankfurt: W. Kramer 1962.

40. DALQUIST, W. W., and S. H. MAMAY: A remarkable concentration of Permian amphibian remains in Haskell County, Texas. J. Geol. **71**, 641–644 (1963).

41. CRAIG, G. Y., and N. S. JONES: Marine benthos, substrate and paleoecology. Palaeontology **9**, 30–38 (1966).

42. VINOGRADOV, A. P.: The elementary chemical composition of marine organisms. Mem. Sears Found. Mar. Res. **2** (1953).

43. LOWENSTAM, H.: Biologic problems relating to the composition and diagenesis of sediments. In: The earth sciences, problems and progress in current research (T. W. DONNELLY, ed.). Chicago: Chicago University Press 1963.

44. MOSS, M. L. (ed.): Comparative biology of calcified tissue. Ann. N.Y. Acad. Sci. **109**, 1–410 (1963).

45. SOGNNAES, R. F. (ed.): Calcification of biological systems.—A symposium. Publ. Am. Assoc. Advan. Sci. **64** (1960).

46. – Mechanisms of hard tissue destruction. Publ. Am. Assoc. Advan. Sci. **75** (1963).

47. KESSEL, E.: Über Erhaltungsfähigkeit mariner Molluskenschalen in Abhängigkeit von der Struktur. Arch. Molluskenk. **70**, 248–254 (1938).

48. FREY-WYSSLING, A., and U. MÜHLETHALER: Ultrastructural plant cytology. Amsterdam: Elsevier 1965.

49. ROELOFSEN, P. A.: The plant cell wall. In: Encyclopedia of plant anatomy, vol. 3/4. Berlin: Borntraeger 1959.

50. SIEGEL, S. M.: The plant cell wall. Oxford: Pergamon Press 1962.

51. BARGHOORN, E. S.: Paleobotanical studies of the Fishweir and associated deposits. Pap. Peabody Fndn. Archeol. **4**, 49–83 (1949).

52. – Degradation of plant remains in organic sediments. Botan. Museum Leaflets, Harvard Univ. **14**, 1–20 (1949).

53. – Degradation of plant tissues in organic sediments. J. Sediment. Petrol. **22**, 34–41 (1952).

54. – Degradation of plant materials and its relation to the origin of coal. In: 2nd Conference on the origin and constitution of coal. Crystal Cliffs, Nova Scotia 1952.

55. –, and R. A. SCOTT: Degradation of the plant cell wall and its relation to certain tracheary features of the Lepidodendrales. Am. J. Botany **45**, 222–227 (1958).

56. VARNER, J. E.: Biochemistry of senescence. Ann. Rev. Plant Physiol. **12**, 245–264 (1961).

57. RICHARDS, A. G.: The integument of arthropods. Minneapolis: Minnesota University Press 1951.

58. SCHÄFER, W.: Fossilisationsbedingungen brachyurer Krebse. Abhandl. Senckenberg. Naturforsch. Ges. **485**, 221–238 (1951).

59. BARTON, L. V.: Seed preservation and longevity. Leonard Hill Books 1961.

60. – Longevity in seeds and the propagules of fungi. Encyclopedia of plant physiology, vol. 15/2. Berlin-Heidelberg-New York: Springer 1965.

61. LIBBY, W. F.: Radiocarbon dates II. Science **114**, 291–296 (1951).

62. – Chicago radiocarbon dates IV. Science **119**, 135–140 (1954).

63. 12th Intern. seed testing convention, Oslo 1959. Proc. Intern. Seed Testing Assoc. 449–497 (1960).

64. BREDER, C. M.: A note on preliminary stages in the fossilization of fishes. Copeia **1957**, 132–135 (1957).

65. PIERCE, W. D.: Fossil arthropods of California, 17, The silphid burying beetles in the asphalt deposits. Bull. S. Calif. Acad. Sci. **48**, 54–70 (1949).

66. KRAUSE, H. R.: Biochemische Untersuchungen über den postmortalen Abbau von totem Plankton unter aeroben und anaeroben Bedingungen. Arch. Hydrobiol. Suppl. **24**, 297–337 (1959).

67. HECHT, F.: Der Verbleib der organischen Substanz der Tiere bei meerischer Einbettung. Senckenbergiana **15**, 165–249 (1933).

68. MÖRNER, C. T., and C. WIMAN: Über einen in Leichenwachs umgewandelten Schweinekadaver aus der Nähe von Göteborg. Göteborgs Kgl. Vetenskaps-Vitterhets Samhäll. Handl. (5) B **6** (9) (1939).

69. MÜLLER, A.: Postmortale Dekomposition und Fettwachsbildung. Zürich 1913.

70. SONDHEIMER, E., W. A. DENCE, L. R. MATTICK, and S. R. SILVERMAN: Composition of combustible concretions of the alewife, *Alosa pseudoharengus*. Science **152**, 221–223 (1966).
71. WIMAN, C.: Über ältere und neuere Funde von Leichenwachs. Senckenbergiana **25**, 1–19 (1942).
72. BERGMANN, W.: Geochemistry of lipids. In: Organic geochemistry (I. A. BREGER, ed.). Oxford: Pergamon Press 1963.
73. ZANGERL, R., and E. S. RICHARDSON: The paleoecological history of two Pennsylvanian black shales. Fieldiana, Geol. Mem. **4**, 1–352 (1963).
74. STOPES, M. C., and D. M. S. WATSON: On the present distribution and origin of the calcareous concretions in coal seams, known as "coal balls". Phil. Trans. Roy. Soc. London, Ser. B, **200**, 167–218 (1907).
75. TAYLOR, E. M.: The decomposition of vegetable matter under soils containing calcium and sodium as replaceable bases. Fuel **6**, 359–367 (1927).
76. WEIGELT, J.: Rezente Wirbeltierleichen und ihre paläobiologische Bedeutung. Leipzig: Max Weg 1927.
77. TOOTS, H.: Sequence of disarticulation in mammalian skeletons. Contrib. Geol. Wyoming Univ. **4**, 37–39 (1965).
78. GUTHRIE, R. D.: Differential preservation and recovery of Pleistocene large mammal remains in Alaska. J. Paleontol. **41**, 243–246 (1967).
79. HOFMANN, J.: Einbettung und Zerfall der Ichthyosaurier im Lias von Holzmaden. Meyniana **6**, 10–55 (1958).
80. HELLER, W.: Organisch-chemische Untersuchungen im Posidonienschiefer Schwabens. In: Advances in organic geochemistry, 1964 (G. D. HOBSON and M. C. LOUIS, eds.). Oxford: Pergamon Press 1966.
81. MOORE, L. R.: The microbiology, mineralogy and genesis of a tonstein. Proc. Yorkshire Geol. Soc. **34**, 235–308 (1964).
82. BOEKSCHOTEN, G. J.: Shell borings of sessile epibiontic organisms as palaeoecological guides (with examples from the Dutch coast). Palaeogeog., Palaeoclimatol., Palaeoecol. **2**, 333–379 (1966).
83. GINSBURG, R. N.: Early diagenesis and lithification of shallow-water carbonate sediments in S. Florida. In: Regional aspects of carbonate deposition (R. J. LE BLANC and J. G. BREEDING, eds.). Spec. Publ. Soc. Econ. Paleont. Mineral. **5**, 80–100 (1957).
84. WETZEL, W.: Die Schalenzerstörung durch Mikroorganismen. Kiel. Meeresforsch. **2**, 255–266 (1938).
85. BYSTROV, A. P.: O razryshenii skeletnykh elementov iskopaemykh zhivotnykh gribami. (On the destruction of skeletal elements of fossil animals by fungi.) Vestn. Leningr. Univ., Ser. Geol. i Geogr. **6**, 30–46 (1956) [Russ.].
86. ROLFE, W. D. I.: The cuticle of some Middle Silurian ceratiocaridid Crustacea from Scotland. Palaeontology **5**, 30–51 (1962).
87. SCHINDEWOLF, O. H.: Parasitäre Thallophyten in Ammoniten-Schalen. Paläontol. Z., Festband H. SCHMIDT, 206–215.
88. BATHURST, R. G. C.: Boring algae, micrite envelopes and lithification of molluscan biosparites. J. Geol. **5**, 15–32 (1966).
89. WOLF, K. H.: "Grain diminution" of algal colonies to micrite. J. Sediment. Petrol. **35**, 420–427 (1965).
90. SHEARMAN, D. J., and P. A. D'E. SKIPWITH: Organic matter in recent and ancient limestones and its role in their diagenesis. Nature **208**, 1310–1311 (1965).
91. CRAIG, G. Y.: Concepts in palaeoecology. Earth-Sci. Rev. **2**, 127–155 (1966).
92. DEGENS, E. T.: Geochemistry of sediments. Englewood Cliffs, N.J.: Prentice-Hall, Inc. 1965.
93. PETTIJOHN, F. J.: Sedimentary rocks, 2nd ed. New York: Harper & Bros. 1957.
94. HALLAM, A.: The interpretation of size-frequency distributions in molluscan death assemblages. Palaeontology **10**, 25–42 (1967).
95. FAGERSTROM, J. A.: Fossil communities in paleoecology: their recognition and significance. Bull. Geol. Soc. Am. **75**, 1197–1216 (1964).
96. CHAVE, K. E.: Carbonate skeletons to limestones: problems. Trans. N.Y. Acad. Sci. (2) **23**, 14–24 (1960).
97. – Factors influencing the mineralogy of carbonate sediments. Limnol. Oceanog. **7**, 218–223 (1962).

98. CHAVE, K. E.: Skeletal durability and preservation. In: Approaches to paleoecology (J. IMBRIE and N. NEWELL, eds.). New York: John Wiley & Sons, Inc. 1964.
99. KLÄHN, H.: Der quantitative Verlauf der Aufarbeitung von Sanden, Geröllen und Schalen in wässerigem Medium. Neues Jahrb. Geol. Palaeontol. Beil.-Bd. **67** B, 313–412 (1932).
100. TAUBER, A. F.: Postmortale Veränderungen an Molluskenschalen und ihre Auswertbarkeit für die Erforschung vorzeitlicher Lebensräume. Palaeobiologica **7**, 448–495 (1942).
101. WEYL, P. K.: The solution alteration of carbonate sediments and skeletons. In: Approaches to paleoecology (J. IMBRIE and N. NEWELL, eds.). New York: John Wiley & Sons, Inc. 1964.
102. BAAS BECKING, L. G. M., I. R. KAPLAN, and D. MOORE: Limits of the natural environment in terms of pH and Eh potentials. J. Geol. **68**, 243–284 (1960).
103. STEVENSON, F. J.: Some aspects of the distribution of biochemicals in geologic environments. Geochim. Cosmochim. Acta **19**, 261–271 (1961).
104. BARGHOORN, E. S., and W. SPACKMAN: Geological and botanical study of the Brandon lignite and its significance in coal petrology. Econ. Geol. **45**, 344–357 (1950).
105. SEN, J.: The organization of structural units in fossil wood. Riv. Ital. Paleontol. **62**, 221–222 (1956).
106. – The chemistry of ancient buried wood. Geol. Foren. Stockholm Forh. **79**, 737–758 (1957).
107. – Fine structure in degraded ancient and buried wood, and in other fossilized plant derivatives II. Botan. Rev. **29**, 230–242 (1963).
108. VAROSSIEAU, W. W., and I. BREGER: Chemical studies on ancient buried wood and the origin of humus. Compte Rendu 3 Congr. Avance. Étud. Strat. Géol. Carbonifère **2**, 637–646 (1952).
109. Geochim. Cosmochim. Acta **28** (10) (1964).
110. EMERY, K. O.: The sea off Southern California. New York: John Wiley & Sons, Inc. 1960.
111. BERNER, R. A.: Distribution and diagenesis of sulfur in some sediments from the Gulf of California. Mar. Geol. **1**, 117–140 (1964).
112. LOVE, L. G., and G. C. AMSTUTZ: Review of microscopic pyrite. Fortschr. Mineral. **43**, 273–309 (1966).
113. BROWN, P. R.: Pyritization in some molluscan shells. J. Sediment. Petrol. **36**, 1149–1152 (1966).
114. EHLERS, E. G., D. V. STILES, and J. D. BIRLE: Fossil bacteria in pyrite. Science **148**, 1719–1721 (1965).
115. IMREH, J., u. N. SURARU: Baryt-Kristalle in Eozän-Versteinerungen. Neues Jahrb. Geol. Palaeontol. Monatsschr. **8**, 441–447 (1963).
116. LOVE, L. G., and J. W. MURRAY: Biogenic pyrite in recent sediments of Christchurch Harbour, England. Am. J. Sci. **261**, 433–448 (1963).
117. MOSEBACH, R.: Mineralbildungsvorgänge als Ursache des Erhaltungszustandes der Fossilien des Hunsrück-Schiefers. Palaeontol. Z. **25**, 127–137 (1952).
118. OPPENHEIMER, C. H.: Bacterial activity in sediments of shallow marine bays. Geochim. Cosmochim. Acta **19**, 244–260 (1960).
119. STOCKS, H. B.: On the origin of certain concretions in the Lower Coal-Measures. Quart. J. Geol. Soc. Lond. **58**, 46–58 (1902).
120. EDWARDS, A. B., and G. BAKER: Some occurrences of supergene iron sulphides in relation to their environments of deposition. J. Sediment. Petrol. **21**, 34–46 (1951).
121. LOVE, L. G.: Micro-organic material with diagenetic pyrite from the Lower Proterozoic Mount Isa Shale and a Coal Measures shale. Proc. Yorkshire Geol. Soc. **35**, 187–202 (1965).
122. GOLDBERG, E. D., and R. H. PARKER: Phosphatized wood from the sea-floor. Bull. Geol. Soc. Am. **71**, 631–632 (1960).
123. ARRHENIUS, G. O. S.: Sedimentation on the ocean floor. In: Researches in geochemistry (P. H. ABELSON, ed.). New York: John Wiley & Sons, Inc. 1959.
124. KRINSLEY, D., and R. BIERI: Changes in the chemical composition of pteropod shells after deposition on the sea floor. J. Paleontol. **33**, 682–684 (1959).
125. BLOKH, A. M.: Rare earths in the remains of Paleozoic fishes of the Russian platform. Geokhimiya **1961**, 404–415 (1962).
126. DODD, J. R.: Environmentally controlled variation in the shell structure of a pelecypod species. J. Paleontol. **38**, 1065–1071 (1964).
127. BOWEN, R.: Paleotemperature analysis. Amsterdam: Elsevier 1966.

128. Jefferies, R. P. S.: The palaeoecology of the *Actinocamax plenus* Subzone (lowest Turonian) in the Anglo-Paris Basin. Palaeontology **4**, 609–647 (1962).
129. Huelsenbeck, P., and J. Beerbower: Paleoecology of Upper Cretaceous (Navesink) beds at Poricy Brook, Monmouth County, New Jersey. Proc. Penn. Acad. Sci. **37**, 175–178 (1964).
130. Einsele, G., and R. Mosebach: Zur Petrographie, Fossilerhaltung und Entstehung der Gesteine des Posidonienschiefers im Schwäbischen Jura. Neues Jahrb. Geol. Palaeont., Abhandl. **101**, 319–430 (1955).
131. Klähn, H.: Die Anlösungsgeschwindigkeit kalkiger anorganischer und organischer Körper innerhalb eines wässerigen Mediums. Zentr. Mineral. Geol., Abt. A **1936**, 328–348, 369–384 (1936).
132. Krejci-Graf, K.: Über Schneckendeckel-Ablagerungen und die Erhaltung von Chitinsubstanz. Senckenbergiana **15**, 22–25 (1933).
133. Matern, H.: Oberdevonische Anaptychen in situ und über die Erhaltung von Chitinsubstanzen. Senckenbergiana **13**, 160–167 (1931).
134. Pfannenstiel, M.: Über Lösungserscheinungen an Gryphäen des Lias. Zentr. Mineral. Geol. **1928** B, 51–61 (1928).
135. Quenstedt, W.: Über Erhaltungszustände von Muscheln und ihre Entstehung. Palaeontographica **71**, 1–66 (1928).
136. Chave, K. E., K. S. Deffeyes, P. K. Weyl, R. M. Garrels, and M. E. Thompson: Observations on the solubility of skeletal carbonates in aqueous solutions. Science **137**, 33–34 (1962).
137. Schindewolf, O. H.: Über Aptychen (Ammonoidea). Palaeontographica A **111**, 1–46 (1958).
138. Cloud, P. E.: Environment of calcium carbonate deposition west of Andros Island, Bahamas. U. S. Geol. Surv., Profess. Papers **350** (1962).
139. Williams, M., and E. S. Barghoorn: Biogeochemical aspects of the formation of marine carbonates. In: Organic geochemistry (I. A. Breger, ed.). Oxford: Pergamon Press 1963.
140. Kennett, J. P.: Foraminiferal evidence of a shallow calcium carbonate solution boundary, Ross Sea, Antarctica. Science **153**, 191–193 (1966).
141. Riedel, W. R.: Siliceous organic remains in pelagic sediments. In: Silica in sediments (H. A. Ireland, ed.). Spec. Publ. Econ. Paleont. Mineral. **7** (1959).
142. Jarke, J.: Beobachtungen über Kalkauflösung an Schalen von Mikrofossilien in Sedimenten der westlichen Ostsee. Deut. Hydrograph. Z. **14**, 6–11 (1961).
143. Heim, A.: Oceanic sedimentation and submarine discontinuities. Eclogae Geol. Helv. **51**, 642–649 (1958).
144. Hollmann, R.: Subsolutions-Fragmente (Zur Biostratinomie der Ammonoidea im Malm des Monte Baldo/Norditalien). Neues Jahrb. Geol. Palaeontol., Abhandl., **119**, 22–82 (1964).
145. Mosebach, R.: Wässerige H_2S-Lösungen und das Verschwinden kalkiger tierischer Hartteile aus werdenden Sedimenten. Senckenbergiana **33**, 13–22 (1952).
146. Richter, R.: Tierwelt und Umwelt im Hunsrückschiefer. Senckenbergiana **13**, 299–342 (1931).
147. Lazar, E.: Ein ungewöhnlicher Erhaltungszustand bei interglazialen Mollusken. Geologie **9**, 308–315 (1960).
148. Sohn, I. G.: Chemical constituents of ostracodes; some applications to paleontology and paleoecology. J. Paleontol., **32**, 730–736 (1958).
149. Taylor, J. H.: Some aspects of diagenesis. Advan. Sci. **20**, 417–436 (1964).
150. Larsen, G., and G. V. Chilingar (eds.): Diagenesis in sediments. Amsterdam: Elsevier 1967.
151. Coombs, D. S.: Some recent work on the lower grades of metamorphism. Australian J. Sci. **24**, 203–215 (1961).
152. Winkler, H. G. F.: Petrogenesis of metamorphic rocks. Berlin-Heidelberg-New York: Springer 1965.
153. Sujkowski, Z. L.: Diagenesis. Bull. Am. Assoc. Petrol. Geologists **42**, 2692–2717 (1958).
154. Bucher, W. H.: Fossils in metamorphic rocks. Bull. Geol. Soc. Am. **64**, 275–300, 997–999 (1953).
155. Abelson, P. H.: Geochemistry of organic substances. In: Researches in geochemistry (P. H. Abelson, ed.). New York: John Wiley & Sons, Inc. 1959.
156. Hudson, J. D.: Pseudo-pleochroic calcite in recrystallized shell-limestones. Geol. Mag. **99**, 492–500 (1962).
157. Fairbridge, R. W.: Diagenetic phases. Bull. Am. Assoc. Petrol. Geologists **50**, 612–613 (1966).
158. Degens, E. T.: Über biogeochemische Umsetzungen im Frühstadium der Diagenese. In: Deltaic and shallow marine sediments (L. M. J. van Straaten, ed.). Amsterdam: Elsevier 1964.

159. SIEVER, R., and R. A. SCOTT: Organic geochemistry of silica. In: Organic geochemistry (I. A. BREGER, ed.). Oxford: Pergamon Press 1963.

160. DEGENS, E. T., G. V. CHILINGAR, and W. D. PIERCE: On the origin of petroleum inside freshwater carbonate concretions of Miocene age. In: Advances in organic geochemistry (U. COLOMBO and G. D. HODSON, eds.). Oxford: Pergamon Press 1964.

161. CHAPPELL, W. M., J. W. DURHAM, and D. E. SAVAGE: Mold of a rhinoceros in basalt, Lower Grand Coulee, Washington. Bull. Geol. Soc. Am. **62**, 907–918 (1951).

162. KRUMBIEGEL, G.: Die tertiäre Pflanzen- und Tierwelt der Braunkohle des Geiseltales. Wittenberg: A. Ziemsen 1959.

163. VOIGT, E.: Fossil red blood corpuscles found in a lizard from the Middle Eocene lignite of the Geiseltal near Halle. Res. Progr. **5**, 53–56 (1939).

164. – Mikroskopische Untersuchungen an fossilen tierischen Weichteilen und ihre Bedeutung für Systematik und Paläobiologie. Z. Deut. Geol. Ges. **101**, 99–104 (1950).

165. GLOB, P. V.: Lifelike man preserved 2,000 years in peat. Nat. Geogr. Mag. **105**, 419–430 (1954).

166. BRADLEY, W. H.: Chloroplast in *Spirogyra* from the Green River Formation of Wyoming. Am. J. Sci. **260**, 455–459 (1962).

167. DARRAH, W. C.: Changing views of petrifaction. Pan-Am. Geol. **76**, 13–26 (1941).

168. BATHURST, R. G. C.: Diagenesis and paleoecology – a survey. In: Approaches to paleoecology (J. IMBRIE and N. NEWELL, eds.). New York: John Wiley & Sons, Inc. 1964.

169. TEODOROVICH, G. I.: Authigenic minerals in sedimentary rocks. New York: Consultants Bureau 1961.

170. FOLK, R. L.: Some aspects of recrystallization. J. Sediment. Petrol. **35**, 14–46 (1965).

171. TUREKIAN, K. K., and R. L. ARMSTRONG: Chemical and mineralogical composition of fossil molluscan shells from the Fox Hills Formation, South Dakota. Bull. Geol. Soc. Am. **72**, 1817–1828 (1961).

172. LOWENSTAM, H. A.: Systematic paleoecologic and evolutionary aspects of skeletal building materials. Bull. Museum Comp. Zool. Harvard Coll. **112**, 287–317 (1954).

173. CURTIS, C. D., and D. H. KRINSLEY: The detection of minor diagenetic alteration in shell material. Geochim. Cosmochim. Acta **29**, 71–84 (1965).

174. GARRELS, R. M., and C. L. CHRIST: Solutions, minerals and equilibria. New York: Harper & Row 1965.

175. SCHMALZ, R. F.: Kinetics and diagenesis of carbonate sediments. J. Sediment. Petrol. **37**, 60–67 (1967).

176. SEILACHER, A.: Biogenic sedimentary structures. In: Approaches to paleoecology (J. IMBRIE and N. NEWELL, eds.). New York: John Wiley & Sons, Inc. 1964.

177. BARGHOORN, E. S., and S. A. TYLER: Micro-organisms from the Gunflint Chert. Science **147**, 563–577 (1965).

178. TASCH, P.: Flora and fauna of the Rhynie Chert. Bull. Univ. Wichita **32**, 1–24 (1957).

179. EICKE, R.: Elektronenmikroskopische Untersuchungen an verkieselten Coniferen. Palaeontographica **97** B, 36–44 (1954).

180. NAGY, B., and J. P. WOURMS: Experimental study of chromatographic-type accumulation of organic compounds in sediments: an introductory statement. Bull. Geol. Soc. Am. **70**, 655–659 (1959).

181. STÜRMER, W.: Achat-Bildungen in Kieselschiefer-Fossilien. Senckenbergiana Lethaea **43**, 335–342 (1962).

182. – Das Wachstum silurischer Sphaerellarien und ihre späteren chemischen Umwandlungen. Palaeont. Z. **40**, 257–261 (1966).

183. SAINT-LAURENT, J. DE: Au sujet du mimétisme de la matière minérale fossilisante. Bull. Soc. Hist. Nat. Afrique Nord **31**, 178–179 (1941).

184. BATHURST, R. G. C.: The replacement of aragonite by calcite in the molluscan shell wall. In: Approaches to paleoecology (J. IMBRIE and N. NEWELL, eds.). New York: John Wiley & Sons, Inc. 1964.

185. SORBY, H. C.: The anniversary address of the president. Proc. Geol. Soc. Lond. **35**, 56–95 (1879).

186. DODD, J. R.: Processes of conversion of aragonite to calcite with examples from the Cretaceous of Texas. J. Sediment. Petrol. **36**, 733–741 (1966).

187. FYFE, W. S., and J. L. BISCHOFF: The calcite-aragonite problem, p. 3 – 13. In: Dolomitization and limestone diagenesis (L. C. PRAY and R. C. MURRAY, eds.). Spec. Publ. Soc. Econ. Paleont. Miner. **13** (1965).
188. STEHLI, F. G.: Shell mineralogy in Paleozoic invertebrates. Science **123**, 1031 – 1032 (1956).
189. NIGGLI, P.: Rocks and mineral deposits. San Francisco: W. H. Freeman & Co. 1954.
190. AMES, L. L.: Volume relationships during replacement reactions. Econ. Geol. **56**, 1438 – 1445 (1961).
191. GARRELS, R. M., R. M. DREYER, and A. L. HOWLAND: Diffusion of ions through intergranular spaces in water-saturated rocks. Bull. Geol. Soc. Am. **60**, 1809 – 1828 (1949).
192. HOLSER, W. T.: Metasomatic processes. Econ. Geol. **42**, 384 – 395 (1947).
193. MURRAY, R. C., and L. C. PRAY (eds.): Dolomitization and limestone diagenesis. Spec. Publ. Econ. Paleont. Mineral. **13** (1965).
194. HOLDSWORTH, B. K.: Dolomitization of siliceous microfossils in Namurian concretionary limestones. Geol. Mag. **104**, 148 – 154 (1967).
195. MURRAY, R. C.: Preservation of primary structures and fabrics in dolomite. In: Approaches to paleoecology (J. IMBRIE and N. NEWELL, eds.). New York: John Wiley & Sons, Inc. 1964.
196. WESTOLL, T. S.: Mineralization of the Permian rocks of South Durham. Geol. Mag. **80**, 119 – 120 (1943).
197. DAVIDSON, C. F.: The origin of some strata-bound sulfide ore deposits. Econ. Geol. **57**, 265 – 274, 1134 – 1137 (1962).
198. FERGUSON, L.: Distortion of *Crurithyris* ... by compaction of the containing sediment. J. Paleont. **36**, 115 – 119 (1962).
199. RUTSCH, R. F.: Die Bedeutung der Fossil-Deformation. Bull. Ver. Schweiz. Petrol.-Geol.-Ingr. **15** (49), 5 – 18 (1949).
200. KLÄHN, H.: Sedimentdruck und seine Beziehung zum Fossil. Jahres.-Ver. Vaterl. Naturk. Württ. **88**, 52 – 80 (1932).
201. THENIUS, E.: Versteinerte Urkunden. Berlin-Göttingen-Heidelberg: Springer 1963.
202. JOHNSON, R. G.: Models and methods for analysis of the mode of formation of fossil assemblages. Bull. Geol. Soc. Am. **71**, 1075 – 1085 (1960).
203. KRUMBEIN, W. C., and R. M. GARRELS: Origin and classification of chemical sediments in terms of pH and oxidation reduction potentials. J. Geol. **60**, 1 – 33 (1952).
204. DURHAM, J. W.: The incompleteness of our knowledge of the fossil record. J. Paleontol. **41**, 559 – 565 (1967).
205. KENNEDY, W. J., and A. HALL: The influence of organic matter on the preservation of aragonite in fossils. Proc. geol. Soc. Lond. **1643**, 253 – 255 (1967).

Addendum

Since this account was written, the valuable review volume "Diagenesis in sediments" [150] has been published. The chapter entitled "Phases of diagenesis and authigenesis" by Fairbridge is especially relevant.

Further discussion of the incompleteness of knowledge of the fossil record, with particular reference to adequate sampling, is available [204].

KENNEDY and HALL [205] have suggested that if conditions are suitable for the preservation of organic content then skeletal aragonite will not be converted to calcite. Amino acids surviving from the organic matrix would form a hydrophobic surface layer, thus preventing the catalytic effect of water essential to the aragonite-calcite transformation.

CHAPTER 9

Introduction to Sedimentology

B. J. BLUCK

University of Glasgow, Glasgow, Scotland

Contents

I. Introduction

The fundamental factors controlling the form and content of sedimentary rocks are the location and extent of tectonism, the processes operating within the contemporary geographical environment, and post-depositional compaction and lithification. Investigations in sedimentology are normally ultimately directed to the elucidation of these factors.

Surfaces on the earth's crust are uplifted and downwarped by forces usually considered to originate within the mantle. The uplifted areas may be regarded in some instances as areas of low entropy [1], and agents such as wind, water, and ice remove materials from the uplifted ground. The materials are transported until they are finally trapped in areas of downwarp in the earth's crust, where they might accumulate to great thicknesses and become converted from loose sediment to indurated rocks.

The sources of sediment – the uplifted areas – comprise previously-formed rocks which may be igneous, sedimentary, or metamorphic, but in any case they are mainly formed of silicate minerals. The source rocks are broken down *in situ* or as they are being transported to the site of sediment accumulation by the transporting agent. In both circumstances the source rocks are open to decay by physical, chemical, and biological means. The physical processes include the freezing and thawing action of ice between grain boundaries. The chemical disintegration involves solution, hydration, oxidation, and hydrolysis. Biological processes involve the action of root-wedging and leaching of elements from the rock.

The weathering and breakdown of the source rocks produces materials which fall into two main groups, solids and solutions. The solids are classified primarily according to size, which might reflect an even more fundamental difference in composition. Particles greater than 2 mm in size are normally rock fragments and the sediments comprised of them are referred to as gravels or, when indurated, conglomerates. Grains whose sizes fall between 2–0.06 mm are usually minerals and are called sand when loose, and make-up sandstone when indurated. Silt refers to particles in the size range 0.06–0.004 mm, and the grains are again usually monomineralic. Particles less than 0.004 mm across are referred to as clay, and these are made up of clay minerals which are hydrous aluminium silicates and which lithify into mudstones and shales, thereby making up the argillaceous rocks.

The materials carried in solution are divided according to their chemical composition, and include silica, iron, sodium, etc. There are naturally-occurring colloids which bridge the gap between solutions and solids.

Sedimentary rocks have originated in a variety of ways, and it is difficult and perhaps not essential to find unity in a classification of them. There are sediments which are residual (i.e., left at the source after transportation of the main mass of detritus) and there are the transported products themselves. The transported sediments are deposited by essentially physical means, thus grains transported in traction and suspension are deposited because the entraining current no longer has sufficient energy to move them. Sediments carried in solution are either deposited chemically or biochemically. But whilst some rocks are evidently chemical deposits, e.g., salt, or biochemical, e.g., coal, many are certainly not chemical or biochemical, and some are demonstrably both. It has been common to subdivide rocks into clastic (transported), non-clastic, and limestones and dolomites (which embody both) [2]. Residual rocks are mostly clays and shales and are classified, unsatisfactorily, with the transported or clastic rocks.

II. Clastic Rocks

Clastic rocks are subdivided primarily according to grain size, since this parameter is of some importance with respect to the strength of depositing currents and proximity to the source. The groupings are conglomerates, sandstones, and shales.

Another factor of some importance is composition. Quartz is the most inert of the common rock-forming minerals. Weathering of the source rocks, chemical and mechanical disintegration of the transported sediment, and local weathering and abrasion within the site of accumulation are events which are continually concentrating quartz by the selective breakdown of the less stable grains. The extent to which quartz is concentrated by these means depends upon intensity of weathering and available time. Time is usually far more important. Also the time a sediment is available for breakdown depends upon how quickly it is yielded from the source, how soon it reaches the site of deposition, and, when there, how rapidly it is covered by new sediment. Evidently the time available for breakdown is controlled by the rate of uplift of the source, the rate of downwarp of the area of accumulation, and the proximity of both these to each other, i.e., location and extent of tectonism.

Processes operating within the environment of deposition will affect the texture of sediments. High energy conditions will sort out the grains according

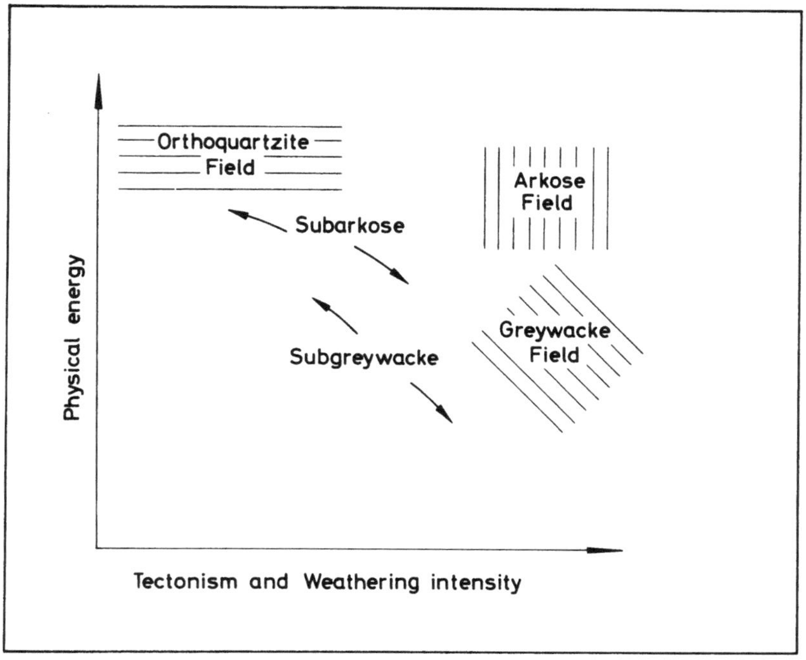

Fig. 1. Illustrating the basis for classifying clastic sedimentary rocks; the example taken is the sandstones. Broadly, rate of uplift and downwarp (tectonic intensity) and the degree of weathering control composition of sediments. Physical energy, again broadly speaking, control texture. The clastic sediments are classified on objective criteria capable of quantification (texture; composition), such that the more fundamental controls are immediately reflected. The other control of sediment properties is the lithological nature of the source, but this is not expressed in the diagram, and is of less importance

to size, shape, and specific gravity; low will not. Deposition from a viscous medium will not sort out sedimentary particles; non-viscous will.

The classification of clastic sediments is given in Fig. 1. The argillaceous rocks are subdivided into residual and transported. The residual shales are further subdivided on the basis of mineralogy of the coarser grains. With increasing amount of silica, iron, etc. shales grade into non-clastic rocks.

III. Non-Clastic Rocks

Non-clastic rocks originate in a number of ways: primary chemical or biochemical precipitation and secondary segregation subsequent to deposition. It is often difficult to distinguish between primary and secondary precipitates. The non-clastic rocks are subdivided on the basis of chemical composition.

A. Siliceous Sediments

The siliceous sediments include radiolarites and diatomites, which are biochemical, and other types of cherts and flints, some of which are chemical. The silica occurs in a variety of forms, opal and chalcedony, but older siliceous sediments of this type recrystallize into quartz. Most naturally occurring waters have silica values of < 50 ppm, where it is in true solution. Silica in the colloidal state may be precipitated by evaporation, cooling, and addition of electrolytes, and the precipitation of dissolved silica may be brought about by organisms (diatoms, radiolarians), adsorption, and reaction with cations [3, 4]. Alexander *et al.* [5], Iler [6], Krauskopf [3], Okamoto [7], and Degens [8] discuss the chemistry of silica more fully.

Organisms such as diatoms have a great capacity for removing silica from sea water [9], and diatomaceous sediments are seen forming today [10], and in the past [11]. Diatoms have been active since mid-Mesozoic times, when they appear to have taken over the role of fixing silica from the sponges [12]. Radiolarians have been active as silica fixers since the Lower Paleozoic when they produced quite thick sequences of chert. Great thicknesses of organic siliceous deposits probably build up where waters rich in silica well-up on the margins of oceans or basins; in the euphotic zone the silica is used to build up some phytoplanktonic skeletons which then settle to the bottom. Some of these skeletons dissolve and finally reprecipitate around the more resistant skeletal remains [11].

Some nodular chert found in limestones may have formed as a chemical gel (and gelatinous silica has been found in a California lake [13]).

B. Iron-Bearing Sediments

Iron-bearing sediments have been studied in some detail and reference may be made to James [14, 15], White [16], Taylor [17], and Garrels [18].

Iron is present in most sedimentary rocks; the average shale has around 6 percent of iron oxides. But at times during the geological past, conditions have been such that great thicknesses of iron-bearing rocks (ironstones make up iron formations which are defined as rocks with greater than 15 percent Fe) have accumulated. It is generally agreed that many ironstones accumulated in shallow-water marine conditions, but some are non-marine and the large Precambrian ore bodies may be lacustrine [19, 20]; some are residual. Ironstones are classified by their chemistry, being split up into sulphides, carbonates, silicates, and oxides. The stability relations of various iron minerals in sea water has been given by Huber [21] and by Garrels and Christ [22].

Iron sulphide usually occurs in the mineral forms pyrite and marcasite, which are the products of sometimes local and sometimes regional reducing conditions. The reducing conditions are known to be either above or below the sediment-water interface. Pyrite forming on the sediment surface has been recorded by Love and Murray [23], and Emery [24], whilst Morreti [25] has described the presence of pyrite in the interior of shells which are enclosed in limestone. Pyrite is believed to have formed in acid or alkaline conditions and marcasite in acid conditions [26, 27]. Some formations are extensive black shales which are rich in pyrite and organic matter, and these must indicate widespread reducing conditions as are found today in the Norwegian fjords and in the Black Sea, where pyrite is actively forming [28, 29].

Siderite, the iron carbonate, is mostly of diagenetic origin. It replaces shells, and precipitates in pore spaces which were probably within the sediment at the time of deposition. Reworked siderite replacements are to be found in the overlying sediments, thus indicating the penecontemporaneous nature of its formation. Chamosite, often found in association with siderite, frequently has an oolitic form, and is believed to be a primary precipitate by some [17, 30] but secondary by others [31]. The oxides of the hematite and limonite group are believed to be both primary and secondary in origin [32, 33].

Whilst great quantities of iron are transported to the sea, a good deal of it is in detrital form and may be transported via clay minerals [34]. Iron is believed to be transported in streams as a stabilized ferric oxide hydrosol. However, the manner of deposition has been a problem, not only a chemical one but also a sedimentological one. Iron formations often lack detrital sediments. In explanation, low-lying land surfaces, drained by sluggish streams, which border the areas of deposition have been proposed.

C. Phosphatic Sediments

Phosphatic sediments are not common, but deposits, usually thin, are found on most continents. Phosphates are normally associated with diastems or unconformities and have as mineral associations calcite, pyrite glauconite, and fluorite. The associated fossils are generally those of creatures which build up phosphatic shells or skeletons, and are commonly inarticulate brachiopods, fish, and conodonts. Phosphates are structurally and compositionally complex, with much isomorphous replacement [35–37].

Phosphate occurs as a crust on surfaces as pebbles, nodules, ovulites, and ooliths. Kazakov [38] divided phosphates into two types, platform and geosynclinal, each with a different association. The formation of phosphate is not certain. It undoubtedly occurs as a replacement of calcareous shells [39, 40] and even wood in recent sediments [41]. But it may also occur as a primary precipitate, and Bluck [42] gives textural evidence which might lead to that conclusion. Phosphates appear to be found in areas of upwelling, e. g., California [24], and also possibly in estuarine conditions without upwelling [43]. Long after deposition, phosphates can suffer much change in elemental composition [8].

D. Evaporites

Evaporites have been studied by Stewart [44], Briggs [45], Braitsch [46], Bersticker *et al.* [47], and many others (see Borchert and Muir [48]). Evaporites refer to minerals which have been concentrated by evaporation (excluding calcium and magnesium carbonate). Although there are numerous evaporite minerals, the predominant varieties include halite, gypsum, and anhydrite. Evaporites are seen forming today, and great thicknesses are known to have formed in the past.

When an original volume of sea water, 3.5 percent salinity, is evaporated to 50 percent original volume, $CaCO_3$ is precipitated, but at 10 percent original volume NaCl begins to crystallize out; below that further sulfates, chlorides, and potash salts crystallize out. The composition of sea water is such that on evaporation 1.5 percent solid is yielded; 78 percent of that solid is halite. Since evaporite deposits are thousands of feet thick, parent columns of water miles deep are implied. This difficulty is overcome by having a continuous supply of water move across the lip of an evaporating area which may be subsiding and which is partly cut off from the sea. A further problem lies in the fact that the proportions of minerals found in evaporite deposits are not the same as would be expected from the straight evaporation of sea water. Many deposits have much calcium sulphate and virtually no sodium chloride. To overcome this problem, it has been suggested that the denser brines from which sodium chloride would precipitate would move out of the basin, along the floor, by a reflux action.

The two dominant evaporite minerals are gypsum and anhydrite. Many investigators have found evidence for the replacement nature of anhydrite [40, 49–53] and some believe that the primary mineral is always gypsum. MURRAY [53] has summarized the gypsum-anhydrite-gypsum replacement relationships observed by numerous workers.

Evaporites form in basins, some with considerable subsidence, which are largely cut off from the sea by organic (reef) growth, or by land or sea movement.

IV. Diagenesis

When the sediments have accumulated on the floor, they are subject to a great number of changes. The word "diagenesis" is intended to cover the changes taking place during this period. The framing of a definition of diagenesis is difficult since on the one hand the sediment is temporarily deposited whilst in transit from source to depositional site and is open to changes which might be regarded as weathering or diagenesis. On the other hand if they suffer great changes after deposition then metamorphism is said to begin. Diagenesis, then, refers to those reactions between minerals, biota, and enclosing fluids, which normally take place within the sediment after its deposition and before its metamorphism. Workers have sometimes found it convenient to split diagenesis into those processes which take place at the sediment-water interface (halmyrolysis) and those taking place after consolidation (epigenesis).

The diagenetic processes involve solution and precipitation using the enclosing fluids as an agent of transfer. The evidence of these happenings are recognised in the rocks by solution pittings, pore-filling cement, and mineral segregations. There are groups of variables which control the course of diagenesis: the composition of the sediment and enclosing fluids, and the heat and pressure supplied by the weight of superincumbent strata. But the diagenetic reactions are slow, and the degree of heat and pressure is continually changing; the system is open and there is much metastability between mineral assemblages [54].

There are a number of rather poorly defined stages in diagenesis which occur after the sediment is laid down and before metamorphism. They are split into three [55], and are considered with reference to different rock types.

A. Early Diagenesis: The Sediment-Water Interface

At the time of sediment deposition, minerals, biota, and enclosing fluids are not in equilibrium. And during this time two major groups of changes take place, biological and chemical.

Biological effects are derived from the burrowing activity of organisms which ingest great quantities of sediment. The main effects, apart from displacement of much of the sediment, is the maceration of grains, solublizing, and changes in the pH. The pH and Eh conditions at the sediment-water interface are largely controlled by organisms in some environments.

Chemical changes at the depositional interface depend not only upon the reagent groups, sediment and water, but also on the rate of sedimentation and the degree of water circulation. Where the rate of deposition is slow and the circulation of water inhibited, the conditions approach those of a closed system. Stagnant water of this type gives rise to a reducing, mildly acid environment, and sulphides might grow at the sediment interface. Hydrogen sulphide is produced by bacteria *Desulfuvibrio* in the Black Sea [28]. In environments with more circulation, mildly oxidizing conditions might favor the growth of glauconite [27], siderite, hematite, and phosphates. Within shells where soft parts are decaying, local reducing conditions may allow for the growth of pyrite [24], and areas of reduction in a surrounding oxidizing environment is thought to favor growth of glauconite [56, 57]. Dolomite is seen to grow on the sediment surface [58] and gelatinous chert is now found in this position [13]. Where conditions for more rapid sedimentation are obtained, the effects of diagenesis on the surface are much reduced by virtue of the short exposure of sediment at any one time, i.e., the surface is quickly covered over.

Clays are particularly susceptible to changes in this position. Experimental work [59] and observations in natural environments show the instability of montmorillonite in sea water, with illite and chlorite as the newly-born clays [60]; but in another environment chlorite is growing from illite [61]. Clays were seen in experiments to release Al_2O_3, SiO_2, and Fe_2O_3 to the adjacent artificial sea water [62].

B. Early Diagenesis: Newly Buried Sediment

Beneath the sediment-water interface the sediment comprises essentially solids with variable interstitial waters. Here the water may have an increase in pH [63, 64], or a decrease in pH [65]. Eh is variable but is regarded as generally negative [66], and silica content of interstitial waters is generally higher than on the surface.

Organic matter exerts a strong influence on the course of diagenesis within the sediment [66]. ZoBELL [67] has shown that the bacterial count changes from 6.3×10^6 gram 2 inches below the surface to 10^3 g at about three feet below the surface (see also LIMBERG-RUBAN [68]). Where water circulation is good, aerobic bacteria may be able to live in the near-surface layers, but anaerobic bacteria may predominate below. The extent of bacterial activity is controlled by the nature of the sediment and the rate of sedimentation. In experiments it has been found that the rate of organic decay is greater in sands than it is in clays [69].

Further considerations of the decay of organic matter in sediments are given by DEGENS [8].

Two processes in clays, which may have an initial porosity of up to 85 percent [2], are the compaction and expulsion of pore water. Muds tend to act as semi-permeable membranes impeding the movement of ions and so concentrating them [70, 71]. Substitutions continue in the clay lattice, and the pore waters may become enriched in certain constituents. Segregations of iron, silica, and calcite in shales, undoubtedly of early burial age, attest to local centers of precipitation. Many sideritic ironstones of the Carboniferous formed in this way, as perhaps did some of the Cretaceous flints.

In sands the changes at this stage are thought to be the solublizing of amorphous (some biochemical) silica and quartz dust. Clays, if present, may replace chert, and then undergoing compaction, and suffer base exchange. In the areas of high pH, brought about by ionic substitutions in the clays, there may be solutioning of quartzes [72]. The dissolved silica may precipitate in areas of low pH, forming overgrowths on quartzes. There is a considerable petrographic literature in this field [40, 72, 73, 74].

During this early burial stage there are many changes in the composition and mineralogy of evaporites [50, 51, 53]; ironstones [17] and pyrite, commonly found filling shells in lithologies which include limestones, owe their origin to this period.

C. Late Stage Diagenesis

Diagenesis in the late stage is induced by the load of superincumbent strata. During this stage the heat and pressure resulting from loading are probably far more influential than pH and Eh. With increase in loading, diagenesis passes into metamorphism. At depths of 20,000 ft the pressures may exceed 20,000 lbs./sq. in. and temperatures may be greater than 150° C [55]. At these depths interstitial waters tend to increase in salinity and chlorinity and may vary from being less than sea water to being ten times as great [75].

The effects of this stage on clays are the expulsion of pore water by compaction and some mineralogical changes. These latter have been studied by BURST [76]. (During this period authigenic minerals such as feldspar may form in shales.)

Sandstones with a good deal of clay (greywackes and subgreywackes) show some of the feldspars going over to muscovite, illite being replaced by chlorite, and the dissolving-out of chert. Clays may begin to replace quartz. Pure quartz sandstones with more pore space, and with quartzes in grain-to-grain contact, show effects of solutioning at the grain boundaries, particularly where small amounts of clay are present [72, 74]. With deep burial the increased temperature and pressure, whilst increasing the solubility of quartz, decrease the solubility of calcium carbonate which then precipitates and replaces both the quartz and the previously-formed quartz overgrowths [73].

V. Carbonate Rocks

Carbonates as discussed here include the minerals calcite, aragonite, and dolomite. Rocks which contain over 50 percent of $CaCO_3$ or $MgCO_3$ are regarded as carbonates and account for about 20 percent of all sedimentary rocks. Dolomites are generally the replacements of limestones. Limestones, although often occurring in areas of tectonic stability where the inclusion of clastics (terrigenous) is inhibited by low-lying surrounding land, can sometimes form on fairly rapidly subsiding areas. Here the clastics are excluded by distance from land, or because they are trapped in periferal areas of downwarp.

For the most part, limestone is made up of whole or fragmented remains of a biological origin of one type or another. But not all $CaCO_3$ is extracted from the enclosing waters (mostly marine) by biochemical means, some is due to direct precipitation. The distribution of lime-rich sediment on the present earth surface has been summarized by RODGERS [78]. There are three main classes of carbonate sediment: deep water oozes (due to the accumulation of $CaCO_3$-fixing pelagic organisms which fall to the sea bed) are not found before the Cretaceous; coral reefs, almost confined to the warm seas of cold latitudes; and finally, associated with the reefs, are the platforms of carbonate accumulation, again most abundant in the warmer regions, but like the coral reefs are found in the colder. By far the most abundant limestone type falls into these two latter categories and attention is therefore confined on these. Over the past decade there has been a considerable advance made in the understanding of carbonate rocks. The recognition that limestone textures might be interpreted in terms of physical energy at the time of deposition, the distinction between recrystallized and pore-filling calcite, and critical investigations of sites of recent carbonate sedimentation have all brought about this advance in understanding.

Essentially limestones are treated as clastic rocks which have occasionally been subject to organic controls, rather than the converse. There are, however, a number of factors which distinguish them from other clastic (terrigenous) rocks.

1. Limestones are endogenetic: the components of which they are made are derived almost wholly from within the very basin of accumulation. In this sense large particles are not deposited near the source in the same way as seen in other clastic rocks. Limestones are made up of the tests of organisms and large organisms can live next to small; both may die at or near the place of their habitation when alive, and become buried to form a limestone.

2. Organisms can bind or hold the sediment so as to reduce the effects of would-be-active currents, or they can build ramparts well above the surrounding sea bed (reefs).

3. Limesand or gravel is highly porous in instances where the component organisms are irregular in shape. The result is that mud or fine sand can percolate through the pores and give the final rock the look of a poorly sorted sediment.

4. The composition of the rock tells a good deal about the environment of accumulation.

Limestones have been subjected to numerous attempts at classification; currently the most successful one is that of FOLK [79]. However, much of the intention of a rock classification is to allow immediate assessment of environment. This might involve more parameters than texture and composition alone. It might involve the types of organisms, the degree of burrowing, etc., and for that reason many European geologists have used the term "microfacies", which takes into account more details of the rock in question.

It is essential to distinguish between limestones which, because of organic growth at the time of deposition, have not behaved as a clastic sediment, and those which have behaved as a clastic sediment. The first are referred to as autochthonous rocks, the second as allochthonous rocks.

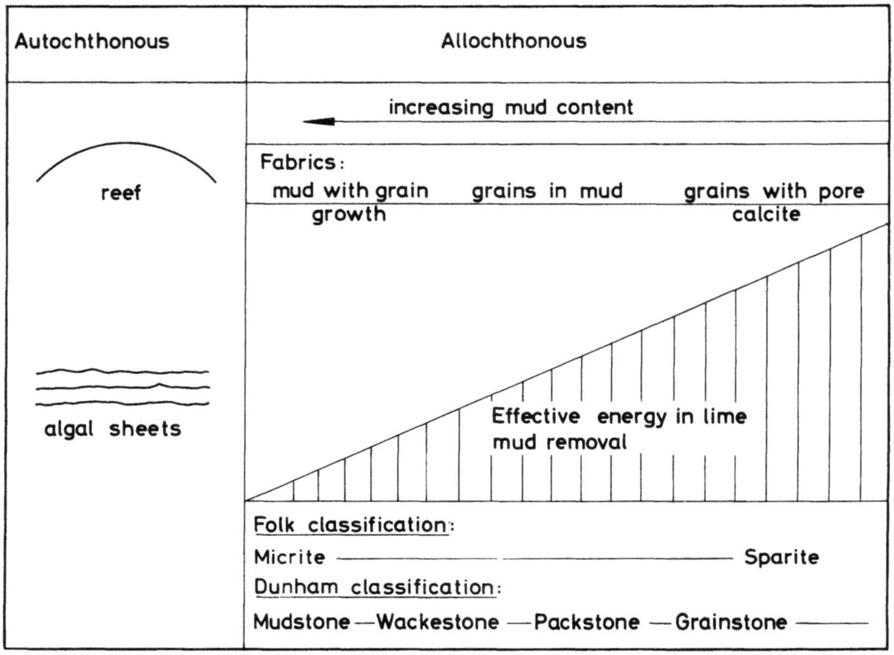

Fig. 2. Illustrating the basis of the classification of limestones. Autochthonous limestones are those where the physical processes of sedimentation are greatly controlled by the biological processes. But the allochthonous limestones have had more physical energy relative to biological activity expended on them. One of the most sensitive components to winnowing currents is lime mud, which therefore becomes an indicator of current activity. Ancient limestones recrystallize easily; but the nature of the calcite between the component grains of the limestone permits the recognition of what was lime mud, and what was a cavity, or pore space free of lime mud

A. Autochthonous

Whilst it might be said with some justification that most limestones have to some degree suffered from organic control over the free movement of their constituent grains on the floor at deposition, there are three types where organic control is very prominent: (1) reefs; (2) mounds and banks; and (3) essentially tabular shaped laminated limestones.

Most of the morphological and biological varieties of reefs which now exist on the earth's surface have been identified in ancient rocks: barrier reefs in the Permian of west Texas [80]; fringing reefs in the Tertiary of the Middle East [81]; and atolls in the Mississippian of Illinois [82]. Coral reefs have been described from the Silurian of Gotland [83] and algal reefs from the Precambrian of South Africa [84]. Reefs are characterized by a framework of organisms capable of resisting normal wave attack, with the framework holding fragments of broken reef materials. The ecology of modern corals and algae has been summarized by

CLOUD [85]. As a generalization it may be said that the corals tend to be the frame-building element and the algae tend to bind the structure and also to bind to the structure those materials which are loose. With an upward growth rate of $\frac{1}{2}$–2 m/100 yrs [86], subsidence is needed to allow the reef to grow upward. But since both the algae and the corals do not live in deep water the subsidence cannot greatly exceed the reef growth rate without the reef dying off. Reefs are therefore prone to grow in areas where the rate of subsidence changes markedly in one direction; for example, when the subsidence is too great it grows towards land.

Mounds and banks are common in the geological past. They may comprise banks of organisms such as oysters [87], crinoids [88], brachiopods, or they may merely be mounds of mud with no diagnostically associated organisms [89–91]. The mounds have characteristic growth forms and oversteepened slopes with sliding of the mud [90]. The binding organisms are not definitely known, but algae have been suggested, and such plants may disintegrate leaving little recognizable evidence of themselves [91, 92].

The essentially planar, or tabular-shaped, organically-bound carbonates are the result of algal activity the remains of which are referred to as stromatolites (although some stromatolites are mound-like). Such laminated carbonates have been recognized in most rocks since the Precambrian. They often contain a moderate amount of organic material [93]. Laminae form by the secretion of a mat of filaments which rapidly cover the sediment surface such as to bind the grains and make them resistant to erosion [94]. Lamination has been described as noctidiurnal, where the organic matter grows faster than the supply of carbonate during the day (forming an organic layer), but slower than the rate of carbonate supply to the mat during the night (forming a carbonate layer) [95].

The lamination takes various shapes, and these have been recently classified [96]. The environment of accumulation very largely controls the shape taken up by the lamination, and nearly all deposits of this kind are found on the areas bordering bodies of water. Recent algal growths have been described from the Bahamas [94, 97], Florida [98], Persian Gulf [93], Australia [99], and Northern France [100]. In ancient sediments stromatolites are recognized by a number of features: alternations of carbonate matter with carbonaceous matter; the growth forms of the laminae; cavities under dome-shaped laminae; associations of growth forms, etc.

The fate of organic matter in stromatolites requires consideration. Algal sediments, in often being porous, are subject to a good deal of leaching. This is accentuated by the fact that stromatolites occupy the zone between land and water which, with fluctuations in water level, is subject to waves of percolating ground water. The relative changes in sea level can be numerous [101]. It is possible that algal growth is accompanied by fungal growth, particularly inside many of the cavities.

B. Allochthonous

The allochthonous rocks are those which are believed to have behaved as clastic sediments. They are classified according to the Folk scheme (Fig. 2) where the basis of subdivision is texture and composition. The limestones with mud between the components are referred to as micrites, and the limestones with clear calcite between the grains are referred to as sparites. It is assumed that the clear calcite is mainly pore-filling, and therefore that the sediment has been well sorted (high energy).

Intraclasts refer to fragments of lithified limestones which have formed in the same area of deposition and at roughly the same times as limestones into which they are incorporated. Most ooliths are believed to form in water not more than a few feet deep, where fragments, thrown up in a solution supersaturated with respect to calcium carbonate, act as nuclei around which precipitation takes place [102, 103]. But the presence of ooliths is not proof of waters of these depths existing at the time and place of the formation of the rocks. The ooliths are commonly swept from very shallow waters into deeper waters. Pellets are roundish or distinctly oval grains of microcrystalline calcite which are believed to be of fecal origin. They sometimes contain a relatively high percentage of organic matter. Fossils refer to the skeletal fragments, and the importance of these in determining the environment of limestone accumulation has already been stressed.

The lime mud, which may form deposits of carbonate mudstone in addition to being incorporated in with other grains, is believed to have formed in a number of ways. The abrasion of skeletal fragments produces fine-grained carbonate, since many organisms which secrete carbonate shells break down into individual crystals. Biological activity, e.g., boring sponges [104], produces fine-grained carbonate, and some organisms themselves disintegrate into mud-sized grains [12]. Chemical precipitation of aragonite is believed to take place on some broad, shallow platforms such as the Bahama Bank. A solution of $CaCO_3$ can remain in a supersaturated state for a long time without crystallizing out. DEGENS [8] believes this is due to the presence of Mg^{++} in sea water. Mainly due to evasion of CO_2, spontaneous crystallization in the form of whitings is thought to take place in the Bahamas [105] and the Persian Gulf [106]. Not everyone is convinced of the spontaneous crystallization of calcium carbonate. LOWENSTAM and EPSTEIN [107] found that aragonite needles on the Bahama Bank have the same isotopic composition as algaly-secreted needles, yet are different from the isotopic composition of the aragonite in ooliths. BROEKER and TAKAHASHI [108] found that some of the aragonite in whitings is thousands of years old, although they do believe in chemical precipitation of aragonite.

The diagenesis of carbonate sediments has received a good deal of attention in recent years. In the early stages aragonite may begin cementing grains on the sea floor [109, 110], aragonite may dissolve or convert to calcite, and high-magnesium calcite may change to low-magnesium calcite [111, 112]. There may be much burrowing [98] and algal boring [113], and dolomite grains may grow on the surface or within the sediment. Where the limestones are near the land, fresh waters may percolate through them and deposit calcite in the pore space so forming beach rocks [114]. There may be reducing conditions below the sediment surface with the development of pyrite; with a slight change in sea level the pyrite may become oxidized. In the later stages of diagenesis the finer (micite or mud) fraction may begin to recrystallize to a coarser mosaic [115, 116], and some of the remaining pores may be infilled with calcite. The calcite presents something of a problem in that in recent environments the cementation by calcite is thought to be achieved in fresh waters [117, 118], and by inference ancient limestones must be cemented in the same way. Those limestones not cemented by pore-filling calcite may suffer solution under pressure.

Glossary

Argillaceous	A clay which has been hardened by recrystallization into an indurated rock.
Authigenetic	Usually refers to minerals formed with or after deposition of the sediment.
Endogenetic	Derived from materials which formed near the site of sediment accumulation.
Entropy	Residual energy.
Euphotic	Within the light zone of the sea, in which plants grow and multiply.
Chert	A rock which is composed of chalcedony and cryptocrystalline quartz.
Diastems	Small breaks in deposition, where time is not represented by sediment at the point in question.
Pelagic	Refers to organisms which are not dependent on the sea floor for their living.
Tectonism	As it is used here, refers to changes in elevation of the earth's surface, which are brought about by forces existing within the earth's interior.

Table 1

Mineral	Composition	Crystal System
Anhydrite	$CaSO_4$	Orthorhombic
Aragonite	$CaCO_3$	Orthorhombic
Calcite	$CaCO_3$	Hexagonal
Chalcedony	SiO_2	None
Chamosite	$Fe_3Al_2Si_2O_{10} \cdot 3\,H_2O$	Monoclinic
Dolomite	$Ca(Mg, Fe)(CO_3)_2$	Hexagonal
Fluorite	CaF_2	Cubic
Glauconite	$KMg(Fe, Al)(SiO_3)_6 \cdot 3\,H_2O$	Monoclinic
Gypsum	$CaSO_4 \cdot 2\,H_2O$	Monoclinic
Halite	NaCl	Cubic
Hematite	Fe_2O_3	Hexagonal
Limonite	$H_2Fe_2O_4(H_2O)_x$	None
Opal	$SiO_2(H_2O)_x$	None
Pyrite	FeS_2	Cubic
Siderite	$FeCO_3$	Hexagonal

References

1. Chorley, R. J.: Geomorphology and general systems theory. U.S. Geol. Surv. Profess. Papers **500**-B (1962).
2. Pettijohn, F. J.: Sedimentary rocks. New York: Harper and Row 1957.
3. Krauskopf, K. B.: The geochemistry of silica in sedimentary environments. Ireland, H. A., ed. Soc. Econ. Paleontologists Mineralogists Spec. Publ. **7**, 4–19 (1959).
4. Mackenzie, F. T., R. M. Garrels, O. P. Bricker, and F. Bicley: Silica in sea water: control by silica minerals. Science **155**, 1404–1405 (1967).
5. Alexander, G. B., W. M. Heston, and R. K. Iller: The solubility of amorphous silica in water. J. Phys. Chem. **58**, 453–455 (1954).
6. Iler, R. K.: Colloid chemistry of silica and silicates. Ithaca: Cornell University Press 1955.
7. Okamoto, G., T. Okura, and K. Goto: Properties of silica in water. Geochim. Cosmochim. Acta **12**, 123–132 (1957).

8. DEGENS, E. T.: Geochemistry of sediments. A brief survey. Englewood Cliffs, N.J.: Prentice-Hall, Inc. 1965.
9. LEWIN, J. C.: The dissolution of silica from diatom walls. Geochim. Cosmochim. Acta **21**, 182–198 (1961).
10. CALVERT, S. E.: Accumulation of diatomaceous silica in the sediments of the Gulf of California. Bull. Geol. Soc. Am. **77**, 569–596 (1966).
11. BRAMLETTE, M. N.: The Monterey formation of California and the origin of siliceous rocks. U.S. Geol. Surv. Profess. Papers **212**, 1–57 (1946).
12. LOWENSTAM, H. A.: Biologic problem relating to the composition and diagenesis of sediments. In: The earth sciences, problems and progress in current research (T. W. DONNELLY, ed.). Chicago: University Press 1963.
13. PETERSON, M. N. A., and C. C. VON DER BORCH: Chert: modern inorganic deposition in a carbonate precipitating locality. Science **149**, 1501–1503 (1965).
14. JAMES, H. L.: Sedimentary facies of iron formation. Econ. Geol. **49**, 235–293 (1954).
15. – Chemistry of iron rich sedimentary rocks. U.S. Geol. Surv. Profess. Papers **440-W** (1966).
16. WHITE, D. A.: The stratigraphy and structure of the Mesabi Range, Minnesota. Minn. Geol. Surv. Bull. **38** (1954).
17. TAYLOR, J. H.: The Mesozoic ironstones of England. Petrology of the Northampton sand ironstone formation. Mem. Geol. Surv. Gt. Brit. (1949).
18. GARRELS, R. M.: Mineral equilibria at low temperature and pressure. New York: Harper and Row 1960.
19. HOUGH, J. L.: Fresh water environment of deposition of Precambrian banded iron formations. J. Sediment. Petrol. **28**, 414–430 (1958).
20. GOVETT, G. J. S.: Origin of banded iron formation. Bull. Geol. Soc. Am. **77**, 1191–1212 (1966).
21. HUBER, N. K.: The environmental control of sedimentary iron minerals. Econ. Geol. **53**, 123–140 (1958).
22. GARRELS, R. M., and C. L. CHRIST: Solutions, minerals, and equilibria. New York: Harper and Row 1965.
23. LOVE, L. G., and J. W. MURRAY: Biogenic pyrite in recent sediments of Christchurch Harbour, England. Am. J. Sci. **261**, 433–448 (1963).
24. EMERY, K. O.: The sea of Southern California. New York: John Wiley & Sons, Inc. 1960.
25. MORETTI, F. J.: Observations on limestones. J. Sediment. Petrol. **27**, 282–292 (1957).
26. EDWARDS, A. B., and G. BAKER: Some occurences of supergene iron sulphides in relation to their environments of deposition. J. Sediment. Petrol. **21**, 34–46 (1951).
27. ANDEL, T. VAN, and H. POSTMA: Recent sediments of the Gulf of Paria. Reports of Orinoco Shelf Expedition, Amsterdam, 1954.
28. CASPERS, H.: Black sea and sea of Azov. Mem. Geol. Soc. Am. **67**, 801–890 (1957).
29. DUNHAM, K. C.: Black shale, oil and sulphide ore. Advan. Sci. **18**, 73 (1962).
30. HALLIMOND, A. D.: Bedded iron ores of England and Wales: petrography and chemistry. Mem. Geol. Survey Gt. Brit. (1925).
31. CAYEUX, L.: Les minerals de fer oolitique de France. II. Minerals de fer secondaires. Paris: Imprimerie Nationale 1922.
32. ALLING, H. L.: Diagenesis of the Clinton hematite ores of New York. Bull. Geol. Soc. Am. **58**, 991–1018 (1947).
33. STOSE, G. W.: Notes on the origin of Clinton hematite ores. Econ. Geol. **19**, 405–411 (1924).
34. CARROL, D.: Role of clay minerals in the transportation of iron. Geochim. Cosmochim. Acta **14**, 1–27 (1958).
35. MCCONNELL, D.: The petrography of rock phosphates. J. Geol. **58**, 16–23 (1950).
36. LOWELL, W. R.: Phosphatic rocks in Deer Creek-Wells Canyon Area, Idaho. U.S. Geol. Surv. Bull. **982-A**, 52 pp. (1952).
37. MCCONNEL, D.: The problem of the carbonate apatites IV structural substitutions involving CO_2 and OH. Bull. Soc. Franc. Mineral. Crist. **75**, 428–445 (1952).
38. KAZAKOV, A. V.: The phosphorite facies and the genesis of phosphorites, in Geological Investigations of Agricultural Ores. Trans. Sci. Inst. Fertilizers and Insecto-fungicides **142**, 95–113 (1937).
39. BUSHINSKY, G. I.: Structure and origin of the phosphorites of the U.S.S.R. J. Sediment. Petrol. **5**, 81–92 (1935).

40. CAROZZI, A. V.: Microscopic sedimentary petrography. New York: John Wiley & Sons, Inc. 1960.
41. GOLDBERG, E. D., and R. H. PARKER: Phosphatized wood from the Pacific sea floor. Bull. Geol. Soc. Am. **71**, 631–632 (1960).
42. BLUCK, B. J.: Petrography of Devonian phosphates of Indiana. Ill. Acad. Sci. Trans. **59**, 43–47 (1966).
43. PEVEAR, D. R.: The estuarine formation of United States coastal plain phosphorite. Econ. Geol. **61**, 251–256 (1966).
44. STEWART, F. H.: Marine evaporites. U.S. Geol. Surv. Profess. Papers **440** (1963).
45. BRIGGS, L. I.: Evaporite facies. J. Sediment. Petrol. **28**, 46–56 (1958).
46. BRAITSCH, O.: Mineral Paragenesis und Petrologie der Stassfurtsalze in Revershausen. Kali Steinsalz **3**, 1–14 (1960).
47. BERSTICKER, A. C., K. E. HOEKSTRA, and J. F. HALL: Symposium on Salt. The Northern Ohio Geol. Soc., Inc. (1963).
48. BORCHERT, H., and R. O. MUIR: Salt deposits. London: D. van Nostrand Co., Ltd. 1964.
49. HOLLINGWORTH, S. E.: Evaporites. Proc. Yorkshire Geol. Soc. **27**, 192–198 (1948).
50. STEWART, F. H.: The petrology of the evaporites of the Eskdale No. 2 Boring, East Yorkshire. Mineral Mag. **29**, 445–475, 557–572 (1951).
51. – Permian evaporites and associated in Texas and New Mexico compared with those of Northern England. Proc. Yorkshire Geol. Soc. **29**, 185–235 (1954).
52. POSNJAK, E.: Deposition of calcium sulphate from sea water. Am. J. Sci. **238**, 559–568 (1940).
53. MURRAY, R. C.: Origin and diagenesis of gypsum and anhydrite. J. Sediment. Petrol. **34**, 512–523 (1964).
54. PACKHAM, G. H., and K. A. W. CROOK: The principle of diagenetic facies and some of its implications. J. Geol. **68**, 392–407 (1960).
55. DAPPLES, E. C.: Behavior of silica in diagenesis (IRELAND, H. A., ed.). Soc. Econ. Paleontologists Mineralogists Spec. Publ. **7**, 36–54 (1959).
56. BURST, J. R.: "Glauconite" pellets: their mineral nature and applications to stratigraphic interpretation. Bull. Am. Assoc. Petrol. Geologists **42**, 310–327 (1958).
57. GALLIHER, E. W.: Glauconite genesis. Bull. Geol. Soc. Am. **46**, 1351–1366 (1935).
58. PETERSON, M. N. A., C. C. VON DER BORCH, and G. S. BIEN: Growth of dolomite crystals. Am. J. Sci. **264**, 257–272 (1966).
59. WHITEHOUSE, U. G., and R. S. MCCARTER: Diagenetic modification of clay mineral types in artificial sea water. Clays and Clay Minerals (Proc. 5th Nat. Conf.), 81–119 (1958).
60. GRIM, R. E., and W. D. JOHNS: Mineral investigations in the Northern Gulf of Mexico. Clays and Clay Minerals (Proc. 2nd Nat. Conf.), 81–103 (1954).
61. POWERS, M. C.: Clay diagenesis in the Chesapeake Bay area. Clays and Clay Minerals (Proc. 2nd Nat. Conf.), 68–80 (1954).
62. CARROL, D., and H. C. STARKEY: Effects of seawater on clay minerals. Clays and Clay Minerals (Proc. 7th Nat. Conf.), 80–101 (1960).
63. EMERY, K. O., and S. C. RITTENBERG: Early diagenesis of California basin sediments in relation to origin of oil. Bull. Am. Assoc. Petrol. Geologists **36**, 735–806 (1952).
64. SIEVER, R., K. C. BECK, and R. A. BERNER: Composition of interstitial waters of modern sediments. J. Geol. **73**, 39–73 (1965).
65. SHEPARD, F. P., and D. G. MOORE: Central Texas coast sedimentation: characteristics of sedimentary environment, recent history and diagenesis. Bull. Am. Assoc. Petrol. Geologists **39**, 1463–1539 (1955).
66. BORDOVSKIY, O. K.: Transformation of organic matter in bottom sediments and its early diagenesis. Marine Geol. **3**, 83–114 (1965).
67. ZOBELL, C. E.: Studies on redox potential of marine sediments. Bull. Am. Assoc. Petrol. Geologists **30**, 477–513 (1946).
68. LIMBERG-RUBAN, YE. L.: The quantity of bacteria in the water and in bottom material in the North Western Pacific. Issled. Dalhevost. Morey SSSR **3** (1952) [in Russian].
69. DESITTER, L. U.: Diagensis of oil field brines. Bull. Am. Assoc. Petrol. Geologists **31**, 2030–2040 (1947).
70. OPPENHEIMER, C. H.: Bacterial activity in marine sediments. Geochemical Symposium, Gostoptekhizdat, Moscow 1960.

71. BREDEHOEFT, J. D. et al.: Possible mechanism for concentration of brines in sub-surface formations. Bull. Am. Assoc. Petrol. Geologists **47**, 257–269 (1963).
72. THOMSON, A.: Pressure solution and porosity (IRELAND, H. A., ed.). Silica in Sediments. Soc. Econ. Paleontologists and Mineralogists. Spec. Publ. **7**, 185 (1959).
73. SIEVER, R.: Siever solubility, 0–200° C and the diagenesis of siliceous sediments. J. Geol. **70**, 127–150 (1962).
74. HEALD, M. T.: Cementation of Simpson and St. Peter sandstones in parts of Oklahoma, Arkansas and Missouri. J. Geol. **64**, 16–30 (1956).
75. COOMBS, D. S.: The Nature and alteration of some Triassic sediments from Southland, New Zealand. Trans. Roy. Soc. New Zealand **82**, 65–109 (1954).
76. CHAVE, K. E.: Evidence on history of sea water from chemistry of sub-surface waters of ancient basins. Bull. Am. Assoc. Petrol. Geologists **44**, 357–370 (1960).
77. BURST, J. F.: Post-diagenetic clay mineral environmental relationships in the Gulf Coast Eocene. Clays Clay Minerals, Proc. 6th Nat. Conf. 327–341 (1959).
78. RODGERS, J.: Distributions of marine carbonate sediments: a review. Regional aspects a carbonate deposition. A Symposium with discussion S. E. P. M. (R. J. LE BLANC and JULIA G. BREEDING, eds.) (1957).
79. FOLK, R. L.: Spectral Subdivision of limestones. Classification of Carbonate rocks. Mem. I. Am. Assoc. Petrol. Geologists, Tulsa **62**, 62–84 (1962).
80. NEWELL, N. D. et al.: The Permian Reef Complex of Guadalupe Mountains Region, Texas and New Mexico. San Francisco: W. H. Freeman & Co. 1953.
81. HENSON, F. R. S.: Cretaceous and Tertiary reef formations and associated sediments in Middle East. Bull. Am. Assoc. Petrol. Geologists **34**, 215–238 (1950).
82. LOWENSTAM, H. A.: Niagran reefs of the Great Lakes Area. J. Geol. **58**, 430–487 (1950).
83. HADDING, A.: The pre-quaternary sedimentary rocks of Sweden, VI, reef limestones. Medd. Lunds Geol. Mineral. Inst. **2**, 37, 137p (1941).
84. YOUNG, R. B.: A comparison of certain stromatolitic rocks in the dolomite series with modern algal sediments in the Bahamas. Trans. Proc. Geol. Soc. S. Africa **37**, 153–162 (1934).
85. CLOUD, P. E.: Facies relationships of organic reefs. Bull. Am. Assoc. Petrol. Geologists **36**, 2125–2149 (1952).
86. KUENEN, PH. H.: Marine geology. New York: John Wiley & Sons, Ltd. 1950.
87. ARKELL, W. J.: The Jurassic System in Great Britain. Oxford 1933.
88. STOCKDALE, P. B.: Bioherms in the Borden Group of Indiana. Bull. Geol. Soc. Am. **42**, 707–718 (1931).
89. LEES, A.: The structure and origin of the Waulsortian (Lower Carboniferous) reefs of W. EIRE. Phil. Trans. Roy. Soc. London Ser. B **247**, 483–531 (1964).
90. SCHWARZACHER, W.: Petrology and structure of some Lower Carboniferous reefs in North Western Ireland. Bull. Am. Assoc. Petrol. Geologists **45**, 1481–1503 (1961).
91. COTTER, E.: Waulsortian-type carbonate banks in the Mississippian lodgepole formation of Central Montana. J. Geol. **73**, 881–888 (1965).
92. WOLF, K. H.: "Grain-diminution" of algal colonies to micrite. J. Sediment. Petrol. **35**, 420–427 (1965).
93. SHEARMAN, D. J., and P. A. D'E SKIPWITH: Organic matter in recent and ancient limestone and its role in diagenesis. Nature **208**, 1310 (1965).
94. BLACK, M.: The algal sedimentation of Andros Island, Bahamas. Phil. Trans. Roy. Soc. London, Ser. B **222**, 165–192 (1933).
95. MONTY, C.: Recent algal stromatolites in the windward lagoon, Andros Island, Bahamas. Ann. Soc. Geol. Belg. Mem. **88**, 269–276 (1964–65).
96. LOGAN, B. W., R. REZAK, and R. N. GINSBURG: Classification and environmental significance of algal stromatolites. J. Geol. **72**, 68–83 (1964).
97. REZAK, R.: Stromatolites of the belt series in Glacier National Park and vicinity, Montana. U.S. Geol. Surv. Profess. Papers **294-D** (1957).
98. GINSBURG, R. N.: Early diagenesis and lithification of shallow-water carbonate sediments in S. Florida. Regional aspects of carbonate deposition S. E. M. P. Spec. Publ. **5**, 80–100 (1957).
99. MAWSON, Sir DOUGLAS: Some South Australian algal limestones in process of formation. Quart. J. Geol. Soc. London **85**, 613–623 (1929).

100. HOMMERIL, P., and M. RIOULT: Etude de la fixation des sediments meubles par deux algues marins: rhodothamniella floridula (Dillwyn) J. Feldm et microcoleus chtonoplastes thur. Marine Geol. **3**, 131 – 155 (1965).
101. BLUCK, B. J.: Sedimentation of Middle Devonian Carbonates, South Eastern Indiana. J. Sediment. Petrol. **35**, 656 – 682 (1965).
102. NEWELL, N. D., E. G. PURDY, and J. IMBRIE: Bahamian oolitic sand. J. Geol. **68**, 481 – 497 (1960).
103. CAROZZI, A. V.: Contribution a l'etude des proprietes geometriques des oolithes – l'example du Grand Lac Sale. Bull. Instn. National Genevois **58**, 1 – 51 (1957).
104. CONRAD, N. A.: Observation on coastal erosion in Bermuda and measurements of the boring rate of the Ponge, Cliono, Campa. Limnol. Oceanog. **2**, 92 – 108 (1966).
105. CLOUD, P. E.: Environment of calcium carbonate deposition west of Andros Island, Bahamas. U.S. Geol. Surv. Profess. Papers **350**, 1 – 138.
106. WELLS, A. J., and L. V. ILLING: Present day precipitation of calcium carbonate in the Persian Gulf. Dev. in Sed. Vol. 1 Deltaic and Shallow Marine Deposits (VAN STRAATEN, ed.) (1964).
107. LOWENSTAM, H. A., and S. EPSTIEN: On the origin of sedimentary aragonite needles of the Great Bahama Bank. J. Geol. **65**, 364 – 375 (1957).
108. BROECKER, W. S., and T. TAKAHASHI: Calcium carbonate precipitation on the Bahama Banks. J. Geophys. Res. **71**, 1575 – 1602 (1966).
109. ILLING, L.: Bahaman calcarious sands. Bull. Am. Assoc. Petrol. Geologists **38**, 1 – 95 (1954).
110. FRIEDMAN, G. M.: Early diagenesis and lithification in carbonate sediments. J. Sediment. Petrol. **34**, 777 – 813 (1964).
111. FYFE, W. S., and J. L. BISCHOFF: The calcite-aragonite problem. Dolomitization and Limestone diagenesis. A Symposium (L. C. PREY and R. C. MURRAY, eds.). S. E. P. M. Spec. Publ. **13**, 3 – 13 Tulsa (1965).
112. GEVIRTZ, J. L., and G. M. FRIEDMAN: Deep sea carbonate sediments of the Red Sea, and their implication on marine lithification. J. Sediment. Petrol. **36**, 143 – 151 (1966).
113. BATHURST, R. G. C.: The replacement of aragonite by calcite in the molluscan Shell wall. Approaches to Paleoecology (J. IMBRIE and N. D. NEWELL, eds.). New York: John Wiley & Sons, Ltd. 1964.
114. STODDART, D. R., and J. R. CANN: Nature and origin of beach rock. J. Sediment. Petrol. **35**, 243 – 273 (1965).
115. BATHURST, R. G. C.: Diagenetic fabrics in some British Dinantian limestones. Liverpool Manchester Geol. J. **2**, 11 – 36 (1958).
116. BEALS, F. W.: Diagenesis in pelleted limestones. Dolomitization and limestone diagenesis. A Symposium (L. C. PAY and R. C. MURRAY, eds.). S. E. P. M. Spec. Publ. **13**, Tulsa (1965).
117. GROSS, M. G.: Variations in the $^{18}O/^{16}O$ and $^{13}C/^{12}C$ ratios of diagenetically altered limestones in the Bermuda Island. J. Geol. **72**, 170 – 194 (1964).
118. BERNER, R. A.: Chemical diagenesis of some modern carbonate sediments. Am. J. Sci. **264**, 1 – 36 (1966).

CHAPTER 10

Organic Matter in Sediments*

D. H. WELTE

Institut für Geologie, Universität Würzburg

Contents

I. Introduction

Organic matter in sediments has to be regarded as the residue of organic life, and is, in the geological sense, a fossil. An exception to this rule may be carbonaceous material produced in early Precambrian times by abiotic processes taking place before the first living cells came into existence. Therefore, organic matter may be found in sediments even older than organized organic life. At present we find it in fossil form in sediments as old as 3×10^9 years. Organic carbon became more important and more abundant with the development and diversification of life.

Phytoplankton and bacteria are and were the main producers of organic matter throughout the history of the earth and thus account for most of the organic matter found in Recent and ancient sediments [1, 2].

II. Incorporation of Organic Matter into Sediments

The preservation of organic matter is almost exclusively restricted to aquatic sediments. Depending on where and how those sediments were formed, they contain varying proportions of materials from pre-existing rocks that were transported somehow to the site of deposition and of other components that were formed *in situ*.

The same principle applies to the organic materials. Sediments may thus contain allochthonous organic matter, that was transported to the site of deposition

* Editors' Note: Dr. WELTE suffered a prolonged illness shortly after preparing the Chapter outline printed here. This illness unfortunately prevented him from writing the complete Chapter.

from elsewhere and they may also contain an autochthonous portion, that originated at the site of deposition.

Autochthonous and allochthonous organic matter differ in composition. Whereas the autochthonous material resembles more closely the primary biological products, the allochthonous fraction is mainly composed of diagenetically formed, secondary reaction products, like humic acids and kerogen. Thus the allochthonous material represents the more stable residual organic matter that has already experienced at least part of the sedimentary cycle.

III. Distribution of Organic Matter in Space and Time

Different environments of sedimentation are represented by different types of sediments with different proportions of autochthonous versus allochthonous organic matter and all the pecularities of the contributing biological realm. Sedimentary rocks formed by chemical precipitation, like certain carbonates and evaporites, rarely contain larger quantities of allochthonous organic matter, as compared to detrital rocks, like sandstones or shales.

Beyond these more basic differences the specific mode of deposition of a certain sediment will result in a variety of geometrical distribution patterns of organic matter within the rock.

This may range from a sheet-like, repeated concentration of organic matter, e.g., in certain oil shales, to a very uniform, finely disseminated pattern as in certain limestones or shales or to the enrichment of coarse organic debris as found in some sandstones.

The amount of organic matter preserved in ancient sediments fluctuates during earth history. It was high during Cambrian time, low in the Silurian, again high in the Carboniferous, low during the Permo-Triassic and rose toward the Tertiary to reach another high [3]. There is good reason to assume that this phenomenon is related, at least in part, to highs and lows in the primary organic production. Such changes, especially in regard to phytoplankton productivity, were shown by TAPPAN and LOEBLICH [2].

Other reasons for changes in organic carbon production and preservation may be looked for among climatic and tectonic events. BITTERLI [4], for instance, states that the formation of bituminous sequences is favored at turning points in the paleographic history, such as those related to orogenic phases and during epeirogenic (or eustatic) oscillations and consecutive transgressions or regressions.

On the other hand major changes in the organic production during earth history may very well have changed the CO_2 and O_2 balance in ocean waters and the atmosphere as suggested by TAPPAN and LOEBLICH [2]. This in turn may influence such processes as the precipitation of carbonates and the development of higher plants and animals.

IV. Role of Organic Matter in Sediment Formation and Compaction

After sedimentary particles have settled to form a sediment they undergo diagenetic changes. Organic matter, especially, being principally unstable under geologic conditions, is subjected to those diagenetic changes. According to the abundance of organic matter a sediment may be more or less influenced during its diagenetic path.

Organic matter supports microbial metabolism. This in turn influences inorganic chemical reactions by changing pH and Eh conditions (production of CO_2 and H_2S). Furthermore, poorly understood complex formation and chelation processes will occur. After microbial activity comes to a halt the mere presence of carboxyl, hydroxyl and amide groups within the organic material will ensure chemical reactions also involving the inorganic matrix.

Finally, physical and mechanical properties of a sediment, such as water retention capacity, porosity and compressibility are to some degree controlled by the amount of organic material within the sediment [5].

V. Conclusions

Organic matter in sediments is not just a component left over from former organic life; it plays an active role in the formation and diagenesis of sediments. Unlike the inorganic components, only small amounts (about 1–2 percent) of organic matter may strongly influence the post-depositional behavior of a sediment.

A careful study of the occurrence and distribution of organic matter in Recent and ancient sediments will be a powerful tool for reconstructing such important aspects of earth history as the evolution of the atmosphere, climatic conditions of the past, and the development of life.

References

1. BORDOVSKIY, O. K.: Accumulation and transformation of organic substances in marine sediments. Marine Geology **3**, 3–114 (1965).
2. TAPPAN, H., and A. R. LOEBLICH JR.: The geobiologic application of fossil phytoplankton evolution and time-space distribution. Special paper by the Geol. Soc. Am. (in press).
3. RONOV, A. B.: Organic carbon in sedimentary rocks (in relation to presence of petroleum). Geochemistry (translation) **5**, 510–536 (1958).
4. BITTERLI, P.: Aspects of the genesis of bituminous rock sequences. Geol. Mijnbouw **42**, 183–201 (1963).
5. WELLER, J. M.: Compaction of sediments. Bull. Am. Assoc. Petrol. Geologists **43**, 273–310 (1959).

CHAPTER 11

Geomicrobiology and Geomicrobiological Attack on Sedimented Organic Matter

L. R. MOORE

University of Sheffield, England

Contents

I. Introduction

Recent developments in the field of geomicrobiology as applied to the organic content of rocks are of fascinating potential. Individual organic compounds and related series of compounds can now be isolated from rocks. The fullest interpretation of the results is dependent upon the existing and rapidly growing knowledge concerning the biological nature of the organic matter contained in these rocks. Such a source of information is largely provided in geological science through micropalaeontological research concerned with the biological contents of sedimentary rocks of all geological ages from the Precambrian to the present day. The most pertinent observations are those concerning remains composed entirely of organic matter, as distinct from studies of remains in which an inorganic protective cover, shell, or test is mainly preserved. Initial investigations were concerned with the isolation of the more resistant total organic entities, e.g., those with a resinous or "chitinous" composition, released by acid maceration processes as "acidic residues". In this manner the studies of palynology and microplanktonology were developed, the former concerned with pollen and spores, the latter with fossil microplankton or with remains which morphologically bear comparison with modern plankton. Recent investigations of the entire organic content of rocks, as released by acid digestion of the inorganic constituents with little or no subsequent differential maceration of the organic matter, have provided additional information concerning the nature of the organic content. In particular, preserved cell structures and tissues and various types of amorphous groundmass organic material including resinous, "humic", and bituminous substances have been examined. The total isolation of organic matter and the relatively non-destructive processes employed in its isolation have demonstrated the presence, nature, and method of attack of geomicrobiological agents concerned with degradation processes. This aspect of study, more analogous to modern micro-biology, is referred to as geomicrobiology or palaeomicrobiology and is principally concerned with the presence and activities of a range of microorganisms referable to as bacteria, actinomycetes, fungi, and algae.

The wellknown preservation of plant remains as carbonized or petrified tissues has unfortunately led to the general view that organic matter is likely to be preserved as carbonized or bituminized residues, which may contain some degree of original structure but with little of its original or near original composition. Micropalaeontologists are increasingly aware of the perfect preservation of the finest structural details in organic remains from fossil sediments, e.g., the preservation of cell walls, cell contents, cells in a state of division, and vegetative and reproductive processes in organic connection. They are also aware of the organisms and processes of biological degradation in geological times. It is therefore not surprising that modern organic geochemistry demonstrates the presence of a wide range of organic substances preserved in rocks, or that chemists believe such compounds may be directly derived from the organic parts of the fossil remains themselves.

The appropriate range of geological conditions for the aggregation of organic matter and the requisite environmental conditions conducive to its preservation in sediments allow organic matter to be represented in a wide variety of sedimentary rock types. The parameters of the physico-chemical and biological environment necessary for this preservation are considered in this volume. They may be indicated by the mineralogical composition or diagenetic changes in the sediment, but the microbial association and organic content itself is further indicative of the environmental conditions. This may be exemplified by the presence of microplankton in a typical marine habitat and the presence of land-derived vegetal debris, spores, and planktonic algae in a freshwater habitat. After initial burial, subsequent geological conditions must be such that the entombed organic matter is preserved by retention within a closed physico-chemical system. In part, this is dependent upon the kind of rock in which the material was originally entombed. Thus, in porous or pervious rocks, e.g., sandstones, some siltstones and limestones, there is a greater opportunity for the introduction, over geological time, of oxygenated solutions which could cause alteration or destruction of some or all of the original organic matter. At the same time, non-original organic compounds and even particulate organic matter may be introduced. Impervious and non-porous rocks, largely composed of finely divided clay minerals and with some degree of carbonate or siliceous inorganic cement, not only reveal a better preservation of the organic matter but are less liable to alteration or addition.

Other geological conditions may profoundly alter the original composition of the organic matter. After completion of the original diagenetic processes, time as such appears to have little effect, as instanced by the preservation of bacterial, fungal, and algal remains in Precambrian rocks and by the presence of viable bacteria in the Permian salt deposits [1]. Of more significance are factors involving depth of burial of the sediment and its association with periods of folding or faulting, with igneous intrusion and extrusion, and with regional crustal movements. Changes of pressure and temperature or physical deformation lead to alteration or metamorphosis of the organic matter. Severe changes ultimately convert the organic material into a carbonized or "graphitic" film, destroying its original composition and detailed structure, but sometimes preserving the original morphological characteristics. The oldest recorded organic remains (3,200+ million years [2] from the Fig Tree Formation of Africa) are so preserved. Not all alterations reach this final stage and organic remains show intermediate changes analogous to rank changes in coal. Thus, a specific miospore, whose characteristics are well known under normal conditions, may be compared with the same miospore preserved in rocks involved in milder metamorphic changes, e.g., differing depths of burial, approach to structural complexity, or igneous intrusion. The changes seen in the spore involve deepening of the color, while a differential color pattern is retained and suggests some progressive chemical alteration. In these cases the total organic matter may not be uniformly altered, for under the prevailing temperature and pressure conditions of metamorphism some parts may be more readily carbonized while others retain their pigmentation. The organic matter may thus provide a sensitive indicator of lower metamorphic grade by organic geochemical analysis.

The organic matter originally available for preservation in sediments will vary for several reasons, for example:

1. The nature of the original organic material which may have changed with geological time as biological evolution proceeded.

2. The source of the organic material, i.e., from plants and animals in mixtures of varying proportions.

3. The derivation of the organic matter, i.e., mainly microplankton in marine enviroments; mainly land-derived vegetation in lacustrine enviroments; the possibility of admixture in varying proportions in "inshore" platform sediments, or in estuaries, deltas and lagoons.

4. The degree to which autochthonous material formed in the given environment is affected by the addition of allochthonous material of a different type admitted by water or wind transport. Such conditions frequently depend on the form and bathymetric nature of the sea floor or water body.

5. The extent to which the organic material has been degraded by microbiological activity and the nature of this activity. The oxidative processes of biological activity may be terminated at various stages and selective attack may concentrate more biologically stable organic residues.

6. The extent of the contribution made by extra-cellular enzymes derived from micro-organisms, either by direct addition to the organic matter or by increasing the proportion of water-soluble or colloidal material.

7. The physico-chemical (oxidation or reduction) and biological parameters of the site of preservation. These control the final possible stage of selective degradation by bottom-dwelling benthon and the indigenous microbiological population, which continues its activities during the diagenetic stages of rock formation.

8. Rocks of the same age but from different sites may illustrate some, many, or even all of the above affects, with differences dependent upon the local biological, physico-chemical, and geological conditions.

The initial chemical composition of the organic matter preserved in a given rock will depend on factors listed under 1–7 above. The final chemical composition will be related to further geological factors, for example:

1b. The properties of the containing rock which may effectively seal off the organic matter, preventing oxidation and renewed biological activity. The remarkable preservation of organic matter in chert is a well known example.

2b. The geological history of the rock. Occurence or exposure within a zone of oxidation, weathering, or erosion can deplete or change the organic content of the rock.

3b. The geological sequence of which the rock forms a part. An impermeable cover protects the rock from percolating solutions and conversely a porous cover provides a means of access for solutions and encourages chemical change. Porous rock may also store gases derived from the underlying organic rich sediment, thus indirectly aiding chemical changes by removing one member of the system.

4b. The total depth of burial beneath younger sediments; depth is reflected by the geothermal gradient and changes in the physico-chemical system.

5b. Geological conditions involving direct temperature changes, as caused for example by the nearness of injected or ejected masses of molten igneous rock.

6b. Conditions involving pressure and temperature changes engendered by geological structural movements. Faults, folds, and more severe structural upheavals may lead to metasomatic or metamorphic alteration of the organic matter and to the formation of bitumenastic or asphaltic substances or to carbon or graphitic films.

In summary, the organic matter in rock of the same age occurring in differing places will differ initially as indicated under conditions 1–7 and in addition, the final chemical composition will depend on factors 1b–6b. The various chemical compounds represented in the initial organic matter will themselves react differently to changes in pressure and temperature.

II. Accumulation of Organic Material

A. The Marine Environment

Considerable pertinent information concerning the nature of organic matter in the marine environment has resulted from studies of recent deposits. Such works as those of BADER *et al.* [3], BORDOVSKY [4], and TRASK [5] are sources of detailed information in this respect. VALLENTYNE [6] discusses the molecular nature of organic matter in lakes and oceans and lists the various chemical compounds known in these and related environments; this work carries comprehensive literature references.

The principal source of organic matter is the autochthonous population of the surface waters and, in particular, the Phytoplankton. Unicellular algae, diatoms, peridineans and dinoflagellates may be present together with zygospores, cysts, protective coverings, and various representative forms of the resting stages in life cycles. Zooplankton remains add their counterpart and other organic detritus of allochthonous origin may contribute. The greater part of the microplankton is consumed by predators; of the remainder, microbiological decomposition accounts for some 70 percent, leaving about 30 percent of microbiologically stable material available for preservation. The principal marine microbiological agents are bacteria which live free in the water or are attached to organic material. Their distribution is uneven for they are particularly abundant only in the surface waters and in the organic detritus of the sea floor. The bacterial biomass shows a parallel relationship to that of the plankton and in certain cases may be similar. KRISS [7] estimated the average bacterial biomass to be 20.8 mg/m^3 for the Caspian Sea, 20 mg/m^3 for the oxygenated zone of the Black Sea and approximately 0.1 mg/m^3 for the Pacific Ocean. BORDOVSKY [4] refers to the chemical composition of marine organisms and in particular stresses the importance of algae and bacteria as the principal contributors. The decomposition of organic matter by autotrophic organisms is in part compensated by synthesis and transformation in the bodies of heterotrophic organisms. Nevertheless, part of the organic matter is discarded and eventually precipitated, and corresponds to the biologically stable portion. Organic matter is present in the seas in true solution, in colloidal suspension or solution, and as particulate suspension of organic detritus of dead and living organisms and their skeletal and exoskeletal remains. The proportions may differ but the dissolved content is often tens or hundreds of times that of all other forms of organic matter. In inland water, colloidal solution may play a more important role. DATSKO [8] cites the following proportions for the Sea of Azov: true solution 66 percent; colloidal solution 22 percent; in suspension 12 percent.

The eventual sedimentation of organic material is accomplished in several ways. The descent of particulate matter is relatively rapid and may amount to 80–1,920 m in a day, the principal factor being particle size. The soluble compounds are partly absorbed by clay minerals and BADER *et al.* [3] account for some 53–62 percent of decomposition products in this way. Coagulation and ageing of colloids form an essential part of the process, and the biochemically stable residual materials (humus) form as condensation products of carbohydrates, proteins and their derivatives. The small proportion of bitumens present, such as the lipids of Phytoplankton, are believed to be resistant to biological oxidation.

Other organic matter includes the particulate allochthonous material and the particulate remains of planktonic or benthonic organisms. Incorporated are the local algal contributions of the green, red, and brown algae and occasionally of higher plants which colonize the coastal regions or shallow water zones of many seas. Bottom-dwelling populations of animals may either extract nutrients from the sea water by filtration and precipitation (seston feeders) or collect detritus from the sea floor (detritus feeders). By their action in working over the sediment these animals bring about chemical changes in the organic matter and add a contribution from their own metabolism. They are in turn the subject of predatory attack. The whole of the organic matter is further worked over by bacterial, actinomycete, and possibly fungal action. In this sphere of activity bacteria are the main contributors to further decomposition and the synthesis of new compounds by enzymic processes. Some 30–40 percent of the carbon content of an organic substrate is converted to new compounds while 60–70 percent is liberated as CO_2 or under particular conditions as hydrocarbons. The availability of assimilable organic matter is the main factor which limits the development of bacteria in bottom deposits. The bacterial biomass is significant in the sediments of some seas; ZHUKOVA [9] found a variation of between 57 and 12,000 million bacterial cells per gram of natural sediment in the upper centimeter of the Northern Caspian Sea, while KRISS [7] recorded a bacterial biomass of 15–30 g/m^3 from the hydrogen sulphide regions of the Black Sea, a figure approaching the biomass of bottom-dwelling organisms.

SPARROW and JOHNSON [10], in reviewing the literature concerning fungi in oceans and estuaries, have remarked on the presence of active fungi at all levels in the water bodies and sediments of ocean basins and shallow seas. The fungi included members of the Mucoraceous Fungi, yeastlike fungi, Actinomycetes, Fungi Imperfecti, Ascomycetes, and Phycomycetes, which are forms essentially parasitic or saprophytic and variously concerned with the breakdown of cellulose, pectic substances and other carbohydrates and, to some extent, lignin. HOHNK [11], in a study of German coastal waters, showed a landward increase in the abundance of Phycomycetes with an accompanying decrease in the frequency of Ascomycetes and Fungi Imperfecti. The precise role of fungi in decomposing organic detritus in the sea is not known; fungi are actively concerned, but their role may be indirect and related to changes in the environment of pH, oxygenation, hydrolysis, nitrification, and ammonification, etc. They may also function in association with bacteria or even algae.

Despite the vast quantity of phytoplankton produced in the oceans and seas, estimated at 9×10^{10} tons of dry matter annually [12], only a small quantity reaches the sea floors. TRASK [5] estimated that 2 percent reached the sea floor in the shallows, as against 0.02 percent in the open sea. USPENSKIY [13] believed that only 0.8 percent of the primary production of organic matter escaped decomposition and was precipitated.

The organic matter in marine sediments is a complex mixture which includes:

1. Original and altered protein material and its decomposition products, including amines, amino acids, and amino complexes. DEGENS [14] has shown some of these compounds are developed during early diagenetic stages. DEGENS, PRASHNOWSKY, EMERY, and PIMENTA [15] describe differences in the relative abundance of amino acids with depth in the sediment which they ascribe to preferential survival in response to differences of pH, Eh and concentration and the action of interstitial water in bringing about solution and redeposition.

2. Carbohydrates. These highly oxidized compounds are rapidly hydrolized enzymatically to carbon dioxide and water, but various carbohydrates, including cellulose and sugars, have been detected in sediments. DEGENS, REUTER, and

SHAW [16] considered carbohydrates to be both the survivors and the products of diagenetic changes.

3. Lignin, although not a normal constituent of marine organisms, may be present as a stable biological entity derived from land organisms.

4. Pigments of various types, include chlorophyll-like compounds, carotenoids, and porphyrins. These compounds have been isolated from sediments and rocks. ORR, EMERY, and GRADY [17] described pigments obtained from recent sediments; these were intermediate in character between chlorophyll and the porphyrins found in crude oil. These authors, HODGSON and PEAKE [18] and HODGSON, PEAKE and BAKER [19] variously refer to the presence of organo-metallic complexes, such as the vanadyl and nickel porphyrins, in recent sediments, in post-glacial lake sediments and in rocks of various geological ages ranging from Devonian to the present day. Porphyrins were not represented in the Precambrian rocks studied [19].

5. Lipids are derived from phytoplankton metabolism and are further synthesized by microbiological activity from carbohydrate and protein sources [4]. The relative stability of fatty acids in sediments and their presence in rocks led HOERING and ABELSON [20], to regard them as of great promise in tracing the biological record back to the earliest known rocks. The presence of solvent-extractable lipid fractions, especially the hydrocarbon content, has been established in many sediments and they have been reported from the most ancient rocks. Lipids normally make up only a small proportion of the organic matter in sediments; TRASK [5] estimated the mean for recent sediments as 0.06 percent. Lipids are of particular interest in relation to the genesis of oil.

6. A further content consists of humic materials and residual unspecified organic matter. DEGENS, EMERY, and REUTER [21] established the presence of benzoic acid derivatives in the humic acid fraction of recent sediments and regard lignin as originating in allochthonous material. The presence of humic matter in deep ocean sediments far removed from land [4] suggests an association with autochthonous marine material. Humic acids may derive in part from proteins. For recent deposits [4] a relationship exists between humic acid concentration and particle size. Humic acid content seems to vary directly with total organic carbon in recent and in ancient sediments. The degree of humification also increases with increasing total organic content. The content of humic matter in different types of sediment in the northwestern Pacific is given as: 33.7 percent in clay and silt clay muds; 28.8 percent in silts; 27.2 percent in fine sands; 22.8 percent in medium and coarse sand; 14.9 percent in gravels [4].

DEGENS, EMERY, and REUTER [21] conclude that organic matter can be preserved under oxidizing as well as under reducing conditions. Under oxidizing conditions the marine organic contributions are largely eliminated, whereas under reducing conditions they are partially preserved, along with continentally derived organic material. Differences in the type of organic matter may be expected under these two environmental conditions, in accord with micropalaeontological evidence, for considerable microfloras and/or microfaunas are preserved in marine deposits from Precambrian to the present day. These collections are derived in the main from the shallower platform accumulations or from inshore localities in past oceans, but a proportion of this fossil organic material has been transported from land. It is significant that when reducing conditions in the environment are deduced from the mineralogical constitution of the rock, organic material is well represented, in a good state of preservation. The most important factors in the preservation of organic matter are those connected with the environment of accumulation and the early diagenetic changes which take place.

B. The Non-Marine-Brackish or Freshwater Environments

These environments cover a range of conditions starting from terrestrial accumulations to the sea as the eventual repository of materials. A series of interconnected environments lie between these extremes. Each of the separate environments is formed under specific physiographic conditions and, dependent upon the requisite geological conditions, each may be preserved and represented at appropriate points in geological time. On the other hand, many such environments are transitory and suffer destruction by erosion. Their sediments, which have been formed under specific conditions, may then be transported by fluvial or marine action into other environments which culminate in inshore or shelf accumulations in the sea. In such circumstances it is important to ascertain for a given rock whether there has been a single cycle of organic accumulation or more than one cycle of accumulation and change. The latter occurrence introduces the strong possibility of further partial biological selectivity under the changed conditions, with the probability that additional organic material may be produced in the new environment. The problem is effectively that of subjecting allochthonous and autochthonous constituents to differing environments but there is particular emphasis on the nature and content of the allochthonous material.

The starting point is terrestrial vegetation which has been preserved *in situ* as peat or has decomposed into litter and its subsequent inclusion into soils and/or water bodies. Water bodies such as lakes may contain sediment with a high proportion of organic ooze composed of the products of biological degradation of a terrestrial flora and, as an allochthonous constituent, the more stable remains of that flora. To this is added autochthonous phytoplankton in which algae are the predominant constituent. Lake deposits, which are governed by physiographic, climatic, and depth factors, may contain alternately organic rich and inorganic rich layers, each pair perhaps a millimeter thick and representing an annual deposit or varve. Under certain circumstances lagoonal, and even inshore gulf sediments may show similar features due to seasonal accumulation [22].

Rivers play the main role in the transport of allochthonous organic matter from inland sources to the sea, via the intermediate depositional environments of estuarine, deltaic and alluvial deposits. They are, in part, responsible for the establishment of extensive mud flat environments. The content and proportion of the various forms of organic matter in river water vary in relation to relief, climate, and other factors. For plains flow the organic matter may reach 40–50 percent of the mineral content, with a predominance of substances in either true or colloidal solution. For rivers in mountain tracts the organic matter may be only 4–5 percent of the mineral content and is mainly carried as detritus. For major rivers SKOPINTSEV and KRYLOVA [23] considered the mean annual organic content to vary between 10 and 43 mg/l, the average being not less than 20 mg/l. Chemically the greater part of this matter is typified by high resistance to biological attack with a wide range of the C/N ratio (8 to 36) indicating the degree of transformation and the depletion of proteinaceous material. Humus derivatives account for a high proportion of the organic matter of river waters and consequently of the sediments associated with fluvial transport.

C. Terrestrial Vegetation and Structural Elements of Plants

Detailed accounts of the organic compounds formed by plants are contained in other chapters. At this stage it is necessary to refer to the major groups of compounds and their relationship to plant structure.

The carbohydrates and allied compounds form a group of which the celluloses are the major constituent of cell walls, accompanied by starch and pectin. Pectin is also a constituent of saps and fibers. Chitin, a related substance, forms resistant cell walls or coverings in some plants, e.g., fungi. The pentosans or plant gums are an essential part of this complex.

Lignin is associated with cellulose fibers in the strengthening of the cell wall and refers to a structural type rather than a single substance. An allied group of compounds, the "lignans", is also present in wood. These compounds possess aromatic nuclei as do the allied plant constituents, the anthocyanins and the catechins.

Proteins and other nitrogeneous compounds, vegetable proteins, both as enzymes and structural units, are readily broken down by degradation processess to amino acids. The green pigment chlorophyll contains nitrogen and is concerned in carbohydrate synthesis. The more stable porphyrin derivatives are present in fossil organic matter. The alkaloids form an important nitrogen-based group of compounds but their contribution to plant structure is of minor significance as is that of the nucleic acids.

Fats, glycerides of fatty acids, and waxes (esters of fatty acids and fatty alcohols) are of vital importance to plant life. Polymerized and crosslinked structures based on fatty alcohols and acids are present in cutin, exine, and suberin, the fundamental units of cuticle, spore exines, and cork. In these materials the long-chain aliphatic compounds exist together with lignin and tannins, and the resultant membranes are resistant to anaerobic oxidation and chemical treatment.

The resins are amongst the plant products with the highest resistance to chemical and biological attack and are capable of polymerization; plant latex, or rubber, is related in function but not chemical structure.

The plant substances referred to above are genetically related, and the complicated conversions in plant life are based on a relatively restricted number of primary reactions which play an important part in the cycles which govern life processes, e.g., the Calvin and Krebs cycles. These processes which occur in the living plant do not terminate with the death of the higher plant. During the degradation of the dead plant material the cycles continue under the influence of the attacking microorganisms.

D. Decomposition of Wood and other Plant Remains

The decomposition of wood has played a major part in the derivation of organic deposits, whether of fossil soils, peats, coals, or of deposits in the non-marine environment. It is therefore of some importance to consider the degradation process involved. Decomposition of wood [24] is almost exclusively a biological process, carried out by a restricted group of organisms. The Hymenomycete fungi, with a few contributors from the ascomycetes and Fungi Imperfecti, are almost the sole organisms concerned. This is accomplished by fungi active within either living or dead wood, by fungi which decompose litter, and by fungi which live symbiotically with the tree, their mycelia forming mycorhiza within the roots. Cell walls in wood contain several layers which either vary in composition, or in the ratio of lignin to cellulose. These layers, according to their composition, are specifically attacked by fungi. Thus, *Merulius lacrymans*, which is responsible for brown or dry rot, digests the cellulose and hemi-cellulose, leaving lignin. *Fomes annosus*, active in white rot, preferentially attacks lignin and hemicellulose, the enzymes responsible being hadromase for lignin breakdown and cellulase for the hydrolysis of cellulose and hemicellulose. GOKSOYR [24], NORD and VITUCCI [25], and NORD and SCHUBERT [26] have considered the chemical nature of these changes, and FLAIG [27] described the chemical changes which follow fungal conversion of lignin to humic substances. Other workers, e.g., SHRIKHANDE [28] and REGE [29], refer to the production of colloidal mucus on biochemical decomposition by fungi and describe it as a mixture of carbohydrates and proteins. Bacteria may work in conjunction with fungi, particularly in the later stages of decomposition when they are capable of breaking down hemicelluloses, cellulose, and pentosans. GOKSOYR [24] referred to aromatic compounds (e.g., pinosylvin) in the heartwood of trees, which are biologically inert and inhibit the action of microorganisms. Nevertheless, some fungi employ oxidative enzymes to detoxify such wood. The mechanism of fungal attack on cellular tissue is specific and largely follows the structural pattern of the wood cells. The active hyphal tip of the fungus produces hydrolytic enzymes which diffuse into and digest the substrate, the hypha growing into the digested region and maintaining contact with the substrate. BAILEY and VESTAL [30], SAVORY [31] and REESE [32] describe the process for cellulose and refer to the direction of attack as being predetermined by the orientation of the cellulose fibrils.

The study of modern peats also illustrates the processes outlined above. KUZNETSOV *et al.* [33] described three principal layers in peat: a surface layer in which oxidizing conditions persist; an intermediate layer of alternating oxidizing and reducing conditions; the lowest layer which is a reducing environment.

The total bacterial population of Russian Berendeyevo swamp peat consisted of 700 million bacteria per gram of dry peat in the surface layer and 25 million at depths of 25–50 cms and 6 meters. NEOFITOVA [34], in a study of the fungal flora from the surface layers of the Gladkoye Boloto peat bog, isolated 74 species of fungi belonging to 24 genera; of these, 54 species were Hyphomycetes and 16 were plant molds. The surface layers of the peat (50 cms.) contained 3 million fungal spores per gram in summer and 170 thousand in winter; no live spores were present below 60 cms. Experimental observations led to the conclusion that fungi were the chief organisms concerned in peat formation and that bacteria further decomposed the tissue of the dead plants.

During sedimentation of allochthonous organic matter in water bodies and as a result of the activities described above, there is a concentration of the more biologically stable material. Thus lignin-type compounds and derivatives, represented as "humus", predominate, accompanied by representatives of the plant products referred to earlier, e.g., fats and waxes, resins and polymers such as cutin, exine, and suberin. A portion of the organic matter consists of carbohydrates and resistant chitinous remains, while derivatives of proteinaceous material and of pigments may be retained. This material may form the organic oozes of lakes in association with inorganic sediment, as exemplified by geological deposits such as the cannel coals. A particular ingredient may be of special importance, e.g., the exine of spores, in which case the geological counterpart is a spore cannel. Where plankton accumulate in large quantities, autolysis of the dead cells is accompanied by release of readily assimilable substances which are mineralized by the bacterial population in the descending order protein, carbohydrate, and hydrocarbons. HELLEBUST [35] showed (for marine phytoplankton) that the 4–16 percent of photoassimilated carbon normally excreted could rise to as much as 17–38 percent at the end of a diatom bloom when large numbers of empty frustules were present. Among the substances excreted were glycolic acid, proteins, amino acids and peptides, glycerol, mannitol, and a chloroform-soluble fraction. Certain algae secrete a single substance, e.g., *Chlorella* sp. gives proline, and *Olithodiscus* sp. gives mannitol. Surface plankton are represented in most lake bottom sediments despite the biological degradation which continues within the upper surface of the ooze. Thus, algae of the type *Botryococcus* form an integral part of the organic matter of cannel coals and other carbonaceous deposits of Carboniferous age, so that some cannels are referred to as algal cannels. The continued activity of bacteria and fungi in the sediments adds further metabolites. In some cases the content of these organisms may be appreciable (e.g., DRATH [36]) but the activity appears to be restricted mainly to the surface layers of the sediment. The data in Table 1 (after KUZNETSOV *et al.* [33]) indicate the change with depth in the composition of the ooze of the freshwater Lake Chernoye at Kosino. VALLENTYNE [37] examined some of the biochemical aspects of limnology and reported that the dispersed organic matter (dissolved and colloidal) exceeds the

Table 1. *Composition of the organic matter in the oozes of Lake Chernoye at Kosino*[a]

Depth from surface of ooze (m)	Ash content (percent dry wt.)	Components (percent) of organic matter			
		alcohol/benzene extract (lipid fraction)	hemicellulose	cellular tissue	humic complex
0.15	49.3	7.4	14.5	7.7	62.0
1.0	49.1	5.8	8.9	5.6	62.0
2.0	47.3	5.4	7.5	4.4	66.8
3.0	48.6	4.6	7.0	4.4	68.7
5.0	46.6	4.9	6.3	6.6	70.6

[a] After KUZNETSOV *et al.* [33].

particulate organic matter by a factor of 5–10. The organic content of lakes, varying in age by up to 10,000 years, included a wide range of amino acids and quantites of free sugars comparable with those found in fresh algae; carotenoids and substances related to chlorophyll were also recorded.

The further aerobic and anaerobic bacterial attack on pre-existing organic material which has been transported to the sea leads to a further increase in the proportion of biologically resistant materials, including humic compounds, resins, cutin, suberin, chitin, and possibly toxic heartwood tissues. These compounds form the allochthonous constituents of estuaries, deltas, and marine inshore environments, and may be preserved in oxidizing environments as the main or sole organic constituent. Where the environment is reducing, they will be preserved alongside localized autochthonous material.

Few attempts have been made to differentiate marine and freshwater sediments by their organic content, but DEGENS and BAJOR [38] studied German Carboniferous sediments of known marine or freshwater origin. The marine sediments contained a high amino acid content, especially of arginine and cystine, and on this basis identification was made.

III. Application of Organic Geochemical Studies

In studying the organic content of rocks, attention should be paid to the geological circumstances surrounding the formation of the rock, for example, whether it was of marine or non-marine origin. This can be variously established by field observation, by the palaeontological content of mega- and microfossils and, in certain circumstances, by the mineralogical composition and the distribution and concentration of minor elements. The conditions prevalent at the time of accumulation, e.g., oxidizing or reducing conditions are of considerable importance with respect to the organic material preserved. Recognition of the biological origin, nature, and relative proportions of the organic material is an important prerequisite for chemical examination. In some rocks concentrations of principally one type of organic material are observed. With modern micropalaeontological techniques several different types can be isolated and selected for individual study. A knowledge of the chemical composition or chemical decay process of single organisms or their remains (or of groups of such) would greatly increase the ease of interpretation. Where samples permit, these organisms or groups of organisms might be studied at various stages of their geological history, or at any given time after different conditions of preservation or physico-chemical change. Thus both the source of the organic material and the nature and degree of the biological degradation are important. A prerequisite to such studies may well be a consideration of modern representatives of the fossil groups and an evaluation of the relative stability of known chemical compounds. It may be unfortunate that chemical procedures for extraction of organic matter bring together compounds which are chemically related but which need not necessarily be related in their biological origin. The term "kerogen" was originally applied to the organic matter of oil shales, torbanites, "Kerosene shales", etc., but these rocks actually carry a biological content similar to that of many other sediments, and examples are known from both marine and non-marine depositional environments. The term has been used to denote total organic matter, both soluble and insoluble, but its more restricted use implies the insoluble organic content of rocks. There has been confusion over this term and interpretations based on these varied uses must thereby differ.

No reference has been made in this account to the content of animal matter, or its degradation products in the organic complex. The contribution is more difficult to assess but must be considerable under favourable circumstances. Animal microfossils are abundant in many sediments.

A. States of Preservation

Reference has been made to the detailed structural (morphological) preservation of plants and other organisms, but chemical composition of the preserved material has received little study. Well-preserved structures include the cell walls of bacteria, fungi, and algae, the latter with such detail that cell contents, nuclei, and organizational details permit taxonomic reference to modern groups. BRADLEY [39] in his detailed study on Green River Shale (Eocene) described a solid *Spirogyra* which still contained its spiral chloroplast. The cell walls and internal structure of plants are preserved in coals and in a variety of rocks. The more resistant cuticles of leaves including stomatal apparatus are known from Devonian times, and the detailed structures of Devonian, Carboniferous, and younger woods have been described. Entire fructifications of plants containing spores or pollen are known and some show the effects of specific parasitic or saprophytic attack. NEVES and TARLO [40] recorded isolated osteocytes from fish scales in the suspension obtained on maceration of a Carboniferous rock. The extended cells and their finely branching processes were represented by unbroken cell walls which the authors considered must have a chemical composition akin to that of cellulose. Cell contents were retained in some cases. NEVES and TARLO [40] also refer to the presence of cellulose-like polysaccharides in fossil dentine.

DOBERENZE and LUND [41] obtained electron micrographs of collagen fibrils in decalcified bone of Jurassic age and noted the similarity to the collagen from Pleistocene bone. The wide range of structural (morphological) detail preserved and the recognition of cellulose, lignin, and chitinous materials in these bones suggests the possibility of identifying further original organic substances. DOMBROWSKI [1] has made the remarkable claim that he has cultured viable bacteria from primary crystalline salt deposits of Precambrian, Middle Devonian, Silurian and Permian ages. In particular he found those of Permian age (*Bacillus circulans*) to be active after a dormancy of 180 m.y. and listed their physiological characteristics. The original chemical constitution is thought to have been preserved by dehydration and reversible denaturing of protein, consequent upon salification. CARLISLE [42] showed that the skeleton of the Cambrian fossil, *Hyolithellus*, includes chitin and a protein structurally related to sclerotin. ABELSON [43] reported a range of amino acids in an Orodovician trilobite *Calymene* (approx. 360+ m.y.) and pointed out the new field of study concerning the composition of the skeletal structures of fossils such as the "chitinous" brachiopods, the arthropods, and the graptolites. HARE [44] considered that the oxidation products of certain amino acids were present in progressively larger amounts in older fossil shells of *Mytilus* and believed oxidation to be one of the possible mechanisms involved in fossilization.

It is likely that considerable revision of the processes involved in fossilization will become apparent once detailed geochemical investigations have been made of separate organisms and distinct groups of organisms.

IV. Techniques of Geomicrobiological Study

The following considerations are most pertinent with regard to actual experimental procedures in geomicrobiological studies:

1. Recognition of the type of rock is important in allowing identification of the conditions of the original preservation process and the subsequent geological history. The rock types may vary widely: cherts, coals and oil shales, siltstones, shales, limestones, salt deposits, sulphide ores, bauxites, tonsteins, and fossil remains, e.g., bone, fossil wood, and amber. From these varied occurrences microbiological remains have been recorded at periods throughout dated geological time.

2. The techniques used depend upon the nature of the rock concerned and care must be taken to ensure that the microorganisms recorded were contained in the rock and do not represent external contamination.

3. Recognition of the microorganisms is often difficult for several reasons including the following: (a) The dimensions of the organisms approach the limit of resolution of the light microscope. They require application of sophisticated electron microscopic techniques (reflected or transmitted electron beam) and the use of replicas. (b) Taxonomic identification poses problems. Despite the variety of forms, there is commonly a lack of known biological affinity between them. Many modern microorganisms are only identifiable on the basis of reproduction processes or characteristic life cycles and form; some characteristics may be physically or chemically dependent on the environment. Identification often rests on morphological similarity with present day forms, which must be vague and can be merely a guide to possible biological affinity. Recognition (but not identification) on the basis of form is possible in the absence of genetic affinity. The use of indeterminate systematic categories for descriptive recognition includes certain groups of organisms, e.g., the Acritarcha, a group which covers most of the microfossils which are apparently unicellular and composed of organic matter. (c) Microorganisms may show distinctive relationship symbiotically with one another or parasitically with a host. Such relationships are difficult to imply from an examination of fossil materials but they have been recorded. Microbiological activity on plant and animal remains results in characteristic lesions. The association of specific bacteria and fungi with particular tissues is recorded for various geological periods (RENAULT [45], MOODIE [46]). A selective effect, following presumed enzymatic ingestion of tissue by fungi, has been demonstrated. (d) Finally, the population of microbiological remains in a specific environment is

often distinctive, e.g., soil populations; similar data have been recorded for rocks [47, 48].

Factors (a) through (d) are considered in the selection of the available techniques. In general they involve: (1) *In situ* study of the microorganisms in the rock by methods which provide either a thin section of the rock for examination by transmitted light, or a specially polished and etched surface which enables reflected light to be used, amyl acetate peels to be taken and replicas to be made for electron microscopy; (2) Physical or chemical action on the inorganic constituents isolating the organic material for direct study following acid digestion, or for chemical maceration and differential oxidation.

A. "In Situ" Studies

One great advantage of *in situ* studies lies in the avoidance of external contamination. It also permits examination of microorganisms in their context with the host and in their association with other organisms and with the organic and inorganic constituents. The organisms are observed substantially free from breakage and alteration resulting from processing. The disadvantages lie primarily in the technical problem of producing a thin section of a heterogeneous inorganic-organic rock satisfactory for study by transmitted light; in addition, the mineral content gives rise to optical interference. There is also the subjective difficulty of identifying sectioned material. The method generally employed involves the cutting of the thin section, impregnation with polyester resins and mounting in a similar medium for grinding. To reach the desired thinness, final grinding may be carried out with mild abrasives only, such as magnesia in glycerine, and finished by hand honing. From such a surface an amyl acetate peel removes a thin film of organic matter, which adheres to the peel particularly if the surface has been mildly etched. The peels may be examined optically and they can be utilized as the basis for the replicas needed for electron microscopy. Specially polished and etched surfaces may be studied with oil immersion and reflected light techniques. Thin sections of rocks which contain a high proportion of lenticular organic matter, particularly when embedded in a clay or carbonate matrix as in oil shales and carbonaceous rocks, may be difficult to prepare. In such cases, small blocks (1 cm side) are subjected to alternate treatment with hydrochloric and hydrofluoric acids for removal of inorganic matter (or treated with nitric acid if pyrites is present) and then dehydrated and placed in phenol for a short period. The softened and spongy block of organic material is washed free of phenol, dehydrated with alcohol, and finally introduced to graded mixtures of xylene/paraffin wax. The wax-impregnated block is then cut on a rotary microtome affording very thin, mineral-free, sections of organic matter.

B. Study of Isolated Organic Material

The microfossil content of rocks is generally studied by dissolving the inorganic matrix of the rock (acid digestion) or alternatively, the rock may be disintegrated by physical techniques.

Chemical Treatment

The acid digestion technique used depends upon the inorganic content of the rock. Thus, acetic and hydrochloric acids dissolve the carbonates of limestones, hydrofluoric acid dissolves silicates, and nitric acid attacks mineral sulphides, e.g., pyrites. The acids are employed singly or in combination as dilute aqueous solutions. The majority of rocks, particularly shales and siltstones, require treatment with hydrochloric and hydrofluoric acids, the latter digestion being carried out in polyethylene containers. Complete digestion may take days or weeks depending on the size of the rock fragments used, the particle sizes of the inorganic minerals, the concentration of the reagents, and the temperature at which the reaction is allowed to take place. The size of rock fragment depends on the nature of the investigation; if isolated remains, e.g., miospores, are being investigated, the rock may be crushed, but the isolation of plant fragments is facilitated by splitting the rock into thin slices along the bedding planes. Slow reaction using dilute acid is preferable, as rapid disintegration tends to break up the fragments of organic matter. Centrifugation, density separation, and flotation techniques are employed for both initial and final separations of the organic and inorganic matter. In certain circumstances, short periods of ultrasonic vibration are effective in separating residues, but they may result in partial disintegration of the organic fragments.

In the study of geomicrobiological remains, plant fragments, and the total organic content, the preferred technique is controlled acid digestion at room temperature. Further mechanical processes are undesirable, and the undried material is mounted in glycerine jelly or "Cellusize" for microscopic study. Some part of this material may be heavily colored or included within the degraded and colloidal groundmass of organic matter. Certain tissues can be "cleared" and separated from the groundmass by differential oxidation of the latter by a maceration process involving nitric acid, fuming nitric acid, or a mixture of potassium chlorate and nitric acid. This destructive process leads to a concentration of waxes, resins, and the more acid resistant structures, and is a technique used in the concentration of miospores. The differential nature of the attack is well known (MOORE [47]) and various stages of breakdown can be obtained. Short periods (e.g., 10 minutes) of acid maceration may be used to "clear" material and longer periods (e.g., 6–12 hours) to destroy the groundmass and release the contained portions of certain organisms and cell structures. The macerated material is treated with dilute base, e.g., potassium or ammonium hydroxide, to dissolve the oxidized humic or ulmic constituents and is then washed and mounted for optical study. At appropriate stages the isolated organic matter can be mounted in paraffin wax for microtome sectioning, or treated with epoxy resins and mounted in Araldite, prior to ultramicrotome sectioning for the electron microscope [49].

Although the extracted organic material is commonly red, brown, and yellow and preserves optical differentiation, it may be stained with water- or alcohol-soluble biological stains. The objectives are increased resolution of optical detail, improved photographic contrast, and differentiation of organic content on the basis of affinity for stains. Little is known concerning the mode of interaction of stains with fossil material, but much of it stains readily. The stains commonly used are gentian violet, safranin, methyl green, basic or acidic fuchsin, and Bismark brown. Cotton blue is used for fungal remains.

Lignites and brown coals being prepared for the extraction and concentration of pollen and spores are processed with a 10 percent solution of potassium hydroxide followed by acetolysis. Extraction of miospores from bituminous coals is achieved by maceration using fuming nitric acid, potassium chlorate, and conc. nitric acid (the Schulze process), or bromine and nitric acid (the Zetsche Kalin process). In recent years the petrographic constituents of coals and their contained structures have been studied mainly by the oblique illumination of polished surfaces under oil immersion, as inaugurated by STACH [50]; the earlier studies of coal by thin section and transmitted light were carried out by THIESSEN [51].

The advantages in studying isolated organic material lie principally in the improved selection and in the better resolution and appearance of microorganisms. The concentration of material thereby facilitates comparison and identification. The disadvantages follow from disassociation of some components and the absence of information concerning the *in situ* association of an organism with organic and inorganic constituents. It is essential to compare both thin section and extracted materials if the problems referred to on p. 278 are to be resolved.

V. The Presence of "Bacteria, Actinomycetes, Fungi, and Algae" in Rocks

Reference has been made to the difficulties inherent in identifying groups of organisms whose genetic affinity is unknown. In the discussion below no such absolute classification is implied, but general similarities in morphological form, a knowledge of reproductive methods, association with other material, and a known habit are all variously used in referring forms in rocks to their modern counterparts. At the present time this range of microorganisms is largely concerned with the destruction of organic matter. ZOBELL [52–54] has referred to more than 100 species of bacteria and related organisms (Bacillus, Bacterium, Mycobacteria, Proactinomyces, Actinomycetes, and Micromonospora) known to attack a wide range of organic substances in soils and sediments. They are tolerant of temperatures between $-10°$ C and $105°$ C, and live forms have been removed from water depths in excess of 5,000 meters and from muds exceeding 100,000 years in age. They are adaptable and are not controlled by environment since their metabolic activity is capable of rapid change. Their biomass may range up to 500 g/m^3 in sediments and soils and constitute from 2–15 percent of the organic matter in soils, sediments and water. Whilst a pH of 6.5–8 is usual, specialized bacteria can exist in highly acidic (pH 1) or in highly alkaline environments. WAKSMAN [55] referred to the ubiquitous distribution of the actinomycetes which form 13–46 percent of the population in the surface layer of an organic-rich soil and which increase to 65 percent at depth. They are present in the waters and deposits of lakes and form 10–45 percent of the microbial population of the water. They are active in the decomposition of chitin, lignin and cellulose and flourish at pH 7 but are inhibited by acid conditions. The action of fungi has been referred to; fungi are less susceptible to variations in pH and can flourish under extremely acid conditions.

A. Bacteria in Rocks

VAN TIEGHEM [56] first recorded fossilized bacteria, present in a Carboniferous shale. Subsequently, examples of the accepted forms (coccoid, bacilloid, filamental, and spirillar) and their distictive arrangements (diplococcoid, staphylococcoid, and streptococcoid, and several bacilloid) have been recorded from every geological system, including the Precambrian, and from rocks as different as peats, coals, carbonaceous shales, flints, limestones, saline deposits, bauxites, fireclays, iron and manganese ores, and phosphate deposits. The preservation of bacteria or their spores varies from the presence of the actual cell wall to mineral replacement of this wall, or to sheath-like coverings of the wall. The association of fossilized bacteria with plant tissues and with animal and insect remains have been described, and in many instances the pathological association with fossil bone or plant tissue was related to characteristic lesions (MOODIE [46], RENAULT [45]). The associations recorded have been referred to autotrophic or chemautotrophic forms, or to groups which include iron bacteria, nitrate- and sulfate-reducing bacteria, calcareous and saprophytic bacteria (PIA [58]). The presence of living bacteria in various modern and recent sediments, including sulfur and iron deposits, tufa and peats, provides a functional analogy with past occurrences. RENAULT [57] identified and described some twenty fossil species, mainly from Carboniferous rocks and related their specific attack to certain cells of plants, to bones, fish teeth, and coprolitic remains. A stratigraphic list of bacterial remains is included in the Swansea Symposium Volume of the Geological Society of London (MOORE [59]). The literature contains frequent reference to what are possibly fossil bacteria, and many workers have noted resemblances to modern forms and referred to *Crenothrix*, *Sphaerotilus* or *Lyngbia*. Recent work, e.g., CLOUD [60], BARGHOORN and TYLER [61], records various forms of bacteria or forms with actinomorphic affinity, from the Precambrian Gunflint Cherts (1.9×10^9 years). SCHOPF, BARGHOORN, MASER, and GORDON [62] reported rod-shaped and coccoid bacteria from this source. Electron microscopic techniques revealed the organically preserved cell wall and its imprint in the rock as a replica. The bacteria were relatively undisturbed and resembled modern bacteria of the genera *Siderocapsa* and *Siderococcus*. Electron microscopic studies allowed SCHOPF, EHLERS, *et al.* [63] to identify sheath bacteria *Sphaerotilus catenulatus*, spiral thread forms *Gallionella pyritica* and segmented microbes allied to *Beggiatoa*, preserved in pyrites from a Carboniferous rock. These bacteria were generically similar to present day forms occurring in a comparable reducing environment and illustrate an ecologic association which has persisted for at least 250 million years.

Reference has already been made to the claim that viable bacteria can be extracted from salt deposits (DOMBROWSKI [1]). The form *Bacillus circulans* from Permian salt crystals was comparable with rare present-day forms inhabiting concentrated brines. KUZNETSOV *et al.* [33] refer to counts of bacteria from Devonian, Carboniferous, and Permian rocks, the numbers ranging from 35–117 million per gram of rock in situations where oil had been developed in the rock. It is doubtful what significance can be attached to these figures, for no bacteria were recorded in associated rocks which did not contain oil.

B. Actinomycetes and Fungi in Rocks

Remains of fungi in rocks are well known and consist of mycelia and fungal spores which are difficult to identify unless connected, as are conidiophores, to hyphal material. The occurrences of the main classes of fossil fungi have been recorded [59], the earliest record of a recognized saprophytic fungus being from the Silurian rocks. There are earlier records of fungi from the Precambrian, but the majority of the remains are known from the Carboniferous and later rocks, particularly the Tertiary, when the higher fungi (Ascomycetes and Basidiomycetes) are well represented. Descriptions of remarkably well-preserved fungi with complete reproductive systems and cell contents, have been given by BERRY [64]. Other remains consisting of fine filaments 0.7–2.0 μ in diameter, which may be genetically related to filamentous bacteria, the actinomycetes, or the Fungi Imperfecti, have received consistent reference in the literature. WEDL [65] described certain remains from animal shells as *Saprolegnia ferox*; KOLLIKER [66] regarded similar organisms as fungi, and RENAULT [57] referred to aseptate filaments in the bark cells of Carboniferous plants as the saprophytic fungus *Phellomyces* and in shales, oil shales and cannel coals as the saprophyte *Anthracomyces cannellensis*. The latter organism was regarded as the agent responsible for the degradation of Carboniferous organic material. KIDSTON and LANG [67] described similar remains from the Devonian and compared them with *Archaeothrix contexta* and *A. oscillatoriformis* and, while they were regarded as possible fungal hyphae, they were thought to be suggestive of the Trichobacteria *Beggiatoa* or *Oscillatoria*. In recent years the Actinomycetes, with affinities placing them between Fungi and Bacteria, have received considerable attention. There seems to be some indecision regarding the systematic position of these organisms, *q.v.* DRECHSLER [68], HENRICI [69], WAKSMAN [55], and BERGEY [70], but their morphological characteristics are well defined. Thus, according to WAKSMAN, the Actinomycetes form elongated, usually filamentous, cells; the filaments are usually 1 μ or less (and do not exceed 1.5 μ) in diameter. They multiply by special spores (oidospores) or by conidia; the former are formed by segmentation or simple division, the latter are formed singly at the ends of simple conidiophores. The saprophyte *Anthracomyces cannellensis*, and the forms *Polymorphyces minor* and *P. major*, [71], *Palynomorphytes diversiformis*, [72], and "*Phellomyces*" (MOORE [47]) constitute a group which have morphologic similarities with the Actinomycetes and may be related to these or to Bacteria or Fungi. Forms such as *Anthracomyces* are common in degraded organic matter whilst *Phellomyces* is present in cellular tissue and resembles a fungus in its action. This group of organisms is widespread in carbonaceous rocks and ubiquitous in distribution. Members of this group have been recorded in the organic matter of the Precambrian Nonesuch Shale [49]. Forms of fungi which attack fossil bones, *Mycelites ossifragus*, are described [46].

Hyphae and other fungal structures are normally well preserved, and HARE and ABELSON [73] refer to the presence of chitin in the cell walls of fungi, yeasts, and green algae. Fungal spores are thick walled and dark in color and frequently appear to retain their internal contents; they are common in organic material which has suffered widespread biological degradation. They are believed to be present in Precambrian rocks. PFLUG [74] named four species of *Polycellaria* with resistant cuticula which he regarded as the conidia of Fungi Imperfecti. MOORE and SPINNER [49] note conidia attached to hyphal threads in the Precambrian Nonesuch Shale. Under favorable conditions fungi produce chitinous schlerotia as resting stages in a life cycle; these schlerotia occur in cellular tissue in coal, particularly Cannel Coals. DRATH [36] described a succession of schlerotia from a Polish Cannel Coal and they have been observed in basin type oil shales and Tertiary coals [75]. The proportion in some coals was sufficient to warrant the establishment of the coal maceral Schlerotinite for classification of these fungal ingradients; TAYLOR and COOK [76] consider Schlerotinite to be discrete fusinised resin bodies and not entirely related to fungal production.

The importance of fungal spores in palynology was also recognized by GRAHAM [77] who suggested a classification for these remains. The Carboniferous fossil fungi from the Silesian Coal Measures were shown to contain mycelial tissues, hyphae and reproductive organs [79]; a further work [75] is concerned with a palaeomycological investigation of Tertiary coals. The authors stress the importance of fungal content and particularly of resting stages, hypnospores, uredospores, teleutospores, chlamidospores, and conidia. They believe fungi play a large part in the coal forming processes and refer to Funginite in the sapropelitic sediments and infer the presence of fungi to be environmentally diagnostic.

C. Presence of Algae in Rocks

The algae constitute a complex group of organisms with a geological history extending from Precambrian times. The oldest known organic remains, *circa* 3,200 million years old [2], contain carbonized remains of filamentous algae. Definite cellular organisms, including *Gunflintia*, *Animikiea*, and *Archaeorestis*, from the Precambrian Gunflint Cherts [60, 61] are related to present day blue-green algae. BARGHOORN and SCHOPF [80] recorded the perfect preservation of septate filaments in Australian Precambrian cherts, the cell walls and contents being indicative of blue-green algae. Resting zygotes and a structure resembling the Oscillatoriaceae, Nostocaceae, and Ulotrichales amongst modern photosynthetic algae were recognized. PFLUG [74] has identified two new species of algae, *Filamentella plurima* and *Fibularis funicula*, in the Precambrian Algonkian of the Canadian Shield Belt Series (1.1×10^9 years); they are preserved in siliceous carbonaceous rocks. A number of workers, including PFLUG [74], TIMOFIEV [81], and ROBLOT [82], have described microorganisms which may be unicellular algae, zygospores, or cysts related to algae. They are either referred to as "sporomorphs" or are included in an indeterminate systematic category which contains most of the apparently unicellular microfossils composed of organic matter, e.g., Acritarchs. The category is not restricted to Precambrian remains but these remains are sufficiently common in the Late Russian Precambrian or Riphean, which covers a time span of 1,000 million years, for Russian workers to have effected subdivisions of the rocks based on this organic microfossil content. It would be impossible in the present work to list individual records and a classification of fossil algae in geological time. The reader is referred to the Swansea Symposium Volume of the Geological Society of London [59] for such detail. The records are continuous throughout geological time and the present main algal subdivisions are recognized. The Lower Cambrian [83] saw new groups of algae appear, including the Dinoflagellates. Filamentous blue-green algae are known from the Ordovician [84] and Devonian; CROFT and GEORGE [85] described three new genera and species of blue-green algae (*Kidstoniella fritschi, Langiella scourfieldii, Rhyniella vermiformis*) from chert. The well-preserved sheaths, cell walls, cell contents and heterocysts enable direct reference to present day classes of algae. By Ordovician times, algae, presumably of the class Protophyceae or Xanthophyceae, had become prominent. *Gleocapsomorpha* is the main organic constituent of Kuchersiti Estonian oil shale, occurring in perfectly preserved, golden brown colonies showing cell structures and as degraded remains. Related algae have continued to be an important constituent of organic residues of lake, swamp and estuarine deposits, especially throughout the Carboniferous and Permian. The present day alga *Botryococcus brauni* Kutzing (BLACKBURN and TEMPERLEY [86]) is well known, and related forms, e.g., *Botryococcus* Pila and Rheinschia, have been recorded from Carboniferous shales, cannel coals, and basin-type oil shales. The boghead cannel coals contain high proportions of these algae amongst their sapropelic constituents. In the later Mesozoic and Tertiary times, the algae have been shown [87] to be increasingly important. In particular, new planktonic groups such as Diatoms, Coccoliths, and Silicoflagellates appeared, and there was a large increase in numbers of the Dinoflagellates and other groups known at the present day. Some of these new groups inhabited both marine and freshwater environments, particularly during Tertiary times. Since that time there have been few significant changes in the algal flora.

D. Presence of Other Fossil Groups in Rocks

There are several important groups of common microfossils which may make significant contributions to the organic content of rocks. Some are undoubtedly related to algae, others are of unknown affinity or are the fragmentary remains of higher plants. Together they are considered as fossil microplankton and are classified on a morphological basis. The principal groups are referred to below.

1. Acritarchs

These are mainly single-celled organisms with a resistant (presumably chitinous) integument and are related to algae or fungi. The majority are considered to be cysts, akinetes, and resting spores of algae and are simple rounded bodies (sporomorphs), spined forms (acanthomorphs), or polygonal shaped (polygonomorphs) and heteromorphic types. Their occurrence in sediments, known to be marine, suggests a phytoplanktonic association. They are present in Precambrian rocks and are the dominant forms of microplankton in the Lower Palaeozoic. Their importance declined during the Upper Palaeozoic.

2. Hystrichospheres

Directly related to the algae, being the cysts of the dinoflagellates [88], they consist of a highly resistant membrane bearing tubular or branching spines. They vary in size between 10–100 microns. The group was relatively unimportant during the Palaeozoic but was an important member of the phytoplankton during Mesozoic, Tertiary and present times.

3. Dinoflagellates

Planktonic marine Protista, these organisms possess a thick organic cell wall often carrying spines, the surface of which is divided by furrows into a system of angular plates. They are largely responsible for plankton blooms and carry pigments. They occur as important fossils in the Tertiary and Mesozoic rocks and have a history which extends back into the Palaeozoic.

4. Coccoliths

Consisting of a spherical, often thick, organic membrane which is covered with minute calcareous plates (coccolith, 2–10 microns), they are dominantly marine, planktonic, unicellular organisms and may assume varied forms. The plates form a high proportion of some limestones and chalks, particularly in Mesozoic and Tertiary rocks.

5. Silicoflagellates

These marine planktonic organisms are related to coccoliths but have a framework of organized silica rods, outside of which lies a layer of cytoplasm. They are known from Mesozoic, Tertiary, and recent deposits and appear to have shown little development.

6. Diatoms

This group contains unicellular or colonial forms of microscopic algae. The skeleton walls are composed of pectin impregnated with silica and they carry a detailed surface sculpture. Many diatoms favor organic rich waters and others are found where extensive decay is taking place. The chromatophores contain a wide range of pigments. Diatoms are known from both marine and freshwater environments and may be planktonic or bottom-living. Fossil diatoms are known from the Mesozoic but were more important elements in the Tertiary and in recent sediments.

7. Chitinozoa

These fossils are of unknown systematic position but resemble the chitinous protozoa and the hydrozoa. They consist of symmetrical, cylindrical, bell, flask, or funnel-shaped chitinous membranes, usually as individuals or joined together and range in size between 0.1–0.3 mms. They were discovered in acid-insoluble residues and are known to range only from the Ordovician to the Devonian.

8. Spores and Pollen

They are derived from terrestrial vegetation and are therefore of greatest significance in freshwater deposits. They also contribute to the organic content of inshore marine sediments.

(a) *Miospores*

The miospores and megaspores of the earlier land plants are characterized by distinctive haptotypic features of triradiate or monolete patterns. They are extremely variable in form and possess a resistant structural wall or exine consisting of cutin, resins and waxes, and chitinous substances. They appeared in the Upper Silurian and were of considerable importance in the Devonian and Carboniferous and of less importance thereafter in Permian and Triassic times.

(b) *Pollen*

Pollen grains have a wall structure with a resistant exine, similar to the miospores, but are generally and broadly characterized by germinal furrows or by colpae on the exine which are either monocolpate, or tricolpate. Acolpate grains do not possess a furrow. Pollen grains are rare in the Carboniferous but become dominant in the Permian and thereafter constitute the principal elements in palynological analyses to the present day.

VI. Geomicrobiological Attack on Specific Fossil Material

The specific material subjected to attack must still be biologically recognizable if the method and nature of the attack is to be evaluated. In general, only part of this attack is represented by the evidence from any given piece of material, but by comparison of related material with differing degrees of degradation it is possible to evaluate the total effect. In the circumstances described below, the attacking organism is preserved *in situ* on, or within, the material.

A. Miospores

The form and structure of spores and of the several layers of the outer wall (intine, exine) are well known, but the chemical composition of this wall is unknown. It has been referred to as "sporopollenin". From the diversity of structure and color differentiation on one and the same spore, the walls are likely to have a variable composition, which may lead to selectivity in attack by microorganisms. The material studied [72] was contained within a chert of Carboniferous age, the organic matter representing the decayed remnants of a terrestrial vegetation, in which the processes of decay were arrested by saturation with siliceous water. The organism *Palynomorphytes diversiformis*, Fig. 1a–h, which was observed to attack miospores, is related to the actinomycetes or fungi and was probably saprophytic in habit. The forms recorded possess a dark, relatively thick, cell wall with hyaline or pale yellow highly refractive contents. The commonest forms are coccoid cells, 0.75–1.5 μ across but they may be ovoid up to 2.0 μ. The cells occur singly, in diplococcoid pairs, or they may form chains of cells in the streptococcal manner. Ovoid or elongated cells provide a linear beaded filament between 2–3 μ and 10–20 μ long, with the ends marked by a highly refractive larger cell; such beaded filaments may be branched. Filaments, with well defined parallel walls

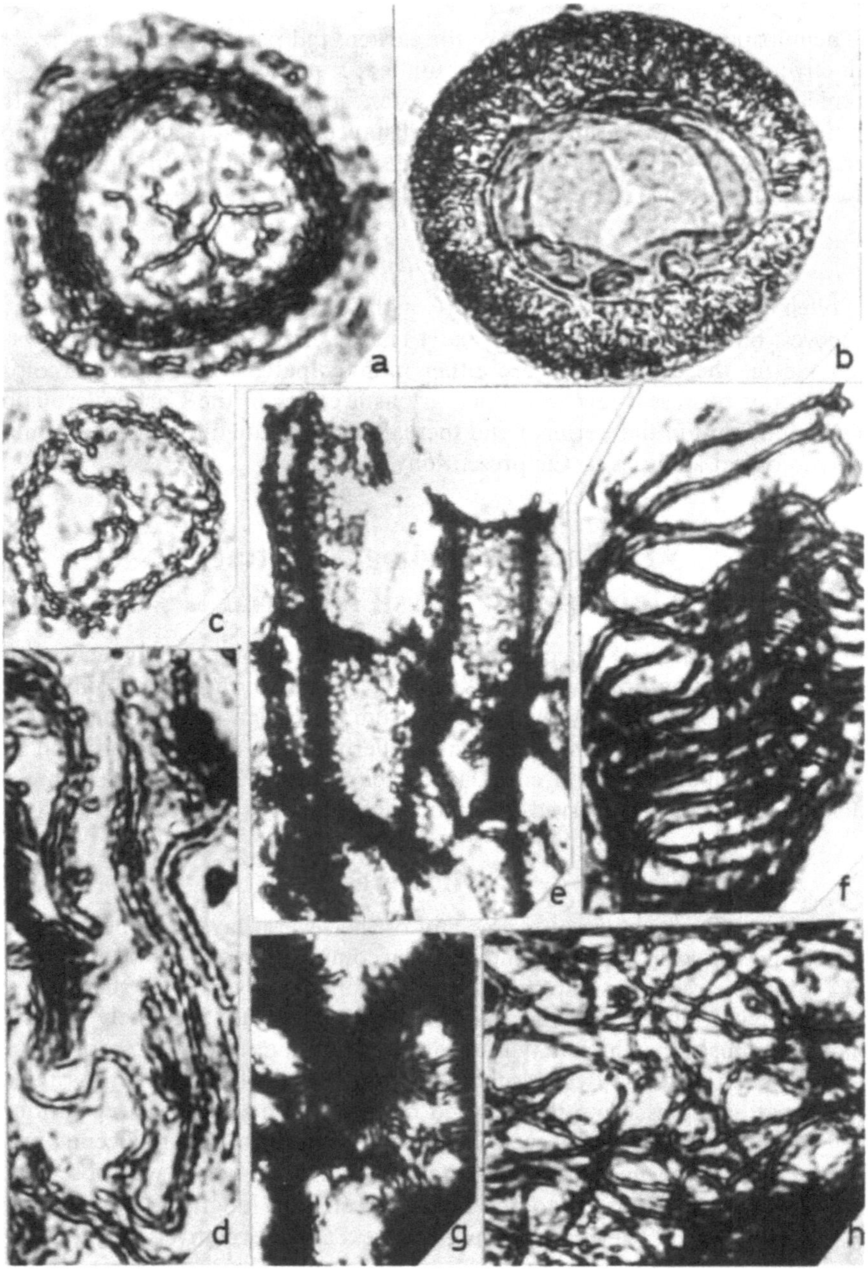

Fig. 1. a) Miospore *Densosporites* sp. attacked by *Palynomorphites diversiformis* on central region and concentric filiments within the annular region. Chert band, Upper Oil Shale Group, Carboniferous, Scotland. (× 1,000). b) Miospore *Endosporites* sp., *Palynomorphites* investing saccus of spore with cocci and fine filaments radially arranged. Erda Tonstein, Upper Carboniferous, Germany. (× 1,000). c) Miospore "*Palynomorphites* pseudomorph" of *Lycospora*. Filaments representing position of equa-

resembling hyphae, vary from 0.75–1.5 μ in width and are straight, flexuous, polygonal, curved, or concentric. Branching is a common feature and is usually associated with a rounded cell, or swollen vesicle. In its branched form the organism resembles the saprophytic fungus, *Anthracomyces cannellensis*, which RENAULT [57] recorded on the surface and within microspores and megaspores. The effects on the spore exine vary with the main structural features of the spores, though all spores of however divergent character, were invested by the organism. The principal effects were the following:

1. In spores with little structural differentiation of the exine, the surface was colonized by cocci or short beaded filaments which often coalesced to enclose circular or hexagonal areas in which the exine was totally destroyed, e. g., on spores such as *Punctatisporites* sp.

2. In spores with a well defined triradiate mark on the central body and a thin equatorial flange of differing composition, the result was to reduce or destroy the flange, then to destroy the exine of the central body and cause a break-up of the spore into segments, e. g., *Lycospora*. When the attack led to the development of filaments, these remained to give a "fungal pseudomorph" of the spore (Fig. 1c).

3. In spores such as *Densosporites* (Fig. 1a) possesing a central area and a wing or cingulum with an inner thickened zone and an outer equatorial flange, the attack is surprisingly different. The inner thickened zone is preferentially selected and filled with an anastomosing mass of fungal hyphae. Thereafter the central area and later the equatorial flange are destroyed. With total destruction of the spore wall there only remains a ring of hyphal filaments to indicate its original presence; the ring in turn breaks up into segments [72].

4. In spores with strong ornamentation and thick rugulae protruding from the surface, the attack is restricted to the thinner exine and follows an irregular pattern, leaving strong hyphae to form a "palynomorphytes skeleton" of the spore, as for *Convolutispora* and *Leiotriletes* [72].

A considerable number of bodies (Fig. 1c) consist solely of an interconnected meshwork of cocci, beaded filaments and filaments of *Palynomorphytes*, representing fungal "pseudomorphs" or "skeletons" of unknown original material. These are additional to the fungal schlerotia which were attached to cellular structures, or were freed and occurred in the degraded humic or sapropelic matter. The organic content was therefore profoundly affected by fungal metabolism and by the presence of fungal remains. ELSIK [89] described the attack of related saprophytic fungi on Tertiary spores and pollen. GOLDSTEIN [90] summarized

torial flange and positions of the triradiate mark; spore exine totally destroyed. Chert band, Upper Oil Shale Group, Carboniferous, Scotland. (× 500). d) Cellular material attacked by hyphae of "*Phellomyces*", wood mainly destroyed. Erda Tonstein, Upper Carboniferous, Germany. (× 1,200). e) Carbonized wood cells. Filaments following cell walls; coccoid cells of bacteria type line the interior of the cells and are isolated in groups within the cells. Erda Tonstein, Upper Carboniferous, Germany. (× 500). f) Vascular cell structure; the form is represented by filaments of "*Phellomyces*". Erda Tonstein, Upper Carboniferous, Germany. (× 1,000). g) Carbonized material, consisting essentially of minute bacteria like cells and short filaments. Erda Tonstein, Upper Carboniferous, Germany. (× 1,000). h) "*Phellomyces*" filaments, in a complex branching pattern forming a "matte" of hyphal remains and organic matter (dark). Erda Tonstein, Upper Carboniferous, Germany. (× 1,000)

the attack of modern aquatic and soil-inhabiting fungi on spores and pollen and drew attention to the effects of Phycomycetes, both filamentous forms, and Chytrids. The latter tend to colonize the surface and to draw their substance from the exine or from within the spore body. Examples of fossil spores with this type of fungal attack occur in carbonaceous shales from the Coal Measures of Staffordshire.

B. Cellular Tissue

MOORE [47] has described the microbial degradation of organic matter in a group of carbonaceous sediments formed in the non-marine environment as shallow water bodies or subhydric soils. They are referred to as tonsteins, rocks which are closely associated with coal seams of various ages. In these rocks recognizable tissue was preserved as host material to the still contained or attached microbiological parasitic or saprophytic organisms. Carbonized cellular tissue (Figs. 1e and g) was invested by chains or groups of cells with thick, dark walls and hyaline contents resembling the bacterium *Micrococcus hymenophagus* Renault. Fragments of apparently carbonised wood consisted entirely of these minute (1 μ) bacterial cells. Although fungal hyphae were occasionally present, bacteria appeared to be the principal agents of degradation.

Non-carbonized cellular material, consisting of tracheids, vascular and parenchymatous tissues, was degraded by the organism *Phellomyces*, which resembles the form *Phellomyces dubius*, described as a saprophytic bark fungus by RENAULT [57]. In its simplest form, the organism consists of straight, aseptate filaments or hyphae, 0.5–1.5 μ wide, which follow, or occur within, the cell walls. The filaments may assume the rectangular or hexagonal pattern of the cell walls, or, by branching, form a more complex pattern. In the tracheid fragment (Figs. 2a and d), filaments are visible within the cell walls, and they emerge where the walls have disappeared to form centers simulating a conidiophore or complex vesicle. Under severe degradation, the cell walls are destroyed and the filaments and coccoid cells remain to form a "*Phellomyces* pseudomorph" of the original structure (Fig. 2d). *Phellomyces* may also form a complex interweaving mass of filaments within the cells, or throughout the entire cellular material (Fig. 1h). This organism, with its resistant chitinous walls, was largely responsible for the breakdown of tissues. Its mode of action has been described [24] and it is presumed that enzymatic processes are involved. Other fungal remains are represented by red brown fungal spores (*Sporonites unionus* Horst), often attached to fragments of disorganized wood (Fig. 2g).

C. Groundmass Organic Matter

This composite matter consisted of partly degraded, but still recognizable tissue and pale yellow amorphous material together with microbiological remains. It is difficult to differentiate between those microbiological remains which are included with, or preserved within, already degraded material and those which have been actively concerned with a later stage of further degradation.

1. In this category, fragments of cell walls invested with *Phellomyces* and pseudomorphs of cell walls are preserved (Figs. 1d and f), together with disorganized material. The commonest ingredient is the saprophytic fungus, *Anthracomyces cannellensis* Renault, with its straight, sinuous or branching filaments, 2.0–2.9 μ long and 1.0 μ wide, ending in a small spherical conidium. The organism produces a felted mass of filaments which spread through the material and form an integral part of the organic content Fig. 2e. Where the organic matter is highly degraded, *Polymorphyces minor* and *P. major* Moore [71] are present. These are rounded, ellipsoidal or polygonal organisms with a thick, dark, smooth or beaded outer wall and a clearer centre containing cells or beaded filaments, possibly of fungal affinity. Small spherical or ellipsoidal brown cells, 0.5–1.0 μ across, referred to bacteria, occur singly, in pairs or with a trifid arrangement and are associated with cellular material or scattered through the groundmass (Fig. 2b); they resemble *Micrococcus hymenophagus* Renault. Larger forms, 1–2 μ across, with thick, dark brown walls and transparent centres, occur only in the groundmass and are referred to *Micrococcus Guignardi* Renault (Fig. 2b). The bacillus form is represented by chains of coccoid cells resembling *Bacillus ozodeus* Renault; others are rod-like with either rounded ends or terminating in a spore or vesicle. They are generally 5–10 μ long × 0.5–1 μ wide with pale brown or hyaline contents and are always associated with highly degraded organic matter in association with *Polymorphyces*. The groundmass commonly shows chains of dark, thick-walled, bacteria-like cells with a microreticulate pattern in which *M. hymenophagous* is prominent; the organization suggests a bacterial colony. In highly-degraded material, colonies of dark red-brown, thick walled cells with spherical vesicles arising directly from the main filament or present on short branches resemble the form of actinomycetes and especially that of the modern genus *Micromonospora*.

2. The pale-yellow amorphous material is generally devoid of contents such as those described above, but does contain small, thick-walled, dark red cells in irregular groups or in concentric colonies. The colony has developed circumferentially leaving within it still further degraded and colorless amorphous material (Fig. 2i); larger colonies of 6 or 8 dark red cells of presumed fungal affinity were also present.

The various organisms described are considered to represent bacteria, actinomycetes and fungi. The known modern environmental conditions under which these organisms flourish allow certain parameters regarding the possible fossil environment to be suggested. The medium was acid (fungal activity, pH 3.5–5.5; actinomycetes, *circa* pH 7) but conducive to a high rate of metabolic activity. The pH was therefore above 4.5 and nearer the range 5.5 to 7.0 (the critical value for bacteria is pH 4.5), with high humidity and temperature and sufficient calcium, magnesium and potassium for biological activity. Such conditions were unfavorable to the continued formation of peat or general organic debris, since microbial activity would occur as rapidly as accumulation of organic matter. The rocks contained 1–2 percent organic matter.

The degraded organic matter then entered an intimately mixed organic-inorganic aluminium silicate colloid or gel, analogous to soil colloids. Separation of these phases led to the *in situ* development of kaolinite macrocrystals of various

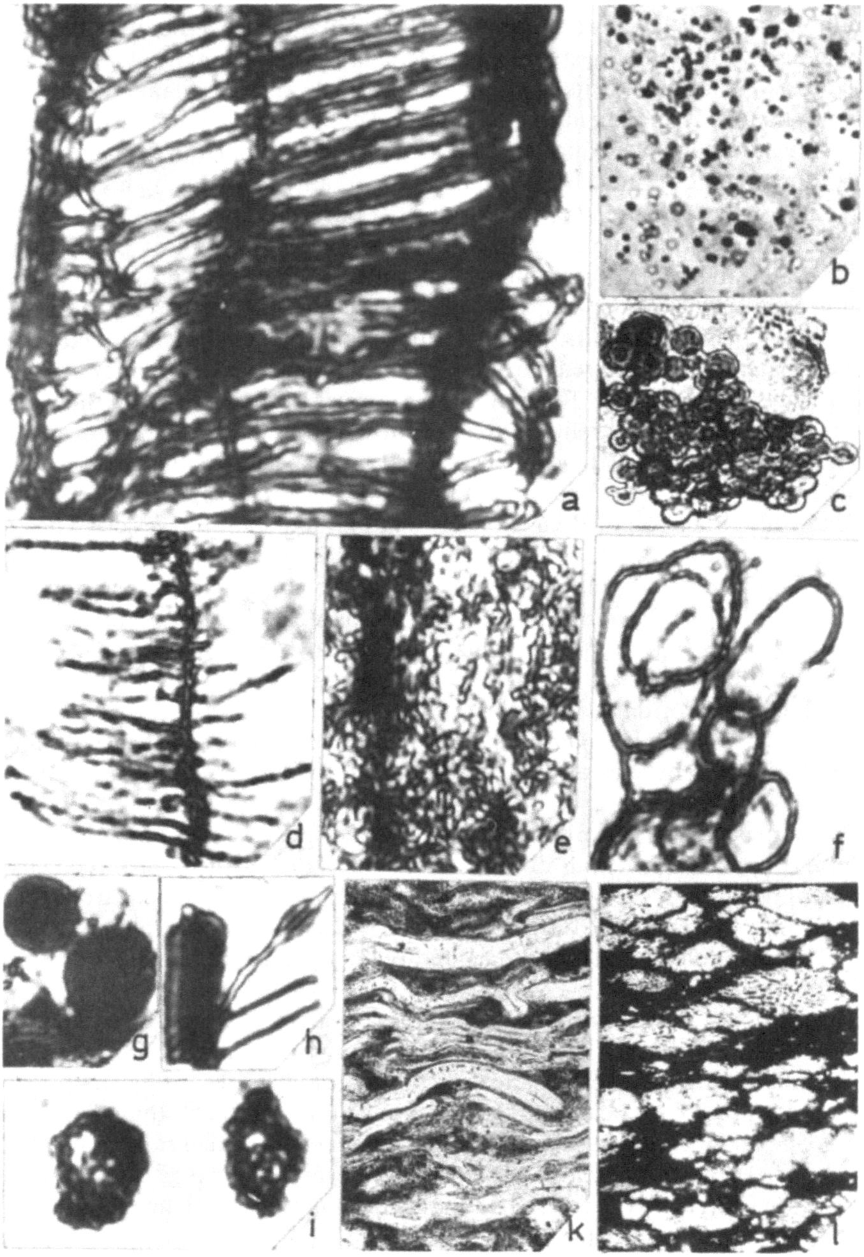

Fig. 2. a) Tracheid degraded by "*Phellomyces*" filaments. To the left filaments predominate, to the right filaments are present within existing cell walls; extreme right filaments converge to a red brown conidiophore. Erda Tonstein, Upper Carboniferous, Germany. ($\times$ 1,200). b) Single coccoid and groups of presumed bacteria cells resembling *Micrococcus hymenophagous* REN. *M. Guignardi* REN. Thin section of Constance Tonstein, Lens Group, Upper Carboniferous, France. ($\times$ 750). c) Colony of thick

forms [47, 48] in which the organic colloidal particles [91] were closely packed by polar adsorption on the surfaces of the orienting mineral particles, which were thereby joined together. Thus, fine organic matter is enclosed within the macrocrystal along lamellar planes while colonies, e.g., of *Anthracomyces cannellensis*, are included within the developed vermiform crystals.

D. Algal Material and Resin Globules

Algal colonies of a type similar to *Botryococcus* were infrequently present in the rocks referred to above and in no instance were they seen to be attacked by microorganisms (Fig. 2c). It is possible that the organisms specific to their attack were not present. On the other hand, the preservation of algal material in rocks which contain highly degraded plant tissues may merely indicate the restricted opportunity for attack on planktonic material reaching a degradation environment which is predominantly humic. DRATH [36] indicated the degree to which boghead algae may be attacked by fungi. The Polish boghead coal described consisted of three main constituents (plus a very small content of what may be considered as humic ingredients, i.e., fusinite and semifusinite, 0.1 percent; vitrinite, 1.0 percent; and exinite, 1.0 percent): (1) alginite (representing algal colonies) with an average content of 23.3 percent and maximum of 33.9 percent; (2) eualginite (heterogeneous matrix material derived from the destruction of algal colonies by fungi), average content 70.4 percent, maximum 86.9 percent; (3) chitinite and semichitinite (representing fungal schlerotia), average content 0.7 percent, maximum 1.7 percent.

Resinous materials, though rare in the above rocks, is often present as deep red, rounded or irregular entities in rocks such as the basin-type oil shales. They are often suspended in the organic ooze and show displacement by entry. They appear to be particularly resistant to attack by the normal microbiological population.

walled resinous algal cells. Erda Tonstein, Upper Carboniferous, Germany. (× 400). d) Complete degradation of wood cells; straight and branching filaments of "*Phellomyces*" produce a pseudomorph of the original cell pattern (see a). Common remains in degraded groundmass organic matter. Erda Tonstein, Upper Carboniferous, Germany. (× 800). e) Completely attacked groundmass organic material; dark wood tissue; cocci, elongate cells and short filaments represent organisms "*Phellomyces*" and *Anthracomyces*. Erda Tonstein, Upper Carboniferous, Germany. (× 1,200). f) "*Phellomyces*" filaments with circular or elliptical habit as seen in groundmass organic matter. Erda Tonstein, Upper Carboniferous, Germany. (× 1,000). g) Red brown cellular bodies attached to wood fragments; similar to *Polymorphites* and regarded as a saprophytic fungus. Erda Tonstein, Upper Carboniferous, Germany. (× 1,000). h) Fungal hyphae arising from wood fragments; one carries swollen vesicle suggestive of conidiophore formation; note detail of preservation. Erda Tonstein, Upper Carboniferous, Germany. (× 1,200). i) Colonies of thick walled red brown cells, found in the organic groundmass; believed to be saprophytic fungi related to *Polymorphites*. Erda Tonstein, Upper Carboniferous, Germany. (× 1,200). k) Thin section of "spore cannel" (right angles to the bedding), showing high concentration of spores (or algae) embedded in the organic detritus. Tasmanite. Permian Age. (× 200). l) Thin section of algal cannel (right angles to bedding) showing high concentration of pale yellow algal colonies related to *Botryococcus* occurring in the organic detritus. Boghead Coal, South Africa. Permo-Carboniferous Age (× 200)

VII. Information Provided by Geomicrobiological Studies

The chronologic and stratigraphic importance of the major groups of microfossils studied in Palynology and Microplanktonology is shown by their effective use in the correlation of rocks and thus of the order of events in earth history. The ubiquity of distribution, the great numbers of specimens available, and the fact that the majority of these remains are of organic constitution, make them particularly significant in the study of organic geochemistry. Changes in the chemical composition of organic matter in rocks may be related to the appearance of different groups of these organisms at a specific geological time. Major environmental differences are indicated by the microfossil content, since certain types of microplankton are entirely marine, while others inhabit freshwater. Land-derived spores, blown or washed out to sea, give rise to an allochthonous content in an otherwise predominantly marine, autochthonous deposit. Partially degraded cellular fragments and the structureless material referred to generally in fossil sediments as "humic constituents", "resin like bodies" or "dark red brown" and "pale yellow" resinous materials, are indicative of the part played by humic constituents. Biological examination of organic matter would assist the evaluation of the proportional representation of the various materials present and the interpretation of the chemical analysis. Separation of the biological ingredients followed by chemical analysis would aid the interpretation of the differing chemical composition of organic matter in sediments. In certain rocks, the organic matter may well be formed by a group of related organisms or substances, e.g., microplankton of the Lower Palaeozoic marine sediments, or microplankton algae in some freshwater lake sediments (Fig. 21).

The distribution over geological time of the more important groups of organisms is well known (see Figs. 3 and 4) and representatives of some groups exist at the present time. It is possible to compare the chemical composition of related organisms at different periods in geological time and to examine the effects of time and method of preservation on their final chamical composition. Reference (p. 267) has been made to the possible chemical changes resulting from preservation under various geological conditions involving heat and/or pressure. Chemical analysis of such material may elucidate these conditions and it may be possible to simulate the conditions under laboratory control. There is also the possibility that organic geochemical analysis would provide a sensitive indicator to the lower metamorphic grade in rocks.

In studies concerned with the microbiological degradation of organic matter, the information is chiefly environmental in character, either in terms of the site of formation or conditions of preservation. Such information is evidenced [92] in the study of tonsteins, where the degraded organic material and microbial population suggest an environment analogous to soils. The selectivity of microbial degradation leads to a progressive concentration of the more biologically stable organic ingredients, particularly in rocks which form a suite bridging across several related environments. Thus, the related suite containing coal, cannel coals, boghead coals, and basin-type oil shales (as distinct from marine) shows changes in organic content which are ascribed to the biological degradation and physicochemical processes affecting a contemporaneous vegetation and resulting peat swamps. The cannel coals, bogheads, and oil shales constitute sapropelic organic accumulations with allochthonous and autochthonous contents formed under shallow water conditions. The environmental controls include: the position of the locus of accumulation in or near a peat swamp; the nature of the drainage and the possibility of wind transport; the potential for supply of the various ingredients; the selection of material which results from physicochemical alteration and microbial degradation, leading to concentration of the more stable remains; the amount of detrital mineral matter available and the state of dilution of the organic ingredients by inorganic matter.

The chemical composition of the preserved organic matter depends on these factors and is particularly influenced by the content of the microorganisms and their metabolic products.

Other, similarly related suites of rock include shales, seatearths, carbonaceous shales and coal, and bauxitic, lateritic, and other types of soil accumulation [12]. The close inter-relationship of organic and inorganic matter in the formation of sediments, or of crystalline matter, is illustrated by the occurrence of a mixed organic-inorganic colloid paste or gel (p. 291). Fungi and algae cause aggregation of mineral particles either mechanically or as a result of impermeable metabolic products [93–95]. Bacteria and algae [96] precipitate crystalline calcium carbonate, the bacterium *Macromonas bipunctata* Bergy acting in symbiotic relationship with the green alga *Synechococcus elongatus* Nag. The flagellate bacterium *Vibrio desulphuricans* [97] was responsible for the initial formation of colloidal sulfur, which, on ageing, resulted in crystallisation. Unusual forms of dolomite crystals with degraded organic matter also enclosed within the crystals, have been noted. EVANS [98] has referred to the solubilization effects which certain organic compounds have on various minerals and on other organic compounds. Under experimental conditions, such relatively water-insoluble minerals as carbonates, phosphates and silicates were brought into solution by A.T.P.

The relationship of microorganisms with sulfide and oxide mineral deposits is well known, but it is not clear whether bacteria are directly and metabolically involved or whether they affect the equilibrium relations between various ions. These matters have been discussed by many workers, particularly EHRLICH [99] and BAAS BECKING [100]. The association of bacteria with Carboniferous pyrite [63] has been established, and LOVE and AMSTUTZ [101] have reviewed evidence for microscopic pyrite from the Devonian Chattanooga Shale and the Rammelsberg Banderz. Of particular interest is the organic matter which is associated with the framboidal or syngenetic pyrite.

FRASER [102] described a modern forest peat in which 3–10 percent copper by dry weight was present as a chelated compound bound up with the organic colloidal complex and associated with 60–80 percent organic matter. LOVERING [103], from experimental evidence, considered the copper to be precipitated as a result of the action of waste products of bacterial metabolism.

The inclusion and preservation of microbiological remains in rocks indicates a lack of serious secondary alteration. Thus their presence within Precambrian cherts and limestones proves the rock to be of primary formation with little or no subsequent modification. Within some cherts and limestones, the otherwise scattered organic fragments are aligned and compressed by their exclusion from the growing crystal lattice e.g., stylolites. In others, the finer organic matter is included within cryptocrystalline material.

The most significant contribution made by geomicrobiology is towards the completion of the fossil record, a matter of intrinsic value in the evolutionary plethora. The effects of microorganisms on all other forms of life, their ubiquity and their power to colonize and to affect every known environment, transcend a wide range of disciplines.

The present hiatus in geomicrobiological studies and the emphasis upon Precambrian studies, follows from the fact that these simple organisms are the earliest known forms of life on Earth. Such information touches upon the origin of life and is the direct concern of the geologist, chemist, biochemist, biologist, and biophysicist. The presence of large and highly complex organic molecules in Precambrian rocks raises considerations of chemosynthesis, but the undoubted presence of bacteria, algae and fungi indicates the action of biological photosynthetic processes and the early establishment of a biological ecology. On present evidence [60], photosynthesis in a biological sense began well over 2,000 million years ago. Most of the presently available geomicrobiological evidence from the Precambrian refers to shallow-water, inshore deposits. HINTON and BLUM [104] favor a terrestrial origin of life in which small and sheltered niches would favor the synthesis of complex organic compounds. The 1965 Symposium on the Evolution of the Earth's Atmosphere touches upon models proposed for the formation of the Earth [105], its atmosphere and early life [106].

The search for identifiable organic remains in ancient sediments and the biochemical and chemical analysis of these remains may give a deeper insight into the evolutionary process.

Finally, in extra-terrestrial matter as represented by the Type I Carbonaceous Chondrites [107], KAPLAN, DEGENS, and REUTER [108] isolated aromatic and aliphatic hydrocarbons, amino acids, and sugars. The absence of pigments, fatty acids and nucleic acids, in conjunction with other biochemical considerations, led these authors to conclude that the organic matter was chemo-synthetic rather than biochemical. The isolation by micropalaeontological techniques (particularly from the Orgueil Type I Carbonaceous Meteorite) of carbonaceous matter having discrete patterns resembling those known in a biological organization has led to the consideration of the existence of extra-terrestrial life. The presumed organic entities have been described by NAGY [109] and the matter though unresolved is one of absorbing interest.

VIII. Geological Distribution of the Main Groups of Other Microfossils

So far, only those microfossils which are entirely or predominantly formed of an organic integument have been described. There are other microfossils which are abundantly represented in sediments of various ages and which, by the decay of their soft parts or by reason of the constitution of their skeletons or protective coverings, may be expected to add significantly to the overall organic content of a sediment. In many cases the remains represent known creatures, in others the biological relationships are entirely unknown. A selection of the more important groups, mainly marine, is referred to below and included in Figs. 3 and 4 which also indicate their geological time range. For fuller information, reference to the following should be made: GLAESNER [110], FRITSCH [111], DOWNIE [112], POKORNY [113], DAVIS [114], Leitfossilien der Mikropalaontologie [115], LINDSTROM [116].

Foraminifera, marine unicellular, chambered protozoa – chitinous inner membranes. Lower Palaeozoic-Recent.

Radiolaria, marine protozoa, pseudochitinous, muscinoid membranes, dense nuclear material, oil and pigments. Ordovician-Recent.

Ostracods, marine and freshwater, bivalved crustaceans, chitinous and calcarreous, profuse in organic-rich sediments. Cambrian-Recent.

Scolecodonts, marine, jaw elements of polychaete worms – chitinous. Ordovician-Devonian – Jurassic-Cretaceous, and Present.

Conodonts, marine, affinity unknown, tooth-like forms, collagen – calcium phosphate with apatite. Cambrian-Trias.

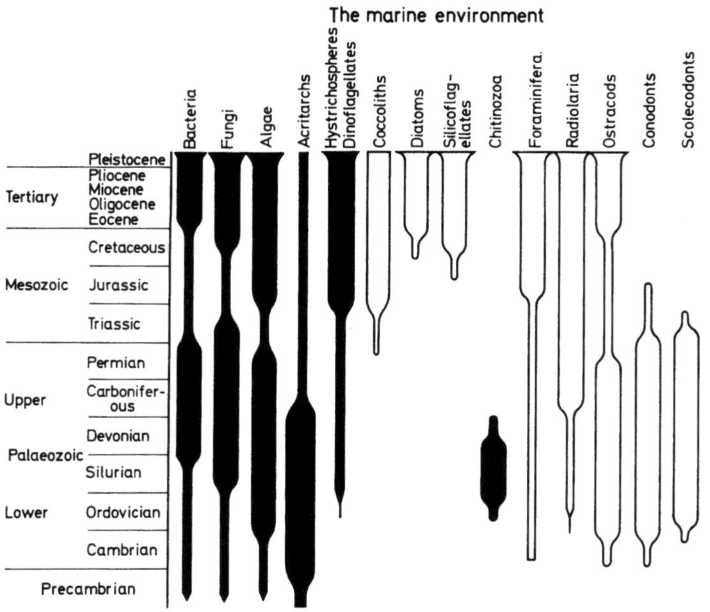

Fig. 3

Figs. 3 and 4 indicate the main groups of microfossils but are not concerned with precise biological subdivisions of each group or with the rarer microfossils. Their intent is to emphasise the possible organic contribution to sediments by expressing the geological range of the organisms and by indicating the prevalent environments in which they occur, e.g., marine and non-marine. In their interpretation the Figures are schematic, for at any one period in geological time and at a given place there may not have existed both marine and non-marine environments; it is likely that the marine sequence may have alternated with non-marine episodes. The Figures do not indicate continuity of the environments but rather the nature of the available material when the environments did occur. No attempt is made to express the relative abundance of the contribution for any group. A narrow area is used to show the appearance of a given group and a wider area

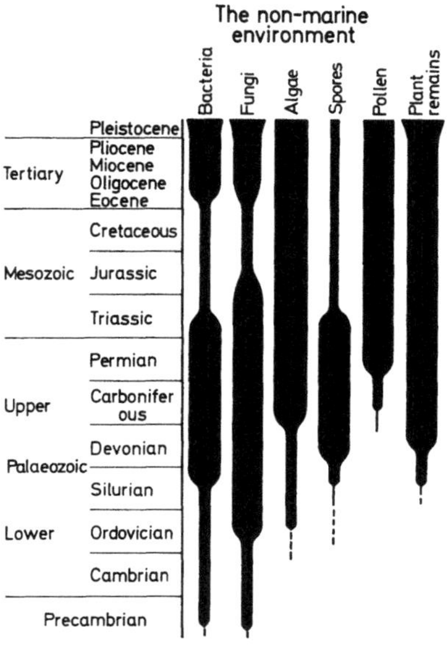

Fig. 4

signifies more common occurrence for that group. The forms which are predominantly organic in constitution are shown in black, while other forms are unshaded.

Fig. 3 is concerned with groups which occur in the marine environment including: Bacteria, Fungi, Algae, Acritarchs, Hystrichospheres, Dinoflagellates, Coccoliths, Diatoms, Silicoflagellates, Chitinozoa, Foraminifera, Radiolaria, Ostracods, Conodonts and Scolecodonts.

Fig. 4 lists the principal forms or sources of organic material available to the non-marine environment including: Bacteria, Fungi, Algae, Spores and Pollen, and Plant remains. Certain animal remains referred to, e.g., Ostracods, also inhabit freshwater environments.

The following glossary refers to microorganisms mentioned in the text and is not intended to represent the complete record of microorganisms known to geological science. The summary attempts to correlate fossil organisms with modern representatives insofar as this has been done by authors recording their geological occurrence. The basis is rarely that of genetic affinity. It is frequently based on either morphological similarity, or an environmental association.

Glossary of Microfossils

Fossil Record	Presumed modern relationship
Bacteria	
Sphaerotilus catenulatus	?*Sphaerotilus natans* Sheathed Eubacteria
Gallionella pyritica	?*Gallionella ferruginea* Stalked Eubacteria
Siderocapsa type	?*Siderocapsa* ssp. ?Siderocapsaceae
Siderococcus type	?*Siderococcus* ssp. ?Siderocapsaceae
Crenothrix	*Crenothrix* ssp.
Bacillus circulans	*Bacillus circulans*
Bacillus ozodeus	*Bacillus* ssp.
Micrococcus hymenophagous	*Micrococcus* ssp.
Micrococcus Guignardii	*Micrococcus* ssp.
Beggiatoa sp.	*Beggiatoa* ssp. ?Bacterial or Algal
Lyngbia type	*Lyngbia* ssp. ?Algal ?Oscillatoriaceae
Fungi, Actinomycetes (Mycelial Eubacteria), Bacteria	
Anthracomyces cannellensis	Fungal, ?Basidiomycetes (Pucciniaceae)
Palynomorphites diversiformis	?Fungal unclassified or Actinomycetes (Mycelial Eubacteria)
Phellomyces dubius	?Fungal unclassified or Actinomycetes (Mycelial Eubacteria)
"*Phellomyces*"	?Fungal unclassified or Actinomycetes (Mycelial Eubacteria)
Archaeothrix contexta	?Fungal or Actinomycetes (Mycelial Eubacteria)
Archaeothrix oscillatiformis	?Fungal or Algal ?Oscillatoriaceae
Mycelites ossifragous	?Fungal
Polycellaria ssp.	Fungal conidia ?Fungi Imperfecti
Polymorphyces minor	?Fungal, Actinomycetes, or Bacteria
Polymorphyces major	?Fungal, Actinomycetes, or Bacteria
Saprolegnia ferox	*Saprolegnia ferox* Fungal, Saprolegniales
Algae	
Kidstoniella fritschi	Algal – Schizophyceae – Stigonemataceae
Langiella scourfieldii	Algal – Schizophyceae – Stigonemataceae
Rhyniella vermiformis	Algal – Schizophyceae – Stigonemataceae
Animikiea septosa	?Algal ?Schizophyceae – Oscillatoriaceae
Gunflintia	Algal ?Ulotrichaceae
Archeorestis	?Algal or problematical
Filimentella plurima	Algal ?Nostocaceae
Fibularis funicula	?Algal
Spirogyra sp.	Algal Zygnemataceae
Botryococcus braunii	*Botryococcus braunii*, Xanthophyta (Heterocontae)
Pila	Algal Xanthophyta (Heterocontae)
Rhienschia	Algal Xanthophyta (Heterocontae)
Gleocapsomorpha	Algal ?Protophyceae

References

1. Dombrowski, H.: Bacteria from palaeozoic salt deposits. Ann. N.Y. Acad. Sci. **108**, 453 (1963).
2. Pflug, H. D.: Structured organic remains from the Fig Tree Series of the Barberton Mountain Land. Univ. Witwatersrand, Econ. Geol. Res. Unit. Information Circular, **28**, 1 – 14 (1966).
3. Bader, R. G., B. W. Hood, and J. B. Smith: Recovery of dissolved organic matter in sea water and and organic sorption by particulate material. Geochim. Cosmochim. Acta **19**, 4 (1960).
4. Bordovsky, O. K.: Accumulation and transformation of organic substance in marine sediments. Parts 1 – 3. Mar. Geol. **3**, 3 – 114 (1965).

5. TRASK, P. D.: Origin and environments of source sediments of petroleum. Houston, Texas: Gulf Publishing Co. 1932.
6. VALLENTYNE, J. R.: The molecular nature of organic matter in lakes and oceans, with less reference to sewage and terrestrial soils. J. Fisheries Res. Board Can. **14**, 1 (1957).
7. KRISS, A. YE.: Marine microbiology (deep water). Izv. Akad. Nauk SSSR, Moscow (1959) [in Russian].
8. DATSKO, V. G.: Organic matter in Soviet Southern waters. Izv. Akad. Nauk SSSR, Moscow (1959) [in Russian].
9. ZHUKOVA, A. I.: The biomass of micro-organisms in bottom sediments of the North Caspian. Mikrobiologiya **24**, 3 (1955) [in Russian].
10. SPARROW, F. K., and T. W. JOHNSON: Fungi in oceans and estuaries. Weinheim: J. Cramer 1961.
11. HÖHNK, W.: Studien zur Brack- und Seewassermykologie I. Veröffentl. Inst. Meeresforsch. Bremerhaven **1**, 115–125 (1952).
12. SKOPINTSOV, B. A.: Organic matter in natural waters (water humus). Trans. Geol. Inst. Akad. Nauk SSSR **17**, 29 (1950) [in Russian].
13. USPENSKIY, V. A.: The carbon balance in the biosphere in relation to the distribution of carbon in the earth's crust. Gostoptekhizdat, Moscow (1956) [in Russian].
14. DEGENS, E. T.: Biogeochemical alterations in the early stages of diagenesis. Proc. 6th Internat. Congr. Sedimentology, Amsterdam-Brussels 1963.
15. – A. PRASHNOWSKY, K. O. EMERY, and J. PIMENTA: Organic materials in recent and ancient sediments. Pt. II. Amino acids in marine sediments of Santa Barbara basin, California. Neues Jahrb. Geol. u. Palaeontol. Monatsh. **8**, 413–426 (1961).
16. – J. H. REUTER, and K. N. F. SHAW: Biochemical compounds in Offshore California sediments and sea waters. Geochim. Cosmochim. Acta **28**, 45–66 (1964).
17. ORR, W. L., K. O. EMERY, and J. R. GRADY: Preservation of chlorophyll derivatives in sediments off Southern California. Bull. Am. Assoc. Petrol. Geol. **42**, 925 (1958).
18. HODGSON, G. W., and E. PEAKE: Metal chlorin complexes in recent sediments as initial precursors to petroleum porphyrin pigments. Nature **191**, 766 (1961).
19. – –, and B. L. BAKER: The origin of petroleum porphyrins. The position of the Athabasca oil sands. Res. Council Alberta (Can.) Inform. Ser. **45**, 75 (1963).
20. HOERING, T. C., and P. H. ABELSON: Fatty acids from the oxidation of kerogen. Ann. Rep. Geophys. Lab. Carnegie Inst. Washington, D. C. 218 (1964–65).
21. DEGENS, E. T., K. O. EMERY, and J. H. REUTER: Organic materials in recent and ancient sediments. Part III. Biochemical compounds in San Diego Trough, California. Neues Jahrb. Geol. Palaeontol. Monatsh. (May 1963).
22. CALVERT, S. E.: Origin of diatom-rich varved sediments from the Gulf of California. J. Geol. **74** (5), 546 (1966).
23. SKOPINTSOV, B. A., and L. P. KRYLOVA: Transport of organic matter by the major rivers of the Soviet Union. Dokl. Akad. Nauk. SSSR **105** (4) (1955) [in Russian].
24. GOKSOYR, J.: Wood decomposing fungi and their adaptation to life in wood. Advan. Sci. **22** (97), 147 (1965).
25. NORD, F. F., and J. C. VITTUCCI: Enzyme studies on the mechanisms of wood decay. Nature **160**, 224 (1947).
26. –, and W. J. SCHUBERT: The biochemical disintegration of wood and the mechanism of lignification. Holzforschung **15**, 1–7 (1961).
27. FLAIG, W.: Effects of micro-organisms in the transformation of lignin to humic substances. Geochim. Cosmochim. Acta **28**, 1523–1535 (1964).
28. SCHRIKHANDE, J. G.: The production of mucus during the decomposition of plant materials. The effects of changes in the flora (mucus producing microorganisms). Biochem. J. **27**, 1563–1574 (1933).
29. REGE, R. D.: Biochemical decomposition of cellulose materials with special reference to the action of fungi. Ann. Appl. Biol. **14**, 1–44 (1927).
30. BAILEY, I. W., and M. R. VESTAL: The significance of certain wood destroying fungi in the study of the enzymatic hydrolysis of cellulose. J. Arnold Arboretum **18**, 196–205 (1937).
31. SAVORY, J. G.: Breakdown of timber by ascomycetes and fungi imperfecti. Ann. Appl. Biol. **41**, 336–347 (1954).
32. REESE, E. T.: Enzymatic hydrolysis of cellulose. Appl. Microbiol. **4**, 37–45 (1955).

33. KUZNETSOV, S. I., M. V. IVANOV, and N. K. LYALIKOVA: Introduction to geological microbiology. New York: McGraw-Hill 1963.
34. NEOFITOVA, V. K.: The microflora of the upper undrained peat-layers and its role in the process of peat formation. Vest. Leningr. Gos. Un-ta, **10** (1953) [in Russian].
35. HELLEBUST, J. A.: Excretion of some organic compounds by marine phytoplankton. Limnol. Oceanog. **10**, 192–206 (1965).
36. DRATH, A.: Boghead coals from Radzionkow. Inst. Geol. (Polona) Biul. **21**, 1 (1939).
37. VALLENTYNE, J. R.: Biochemical limnology. Science **119**, 605–606 (1954).
38. DEGENS, E. T., and M. BAJOR: Distribution of amino acids in fresh water and marine shales of the Pennsylvanian (Ruhr District) Germany. Fortschr. Geol. Rheinland-Westfalen **3**, 429–440 (1962).
39. BRADLEY, W. H.: Chloroplast in spirogyra from the Green River formation of Wyoming. Am. J. Sci. **260**, 455–459 (1962).
40. NEVES, R., and L. B. H. TARLO: Isolation of fossil osteocytes. J. Roy. Microscop. Soc. **84**, 217–219 (1954).
41. DOBERENZ, A. R., and R. LUND: Evidence for collagen in a fossil of the Lower Jurassic. Nature **212**, 1502–1503 (1966).
42. CARLILSE, D. B.: Chitin in a Cambrian fossil *Hyolithellus*. Biochem. J. **90**, p. 1 (1964).
43. ABELSON, P. H.: Amino acids in fossils. Science **119**, 576 (1954).
44. HARE, P. E.: Amino acid composition of the organic matrices of the structural shell units of *Mytilus Californianus*. Science **139**, 3551, 216–217 (1963).
45. RENAULT, B.: Recherches sur les bacteriacees fossiles. Ann. Sci. Nat., VIII. Ser. Bot. **275** (1896).
46. MOODIE, R. L.: Palaeopathology, an introduction to the study of ancient evidence of disease. Urbana: University of Illinois Press 1923.
47. MOORE, L. R.: The microbiology, mineralogy and genesis of a tonstein. Proc. Yorkshire Geol. Soc. **34**, 235–291 (1964).
48. – The "in situ" formation and development of some kaolinite macrocrystals. Clay Minerals Bull. **5** (31), 338–352 (1964).
49. –, and E. G. SPINNER: A further study of the geomicrobiology of a Precambrian shale. Proc. Yorkshire Geol. Soc., in press (1969).
50. STACH, E.: Die Auschliff sporendiagnose des Ruhr-Kohlenflözes Baldur. Palaeontographica, Abt. B **102**, 71–95 (1957).
51. THIESSEN, R.: Structure of palaeozoic bituminous brown coals. U.S. Bur. Mines Bull. **117**, 1–256 (1920).
52. ZOBELL, C. E.: Action of microorganisms on hydrocarbons. Bacteriol. Rev. **10**, 1 (1946).
53. – Soil and water microbiology. Fourth Intern. Congr. Microbiol. Copenhagen, Sect. VII, 453 (1947).
54. – Geochemical aspects of the microbial modification of carbon compounds. Scripps Inst. Oceanog., Contrib. 1703, **34**, 1653 (1964).
55. WAKSMAN, S. A.: The actinomyces. 1st ed. XVIII, 230 pp. Waltham, Mass.: Chronica Botanica 1950.
56. TIEGHEM, P. VAN: Sur le ferment butrique (*Bacillus amylobacter*) à l'epoque de la houille. Compt. Rend. **89**, 1102 (1879).
57. RENAULT, B.: Sur quelques microorganisms des combustibles fossiles. Bull. Soc. Ind. Min. T. **13–14** (1900).
58. PIA, J.: Die vorzeitlichen Spaltpilze und ihre Lebensspuren. Palaeobiologica **1**, 457–474 (1928).
59. MOORE, L. R.: Fossil bacteria. The Fossil Record. Swansea Symposium Volume, Geol. Soc. London, Pt. II, pp. 164–180 (1967).
60. CLOUD, P. E.: Significance of the gunflint (Precambrian) microflora. Science **148**, 27 (1965).
61. BARGHOORN, E. S., and S. A. TYLER: Microorganisms from the gunflint chert. Science **147**, 563 (1965).
62. SCHOPF, J. W., E. S. BARGHOORN, M. D. MASER, and R. O. GORDON: Electron microscopy of fossil bacteria two billion years old. Science **149**, 1385 (1965).
63. – E. G. EHLERS, D. V. STILES, and J. D. BIRLE: Fossil iron bacteria preserved in pyrite. Proc. Am. Phil. Soc. **109**, 288–308 (1965).
64. BERRY, E. W.: Remarkable fossil fungi. Mycologia **8**, 73 (1916).

65. WEDL, C.: Über die Bedeutung der in den Schalen von manchen Acephalen und Gasteropoden vorkommenden Canale. Sitzber. Kais. Akad. Wiss. Wien **33**, 451 (1858).
66. KOLLIKER, A. W.: On the frequent occurrences of vegetable parasites in the bone structures of animals. Proc. Roy. Soc. (London), **10**, 95–100 (1859).
67. KIDSTON, R., and W. H. LANG: On Old Red Sandstone plants showing structure from the Rhynie Chert bed, Aberdeenshire. Pt. V. The thallophyta. Trans. Roy. Soc. (Edinburgh), **52**, 831–855 (1921).
68. DRECHSLER, C.: Morphology of the genus *Actinomyces*. Bot. Gaz. **67**, 147–164 (1919).
69. HENRICI, A.: Molds, yeasts and actinomycetes, 2nd ed. New York: John Wiley & Sons, Inc. (1947).
70. BERGEY's manual of determinative bacteriology, 6th ed. Baltimore: Williams & Wilkins Co. 1948.
71. MOORE, L. R.: On some microorganisms associated with the scorpion Gigantoscorpio Willsi STØRMER. Skrifter Norske Videnskaps-Akad. (N. S.) **9**, 1–14 (1963).
72. — Microbiological colonisation and attack on some Carboniferous miospores. Palaeontology **6**, 349–372 (1963).
73. HARE, P. E., and P. H. ABELSON: Amino acid composition of some calcified proteins. Ann. Rep. Geophys. Lab. Carnegie Inst. Washington, D. C. 223 (1965).
74. PFLUG, H. D.: Organische Reste aus der Belt-Serie (Algonkium) von Nordamerika. Palaeont. **39**, 10–25 (1965).
75. BENES, K., and J. KRAUSSOVA: Paleomycological investigation of Tertiary coals of some basins in Czechoslovakia. Sb. Geol. Ved. **6**, 149–168 (1963).
76. TAYLOR, G. H., and A. C. COOK: Selerotinite in coal: Its petrology and classification. Geol. Mag. **99**, 1, 41–52 (1962).
77. GRAHAM, A.: The role of fungal spores in palynology. J. Palaeont. **36**, 60–68 (1962).
78. MOORE, L. R.: The microbiology of some tonsteins. Cinquieme Congres Int. de Stratigraphie et de Geol. du Carbonifere, Paris, 587 (1964).
79. BENES, K., and J. KRAUSSOVA: Carboniferous fossil fungi from the Upper Silesian basin. Sb. Geol. Ved. Fada **4**, Praha (1963).
80. BARGHOORN, E. S., and J. W. SCHOPF: Microorganisms from the late Precambrian of Central Australia. Science **150**, 337 (1965).
81. TIMOFIEV, B. V.: The earliest Baltic flora and its stratigraphic importance. Trans. V.N.I.G.R.I. **129**, 1–320 (1959).
82. ROBLOT, M. M.: Sporomorphes du Precambrian Normand. Rev. Micropal. **7**, 2, 153–156 (1964).
83. TIMOFIEV, B. V.: Precambrian spores. Int. Geol. Congr. XXI Session. Repts. Soviet geologists problem IX, Precambrian stratigraphy and correlations. Acad. Sci. USSR, Moscow, Leningrad, 138–714 (1960).
84. STARMACH, K.: Blue-green algae from the Tremadocian of the Holy Cross Mountains (Poland). Acta Palaeontol. Polon. **8** (4), 451–460 (1963).
85. CROFT, A. N., and E. A. GEORGE: Blue green algae from the Middle Devonian of Rhynie, Aberdeenshire. Bull. Brit. Museum Geol. **3** (10), 341–353 (1959).
86. BLACKBURN, K. B., and B. N. TEMPERLEY: *Botryococcus* and the algal coals. Pts. I and II. Trans. Roy. Soc. Edinburgh **58** (3), 841 (1936).
87. JOHNSON, J. H.: Bibliography of the fossil algae 1942–1955. Quart. Colo. School Mines **52** (2), 1–92 (1957).
88. DOWNIE, C., W. R. EVITT, and W. A. S. SARJEANT: Dinoflagellates, hystrichospheres and the classification of the acritarchs. Stanford Univ. Publ. Univ Ser. Geol. Sci. **7**, 3 (1963).
89. ELSIK, W. C.: Biologic degradation of fossil pollen grains and spores. Micropalaeontology **12**, 515–518 (1966).
90. GOLDSTEIN, S.: Degradation of pollen by phycomycetes. Ecology **41**, 543–545 (1960).
91. BAVER, L. D.: Soil physics, 3rd ed. London: Chapman and Hall 1956.
92. MOORE, L. R.: Coal and coal bearing strata (D. G. MURCHISON and T. S. WESTOLL, eds.). Edinburgh: Oliver and Boyd 1968.
93. GEOHEGAN, M. J.: Trans. 4th Int. Congr. Soil Science **1**, 198 (1950).
94. MCCALLA, T. M.: Proc. Soil Sci. Soc. Am. **7**, 209 (19).
95. HAMMERIL, P., and M. RIOULT: Etude de la fixation des sediments meubles par deux algues marines *Rhodothamniella floridula* and *Microcoleus chtonoplastes*. Marine Geology **3** (1/2), 131 (1965).

96. MASON-WILLIAMS, A.: Biological aspects of calcite deposition. Symp. Internat. di Speleologia Varenna 1 (1961).
97. SUBBA RAO, M. B., K. K. IYA, and M. SREENIVASAYA: Microbiological formation of elemental sulphur in coastal area. Fourth Intern. Congr. Microbiol. Copenhagen, Sect. VII, 494 (1947).
98. EVANS, W. D.: The organic solubilization of minerals in sediments. Advances in organic geochemistry. Proc. Intern. Meeting, Milan (1962), 1–8. New York: Pergamon Press, Inc. 1964.
99. EHRLICH, H. L.: Observation on microbial association with some mineral sulphides. Proc. Nat. Sci. Found. Symp. Yale 153 (1962).
100. BAAS BECKING, L. D. M.: Geology and microbiology. New Zealand Oceanograph. Inst. Mem. **3**, 48 (1959).
101. LOVE, L. G., and G. C. AMSTUTZ: Review of microscopic pyrite from the Devonian Chattanooga shale and Rammelsberg Banderz. Fortschr. Mineral. **43** (2), 273–309 (1966).
102. FRASER, D. C.: Organic sequestration of copper. Econ. Geol. **55** (2), 402–414 (1961).
103. LOVERING, T. S.: Organic precipitation of metallic copper. Bull. Geol. Soc. U.S. **785-C**, 45–52 (1927).
104. HINTON, H. E., and M. S. BLUM: Suspended animation and the origin of life. New Scientist, **Oct.** 270 (1965).
105. RINGWOOD, A. E.: Chemical evolution of the terrestrial planets. Geochim. Cosmochim. Acta **30**, 41 (1966).
106. FISCHER, A. G.: Fossils, early life and atmospheric history. Proc. Nat. Acad. Sci. U.S. **53** (6), 1205 (1965).
107. MASON, B.: The "enstatite chrondites". Geochim. Cosmochim. Acta **30**, 23–41 (1966).
108. KAPLAN, I. R., E. T. DEGENS, and J. H. REUTER: Organic compounds in stony meteorites. Geochim. Cosmochim. Acta **27**, 805 (1963).
109. NAGY, B.: Investigations of the Orgueil carbonaceous meteorite. Geol. Fören. Stockholm Forh. **88**, 235–272 (1966).
110. GLAESSNER, M. F.: Principles of micropalaeontology. Victoria: Melbourne University Press 1945.
111. FRITSCH, F. E.: The structure and reproduction of the algae, vol. 1. New York: Cambridge University Press 1948.
112. DOWNIE, C.: The geological history of the microplankton. Rev. Palaeobot. Palynol. **1**, 269–281 (1967).
113. POKORNY, V.: Principles of zoological micropalaeontology, vol. II. New York: Pergamon Press, Inc. 1965.
114. DAVIES, C. C.: The marine and freshwater plankton. East Lansing: Michigan State University Press 1955.
115. Leitfossilien der Mikropalaeontologie. Berlin-Nikolassee: Gebrüder Borntraeger 1962.
116. LINDSTROM, M.: Conodonts. Elsevier Publ. Co (1964).

CHAPTER 12

Biogeochemistry of Stable Carbon Isotopes *

E. T. DEGENS

Department of Chemistry,
The Woods Hole Oceanographic Institution,
Woods Hole, Mass.

Contents

I. Introduction

A. Historical Background

In 1947, in his classical paper on the thermodynamic properties of isotopic substances, H. C. UREY [1] laid the foundation of modern isotope geochemistry. At the same time, A. O. NIER [2] designed a new mass spectrometer which allowed the measurement of small differences in isotope abundance ratios. A modification in the Nier-type mass spectrometer and a refinement in instrumentation techniques by McKINNEY *et al.* [3] finally initiated stable isotope studies of the type that will be discussed in this review.

* This research represents part of a program at the Woods Hole Oceanographic Institution concerned with the stable isotope distribution in geological and biological systems, headed by J. M. HUNT. His advice and encouragement in the course of this endeavour are gratefully acknowledged. I wish to express sincere appreciation to my colleagues at the Institute, in particular, J. H. RYTHER, H. G. TRÜPER, S. W. WATSON, J. A. HELLEBUST, W. G. DEUSER, M. BLUMER, and R. R. L. GUILLARD who stimulated and guided the research efforts in discussions and by active participation. To D. W. SPENCER, I extend my gratitude for critically reading the manuscript, and to J. MATHEJA for advice in the presentation of the data.

The work was sponsored by the National Aeronautics and Space Administration and by a grant from the Petroleum Research Fund administered by the American Chemical Society. Grateful acknowledgement is hereby made to NASA and the donors of said fund.

The manuscript was submitted for publication on Febr. 15, 1967.

Woods Hole Oceanographic Institution contribution No. 2020.

The early work done in the field of stable carbon isotope biogeochemistry can be described as a reconnaissance survey, because little was actually known at that time of the ^{13}C content in the various biological and geological materials. This introductory chapter of isotope biogeochemistry was largely written by BAERTSCHI [4], CRAIG [5], LANDERGREN [6], NIER and GULBRANSEN [7], RANKAMA [8], WEST [9], and WICKMAN [10].

A significant breakthrough in isotope research started about 1960. PARK and EPSTEIN [11, 12] and ABELSON and HOERING [13] discovered that photosynthetic carbon dioxide reduction may label the various metabolic products to different extents. Both groups of authors realized the significance of this observation for studies on the biosynthetic pathways of carbon and the geological carbon cycle.

Applying data by BIGELEISEN [14] on the relative velocities of reactions undergone by isotopically-labeled molecules, SILVERMAN [15] and SACKETT *et al.* [16] showed that thermal cracking of carbon-carbon bonds in various organic materials may introduce isotope fractionation.

The present trend is to analyze pure and chemically well-defined carbon compounds and to study the isotope relationships between co-existing organic phases in sediments or in metabolically related compounds. This feature is analogous to the approach of the geochemist, who is generally more interested in the element or isotope composition of individual minerals than in the chemistry of a homogenized rock powder.

B. Scope of Studies

There are a great number of specific problems, both in the field of biology and geology, where carbon isotope data can be used profitably. The significant application to biology lies in the delineation of biosynthetic pathways through a comparison of isotope ratios in discrete biochemicals. Inasmuch as organisms can be studied in their natural habitat, the advantage of stable carbon isotope methods over radiocarbon techniques becomes evident.

In geology on the other hand, we are principally dealing with two sets of problems. The first one is concerned with the diagenetic fate of organic matter through geologic time; the second one tries to gain more insight into the nature of pre-biological carbon in terrestrial and extraterrestrial bodies.

During our discussion, there will be ample opportunity to see the intimate relationship between carbon-containing compounds from living and fossil organisms. Furthermore, it will be shown that information on the distribution of carbon isotopes in biological and geological material is essential to understand the complex biogeochemical cycle of carbon through time and space.

II. Methods of Analysis

A. Combustion System for Organic Matter

The working gas of carbon isotope ratio mass spectrometers is usually carbon dioxide. Thus, the first step in sample preparation requires conversion of organic carbon to carbon dioxide, and purification of the carbon dioxide from any contaminating gas contained in the system. This is accomplished following a procedure outlined by CRAIG [5].

The equipment required for the combustion of the organic matter is illustrated in Fig. 1. The operation starts by placing the samples, if a solid, in an alundum boat in quantities sufficient to provide about 10–15 cc at N.T.P. of carbon dioxide. Carbonate-containing samples require acidification prior to combustion to release the inorganic carbon as CO_2. The boat is subsequently placed in a quartz combustion tube which is half-filled with copper oxide. After the system is closed and pumped down to high vacuum, oxygen is admitted and combustion carried out to completion at temperatures in the neighborhood of 900° C. During combustion, the generated gases are continuously recycled via an electrically operated Toepler pump to secure a total conversion of carbon monoxide to carbon dioxide; a liquid-nitrogen-cooled trap provides a simultaneous collection of carbon dioxide.

Following a suggestion of SACKETT and THOMPSON [17], the CO_2 is passed over hot (500° C) manganese dioxide and copper to remove contaminating sulfur and nitrogen oxides. A set of cooled

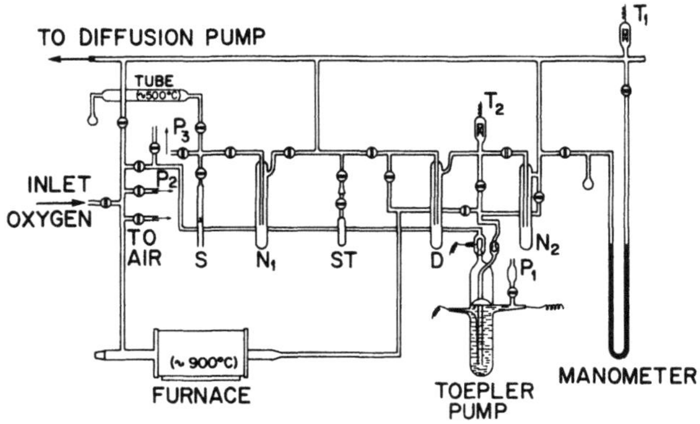

Fig. 1. Carbon combustion apparatus [5, 16]. Legend: D – acetone-dry ice trap; N_1, N_2 – liquid nitrogen traps; P_1, P_2, P_3 – to mechanical pump; S – tube equipped with break seal for thermal cracking studies; ST – carbon dioxide transfer tube; T_1 T_2 – thermocouple pressure gauge

traps, filled with acetone-dry ice and liquid nitrogen, finally separates the carbon dioxide from any water or gas contaminant. The carbon dioxide yield can subsequently be measured in a calibrated manometer, after which the gas is transferred to a sample tube for mass spectrometric analysis.

A variety of techniques have been proposed for the extraction of the carbon compounds from liquid or gas samples such as petroleum or methane [5, 18–20].

B. Mass Spectrometer Analysis

The basic details of the mass spectrometer commonly used in stable isotope work have been described by NIER [2]. Improvements in design [3] have increased the precision of the Nier instrument by more than 50 percent.

The instrument is a 60° sector-type mass spectrometer and contains a double collecting system. The analysis starts by feeding the CO_2 sample via a small gas leak into the source of the mass spectrometer where the CO_2 is ionized. The ions, after being accelerated in an electrostatic field and focused into a single beam, have to pass through a magnetic analyzer where they are separated according to their mass. The resolved ion-beams upon hitting a collector become neutralized. The electric current thereby released is electronically amplified and recorded on a potentiometric recorder.

The $^{13}C/^{12}C$ ratio is determined by comparing mass 45 ($^{13}C^{16}O^{16}O$) with mass 44 ($^{12}C^{16}O^{16}O$). Variations in mass 45/44 ratio are measured relative to an isotopically known standard gas. An inlet system incorporating a magnetically-operated valve allows a switchover from the standard CO_2 to the unknown CO_2 gas in a matter of seconds permitting a rapid comparison of carbon isotope ratios of both gases under the same mass spectrometric conditions.

The complete $\delta^{13}C$ procedure requires an amount of sample typified by about 20 mg of dry marine phytoplankton and about two hours for combustion, collection, and mass spectrometry.

C. Isotope Standards

In the literature, the $^{13}C/^{12}C$ ratios are reported in a number of different ways. This is unfortunate because it makes it more difficult to compare the results of the various authors. The reference scales most frequently used include the Stockholm, NBS (National Bureau of Standards, Nos. 20, 21 and 22), Wellington, Nier-Solenhofen limestone, Basel, and PDB-Chicago standards. In order to be consistent throughout this survey, it was decided to select the most commonly used standard, i.e., the PDB-Chicago standard, as a reference scale. All data reported differently have been adjusted to the PDB standard using the appropriate conversion factors [21, 15] in order to facilitate comparison of published information.

The carbon isotope data are reported as δ-values, which are deviations in parts per thousand (per mil) of the $^{13}C/^{12}C$ ratios of the samples from that of the CO_2 obtained from the belemnite standard (Peedee Formation, Upper-Cretaceous, South Carolina), abbreviated PDB-standard, used by the University of Chicago group [5, 21]. δ-values are defined by the formula:

$$\delta^{13}C=\left(\frac{R}{R_s}-1\right)\times 1{,}000$$

where

$$R={}^{13}C/{}^{12}C \text{ ratio in the sample}$$

$$R_s={}^{13}C/{}^{12}C \text{ ratio in the standard.}$$

A negative $\delta^{13}C$ implies that the sample is depleted in ^{13}C relative to the chosen PDB standard. In contrast, a positive $\delta^{13}C$ means an enrichment in ^{13}C relative to the PDB standard.

Appropriate corrections are commonly applied for the ^{17}O contributions to the signal for mass 45, the error due to mixing of sample and standard at the analyzer tube inlet, and the tailing of the mass 44 peak under the mass 45 peak [21]. The precision of the reported analyses is ± 0.2 per mil or better. At present most laboratories engaged in isotope work achieve a precision of ± 0.05 to 0.1 permil depending on type of sample material.

D. Sample Material

This review is based on more than 5,000 carbon isotope data which have been largely obtained from the literature. Unpublished results of more than 500 analyses are also included in the figures: they have kindly been made available by M. Blumer, W. G. Deuser, K. O. Emery, R. L. Guillard, J. A. Hellebust, J. M. Hunt, W. M. Sackett, H. G. Trüper, and S. W. Watson.

The individual samples have been systematically grouped under class of compound and have been plotted in the form of cumulative frequency diagrams (2 sigma range) to summarize the information in a comprehensive form (Figs. 15–21).

III. Isotope Fractionation by Living Matter

A. Photosynthesis

Substantially all living matter is a product of photosynthesis directly as green plants or indirectly as animals and organotrophic (i.e., not directly photosynthetic) plants. Let us, therefore, briefly consider the main avenues of carbon dioxide reduction during this process and ear-mark possible isotope fractionation barriers. For a comprehensive account of photosynthesis one may consult Calvin and Bassham [22], Bassham [23], and Bonner and Varner [24].

1. Environmental Effects

Plants extract carbon from two sources, atmospheric CO_2, and molecular CO_2 and HCO_3^- in the hydrosphere. Thus, the isotope composition of an organism will be influenced by the $\delta^{13}C$ of the carbon source utilized during photosynthesis.

It has long been recognized that atmospheric CO_2 is depleted in ^{13}C relative to the inorganic carbon pool of the oceans which contains about 98 percent of all the non-biological carbon in the atmosphere-hydrosphere system. The equilibrium fractionation in the exchange:

$$^{13}CO_2 + H^{12}CO_3^- \rightleftharpoons {}^{12}CO_2 + H^{13}CO_3^-$$

has recently been determined [25]; the carbon isotope fraction between gaseous CO_2 and HCO_3^- decreases from 9.2 to 6.8 permil over the temperature range 0° to 30° C, the lighter carbon species being enriched in the CO_2. This fractionation occurs in the hydration stage, not in the passage of atmospheric CO_2 through the air-water interface. Consequently, in isotopic equilibrium, the molecular CO_2 present in any natural water body should have the same $\delta^{13}C$ value as the atmosphere [26], whereas the HCO_3^- should be enriched in ^{13}C by about 7 to 9 permil relative to atmospheric CO_2. The ratio of the carbonic acid substances in solution is principally pH and temperature controlled. For example normal sea water (pH 8.5) contains more than 99 percent of its dissolved inorganic carbon as HCO_3^-, whereas fresh waters are usually acidic (pH 5–7) and thus have predominantly molecular CO_2. In turn, the carbon isotope differences observed between fresh waters (light) and marine waters (heavy) can no longer be interpreted to be exclusively a result of the large biogenic CO_2 contributions to rivers and lakes [27, 28].

It is known that for gas molecules the velocities of isotopic species are proportional to the inverse square root of the molecular weights. For carbon dioxide one can write:

$$\frac{\text{Velocity}\,(^{12}C^{16}O^{16}O)}{\text{Velocity}\,(^{13}C^{16}O^{16}O)} = \sqrt{\frac{45}{44}} = 1.011$$

This implies that collisions of CO_2 of mass 44 with a photosynthesizing leaf are 1.1 percent more frequent than those of CO_2 of mass 45 [29]. In consequence land plants are expected to be substantially depleted in ^{13}C relative to marine plants because the latter utilize dissolved and not gaseous CO_2 during photosynthesis.

Preliminary data on the carbon isotope distribution in marine and terrestrial organisms seemed to support this inference [5]. Recent experiments with marine phytoplankton populations, however, cast doubt on the reality of the postulated large kinetic effect [30]. These experiments indicate that the observed carbon isotope fractionation is principally determined by pH, water temperature, concentration of carbonic acid species, and growth rate of the organisms. As a general rule, maximum fractionation is achieved when pH and water temperature are low, the dissolved CO_2 concentration is high, and the growth rate is moderate (Fig. 2). The experiments further indicate that molecular CO_2 is the exclusive carbon source during photosynthesis. If molecular CO_2 is highly abundant and in isotopic equilibrium with the large HCO_3^- pool in the sea, the observed fractionation between cell carbon and HCO_3^- may be as high as −28 permil. As the molecular CO_2 pool becomes effectively drained due to a number of environmental or biological circumstances, the isotopic equilibrium between molecular CO_2 and the pool of ionized carbonic acid species is broken down; ^{13}C enriched molecular CO_2 is released from the HCO_3^- and is subsequently taken up by the plants. This

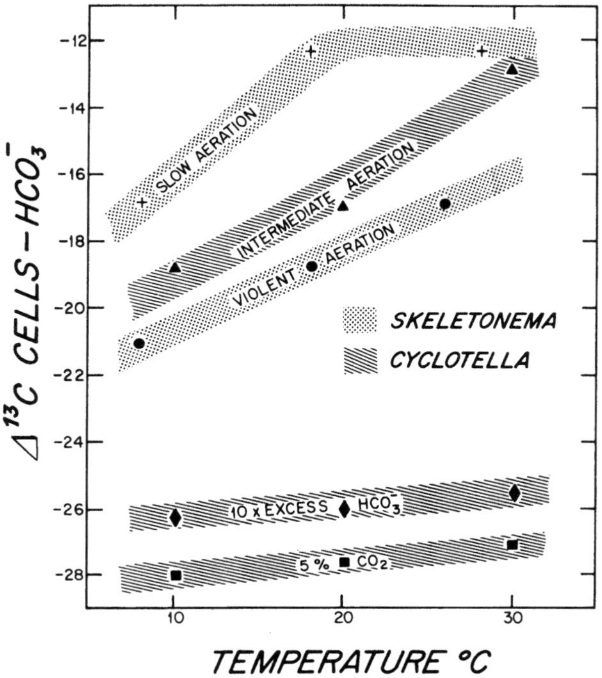

Fig. 2. Temperature and CO_2 concentration effects in laboratory cultured marine plankton [31]. $\Delta^{13}C$ refers to the difference between two $\delta^{13}C$ values, that for the cell carbon and that for HCO_3^-

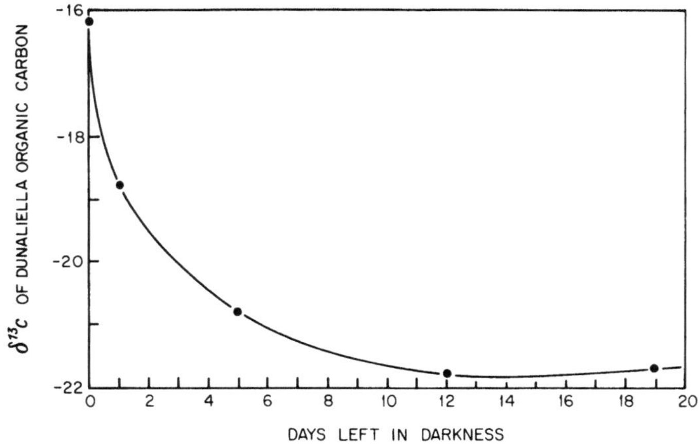

Fig. 3. Respiration effects in laboratory cultured marine plankton [31]

will result in a lowering of the observed fractionation between CO_2 and HCO_3^-; in the present case a minimum plateau value of about -12 is achieved (Fig. 2).

Recent data [31] which showed pronounced differences for marine plankton from warm water (low latitudes) and cold water (high latitudes) can now be reasonably explained in terms of changes in water temperature or other environmental characteristics. The total spread in $\delta^{13}C$ between the various natural plankton samples so far examined is about 15 permil (−12 to −27) which is exactly the range predicted from the aformentioned laboratory-cultured phytoplankton population.

Phytoplankton cultures kept in darkness for 12 days show a depletion in ^{13}C by as much as 5 permil relative to the ^{13}C-content at the beginning of the period of dark respiration (Fig. 3). Thus isotope fractionations in living material occur in both synthetic and degenerative processes [30].

2. Metabolic Effects

The first reaction of carbon dioxide reduction in photosynthesis is a carboxylation reaction involving ribulose-1,5-diphosphate (RuDP) and 3-phosphoroglyceric acid (PGA). From here a stepwise formation of various sugars takes place and the monosaccharide conversion via phosphorylated derivatives will proceed. The available data suggest, that a subsequent polysaccharide synthesis from mixtures of sugar nucleotides will not result in a significant rearrangement of the individual carbon skeletons.

An examination of the carbon isotope relationship in this portion of the photosynthetic carbon reduction cycle is of interest. PARK and EPSTEIN [11, 12] have shown that the enzymatic fixation of CO_2 leading to PGA is accompanied by a maximum 17 permil depletion in ^{13}C. They concluded that this carboxylation step is the major metabolic event controlling the isotopic carbon composition of the plant as a whole. It was further suggested that except for the chloroform-extractable lipids, no other major biochemical compound is labeled isotopically differently from the total plant. However work by DEGENS *et al.* [32] indicates that different carbohydrates in an organism show different degrees of ^{13}C-depletion relative to the starting CO_2. Namely individual sugars are spread over a wide $\delta^{13}C$ range whereas the total sugar fraction is isotopically identical to the whole plant.

The biosynthesis of amino acids follows a rather complex pattern. Principally, intermediates of the glycolytic pathway (e.g., phosphoenolpyruvate, pyruvate) and the tricarboxylic acid (TCA) cycle (e.g., α-ketoglutarate, oxalacetate, glyoxalate) supply the necessary carbon skeletons from which by reductive amination or transamination amino acids arise. In view of the numerous ways in which carbon skeletons can be arranged in the TCA–cycle, there is ample opportunity for carbon isotope fractionation.

Differences in $\delta^{13}C$ between the various amino acids in a protein hydrolyzate are vividly displayed in a study by ABELSON and HOERING [13]. The internal $\delta^{13}C$ variations cover a range of about 17 permil. Most enriched in ^{13}C are serine, threonine, glycine and aspartic acid, whereas the leucines and aromatic amino acids are most depleted in ^{13}C. Of further significance is the observation that the carboxyl functions of amino acids are generally enriched in ^{13}C by as much as 20 permil relative to the remainder of the molecule. This characteristic has great importance in the evaluation of isotope data from ancient sediments. The internal isotope fluctuations within and between amino acids are not reflected in the $\delta^{13}C$ of the total protein which actually exhibits the same $\delta^{13}C$ content as the associated carbohydrate fraction. This implies that under conditions of steady photosynthesis, the material balance of carbon isotopes will always remain constant; in most instances, the constant will be determined at the RuDP-PGA barrier.

Lignin is a major biochemical compound in higher plants, but simple intermediates, e.g., hydroxybenzoic acids, are known from many primitive organisms. The theory has been advanced that certain quinoid derivatives are involved in the electron transport of cell respiration, instead of being only physical impregnations of the maturing cell. The starting point of all phenylpropanoid compounds is either phenylalanine or tyrosine. Since these two amino acids belong to the ones most depleted in ^{13}C, this characteristic will be reflected in the lignin fraction. Of course, if the lignin fraction of a plant represents 50 percent or more of the total organic matter, the isotope difference relative to the other biochemicals present will grow smaller for reasons of material balance.

The biosynthesis of lipids involves the oxidation of pyruvate to acetyl coenzyme *A* and carbon dioxide. From acetyl CoA the formation of the various lipids proceeds. It has been demonstrated by many investigators [11, 12, 15] that lipid materials tend to be depleted in ^{13}C by several permil relative to the plant organic matter as a whole. PARKER [33] has extended this knowledge by showing that the maximum fractionation between the individual fatty acids of an organism can amount to 4 permil, and that different organisms display various degrees of ^{13}C depletion in lipids, i.e., about 4 to 15 permil relative to the total plant. Another interesting aspect of the isotope chemistry of lipids is worth mentioning. It was observed [11, 31, 32] that an increase in lipid content results in a lowering of the $\delta^{13}C$ difference between the lipid fraction and the total plant. This is probably linked to the glyoxylate cycle which is responsible for the conversion of lipids to carbohydrates.

The major steps of carbon isotope fractionation during photosynthesis are enumerated below. The Roman numerals I to VII indicate the positions where the principal fractionation barriers exist (numbers in parenthesis refer to the maximum fractionation in terms of $\delta^{13}C$):

I.	Inorganic-organic boundary effects (uptake of CO_2 by the plant cytoplasm)	(7)
II.	Carboxylation at the RuDP-PGA-level	(17)
III.	Sugar interconversions	(11)
IV.	Amino acid interconversions	(15)
V.	Respiration (decarboxylation reaction)	(20)
VI.	Fatty acid interconversions	(4)
VII.	Lipid-carbohydrate conversions	(18)

The reference body is dissolved or atmospheric CO_2. A change in $\delta^{13}C$ at Step I is kinetically controlled as CO_2 passes from the atmosphere into the plant cytoplasm during photosynthesis. However, it was mentioned earlier that there is some doubt on the reality of the postulated large kinetic effect.

The total $\delta^{13}C$ range of Step II is 17 permil. This represents the maximum fractionation at the ribulose-1,5-diphosphate/3-phosphoroglyceric acid level [11]. A brief outline of the photosynthetic carbon fractionation at Steps III to VI has already been presented in our discussion. It is noteworthy that the $\delta^{13}C$ in cellulose and lignin will remain unaltered since these compounds do not participate actively in the metabolic turnover of carbon. The glyoxylate cycle (Step VII), counteracts the ^{13}C depletion introduced during respiration (Step V) and lipid formation, by converting fats into carbohydrates. A further isotope "homogenization" may be accomplished via the glycolytic pathway which involves the degradation of hexose to pyruvate (oxidative pentose phosphate cycle), or via the tricarboxylic acid cycle which acts as an incinerator for carbon skeletons. As a consequence of these effects, $\delta^{13}C$ differences between coexisting biochemical compounds become less pronounced.

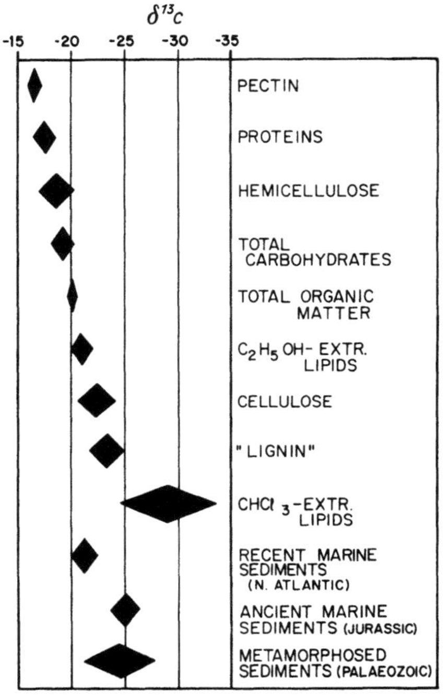

Fig. 4. $\delta^{13}C$ in various biochemical constituents isolated from marine plankton [32]. For comparison, the $\delta^{13}C$ in representative Recent and ancient sediments is included. Diamond-shaped figures represent 1-sigma ranges

Finally, the sum of the negative changes in $\delta^{13}C$ across the various barriers must be matched by an increase in ^{13}C at some other points. Zero potential relative to the inorganic carbon pool utilized during carbon fixation is obtained under steady state photosynthesis by translocation of CO_2 during assimilation, and the removal of CO_2 during respiration. The CO_2 hereby released from the system is enriched in ^{13}C to the same extent as the cell carbon is depleted in ^{13}C.

The isotopic compositions in various types of living matter have been summarized in Figs. 4 and 5. The diamond-shaped figures represent the I sigma values and are obtained from the more detailed diagrams in the appendix of this chapter.

For comparative purposes, isotope determinations on cultures of the marine chemoautotrophic bacterium *Nitrosocystis oceanus* are included. This bacterium assimilates CO_2 via the reductive pentose phosphate cycle [34]. The observed carbon isotope fractionation between HCO_3^- and cell carbon is about twice that commonly found in marine phytoplankton. The pH at the start of the *Nitrosocystis* growth is about 7.6, but this is rapidly lowered to pH 5.5 by oxidation of ammonia. Under those conditions, most of the dissolved carbon is present as molecular CO_2. Phytoplankton grown under such circumstances would exhibit a fractionation (cell carbon-HCO_3^-) of about −28 permil, rather than −35 permil shown by the bacterium. While it is possible that the high fractionation of *Nitrosocystis* results from efficient fractionation in an enzymatic carboxylation at the ribulose-1,5-diphosphate level, it seems more likely that it is due to the greater abundance (25–30 percent) of ^{13}C-depleted lipid compounds in *Nitrosocystis*.

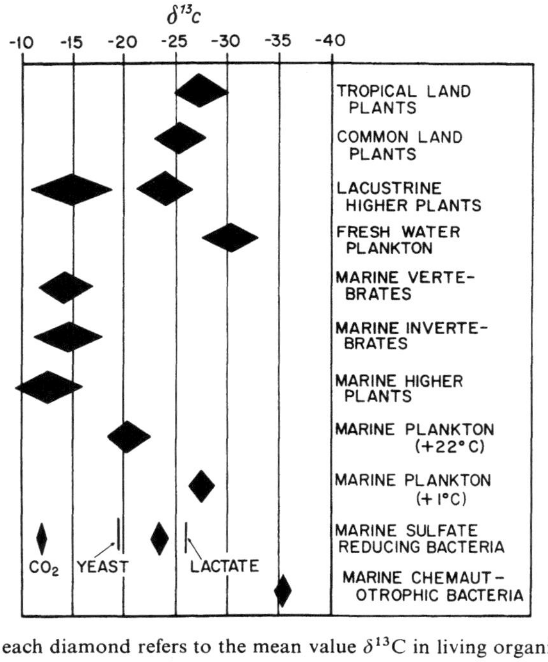

Fig. 5. The center of each diamond refers to the mean value $\delta^{13}C$ in living organisms. Two diamonds are shown for lacustrine higher plants because of differences in their photosynthesis behavior

In contrast, marine sulfate reducers, grown on yeast extract and sodium lactate, are isotopically only slightly different from their carbon substrate (Fig. 5). Kaplan and Rittenberg [35], in relating their observed fractionation effects in sulfate reducing bacteria to the participation of the carboxyl function of lactate, observe a depletion in ^{13}C by a few permil.

IV. Isotope Distribution in Sediments

A. Recent Sediments

Organic matter in Recent sediments has about the same isotopic composition as the organisms living in the environment of deposition. The mean $\delta^{13}C$ of fresh water sediments amounts to −25 permil, whereas marine muds are generally

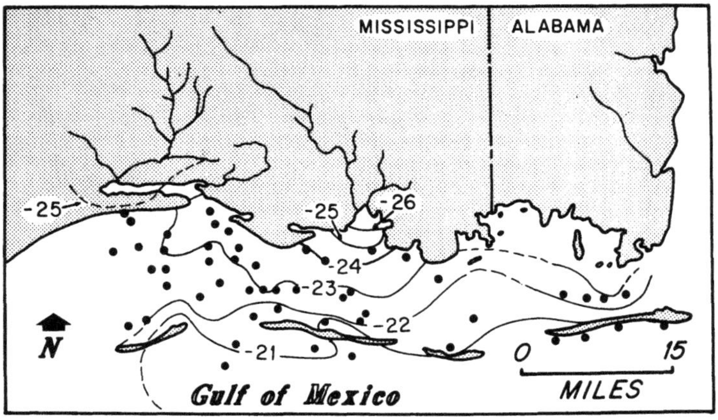

Fig. 6. Isotopic composition of organic carbon in Recent Gulf Coast sediments [17]

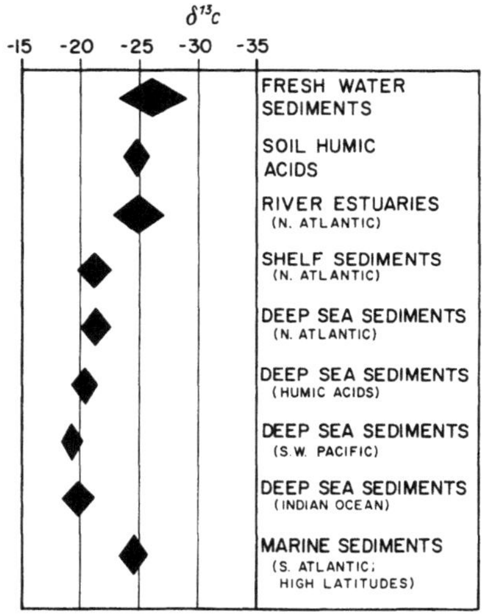

Fig. 7. Isotopic composition of organic carbon in Recent sediments. (Unpublished data, J. M. HUNT, K. O. EMERY)

5 permil heavier, i.e., -20 permil. Only the sediments deposited in river estuaries show a terrestrial influence and $\delta^{13}C$ gradations can frequently be observed. To illustrate the type of gradation pattern that may occur, the results from near-shore sediments from the Mississippi Sound area are presented in Fig. 6. The systematic increase in ^{13}C with increasing distance from the shore line can either be attributed to the lesser influence of land derived organic detritus ($\delta^{13}C \sim -25$ permil), or the diminishing effect of light fresh water bicarbonate ($\delta^{13}C \sim -5$ to -10 permil).

The similarity in $\delta^{13}C$ values in deep sea sediments and shelf deposits is striking and can only mean that contributions of organic debris from land are of minor significance as soon as we leave the immediate environs of the river inlets (Fig. 7).

Bacterial populations and burrowing organisms will not grossly modify the $\delta^{13}C$ of recent sediments. Of course, their activities will result in metabolic waste products, some of which subsequently give rise to so-called humic materials. These compounds are similar in their ^{13}C content to the original biochemicals. During the process of fermentation and decomposition, carbon dioxide and methane of widely different isotopic composition are produced [36], the isotope relationships displayed in such a system are illustrated in Fig. 17 (see appendix, this chapter).

In conclusion, the $\delta^{13}C$ data for Recent sediments are compatible with the general biological and geological viewpoint that the bulk of the organic debris contained in sedimentary deposits is derived from the local biomass. Contributions of fresh water organic matter to marine environments are negligible, and substantially restricted to the small amounts of adsorbed organics on detrital clay minerals.

B. Ancient Sediments

After the major nutrients such as proteins and non-cellulose carbohydrates have been largely eliminated from sediments by action of, *inter alia*, microorganisms and burrowing animals, the further diagenetic degradation of organic matter is that of a slow non-biological maturation. This process essentially involves the elimination of functional groups, breaking of carbon-carbon bonds, and a reorganization of the resulting reaction products. Given sufficient time, the carbonaceous fossil organic matter will eventually end up as CO_2, methane, and graphite. The question immediately arises: what does this trend mean for the carbon isotope composition in the various intermediary stages?

It is well established [14] that the substitution of a heavy for a light isotope lowers the vibrational frequencies and the zero point energy of a chemical bond. Consequently, to break a ^{12}C–^{12}C bond should require less energy than to part a ^{13}C–^{12}C bond. Since energy differences become more effective as the carbon-carbon bond dissociation energy in organic molecules decreases, the extent of carbon isotope fractionation will depend on the terminal bond dissociation energies i.e., those resulting in the loss of single carbon atoms as methane, and therefore, vary accordingly.

Other factors important in the diagenetic fractionation of carbon isotopes include the preferential elimination in the early stages of diagenesis of ^{13}C-rich compounds such as proteins or carbohydrates. Furthermore, a decarboxylation of a molecule, where the carboxyl carbon is ^{13}C enriched, will cause a lowering of the ^{13}C-content in the organic residue, while a ^{13}C-enriched CO_2 is simultaneously released. And finally, isotopic exchange reactions, for instance, between carbon dioxide and methane, might produce fractionation effects, although no conclusive laboratory experiments bearing on this phenomenon have been made. The isotopic equilibrium constants for such simple systems can be calculated by means of spectroscopical data and the techniques of statistical mechanics [5, 14, 37].

1. Coal

Coal is a product of plant debris that has undergone severe physical and chemical alteration throughout geological history. Because of its structural

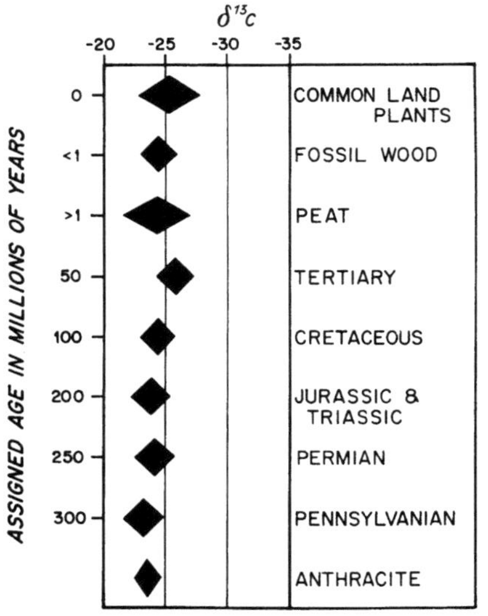

Fig. 8. $\delta^{13}C$ in coals of various geologic ages [5, 38, 39] and for comparison, the $\delta^{13}C$ in land plants, fossil wood and peat

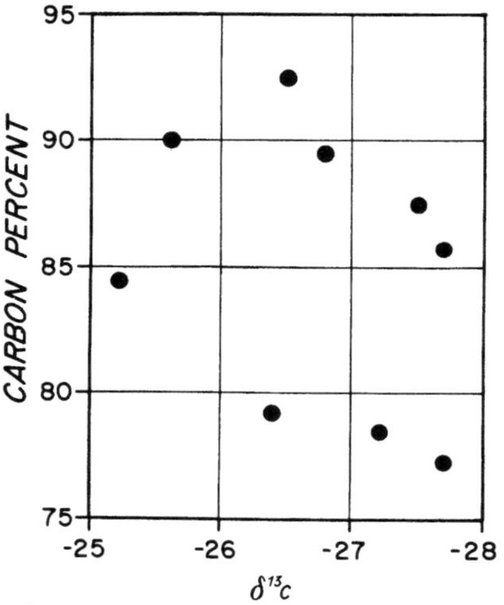

Fig. 9. Relationship between $\delta^{13}C$ and carbon content in partly metamorphosed Tertiary lignites [38]

character, lignin is regarded as the predominant biochemical starting material for all coals. In essence, the bulk of the carbon is organized in aromatic nuclei as in substituted benzenes, naphthalenes, diphenyls, and phenanthrenes, or their multiples. A substantial fraction of the non-aromatic carbon is arranged in hydroaromatic rings.

In view of the extensive diagenetic alteration, it is quite remarkable that this process has not left any recognizable imprint on the $\delta^{13}C$ of various coals. There is no correlation between isotopic composition, degree of coalification, and geologic age of the coals (Fig. 8). The data fall close to the mean of wood $\delta^{13}C$ of -25 permil [5, 38, 39]. It appears, therefore, reasonable to assume that little isotopic fractionation occurs during diagenesis, and that the former land plants had essentially the same isotopic composition as modern wood specimens. To further underline this concept, $\delta^{13}C$ data on thermally metamorphosed Tertiary coals are presented in Fig. 9.

Gases generated during the coalification include principally methane and carbon dioxide. Relative to coal, methane is highly depleted in ^{13}C, in contrast to CO_2 which contains more ^{13}C. In view of the uniform carbon isotope distribution in coals of all ranks, types, and ages, one has to assume that the isotopic composition of the aromatic coal structure is inherited without isotopic fractionation from former lignin precursors. It also implies that whatever happens during coalification, particularly in terms of CO_2 and methane production, the enrichment in ^{13}C in one compound must be matched by a decrease in ^{13}C in the other.

2. Petroleum and Gases

Most crude oil source rocks are sediments which were deposited under marine conditions. This was generally interpreted to mean that marine organisms represent the principal source for petroleum. With the advance of carbon isotope biogeochemistry, this geologically reasonable interpretation was partly abandoned in favor of a new hypothesis which assumes large contributions of organic matter from land. In the following discussion, however, isotope information is presented that agrees with well-rooted principles of geology, i.e., $\delta^{13}C$-values are consistent with a marine origin for most petroleum occurrences.

In this connection, a comparison of present-day marine plankton and crude oil is quite revealing. Analogous to the coal-wood relationships, there is no $\delta^{13}C$ difference established between petroleum and its proposed biological source (Fig. 10). The range for crude oils is similar to the range observed for marine plankton. This is particularly so if one considers lipid compounds as a major precursor of hydrocarbons. Values for Recent biological hydrocarbons and fatty acids obtained from marine plankton fall in the $\delta^{13}C$ range of -25 to -26 permil, which is exactly the mean for all crude oils.

An interesting age relationship can be recognized which may largely be due to fluctuations in environmental conditions throughout the earth's history (Fig. 10). We have previously seen that the $\delta^{13}C$ in marine phytoplankton is predominantly environmentally controlled, i.e., pH and water temperature excercise a major influence on the final $\delta^{13}C$ in the cell carbon by determining the ratio and amounts of the dissolved carbonic acid species in the sea. The isotopic variations displayed in Fig. 10 may thus simply reflect changes in these environmental parameters in the ancient ocean. For example, the rather negative $\delta^{13}C$ values in the Triassic and pre-Devonian crude oils may have been caused by a slight increase in the atmospheric CO_2 pressure. This phenomenon would cause a lowering of the pH in the former sea but simultaneously an increase in the molecular CO_2 content. Under such circumstances, the phytoplankton population may utilize a CO_2 which is in isotopic equilibrium with atmospheric CO_2 and aqueous bicarbonate, and a $\delta^{13}C$ of about -28 permil for the cell carbon is expected. In contrast, if the isotopic equilibrium between the molecular CO_2 and the pool of ionized carbonic acid species is broken down due to *inter alia* – a decrease in atmospheric CO_2 pressure or an increase in water temperature – $\delta^{13}C$ values as low as –12 permil may be obtained for the plant material.

It was already pointed out that diagenesis will preferentially eliminate the ^{13}C-enriched carbohydrate/protein fraction, thus concentrating the light carbon isotope in the organic residue. On the average, the fossil biomass in marine sediments is about 6 permil lower in $\delta^{13}C$ compared to Recent marine sediments (–26 vs. –20 permil). A crude oil generated from a –12 permil plankton would thus be expected to have a $\delta^{13}C$ of –18; and a crude oil formed from –28 permil plankton would yield a $\delta^{13}C$ of –34. In comparing the actually measured $\delta^{13}C$ in petroleum from various geologic ages (Fig. 19), there is a good agreement between predicted and analytically determined $\delta^{13}C$ values.

The primary factors that control the isotopic carbon variation in naturally occuring gases liberated from organic matter during diagenesis are determined by the strength of terminal carbon-carbon bonds, the diagenetic temperature, the geologic age, and the maturity* of the parent material [16]. At temperatures below 100° C the extent of isotopic fractionation in methane is considerable; namely $\delta^{13}C$ values as low as –60 to –80 permil may be produced from organic matter with a $\delta^{13}C$ of –25 permil. These values are in the same range as the bacterial CH_4 fractionation data obtained during the fermentation of methanol [40].

Interpretations of carbon isotope values for methane, however, are complicated by the fact that the gas might equilibrate isotopically with CO_2 which commonly is highly enriched in ^{13}C. The isotope exchange reaction can be written:

$$^{12}CO_2 + {}^{13}CH_4 \rightleftharpoons {}^{13}CO_2 + {}^{12}CH_4 .$$

Following equations developed by CRAIG [5], and assuming that isotopic equilibrium is established in this system, effective temperatures can be computed. Since the kinetics of this exchange are unknown, equilibrium considerations are very difficult to make [18, 41, 42].

In conclusion, the carbon isotope data in crude oils are consistent with the geological consensus that the biological source material for petroleum is predominantly of marine origin. Age differences are attributed to a number of influences which include fluctuations in water temperature, photosynthetic CO_2 fixation mechanisms, or pH effects. The general trend of increasing paraffinicity with depth of burial or time, may also be responsible for some of the variation in $\delta^{13}C$. Fluctuations in $\delta^{13}C$ between hydrocarbons of a single oil are related to carbon number, odd/even characteristics, or molecular structure (e.g. aromatic vs. chain hydrocarbon). The $\delta^{13}C$ in methane and CO_2 liberated in the course of formation of petroleum, coal, or kerogen, is dependant on differences in reaction rates and isotope exchange equilibria.

* Since the maturity of fossil organic matter will determine its observed $^{13}C/^{12}C$ fractionation to methane at a given temperature, one may utilize this information for studies concerned with the recognition of petroleum source rocks. This will principally involve a thermal degradation of a sediment sample at a given temperature, e.g., 200° C, and for a given time, and the analysis of the methane liberated. The degree of isotopic fractionation between the starting organic matter and the produced methane may eventually be used as an index of diagenetic maturity of the sediment organic matter. The smaller the difference in $\delta^{13}C$ between the two organic phases, the more matured the organic matter should be. Consequently a sediment which exhibits a large fractionation is less likely to be considered as a petroleum source rock.

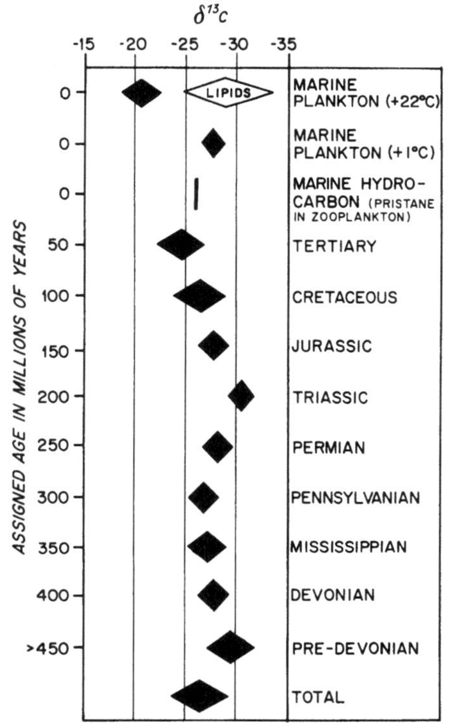

Fig. 10. Carbon isotope distribution in crude oils of various geologic ages and, for comparison, organic material and compounds of Recent origin

3. Kerogen

Most fossil organic matter is found in shales, slates, and schists. The relative quantity present in sandstones and limestones is comparatively small. In connection with unmetamorphosed sediments, the term kerogen is frequently used for the finely disseminated organic matter insoluble in organic solvents; in metamorphosed sediments, the organic material is more commonly named "graphite".

Chemically, kerogen appears to be a denatured form of humic acid. The loss in oxygen and nitrogen and the gain in carbon when compared to humic acids can essentially be linked to dehydration, decarboxylation, loss of methoxyl and carbonyl groups and deamination phenomena. Due to the resulting increase in aromatic structures, kerogen becomes similar to coal, and in the final stage, the organic residue will resemble graphite.

The carbon isotope composition of all ancient sediments of fresh water and marine origin is close to -26 permil, except for early Paleozoic and in particular Precambrian samples that show more negative $\delta^{13}C$ values (Fig. 11). This relationship may mean that environmental conditions in the older Paleozoic and Precambrian time were somewhat different from those established during later geological periods. In this connection, it is noteworthy that the $\delta^{13}C$ in marine limestones ($CO_3^=$) does not show a systematic trend throughout earth history.

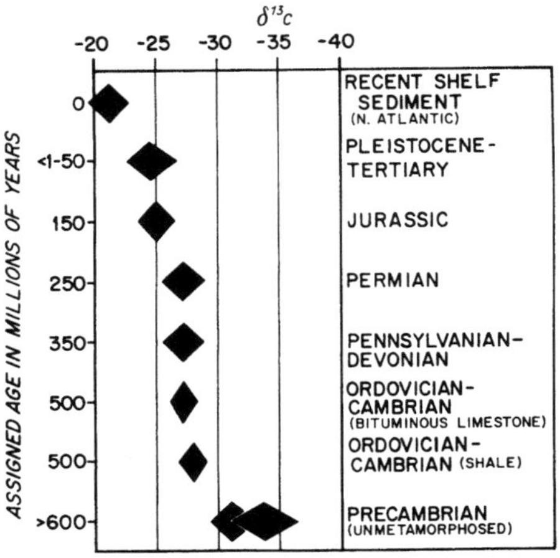

Fig. 11. Carbon isotope distribution in ancient sediments and, for comparison, Recent shelf sediments

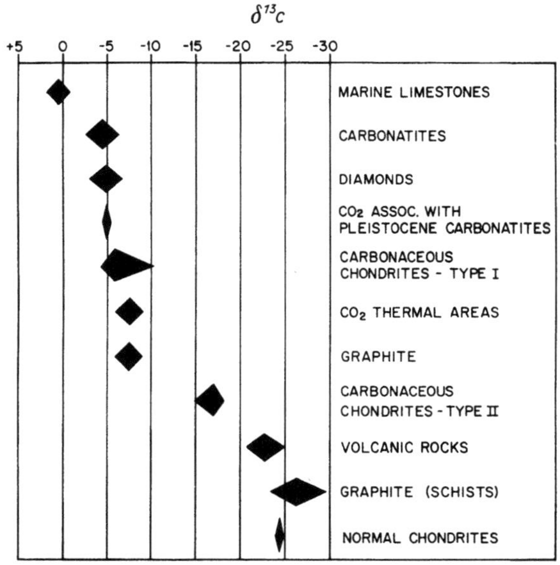

Fig. 12. Carbon isotope distribution in various terrestrial and extraterrestrial materials

From the more than 4,000 limestone samples reported in the literature the majority have a $\delta^{13}C$ close to zero with a I sigma range of $+2$ to -1 (Fig. 12). One may thus safely assume that the $\delta^{13}C$ of the bicarbonate source in the sea stayed fairly uniform through time. The isotopically light organic matter in the Precambrian sediments is, therefore, most likely a consequence of a higher abundance of molecular CO_2 in the Cambrian/Precambrian sea as a function of a slightly lower pH relative to present day conditions or colder temperatures. This would result in a maximum fractionation for green plants of -28 permil. Since a lowering of pH or a change in temperature will significantly affect the solubility characteristics of $CaCO_3$, perhaps, the appearance of shell-forming organisms at the Cambrian/Precambrian boundary can be linked to it.

It was mentioned previously that the $\delta^{13}C$ values of marine lipids are spread over a wide range. However, significant $\delta^{13}C$ differences from the organism as a whole are only found where the lipid content is small. Since in general marine plankton has a high fat content (about 20 to 40 percent) isotopic differences from the total plankton become small for reasons which have previously been outlined. In this manner, the observation of ECKELMANN *et al.* [43] and KREJCI-GRAF and WICKMAN [44], may find its explanation; they found that crude oils and their associated shales have essentially the same isotopic composition.

In a similar fashion to coals, the finely disseminated organic matter in sediments, after the early microbiological stage has passed, appears to be isotopically unaffected by diagenesis. Even metamorphism does not significantly change the isotope pattern of the carbonaceous material; this was clearly demonstrated in studies on metamorphosed and unmetamorphosed sediments from otherwise identical geological settings [6, 38, 45].

V. Primordial Carbon

A. Igneous Rocks

There are several ways one might conceivably obtain information on the $\delta^{13}C$ of primordial (prebiological) terrestrial carbon. One approach is a study of carbon associated with igneous rocks. The other alternative involves a comparison of the $\delta^{13}C$ in gases associated with thermal areas. In doing so, a controversial picture is obtained (Fig. 12). Namely, one set of data suggests a $\delta^{13}C$ of about -4 to -8, whereas another set of data indicates a $\delta^{13}C$ of about -23 for the primordial carbon (Fig. 21). However, the basalts, on which the figure of -23 permil was obtained, are almost certainly contaminated with biogenic carbon [5]. Therefore, an independent check by another method is advisable.

In Table 1, the carbon content in various biological and geological materials is presented together with their mean $\delta^{13}C$. It is apparent that limestones and shales represent the only major exogenic sinks which take care of magmatic carbon released continuously during the history of the earth. Taking the relative proportions of carbonate and shale carbon into consideration, one can estimate the $\delta^{13}C$ of their common source. A $\delta^{13}C$ value of about -6 relative to PDB standard is obtained. In view of the close agreement between geological estimates which are based on the relative abundance of shales, limestones and sandstones, and the geochemical carbon balance (Table 1), a $\delta^{13}C$ of -23 for magmatic carbon as inferred from elemental carbon inclusions in basalts becomes rather unlikely. Instead, carbonatites, certain graphites, diamonds, and carbon dioxide from thermal areas appear to be the closest representatives of magmatic carbon (Fig. 12).

Unfortunately, little is known of the carbon isotope distribution in magmatic rocks, and studies in this direction are worth pursuing. For instance, granites are supposed to have formed by crustal melting (anatexis); if sediments have been granitized this way, this characteristic may still be reflected in the $\delta^{13}C$ of carbon in granite.

Table 1. *Biogeochemical balance of carbon* [46–48]

	Carbon g/cm^2 [a]	$\delta^{13}C$
Carbonates (calcite + dolomite)	2,340 [b]	0
Shales and Sandstones	633	−26
Coals	1.1	−25
Petroleum	0.035	−26
Ocean ($HCO_3^- + CO_3^{-2}$ + dissolved CO_2)	7.5	0
Atmospheric CO_2	0.125	−7
Living matter (land)	0.054	−25
Living matter (marine)	0.00005	−20
Dissolved organic matter (marine)	0.53	−27
Mean (magmatic carbon)		~ −6

[a] Gram per square centimeter earth surface.
[b] Corresponding to a layer ~100 m thick over the whole surface of the earth.

B. Meteorites

Meteoritic carbon falls into three discrete groups, yet a certain trend is noticeable (Fig. 13), i.e., the materials most enriched in carbon have in general the highest $\delta^{13}C$ values [50, 51]. Carbonaceous chondrites (Type I) have a mean $\delta^{13}C$ that corresponds with terrestrial magmatic carbon, whereas normal chondrites fall into the range of land plants.

In following a concept developed by MASON [49] and others, it is conceivable that carbonaceous chondrites (Type I) represent primitive materials, perhaps aggregates of dust from the primordial solar nebula, from which other chondrites were formed by a thermal metamorphism. This would result in partial or complete

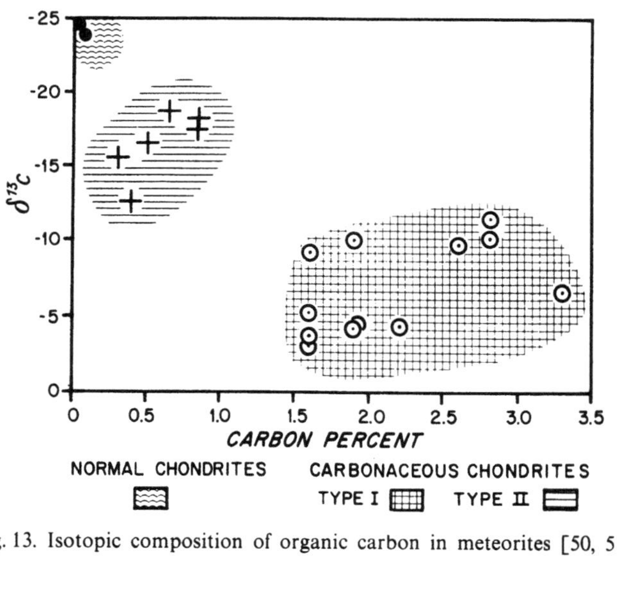

Fig. 13. Isotopic composition of organic carbon in meteorites [50, 51]

loss of volatile compounds and a lowering of the carbon content. Such a loss may be accompanied by isotope fractionation of the kind observed in Fig. 13.

In this context, the observation by CLAYTON [52], showing an about 60 permil enrichment in ^{13}C in meteoritic dolomites over the associated organic carbon, is quite revealing. This fractionation is in the right direction, if we assume that a decarboxylation mechanism yielded a CO_2 enriched in $\delta^{13}C$ which subsequently was used as a source in the dolomite formation. One may also propose that, different from earth, the production of organic carbon exceeded by far the formation of carbonates, and that this stage actually came late in the history of the carbonaceous chondrites. That is at a time when the residual carbon pool may already have been highly enriched in ^{13}C, due to a slight preference of ^{12}C-extraction during the inorganic synthesis of carbon-containing molecules. So far, dolomite is known from carbonaceous chondrites but not chondrites, and in all probability a cause and effect relationship between organic matter and carbonates may be anticipated.

Several papers have appeared recently which assume a biogenic origin for the organic matter contained in carbonaceous chondrites. The idea has even been put forward that photosynthetic organisms were principally involved in the production of this extraterrestrial organic material. Here is not the place to review these claims critically, but the close agreement between the $\delta^{13}C$ in terrestrial magmatic carbon and in the Type I carbonaceous chondrite should make everybody aware that isotope data do by no means support such inference, but that they are more in line with an abiotic origin of meteoritic carbon.

VI. Summary and Conclusions

All isotope information pertinent for the reconstruction of the carbon cycle in nature has been summarized in Fig. 14. A $\delta^{13}C$ of -6 for the magmatic carbon source is postulated. This is close to the isotopic composition now present in the atmosphere. In this way, significant carbon contributions by volcanic gases will not change the isotope composition of atmosphere and hydro-

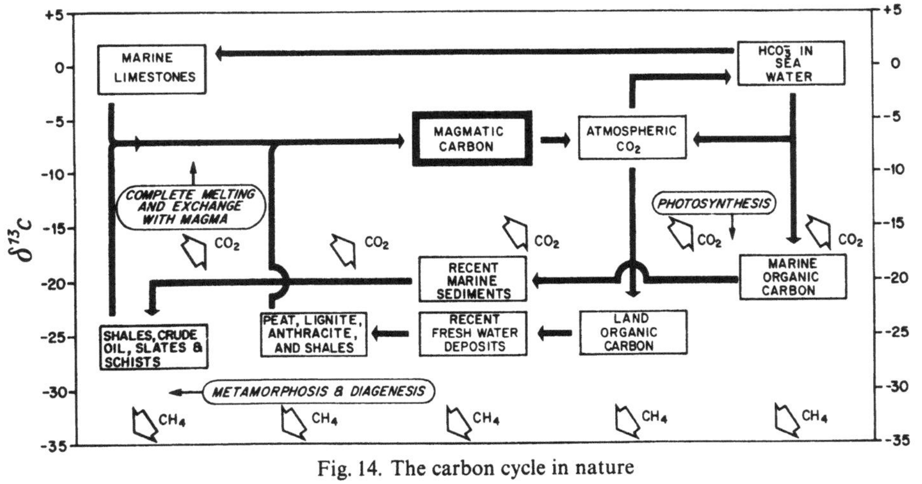

Fig. 14. The carbon cycle in nature

sphere. Only the partial CO_2 pressure and eventually the pH in sea water will be affected.

The uniform isotope ratios for all post-Precambrian carbonates and organics can only mean that we are dealing with a rather complex biochemical system, where the production of organic matter and the formation of limestones is in a well-balanced state. From the wider range in the isotope composition of Precambrian organic matter, we may infer that the carbon cycle was still in a state of flux.

Diagenesis and metamorphism does not alter the isotope composition of the organic matter to a large extent. A complete melting of the whole rock and exchange with magmatic carbon is required to erase the isotope record, corresponding to a lowered content of ^{13}C which was imprinted during photosynthesis.

Appendix

This chapter is based on a collection of all carbon isotope data which have been used in the preceding discussion. More than 5,000 individual analyses have been grouped in such a way as to facilitate a convenient comparison of the various samples (Figs. 15–21). The list of publications from which the isotope data were

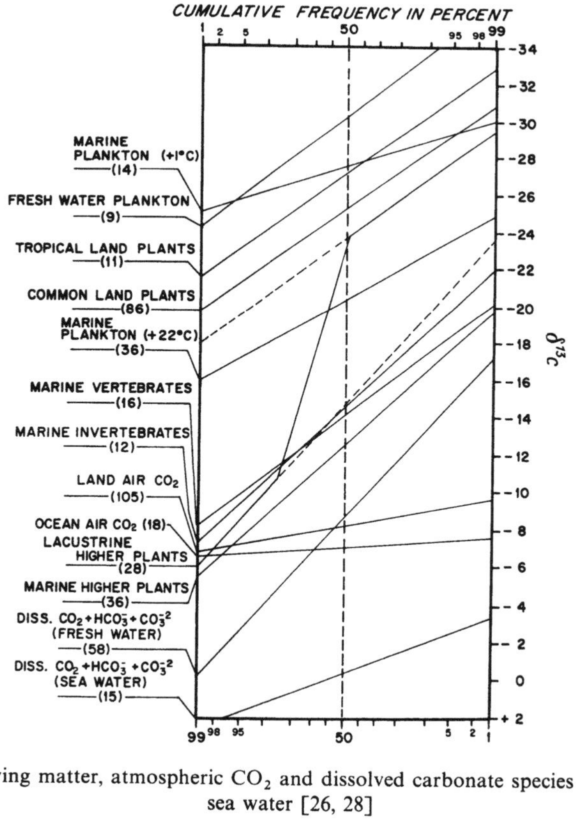

Fig. 15. $\delta^{13}C$ in living matter, atmospheric CO_2 and dissolved carbonate species in fresh water and sea water [26, 28]

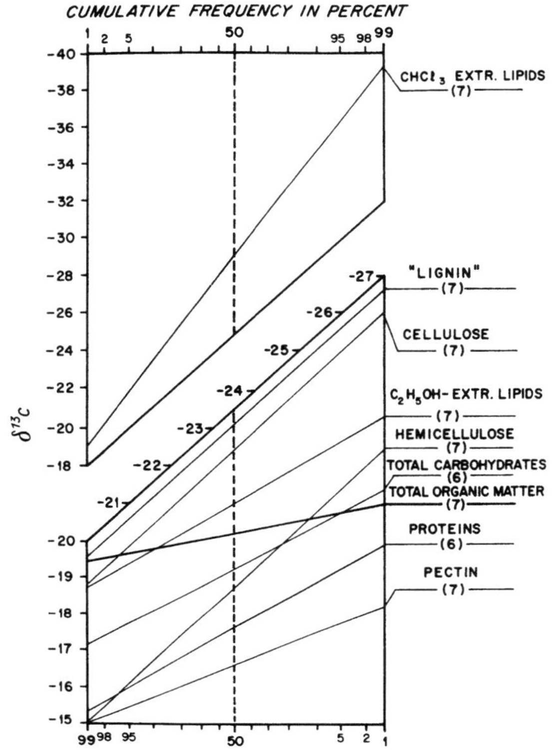

Fig. 16. $\delta^{13}C$ in various biochemical fractions extracted from marine plankton [32]

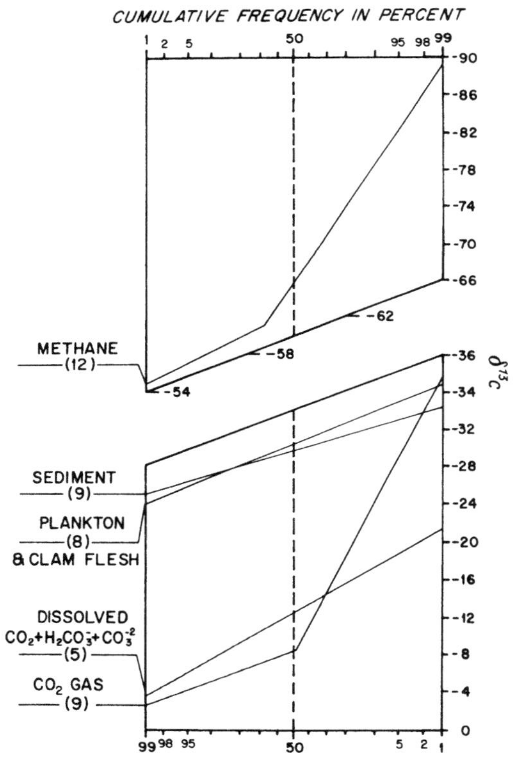

Fig. 17. $\delta^{13}C$ relationships in coexisting carbon-containing compounds from Recent lacustrine environments [36]

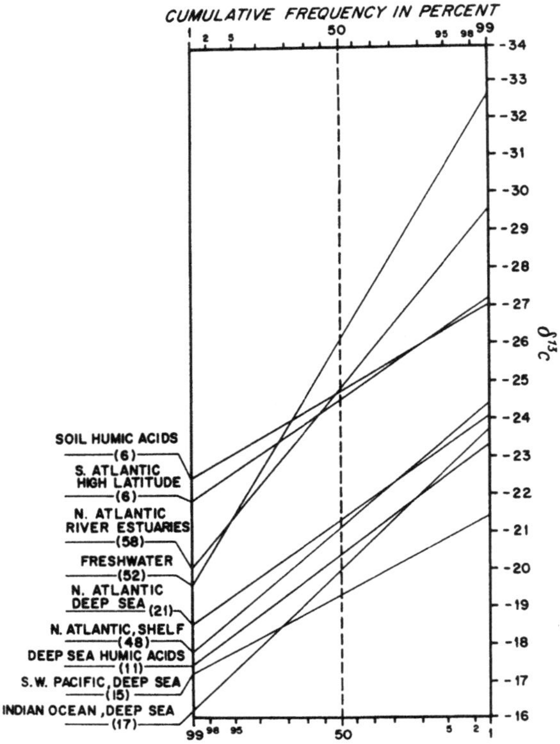

Fig. 18. $\delta^{13}C$ in Recent sediments

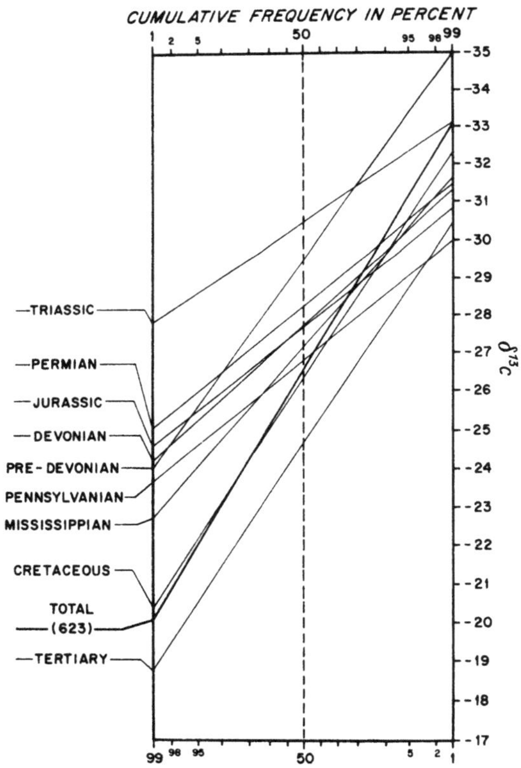

Fig. 19. $\delta^{13}C$ in crude oils from various geologic ages

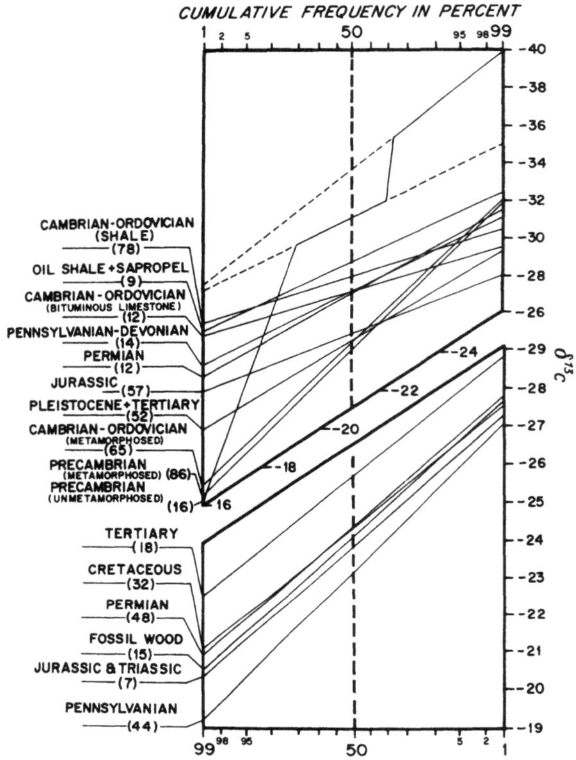

Fig. 20. $\delta^{13}C$ in ancient sediments (top diagram) and coals (bottom diagram)

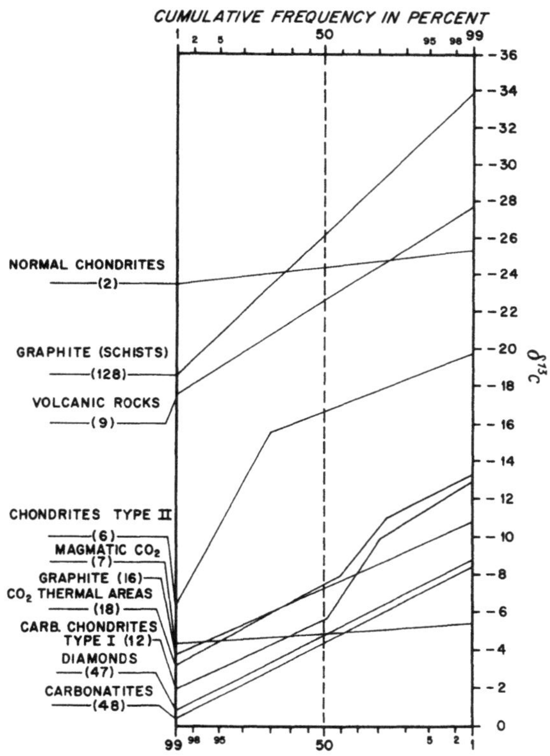

Fig. 21. $\delta^{13}C$ in various igneous, metamorphic, and meteoritic samples

obtained include about 500 titles. The author regrets that due to lack of space no proper credit can be given here to all sources of information.

Terminology: 2 sigma (σ) range: all data came within the 99 percent probability range, i.e., out of 100 measurements 99 came in the specified range. 1 sigma range: all data came within the 66 percent probability range.

References

1. Urey, H. C.: The thermodynamic properties of isotopic substances. J. Chem. Soc. **1947**, 562–581 (1947).
2. Nier, A. O.: A mass spectrometer for isotope and gas analysis. Rev. Sci. Instr. **18**, 398–411 (1947).
3. McKinney, C. R., J. M. McCrea, S. Epstein, H. A. Allen, and H. C. Urey: Improvements in mass spectrometers for the measurement of small differences in isotope abundance ratios. Rev. Sci. Instr. **21**, 724–730 (1950).
4. Baertschi, P.: Die Fraktionierung der natürlichen Kohlenstoffisotopen im Kohlendioxydstoffwechsel grüner Pflanzen. Helv. Chim. Acta **36**, 773–781 (1953).
5. Craig, H.: The geochemistry of stable carbon isotopes. Geochim. Cosmochim. Acta **3**, 53–92 (1953).
6. Landergren, S.: On the relative abundance of the stable carbon isotopes in marine sediments. Deep-Sea Res. **1**, 98–120 (1954).
7. Nier, A. O., and E. A. Gulbransen: Variation in the relative abundance of the carbon isotopes. J. Am. Chem. Soc. **61**, 697–698 (1939).
8. Rankama, K.: New evidence of the origin of pre-Cambrian carbon. Bull. Geol. Soc. Am. **59**, 389–416 (1948).
9. West, S. S.: Relative abundance of the carbon isotopes in petroleum. Geophysics **10**, 406–420 (1945).
10. Wickman, F. E.: Variations in the relative abundance of the carbon isotopes in plants. Geochim. Cosmochim. Acta **2**, 243–254 (1952).
11. Park, R., and S. Epstein: Carbon isotope fractionation during photosynthesis. Geochim. Cosmochim. Acta **21**, 110–126 (1960).
12. – – Metabolic fractionation of ^{13}C and ^{12}C in plants. Plant Physiol. **36**, 133–138 (1961).
13. Abelson, P. H., and T. C. Hoering: Carbon isotope fractionation in formation of amino acids by photosynthetic organisms. Proc. Nat. Acad. Sci. U.S. **47**, 623–632 (1961).
14. Bigeleisen, J.: The relative reaction velocities of isotopic molecules. J. Chem. Phys. **17**, 675–678 (1949).
15. Silverman, S. R.: Investigations of petroleum origin and evolution mechanisms by carbon isotope studies. In: Isotopic and cosmic chemistry (H. Craig, S. L. Miller, and G. J. Wasserburg, eds.). Amsterdam: North-Holland Publ. Co. 1964.
16. Sackett, W. M., S. Nakaparksin, and D. Dalrymple: Carbon isotope effects in methane production by thermal cracking. Third Intern. Meet. Org. Geochem., London, Sept., 1966.
17. –, and R. R. Thompson: Isotopic organic carbon composition of recent continental derived clastic sediments of eastern Gulf Coast, Gulf of Mexico. Bull. Am. Assoc. Petrol. Geologists **47**, 525–528 (1963).
18. Zartman, R. E., G. J. Wasserburg, and J. H. Reynolds: Helium, argon, and carbon in some natural gases. J. Geophys. Res. **66**, 277–306 (1961).
19. Silverman, S. R., and S. Epstein: Carbon isotopic compositions of petroleums and other sedimentary organic materials. Bull. Am. Assoc. Petrol. Geologists **42**, 998–1012 (1958).
20. Keeling, C. D.: The concentration and isotopic abundances of atmospheric carbon dioxide in rural areas. Geochim. Cosmochim. Acta **13**, 322–334 (1958).
21. Craig, H.: Isotopic standards for carbon and oxygen and correction factors for mass spectrometric analysis of carbon dioxide. Geochim. Cosmochim. Acta **12**, 133–149 (1957).
22. Calvin, M., and J. A. Bassham: The photosynthesis of carbon compounds. New York: W. A. Benjamin, Inc. 1962.
23. Bassham, J. A.: Photosynthesis: The path of carbon. In: Plant biochemistry (J. Bonner and J. E. Varner, eds.). New York: Academic Press 1965.

24. Bonner, J., and J. E. Varner: Plant biochemistry. New York: Academic Press 1965.
25. Deuser, W. G., and E. T. Degens: Carbon isotope fractionation in the system CO_2(gas)–CO_2 (aqueous)–HCO_3^-(aqueous). Nature **215**, 1033–1035 (1967).
26. Keeling, C. D.: Concentration and isotopic abundances of carbon dioxide in rural and marine air. Geochim. Cosmochim. Acta **24**, 277–298 (1961).
27. Vogel, J. C., and D. Ehhalt: The use of the carbon isotopes in groundwater studies. In: Radioisotopes in hydrology, p. 383–395. Vienna: Intern. Atomic Energy Agency 1963.
28. Münnich, K. O., and J. C. Vogel: Untersuchungen an pluvialen Wässern der Ost-Sahara. Geol. Rundschau **52**, 611–624 (1962).
29. Craig, H.: Carbon-13 in plants and the relation between carbon-13 and carbon-14. J. Geol. **62**, 115–149 (1954).
30. Degens, E. T., R. L. Guillard, W. M. Sackett, and J. A. Hellebust: Metabolic fractionation of carbon isotopes in marine plankton, Part I. Deep Sea Res. **15**, 1–9 (1968).
31. Sackett, W. M., W. R. Eckelmann, M. L. Bender, and A. W. H. Be: Temperature dependence of carbon isotope composition in marine plankton and sediments. Science **148**, 235–237 (1965).
32. Degens, E. T., M. Behrendt, B. Gotthardt, and E. Reppmann: Metabolic fractionation of carbon isotopes in marine plankton, part II. Deep-Sea Res. **15**, 11–20 (1968).
33. Parker, P. L.: The isotopic composition of the carbon of fatty acids. Ann. Rep. Dir. Geophys. Lab. Carnegie Inst. Wash. Year Book **61**, 187–190 (1961–1962).
34. Campbell, A. E., J. A. Hellebust, and S. W. Watson: Reductive pentose phosphate cycle in *Nitrosocystis oceanus*. J. Bacteriol. **91**, 1178–1185 (1966).
35. Kaplan, I. R., and S. C. Rittenberg: Carbon isotope fractionation during metabolism of lacate by *Desulfovibrio desulfuricans*. J. Gen. Microbiol. **34**, 213–217 (1964).
36. Oana, S., and E. S. Deevey: Carbon-13 in lake waters and its possible bearing on paleolimnology. Am. J. Sci. **258-A**, 253–272 (1960).
37. Craig, H.: The isotopic geochemistry of water and carbon in geothermal areas. In: Nuclear geology on geothermal areas. Consiglio Nazionale Delle Richerche, Pisa, p. 17–53 (1965).
38. Wickman, F. E.: The cycle of carbon and the stable carbon isotopes. Geochim. Cosmochim. Acta **9**, 136–153 (1956).
39. Compston, W.: Carbon isotopic compositions of certain marine invertebrates and coals from the Australian Permian. Geochim. Cosmochim. Acta **18**, 1–22 (1960).
40. Rosenfeld, W. D., and S. R. Silverman: Carbon isotope fractionation in bacterial production of methane. Science **130**, 1659 (1959).
41. Wasserburg, G. J., E. Mazor, and R. E. Zartman: Isotopic and chemical composition of some terrestrial natural gases. In: Earth science and meteoritics (J. Geiss and E. D. Goldberg, eds.), p. 219–240. Amsterdam: North-Holland Publ. Co. 1963.
42. Nakai, N.: Carbon isotope fractionation of natural gas in Japan. J. Earth Sci. **8**, 174–180 (1960).
43. Eckelmann, W. R., W. S. Broecker, D. W. Whitlock, and J. Allsup: Implications of carbon isotope composition of total organic carbon of some recent sediments and ancient oils. Bull. Am. Assoc. Petrol. Geologists **46**, 699–704 (1962).
44. Krejci-Graf, K., and F. E. Wickman: Ein geochemisches Profil durch den Lias *alpha* (zur Frage der Entstehung des Erdöls). Geochim. Cosmochim. Acta **18**, 259–272 (1960).
45. Gavelin, S.: Variations in isotopic composition of carbon from metamorphic rocks in Northern Sweden and their geological significance. Geochim. Cosmochim. Acta **12**, 297–314 (1957).
46. Borchert, H.: Zur Geochemie des Kohlenstoffs. Geochim. Cosmochim. Acta **2**, 62–75 (1951).
47. Ryther, J. H.: Organic production by plankton algae and its environmental control. In: The ecology of algae. Symposium held at the Pymatuning Laboratory of Field Biology, University of Pittsburgh, Spec. Pub. No. 2, p. 72–83 (1959).
48. Weeks, L. G.: Habitat of oil and factors that control it. In: Habitat of oil (L. G. Weeks, ed.), Tulsa Oklahoma. Bull. Am. Assoc. Petrol. Geologists **42**, 1–61 (1958).
49. Mason, B.: The carbonaceous chondrites. Space Sci. Rev. **1**, 621–646 (1962–1963).
50. Boato, G.: The isotopic composition of hydrogen and carbon in the carbonaceous chondrites. Geochim. Cosmochim. Acta **6**, 209–220 (1954).
51. Briggs, M. H.: Evidence of an extraterrestrial origin for some organic constituents of meteorites. Nature **197**, 1290 (1963).
52. Clayton, R. N.: Carbon isotope abundance in meteoritic carbonates. Science **140**, 192–193 (1963).

CHAPTER 13

Hydrocarbons – Saturated, Unsaturated and Aromatic

W. G. MEINSCHEIN

Indiana University, Bloomington, Indiana

Contents

I. Introduction

Hydrocarbons are compounds that are composed solely of carbon and hydrogen atoms. Saturated, unsaturated, and aromatic hydrocarbons are the three principal classes of naturally occurring hydrocarbons. These classes differ in their chemical and metabolic* activities and relative abundances in organisms and sedimentary deposits. Saturated hydrocarbons are called paraffins, a name derived from the Latin *parum affinis*, which means slight affinity. Paraffins or alkanes are trace constituents of biological lipids, but alkanes are the most stable and abundant hydrocarbon constituents of terrestrial rocks. Unsaturated hydrocarbons are commonly referred to as olefins because ethylene, a gaseous unsaturated hydrocarbon, was found to react with chlorine and bromine to yield oily products. Its trivial name "olefiant gas" or oil-forming gas gave rise to the term olefin which has come to denote compounds related to ethylene. Olefins are more abundant than alkanes in biological lipids. Squalene, a polyolefinic triterpene, is an anabolic* intermediate of plant and animal steroids which may be the precursors of some aromatic hydrocarbons in terrestrial rocks [1]. Aromatic hydrocarbons contain at least one six-carbon ring structure with alternate single and double bonds, a benzenoid ring system, and they owe their family name to the fact that certain of

* Metabolism is the sum of processes concerned with the synthesis (anabolism) and destruction or degradation (catabolism) of protoplasm and its molecular constituents. Metabolic, anabolic, and catabolic, as used in this chapter, pertain to the metabolism, anabolism, and catabolism of biological compounds respectively.

these compounds have pleasant odors. Aromatic hydrocarbons are far more abundant than olefins but less abundant than alkanes in sedimentary deposits. The concentrations of aromatic hydrocarbons in organisms are negligible. Most aromatic hydrocarbons are metabolically and chemically less active than olefins.

Scientific interest in the origin of hydrocarbons was enhanced by the discovery and commercial utilization of petroleum. Hydrocarbons are principal constituents of crude oils, and studies of the origins of hydrocarbons are commonly viewed as an integral part of studies of the origin of petroleum [2]. A consequence of these studies is the discovery that certain hydrocarbons in petroleum resemble biological compounds [3]. This discovery has led to the widespread use of hydrocarbons in paleobiological, exobiological, and geological investigations [4], but heated controversies about meteoritic hydrocarbons emphasize the need for a careful appraisal of diverse theories and relevant data on the origins of hydrocarbons [5]. In this chapter, we will consider the pertinence of the chemistry of carbon, evolutionary concepts, metabolic processes, and natural environments to the compositions, distributions, and scientific potentials of terrestrial hydrocarbons and their possible precursors.

Organic compounds generally may be made by either abiotic or anabolic reactions. Carbon atoms are chemically unique in their abilities to combine with one another and with atoms of other elements in an infinite number of ways. The infinite variability of organic reactions makes it impossible to prove that a specific compound was produced by a specific reaction or sequence of reactions. Synthetic chemicals, such as plastics, may be made from abiotic or biological intermediates, and any biological compound may theoretically be made by abiotic reactions. Whereas the chemical versatility of carbon precludes absolute determinations of the origins of organic compounds solely on the basis of their compositions, certain highly selective processes do apparently yield organic compounds for which origins may be reliably deduced, and if we may accept the universality of chemical and physical laws, the same criteria may be used for evaluating the origins of terrestrial and extraterrestrial organic materials.

Most highly selective reactions are directly or indirectly associated with life. Active catalysts and reactive intermediates may produce high yields of organic products of specific structures, and if these products are complex, they can usually be recognized as biological compounds or artifacts. Reactions carried out in attempted imitation of molecular evolution may also duplicate some biological compounds. Fox believes that such duplication may limit the value of chemical fossils [6], but as Wald suggests, the directive influences of scientists and of ubiquitous biological materials may be confused with abiotic directive influences [7]. Although most scientists agree that the molecular constituents of the first organisms were formed abiotically, Morrison explains why the continuity between these molecules and organisms was probably neither quantitatively important nor qualitatively persistent. He logically proposes that life arose from the "rare but interesting deviates" in an abiotic assemblage of molecules in primordial waters [8], and he notes that in the subsequent period of biological evolution a discontinuity between the free energy distribution curves of the abiotic and self-replicating molecules must have developed either instantaneously or rapidly. This discontinuity and the sparsity of the rare molecular deviates directly involved in the origin of life, which MORRISON describes, would ensure that biological molecules are analytically distinguishable from any natural assemblage of abiotic compounds.

Additional confirmation of the compositional distinctiveness of biological molecules may be indirectly deduced from the inadequacies of mathematical considerations of reproduction. WIGNER shows the quantum mechanical impossibility of keeping genetic information from growing increasingly disordered as it is transmitted [9], but WALD points out that some disorder is a prerequisite to evolutionary processes. Variations are needed if natural selection is to be operative, and the survival of the fittest apparently serves to reduce the disorder which must result from information transfers during reproduction. WALD states, "Order in living organisms is introduced not beforehand, by preconceived design, but after the fact – the fact of random mutation – by a process akin to editing" [10].

In essence, biological evolution apparently preserves and enhances the order of transferred genetic information which in turn controls the production, combination, and degradation of metabolic intermediates in organisms. Metabolic reactions, unlike abiotic reactions which are defined by quantum mechanics, follow well-defined and numerically restricted pathways. Organisms are largely isolated systems with directed metabolisms. Plants and animals, in widely different environments, make and use many of the same compounds. A general uniformity in the biological code is clearly indicated [11], and the conformity between morphologically and molecularly determined phylogenies effectively establishes that life may be characterized at the molecular as well as at the cellular level [12].

Admittedly, the enigma of life may cast some doubt on the adequacy of any theoretical explanation of evolution. The discussions above are based on the premise that organic substances obey the same chemical laws as inorganic substances, but FOX suggests that complex carbon compounds are operationally easier to make than theoretical considerations permit [6]. Although substantial data indicate that thermodynamical laws govern the reactions of carbon, the controversies about whether or not molecular remnants of pre-existent organisms can be analytically distinguished from products made solely by abiotic reactions may not be resolved completely by theoretical considerations.

Experimental data acquired on natural samples may provide an insight into the compositional controls that metabolic and abiotic reactions exert on organic substances. Conditions that exist in various regions of the Earth apparently insure the availability of different carbon compounds whose compositions are determined mainly by either biological or abiotic processes. Metabolic and abiotic reaction rates have markedly different temperature and environmental dependencies. Metabolic activity is greatest under moderate temperatures and oxidizing conditions, whereas abiotic syntheses of complex organic compounds requires either high thermal or irradiation energies and reducing conditions [13]. Thus, at the Earth's surface, organisms dominate the production and utilization of carbon compounds, and fossil records indicate that this biological dominance has persisted for more than three billion years [14]. Estimates of biological productivity and fossil fuel reserves suggest that former life has made 10^4 to 10^8 times more organic materials than are retained in ancient rocks [15]. Compositely, the antiquity and abundance of life seemingly establish that the remnants or elimination products of pre-existent organisms were effectively the sole sources of carbonaceous materials that were initially incorporated into the sediments which formed extant sedimentary rocks.

Sedimentation subjects the remnants of former life to conditions that change progressively. At relatively shallow depths beneath sediment surfaces, organisms frequently deplete the supply of atmospheric oxygen, and anaerobes commonly replace aerobes as the principal forms of life. Biological activity, food substances, and available oxygen normally decrease with increasing depth in a sedimentary column, and sediments beneath sufficient overburden may undergo lithification. The temperatures and pressures of rocks and their organic constituents generally increase with depth. Overall the sedimentation process gradually moves organic materials from biologically to abiotically active regions.

It is suggested that abiotic reactions produce, as well as alter, carbonaceous substances in ancient rocks [16]. Certainly, abiotic alternations can destroy all structural and distributional characteristics which may distinguish biological remnants from abiotic products, and the resemblances between organic materials in rocks and abiotic substances may be both real and deceptive. The principal deficiency in most abiotic theories of the origin of carbon compounds in rocks appears to be the failure of these theories to propose a plausible source of abiotic reagents and catalysts. It seems unlikely that conditions within a specific rock or magma volume may release abiotic carbon, hydrogen, and catalysts in reactive forms and intimate associations at temperatures and pressures which permit their combination as complex hydrocarbons. Furthermore, the rocks with mild thermal histories and low permeabilities which contain the great preponderance of hydrocarbons apparently are better suited for the retention of altered and unaltered biological remnants than for the production or incorporation of completely abiotic carbon compounds.

Disparities in the compositions of organisms and fossil organic substances may be traced mainly to variations in the catabolic activities, stabilities, and solubilities of biological remnants and their alteration products. It is not in the best interest of life that carbon and energy be wasted in the production of organic materials which are retained for geologic time in rocks. The remarkable efficiency of the carbon cycle attests to the ability of organisms to conserve carbon and energy. This ability apparently results from the great rates of metabolic reactions, for otherwise the destinies of active biological molecules are left to chance or abiotic reactions. Life continues because the speed of biological processes permits organisms to control the structures and abundances of their molecular constituents. We may expect that this control is accomplished by a balance between anabolic and catabolic reactions. Organisms may not maintain high metabolic rates by producing compounds in large quantities which are not utilized in large quantities, and thus only minor anabolic products may have low catabolic activities.

An ambiguity may develop about the origins of minor biological products. The dominance that metabolic processes exert on organic substances at the Earth's surface apparently does not extend to the organic compounds which escape the carbon cycle, and the role of abiotic reactions in this escape may be significant. Many metabolic reactions are reductions, and in the reducing micro-environments within cells, abiotic reactions which are prohibited by atmospheric oxygen may occur. It is possible that minor anabolic products with low catabolic activities are in reality abiotic attritions, but this possibility would be difficult either to prove or to disprove.

Irrespective of whether or not the minor and catabolically least active constituents of organisms are anabolic products that may have undergone minor abiotic alterations within living cells, the greatest resemblance between cellular and preserved carbonaceous substances is observed for relatively large, chemically stable compounds that are present in fossil organic materials in higher concentrations than in organisms. Low water solubilities and low catabolic and chemical activities are commonly shared characteristics of ancient organic compounds which are structurally and distributionally similar to compounds in plants and animals.

II. Gross Compositions and Distributions of Natural Hydrocarbons

Organic geochemists seek to detect the existence, kind, and antiquity of life, to determine paleoenvironmental conditions and to define the origins of economically important organic materials. All these objectives may be achieved by recognizing and tracing processes associated with the production and alteration of biological compounds. Basically, the requirements for the recognition and definition of processes are much the same. Only ancient processes that have changed systems are detectable, and information about the initial, intermediate, and final states of systems are needed to trace such processes.

Carbon compounds are remarkably constituted for defining natural processes that occur at moderate levels of energies. Both homologous and chemically dissimilar organic compounds may undergo structural and/or distributional changes which precisely define conditions that exist in most of the accessible regions of the Earth's crust. Because organisms produce select arrays of carbon compounds, biological remnants may serve as indicators of evolutionary as well as of physical and/or chemical processes. Abiotic products are geochemically less useful than certain biological compounds, for the composition of a randomly synthesized carbon compound neither defines the reactants nor indicates the sequence of reactions that formed this compound. Likewise, the geochemical usefulness of a simple or an extensively-altered molecular remnant of life is limited, since it may not be distinguished from an abiotic compound.

Certain hydrocarbons apparently have great potentials in geological, paleobiological and exobiological investigations. Many complex hydrocarbons found in nature are sufficiently stable to retain the structural order that may characterize biological remnants [17], and these hydrocarbons or their precursors may have been important in biologic evolution. Hydrocarbons are ubiquitous but minor products of plants and animals [18]. Extensive analytical results indicate close structural and genetic relationships between the non-hydrocarbon and hydrocarbon constituents of biological lipids.

BLOCH lucidly discusses the role of lipids in evolutionary development. He writes, "Biochemical unity, in the broadest sense, prevails for many if not all of the processes that are common to all organisms and constitute the essential and minimal manifestation of life. Biochemical diversity, on the other hand, is the molecular expression of cellular differentiation and specialization of function". Lipids probably form structural components of cytoplasmic and intercellular membranes. BLOCH suggests that differences in the lipid patterns of various phylogenetic groups show that the invention of certain lipid molecules were decisive events in evolutionary diversification. He notes the absence of sterols and polyunsaturated acids in procaryotes and the presence of these compounds in eucaryotic* cells as evidence in support of his suggestion [19].

It is not an easy task to establish the specific genetic ties that may exist between the hydrocarbons and the acids and alcohols in biological lipids. The low metabolic activities and abundances of many hydrocarbons apparently contribute to a general lack of interest in their biosynthesis, and we lack direct information about the anabolism of most hydrocarbons, except terpenes, in organisms. Also, differences in the compositions of lipids and the catabolic activities of individual lipid compounds make it difficult to ascertain which compositional and concentrational variations in naturally occurring hydrocarbon fractions are attributable to biological fluctuations, abiotic alterations, and physical redistributions, respectively.

In our laboratory, we have gathered analytical data on biological and sedimental lipids for the purpose of determining the controls that different natural processes may exert on hydrocarbon compositions. We have arbitrarily concentrated our attentions on C_{15} and larger ($>C_{15}$) organic compounds. This restriction simplifies sample recovery and reduces ambiguities about the origins of compounds, because: (1) $>C_{15}$ lipids can be semi-quantitatively recovered, by low temperature evaporative processes, from their solutions in the volatile solvents which are used in extractions and chromatographic separations, and (2) the probability that uncontrolled reactions may yield a specific carbon compound is many times less for a $>C_{15}$ compound than for a compound in the C_1–C_{15} range. Our analyses have been mainly spectrometric and spectroscopic analyses of complex liquid-solid chromatographic fractions of organic extracts and their hydrogenated or dehydrogenated products; however, many capillary gas-liquid chromatograms (GLC) and tandem GLC-mass spectra (GLC-MS)** have also been obtained of selected fractions.

Comparisons of the structural types and distributions of $>C_{15}$ hydrocarbons, acids, and alcohols from organisms, fecal materials and sedimentary deposits indicate that: (1) the concentrations of olefinic hydrocarbons, principally terpenes, in organisms greatly exceed the concentrations of these compounds in fossil organic matter; (2) the alkanes in organisms and sediments are largely isoprenoids, steranes, and n-paraffins which structurally resemble biological acids and alcohols,

* *Procarytoes* (bacteria and blue-green algae) have only one structurally defined membrane, the cytoplasmic membrane; while in *eucaryotic cells* the membrane-bound cytoplasm contains organelles (nuclei, mitochondria, etc.) which are also enclosed by membranes.

** A tandem GLC-MS or GC-MS consists of a gas-liquid chromatographic column connected in series with an inlet of a mass spectrometer. Alkanes eluted from the GLC column as chromatographic peaks pass directly into the mass spectrometer, and mass spectra are obtained of the individual peaks. The retention times and mass spectra thus obtained are used in determining the structures of the alkanes.

but the relative abundances of alkanes of specific carbon numbers differ significantly from the relative abundances of analogous acids and alcohols within most extracts; (3) the concentrations of alkanes in fecal extracts and those of Recent sediments are comparable, and these concentrations are an order of magnitude greater than the concentrations of alkanes in the average biological lipid; (4) aromatic hydrocarbons are present in negligible or trace quantities in biological lipids, and the aromatic hydrocarbons in Recent marine sediments are most commonly detected in sediments which are buried at depths in excess of a few feet; and (5) non-alkyl substituted phenanthrene, pyrene, chrysene, fluoranthene, triphenylene, 1,2-benzanthracene, perylene, 1,12-benzperylene, and coronene are the prominent polycyclic aromatic constituents of Recent marine sediments, whereas alkyl substituted polycyclic aromatic hydrocarbons predominate in the aromatic fractions from ancient rocks and crude oils [3, 20].

A permitted interpretation of the above analyses is: (1) the low concentrations of biologically abundant terpenes in sedimentary deposits suggest that these compounds are catabolically and chemically too active to escape the carbon cycle and to accumulate in appreciable quantities in sediments; (2) the low catabolic and chemical activities of alkanes make it possible for these compounds to be preferentially concentrated and preserved among the remnants of former life; and (3) most aromatic hydrocarbons are not formed within living cells, but the simplicities and locations of aromatic fractions in some Recent marine sediments suggest that anaerobes participate in the production of certain aromatic hydrocarbons. Sterols and isoprenoids are the probable biological precursors of many aromatic hydrocarbons. Abiotic production and alteration as well as physical redistribution may cause the compositional and concentrational differences between the aromatic compounds in Recent sediments and those in ancient rocks [3, 20].

Students of the origins of hydrocarbons disagree mainly about the origins of alkanes. Most scientists do not concur with the view that $>C_{15}$ alkanes in sediments are largely products of enzymatically controlled reactions. HUNT, PHILIPPI, BRAY and EVANS, and other investigators believe that the production of alkanes in "source rocks" is indicated by compositional and concentrational variations of alkanes in Recent and ancient sedimentary deposits, and they cite substantial analytical evidence in support of their belief [21]. The analyses of HUNT, presented in Table 1, show conclusively that many Cenozoic and Paleozoic rocks contain much higher concentrations of hydrocarbons than do Recent sediments, and the general compositional differences between crude oils and the organic extracts of "source rocks", as determined by HUNT, are shown in Fig. 1. Extensive data on the compositional and concentrational variations of hydrocarbons in Recent and ancient sedimentary deposits are also discussed by PHILIPPI and BRAY and EVANS [21].

Some of the disagreement on the role of "source rocks" in the formation of alkanes may be traced to problems associated with sampling methods and with distinguishing between compositional and concentrational changes produced by chemical or physical processes. Additional attention will be paid to the origins of alkanes after the identifications, quantitative assays, and biosynthesis of individual and homologous hydrocarbons have been considered.

Table 1. *Distribution of hydrocarbons and associated organic matter in Recent and ancient non-reservoir sediments* [21]

	No. of samples	Average composition and ranges (ppm)[a]		
		Hydrocarbon	Asphalt	Kerogen
Recent sediments[b]				
Mediterranean Sea	1	29	461	9,000
Gulf of Mexico	10	32 (12– 63)	275 (113– 790)	6,400 (3,800– 9,100)
Gulf of Batabano, Cuba	10	40 (15– 85)	575 (136–1,023)	17,100 (2,900–34,900)
Orinoco Delta, Venezuela	10	60 (27–110)	555 (283–1,355)	10,500 (7,500–14,700)
Lake Maracaibo, Venezuela	8	68 (24–116)	1,060 (266–1,600)	26,900 (13,700–41,500)
Cariaco Trench, Venezuela	16	105 (56–352)	1,250 (224–2,600)	25,000 (8,100–31,200)
Ancient sediments				
Shales	791	300	600	20,100
Carbonates	281	340	400	2,160

[a] Ranges for Recent sediments are shown in parentheses. In this and subsequent tables the hydrocarbon is defined as the organic matter extracted from the rock that contains only the elements, carbon and hydrogen. The asphalt is the soluble non-hydrocarbon material and the kerogen is the insoluble organic matter remaining in the rock.

[b] All samples are clay muds except those from the Gulf of Batabano, which are carbonate.

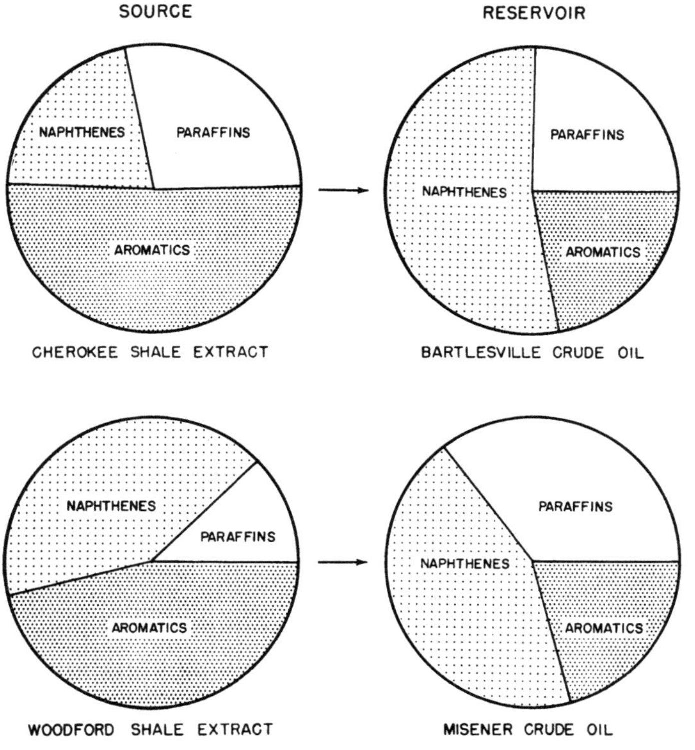

Fig. 1. Compositions of oils extracted from reservoir sands and from the presumed source rock

III. Hydrocarbons as Biological, Environmental, and Chemical Indicators

HILL and WHITEHEAD note that triterpenoids as well as steroids display a phylogenetic pattern. They report a progressive change in the steroid and triterpenoid contents of plants. Bacteria, possibly the simplest plant forms, contain neither steroids nor pentacyclic triterpenoids [22]. Blue-green algae, the most primitive form of algae which rival bacteria in their simplicity, are also devoid of sterols, whereas as stated by BLOCH all representatives of major algal divisions synthesize sterols [23]. Angiosperms, the dominant type of modern flora, produce a great variety of sterols and pentacyclic triterpenoids [22]. Based on the distribution of triterpanes within the plant kingdom, HILL and WHITEHEAD propose that the triterpanes in a Nigerian crude oil were more likely derived from land plants than marine organisms. They suggest simple plant forms or perhaps protozoa as sources of gammacerane in the Green River oil shale bitumen. Gammacerane is the parent hydrocarbon of tetrahymanol, the first pentacyclic triterpenoid alcohol isolated from an animal [24].

CLARK comprehensively reviewed the analyses of n-paraffins in biological materials [25]. The concentrations of odd-carbon number n-paraffins substantially exceed the concentrations of even-carbon number n-paraffins in the lipids of many organisms. This odd-carbon preference is usually most pronounced over specific carbon number ranges. n-Paraffins from many higher order plants display a strong odd-carbon preference in the C_{23} to C_{36} range, whereas neither n-pentadecane (n-C_{15}) nor n-heptadecane (n-C_{17}) and rarely n-nonadecane (n-C_{19}) are singly abundant in algal alkanes. EGLINTON *et al.*, propose that n-paraffins and other alkanes may be useful in chemical taxonomy [26], and CLARK and BLUMER [27] discuss n-paraffin distributions in marine organisms and marine sediments. The latter authors note that benthic algae may be grouped into two categories on the basis of their hydrocarbon distributions. n-C_{17} predominates in red and green algae, but n-C_{15} predominates in brown algae. A moderate predominance of an odd-carbon n-paraffin, n-C_{17}, was observed in only one planktonic algae, *Syracosphaera carterae*.

Limited data indicate that n-paraffin concentrations show environmental as well as taxonomic differences. CLARK and BLUMER observe that a sample of *Chondrus crispus* and a sample of the *Fucus* species of seaweed from Falmouth, Massachusetts, contained four times as much n-C_{19} relative to other n-paraffins as any other algae sample from another location. They suggest that this variation in the n-paraffin distributions of these Falmouth samples may be due to environment [27]. Results obtained in our laboratory and by other investigators are also indicative of an environmental variation in n-alkane distributions. We analyzed the lipids of cow manure from New Jersey to test the hypothesis that the catabolically less-active alkanes in plant lipids would be concentrated relative to fats and oils in the manure lipids. We found the n-paraffins from the manure were distributionally similar to plant n-paraffins, and n-nonacosane (n-C_{29}) was the most abundant alkane in our samples [20]. ORO *et al.* tested our hypothesis directly by analyzing pasture plants consumed and manure produced by cows in Texas. They observed a marked similarity in n-paraffin distributions from the plants and manure, but n-hentricontane (n-C_{31}) was the most abundant n-paraffin in their samples [28]. Similar variations have been observed in the concentrations of n-paraffin in Recent sediments. Data obtained in our laboratory show that n-C_{29} or n-C_{31} is present in the highest concentration in alkanes from various Gulf of Mexico sediments, whereas n-pentacosane (n-C_{25}), or n-heptacosane (n-C_{27}), is found in the highest concentrations in alkanes from lake sediments in Canada and the northern United States.

BLUMER, MULLIN and THOMAS [29] present a significant study of pristane, 2,6,10,14-tetramethylpentadecane, in the marine environment. They review the distribution of pristane in marine organisms. Their analysis of zooplankton, shows that the genus *Calanus*, which lives predominately in cold waters, contains ten times more pristane than other genera. Metabolic studies of *Calanus hyperboreus* under starvation conditions at four degrees Centigrade reveal that lipids serve as the main metabolic substrate. Because pristane is less utilized for energy than other lipid components, pristane concentrations increase in the lipid fractions of starving *Calanus*. BLUMER *et al.* believe that the metabolism of pristane by *Calanus* has assumed special importance. This buoyant hydrocarbon (density – 0.78) apparently permits *Calanus* to conserve energy that they would otherwise waste in swimming to maintain their position in the water column during periods of "diapause", and the key position of *Calanus* in the marine food chain suggests that *Calanus* may be a major primary source of pristane in the marine environment. They propose that the phytol group in the chlorophyll molecule, which *Calanus* obtain from ingested phytoplankton, is converted into pristane in three steps: hydrogenation of the double bond, oxidation of the alcohol to acid, and decarboxylation to the saturated hydrocarbon [29]. These reactions are:

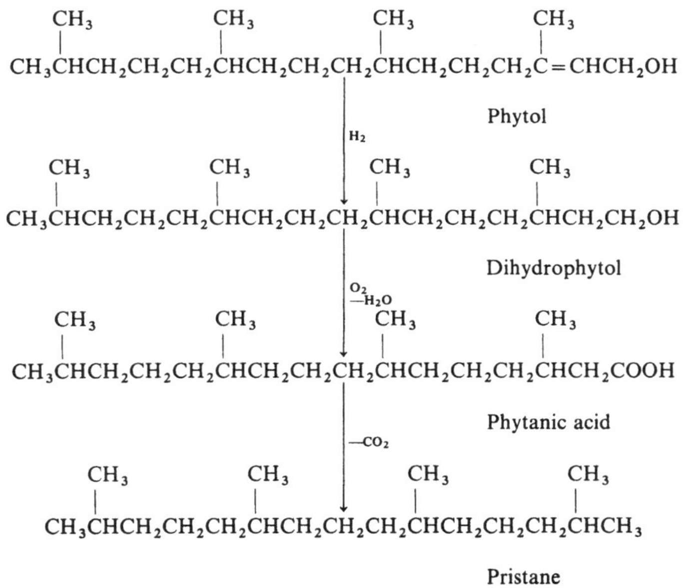

It may be noted that a conditionally and energetically favored conversion of phytol to pristane could be accomplished by hydrogenation and hydrogenolysis as follows [20]:

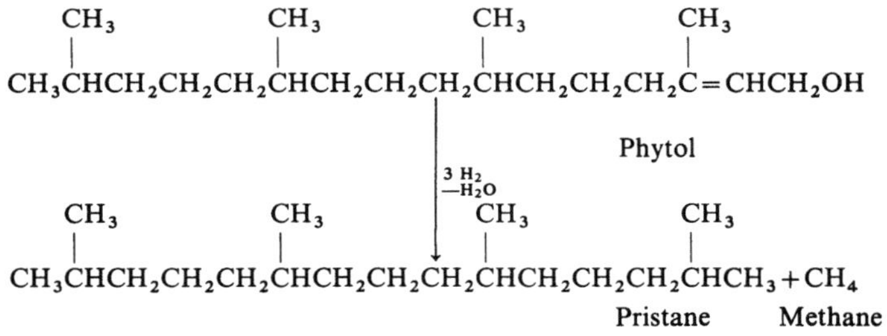

Table 2. *Biosynthesis of terpenoids and steroids* [30]

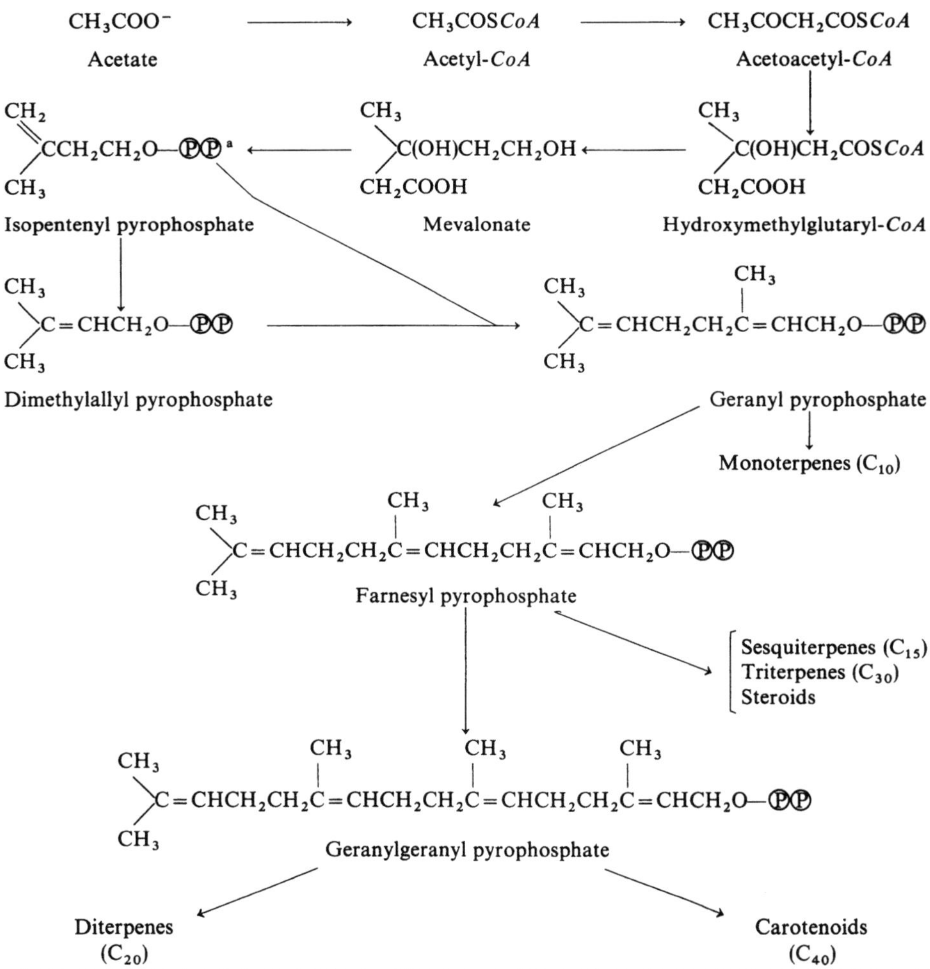

[a] O—ⓅⓅ indicates pyrophosphate group.

IV. Biosynthesis of Hydrocarbons and Related Compounds

Biological acids, alcohols, isoprenoid hydrocarbons, and presumably alkanes are biosynthesized from acetate units. The greater part of biological branched-chain and methyl- or isopropyl-substituted cyclic acids, alcohols, and olefins are either terpenoids or compounds derived from terpenoids, such as steroids, sapogenins, and plastoquinones. A number of biosynthetic steps are common to the formation of all terpenoids and genetically related compounds. Mevalonate, the coenzyme A, *CoA*, derivative of dihydroxy-3-methylpentanoic acid, is a key intermediate in terpene synthesis. Common metabolic intermediates are also involved in the biological production of linear acids and alcohols and probably n-alkyl-substituted cyclic acids and alcohols. Malonate, 1,3-propandioate, is a

Table 3. *Biosynthesis of fatty acids (animal tissues)* [31]

No. malonates reacting	No. carbon atoms per acid molecule	*de novo* synthesis						
x	$2x+2$	Acetyl-$CoA + x$(Malonyl-CoA) ↓						
7	C_{16}	Palmitic	$\xrightarrow{DS}$	Δ^9 (Palmitoleic)				
		↓ *de novo* or E		↓ E				
				Δ^{11} (cis-Vaccenic)				
8	C_{18}	Stearic	$\xrightarrow{DS}$	Δ^9 (Oleic)	$\xrightarrow{DS}$	$\Delta^{6,9}$		
		↓ E		↓ E		↓ E		
9	C_{20}	Arachidic		Δ^{11}		$\Delta^{8,11}$	$\xrightarrow{DS}$	$\Delta^{5,8,11}$
		↓ E		↓ E				
10	C_{22}	Behenic		Δ^{13}				
		↓ E		↓ E				
11	C_{24}	Lignoceric		Δ^{15}				
		Saturated acids		Monoenoic acids		Dienoic acids		Trienoic acids

E = elongation mechanism by addition of 2C from acetyl-*CoA* to preformed acids.
DS = desaturation mechanism by removal of 2H from preformed acids.

key intermediate in the latter syntheses. The two major reaction pathways outlined here indicate how terpenoid and linear compounds are made in organisms.

The biosynthetic pathways leading to fatty acids in plants and animals are similar. Palmitic (n-hexadecanoic) acid and palmitoleic (Δ^9 n-hexadecenoic) acid are the dominant acids in animal tissues, whereas stearic (n-octadecanoic), oleic (Δ^9 n-octadecenoic) and other C_{18} acids are dominant in plant tissues. The reason for this variation is not known, but it can be attributed to the specificity of the fatty acid synthetase systems. Animal synthetase releases free palmitic acid from the coenzyme A linkage by some deacylating mechanism, whereas plant synthetase releases stearic acid to the medium [31].

Terpenes are the most abundant biological hydrocarbons. Terpenes contain an integral number of isoprene $\left(\begin{matrix}C\\C\end{matrix}\!\!>\!C{-}C{=}C\right)$ skeletal units, but compounds composed of five and seven isoprene units are not found in organisms. Monoterpenes (C_{10}), sesquiterpenes (C_{15}), diterpenes (C_{20}), triterpenes (C_{30}), and tetraterpenes (C_{40}) are distributed through the plant kingdom.

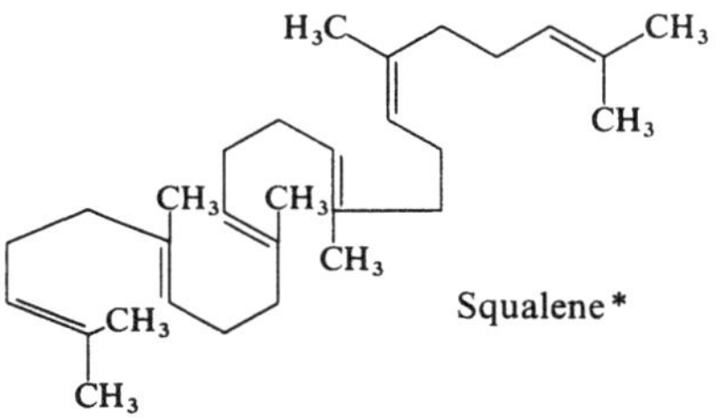

Squalene*

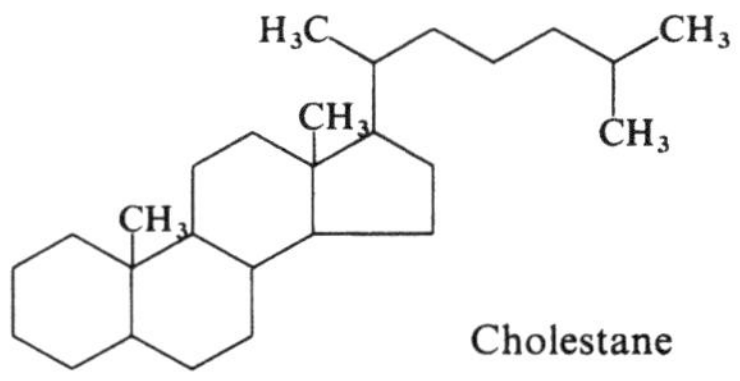

Cholestane

Squalene is the triterpene which is most commonly produced by animals. Steroids are biosynthesized in plants and animals from squalene. Cholesterol (a C_{27} steroid), the C_{24}–C_{27} bile acids, and the C_{18}–C_{19} sex hormones comprise the principal stereoidal constituents of animals. Plants contain a variety of C_{27} to C_{29} steroids, and the C_{29} steroids are the dominant type. Some terpenoids and steroids display strong physiological activities. Carotenoids (tetraterpenes) are photochemically active precursors of Vitamin A, a diterpene. Although olefinic hydrocarbons are a major fraction of the mono- and tetraterpenoids in plants, the preponderance of biological isoprenoids are alcohols, acids and multifunctional derivatives. A vast body of information is available on the biosynthesis and structures of steroids and isoprenoids [30].

n-Paraffins are the most abundant biological alkanes. Bacteria produce great quantities of methane, frequently called "marsh gas", and n-paraffins in the C_1 to C_{62} range are reported in organisms [32]. Published analyses indicate that C_2 through C_6 and C_{36} or larger n-alkanes are biologically less abundant than their C_1 and C_7 to C_{36} homologs. Pristane, an isoprenoid type alkane, is abundantly present in a variety of unicellular and multicellular marine and terrestrial animals [29]. Non-isoprenoid branched chain alkanes and olefinic and acetylenic hydrocarbons are found in plant waxes and seed lipids. Rose petal wax contains isoparaffins**. Anteiso** – in addition to isoparaffins – are constituents of tobacco leaf, wool, and sugar cane waxes [33]. Ethylene is released by ripening fruit. Coffee beans yield 1,3-butadiene, and polyolefinic linear hydrocarbons in the C_{13} to C_{28} carbon number range appear in olive oils. Dahlia, mugwort, dog grass and other members of the Compositae family produce allenes and acetylenes [18].

Metabolic studies of hydrocarbons are mainly limited to investigations of the structures, abundances, and isotopic compositions of hydrocarbons in organisms and to identifications of metabolites which are produced by organisms from hydrocarbon substrates. Most lines of evidence gathered from these studies suggest that hydrocarbons and nonhydrocarbons in biological lipids are bio-

* Squalene structure is drawn in this conformation to resemble a sterol configuration.

** Isoparaffins and anteisoparaffins are 2-methyl and 3-methyl substituted branched chain paraffins respectively.

synthesized and catabolized by similar enzyme systems. Isotopically, hydrocarbons and other lipid components have lower ^{13}C contents than the non-lipid components of organisms [34], and the metabolites of hydrocarbons are generally organic acids and alcohols. Normal and branched-chain alkanes and olefins are commonly converted to fatty acids by ω-oxidation. Dibasic acids, dihydric compounds, and hydroxy acids are normally products of biodegradations of cyclic hydrocarbons [35].

Investigations of the biosynthesis of alkanes are limited in scope. KLEIN finds that the non-saponifiable and non-steroidal or "hydrocarbon" fractions of lipids produced from 1-^{14}C-acetate, in a cell-free system from yeast, are radioactive [34].

Table 4. *Schematic representation of biosynthetic pathways for C_{29} compounds in Brassica oleracea* [36]

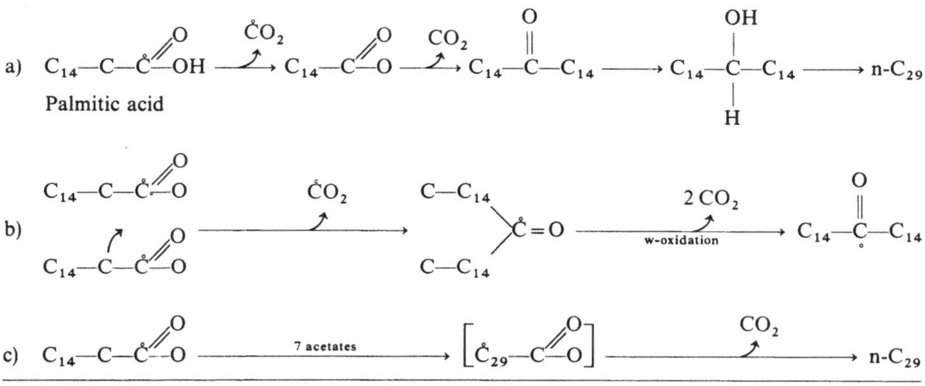

° = site of ^{14}C label.

Calculations, based on his analyses, show that the radioactivities of the hydrocarbon and nonhydrocarbon fractions of these lipids are equal. KOLATTUKUDY reviews the attempts to incorporate acetate into biological paraffins and presents a detailed study of the biosynthesis of n-nonacosane (n-C_{29}), 15-nonacosanone, and 15-nonaconsanol in cabbage and broccoli leaves, *Brassica oleracea*, in which these compounds are the principal constituents of the cuticular wax [36]. Three major pathways, which are shown in Table 4, have been proposed for the biosynthesis of n-C_{29} and its derivatives in this plant [36].

KOLATTUKUDY [36] employed a variety of ^{14}C-labeled intermediates and labeled 15-nonacosanone as well as inhibitors to investigate the precursor-product relationships and biosynthetic pathways of the C_{29} compounds in choppings and disks of broccoli leaves. The specific activities of the C_{29} compounds which were isolated from the leaves after administration of carboxyl (1-^{14}C) labeled acetate decreased from n-nonacosane to 15-nonacosanone to 15-nonacosanol, and labeled 15-nonacosanone was not metabolized by leaf disks. (1-^{14}C) C_{10} through C_{18} fatty acids were individually incorporated into the C_{29} compounds of leaf cuttings, and a sharp increase in the percent of label incorporated was observed on going from cuttings administered C_{10} to C_{14} fatty acids to those administered C_{16} to C_{18} fatty acids. Uniformly (U-^{14}C) and carboxyl (1-^{14}C)

labeled palmitic acid yielded C_{29} compounds of equivalent activities, and palmitic acid was more rapidly incorporated into the C_{29} fraction by the leaf cuttings than was n-pentadecanoic acid. Imidazole, which inhibits the α-oxidation of fatty acids, did not inhibit acetate incorporation into the C_{29} compounds of the leaves. 3-(4-chlorophenyl)-1,1-dimethylurea (CMU), apparently a potent inhibitor of photophosphorylation, limited the acetate incorporation into fatty acids more severely than acetate incorporation into the C_{29} compounds of broccoli leaves [36].

On the basis of these experimental results, pathways a and b of Table 4 are effectively ruled out, and KOLATTUKUDY presents the working hypothesis for the compartmentized biosynthesis of C_{29} compounds in *B. oleracea* which is shown diagrammatically in Table 5.

Table 5. *Biosynthesis of n-nonacosane in broccoli leaves* [36]

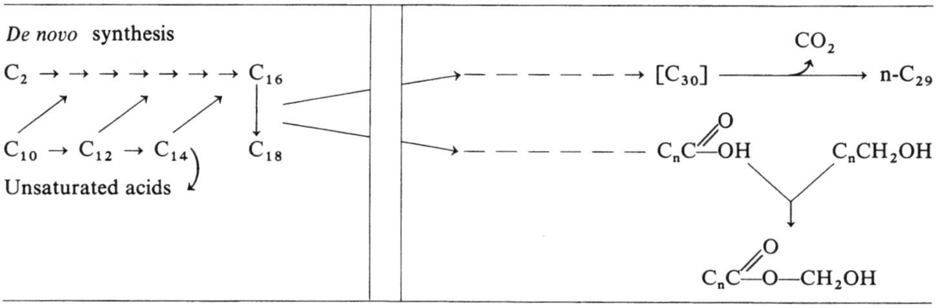

Although it may be logically assumed that alkanes generally are biosynthesized, as is n-nonacosane, in much the same ways as are structurally, isotopically, and metabolically related acids and alcohols, the findings of KOLATTUKUDY and analytical results discussed above clearly suggest certain non-equivalences in the anabolisms of alkanes and of acids and alcohols in biological systems. CLARK and BLUMER observe that the high concentrations of n-pentadecane and n-heptadecane cannot be correlated with the relative abundances of palmitic and stearic acids in different benthic algae [27] and I have previously reported marked distributional differences between alkanes in biological and sedimental samples and acids and alcohols in organisms [20]. These differences seemingly confirm that biological reasons exist for the production of alkanes, but our knowledge of why organisms make one homolog in preference to another is usually more incomplete than our knowledge of how the compound is made. In the case of alkanes, our dependencies on metabolic studies of acids and alcohols for an understanding of how and why alkanes are biosynthesized may be excessive.

The purpose here is not to deny that we may benefit from the insight which the vast body of information on the biosynthesis of acids and alcohols provides about the anabolic production of hydrocarbons, but rather to indicate the need for data that pertain to the metabolism of alkanes. This need may be most apparent for the non-isoprenoid type of cycloalkane. Only mass spectral type-analyses indicate the presence of n-alkyl substituted cyclopentanes and cyclohexanes in biological lipids [20]. These analyses, however, do suggest great structural resemblances between alkanes from organisms and Recent sediments.

Structural ties are also demonstrated for all types of alkanes in Recent and ancient sedimentary deposits [20], and identifications establish the presence of n-alkylcyclohexanes in ancient rocks [33]. Ample analytical evidence proves the widespread distributions of cyclopentyl and cyclohexyl hydrocarbons in petroleum, but seed oils of the family Flacourtiaceae are as yet held to be an exclusive source of cyclopentenyl acids in organisms.

JOHNS *et al.* recognize the dilemma concerning the origins of cyclohexyl substituted n-alkanes in Precambrian rocks. They suggest that mono-olefinic acids with double bonds in the 6 (Δ^6) or 7 (Δ^7) position may undergo cyclization and decarboxylation to form n-alkylcycloalkanes [33]. Their suggestion is noteworthy in that it directs attention to a testable hypothesis of hydrocarbon genesis.

BLOCH discusses the reactions that form unsaturated acids in organisms. Olefinic compounds are produced metabolically by dehydrogenation and dehydration of saturated compounds. Aerobic desaturation or dehydrations of long chain fatty acids normally yield Δ^9 mono-olefinic acids. Exceptions to the 9,10-specificity of oxidation desaturations are noted in *Mycobacterium phlei* and *Bacillus megatherium* which make Δ^{10} and Δ^5 monoenoic acids respectively. Anaerobic bacteria biosynthesize monoolefinic acids by successive additions of C_2 units to C_{10}- and C_{12}-β-hydroxy acids that are derived in the conventional way from acetyl-*CoA* and malonyl-*CoA*. Experimental results verify the pathways of the anaerobic synthesis of unsaturated acids. It is demonstrated that the successive elongations of 3-decenoate and 3-dodecenoate maintain the double bonds in the same positions relative to the methyl terminals of the fatty acid chains. Thus, 3-decenoate yields the Δ^7-C_{14}, Δ^9-C_{16}, and Δ^{11}-C_{18} acid series, whereas 3-dodecenoate forms the Δ^7-C_{16} and Δ^9-C_{18} acids [37]. WOLFF reports Δ^3-mono-olefinic acids in plants of the Compositae family [38].

As only Δ^7-C_{14} and Δ^7-C_{16} acids and no Δ^6-monoenoic acid appear commonly in biological lipids, polyolefinic acids may serve more suitably than mono-olefinic acids as non-isoprenoid cycloalkane precursors. One reason for favoring polyolefinic acids is that such precursors are more reactive and versatile than mono-olefinic acids, but bond systems of the divinyl-methane type found in most polyunsaturated fatty acids would not be expected to undergo cyclization reactions as do the conjugated bond systems in terpenoids. Our knowledge of the biosynthesis or biological precursors of n-alkyl substituted cycloalkanes remains sadly inadequate.

V. Origin and Preservation of Hydrocarbons

As suggested already, carbon compounds may provide the best means of tracing many natural processes which have occurred within most of the accessible regions of Earth during the geological past. The great potential of carbon compounds in paleoenvironmental and/or paleobiological investigations results from the unique chemical flexibility of carbon and the systematic variations in the chemical and physical properties of carbon compounds. At moderate temperatures and redox potentials, carbon compounds may be found that will undergo structural or distributional changes which precisely define the chemical or physical processes which are operative on these compounds. Also, because biochemical unity is a minimal manifestation of life, stable molecular remnants of life may retain their identities as biological products. It has been noted as well that fossil records and the conditional dependencies and relative efficiencies of metabolic and abiotic reactions leave little doubt that pre-existent organisms were effectively the sole source of the organic materials initially emplaced in sediments which have formed extant sedimentary rocks.

It seems generally agreed that biological lipids are the principal source of the C_{15} and larger hydrocarbons or their precursors. The evidence supporting this agreement is substantial, but the issue of how alkanes are derived from lipids is only partially resolved. Although the scarcity of fossil olefinic hydrocarbons and a consensus that aromatic hydrocarbons are derived from isoprenoids or steroids has removed major motivations for debate on the origins of these compounds, considerations of the geochemical significance of olefinic and aromatic concentrations in natural systems may provide some insight into the origins of alkanes.

In dealing with the general problem of the origins of fossil organic materials, one is forced to recognize certain limitations. Many biological molecules are either metabolically too reactive or chemically too unstable to be preserved in sediments at concentrations that exceed analytical blanks in laboratories or biological blanks in rocks. ABELSON has indicated the necessity for appraising molecular stabilities [39], and CLARK and BLUMER justifiably point to contamination as a source of error in organic geochemical research [27]. It is in regard to the stabilities or activities and minimal concentrations of geochemically important organic compounds that we may now consider olefinic and aromatic hydrocarbons.

Of the major types of hydrocarbons in organisms, olefins are the most abundant and aromatic hydrocarbons are the least abundant. Although abundance alone does not define the metabolic rates of biological compounds, we may accept for the purposes at hand that it does. Thus, for biological molecules generally, we may assume that the series carbohydrates, proteins, fatty acid, alcohols, olefins, alkanes, and aromatic hydrocarbons is arranged in order of decreasing metabolic activity. Without restating previous discussions, the low concentrations of olefins in fossil organic matter suggest that the geochemical potentials of compounds which are metabolically (catabolically) and chemically more reactive than olefins are limited, whereas the structural ties which relate the relatively abundant aromatic hydrocarbons in sediments to isoprenoids and steroids suggest that abiotic alterations that do not destroy the biological identities of biological compounds may enhance their geochemical potentials. Because alkanes are catabolically and chemically less reactive than olefins, alkanes may have the greatest geochemical potential of all biological molecules.

Returning to the issue of whether $>C_{15}$ alkanes in ancient rocks are mainly altered or unaltered molecular remnants of ancient life, we may assume that the ubiquity, distributions, ^{14}C contents, physical properties, and structures of alkanes in organisms, soils, and Recent marine sediments establish beyond reasonable doubt that some biosynthetic alkanes escape the carbon cycle. Furthermore the energies of degradations of specific biological alkanes, determined by

ABELSON [39], as well as their presences in rocks of essentially all ages confirm that a major portion of these hydrocarbons can retain their structural integrities in most sedimentary environments. Thus, *the issue here is* apparently not whether some alkanes are molecular remnants of former life but rather *how many or what proportion of the alkanes in ancient rocks were derived as alkanes from biological lipids.*

In retrospect, it seems apparent that analytical capabilities have strongly influenced our thinking on the origins of hydrocarbons generally and alkanes specifically. The "source rock" concept was developed prior to 1920, many years before analytical techniques and methods were available for the detection and identification of alkanes in most rocks. This concept logically presumed that the vast quantities of hydrocarbons found in petroleum reservoirs were formed in the only other rocks then known to contain hydrocarbons, namely organic-rich shales or "source rocks". Even today, it is not possible to refute the argument that biological remnants are converted into $>C_{15}$ alkanes within sedimentary deposits. Certainly the aromatic hydrocarbons must be derived mainly in this manner, and disproportionation reactions which may provide the energetically favored ways of changing terpenoids and steroids into aromatic hydrocarbons would also yield cycloalkanes [20, 40].

Nonetheless, I believe that the emphasis on the chemical conversions of acids and alcohols into alkanes and the alteration or maturation of alkanes is excessive. Alkanes in organisms and in rocks are effectively shielded from chemical reagents by other organic compounds which are both more reactive and abundant than alkanes. Many of the analytical data on the hydrocarbon constituents of rocks which have been cited as evidence of the production or alteration of alkanes may be interpreted equally as well in terms of the physical redistribution of these compounds. JOHNS *et al.* note, for example, that the decrease in the odd-carbon preference of n-paraffins which BRAY and EVANS present as an indicator of "source rocks" may merely reflect the variations in n-paraffin distributions in biological sources [33]. CLARK and BLUMER show that n-paraffins from cultured plankton are distributed similarly to n-paraffins in some crude oils [27].

Some deductions concerning the origins of alkanes may be drawn from a study of the gas-liquid chromatograms shown in Figs. 2–9. Fig. 2 presents a GLC of alkanes from bat guano. It is assumed that the latter alkanes are largely a composite of paraffins from insects which the bats consumed. The GLC in Fig. 2 is markedly similar to the GLC's of alkanes from pasture plants and cow manure [28], but it does not closely resemble the GLC of bacterial alkanes from *Vibrio ponticus* shown in Fig. 3. Comparisons of the chromatographic peaks in Figs. 2–9 with those of reference n-paraffins, pristane, and phytane in Fig. 9 indicate that n-paraffins are commonly the most abundant type of alkanes in organisms and sedimentary deposits and that pristane and phytane are usually present in alkanes from ancient rocks or crude oils. Pristane and phytane peaks appear, also, in Figs. 3 and 4, however it should be noted that contaminants may be the source of the phytane peaks in the alkanes from *Vibrio ponticus* and Recent sediments. CLARK and BLUMER note that phytane is not detected in alkanes which they have isolated from algae and Recent sediments. As they state, the apparent absence of phytane in their samples is geochemically significant [27]. The ubiquity of phytane

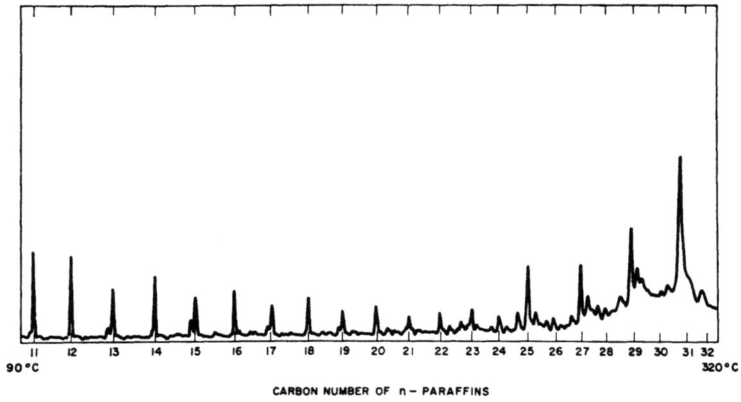

Fig. 2. Gas-liquid chromatogram of alkanes from bat guano. Age $\cong 5 \times 10^3$ years. Apiezon L-capillary column

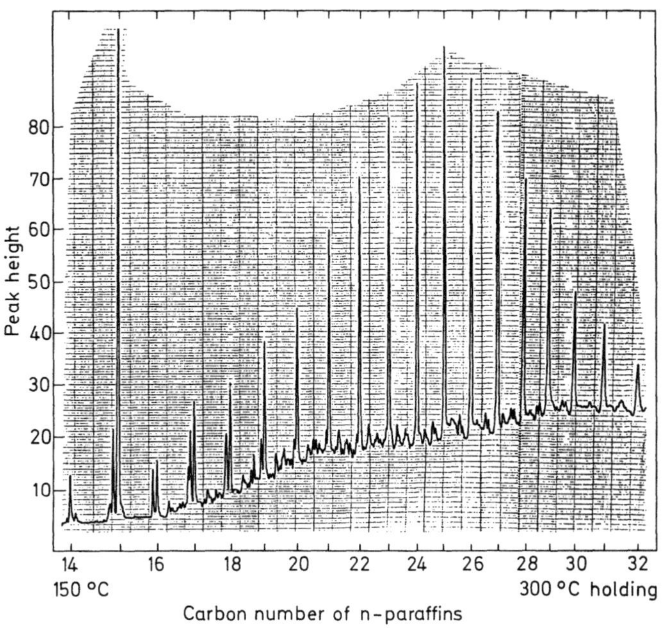

Fig. 3. Gas-liquid chromatogram of alkanes from *Vibrio ponticus*. Age <20 years. Apiezon L-capillary column. Sample supplied by T. S. OAKWOOD, Pennsylvania State University

in ancient rocks and its established absence in organisms would confirm either that phytane is produced in sediments or that extant organisms do not make phytane as did some of their progenitors.

Although contamination may explain the "phytane" peaks in Figs. 3 and 4, these GLC's indicate much the same distributions of n-paraffins that CLARK and BLUMER found for n-paraffins from some algae and Recent sediments [27], and a comparison of these distributions with those of the n-paraffins from the Eocene sediments and crude oils in Figs. 5–7 show, as JOHNS *et al.* have noted [33], that

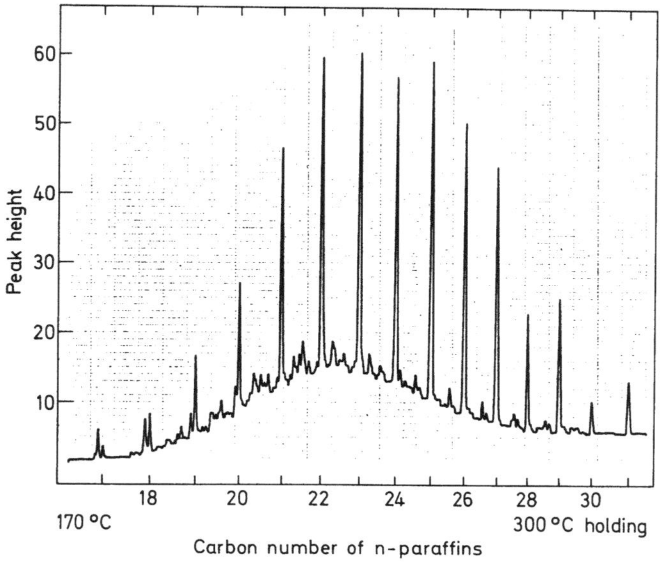

Fig. 4. Gas-liquid chromatogram of alkanes from Recent marine sediment. Age $\cong 5 \times 10^3$ years. Apiezon L-capillary column. Sample supplied by Jersey Production Research Company, Tulsa, Oklahoma

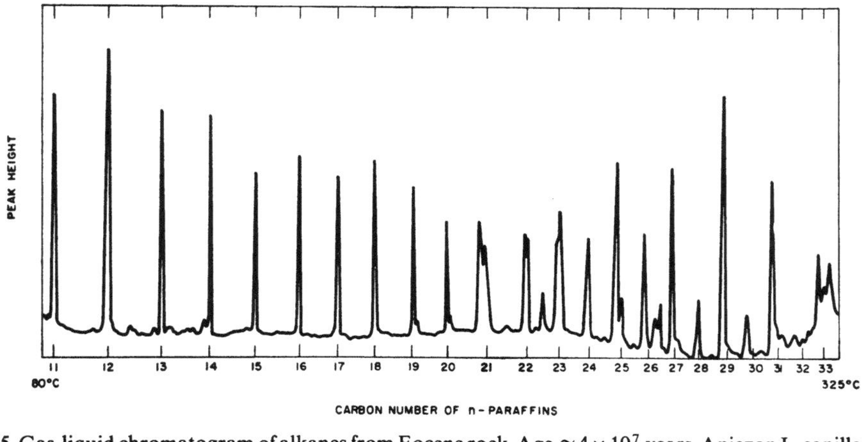

Fig. 5. Gas-liquid chromatogram of alkanes from Eocene rock. Age $\cong 4 \times 10^7$ years. Apiezon L-capillary column

n-paraffin distributions may not serve as reliable indicators of "source rocks". This comparison shows specifically that the distributions of $> C_{15}$ n-paraffins in *Vibrio ponticus* and crude oils are similar, whereas the distributions of n-paraffins from bat guano resemble those of n-paraffins from Eocene rocks more closely than those of n-paraffins from some Recent sediments.

Mass spectral type analyses of biological and sedimental alkanes are presented in Table 6. These analyses indicated approximately the same mean compositions for $> C_{15}$ alkanes from organisms and sediments generally, but paraffins and mono- and dicycloalkanes appear more concentrated in the alkanes from organ-

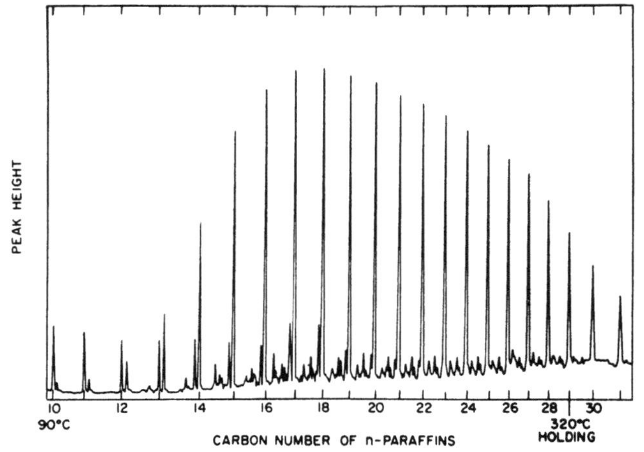

Fig. 6. Gas-liquid chromatogram of alkanes from Cretaceous crude oil. Age $\cong 1 \times 10^8$ years. Apiezon L-capillary column

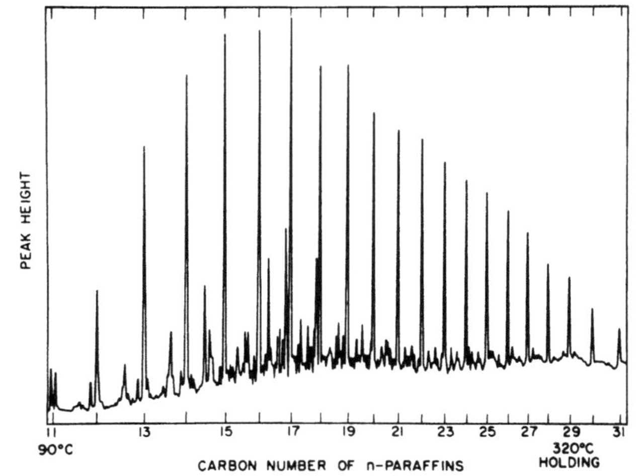

Fig. 7. Gas-liquid chromatogram of alkanes from Cambrian crude oil. Age $\cong 5 \times 10^8$ years. Apiezon L-capillary column

isms and other sedimentary deposits. PHILIPPI discusses the systematic changes in compositions and concentrations of alkanes that result from the oil generation process [21]. The large standard deviations recorded in Table 6 for the analyses of alkanes from various sources and the GLC's in Figs. 2–8 may partially explain why a diversity of opinions exist about the origins of alkanes. It seems that either chemical alterations or physical redistributions of biological products may be indicated by different portions of the data under consideration. I prefer to stress the affects of physical processes rather than chemical alterations on alkane compositions because I believe hydrocarbon movements exert a greater control than do chemical reactions on the concentrations and distributions of $> C_{15}$ alkanes in rocks.

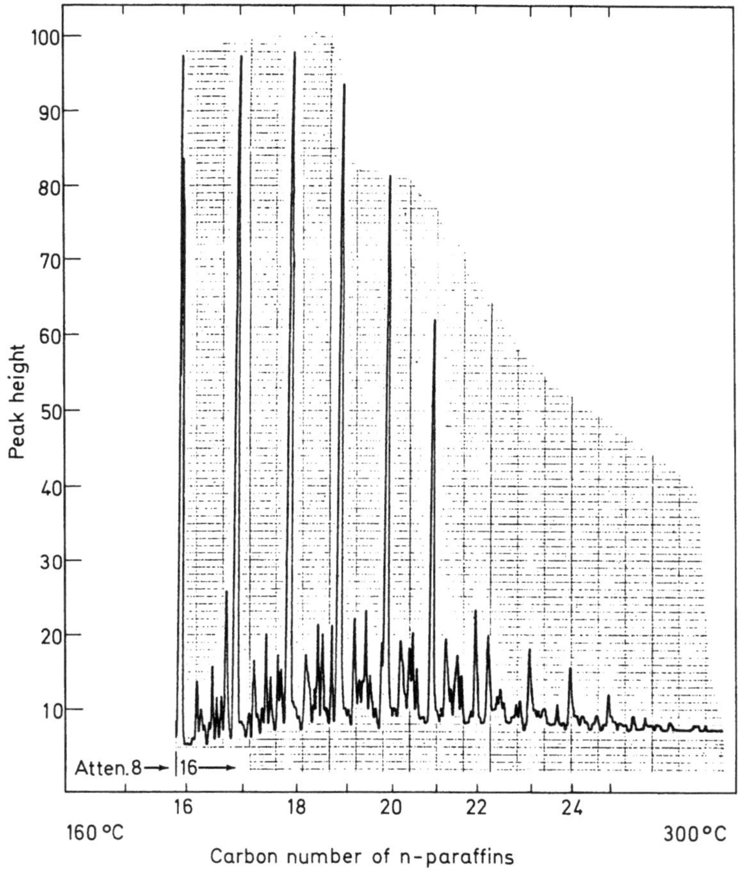

Fig. 8. Gas-liquid chromatogram of alkanes from Precambrian crude oil. Age $\cong 1 \times 10^9$ years. Apiezon L-capillary column

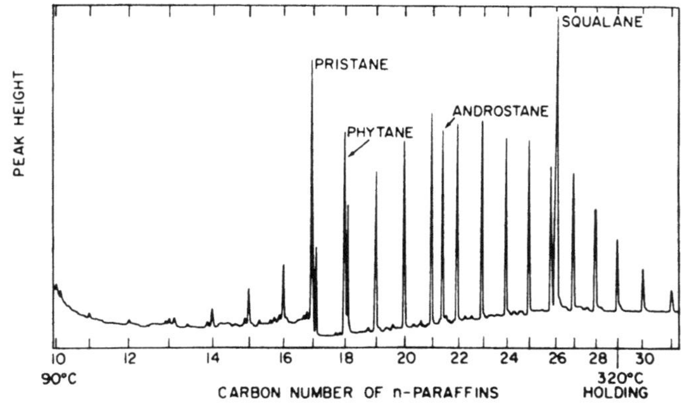

Fig. 9. Gas-liquid chromatogram of reference n-paraffins, pristane, phytane, androstane, and squalane. Apiezon L-capillary column

Many of the data that are cited as evidence for the generation of $>C_{15}$ alkanes in "source rocks" may also be indicative of hydrocarbon migrations. It is not unreasonable to assume that the redistributions of hydrocarbons which lead to the emplacement of natural gas and petroleum in reservoir rocks may enhance the concentrations of some hydrocarbons in organic-rich shales, and one would expect the compositions of the hydrocarbons absorbed on these organophilic rocks to resemble the mobile hydrocarbon-rich or petroleum phases in sedimentary deposits. Neither is the fact that the concentrations of $>C_{15}$ alkanes in early Quaternary to Paleozoic rocks are several times greater than the concentrations of alkanes in Recent sediments conclusive evidence of the production of alkanes

Table 6. *Mass spectrometric type analyses of the saturated hydrocarbons (n-C_7 eluates from silica gel columns) from crude oils, organisms and Recent and ancient sediments*

Sample	Percents of saturated hydrocarbon types[a]						
	Paraffins	Cycloalkanes					
	σ	1-ring σ	2-ring σ	3-ring σ	4-ring σ	5-ring σ	6-ring σ
Organisms (11)	24.9 (6.2)	27.5 (2.8)	14.8 (1.4)	11.1 (2.5)	10.0 (2.1)	5.8 (2.6)	5.9 (2.6)
Sediments:							
Recent marine (12)	21.6 (6.2)	23.6 (2.5)	17.3 (2.4)	16.7 (1.1)	11.6 (1.5)	5.9 (1.0)	3.3 (0.6)
Ancient (19)	20.6 (6.0)	22.2 (3.6)	18.5 (2.9)	14.7 (3.2)	11.8 (3.2)	6.2 (1.2)	6.1 (2.7)
Crude oils (5)	26.1	28.5	20.2	13.0	7.6	3.5	2.8

[a] Percentages given for samples are averages of the number of samples shown in parentheses to the right of the sample designations. Standard deviations in the percents of different types of hydrocarbons in the various samples are presented to the right of percentage values.

in these rocks. The sedimentation cycle causes a substantial turnover of terrestrial rocks, and the movements of hydrocarbons from rocks subjected to high temperatures as well as the migrations of hydrocarbons during petroleum-forming processes may enhance the alkane concentrations of many rocks of intermediate geological ages. Evidence that certain very old rocks have lost alkanes is provided by the low alkane contents of many Precambrian rocks [41].

It is obvious that some alkanes are produced or altered in ancient rocks. Adamantane, tricyclo $(3,3,1,1^{3,7})$-decane, is not found in organisms, but it is widely distributed in crude oils. BOGOMOLOV shows that heating petroleum with clay increases the adamantane content of the oil. He proposes that adamantane is a rearrangement product of monoterpenes [42]. Undoubtedly there are other alkanes in ancient rocks that are products of rearrangements or degradations of terpenes. All the C_{16} through C_{22} hydrocarbons in the same isoprenoid series as pristane are found in ancient rock samples. Pristane (C_{19}) and phytane (C_{20}) are commonly the most abundant and 2,6,10-trimethyltetradecane (C_{17}) is usually the least abundant of the C_{16}–C_{22} isoprenoid alkanes in rocks. JOHNS *et al.* [33] and MCCARTHY and CALVIN [43] suggest that the C_{16} to C_{19} isoprenoid alkanes in rocks may be formed by cleavages of bonds in compound with the skeletal structure of phytol. If these cleavages occur near the skeletal end which contains the hydroxyl group in phytol, two cleavages are required to form the C_{17} isoprenoid as shown:

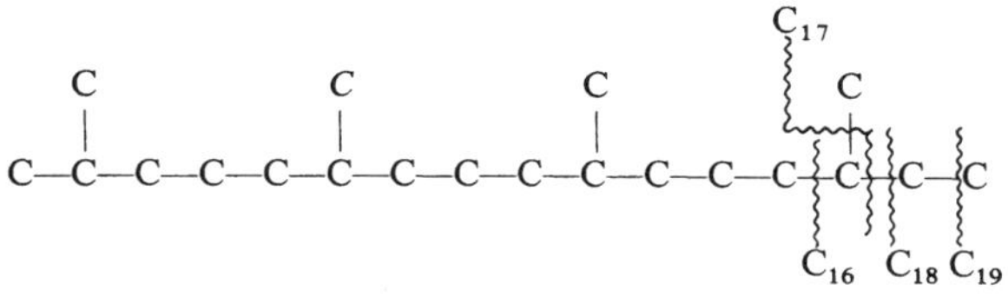

Since the C_{16}, C_{18}, and C_{19} isoprenoids require only one cleavage, JOHNS *et al.* [33] and MCCARTHY and CALVIN [43] suggest that their abundances in rocks may exceed the abundance of the C_{17} isoprenoid because it is more likely that one bond should be broken than two.

Irrespective of the mode of formation of the C_{16} to C_{22} isoprenoids in rocks, the low or negligible concentrations of the C_{17} isoprenoid in fossil alkanes clearly indicates that pristane, phytane, and the C_{21} isoprenoids have not been extensively degraded to the C_{17} isoprenoid because the cleavage of one or more single bonds in any of the C_{18} to C_{22} isoprenoids could yield the C_{17} isoprenoid. Similar and complementary evidence of the stabilities of alkanes in sedimentary environments is provided by the analyses of MARTIN, WINTERS, and WILLIAMS. Their data show that certain Ordovician crude oils, which are among the oldest oils produced commercially, contain much higher concentrations of odd-carbon number than even-carbon number C_{10} to C_{20} n-paraffins [44]. Since degradations of n-paraffins would statistically produce approximately equivalent amounts of odd- and even-carbon number products, the strong preponderance of odd-carbon number n-paraffins in the Ordovician oils suggests that degradations have not destroyed the apparently biologically controlled distributional pattern of the n-paraffins in these 4×10^8 year old samples.

It is important to note that the retentions of distributional patterns of specific alkanes which appear in approximately the same concentrations in rocks of different geologic ages places certain limitations either on the kinds or amounts of alkanes formed in ancient sedimentary basins. As suggested previously in this chapter, the great compositional differences between the principal organic constituents of rocks and organisms probably results mainly from the extensive abiotic alterations that most biological remnants undergo in sediments. Thus, these compositional differences may be viewed as evidence of the inability of abiotic reactions to duplicate biological products in natural systems. If we may accept this inability for the relatively stable alkanes as well as for less stable biological compounds, the general distributional and concentrational uniformities observed for specific alkanes in rocks of different geologic ages could not pertain if abiotic reactions were forming or altering a substantial portion of fossil alkanes.

In conclusion, the potentials of alkanes in chemical taxonomy, paleobiology, and geological research cannot be fully realized without extensive additional research. As EGLINTON and HAMILTON show by their investigation of leaf epicuticular wax, the development of these potentials will require interaction between students of biology, biochemistry, chemistry, physics, and geology [45]. Alkanes are compounds with a bright future and apparently a definable history.

References

1. MAIR, B. J., and J. L. MARTINEZ-PICO: Composition of the trinuclear aromatic portion of the heavy gas oil and light lubricating distillate. Proc. Am. Petrol. Inst. **42**, 173 (1962).
2. HEALD, H. C.: Fundamental research on the occurrence and recovery of petroleum 1952–1953. Baltimore: The Lord Baltimore Press.
   LINK, F. A.: Whence came the hydrocarbons. Bull. Am. Assoc. Petrol. Geologists **41**, 1387 (1957).
3. MEINSCHEIN, W. G.: Origin of petroleum. Bull. Am. Assoc. Petrol. Geologists **43**, 925 (1959).
   – Significance of hydrocarbons in sediments and petroleum. Geochim. Cosmochim. Acta **22**, 58 (1961).
4. –, E. S. BARGHOORN, and J. W. SCHOPF: Biological remnants in a Precambrian sediment. Science **145**, 262 (1964).
   EGLINTON, G., P. M. SCOTT, T. BELSKY, A. L. BURLINGAME, and M. CALVIN: Hydrocarbons of biological origin from a one-billion-year-old sediment. Science **145**, 263 (1964).
   BARGHOORN, E. S., W. G. MEINSCHEIN, and J. W. SCHOPF: Paleobiology of Precambrian shale. Science **148**, 461 (1965).
   ORO, J., D. W. NOONER, A. ZLATKIS, S. A. WIKSTROM, and E. S. BARGHOORN: Hydrocarbons in a sediment of biological origin about two billion years ago. Science **184**, 77 (1965).
   CLOUD, P. E., JR., J. W. GRUNER, and H. HAGEN: Carbonaceous rocks of the Soudan Iron Formation. Science **148**, 1718 (1965).
   BELSKY, T., R. B. JONES, E. D. MCCARTY, A. L. BURLINGAME, W. RICHTER, and M. CALVIN: Evidence of life processes in a sediment two and a half billion years old. Nature **206**, 446 (1965).
   BURLINGAME, A. L., P. HAUG, T. BELSKY, and M. CALVIN: Occurrence of biogenic steranes and pentacyclic triterpanes in an Eocene shale and in an Early Precambrian shale. Proc. Nat. Acad. Sci. U.S. **54**, 406 (1965).
   EGLINTON, G., and M. CALVIN: Chemical fossils. Sci. Am. **216**, 32 (1967).
5. NAGY, B., W. G. MEINSCHEIN, and D. J. HENNESSY: Mass spectroscopic analysis of the Orgueil meteorite: evidence for biogenic hydrocarbons. Ann. N.Y. Acad. Sci. **93**, 25 (1961).
   Meteorite hydrocarbons and extraterrestrial life. Ann. N.Y. Acad. Sci. **93**, 658 (1962).
   ANDERS, EDWARD: Meteorite hydrocarbons and extraterrestrial life. Ann. N.Y. Acad. Sci. **93**, 649 (1962).
   MEINSCHEIN, W. G., B. NAGY, and D. J. HENNESSY: Evidence in meteorites of former life. Ann. N.Y. Acad. Sci. **108**, 553 (1963).
   BRIGGS, M. H., and G. MAMIKUNIAN: Organic constituents of the carbonaceous chondrites. Space Sci. Rev. **1**, 647 (1963).
   MEINSCHEIN, W. G.: Hydrocarbons in terrestrial samples and Orgueil meteorite. Space Sci. Rev. **2**, 653 (1963).
   UREY, H. C.: Biological material in meteorites; a review. Science **151**, 157 (1966).
6. FOX, S. W.: Biology and the exploration of mars (C. S. PITTENDRIGH, WOLF VISHNIAC, and J. P. T. PEARMAN, eds.). Publ. No. 1296. Washington, D.C.: National Academy of Sciences National Research Council 1966.
7. WALD, GEORGE: Origin of optical activity. Ann. N.Y. Acad. Sci. **69**, 352 (1957).
8. MORRISON, P.: A thermodynamic characterization of self-reproduction. Rev. Mod. Phys. **36**, 517 (1964).
9. WIGNER, E. P.: The logic of personal knowledge (Festschrift for Michael Polanyi). London: Routledge and Kegan Paul 1961.
10. WALD, GEORGE: The origins of life. Proc. Nat. Acad. Sci. U.S. **52**, 595 (1965).
11. JUKES, T. H.: Some recent advances in studies of the transcription of the genetic message. Advanc. Biol. Med. Phys. **9**, 1 (1963).
12. FITCH, W. M., and E. MORGOLIASH: Construction of phylogenetic trees. Science **155**, 279 (1967).
13. OPARIN, A. I.: The origin of life. New York: The Macmillan Co. 1938.
   – Origin of life on earth. London: Pergamon Press 1959.
   MILLER, S. L.: A production of amino acids under possible primitive earth conditions. Science **117**, 528 (1953).
   –, and H. C. UREY: Organic compound synthesis on the primitive earth. Science **130**, 245 (1959).
14. BARGHOORN, E. S., and J. W. SCHOPF: Microorganisms three billion years old from the Precambrian of South Africa. Science **152**, 758 (1966).

ORO, J., and D. W. NOONER: Aliphatic hydrocarbons in Precambrian rocks. Nature **213**, 1082 (1967).
15. BARGHOORN, E. S.: Degradation of plant materials and its relation to the origin of coal. Second Conference on the Origin and Constitution of Coal, Crystal Cliff, Nova Scotia **181** (1952).
VALLENTYNE, J. R.: Solubility and the decomposition of organic matter in nature. Arch. Hydrobiol. **58**, 423 (1962).
16. ROBINSON, ROBERT: Duplex of petroleum. Nature **199**, 113 (1963).
PONNAMPERUMA, CYRIL, and K. PERING: Possible abiogenic origin of some naturally occurring hydrocarbons. Nature **209**, 979 (1966).
17. MEINSCHEIN, W. G.: Origin of petroleum. Paper presented before Southwest Regional Meeting Am. Chem. Soc., 4–6 December, 1957, Tulsa, Oklahoma.
Preliminary Proposal for Government Contract Research on Development of Hydrocarbon Analyses as a Means of Detecting Life in Space, Esso Research and Engineering Company, Submitted to NASA, 31 January 1962.
18. GERARDE, H. W., and D. F. GERARDE: The ubiquitous hydrocarbons. Association of Food and Drug Officials of the United States, vols. XXV and XXVI, 1961 and 1962.
19. BLOCH, KONRAD: Evolving of genes proteins. Symposium Rutgers State University 1964 (1965).
20. MEINSCHEIN, W. G., and G. S. KENNY: Analyses of a chromatographic fraction of organic extracts of soils. Anal. Chem. **29**, 1153 (1957).
– Living things – major producer of petroleum hydrocarbons. Annual Meetings Geological Society of America, Nov. 2, 1960.
Origin of petroleum, in: Enciclopedia del petrolia e dei gas Naturali. Rome: Press of L. Instituto Chimico dell'Università Roma (in press).
Carbon compounds in terrestrial samples and the Orgueil meteorite. Science and Space Research III, 165 (1965).
Quarterly Reports, Contract No. NASw-508, April 1 and July 1, 1963.
Benzene extracts of the Orgueil meteorite. Nature **197**, 833 (1963).
21. HUNT, J. M.: Distribution of hydrocarbons in sedimentary rock. Geochim. Cosmochim. Acta **22**, 37 (1961).
PHILIPPI, G. T.: On the depth, time and mechanism of petroleum generation. Geochim. Cosmochim. Acta **29**, 1021 (1965).
BRAY, E. E., and E. D. EVANS: Hydrocarbons in non-reservoir-rock source beds. Bull. Am. Assoc. Petrol. Geologists **49**, 248 (1965).
22. HILLS, I. R., and E. V. WHITEHEAD: Triterpanes in optically active petroleum distillates. Nature **209**, 977 (1966).
Triterpanes in petroleum. Paper presented at Summer Meeting of The American Petroleum Institutes Research Project No. 60, Laramie, Wyoming, July 1966.
HILLS, I. R., E. V. WHITEHEAD, D. E. ANDERS, and W. E. ROBINSON: An optically active triterpane, grammacerane in Green River, Colorado, oil shale bitumen. Chem. Comm. **1966**, 752.
23. STANIER, R. Y., and C. B. VAN HEIL: The concept of a bacterium. Arch. Microbiol. **42**, 17 (1963).
24. TSUDA, Y., A. MORIMOTO, T. SANO, Y. INUBUSHI, F. B. MALLORY, and J. T. GORDON: Synthesis of tetrahymanol. Tetrahedron Letters **1965**, 1427.
25. CLARK, R. C., JR.: Occurrence of normal paraffin hydrocarbons in nature. Technical Report Reference No. 66–34, Woods Hole Oceanographic Institution, Woods Hole, Massachusetts, 1966.
26. EGLINTON, G., A. G. GONZALES, R. J. HAMILTON, and R. A. RAPHAEL: Hydrocarbon constituents of the wax coatings of plant leaves. Phytochemistry **1**, 89 (1962).
27. CLARK, R. C., JR., and M. BLUMER: Distribution of paraffins in marine organisms and sediment. Limnol. Oceanog. **12**, 79 (1967).
28. ORO, J., D. W. NOONER, and S. A. WIKSTROW: Paraffinic hydrocarbons in pasture plants. Science **147**, 870 (1965).
29. BLUMER, M., M. M. MULLIN, and D. W. THOMAS: Pristane in the marine environment. Helgolaender Wiss. Meeresuntersuch. **10**, 187 (1964).
30. GOODWIN, T. W.: Biosynthetic pathways in higher plants (J. B. PRIDHAM and T. SWAIN, eds.). London and New York: Academic Press 1965.
WOLSTENHOLME, G. E. W., and MAEVE O'CONNOR: Ciba Foundation Symposium on The Biosynthesis of Terpenes and Sterols. London: J. & A. Churchill, Ltd. 1959.

31. WAKIL, S. J., and P. K. STUMPF, in: Metabolism and physiological significance of lipids. London and New York: John Wiley & Sons, Ltd. 1964.
32. KOONS, C. B., G. W. JAMIESON, and L. S. CIERESZKO: Normal alkane distribution in marine organisms; possible significance to petroleum origin. Bull. Am. Assoc. Petrol. Geologists **49**, 301 (1965).
33. JOHNS, R. B., T. BELSKY, E. D. MCCARTHY, A. L. BURLINGAME, PAT HAUG, H. K. SCHNOES, W. RICHTER, and M. CALVIN: The organic geochemistry of ancient sediments, 1–55. Part II, Technical Report on NsG 101-61, Series No. 7, Issue No. 8. Berkeley, California: University of California Press 1966.
34. PARK, R. B., and S. EPSTEIN: Carbon isotope fractionation during photosynthesis. Plant Physiol. **36**, 133 (1961).
KLEIN, H. P.: Some observations on a cell free lipid synthesizing system from Saccharomyces cerevisiae. J. Bacteriol. **73**, 530 (1957).
35. MCKENNA, E. J., and R. E. KALLIO: The biology of hydrocarbons. Ann. Rev. Microbiol. **19**, 183 (1965).
JOHNSON, M. J.: Utilization of hydrocarbons by micro-organisms. Chem. Ind. **1964**, 1532.
36. KOLATTUKUDY, P. E.: Relation of fatty acids to wax in Brassica oleracea. Biochemistry **5**, 2265 (1966).
37. BLOCK, KONRAD: The control of lipid metabolism (J. K. GRANT, ed.). London and New York: Academic Press, Inc. 1963.
38. WOLFF, I. A.: Seed lipids. Science **154**, 1140 (1966).
39. ABELSON, P. H.: Researches in geochemistry (P. H. ABELSON, ed.). London and New York: John Wiley & Sons, Inc. 1959.
40. FIESER, L. F., and MARY FIESER: Organic chemistry. Boston: D. C. Heath & Co. 1944.
41. MEINSCHEIN, W. G.: Soudan formation: organic extracts of Early Precambrian rocks. Science **150**, 601 (1965).
42. BOGOMOLOV, A. I.: Correlation of the constituents of petroleum with their organic origin. Tr. Vses. Neftegaz. Nauch.-Issled. Geologorazved. Inst. **1964**, 10.
43. MCCARTHY, E. D., and M. CALVIN: The isolation and identification of the C_{17} isoprenoid saturate hydrocarbon, 2,6,10-trimethyltetradecane: squalane as a possible precursor. Abstract for ACS Meeting, Petroleum Chemistry Division, New York, September 1966.
44. MARTIN, R. L., J. C. WINTER, and J. A. WILLIAMS: Distributions of n-paraffins in crude oils and their implications to origin of petroleum. Nature **199**, 110 (1963).
45. EGLINTON, G., and R. J. HAMILTON: Leaf epicuticular waxes. Science **156**, 1322 (1967).

CHAPTER 14

Fatty Acids and Alcohols

P. L. PARKER
University of Texas,
Marine Science Institute at Port Aransas,
Port Aransas, Texas

Contents

I. Introduction

The fatty acids continue to be one of the most rewarding types of organic compounds studied by geochemists. Almost without exception every research group concerned with fatty acid geochemistry has produced qualitatively new data and relationships. Several factors make fatty acids favorable organic molecules for geochemical study; they are major components of most organisms, many of them have the intrinsic chemical stability to persist for geologically long periods of time, and sufficient chemically-different types of carboxylic acid occur naturally to carry useful information about their origin and transformations.

Very little is known about the geochemical occurrence of fatty alcohols. The straight chain alcohols occur less frequently and generally in much lower concentrations in the biosphere than do the corresponding acids. Nevertheless, they should have an interesting geochemistry, if for no other reason than that they are relatively rare and may be indicators of special environments, such as lakes receiving forest runoff. Because so little data is available on the alcohols, this chapter will be concerned mostly with the acids although the concepts discussed may apply to the alcohols. The reader may want to refer to Chapters 13, 19 and 24 for further information on the fatty alcohols. There has been considerable geochemical interest in phytol, the branched chain alcohol which is a side chain of the chlorophyll molecule. This interest, centered on likely degradation products of phytol rather than phytol itself, will be discussed directly.

It has been known for a long time that fatty acids occur in materials of geological interest. SCHREINER and SHOREY [1], investigating soil chemistry in 1908, found dihydroxystearic acid in soils of low fertility. Since then geochemical interest in the organic acids has steadily grown as new tools and broader interests evolved. In a very lucid manner BERGMANN [2] reviewed much of the early work on fatty acid geochemistry. Research on the organic acids associated with petroleum, in particular the napthenic acids, was summarized by LOCHTE and LITTMANN [3], a good deal of the research being the work of LOCHTE and his graduate students. VALLENTYNE [4] includes fatty acids in his general review of organic matter in lakes and oceans. DEGENS [5] devotes a large part of his recent book to organic geochemistry with some mention of fatty acids. Instead of an exhaustive review of the literature an attempt has been made here to critically examine and relate several recent studies. At the same time it is hoped that enough factual background material has been included to allow the reader to construct a conceptual framework in which to evaluate fatty acid geochemistry and perhaps to some extent organic geochemistry as a whole. A few of the elements of such a framework can be stated.

Almost without exception all the organic matter found on the surface of the earth is derived from once living organisms. This is certainly true for the fatty acids. In discussing fatty acid geochemistry it is useful to consider a model. The simplest model is organic carbon flowing from organism to Recent sediment and hence to ancient organic deposit. To elaborate on this model let us consider a marine bay, although other environments would do equally well. A marine bay is an ecological system – a community of organisms, never a single organism or even a single species. Most of the inorganic carbon (e.g., CO_2) converted to organic matter by the plants is quickly oxidized back to inorganic carbon (CO_2) again by the community. A small amount of organic matter is not destroyed but is trapped in the top sediment. This material is further modified by bacteria and detritus-feeding animals (Chapter 11). As the top sediment is buried under more sediment, biological activity gradually ceases and the organic matter of the Recent sediment, already highly modified, begins the slow chemical and physical transformations which will eventually change it into the organic matter of an ancient sediment or perhaps into petroleum. Such a model allows many questions. For fatty acids we can ask "whose" fatty acids are finally preserved? Are such "fossil acids" those of the most abundant member of the community, those of the major photosynthetic producers, or those of bacteria? Are the fatty acids, possibly of unusual structure, present in minor members of the community ever preserved? There is much to be learned about the origin, transformation and eventual fate of fatty acids, but substantial data and important generalizations are already appearing.

The comparative biochemistry of the fatty acids is the first consideration necessary for understanding fatty acid geochemistry. Fortunately the comparative occurrence and distribution of fatty acids in organisms has been discussed in other Chapters. Both HILDITCH and WILLIAMS [6] and MARKLEY [7] give extensive surveys of the comparative literature as well as useful introductory material on the nomenclature, and physical and chemical properties of the fatty acids. Shorter reviews by SHORLAND [8, 9] give additional comparative data and discuss biosynthetic pathways to various acids. An article by MEAD *et al.* [10] provides a concise introduction to the fatty acids, long chain alcohols and waxes, and may be of use to the hurried reader. The comparative literature is weighted toward acids of plants and animals of economic importance to the fats and oil industry. However, a balanced picture of likely fatty acid patterns for typical environments will result from critical reading.

Table 1. *Percent composition of the total fatty acids of selected organisms*

Acid	Ryegrass[a]	Mixed plankton[b]	Bacteria isolate[c]	Blue-green algae[d]
i 12:0		0.03		
12:0	0.4	0.77		2.2
12:1	0.2			
13:0		0.06		
i 14:0		0.17	2.7	
14:0	1.4	9.60	3.4	21.0
14:1	0.5	0.95		tr
i 15:0		0.17	57.0	
15:0		1.00		1.4
15:1				
i 16:0		0.14		
16:0	10.6	26.00	15.0	17.0
16:1	4.1	13.00	11.0	3.7
16:2				
i 17:0			5.4	
17:0		3.0		
17:1				
i 18:0		0.2		
18:0	1.5	6.3	2.0	2.6
18:1	4.6	8.0	4.1	2.8
18:2	11.6	3.2		4.2
18:3	62.8	4.3		19.0
20:0	0.4	5.8		
>20:	1.9	16.0		

Note—12:0 means n-dodecanoic acid and 12:1 means dodecenoic acid, etc.

[a] SHORLAND [8].

[b] P. L. PARKER [20] and P. L. PARKER, unpublished data; plankton tow made in Arcansas Bay, Texas. Sample includes phytoplankton, zooplankton and detritus; "i" indicates iso plus anteiso acids; data based on GLC of methyl esters, peak area proportional to concentration.

[c] PARKER, VAN BAALEN, and MAURER [11]. This bacteria is one of four species isolated from the top of Baffin Bay, Texas. Data based on GLC of methyl esters.

[d] PARKER, VAN BAALEN, and MAURER [11]. The sample was a natural bloom of *Trichodesmium erythaeum*. Contained 27 percent 10:0. Data based on GLC of methyl esters.

The fatty acid compositions of several organisms are shown in Table 1. The organisms selected include both common and rare species and some of the fatty acid patterns may at first appear strange. The saturated, straight chain, even carbon number acids are major acids, with palmitic acid (I) being the first or second most abundant acid, of most organisms. Saturated, straight-chain, odd carbon-number acids are present in most organisms but often only in small amounts. Straight-chain, even carbon-number acids with one or more double bonds are the

$CH_3(CH_2)_{14}COOH$

(I) $C_{16:0}$ or Palmitic Acid

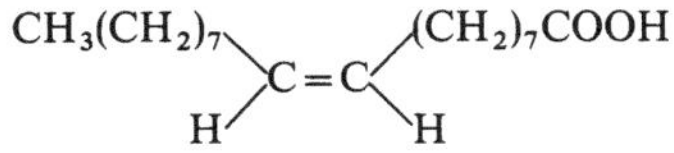

(II) $C_{18:1}$ or Oleic Acid

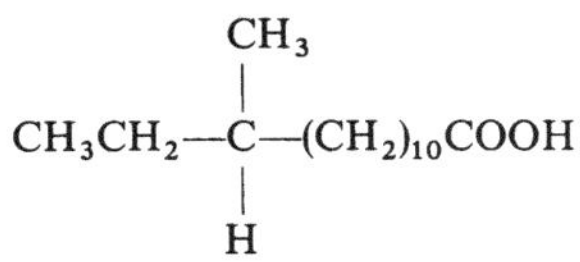

(III) *anteiso*-C_{15} or 12-methyltetradecanoic acid

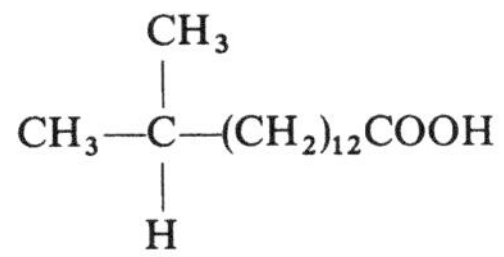

(IV) *iso*-C_{16} or 14-methylpentadecanoic acid

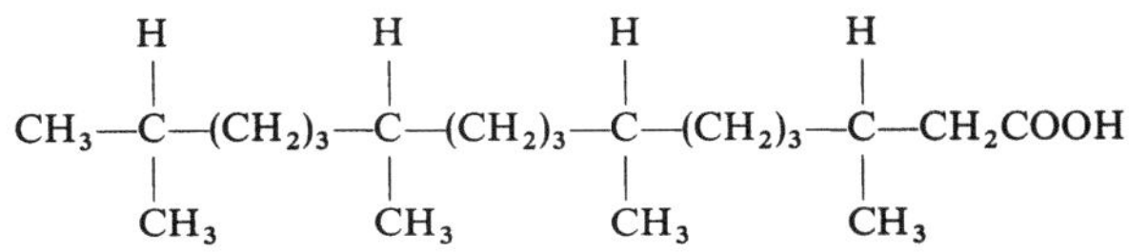

(V) phytanic acid or 3,7,11,15-tetramethylhexadecanoic acid

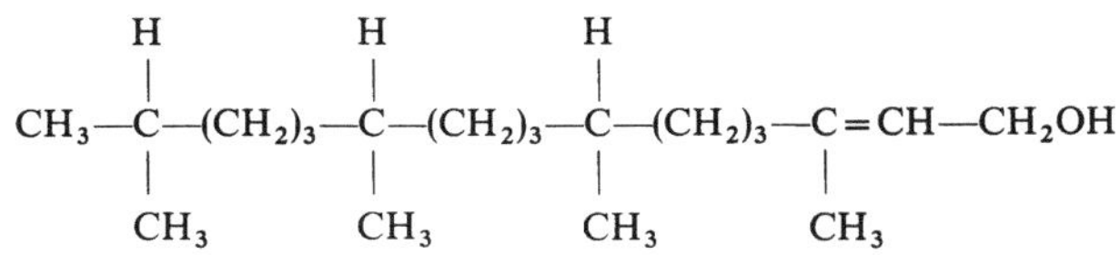

(VI) phytol or 3,7,11,15-tetramethylhexadec-2-enol.

Fig. 1. Structural formulas of fatty acids

second major type of naturally occurring fatty acid. Oleic acid (II) is widely distributed, often in higher concentrations than palmitic acid. Polyunsaturated acids with two, three, and sometimes more double bonds are common constituents of natural lipids, molecules with 16, 18, 20, 22 and 24 carbon atoms being favored. Saturated acids with single and multiple methyl branched-chains of both even and odd carbon number are of special interest to organic geochemists, partly because they are atypical and rare in most organisms. *Anteiso*-C_{15} (III), *iso*-C_{16} (IV) and phytanic acid (V) are representative of the branched-chain acids. Space does not permit mention of the variety of other naturally occurring fatty acids, although any one of them may some day be shown to be of geochemical significance. Nor has any mention been made of the chemical combinations of fatty acids in the lipids of organisms, for example, phospholipids and glycerides. The trivial names of the common fatty acids will be used. Only when necessary will both systematic names and trivial names be given. An abbreviation system will be adopted. For example, $C_{16:0}$ refers to the straight-chain 16 carbon acid with zero double bonds.

II. Geochemical Occurrence of Fatty Acids

In 1961 WILLIAMS [12] reported dissolved fatty acids in Pacific Ocean waters. Similar results were obtained by SLOWEY *et al.* [13] for water from the Gulf of Mexico. The concentrations were small but considering the vastness of the sea, they have biological and geochemical significance. Adsorption of dissolved material on clay is one mechanism for transferring organic matter from solution into a sediment. Typical data for fatty acids dissolved in sea water are given in Table 2. The acid patterns qualitatively are like that shown in Table 1 for plankton, but quantitatively they are more complex. Several of WILLIAMS' samples [12] contained high concentrations of three unidentified acids which he suggests may be $C_{15:0}$, $C_{17:0}$, and $C_{19:0}$, based on gas chromatographic retention times. If so, it would be a very interesting observation and is good reason for further study of acids in sea water. WILLIAMS [14] later reported ten-fold lower fatty acid concentrations for water around Vancouver Island, which would be 100-fold lower than the water from the Gulf of Mexico. Unfortunately, total dissolved organic matter was not measured for any of the samples so that there is no basis for meaningful comparison of concentrations. JEFFREY *et al.* [15], in a more detailed study of lipids in sea water from the Gulf of Mexico, found higher concentrations than WILLIAMS. Of more interest, they report relatively high $C_{15:0}$, $C_{17:0}$ and

Table 2. *Fatty acids dissolved in sea water*[a]

Acid	Pacific Ocean[b] µg/l	Gulf of Mexico[c] µg/l	Vancouver Island[d] µg/l	Gulf of Mexico	
				Free acids percent distribution	Sterols percent
10:0		21		0.09	
12:0	0.4	329		0.32	
13:0					3.95
14:0	1.8	14	1.6	3.91	4.36
14:1		7		0.26	
15:0	6.6			3.13	1.65
16:0	14.6	91	1.53	17.14	21.48
16:1		161	0.41	0.16	7.67
16:2			0.5	0.32	3.38
16:3				4.52	
17:0	15.5				4.95
18:0	22.2	49	0.80	10.0	5.95
18:1				52.68	14.06
18:2			0.25		
18:3				1.42	3.95
19:0	14.1				21.14
>19			1.97	2.36	7.4
Total	89.7	700	7.07		

[a] Filtered through 0.45 µ Type Millipose filters.

[b] WILLIAMS [12]; µg/l; 15:0, 17:0 and 19:0 were very tentatively identified; 16:1 and 18:1 were present in some samples; water taken at 1,200 m depth. Total acids.

[c] SLOWEY *et al.* [13]; µg/l; water taken at 50 m. Total acids.

[d] WILLIAMS [14]; µg/l; water taken at 20 m. Total acids.

[e] JEFFREY *et al.* [15] and JEFFREY [47]; percent composition of fatty acids.

$C_{19:0}$ concentrations, especially for the acids esterified by the sterol fraction of the total dissolved lipids. Both groups of workers noted non-systematic changes in fatty acid pattern with water depth and location. Detection and conclusive identification of dissolved material at these low concentrations is extremely difficult and it may be some time before a clear picture emerges. The data thus far available suggests that detailed study of the dissolved acids with respect to time of year, water depth and location will produce many interesting results.

Fatty acids have been found in Recent and ancient sediments by several investigators. The results of these studies, although carried out independently, overlap and will be discussed so as to bring out their relationship to each other.

Table 3. *Fatty acids in Recent and ancient sediment*

Fatty acid	Recent sediment		Ancient sediment	
	Baffin Bay[a] µg/g	San Nicholas[b] µg/g	Eagle Ford[b] µg/g	Green River[c] percent
12:0	0.73			2.5
13:0	0.21			7.0
i 14:0	0.97			
14:0	2.6	1.05	0.18	3.0
14:1	0.5			
ia 15:0	3.1			
15:0	1.3	0.74	0.20	4.0
15:1	+			
i 16:0	1.3			
16:0	11.0	17.02	0.50	7.0
16:1	5.5			
i 17:0	1.1			
17:0	1.1	0.70	0.32	5.0
17:1	0.8			
i 18:0	0.84			
18:0	3.0	3.83	0.38	7.5
18:1	7.5			
19	+	0.70	0.18	5.0
20	+	2.36	0.14	5.2
21		0.65	0.12	4.0
22		2.49	0.10	6.5
23		0.75	0.08	5.0
24		4.00	0.08	5.6
25		0.81	0.08	3.0
26		3.71	0.08	4.1
27		0.47	0.06	2.0
28		3.11	0.06	3.7
29		0.24	0.04	2.3
30		1.39	0.08	4.6
$\frac{C\,16+C\,18}{2\,C\,17}$	6.8	14.9	1.71	1.5

[a] P. L. PARKER [20]; average of 9 surface grab samples; µg fatty acid per g dry sediment; "i" is *iso* and "ia" is *iso* plus *anteiso* acid; method used would not detect molecules with more than 20 carbon atoms. + indicates trace amounts.

[b] COOPER [16]; µg/g; method excluded unsaturated and branched-chain acids. Age Upper Cretaceous.

[c] LAWLOR and ROBINSON [18]; percent composition of total fatty acids; acids were 1 percent of the total organic matter; acids with 10 and 11 and 31–35 carbon atoms were also found. Age Eocene.

In his study of the straight-chain, saturated acids in Recent and ancient sediments, COOPER [16] brought several important problems in fatty acid geochemistry into sharp focus. Some of his results are given in Table 3 along with some additional data for comparison. Allowing for the fact that his method excluded branched-chain and unsaturated acids, COOPER's data for the acids of recent sediments (Table 3) are similar to those for the acids from marine plankton (Table 1). However, the fatty acid patterns of his ancient sediments are strikingly different from those of Recent sediments and of living organisms in that the ancient materials are relatively enriched in odd carbon number fatty acids. The ratio, $C_{16}+C_{18}/2\,C_{17}$, a measure of the even-odd relative concentration, is 14.9 for the Recent sediment and 1.71 for the ancient sediment. COOPER [17] accepted the concept of high even-odd ratios for Recent material and low even-odd ratios for ancient sediments and proposed a mechanism to explain it. The mechanism postulates that each fatty acid can lose CO_2 to form an unstable radical which reacts to form either a fatty acid of one less carbon atom or a normal paraffin of one less carbon atom.

$$RCH_2{-}C(=O)OH \rightarrow RCH_2^{\bullet} + CO_2 + H^{\bullet}$$

$$RCH_2^{\bullet} + R'H \rightarrow RCH_3 + R'^{\bullet}$$

$$RCH_2^{\bullet} + (O) \rightarrow RC(=O)OH + H_2O.$$

If this process continued for a long time a sediment with a high even-odd ratio would be transformed into one with a low even-odd ratio and the overall chain length would progressively decrease. This hypothesis was especially useful for petroleum geochemistry. Normal fatty acids have long been considered a likely source material for the normal paraffin fraction of petroleum. KVENVOLDEN [19], like COOPER, found low even-odd ratios for the fatty acids extracted from similar Cretaceous sediments. However, his even-odd ratios for both Recent and ancient sediments showed a broader range than COOPER's. PARKER [20] has reported ratios as low as 2.3 for Recent sediments. The biological and geochemical processes that determine the fatty acids in a given sediment are certainly complex and may differ significantly from case to case. It is clear that the chemical nature of the biological source material varies with the individual environment and, further, that post-depositional chemical transformations play a significant part. KVENVOLDEN's data indicate that the fatty acid concentration in Recent sediments is adequate for the generation of the amount of n-alkanes found in ancient sediments. Experimental evidence for such changes is also known and is cited below.

JURG and EISMA [21] have presented experimental evidence for the production of hydrocarbons from fatty acids. Behenic acid ($C_{22:0}$) was heated with bentonite clay at 200° C. The normal C_{21} alkane, which would be formed by decarboxylation of the fatty acid, was the chief reaction product. But significant amounts of normal C_{24} to C_{36} alkanes were produced indicating that simple decarboxylation is not the only reaction taking place. When the experiments when carried out at 300° C, C_{23} and C_{24} fatty acids as well as acids with less than 22 carbon atoms could be isolated [22] from the product. Details of these experiments are given in Chapter28 of this book.

At the present time the origin of the high molecular weight (20–36 carbon atoms) fatty acids found in Recent and ancient sediment is not known. MEINSCHEIN and KENNY [23] found this material in soil where it can be explained as being derived from plant waxes. However, plant waxes are not generally abundant in marine environments. HOERING (personal communications, 1967) has obtained high molecular weight (C_{20} to C_{35}) dicarboxylic acids by mild oxidation of sediments. It may well be that certain marine environments receiving runoff from highly-productive land areas would accumulate plant waxes. Other marine environments not receiving organic rich runoff might not accumulate waxes. If so, perhaps a detailed study of both types of recent environments would give some insight into ancient environments. The origin of these long chain acids may be solved soon.

ABELSON and PARKER [24] and ABELSON *et al.* [25] examined several Recent and ancient sediments for fatty acids. They were able to detect even carbon-number saturated acids in the 500-million-year-old Alun shale, thus giving some hope that fatty acids might be found in still older sediments. HOERING and ABELSON [26] found that mild chromic acid oxidation of a sediment, which had previously been freed of fatty acids by exhaustive solvent extraction, produced a new suite of saturated, straight-chain even and odd carbon-number acids. They suggest that oxidation frees long-chained aliphatic structures which are bound somehow into the insoluble organic matter (kerogen) of the sediment. They have used this "kerogen-oxidation" technique on the Precambrian McMinn shale (1,600 m.y.) and obtained a complete suite of saturated, straight-chain acids at each carbon number from C_{11} to C_{21}. These acids may represent portions of molecules laid down by organisms at the time sediment was formed.

LAWLOR and ROBINSON [18], in their detailed investigation of the Green River shale (Eocene) from Wyoming, have studied the fatty acids present in various samples of the shale. They found a straight-chain, saturated acid at each carbon number from C_{10} to C_{35} with even carbon-numbered acids slightly predominating. They point out that their data indicate that paraffins are being formed from acids. HOERING and ABELSON [26] found very similar results from the Green River shale. RAMSAY [27] studied the distribution of "free" fatty acids in two samples taken from 1100-ft. and 1900-ft. levels of the Green River formation. He found the same C_{10} to C_{30} series already mentioned. He makes two observations concerning the acids: (1) for both depths the even-odd ratio is the same with respect to the lower acids (C_{12}–C_{18}), but shows a definite decrease at the deeper level for the longer chain acids (C_{19}–C_{29}) suggesting that the two groups of acids may have different histories; (2) the total quantity of acid isolated was five times less at the deeper level. This, along with a corresponding increase observed in the hydrocarbon content and even-odd ratio, is further evidence that fatty acids may give rise to hydrocarbons. It is reassuring to note that so long as comparable experimental procedures are used there seems to be fair agreement among investigators on kinds and amounts of fatty acids found in the Green River shale.

The importance of experimental technique, especially the isolation procedure prior to instrumental analysis, is well illustrated in the case of the branched-chain fatty acids. LEO and PARKER [28], using a chemical procedure designed not to exclude any type of acid, found a homologous series of *iso*-(IV) and *anteiso*-(III) branched-chain acids in Recent and ancient sediments. The C_{15} branched-chain acids were especially abundant. Higher organisms and marine plankton do not, insofar as present surveys go, contain enough branched-chain acid to account for the observed concentrations without postulating selective preservation. Bacterial lipids are a possible source and PARKER, VAN BAALEN and MAURER [11] have recently isolated bacteria from Recent sediments and shown that two of the four species isolated have the *iso* and *anteiso* C_{15} acids as their major fatty acids. PETERSON [29], in a thorough study of fatty acid distributions in Recent sediments off the coast of Washington state, found similar patterns of branched-chain acids. He suggests that the methyl-branched C_{15} and normal $C_{15:0}$ are sensitive indicators of depositional environment.

EGLINTON *et al.* [30] and RAMSAY [27] have recently discovered isoprenoid fatty acids in the Green River shale. Besides phytanic acid (V), they found isoprenoid acids with 14, 15, 16, 17, 19, 20 and 21 carbon atoms. Earlier investigators did not report these acids. The reason was that they used the urea adduction method in their purification procedure and this method excludes molecules with branched chains. The high concentrations of isoprenoid acids found in the Green River shale indicate that these compounds will be useful biological tracers in future studies of Recent and ancient materials. CASON and GRAHAM [31] had previously isolated the C_{14}, C_{15}, C_{19} and C_{20} isoprenoid acids from a California petroleum. The key to understanding the origin of these related acids is consideration of phytanic acid. Phytanic acid is found in a variety of organisms in small amounts and its geochemical route may be direct from organism to sediment. No single abundant source of this acid has yet been reported. However, as pointed out by RAMSAY, the structural similarity between phytanic acid (V) and phytol (VI) strongly suggests that phytol is the precursor. Chlorophyll, which contains phytol as an ester-linked side-chain, is an abundant organic constituent of Recent sediments, so there is little doubt that enough material is available. If phytanic acid is indeed derived from phytol, then the other isoprenoid acids may have been formed by more extensive decomposition of phytol. The C_{21} acid requires a different origin. Other naturally occurring isoprenoid compounds which could serve as source material include a C_{20} isoprenoid di-ether found by KATES, YENGOYAN and SASTRY [32] in halophyllic bacteria and the C_{15} isoprenoid alcohol farnesol which replaces phytol in certain bacterial chlorophylls. Whatever the ultimate source of these isoprenoid units, their geochemical occurrence is widespread. BLUMER [33] has done extensive work on phytol-related hydrocarbons in recent ecological systems. BENDORAITIS, BROWN, and HEPNER [34] isolated pristane, and DEAN and WHITEHEAD [35] phytane from petroleum. ROBINSON, CUMMINS, and DINNEEN [36] systematically studied the isoprenoid hydrocarbons in the Green River shale. The reader is referred to the chapter on hydrocarbons (Chapter 13) for a more detailed discussion of these results and for a discussion of the significance of isoprenoid hydrocarbons as biological markers in Precambrian sediments. Here it is sufficient to point out that fatty acids and alcohols, especially phytol, are probably important precursors of isoprenoid hydrocarbons.

The unsaturated fatty acids merit special consideration because taken together they account for more than half of the fatty acids of most organisms. Most investigators have used chemical procedures, such as urea adduction, which excluded unsaturated acids so that there is not much data available. Polyunsaturated acids are generally supposed to be too unstable and too reactive toward other components of cellular debris to persist in a sediment for geologically long periods of time. ABELSON [37] demonstrated this by heating Chlorella cells at 190° C for varying periods of time in the absence of oxygen and measuring the fatty acids left. He found that the polyunsaturated acids disappeared quickly but that oleic acid (II) was relatively stable. PARKER and LEO [38] observed a similar change in fatty acid patterns of successive layers of living and recently-living blue-green algal mats. The uppermost living layer was rich in unsaturated acids while in the lower layers the unsaturated acids were absent. ROSENFELD [39] measured the decrease in iodine number with depth in sediment and arrived at the same conclusion. His sample was a 48 inch core of recent marine mud estimated to be 440 years old at the base. The fact that extractable polyunsaturated acids are rare in sediments does not necessarily mean that these molecules have been destroyed but rather that in undergoing partial reaction, such as hydrogenation, addition, etc., they have lost their biochemical identity. The carbon skeleton of these reactive molecules may well serve as a major building block for kerogen formation, in which case the acids not directly detected could be geochemically the most significant ones; information relative to the structure of kerogen is urgently needed.

Unsaturated fatty acids are not completely absent in geological materials. Both $C_{16:1}$ and $C_{18:1}$ are present in Recent sediment in significant amounts. LEO [40] detected $C_{16:1}$ and $C_{18:1}$ in both the Green River shale and the Thermopolis shale (Cretaceous). RAMSAY [27] found $C_{18:1}$ in the D'Arcy oil and tentatively identified it as oleic acid (II). If this is indeed oleic acid and if it is the originally deposited material, then oleic acid is very stable, for RAMSAY gives the age of the D'Arcy oil as Carboniferous (250–300 million years). PETERSON [29] reported the highly unsaturated $C_{20:5}$ to be present in a 2 meter core from a highly reducing Recent sediment. He indicates that such occurrences were not general in his work and suggests that they may be due to relatively unaltered organic matter. The occurrence of such highly unsaturated acids in sediment demonstrated the importance of source material and specific depositional environment.

NAGY and BITZ [41] found that the Orgueil meteorite contained straight-chain, saturated acids with from 14 to 28 carbon atoms and with little even-odd preference. If these acids are actually an original part of the meteorite, then the organic geochemistry of other non-terrestrial materials should be very interesting in years to come.

Fatty acids have been detected in a variety of other geological materials. While sea water has been discussed at some length in this article, the literature contains little mention of freshwater although GORYUNOVA [42] reports high concentrations of high molecular weight fatty acids in lake water. LEO [40] analyzed two samples of fossil bog butter (see BERGMANN [2]) and found that relative to modern butter the polyunsaturates had almost vanished and about one-half of the triglyceride bonds had been broken. The bog butter was approximately 300 years old by ^{14}C dating. VISWANATHAN, BAI and PALLAI [43] studied the fatty acids in raw and treated sewage. They found less fatty acid in aerobic effluent than in anaerobic effluent which is consistent with the geochemical experience of the reactivity of organic matter.

III. Concluding Remarks

Organic geochemistry as an active field of research is new. In the case of the fatty acids the answers are not yet in, but the tools have been sharpened. The fact that at least four distinct and structurally different chemical types of fatty acids are found in living organisms means that by measuring ratios of these to each other and to hydrocarbons, insight into the origin, changes and fate of organic matter can be gained. There are many types of Recent environments which have not been studied with regard to fatty acids. New investigators might profitably study environments close to their laboratories. BRADLEY [44], in a search for modern analogues of the Eocene lakes that gave rise to the Green River shale, has demonstrated how very interesting and fruitful a study of the organic matter in a present day lake can be.

IV. Experimental Procedures

As pointed out earlier, recent advances in organic geochemistry are closely related to advances in experimental technique. Recent progress is due to the rapid development of new instruments and apparatus which has taken place in the last two decades. The following brief account will serve to introduce the non-geochemist to some of the current experimental procedures. The Chapter of this book devoted to techniques should be consulted for details.

A. Extraction of Fatty Acids from the Sample

Fatty acids are readily extracted from organisms with solvent mixtures such as methanol-benzene or methanol-chloroform. Either a Soxhlet apparatus or sonication serves to speed the extraction. Care must be taken to avoid direct heating of the sample if the unsaturated acids are of interest.

It is difficult to extract all the fatty acids from recent and ancient sediments. A fatty acid, labeled with ^{14}C and added to a sediment, is difficult to recover in high yields. Fig. 2 illustrates a procedure for the extraction and purification of the total soluble fatty acids from a Recent sediment. EGLINTON [30] *et al.* have found that slightly more acid can be recovered if the sediment sample is partially demineralized by hydrofluoric/hydrochloric acid (1:1) prior to extraction with organic solvent. They also use sonication to speed the extraction. COOPER and BRAY [16, 17] extracted the acids as the potassium salts from air-dried sediment. In their procedure the sediment was refluxed with methanol-KOH and the fatty acids extracted (CCl_4) after acidification with HCl. One might expect that the "methanol-KOH method" would give slightly higher yields. Hydrolysis of ester bonds linking fatty acid residues to the insoluble kerogen matrix could free additional acids. However, no such increase has been noted for Recent sediments in my laboratory. HOERING and ABELSON [26] have found that additional fatty acid is rendered extractable if the sediment is subjected to a mild chromic acid oxidation. There has been no systematic comparison of different methods by different workers on a standard sample.

B. Purification

It is usually necessary to separate the fatty acid fraction from the rest of the lipid extract. This separated fatty acid fraction may or may not require further purification depending on the design of the total experiment. Separation and purification were accomplished (Fig. 2) by solvent extraction. Inorganic salts and

Frozen recent sediment

|

Treat with dilute HCl to remove carbonate
Wash with distilled water to remove inorganic salts
Recover sediment by filtration

|

Extract moist sediment with methanol with stirring and ultrasonic vibrations for 30 min
Filter, save both fractions

|

Extract moist sediment with chloroform with stirring and ultrasonic vibrations for 30 min
(Filter, dry and weigh sediment)

|

Reduce volume of combined methanol and chloroform on rotating evaporator at 50° C

|

Wash organic phase with water to remove inorganic salts
(add chloroform if necessary to cause phases to separate)

|

Evaporate organic phase to near dryness under stream of nitrogen
Hydrolyze with KOH-methanol (3 percent) on steam bath for 1 hr.

|

Extract basic solution with hexane to remove non-acidic lipids. Save KOH-methanol phase

|

Acidify KOH-methanol with HCl and extract fatty acids into a few ml. of hexane

|

Prepare methyl esters by BF_3-methanol method

|

Gas-liquid chromatography
Infra-red Spectrometry
Mass Spectrometry *etc.*

Fig. 2. Flow-chart of the extraction and purification of total fatty acids from sediment (LEO and PARKER [28])

water-soluble organic compounds were removed in a water-chloroform system. Non-acidic lipids such as hydrocarbons were removed by a KOH-methanol vs. hexane partition. COOPER and BRAY [16, 17] included a similar acid-base extraction procedure in their extraction and separation method as did LAWLOR and ROBINSON [18]. EGLINTON *et al.* [30] recovered "free fatty acids" from the total lipid extract by passing the total extract through a column containing potassium hydroxide supported on silicic acid. In their method the total soluble fatty acids would be obtained by hydrolysis of the total lipid extract prior to the column separation.

Further purification of the fatty acid fraction has been carried out by several investigators. Urea adduction of the fatty acids or their methyl esters was used by COOPER and BRAY [16], LAWLOR and ROBINSON [18], and HOERING and ABELSON [26]. In this method straight-chain, saturated acids form a clathrate structure with urea. By decomposing the urea clathrates with water, the straight-chain acids are recovered. Column and preparative thin layer chromatography have been used to further purify methyl esters by RAMSAY [27] and PETERSON [29].

C. Identification and Quantitative Determination

Gas-liquid chromatography (GLC) is the single most useful tool for the qualitative and quantitative determination of fatty acids. It can readily be combined with chemical modifications and other physical tools. Pure standards of the material sought are generally needed. WOODFORD and VAN GENT [45] have shown that GLC retention times plotted against carbon number yield a family of straight lines; one for each homologous series. This serves as an internal check on the identification of peaks.

Identification of GLC peaks can be confirmed in a number of ways. GLC on a second liquid phase which shifts the relative retention time of the unknown peak is easy and useful. Unsaturation in a fatty acid can be detected by bromination or hydrogenation of the sample. The GLC peak due to an unsaturated acid will not appear in the resulting GLC chromatogram. The GLC chromatogram shown in Fig. 3 illustrates the complex pattern of acids found in a highly purified extract from a sediment.

Fatty acids and their esters have fairly characteristic infrared spectra. O'CONNOR [46] has summarized many of the useful correlations between IR spectra and structure. As in the case of GLC, IR can be used as a fingerprint if standards are available. The identification of individual GLC peaks can be confirmed by trapping the substance in a glass tube as it emerges from the GLC instrument. This is then used to obtain an IR spectrum. Fig. 4 illustrates how IR can be used to confirm an identification. The observed splitting in the region 1360 to 1380 cm^{-1} is characteristic of methyl iso-branched acids.

The mass spectrometer is a very powerful tool. The mass spectra of trapped GLC peaks yield especially valuable structural information. Fig. 3 illustrates the use of GLC-mass spectrometry in the identification of two compounds which had not previously been reported in sediment. The reader is advised to turn to Chapter 4 for a discussion of this powerful method.

The isolation of dicarboxylic acids with 12, 13, 15, 16, 17 and 18 carbon atoms from the Green River shale provides geochemists with yet another type of organic acid which will doubtless be looked for in other sediments and which will add to our understanding of the chemical transformations of these materials in geological time [48]. A bacterial origin was suggested for these acids.

Recently BLUMER and COOPER [49] have shown that several of the isoprenoid acids found in ancient sediment are present in Recent sediment. They suggest that phytol is converted to a few isoprenoid acids, mostly phytanic and pristanic acids, by organisms, then these decompose very slowly to give the other isoprenoid acids found in ancient sediment.

This brief mention of experimental methods is meant to point out some techniques in general use. The critical reader will also take note of our need for new and better techniques. Contamination during laboratory operations becomes

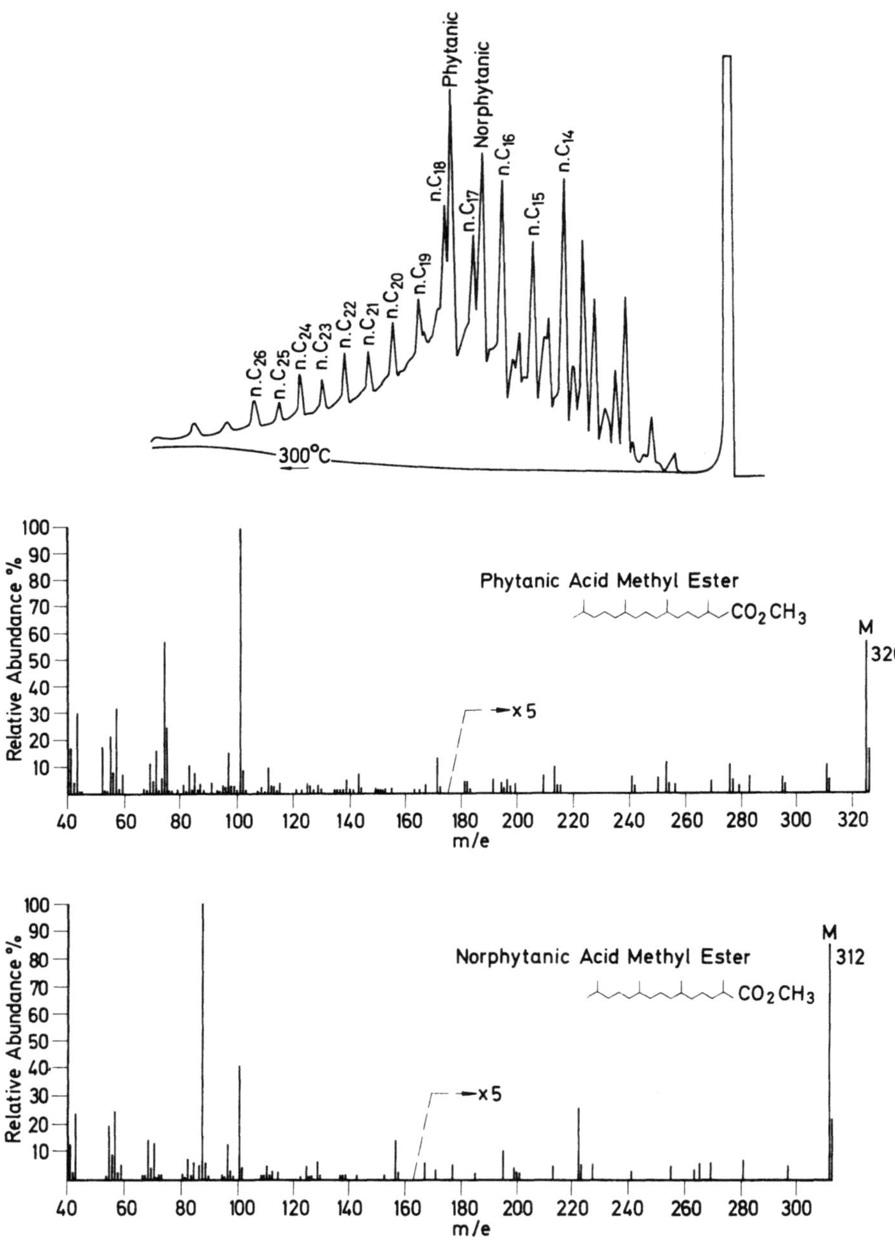

Fig. 3. (EGLINTON *et al.* [30].) a) Gas-liquid chromatogram of the free fatty acid fraction (as methyl esters) of a sample from the Green River shale. Conditions: 3 m by 3 mm column containing 1 percent SE-30 on Gas Chrom P, temperature programmed from 150° to 300° C at 8°/min. Flow rate, 20 ml/min. The base-line signal for injection of solvent alone is also shown. b) Mass spectra of isoprenoid fatty acids (as methyl esters) isolated from the Green River shale. Spectra obtained by combined gas chromatography-mass spectrometry (2 m by 4 mm column containing 6 percent SE-30 on Gas Chrom Q; temperature, 190° C, and scan time 2 seconds) on fractions trapped from a Versamid column (3 m by 6 mm column containing 3 percent Versamid on Gas Chrom Q)

an increasing hazard as the scale of the experiments diminishes. Skin lipids, soaps, detergents, bacterial action are ever-present risks.

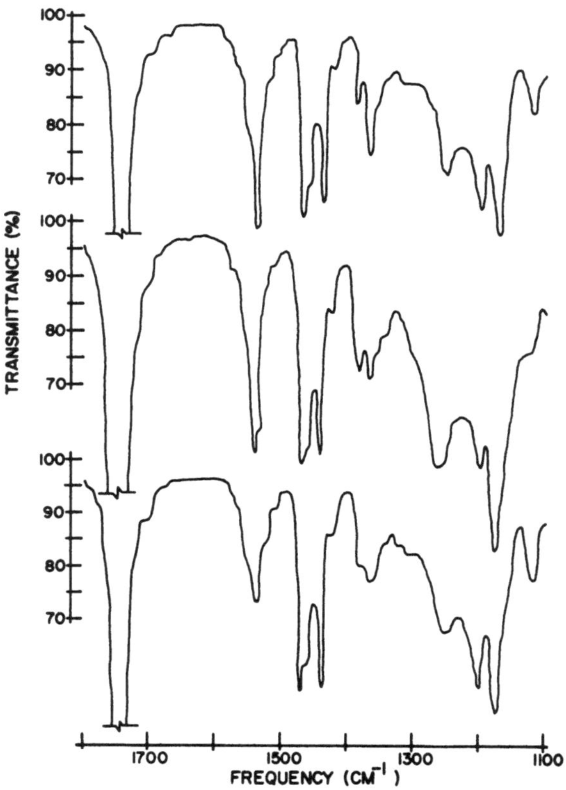

Fig. 4. (LEO [40].) Infrared spectra of fatty acids (esters of). Top: Methyl isotetradecanoate (i-C_{14}), standard. Middle: Methyl isopentadecanoate (i-C_{15}), sample peak collected from GLC. Bottom: Methyl pentadecanoate ($C_{15:0}$), standard. Note: All the above were in solution in carbon tetrachloride. The substance whose spectrum appears in the middle figure was not pure but contained a-C_{15} as well. Iso acids can be distinguished from their corresponding anteiso and normal forms in that they show a doublet in the range from 1360 to 1380 cm^{-1}

References

1. SCHREINER, O., and E. C. SHOREY: The isolation of dihydroxystearic acid from soils. J. Am. Chem. Soc. **30**, 1599–1607 (1908).
2. BERGMAN, W.: Geochemistry of lipids. In: Organic geochemistry (I. BREGER, ed.), p. 503–542. New York: The Macmillan Co. 1963.
3. LOCHTE, H. L., and E. R. LITTMANN: The petroleum acids and bases. New York: The Chemical Publishing Co., Inc. 1955.
4. VALLENTYNE, J. R.: The molecular nature of organic matter in lakes and oceans, with lesser reference to sewage and terrestrial soils. J. Fisheries Res. Board Can. **14** (1), 33–82 (1957).
5. DEGENS, E. T.: Geochemistry of sediments. Englewood Cliffs, N. J.: Prentice-Hall, Inc. 1965.
6. HILDITCH, T. P., and P. N. WILLIAMS: The chemical constitution of natural fats. 4th ed. New York: John Wiley & Sons, Inc. 1964.
7. MARKLEY, K. S.: Fatty acids: their chemistry, properties, production and uses, Part 1 (K. S. MARKLEY, ed.). New York: Interscience Publishers 1960.

8. SHORLAND, F. B.: The distribution of fatty acids in plants lipids. In: Chemical plant taxonomy (T. SWAIN, ed.), p. 253–311. New York: Academic Press, Inc. 1963.
9. – The comparative aspects of fatty acid occurrence and distribution. In: Comparative biochemistry, Vol. 3/A (M. FLORKIN and H. S. MASON, eds.), p. 1–92. New York: Academic Press, Inc. 1962.
10. MEAD, J. F., D. R. HOWTON, and J. C. NEVENZEL: Fatty acids, long chain alcohols, and waxes. In: Comprehensive biochemistry, Vol. 6 (M. FLORKIN and E. H. STOTZ, eds.), p. 1–49. Amsterdam: Elsevier Publishing Co. 1965.
11. PARKER, P. L., C. VAN BAALEN, and L. MAURER: Fatty acids in eleven species of blue-green algae: geochemical significance. Science **155**, 707–708 (1967).
12. WILLIAMS, P. M.: Organic acids in Pacific Ocean waters. Nature **188**, 219–220 (1961).
13. SLOWEY, F. J., L. M. JEFFREY, and D. W. HOOD: The fatty-acid content of ocean water. Geochim. Cosmochim. Acta **26**, 607–616 (1962).
14. WILLIAMS, P. M.: Fatty acids derived from lipids of marine origin. J. Fisheries Res. Board Can. **22** (5), 1107–1122 (1965).
15. JEFFREY, L. M., B. F. PASBY, B. STEVENSON, and D. W. HOOD: Lipids of ocean water. In: Advances in organic geochemistry (U. COLOMBO and G. D. HOBSON, eds.), p. 175–214. New York: The Macmillan Co. 1964.
16. COOPER, J. E.: Fatty acids in recent and ancient sediments and petroleum reservoir waters. Nature **193**, 744–746 (1962).
17. –, and E. E. BRAY: A postulated role of fatty acids in petroleum formations. Geochim. Cosmochim. Acta **27**, 1113–1127 (1963).
18. LAWLOR, D. L., and W. E. ROBINSON: Fatty acids in Green River Formation oil-shale. Div. Petrol. Chem. Am. Chem. Soc., Abstracts, Detroit Meeting (1965).
19. KVENVOLDEN, K. A.: Molecular distributions of normal fatty acids and paraffins in some lower Cretaceous sediments. Nature **209**, 573–577 (1966).
20. PARKER, P. L.: Fatty acids in recent sediments. Publ. Inst. mar. Sci. Univ. Tex. (in press).
21. JURG, J. W., and E. EISMA: Petroleum hydrocarbons: generation from fatty acids. Science **144**, 1451–1452 (1964).
22. – – On the mechanism of the generation of petroleum hydrocarbons from a fatty acid. Preprint, 3rd Internat. Meeting on Organic Geochemistry, London (1966).
23. MEINSCHEIN, W. G., and G. S. KENNY: Analyses of a chromatographic fraction of organic extracts of soils. Anal. Chem. **29**, 1153–1161 (1957).
24. ABELSON, P. H., and P. L. PARKER: Fatty acids in sedimentary rocks (Annual Rept. Director Geophys. Lab.). Carnegie Inst. Wash. Yr. Bk. **61**, 181–184 (1962).
25. – T. C. HOERING, and P. L. PARKER: Fatty acids in sedimentary rocks. In: Advances in Organic Geochemistry (U. COLOMBO and G. D. HOBSON, eds.), p. 169–174. New York: The Macmillan Co. 1964.
26. HOERING, T. C., and P. H. ABELSON: Fatty acids from the oxidation of kerogen. (Annual Rept. Director Geophys. Lab.) Carnegie Inst. Wash. Yr. Bk. **64**, 218–223 (1965).
27. RAMSAY, J. N.: Organic geochemistry of fatty acids. M.S. Thesis, University of Glasgow (1966). Also A. G. DOUGLAS, K. DOURAGHI-ZADEH, G. EGLINTON, J. R. MAXWELL, and J. N. RAMSEY, Fatty acids in sediments including the Green River shale (Eocene) and Scottish Torbanite (Carboniferous). Preprint, 3rd Internat. Meeting on Organic Geochemistry, London (1966).
28. LEO, R. F., and P. L. PARKER: Branched-chain fatty acids in sediments. Science **152**, 649–650 (1966).
29. PETERSON, D. H.: Fatty acid composition of certain shallow-water marine sediments. Ph. D. Thesis, University of Washington (1967).
30. EGLINTON, G., A. G. DOUGLAS, J. R. MAXWELL, J. N. RAMSAY, and S. STALLBERG-STENHAGEN: Occurrence of isoprenoid fatty acids in the Green River shale. Science **153**, 1133–1135 (1966).
31. CASON, J., and D. W. GRAHAM: Isolated of isoprenoid acids from a California petroleum. Tetrahedron **21**, 471–483 (1965).
32. KATES, M., L. S. YENGOYAN, and P. S. SASTRY: A diether analog of phosphatidyl glycerophosphate in *Halobacterium cutirubrum*. Biochim. Biophys. Acta **98**, 252–268 (1965).
33. BLUMER, M.: Organic Pigments: their long-term fate. Science **149**, 722–726 (1965).

34. Bendoraitis, J. G., B. L. Brown, and L. S. Hepner: Isoprenoid hydrocarbons in petroleum. Isolation of 2,6,10,14-tetramethylpentadecane by high temperature gas-liquid chromatography. Anal. Chem. **34**, 49–53 (1962).
35. Dean, R. A., and E. V. Whitehead: The occurrence of phytane in petroleum. Tetrahedron Letters **21**, 768–770 (1961).
36. Robinson, W. E., J. J. Cummins, and G. U. Dinneen: Changes in Green River oil-shale paraffins with depth. Geochim. Cosmochim. Acta **29**, 249–258 (1965).
37. Abelson, P. H.: Thermal stability of algae. (Annual Rept. Director Geophys. Lab.) Carnegie Inst. Wash. Yr. Bk. **61**, 179–181 (1962).
38. Parker, P. L., and R. F. Leo: Fatty acids in blue-green algal mat communities. Science **148**, 373–374 (1965).
39. Rosenfeld, W. D.: Fatty acid transformations by anaerobic bacteria. Arch. Biochem. Biophys. **16**, 263–273 (1948).
40. Leo, R. F.: The geochemistry of fatty acids in recent marine sediments. M.A. Thesis, University of Texas (1966).
41. Nagy, B., and Sister Mary C. Bitz: Long-chain fatty acids from the Orgueil Meteorite. Arch. Biochem. Biophys. **101**, 240–248 (1963).
42. Goryunova, S. V.: Characterization of dissolved organic substances in water of Gluboke Lake. Chem. Abstr. **47**, 8293 h (1953).
43. Viswanathan, C. V., B. M. Bai, and S. C. Pillai: Fatty matter in aerobic and anaerobic sewage studies. J. Water Pollution Control Federation **34**, 189–194 (1962).
44. Bradley, W. H.: Tropical Lakes, copropel, and oil shale. Geol. Soc. Am. Bull. **77**, 1333–1338 (1966).
45. Woodford, F. P., and C. M. van Gent: Gas-liquid chromatography of fatty acid methyl esters: the "carbon number" as a parameter for comparison of columns. J. Lipid Res. **1**, 188–190 (1960).
46. O'Connor, R. T.: Spectral properties, Part 1 (K. S. Markley, ed.), p. 379–498. New York: Interscience Publishers 1960.
47. Jeffrey, L. M.: Lipids in sea water. J. Am. Oil. Chem. Soc. **43**, 211–214 (1966).
48. Haug, P., H. K. Schnoes, and A. L. Burlingame: Isoprenoid and dicarboxylic acids isolated from Colorado Green River shale (Eocene). Science **158**, 772–773 (1967).
49. Blumer, M., and W. J. Cooper: Isoprenoid acid in recent sediments. Science **158**, 1463–1464 (1967).

CHAPTER 15

Fossil Carbohydrates *

F. M. SWAIN

University of Minnesota, Minneapolis, Minnesota

Contents

I. Introduction

Although fossil carbohydrates were separated from Tertiary and Cretaceous materials as early as 1922 by GOTHAN [1] and other workers (see VALLENTYNE [2], for review of early literature) it was only recently that active work has been done on these substances. PALACAS [3] seems to have been the first to identify monosaccharides from Paleozoic and from pre-Cretaceous Mesozoic rocks (Table 1). In addition to several monosaccharides which he characterized chromatographically, D-glucose was identified chemically in Devonian Marcellus Black Shale. Subsequently ROGERS [4] identified additional monosaccharides from Devonian and other rocks mainly employing chromatographic and enzymatic methods. He repeated PALACAS' chemical determination of D-glucose from Marcellus Shale

* The following assisted or provided advice and information for the preparation of this chapter: JUDY G. BRATT, GUNTA V. PAKALNS, SAMUEL KIRKWOOD, WILL SALO, M. ALAN ROGERS, JAMES G. PALACAS and MICHAEL FLEMING. Some of the unpublished experimental work recorded herein was supported by a grant (NGR-24-005-054) from the National Aeronautics and Space Administration, by a grant from the National Science Foundation (GB 18856), and by a grant from the Graduate School, University of Minnesota. The writer expresses sincere thanks for this assistance and support.

and further characterized D-xylose from the same black shale by crystallization and chemical methods. D-glucose was characterized by converting the crystals into N-p-nitrophenyl-β-D-glucosylamine—m.p. and mixed m.p. 184° C and $[\alpha]_D^{22}$–200° (approx.) in pyridine (c. 0.007). D-xylose was characterized by converting the crystalline residue into di-O-benzylidene-D-xylose dimethylacetal, [m.p. and mixed m.p. 211–212° C and $[\alpha]_D^{23}$–7° (approx.) in chloroform (c. 0.015)].

Other studies [5–9] have shown that carbohydrate materials are widely distributed in rocks of many ages, and bear a relationship to source materials and environments of deposition of the sediments.

Table 1. *Carbohydrate components of sedimentary rocks* [10]

Formation	Age	Component sugars				
		glucose	galactose	arabinose	xylose	rhamnose
Stonehenge	Ordovician	+		(?+)	(?+)	
Simpson	Ordovician	+	(?+)	+	+	
Upper Chambersburg	Ordovician	+		(?+)		
Marcellus	Devonian	+	(?+)	+	+	+
Woodford	Upper Devonian	+		+	+	
Des Moines	Pennsylvanian	+		+	+	
Missouri	Pennsylvanian	+				(?+)
Leonard	Permian	+	(?+)		+	
Schuler	Jurassic	+		+	+	
Green River	Eocene	+	(?+)		(?+)	
Elko	Miocene	+				

The rock was treated with hot 0.5 N sulphuric acid and the sugars identified chromatographically (indicated by +).

The organic material from which these carbohydrate residues were obtained was finely divided and disseminated through the rocks and was not referable to any recognizable fossil species. Water extraction yielded, as monosaccharide, only 10–15 percent of the total carbohydrate content of the rock as determined by a phenol-sulfuric acid test cited below. For this reason, in the studies noted above, hot dilute sulfuric acid was used to extract the monosaccharides from the rocks. Presumably the bulk of the residual carbohydrates occur as polymers, or are complexed with humic substances in the rock [10, 11].

The general range of values of total carbohydrates of sedimentary rocks is given in Table 2. The determinations of total carbohydrates are based on a phenol-sulfuric acid test for furfurals in which the absorption maximum of the colored product is compared to that obtained for D-glucose.

The carbohydrates and related substances that have been recorded from geologic samples are shown in Figs. 2–9.The hexoses (Fig. 1) have been found in sediments [3, 8–10, 12–15]; the configuration, however, is known only for glucose and galactose. The pentoses, of unknown configuration (Fig. 2), have also been found in sediments, rocks and fossils (see references cited for hexoses) but in somewhat smaller amounts. Uronic acids (Fig. 3) have been reported in sediments and rocks [4, 12]. Sugar alcohols (Fig. 4) have been questionably identified from Devonian rocks in unpublished studies by the writer. Amino sugars (Fig. 5) have

Table 2. *Total carbohydrate content of some geologic samples* [12]

Geologic formation	Age and location	Description	Carbohydrate content; range in parentheses (ppm)	Absorption maximum of colored product (mμ)
Coutchiching (3 samples)	Early Precambrian, Ontario	Dark green biotite slate and phyllite (basinal)	11 (6–14)	490
"Carlim-Lowville" (1 sample)	Early Middle Ordovician, Pennsylvania	Dense, dark gray, carbonaceous, pyritic stylolitic limestone; graphitic residue on stylolites (neritic)	37	n. d.
Burket Shale (1 sample)	Late Devonian, Pennsylvania	Black fissile shale, with *Paracardium doris*	17	n. d.
Pocono Sandstone, Burgoon Member (1 sample)	Early Mississippian, Pennsylvania	Coal (paludal)	126	488
Twin Creek Limestone (4 samples)	Middle Jurassic, Wyoming	Light gray thin-bedded limestone (neritic)	31 (14–55)	490
Twin Creek Limestone (9 samples)	Middle Jurassic, Utah	Pale gray-brown sublithographic limestone (neritic)	23 (6–45)	490
Mariposa Shale (3 samples)	Late Jurassic, California	Dark gray micaceous argillite (basinal)	18 (10–30)	n. d.
Twist Gulch Formation (7 samples)	Late Jurassic, Utah	Pale gray soft, calcareous shale (alluvial)	48 (14–88)	484–487
Arapien Shale (1 sample)	Middle Jurassic, Utah	Pale gray calcareous carbonaceous gypsiferous shale (alluvial)	24	486
Smackover Formation (1 sample)	Late Jurassic Louisiana, Waller B-1, 10, 630–10, 635 feet	Dark grayish brown, fine to medium grain, slightly calcareous argillaceous bituminous sandstone (neritic)	55	482
Schuler Formation (1 sample)	Late Jurassic, Louisiana, Greene 9085–9086 feet	Medium to light gray silty shale or siltstone (neritic)	21	482
Smackover Formation (5 samples)	Late Jurassic, Louisiana, Bolinger Well	Medium light gray pisolitic algal colitic limestone (littoral)	20 (0–66)	482–490
La Casita Shale (1 sample)	Late Jurassic, Mexico	Sooty black fissile shale (basinal)	163	486
Smackover Formation (2 samples)	Late Jurassic, Louisiana, Timmons Nol, 10,579.5–11,267 feet	Light gray oolitic limestone (littoral reef)	1–11	484–488

Aldohexoses

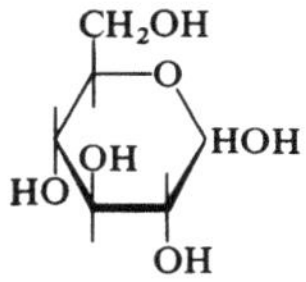
D-glucose

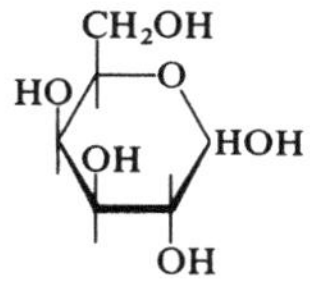
D-galactose

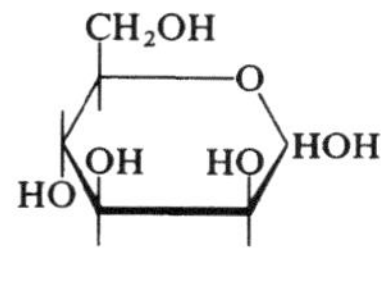
D-mannose

Ketohexose

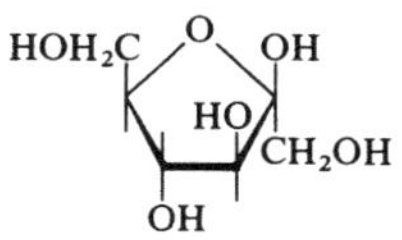
β-D-fructose

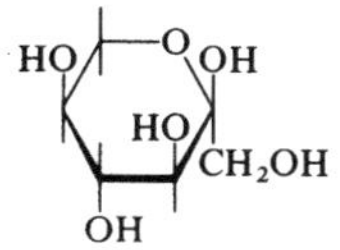
α-L-sorbose

Fig. 1. Hexoses that have been found in geological samples

Pentoses

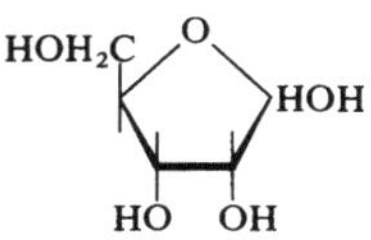
D-ribose

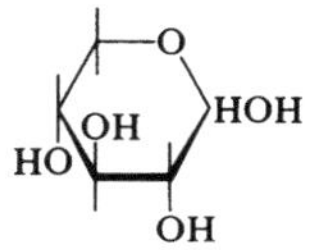
D-xylose

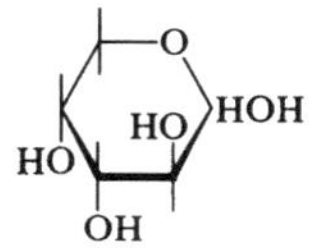
D-arabinose

Methyl pentoses

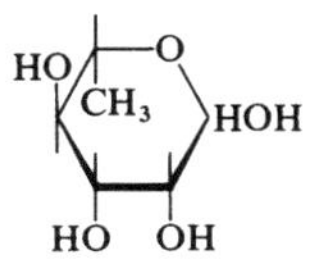
L-rhamnose

Fig. 2. Pentoses that have been found in geological samples

Uronic acids

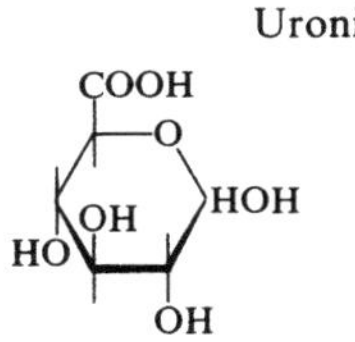
D-glucuronic acid

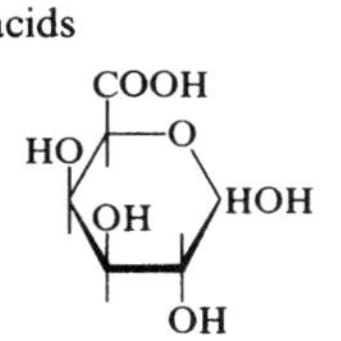
D-galacturonic acid

Fig. 3. Uronic acids that have been found in geological samples

been noted in lake and bog sediments in the course of analyses for amino acids (Fig. 6).

Oligosaccharides (Fig. 7) have not been definitely identified as such in geological materials, but cellobiose, cellotriose, laminaribiose, etc., were formed as enzymatic products of cellulase- and laminarase-treated Devonian shales [4].

Polyhydroxy alcohols

Mannitol D-sorbitol (D-glucitol)

Fig. 4. Polyhydroxy alcohols that have been found in geological samples

Amino sugars

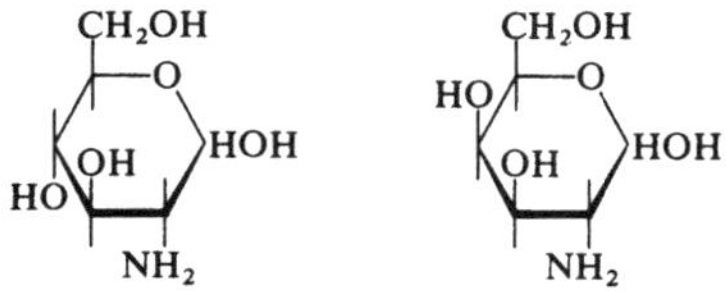

D-glucosamine D-galactosamine

Fig. 5. Amino sugars that have been found in geological samples

Several unidentified, possibly oligosaccharide, compounds having low R_f values were noted on paper chromatograms of acids hydrolyzates of Devonian rock samples.

Starch, cellulose, mannan and xylan (Fig. 8) have been found in sediments [12] and cellulose and possibly starch have been recorded from fossils [2]. Laminarin (Fig. 9) has been recorded from Devonian shales [4] and chitin from fossil insects [16]. Bacterial polysaccharides are questionably recorded from sediments [17].

II. Methods of Analysis of Fossil Carbohydrates

The aims of carbohydrate analyses of geological samples, including Recent sediments, rocks, and fossils are the following: first, to learn whether the sample contains carbohydrate material in any form; second, the total amounts of carbohydrates present; third, what monosaccharides are present and their amounts; and fourth, what polymeric material including polysaccharides may occur and their amounts. Finally, an attempt is made to understand the sources, conditions of sedimentation and diagenesis, and possible evolutionary implications of the carbohydrate residues.

A recommended procedure for study of geological samples suspected to contain carbohydrates is as follows: (1) test the sample for total carbohydrates by phenol-sulfuric acid technique, or less desirably, by distillation of the sample with acid; (2) if carbohydrates are present, analyze a sample for free sugars by water extraction and a chromatographic method or for combined sugar units by acid extraction and chromatography; (3) analyze a sample for glucose, galactose and possibly other monosaccharides by an enzymatic method which provides

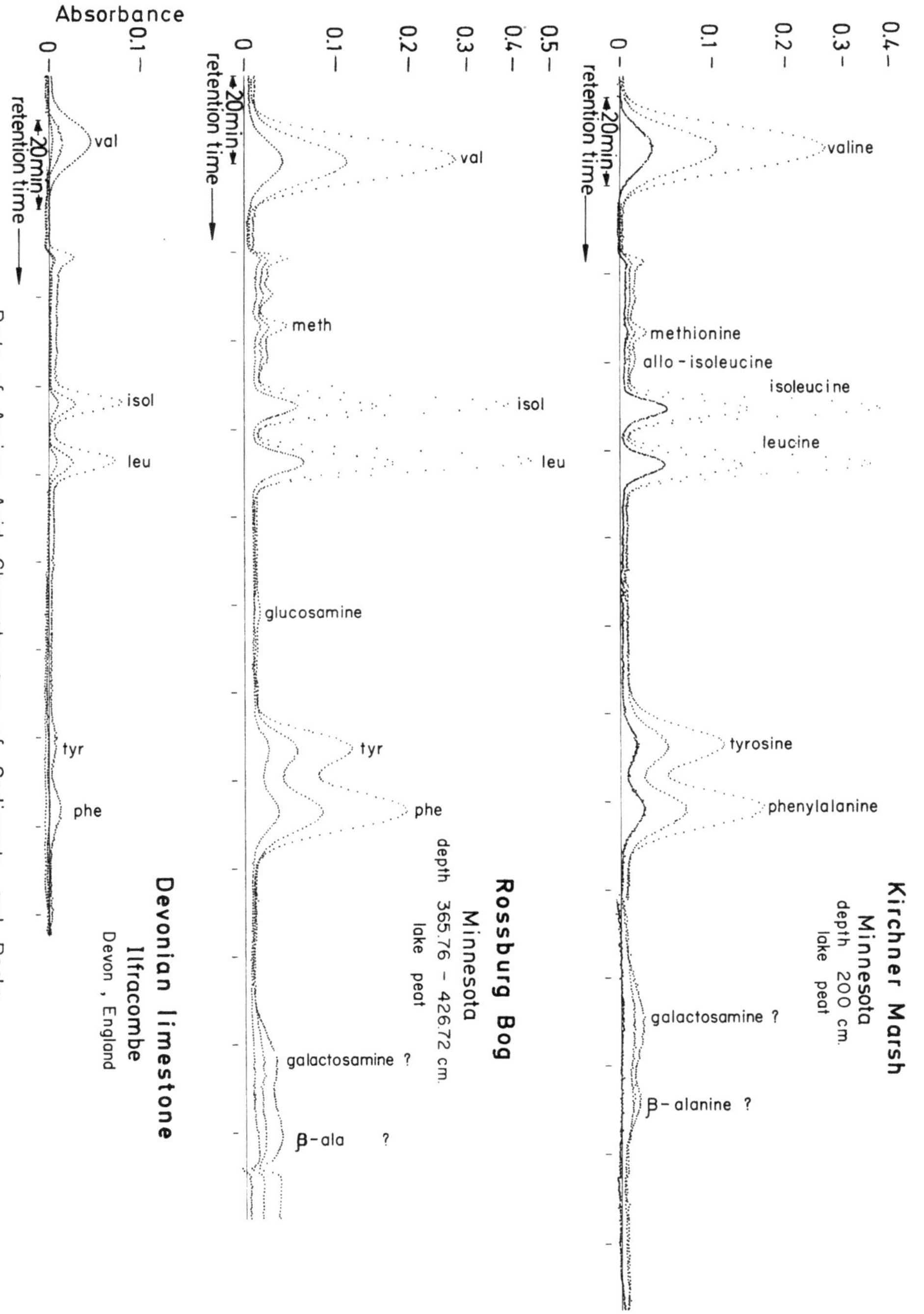

Fig. 6. Portions of amino acid chromatograms (from Moore-Stein-type analyzer) showing glucosamine and galactosamine in acid hydrolyzates of peat

Oligosaccharides

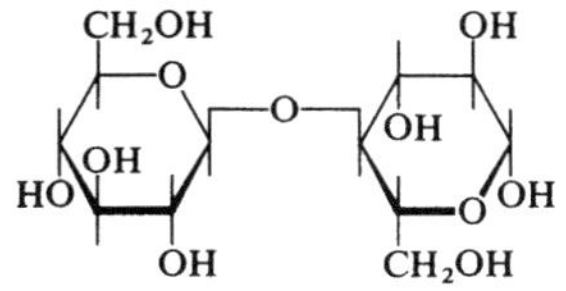

Cellobiose
4-0-β-D-glucopyranosyl-D-glucose
also cellotriose

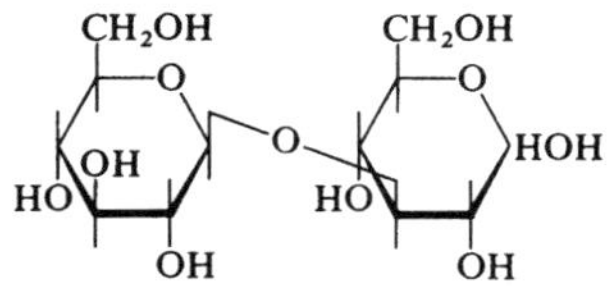

? Laminaribiose
3-0-β-D-glucopyranosyl-D-glucose
? also laminaritriose, -tetraose, -pentaose, -hexaose

Fig. 7. Oligosaccharides that have been detected as degradation products of glucosan- and laminaran-polysaccharides in enzymatic analyses of geological samples

Polysaccharides

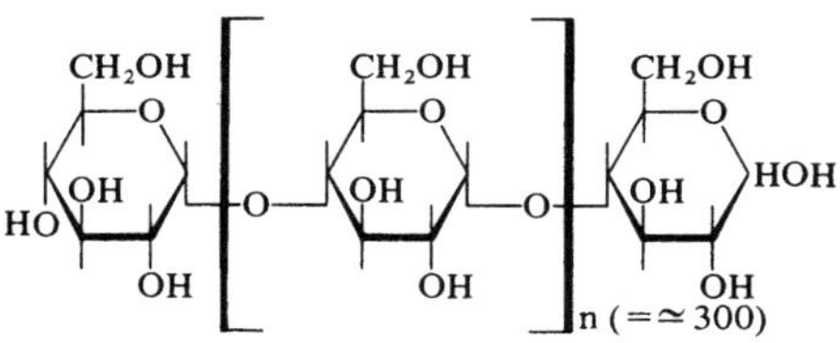

Amylose

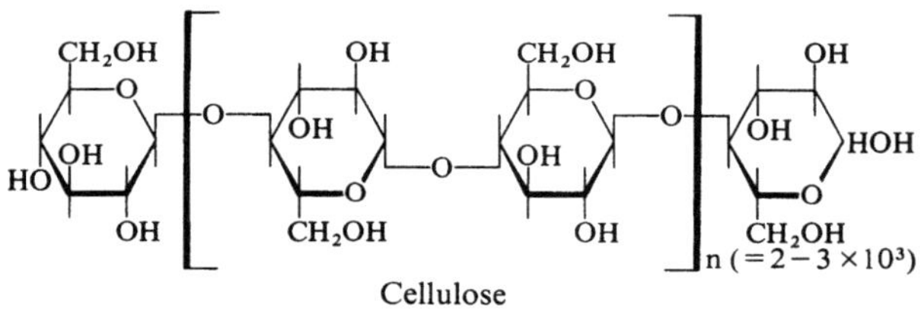

Cellulose

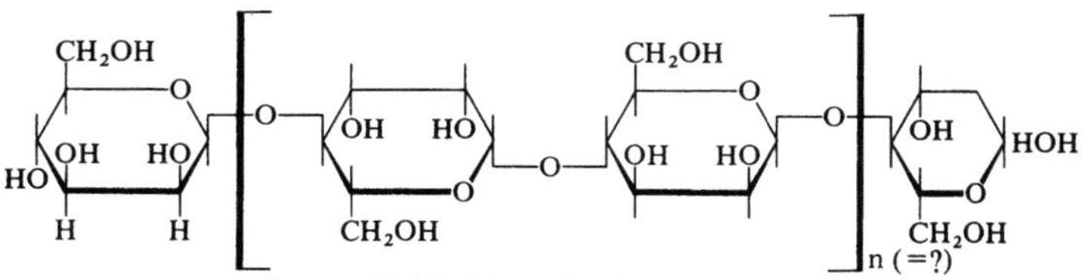

Mannan (1,4-); 1,3- and other links possible

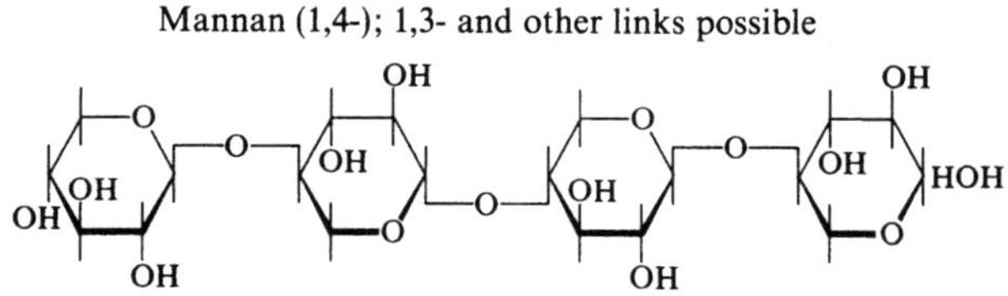

Xylan (1,4-); 1,3- and other links possible (e.g. algal xylans)

Fig. 8. Some polysaccharides that have been found in geological samples

Polysaccharides

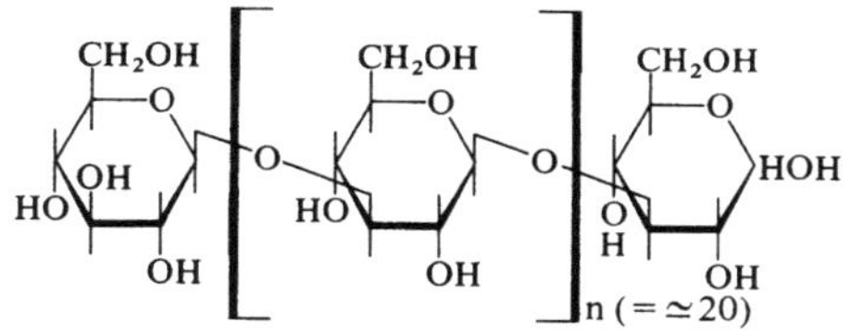

Laminaran

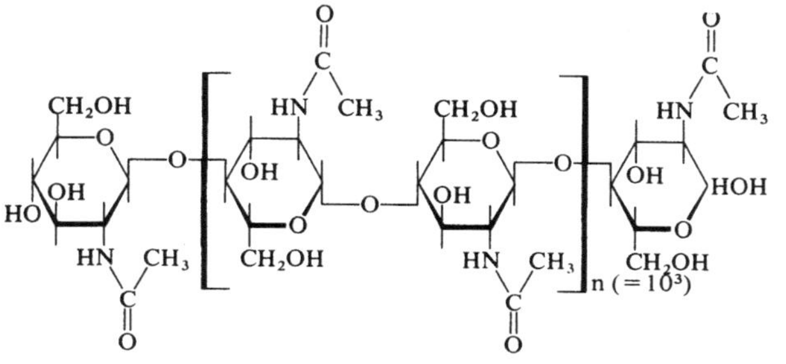

Chitin

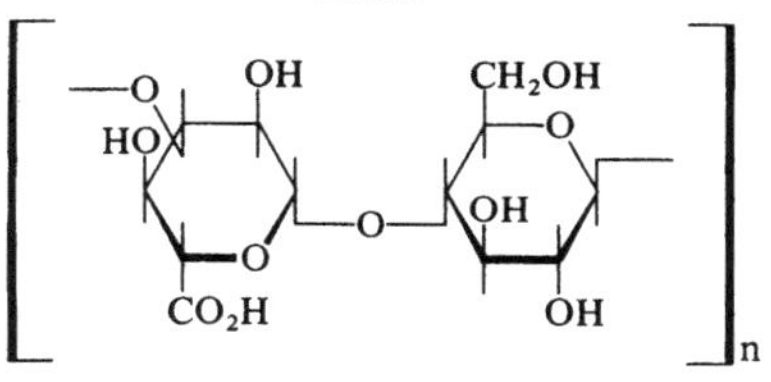

Bacterial polysaccharides
(Type III, Pneumococcus)

Fig. 9. Additional polysaccharides that have been found in geological samples

information as to configuration; (4) analyze a sample for polysaccharides by enzymatic methods if enough original material is available; and (5) compare analytical results with analyses of modern counterparts of organic geochemical material.

Rock and fossil samples are prepared by cleaning the exterior surfaces of dirt, plant growths, and weathered crusts; washing the surface with concentrated chromic acid solution to destroy contaminating organic material; crushing and powdering the sample under sterile conditions; and storing the powders in glass jars.

Analytical reagents are used as "controls" and blank runs are obtained for each procedure described here. Depending on the procedure used, the results on geological samples are considered to be valid to 1–5 ppm for most of the colorimetric and chromatographic analyses, and 0.1–1 ppm for enzymatic tests and for some of the gas chromatographic tests.

III. Total Carbohydrates

A. Distillation Techniques

Where only trace amounts occur the presence of carbohydrate materials and a general idea of their contents in geological samples may be detected by the reaction of pentoses and hexoses with 12 percent hydrochloric acid to form furfural and 5-hydroxymethylfurfural, respectively (Fig. 10). The procedure is complicated by the inclusion of hydrocarbons, heterocyclic compounds and other substances in the HCl distillates of geological samples [7] (Fig. 11).

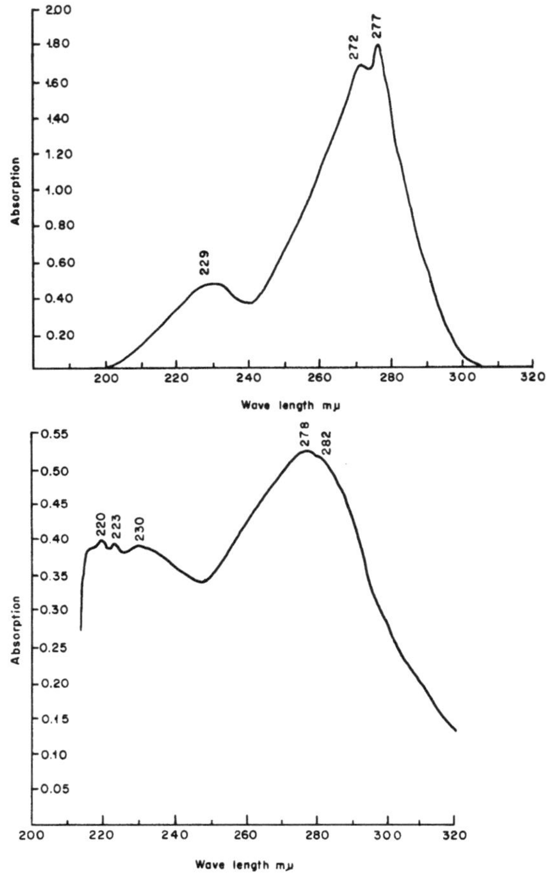

Fig. 10. Ultraviolet absorption spectra of furfural and 5-hydroxymethylfurfural obtained by hydrochloric acid treatment of arabinose and glucose respectively; pH of spectral solution 3 [6]

B. Phenol-Sulfuric Acid Technique

A method of DUBOIS *et al.* [18] is useful for total carbohydrates in geological samples as it is reasonably quantitative at low concentrations. Simple sugars, oligosaccharides, polysaccharides and their derivatives, including methyl ethers having free or potentially free reducing (–CHO) groups, will produce an orange-yellow color when treated with phenol and sulfuric acid.

IV. Monosaccharides

A. Water Extraction of Free Monosaccharides

Pulverized samples of 1–5 g of modern sediment or 500–1,000 g of rock are extracted for 8–10 hours with distilled water under reflux at the boiling temperature of the mixture. The mixture is centrifuged and desalted first by ethanolic precipitation and then by use of electric-desalting equipment or by ion-exchange resins. The desalted solution is concentrated to a standard volume. Desalting is done on Dowex 50(H+) 8 percent cross-linked, or similar cation-exchange resin followed by Duolite A-4(OH^-) or equivalent weak anion exchange resin and finally by Dowex 50 or Amberlite IR-20(H+) resin again. The three columns are mounted one above the other.

B. Alcohol Extraction of Free Monosaccharides

Ethyl alcohol may be used in place of water for extraction of free sugars. The solution is desalted on ion-exchange resins and concentrated to a standard volume as above.

C. Acid Extraction of Monosaccharides

The purpose of this method is to completely hydrolyze all the carbohydrate polymers in the sample and to identify the combined sugar units. Amounts of samples comparable to those above are used for acid hydrolysis. The alkaline constituents of rock samples are first determined by preliminary acid extraction of a 1-gram sample of powdered rock and subsequent back titration with NaOH in the presence of phenolphthalein. The pulverized samples are boiled under reflux for 8–10 hours with 0.5 N sulfuric acid; alternatively, the sample is pretreated with cold concentrated (72 percent) sulfuric acid for 2–4 hours, the solution is diluted to 0.5 N and extracted as above. The hydrolyzate is centrifuged and the solution neutralized with $BaCO_3$, followed by desalting and concentration as described above. During neutralization and concentration the pH is maintained in the 5–7 range and alkalinity, which tends to degrade the carbohydrate materials, is avoided.

D. Infrared Spectra of Monosaccharides

A preliminary idea of the makeup of the carbohydrate-bearing sample may be gained from infrared spectral analysis [19]. In addition to the C–H stretching region (2,800–3,000 cm^{-1}) and the CH_2 symmetrical bending vibration (1,400–1,460 cm^{-1}) of cellulose, the following absorption bands of the pyranose ring are often prominent though not definitive: 967, 917, 891, 880, 867, 844, and 770 cm^{-1}. The principal use of infrared, as well as of nuclear magnetic resonance spectra [20] of carbohydrates is in structural studies.

E. Paper Chromatography of Monosaccharides

Separation of carbohydrate extracts containing about 5 ppm or more monosaccharides is carried out on Whatman No. 1 filter paper, general solvents: (1) butanol:acetic acid:water (4:1:5) or (2) pyridine:ethyl acetate:water (1:1:3 or

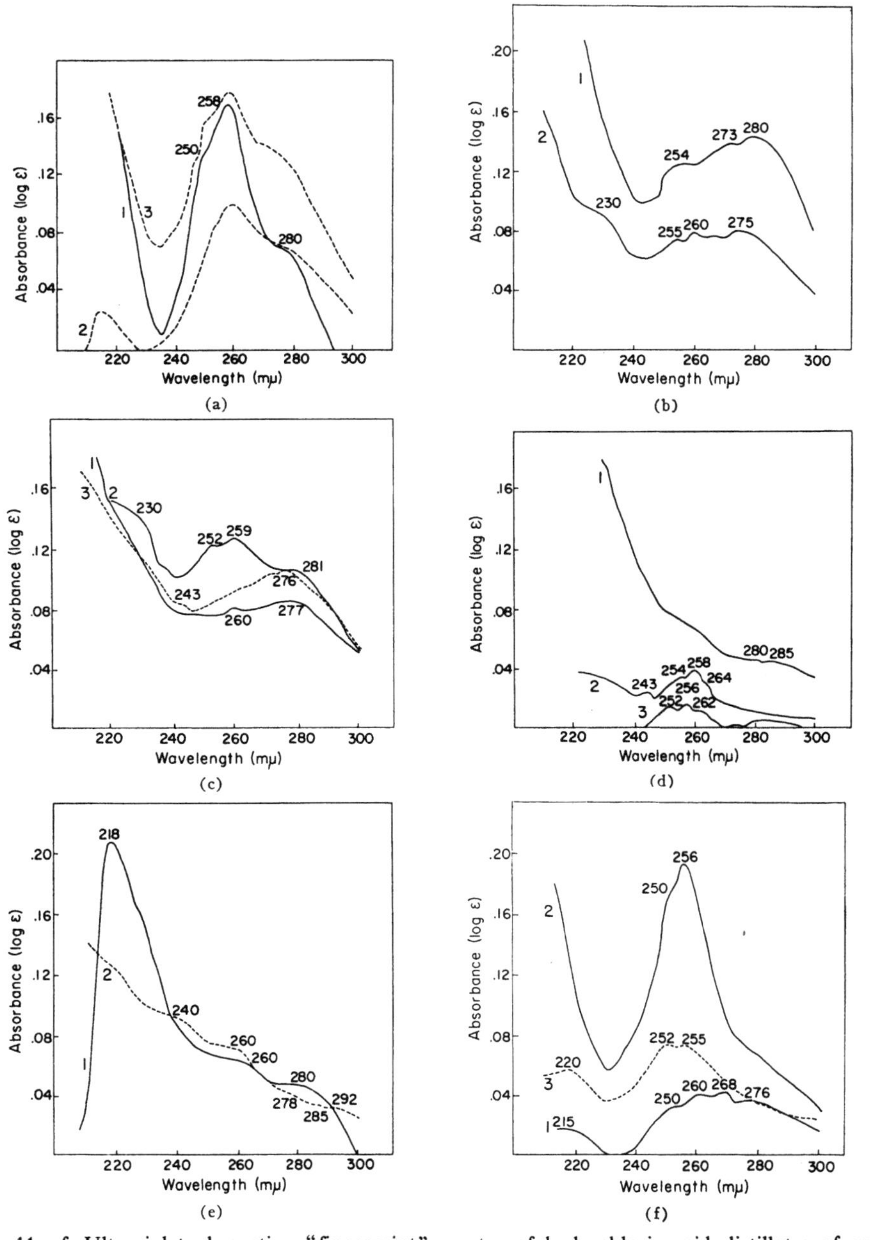

Fig. 11 a–f. Ultraviolet absorption "fingerprint" spectra of hydrochloric acid distillates of some Jurassic geochemical samples showing substances that may interfere with determination of furfurals. a) 1 – Schuler prodeltaic sandstone, Nat. Gas and Oil No. 1 Napper, 9,770–9,774 feet (may represent mixture of furan and anthracene); 2 – Schuler littoral sandstone (same as 1), 9,885 feet (may represent anthracene); 3 – Schuler littoral sandstone (same as 1), 10,184–10,187 feet (may represent mixture of furan and anthracene). b) 1 – Smackover neritic limestone, Ohio No. B-1 Waller, 10,491–10,535 feet (may represent mixture of furan, furfural and purine such as adenine); 2 – Smackover oölitic littoral limestone, Charter No. 1 Bolinger, 10,574–10,575.5 feet (may represent mixture of furan, furfural and pyridine. c) 1 – Smackover oölitic, stromalolitic, littoral limestone, Carter No. 1 Bolinger, 10,568.5–

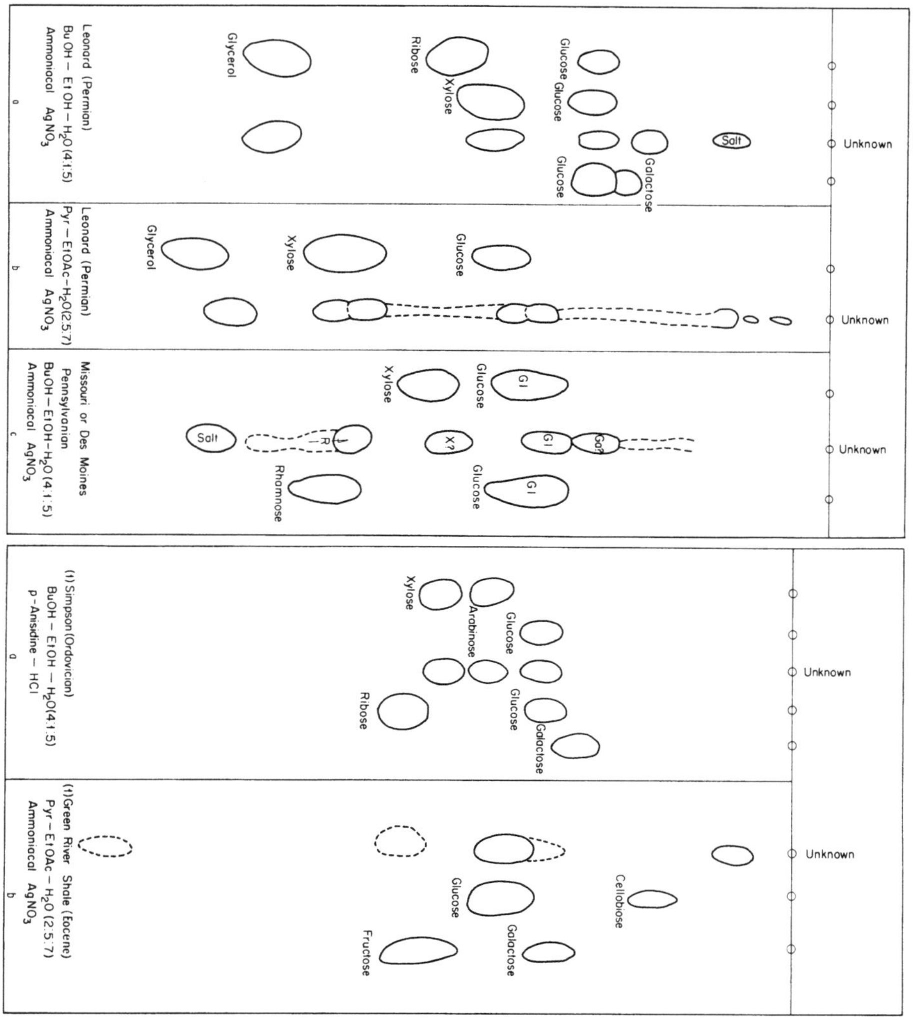

Fig. 12a and b. Carbohydrate paper chromatograms of geochemical samples. a) Leonard Series limestone, Middle Permian, Word County Texas, two samples developed in different solvents. Middle Pennsylvanian Missouri or Des Moines Series limestone, Tom Green County, Texas. b) Simpson limestone, Middle Ordovician, Eaton County, Texas. Green River Oil Shale, Lower Eocene (Mahogany Ledge), Uinta County, Utah [3]

10,569.5 feet (may represent mixture of furan and purine); 2 – (same as 1), 10,560.5–10,561.5 feet (may represent furfural); 3 – (same as 1), 10,603.5–10,605.5 feet (may represent mixture of furan and purine). d) 1 – Smackover oölitic, stromalolitic limestone, Placid No. 1 Odom, 10,250–10,255 feet (may represent mixture of furan and a little furfural); 2 – (same as 1), 10,299.5 feet (may represent pyrazine); 3 – (same as 1), 10,570 feet (may represent aniline). e) 1 – Smackover algal oölitic limestone, Carter No. 1 Bolinger, 10,591.5–10,592.5 feet (may represent naphthalene); 2 – Schuler neritic gray shale, Carter No. 3 Davis, 12,964 feet (may represent mixture of furan and pyrimidine). f) 1 – Schuler littoral pink siltstone, Nat. Gas and Oil No. 1 Napper, 9,265–9,272 feet (may represent mixture of quinolinic acid and furfural); 2 – Buckner littoral red mudstone and anhydrite, Carter No. 1 Hope Fee, 5,770.5 feet (may represent mixture of furan and anthracene); 3 – Schuler alluvial red-green mudstone, Jackson No. 1 Green, 9,098–9,115 feet (may represent pryidine) [7]

2:5:7); from 1 to 80 μl of sample is spotted on chromatogram; development is for about 24 hours; general sprays: (1) aniline: phthalic acid: water-saturated butanol (0.9 g:1.6 ml:100 ml) [21] or (2) a 3 percent solution of p-anisidine:HCl in aqueous n-butanol [22] or (3) a mixture of equal volumes of 2 N $AgNO_3$ and concentrated ammonia.

A solution containing a series of standard monosaccharides is co-chromatogrammed with several samples to be analyzed (Fig. 12). The developed and stained chromatogram is cut into strips and scanned in a densitometer. If a manual densitometer is used, quantitative estimations are made using an area × density method.

An alternate method is the use of a recording densitometer and integrator which gives more reproducible results than the preceding method. Paper chromatography of sugars, however, does not have a reproducibility of more than 10–15 percent.

F. Automated Carbohydrate Analyzer

An analyzer for carbohydrates similar to the automatic analyzers used for amino acids has recently been described [23]. Mixtures of sugars are fractionated as borate complexes on strongly basic anion exchange resin in borate form by elution with borate buffers. The eluate is reacted continuously with a phenol-sulfuric acid mixture and the absorbancy of the colored reaction products is monitored and recorded.

G. Thin-Layer Chromatography of Carbohydrates

The extraction and concentration of free or polymeric carbohydrates is carried out by the methods described above. Thin-layer plates are prepared using silica gel and calcium sulfate, Kieselguhr G, powdered cellulose, or commercially-prepared plates such as Eastman K 301 R can also be used for separation of the sugars. The chromatograms are developed in BuOH-HOAc-H_2O (4:1:5), PrOH-EtOAc-H_2O (3:2:1), or EtOAc-pyridine-H_2O (1:1:3) and are stained with one of the sprays described above (Fig. 13) [24].

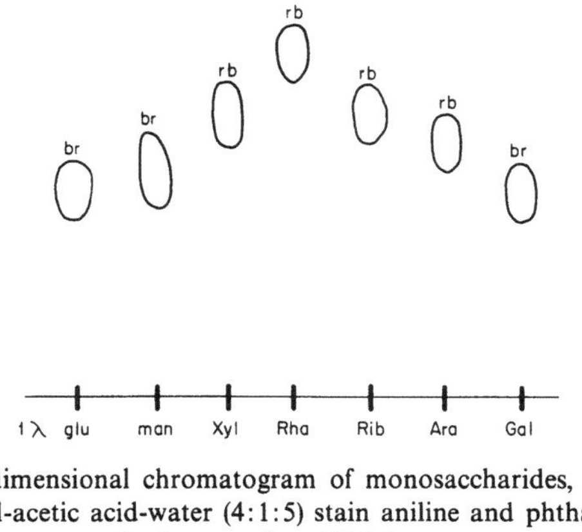

Fig. 13. Thin layer one-dimensional chromatogram of monosaccharides, silica gel plate (Eastman K 301 R), solvent butanol-acetic acid-water (4:1:5) stain aniline and phthalic acid; colors of spots: br, brown; rb, reddish-brown

H. Gas Chromatography of Carbohydrates

Individual sugars may be separated on gas-chromatographic columns as their methyl ethers, acetates or as their trimethylsilyl or chloromethyldimethylsilyl derivatives.

Fully methylated and partially methylated glycosides [25, 26] are obtained by refluxing methylated sugars with methanolic HCl for several hours. The methyl glycopyranosides are separated on Apiezon M on Celite 545, or similar gas chromatographic columns, using thermal conductivity detectors. An alternate method uses flame ionization detectors. One of the major problems in this and following methods of preparation of carbohydrate derivatives is the formation of anomers.

Monosaccharide acetates are prepared [27] by treating 1 to 5 mg of individual sugars with anhydrous sodium acetate (10 mg) and acetic anhydride (1 ml). The crystalline sugar acetate is dissolved in CCl_4, acetone or ethyl acetate, and a 1–2 μl aliquot is injected onto a 5–10 foot gas chromatographic column (Chromo-

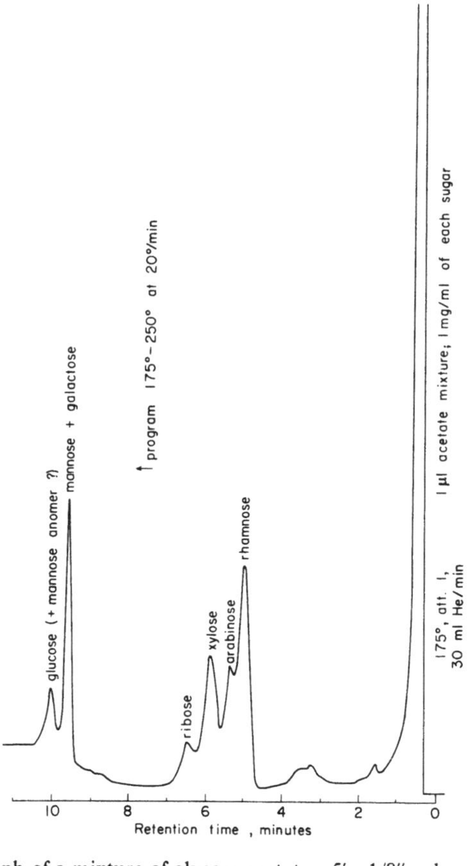

Fig. 14. Gas chromatograph of a mixture of glycose acetates; 5′ × 1/8″ column packed with SE-30 on 60/80 mesh chromosorb W; programmed 175–275° C, att. 1, injector and detectors temps. 300° C, helium flow rate 25 ml/min. 0.2 μl of pentacetates of rhamnose, arabinose, xylose and ribose and hexacetates of glucose, galactose and mannose in EtOAc, 1 mg/ml; hexoses did not separate under these conditions

sorb W, 80–100 mesh and XE-60, ECNSS-M, or other suitable adsorbent) with a flame ionization detector (Fig. 14). The method is relatively freer of anomer formation than the other derivative preparations, but the α- (or β-) anomer of mannose typically is formed.

Alditol acetates are prepared [28] by reducing the glycose mixture with sodium borohydride for three hours; excess borohydride is neutralized with acetic acid and the solution evaporated to dryness; the dry mixture is refluxed for four hours with a mixture containing equal amounts of acetic anhydride and pyridine (*ca.* 1 cc/100 mg of sugar); the solution is cooled and directly injected into the gas chromatograph for analysis.

Trimethylsilyl ethers are prepared [29] by treating 10 mg of carbohydrate with 1 ml of anhydrous pyridine, 0.2 ml of hexamethyldisilazane and 0.1 ml of trimethylchlorosilane for five minutes or longer. From 0.1 to 0.5 μl of the resulting reaction mixture is injected on a gas chromatographic column, 6 feet 0.25 inches packed with 3 percent SE-52 or similar absorbent and equipped with a flame-ionization detector.

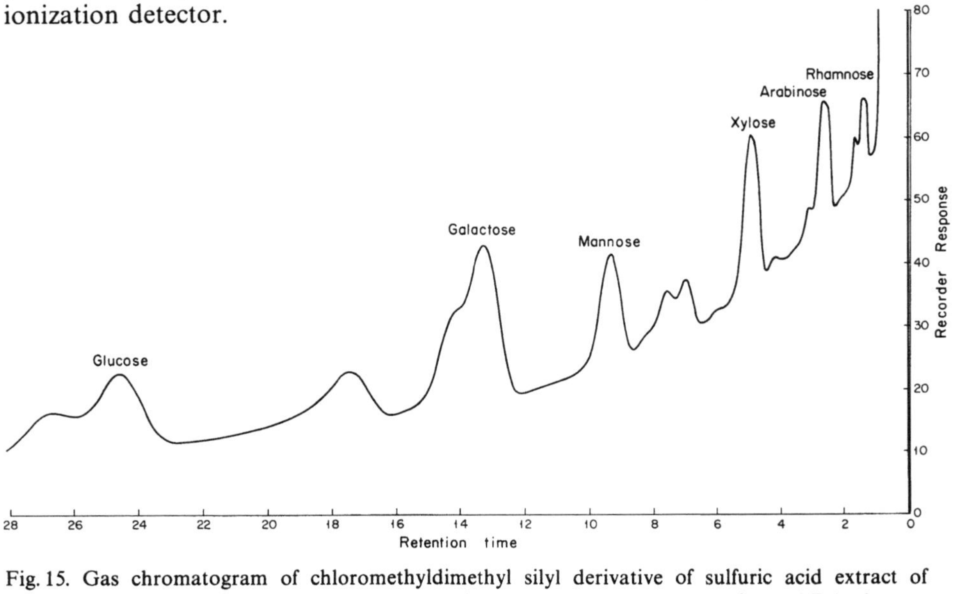

Fig. 15. Gas chromatogram of chloromethyldimethyl silyl derivative of sulfuric acid extract of *Trimerophyton robustius* (DAWSON), Lower Devonian, Cap Aux Os, Quebec; column QF-1, electron capture detector, operating temperature 170° C, He flow rate 30 ml/min, attenuation 8, 0.1 μl of solution added to column

Chloromethyldimethylsilyl ether derivatives of sugars are prepared [30] by reacting 5 mg of each sugar with 1 ml of dichloromethyltetramethyldisilazane and chloromethyldimethylchlorosilane* in either pyridine or dimethylformamide for approximately 30 minutes at room temperature. Microgram quantities are injected directly on the gas chromatography column where flame ionization detectors are used (Fig. 15). When nanogram (10^{-9}) quantities are to be injected, 1 μl of the reagent solution is diluted with 1 ml of n-hexane; in this case an electron capture detector is used.

* Available from Applied Science Laboratory, Inc., State College, Pennsylvania.

I. Enzymatic Detection of Monosaccharides in Geologic Samples

Enzymatic methods have been developed for the analysis of glucose and galactose in geologic samples and for the determination of water-soluble polysaccharides [4].

D-glucose and D-galactose are detected by enzyme-oxidase preparations* The use of these oxidases is discussed by ROGERS [4, 11] and SWAIN and ROGERS [8].

The glucose-oxidase technique is a coupled enzyme system based on the following scheme of reaction:

$$\text{D-glucose} + O_2 + H_2O \xrightarrow{\text{glucose-oxidase}} H_2O_2 + \text{gluconic acid}$$

$$\underset{\text{(o-dianisidine)}}{H_2O_2 + \text{reduced chromogen}} \xrightarrow{\text{peroxidase}} \text{oxidized chromogen}$$

The purification and properties of the enzyme, its commercial applications and specificity are described by BENTLEY [31], UNDERKOFLER [32] and MCCOMB, YUSHOK and BATT [33]. The method is specific for D-glucose with the exception that 2-deoxy-D-glucose is oxidized at 12 percent of the rate for glucose. See ROGERS [4, 11] for protocol in the glucose-oxidase technique.

The galactose-oxidase determinations are based on the following schematic reactions:

$$\text{D-galactose} + O_2 \xrightarrow{\text{galactose-oxidase}} \text{D-galacto-hexodialdose} + H_2O_2$$

$$H_2O_2 + \text{reduced chromogen} \xrightarrow{\text{peroxidase}} \text{oxidized chromogen}.$$

In contrast to the glucose-oxidase enzyme, the galactose-oxidase enzyme also attacks D-galactosamine and other monosaccharides, galactosides and oligosaccharides [34]. ROGERS [4, 11] presents the protocol for the galactose-oxidase techniques and discusses its geochemical applications.

V. Enzymatic Detection of Polysaccharides

The analysis of geological samples for polysaccharides was discussed by ROGERS [4, 11] and by CARLISLE [35]. Most of the carbohydrate materials in rock samples are believed to occur in polymeric form [4, 10, 11] and are separable from the rock matrix only by extraction of the rock sample with acid. This process destroys the glycosidic bonds of polysaccharides to yield monosaccharides, fragmentary polymers, furfurals and other products.

According to ROGERS [4, 11] however, some of the material extractable from rocks with hot distilled water includes enzymatically-detectable polysaccharides, specifically β-1,3 glucans (such as laminarans and related sugars) and β-1,4 glucans (such as celluloses and related sugars); other mixed-linkage polysaccharides that might be detected are cereal glucans, lichenen (from lichens), callose and amylose (from higher plants).

Chitin has been detected by enzymatic analysis in a Cambrian *Hyolithellus* [35] (CARLISLE, 1963). The chitinase enzyme preparation was obtained by treating mushrooms (100 g) with 35 percent (w/v) sodium chloride solution (100 µl) at 4° C

* Available from Worthington Biochemical Corporation, Freehold, New Jersey.

for 48 hours followed by purification on Sephadex. Fragments of fossil material and of locust wing, as a control, were digested with the enzyme preparation for 72 hours and the samples were chromatographed using 1-butanol:ethanol:water (4:1:5) as solvent. N-Acetylglucosamine was shown to be present with ELSON and MORGAN reagents which convinced the author that his determinations of the presence of chitin were valid.

VI. Carbohydrate Components of Some Living Materials of Geologic Importance

Certain carbohydrates are characteristic of, although not in all cases restricted to, certain classes of organisms. For example in the carageenan of Irish Moss (*Chondrus crispus*-Rhodophyta) a 3,6-anhydro-D-galactose residue is joined by a β-1,4 link to a D-galactose 4-sulfate residue in the form of a chain of alternating links, together with other modifications [36]; guluronic and mannuronic acids and 6-deoxy-L-galactose (fucose) are typically found in the Phaeophyta [36], galactosamine is common in the cartilage of animals [36]. The reader is referred to PIGMAN [36] or other treatises on carbohydrates for further information on the restricted occurrences of certain sugars.

As a basis for comparison of fossil carbohydrate residues with those of Recent organisms, analyses of Recent specimens are herein summarized from studies by FOGG [37], HOUGH *et al.* [38], ROGERS [12, 39], SWAIN [17] and SWAIN *et al.* [14, 15].

The acid-extractable monosaccharides of freshwater algae, in addition to glucose and galactose (Fig. 16), include arabinose and xylose. Arabinose has been recorded in gums obtained by aqueous extraction of *Nostoc* and other blue-green algae [37], and as part of an amorphous polysaccharide complex in *Anabaena* [37]. HOUGH *et al.* [38] suggested that the xylose in several blue-green algae they studied is part of a mucilaginous complex.

Among marine algae (Fig. 16), the Chlorophyta had relatively large amounts of rhamnose, the Pheophyta (Fig. 16) are rich in ribose, and the Rhodophyta have abundant galactose, all in addition to glucose. Galactose is a prominent component of the algal polysaccharides agar and carageenan. HOUGH *et al.* [38] and FOGG [37] found rhamnose as a minor but widespread component of mucilaginous polysaccharides in algae they studied.

The carbohydrate content of a typical species of lichen is shown in Fig. 16. It consists almost entirely of hexoses but small amounts of pentoses are present in several species and *Parmelia phycodes* has a considerable content of xylose and rhamnose.

The horsetail *Equisetum arvense* has appreciable amounts of pentoses including arabinose (Fig. 16).

The principal carbohydrates of *Phylloglossum*, a rare cryptogram thought to be a lycopod, were found [41] to consist of sucrose, glucose, and fructose.

The species of bryophytes studied also contain important amounts of pentoses in addition to the generally predominant hexoses (Fig. 16). *Polypodium vulgare* is notable for its high content of mannose as well as of pentoses.

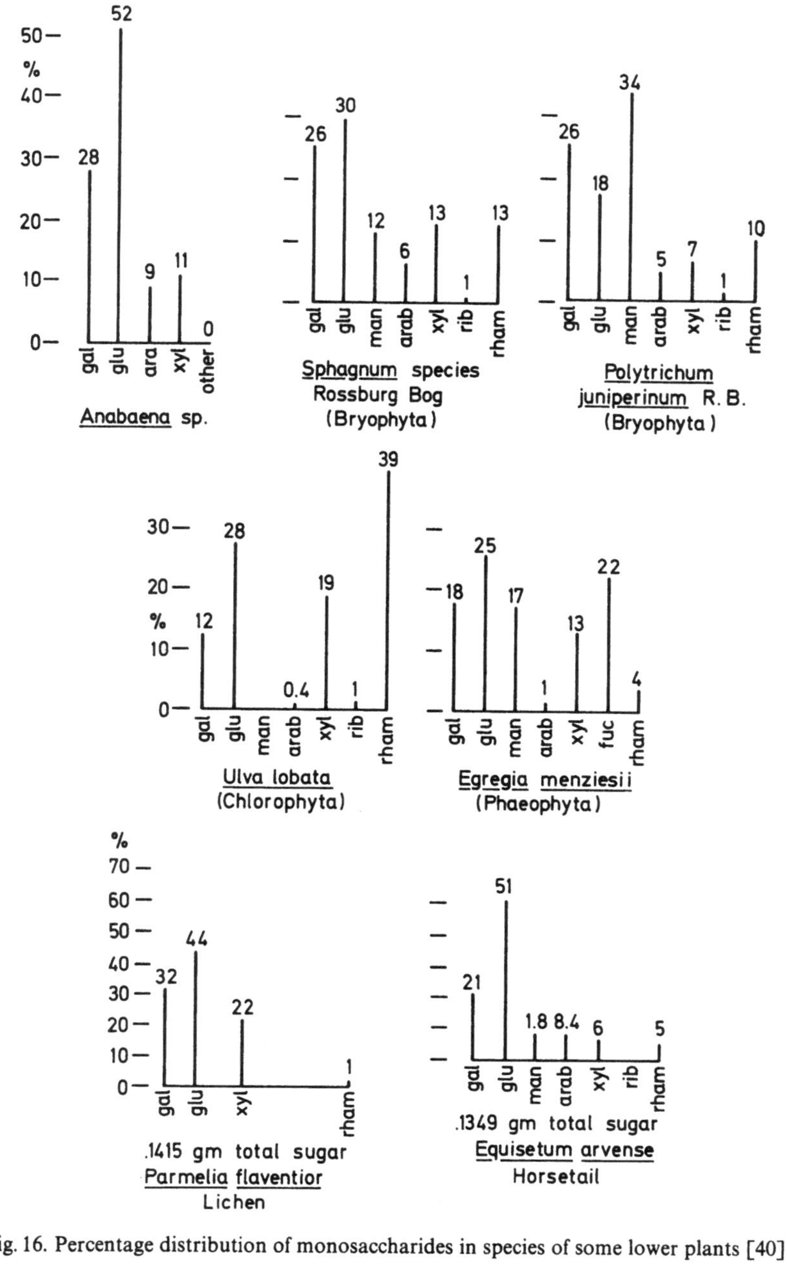

Fig. 16. Percentage distribution of monosaccharides in species of some lower plants [40]

Other species of higher aquatic plants studied by SWAIN, PAKALNS and BRATT [15] contain significant amounts of xylose or arabinose or both, in addition to glucose and galactose (Fig. 17).

The carbohydrates of several species of marine animals (Fig. 17) are mainly glucose and galactose, with lesser amounts of ribose in some species.

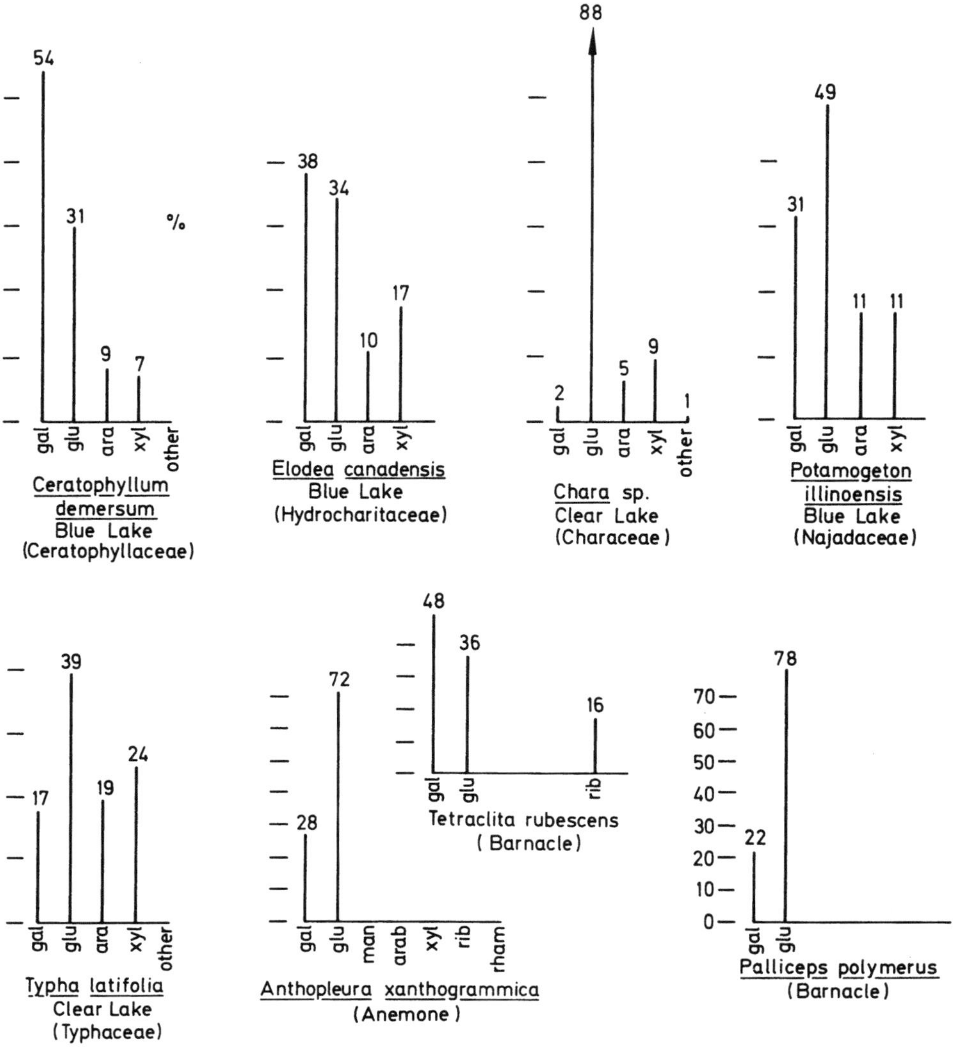

Fig. 17. Percentage distribution of monosaccharides in species of some higher plants and of some invertebrates [40]

VII. Carbohydrate Components of Rock Samples

Precambrian rock samples studied by SWAIN *et al.* [15] contained carbohydrates in small amounts (Table 3). D-glucose and D-galactose were verified enzymatically in several samples and those two sugars as well as arabinose were tentatively separated chromatographically (Table 4). The sources of the Precambrian carbohydrates, if they are indigenous to the rock, could have been Protista or Thallophyta of primitive seaweed type. Remains of Protista have been recorded from the sequence of Precambrian formations from which the carbo-

Table 3. *Total carbohydrate contents of Precambrian rocks* [15], *using phenol-sulfuric techniques*

Formation	Locality	Type of rock	Total Carbohydrate (ppm)	Absorption maximum (mμ)
Coutchiching	Ontario	Chlorite schist	6–14	488
Soudan	Minnesota	Carbonaceous schist	7–8	488
Thomson	Minnesota	Argillite	25[a]	486
Cuyuna	Minnesota	Argillite	16	486
Biwabik	Minnesota	Argillite	0–tr	488
Biwabik	Minnesota	Algal chert	tr	490
Rove	Minnesota	Argillite	0–tr[a]	490
Rove	Ontario	Carbonaceous argillite	15[a]	484
Rove	Minnesota	Anthraxolite	0–tr	490
Wynniatt	Victoria Island	Argillite	9	488
Killiam	Victoria Island	Argillite	0	–

[a] Presence of D-glucose verified by glucose-oxidase test. tr = trace.

Table 4. *Monosaccharides in Precambrian samples (ppm)* [15]

Formation	Galactose	Glucose	Arabinose	Unknown	Total
Coutchiching	–	–	–	a[a]	–
Thomson	0.19 a	0.19 a	0.09 a	–	0.47 a
	–	0.41 b[b]			
Rove, Gunflint	–	0.29 b	–	–	0.29 b
Rove (Anthraxolite)	–	0.87 a	–	–	0.87
		0.30 b	–	–	

[a] Paper chromatographic analysis denoted by "a".
[b] Enzymatic analysis denoted by "b".

hydrate-bearing samples were obtained [42, 43]. Metazoans, the remains of which might also have yielded the above monosaccharides, have not been found in any direct form of fossil remains in the Precambrian sequence.

The total carbohydrates of Paleozoic and younger sedimentary rocks studied [4, 6–10, 14, 15] range from traces to over 600 ppm. Values obtained for various facies of the Onondaga Beds, Middle Devonian of the eastern United States were typical (Table 5). The rock samples richest in carbohydrates are gray shales and limestones of predominantly nearshore origin or where offshore plankton accumulations have occurred. In the case of the Paleozoic rocks [8, 15], thallophytic vegetation and primitive higher plants of both marine and nearshore terrestrial origin are believed to account for most of the residues studied. Furfural residues, presumably derived from carbohydrates were found [7, 9] to be more abundant in littoral Paleozoic and Mesozoic rocks than in marine basinal or alluvial facies of the same ages, presumably owing to a more plentiful supply of plant sources in the littoral environment.

The ribose content [4] of Devonian dark shales (Table 6) may have been contributed by phaeophyte algae or by micro-organisms acting on settled plankton or both. Mannose residues of some of the Devonian dark shales could have been supplied by marine yeasts under nearshore conditions [44]. Glucuronic acid-

Table 5a. *Stratigraphic distribution of glucose equivalent measurements* [8] *in Lower Onondaga formation, Devonian, of Pennsylvania*

Subdivision	Environmental type	Glucose equivalent (ppm)
Esopus shale, eastern Pennsylvania (2 samples)	Prodeltaic and deltaic shale	12 (8–16)
Huntersville chert, western and southern Pennsylvania (8 samples)	Neritic depressional shale bordering low-lying source area	52 (0–238)
Beaverdam Run shale, southern and southwestern Pennsylvania (9 samples)	Neritic depressional black shale	49 (0–125)

Table 5b. *Stratigraphic distribution of glucose equivalent measurements* [8] *in Middle Onondaga formation, Devonian, of Pennsylvania*

Subdivision	Environmental type	Glucose equivalent (ppm)
Scoharie shale, eastern Pennsylvania (4 samples)	Inner neritic to prodeltaic fine clastic shelf	11 (5–18)
Nedrow limestone northwestern Pennsylvania (6 samples)	Inner neritic carbonate shelf bordering low-lying source area	185 (10–392)
Needmore shale central and southern Pennsylvania (11 samples)	Inner neritic carbonate shelf bordering low-lying source area	31 (0–70)

Table 5c. *Stratigraphic distribution of glucose equivalent measurements* [8] *in Upper Onondaga formation, Devonian, of Pennsylvania*

Subdivision	Environmental type	Glucose equivalent (ppm)
Buttermilk Falls, limestone and chert, eastern Pennsylvania (4 samples)	Inner neritic carbonate shelf bordering fine clastic shelf source area	34 (3–67)
Moorhouse limestone, northern Pennsylvania (3 samples)	Inner neritic carbonate shelf, bordering low-lying source area	54 (13–128)
Selinsgrove, limestone, and shale, central Pennsylvania (21 samples)	Inner neritic fine clastic shelf with local carbonate banks	36 (3–70)

Table 6. *Monosaccharide compounds of Paleozoic rocks* [4, 14], *based on paper chromatographic separation*

Geol. unit	Age	Gal	Glu	Man	Rha	Ara	Xyl	Rib	Fru	Gluc acid	Unkn
Antarctic sample	Cambrian		×	×	×			×			
Onondaga Gr.	Devonian		×		×	×	×		×		×
Marcellus Fm.	Devonian	×	×	×	×	×	×	×	×	×	×
Chemung Gr.	Devonian	×	×	×		×	×				
New Albany Fm.	Devonian	×	×			×	×				×

xylose associations in some of the Devonian samples studied may have originated in the hemicelluloses of larger land plants, judging from present day occurrences. Such associations have not yet been found in pre-Devonian samples studied. Glycerol residues [10] which appear to be associated with glucose in some marine Ordovician rocks may have originated in the polyhydroxy-alcohol sorbitol which occurs in some modern seaweeds in the form of food reserve glycosides in the Rhodophyta, or in bacteria.

VIII. Carbohydrate Components of Fossil Specimens

The total carbohydrate contents of a set of 38 specimens of fossils (Table 7) ranged from 4 to 900 ppm. Absorption spectra of the colored products obtained by the phenol-sulfuric acid tests for total carbohydrates indicated that both hexose residues (λ max 490 mμ) and pentoses (λ max 480 mμ) were present in the specimens [1].

Table 7. *Total carbohydrates of Paleozoic fossils* [15] *based on phenol-sulfuric acid technique*

Phylum	Age	*Specimen*, (ppm) of powdered sample	*Matrix*, (ppm) of powdered sample
Foraminiferida	Pennsylvanian	300	–
Coelenterata	Silurian	55	–
Bryozoa	Ordovician	15	104
Brachiopoda	Ordovician-Pennsylvanian	4–383[a]	47–160
Annelida	Ordovician-Devonian	11–163	7–126
Ostracoda	Ordovician-Silurian	153–347	36–132
Trilobitomorpha	Devonian	14–20	15
Plantae	Ordovician-Permian	22–260[a]	50–72

[a] Presence of D-glucose verified by glucose-oxidase test.

In the fossil specimens studied [15] the supposed neritic types are slightly higher in total carbohydrates than the littoral types, the reverse from that observed in the associated rock samples. Paleozoic fossil specimens from geosynclinal rocks studied typically yielded more carbohydrates than the associated matrix; in shelf-environment samples, however, Paleozoic fossils contained less carbohydrates than the matrix. Mesozoic and Cenozoic fossil specimens and associated matrix were more nearly equal in amount.

Only small amounts of monosaccharides have so far been extracted from individual species of pre-Recent fossil plants and animals (Table 8). Glucose has been identified chromatographically in a Devonian brachiopod, in a Mississippian *Lepidodendron* and a Pennsylvanian *Cordaites*. These three fossils also gave positive glucose-oxidase tests for D-glucose. Small amounts of D-galactose have also been detected enzymatically in Devonian, Mississippian and Permian plant fossils. Other chromatographically-determined monosaccharides in the *Lepidodendron* fossil include xylose and rhamnose. HOUGH *et al.* [38] recorded a mucilaginous polysaccharide from the algae *Nostoc* that contained 10 percent rhamnose,

25 percent xylose and the rest hexosans or hexuronic acids. At present it is uncertain whether the rhamnose and xylose of the *Lepidodendron* was part of the woody tissue of the plant or whether a mucilaginous polysaccharide of microbiologic origin may have formed on the decaying *Lepidodendron*.

Except for several unresolved, possibly oligosaccharide, compounds having low R_f values in paper chromatograms, the only polymeric carbohydrates so far found in Paleozoic rock samples studied are those that ROGERS [4, 11] separated from a Devonian dark shale. He characterized a β-1,3-linked polysaccharide by enzymatic means, and also tentatively identified β-1,4-linked polysaccharides. In the enzymatic identification of these polysaccharides ROGERS [4, 11] identified, as enzymatic products, a number of oligosaccharides, such as laminaribiose, cellobiose, etc. which helped to bear out the β-1,3- and β-1,4-linkages.

Table 8. *Monosaccharide components of some Paleozoic fossils (ppm)* [15]

Species	Age	Gal	Glu	Xyl	Ara	Rha	Unkn	Sum
Echinocoelia sp.	Devonian	tr-b[b]	0.31 a[a], b				a	0.31 a
Zosterophyllum sp.	Devonian	6.3 b	20 b					
Callixylon sp.	Devonian	26.2 b	0 b					
Lepidodendron sp.	Mississippian	0.18 a	0.47 a, b	95 a		0.21 a		1.63 a
Lepidophloios sp.	Pennsylvanian	1.17 b	1.34 b					
Cordaites sp.	Pennsylvanian	0.07 a, b 0.005 b	0.09 a, b 0.33 b	0.05 a	0.01 a			0.22 a
Sigillaria sp.	Permian	0–0.02 b	0–3.5 b					

[a] Paper chromatographic analysis denoted by "a".
[b] Enzymatic analysis denoted by "b".

The reports of earlier workers on occurrences of cellulose, hemicellulose, and several individual sugars in Mesozoic and Cenozoic plant fossils and lignite have been summarized by VALLENTYNE [2]. Many of these reports are based on colorimetric and solubility tests, but in a few instances X-ray powder diagrams were taken of the cellulose fraction [45] and D-glucose and other sugars characterized chemically [46, 47]. Glucosamine was isolated from Miocene insect wings and was firmly characterized by crystallization and measurement of specific rotation of the phenylhydantoin derivative [16].

The presence of chitin in a Cambrian Pogonophora, *Hyolithellus micans* Billings was detected by enzymatic analysis [35], the analysis was based on separation of N-acetylglucosamine by paper chromatography of the chitinase-treated fossil material.

The results of these studies seem to confirm existing paleontologic evidence that microscopic Thallophyta and possibly a few larger algae were the principal Precambrian plants. The wide variety of monosaccharide residues in Paleozoic rocks and fossils is an indication that the principal polysaccharides of modern aquatic and perhaps terrestrial organisms were in existence by Devonian time.

The stratigraphic value of the study of carbohydrate residues in rocks appears to rest mainly in their indications as to source-organisms, provenance and environment of deposition of the carbohydrate material.

IX. Thermal Studies of Carbohydrate Residues

When crushed Devonian shale samples were heated in air [4] and in nitrogen [14] and then tested for total carbohydrates, yields were similar or slightly reduced up to about 150–180° C in which range an increase of up to 30 percent in air and 275 percent in nitrogen were obtained. Above 200° C yields became lower, and in the sample heated under nitrogen, yield was zero at 225° C. The clay minerals of the shale appear to act as a protective medium for the carbohydrates up to about 200° C, above which the protective mechanism breaks down. Pre-heating of the sample up to about 150–180° C seems to increase the yield of carbohydrate materials by breaking the glycosidic bonds of polymeric materials.

A sample of Devonian shale was analyzed by the acid-hydrolysis and chromatographic methods described above [4]. A duplicate sample was heated in air at 150° C for 24 hours before analysis. The unheated sample yielded: galactose 0.39 μg/g; mannose 0.58 μg/g; glucose 0.82 μg/g; and fructose 0.78 μg/g. After thermal treatment only 0.30 μg/g of galactose was present.

X. Suggested Additional Research

Some of the problems in the geochemistry of carbohydrates that need further study and for which analytical methods have been developed are the following: (1) further documentation should be made of monosaccharide, oligosaccharide and polysaccharide components of modern plants and animals that have geologic importance; (2) thermal degradation of carbohydrate compounds should be studied, both of isolated compounds and of organic and inorganic mixtures, and thermal stability series established for carbohydrates; (3) geochronologic series of fossil animal and plant remains should be examined for their carbohydrate components to find out whether any recognizable evolutionary changes may have occurred in these compounds with the progress of time; (4) examination of amino sugars and sugar alcohols of modern organisms and their geochemical equivalents may provide valuable information as to metabolic conditions in primitive organisms; and (5) carbohydrate analyses of selected rock facies will enable the geologist to interpret paleoenvironmental conditions to a greater extent.

Several improvements need to be made in the separation and the characterization of carbohydrate residues in fossils and rocks. Acid extraction is apparently effective in separating only 30–60 percent of the polymeric carbohydrate residues in rocks. Gas chromatographic analysis of esters of carbohydrate residues seems at the present time to offer the best possibilities for geological samples, but many improvements need to be made. Enzymatic tests for certain monosaccharides are very useful but need to be developed for more compounds. Polysaccharide analysis by enzymatic means, also very useful, hopefully will be improved and its coverage extended to more compounds.

References

1. Gothan, W.: Neue Arten der Braunkohlenuntersuchung, IV. Braunkohle **21**, 400–440 (1922).
2. Vallentyne, J. T.: Geochemistry of carbohydrates. In: Organic Geochemistry (I. A. Breger, ed.). Oxford: Pergamon Press 1963.

3. PALACAS, J. G.: Geochemistry of carbohydrates. Unpublished doctoral dissertation, University of Minnesota 1959.
4. ROGERS, M. A.: Organic geochemistry of some Devonian black shales from eastern North America: Carbohydrates. Unpublished doctoral dissertation, University of Minnesota 1965b.
5. SWAIN, F. M.: Organic materials of early Middle Devonian, Mt. Union area, Pennsylvania. Bull. Am. Assoc. Petrol. Geologists **42**, 2858–2891 (1958).
6. – Stratigraphic distribution of furfurals and amino compounds in Jurassic rocks of Gulf of Mexico region. Bull. Am. Assoc. Petrol. Geologists **45**, 1713–1720 (1961).
7. – Stratigraphic distribution of some residual organic compounds in Upper Jurassic. Bull. Am. Assoc. Petrol. Geologists **47**, 777–803 (1963).
8. –, and M. A. ROGERS: Stratigraphic distribution of carbohydrate residues in Middle Devonian Onondaga beds of Pennsylvania and western New York. Geochim. Cosmochim. Acta **30**, 497–509 (1966).
9. – Distribution of some organic substances in Paleozoic rocks of central Pennsylvania. In: Coal Science, Advan. Chem. Ser. **55**, 1–21 (1966a).
10. PALACAS, J. G., F. M. SWAIN, and FRED SMITH: Presence of carbohydrates and other organic substances in ancient sedimentary rocks. Nature **185**, 234 (1960).
11. ROGERS, M. A.: Enzymatic detection of polysaccharides in rocks. Submitted for publication to Geochim. Cosmochim. Acta (1967).
12. SWAIN, F. M., M. A. ROGERS, R. D. EVANS, and R. W. WOLFE: Distribution of carbohydrate residues in some fossil specimens and associated sedimentary matrix and other geologic samples. J. Sediment. Petrol. **37**, 12–24 (1967).
13. ROGERS, M. A.: Carbohydrates in aquatic plants and associated sediments from two Minnesota Lakes. Geochim. Cosmochim. Acta **29**, 183–200 (1965a).
14. PRASHNOWSKY, A., E. T. DEGENS, K. O. EMERY, and J. PIMENTA: Organic materials in recent and ancient sediments. Part 1. Sugars in marine sediments of Santa Barbara Basin, California. Neues Jahrb. Geol. Palaeontol. **8**, 400–413 (1961).
15. SWAIN, F. M., G. V. PAKALNS, and J. G. BRATT: Possible taxonomic interpretation of some Palaeozoic and Precambrian carbohydrate residues. In: Adv. in Org. Geochem. 1966, Oxford: Pergamon Press, 461–483 (1968).
16. ABERHALDEN, E., u. K. HEYNS: Nachweis von Chitin in Flügelresten von Coleopteran des oberen Mitteleocans (Fundstelle Geiseltal). Biochem. Z. **259**, 320–321 (1933).
17. SWAIN, F. M.: Stratigraphy and biochemical paleontology of Rossburg peat, north-central Minnesota. In: Essays in stratigraphy (C. TEICHERT and E. YOCHELSON, eds.). Lawrence, Kansas: University of Kansas Press 445–475 (1967).
18. DUBOIS, M., K. A. GILLES, J. K. HAMILTON, P. D. REBERS, and FRED SMITH: Colorimetric method for determination of sugars and related substances. Anal. Chem. **28**, 350–356 (1956).
19. SPEDDING, H.: Infrared spectroscopy and carbohydrate chemistry. Advan. in Carbohydrate Chem. **19**, 23–49 (1964).
20. HALL, L. D.: Nuclear magnetic resonance. Advan. in Carbohydrate Chem. **19**, 51–93 (1964).
21. LEDERER, E., and M. LEDERER: Chromatography. New York: Elsevier 1954.
22. HOUGH, L., J. K. N. JONES, and W. E. WADMAN: Quantitative analysis of mixtures of sugars by the method of partition chromatography, V. Improved methods for the separation and detection of their methylated derivatives on the paper chromatogram. J. Chem. Soc. 1702–1706 (1950).
23. GREEN, J. C.: Automated carbohydrate analyzer: experimental prototype. Nat. Cancer Inst. Monograph. **21**, 447–461 (1966).
24. RANDERATH, KURT: Thin-layer chromatography (D. D. LIBMAN, trans.). New York and London: Academic Press 1964.
25. McINNES, A. G., D. H. BALL, F. P. COOPER, and C. T. BISHOP: Chromatog. **1**, 556 (1958).
26. BISHOP, C. T.: Gas-liquid chromatography of carbohydrate derivatives. Advan. in Carbohydrate Chem. **19**, 95–147 (1964).
27. CASON, JAMES, and HENRY RAPPAPORT: Laboratory text in organic chemistry. Englewood Cliffs, N. J.: Prentice-Hall, Inc. 1950.
28. SAWARDEKER, J. S., J. H. SLONAKER, and ALLENE JEANES: Quantitative determination of monosaccharides as their alditol acetates by gas-liquid chromatography. Anal. Chem. **37**, 1602–1604 (1965).

29. SWEELEY, C. C., RONALD BENTLEY, M. MAKITA, and W. W. WELLS: Gas liquid chromatography of trimethylsilyl derivatives of sugars and related substances. J. Am. Chem. Soc. **85**, 2497–2507 (1963).
30. SUPINA, W. R., R. S. HENLY, and R. F. KRUPPA: Silane treatment of solid supports for gas chromatography. J. Am. Oil Chemists Soc. **43**, 202 A – 204 A, 228 A – 232 A (1966).
31. BENTLEY, R.: Glucose aerodehydrogenase (glucose oxidase). Methods in enzymology (S. P. COLOWICH and N. O. KAPLAN, eds.), vol. 1 (1955).
32. UNDERKOFLER, L. A.: Properties and applications of the fungal enzyme glucose oxidase. Proc. Intern. Symp. Enzyme Chem. (1957).
33. MCCOMB, R. B., W. D. YUSHOK, and W. G. BATT: 2-deoxy-D-glucose, a new substrate for glucose oxidase. J. Franklin Inst. **263**, 161 (1957).
34. AVIGAD, G., D. AMARAL, C. ASENSIO, and B. L. HORECKER: The D-galactose oxidase of *Polyporus arcinatus*. J. Biol. Chem. **237**, 2736 (1962).
35. CARLISLE, D. B.: Chitin in a Cambrian fossil, *Hyolithellus*. Biochem. J. **90**, 1 c, 2 c (1964).
36. PIGMAN, W. *et al.*: The carbohydrates. New York: Academic Press, Inc. 1957.
37. FOGG, G. E.: The comparative physiology and biochemistry of the blue-green algae. Bacteriol. Rev. **20**, 148–165 (1956).
38. HOUGH, L., J. K. N. JONES, W. E. WADMAN: An investigation of the polysaccharide components of certain fresh-water algae. Chem. Soc. 3393–3399 (1952).
39. ROGERS, M. A.: Carbohydrates in aquatic plants and associated sediments from two Minnesota Lakes. Unpublished Master of Science thesis, University of Minnesota (1962).
40. In part after ROGERS [12] and SWAIN *et al.* [14, 15], in part unpublished studies under NASA Grant NGR-24-005-054.
41. WHITE, ELEANOR, AIDA TSE, and G. H. N. TOWERS: Lignin and certain other chemical constituents of *Phylloglossum*. Nature **1967 I**, 285, 286.
42. GRUNER, J. W.: The origin of the sedimentary iron formations: the Biwabik formation of the Mesabi Range. Econ. Geol. **17**, 407–460 (1922).
43. BARGHOORN, E. S., and S. A. TYLER: Microorganisms from the Gunflint Chert. Science **147**, 563–577 (1965).
44. UDEN, N. VAN, and I. TAYSI: Occurrence and population densities of yeast species in an estuarine-marine area. Limnol. Oceanog. **9**, 42–45 (1964).
45. OPFERMANN, E., u. G. RUTZ: Über den Feinbau der Holztracheiden nach Beobachtungen an dem Fasermaterial von fossilem Holz. Papier-Fabr. (Tech.-Wiss. Teil) **28**, 780–786 (1930).
46. KOMATSU, S., and H. UEDA: Chemistry of Japanese plants. II. Composition of fossil wood. Mem. Coll. Sci. Kyoto Univ., Ser. A, **7 A**, 7–13 (1923).
47. BRASCH, J. D., and J. K. N. JONES: Investigation of some ancient woods. Tappi **42**, 913 (1959).

Addendum

Since this chapter was submitted several other studies of fossil carbohydrates have been completed, including examination of free sugars and polysaccharides as well as of acid extractable sugars (A 1, A 2, A 3, A 4). Free sugars have been separated from several species of Paleozoic plants (*Rhynia*, *Calamites*, and *Lepidophloios*); galactose and glucose are generally the most common in the specimens but large amounts of ribose and rhamnose in several specimens suggest a bacterial origin (A 4). The latter may represent contaminating material from either Holocene or older geological sources. The same explanation may hold for those specimens in which free sugars considerably exceed combined sugars (A 1–A 3).

By means of enzymatic analyses using α-amylase, β-amylase and purified cellulase preparations, traces of linear α-1→4 glucopyranose units (starch) and linear β-1→4 glucopyranose (cellulose) were detected in Devonian *Rhynia*, and Pennsylvanian *Calamites* (A 4).

The study of thermal degradation of simple reducing sugars such as glucose is complicated by the fact that dimer or polymer substances (A 5) are formed on heating the sugar. Glucose alone was found to degrade with an activation energy of approximately 22 Kcal/mole in the temperature range 180° to 250° C. In artificial association with montmorillonite, the activation energy of degradation is about 21 Kcal/mole. In artificial association with Devonian black shale, however, the activation energy of degradation is about 25.4 Kcal/mole. The preservation of carbohydrates in ancient geological materials indicates that other factors may be involved in the occurrence of sugars in ancient rocks or perhaps that much of the carbohydrate geochemical material may not be in its original form.

References to Addendum

A 1. SWAIN, F. M., J. M. BRATT, and S. KIRKWOOD: Carbohydrate components of some Paleozoic plant fossils. J. Paleontol. **41**, 1549–1554 (1967).

A 2. – – – Possible biochemical evolution of some Paleozoic plants. J. Paleontol. **42**, 1078–1082 (1968).

A 3. – – – Carbohydrate components of Upper Carboniferous plant fossils from Radstock, England. J. Paleontol. **43**, in press (1969).

A 4. – – –, and P. TOBBACK: Carbohydrate components of Paleozoic plants. Adv. in Organic Geochemistry 1968. Amsterdam: Verweig & Sohn in press (1969).

A 5. SUGISAWA, H., and H. EDO: The thermal degradation of sugars I. Thermal polymerization of glucose. Chem. and Ind. (London) **1**, 892 (1964).

CHAPTER 16

Terpenoids – Especially Oxygenated Mono-, Sesqui-, Di-, and Triterpenes

M. STREIBL and V. HEROUT

Czechoslovak Academy of Sciences,
Institute of Organic Chemistry and Biochemistry,
Prague, Czechoslovakia

Contents

This chapter attempts to review the geochemistry of isoprenoids*. The importance of this field is being increasingly appreciated. Some of this data has previously been summarized in the monographs written by BREGER [1, 2], and articles by DOUGLAS and EGLINTON [3], KROEPELIN [4] and others. Isoprenoid compounds are regarded as significant biological markers because their degree of structural specificity is indicative of biogenesis. The chapter is principally concerned with those isoprenoids containing oxygen functions, although in some cases some attention must also be given to the related hydrocarbons.

* Throughout this chapter the stereochemistry has been indicated where possible. The stereochemistry must be defined for the identification and complete characterization of organic compounds possessing asymmetric centers.

I. Structure and Stability of Terpenoids from the Geochemical Viewpoint

Terpenoids may be characterized as natural products which are built up of 5-carbon subunits arranged according to the isoprene rule [5, 6]. The molecule only rarely consists of one isoprene unit of five carbon atoms (hemiterpenes); frequently they are composed of ten (monoterpenes), fifteen (sesquiterpenes), twenty (diterpenes), thirty (triterpenes), or forty carbon atoms (tetraterpenes or commonly called carotenoids). Polyterpenic compounds are not scarce in nature (caoutchouc, balata, etc.). The terpenes may be hydrocarbons, alcohols and their esters, aldehydes, ketones, oxides or rarely peroxides, but more frequently are acids and lactones. The terpene classes consist mostly of isomeric compounds having a characteristic molecular formula for each group: $C_{10}H_{16}$ for monoterpenes, $C_{15}H_{24}$ for sesquiterpenes including even more highly unsaturated substances. The latter are predominately carotenoids, typical representatives of which are polyenes containing conjugated double bonds. On the other hand, the special case of a relatively highly saturated terpenic alcohol is represented by phytol (I), the alcoholic moiety of an ester of the most important and widely distributed plant pigment, chlorophyll.

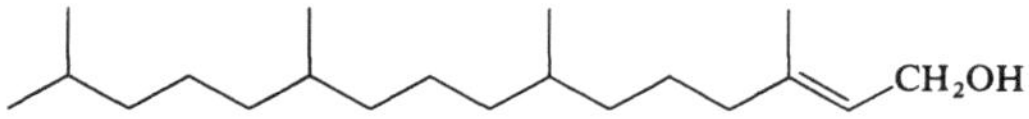

I (Phytol)

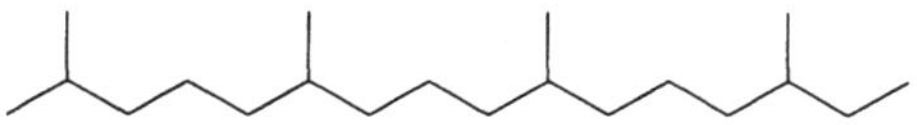
VIII (Phytane)

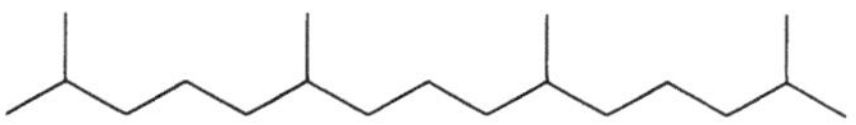
IX (Pristane)

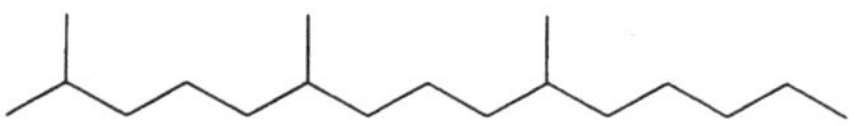
X

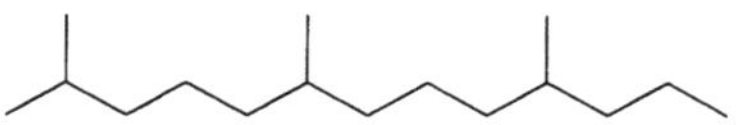
XI

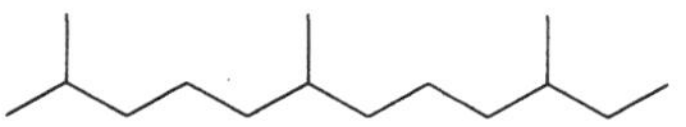
XII (Farnesane)

In general, the occurrence of terpenoids is confined to the plant kingdom; apart from steroids, terpenic derivatives occur in the animal kingdom only rarely and then in small amounts. The latter substances, are derived mostly from plant material which is later modified (pristane, carotenoids, vitamin A, and others). From this point of view, the existence of squalene (originally found in shark liver) is not really typical. From the geochemical point of view, we may deduce that most of the terpenoids and their derivatives are likely to be of plant origin.

Because of the unsaturated character the low-molecular weight terpenes are often unstable and quite readily undergo changes such as polymerization, oxidation, or reduction. However, if such a process results in saturated cyclic or aliphatic derivatives, or if cyclic compounds become aromatized, the final products may be remarkably stable and possibly even survive for millions or hundreds of millions of years in various geological formations [7–10, 10a].

Hydrocarbons are the final residue from the changes affecting isoprenoids during fossilization, particularly because microorganisms do not attack them to any degree. One example of the changes of monoterpenic hydrocarbons, originally preserved as turpentine oil in wood stumps of some species of *Pinus*, was demonstrated by SKRIGAN [11]. During fossilization the principal original component of turpentine, α-pinene (II), which is still partly preserved in the younger layers of wood stumps covered by peat, gradually changes entirely. In the lower layers of fossil stumps of such peat only p-cymene (III) and p-menthane (IV) are present. We may conclude that isomerization and a process analogous to disproportionation on metal catalysts occurred there.

MULLIK and ERDMAN [12] demonstrated to what extent increased temperature may bring about changes, particularly in carotenoids. The increase in temperature and pressure in moist sediments rich in organic material from the sea floor was responsible for the production of large amounts of low-molecular weight aromatic hydrocarbons (benzene, toluene, m-xylene, and others). This fact suggests a direct possibility of deriving crude oil from isoprenoid substances.

Acids from the geochemical viewpoint are interesting because of their abundant occurrence, particularly in the exudates of conifers as the so-called "resin acids". Along with other diterpenes, they constitute the essential part of resins which very often may be found in the fossil or subfossil state (amber, kauri copal, and others). When they become fossilized, decarboxylation is typical; consequently, the hydrocarbon produced is later subjected to changes analogous to those mentioned for pinene. The chemistry of the changes may best be demonstrated by diterpenic acids of the abietane type (V) from various *Pinus* sp., which are converted at the same time to both retene (VI) and fichtelite (VII). The above process which takes place in wood stumps sunk in peat for different lengths of time has been well described by SKRIGAN [13].

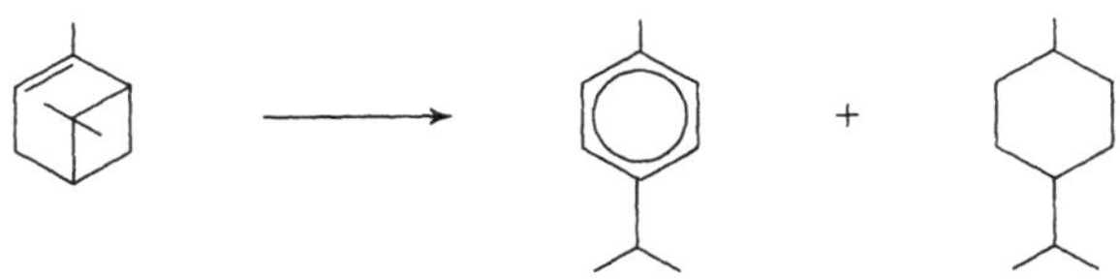

II III IV

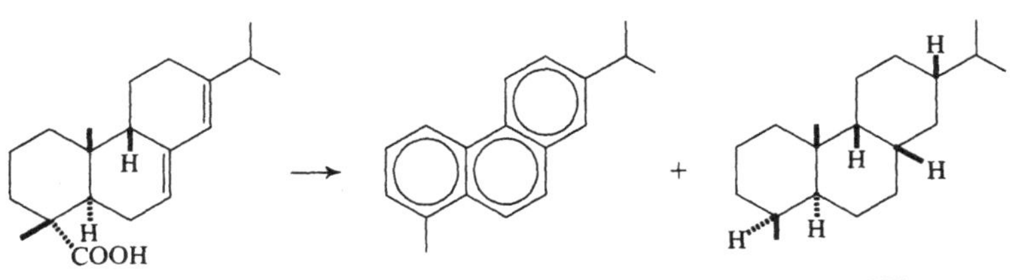

V (Abietic acid) VI VII

Other oxygen-containing terpenoids undergo analogous changes. For instance, the change of triterpenic substances from North Bohemian brown coal has largely been demonstrated in conjunction with the analysis of montan wax [14–19]. (For a review of all the triterpenic substances identified and the products of gradual aromatization and alteration into hydrocarbons, see pp. 410–412, 419.)

Without doubt, the most common organic compounds in various geological formations are products derived from phytol (I). Examples of their occurrence are observed in Precambrian sediments [7–9], chert [10], oil shale [20, 21], crude oil [22–24, 24a] and its fractions [25], low temperature carbonization tars [26, 27], and recent sediments [28]. Several theories on the mechanism of derivation have been developed (cf. BOGOMOLOV [22]), but the essential fact is that the final products are especially the hydrocarbons 2,6,10,14-tetramethylhexadecane (phytane VIII), 2,6,10,14-tetramethylpentadecane (pristane IX), 2,6,10-trimethylpentadecane (X), 2,6,10-trimethyltridecane (XI) and 2,6,10-trimethyldodecane (farnesane XII). The content of these hydrocarbons in Green River oil shale has been studied by ROBINSON *et al.* [21]. From a core taken at various depths ranging from 1,036 to 1,923 feet all the above products of phytol degradation have been isolated; it is worth noting that the relative amount of phytane decreased with depth, whereas that of C_{15}, C_{16}, C_{18}, and C_{19}-hydrocarbons tended to increase. Once formed, phytane appears to be extremely stable as a rule, as can be seen by its presence in the oldest sediments yet investigated [9, 10]. The above relative decrease in quantity might be explained by different mechanisms of changes of phytol in various geological layers rather than by a stepwise degradation of phytane previously formed.

In a large number of cases, it is impossible to follow the degradation of oxygen-containing terpenoids. From SKRIGAN'S paper [11] there is also a probable change of monoterpenic alcohols from turpentine into substances of a less well defined phenolic character; the occurrence of isoprenoid acids in crude oil [29] or shale [29a] may be regarded as one of the final stages in decomposition of phytol or its degradation products. Identification of the final degradation product in a geological formation does not mean it is always possible to ascertain the parent material. For example, the occurrence of polycyclic aromatic hydrocarbons (partly carcinogenic) in soils [30] may be connected with the origin of triterpenoids only in a general way. The existence of a remarkably interesting hydrocarbon, adamantane (XIII), which was isolated for the first time by LANDA and MACHACEK from Hodonin crude oil [31], or its thio analogue, thioadamantane (XIV) [32], suggests a terpenoid origin [22]; however, see addendum.

One of the factors effective in the conversion of terpenoids may be the presence of sulfur. DOUGLAS and MAIR [33] proved by their experiments that even at very low temperatures (up to 150° C) both terpenoids and steroids react in the presence of sulfur to form aromatic hydrocarbons; principally, this process is dehydrogenation. The composition of certain natural sulfur compounds may be derived from the products of cleavage of terpenoids; for instance, 1,4,4-trimethyl-4,5,6,7-tetrahydroisothianaphthene (XV) was obtained [34] from a concentrate of the kerosene fraction of Middle East crude oil.

The fate of volatile terpenoids, which evaporate in enormous amounts into the atmosphere from living plants, was considered by WENT [35]. Since there are no common mechanisms responsible for the degradation of these substances, volatile terpenoids accumulate after rainfalls in river deltas and their sediments are altered to bitumens or asphalts. This suggests a possible relation to petroleum formation.

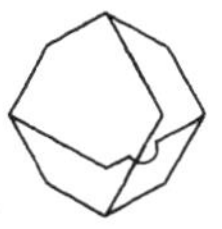
XIII

XIV

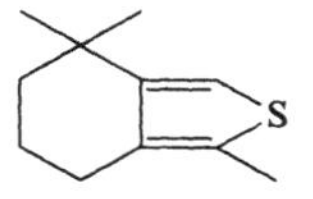
XV

II. Mono- and Sesquiterpenoids

Occurrences of monoterpenic substances are not mentioned to any large extent in the literature. This may be due to a relatively high volatility, as well as instability. Therefore, their occurrence in older formations is something of a rarity.

Terpin-hydrate $C_{10}H_{20}O_2$ (XVI) was found as a crystalline substance, m.p. 120–121° C, in pine logs buried for at least 500 years. Originally called Flagstaffite [36], the identity of this very common terpenic derivative (produced, e.g., by a hydration of pinene) was proved mainly by melting point and mixed melting point determinations.

The hydrocarbons p-cymene, $C_{10}H_{14}$ (III) and p-menthane, $C_{10}H_{20}$ (IV), have been identified in a similar material, as described above [11]. In this case, the age of the buried pine logs has been estimated to be 6–8 thousand years. Hydrocarbons were steam distilled from the logs and the resulting fossil turpentine separated by fractional distillation.

Instead of α-pinene, which is typical of the light fraction of a similar distillate from recent pines or from the wood stumps covered by peat of age 1.5 up to 3 thousand years, the aromatic p-cymene was obtained along with the fully saturated hydrocarbon.

4-Isopropylidene cyclohexanone $C_9H_{14}O$ (XVII) in lignite resins [37] from Central German brown coal, is also undoubtedly the degradation product of some monoterpenic derivative contained in the original resin.

XVI

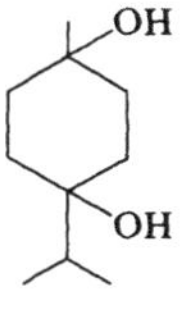

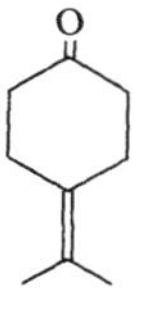

XVII

Adamantane, $C_{10}H_{16}$ (XIII), a crystalline hydrocarbon (m.p. 268° C sealed tube), derived from crude oil (Hodonin, Czechoslovakia), is also related to substances of terpenic character [22, 31]. It was isolated from the light fraction of this high-boiling crude oil in the form of crystals (d 1.07), specifically of the cubic system in accordance with the symmetry of the molecule.

In connection with adamantane, *thiaadamantane* (XIV), $C_9H_{14}S$, (m.p. 320° C sealed tube) has been identified [32] in Middle East oil from Agha Jari, South Iran, and crystallized from the distillate fraction boiling at 78°/3 mm.

Sesquiterpenes have not been widely reported, obviously due to their instability. The several reports do not permit a decision on whether the native terpenic derivative was a sesquiterpene or another polyprenoid, since the fossil substance in question is now a high polymer. One of such statements [38], for example, describes the isolation of a high molecular weight unsaturated hydrocarbon $(C_{15}H_{28})_n$ from peat wax separated by deresinification in organic solvents followed by chromatography on an alumina column. A white crystalline compound, m.p. 230° C, is thought to be related to a hydrogenated bicyclic sesquiterpene. Among the sesquiterpenoid derivatives there are, for instance, hydrocarbons of molecular formula $C_{15}H_{20}$, $C_{15}H_{26}$, and $C_{15}H_{28}$, extracted with alcohol from Italian brown coal [39–41] in the environs of Fognano, or the hydrocarbon $C_{15}H_{26}$, isolated from the brown coal from Steyer, Piberstein [42]. Two similar hydrocarbons of molecular formula $C_{15}H_{26}$ were isolated from North Bohemian coal [43]; it has been proved by the use of mass spectroscopy that these hydrocarbons are tricyclic and substituted with methyl or ethyl-groups; neither mass- nor IR-spectra resemble any of the large number of the saturated sesquiterpenes described so far. In some cases (cf. ROBINSON *et al.* [21] and KOCHLOEFL *et al.* [26]), the hydrocarbon farnesane, $C_{15}H_{32}$ (XII), along with the products of degradation of phytol, was also obtained. Thus, it is not clear whether in the above case we are concerned with a product of the full hydrogenation and/or dehydrogenation of the sesquiterpenes, farnesene or farnesol, or a product derived from phytol.

Destructive distillation of a fossil resin [44] from a bed of lignite deposits in Warkalay, India, produced an oil which on fractionation yielded blue-colored fractions. These apparently contained *azulenes*, typical products of dehydrogenation of a large number of cyclic sesquiterpenoid compounds.

Also from a crude oil there were often isolated alkylnaphthalenes, which may be regarded as products of dehydrogenation of some sesquiterpenoid substances. These compounds were established for the first time in Rumanian crude oil [45], and characterized in more detail in an oil of Trinidad origin [46]. From the fraction (b.p. 267–278° C), alkyl naphthalenes were extracted selectively with furfuraldehyde; from this enriched fraction picrates were formed in ethanol. By treatment with alkali an oil was regenerated from the latter, and this fraction, when cooled, yielded crystals of 2,3,6-trimethylnaphthalene. After the mother liquors were distilled, the fractions were transformed into picrates and/or styphnates, and when subjected to another crystallization, respective derivatives were isolated. Further treatment with base converted them into crystalline 1,2,5-trimethylnaphthalene (m.p. 31° C) and liquid 1,2,7-trimethylnaphthalene.

III. Diterpenoids

Diterpenes are the most common isoprenoids in fossils. In addition to low volatility, this fact may be due to their relatively wide occurrence in the plant kingdom, essentially in a concentrated state. Resins found in trees, particularly conifers, are composed mainly of diterpenoids. The outpours form drops of considerable size, which are quite readily preserved since they were formed by ample flora and frequently have accumulated for long periods of time. This is why at present some of the fossil resins are being extracted for technical purposes (kauri copal, Manila copal or amber). Deposits of amber were formed, e.g., in Baltic area [47–49], during the Lower Oligocene under the climatic conditions favorable for an abundant growth of the *Pinus silvatica*, *P. baltica* and *P. cembrifolia*. For information on their composition and transformation, see Chapter 25.

Fichtelite $C_{19}H_{34}$ (VII) is one of the longest-known and chemically best examined fossil diterpenes. Remains of coniferous trees in peat and lignite beds often contain white or yellowish paraffin-like deposits either crystalline or amorphous in appearance. These deposits have been referred to as "earth resins" (see STERLING and BOGERT [50]) and given different names by various investigators (e.g., fichtelite, phylloretine, tekoretine, scheererite, branchite, hartite, hatschetite, etc.). Nearly all such naturally occurring materials are a mixture of hydrocarbons, the chief constituents being fichtelite (VII) and retene (VI).

The method of isolating pure fichtelite (m.p. 46.5° C, ethanol) and the determination of its constitution have been the topic of several papers; for the reviews, see SIMONSEN and BARTON [51]. Quite commonly it has been isolated by fractional crystallization from organic solvents, but there is no doubt that modern chromatographic methods, especially adsorption column chromatography on silica gel and primarily preparative gas chromatography, would lead to the same compound. For the purification of fichtelite by chromatography on alumina, the reader is referred to SKRIGAN [13], RUZICKA and his co-workers [52, 53], and others who determined its structure, shown by formula (VII). The absolute configuration has only recently been established [54].

Retene $C_{18}H_{18}$ (VI) is the product associating (see the scheme above) the origin of fichtelite with resin acids, particularly of the abietane type. Therefore, it is a widely distributed hydrocarbon occurring predominantly in peats, lignites, and brown coals. To separate it from fichtelite and other impurities, chromatography is again recommended; to characterize this aromatic hydrocarbon (m.p. 99° C) the picrate (m.p. 124° C) can be used, unless it is preferable to employ spectral methods (IR, UV, NMR and mass spectrometry).

Iosene $C_{20}H_{34}$ (XVIII), along with fichtelite, is another hydrocarbon frequently found in various brown coal deposits, as indicated by the large number of synonyms for this fossil diterpene, such as hartite, hofmanite, branchite, rhetenite, krantzil and bombiccite. The exact formula, however, was later recognized by BRIGGS [56], who identified it as α-dihydrophyllocladene (XVIII), which still occurs in different species of recent conifers such as *Phyllocladus*, *Dacrydium*, *Podocarpus*, *Sciadopitys*, *Cryptomeria*.

Iosene has been isolated by extraction from coal samples, using carbon disulphide and crystallization from ethanol. As in the case of fichtelite, chromatographic methods would again be of great assistance in isolating iosene. The pure hydrocarbon melts at 73.4° C, $(\alpha)_D + 23.75°$.

Agathene dicarboxylic acid (XIX) is contained in some recent fossil resins (kauri copal, Manila copal) as demonstrated by RUZICKA and HOSKING [57].

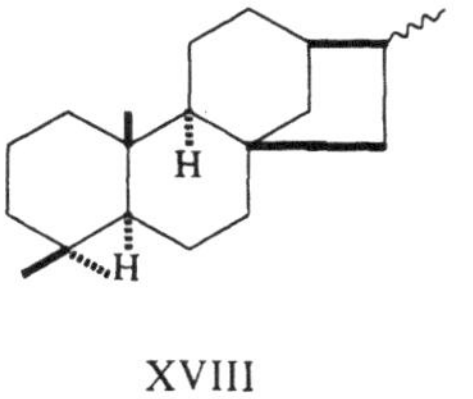

XVIII

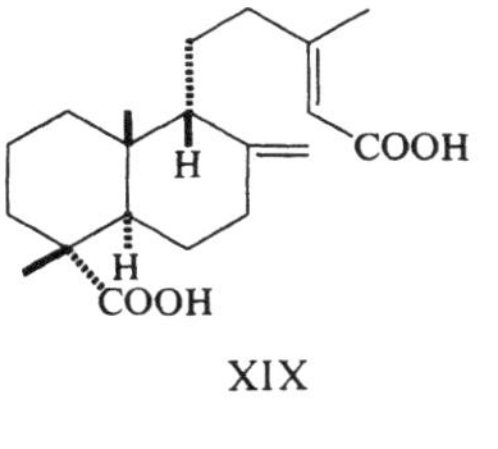

XIX

Phytol (I) undergoes very rapid degradation and thus cannot be found in the fossil formations. It changes to a variety of derivatives occurring abundantly as hydrocarbons or acids, as has already been shown above. Principles for the isolation of substances of analogous character are given in detail in Chapter 1; only one example is provided of the processes aimed at detecting imperceptible amounts of aliphatic hydrocarbons in chert (see ORO *et al.* [10]). The finely powered (100 mesh) siliceous sample was either demineralized by hydrofluoric acid or extracted in a Soxhlet extractor with benzene and methanol (3:1). The hydrocarbon fraction was eluted with heptane from a silica column and the hydrocarbons isolated analyzed by programmed gas chromatography on a capillary column coated with Apiezon L in combination with mass spectroscopy. The whole procedure proved the presence of n-paraffinic hydrocarbons, pristane, $C_{19}H_{40}$ (IX) and phytane, $C_{20}H_{42}$ (VIII).

In another case of isolating isoprenoid acids from crude oil (San Joaquin Valley, California), the following methods [29] have been applied: the concentrate of naphthenic acids was fractionated by distillation, the products were concentrated by thiourea adduction and then analyzed by gas chromatography. A subsequent crystallization yielded altogether four acids of isoprenoid character as follows: 2,6,10-trimethylundecanoic, $C_{14}H_{28}O_2$ (XX), 3,7,11-trimethyldodecanoic, $C_{15}H_{30}O_2$ (XXI), 2,6,10,14-tetramethylpentadecanoic, $C_{19}H_{38}O_2$ (XXII), and 3,7,11,15-tetramethylhexadecanoic, $C_{20}H_{40}O_2$ (XXIII) acids.

2,2,6-Trimethylcyclohexylacetic acid, $C_{11}H_{40}O_2$ (XXIV), was isolated from the last mentioned crude oil and its identity verified by synthesis [58]; we may presume, however, that this product of decomposition was not obtained from diterpenic derivatives but rather from some of the carotenoids.

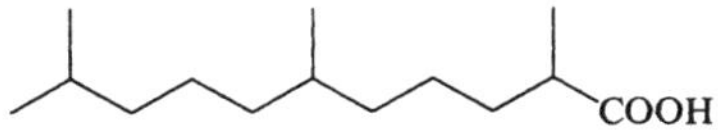

XX

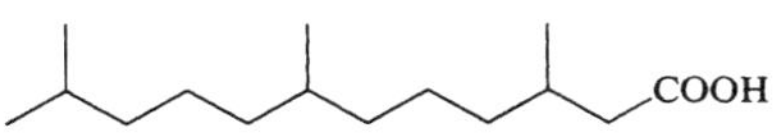

XXI

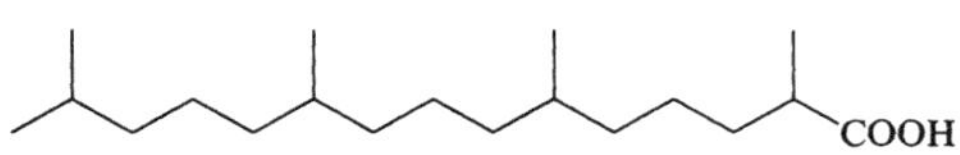

XXII

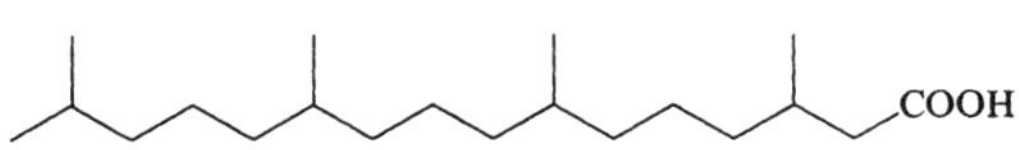

XXIII

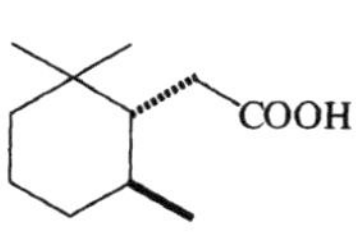

XXIV

IV. Triterpenoids

Triterpenoids constitute a large group, the common precursor of which is the aliphatic isoprenoid, squalene (XXV) [59]. Having in general complicated chemical structures, these compounds usually contain four or five carbon rings. The recent triterpenes are widely distributed in higher plants and, on the basis of their

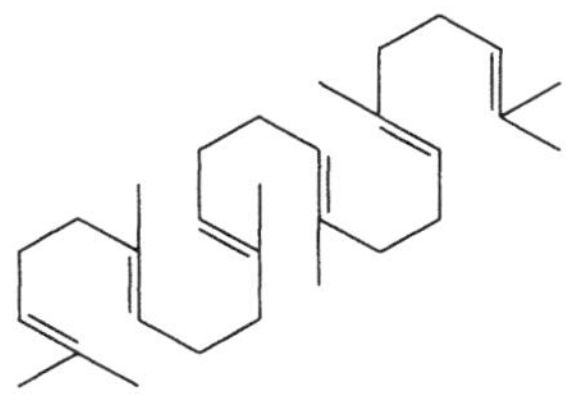

XXV Squalene

chemical structure, are divided into several groups; for more detailed description, see monographs [59, 60]. The large number of asymmetric centers in cyclic triterpenes gives rise to several stereoisomers of high optical activity, which are often preserved even in fossil triterpenes or substances produced later with time [61]. In peats and brown coals, derivatives of betulin (XXVI) have very often been identified; betulin is assigned to the lupeol series present to a large extent in the outer bark of white birch [62]. Pentacyclic triterpenes in the pure state are highly crystalline with high melting points and, considering their chemical structure, exhibit remarkable stability. They can thus be regarded as a source of organic matter remaining from the original biomass in all fossil formations – whether in recent sediments in the original form or in ancient ones in a transformed state. They are presumed to be the source of crude oil hydrocarbons with both aromatic as well as nonaromatic nuclei [63], and thus modified, constitute a component of the optically active material in petroleum [61].

In earlier work on the composition of organic matter in sediments, many effective working methods, particularly chromatography, were not available to produce satisfactory results both in the isolation and identification of triterpenic compounds. In spite of the difficulties mentioned above, RUHEMANN and RAUD [37], having conducted laborious crystallizations from the extract of Central German brown coal, succeeded in isolating a major amount of individual organic compounds and some triterpenes, namely betulin (XXVI), allobetulin (XXVII) and other unidentified compounds with thirty carbon atoms. The same substances, accompanied by some others, were obtained by MUENCH [64] from another German brown coal. Somewhat later LAHIRI [65] used chromatographic methods to separate extracts from Mitchell Main coal, and spectroscopic analysis (IR and UV) of some of the chromatographic fractions ascertained the presence of aromatic and triterpenic compounds. More detailed reviews of earlier papers dealing with triterpenes in fossils were written by BERGMANN [66], JAROLIM and co-workers [15].

Using column chromatography, CARRUTHERS and COOK [67] isolated in high-boiling fractions from an American petroleum, a high-melting crystalline substance of the approximate composition $C_{30}H_{46}O_2$, which was later identified by BARTON and co-workers [68] as oxyallobetul-2-ene (XXVIII) by means of comparison with an authentic specimen. Its presence supports the idea of at least a partial plant origin of the petroleum mentioned above. IKAN and MCLEAN [69], using chromatographic analysis for an unsaponifiable fraction of the extract from Scottish lignite, isolated in the sterol fraction a triterpenic ketone, friedelin (XXIX) and the corresponding alcohol, friedelan-3β-ol (XXX). The triterpenic alcohol (XXX), accompanied by a mixture of sterols, was found in Scottish peat [70] from Cumbernauld. In Israel "young peat", IKAN and KASHMANN [71] identified the same triterpenes (XXIX, XXX) and sterols, employing chromatographic separation on activated alumina and thin layer chromatography.

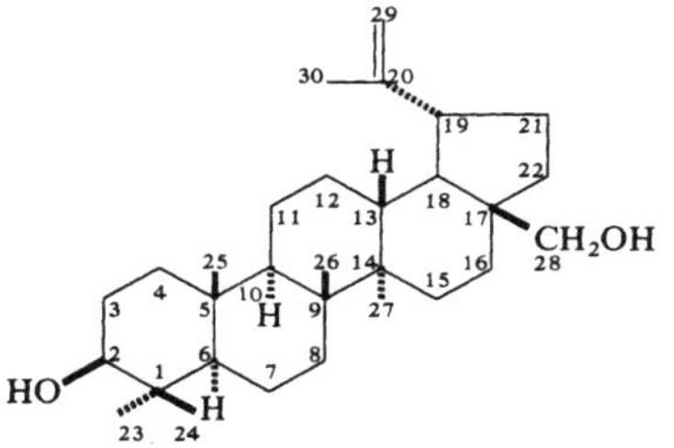

XXVI Betulin

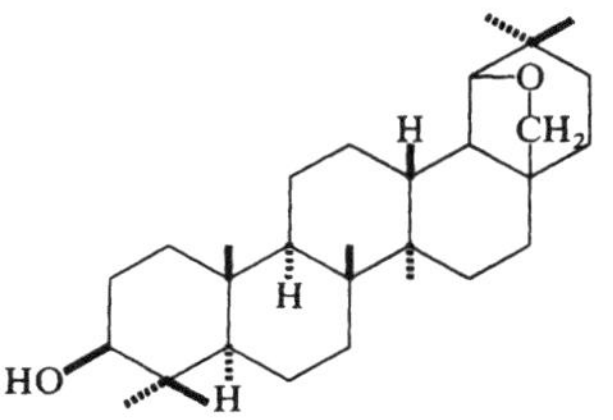

XXVII Allobetulin

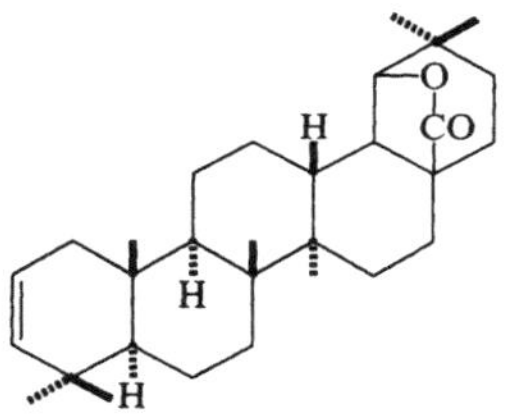

XXVIII Oxyallobetul-2-ene

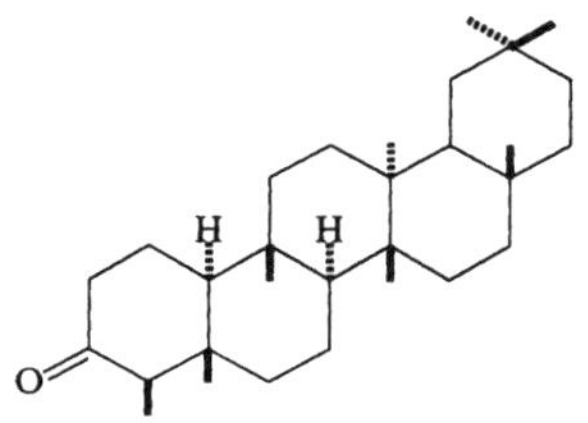

XXIX Friedelin

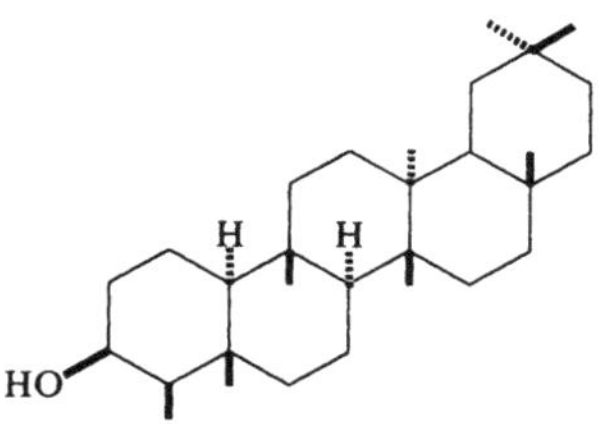

XXX Friedelan-3β-ol

An extensive and so far unfinished study of the chemical composition of the extract from North Bohemian brown coal, the age of which is estimated stratigraphically to be 30–50 × 10^6 years, has recently been initiated by Czechoslovak chemists ŠORM and co-workers [14–19].

The benzene extract of this wax coal from the mine "Josef-Jan", which is industrially produced as montan wax, was first separated by the authors into wax, resin, and asphalt fractions on the basis of solubility in 2-propanol.

More than 30 individual compounds were isolated by the authors by crystallization and various types of chromatography. Employing IR, UV, mass spectroscopy and NMR, and also by comparison with naturally extant triterpenes, they identified more than twenty triterpenic and aromatic compounds; for some other isolated compounds they determined the partial structure, which was established on the basis of spectroscopic measurements, physical constants and chemical reactions. For the review of isolated compounds see Table 2 in Experimental.

Such a complete analysis of one fossil material makes it possible to follow and consequently determine some of the chemical processes that take place during coalification of the deposited material. It may be inferred, for instance, that under conditions enabling dehydrogenation a gradual aromatization of triterpenoid rings took place (XXXIII → XXXVII → XXXIX → XXXV) in the A ring and proceeded through B, C, and D rings up to E rings, with the angular methyl groups being commonly eliminated at the same time. As may be demonstrated by the examined substances, some triterpenic compounds became completely aromatized, forming methylated picenes (XXXVI), and the path was proved experimently in the laboratory by the authors. The other materials, too, showed that chemical transformation of triterpenic substances into aromatic polycondensed hydrocarbons might be regarded as quite common [11, 13, 72–74]. Even from the paleobotanic point of view, such studies are of interest, since some

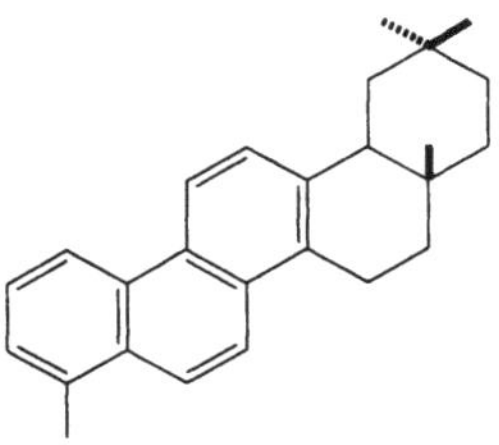

XXXI Octahydrotetramethylpicene

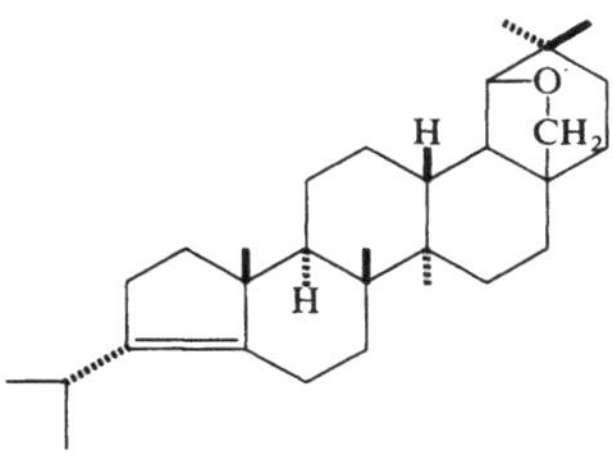

XXXII α-Apoallobetulin

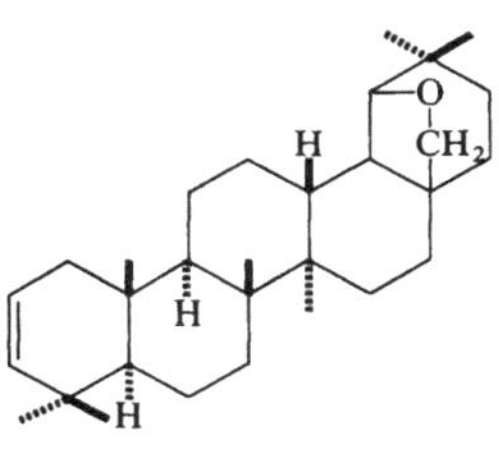

XXXIII Allobetul-2-ene

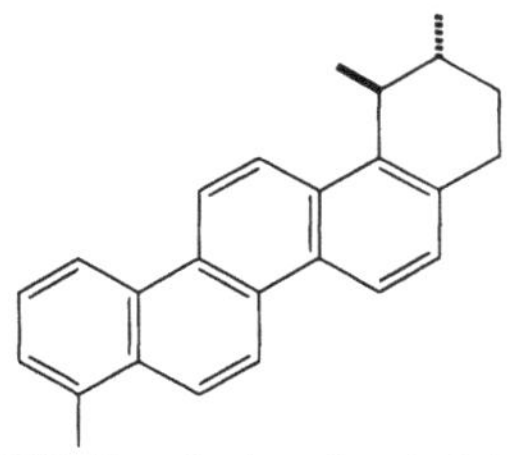

XXXIV Tetrahydrotrimethylpicene

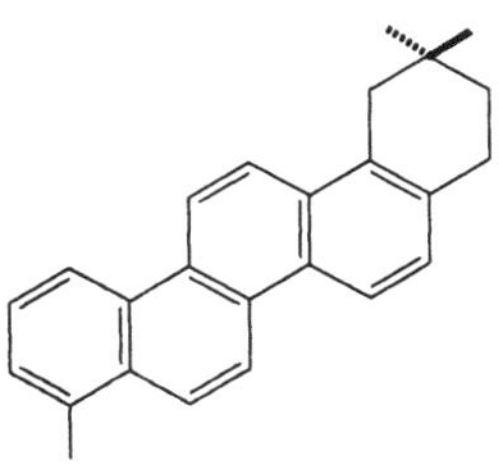

XXXV Tetrahydrotrimethylpicene

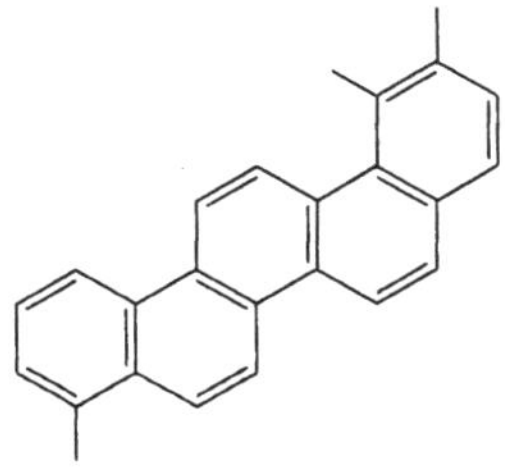

XXXVI 1,2,9-Trimethylpicene

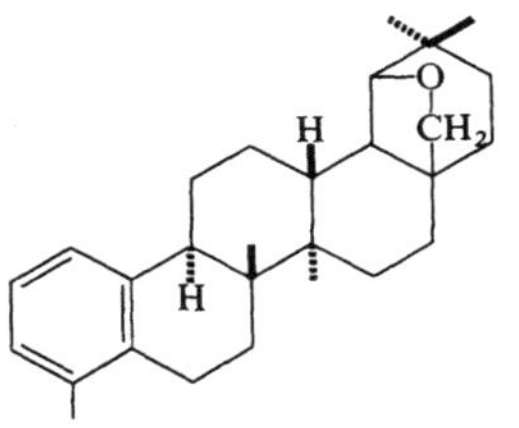

XXXVII Bisnormethyl-2-desoxyallobetul-triene

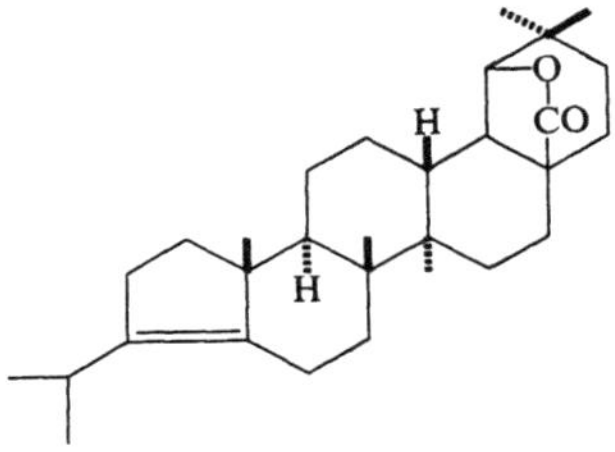

XXXVIII α-Apooxyallobetul-3-ene

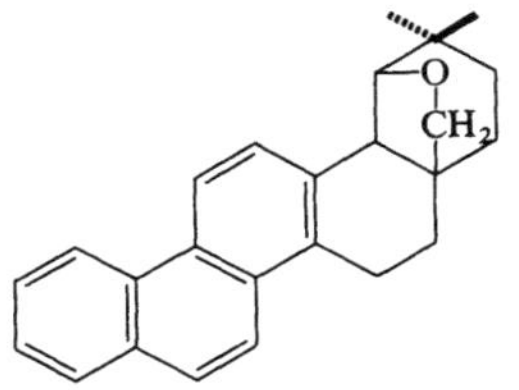

XXXIX Pentanormethyl-2-desoxyallobetul-heptaene

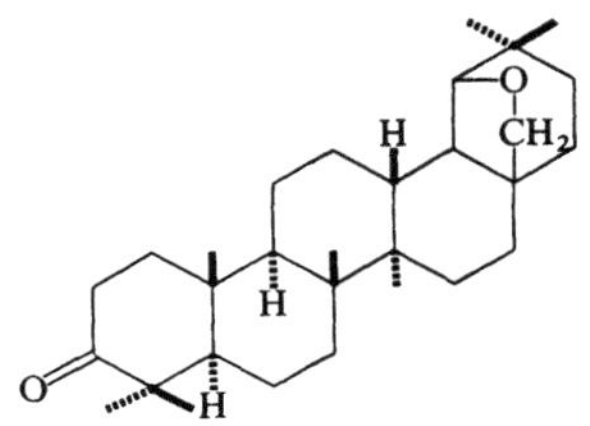

XL Allobetulone

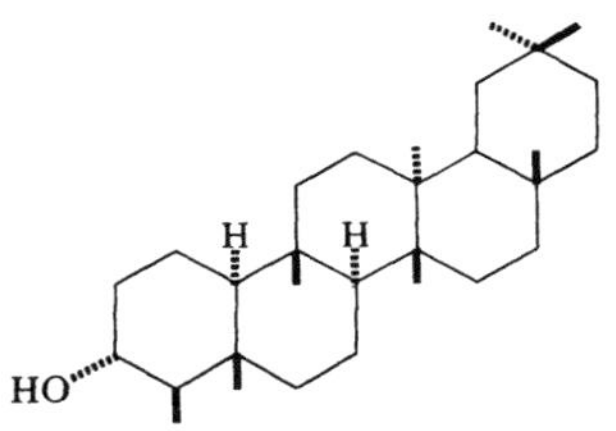

XLI Friedelan-3α-ol

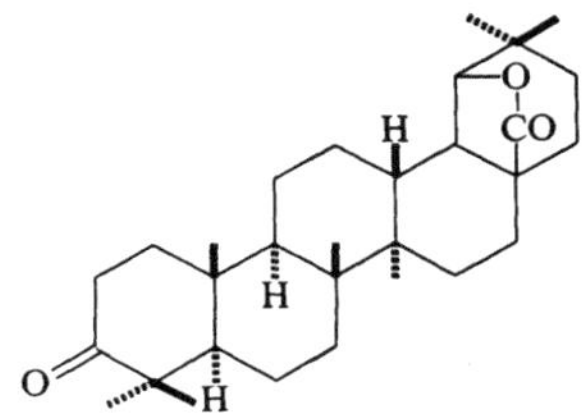

XLII Dehydro-oxyallobetulin

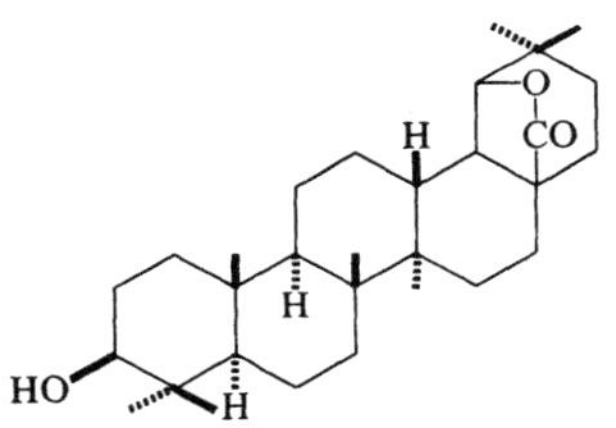

XLIII Oxyallobetulin

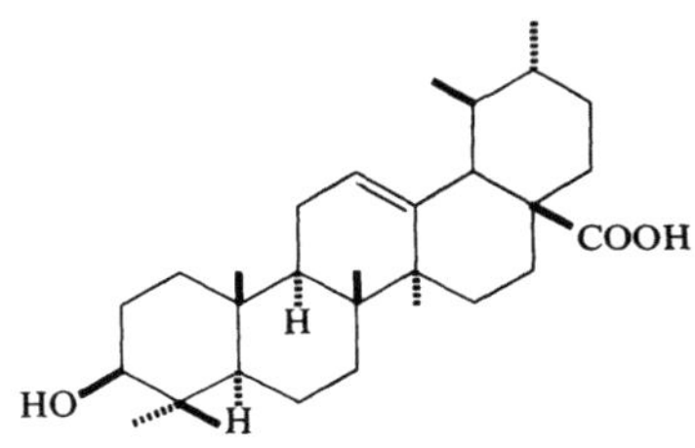

XLIV Ursolic acid

of the discovered triterpenes also occur in recent plants. If sufficient analogous results are available, certain conclusions might possibly be drawn about the flora present in the locality during a given age. Additionally, use might also be made of the results obtained by other methods such as pollen analyses, paleobotanic examination of plant debris and the like.

Using modern experimental microtechniques and physicochemical measurements made on minute quantities of substances, the research workers of Professor CALVIN'S group [75] from California proved the occurrence of biogenic pentacyclic triterpanes in an Eocene and Early Precambrian shales. In the reducing medium of these sediments, the oxygen functions in the original triterpenic and steroidal compounds were gradually removed. The authors believe that the reduction of the original oxygenated isoprenoids was a non-stereospecific reaction, not brought about by microorganisms. Recently HILLS *et al.* [61] isolated three pentacyclic triterpenes from Nigerian crude oil. It was found that the hydrocarbon (m.p. 292–297° C) previously isolated from Green River oil shale bitumen [20] is according to physical data identical with gammacerane, $C_{30}H_{52}$ (XLIV a). Thus this represents a further optically active and exactly defined triterpenic derivative obtained from fossil materials in addition to the substances listed in Table 2.

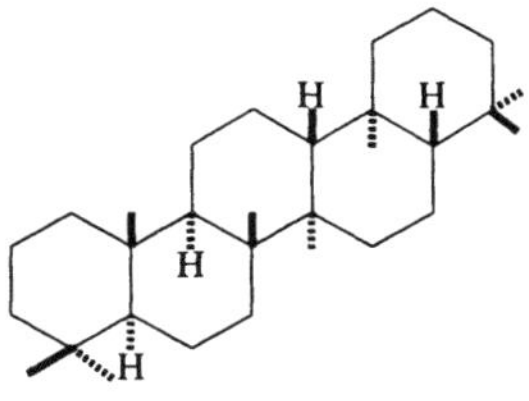

XLIV a

V. Steroids

At present steroids are classified as terpenoids. The former originate from the basic structural unit by a biochemical synthesis through a triterpenoid intermediate type (C_{30}) of squalene, $C_{30}H_{50}$ (XXV), [59] by the elimination of several carbon atoms. From the geochemical point of view, the steroid alcohols containing an aliphatic chain, the so-called sterols (XLV, R = H, $-CH_3$, $-C_2H_5$), are of special interest. All of them possess Δ^5-unsaturation and OH-group in position 3. Owing to several asymmetric centres in the molecule, they can exist as several stereoisomers. The sterols are thus optically active, and the compounds derived from sterols of triterpenes are thought to be responsible for the optical activity of crude oils [63]. Oxidation or, more properly, dehydrogenation reactions, which often participate in geochemical changes of organic substances, give rise to derivatives of aromatic hydrocarbons, particularly phenanthrene, $C_{14}H_{10}$ (XLVI), and probably of anthracene, $C_{14}H_{10}$ (XLVII), in the case of a rearrangement [76, 77].

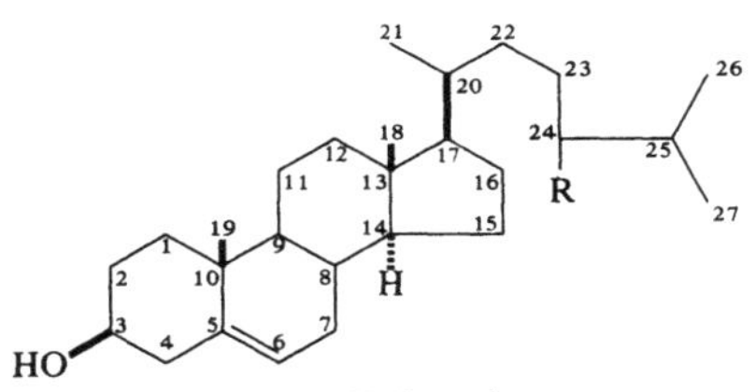

XLV Sterol

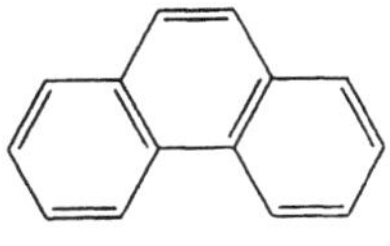

XLVI Phenanthrene

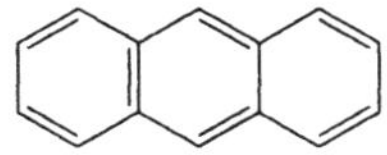

XLVII Anthracene

The sterols are widely distributed as free sterols or esters of higher aliphatic acids in nature. They are classified on the basis of occurrence as zoosterols, phytosterols, mycosterols, and marine sterols. Cholesterol (XLV, R = H) is the principal sterol in vertebrates, but invertebrates have a wide diversity of other sterols occurring in significant quantities. The sterols, common components of the unsaponifiable fractions of lipids, are crystalline in the pure state. From the geological point of view, it is important to recall that they are relatively stable with regard to their chemical structure. Thus they may be preserved either as the original or transformed compound, forming essential constituents of the geological biomass. No wonder they are present in most special geological formations containing organic matter.

For example, recently their presence has been established in chromatographic fractions of lipids from several samples of sea water by the use of the Liebermann-Burchard color reaction [78]. In an analogous way, Liebermann-Burchard visible spectra indicate that sterols other than cholesterol are present in both recent marine and fresh-water sediments [79]. TLC made it possible to confirm the presence of 17-keto-steroids in some medicinal muds [80].

Many papers were devoted to the isolation and identification of sterols from soils, in as much as steroids of various types undoubtedly occur in soil in a practically non-transformed state. It can be presumed that earlier work on this subject (until about the 1950's) is not satisfactorily convincing because the technical standard at that time was insufficiently exact. In accordance with BERGMANN'S [66] opinion, phytosterol isolated by SCHREINER and SHOREY [81], and notably "agrosterol" [82] (m.p. 237° C), are not likely to have been individual sterols but a mixture related to the triterpenic betulins; moreover, the same substances, which were later isolated from montan wax [83] and lignite [84], were certainly rearranged triterpenes. The other earlier works, which tried to prove the existence of sterols in various fossil materials (petroleum, black shales, asphalts, muds, guano, lignite, calcinated shells) by means of unsatisfactory chemical reactions, should at present be evaluated with some reservation (for review, see BERGMANN [66]).

With the help of colorimetric analysis TURFITT [85] studied some samples of English soils for their sterol content and found that cholesterol is rapidly destroyed biologically in aerated soils, but is more stable in wet, acidic conditions in the absence of oxygen. In conjunction with hydrogenation experiments, effective techniques of column chromatography, IR-, and particularly mass spectrometry were helpful in proving the presence of sterols and probably penta- and hexacyclic triterpenes in several subsoils [86]. Recently, chromatographic separation was used for analyzing unsaponifiable fractions of the extracts from various peats and coals [87–89]. In many cases, isolation of individual sterols was carried out successfully. One of the most commonly occurring sterols is β-sitosterol, $C_{29}H_{50}O$ (XLV, R = —C_2H_5), often accompanied with β-sitostanol, $C_{29}H_{52}O$ (XLVIII).

The last-mentioned sterol was isolated from Scottish peat moss samples [87]. It was obtained, along with β-sitostanol, from Canadian peat moss and both compounds were identified on the basis of oxidation and catalytic hydrogenation experiments [88]. IKAN and MCLEAN [69] obtained these compounds from

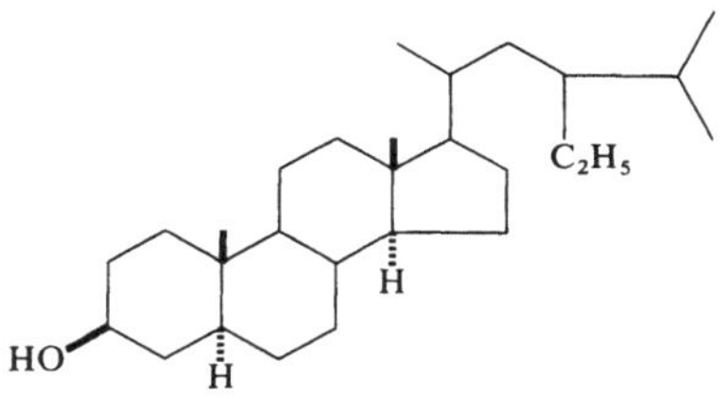

XLVIII β-Sitostanol

Scottish lignite belonging to the Oligocene period, and later also from "young" Israel peat [71]. β-Sitosterol was also identified in a Russian peat [89] showing a high degree of decomposition by a comparison of physical constants, chromatographic values and IR-spectra of isolated and authentic specimens and their esters. From this viewpoint it follows that the occurrence of unchanged sterols in recent fossils may be quite common, which is in accord with the opinion that peats, lignites, and coals are related to each other in having been formed from plant material such as mosses, marine and marsh plants, grasses, and tree parts [71].

The occurrence of biogenic steranes (hydrocarbons possessing a steroidal carbon skeleton without any oxygen functions) in ancient sediments has recently been established [75] as well as triterpanes in non-marine Green River (Colorado) shale of Eocene age and in Soudan Iron shale (Minnesota) of Early Precambrian age, the latter being of marine origin. The steranes C_{27}, C_{28} and C_{29} were separated from the branched-cyclic alkane fraction by preparative GLC and their tetracyclic structures elucidated from their mass spectra by comparison with those of authentic steranes and triterpanes. The authors assume the occurrence of several isomers of each homologous sterane, which may be interpreted in terms of abiogenic reduction of natural precursors (ergosterol, etc.). Mass spectrometry also detected sterane skeletons in other recent marine sediments [90] and petroleum [91]. In addition, MEINSCHEIN [90] refers to the presence of sterenes, unsaturated hydrocarbons with a sterane skeleton, in some sediments and crude oils. Having isolated this group of substances with the use of column chromatography on alumina, he established that their mass spectrometric cracking patterns differed from those of the steranes. He assumes that the unsaturation occurs both in the alkyl group located in the 17-position and in the sterol ring system.

It is noteworthy that the mass spectra of the organic matter of the Orgueille meteorite contain peaks suggesting a steroid skeleton; analogously, neither IR- nor UV-spectra exclude the presence of saturated steroid nuclei (for a review, see MURPHY and NAGY [92]).

VI. Polyterpenoids

The longest isoprenoid molecules are obviously those of natural rubbers whose aliphatic chains are composed of many isoprene units (IL). They have been

$$[-CH_2-\underset{\displaystyle CH_3}{\underset{|}{C}}=CH\cdot CH_2(CH_2-\underset{\displaystyle CH_3}{\underset{|}{C}}=CH\cdot CH_2)_x CH_2-\underset{\displaystyle CH_3}{\underset{|}{C}}=CH\cdot CH_2-]$$

IL

preserved almost in the original state in certain German lignites and are named "monkey hair"; they are presumed to have been derived from latex-bearing plants or trees. In the course of the coalification process the sulfur present caused them to become partially vulcanized [93] (cf. Bergmann [66]).

VII. Taxonomic Importance of Terpenoids – "Biological Tracers"

From the geochemical viewpoint, the presence of isoprenoid compounds may be regarded as especially significant. With regard to the high degree of structural specifity which is indicative of biogenesis, some of them even deserve the denotion "biological markers" and/or "tracers". This term includes especially hydrocarbons such as phytane, pristane, farnesane, and others. If the secondary contamination of a geological substance is excluded, there can be no doubt that living organisms also participated in forming a sediment containing the above mentioned isoprenoids. This fact is clearly demonstrated in this Chapter.

When trying to make another use of the material gathered on fossil isoprenoids, which would be analogous to the chemotaxonomic conclusions on recent flora, we are faced with considerable difficulties. This chapter only provides proof that in earlier sediments there occur primarily the changed products of the original isoprenoids, making it almost impossible to identify the original chemical structure. Fichtelite (VII), for instance, may be derived from any resin acid of the abietane type, whereas retene (VI) may be derived from an even larger group of tricyclic diterpenoids. When studying fossil sediments, we are merely restricted to the analysis of a deposit representing a mean value of material "*in situ*" or the material transported to this place. Unlike the chemotaxonomic studies of the recent flora, there is seldom the chance in geochemical studies to examine material of an undoubtedly uniform origin. Iosene (XVIII), for example, somehow indicated that it possibly originated from some conifers related to those in which at present phyllocladene is found (see p. 407); a paleobotanical proof of much greater value is furnished by the discovery of a fossil that can be identified as the part of the conifer in question. Abundant occurrence of the derivatives of betulin in brown coals [14–19, 37, 64], bears evidence that the source of this fossil material is very probably connected with *Betula* or related plants, which typically contain triterpenes of this type. However, the same cannot be unequivocally said about the origin of crude oil, although it has been confirmed that a derivative of betulin is also present there. The conclusion of a partial plant origin for the crude oil under consideration is possibly very tenable.

VIII. Experimental Procedure

Flow diagram I presents the procedure of the isolation of terpenic substances from North Bohemian brown coal [14–19]. The resin fraction (1,500 g) was chromatographed on 15,000 g neutral alumina with the use of a special eluotropic series (see Table 1). A–E portions obtained by using a solvent or mixtures of solvents were rechromatographed on a sufficient excess of neutral alumina by means of the same solvents stated above. Thus, portion A (225 g) was chromatographed on 7,000 g of alumina and separated into 232 fractions which gradually crystallized to yield some compounds. Analogously, portion B (390 g) was chromatographically separated into 667 fractions, portion C (200 g) into 275 fractions, portion D (90 g) into 170 fractions, and portion E (570 g) into 183 fractions.

Table 1. *Chromatography of the resin fraction of Bohemian montan wax*

Part	Solvent	Total volume litres	Total weight grams	Description
A	Petroleum ether	450	225	fatty consistence yellow-brown
B	Benzene	350	390	semisolid red-brown
C	Chloroform	350	200	semisolid red-brown
D	Chloroform-alcohol (1:1)	300	90	solid red-brown
E	Chloroform-alcohol (1:1)+5 percent acetic acid	300	570	solid black

As a further example of the procedure, the separation of portion B is described in more detail. Portion B was first adsorbed in a chloroform solution on unactivated alumina (500 g) and after evaporation of the solvent it was transferred, together with this alumina, to the top of a column of neutral alumina (9,000 g) of activity (III). Elution was carried out subsequently by means of petroleum ether, chloroform and ethanol, or, respectively, their mixtures, and 1.5 l fractions were collected. After evaporation of the solvent the dry residues were dissolved in a mixture of benzene and chloroform and allowed to crystallize. The crystalline mixtures were separated further by fractional crystallizations until individual compounds with constant melting points were obtained. Thin layer chromatography served as a control for the determination of the contents of chromatographic fractions as well as a criterion of the purity of the isolated components. Fig. 1 shows the R_f values of some oxygenated triterpenes isolated in this way. From the fractions 40–163, which were eluted by means of petroleum ether, the substance (XXIX) (m. p. 253–255° C) was thus crystallized. The IR spectrum (Fig. 2) showed a peak characteristic for a keto group in a six-membered ring (1,709 cm^{-1}) which was identical with the spectrum of an authentic specimen of friedelin. In the fractions 225–258 eluted by means of a petroleum ether-benzene mixture, a substance (m. p. 285–286° C) crystallized which was identified, according to the mixed melting point with its acetate and by the comparison of the IR-spectrum (Fig. 2) with that of an authentic standard, as friedelan-3β-ol (XXX). This proof was also confirmed by the fact that oxidation of this substance with chromium trioxide in pyridine yielded the ketone friedelin (XXIX). In the same way the structure of substance (XLI) was also confirmed.

Other physical methods were also utilized for the determination of some structures of the isolated substances. For example, the structure of substance (XXXI) isolated from eluate A was determined as follows: the mass spectrum of this substance contained a molecular peak at mass m/e 342; thus its elementary composition was corrected to $C_{26}H_{30}$, which was not possible to establish

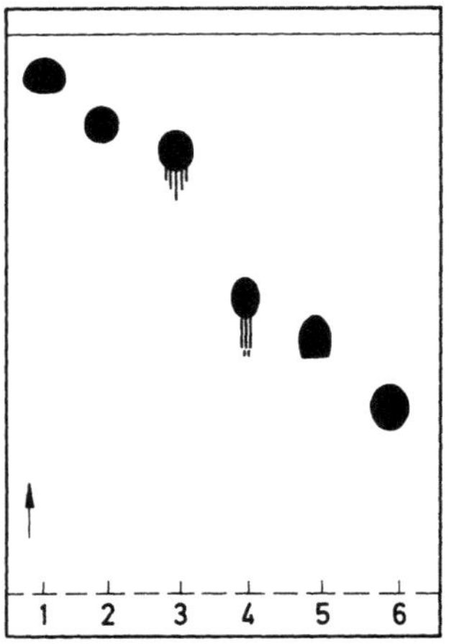

Fig. 1. T.l.c. diagram of some oxygenated triterpenoids isolated from brown coal: 1, friedelin; 2, allobetulone; 3, friedelan-3β-ol; 4, friedelan-3α-ol; 5, oxyallobetulin; 6, betulin. Kieselgel G, chloroform with ether (8 percent)

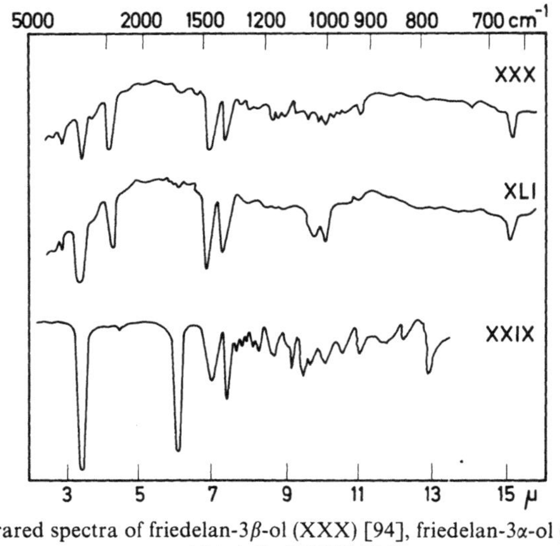

Fig. 2. Infrared spectra of friedelan-3β-ol (XXX) [94], friedelan-3α-ol (XLI) [94] and friedelin (XXIX) [15]

previously on the basis of the elementary analysis alone. The UV- and especially the IR-spectrum contained peaks for an aromatic system of the phenanthrene type and, moreover, the IR-spectrum proved the presence of the geminal dimethyl group $(CH_3)_2C<$, which was also found in the partially hydrogenated product (1,368; 1,380; 1,387 cm^{-1}). By dehydrogenation with selenium at 350° C, either 2,9-dimethylpicene or the substance (XXXV), which was also isolated from brown coal, were obtained. The NMR spectrum detected three methyl groups always attached to a fully substituted carbon atom (τ 8.94; 4.05; 4.11) and one methyl group linked to an aromatic ring (τ 7.28). Only the composition of octahydrotetramethylpicene corresponds in all points to these findings for

Table 2. *Triterpenoic and related compounds isolated from Bohemian montan wax*

Number	Compound	M. P. (°C)	Formula	$[\alpha]_D^{20}$	Functional groups
	?	227	$C_{30}H_{52}$	−28.8	hydrocarbon
XXXI	Octahydro-2,2,4a,9-tetramethylpicene	233–235	$C_{26}H_{30}$	−34.2	aromatic hydrocarbon
XXXII	α-Apoallobetulin	203–206	$C_{30}H_{48}O$	+65.5	unsaturated ether
XXXIII	Allobetul-2-ene	249–250.5	$C_{30}H_{48}O$	+47.6	unsaturated ether
XXXIV	Tetrahydro-1,2,9-trimethylpicene	230–231.5	$C_{25}H_{24}$	+50.0	aromatic hydrocarbon
	?	275–275.5	$C_{25}H_{24}$	+146.1	aromatic hydrocarbon
	?	204–205	$C_{25}H_{24}$		aromatic hydrocarbon
XXXV	Tetrahydro-2,2,9-trimethylpicene	251–252	$C_{25}H_{24}$	0.0	aromatic hydrocarbon
XXIX	Friedelin	253–255	$C_{30}H_{50}O$	−21.6	ketone
XXXVI	1,2,9-Trimethylpicene	277	$C_{25}H_{20}$		aromatic hydrocarbon
XXXVII	23,25-Bisnormethyl-2-desoxyallobetul-1,3,5-triene	261	$C_{28}H_{40}O$	+8.4	aromatic ether
XXXVIII	α-Apooxyallobetul-3-ene	289–291	$C_{30}H_{46}O_2$	+70.3	unsaturated lactone
XXVIII	Oxyallobetul-2-ene	363	$C_{30}H_{46}O_2$	+75.2	unsaturated lactone
XXXIX	23,24,25,26,27-Pentanormethyl-2-desoxyallobetul-1,3,5,7,9,11,13-heptaene	250	$C_{26}H_{28}O$	+150.8	aromatic ether
XXX	Friedelan-3β-ol	285–286	$C_{30}H_{52}O$	+20.7	alcohol
XL	Allobetulone	228–229	$C_{30}H_{48}O_2$	+84.4	ketoether
XLI	Friedelan-3α-ol	310	$C_{30}H_{52}O$	+24.2	alcohol
XLII	3-Dehydrooxy-allobetulin	338	$C_{30}H_{46}O_3$	+91.3	ketolactone
XXVII	Allobetulin	266	$C_{30}H_{50}O_2$	+50.7	hydroxyether
XLIII	Oxyallobetulin	347	$C_{30}H_{48}O_3$	+46.0	hydroxylactone
XXVI	Betulin	254–255	$C_{30}H_{50}O_2$	+26.4	unsaturated diol
	?	330			trisubst. double bond, polyalcohol
	?	267–269	$C_{30}H_{50}O_2$		hydroxyketone
	?	196–198	$C_{30}H_{52}O_4$		
XLIV	Ursolic acid	285–287	$C_{30}H_{48}O_3$	+63.3	carboxylic acid
	?	253–256	$C_{30}H_{50}O_3$	+83.8	polyalcohol

this substance. The results of the analyses of substance (XXXV) also facilitated the determination of its structure in a similar manner.

At present thin-layer [95, 96] and gas chromatography [97–99] of triterpenic compounds have been elaborated in detail and new papers on mass spectroscopy [100] of these compounds are being published.

Flow diagram 1

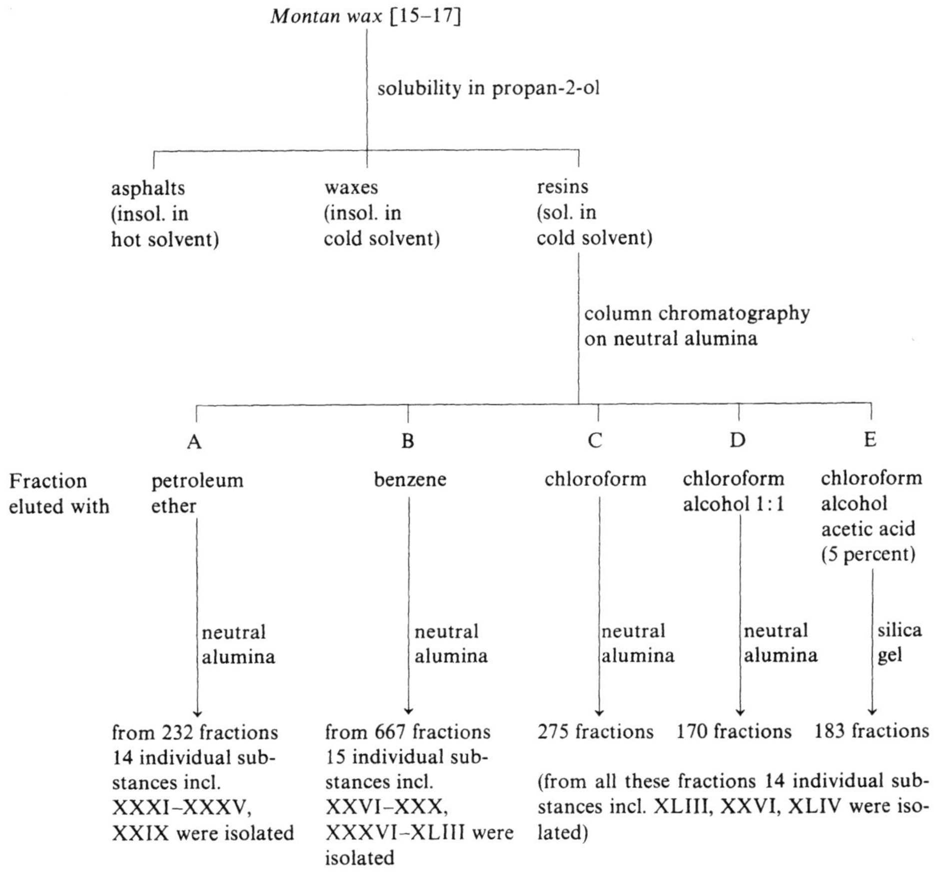

References

1. BREGER, I. A.: Organic geochemistry. Oxford: Pergamon Press 1963.
2. – Geochemistry of lipids. J. Am. Oil Chemists' Soc. **43**, 197–202 (1966).
3. DOUGLAS, A. G., and G. EGLINTON: The distribution of Alkanes. In: Comparative phytochemistry (T. SWAIN, ed.), p. 57–77. London: Academic Press 1966.
4. KROEPELIN, H.: Ergebnisse, Methoden und Probleme der organischen Geochemie. Fortschr. Mineral. **43**, 22–46 (1966).
5. RUZICKA, L.: History of the isoprene rule. Proc. Chem. Soc. **1959**, 341–360.
6. – The isoprene rule and the biogenesis of terpenic compounds. Experientia **9**, 357–367 (1953).
7. MEINSCHEIN, W. G., E. E. BARGHOORN, and J. W. SCHOPF: Biological remnants in a Precambrian sediment. Science **145**, 262–263 (1964).
8. EGLINTON, G., P. M. SCOTT, T. BELSKY, A. L. BURLINGAME, and M. CALVIN: Hydrocarbons of biological origin from a one-billion-year-old sediment. Science **145**, 263–264 (1964).
9. BELSKY, T., R. B. JOHNS, E. D. MCCARTHY, A. L. BURLINGAME, W. RICHTER, and M. CALVIN: Evidence of life processes in a sediment two- and one-half-billion-years-old. Nature **206**, 446–447 (1965).
10. ORÓ, J., D. W. NOONER, A. ZLATKIS, S. A. WIKSTRÖM, and E. S. BARGHOORN: Hydrocarbons of biological origin in sediment about two billion years old. Science **148**, 77–79 (1965).

10a. BLUMER, M.: Organic pigments, their long-term fate. Science **149**, 722–724 (1965).

11. Skrigan, A. I.: Composition of turpentine from a swamp rosin 1,000 years old. Dokl. Akad. Nauk SSSR **80**, 607–609 (1951); Chem. Abstr. **46**, 5337 (1952).
12. Mullik, J. D., and J. G. Erdman: Genesis of hydrocarbons of low molecular weight in organic-rich aquatic systems. Science **141**, 806–807 (1963).
13. Skrigan, A. I.: Preparation and utilisation of fichtelite and retene. Tr. Vses. Nauchn.-Tekhn. Soveshch. Gorki **1963**, 108–115 (1964); Chem. Abstr. **62**, 10664 (1965).
14. Jarolím, V., M. Streibl, M. Horák, and F. Šorm: Isolation of triterpenes from North Bohemian brown coal. Chem. Ind. (London) **1958**, 1142–1143.
15. – – K. Hejno u. F. Šorm: Über einige Inhaltsstoffe des Montanwachses. Collection Czech. Chem. Commun. **26**, 451–458 (1961).
16. – K. Hejno, M. Streibl, M. Horák u. F. Šorm: Über weitere Inhaltsstoffe des Montanwachses. Collection Czech. Chem. Commun. **26**, 459–465 (1961).
17. – – u. F. Šorm: Über einige weitere Inhaltsstoffe des Harzanteils des Montanwachses. Collection Czech. Chem. Commun. **28**, 2318–2327 (1963).
18. – – – Struktur einiger aus Montanwachs isolierter triterpenischer Verbindungen. Collection Czech. Chem. Commun. **28**, 2443–2454 (1963).
19. – – F. Hemmert u. F. Šorm: Über einige aromatische Kohlenwasserstoffe des Harzanteils des Montanwachses. Collection Czech. Chem. Commun. **30**, 873–879 (1965).
20. Cummins, J. J., and W. E. Robinson: Normal and isoprenoid hydrocarbons isolated from oil shale bitumen. J. Chem. Eng. Data **9**, 304–306 (1964).
21. Robinson, W. E., J. J. Cummins, and G. U. Dinneen: Changes in Green River oil-shale paraffins with depth. Geochim. Cosmochim. Acta **29**, 249–254 (1965).
22. Bogomolov, A. I.: Correlation of the constituents of petroleum with their organic origin. Tr. Vses. Neft. Nauchn.-Issled. Geologorazvěd. Inst. **1964**, 10–20; Chem. Abstr. **62**, 7549 (1965).
23. Dean, R. A., and E. V. Whitehead: The occurrence of phytane in petroleum. Tetrahedron Letters **1961**, 768–770.
24. Bendoraitis, J. G., B. L. Brown, and L. S. Hepner: Isolation of 2,6,10,12-tetramethylpentadecane by high temperature gas-liquid chromatography. Anal. Chem. **34**, 49–53 (1962).
24a. Hoeven, W. van, P. Haug, A. L. Burlingame, and M. Calvin: Hydrocarbons from Australian oil, two hundred million years old. Nature **211**, 1361–1365 (1966).
25. Mair, B. J., N. C. Krouskop, and T. J. Mayer: Composition of the branched paraffin-cycloparaffin of the light-gas-oil fraction. J. Chem. Eng. Data **7**, 420–426 (1962).
26. Kochloefl, K., P. Schneider, R. Řeřicha u. V. Bažant: Isoparaffinische und cycloparaffinische Kohlenwasserstoffe. Collection Czech. Chem. Commun. **28**, 3362–3381 (1963).
27. Bricteux, J., and M. Neuray: Evidence of isoprenoid olefins in low-temperature carbonisation tars. Ann. Mines Belg. **1965a**, 1175–1182.
28. Blumer, M., and W. D. Snyder: Isoprenoid hydrocarbons in recent sediments. Presence of pristane and probable absence of phytane. Science **150**, 1588–1590 (1965).
29. Cason, J., and D. W. Graham: Isolation of isoprenoid acids from a California petroleum. Tetrahedron **21**, 471–475 (1965).
29a. Eglinton, G., A. G. Douglas, J. R. Maxwell, J. N. Ramsay, and S. Stalberg-Stenhagen: Occurrence of isoprenoic fatty acids in the Green River shale. Science **153**, 1133–1135 (1966).
30. Borneff, J., and T. Kunte: Carcinogenic substances in water and soils. XIV. Polycyclic aromatic hydrocarbons in soil samples. Arch. Hyg. Bakteriol. **147**, 401–409 (1963); Chem. Abstr. **60**, 2280 (1964).
30a. Blumer, M.: Benzpyrenes in soil. Science **134**, 474–475 (1961).
31. Landa, S., et V. Macháček: Sur l'adamantane, nouvel hydrocarbure extrait du naphte. Collection Czech. Chem. Commun. **5**, 1–5 (1935).
32. Birch, S. F., T. V. Cullum, R. A. Dean, and R. L. Denyer: Thiaadamantane. Nature **170**, 629–630 (1952).
33. Douglas, A. G., and B. J. Mair: Sulfur: role in genesis of petroleum. Science **147**, 499–501 (1965).
34. Birch, S. F., T. V. Cullum, R. A. Dean, and D. G. Redford: Sulphur compounds in the kerosene boiling range of middle east distillates. Occurrence of a bicyclic thiophene and thienyl sulphide. Tetrahedron **7**, 311–318 (1959).
35. Went, F. W.: Organic matter in the atmosphere, and its possible relation to petroleum formation. Proc. Nat. Acad. Sci. U.S. **46**, 212–221 (1960).
36. Guild, F. N.: The occurrence of terpin-hydrate in nature. J. Am. Chem. Soc. **44**, 216 (1922).

37. RUHEMANN, S., u. H. RAUD: Über die Harze der Braunkohle I. Die Sterine des Harzbitumens. Brennstoff-Chem. **13**, 341–345 (1932).
38. BEL'KEVICH, P. E., F. C. KAGANOVICH i E. A. YURKEVICH: Khimija i genezis torfa i sapropelei. Akad. Nauk Belorussk. SSR, Inst. Torfa **1962**, 173–176; Chem. Abstr. **58**, 8817 (1963).
39. CIUSA, R., and A. GALIZZI: Constituents of lignites. Gazz. Chim. Ital. **51** (I), 55–60 (1921).
40. –, and M. CROCE: Constituents of lignites II. Gazz. Chim. Ital. **52** (II), 125–128 (1922).
41. –, and A. GALIZZI: Some constituents of lignites (III). Ann. chim. applicata **15**, 209–214 (1925).
42. SOLTYS, A.: Iosene, a new hydrocarbon from steyrian lignite. Monatsh. Chem. **53–54**, 175–184 (1929).
43. STREIBL, M., u. F. ŠORM: Über weitere, namentlich ungesättigte Kohlenwasserstoffe des Montanwachses. Collection Czech. Chem. Commun. **31**, 1585–1595 (1966).
44. VARIER, N. S.: Fossil resin from the lignite beds of warkalay. Bull. Central Res. Inst. Univ. Travancore, Trivandum, Ser. A **1**, 117–118 (1950); Chem. Abstr. **46**, 10581 (1952).
45. GAVAT, I., u. I. IRIMESCU: Aromatische Kohlenwasserstoffe aus rumänischen Erdölen. I. Mitt. Isolierung von Dimethylnaphthalinen aus Gasöl. Chem. Ber. **74**, 1812–1817 (1941).
46. CARRUTHERS, W., and A. G. DOUGLAS: Trimethylnaphthalenes in a Trinidad oil. J. Chem. Soc. **1955**, 1847–1850.
47. ROZHKO, E. E.: Amber and amber-bearing sediments. Uch. Zap. Leningr. Gos. Ped. Inst. 267, 193–203 (1964); Chem. Abstr. **64**, 9458 (1966).
48. SERGANOVA, G. K., and S. R. RAFIKOV: Structure and properties of Baltic amber. Zh. Prikl. Khim. **38**, 1813–1821 (1965).
49. TROFIMOV, V. S.: Amber placer deposits and their origin. Geol. Rossypei, Akad. Nauk SSSR, Otd. Nauk o Zemle **1965**, 77–97; Chem. Abstr. **63**, 17729 (1965).
50. STERLING, E. C., and M. T. BOGERT: Synthesis of 12-methylperhydroretene (≡abietane) and its non-identity with fichtelite. J. Org. Chem. **4**, 20–28 (1939).
51. SIMONSEN, J. L., and D. H. R. BARTON: The terpenes, Vol. III, p. 337. London: Cambridge University Press 1952.
52. RUZICKA, L., F. BALAŠ u. H. SCHINZ: Zur Kenntnis des Fichtelits und der Stereochemie hydrierter Phenanthrensderivate. Helv. Chim. Acta **6**, 692–697 (1923).
53. –, u. E. WALDMANN: Zur Konstitution des Fichtelits. Helv. Chim. Acta **18**, 611–612 (1935).
54. BURGSTAHLER, A. W., and J. N. MARX: The synthesis and stereochemistry of fichtelite. Tetrahedron Letters **1964**, 3333–3338.
55. SOLTYS, A.: Three compounds extracted from steyrian lignite. Monatsh. Chem. **53–54**, 185–186 (1929).
56. BRIGGS, L. H.: The identity of α-dihydrophyllocladene with iosene. J. Chem. Soc. 1035–1036 (1937).
57. RUZICKA, L., u. J. R. HOSKING: Über die Agathen-disäure, die krystallisierte Harzsäure $C_{20}H_{30}O_4$ des Kaurikopals, des Hart- und des Weichmanilakopals. Ann. Chem. **469**, 147–192 (1929).
58. CASON, J., and K. L. LIAUW: Characterization and synthesis of a monocyclic eleven-carbon acid isolated from a California petroleum. J. Org. Chem. **30**, 1763–1766 (1965).
59. BOITEAU, P., B. PASICH, et A. RAKOTO RATSIMAMANGA: Les Triterpenoides en physiologie végétale et animale. Paris: Gauthier-Villars 1964.
60. SIMONSEN, J., and W. C. J. ROSS: The terpenes, Vol. IV. London: Cambridge University Press 1957.
61. HILLS, I. R., and E. V. WHITEHEAD: Triterpanes in optically active petroleum distillates. Nature **209**, 977–979 (1966).
62. SIMONSEN, J., and W. C. J. ROSS: The terpenes, Vol. IV, p. 289. London: Cambridge University Press 1957.
63. MAIR, B. J.: Terpenoids, fatty acids and alcohols as source materials for petroleum hydrocarbons. Geochim. Cosmochim. Acta **28**, 1303–1321 (1964).
64. MÜNCH, W.: Über die Harze der Braunkohle. Die Sterine des Harzbitumens. Oel und Kohle **2**, 564–567 (1934).
65. LAHIRI, A.: Chromatographic analysis of coal bitumen. Fuel **24**, 66–73 (1945).
66. BERGMANN, W.: Geochemistry of Lipids. In: Organic Geochemistry, ed. I. A. BREGER, p. 503–542. Oxford: Pergamon Press 1963.
67. CARRUTHERS, W., and J. W. COOK: The constituents of high-boiling petroleum distillates. I. Preliminary Studies. J. Chem. Soc. **1954**, 2047–2052.

68. BARTON, D. H. R., W. CARRUTHERS, and K. H. OVERTON: Triterpenoids (XXI). A triterpenoid lactone from petroleum. J. Chem. Soc. **1956**, 788.
69. IKAN, R., and J. MCLEAN: Triterpenoids from lignite. J. Chem. Soc. **1960**, 893 – 894.
70. MCLEAN, J., G. H. RETTIE, and F. S. SPRING: Triterpenoids from peat. Chem. Ind. (London) **1958**, 1515 – 1516.
71. IKAN, R., and J. KASHMAN: Steroids and triterpenoids of Hula peat as compared to other humoliths. Israel J. Chem. **1**, 502 – 508 (1963).
72. SAKABE, T., and R. SASSA: An aromatic hydrocarbon isolated from coal bitumen. Bull. Chem. Soc. Japan **25**, 353 – 355 (1952).
73. OUCHI, K., and K. IMUTA: The analysis of benzene extracts of Yubari coal. I. Chromatographic fractionation of the petroleum ether-methanol-soluble portion and UV- and IR-spectra of the fractions. Fuel **42**, 25 – 35 (1963).
74. ERDMAN, J. G.: Some chemical aspects of petroleum genesis as related to the problem of source bed recognition. Geochim. Cosmochim. Acta **22**, 16 – 36 (1961).
75. BURLINGAME, A. L., P. HAUG, T. BELSKY, and M. CALVIN: Occurrence of biogenic steranes and pentacyclic triterpanes in an Eocene shale (52 million years) and in an early Precambrian shale (2.7 billion years): A preliminary report. Proc. Nat. Acad. Sci. U.S. **54**, 1406 – 1412 (1965).
75a. HILLS, I. R., E. V. WHITEHEAD, D. E. ANDERS, J. J. CUMMINS, and W. E. ROBINSON: An optically active triterpane, gammacerane in Green River, Colorado, oil shale bitumen. Chem. Communs. **1966**, 752 – 754.
76. BURGSTAHLER, A. W.: A contribution to the anthrasteroid problem. J. Am. Chem. Soc. **79**, 6047 – 6050 (1957).
77. SLATES, H. L., and N. L. WENDLER: Realdoltransformations of 5β-methyl-9α,10α-dihydroxy-19-norcholestane-3,6-dione. Experientia **17**, 161 (1961).
78. JEFFREY, L. M.: Lipids in sea water. J. Am. Oil Chemists' Soc. **43**, 211 – 214 (1966).
79. SCHWENDINGER, R. B., and J. G. ERDMAN: Sterols in recent aquatic sediments. Science **144**, 1575 – 1576 (1964).
80. CURRI, S. B.: Identification of lipids and steroids and determination of fatty acids in the peloids (mud) of the Euganean basin. Anthol. Med. Santoriana **69**, 100 – 104 (1963); Chem. Abstr. **60**, 14335 (1964).
81. SCHREINER, O., and E. C. SHOREY: Cholesterol bodies in soils: phytosterol. J. Biol. Chem. **9**, 9 – 11 (1911).
82. – – The presence of cholesterol substance in soils: agrosterol. J. Am. Chem. Soc. **31**, 116 – 118 (1909).
83. PSCHORR, R., u. J. K. PFAFF: Beitrag zur Kenntnis des Montanwachses mitteldeutscher Schwellkohle. Chem. Ber. **53**, 2147 – 2162 (1920).
84. ZECHMEISTER, L., and O. FREHDEN: A chromatographic study of lignite. Nature **144**, 331 – 335 (1939).
85. TURFITT, G. F.: The microbiological degradation of steroids. Biochem. J. **42**, 376 – 383 (1943).
86. MEINSCHEIN, W. G., and G. S. KENNY: Analyses of a chromatographic fraction of organic extracts of soils. Anal. Chem. **29**, 1153 – 1161 (1957).
87. BLACK, W. A. P., U. J. CORNHILL, and F. N. WOODWARD: A preliminary investigation on the chemical composition of Sphagnum moss and peat. J. Appl. Chem. (London) **5**, 484 – 492 (1955).
88. IVES, A. J., and A. N. O'NEILL: The chemistry of peat. I. The sterols of peat moss (Sphagnum). Can. J. Chem. **36**, 434 – 439 (1958).
89. BEL'KEVICH, P. I., G. P. VERKHOLETOVA, F. L. KAGANOVICH, and I. V. TORGOV: β-Sitosterine from peat wax. Izv. Akad. Nauk SSSR, Otd. Khim. Nauk **1963**, 112 – 115; Chem. Abstr. **58**, 10011 (1963).
90. MEINSCHEIN, W. G.: Origin of petroleum. Bull. Am. Assoc. Petrol. Geologists **43**, 925 – 944 (1959).
91. SCHISSLER, D. O., D. P. STEVENSON, R. J. MOORE, G. J. O'DONNELL, and E. E. THORPE: Steranes in petroleum. ASTM E-14 Meeting, New York 1957.
92. MURPHY, Sister M. T. J., and B. NAGY: Analysis for sulfur compounds in lipid extracts from the Orgueille meteorite. J. Am. Oil Chemists' Soc. **43**, 189 – 196 (1966).
93. KINDSCHER, E.: Über ein Vorkommen von Kautschuk in mitteldeutschen Braunkohlenlagern. Chem. Ber. **57**, 1152 – 1157 (1924).
94. NONOMURA, S.: Friedelin and epifriedelinol from Salix japonica Thumb. J. Pharm. Soc. Japan **75**, 1303 – 1304 (1955).

95. TSCHESCHE, R., F. LAMPERT u. G. SNATZKE: Dünnschicht- und Ionenaustauscherpapierchromatographie von Triterpenoiden. J. Chromatog. **5**, 217–224 (1961).
96. – I. DUPHORN, and G. SNATZKE: Thin-layer and ion exchange chromatography of triterpenoid compounds. In: Methods for the detection and estimation of radioactive compounds separated by g.l.c. (A. T. JAMES and L. J. MORRIS, eds.), p. 247–259. New York: D. Van Nostrand 1964.
97. EGLINTON, G., R. J. HAMILTON, R. HODGES, and R. A. RAPHAEL: Gas-liquid chromatography of natural products and their derivatives. Chem. Ind. (London) **1959**, 955–957.
98. IKEKAWA, N., S. NATORI, H. ITOKAWA, S. TOBINAGA, and M. MATSUI: Gas chromatography of triterpenes. I. Ursane, oleanane and lupane groups. Chem. Pharm. Bull. (Tokyo) **13**, 316–319 (1965).
99. – – H. AGETA, K. IWATA, and M. MATSUI: Gaschromatography of triterpenes. II. Hopane, zeorinane and onocerane groups. Chem. Pharm. Bull. (Tokyo) **13**, 320–325 (1965).
100. COURTNEY, J. L., and J. S. SHANNON: Triterpenoids. Structure assignment to some friedelane derivatives. Tetrahedron Letters **1963**, 13–20.

Addendum

The paper of NOMURA, M., P. VON R. SCHLEYER, and A. A. ARZ [Alkyladamantanes by rearrangement from diverse starting materials. J. Am. Chem. Soc. **89**, 3657–3659 (1967)] presents the following information: a variety of substances, including isoprenoids, such as cholesterol, cholestane, abietic acid, cedrene, caryophyllene, camphene, and squalene, formed different alkyladamantane mixtures on reaction with $AlBr_3$ or $AlCl_3$ "sludge" catalyst (110°–130° C; 2–5 days). E.g. methyl-(1 %), dimethyl-(16 %), trimethyl-(26 %), tetramethyl-(35 %), methylethyl-(traces), dimethylethyl-(3 %), trimethylethyl-(12 %) and other alkyl-adamantanes were produced from cholesterol. This fact may explain how the adamantane derivatives came to be present in petroleum.

BURLINGAME, A. L., and B. R. SIMONEIT [Isoprenoid fatty acids isolated from the kerogen matrix of the Green River formation (Eocene). Science **160**, 531–533 (1968)] demonstrated by oxidation the presence of a series of isoprenoid acids [XXI, 4,8,12-trimethyltridecanoic, 5,9,13-trimethyltetradecanoic (norphytanic), XXIII (phytanic acid) and 4,8,12,16-tetramethylheptadecanoic acid] bound to the insoluble kerogen matrix. This confirms the indigenous nature of the original insoluble material. A similar series of free acids and dicarboxylic acids were found in the same material [HAUG, P., H. K. SCHNOES and A. L. BURLINGAME: Isoprenoid and dicarboxylic acids isolated from Colorado Green River shale (Eocene). Science **158**, 772–773 (1967)]. However, similar isoprenoid acids have also been detected in quite recent marine sediments [BLUMER, M., and W. I. COOPER: Isoprenoid acids in Recent sediments. Science **158**, 1463–1464 (1967)].

HENDERSON, W., V. WOLLRAB, and G. EGLINTON [Identification of steroids and triterpenes from a geological source by capillary gas-liquid chromatography and mass spectrometry. Chem. Commun. (1968) 710–712] presented the utility of gas-liquid chromatography coupled with mass spectrometry for identification of traces of steroids and triterpenoids in fossils, e.g. cholestanes, stigmastane, hopane, gammacerane, etc.

MURPHY, Sister M. T. J., A. MCCORMICK, and G. EGLINTON [Perhydro-β-carotene in the Green River shale. Science **157**, 1040–1042 (1967)] demonstrated the surprising and unique finding of carotane, though the instability of the original β-carotene (cf. [12]) is well known.

CHAPTER 17

Carotenoids

RICHARD B. SCHWENDINGER

Esso Agricultural Products Laboratory,
Esso Research & Engineering Company,
Linden, New Jersey

Contents

I. Introduction

The occurrence of carotenoids in recent sediments is typical of the geochemistry of any class of organic substances: it is primarily an analytical problem. Analysis of organic compounds in recent sediments is most challenging, and at the same time most frustrating. The difficulty of attempting to identify qualitatively or quantitatively trace quantities of organic material within a fantastically complicated matrix of other organic materials (most of whose make-up is still largely a mystery) is compounded by the fact that most often this organic matrix is itself a very minor component of an essentially inorganic system. Add to these purely chemical problems such insoluble questions as how to obtain a representative laboratory sample from an inhomogeneous sedimentary system many cubic kilometers in extent, or how to obtain from the field a sample sufficiently undisturbed so as to prevent chemical and especially biochemical changes. Pondered coldly and logically, such an analytical problem becomes insuperable and ought never be attempted. Yet, as a matter of fact, such attempts are made and, more surprisingly, often are made successfully. This chapter, then, will not simply enumerate the kind and quantity of carotenoids found in various sediments, but will try to investigate critically the literature of the subject and interpret it in light of the analytical problems involved and the techniques presently available.

KARRER and JUCKER [1] have defined carotenoids as "yellow to red pigments of aliphatic or alicyclic structure, composed of isoprene units (usually eight) linked so that the two methyl groups nearest the center of the molecule are in positions 1:6 while all the other lateral methyl groups are in positions 1:5; the

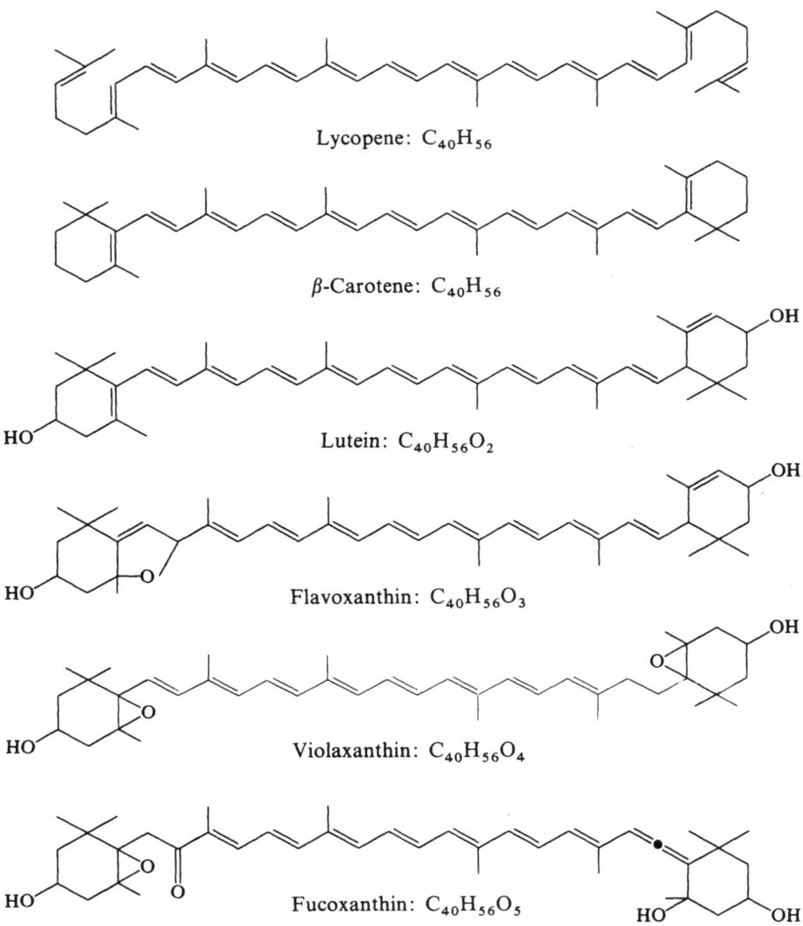

Fig. 1. Structures of some typical carotenes and xanthophylls

series of conjugated double bonds constitutes the chromophoric system ...". In Fig. 1 are listed the structures of a few common carotenoids which specifically demonstrate some of the types of oxygen-containing functional groups that occur. The chemistry and biochemistry of this class of pigments has been considered in detail elsewhere [1–3].

Carotenoids are generally subdivided into two classes: carotenes, which are hydrocarbons, and xanthophylls, which contain oxygen. This division historically has been made on the basis of the partition of the carotenoids between hexane and 90 percent methanol in a two-phase system. A h/e ratio is then commonly used to represent the proportion of xanthophylls (hypophasic) to carotenes (epiphasic). In almost all cases this separation is remarkably complete. However, carotenoids possessing a single hydroxyl group as well as some with two hydroxyl groups show intermediate solubility [4], a most important point which will be discussed later.

II. Literature

A. Carotenoids in Living Organisms

Carotenoids are ubiquitous in nature [5, 6]. They occur in all green plant tissue as integral constituents of the photosynthetic process and also can be found in some nonphotosynthetic bacteria and fungi. These organisms synthesize their own carotenoids and provide the source of carotenoids which are found in most of the animals. Mammals need carotenoids for Vitamin A production but their function in the lower animals is often vague although none the less essential. The occurrence of these red pigments in almost all organisms makes them prime objects for evolutionary study [5, 6]; results have not been overly promising yet.

The principle carotenoids in the chloroplasts of higher plants [7] are β-carotene, lutein, violaxanthin, and neoxanthin. Flowers and fruits contain many rather exotic carotenoids. These latter represent very minor percentages of the total found in phanerogams but they are sometimes important for their nutritional value. It is in the cryptogams, especially the algae, that the diversity of nature is clearly shown by the qualitative [5, 6] and quantitative [8] proliferation of xanthophylls. Table 1 shows the major carotenoids to be found in algal classes; minor components are not shown but demonstrate even greater variation.

Table 1. *Qualitative and quantitative distribution of the major carotenoids of the algal classes*[a]

Algal class	Carotenes		Xanthophylls		h/e[b]
	dominant	mg/100 g	dominant	mg/100 g	
Predominantly fresh-water					
1. Chlorophyceae	β-carotene	10–50	lutein, violaxanthin, neoxanthin	70–350	7
2. Xanthophyceae	β-carotene		zeaxanthin-like unknowns		
3. Chrysophyceae	β-carotene	10–40	fucoxanthin	250–320	11
4. Cryptophyceae	α-carotene		zeaxanthin, diatoxanthin		
5. Charophyceae	β-carotene		lutein		
6. Euglenineae	β-carotene		lutein, antheraxanthin, neoxanthin		
7. Cyanophyceae	β-carotene		echinenone, myxoxanthophyll		
Predominantly marine					
8. Diatomophyceae	β-carotene	2–30	fucoxanthin	65–550	19
9. Dinophyceae	β-carotene	9	sulcatoxanthin (peridinin)	70	8
10. Phaeophyceae	β-carotene	3–30	fucoxanthin	11–180	6
11. Rhodophyceae	β-carotene	2–30	lutein, violaxanthin, neoxanthin taraxanthin	8–180	6

[a] Adapted from GOODWIN [5, 6], PARSONS *et al.* [8], and various literature sources.
[b] Ratio of xanthophylls (hypophasic) to carotenes (epiphasic).

One interesting fact not explicitly shown in Table 1 is that fucoxanthin is the major xanthophyll of the algae (diatoms) that make up approximately 85 percent by weight of the plant biomass [9] of the oceans. Assuming that 1.9×10^{19} tons of organic carbon are produced in the oceans per year, that algae are 45 percent carbon by dry weight, and that fucoxanthin constitutes 0.3 percent by dry weight, then 6.8×10^{16} tons, or 200 mg/sq.m. of ocean, are synthesized each year. If fucoxanthin is 75 percent of the total xanthophylls in the algae in which it occurs, if the h/e ratio is 7, and if the other 15 percent of the algae have about the same carotenoid content, the yearly production of carotenoids in the ocean is 1.2×10^{17} tons or 360 mg/sq.m. The total pigment production on land is about the same order of magnitude but here no one single carotenoid predominates. Since fucoxanthin is the carotenoid produced in the greatest quantity on a world-wide basis, it is interesting that its structure only lately has been established [10].

B. Carotenoids in Sediments

The occurrence of carotenoids in lake sediments (which is often overlooked) was first reported by LYUBIMENKO [11] in 1923 and in marine sediments by TRASK and WU [12] in 1930. Since that time, carotenoids have been shown to occur in the recent sediments of all environments studied: a cave, lacustrine deposits, sapropels, a river estuary, an intertidal sand, and marine sediments. Historical reviews to 1960 have been given most adequately by VALLENTYNE. DUNNING [15] also reviewed the subject in 1963. ERDMAN [16, 17] has speculated on the role sediment carotenoids may play as progenitors of petroleum. He pointed out that, although recent sediments contain most of the hydrocarbons normally present in crude oil, the low molecular weight aromatics are unique in being absent. Carotenoids were shown both theoretically and experimentally [18, 19] to be probable precursors of such aromatics as toluene and xylenes, and this fact was used to argue for the necessity of diagenesis for formation of petroleum [17].

Although about twenty different carotenoids have been tentatively identified in sediments, only six have been unequivocally confirmed. VALLENTYNE's [20] identification of β-carotene in Searles Lake sediments is a classic example of one of these six carotenoids, as can be judged by the spectral data published later [15]. Generally, carotenoids disappear rapidly in sediments with age, the two oldest coming from a Searles Lake sediment [20] estimated to be 20,000 years old and an interglacial gyttja [21] estimated to be 100,000 years old.

It is not surprising that many different carotenoids have been found in sediments; it is surprising that fucoxanthin has been reported [22] only once in recent marine sediments. A plausible explanation turns on some observations by HARVEY [9]. He believes that only minute amounts of dead plants ever reach the ocean floor. Phytoplankton are eaten by herbivorous zooplankton which in turn are eaten by omnivorous and carnivorous zooplankton, some of which are notably voracious and void much undigested food. Whatever escapes these organisms may well provide food for the bottom fauna which tend to be scavengers. Just as the major weight of marine algae are diatoms, so the main weight of marine zooplankton are copepods [9] whose characteristic xanthophyll is astaxanthin. Thus the chances are not great that large quantities of undecomposed fucoxanthin sediment reach the ocean floor. The greatest possibility of this happening ought to be an area where phytoplankton production is high and bottom decomposition is low. Some of the anaerobic basins off the coast of Southern California fit this description, and it is here where fucoxanthin might best be sought in recent sediments.

Even though fucoxanthin has not lived up to prior expectations as an estimator of past sedimentary history, at least one successful application of carotenoids as paleobiological indicators has been made. Since myxoxanthophyll is unique to the blue-green algae (Cyanophyceae), and since these algae are responsive to pollution, ZÜLLIG [23, 24] used the changes in myxoxanthophyll content with depth of sediment to determine the past history of several Swiss lakes. It also has been shown that the carotenoid content of recent marine sediments is appreciably higher than recent terrestrial sediments [25] when expressed as a function of total organic carbon. No explanation of the phenomenon was advanced. The carotenoids still show promise as indicators of past history above and within sediments [26] but the difficulties encountered in their analysis is the most probable reason for their neglect compared to amino acids, sugars, and hydrocarbons.

III. Experimental Techniques

A. Sampling

For years, agricultural science has used statistics to good effect in sampling problems. The geochemist recently has begun to look to statistics for design and analysis of prospecting surveys [27, 28]. But little work has been done on the problems of sampling as they effect the determination of trace organic compounds in inhomogeneous sediments. At least three basic approaches are possible. A single sample may be taken and this used on the assumption that it is "representative". Since organic geochemical analyses are time consuming, each one involving many days or weeks of work, this is not a difficult assumption to make. Another method may be to take some number of samples, pool and homogenize them, and analyze an aliquot. This will provide an answer with a built-in "average" but leaves the question of confidence limits unanswered. Still a third approach would be to make analyses on each of a number of separate samples.

Unpublished work by the author [29] demonstrates the results obtained when carotenoids were analyzed [25] by a refinement of the third method enumerated above. A number of cores were taken at each of two locations within each of three environments. Cores within each location were about one meter apart, cores at different locations within an environment were tens of meters to kilometers apart (cf. ERDMAN [17] for maps). Carotenoid contents and h/e ratios are shown in Table 2. While the variation between cores within each location is, as expected *a priori*, generally less than between cores in different locations within each environment, it is surprising that these differences are not greater. It means that the sediment systems studied here not only varied widely at great distances but also at quite small distances. Most organic geochemists would be satisfied that the differences among the three environments are real; most statisticians and chemists would vehemently deny that any conclusions could be drawn from data with such wide variation. Until someone spends a few years doing a statistical study of the variance normally found in such systems and of the confidence limits needed to draw valid conclusions from the data, a large amount of faith and intuition are a prerequisite for a worker in this field.

Table 2. *Carotenoids in the 0–15 cm. horizon of Recent sediments* [29]

	Mg. carotenoids/1,000 g org. carbon				h/e ratios			
	duplicate determinations[b]			average	duplicate determinations[b]			average
Bellefontaine Marsh (brackish)								
Location 1	25	28		27	3.3	2.7		3.0
Location 2D	30	21		26	7.0[a]	3.5		3.5
	Environmental average			27	Environmental average			3.3
Mississippi Sound (marine lagoon)								
Location 1	127	140	193	153	3.8	3.3	3.2	3.4
Location 4	318	264		291	2.1	3.0		2.6
	Environmental average			222	Environmental average			3.0
Gulf of Mexico (shallow marine)								
Location 3A	217	233	300	250	11.7[a]	3.7	3.5	3.6
Location 4	350	317	400	356	2.5	4.1	3.8	3.5
	Environmental average			303	Environmental average			3.6

[a] These values are so high as to justify discarding on the basis of experimental error.

[b] Duplicates are not aliquots. Each was determined on different core samples from each location within the environment.

B. Hypophasic Epiphasic (h/e) Ratios

The real enigma of sedimentary carotenoids is the report that their h/e ratios are unity or lower [30] even though values of 3–10 are usual for the higher plants and 4–9 for algae. This has been attributed to preferential oxidation of the xanthophylls, even though there is some question of whether this chemical explanation is in fact true, especially since sediments are very often oxygen deficient. Besides, a careful review of the literature together with some simple calculations shows that an almost equal number of papers report an h/e ratio in the range of plants as express the opposite view. The most extensive experimental survey of h/e ratios as a function of depositional environment [25] found values of 1.3–3.6: low compared to plants but much higher than those claimed to be so characteristic of sediments. An inquiry into these discrepancies is interesting because of some of the basic analytical problems it emphasizes.

It has been mentioned that, although carotenoids are generally separated into carotenes and xanthophylls on the basis of their distribution between hexane and 90 percent methanol, the monohydroxy and some of the dihydroxy carotenoids are appreciably soluble in both phases. This is a well-known phenomenon in the case of the monohydroxy compounds and is not especially disturbing since this class of carotenoids is known to constitute a very small percentage of naturally occurring mixtures. That the dihydroxy carotenoids, lutein and zeaxanthin, are appreciably soluble in the epiphase of a hexane: 90 percent methanol system is a published [4] but mostly ignored fact, a fact that becomes important since one or both of these compounds is usually a major percentage of most plant xanthophyll fractions. SCHWENDINGER and ERDMAN [25] used a hexane: 95 percent methanol system to circumvent this particular problem and obtained consistently high h/e ratios of 1.3–3.6 for aqueous recent sediments ranging from fresh water through brackish to marine. These are much closer to the values of 3–9 expected from such probable source material as plants. Lest such a logical and simple explanation seems to solve this particular problem once and for all, the work of WHITE and ZSCHEILE [31] ought to be mentioned; they found that partition coefficients of carotenoids were dependent on the carotenoid components of the mixture being studied. Besides, it is certain that other materials (essentially colorless) come through the sediment extraction-purification scheme, but equally uncertain what effect these unknown substances also may have on the distribution of carotenoids between hexane:95 percent methanol in so far as they change the h/e ratios. To further confuse this initially simple solution to what really is a vexing problem, it is certain that at least two other possible sources of error exist in the analyses normally used. Before discussing these, some description of the analytical procedures are necessary.

C. Spectrophotometry

The analytical method used almost universally to determine the gross quantity of carotenoids in sediments is some variation of the plant assay methods enumerated by GOODWIN [32]. In outline form, the material to be tested is homogenized and extracted with a suitable solvent, this solution saponified with base and carotenoid pigments extracted with ether. These pigments can be transferred to a more suitable solvent and their quantity determined using a colorimeter with a suitable filter or a spectrophotometer set at, say 452 mμ. The h/e ratio can be obtained similarly after distribution of the carotenoids in the recommended two-phase system, using a colorimeter or a spectrophotometer set at 452 mμ for the carotenes and 445 mμ for the xanthophylls. Modifications of solvents (absolute methanol, acetone, aqueous acetone) or of some of the manipulations (nitrogen atmosphere, sonic dispersers, choice of base) depend on preference of the investigator. Assuming these preferences produce only second-order corrections, the key to all these methods lies in the proper use of an instrument. This requires some explanation.

Underlying present use of the automatic electronic gear no laboratory can now afford to be without, is the "Principle of Modern Instrument Infallibility" which can be stated as follows: If an electronic instrument is utilized somewhere in the analytical scheme, the results must be correct. A most important corollary is: the more expensive the electronic apparatus used, the more exact is the data produced. This is in no way meant to disparage these marvelous devices, it is mentioned simply to emphasize that no amount of automation will ever replace the need for the analyst's good technique, precise judgment, and intelligent interpretation. What, then, can a colorimeter or a spectrophotometer do and, more important, what can they not do?

Both colorimeters and spectrophotometers operate in the visible light region on the same basic principles: light from an integral source is directed through the solution to be analyzed and detected in some way so that an electrical signal which may be amplified is generated sufficient to operate a visual indicator. The amount of light absorbed by the solution is then assumed to be in some manner proportional to the material to be analyzed. A colorimeter passes all of the source light or, at the very least, some large percentage of it (e.g., mostly red or mostly blue by the use of glass filters) through the solution, thus automatically integrating the area under the spectral curve of the compound(s) in solution. However, it is incapable of discriminating between wavelengths except through the use of filters, which is to say, inexactly. Assuming that the compound(s) to be analyzed have well known spectra and that the spectra of possible impurities are equally well known and are sufficiently different so that a filter can discriminate between the two sets of spectra, a colorimeter can produce results of almost any desired accuracy.

A spectrophotometer disperses the source light into a spectrum and passes parts of this spectrum through the solution. The narrowness of the band passed through the solution is essentially directly proportional to the expense of the basic instrument. Thus a spectrophotometer discriminates between wavelengths but integrates only a very small area beneath the spectral curve, and is useful for determining a spectral curve or for analyzing a compound once its spectrum is known. Recording and double-beam instruments enhance the usefulness but do not alter the basic operating principles of spectrophotometers. This ability to discriminate between wavelengths is extremely useful for the analysis of a single compound containing a sharp spectral peak but can be most unwelcome when simultaneously analyzing for a family of compounds with similar spectra peaking at various wavelengths. Application of these remarks to the analysis of the families of compounds making up the carotenoids, xanthophylls and carotenes is not quite so obvious as it seems.

The spectra of carotenoids are greatly influenced by solvent but it is fair to say that major absorption peaks will be somewhere between 400 and 575 mμ no matter what the solvent with the range 420–525 mμ including most of the maxima. This is the reason for the use of a filter when analyzing with a colorimeter. Since the maximum for β-carotene is 425 mμ in hexane, and this compound is a major component of most higher plant carotenoid arrays, this is the wavelength often chosen when analyzing carotenoids and carotenes with a spectrophotometer, with the 445 mμ setting for xanthophylls being a more or less empirically found value also derived from work with higher plants. The use, then, of these particular wavelengths depends on the rather fair assumptions that the carotenoids of higher plants are essentially very similar mixtures and do not contain significant amounts of interfering substances when isolated by the recommended extraction-separation scheme. But the application of this same reasoning to the analysis of sediment carotenoids is not quite so logical. Algae in many cases are known to contain xanthophyll arrays quite different from higher plants [5, 6] and they are a possible major source material for sediment carotenoids. In this case, the empirical setting of 445 mμ no longer necessarily holds and even integration of areas under curves, either using a colorimeter or the area under the experimentally determined curve [25], is still rather unsatisfying in the absence of more pertinent standards. The difficulty is not only with spectral maxima but also with extinction coefficients since fucoxanthin, the major xanthophyll in brown algae, has an extinction coefficient approximately half as large as the "normal" xanthophyll extinction coefficients. However, in the absence of absolute standards, integration methods are more likely to provide mutually comparable results. This could account, in part, for the often found low h/e ratios for sediments and for the more nearly normal values found when a curve-integration method was used [25].

Spectra of pure β-carotene samples and carotenoids extracted from recent sediments [29] are shown in Fig. 2. Despite the fact that absorption maxima of carotenoids shift with change in type of solvent, the spectra of β-carotene in

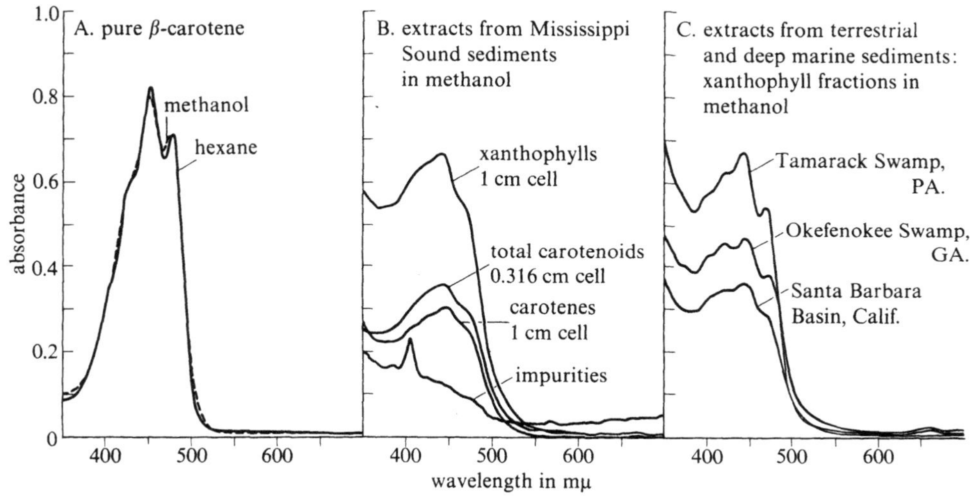

Fig. 2. Typical spectra of pure β-carotene and carotenoid extracts from sediments

hexane and absolute methanol are almost superimposable (Fig. 2a). Absolute methanol was the somewhat unusual solvent chosen [25, 29] both for extraction of sediments and spectral work. The saponified extract from a Mississippi Sound sediment [17, 25] demonstrates the appearance of typical spectral curves (Fig. 2b). It is interesting that the three spectra of total carotenoids, carotenes, and xanthophylls are quite similar except that the absorption maximum is displaced toward higher wavelengths in the case of the carotenes and lower wavelengths in the case of the xanthophylls as compared to the total carotenoids. The spectrum of the xanthophyll fraction of Mississippi Sound, a marine lagoonal environment, is similar to the spectrum of the same fraction from Santa Barbara Basin, a deep marine environment (Fig. 2c). Both differ from the spectra of the xanthophyll fractions of two terrestrial aquatic sediments: Tamarack and Okefenokee Swamps (Fig. 2c). All these mixtures are far too complex to allow identification of components by position of spectral maxima.

Vallentyne [14] warned of the difficulties inherent in listing absorption maxima for carotenoids without specifying solvent. Even more basic are the generally ignored difficulties inherent in reporting spectral data without specifying the limits of error of the instrument used [15]. The author's laboratory [29] used two spectrophotometers: a manual Beckman DU and a recording Beckman DK-1. The Beckman DU had been specially rebuilt at the factory and wavelengths were known to be accurate; it was used to obtain exact absorption maxima of spectra of special interest as they were observed on the recording instrument. Although the Beckman DK-1 had been reconstructed in this laboratory, absolute wavelengths could be reproduced to ± 2 mμ at best, yet this was far better than most other similar instruments that were critically investigated. Such factors as gear lash, reproducibility of initial instrument settings, and inaccuracies of the chart paper (to name but a few) must be taken into account when interpreting wavelength positions from any recording spectrphotometer. Further, rising background absorption can shift spectral peaks a number of mμ and recording speed may have a great influence on peak height. Even such basic preliminaries as checking instrument base line, cell match, and noise level are all too often bypassed.

Spectrophotometers seldom suddenly cease functioning; they usually fail over a period of time. Unless a regular maintenance and calibration program is followed assiduously, the point in time at which the instrument was no longer reliable cannot be traced and the appropriate spectra rerun. Such malfunctioning of instruments unfortunately is not rare and too strict adherence to the "Principle of Modern Instrument Infallibility" can produce strange results.

The presence of interfering substances is another possible source of error. Saponification of the initial extract is performed to remove neutral fat and to remove the chlorophylls which interfere in a number of ways with carotenoid analysis [32]. Procedures for pigment analyses in plants are well established. But mention has been made previously of the complex nature of sedimentary organic matter: a product not only of plants and animals living above the sediment but also of biological, biochemical and chemical activities within the sediment. To apply analytical techniques developed for plant analyses to sediments without investigating their applicability is an invitation to error. Here the error is to assume that, in sediments as in plants, saponification will remove any tetrapyrrole pigments present. The author has found, often to his dismay, that this last assumption need not be true. Since it was routine to use a double-beam recording spectrophotometer for all scanning as well as analytical work, sediment extracts many times were found to contain unsaponifiable tetrapyrrole pigments as shown by absorption peaks between 400–420 mμ and 550–670 mμ (Figs. 2b and c). Whether these are of some obscure microbial origin or a degradation product of chlorophyll is a matter for conjecture. What is important is that these pigment impurities often were present in sufficient quantities to cause gross errors unless suitably corrected in some way; in this case by integrating the curve between 420 and 500 mμ to avoid the Soret peak. Using a colorimeter for analyzing such extracts without previously checking the spectral curve could have been a disaster.

One last observation ought to be made about spectral curves of sediment organic extracts. No matter how carefully these extracts are "cleaned up", a background absorption from minute quantities of unknown materials is always present. This normally does not appear to be too intense in the visible region (400–700 mμ) but does intensify rapidly as the wavelength gets shorter so that most extract spectral curves go offscale by about 250–230 mμ in the ultraviolet region. The lower part of the visible region (400–500 mμ) where the carotenoids absorb is a borderline area as far as this background absorption is concerned (Figs. 2b and c). However, it would be interesting to have something more definitive on its nature and intensity so as to be sure that this is not still another source of error.

D. Chromatography and Countercurrent Distribution

An obvious question by now ought to be: why not use chromatography to separate the red and green pigments and whatever other impurities are present, quantitatively determine the carotenoids one by one, and get a complete picture of their occurrence in sediments? This would certainly circumvent most of the difficulties mentioned so far. TSWETT first demonstrated the use of chromatography by separating plant pigments on an alumina column over half a century ago. STRAIN [7], among others, has shown how useful column chromatography can be for investigating chloroplast pigments. Procedures for separation and identification of the carotenoids from plants have been well worked out. These have been applied to sediments with some successes [20, 22] but they are far outweighed by the failures which is one of the reasons for the many tentative but few conclusive identifications of carotenoids in recent sediments. Again, the explanation turns on the fantastic complexity of the mixture in question, on the minute quantities that can be extracted from large volumes of sediment only after much tedious work, and on the indeterminate amount of material that always is lost through decomposition or other means on the chromatographic substrate. It is obvious that the procedures used to separate carotenoids from plant extracts using column chromatography cannot be applied to sediment extracts without gross modifications.

Thin layer chromatography of carotenoids [33] is a technique that shows some promise. However, over the last five years, paper chromatography of chloroplast pigments [34, 35] has been developed into an unusually powerful tool. Methods for quantitative estimation of carotenoids by paper chromatography also have been developed [36, 34].

ANDERSEN and GUNDERSEN [21] separated gyttja carotenoids using paper chromatography and were able to identify β-carotene and suggest that fucoxanthin and lutein or related compounds also were present. Most of the red pigments seemed to be contaminated by tetrapyrrole impurities with the likelihood that shifting of xanthophyll peaks could have been serious. ZÜLLIG [24] extended paper chromatographic methods successfully to the quantitative estimation of myxoxanthophyll in lake sediments. Personal experience of the author was that paper chromatography was applicable to separation of sedimentary carotenoids but that unidentified lipids extracted also from the sediments interfered with, and often reversed, the usual positions of the individual xanthophylls. Decomposition of the carotenoids on the paper was common despite precautions. Elution of sufficient material for spectrophotometric identification was most difficult at best, impossible normally.

There is one method that should be especially applicable to this problem. Using a 200 tube countercurrent extractor, CURL [37] was able to resolve naturally occurring mixtures of carotenoids into carotenes, monols and polyols and the further separate the polyols into dihydroxy, dihydroxy monoepoxide, dihydroxy diepoxide, and higher polyol carotenoids. Each group then was passed through a magnesium oxide column to effect the final separation into individual substances. Exploratory work by the author has convinced him that such a system, with minor modifications, would work well for the analysis of carotenoids in sediments but this method was never pursued beyond confirming that the h/e ratios of the sediments being studied were indeed well above unity.

IV. Conclusion

If this chapter has sounded pessimistic, almost hopeless, the impression created is incorrect. Much is known about the kind and quantity of carotenoids in a number of specific sediments. More needs to be done to obtain some picture of how carotenoids are distributed through sediments, how long they last, and what their ultimate fate is. Some of the analytical problems that must be involved in such an undertaking were reviewed and the pertinence of such commentary to the interpretation of past work was stressed. That an organic geochemist must also be a geologist, in the sense expressed by RUSSELL [38], now ought to be clear. Earlier it was stated that the organic geochemist needed a large faith and a fine intuition; to this should be added an infinite patience and a good sense of humor.

References

1. KARRER, P., and E. JUCKER: Carotenoids. Translated and revised by E. A. BRAUDE. New York: Elsevier 1950.
2. WEEDON, B. C. L.: Chemistry of the carotenoids. In: Chemistry and biochemistry of plant pigments (T. W. GOODWIN, ed.). New York: Academic Press 1965.
3. GOODWIN, T. W.: The comparative biochemistry of carotenoids. London: Chapman & Hall 1952.
4. PETRACEK, F. J., and L. ZECHMEISTER: Determination of partition coefficients of carotenoids as a tool in pigment analysis. Anal. Chem. **28**, 1484–1485 (1965).
5. GOODWIN, T. W.: Carotenoids: structure, distribution, and function. In: Comparative biochemistry, vol. IV, part B (M. FLORKIN and H. S. MASON, eds.). New York: Academic Press 1962.
6. – Distribution of carotenoids. In: Comparative biochemistry, vol. IV, part B (M. FLORKIN and H. S. MASON, eds.). New York: Academic Press 1962.
7. STRAIN, H. H.: Chloroplast pigments and chromatographic analysis. University Park, Pa.: Pennsylvannia State University 1959.
8. PARSONS, T. R., K. STEPHENS, and J. D. H. STRICKLAND: On the chemical composition of eleven species of marine phytoplankters. J. Fisheries Res. Board Can. **18**, 1001–1025 (1961).
9. HARVEY, H. W.: Chemistry and fertility of sea waters, 2nd ed. New York: Cambridge University Press 1957.
10. BONNETT, R., A. K. MALLAMS, J. L. TEE, B. C. L. WEEDON, and A. MCCORMICK: Fucoxanthin and related pigments. Chem. Comm. No. 15, 515–516 (1966).
11. LYUBIMENKO, V. N.: Chlorophyll in deposits of lake silts. Zh. Russk. Bot. Obshch. **6**, 97–105 1923).
12. TRASK, P. D., and C. C. WU: Does petroleum form in sediments at time of deposition? Bull. Am. Assoc. Petrol. Geologists **14**, 1451–1463 (1930).
13. VALLENTYNE, J. R.: Fossil pigments: the fate of carotenoids. In: Symposia on comparative biology, vol. I. Comparative biochemistry of photoreactive systems. (M. B. ALLEN, ed.). New York: Academic Press 1960.
14. – The molecular nature of organic matter in lakes and oceans, with lesser reference to sewage and terrestrial soils. J. Fisheries Res. Board Can. **14**, 33—82 (1957).
15. DUNNING, H. N.: Geochemistry of organic pigments: carotenoids. In: Organic geochemistry (I. A. BREGER, ed.). New York: MacMillan Co. 1960.

16. ERDMAN, J. G.: Some chemical aspects of petroleum genesis as related to the problem of source bed recognition. Geochim. Cosmochim. Acta **22**, 16–36 (1961).
17. – Petroleum–its origin in the earth. Fluids in subsurface environments–a symposium. Am. Assoc. Petrol. Geol., Tulsa 1965.
18. MILIK, J. D., and J. G. ERDMAN: Genesis of hydrocarbons of low molecular weight in organic rich aquatic sediments. Science **141**, 806–807 (1963).
19. DAY, W. C., and J. G. ERDMAN: Ionene: a thermal degradation product of β-carotene. Science **141**, 808 (1963).
20. VALLENTYNE, J. R.: Carotenoids in a 20,000 year old sediment from Searles Lake, California. Arch. Biochem. Biophys. **70**, 29–34 (1957).
21. ANDERSEN, S. T., and K. GUNDERSEN: Ether soluble pigments in interglacial gyttja. Experienta **11**, 345–348 (1955).
22. FOX, D. L., D. M. UPDEGRAFF, and D. G. NOVELLI: Carotenoid pigments in the ocean floor. Arch. Biochem. Biophys. **5**, 1–23 (1944).
23. ZÜLLIG, H.: Sediments as indicators of the condition of water. Schweiz. Z. Hydrol. **18**, 5–143 (1956).
24. – The determination of myxoxanthophyll in core profiles as an indicator of past activity of blue-green algae. Verh. Intern. Ver. Limnol. **14**, 263–270 (1961).
25. SCHWENDINGER, R. B., and J. G. ERDMAN: Carotenoids in sediments as a function of environment. Science **141**, 808–810 (1963).
26. FOGG, G. E., and J. H. BELCHER: Pigments from the bottom deposits of an English lake. New Phytologist. **60**, 129–138 (1961).
27. MIESCH, A. T., and J. J. CONNOR: Investigation of sampling-error effects in geochemical prospecting. U.S. Geol. Surv. Profess. Paper **475-D**, 84–88 (1964).
28. MACKOWSKI, M. P.: Statistical design and analysis of geochemical exploration surveys. Mining Congr. J. **50** (5), 56–59 (1964).
29. SCHWENDINGER, R. B.: Unpublished data. We are indebted to the Multiple Fellowship on Petroleum sponsored by the Gulf Research and Development Company at Mellon Institute for permission to publish these data.
30. FOX, D. L., and C. H. OPPENHEIMER: The riddle of sterol and carotenoid metabolism in muds of the ocean floor. Arch. Biochem. Biophys. **51**, 323–328 (1954).
31. WHITE, J. W., and F. P. ZSCHEILE: Studies of the carotenoids. III. Distribution of pure pigments between immiscible solvents. J. Am. Chem. Soc. **64**, 1440–1443 (1942).
32. GOODWIN, T. W.: Carotenoids. In: Modern methods of plant analysis, vol. III (K. PAECH and M. V. TRACEY, eds.). Berlin-Göttingen-Heidelberg: Springer 1955.
33. BOLLINGER, H. R.: Vitamins. Thin layer chromatography: a laboratory handbook (E. STAHL, ed.). New York: Academic Press 1965.
34. SESTAK, Z.: Paper chromatography of chloroplast pigments (chlorophylls and carotenoids) (M. LEDERER, ed.). Chromatographic reviews, vol. 7, p. 65–91. New York: Elsevier 1965.
35. BOOTH, V. H.: Mapping plant lipids by paper chromatography. Chromatographic reviews, vol. 7, p. 98–118. New York: Elsevier 1965.
36. JENSEN, A., and S. L. JENSEN: Quantitative paper chromatography of carotenoids. Acta Chem. Scand. **13**, 1863–1868 (1959). (See also JENSEN and STAHL in: Carotine and carotonide. Darmstadt: Dr. Dietrich Steinkopff 1963).
37. CURL, A. L.: Application of countercurrent distribution to Valencia orange juice carotenoids. J. Agr. Food Chem. **1**, 456–460 (1953).
38. RUSSELL, R. D.: The geologist–a scientist? an interpreter? Geotimes **10** (6), 18–20 (1966).

CHAPTER 18

Geochemistry of Proteins, Peptides, and Amino Acids

P. E. HARE

Geophysical Laboratory, Washington, D. C.

Contents

I. Introduction

Proteins account for most of the nitrogen compounds in living systems and in fact make up a sizable fraction of the bulk weight in many organisms, often over 50 percent on a dry-weight basis [1]. Peptides and free amino acids, though present in all living cells as metabolic intermediates and products, make up only a relatively small fraction of the nitrogen compounds present [2, 3].

As enzymes, proteins catalyze a tremendous variety of complex biochemical reactions. Antibodies, many bacterial toxins, and numerous plant products are proteins [2]. As structural components, proteins constitute such diverse materials as muscle, silk, and sponge. Combined with other materials proteins form a number of important compounds such as the various hemoglobins and cytochromes.

Proteins function in biological mineralization processes, forming organic matrices intimately associated with and in large part protected by the mineral phase involved [4–7]. A study of these mineralized proteins is of particular significance to paleontological and geological work because of the very nature of the fossil record [8] and the possible interpretation of past biochemical processes. Mineralized proteins also provide a useful system for the study of protein diagenesis, since potentially destructive agents such as microorganisms, moisture, and oxygen must penetrate the mineral phase before reaching the organic matter within [9].

Although the various proteins in living systems exhibit a wide variety of properties and functions, they are all made up of the same relatively few amino acids. Only about twenty amino acids are found commonly as hydrolysis products

Table 1. *L-amino acids isolated from acid hydrolysates of proteins*

Group	L-amino acid	Formula 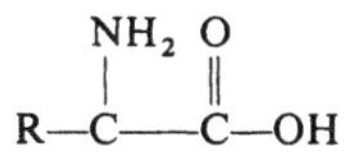 (* asymmetric carbon atom)	Abbreviation
Neutral	Glycine	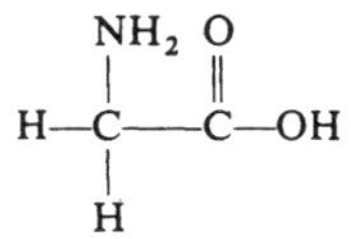	GLY
	Alanine	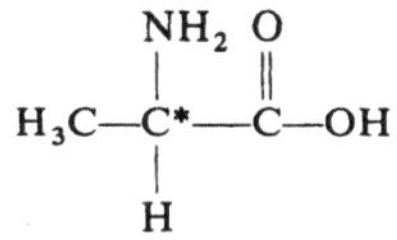	ALA
	Valine	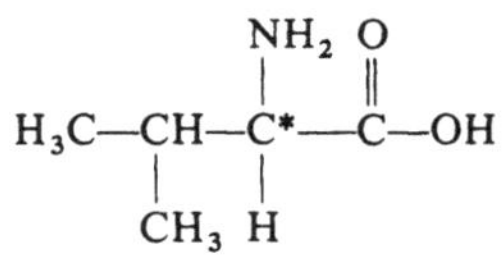	VAL
	Isoleucine	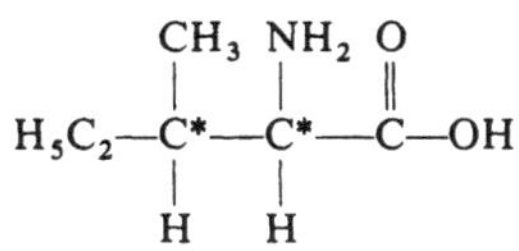	ISO
	Leucine	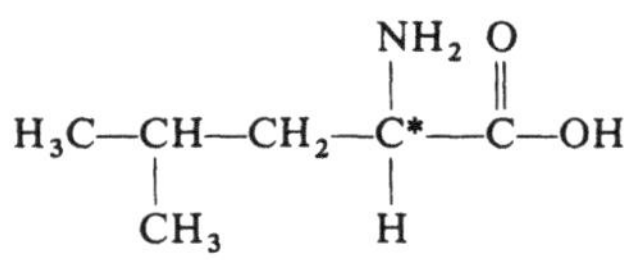	LEU
Imino	Hydroxyproline	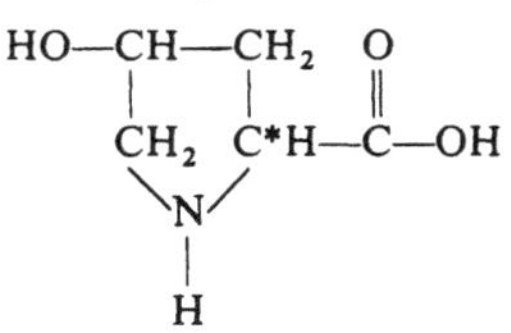	HPR
	Proline	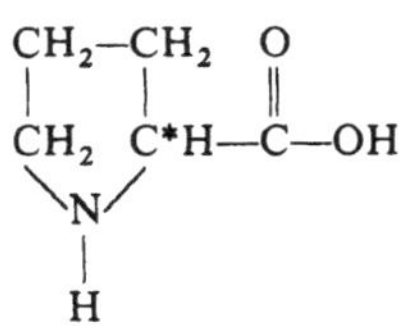	PRO

Table 1. (Continued)

Group	L-amino acid	Formula 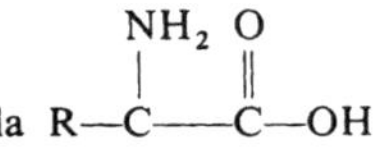 (* asymmetric carbon atom)	Abbreviation
Hydroxyl	Serine	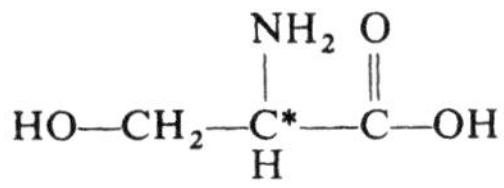	SER
	Threonine	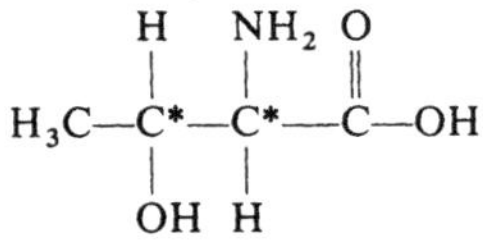	THR
Acidic	Aspartic acid	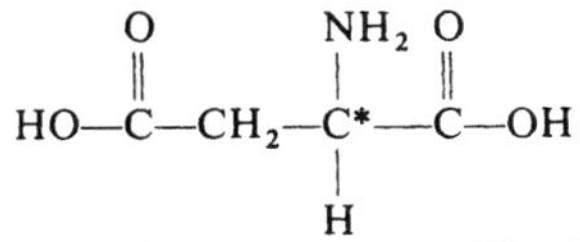	ASP[a]
	Glutamic acid	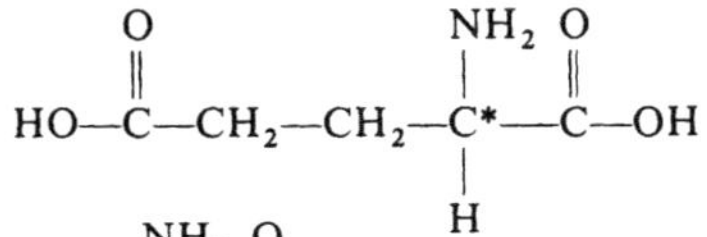	GLU[a]
Sulfur	Cystine	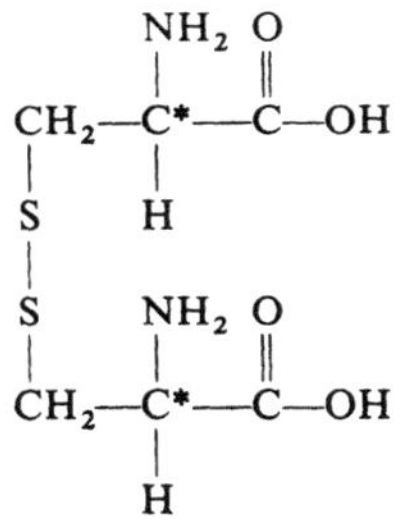	CYS
	Methionine	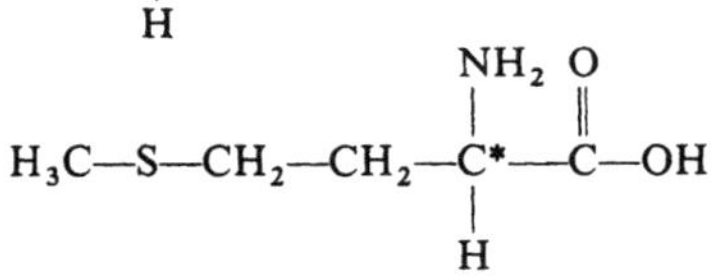	MET
Aromatic	Tyrosine	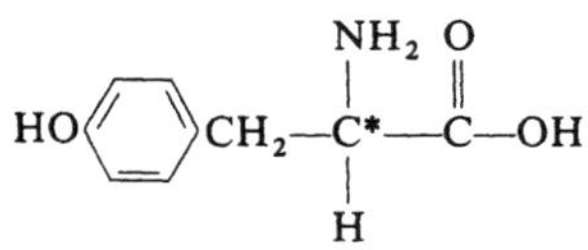	TYR
	Phenylalanine	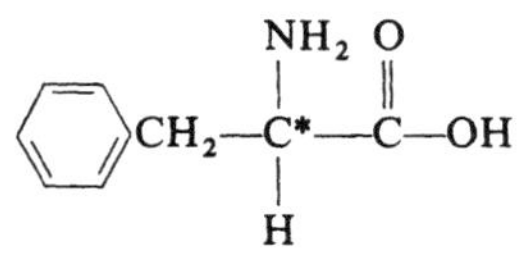	PHE
Basic	Hydroxylysine	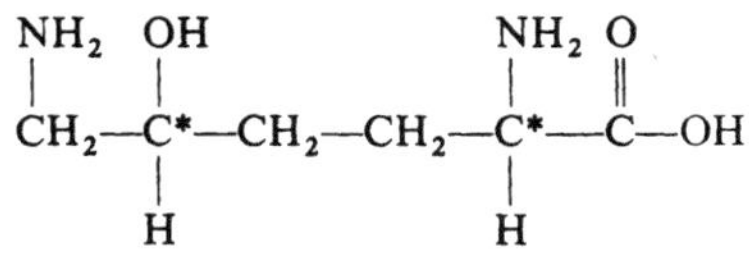	OH-LYS

[a] The amino acids asparagine and glutamine are converted by acid hydrolysis respectively to ASP and GLU.

Table 1. (Continued)

Group	L-amino acid	Formula 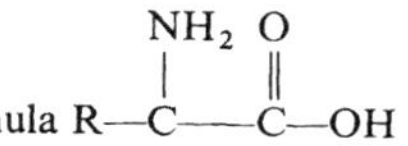 (* asymmetric carbon atom)	Abbreviation
Basic	Lysine	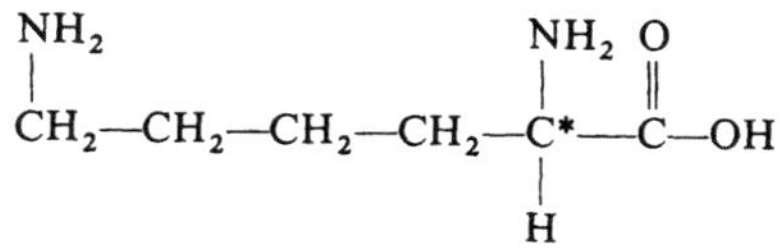	LYS
	Histidine	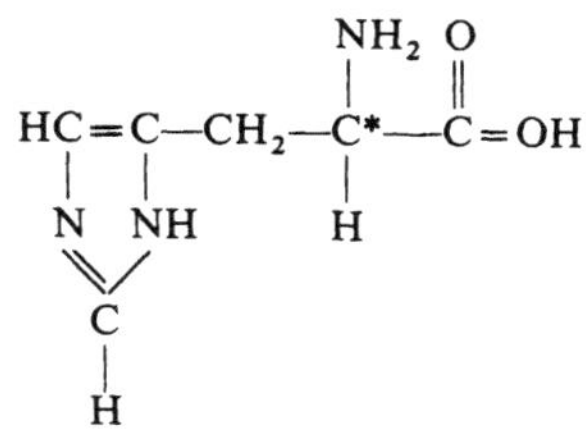	HIS
	Tryptophane	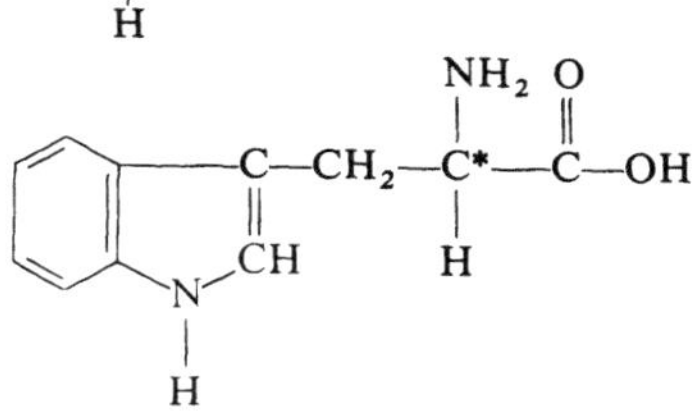	TRY
	Arginine	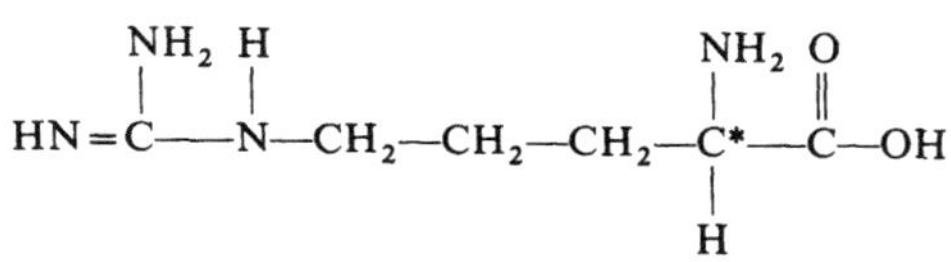	ARG

Table 2. *Other amino acids and some ninhydrin-positive compounds found in geologic materials*

Name	Formula	Probable source
D-amino acid	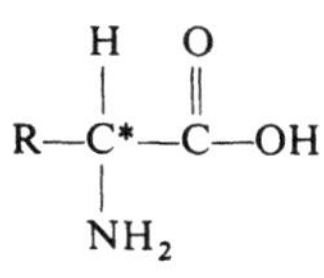	Racemization of protein L-amino acids (Fig. 1) Bacteria and other organisms
D-alloisoleucine	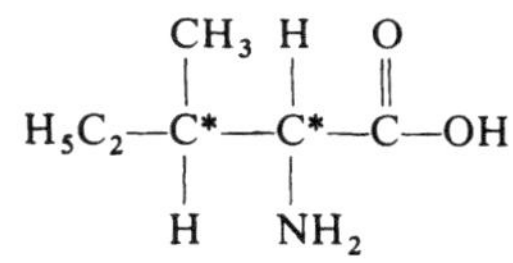	L-isoleucine
Ornithine (ORN)	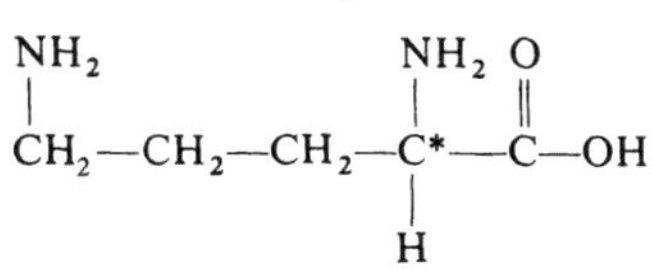	Arginine
Citrulline (CIT)	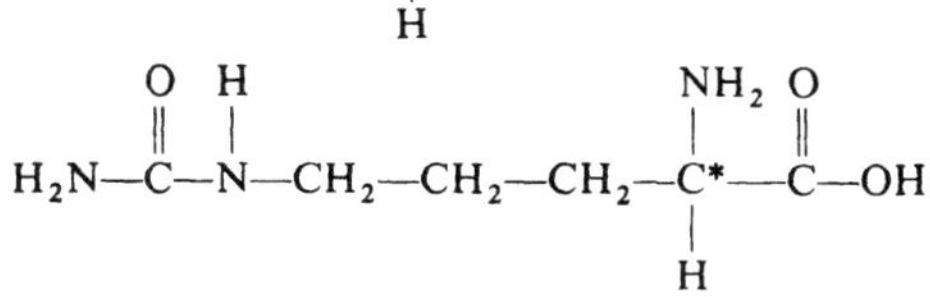	Arginine

Table 2. (Continued)

Name	Formula	Probable source
α-aminobutyric acid	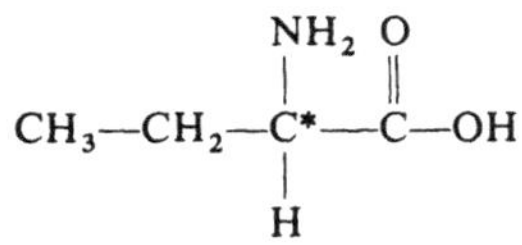	Glutamic acid
γ-aminobutyric acid	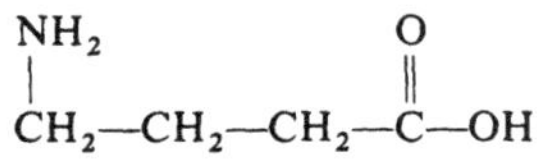	Glutamic acid
β-alanine	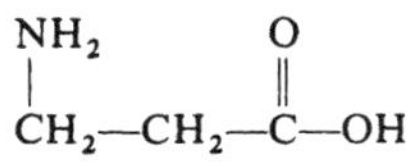	Aspartic acid
Cysteic acid	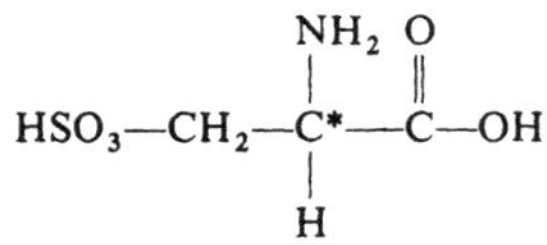	Cysteine and cystine
Methionine sulfone (MET SUL; sulfoxide has one less oxygen on the sulfur atom)	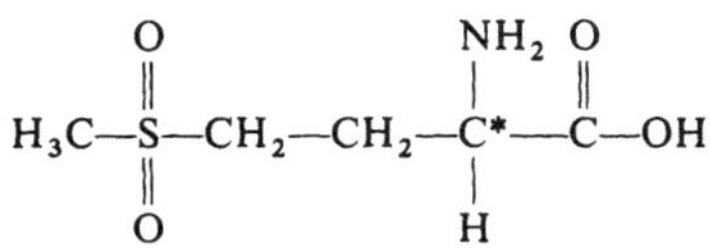	Methionine
Diamino pimelic acid	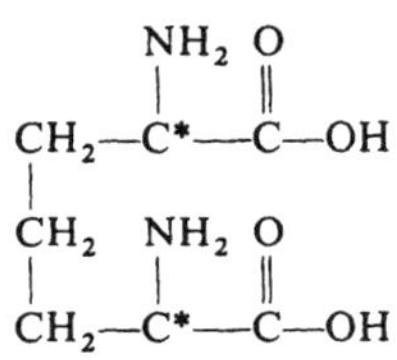	Bacteria
Taurine	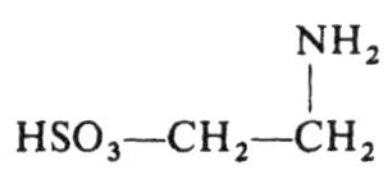	Cysteic acid
Urea	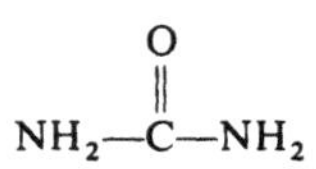	Arginine
Ammonia	NH_3	Deamination of amino acids
Amines	$R—NH_2$	Decarboxylation of amino acids
Glucosamine	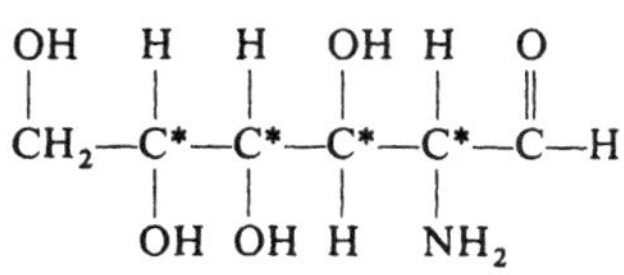	Chitin and shell organic matrix
Galactosamine (=chondrosamine)	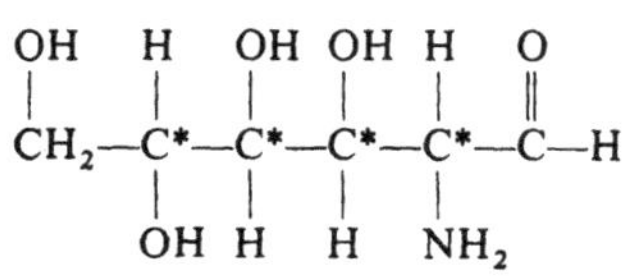	Shell organic matrix

of proteins (Table 1). Some of the nonprotein amino acids found in geologic environments are listed in Table 2.

The primary structure of proteins is the linear sequence of amino acids in peptide linkage. In protein synthesis by living cells this sequencing of amino acids is accomplished by the purine and pyrimidine base sequence in the nucleic acids with a triplet of bases coding for a specific amino acid [10]. It is this genetic code that limits the number of amino acids found in proteins. This code, with its associated enzyme systems, further selects only a single optical isomer of each of the amino acids having one or more asymmetric carbon atoms (Table 1).

Of the protein amino acids, only glycine lacks at least one asymmetric carbon atom. Other amino acids isolated from acid hydrolysates of proteins are optically active, a property apparently restricted to materials of biologic origin. The usefulness of optical activity as an aid in determining the mode of origin of amino acids in geologic materials will be considered later in the chapter.

The complete characterization of a protein would include its amino acid sequence as well as its three dimensional configuration (secondary and tertiary structure). This data is difficult and time consuming to accumulate and is available at the present time only on a very few soluble proteins [11], none of which are apparently preserved in geologic environments. The amino acid composition of protein material, on the other hand, is relatively easy to determine. Although giving only limited characterization of protein material, it may still provide useful data in evaluating the effects of geologic environments on proteins.

After a short discussion on the nitrogen cycle and a brief review on the reported occurrences of amino acids in the geologic record, we shall consider the changes in the amino acid composition (including the D and L configuration) in the shell proteins of *Mercenaria* from a series of fossil shells and in a series of recent shells which were treated to simulate the geologic environment. Application of the data will be made to geochronology, geothermometry, and to the mode of origin of amino acids found in various geologic environments.

II. Nitrogen Cycle

The path of nitrogen from the atmosphere through the biosphere and back to the atmosphere involves a number of steps in which microorganisms play a vital role [12, 13]. The nitrogen cycle has no doubt differed in kind and degree in the geologic past as life itself has changed with the environment. The most primitive nitrogen cycle possibly consisted in both synthesis and decomposition via electric discharge, ultraviolet, and thermal mechanisms with a resulting steady state reservoir of nitrogen-containing carbon compounds.

With the "invention" of photosynthesis and nitrogen fixation, organisms utilized atmospheric materials to manufacture their own supplies of essential organic compounds. Organisms filling ecological niches at various parts of the nitrogen cycle obtained food and energy in return for nitrogen fixation or denitrification. In present day soils the total quantity of living microorganisms within the zone of nitrogen exchange is of the order of one ton per acre [12].

The part of the nitrogen cycle involved in releasing nitrogen from the decomposition of plant and animal material is of more direct concern to the problems of this chapter. It seems probable that over geologic time biologic processes have synthesized an amount of organic material comparable to the total mass of the earth [14]. Compared to the amount of organic matter in sedimentary rocks [15] and in the biosphere, this indicates that the efficiency of the nitrogen cycle in material balance approaches 100 percent.

The carbon to nitrogen ratios in organic material of various ages should give some indication of the relative stability of organic nitrogen compounds to other non-nitrogenous organic material. Studies of the carbon to nitrogen ratios in present day soils [16, 12] and in sediments and rocks of various lithologies and ages [17–19] indicate that the extreme values differ only by one or at most two orders of magnitude. Soils and Recent sediments have C/N values from five to around fifteen, or about the same as the range found in microorganisms [13]. In sedimentary rocks C/N ratios vary from near that found in soils to values of over 100.

In a series of six Paleozoic shales STEVENSON [19] found the C/N ratio varied from less than 10 to nearly 40. Subtracting out the fixed ammonium nitrogen, the variation in the C/N ratio in the six samples was only from 35 to 43. Low values of the C/N ratios in rocks appear to be due to the presence of fixed ammonium.

In a series of igneous rocks STEVENSON showed that more than half to over 95 percent of the total N was present as fixed ammonium. In none of the rock samples (sedimentary or igneous) is there any trend with the geologic age of the sample. There does seem to be a correlation with the type of organic matter originally deposited and in the lithology involved.

While most of the nitrogen is in the form of ammonium, at least a small fraction of the nitrogen appears to be in the form of carbon-containing compounds. Depending on environmental conditions organic nitrogen compounds sometimes escape the fate of the nitrogen cycle and are preserved to some degree in sedimentary materials. To what extent these materials can move around in the geologic column is not clear, but it may be that some of these organic nitrogen compounds are in a kind of steady state concentration transported by ground water from more Recent soils and other zones of biologic activity into more ancient sediments.

Since microorganisms are themselves made up of appreciable quantities of protein material and since they also play an important role in nitrogen metabolism [20], a consideration of the distribution of microorganisms should be helpful in evaluating the geological occurrences of amino acids and other organic nitrogen compounds.

The distribution of microorganisms is not easy to determine. Because of their ubiquitous presence in the surficial zone of the earth, it is not always possible to know whether the observed presence of certain living organisms in samples brought from the depths of the earth were actually living there or were introduced during the collection of the sample. Enough data have been assembled, however, to indicate that there are few places on or in the crust of the earth where living organisms cannot live [21–23].

Where living organisms do exist, and this includes anerobic as well as aerobic environments, organic material is being synthesized and degraded. Therefore, *when* a particular organic-rich sediment was formed may not be as important to the preservation of the organic material as the question of how much alteration has been or is being accomplished by living organisms as a part of the nitrogen cycle.

III. Geologic Occurrences of Amino Acids (Free and Combined)

A. Fossils

ABELSON first pointed out the significance of the amino acids found in fossil shells [24–26]. He showed that the thermally more stable amino acids like alanine, glycine, valine, proline, and the leucines were present in older fossil materials, while the thermally less stable amino acids were either absent entirely or present only in trace amounts.

A dramatic difference in the amount of peptide-bound amino acid material was found between fossil and Recent shells of *Mercenaria*. In the Recent shell sample nearly all of the amino acids were peptide-bound as insoluble protein components (insoluble in trichloroacetic acid). Less than one percent of the total amino acids recovered were free. In the Pleistocene sample less than 40 percent of the amino acids were protein-bound (TCA insoluble), while the free amino acid fraction amounted to 20 percent of the total. In the Miocene shell none of the amino acids were detected in peptide linkage.

FLORKIN *et al.* [27, 28] showed that the organic matter in a series of fossil nacreous shell structures was physically preserved so well that the distinctive pattern observed for Recent Cephalopod Molluscan classes can readily be recognized in electron micrographs of the fossil structures. Amino acid compositions of these preserved matrices, however, showed that the original composition had been modified. The amount of nitrogen in the nacreous shell structure of *Nautilus* dropped from 0.4 percent in the Recent specimen to 0.01 percent in the Eocene specimen. Glycine and alanine are the most abundant amino acids in the Recent *Nautilus*, while serine is the most abundant in the fossil specimen.

In a study of the early diagenesis of the insoluble amino acid-containing components in a series of radiocarbon-dated *Mytilus* shells [29], it was found that the amino acid composition of this insoluble fraction changed progressively with time, as shown in Fig. 1. In Recent samples the glycine and alanine concentrations are about equal and together make up nearly 60 percent of the total amino acids from the insoluble fraction. In older shells the glycine progressively decreases relative to alanine so that in an Upper Pleistocene sample from a marine terrace (^{14}C age > 30,000 years) the alanine to glycine ratio was 15 [30]*.

A later study by DEGENS and LOVE [31] of a sequence of fossil shells of *Gyraulus* from the Tertiary of Steinheim, Germany, showed a remarkable similarity in the amino acid ratios of the fossil specimens compared to a Recent *Planorbis*. The

* Treatment of the insoluble protein from a Recent shell with 0.1 N HCl at 110° C for a few hours yielded insoluble residues with amino acid compositions parallel to that found in the fossil series [30]. Analysis of the soluble fraction from these treatments showed that little, if any, of the amino acid material was being destroyed, but rather some components of the insoluble fraction were being taken selectively into solution.

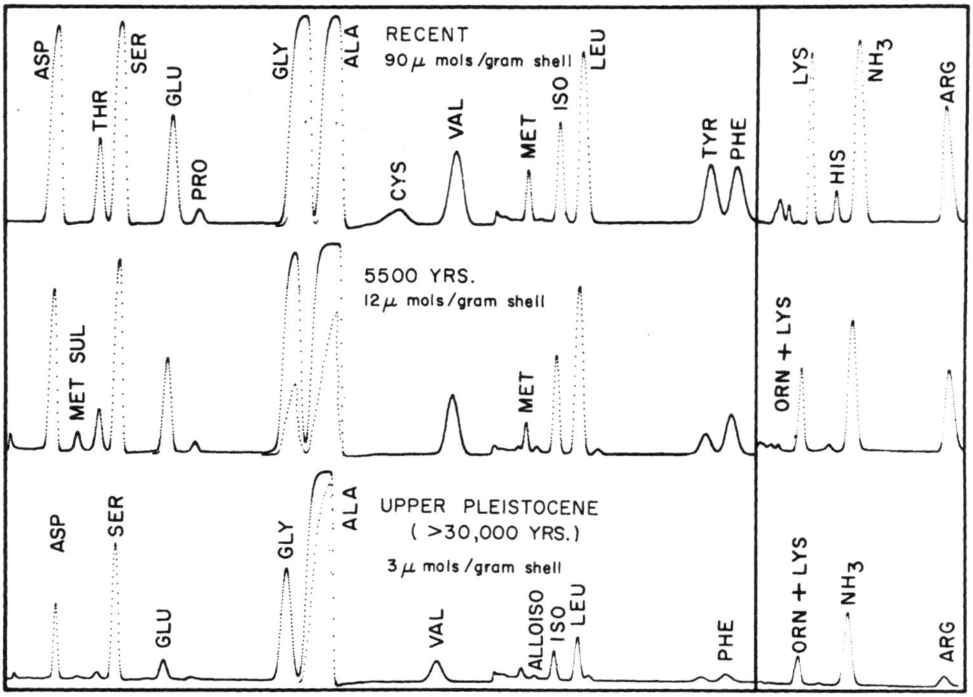

Fig. 1. Early diagenetic changes in the shell proteins of *Mytilus*. Amino acids from the acid hydrolysates of the insoluble fraction in *Mytilus* shells. Micromoles of amino acids found in this fraction are indicated per gram of shell [29, 30]

fossil series had less than one tenth of the total amino acids as the Recent sample, but such presumably unstable amino acids as threonine, serine, and cystine were reported present. No trends with time were apparent, and differences were interpreted as reflecting environmental or evolutionary changes.

A recent study by JOPE [32, 33] on the amino acids in fossil brachiopod shells shows little similarity with shells of their nearest Recent relatives. The ages of the fossils ranged from Cretaceous to Silurian and in all cases showed the presence of such presumably unstable amino acids as serine, threonine, and arginine.

FOUCART *et al.* [34] have made a study of the combined amino acids in some Paleozoic graptolites and have compared them to the amino acids in the surrounding rock. The level of amino acids in the rock is about one fifth of that in the graptolite fragments. The amino acid ratios, however, are not greatly different. Most of the amino acid fraction in both the graptolite and the rock is made up of serine, glycine, alanine, glutamic acid, and aspartic acid.

Fossil bones and teeth have been the subject of a number of amino acid studies [24, 35–38]. In addition, reports on the electron microscopy of the organic matrix of fossil bones are of significance [39, 40]. In general the results parallel the work described on shells. The organic matrix of tooth dentine and bone is largely made up of the protein collagen. Collagen is characterized by a glycine content of about one third of its total amino acid residues. In addition, collagen contains hydroxylysine and hydroxyproline, a characteristic which is often used to distinguish collagen from other proteins. Collagen also has a distinctive pattern in the electron microscope.

The structural preservation of collagen in bone samples of Pleistocene and Miocene ages and even in Devonian and Triassic samples has been reported [41, 38]. Amino acid ratios from Late Pleistocene bone, while showing a number of similarities to Recent collagen, also reveal a number of significant changes indicating that the chemical integrity of the organic matrix has been modified during fossilization [37].

A recent study by ARMSTRONG and TARLO [38] includes data on a fossil Pliosaur tooth and bone and some conodonts. A bone of Upper Jurassic age and the rock material surrounding the bone were both analyzed for amino acids. The rock matrix was found to have all the amino acids found in the bone but in even greater concentrations. HELLER [42] has found concentrations of free amino acids in rocks near fossil bones and shells and has interpreted this as indicating organic matter coming out of the fossil into the surrounding rock. Perhaps the organic matter originally present around the bone provided an environment for anerobic microorganisms to flourish, and the remains of this system have been superimposed on the original *in situ* deposit. More data are clearly needed to interpret the mode of origin of many of the amino acid occurrences in fossils.

B. Sediments and Rocks

ERDMAN *et al.* [43] recovered several amino acids from an Oligocene marine sediment and compared them to the amino acids in a Recent sediment. As in the earlier study of shells by ABELSON [24], the more thermally stable amino acids were relatively more abundant in the older sediment. Amino acids have been reported from sediments as old as Precambrian. HARINGTON reported [44, 45] the presence of several amino acids in crocidolite asbestos of Precambrian age from South Africa; serine was the most abundant amino acid. Several amino acids, including serine, have been reported [46] recently from the Mountsorrel Bitumen, a Carboniferous mineralization associated with some dolerite dikes. The authors considered the possibility of an abiogenic origin for the amino acids. Previously an abiotic origin for the hydrocarbon fraction had been postulated [47].

SELLERS [48] in a systematic study on the stabilities of the various amino acids in sediments found that most of the amino acids present in a sample of a Recent marine sediment were also found in a sample of Miocene mudstone. The Miocene sample, however, had less than 10 percent of the total amino acid concentration of the Recent marine sediment. Heating of both Recent and Miocene samples in sealed bombs resulted in the appearance of some additional ninhydrin-positive materials, one of which was tentatively identified as alloisoleucine, the diastereoisomer of isoleucine. The absence of alloisoleucine in the unheated samples seems significant in view of some recent findings discussed later in the chapter.

Studies on the amino acids in marine [49] and lake sediment [50] cores show generally decreasing total amounts with depth but highly irregular distributions with respect to many individual amino acids. The water-sediment interface shows the highest concentration of amino acid material with the sediment surface containing as much as three to four orders of magnitude more amino acid material (free and combined) than the overlying water. This implies that biologic activity is responsible for the bulk of the sediment surface amino acids, and as the microbiologic activity decreases with increasing depth of sediment so does the total amount of amino acid material.

C. Soils

Though not as extensive as the sediment-ocean interface, the soil represents the sediment-atmosphere counterpart and shows a number of similarities. Microbiological activity is again a maximum near the surface and diminishes with depth [13]. The area distribution is heterogeneous with samples only a centimeter apart showing vastly differing biologic activity. Arctic tundra soils that are frozen most of the year may have over a million bacteria per gram. Even desert soils with near oven-dry conditions have considerable numbers of microorganisms present [13].

Work on the nitrogen content of soils shows that much of the nitrogen is present as ammonium ions. The fraction present as fixed nitrogen increases with depth and presumably is an integral part of the clay structure [51].

The amino acid distribution in soils shows a marked correlation with microbiologic activity. Roots of plants are known to excrete free amino acids and even peptides [12, 52]. Disintegration and decomposition of dead microorganisms produce a variety of free amino acids and peptides. STEVENSON [53] has found over 30 ninhydrin-reactive materials in soil samples including all the protein amino acids, some degradation products like ornithine, and amino sugars. He also identified diaminopimelic acid, which is presumably restricted to bacteria.

A recent study by SOWDEN and IVERSON [54] on the free amino acids in soils has shown that although the water-extractable free amino acid concentration is quite low (0.01 to 1 microgram per gram of soil), there is another fraction of free amino acids extractable with carbon tetrachloride amounting up to 100 times as much. This appears to be the result of injury to microbial cells releasing free amino acids within the cells. Compared to the total amino acids obtained by hydrolysis, the free amino acid fraction made up one percent or less.

D. Meteorites

Occurrences of amino acids have been reported [55–57] from several samples of meteorites including both carbonaceous chondrites and stony chondrites. Although in lower concentrations, the amino acids in stony chondrites have a somewhat similar pattern to that in the carbonaceous meteorites. Some of the thermally less stable amino acids such as serine, threonine, and aspartic acid are present. The origin of the amino acids in meteorites has been variously proposed to be 1) biologic *in situ*, 2) contamination, or 3) abiotic.

E. Hydrosphere

Several reports on the organic matter in sea water [58, 59] indicate that most of the common amino acids are present. JOHANNES and WEBB [60] have shown that living zooplankton release appreciable amounts of dissolved amino acids.

The difficulties in quantitatively extracting amino acids from sea water are many, and the extraction process itself often results in the loss of some amino acids originally present. The concentration level of amino acids in sea water is somewhat more uniform but much lower than in the sediments near the water-sediment interface [61].

Trapped sea water or connate water in Paleozoic and Tertiary sediments have been shown [61] to be fairly rich in amino acids including the less stable amino acids serine, threonine, and aspartic acid. Because of the low level of bacterial contamination found, it was concluded this was probably not a significant source of the amino acids.

Ground water is included in the hydrosphere and is of extreme importance because of its possible role in transporting and redistributing organic matter throughout the geologic column. Only indirect evidence is available on the presence of amino acids in ground waters. Carbon to nitrogen ratios have been reported [62, 63] from a number of samples of ground water. The values seem to be independent of depth and run from near 5 to nearly 40. Most of the values are near 10, a value similar to that found in soil organic matter.

F. Atmosphere

Although the atmosphere is usually not considered as a major source of biochemical material, recent studies [64, 65] have shown measurable quantities of organic matter are present in the atmosphere. The atmosphere is an effective agent in the transportation of surficial material of the earth including both living and fossil organisms. The recent study of DELANY *et al.* [66] has indicated that much of the dust collected on the island of Barbados had been transported over 5,000 km. The quantity of dust collected indicates that a sizable fraction of the deep sea clay is probably wind transported. The organic constituents of the dust included remains of both marine and fresh water organisms. Obviously amino acid-containing material is present. In other studies [55, 67] it has been found that dust settling out from the atmosphere contains appreciable quantities of combined amino acid material. Rain collected at the earth's surface often has appreciable amounts of organically combined nitrogen [68], much of it in the form of pollen, spores, bacteria, and dust, all originally of terrestrial origin.

G. Amino Acids from Nonbiological Sources

Before WOHLER's synthesis of urea in the early 19th century all organic compounds were assumed to be products of biological processes. The nonbiological synthesis of organic compounds has progressed from WOHLER's relatively simple synthesis of urea to recently constructed macromolecules like insulin with 51 amino acids in peptide linkage [69, 70].

It might be argued that these nonbiologic syntheses are in fact biological since human intervention and direction are involved. The work of MILLER [71] and others, however, has shown that even under simulated primitive earth conditions, it is possible to synthesize a large number of biologically important compounds including amino acids and polypeptides [71–76]. The amino acids so synthesized are not optically active but are racemic mixtures of both the D and L amino acid configurations.

It is conceivable that there may be amino acid deposits or occurrences which have been synthesized by abiotic processes during the geologic history of the earth. Problems in the use of optical activity as an indicator of mode of origin will be considered later.

IV. Problems of Contamination

There is little doubt that both combined and free amino acids are present in most geological materials. Are these *in situ* occurrences of the amino acids and hence essentially as old as the host rock from which they are extracted, or is it

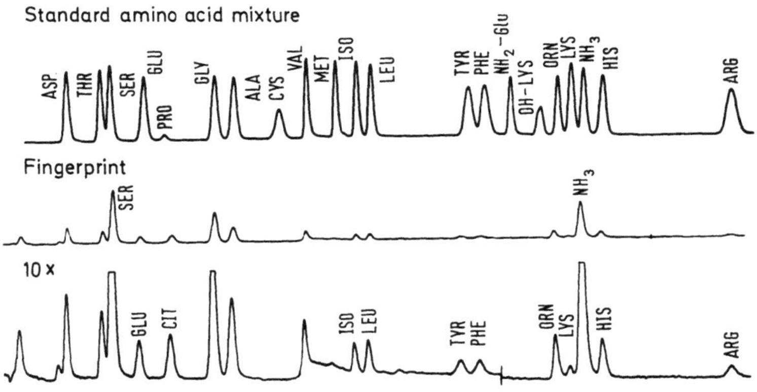

Fig. 2. Amino acids in a single human fingerprint compared to standard amino acid mixture containing 0.01 micromoles of each amino acid [67]. Third line is a 10X recorder amplification of the center chromatograph. NH_2-Glu in the standard is glucosamine

possible that samples have been contaminated by more recent material before or after collection of the sample? Running a laboratory blank may reveal some of the possible sources of laboratory contamination but will reveal nothing about *in situ* contamination of the sample before collection.

A possible and a potentially serious source of contamination is found in the fingerprints of persons who handle samples [67, 77, 78]. This surface contamination is not easily eliminated even by extensive acid-prewashing treatments. Amino acids have been detected in fingerprints made on paper twelve years previously [79]. The pattern of amino acids found in human fingerprints shown in Fig. 2 is characterized by rather large amounts of some of the thermally less stable amino acids like serine, threonine, and aspartic acid. Cystine is apparently absent.

Reagents often used in extracting amino acids have sometimes enough amino acids in them to cause problems in interpretation. Even reagent-grade chemicals, such as hydrochloric acid which is often used in dissolving samples and in peptide hydrolysis, have sometimes been found to contain appreciable amounts of amino acids [67]. The amino acid pattern is very much like that of the fingerprint in Fig. 2.

Other possible sources of contamination include paper and cloth used in wrapping specimens, saliva from sneezing, and dust from the atmosphere. Although it may not be realistic to eliminate the problem, a knowledge of the unavoidable background of possible contamination from these sources is critical for the appraisal of the occurrences of low-level amounts of amino acids in geologic environments.

V. Nonprotein Amino Acids in Fossil Shells

Some amino acids not normally present in proteins have been found in a series of fossil shells of *Mercenaria* [80]. These nonprotein amino acids appear to be degradation and racemization products of the original amino acid constituents of the shell proteins. We have studied the diagenesis of shell proteins in a sequence of progressively older *Mercenaria* shells extending from the shells of living animals to as old as Miocene. Shells with radiocarbon ages from 1,000 to 33,000 years are included in the study.

Many molluscan shell structures appear to be fair approximations to closed systems, so that the amino acids found in fossil shell structures should be representative of *in situ* diagenesis. Experiments of heating shell fragments of *Mercenaria* both in water and in concentrated solutions of amino acids indicate little exchange of material within the shell structure itself [81].

The shells of living mollusks contain a proteinaceous organic matrix characteristic of the type of shell structures present [7, 82]. Most shells have more than one type of shell structure and therefore more than one type of organic matrix. These differ in amino acid composition as well as in several physico-chemical properties. The largely insoluble nature of the organic matrix of most recent molluscan shells has made it difficult to study the intact proteins for molecular weight, amino acid sequence, etc. Amino acid composition is relatively easy to determine and can be a useful parameter in determining organic matrix types for recent mollusk shells.

In fossil shells the organic matrix is progressively more soluble with age. The amino acid composition is readily determined, and comparison with its recent counterpart reveals changes in the amino acid pattern due primarily to *in situ* diagenesis of the organic matrix.

A. Sample Preparation

Fragments of shell structures of a few milligrams to a gram, depending on the amino acid content, are placed in a Pyrex test tube 150 × 15 mm. Polyethylene tubes 13 × 100 mm with a lip and an 0.006″ hole drilled in the bottom are placed in each sample-containing Pyrex tube. Twice the stoichiometric amount of 12 N HCl plus 1/2 ml of 6 N HCl is added to the polyethylene tube. These are then placed in a vacuum centrifuge, where the acid rapidly dissolves the sample and gives a final solution 6 N in HCl. The centrifuging keeps frothing to a minimum and the vacuum enables the reaction to be completed in three to four minutes at near 0° C to break down proteins and peptides to free amino acids.

After hydrolysis the samples are transferred to polyethylene tubes and twice the stoichiometric amount of 48 percent HF is added to precipitate Ca^{++} as CaF_2. After centrifuging, the supernatent is transferred to a second polyethylene tube and evaporated to dryness in the vacuum centrifuge over NaOH.

Standard amino acid mixtures carried through the same procedure show negligible losses and little or no racemization. Earlier attempts at ion-exchange desalting resulted in significant losses of several amino acids.

B. Amino Acid Analysis

To determine quantitatively submicrogram amounts of amino acids, a sensitive instrument (see Fig. 3) based on the ion-exchange method of SPACKMAN, MOORE, and STEIN has been developed [83, 84]. The system uses $1\frac{1}{2}$ mm inside diameter Teflon columns and regulated nitrogen pressure to force the buffers through the system. Pressure is also used to force the ninhydrin reagent into the mixer and heating coil. The system is highly stable, sensitive, and capable of greater than one percent reproducibility at the nanomole level on consecutive runs.

A system of automated sample injection and multiple-column analysis allows the rapid loading of several samples followed by the continuous unattended analysis of each individual sample.

The basic amino acids are analyzed on a single ion-exchange column using pH 4.4 citrate buffer at 50° C. This allows for the resolution of ornithine from lysine, and glucosamine from galactosamine with a total analysis time of $1\frac{1}{2}$ hours for each sample.

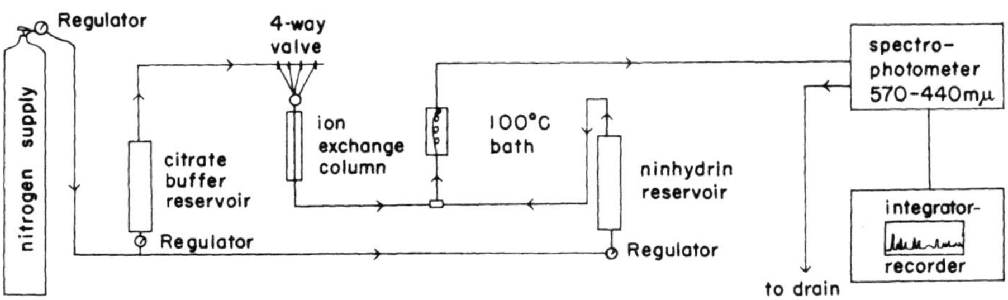

Fig. 3. Flow diagram of amino acid analyzer

The samples are drawn in with a syringe into 100 cm lengths of 32 gauge Teflon tubing, which are then connected through an automatic switching valve to the top of the column. A timer programs the flow consecutively through each sample tube injecting the sample into the column ahead of the buffer.

For the analysis of the acidic and neutral amino acids separate columns for each sample are used and are connected to a second automatic switching valve, which is programmed to allow flow consecutively through the different columns. At the high pressure side there are separate lengths of tubing which contain just enough of the pH 3.25 and 4.25 buffers, as well as 0.2 N NaOH for regeneration, for each analysis. This eliminates the need for a timed buffer change during each run.

The ninhydrin is added to the effluent stream by pressure applied to a Teflon coil containing ninhydrin reagent to produce a two-to-one mixture of buffer to ninhydrin reagent. The entire effluent line leading from the column through the heating coil to the colorimeter is of 32 gauge Teflon tubing. A chromatogram from a standard amino acid mixture is shown in Fig. 2 showing the sensitivity of the instrument to 0.01 micromoles of each amino acid.

C. Results

In a Recent *Mercenaria* shell there is considerable variation both in the amount and in the composition of organic matrix from one structural layer to another. The shell is made up largely of three calcified structural units laid down by different parts of the mantle. The total amino acid composition, as well as the per cent of amino acids per gram of shell weight, for each layer is shown in Fig. 4. The inner and outer layers both have over twice as much organic matrix per gram of shell as does the middle layer. The middle layer has relatively more aspartic acid and less glycine than either of the other layers. In the series of shells used in this study, only a single structural unit, the inner layer, was used.

Fig. 5 shows the amino acid patterns for three of the samples: Recent, Upper Pleistocene, and Miocene *Mercenaria* shells [85]. The total amino acids recovered

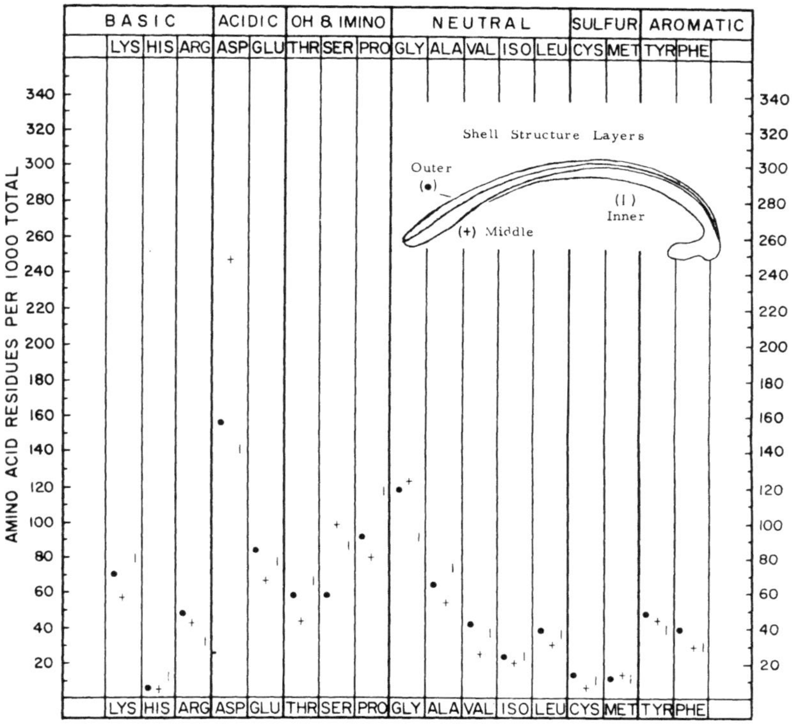

Fig. 4. Amino acid composition of the individual shell layers in *Mercenaria* [81]. Total amino acids recovered: outer shell layer – 21 micromoles per gram; middle shell layer – 8 micromoles per gram; inner shell layer – 16 micromoles per gram

decreases progressively from 16 micromoles per gram of shell to one micromole per gram of shell. In addition to this overall decrease the amino acid ratios change in a systematic way with progressively older shells. Threonine, serine, cystine, histidine, and arginine decrease to very low amounts or are absent entirely in the Miocene sample. Aspartic acid is the most abundant amino acid in the Recent shell but drops to the eleventh most abundant in the Miocene. Alanine, glutamic acid, proline, valine, and the leucines make up over 80 percent of the total amino acids found.

Of particular interest and significance is the presence of nonprotein amino acids in fossil specimens. One example is ornithine, derived from the protein amino acid arginine. The ornithine-to-arginine ratio increases from zero for the Recent shell to essentially infinity with the absence of arginine in the Miocene shell.

Another example, alloisoleucine derived from isoleucine, increases from essentially zero to an equilibrium value of approximately 1.3 times the isoleucine value. In a 1,000-year-old shell the alloisoleucine-to-isoleucine ratio is approximately 0.1. Samples of isoleucine when heated in water in sealed tubes yield nearly identical equilibrium values. An equilibrium mixture of alloisoleucine and isoleucine different from a 50:50 mixture indicates a difference in the free energy of formation of the two diastereoisomers.

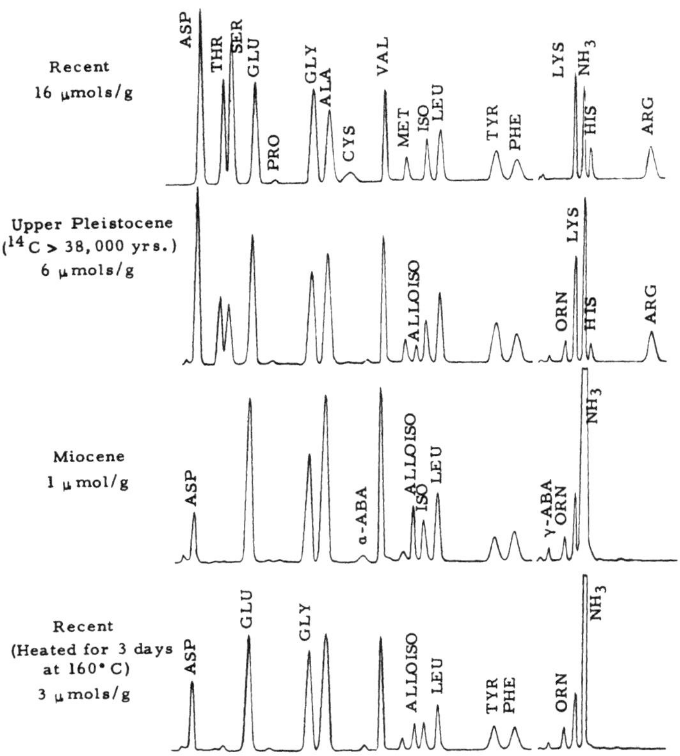

Fig. 5. Amino acids from the inner layer of *Mercenaria* shells [81, 85]. Total amino acids recovered are indicated in micromoles per gram of shell. Bottom pattern obtained by heating a fragment of a Recent shell in water in a sealed tube for three days at 160° C. α-ABA and γ-ABA are α and γ amino butyric acids

Allothreonine, the diastereoisomer of threonine, was not resolved in the present experiments. Some evidence of its presence, however, comes from a marked broadening of the threonine peak indicating the possible presence of allothreonine. Threonine is nearly absent in the older fossil specimens.

D. Laboratory Simulation of Geologic Environments

Fragments of Recent *Mercenaria* shells heated to temperatures between 150 and 200° C in the presence of water in sealed tubes yield amino acid patterns closely resembling the patterns found in fossil shells of *Mercenaria* [81]. The amount of water does not seem to be critical as long as some excess is present relative to the amount of organic matter. Apparently decomposition of amino acids without an excess of water present, results in amino acid patterns very different from that found in fossil materials [50].

pH can be expected to play a significant part in the kinetics of amino acid decomposition. We have found that pure arginine is relatively stable in acid solutions but in basic solutions decomposes readily to ornithine and urea. Similarly the reaction of isoleucine to alloisoleucine is accelerated in alkaline solutions. The pH dependence, although a complicating factor, may not be a serious problem for carbonate shell amino acids, since any water entering the shell would tend to be buffered by the $CO_3^=$ of the shell. By taking actual shell fragments in the heating experiments presumably the pH of the natural environment would be approximated.

Fig. 5 also shows a chromatogram of a shell fragment from a Recent *Mercenaria* heated at 160° C in water in a sealed tube for three days. Aspartic acid decrease relative to glutamic acid. Threonine and serine decompose relatively quickly. The ornithine-to-arginine ratio and also the alloisoleucine-to-isoleucine ratio change in a similar way to that found in the series of fossils. The equilibrium ratio of alloisoleucine to isoleucine is virtually identical in the prolonged heating experiments to the Miocene fossil specimen. The resemblance of the amino acid pattern to that in the series of fossils suggests a close approximation of the laboratory treatment to the natural environment.

There is no single heating experiment that will duplicate the results in a particular fossil during a single run. This is because of the very different activation energies for the various reactions. For example, to duplicate the alloisoleucine to isoleucine ratio found in the Upper Pleistocene sample it is necessary to heat the shell in water at 140° C for approximately six days. To get a similar ratio for ornithine to arginine, slightly less than one day at 140° C duplicates the Upper Pleistocene ratio. To duplicate the threonine and serine values found in the Upper Pleistocene sample, only around eight hours of heating are necessary.

E. Potential Application to Geothermometry and Geochronology

Each amino acid has its characteristic activation energy, which can be determined by a study of the kinetics of decomposition at a series of temperatures. Time and temperature are the main variables. With two unknowns (time and temperature) for each equation, it should be possible to define both time *and* temperature by the use of two or more amino acid reactions, e.g., arginine-to-ornithine and isoleucine-to-alloisoleucine.

A knowledge of the temperature dependence of the rate of chemical reactions can be useful in predicting rates of reaction at lower temperatures, where the reaction rate is too slow for laboratory study. Preliminary work at our laboratory suggests that decomposition studies at several temperatures between 88° and 225° C may be useful in predicting reaction rates at temperatures between 0° and 30° C and thereby be of potential use in both geochronologic and geothermometric studies. Comparison of results with data from older fossils indicates that the method might be useful for shells at least as old as Miocene.

The amino acid pattern of fossil shell structures should also be useful for stratigraphic correlation. Samples of 100 mg or even less are usually sufficient. The presence of reworked older fossils in a younger deposit should be readily detected by a comparison of the amino acid patterns of the shells in a fossil assemblage.

VI. Optical Configuration of Amino Acids in Fossils

To gain further insight into the possible mode of formation of fossil amino acid occurrences we have made a preliminary study of the optical configuration of the amino acids in fossil shells [85].

Table 3. *Structural formulae for the isomers of isoleucine*

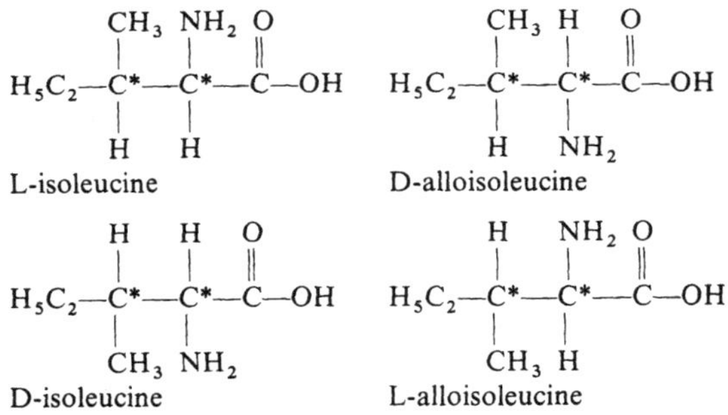

As indicated earlier in the chapter, most of the amino acids have only one asymmetric carbon atom with two possible optical isomers designated as D and L amino acid. Four protein amino acids (hydroxyproline, hydroxylysine, threonine, and isoleucine) have a second asymmetric carbon atom (see Table 1). In addition to the D and L optical isomers, these each have two diastereoisomers designated as the D and L allo amino acids. This is illustrated in Table 3 for isoleucine.

Amino acids synthesized by total inorganic synthesis are racemic mixtures of optical isomers and diastereoisomers and therefore show no optical activity. In contrast the amino acids isolated from acid hydrolysates of biological proteins are optically active. Alkaline hydrolysis leads to extensive racemization of amino acids. It would seem that the property of optical activity would be useful for distinguishing the mode of origin of amino acid mixtures found in various geologic environments.

To determine the optical configuration of the amino acids we have utilized the very specific enzyme L-Amino Acid Oxidase, which destroys only the L-configuration of the amino acids leaving the D-amino acid intact [86]. As a confirmatory check the enzyme D-Amino Acid Oxidase was used as well as some of the specific decarboxylases. Aliquots of the amino acids isolated from the shells were treated with the enzymes, after which the enzymes were precipitated with trichloroacetic acid and the solution placed on the amino acid analyzer. Untreated aliquots were run for comparison. The difference between the two runs reflects the amount of L or D amino acid acted on by the enzyme. Not every amino acid is acted on by the enzymes. The amino acids listed in Table 4 are reactive and have been calibrated by standard solutions of the D, L, and DL amino acid configurations.

The results show the Recent shell contains only L-amino acids, while in the Miocene sample the amino acids are virtually racemized to equal amounts of D and L amino acids.

The Upper Pleistocene sample (age approximately 70,000 years) shows appreciable racemization with approximately 25 percent of the D-amino acids present. The insoluble residue from the Upper Pleistocene shows that some racemization has occurred while the amino acids are presumably still in peptide linkage.

It appears that the L-amino acids are not stable but eventually racemize to form an optically inactive racemic mixture of D and L amino acids indistinguishable from the amino acids synthesized in the laboratory. An exception, however,

Table 4. *Percentage of D-amino acid isomers in the inner layer of shells of Mercenaria* [85]

	Recent, total	Upper pleistocene, total	Upper pleistocene, insoluble	Miocene, total
Glutamic acid	<5	22	11	47
Proline	0	31	19	52
Alanine	0	40	10	51
Valine	0	28	5	52
Alloisoleucine	–	>98	100	>95
Isoleucine	0	0	0	<5
Leucine	0	26	5	48
Tyrosine	0	25	16	50
Phenylalanine	0	25	8	49

has been found for isoleucine. From the data in Table 4 it appears that L-isoleucine does not racemize to D-isoleucine but rather to D-alloisoleucine. The second asymmetric carbon atom (Table 3) apparently does not easily racemize, if at all, since there are no readily ionizable groups on the atom nor is it adjacent to a carboxyl group for enolization. Thus L-isoleucine from biologically produced proteins racemizes to D-alloisoleucine (to racemize to D-isoleucine would require simultaneous racemization of both asymmetric carbon atoms). In the Miocene shells of *Mercenaria* only D-alloisoleucine and L-isoleucine were found, whereas all other amino acids were present in a racemic mixture. In Recent shell fragments prolonged heating in water in sealed tubes yielded racemic mixtures of most amino acids but only L-isoleucine and D-alloisoleucine. Although these relationships were discovered in shell amino acids, they are of general application to biological proteins and their diagenesis in geologic environments.

A. Criteria for Mode of Origin

It would appear that the isomeric composition of isoleucine, because of its very stable second asymmetric carbon atom, should be useful in distinguishing the mode of origin of specific geologic occurrences of amino acids. Isoleucine is always present in protein hydrolysates. Recent samples should have largely L-isoleucine, while in older samples progressively increasing amounts of D-alloisoleucine should be present to an equilibrium ratio of alloisoleucine to isoleucine. Isoleucine of non-biological origin would be expected to be racemic mixtures of both D, L-isoleucine and D, L-alloisoleucine. The presence of only L-isoleucine in very ancient rocks would suggest contamination by recent biological material. These criteria for mode of origin are summarized in Table 5.

Table 5. *Criteria for mode of origin of amino acids based on isomeric composition of isoleucine*

Isoleucine Isomers Present	Significance
1. Only L-isoleucine	Recent biological origin
2. L-isoleucine and D-alloisoleucine	Biological origin (probably older than Pliocene if isomers are in equilibrium)
3. D, L-isoleucine and D, L-alloisoleucine	Abiotic origin

VII. Discussion and Summary

Although work is just beginning on the optical configuration of isoleucine and other amino acids in problems of geologic interest, it is possible to make a few preliminary observations on some of the reported geologic occurrences of amino acids with regard to the presence or absence of alloisoleucine. In a series of fossil carbonate shells we have found progressively increasing amounts of alloisoleucine relative to isoleucine. This seems to be the result of *in situ* diagenesis on the original shell proteins, which contained only L-isoleucine. The rock matrix around the fossil shells has no appreciable amounts of alloisoleucine, although there are significant amounts of isoleucine and other amino acids [30]. The absence of alloisoleucine suggests that the amino acid material in the rock matrix is biologically recent and not *in situ* as it is in the shell. The absence of alloisoleucine reported [48] in unheated Miocene sediments, but its probable presence in heated samples, also seems to indicate that the amino acids in certain Miocene sediments, at least, are of recent biological origin.

Chromatograms of amino acids from some meteorite samples [55, 57], as well as from a Precambrian asbestos sample [44], show little or no alloisoleucine. This seems to be consistent only with a recent biological origin for the amino acid fraction. From these few examples it appears that some geologic occurrences of amino acids may not be as old as the host rock but rather of more recent biological origin.

FLORKIN [28] and GREGOIRE [87] and their group have made a number of studies relative to the preservation of the organic matrix in fossil mollusk shells. From electron micrographs the physical preservation of organic matrix tissues has been demonstrated for Paleozoic shell structures. It seems possible, however, that physical preservation may not necessarily imply that the original chemical integrity has also been preserved. Although direct comparisons are difficult because Paleozoic forms are largely extinct, enough is known on present-day shell structure types to make reasonable estimates of the original amino acid composition of many extinct forms. In general the amino acid composition reported for these very ancient shells is totally unlike any recent shell structures that have been reported. The amino acids in these Mesozoic and Paleozoic shells seem to be greatly altered from the original composition and possibly consist of a mixture of *in situ* amino acid material with material of more recent origin.

One of the most interesting aspects of the very thorough and systematic work reported by GREGOIRE [87] has been the evidence for fossil peptide-bound amino acid material. Based on the biuret reaction for the peptide bond, positive reactions have been obtained for fossil nacreous shell structures as old as Ordovician. These results are intriguing. If peptides of the original shell structures still persist in material this old, then the possibility exists of extracting the amino acid sequences of at least part of the original proteins involved. Comparison with similar data of Recent shell proteins should then reveal important phylogenetic and evolutionary relationships as well as information on early mechanisms of shell formation in mollusks.

Our data, however, leads us to believe that peptide bonds are hydrolyzed in geologically short times of 10^4 to 10^6 years [89]. In older fossils the total amino acid concentration is essentially equivalent to the free amino acid fraction. Further data are necessary to resolve the obvious differences in the results of these studies on peptide bond stability. It is possible that amino acids and peptides may form complexes in sedimentary materials that would increase their resistance to decomposition [13]. If simple heating experiments of the amino acid-containing samples in water cause changes to occur in the peptide and amino acid fractions, it seems likely that geologic time alone would accomplish similar results at ambient temperatures. Our data on shell structures strongly suggest this, and experiments with proteins and amino acids in other pH environments suggest general validity of these results for other geologic materials.

In summary, there have been a number of studies of the amino acids in generally random samples of geologic interest. Samples representing most of the geologic column including the Precambrian have been reported to contain several amino acids including some relatively unstable ones. This somewhat confusing picture will only become clearer when more systematic studies are made and more of the diagenetic reactions of the amino acids are learned. The ability to simulate geologic environments in the laboratory should make it possible to better understand the reactions in nature and to be able to say something about their rates at various temperatures in the earth. The application of enzyme techniques to the determination of optical configuration of amino acids in geologic environments promises to be a significant aid in the elucidation of the age, as well as the mode of origin, of many geologic occurrences of amino acids.

References

1. Florkin, M.: Biochemical evolution. New York: Academic Press, Inc. 1949.
2. Fruton, Joseph S., and Sofia Simmonds: General biochemistry. New York: John Wiley and Sons, Inc. 1958.
3. Fox, Sidney W., and Joseph F. Foster: Introduction to protein chemistry. New York: John Wiley and Sons, Inc. 1957.
4. Wilbur, Karl M.: Shell structure and mineralization in molluscs. In: Calcification in biological systems (Reidar F. Sognnaes, ed.). Publ. Am. Assoc. Advan. Sci. No. 64, 15–40 (1960).
5. Glimcher, Melvin J.: Specificity of the molecular structure of organic matrices in mineralization. In: Calcification in biological systems (Reidar F. Sognnaes, ed.). Publ. Am. Assoc. Advan. Sci. No. 64, 421–487 (1960).
6. Hare, P. Edgar: Amino acids in the proteins from aragonite and calcite in the shells of *Mytilus californianus*. Science **139**, 216–217 (1963).
7. Hare, P. E., and P. H. Abelson: Amino acid composition of some calcified proteins. Carnegie Inst. Wash. Year Book **64**, 223–232 (1965).
8. Lowenstam, H. A.: Biologic problems relating to the composition and diagenesis of sediments. In: The earth sciences (Thomas W. Donnelly, ed.). Chicago: Chicago University Press 1963.
9. Abelson, Philip H.: Geochemistry of organic substances. In: Researches in geochemistry (Philip H. Abelson, ed.), p. 79–103. New York: John Wiley & Sons, Ltd. 1959.
10. Jukes, Thomas H.: The genetic code, II. Am. Scientist **53**, 477–487 (1965).
11. Phillips, David C.: The three-dimensional structure of an enzyme molecule. Sci. Am. **215**, 78–90 (1966).
12. Mortensen, James L., and Frank L. Himes: Soil organic matter. In: Chemistry of the soil (Firman E. Bear, ed.), p. 206–241. New York: Reinhold 1964.
13. Alexander, Martin: Introduction to soil microbiology. New York: John Wiley & Sons, Inc. 1961.

14. MASON, BRIAN: Principles of geochemistry, p. 215 – 238. New York: John Wiley & Sons, Inc. 1958.
15. RUBEY, W. W.: Geologic history of sea water. Bull. Geol. Soc. Am. **62**, 1111 – 1148 (1951).
16. STEVENSON, F. J.: Carbon-nitrogen relationships in soil. Soil Sci. **88**, 201 – 208 (1959).
17. BADER, RICHARD G.: Carbon and nitrogen relations in surface and subsurface marine sediments. Geochim. Cosmochim. Acta **7**, 205 – 211 (1955).
18. FORSMAN, J. P., and JOHN M. HUNT: Insoluble organic matter (kerogen) in sedimentary rocks. Geochim. Cosmochim. Acta **15**, 170 – 182 (1958).
19. STEVENSON, F. J.: Chemical state of the nitrogen in rocks. Geochim. Cosmochim. Acta **26**, 797 – 809 (1962).
20. MCLAREN, A. D.: Biochemistry and soil science. Science **141**, 1141 – 1147 (1963).
21. STONE, ROBERT W., and CLAUDE E. ZOBELL: Bacterial aspects of the origin of petroleum. Ind. Eng. Chem. **44**, 2564 – 2567 (1952).
22. BEERSTECHER, ERNEST, JR.: Petroleum microbiology. Houston, Texas: Elsevier 1954.
23. KUZNETSOV, SERGEY IVANOVICH, MIKHAIL VLADIMIROVICH IVANOV, and NATAL'YA NIKOLAYEVNA LYALIKOVA: Introduction to geological microbiology. New York: McGraw-Hill Book Co. 1963.
24. ABELSON, P. H.: Organic constituents of fossils. Carnegie Inst. Wash. Year Book **53**, 97 – 101 (1954).
25. – Organic constituents of fossils. Carnegie Inst. Wash. Year Book **54**, 107 – 109 (1955).
26. – Paleobiochemistry. Sci. Am. **195**, 83 – 92 (1956).
27. FLORKIN, M., CH. GRÉGOIRE, S. BRICTEUS-GRÉGOIRE, and E. SCHOFFENIELS: Conchiolines de nacres fossiles. Compt. Rend. **252**, 440 – 442 (1961).
28. — A molecular approach to phylogeny, p. 133 – 156. London: Elsevier 1966.
29. HARE, P. E.: The amino acid composition of the organic matrix of some west coast species of *Mytilus*. Ph. D. thesis California Institute of Technology 1962, 109 p.
30. – Unpublished data.
31. DEGENS, EGON T., and STEVEN LOVE: Comparative studies of amino-acids in shell structures of *Gyraulus trochiformis*, Stahl, from the Tertiary of Steinheim, Germany. Nature **205**, 876 – 878 (1965).
32. JOPE, MARGARET: The protein of brachiopod shell – I. Amino acid composition and implied protein taxonomy. Comp. Biochem. Physiol. **20**, 593 – 600 (1967).
33. – The protein of brachiopod shell – II. Shell protein from fossil articulates: Amino acid composition. Comp. Biochem. Physiol. **20**, 601 – 605 (1967).
34. FOUCART, M. F., S. BRICTEUX-GRÉGOIRE, CH. JEUNIAUX, and M. FLORKIN: Fossil proteins of graptolites. Life Sci. **4**, 467 – 471 (1965).
35. SINEX, F. M., and B. FARIS: Isolation of gelatin from ancient bones. Science **129**, 969 (1959).
36. WYCKOFF, RALPH W. G., WILLIAM F. MCCAUGHEY, and ALEXANDER R. DOBERENZ: The amino acid composition of proteins from pleistocene bones. Biochim. Biophys. Acta **93**, 374 – 377 (1964).
37. HO, TONG-YUN: The amino acid composition of bone and tooth proteins in Late Pleistocene mammals. Proc. Natl. Acad. Sci. U.S. **54**, 26 – 31 (1965).
38. ARMSTRONG, W. G., and L. B. HALSTEAD TARLO: Amino-acid components in fossil calcified tissues. Nature **210**, 481 – 482 (1966).
39. WYCKOFF, RALPH W. G., ESTELLE WAGNER, PHILIP MATTER, III, and ALEXANDER R. DOBERENZ: Collagen in fossil bone. Proc. Natl. Acad. Sci. U.S. **50**, 215 – 218 (1963).
40. DOBERENZ, ALEXANDER R., and RALPH W. G. WYCKOFF: Fine structure in fossil collagen. Proc. Natl. Acad. Sci. U.S. **57**, 539 – 541 (1967).
41. ISAACS, W. A., K. LITTLE, J. D. CURREY, and L. B. H. TARLO: Collagen and a cellulose-like substance in fossil dentine and bone. Nature **197**, 192 (1963).
42. HELLER, WOLFGANG VON: Tonmineralien bituminöser Schiefer als natürliche Systeme der Verteilungschromatographie. Erdoel, Kohle, Erdgas, Petrochem. **19**, 557 – 561 (1966).
43. ERDMAN, J. G., EVERETT M. MARLETT, and W. E. HANSON: Survival of amino acids in marine sediments. Science **124**, 1026 (1956).

44. HARINGTON, J. S.: Natural occurrence of amino acids in virgin crocidolite asbestos and banded ironstone. Science **138**, 521 – 522 (1962).

45. –, and J. J. LE R. CILLIERS: A possible origin of the primitive oils and amino acids isolated from amphibole asbestos and banded ironstone. Geochim. Cosmochim. Acta **27**, 411 – 418 (1963).

46. AUCOTT, J. W., and R. H. CLARKE: Amino-acids in the Mountsorrel bitumen, Leichestershire. Nature **212**, 61 – 63 (1966).

47. PONNAMPERUMA, CYRIL, and KATHERINE PERING: Possible abiogenic origin of some naturally occurring hydrocarbons. Nature **209**, 979 – 982 (1966).

48. SELLERS, G. A.: Hydrothermal experiments on the thermal stability of amino substances in sediments. Ph. D. thesis California Institute of Technology. University Microfilms, In., No. 662197, Ann Arbor, Michigan 1966.

49. RITTENBERG, S. C., K. O. EMERY, JOBST HULSEMANN, E. T. DEGENS, R. C. FAY, J. H. REUTER, J. R. GRADY, S. H. RICHARDSON, and E. E. BARY: Biogeochemistry of sediments in experimental Mohole. J. Sediment. Petrol. **33**, 140 – 172 (1963).

50. JONES, J. D., and J. R. VALLENTYNE: Biogeochemistry of organic matter – I. Polypeptides and amino acids in fossils and sediments in relation to geothermometry. Geochim. Cosmochim. Acta **21**, 1 – 34 (1960).

51. STEVENSON, F. J., and A. P. S. DHARIWAL: Distribution of fixed ammonium in soils. Soil Sci. Soc. Am. Proc. **23**, 121 – 125 (1959).

52. STEVENSON, I. L.: Biochemistry of soil. In: Chemistry of the soil (FIRMAN E. BEAR, ed.), p. 242 – 291. New York: Reinhold 1964.

53. STEVENSON, F. J.: Isolation and identification of some amino compounds in soils. Soil Sci. Soc. Am. Proc. **20**, 201 – 204 (1956).

54. SOWDEN, F. J., and K. C. IVERSON: The "free" amino acids of soil. Can. J. Soil Sci. **46**, 109 – 120 (1966).

55. VALLENTYNE, J. R.: Two aspects of the geochemistry of amino acids. In: The Origins of Prebiological Systems (SIDNEY W. FOX, ed.), p. 105 – 125. New York: Academic Press 1965.

56. KAPLAN, I. R., E. T. DEGENS, and J. H. REUTER: Organic compounds in stony meteorites. Geochim. Cosmochim. Acta **27**, 805 – 834 (1963).

57. DEGENS, EGON T.: Genetic relationships between the organic matter in meteorites and sediments. Nature **202**, 1092 – 1095 (1964).

58. PARK, KILHO, W. T. WILLIAMS, J. M. PRESCOTT, and D. W. HOOD: Amino acids in deep-sea water. Science **138**, 531 – 532 (1962).

59. DEGENS, EGON T., JOHANNES H. REUTER, and KENNETH N. F. SHAW: Biochemical compounds in offshore California sediments and sea waters. Geochim. Cosmochim. Acta **28**, 45 – 66 (1964).

60. JOHANNES, R. E., and K. L. WEBB: Release of dissolved amino acids by marine zooplankton. Science **150**, 76 – 77 (1965).

61. DEGENS, EGON T.: Geochemistry of Sediments, p. 202 – 280. Englewood Cliffs, N. J.: Prentice-Hall 1965.

62. BARS, E. A., and S. S. KOGAN: Some rules of variation in the nature of the organic matter dissolved in the formation water of the Volga region oil fields. In: The geochemistry of oil and oil deposits (L. A. GULYAEVA, ed.), p. 179 – 191. New York: Daniel Davey 1964.

63. –, and L. N. NOSOVA: Dissolved organic matter in Cretaceous and Jurassic formation waters from the middle part of the Ob'-Irtysh Basin. In: The geochemistry of oil and oil deposits (L. A. GULYAEVA, ed.), p. 129 – 209. New York: Daniel Davey 1964.

64. WENT, F. W.: Organic matter in the atmosphere, and its possible relation to petroleum formation. Proc. Natl. Acad. Sci. U.S. **46**, 212 – 221 (1960).

65. BEAR, I. J., and R. G. THOMAS: Genesis of petrichor. Geochim. Cosmochim. Acta **30**, 869 – 879 (1966).

66. DELANY, A. C., AUDREY CLAIRE DELANY, D. W. PARKIN, J. J. GRIFFIN, E. D. GOLDBERG, and B. E. F. REIMANN: Airborne dust collected at Barbados. Geochim. Cosmochim. Acta **31**, 885 – 909 (1967).

67. HARE, P. E.: Amino acid artifacts in organic geochemistry. Carnegie Inst. Wash. Year Book **64**, 232–235 (1965).

68. McKEE, H. S.: Nitrogen metabolism in plants. Oxford: Clarendon Press 1962.

69. KATSOYANNIS, PANAYOTIS G., KOUHEI FUKUDA, ANDREW TOMETSKO, KENJI SUZUKI, and MANOHAR TILAK: Insulin peptides X. The synthesis of the B-chain of insulin and its combination with natural or synthetic A-chain to generate insulin activity. J. Am. Chem. Soc. **86**, 930–932 (1964).

70. ZAHN, HELMUT: Chemische Synthese von Proteinen. Naturwissenschaften **54**, 396–402 (1967).

71. MILLER, A. L.: Production of some organic compounds under possible primitive earth conditions. J. Am. Chem. Soc. **77**, 2351–2361 (1955).

72. ABELSON, P. H.: Paleobiochemistry: inorganic synthesis of amino acids. Carnegie Inst. Wash. Year Book **55**, 171–174 (1956).

73. GROSSENBACHER, KARL A., and C. A. KNIGHT: Amino acids, peptides, and spherules obtained from primitive earth gases in a sparking system. In: The origins of prebiological systems (SIDNEY W. FOX, ed.), p. 113–186. New York: Academic Press 1965.

74. HARADA, KAORU, and SIDNEY W. FOX: The thermal synthesis of amino acids from a hypothetically primitive terrestrial atmosphere. In: The origins of prebiological systems (SIDNEY W. FOX, ed.), p. 187–201. New York: Academic Press 1965.

75. ORÓ, J.: Stages and mechanisms of prebiological organic synthesis. In: The origins of prebiological systems (SIDNEY W. FOX, ed.), p. 137–171. New York: Academic Press 1965.

76. ABELSON, P. H., and P. E. HARE: Action of 2537 Å radiation on HCN solutions. Carnegie Inst. Wash. Year Book **65**, 358–360 (1966).

77. ORÓ, J., and H. B. SKEWES: Free amino-acids on human fingers: the question of contamination in microanalysis. Nature **207**, 1042–1045 (1965).

78. HAMILTON, PAUL B.: Amino-acids on hands. Nature **205**, 284–285 (1965).

79. ODEN, S., and B. VON HOFSTEN: Detection of fingerprints by the ninhydrin reaction. Nature **173**, 449–450 (1954).

80. HARE, P. E., and R. M. MITTERER: Nonprotein amino acids in fossil shells. Carnegie Inst. Wash. Year Book **65**, 362–364 (1966).

81. MITTERER, R. M., and P. E. HARE: Unpublished data.

82. AKIYAMA, MASAHIKO: Conchiolin-constituent amino acids and shell structures of bivalved shells. Proc. Japan Acad. **42**, 800–805 (1966).

83. HARE, E.: Automatic multiple column amino acid analysis – the use of pressure elution in small bore ion-exchange columns. Federation Proc. **25**, 709 (1966).

84. SPACKMAN, D. H., W. STEIN, and S. MOORE: Automatic recording apparatus for use in the chromatography of amino acids. Anal. Chem. **30**, 1190–1206 (1958).

85. HARE, P. E., and P. H. ABELSON: Racemization of amino acids in fossil shells. Carnegie Inst. Wash. Year Book **66**, in press (1967).

86. GREENSTEIN, JESSE P.: The resolution of racemic α-amino acids. In: Advances in Protein Chemistry (M. L. ANSON, KENNETH BAILEY, and JOHN T. EDSALL, eds.), p. 121–202. New York: Academic Press 1954.

87. GRÉGOIRE, CHARLES: On organic remains in shells of paleozoic and mesozoic cephalopods. Inst. Roy. Sci. Natl. Belg. Bull. **42**, No. 39 (1966).

Addendum

Amino acids isolated from samples of the Precambrian Gunflint Chert include substantial amounts of the amino acids serine and threonine and little or no alloisoleucine [1, 2]. This pattern of amino acids seems consistent only with a recent biological origin for at least some of the amino acids found. This would imply *in situ* contamination of the chert with recent organic materials.

An alternative possibility is that the chert matrix has protected the amino acids against normal diagenetic reactions. To check this possibility, samples of Gunflint Chert were heated in the presence of water at 165° C for periods of 1 day and longer. If the chert stabilized the amino acids it should be evident in such an experiment. Instead, at the end of the experiment serine and threonine had been destroyed and significant amounts of alloisoleucine had been formed, a result almost identical to the heating experiments of shells [3] which in turn compared closely with the amino acids isolated from a natural series of fossil shells. There seems to be no stabilizing effect on the amino acids by the chert matrix.

These results emphasize the need to critically evaluate the occurrences of amino acids in geologic environments. It appears that there is a ubiquitous background level of recent biologically produced amino acids that makes it difficult to interpret the results of amino acids found in very ancient materials.

1. SCHOPF, J. WILLIAM, KEITH A. KVENVOLDEN, and ELSO S. BARGHOORN, Amino Acids in Precambrian Sediments: An Assay. Proceedings of the National Academy of Sciences, **59**, No. 2, 639–648 (1968).

2. ABELSON, P. H., and P. E. HARE, Recent Amino Acids in the Gunflint Chert. Carnegie Inst. Wash. Year Book **67**, 208—210 (1969).

3. HARE, P. E., and R. M. MITTERER, Laboratory Simulation of Amino Acid Diagenesis in Fossils. Carnegie Inst. Wash. Year Book **67**, 205–208 (1969).

CHAPTER 19

Porphyrins

EARL W. BAKER

Mellon Institute,
Carnegie-Mellon University,
Pittsburgh, Pennsylvania

Contents

I. Introduction

The history of life is encoded in the variety of the chemical structures of organic materials which have survived to the present time. Crude petroleum and bitumens represent the organic residue of some of the oldest living matter. In particular, petroporphyrins, are easily traceable artifacts of living systems from as old as a billion years upto the present. Even though the porphyrins account for only a trace of the carbon present in petroleum, they have great geochemical significance. This fact was immediately recognized by A. TREIBS in 1934 when he discovered that a wide variety of petroleum and bitumens contained porphyrins. He stated "The findings compel extensive geological conclusions. The demonstration of porphyrin is just as sure and exact as the spectroscopic detection of an element ... The proof that chlorophyll-bearing plants played a decisive part in the formation of bitumens and petroleums of various origins and of all geological ages is brought out with full certainty" [1]. Following this announcement and after working with a very wide sampling of minerals, shales, bitumens and crudes, TREIBS put forward his famous postulate for the transformation of chlorophyll into desoxophylloerythroetioporphyrin [2].

The implications of these ideas are extremely far reaching and to a great extent the methods and conclusions put forward in that seminal paper are the basis of organic geochemistry. The method of selecting genetically related, starting material-product pairs and connecting them with feasible geochemical reactions has been extended to many classes of compounds. The assumption made by TREIBS that the overall biochemistry of past organisms is similar or identical to present ones is implicit in all such studies.

A highly refractory aromatic nucleus combined with peripheral groups of variable stability make chlorophyll chemically unique. The aromatic nucleus ensures geologic survival and the peripheral groups of its chemical progeny rather accurately reflect the environment to which the material has been exposed.

Thus, because the porphyrins were the premier organic geochemical tracers and secondly because their potential information content is so great, these materials continue to intrigue the geochemist. Recent refinements in separation techniques and methods of spectrometric identification have made the studies especially fruitful.

A certain jargon usually pertains to each speciality and the field of the chemistry of tetrapyrrole pigments is not excepted. The term "petroporphyrin" to describe the plurality of porphyrins found in petroleum was coined because of indications that at least some of the components were not identical with known porphyrins. The indications have been confirmed and these materials do constitute a new class of naturally occurring compounds. The designation of this class of porphyrins as "petroporphyrins" seems especially appropriate [3, 4].

The terms "complex" and "chelate" have been used interchangeably to designate the compounds formed from the reaction of metal ions and porphyrins. However, with an increasing number of studies of the coordination of the metal to additional ligands beyond the four porphyrin nitrogens, new terminology is required. In line with suggestions by a number of workers in the field, the term "chelate" will be used to designate the "normal" square planar metalloporphyrins, while the term "complex" is reserved for compounds where additional ligands are present to form square pyramidal or octahedral coordination spheres.

II. Concentration and Purification of the Pigments

An old English recipe for rabbit stew begins, "Catch a rabbit". And so it is with the study of the fossil porphyrin pigments. Since these pigments in both contemporary and ancient sources are minor constituents of the matrix in which they occur and because of their complex, high molecular weight structures are subject to alteration, their isolation has continued to challenge the chemist. For convenience we shall divide the methods used for obtaining the pigments into three groups: isolation, fractionation, and identification.

By isolation we shall mean procedures which reject the bulk of the matrix to yield a pigment concentrate. The main requirements for a suitable procedure are first, that the integrity of pigments be preserved, second, that little or no loss of the pigment in the rejected matrix occurs and, finally, that the method produces utilizable amounts of material in a reasonable period.

Fractionation is defined as a means of segregation of the porphyrin aggregate obtained from the initial isolation into homogeneous components as determined by some criterion, usually based on a specific physical method or methods and does not imply that each fraction is a pure compound. Since the quantity of material here is 2 or 3 orders of magnitude lower than that prior to the isolation, more sophisticated and precise methods should be called into play.

Identification takes on various aspects but in general, means assigning the porphyrins to a specific spectral group or to a chemical class by establishing the presence or absence of functional groups.

The following sections present selected methods for these operations. Some traditional ones are included as well as some taken from the current literature.

A. Isolation of Porphyrins

The porphyrins in bitumens are present predominantly as vanadyl and nickel chelates. These chelates are in many of their physical properties very similar to the materials with which they are associated in the bitumens, petroleums or sediments. For this reason, there is no single method of isolation suitable for all cases. The most common and convenient method, demetallation by strong acids, warrants the most detailed discussion. However, it does have the disadvantage that the distinction between the porphyrins chelated with nickel and those chelated with vanadium is lost. Such methods as adsorption chromatography, liquid-liquid extraction and a recent addition, gel permeation chromatography, preserve the metallo-chelates for study.

It should be noted that the recovery of porphyrin by these procedures is not quantitative and all probably discriminate against the carboxylic acid porphyrins and against those of high molecular weight.

1. Acid Extraction Methods

The original and still most common method of isolation is based on demetallation. Displacement of the chelated metal converts the metalloporphyrin, which is essentially neutral, to a sufficiently strong base ($pK_b \approx 5$) to be easily extracted from the bituminous residue with dilute aqueous acid.

Naturally, the choice of demetallating agent is dictated by the most refractory chelate present. In petroleum samples, this is the vanadyl and only very strong acids can displace the metal. The strength of the acid needed is shown by the fact that HCl-acetic acid will not demetallate vanadyl porphyrins even at elevated temperatures, while HBr-acetic acid demetallates vanadyl porphyrins slowly at room temperature but rather quickly at 50° C [5].

Table 1. *Acidic reagents for extraction of porphyrins from bitumens*

Reagent	Reference
HBr-acetic acid	TREIBS [6], GROENNINGS [7], COSTANTINIDES *et al.* [8]
H_2SO_4	TREIBS [9], DEAN and GIRDLER [20]
HBr-formic acid	SUGIHARA and GARVEY [10]
Methanesulfonic acid	ERDMAN [11], BAKER [3]
H_3PO_4	CORWIN [12]

Table 1 lists other reagents which have been successfully used for opening vanadyl porphyrins. Of these, the most widely used reagent for the demetallation of porphyrin aggregates in petroleum and petroleum residues is HBr-acetic acid.

Its use was originated by TREIBS [6]. After some improvements by GROENNINGS [7], it became the standard method for quantitative measurements of porphyrin contents. That the method still had certain shortcomings was recognized, so that COSTANTINIDES *et al.* critically reviewed the experimental details of the procedure and incorporated several further improvements into the method given below [8]. That any free bromine in the HBr-AcOH reagent will produce artifacts has been mentioned by many workers and the nature of some of the artifacts are now known. Treatment of deutero- or protoporphyrin in HBr-AcOH leads to a compound of the chlorin class [8a] when bromine is present.

a) Method for the Quantitative Determination of Porphyrins in Petroleum Products

A five gram sample is dissolved in 15 ± 1 ml of toluene and poured into an ampoule with 40 ± 1 ml of reagent, which consists of a solution of a 30 ± 2 weight percent HBr in anhydrous acetic acid (< 0.2 percent H_2O). For the preparation of the HBr-acetic acid reagent a fritted glass inlet tube for HBr gas is recommended since HBr dissolves with difficulty in acetic acid. The ampoule is cooled with dry ice, sealed and immersed for 48 hrs. in a thermostatic bath at $50 \pm 1°$ C and shaken frequently.

The ampoule is opened and the reaction mixture is poured into a separatory funnel containing 50 ml of 33 percent by volume aqueous acetic acid. The funnel is vigorously shaken. The acid phase is drawn off and the oil phase washed three times with 20 ml of 50 percent acetic acid. The combined acid phases are boiled to remove HBr and toluene, then filtered, and brought to a suitable volume for optical density measurement. 50 percent acetic acid is used for dilution and the spectrum of the acid solution is determined in the 330–450 mμ range. The measure of the intensity of the Soret band is made by integrating the area defined by the base line and the absorption curve. An integral absorption coefficient of 65×10^8 mole^{-1} cm^2 mμ is used to calculate the concentration of porphyrin aggregate in the sample. Duplicate determinations of porphyrin content on the same sample give results which do not differ by more than 3 percent.

If a preparative rather than analytical method is required, the first portion of the method is used and the acid phase is fractionated by the method described in the section on isolation of porphyrin from oil shale.

A number of adaptations of the basic method have been reported. An open container rather than a sealed ampoule can be used [13] and the digestion time reduced from 2 days to 2 hrs. by continuous shaking [14].

For solid bitumens somewhat different methods are required. A very satisfactory method for a preliminary concentration of the pigment containing fraction of gilsonite was given by SUGIHARA and MCGEE [15].

b) Solvent Extraction of Gilsonite

Gilsonite (200 g) was placed in a Waring blender with ethyl acetate (500 ml) and agitated at high speed for 1 min. The suspension was allowed to settle and the liquid decanted. A second 500 ml portion of solvent was added, agitated, and decanted. Removal of the ethyl acetate-soluble compounds was essentially complete after eight extractions. Combination of the extracts and evaporation of the solvent gave 53.4 g of black tar.

The tar may then be treated by any of the acid extraction methods to yield demetallated porphyrins or can be used as the starting point for a chromatographic concentration procedure.

In shales and kerogens, still different techniques are required for the initial opening of the matrix. The method given by TREIBS for the extraction of porphyrins from oil shale is suitable for a wide variety of such solid samples.

c) Isolation of Porphyrin from Oil Shale According to TREIBS [6]

A total of 12 kg of shale (pulverized in a ball mill) is added in portions of 3–4 kg to equal amounts of acetic acid, heated on a steambath for two days, cooled, filtered and thoroughly washed. The treatment is repeated 6 times so that exhaustive extraction is approached.

The acetic acid extract is vacuum distilled nearly to dryness, the residue of salts dissolved in water, from which the complex salts together with extracted oil are quantitatively collected in a moderate amount of chloroform. The residue on evaporation amounts to about 100 g of dark brown oil, from which the odor of sulfur compounds arises. After further drying under vacuum at steambath temperature, the oil is treated in 4 portions with 100 g of hydrogen bromide-acetic acid for 4 days at 50°; the sealed ampoule is shaken frequently. The crude porphyrin mixture is transferred to ether in the usual way by addition of sodium acetate. By repeated extraction with 10 percent hydrochloric acid and finally 20 percent hydrochloric acid, the quantity of interest (porphyrin) is mostly separated from the oily impurity; however, complete removal occurs only in the course of the following acid fractionation.

The separation yields several different basic fractions and each is subjected to further acid fractionation for complete separation. Treatment of the ether solution with hydrochloric acid of 2.5, 5, 10, and 20 percent, yields four fractions with the 2.5 percent fraction containing the main part of the porphyrins. The acid porphyrins (those with carboxyl functions) can be separated with dilute sodium hydroxide from another solution while the upper phase contains all of the etioporphyrins.

The use of methanesulfonic acid (MSA) as a demetallating agent was discovered by ERDMAN [11] and exploited by BAKER and collaborators for the preparation of petroporphyrins from asphaltenes [16].

d) MSA Extraction of Petroporphyrins

Asphaltene (40.0 g), obtained by pentane precipitation from crude petroleum was treated with MSA (200 ml) in a ball mill (0.3 gallon) containing ceramic balls (usually 10 balls, 1 inch diameter). After stirring to break up lumps, the mill was closed and placed on rollers at slow speed. Heat was supplied with a bank of infrared lamps or a heat gun. The heat input was adjusted by trial and error to give a temperature at the end of a four-hour run of $105 \pm 5°$ C. After four hours, the mill was opened, and 200 g of ice with 200 ml of water was added with stirring. The solution (*ca.* 50° C) was filtered with suction through an 11 cm filter funnel.

The volume of the wine colored extract was measured and the yield assayed by volumetric dilution and the optical density at 546 mμ due to the di-cation was determined. An ε value of 17×10^3 l mole^{-1} cm^{-1} was used to calculate the yield after subtraction of a small background correction.

The porphyrins were then transferred from the aqueous acid to methylene chloride [17] (*ca.* 250 ml) in a liquid-liquid extractor [18]. The methylene chloride layer was then withdrawn from the extractor and the porphyrin converted from the di-cation to the free base by treatment with sodium acetate solution (100 ml, 5 percent w/v). The methylene chloride layer was transferred to a rotary evaporator and the solvent removed. The oily black residue was taken up with a minimum of benzene and applied to a silica gel chromatographic column (40 mm diameter; 12 cm of silica gel) which had been prepared with cyclohexane. The column was developed with cyclohexane/benzene, benzene, and finally with benzene containing 1–5 percent ether. The purified porphyrin was finally recovered by evaporation of the solvent.

The use of phosphoric acid has been mentioned briefly by CORWIN [12].

e) Phosphoric Acid Treatment of Petroleum Samples

The sample of crude petroleum to be treated is mixed with three times its weight of 85 percent phosphoric acid and heated to 180° for an hour. It is then poured onto cracked ice and treated with an equal volume of 5 percent HCl. The mixture is allowed to stand overnight, whereupon the phases separate. The porphyrins will be found in the aqueous phase. When nickel or vanadyl etioporphyrin II is treated by this method, the removal of metal is quantitative.

Some confusion has arisen on the use of sulfuric acid as an extractive agent for porphyrins. Some references purport that vanadyl porphyrins are not demetallated by this reagent [19] and hence cannot be extracted from bitumens or petroleum.

That vanadyl porphyrins are demetallated, at least in part, by 90 percent H_2SO_4 at room temperature was demonstrated by DEAN and GIRDLER [20]. Actually, sulfuric acid was used by TREIBS in some of his earlier work as an extractive agent of oil shale and later by CORWIN *et al.* [21]. However, its use cannot be recommended because, as noted by TREIBS, at times the porphyrin may be reabsorbed on the bitumen when the acid is diluted or at other times inexplicably destroyed.

SUGIHARA and GARVEY have proposed the substitution of formic acid for acetic as the solvent for HBr in the extraction of porphyrins from a heavy crude oil [10]. With this method, a sealed ampoule cannot be used and a continuous flow of HBr through the reaction vessel was used instead. Further mention will be made of the use of formic acid in Section II, C-1.

2. Adsorption Chromatography

Probably the most convenient method of obtaining metallochelate concentrates is column adsorption chromatography. As pointed out earlier, the distinction between those porphyrins chelated with nickel and those chelated with vanadium is not lost in this treatment. Further, the ever present danger of creating artifacts under the rigorous conditions of demetallation is avoided. It is not difficult to obtain a porphyrin concentrate of say 5–10 percent purity, sufficient for some types of spectral studies. However, column chromatography is in general not sufficiently selective to reject tightly clinging fluorescent impurities which are present. Therefore, one often resorts to a combination of chromatographic concentration followed by acid demetallation of the concentrate. Demetallation of such concentrates will be discussed in a later section.

a) Isolation of Metalloporphyrin Aggregates from Crude Petroleum [22]

Equal weights of crude oil and isooctane are mixed and shaken with an equal weight of alumina in an Erlenmeyer flask. The supernatant liquid is decanted and the alumina washed with 20/80 benzene/isooctane to remove colored non-porphyrin impurities. The washing is continued until the washings are nearly colorless. The metalloporphyrins are desorbed by washing with 30/70 ethylene dichloride/benzene until the wash liquid contains no appreciable porphyrin. A second fraction containing some porphyrin and other colored impurities can be recovered with chloroform containing some methanol. Reduction of the combined ethylene dichloride washings to dryness yields a porphyrin-rich residue which is then chromatographed on alumina with the same solvents as above to give a metalloporphyrin concentrate.

Careful development of a chromatographic column as in the above procedure, yields a porphyrin concentrate of, perhaps, 10 percent purity. The difficulties increase rapidly, however, when a high degree of purity of the separated porphyrin aggregate is required, i.e., as one passes from the isolative step to the purification. The non-porphyrins which have been carried through the separation scheme this far must have chromatographic properties very similar to the porphyrins. Improvement in purity beyond 50 percent is achieved with increasing difficulty and generally at the sacrifice of some of the porphyrins. Persistent and repeated application of a combination of column and partition chromatography will produce samples of reasonably good purity. Sufficient standardization has not yet been achieved in this difficult area that a broadly useful method can be set down. For this reason, it is best to refer to the original literature for details of this work. The particular combination is highly individualistic and not necessarily repeatable on a material from a different source. Useful guidelines may be taken from the successful isolation and purification of the vanadyl porphyrins from Athabasca oil sands by MILLSON, MONTGOMERY and BROWN [23].

A promising development is the system of SNYDER for prediction of the separation factors of a large number of compounds found in petroleum. This system relates the compound retention volume to adsorbent, eluent and sample parameters [24]. The parameters for a number of synthetic porphyrins and their vanadium chelates have been reported [25]. As these parameters are confirmed experimentally and the listing is extended to other metallochelates (especially nickel), these parameters should be increasingly useful in selecting the optimum eluant for a required separation or purification of the petroporphyrins.

3. Liquid-Liquid Extraction

Because of its ease and simplicity, solvent extraction has been often used as a technique for concentration of the metalloporphyrin pigments. The list of solvents employed is long and no attempt will be made to list them. As is expected, the most success is achieved with polar, coordinating solvents.

a) Aqueous-pyridine Extraction of Vanadyl Porphyrins [26]

50 ml of oil or residuum is treated with 50 ml pyridine and 10 ml of water in an extraction funnel. 6–20 ml of xylene may be added as an aid to separation of the layers. The partition coefficient for vanadyl porphyrin is *ca.* 0.55 to 0.65 so that essentially all porphyrins are removed after 10 extractions.

Other successful solvent pairs include aniline-kerosene, 3,3′-oxydipropionitrile-hydrocarbon [23], and dimethylformamide-cyclohexane [4].

Countercurrent extraction by the Craig method allows close control of a number of variables not possible in simple extraction methods. Details of the use of this method on petroporphyrin concentrates have not been reported; however, COSTANTINIDES and ARICH so fractionated a petroleum residuum.

b) Countercurrent Extraction of a Petroleum Residuum [27]

50 g of residuum was fractionated in an 8-stage Craig extraction unit with 300 ml decalin and 300 ml DMF in each stage. Before addition to the extractor, the solvents were co-saturated and steps taken to avoid moisture absorption by the DMF. Emulsions were broken by centrifugation. Partition coefficients determined on synthetic chelates are: vanadyl etioporphyrin $K = 10.9$ and nickel etioporphyrin $K = 1.91$.

The pigments in recent sediments, perhaps in keeping with the less fossilized nature of the matrix and because of their retention of solubilizing functional groups (i.e., keto, carboxyl), are easily recovered by solvent extraction. A typical procedure uses aqueous acetone.

c) Quantitative Extraction of Pigments from Sediments [28]

Samples of wet sediment (*ca.* 20 g) were extracted at least three times with aqueous acetone, 10 percent v/v (40 ml). The suspension was shaken for 5 min, centrifuged and the solvent decanted. The pigment concentration was determined from the optical density of the absorption max in the 670 mμ region using a coefficient of 4×10^4 l mole^{-1} cm^{-1} for pheophytin-like compounds.

It should be noted that pheophytin in 80 percent acetone has a molar extinction coefficient of 5.7×10^4 l mole^{-1} cm^{-1} at 667 mμ. The values for pigments in sediments are, at best, estimates and may in some cases over-estimate the amount of pigment present.

4. Gel Permeation Chromatography

A newer method, gel permeation chromatography, promises to be among the most useful of the chromatographic techniques. Under proper conditions, this method separates according to molecular size. It was developed by biochemists under the name of gel filtration. Cross-linked polystyrene gels were prepared and found to be suitable for use in organic solvents [29, 30, 54]. A derivative of sephadex, called Sephadex LH-20, is also suitable for use in organic solvents [31, 54]. Two typical isolation procedures for metalloporphyrins follow.

a) Isolation of Metalloporphyrins from Petroleum Residues [30]

A glass column (7 × 112 cm) was packed with a polystyrene cross-linked with divinyl-benzene gel (Dow Chemical Co.). The gel was loaded into the column as a thin slurry in benzene and allowed to settle while benzene was passed through the column. Samples were placed on the column in as small a volume as possible without disturbing the gel and eluted with benzene at 0.6 ml/min/cm^2 of cross-sectional area. About 5 g of oil could be handled without overloading. Fractions rich in metalloporphyrins follow the elution of molecules too large to enter the pores (asphaltenes). The elution of the metalloporphyrins may be monitored by spectrophotometric measurement of the visible bands. Combination of these cuts and rechromatography over alumina and silica gel produce a material of good purity.

b) Isolating a Metalloporphyrin Fraction from Asphaltenes [31]

A column (2 × 25 cm) was packed with Sephadex LH-20 (Pharmacia Fine Chemicals, Inc.) which was swollen overnight in THF. About 100 mg of asphaltene dissolved in THF (*ca.* 10 ml) is placed on the column and the flow rate of THF is adjusted to about 1.5 ml/min. The metalloporphyrin band develops behind the main asphaltene band.

B. Fractionation

Many of the methods of isolation of the petroporphyrins are applicable with some refinements to their fractionation. Handling problems are simplified since the quantity of material is much reduced.

1. Base Fractionation

The simplest fractionating agent is dilute alkali. TREIBS extracted ether solutions of porphyrins with 10 percent NaOH to remove the carboxylated porphyrins and leave in the ether layer the alkylporphyrins. The carboxylated porphyrins are then recovered by transfer into ether by acidification of the alkaline layer [1].

2. Chromatographic Methods

Carboxylated porphyrins may also be readily fractionated from the bulk of the porphyrins by chromatography over a weak adsorbent such as cellulose powder.

Column chromatography may also be used to fractionate the porphyrin aggregate obtained by demetallation into fractions having rhodo-, etio-, and desoxophyllo-type visible spectra [3].

a) Chromatographic Fractionation of Boscan Petroporphyrin

50 mg of porphyrin aggregate obtained by demetallation (see Section A-1) is taken up in a minimum of benzene and applied to a silica gel (Davison No. 923) (bed size 4 × 12 cm) which had been prepared with cyclohexane. Slow development of the column to the level of 1:1 cyclohexane/benzene, benzene and finally 50:1 benzene/ether, yields fractions b, d and f with visible spectra as shown in Fig. 1. The visible spectrum of the porphyrin aggregate is shown in a and unresolved mixtures in c and e.

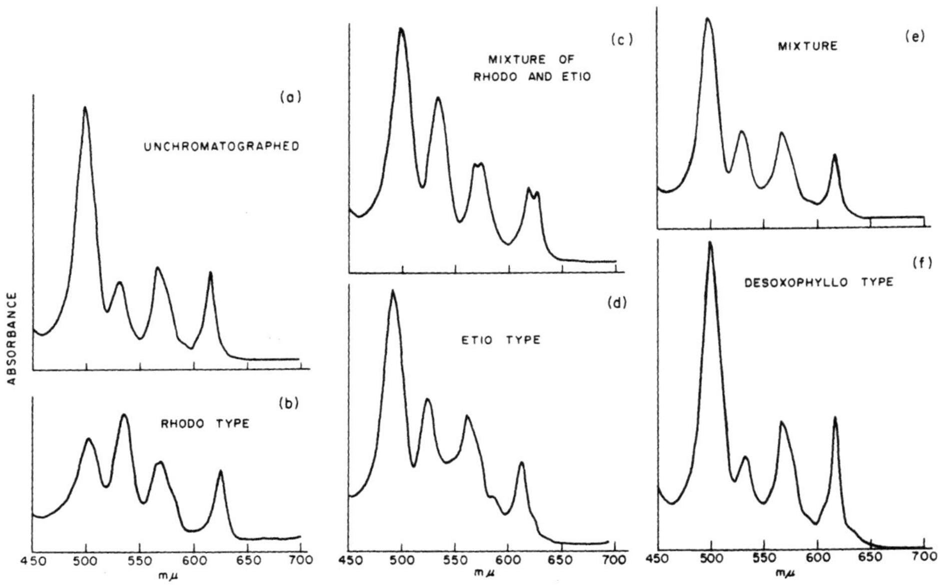

Fig. 1. Chromatographic separation of petroporphyrin (see text for details)

A slightly different chromatographic procedure which accomplishes the same fractionation has been given by HOWE [32].

Because vanadyl porphyrins are considerably more polar than their nickel counterparts, it is relatively easy to make a separation into fractions on this basis.

b) Column Chromatographic Fractionation of Metalloporphyrin Aggregates [28]

The metalloporphyrin aggregate (see Section II, A-2,3) is taken up in 1:1 benzene/n-hexane solution. Development of the silica gel column with the same mixed solvent elutes the nickel porphyrins and subsequent development with 1:1 benzene/chloroform elutes the vanadyl porphyrins.

3. The Hydrochloric Acid Number

WILLSTÄTTER and MIEG [33] devised a very useful method for separation and determination of chlorophyll derivatives, which depends on the different distribution of these pigments between ether and dilute hydrochloric acid. The name

"HCl Number" is given to the percentage content of acid that extracts *ca.* two-thirds of the dissolved substance from an equal volume of ether. However, since the ratio in which the pigment is distributed between the ether and the aqueous acid is extraordinarily sensitive to change in acid concentration, the "HCl Number" may be measured with sufficient accuracy in a test tube. A variation in acid concentration of a few percent (for example, use of 6 percent HCl in place of 3 percent) shifts the ratio from almost zero to nearly infinity. Thus, the "HCl Number" is most useful as an indicator in laboratory workup procedures but is less useful as a constant of characterization.

A method for the acid fractionation of a porphyrin aggregate from oil shale has been cited (Section II, A-1). It was noted that the 2.5 percent fraction contained the bulk of the porphyrin. This is certainly easily rationalized when compared with an "HCl Number" of 2.5 for desoxophylloerythroetioporphyrin and 3.0 for etioporphyrin [1].

C. Purification

Preparation of purified samples of metallochelates by any of the procedures cited is tedious. One detects the impurities either by their fluorescence or by noting a high background in the absorption spectrum in the near ultraviolet in the range of 350 mμ, i.e., just below the Soret peak. The most straightforward way of dealing with these impurities is an acid demetallation which quite effectively separates the porphyrins from the nonporphyrin materials. Procedures by which vanadyl porphyrins can be demetallated are considered because of their refractory nature. The nickel chelates are more labile and generally present no unusual difficulties in demetallation.

1. Demetallation of Vanadyl Porphyrins

It sometimes may be advantageous to obtain free base porphyrins from concentrates rather than from the raw matrix. But in this regard, it has been pointed out by Millson, Montgomery, and Brown [33] that while the use of the Groennings [7] method gives satisfactory results with crude oils and asphaltenes, attempts to demetallate purified fractions of the porphyrin pigments results on the other hand in almost complete destruction of the porphyrin. They found that the addition of a small amount of 4,4'-methylenebis (2,6 di-*t*-butylphenol) to the reaction mixture gave a good yield of porphyrin. It has generally been assumed that the destruction of the porphyrin was caused by free bromine in the HBr reagent*. Complete exclusion of free bromine from HBr is difficult and the use of a bromine scavenger (hindered phenol) would be expected to improve the yield, as was observed. In the cases where crude oil and asphaltenes are present, these materials certainly contain potential bromine scavengers, and, hence, an explanation for the success of the Groennings method is apparent.

Extraction of asphaltene with methanesulfonic acid was reported to give satisfactory yields [16]. Interestingly, attempts to demetallate either purified vanadyl petroporphyrin fractions or vanadyl porphyrins with this reagent lead

* The structure of the bromination product has been worked out for some cases and is of the chlorin class [8a].

to destruction of the porphyrin. However, the addition of hydrazine sulfate to the reaction mixture produces essentially quantitative yields of free base porphyrin.

It appears then, that the presence of free bromine is not the sole cause for loss of porphyrin in the demetallation. The destruction of the porphyrin is observed in both H_2SO_4 and MSA, while the addition of a reducing agent (hydrazine) gives high recoveries of the porphyrin. Hindered phenols are good reducing agents and it is reasonable that the efficacy of this reagent is as much due to its reducing powers as its ability to scavenge free bromine. Likewise, the improvement in yield shown by using formic acid [10] in place of acetic acid may also depend on the reducing potential of the former.

a) Demetallation of Vanadyl Porphyrins [31]

Vanadyl mesoporphyrin IX (5.0–5.5 mg) and high purity MSA (9–10 ml) together with a few crystals of hydrazine sulfate is heated and stirred at 50–60° C in a 100 ml 3-neck flask under a N_2 atmosphere. Demetallation of the porphyrin is essentially complete after 1–1.5 hrs. as indicated by loss of the vanadyl mesoporphyrin IX spectrum and the appearance of spectral peaks corresponding to the porphyrin di-cation.

b) Alkaline Demetallation

An alkaline reductive method for the removal of various chelated metals including vanadyl from porphyrins has been reported [34]. The method involves the use of lithium in ethylenediamine to convert the metalloporphyrin to the di-lithio complex which is then decomposed by water to give the free base porphyrin. The reagent has, until now, found only limited use and no procedure for the recovery of porphyrin from petroleums or bitumens has been reported; however, it has been shown to reduce the vanadyl porphyrin content in Boscan asphaltene fractions.

2. Miscellaneous Purification Methods

A number of procedures have been used for the final rejection of non-porphyrin materials following the demetallation in order to prepare porphyrin samples suitable for characterization and elemental analysis.

a) Crystallization of Porphyrins

Crystallization by slow evaporation of a benzene solution of the porphyrin was used by TREIBS. Use of a mixed solvent system such as benzene-methanol may be necessary if the amount of sample is very small.

It has been mentioned by VANNOTTI that the dihydrochloride of protoporphyrin is soluble in chloroform and a few other selected chlorinated hydrocarbons [17]. This rather unusual property serves as the basis of a useful purification method already described, whereby the porphyrin di-cation from MSA extraction is transferred from aqueous acid to methylene chloride solution (see Section II, A-1). Crystallization of the porphyrin di-cation from the methylene chloride is an effective final purification step.

Thin layer chromatography (TLC) has been found effective for the purification of porphyrins from acid extracts. THOMAS and BLUMER used Silica Gel G with benzene as the eluant to obtain eight fractions of a complex porphyrin mixture from oil shale [35]. TLC can sometimes be used as a combination isolation-purification procedure if only small quantities of material are needed.

b) Thin Layer Chromatography of Tetraphenylporphine [36]

Kiesel Gel activated for 1 hr. at 115° and developed with 1 percent acetone in benzene, quantitatively separates tetraphenylporphine from complex reaction mixtures. The porphyrin band on the TLC plate is located under a u.v. lamp and removed with a microspatula. Benzene extraction of the silica yields the purified porphyrin.

The related method of paper chromatography was applied extensively by DUNNING to the petroporphyrins [37, 38]. While this type of chromatography is highly selective, it has the disadvantage that only very small quantities of resolved material are available for further study.

Purification methods such as sublimation and molecular distillation which depend on the low but definite vapour pressure of the porphyrins are useful in some cases [39a]. However, even under relatively hard vacuum (*ca.* 10^{-6} mm Hg) the temperatures required for volatilization of porphyrins with carboxylic acid groups is sufficiently high that pyrolytic reactions such as dehydration and decarboxylation occur. Hence, these methods are probably applicable only to the alkylporphyrins and their metal chelates.

In 1962, KLESPER, CORWIN and TURNER reported the separation of a metalloporphyrin mixture by a method analogous to gas chromatography except that the carrier gas, dichlorodifluoromethane, was held above its critical point [40]. This chromatographic method, since designated hyperpressure gas chromatography, promises to be among the most powerful yet devised. All of the usual advantages of gas chromatography such as speed and versatility appear to accrue, together with a separation power not previously available to the porphyrin chemist. In this way, separation of porphyrin homologues seems to be a real possibility [41].

III. Identification and Characterization

Specialized methods are required for the identification of the petroporphyrins and related materials. Their properties deny the application of many of the more conventional methods. For example, melting points have little meaning in most cases, because the materials do not liquify below 300° and almost always decompose rather than melt reversibly. Likewise, their inertness generally makes the preparation of derivatives unrewarding and finally the interpretation of the data from elemental analysis is troublesome because of the similarities of the analysis calculated for differing structures, even if the many component mixtures can be satisfactorily resolved.

As is often the case, those very properties which preclude characterization by one means open the door to others. The chemical inertness of the porphyrin system is one reflection of its aromaticity, as also is its rich visible spectrum. The uses of the visible spectrum for identification of the pigments from the time of BERZELIUS and HOPPE-SEYLER down to the present are voluminous. Over the years, the porphyrin chemists were quick to take advantage of improvements in optical spectrometry which would aid in extracting structural information from the electronic spectrum. In particular, they seized upon the ratio-recording spectrophotometer and made it a major tool in their research.

In contrast to this, appreciation of mass spectrometry in the field of the tetrapyrrole pigments has been slow in coming, but as the following sections will show, this method may be the most powerful yet applied. The porphyrins produce nearly ideal spectra when inserted directly into the ionizing beam, a simple and straightforward procedure on modern spectrometers. The many component mixtures which yield in most spectral presentations an uninterpretable picture, show at low voltage a mass distribution which is easily understood. The interpretation is straightforward and unambiguous. Fortuitously, the information obtained in this way nicely complements that from the electronic spectrum.

A. Electronic Spectra

The richness of the electronic spectra of the porphyrins and chlorins has long intrigued chemists and certainly in no small way contributed to their discovery in bitumens, etc. The distinctive sharp bands in the visible region provide a very sensitive indicator of their presence. Since the electronic spectra have been so useful to the chemist, a brief mention of nomenclature and empirical rules which aid in their interpretation seems in order.

1. Porphyrins

In an oversimplified, but useful way, the electronic spectra of the porphyrins can be separated into two parts, visible and near u.v. The metallochelates are characterized by a two-banded visible spectrum, while the free base porphyrins have four bands. The near u.v. band is extremely intense and named after its discoverer, SORET [42]. The visible part of the spectrum is the most useful, first because it is more sensitive to structural changes and second, because it is more easily observed. Hence, most of the subsequent discussion will be directed to the 450 to 700 mμ portion of the spectrum.

The theoretical basis for the electronic spectrum of the porphyrins has been much discussed and is not of direct interest here. However, certain generalizations concerning the relative intensities of the visible bands are useful in assigning structural types. It was noted very early that the relative intensities of the four main visible bands, numbered I-IV starting from the long wavelength, varied markedly with substituents [42a]. Some of these types are so typical that they have been given specific names. Three common types designated as phyllo, etio and rhodo are shown in Fig. 2a, b and c, together with an additional type, of particular interest here, as shown in Fig. 2d. It will be called the DPEP type after the parent compound in the series, desoxophylloerythroetioporphyrin. The structures of the parent compounds for each of these four series are given in Fig. 3.

The etio-type spectrum Fig. 2b where IV > III > II > I is characteristic of all porphyrins in which 6 or more β positions carry alkyl side chains, regardless of the combination or orientation around the nucleus. However, there is a diminution of the intensity of band III with an increasing number of open β positions. Deuteroporphyrin IX has a considerably less intense III band than its 2,4 diethyl analog, mesoporphyrin IX.

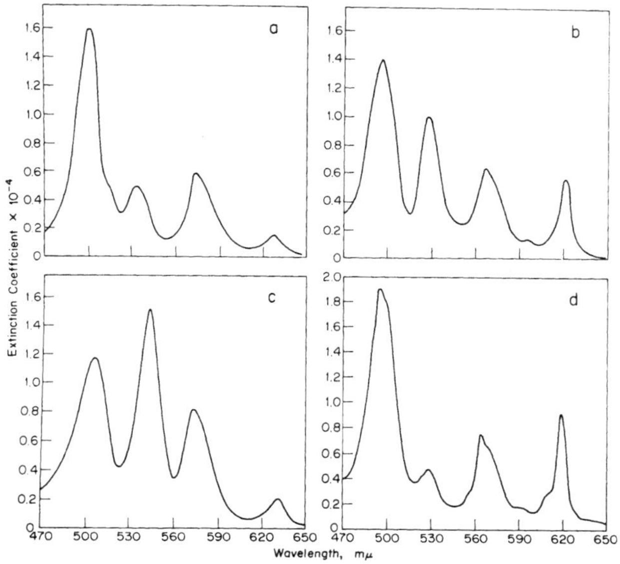

Fig. 2a – d. Visible spectra of porphyrins. (a) γ-phylloporphyrin XV; (b) etioporphyrin III; (c) rhodoporphyrin XV; (d) desoxophylloerthroetioporphyrin (for structural formulae, see Fig. 3)

Compound	Substituents[a]												Spectral type
	R_1	R_2	R_3	R_4	R_5	R_6	R_7	R_8	α	β	γ	δ	
Etioporphyrin III	M	E	M	E	M	E	E	M	H	H	H	H	Etio
γ-phylloporphyrin XV	M	E	M	E	M	H	P	M	H	H	M	H	Phyllo
Rhodoporphyrin XV	M	E	M	E	M	C	P	M	H	H	H	H	Rhodo
Desoxophyllo-erythroetioporphyrin	M	E	M	E	M	CH_2CH_2	E	M	H	H		H	DPEP

[a] $M = CH_3$; $E = CH_2CH_3$; $P = CH_2CH_2COOH$; $C = COOH$.

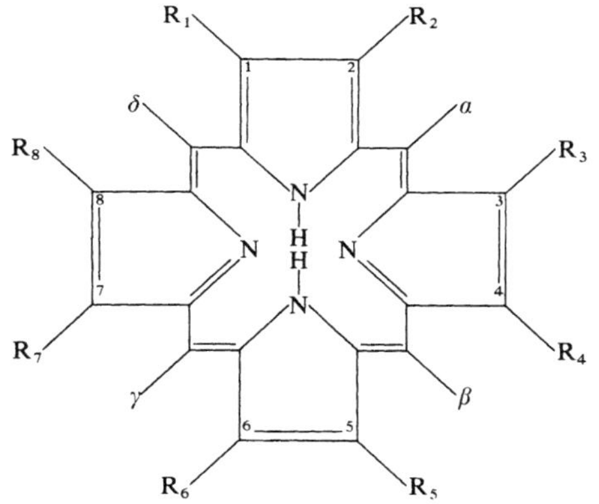

Fig. 3. Structures of parent porphyrins of the four spectral series

Isolated carboxyl groups such as those in acetic or propionic acid side chains behave nearly the same as alkyl groups but, depending on the number of intervening methylenes in each chain and the number of carboxyl groups, may have a barely perceptible effect on the spectrum. Comparison of coproporphyrin (tetramethyl, tetrapropionic acid porphin) with uroporphyrin (tetraacetic acid, tetrapropionic acid porphin) shows that the absorption bands are moved about 5 mμ to the red in uroporphyrin or *ca.* 1 mμ per isolated carboxyl group.

The above example of substituent effects on band positions shows that they are equally as useful as the intensities for identifications. Likewise, by extending the length of the conjugation, a vinyl group shifts the bands *ca.* 6 mμ to the red as compared to its ethyl analog without appreciably altering the intensities.

The phyllo-type spectrum Fig. 2a (IV, II, III, I) appears to arise in either of two ways. It is commonly associated with substitution at a bridge carbon such as γ-phylloporphyrin from which the type name arose. However, as mentioned earlier, there is a diminution of the intensity of band III with an increasing number of open β positions. Therefore, the phyllo-type spectrum may also arise when there are four or more open positions.

Strongly electron-withdrawing groups such as carbonyl or carboxyl conjugated with the porphyrin, cause band III to become most intense and produce the rhodo-type spectrum Fig. 2c. It is apparent that such conjugation must also be associated with a red shift of the absorption band [43].

A fourth type, the DPEP type named after the parent compound, desoxophylloerythroetioporphyrin is shown in Fig. 2d. This spectral type is related to the phyllo-type except that peak I is considerably intensified. Spectra of this type are apparently restricted to those porphyrins having an isocyclic ring, and as such are of particular interest here. There has been some confusion of this type with the phyllo-type spectrum but comparison of Fig. 2a and d shows the differences clearly*. The intensity order for the DPEP type is IV, I, II, III while for the phyllo-type it is IV, II, III, I.

A qualitatively different spectral type in which the long wavelength peak is shifted to the red (650–680 mμ) and dominates the visible spectrum is generally identified with the chlorins (dihydroporphyrins). In these green pigments, the intensities of the absorption bands in the central part of the visible spectrum are reduced and appear as small bumps rather than strong discrete bands.

It is customary to refer to any green tetrapyrrole pigment as a chlorin; however, such a spectrum also arises by oxidation as well as reduction of the porphyrin nucleus. For example, alkyl porphyrins may be oxidized with OsO_4 to compounds having a chlorin-type spectrum and characterized as 1,2-dihydroxyporphyrins [44]. It will be noted that the dihydroxyporphyrins possess the same conjugated system as the chlorins and it is therefore quite reasonable that the spectra are similar. A further and germane example of the formation of green pigments under oxidizing conditions was mentioned earlier in regard to demetallation of metallo-porphyrins. The presence of free bromine in an HBr-acetic acid reagent yields in certain cases

* For example the spectrum given by HODGSON *et al.* is the phyllo-type rather than the DPEP type [43a]. The confusion has probably arisen because it was incorrectly assumed that petroporphyrins from Athabasca Tar sands were of the DPEP type. Actually they are mixtures of etio- and DPEP-types leading to a phyllo-type spectrum. See Fig. 1 and [4] and [16].

a green pigment [8a]. Thus, the assumption that a chlorin-type spectrum is indicative of reducing conditions can obviously lead to erroneous conclusions.

In contrast to the four-banded visible spectrum of the porphyrins, the metallochelates of interest here have a two-banded spectrum. The nickel chelate is an example of such a simple square-planar chelate. The two bands, usually called α and β, are found in nickel etioporphyrin in neutral solvents at 550 and 514 mμ with an α/β intensity ratio of *ca.* 3. Coordination with additional ligands shifts the bands to longer wavelengths and reduces the α/β ratio. The octahedral complex of nickel mesoporphyrin has absorption maxima at 574 and 543 mμ [45].

Vanadyl porphyrins also have the normal two-banded spectrum, with the maxima for vanadyl etioporphyrin found at 570 and 531 mμ with an α/β ratio of *ca.* 2. This shift to longer wavelengths and reduced α/β ratio as compared to the square planar nickel chelate can be simply considered as additional coordination with the oxygen ($V=O$) to yield a pentacoordinate complex.

The position and intensity of the metallo-chelate absorption bands are, in general, much less sensitive to alterations by substituents than are the spectra of the free base porphyrins. Hence, they are less useful and less used than the porphyrin spectra.

A different type of two-banded spectrum than that of the simple metallochelates is shown by the porphyrin di-cations. For the simple alkylporphyrins, i.e., etio, the bands are observed at *ca.* 550 and 590 mμ but here the β band is predominant with the α/β ratio *ca.* 1/3.

2. Metallochlorin Spectra

The changes in the spectrum produced in going from a chlorin to a metallochlorin are much less than the corresponding changes in the porphyrins. In the most thoroughly studied case, chlorophyll a *vs.* pheophytin a, a shift of only 5 mμ (from 662 to 667 mμ) in the main red band is observed. The intensities of the less intense bands in the central part of the visible spectrum are "rearranged" and the Soret peak shape changed. However, all of these are dependent to some extent on the particular solvent used.

B. Mass Spectra

Use of mass spectrometry in the field of the tetrapyrrole pigments is quite recent and, in some of the literature reports, only an outline of the experimental procedure is given. Because the experimental parameters can influence the quantitative and even the qualitative features of the spectrum, it is felt that some detail in this area is needed. However, only those details which are required for the application of mass spectrometry of the tetrapyrrole pigments will be included.

The low volatility of the porphyrins in general precludes introduction through a normal heated inlet system. Attempts to do this generally are frustrated by insufficient sample to obtain a readable spectrum or upon further increase in temperature, only pyrolysis peaks are obtained. Thus, a spectrometer equipped with a solid sample injection probe so that the sample is introduced directly to the ionizing beam is nearly essential.

1. Experimental Parameters

In dealing with many-component porphyrin mixtures such as those found in petroleum, bitumens and sediments, two parameters are of especial importance. These are probe temperature and beam energy.

BAKER *et al.* determined the suitable temperature range for obtaining mass spectra of porphyrin mixtures in the following way [16]: a series of five samples of petroporphyrin was introduced to the spectrometer at temperatures ranging from 195 to 280° C. In Table 2 is shown the mean mass of the petroporphyrins as a

Table 2. *Mean mass of petroporphyrin as a function of probe temperature*

Probe temperature	Mean mass of etio series of Boscan petroporphyrin[a]
195	459
213	463
233	468
259	468
265	468

[a] Calculated according to $\frac{\Sigma IM}{\Sigma I}$, where I is the intensity of mass peak M.

function of the indicated temperature of the probe. The center of mass of the porphyrin envelope moves up as a function of temperature until about 230° C and then remains relatively constant, thus setting a lower level on the probe temperature, while at temperatures somewhat above 300° C pure alkyl porphyrin begins to produce thermal fragment peaks. Thus, a probe temperature of *ca.* 260° C was selected.

Since in analysis of mixtures one wishes to observe as far as possible only parent molecular ions without the complications of fragment ions, it is necessary to select an appropriate beam energy. To be certain that only parent ions were observed, the energy of the beam would have to be somewhat above the ionization potential of the molecule and below the appearance potential of the first fragment ion. Taking these factors into consideration, a compromise value of 12 eV is recommended [16].

2. Interpretation of the Mass Spectra of the Porphyrins

The molecular weight of porphin is 310 ($C_{20}H_{14}N_4$) and alkyl substituted porphyrin must fall in the series 310 + 14 n, where n is an integer. Likewise, porphyrins with an isocyclic ring (desoxophylloerythro series) must have molecular weights of 308 + 14 m, where m is an integer 2 or greater. Table 3 shows the masses of these two series of porphyrins. Fig. 4 shows a mass spectrum of a typical petroporphyrin. The peaks corresponding to the two major homologous series are enclosed in dotted envelopes. In addition to the two major series of alkylporphyrins, a minor series (amounting to 1–5 percent of the porphyrin present) exhibiting a rhodo-type visible spectrum has been found in a variety of petroleums and bitumens. The small peaks between the major porphyrin peaks in Fig. 4 at 458 +

Table 3. *Molecular weights of alkyl-substituted porphyrins*

m or n [a]	Etio and phyllo series	Desoxophyllo-erythro series
8	422	420
9	436	434
10	450	448
11	464	462
12	478	476
13	492	490
14	506	504
15	520	518
16	534	532
17	548	546
18	562	560

[a] Number of methylene groups attached to the porphyrin nucleus.

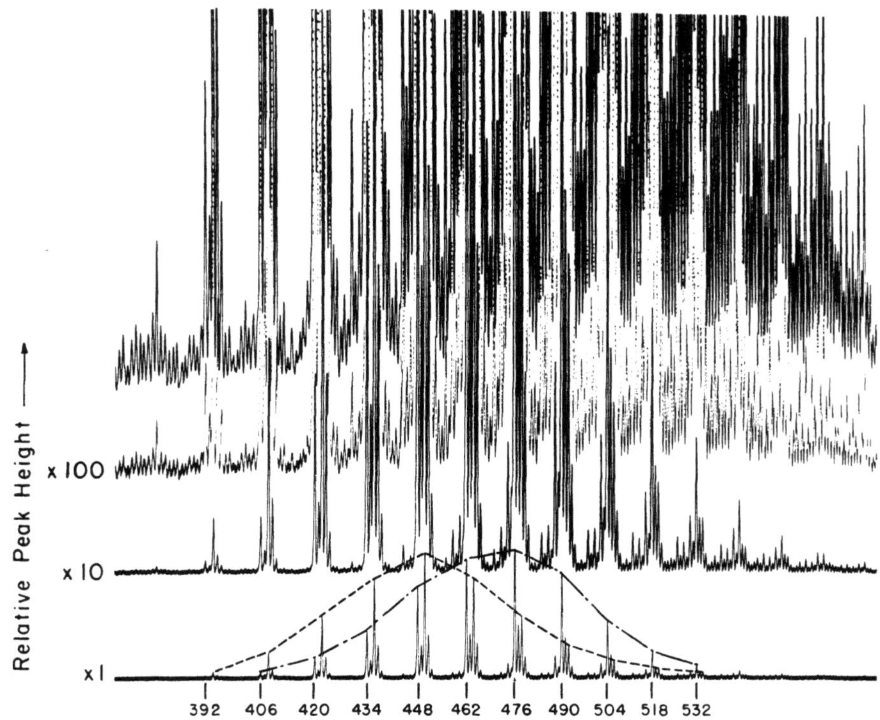

Fig. 4. Partial low voltage mass spectrum of a typical petroporphyrin of marine origin

14 n and 456 + 14 n show the presence of these series [16]. Further investigations of the fossil porphyrins may well reveal that other minor homologous series of porphyrins are present.

The components in an unknown series of porphyrins may be assigned to one or the other of the major series simply on the basis of nominal masses. High resolution accurate mass measurements of the major peaks may then be used to determine: (1) that all components of a peak are isomers, and (2) that only C, H and N are present. From these observations, in conjunction with electronic spectral data, which show the presence of the porphyrin nucleus, an outline of the structures of the porphyrins in the major series can be deduced. For example, the porphyrin component of the DPEP series and a mass of 420 must have at least one unsubstituted β-position since only 8 methylene units are present and one is attached to the γ bridge as part of the isocyclic ring.

As can be seen from the above, the low voltage mass spectrum is the heart of the characterization technique of a petroporphyrin mixture. No other method deals so effectively with the complex mixtures of compounds present. Used in conjunction with low voltage mass spectra, electronic spectral data and chemical evidence do greatly extend the amount of structural information of any specific group of porphyrins, but in general, they must be viewed as secondary rather than primary methods. A second mass spectroscopic technique, high resolution accurate mass measurement, has already been mentioned. The knowledge of the exact mass and hence the molecular composition of a particular mass peak, of course, greatly enhances the evidence from the low voltage spectrum.

The high voltage (70 eV) mass spectra of these materials should also be mentioned. Since the fragmentation patterns of a number of model porphyrins are known, it should be possible to deduce from the fragmentation patterns of the petroporphyrin mixtures some structural details of the substituents. Up to the present, this approach has been thwarted by the low intensity of the fragment ions and the complexity of the mixtures.

3. Other Spectral Methods

Nuclear magnetic resonance (NMR) and infra-red absorption (IR) spectra have until now been much less useful in characterizing fossil tetrapyrrole pigments than electronic and mass spectra. The main constraint on both hinges on the complexity of the spectra and sensitivity to small amounts of impurities. Especially with reference to NMR spectra, this statement would seem unwarranted because most NMR spectra may be taken with little regard to sample purity. Extraneous materials present in an amount of 5 or even 10 percent are often buried in the normal noise and can be ignored. In contrast, with materials of molecular weight of 500 or more even as much as 1–2 percent of a low molecular weight material can produce a proton resonance signal equal in size to that of the porphyrin. Thus, sample purity becomes very nearly the limiting factor in the application of NMR to the petroporphyrins. To further complicate matters, the complex mixtures (10–50 compounds in several homologous series) normally found in petroleums and bitumens present an unresolvable and uninterpretable maze at the present. However, as the separation and purification techniques are refined and high resolution NMR instrumentation becomes more widely available, this spectroscopic method will undoubtedly become a major tool. For a lucid description of the application of the technique to the parallel and equally difficult field of chlorophyll chemistry, as well as a review of the work on pure porphyrins, see KATZ, DOUGHERTY, and BOUCHER [46].

Many of the same difficulties are met in the application of infra-red absorption spectroscopy to the petroporphyrin mixtures. The IR spectrum of porphyrins is sufficiently complex that even with pure single compounds recognition of extraneous bands caused by moisture, contaminants and solvent residues is difficult. This, together with observed shifts in shapes and positions of bands with different sample preparative techniques [47], has caused reluctance to place much reliance on the method as a means of recognition of functional groups in a complex mixture of porphyrins. The main uses described so far for IR spectra in this field are to indicate the absence of a carbonyl group by noting the absence of the 5.7 μ carbonyl stretch in a vanadyl petroporphyrin sample [48] and similarly in a rhodo-type porphyrin extract [49].

IV. Geochemistry

A. The Scheme of Treibs

Treibs identified the basic porphyrin fraction from bitumens as spectroscopically identical with desoxophylloerythrin and concluded from elemental analysis that the material was DPEP [1, 2]. Making the explicit assumption that the precursor was chlorophyll *a*, Treibs explained that vanadyl DPEP (see Fig. 5) resulted from a sequence of degradative steps as follows: (1) magnesium loss,

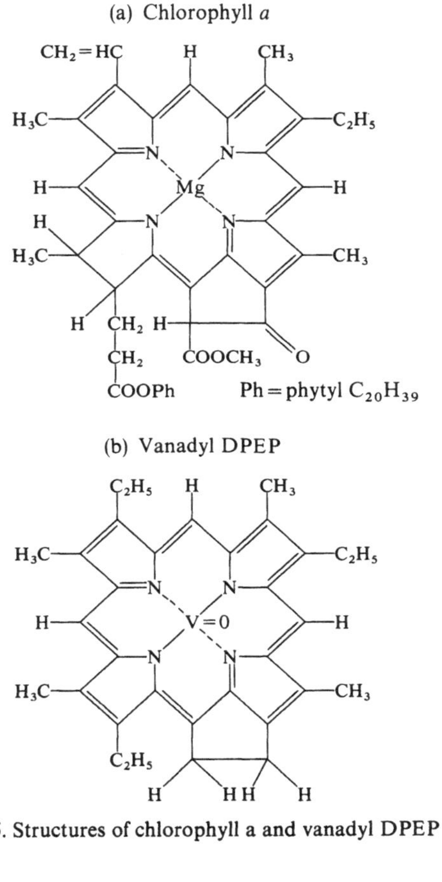

Fig. 5. Structures of chlorophyll a and vanadyl DPEP

(2) ester hydrolysis, presumably under acidic conditions, since alkaline hydrolysis would also open the isocyclic ring, (3) hydrogenation of the vinyl group, (4) dehydrogenation of the green (chlorin) system to the red (porphyrin) system, (5) reduction of the 9-carbonyl group to CH_2, (6) decarboxylation, and (7) chelation.

It is apparent and was so noted by TREIBS that if the above reactions were operative on chlorophyll b or bacteriochlorophyll, the same end product would result. By noting laboratory conditions under which these reactions would proceed, inferences were drawn on conditions of petroleum formation. For instance, it would appear from reactions 2 and 5 that acidic reducing environments applied at least initially. The survival of the porphyrins establishes an upper limit of the thermal history of the bitumens. Conversely, the observed decarboxylation of the propionic acid group places some lower limit on the integrated thermal environment. The observation of some acidic (undecarboxylated) materials perhaps further narrows the range of possible conditions.

As will be noted subsequently, some of the conclusions reached by TREIBS concerning porphyrins of the etio series may be in doubt because the second metallochelate present is nickel, not iron as he supposed. But, in retrospect, one must contemplate with awe the insight provided by TREIBS. Only recently has it been possible to extend his conclusions significantly and to prove that the situation is more complex than he supposed.

B. The Starting Materials

Following TREIBS' lead, we shall assume that fundamental biosynthesis has not changed since the times of petrogenesis and we shall now discuss the structures of the possible precursors of the fossil porphyrins.

Because of discoveries of many new plant pigments, we can call upon a much wider range of possible starting materials than could TREIBS. At the time of TREIBS' proposal, the structure of hemin had just been proved and that of chlorophyll proposed but not yet confirmed in all details. It is natural and correct that he should restrict himself to those known structures as the cornerstone of his theory. The fact that both the plant and animal kingdom would provide a wide number of biological variations on the basic skeleton as starting materials for the diagenetic transformations has become known only in the course of time. That it has taken so long to require changes in the basic scheme stands as a tribute to the original postulate.

For the purposes of the discussion here, it will be useful to designate the tetrapyrrole pigments as belonging to the heme or to the chlorophyll family. This distinction is made not on a "green" or "red" basis, but on the presence or absence of the isocyclic ring, because this structural difference is preserved while the 7,8 hydrogens are lost.

1. Types of Chlorophylls

In recent years, there has been a vast expansion of our knowledge of chlorophyll. From the geochemists' point of view, the most interesting aspect of these advances has been the demonstration that the carbon skeletons of all chlorophylls are not identical. In fact, as shown in Fig. 6, the variations in chlorophyll structures discovered to date are sizeable, and even further examples of non-homogeneity may yet be found.

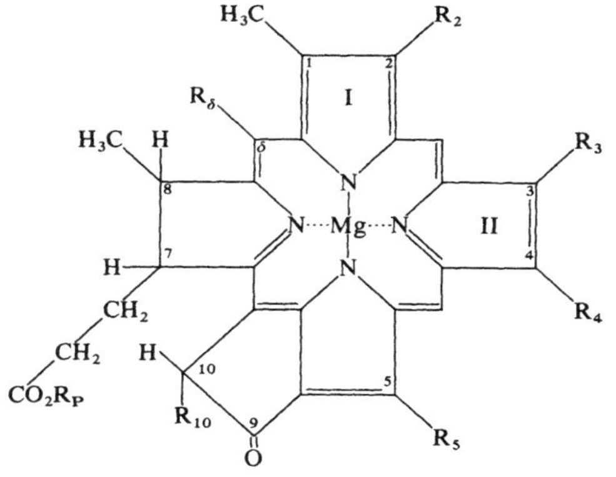

Designation	Substituents[a, b]							C_N[c]
	R_2	R_3	R_4	R_5	R_{10}	R_δ	R_p	
Chlorophyll								
a	V	M	E	M	C	H	Ph	32
b	V	F	E	M	C	H	Ph	32
d	F	M	E	M	C	H	Ph	31
Chlorobium chlorophyll 650								
Fraction 1	D	M	B	E	H	H	Fr	35
Fraction 2	D	M	P	E	H	H	Fr	34
Fraction 3	D	M	B	M	H	H	Fr	34
Fraction 4	D	M	E	E	H	H	Fr	33
Fraction 5	D	M	P	M	H	H	Fr	33
Fraction 6	D	M	E	M	H	H	Fr	32
Chlorobium chlorophyll 660								
Fraction 1	D	M	B	E	H	E	Fr	37
Fraction 2	D	M	B	E	H	M	Fr	36
Fraction 3	D	M	P	E	H	E	Fr	36
Fraction 4	D	M	P	E	H	M	Fr	35
Fraction 5	D	M	E	E	H	M	Fr	34
Fraction 6	D	M	E	M	H	M	Fr	33
Bacteriochlorophyll								
a	A	M, H	E, H	M	C	H	Ph	32

[a] The subscripts to the R refer to the substituent positions on the standard phorbin nucleus.
[b] $M=CH_3$; $E=CH_2CH_3$; $V=CH=CH_2$; $F=CHO$; $C=COOCH_3$; $D=CHOHCH_3$; $B=CH_2CH(CH_3)_2$; $A=COCH_3$; $P=CH_2CH_2CH_3$; Ph = Phytyl; Fr = Farnesyl.
[c] C_N is the number of carbons in the molecular skeleton excluding saponifiable alcohols and carboxyl carbons. It therefore corresponds to the carbon numbers of the porphyrins derived by a standard Treibs' scheme.

Fig. 6. Structures of chlorophylls

However, it should be recognized that all of the chlorophylls other than the *a* type are of limited distribution and thus must be considered as unlikely sources of widely distributed petroporphyrins. In particular, the *Chlorobia* are chemo-autotrophic, and with this requirement they must have a very restricted distribution.

2. The Heme Family

In Fig. 7 is shown the structure of some representative porphyrins of the heme family. Reduction and decarboxylation yields etioporphyrins. It is seen that it is possible from a strictly structural chemical point of view for a mixture of fossil porphyrins to arise.

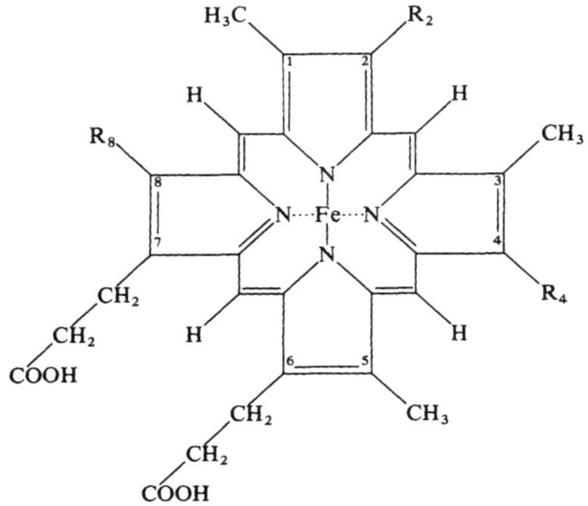

Name	Substituents[b]			
	R_2	R_4	R_8	C_N[c]
Protoheme IX	V	V	M	32
Chlorocruoroheme	F	V	M	31
Heme c	V_c	V_c	M	32
Heme a	V_f	V	F	47
Deuteroheme	H	H	M	30

[a] See General Reference [4] and references therein for distribution and structure proofs of these hemes.

[b] $V = CH{=}CH_2$; $M = CH_3$; $F = CHO$; $V_c = CH(CH_3)SCH_2CH(NH_2)COOH$;
$V_f = CHOHCH_2(CH_2CH{=}\underset{\displaystyle CH_3}{\underset{|}{C}}{-}CH_2)_3H$.

[c] C_N is the number of carbons in the molecular skeleton excluding saponifiable alcohols and carboxyl carbons. It therefore corresponds to the carbon numbers of the porphyrins derived by a standard TREIBS' scheme.

Fig. 7. Structures of some of the known hemes[a]

However, as CORWIN has pointed out, the chances of finding porphyrins of animal origin is low since the relative amount of material from plant sources is so overwhelming, perhaps 100,000 to 1 [4]. This makes it likely that minor pigments of the heme type in the plant sources contribute more to the overall store of the petroporphyrins than the major pigments of animal origin.

C. Fossil Porphyrins

1. Types of Pigments in Recent Sediments

For the most part, the detection and study of the types of pigments in recent sediments has depended on the visible spectral evidence. Certain of the chemical alterations are known to proceed very rapidly on the geologic time scale, in fact, on any time scale. The loss of magnesium probably occurs immediately at the time of burial or it has been suggested that in marine sediments it occurs even in the water column [50]. Likewise, the hydrolysis of the phytol is so facile that its attachment to the tetrapyrrole nucleus for any length of time is unlikely. The first product that would be expected to survive long enough to be isolated from sediments then would be pheophorbide. Since the determinations have been made by optical spectra and since pheophorbide and pheophytin have identical visible spectra, the evidence on this point is not compelling but seems a reasonable possibility. Substances having a visible spectrum corresponding to the nickel and vanadyl chlorins have been reported in recent sediments [23, 51, 52]. Thus, it appears that re-chelation occurs at a relatively early stage but at just what stage in the degradation cannot be stated since the details of the structures of the materials are not known.

2. Petroporphyrins in Ancient Sources

Treibs identified the basic porphyrin fraction from bitumens as spectroscopically identical with desoxophylloerythrin and concluded by elemental analysis that it was DPEP. There have been over the years, a number of indications that the petroporphyrins were not a single species [3]. These indications came from many sources. For example, chromatographic results seemed to be best explained on the basis of a group of isomers rather than a single compound. X-ray powder patterns and volatility studies were similarly indicative of inhomogeniety. It remained, however, for the application of mass spectrometry to confirm and measure the degree of inhomogeniety. The porphyrins from an oil shale of Swiss origin were shown to contain pigments ranging from C_{27} to C_{34} and it was also reported [35, 53] that the porphyrin in Colorado (Green River) shale had a similar distribution.

Fig. 4 shows a mass spectrum of the petroporphyrins of typical crude petroleum of marine origin [16]. Those peaks which fall at 420, 434, 448, etc., are porphyrins of the desoxophylloerythro series (intact isocyclic ring). Approximately a dozen different porphyrins of this series are present, encompassing those with as few as eight methylene groups to as great as 18 or even 19. Generally, DPEP itself (12 CH_2 groups) is the largest component of the mixture. The quantities of the other porphyrins of this series fall off in a symmetrical way with increasing or decreasing molecular weights. Those porphyrins with molecular weights of 422, 436, 450, etc., must be of the etio or phyllo series and the distribution is similar to the DPEP series.

Table 4 gives the low voltage mass spectra of a number of petroporphyrins. Qualitatively, it can be seen that there are fairly wide differences in molecular weight ranges and that in some, one series predominates while in others nearly

Table 4. *Low voltage mass spectra of petroporphyrins*[a]

Name/Country	Peak mass																									
	392	394	406	408	420	422	434	436	448	450	462	464	476	478	490	492	504	506	518	520	532	534	546	548	560	562
Agha Jari, Iran	3		12		27		62		*91*		88		70		25		10		4							
		12		44		80		*100*		78		50		24		11		11		4						
Athabasca Tar Sand, Canada			7		16		36		72		*100*		96		55		29		19							
				17		36		62		*67*		55		38		24		19		12						
Baxterville, U.S.A.	2		5		11		18		*22*		19		14		10		7		5		4		3			
		16		49		86		*100*		79		50		26		17		11		8		5		3		
Belridge, U.S.A.			5		16		40		64		*100*		64		15		8									
				11		26		32		34		*40*		12		5		3								
Boscan, Venezuela					6		16		49		74		*100*		82		54		30		16		10		6	
						20		50		76		*79*		60		37		22		12		9		6		4
Burgan, Kuwait			6		18		39		60		*70*		60		42		30		21		15		12		9	
		11		18		48		83		*100*		92		70		45		30		21		18		12		9
Colorado Oil Shale, U.S.A.			1		2		7		23		*100*		73		9		5									
				2		4		5		8		*12*		9		2		1								
Gilsonite, U.S.A.							2		6		26		*100*		73		12									
								10		31		*57*		45		16		4								
Mara, Venezuela			7		18		39		71		93		*100*		82		46		25		14		5			
		4		21		50		78		*100*		80		50		28		16		11		7		2		
Melones, Venezuela			3		10		31		63		88		*100*		54		29		12							
				11		38		79		*100*		92		60		33		15		7						
Rozel Point, U.S.A.							4		10		28		*100*		33		4									
								3		3		4		*9*		6		5								
Santiago, U.S.A.			3		9		30		59		*100*		73		6											
				5		7		*25*		*25*		16		7												
Wilmington, U.S.A.	4		16		36		56		*92*		72		68		28		20									
		20		52		76		*100*		96		64		36		20		12								

[a] Spectra obtained using solid sample inlet on AEI MS-9 using an ionizing voltage of 12 eV; essentially only parent peaks are observed. Peak intensities are normalized to 100 for the most intense peak. Intensities given in the first line for each sample are for the DPEP series; those in the second line, for the Etio series.

equal amounts are present. These observations are quite consistent with those of Treibs who noted on the basis of qualitative visible spectra that the relative amounts of DPEP and Etio varied from petroleum to petroleum [1].

An analysis of the visible spectra of the porphyrins of the 478 series shows that these porphyrins have a relatively low intensity band III. Such a low intensity of band III is compatible with either incomplete β-substitution (Section III, A-1) or bridge substitution. Probably both types of porphyrins are present, with incomplete β-substitution occurring in porphyrins of lower molecular weight (5–7 methylene groups) and bridge substitution in those of higher molecular weight (13–17 methylene groups).

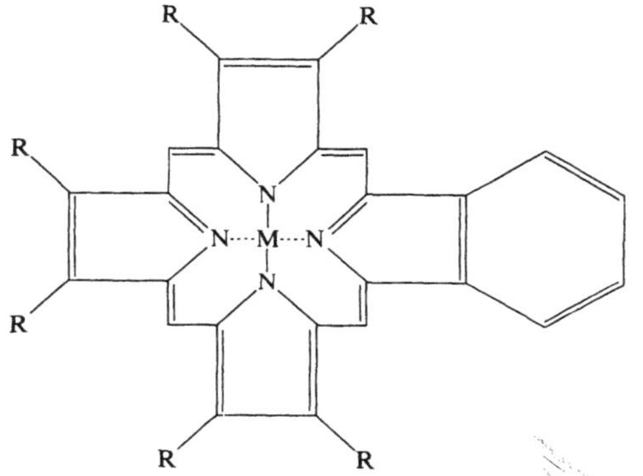

Fig. 8. Proposed structure for rhodo-type petroporphyrin

In passing, it might be added that the basis for Treibs elemental analysis corresponding to DPEP can now be seen (Section IV, A). Obviously, the combustion of a mixture whose mean molecular weight is almost equal to DPEP will give C, H, and N analysis corresponding to DPEP.

In addition to the major series of porphyrins, the presence of a minor series (or perhaps two) has been mentioned (Section III, B-2). An alkylbenzporphyrin structure (Fig. 8) has been proposed for these materials and an explanation for their electronic spectra advanced [16]. The implications of such a structure are rather far reaching. No chlorophylls with such a carbon skeleton or even readily convertible to it have been found. Therefore, the likely source of these materials appears to lie in diagenesis. If the diagenetic formation of one fused aromatic ring is admitted there seems no reason to reject structures which include other similar major carbon skeleton alterations. Indeed, there are indications that in Triassic Oil Shale, materials with a vanadyl porphyrin-type visible spectrum extend into the very high molecular weight range [54]. Confirmation of this result on a broader range of bitumens and asphaltenes would lead one directly to the conclusion that the porphyrin has become incorporated into the aromatic asphaltene sheet*.

* Extracts from a shale of Wyoming origin, Minnelusa formation Pennsylvania age, have been examined by gel permeation chromatography and have been found to contain porphyrin of high molecular weight similar to the distribution found in Serpiano shale [54a].

3. Metallochelates

From a Swiss marl of exceptionally high porphyrin content (0.4 percent), TREIBS isolated and crystallized a metalloporphyrin, spectroscopically and chemically distinct from any known at that time [9]. The chelate with absorption bands at 573 and 534 mμ and unusual stability toward acid cleavage was found to contain vanadium. Further investigation showed that an extremely wide variety of shales and petroleum contained this same chelate. TREIBS had some difficulty in formulating the chelate as his analysis even on synthetic vanadium chelates did not indicate the oxidation level of the vanadium nor distinguish the type of metal oxide bond present. Some possibilities were suggested: vanadyl, VO; its hydrate $V(OH)_2$; or VO_2; the latter two of which might form dimers. The question was not settled until some 20 years later when ERDMAN *et al.* showed that the chelate was a monomer and was correctly formulated as vanadyl on the basis of the vanadyl stretching frequency at 990 cm^{-1} [39].

From the same porphyrin-rich marl, TREIBS isolated a small amount of a second chelate. With some reservation, mainly because the α band appeared at 552 instead of 547 mμ, he proposed that this subsidiary chelate was iron. Fifteen years later, GLEBOVESKAIA and VOL'KENSHTEIN, in a careful reinvestigation, identified the second major metallic component as nickel [57].

In some sources, such as gilsonite, only the nickel chelate is present, while in others, notably Serpiano shale, the vanadyl chelate predominates, and examples can be found which range in between.

Again, to return to the iron question, it is quite apparent from TREIBS' data that he indeed had the nickel, not the iron chelate. Other cases have been cited where the extracts from oil shale appear to contain the iron chelate [58]; however, the chelate has not been isolated and its importance as a geological stabilizer of the porphyrins remains in doubt.

The presence of metallochlorins in ancient sediments has been reported [55]. It was suggested that these pigments were the result of a reversible redox system at equilibrium [56].

D. Conclusions

1. Geochemical Reactions

One is here concerned with those reactions by which the chemical substances in living matter are converted to chemical fossils. The reactions postulated by TREIBS (Section IV, A) will be examined in light of recent knowledge, suggestions by later workers reviewed, and some speculations offered to account for the observed products.

In the conversion of chlorophyll to porphyrin, intermediate compounds of varying degrees of chemical stability are formed. We will discuss the reactions connecting these stability levels in groups which correspond roughly to the severity of required conditions and perhaps coincidently to the reaction rate.

Reactions leading to the first level comprise loss of magnesium, ester hydrolysis and decarboxylation of the β keto acid, and probably chelation. The loss of magnesium is so facile that while it is not an unimportant step, it does little to define the conditions of preservation of the pigments [50]. Hydrolysis of both ester groups (see Fig. 5) occurs very quickly under mild conditions. After hydrolysis of the phytol, a stable propionic acid group remains. In contrast, after hydrolysis of the 10-carbomethoxyl group, an unstable β keto acid remains and decarboxylation occurs quickly.

The argument for chelation with nickel and vanadium at an early stage is based on somewhat scattered evidence. CORWIN and co-workers first showed that vanadyl chelates of phorbides could be produced in the laboratory under conditions analogous to those existing in deposition beds [21]. HODGSON and PEAKE found nickel chlorin pigments in muds from lake beds [52], and, finally, it was demonstrated by CORWIN that chelation with either nickel or vanadyl stabilizes the porphyrin [4]. Such stabilization is nearly a requirement if the porphyrin is to survive through geological time. Intuitively, one tends to regard the square planar chelates as more thermodynamically stable than the free base porphyrins. Studies by ROSSCUP and BOWMAN have indicated the degree of stabilization and suggested an explanation [59]. These workers propose that in these compounds the thermal stability is related to the energy of the longest wavelength (lowest energy) electronic absorption band. It is quite plausible that the energy of the first excited state is roughly proportional to the energy of dissociation, and may therefore be used as an indicator of their stability. For nickel etioporphyrin, the energy of the lowest excited state is 52 kcal/mol, for vanadyl, 50 kcal/mol and a factor of about 1.7 in the first order rate constants for unimolecular decomposition is observed. On this basis, one expects both to be considerably more stable than etioporphyrin where the energy of the lowest excited state is *ca.* 46 kcal/mol.

A second level of stability is achieved by aromatization (oxidation) of the nucleus and reduction of the peripheral functional groups. The dehydrogenation of the chlorin system to the porphyrin system at first appears as a simple "aromatization", analogous to the conversion of cyclohexadiene to benzene. Indeed, simple chlorins are readily dehydrogenated (oxidized) to porphyrins, and chlorin itself is so susceptible to dehydrogenation that attempted preparation of the copper chelate in acetic acid leads only to copper porphine. However, this is not so for chlorophyll itself or its derivatives. Surprisingly, these substances can be aromatized only under severe conditions. The explanation for this behavior was advanced by WOODWARD [60]. The lower periphery of porphyrins (as the formula is usually drawn) derived from chlorophyll (such as DPEP) is so heavily laden with substituents that there is not room for all of them to lie in the plane of the ring. Hence, there is considerable distortion of bond angles and lengths. Removal of hydrogen atoms from the 7 and 8 positions of a 6, γ, 7 substituted chlorin (for example, chlorophyll) transforms carbons 7 and 8 from tetrahedral to trigonal hybridization. In the trigonal hydridization, substituents are forced into the plane of the ring with resultant distortion of bond angles and lengths. Conversely, there is a strong steric factor which favors the conversion of trigonal carbons, 7 and 8 to tetrahedral ones. Said another way, this means that such porphyrins (i.e., DPEP) are easily hydrogenated. On the other hand, etioporphyrins are reduced to chlorins only under severe conditions. It is then clear that the ease of dehydrogenation of the chlorins to the porphyrins depends largely on the presence or absence of a γ substituent or isocyclic ring. Thus, the direct observation of porphyrins of the

DPEP series in crude petroleum places a limit on the reduction potential present, otherwise chlorins rather than porphyrins would be the major material present.

On the other hand, quite strong reducing potentials were required for the reduction of the functional groups of pyropheophorbides under Wolff-Kishner conditions (hydrazine hydrate in methanol-$NaOCH_3$, 120°, 8 hrs.) and the order of reduction is vinyl, 3-formyl and 9-keto [61]. Steric arguments similar to those advanced above, probably also explain the ease of reduction of the 9 carbonyl group to CH_2. The planar oxygen substituent is replaced by two *non-planar* protons thus alleviating the peripheral crowding. At the same time the carbonyl carbon is converted from trigonal (120°) to tetrahedral (109°) hybridization, with the concomitant approach to the unstrained 108° interior angle of a five-membered ring.

A still further level of stability is achieved by decarboxylation of the propionic acid group. The ratio of the carboxylated porphyrins to the decarboxylated ones has been cited very often as a measure of the thermal environment of the petroleum or bitumen matrix. Porphyrin carboxylic acids, when they are found at all, are found only in small quantities. Their absence in petroleums though may not indicate a similar paucity in the producing sands for selective absorption may play a role. Thus, about the most that can be said is that decarboxylation has occurred to some indeterminate degree.

TREIBS considered the decarboxylation of the porphyrin acid to the corresponding etioporphyrins as thermally equivalent to the decarboxylation of aliphatic acids [2]. This comparison is not quite accurate and a more complete picture is possible if one considers the reaction from a kinetic point of view. Activation energies for decarboxylation depends on the substituents near the carboxyl group and range from 30–50 kcal/mol. Non-catalytic carbon-carbon bond breakage requires an activation energy of *ca.* 58 kcal/mol. Since the carboxyl group of the propionic acid side-chain is effectively insulated from the aromatic system, uncatalyzed decarboxylation probably has an activation energy close to 50 kcal/mol [62]. However, the breakage of the C-C bond β to the porphyrin ring ($P\text{-}CH_2 \cdot\cdot CH_2COOH$) is also a possibility. This bond is doubly activated being β to both the aromatic system and the carboxyl group. The activation energy for this reaction may be sufficiently low as to make the two-carbon-loss the preferred reaction.

Other mechanisms involving metal ion catalysis have been suggested [63]. Such mechanisms would be expected to have much lower activation energies but still might not dominate because they are bimolecular and the concentrations of the porphyrins are generally low.

In either case, the expected products would be a mixture of the 7 H, 7 Me homologs of DPEP and DPEP itself, if no reactions with the matrix occur. However, interaction with the media must be considered as likely and will be discussed in a later section.

We conclude from this discussion of the reactions proposed by TREIBS that: (1) the reactions are neither as selective or as simple as TREIBS supposed and therefore even if only those reactions were operative, many porphyrin homologues of DPEP would be expected to be present, and (2) additional reactions must be involved because the multiplicity of porphyrins found is too extensive to be accounted for, even though the non-selectivity of the TREIBS reaction is recognized.

A number of reactions which might reasonably occur have been suggested, such as devinylation, transalkylation, polymerization and pyrolysis.

The devinylation of heme porphyrins proceeds very readily in the presence of fused resorcinol (complete in 0.5 hr. at 160° C). By analogy one might suspect that any medium rich in hydroxy groups (i.e., cellulose, starch) could act as a promoter for the reaction. TREIBS apparently only considered this reaction important in coals where he reportedly found deuteroporphyrin [64]. We view the loss of vinyl groups (or their chemical equivalent, α hydroxy ethyl) as competitive with their reduction. Such a devinylation reaction plus the TREIBS reactions on chlorophyll *a* would yield 2-desethyl DPEP. Alternately, condensation reactions with other aromatic materials (e.g., asphaltenes) at the vinyl group, one of the most reactive groups in the molecule, would produce high molecular weight products.

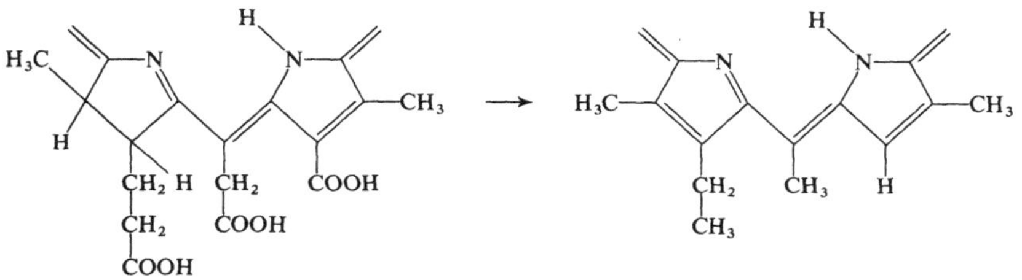

Fig. 9. Partial structures of chlorin e_6 and its C-31 decarboxylated derivative

The "crossover" from the DPEP series to the Etio series by opening of the isocyclic ring was not considered by TREIBS. However, the fact that the porphyrins from common marine sources contain nearly equal amounts of porphyrins of the 310+14n (Etio) series and the 308+14m (DPEP) series leads to the conclusion that opening of the isocyclic ring is an important reaction. Several mechanisms suggest themselves.

Under strong alkaline conditions, the isocyclic ring of chlorophyll is opened to give chlorin e_6 (Fig. 9). Such strongly alkaline conditions are not likely in the deposition beds. Complete decarboxylation and aromatization would yield a C-31 phyllo-type porphyrin. Under acidic conditions, the 10-$COOCH_3$ group is lost (by hydrolysis and decarboxylation of the β keto acid) leading generally to phylloerythrin and preservation of the isocyclic ring.

Sufficiently rigorous thermal conditions will open the ring and probably also expel γ substituents. For example, the porphyrins from retorted shale oil contained a much larger proportion of the Etio series than porphyrin recovered from the unretorted shale [53].

A transalkylation (possibly acid-catalyzed) has been proposed to account for the Gaussian-like distribution of molecular weights observed for the petroporphyrins [3]. The observed symmetry of the molecular weight envelopes is rationalized by this mechanism because alkyl groups subtracted from one presumably are added to the other. Transalkylation need not have occurred between pairs of porphyrins but could equally well have been between porphyrins and other aromatic compounds. Asphaltenes, which are condensed aromatics with alkyl side-chains, 1 to 4 carbon atoms in length, seem especially likely because of their concentration. Thus, the envelope may represent the porphin nucleus plus the average petroleum alkyl chain.

The symmetry pattern might also be explained by radical cleavage of the bonds β to the porphyrin ring and subsequent 'scrambling' of the alkyl groups during the termination step. The comments above concerning the source and length of the alkyl chains in the transalkylation apply equally here. Indeed, the difference in the two mechanisms is only whether the locus of the reaction is the bond α to the porphyrin ring or the β bond. Facile cleavage of the β bonds is plausible on the basis of the double bond rule: single bonds to a double-bonded carbon are strengthened by the presence of the double bond; bonds once removed from the double bond are weakened. Fragmentation patterns in the mass spectra show that β bond cleavage is the predominant mode of decomposition under electron impact.

The possibility that monomeric porphyrins may be regenerated from polymers with asphaltenes or kerogen has been mentioned [54]. At present, such suggestions are speculative.

2. Biogenesis or Diagenesis?

If a strict interpretation of the term biogenesis would be taken, it is apparent from the structural information available that the fossil porphyrins are the result of diagenesis. There are manifest differences in the chemical structures of the buried porphyrins and the biologically active tetrapyrrole pigments. It is, however, pertinent to ask which portions of the petroporphyrin structure are due to biosynthesis and which arose by diagenesis.

Major carbon skeletal alterations are required for the formation of benzporphyrins (Fig. 8) from the biosynthetic pigments. Here the evidence for diagenesis is compelling. The diagenetic formation of one fused ring suggests that other structures with multiple fused rings are also present. Indication of the presence of very high molecular weight porphyrins [54] provides some experimental support for the suggestion. However, none of these classes of porphyrins are very abundant and altogether account for only a small fraction of the petroporphyrins.

In the major porphyrins series (DPEP and Etio) the evidence that diagenesis was the major factor in fixing the carbon skeleton is less rigorous. The possibility cannot be dismissed that there was at one time a group of photosynthetic bacteria where each member possessed a distribution of chlorophylls centered at higher or lower masses than those of the chlorobium chlorophylls. The molecular weight distribution observed for the various petroporphyrins would then represent the remnants of a distribution of chlorophylls. Environmental requirements severely limit the distribution of the chlorobia, and if it is assumed that similar environmental restrictions hold for other producers of non-homogeneous chlorophylls, then these are unlikely sources of the ubiquitous petroporphyrins.

On the other hand, if one assumes a constancy of chlorophylls through the geologic ages, chlorophyll *a* must be the major source of the nucleus of the petroporphyrins and their carbon skeletons are a reflection of the "storage" conditions. Final proof of this hypothesis awaits experimental determination of the sequence of the alkyl groups on any of the higher molecular weight (C-34 or greater) petroporphyrins.

References

1. TREIBS, ALFRED: Chlorophyll- und Häminderivate in bituminösen Gesteinen, Erdölen, Erdwachsen und Asphalten. Ann. Chem. **510**, 42 (1934).
2. – Chlorophyll- und Häminderivate in organischen Mineralstoffen. Angew. Chem. **49**, 682 (1936).
3. BAKER, E. W.: Mass spectrometric characterization of petroporphyrins. J. Am. Chem. Soc. **88**, 2311 (1966).
4. CORWIN, A. H.: Petroporphyrins. Paper V-10, 5th World Petroleum Congress New York 1959.
5. ERDMAN, J. G., J. W. WALTER, and W. E. HANSON: The stability of the porphyrin metallo complexes. Preprints, Div. of Pet. Chem., Am. Chem. Soc. **2**, 259 (1957).
6. TREIBS, ALFRED: Über das Vorkommen von Chlorophyllderivaten in einem Ölschiefer aus der oberen Trias. Ann. Chem. **509**, 103 (1934).
7. GROENNINGS, SIGURD: Quantitative determination of the porphyrin aggregate in petroleum. Anal. Chem. **25**, 938 (1953).
8. COSTANTINIDES, G., G. ARICH, and C. LOMI: Detection and behaviour of porphyrin aggregates in petroleum residues and bitumens. Paper V-11, 5th World Petroleum Congress New York 1959.

8a. CHANG, Y., P. S. CLEZY, and D. B. MORELL: The chemistry of pyrrolic compounds. VI. Chlorins and related compounds. Australian J. Chem. **20**, 959 (1967).

9. TREIBS, ALFRED: Chlorophyll- und Hämin-Derivate in bituminösen Gesteinen, Erdölen, Kohlen, Phosphoriten. Ann. Chem. **517**, 172 (1935).
10. SUGIHARA, J. M., and R. G. GARVEY: Determination of vanadyl porphyrins by demetalation with hydrogen bromide-formic acid. Anal. Chem. **36**, 2374 (1964).
11. ERDMAN, J. G.: Process for removing metals from a mineral oil with an alkyl sulfonic acid. U.S. Patent 3, 190, 829 (June 22, 1965).
12. CORWIN, A. H.: Porphyrins in petroleum. A report to Petroleum Research Fund of Am. Chem. Soc., Jan. 1959.
13. SANIK, JOHN, JR.: Determination of porphyrins in crude petroleum. Preprints, Div. of Pet. Chem., Am. Chem. Soc. **3**, No. 1, 307 (1958).
14. MOORE, J. W., and H. N. DUNNING: Metal-porphyrin complexes in an asphaltic midcontinent Crude oil. Bureau of Mines, Report of Investigations 5370 (Nov. 1957).
15. SUGIHARA, J. M., and L. R. MCGEE: Porphyrins in gilsonite. J. Org. Chem. **22**, 795 (1957).
16. BAKER, E. W., T. F. YEN, J. P. DICKIE, R. E. RHODES, and L. F. CLARK: Mass spectrometry of porphyrins II. Characterization of petroporphyrins. J. Am. Chem. Soc. **89**, 3631 (1967).
17. VANNOTTI, A.: Porphyrins. London: Hilger & Watts Ltd. 1954.
18. WIBERG, K. B.: Laboratory Technique in Organic Chemistry. New York: McGraw-Hill Book, Inc. 1960.
19. DUNNING, H. N., and N. A. RABON: Porphyrin-metal complexes in petroleum stocks. Ind. Eng. Chem. **48**, 951 (1956).
20. DEAN, R. A., and R. B. GIRDLER: Reaction of metal etioporphyrins on dissolution in sulphuric acid. Chem. Ind. (London) 100 (Jan. 23, 1960).
21. CORWIN, A. H., W. S. CAUGHEY, A. M. LEONE, J. E. DANIELEY, and J. F. BAGLI: Petroporphyrins. Preprints, Am. Chem. Soc., Div. of Pet. Chem. **2**, A-35 (1957).
22. – Private communication.
23. MILLSON, M. F., D. S. MONTGOMERY, and (in part) S. R. BROWN: An investigation of the vanadyl porphyrin complexes of the Athabasca oil sands. Geochim. Cosmochim. Acta **30**, 207 (1966).
24. SNYDER, L. R.: Chromatography. 2nd ed. (E. HEFTMAN, ed.), Chap. 4. New York: Reinhold 1967.
25. –, and B. E. BUELL: Compound type separation and classification of petroleum by titration, ion exchange, and adsorption. J. Chem. Eng. Data **11**, 545 (1966).
26. BEACH, L. K., and J. E. SHEWMAKER: The nature of vanadium in petroleum. Ind. Eng. Chem. **49**, 1157 (1957).
27. COSTANTINIDES, G., and G. ARICH: Research on metal complexes in petroleum residues. Paper V-11, 6th World Petroleum Congress Frankfurt, Germany 1963.
28. HODGSON, G. W., N. USHIJIMA, K. TAGUCHI, and I. SHIMADA: The origin of petroleum porphyrins: pigments in some crude oils, marine sediments and plant material of Japan. Sci. Rept. Tohoku Univ., 3rd Ser., **8**, No. 3, 483 (1963).
29. MOORE, J. C.: Gel permeation chromatography. I. A new method for molecular weight distribution of high polymers. J. Polymer Sci. **2**, 835 (1964).

30. ROSSCUP, R. J., and H. P. POHLMANN: Molecular size distribution of hydrocarbons and metalloporphyrins in residues as determined by gel permeation chromatography. Preprints, Div. of Pet. Chem., Am. Chem. Soc. **12**, (2), 103 (1967).

31. BAKER, E. W., and L. F. CLARK. Unpublished data.

32. HOWE, W. W.: Improved chromatographic analysis of petroleum porphyrin aggregates and quantitative measurement by integral absorption. Anal. Chem. **33**, 255 (1961).

33. WILLSTÄTTER, R., u. W. MIEG: Über eine Methode der Trennung und Bestimmung von Chlorophyllderivaten. Ann. Chem. **350**, 1 (1906).

34. EISNER, ULLI, and M. J. C. HARDING: Metalloporphyrins. Part I. Some novel demetallation reactions. J. Chem. Soc. 4089 (1964).

35. THOMAS, D. W., and M. BLUMER: Porphyrin pigments of a Triassic sediment. Geochim. Cosmochim. Acta **28**, 1147 (1964).

36. BALEK, R. W., and A. SZUTKA: The quantitative separation of tetraphenylporphines by thin-layer chromatography. J. Chromatog. **17**, 127 (1965).

37. DUNNING, H. N., and J. K. CARLTON: Paper chromatography of a petroleum porphyrin aggregate. Anal. Chem. **28**, 1362 (1956).

38. FISHER, L. R., and H. N. DUNNING: Chromatographic resolution of petroleum porphyrin aggregates. Anal. Chem. **31**, 1194 (1959).

39. ERDMAN, J. G., V. G. RAMSEY, N. W. KALENDA, and W. E. HANSON: Synthesis and properties of porphyrin vanadium complexes. J. Am. Chem. Soc. **78**, 5844 (1956).

39a. ERDMAN, J. G., V. G. RAMSEY, and W. E. HANSON: Volatility of metallo-porphyrin complexes. Science **123**, 502 (1956).

40. KLESPER, E., A. H. CORWIN, and D. A. TURNER: High pressure gas chromatography above critical temperatures. J. Org. Chem. **27**, 700 (1962).

41. KARAYANNIS, N. M., A. H. CORWIN, E. W. BAKER, E. KLESPER, and J. A. WALTER: Hyperpressure gas chromatography. I. An apparatus for gas chromatography of non-volatile compounds. Anal. Chem. **40**, 1736 (1968).

42. SORET, J. L.: Sur le spectre d'absorption du sang dans la partie violette et ultra-violette. Compt. Rendus **97**, 1269 (1883).

42a. STERN, A. and associates: The light absorption of porphyrins. Z. Phys. Chem. A-**170**, 337 (1934); A-**174**, 81, 321 (1935); A-**175**, 405 (1935); A-**176**, 81 (1936); A-**177**, 40, 165, 365, 387 (1936); A-**178**, 161 (1937); A-**179**, 275 (1937); A-**180**, 131 (1937).

43. FALK, J. E.: Porphyrins and metalloporphyrins. New York: Elsevier Publ. Co. 1964.

43a. HODGSON, G. W., B. L. BAKER, and E. PEAKE: Geochemistry of porphyrins. In: Fundamental aspects of petroleum geochemistry (B. NAGY and U. COLOMBO, eds.). New York: Elsevier Publ. Co. (1967), and HODGSON, G. W., and B. L. BAKER: Chem. Geol. **2**, 187 (1967).

44. FISCHER, HANS, and H. PFEIFFER: Oxydation von Porphyrinen und Chlorinen mit Osmiumtetroxyd. Ann. Chem. **556**, 131 (1944).

45. BAKER, E. W., M. S. BROOKHART, and A. H. CORWIN: Piperidinate complexes of nickel and copper mesoporphyrin IX. J. Am. Chem. Soc. **86**, 4587 (1964).

46. KATZ, J. J., R. C. DOUGHERTY, and L. J. BOUCHER: Infrared and NMR spectroscopy of chlorophyll. In: The chlorophylls (L. P. VERNON and G. R. SEELY, eds). New York: Academic Press 1966.

47. SCHWARTZ, S., M. H. BERG, I. BOSSENMAIER, and H. DINSMORE: Determination of porphyrins in biological materials. Methods Biochem. Anal. **8**, 221 (1960).

48. BAKER, E. W., and A. H. CORWIN: Structure studies on petroporphyrins. Preprints, Div. of Pet. Chem., Am. Chem. Soc. **9**, (1), 19 (1964).

49. FISHER, L. R., and H. N. DUNNING: Chromatographic resolution of petroleum porphyrin aggregates. Bureau of Mines Report of Investigation 5844 (1961).

50. ORR, W. L., K. O. EMERY, and J. R. GRADY: Preservation of chlorophyll derivatives in sediments off Southern California. Bull. Am. Assoc. Petrol. Geologists **42**, 925 (1958).

51. HODGSON, G. W., B. HITCHON, R. M. ELOFSON, B. L. BAKER, and E. PEAKE: Petroleum pigments from Recent fresh-water sediments. Geochim. Cosmochim. Acta **19**, 272 (1960).

52. –, and E. PEAKE: Metal chlorin complexes in Recent sediments as initial precursors to petroleum porphyrin pigments. Nature **191**, 766 (1961).

53. MORANDI, J. R., and H. B. JENSEN: Comparison of porphyrins from shale oil, oil shale and petroleum by absorption and mass spectrometry. J. Chem. Eng. Data **11**, 81 (1966).

54. BLUMER, M., and W. D. SNYDER: Porphyrins of high molecular weight in a Triassic oil shale: Evidence by gel permeation chromatography. Chem. Geol. **2**, 35 (1967).
54a. –, and M. RUDMAN: Unpublished data.
55. –, and G. S. OMENN: Fossil porphyrins: uncomplexed chlorins in a Triassic sediment. Geochim. Cosmochim. Acta **25**, 81 (1961).
56. BLUMER, MAX: Organic pigments: their long-term fate. Science **149**, 722 (1965).
57. GLEBOVSKAIA, E. A., and M. V. VOL'KENSHTEIN: Spectra of bitumen and petroleum porphyrine. Zh. Obshch. Khim. **18**, 1440 (1948).
58. MOORE, J. W., and H. N. DUNNING: Interfacial activities and porphyrin contents of oil-shale extracts. Ind. Eng. Chem. **47**, 1440 (1955).
59. ROSSCUP, R. J., and D. H. BOWMAN: Thermal stabilities of vanadium and nickel petroporphyrins. Preprints, Div. of Pet. Chem., Am. Chem. Soc. **12** (2), 77 (1967).
60. WOODWARD, R. B.: The total synthesis of chlorophyll. Pure Appl. Chem. **2**, 383 (1961).
61. FISCHER, H., u. H. GIBIAN: Übergang von der Chlorophyll b in die a-Reihe. Ann. Chem. **552**, 153 (1942).
62. ABELSON, P. H.: Organic geochemistry and the formation of petroleum. Paper I-41, 6th World Petroleum Congress Frankfurt, Germany 1963.
63. COOPER, J. E., and E. E. BRAY: A postulated role of fatty acids in petroleum formation. Geochim. Cosmochim. Acta **27**, 1113 (1963).
64. TREIBS, ALFRED: Porphyrine in Kohlen. Ann. Chem. **520**, 144 (1935).

General References

DUNNING, H. N.: Geochemistry of organic pigments. In: Organic geochemistry (I. A. BREGER, ed.). New York: Macmillan 1963.
EGLINTON, G., and M. CALVIN: Chemical fossils. Sci. Am. **216**, 32–43 (1967).
FALK, J. E.: Porphyrins and metalloporphyrins. New York: Elsevier, Publish. Co. 1964.
VERNON, L. P., and G. R. SEELY: The chlorophylls. New York: Academic Press 1966.

CHAPTER 20

Fossil Shell "Conchiolin" and Other Preserved Biopolymers

MARCEL FLORKIN*

University of Liège, Liège, Belgium

Contents

I. Introduction

Organic molecules have been detected in many sediments and shales, but structurally preserved biopolymers kept *in situ* in their anatomical location in fossils, with a lesser possibility of being introduced from outside, should constitute the most reliable material for studies in paleobiochemistry, or the biochemistry of fossils**.

Until recently, it has long been accepted that the only organic material remaining in fossils is present in a state approaching complete carbonisation. Since free amino acids were first recognized by ABELSON [1–4] in fossils of various ages, it has been apparent that organic biopolymers may be preserved for about 100,000 years, which was the approximate age of the Pleistocene shell of *Mercenaria mercenaria* studied by JONES and VALLENTYNE [5]. They detected in this shell a "conchiolin" which liberated free amino acids by hydrolysis. Since their report

* The author has to thank Dr. CHARLES GRÉGOIRE, who is in charge of the electron microscopy section in his department, for providing the micrographs reproduced in this chapter, and for reading the manuscript and making valuable suggestions for its improvement.

** The use of the word "paleobiochemistry" as meaning a comparative approach to living organisms along the lines of phylogeny can only lead to confusion.

in 1960, data showing that animal proteins, identified by the liberation of their component amino acids by hydrolysis, may be preserved in fossils for several geologic ages, have been produced [6] and have been followed by a series of demonstrations [7–11] of other such "paleoproteins". The presence of chitin has also been detected in a Cambrian fossil [12].

II. Modern Molluscan Conchiolin

The exoskeleton of molluscs is typically represented by the shell. Several kinds of shell architectures are found in molluscs. But whether external or internal, whether used as a protecting armor, a moisture-keeping device, a swimming apparatus, or in any other adaptation, the typical molluscan shell is generally made of three layers secreted by specialized regions of the mantle: the outermost thin layer named cuticle or, better still, periostracum is often uncalcified; the prisms' layer is laid down under the periostracum and is characterized by the columnar arrangement of crystals; the innermost layer has a lamellar structure and is named "nacre" or "mother-of-pearl" if the crystalline form of calcium carbonate is aragonite, whereas it is named "calcitostracum" if the crystals are calcite. The cuttle bone of *Sepia* still shows oblique calcareous particles similar to the septa of the chamber skeleton of *Nautilus*, but it has acquired a secondary structure in relation to its function in density regulation. The organic residue from the decalcification of a molluscan shell has been called conchiolin by Frémy [13].

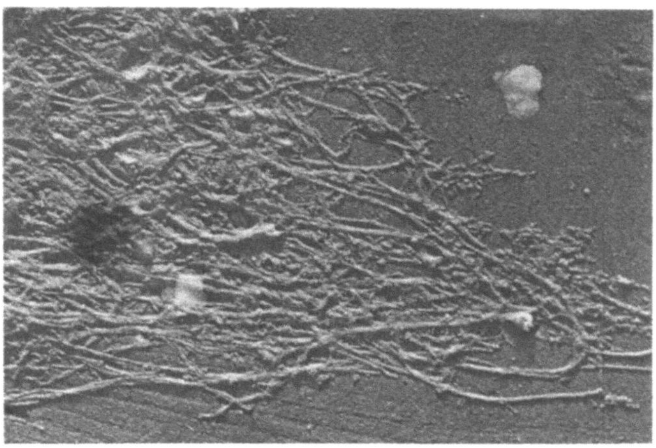

Fig. 1. *Pinctada margaritifera* Linné. Nacroin fibers. Shadow cast with palladium. × 28,000 [14]

In nacre, the mother-of-pearl layer, the organic material is composed of fibres of nacroin associated with a protein component (nacrin). The amino acid components of nacroin are mostly alanine and glycine, the nitrogen of these two amino acids amounting to 83.6 percent of total N in the substance in *Pinctada margaritifera* (Fig. 1). A recent investigation in our laboratory revealed however the presence of chitin, a high polymer of β-N-acetyl-D-glucosamine, in the nacroin residue of the mother of pearl for all the species so far studied (*Nautilus*, *Nucula*,

Pinctada, Pteria, Mytilus). The chitin amounts to from 0.5 to 6 percent of the dry weight of the nacroin [15]. It thus seems convenient to consider the nacroin as a mucopolysaccharide. Nacroin constitutes various proportions of the organic skeleton of mother-of-pearl; in the nacre of *Nautilus*, this N makes up for about 5 percent of total N. When the protein constituent of the nacre of *Nautilus* is decalcified and hydrolyzed, the amino acids forming the largest percentage of the total are glycine, alanine, serine, aspartic acid and glutamic acid. The fact that this pattern of the "conchiolin of nacre" differs from that of nacroin is due to the predominance of these amino acids in nacrine, the protein component associated

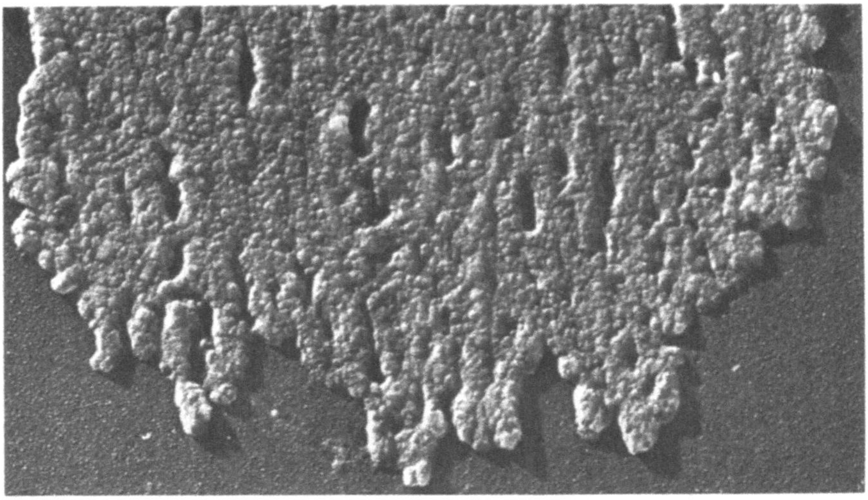

Fig. 2. *Nautilus pompilius* LAMARCK. Conchiolin from mother-of-pearl. Nautiloid pattern of structure. × 50,000 [16]

with the filament of nacroin. This protein is partly in water-soluble form, partly in insoluble form. Without taking any position with respect to the existence of two proteins, we can call these fractions soluble nacrine and insoluble nacrine. The insoluble nacrine is 2–3 times more abundant than the soluble nacrine.

When the calcium carbonate has been dissolved away, the complex constituting the "conchiolin of nacre" appears as soft iridescent membranes. Ultrasonic waves dissociate these into thin membranes, permeable to electrons. Their structure can therefore be studied directly by the method of electron microscopy. The study of the fragments or leaflets of nacreous conchiolin has shown that the membranes are perforated by openings or pores and have a reticulated lace-like texture. Three patterns of structure based on differences in the size, shape and frequency distribution of the pores in the three classes of molluscs we have studied, were recognized in nacreous conchiolin (Figs. 2–4).

In the lace-like structures described above, the strands forming the reticulated sheets of these structures are knobbly cords, studded with hemispherical protuberances of various sizes giving to the trabeculae the appearance of "iris rhizomes" [18]. The conchiolin of nacre appears to consist of a core of muco-

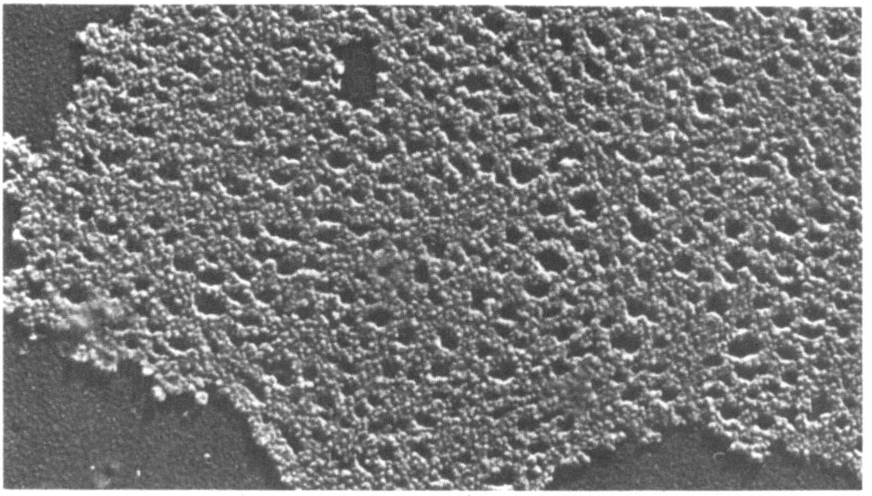

Fig. 3. *Angaria delphinus* LINNÉ. Conchiolin from mother-of-pearl. Gastropod pattern of structure. × 50,000 [17]

Fig. 4. *Mytilus crenatus* LINNÉ. Conchiolin from mother-of-pearl. Pelecypod pattern of structure. × 50,000 [17]

polysaccharide fibrils, surrounded by a double coating of insoluble protein and of soluble protein. The studies described above have established the existence of the pores previous to decalcification and to treatment with ultrasonic waves or electron beams.

Subsequent studies of mother-of-pearl, using replicas of inner surfaces of normal shells, fragments of cleavage produced by fracture, and polished surfaces before and after etching [18], have shown that the reticulated structures belong to the system of organic membranes or sheets which run as continuous formations in between adjacent stratified aragonite lamellae. Recently, intra-crystalline systems of conchiolin have been detected [19] in ultra-thin sections of shells.

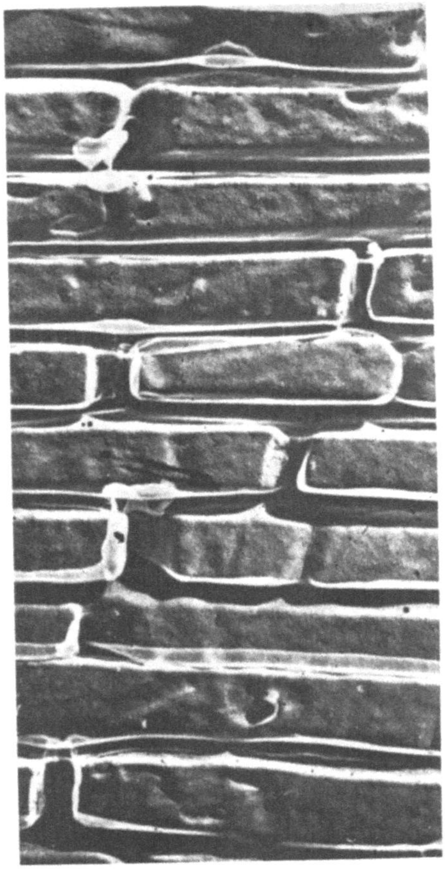

Fig. 5. *Anodonta cygnaea* LINNÉ. Carbon replica of a polished and etched transverse section of mother-of-pearl. Brick wall appearance. Conchiolin membranes wrap the crystals of aragonite. × 25,000 [18]

Thus, mother-of-pearl is typically composed of lace-like protein leaflets associated with tabular crystals of aragonitic calcium carbonate as shown in Fig. 5. However, several species of the class Pelecypoda (bivalves), the internal layer of the shell is in the nature of pseudo-nacre or calcitostracum, in which the mineral constituent, calcium carbonate, is in the mineralogic, prismatic form of calcite, and the organic component is constituted of vitreous, granular or fibrillar membranes without the lace-like structures characteristic of nacre [14, 20].

The calcitostracum replaces the nacre in *Ostreidae*, *Pectinidae*, *Anomiidae*, *Ledidae* and *Arcidae*. Prism conchiolin surrounds each prismatic mineral structure. The prism conchiolin is, as is the case for nacreous conchiolin, built on a fibrillar web (Fig. 6), but the protein combined to the latter has a very dense structure unlace-like in appearance, except in *Mytilus*. If it is true that the lace-like structure of the nacreous conchiolin is always embedded in a mineral structure made of aragonite crystals, it appears that aragonite is sometimes found in prisms, surrounded by their protein structure. As is the case of some gastropods and

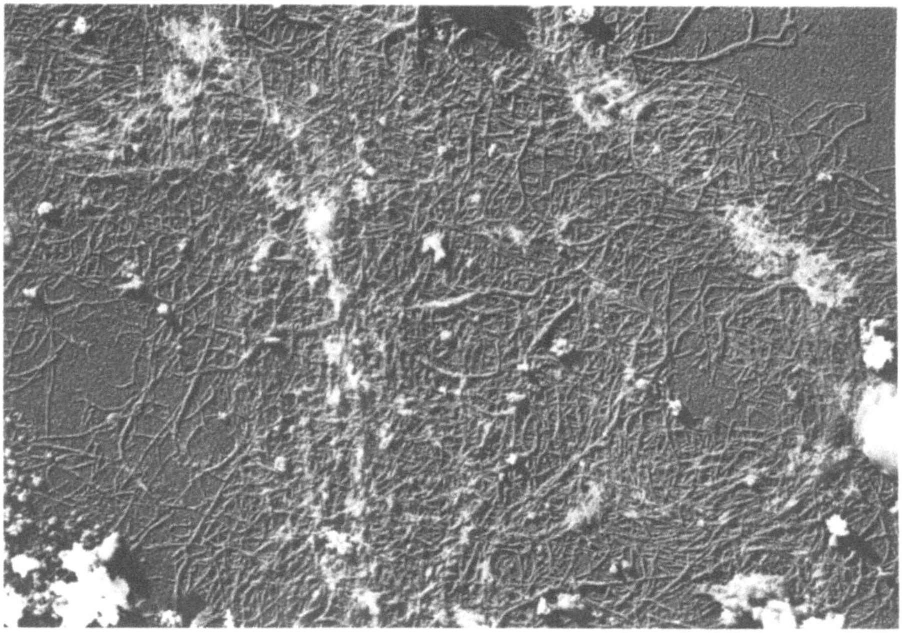

Fig. 6. *Vulsella vulsella*. Fibrillar residue of decalcification of a calcitic prism. × 16,500 [21]

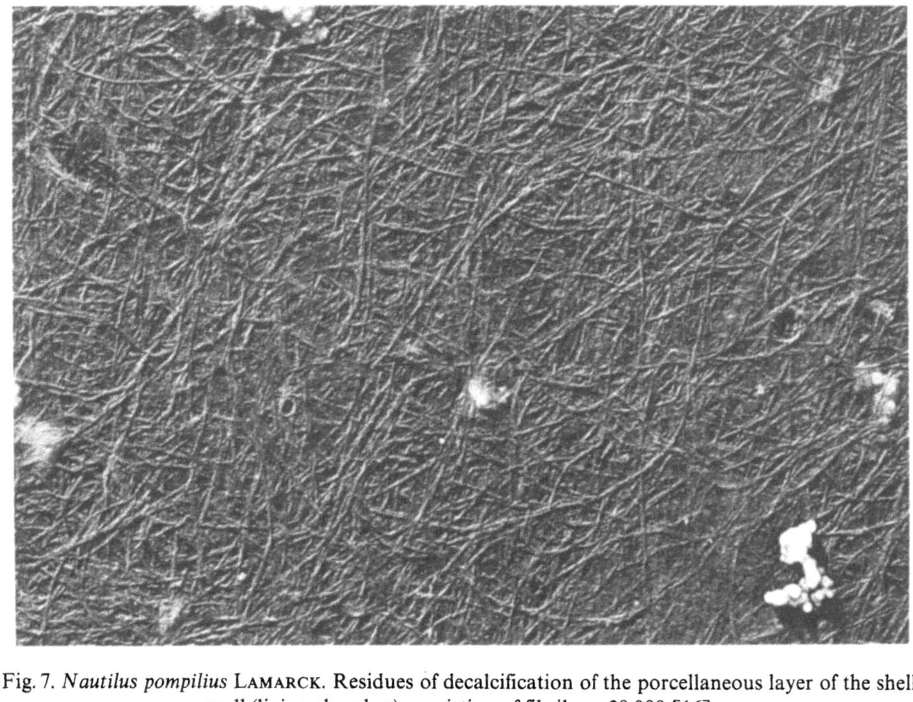

Fig. 7. *Nautilus pompilius* LAMARCK. Residues of decalcification of the porcellaneous layer of the shell wall (living chamber) consisting of fibrils. × 38,000 [16]

Nautilus, it appears that, in the external layer of the shell, the prismatic layer is replaced by a very dense mineral structure, the porcelain component. This is deposited on a system of fibrils as shown in Fig. 7.

III. Fossil Molluscan "Conchiolins"

GRÉGOIRE [22, 23] has identified in fossil shells, lace-like structures which are similar to those described above at the level of nacreous conchiolin. When we were able to obtain enough material of such nacreous remains, we proceeded with a chemical study of them [6] and showed that the lace-like structures observed in fossils by the study of replicas and of decalcified material was actually conchiolin. The remains still give the amino acid pattern of nacreous conchiolin. But even more striking is the fact that the chemical composition typical of the conchiolin of mother-of-pearl has been found in fossil shells from the Eocene period (60 million years old) [6].

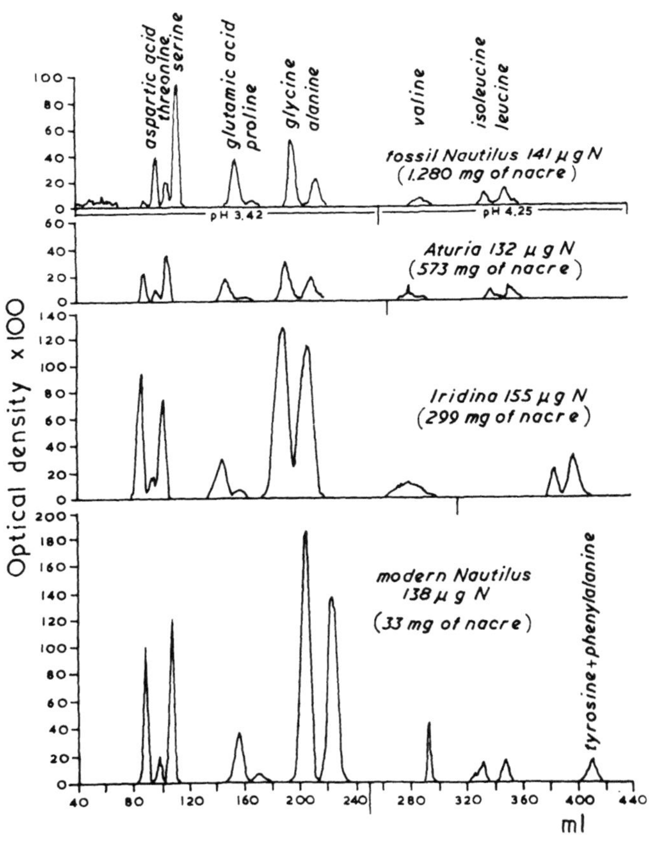

Fig. 8. Separation by chromatography on Dowex 50, using the Moore and Stein technique, of neutral amino acids and acids resulting from the hydrolysis of the residue of nacre after washing out the free amino acids, decalcification and dialysis, of modern *Nautilus*, fossil (Eocene) *Nautilus*, Recent (Holocene) *Iridina* and fossil (Oligocene) *Aturia* [6]. The ultrastructures of the samples used are shown in Fig. 9

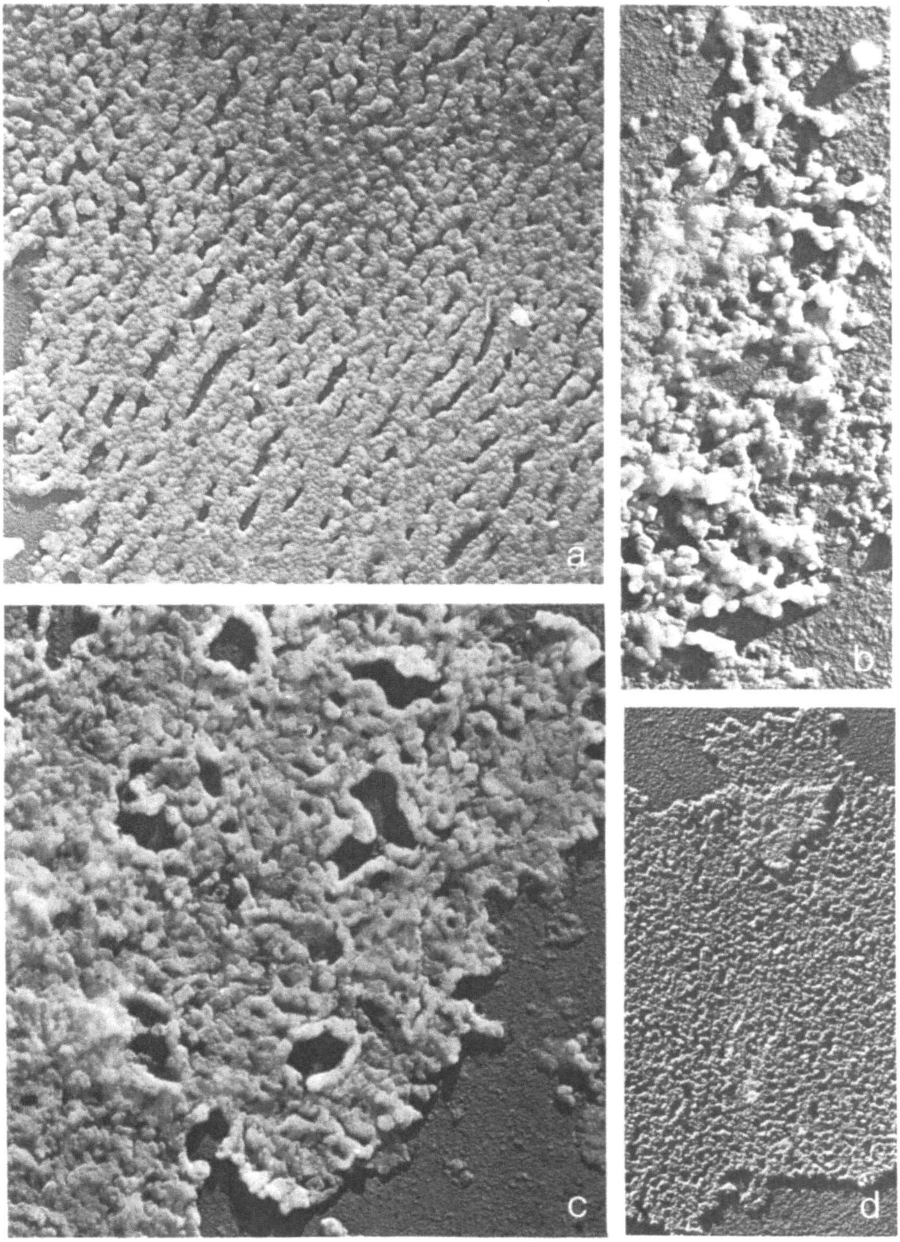

Fig. 9. Electron micrographs of nacreous organic remnants after decalcification with chelating agents. Fragments of lace-like reticulated sheets of conchiolin. a) *Nautilus macromphalus* SOW. (Cephalopoda, Nautiloida.) Recent (Nautiloid pattern). × 36,000. b) *Nautilus* sp. (Cephalopoda. Nautiloida.) Eocene (60 million years). Nautiloid pattern still recognizable in some regions. × 36,000. c) *Aturia* sp. (Cephalopoda. Nautiloida.) Oligocene (40 million years). Nautiloid pattern still recognizable. × 36,000. d) *Iridina spekii* WOODWARD. (Pelecypoda. Mutelida.) Holocene (< 10,000 years). Pelecypod pattern [23]. × 36,000

Table 1. "*Conchiolin*" *of mother-of-pearl. Amino acid composition: molecular fraction per 100 molecules of amino acids* [6].

	Nautilus Eocene (60 million years)	*Aturia* Oligocene (40 million years)	*Iridina* Holocene (> 10,000 years)	*Nautilus* Modern
Aspartic acid	8.7	9.0	10.1	9.0
Threonine	5.6	3.8	1.6	1.5
Serine	24.0	16.7	7.8	10.9
Glutamic acid	15.6	11.9	5.0	5.5
Proline	3.7	4.8	2.0	1.8
Glycine	20.8	23.3	29.8	35.7
Alanine	9.7	15.2	28.2	27.2
Valine	3.1	5.7	4.0	2.2
Isoleucine	3.1	3.3	2.4	1.8
Leucine	5.6	6.2	4.0	1.9
Tyrosine	0	0	0	2.4
Phenylalanine	0	0	0	2.4
Histidine	0	–	0	0
Lysine	0	–	2.3	0
Arginine	0	–	1.9	0

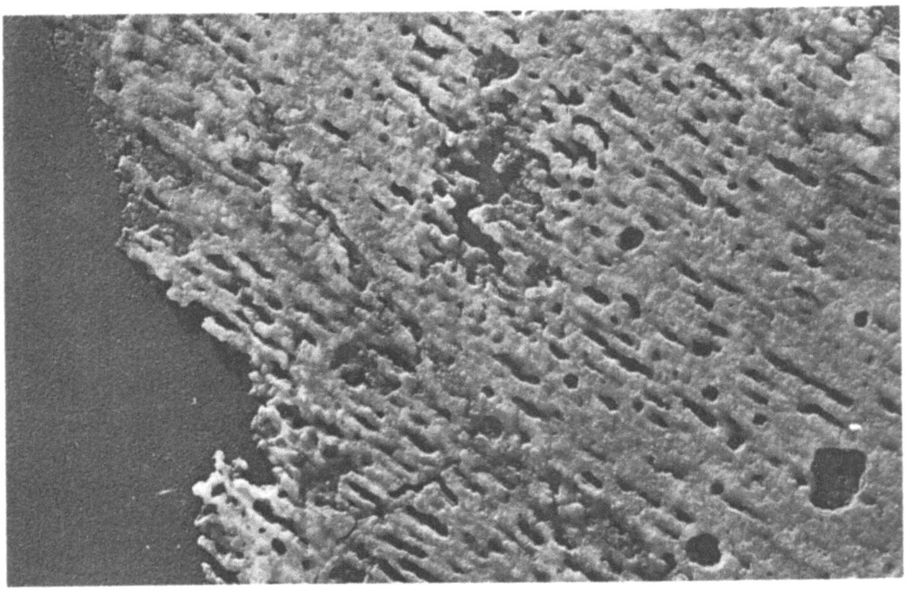

Fig. 10. Conchiolin membrane from a large unidentified Pennsylvanian *Nautilus* (about 300 million years). × 29,000 [24, 25]

The chromatograms shown in Fig. 8 are those for similar weights of nitrogen remaining in fragments of mother-of-pearl, the ultrastructures of which are shown in Fig. 9. The free amino acids were first washed out of the ground material, which was then decalcified and the subsequent hydrolysate subjected to dialysis. In each case, the chromatogram shows a distribution of neutral amino acids close

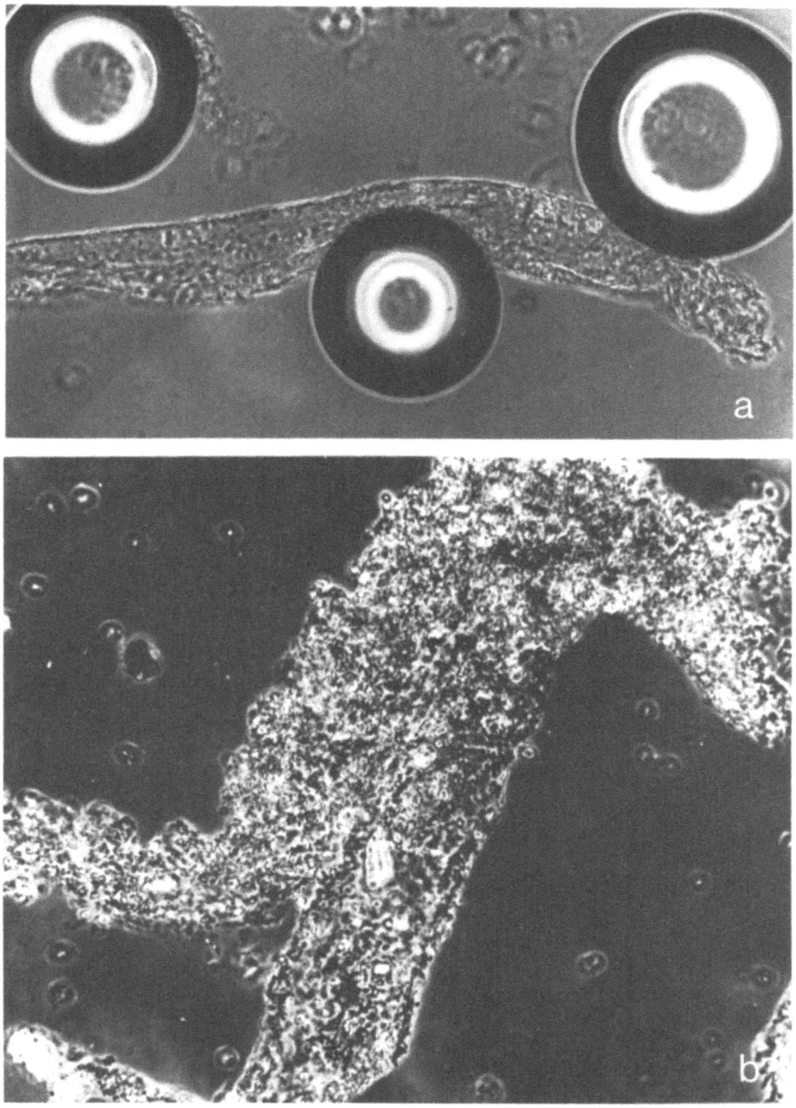

Fig. 11. a) *Atrina (Pinna) nigra* DURH (Recent). End stage of decalcification of a single prism. In the sheath shown here, shrinkage and wrinkling conceal the transverse striation. (From GRÉGOIRE [21].) Phase contrast ×300. b) *Pinna affinis* (London, Clay, Lower Eocene, 60 million years). Elongated, folded and wrinkled fragments of tubular prism sheaths. The transverse striation is concealed in part by small mineral fragments resisting demineralization by the chelating agent used, and, as in the living sample (see Fig. 11 a) by shrinking of the fragments. Phase contrast ×400 [11]

to that of the conchiolin of mother-of-pearl. On this evidence one might infer the presence of this protein. Another possibility is the conservation *in situ*, after a destruction of the conchiolin, of free amino acids bound in the form of carbamates to the mineral portion. Nevertheless, the washings and dialysis after decalcification

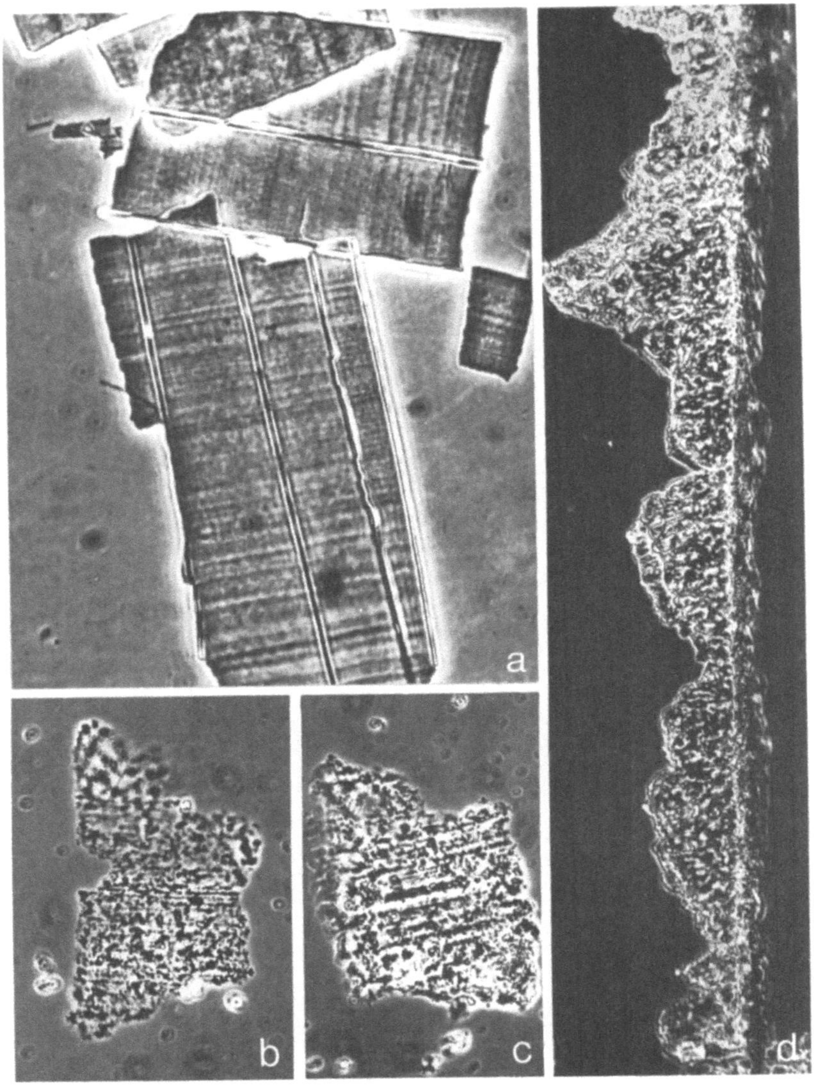

Fig. 12. a) *Atrina (Pinna) nigra* DURH (Recent). Sheath fragments from decalcified prisms showing a transverse striation on rectangular facets. (From GRÉGOIRE [21].) Phase contrast × 800. b) and c) *Pinna affinis* (London Clay, Lower Eocene, 60 million years). Debris of prism sheath facets showing small mineral granules disposed along the transverse striation (see Fig. 12a). Phase contrast × 400. d) *Inoceramus* sp. (Gault, Cretaceous, 135 million years). Elongated sheath shred left by decalcification of a prism. As in Fig. 11b, abundant microcrystals, escaping demineralization, subsist, either attached to or embedded in the substance of the sheath. Phase contrast × 400 [11]

would have eliminated some compounds. On the other hand, in the decalcified residue of mother-of-pearl, particles are observed with the ordinary microscope which show a positive biuret reaction, characteristic of the peptide. The sum of these observations brings us to the conclusion that proteins are present in the structures studied. We can compare (Table 1) the compositions of the "conchiolin" as revealed by the amounts of amino acids liberated by hydrolysis in boiling 6N HCl. A marked similarity is apparent between the conchiolin of the mother-of-pearl of modern *Nautilus* and of the "conchiolin" of the *Iridina* from the Holocene. The "conchiolins" of *Aturia* from the Oligocene and of *Nautilus* from the Eocene contain less glycine and alanine. As the contents in serine and glutamic acid are higher in these samples of fossil mother-of-pearl, we must consider the possibility of a complete or partial loss of constituent proteins which are richer in alanine and glycine.

Using the electron microscope, morphological structures similar to those found in the modern and fossil examples described above, have been detected in nacreous layers of fossil shells 500 years to 430 million years old (Fig. 10) [24, 26, 27].

The submicroscopic structure of the conchiolin of the prisms in molluscan shells has been studied by GRÉGOIRE [21, 28, 29]. The inorganic part of a prism is made up of calcite crystals piled one upon the other. The pile is wrapped inside a conchiolin sheath, comprised of a fibrillar structure covered in turn by a very dense protein component.

BRICTEUX-GRÉGOIRE *et al.* [11] have isolated individual prisms from fossil shells, namely those of *Pinna affinis* (London Clay, Lower Eocene) and *Inoceramus* sp. (Gault, Cretaceous) and have decalcified these structures. Demineralization of the prisms, performed in saturated aqueous solutions of EDTA (ethylene-

Table 2. *Comparison of amino acid compositions for "conchiolins" of prisms from modern and fossil shells* [11]

	Modern						Fossil					
	Atrina nigra			*Pinna nobilis*			*Pinna affinis*			*Inoceramus*		
	µg/g	per-cent N	mol. fr.	µg/g	per-cent N	mol. fr.	µg/g	per-cent N	mol. fr.	µg/g	per-cent N	mol. fr.
Lysine	145	0.7	0.8	110	0.5	0.6	tr			1.58	0.7	4.8
Histidine	70	0.3	0.4	89	0.4	0.5	tr			1.13	0.5	3.2
Arginine	252	1.1	1.1	365	1.5	1.7	tr			0.96	0.4	2.4
Aspartic acid	1,750	9.6	10.4	3,920	20.5	23.3	23.4	2.6	11.1	3.03	1.6	10.0
Threonine	173	1.1	1.1	231	1.4	1.5	8.8	1.1	4.7	1.18	0.7	4.4
Serine	530	3.7	4.0	617	4.1	4.6	27.5	3.8	16.5	3.88	2.5	16.4
Glutamic acid	330	1.6	1.8	348	1.7	1.9	31.8	3.1	13.6	5.13	2.4	15.5
Proline	314	2.0	2.2	316	1.9	2.2	tr			tr		
Glycine	4,390	42.4	46.2	3,500	32.4	36.9	26.8	5.2	22.5	3.60	2.3	21.3
Alanine	530	4.3	4.7	640	5.0	5.7	20.0	3.3	14.1	1.62	1.2	8.1
Valine	930	5.8	6.3	833	4.9	5.6	9.2	1.1	5.0	1.30	0.8	4.9
Isoleucine	334	1.8	2.0	558	3.0	3.4	7.9	0.9	3.8	0.99	0.5	3.4
Leucine	1,220	6.8	7.4	740	3.9	4.5	12.2	1.4	5.9	1.69	0.9	5.7
Tyrosine	2,130	8.6	9.3	1,200	4.6	5.3	tr			tr		
Phenylalanine	472	2.1	2.3	508	2.1	2.4	7.6	0.7	2.9	tr		
		91.9	100.0		87.9	100.1		23.2	100.1		15.5	100.1
Amino nitrogen	1,930			2,020			96			20		
Ammonia	70	3.0		90	3.7		131	113		14.1	56.7	

diamine tetraacetic acid, disodium salt, or Titriplex III), and in 2 percent followed by 25 percent hydrochloric acid solutions, leaves transparent, glassy, brittle, sometimes tubular, fragments of substance which are the remains of the prism sheaths. Figs. 11 and 12 show the structure of this debris, recorded photographically under the phase contrast microscope using aqueous suspensions held between slide and coverglass. In several shreds, the characteristic transverse striation, along which small mineral particles are aligned, is still distinctly recognizable. After being washed to remove free amino acids, the isolated sheaths from prisms isolated from both modern and fossil specimens have been hydrolyzed and the products analysed (Table 2). The results confirm the protein nature of the material.

IV. Graptolites

FOUCART *et al.* [7] have carried out analyses on three different samples of fossils belonging to the order of Graptoloidea (Silurian *Pristiograptus gotlandicus*, *Pristiograptus dubius*; Silurian *Monograptidae, gn. sp.*, contaminated with a few *Retiolitidae*; Ordovician *Climacograptus typicalis*, provisional identification).

The calcareous rocks containing the fossils have been dissolved with cold HCl (0.5 N). The graptolites were separated with micropipettes and washed repeatedly with cold HCl and distilled water in order to remove the soluble components such as free amino acids. As shown in Table 3, in addition to high amounts

Table 3. *Amino acids in hydrolysates of Graptolites, after decalcification and washings*[a] [7]

	Pristiograptus gotlandicus and *P. dubius* (Silurian)		Monograptidae *gn. sp.* (Silurian)		*Climacograptus typicalis* (Ordovician)	
	μg/g	mole-fraction per 100	μg/g	mole-fraction per 100	μg/g	mole-fraction per 100
Aspartic acid	218	9	560	8.6	68	10
Threonine	108	4.9	290	4.9	26	4.3
Serine	214	11	550	10.6	122	22.8
Glutamic acid	380	13.9	1,100	15.3	96	12.8
Proline	(83)	(3.9)	(340)	(6)	trace	
Glycine	280	20.1	760	20.8	89	23.4
Alanine	103	6.3	410	9.5	40	8.9
Valine	116	5.3	(240)	(4.1)	trace	
Isoleucine	97	4	230	3.6	(15)	2.3
Leucine	190	7.8	390	6.1	(13)	(2)
Tyrosine	(42)	(1.2)	(90)	(1.1)	–	
Phenylalanine	(87)	(2.9)	(110)	(1.3)	–	
Lysine	96	3.5	(310)	(4.4)	70	9.4
Histidine	60	2.2	(84)	(1.1)	29	3.7
Arginine	128	4	(140)	(2.2)	trace	
Ammonia	223					

[a] The amounts indicated in brackets could only be calculated approximately on account of their low values.

of ammonia, amino acids are present in the hydrolysates of the three samples studied (*Pristiograptus* – 2,202 μg/g; *Monograptidae* – 5,604 μg/g; *Climacograptus* – 568 μg/g). As soluble material had been previously removed by repeated treatments, and as possible sources of contamination have been avoided, we consider that the amino acids contained in the hydrolysates are of protein origin. An identification of peptide linkages by means of the biuret reaction, as used by FLORKIN *et al.* [6] in the case of "conchiolin" remnants, was not possible because of the dark color of the graptolite fragments.

The question can be asked as to whether these proteins were contained in the graptolites themselves or whether they were present in the sediment from which the graptolites had been removed. The analysis of one graptolitic rock, when compared with that of the graptolites extracted from the same rock, allowed us to eliminate the latter hypothesis. The results (Table 4) show that the amino acid

Table 4. *Comparison between the amounts of amino acids of protein origin in Graptolites removed from a rock with those found in the rock itself* [7]

	μg/g		Mole-fraction	
	graptolites [a]	rock [b]	graptolites	rock
Aspartic acid	68	12.8	10	7.4
Threonine	26	8	4.3	5.1
Serine	122	45.4	22.8	33.4
Glutamic acid	96	9.6	12.8	5.1
Proline	tr	–	tr	–
Glycine	89	30	23.4	30.7
Alanine	40	12	8.9	10.4
Valine	tr	–	tr	–
Isoleucine	(15)	5.6	(2.3)	3.3
Leucine	(13)	7.4	(2)	4.3
Tyrosine	–	–	–	–
Phenylalanine	–	–	–	–
Lysine	70	–	9.4	–
Histidine	29	–	3.7	–
Arginine	tr	–	tr	–
Total	568	130.8	99.6	99.7

[a] *Climacograptus typicalis.*
[b] Ordovician limestone containing *Climacograptus typicalis* (from Ohio) after decalcification and removal of the Graptolite fragments.

concentrations are much lower than those of the graptolites. On the other hand, some amino acids present in the graptolites have not been detected in the rock. The organic composition of the graptolite remnants is thus quantitatively and qualitatively distinct from that of the kerogen of the parent sedimentary rock. Contamination of the graptolites themselves is also eliminated. Observation of the fragments with the electron microscope did not reveal appreciable proportions of typical contaminants such as bacteria, algal filaments, "conchiolin" remnants and so on. On the other hand, parallelism between the results obtained for the three samples of different origin and geological age could hardly be consistent with the possibility of a contamination by exogenous organic material.

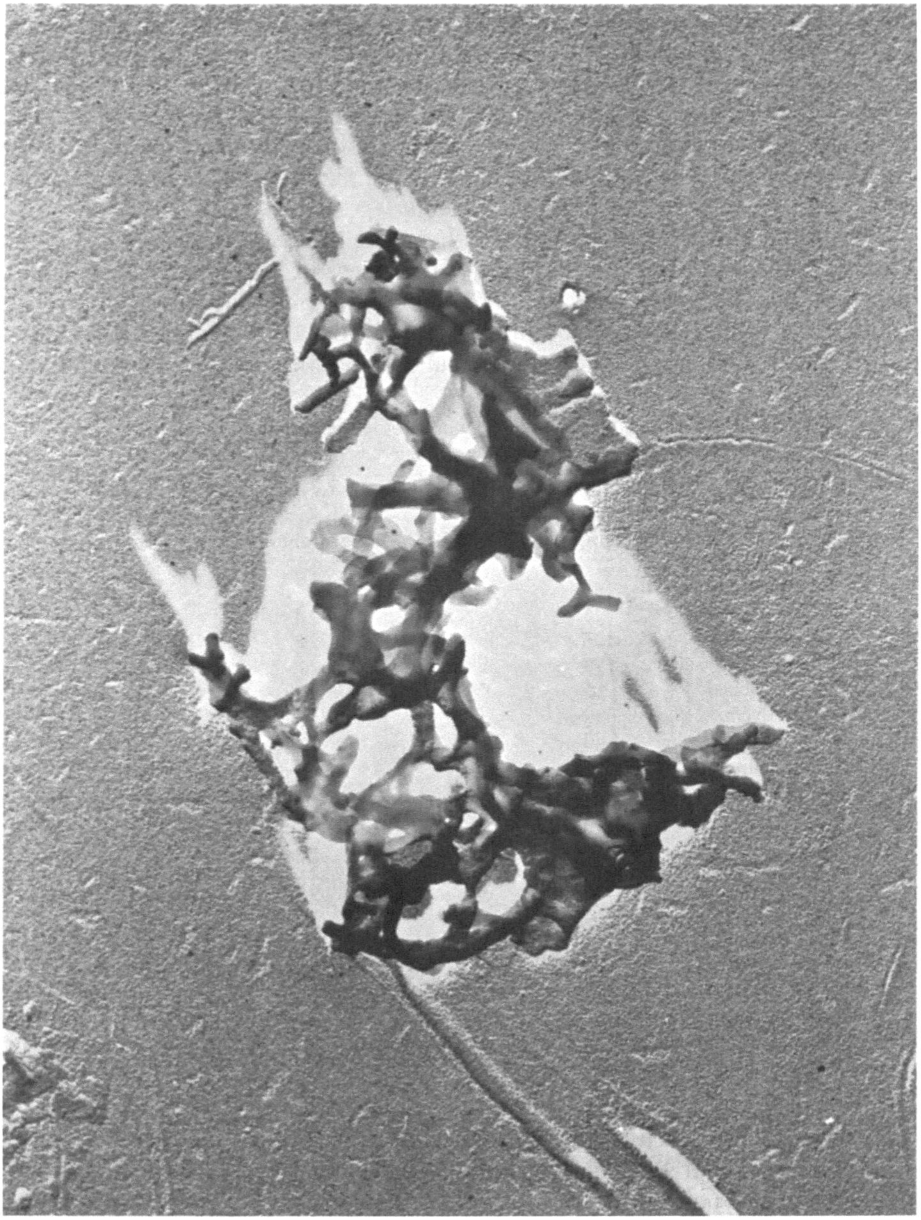

Fig. 13. Fragments of monograptidae and of retiolitidae (Silurian). × 52,000 [10]

The amino acids observed in the hydrolysates of washed and decalcified fragments of graptolites can certainly be considered as originating in remnant fossil proteins belonging to the test of these animals. The three species of graptolites so far examined, show a similar amino acid pattern, characterized by high amounts

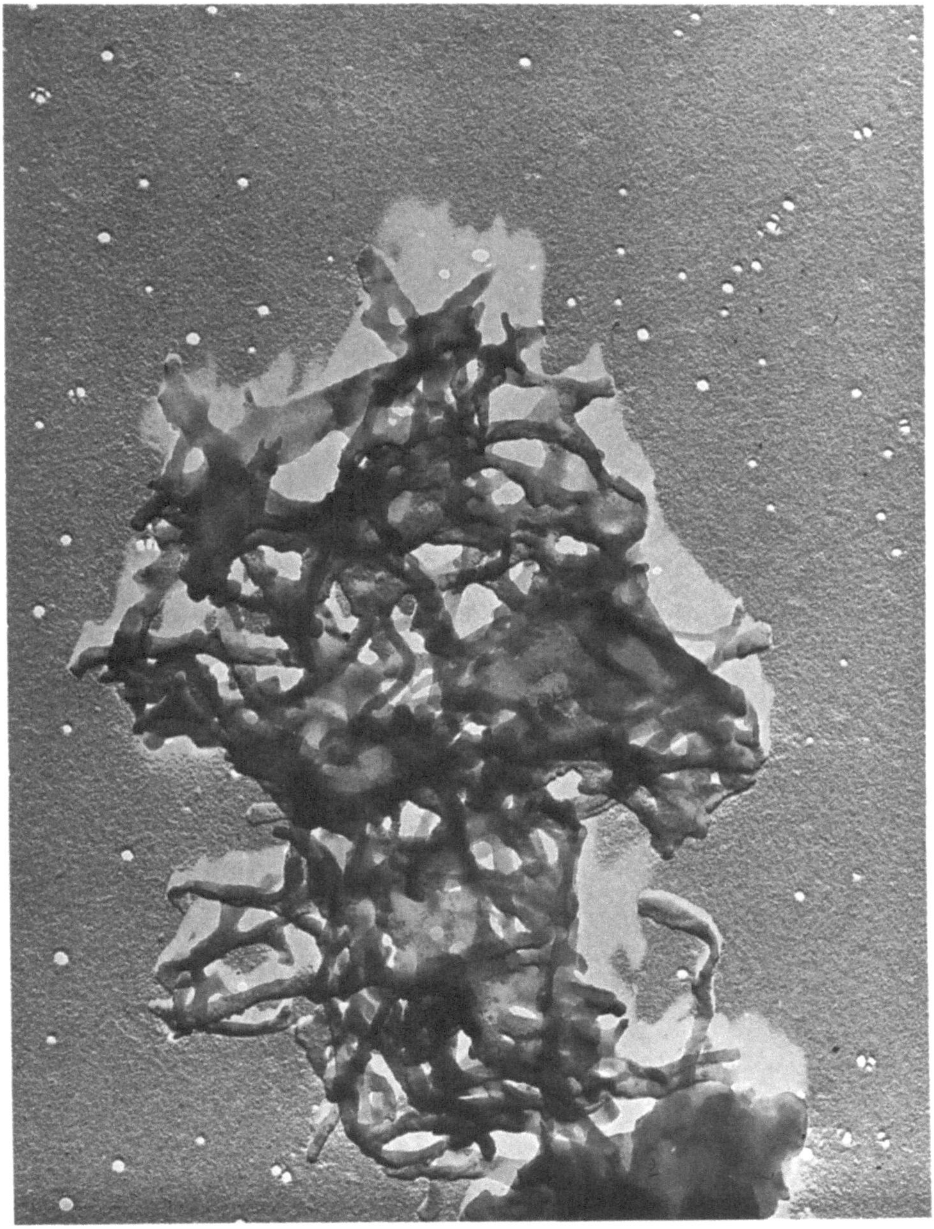

Fig. 14. Fragments of monograptidae and of retiolitidae (Silurian). ×52,000 [10]

of serine (mole-fraction: 10.6 to 22.8), alanine (6.3 to 9.5), glycine (20.1 to 23.4), aspartic acid (8.6 to 10) and glutamic acid (12.8 to 15.3). Such a composition, particularly the high total amount of glycine, serine and alanine, suggests that the graptolitic proteins are of a scleroprotein nature. The study of this material

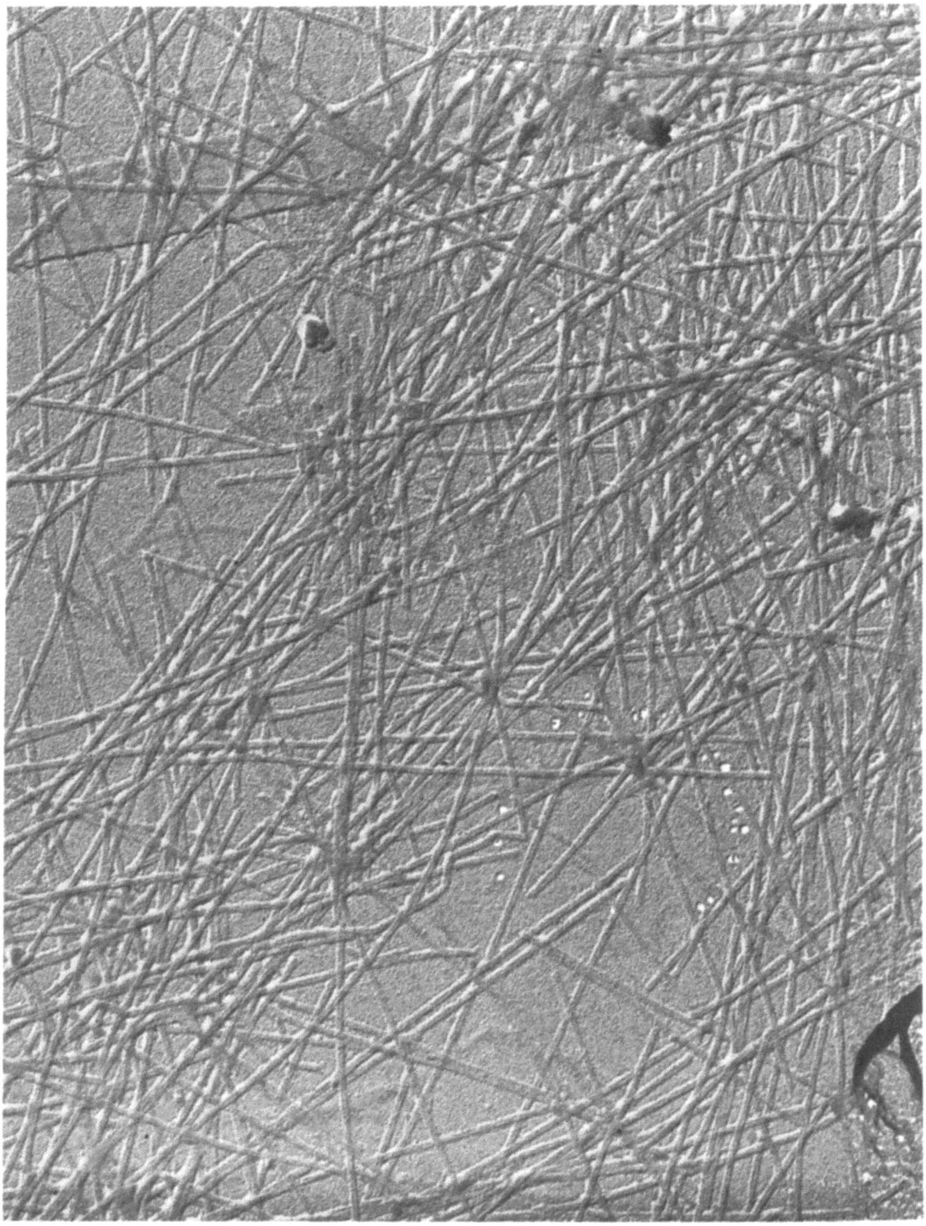

Fig. 15. *Cephalodiscus inaequatus* (ANDERSSON): fragments of coenecium. × 20,500 [10]

under the electron microscope is being continued by Dr. C. GRÉGOIRE in our laboratory.

In contrast to the general belief in the chitinous nature of graptolite tests [30, 31], the three samples examined did not contain any trace of chitin; this polysaccharide was not revealed by the specific enzymatic method of JEUNIAUX [32] and the chromatograms of the hydrolysates did not disclose the presence of glucosamine. Furthermore, the residues of the graptolite samples did not reveal any trace of cellulose. The lack of chitin in graptolites contra-indicates a systematic affinity between Graptolites and the Bryozoa, Hydrozoa or other Cnidaria, all of which are provided with a chitinous test.

The ultrastructures of the protein of Graptolites shown in Figs. 13 and 14 may be described as meshes of bandlike material, the network being more loose or more tight according to the parts considered. The type of ultrastructure is different from that shown in the case of *Cephalodiscus inaequatus* (Fig. 15), a species of a genus belonging to the class of Pterobranchia which has been considered as showing systematic affinities with Graptolites. As well as Graptolites, the organic exoskeleton (or coenecium) of *Cephalodiscus* (a living Pterobranch), is also lacking chitin [33]. Its proteins show a high percentage of glycine [33] as is the case in proteins of Graptolites. On the other hand, the coenecium of *Cephalodiscus* and the tests of Graptolites differ in their ultrastructure.

V. Fossil Bones, Teeth, and Eggs

In electron microscopy, typical collagen fibrils can still be seen in fossil bones and teeth [34–37]. These collagens have been recognized as such by the similar composition of fossil and modern specimens [34–40].

As shown in this laboratory by VOSS-FOUCART [41] the decalcification of Dinosaur egg shells (very likely *Megalosaurus*) liberates two different structures: one composed of white ribbons forming an external and an internal net joined by thin hollow tubes which correspond to the air channels, the other consisting of thin brown sheets superposed in the mass of the shell. These two structures still contain proteins, the ultrastructure of which has been studied with the electron microscope (Figs. 16 and 17). The determination of their amino acid composition has been performed by ion exchange column chromatography after hydrolysis and has shown similarities to the amino acid composition of the shell matrix of hens' eggs.

VI. Diagenetic Change

When comparing paleoproteins with the corresponding native proteins of modern organisms, as in the comparison of nacre conchiolin of modern and fossil molluscan shells [6], the harder protein layer of modern and fossil Brachiopod shells [9], modern and fossil Gastropod shells, modern and fossil prism conchiolins [11] and modern and fossil collagens, differences have repeatedly been observed in the patterns of amino acid composition [38–40, 42]. These differences may of course reflect differences in the pattern of amino acids in the native state of both kinds of proteins. But this would be a rather naive conclusion without a previous critical examination of possible alternative explanations. These include alterations in the structure and composition of the paleoprotein after the death of the animal containing it, during the periods of diagenesis of the sediment concerned and during any metamorphism of the sedimentary rock in which the

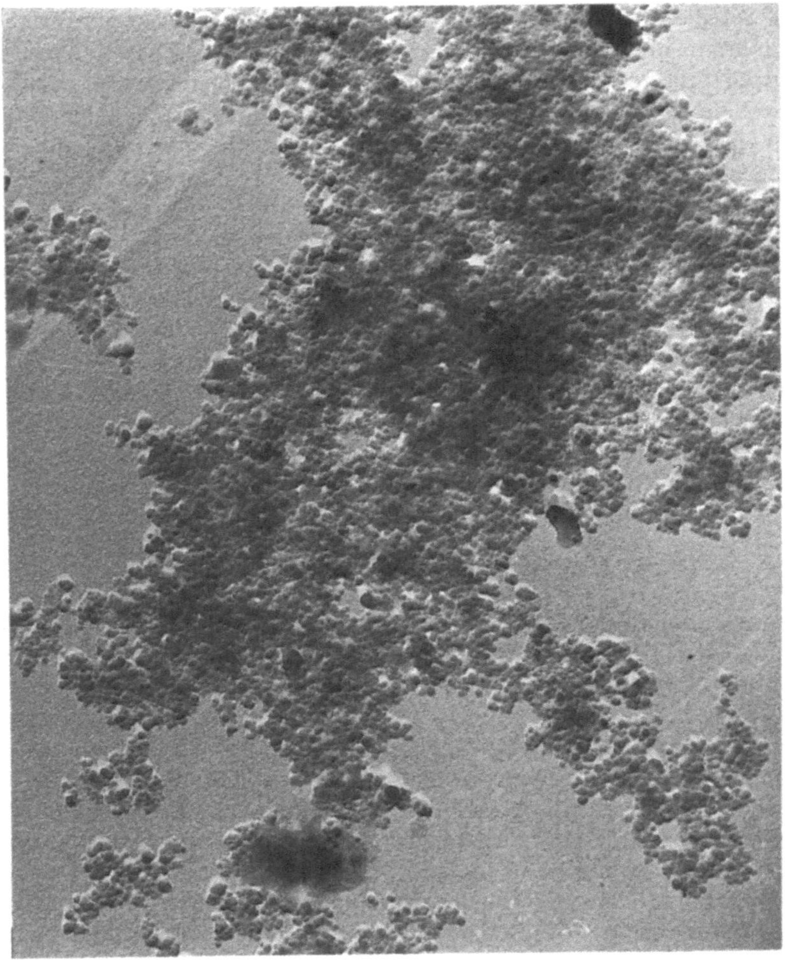

Fig. 16. Fragments of decalcified Dinosaur egg shells (very likely *Megalosaurus*). Corpuscles of the thin stratified sheets. ×42,000 [10]

fossil was imbedded. We may denote the chemical events in the time during which a native protein in a living animal has become a paleoprotein recovered from a fossil by the term *paleization*. This term encompasses post mortem and diagenetic changes and implies no commitment concerning either the nature of possible alterations or their chronology. Paleization may also be accompanied by ultrastructural alterations. As mentioned by GRÉGOIRE [26], many fossil molluscan shells possess "conchiolin" of altered ultrastructure as revealed by the electron microscope. Several kinds of alterations are described by GRÉGOIRE "in the form of discs, lenticular or spheroidal, pebble-shaped bodies, corpuscles in the form of knobs, perforated membranes". GRÉGOIRE has exposed fragments of modern *Nautilus* shells to a few of the factors involved in paleization processes, such as heat and pressure. He has concluded that "heat alone in open air, or associated

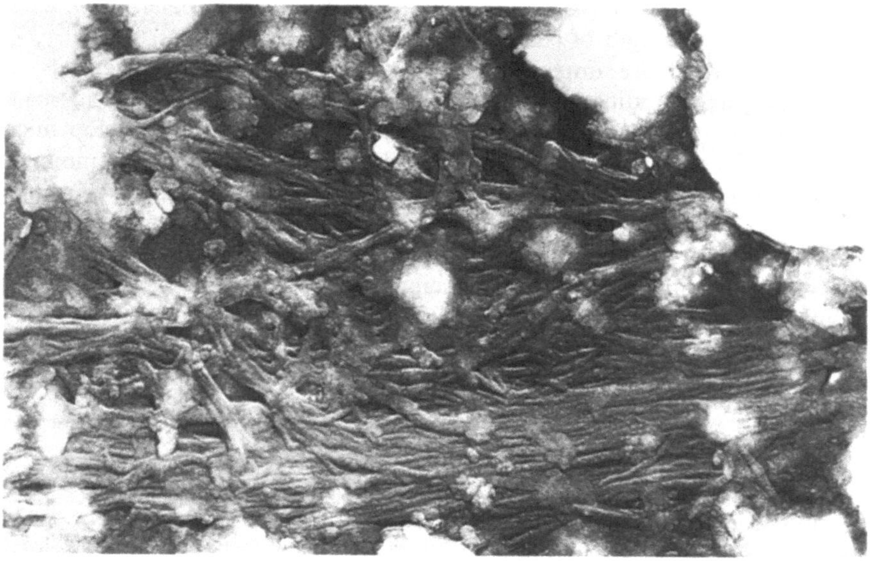

Fig. 17. Fragments of decalcified Dinosaur egg shells (very likely *Megalosaurus*). Fibres of the white ribbons forming the superficial networks and the walls of the air channels (negative ×48,000) [10]

with pressure produced various stages of ultrastructure degradation of conchiolin, similar or identical on the electron microscope scale to those observed consistently in remnants of decalcification of Paleozoic and Mezozoic nautiloid and ammonoid shells" [44].

On the other hand, in comparing the compositions of fossil proteins and the corresponding proteins of living organisms, attention must be paid to possible contamination due to an invasion of the structure by fossil organisms which were contemporary with the organism concerned. It is therefore important to rule out, by thorough control with the electron microscope, the possibility of contamination by epizoans, parasites and especially boring predators [26].

We must also bear in mind that, when dealing with such structures as conchiolins, or other scleroproteins, we are dealing with mixtures of protein species. The pattern of "nacre conchiolin" [6], for instance, is the pattern of a composite mixture separable into several components with different amino acid patterns [14]. In the course of paleization protein fabrics may undergo changes due to the different reactivities of their components, or of alterations resulting from the removal of end segments or side chains of polypeptides. From these changes a modified amino acid composition may result, to which of course no phylogenetic meaning should be conferred.

VII. Phylogenetic Significance

The above pitfalls are inherent in the comparison of paleoproteins and corresponding proteins of living forms. However, according to our present knowledge, the primary structure (i.e., the amino acid sequence), which is the only reliable test of homology [43] is unlikely to be modified in paleization, except with respect to the length of chains.

It appears therefore imperative to study, by experimental methods, the phenomena corresponding to protein paleization. There is a need to isolate definite chemical species from composite paleoproteins and to develop micromethods for the determination of amino acid sequences in the small amounts of material which can be isolated. Our aim must be to define the aspects of primary protein structures (amino acid sequences) which can be compared in homologous protein chains of fossil and modern organisms. Major contributions to our knowledge of the phylogeny of biopolymers should emerge as this area of research develops.

References

1. ABELSON, P. H.: Paleobiochemistry. Organic constituents of fossils. Carnegie Inst. Wash. Yearbook **54**, 107–109 (1955).
2. – Paleobiochemistry. Sci. Am. **195**, 83–92 (1956).
3. – Some aspects of paleobiochemistry. Ann. N.Y. Acad. Sci. **69**, 276–285 (1957).
4. – Organic constituents of fossils. Geol. Soc. Am. Mem. **67**, 87–92 (1957).
5. JONES, J. D., and J. R. VALLENTYNE: Biogeochemistry of organic matter. I. Polypeptides and amino acids in fossils and sediments in relation to geothermometry. Geochim. Cosmochim. Acta **21**, 1–34 (1960).
6. FLORKIN, M., C. GRÉGOIRE, S. BRICTEUX-GRÉGOIRE et E. SCHOFFENIELS: Conchioline de nacres fossiles. Compt. Rend. **252**, 440–442 (1961).
7. FOUCART, M. F., S. BRICTEUX-GRÉGOIRE, C. JEUNIAUX, and M. FLORKIN: Fossil proteins of graptolites. Life Sci. **4**, 467–471 (1965).
8. DEGENS, E. T., and S. LOVE: Comparative studies of amino acids in shell structure of *Gyraulus trochiformis* STAHL. from the Tertiary of Steinheim, Germany. Nature **205**, 876–878 (1965).
9. JOPE, M.: The protein of Brachiopod shell. II. Shell protein from fossil articulates: amino acid composition. Comp. Biochem. Physiol. **20**, 601–605 (1967).
10. GRÉGOIRE, C.: Unpublished observations.
11. BRICTEUX-GRÉGOIRE, S., M. FLORKIN, and C. GRÉGOIRE: Prism conchiolin of modern or fossil molluscan shells. An example of protein paleization. Comp. Biochem. Physiol. **24**, 567–572 (1968).
12. CARLISLE, D. B.: Chitin in a Cambrian fossil *Hyolithellus*. Biochem. J. **90**, 1C–2C (1964).
13. FRÉMY, E.: Recherches chimiques sur les os. Ann. Chim. (Paris) **43**, 96–99 (1955).
14. GRÉGOIRE, C., G. DUCHÂTEAU et M. FLORKIN: La trame protidique des nacres et des perles. Ann. Océanog. (Paris) **31**, 1–36 (1955).
15. GOFFINET, M., and C. JEUNIAUX: Composition chimique de la fraction "nacroine" de la conchioline de nacre de *Nautilus pompilius* LAMARCK. Comp. Biochem. Physiol. **29**, 277–282 (1969).
16. GRÉGOIRE, C.: On submicroscopic structure of the *Nautilus* shell. Inst. Roy. Sci. Natl. Belg. Bull. **38**, 1–71 (1962).
17. – Further studies on structure of the organic components in mother-of-pearl, especially in Pelecypods (Part I). Inst. Roy. Sci. Natl. Belg. Bull. **36**, 1–22 (1960).
18. – Topography of the organic components in mother-of-pearl. J. Biophys. Biochem. Cytol. **3**, 797–808 (1957).
19. WATABE, N.: Studies on shell formation. XI. Crystal matrix relationships in the inner layer of Mollusk shells. J. Ultrastruct. Res. **12**, 351–370 (1965).
20. GRÉGOIRE, C.: Sur la structure, étudiée au microscope électronique des constituants organiques du calcitostractum. Arch. Intern. Physiol. Biochim. **66**, 658–661 (1958).
21. – Sur la structure submicroscopique de la conchioline associée aux prismes des coquilles de mollusques. Inst. Roy. Sci. Natl. Belg. Bull. **37**, 1–34 (1961).
22. – Essai de détection au microscope électronique des dentelles organiques dans les nacres fossiles (ammonites, céphalopodes, gastéropodes et pélécypodes). Arch. Intern. Physiol. Biochim. **66**, 674–676 (1958).

23. GRÉGOIRE, C.: A study on the remains of organic components in fossil mother-of-pearl. Inst. Roy. Sci. Natl. Belg. Bull. **35**, 1–14 (1959).
24. – Conchiolin remnants in mother-of-pearl from fossil Cephalopoda. Nature **184**, 1157–1158 (1959).
25. –, and C. TEICHERT: Conchiolin membranes in shell and cameral deposits of Pennsylvania Cephalopods, Oklahoma. Oklahoma Geol. Notes **25**, 175–201 (1965).
26. – On organic remains in shells of Paleozoic and Mesozoic Cephalopods (Nautiloids and Amonoids). Inst. Roy. Sci. Natl. Belg. Bull. **42**, fasc. 39, 1–36 (1966).
27. GRANDJEAN, J., C. GRÉGOIRE, and A. LUTTS: On the mineral components and the remnants of organic structures in shells of fossil molluscs. Bull. Classe Sci., Acad. Roy. Belg. **50**, 562–595 (1964).
28. GRÉGOIRE, C.: Sur les fourreaux organiques des prismes de *Mytilus edulis* L. Arch. Intern. Physiol. Biochim. **68**, 836–837 (1960).
29. – Structure of the conchiolin cases of the prisms in *Mytilus edulis* LINNÉ. J. Biophys. Biochem. Cytol. **9**, 395–400 (1961).
30. KRAFT, P.: Ontogenetische Entwicklung und Biologie von Diplograptus und Monograptus. Paleont. Z. **7**, 207–249 (1926).
31. MANSHAYA, S. M., and T. V. DROZDOVA: Transformation of organic compounds in sedimentary rocks and the organic matter of graptolites of dictionemae shales. Geokhimiya **11**, 952–962 (1962).
32. JEUNIAUX, C.: Chitine et chitinolyse. Un chapitre de biologie moléculaire. Paris: Masson & Cie. 1963.
33. FOUCART, M. F., et C. JEUNIAUX: Paléobiochimie et position systématique des Graptolithes. Ann. Soc. Roy. Zool. Belg. **95**, 39–45 (1965).
34. WYCKOFF, R. W. G., E. WAGNER, P. MATTER, III, and A. R. DOBERENZ: Collagen in fossil bone. Proc. Natl. Acad. Sci. U.S. **50**, 215–218 (1963).
35. SHACKLEFORD, J. M., and R. W. G. WYCKOFF: Collagen in fossil teeth and bones. J. Ultrastruct. Res. **11**, 173–180 (1964).
36. WYCKHOFF, R. W. G., and A. R. DOBERENZ: The electron microscopy of Rancho La Brea bone. Proc. Natl. Acad. Sci. U.S. **53**, 230–233 (1965).
37. – – Le collagène dans les dents Pleistocènes. J. Microscop. **4**, 271–274 (1965).
38. WYCKOFF, R. W. G., W. F. MCCAUGHEY, and A. R. DOBERENZ: The amino acid composition of proteins from Pleistocene bones. Biochim. Biophys. Acta **93**, 374–377 (1964).
39. HO, T. Y.: The amino acid composition of bone and tooth proteins in Late Pleistocene mammals. Proc. Natl. Acad. Sci. U.S. **54**, 26–31 (1965).
40. – The isolation and amino acid composition of the bone collagen in Pleistocene mammals. Comp. Biochem. Physiol. **18**, 353–358 (1966).
41. VOSS-FOUCART, M.-F.: Paléoprotéines des coquilles fossiles d'œufs de Dinosauriens du Crétacé supérieur de Provence. Comp. Biochem. Physiol. **24**, 31–36 (1968).
42. ARMSTRONG, W. G., and L. B. H. TARLO: Amino-acid components in fossil calcified tissues. Nature **210**, 481–482 (1966).
43. FLORKIN, M.: A molecular approach to phylogeny. Amsterdam: Elsevier Publ. Co. 1966.
44. GRÉGOIRE, C.: Thermal changes in the *Nautilus* shell. Nature **203**, 868–869 (1964).

Addendum

Since the completion of the manuscript, new studies on the effect of thermal changes on modern shell proteins have been performed in the author's laboratory. GRÉGOIRE [b] has uniformly detected biuret-positive shreds among the residual matrices of the pyrolysed modern samples, including those heated to 900° C, as was the case for fossil material of various ages (FLORKIN *et al.* [6]; GRÉGOIRE [a]). The fact that polypeptide assemblages were still present in the samples heated to 900° illustrate the remarkable resistance of conchiolin to pyrolysis. This has been tentatively explained (GRÉGOIRE [b]) by a kind of stabilization caused by the surrounding mineral matter. According to TOWE and HAMILTON [c], the intracrystalline matrices are deformation artefacts produced by ultramicrotomy. As already pointed out by FLORKIN *et al.* [6], AKIYAMA [d] and MITTERER [e] have observed a reduction in the amount of proteins preserved, proportional to the age of the fossil studied.

Concerning fossil collagen, PAWLICKI *et al.* [f] have detected, using electron microscopy, the typical striation pattern of collagen in the walls of blood vessels in fossil Dinosaur bone. DOBERENZ and WYCKOFF [g] have made similar observations on Pleistocene bones.

A study of amino acids of bone and dentine collagen in Pleistocene mammals has been accomplished by HO [h]. He has observed that the fossil collagen studied generally contains less Leu, Phe and Tyr amino nitrogen and more Gly amino nitrogen. In Dinosaur bones (Cretaceous and Jurassic) MILLER and WYCKOFF [i] have found amino-acid proportions differing from those of Pleistocene bones.

References to Addendum

[a] GRÉGOIRE, C.: Experimental diagenesis of the *Nautilus* shell. Third Intern. Meeting on Organic Geochemistry, London, 26–28 Sept. 421–433 (1966).

[b] – Experimental alteration of the *Nautilus* shell by factors involved in diagenesis and metamorphism. Part I. Thermal changes in conchiolin matrix of mother-of-pearl. Bull. Inst. Roy. Sci. Nat. Belg. **44** (25) 69 pp. (1968).

[c] TOWE, K. M., and G. H. HAMILTON: Ultramicrotome-induced deformation artifacts in densely calcified material. J. Ultrastruct. Res. **22**, 274–281 (1968).

[d] AKIYAMA, M.: Quantitative analysis of the amino acids included in Japanese fossil scallop shells. J. Geol. Soc. Japan **70**, 508–516 (1964).

[e] MITTERER, R. M.: Amino acid and protein geochemistry in Mollusks shells. Florida State Univ. Ph. D. Geology (1966).

[f] PAWLICKI, R., A. KORBEL, and H. KUBIAK: Cells, collagen fibrils and vessels in Dinosaur bone. Nature **211**, 655–657 (1966).

[g] DOBERENZ, A. R., and R. W. G. WYCKOFF: Fine structure in fossil collagen. Proc. Nat. Acad. Sci. U.S. **57**, 539–541 (1967).

[h] HO, T. Y.: The amino acids of bone and dentine collagen in Pleistocene mammals. Biochim. Biophys. Acta **133**, 568–573 (1967).

[i] MILLER, M. F. II, and R. W. G. WYCKOFF: Proteins in Dinosaur bones. Proc. Nat. Acad. Sci. U.S. **60**, 176–178 (1968).

CHAPTER 21

Organic Compounds in the Gas-Inclusions of Fluorspars and Feldspars

REIMAR KRANZ

Institut für Physikalische Chemie der Kernforschungsanlage Jülich GmbH, Jülich, West Germany

Contents

I. Introduction

The remnants of a former crystallization have been shown to remain trapped in the gas and liquid inclusions of minerals. It is therefore understandable that these inclusions have aroused increasing interest. One hopes, from the analysis of these inclusions, to gain new insight into the chemistry of the gases and solutions which are present during mineral formation, and to gain new knowledge of fundamental significance for the investigation of the conditions of formation of minerals and rocks.

From Recent investigations, especially those of WAHLER [1] and GOGUEL [2], we know the main components of the liquid and gas inclusions of the minerals of some granites and pegmatites. Some disagreements about the presence of hydrogen and carbon dioxide have also been cleared up. However, from these investigations the question of trace impurities arose, especially with regard to the distribution of organic compounds in the gas contents of this type of inclusion. Whereas some statements can be made about the methane content of these minerals, the occurrence of higher molecular weight hydrocarbons or of their derivatives must be considered improbable on the basis of the assumed "magmatic" origin and of the high temperature stress present at the genesis of these minerals.

The analysis of the gases driven out by laboratory pyrolysis of the mineral and rock samples has been carried out, up to now, mainly by pressure–volume methods through fractional freezing-out, absorption, and combustion. Using these methods, GOGUEL reports the limit of sensitivity for hydrocarbons in the presence of carbon monoxide to be a few mm^3 STP/gram. He was only able to find traces of volatile organic compounds. For the first time, JEFFERY and KIPPING [3] found some low molecular weight saturated and unsaturated hydrocarbons in inclusions. They used pyro-gas chromatography and at the time, these measurements were difficult to reproduce. The authors assumed that the hydrocarbons were formed (as is carbon monoxide) during the pyrolysis of the material, by a reduction of carbon dioxide to carbon monoxide by iron, and from the further catalytic reduction of the CO by hydrogen. The authors asserted that, except for these compounds, it was impossible to extract other organic material from the rock.

The mass spectrometer is a very sensitive instrument for determining gases down to a partial pressure of about 10^{-9} Torr, even in the presence of a great excess of inorganic gases. Thus with this method one should be able to establish to what extent hydrocarbons are involved in filling the gas inclusions. Fluorite was chosen as the first substance for investigation; its deep-violet, generally uraniferous varieties give off sharp-smelling gases when they are struck. This unpleasant smell is generally ascribed to free halogen, but conclusive evidence for this has not been presented. The second substance studied was feldspar. Gas inclusions in feldspars (especially potassium-feldspars, some of which are called "boiling" feldspars) have given rise to the disagreeable property of evolving large amounts of nitrogen during the process of ceramic manufacture. Numerous gas bubbles appear in the ceramic glazing and lead to production losses.

These investigations [4, 5] are limited to minerals of the Oberpfalz district (about 60–100 km north of Regensburg, West Germany). For comparison, some light-colored fluorites of the Swiss Alps, some potassium-feldspars of Norway and South Africa and, as a genetic counterpart, an extraordinarily fine-grained dark-violet fluorite sample which was removed from silicified wood were investigated. Microscopic examination of the material investigated showed inclusions of the order of magnitude of 10 microns, with isolated inclusions of up to 100 microns being found. The degree of filling is, as far as is discernable, quite variable and goes up to about 75 percent. With the deep-violet fluorite crystals, the so-called "stinkspate", the microscopic examination was made very difficult as a result of the almost completely opaque character of the samples. The inclusions, often found oriented in growth or color zones, could be observed only in very thin sections. It is assumed that the gas inclusions in these "stinkspate" are significantly larger and/or more numerous than those in the light fluorite varieties, since larger gas quantities are released from a small piece upon striking.

The preparation of the samples was carried out in the same manner as described earlier [2, 4]. Thoroughly cleaned granules of the sample were ground up under vacuum, the extracted gases were fractionally frozen out and the individual fractions were separately pumped into the storage container of a 60°-single-focusing mass-spectrometer.

II. Analytical Results

In Fig. 1 the individual spectra of the fractionated products of a "stinkspate" sample are summarized. The ionization, obtained by electron impact, causes, as a result of the severe fragmentation of gaseous organic molecules, extraordinarily

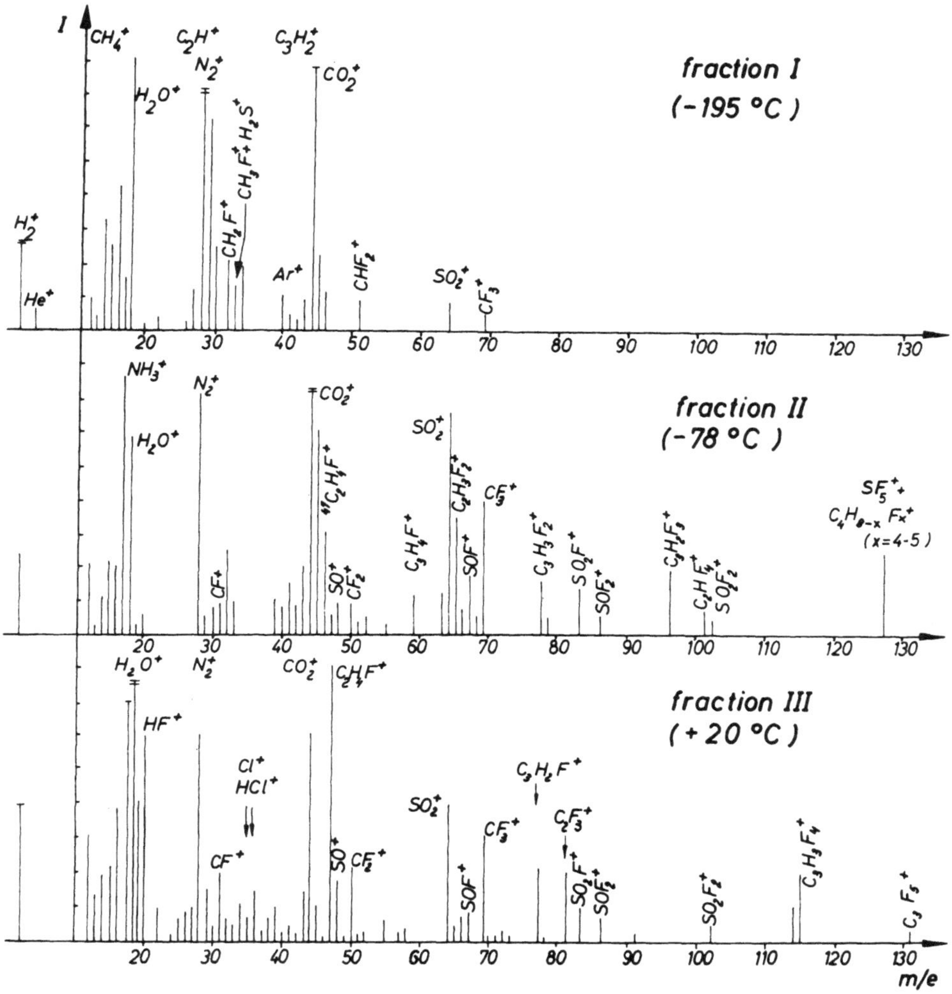

Fig. 1. Mass spectra of the gases extracted from a "stinkspate"; gas fractionation by freezing-out; Fluorite, Johannesschacht (T = ion current × 10, ‡ = ion current × 100). (: × 10 means signal reduced by this factor)

complex spectra. The evaluation and satisfactory assignment of such spectra are difficult, especially in the ppm region, and for trace impurities only the most abundant fragment-ions can be identified. Rough assignments according to compound type are possible. Only tentative structures can be advanced for individual components, since exact mass determinations are not easily carried out with a single-focusing instrument. A final identification was often possible only by a combination of gas chromatography and mass spectrometry.

The mass spectrum of the first fraction is still relatively simple to examine. Nitrogen is present in large quantities, in addition to hydrogen and the noble gases helium and argon, as are the unfrozen remnants of water, carbon dioxide, and sulfur dioxide. Of the hydrocarbons, methane, ethane and propane are present.

The presence of ethylene and perhaps also propylene can be surmised from gas chromatographic investigations. In addition, the presence of mono-, di-, and tri-fluoromethane is demonstrated through the ions of 33, 34, 51, and 69 amu*. Traces of hydrogen sulfide are indicated from the ratio of the ion currents of masses 33 and 34, which would be equally intense for fluoromethane (CH_3F) on the instrument used.

In the spectrum of the second fraction, more highly fluorinated and higher molecular weight compounds appear. The distinct peaks at mass numbers 69 (CF_3^+), 50 (CF_2^+), and 31 (CF^+) indicate that a large number of fluorinated ions are distributed in this spectrum. In addition to the strong content of thionyl fluoride and sulfuryl fluoride in this fraction, the ions of masses 69 and 127 can be assigned. Under the assumption that these peaks are due to an aliphatic compound composed solely of C, H and F, one can calculate an ion $C_4H_3F_4^+$. An unsaturated compound with so high a degree of fluorination seems unlikely. It must therefore be assumed that this olefin-ion is first stabilized by additional splitting off of HF at the time of ionization. Since volatile amines and ammonia are also present in the contents of the gas inclusions, a further possible interpretation of the mass 127 is as the nitrogen-containing ion $C_3NHF_4^+$. With a high resolution scan this peak is resolved into a doublet, so that a final decision cannot, for the time being, be made as to which of the previously mentioned fragments we are dealing with. Ions of this mass number are also preferentially formed from fluoroalkylbenzenes ($C_7H_5F_2^+$), but these compounds would be expected in the higher-boiling gas fraction. Again SF_6 would form a strong ion current of this mass number, and one of the two peaks probably belongs to an ion SF_5^+, as a preliminary splitting of the gas mixture in a gas chromatographic column indicates.

In the third fraction, further fluorocompounds occur. They may include more highly fluorinated species, as for example the ion $C_3F_5^+$. The remaining fluorinated ions are certainly fragments of heavier molecules. The ions of fluorine and hydrogen fluoride also appear in this fraction. Under increased resolving power a large number of double- and triple peaks at single mass numbers appear and unambiguous assignments are impossible.

In Fig. 2, the mass spectrum of the gas evolved from a light green fluorite is presented for comparative purposes. In addition to the known inorganic compounds, the main components are hydrocarbons accompanied by smaller amounts of the fluoro-compounds. The content of ammonia is relatively high, as is also, contrary to all expectation, that of free fluorine and hydrogen fluoride.

The main components of the gases were quantitatively determined and the following amounts found (in mm^3 STP/g): H_2, 7–48; He, 0.1–6.0 (in dark uraniferous minerals only); N_2, 12–63; Ar, 0.05–0.3; CH_4, 1–10; H_2S, 0.1–4.0; CO_2, 7–83; and a very large amount of water vapor (a few ml). The dark "stinkspates" always contain larger quantities of gas than do the light colored fluorites. Using the methane content as reference, the volume percent of the remaining trace compounds can be roughly estimated, assuming their total concentration is of the same order of magnitude, at about 0.1 to 1.0 percent. Of this total, at least half is due to the compounds C_2H_6, C_3H_8, SO_2, SOF_2, SO_2F_2, and NH_3, so

* 1 amu = 1/12 of the mass of a neutral ^{12}C atom, 1 amu = 1.000317917 ME (^{16}O = 16.000) = 1.66×10^{-27} kg = 0.931441 MeV.

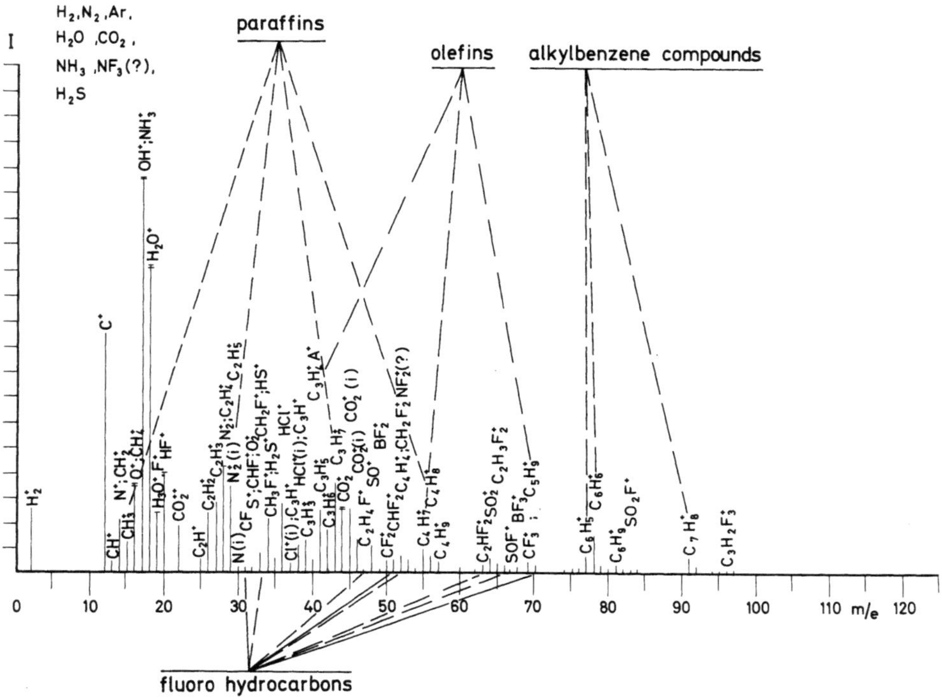

Fig. 2. Mass spectrum of the gas extracted from a green fluorite sample (ERIKA MINE)

that each of the remaining volatile organic compounds was present in the order of 10 to 100 vpm, up to a maximum of 1,000 vpm*. According to a type-analysis of light hydrocarbons [6], as is frequently applied in the petroleum industry, the ions noted in the table are typical for hydrocarbons. A similar pattern for the

Table. *Type analysis of light hydrocarbons*

Compound-type	Characteristic ions	Mass numbers
Paraffins	$C_nH_{2n+1}^+$	43, 57, 71, etc.
Cycloparaffins and monoolefins	$C_nH_{2n-1}^+$	41, 55, etc.
Cycloolefins, Diolefins, Acetylene } "coda" group	$C_nH_{2n+3}^+$, C_nH_{2n-2}	67, 68, 81, 82, etc.
Alkylbenzene compounds	$C_6H_5^+$, $C_6H_6^+$, $C_6H_5CH_2^+$, $C_6H_5CH_3^+$ etc.	77, 78, 91, 92, etc.

fluoro-carbons can be set up qualitatively, taking into consideration the frequently occuring ions which are found in these spectra: CF^+ (31), CF_2^+ (50), CF_3^+ (69), $C_2F_3^+$ (81), $C_3F_5^+$ (131), CH_2F^+ (33), $C_2H_4F^+$ (47), CHF_2^+ (51), $C_2H_3F_2^+$ (65), $C_3H_2F_3^+$ (95), and $C_4H_3F_4^+$ (?) (127).

* vpm = volume-parts per million.

By summation of the ion currents of the individual groups, one obtains dimensionless numbers. These values, although undoubtedly qualitative, can be compared with one other, since the same compound types are always present in the gases trapped in fluorites. The dark violet fluorspars are uranium-bearing as evinced by the measurable helium content. A plot of the helium concentration as a function of the radioactivity of the sample, is a straight line (Fig. 3). Similar plots of the ion currents of the CF_3^+ ion turn out to be exactly parallel with this

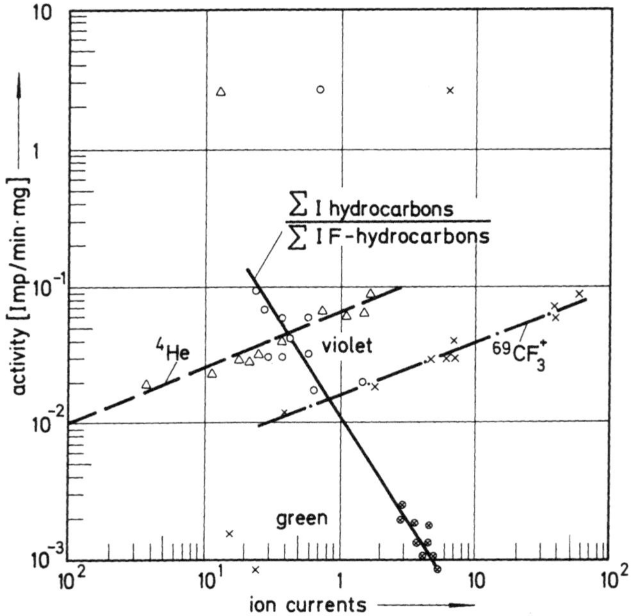

Fig. 3. Gas composition as a function of the radioactivity of the fluorite samples (abscissa in arbitrary units). (× = violet-colored fluorite, ○ = light-colored fluorite)

line. If one calculates the ratios Σ I (hydrocarbons)/Σ I (fluoro-hydrocarbons) for the analysis described, the higher fraction of fluorinated compounds in the violet, uranium-bearing fluorites is again recognizable. The marked scatter of measured values is understandable when one considers that, in nature, in addition to pressure and temperature, many other factors, such as the solvent partner, play a role. The value for one sample deviated extraordinarily from the usual values. In the thin section of this sample, clusters of pyrite-pitchblende were clearly recognizable; the uranium of this sample was thus not homogeneously distributed in the fluorite lattice.

According to TOORKS [7] a very strongly fluorescing, uraniferous synthetic calcium fluoride is obtained when one heats CaF_2 with 6 to 25 mole percent CaO and 0.004 to 0.5 mole percent UO_2. KRÖGER [8] holds that a solid solution of $CaUO_4$ in CaF_2 or a diadochie (Ca_2F_4) $(CaUO_4)$ is present. RECKER [9] reports the same thing for syntheses effected without the addition of CaO, but in an oxidizing atmosphere; in a reducing atmosphere, on the other hand, a replacement of parts of the CaF_2 lattice by UO_2 groups is conceivable. In both cases the uranium ions are incorporated into the fluorite structure in place of Ca^{2+}, and certainly simple individual $(CaF_8)^{-6}$ groups may be replaced by $(UO_4F_4)^{-6}$ ions (in an oxidizing atmosphere) or by $(UO_2F_6)^{-6}$ ions (in a reducing atmosphere); in either case the correct valencies are maintained [10]. In Fig. 4 the same relationships are represented in a triangular diagram. Here also all the measured points lie approximately on a straight line between the apex "fluorinated hydrocarbons" and the side connecting "paraffins" and "olefins".

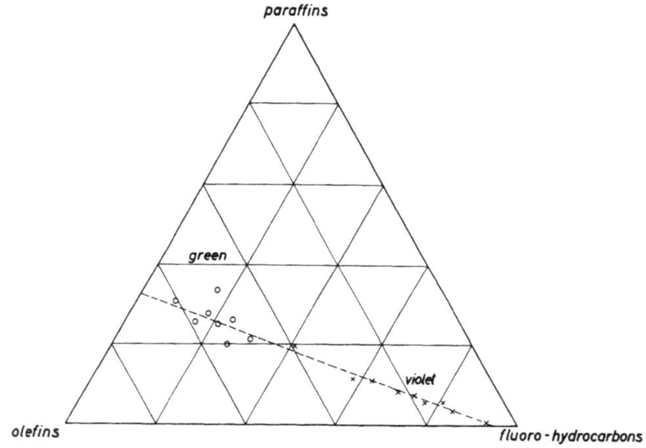

Fig. 4. Results of the compound-type analysis (× = violet colored fluorite, ○ = light-colored fluorite)

The mass-spectra of the gases extracted from feldspars look quite different (Fig. 5). The following volatile compounds could be detected in the gas inclusions:

nitrogen	carbon dioxide	methane	methylamine
hydrogen	carbon monoxide	ethane	dimethylamine
helium	water	propane	ethylamine
argon	ammonia	n-butane	fluoromethane (CH_3F)
		iso-butane	
		ethylene	

Furthermore, correlations could be observed between the extractable gases and the technological properties (in spite of good or bad melting properties):

1. Technologically good and glassy smooth-melting feldspars have a low gas content; the very dark potassium feldspars, which include an appreciable quantity of hydrocarbons, are exceptions.

2. Potassium feldspars contain, beside hydrocarbons, low molecular weight amines and ammonia, which are present to a marked degree in the so-called "boiling" feldspars.

3. The appearance of low molecular weight amines is dependent on the content of uranium, radium and thorium in the minerals.

In discussing the chemical state of nitrogen and of nitrogen compounds in silicate minerals, it is to be emphasized that a very large proportion of the nitrogen is fixed in the lattice in the form of ammonium ions [11]. One can study this occurrence in relation to ion exchange. The mineral can be regarded as a huge, well-organized polymer of which only the outer surfaces are initially accessible and in equilibrium with an aqueous solution. Some years ago MARSHALL [12] demonstrated the strong fixation of ammonium ions in the feldspar lattice.

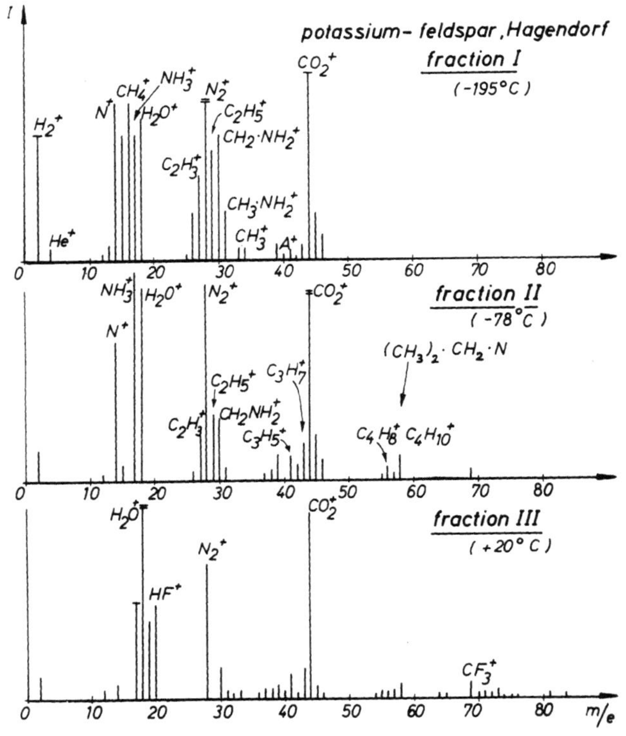

Fig. 5. Mass spectra of the gases extracted from a sample of potassium feldspar (HAGENDORF-MINE)

Investigations by the present author [5] lead to the conclusion that ion exchange is greatly increased by low concentrations of amines. The oxygen-silicon bonds that hold the structure together are very much weaker than the carbon-carbon bonds of organic polymers. Hence a certain low concentration of silicic acid exists in solution in equilibrium with the silica framework. In this way reactions of ammonia with the silica framework become possible. The nitrogen inclusions so formed are responsible for the release of nitrogen during the ceramic combustion which is brought about at high temperatures. High pressure investigations confirmed this opinion. The presence of low molecular weight amines, amino acids or nucleic acids resulted in the introduction of nitrogen into the feldspar lattice. Little can be said about the mechanism of the reaction, but it is to be expected that nitrogen enters the lattice by forming Si – N – Si boundaries or $Si-NH_2$ compounds like water does at distorted Si – O – Si boundaries.

III. Discussion

It is impossible at this stage in the investigation, to give an answer to the question of the ultimate origin of this much altered organic material. That the material was strongly altered by radiochemical reactions during the course of geologic time (the age of the Wölsendorf formation is about 185 million years) is proved by the results of the investigation. Through the friendly cooperation of Professor Dr. F. K. DRESCHER-KADEN, I obtained a sample of silicified wood, the core of which was filled up with very fine-grained violet fluorite. This "wood" contains the same gas inclusions as the other "stinkspate" samples, especially those characterized by the mass numbers 69 and 127 and also by sulfuryl and thionyl fluoride.

These results support the hypothesis that the substances found are the decomposition products of high molecular weight substances. Volatile amines and ammonia which are typical decomposition products of biogenic material are present. A large portion of the bound nitrogen is split off as amino acids by hydrolysis of the "humic acids", formed by bacterial decomposition of cellulose and lignin. Progressive decomposition of these substances, in conjunction with radiochemical reactions and partial fluorination, would lead inevitably to the above compounds. This supposition is also in agreement with the already-observed high hydrogen content, hydrogen being produced by radiochemically induced dehydrogenation of long chain aliphatic compounds.

If one accepts the biogenic origin of the organic material, then one could discuss the influence of these compounds on the solubility and recrystallization of mineral material. Plants can assimilate with great ease slightly soluble or, under the conditions prevailing in the soil, nearly insoluble inorganic substances, which then recrystallize as incrustations on cell walls or within the cells. A related phenomenon is reflected by silicified wood, which clearly shows the replacement of plant-like cells by silica, but with retention of the wood's structure. Only in the inner core has the fiber structure of the wood disappeared and made room for the finely crystalline fluorite. Many examples of this type are known in botany, and all lead to the recognition that solution mechanisms must play a role in plant material balance, a role for which the aqueous solubility of the various minerals alone does not offer a satisfactory explanation. For this reason, the chemical composition of hydrothermal solutions is spiritedly discussed.

NEUBERG, MANDL and GRAUER [13, 14], as well as EVANS [15, 16] found a considerable increase in the solubility of various rock-forming minerals and ores in very dilute aqueous solutions of amino acids, nucleic acids, and the so-called "humic acids". These results were confirmed by the author's investigations on feldspars [5]. SZALAY [17] found a preferential adsorption of uranyl and thorium ions, as well as those of the rare earths, on such humic acids, and thereby explained the high uranium content of biogenic sediments. Enrichment factors of up to 10,000:1 were found. Some model experiments prove the high solubility of mineral matter in very dilute solutions of amines, amino-acids and nucleic-acids [18]. The participation of high molecular weight, apparently biogenic, organic compounds in the transport processes of inorganic materials in the upper crust of the earth is consequently quite conceivable. The influence of such transport reactions on the distribution of radioactive nuclides of the uranium and thorium series and on age determination by lead-lead and uranium-lead methods should be mentioned.

One can object to the proposed biogenic origin of these organic compounds in the gas inclusions on the grounds that large quantities of ammonia are also present, in addition to hydrocarbons, in the gases extractable from light fluorites. It is generally assumed that the nitrogen in minerals is present as NH_4^+ ions [11]. Further significant portions are present as trapped elemental nitrogen. In silicate minerals, the additional possibility of the inclusion of nitrogen in nitride form exists, as MULFINGER and MEYER [19] have shown in glasses. The introduction of nitrogen in feldspars has been shown in this investigation. VINOGRADOV and co-workers [20] have succeeded in proving that an ammonia synthesis is quite possible in nature; this was done by the circulation of a stoichiometric gas mixture over various stone and mineral powders at 350° C, in the presence of water vapor.

Without going into the sharply disputed question concerning the origin of petroleum, one must mention here that the possibility of a non-biogenic synthesis of relatively simple organic compounds by means of a radiochemically induced polymerization must be considered. Thucolite is cited as a specific example of one such high molecular weight organic substance. "Thucolite" is an asphaltic material of pneumatolytic-hydrothermal origin which contains, in addition to the predominant hydrocarbons, U, Th, rare earths, SiO_2 and H_2O. The material occurs commonly as cubic crystals which are, no doubt, pseudomorphs of uranium pitchbende and whose genesis is completely unexplained at the present time. Radiochemical reactions probably play a prominent role in the formation of these compounds. In the case of feldspars, a direct relationship could be detected between the content of organic compounds, especially of amino compounds, and the uranium and thorium content. Furthermore, in a recent investigation [18], amino acids (glycine, alanine and aminobutyric acid) were isolated from highly uraniferous samples of fluorite, while in the light-colored, uranium-free specimens none of these compounds could be detected. These results make plausible the hypothesis that the substances found are radiochemically derived from the highly volatile hydrocarbons, ammonia and water trapped inside the mineral; even amino acids could be formed in this way during geological time. The occurrence of amino acids in rock-forming minerals is therefore a dubious proof of the previous existence of life, since an abiogenic, radiochemical origin of these compounds is conceivable. Further investigations which will certainly raise many more questions in this direction, must be carried out. At any rate, systematic work in this area will yield answers to many previously unsolved problems on the genesis of minerals and mineral deposits.

A hypothesis advanced in 1863 by SCHÖNBEIN and in 1866 by WYROUBOFF [20], according to which the coloration of fluorites was caused by organic substances, has been out of favor since HEINRICH [21] recognized the connection between the uranium content of the WÖLSENDORF deposits and the color and odor of the "stinkspate". The present investigation shows that SCHÖNBEIN'S hypothesis deserves renewed consideration.

Light-colored fluorite becomes blue upon gamma- or X-ray-irradiation. With short irradiations, samples become colored in zones, but after long irradiation, they took on a homogeneous coloration. Careful preparation of these zones followed by individual investigation of their gas contents, showed that the zones which became colored early in the irradiation were those of the highest organic-gas content. One specimen of colorless fluorite became blue-colored in a single layer, when irradiated with 1×10^5 R of 200 kV X-rays; the rest of the crystal remained colorless, even after a long irradiation with more than 5×10^6 R. These colorless regions were found to contain no measurable quantities of organic inclusions.

IV. Experimental Procedures

The preparation of the samples was carried out mainly in the same manner as that described by GOGUEL [2]. The fraction of the sample-containing granules of 1–3 mm diameter was washed with freshly-distilled benzene, double-distilled water, and methanol, freed of adsorbed gases and solvent remnants under high vacuum ($p = 1.10^{-6}$ Torr), and finally ground up under vacuum in a steel grinding-container with two Widia balls, held on a vibratory mill. The grinding of the sample took place at room temperature. Grinding at low temperatures in order to minimize possible thermal reactions was not necessary.

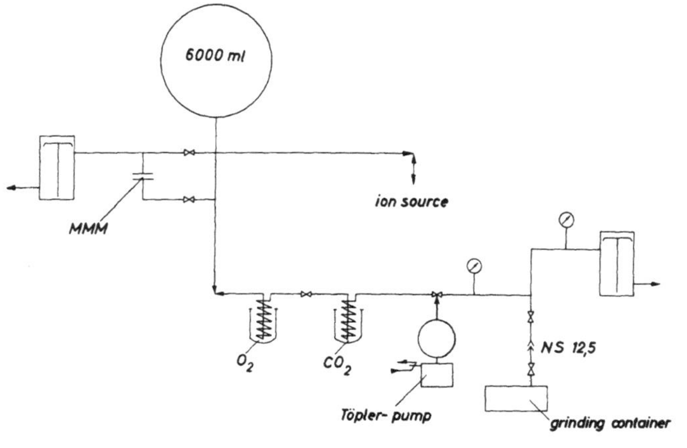

Fig. 6. Gas inlet section of the mass spectrometer. (MMM = micro-membrane-manometer, NS = standard taper joint)

The extracted gases were pumped, with the help of a Toepler pump, into the inlet system of the mass spectrometer, fractionally frozen out with liquid nitrogen and dry-ice, and the individual fractions separately let into the storage container of the instrument. The occurrence of a higher boiling compound in a lower boiling fraction is minimized for trace impurities, as a result of the final vapor pressure of the substance in question and the large working volume of the gas inlet section (Fig. 6).

The mass spectrometer used was a single-focusing 60° Atlas-CH 4 Spectrometer.

Measurement Conditions

Ion source "AN_4" for gas-inlet; ionization by electron impact (70 eV electrons); electron current, 60 μA; slit width, ion source, 0.1 mm; ion detector, 0.3 mm; mass resolution, $A_{(1\,percent)} = 800$; ion detector, secondary electron multiplier at 1.6 kV and 1×10^6 ohms resistance, corresponding to a multiplication factor of 2×10^6. For the more accurate identification and separation of ions of the "same" mass number (m/e ratio), but of different elemental composition, it was necessary, in the course of further experiments, to increase the resolving power of the mass spectrometer considerably. By using an entry-slit width of 0.01 mm and an exit-slit width of 0.02 mm and optimal ion focusing, a resolving power of $A_{(1\,percent)} =$ 1,800 could be attained. This means that in the intermediate mass region at roughly mass number 50, two masses with the mass difference $M = 0.025$ amu are still clearly separated, and with $M = 0.01$ amu are still recognizable as separate masses. The sharp decrease in detection sensitivity was largely compensated by increasing the amplification factor.

Further experiments were carried out with a combined gas chromatograph-mass spectrometer. Coupling of gas chromatography and mass spectrometry is extremely useful in the study of natural products where one may encounter complex mixtures which contain trace components that are difficult to isolate and to identify. Packed columns with various cross-sections were tested; a column diameter of 2.5 mm was chosen as optimal and a helium flow rate of 8 ml per minute was used. For separation of the various compounds extracted from mineral inclusions, three different types of columns (4-meter in length) were necessary: charcoal, dimethylsulfolane and polypropyleneglycol, both 5 percent on acid-washed and silanized 100–120 mesh Chromosorb. A Bieman-Watson type pressure reduction system produced pressure drop from 1 atm to 10^{-5} mm Hg.

References

1. WAHLER, W.: Über die in den Kristallen eingeschlossenen Flüssigkeiten und Gase. Geochim. Cosmochim. Acta **9**, 105 (1955).
2. GOGUEL, R.: Die chemische Zusammensetzung der in den Mineralien einiger Granite und ihrer Pegmatite eingeschlossenen Gase und Flüssigkeiten. Geochim. Cosmochim. Acta **27**, 155 (1963).
3. JEFFERY, P. G., and P. I. KIPPING: The determination of constituents of rocks and minerals by gas chromatography. Analyst **88**, 266 (1963).
4. KRANZ, R.: Organische Fluor-Verbindungen in den Gasschlüssen der Wölsendörfer Flußspäte. Naturwissenschaften **53**, 593 (1966).
5. – Stickstoffverbindungen in Silikatmineralien und ihre Bedeutung für das technologische Verhalten der Feldspäte. Ber. Deut. Keram. Ges. (in press).
6. BROWN, R. A.: Compound types in gasoline by mass spectrometer analysis. Anal. Chem. **23**, 430 (1951).
7. TOORKS, W. P.: USA-Patent 2.323284. Z. Angew. Chem. **73**, 40 (1961).
8. KRÖGER, F. A.: The incorporation of uranium in calcium fluoride. Physica **14**, 488 (1948).
9. RECKER, K.: Über den Einbau von Uran in CaF_2. Z. Angew. Chem. **73**, 40 (1961).
10. STRUNZ, H.: Die Uranfunde in Bayern von 1804 bis 1962. Naturwiss. Verein Regensburg 1962.
11. STEVENSON, F. J.: Chemical state of nitrogen in rocks. Geochim. Cosmochim. Acta **26**, 797 (1962).
12. MARSHALL, C. E.: Reactions of feldspars and micas with aqueous solutions. Econ. Geol. **57**, 12/9 (1962).
13. MANDL, I., A. GRAUER, and C. NEUBERG: Solubilization of insoluble matter in nature. Biochim. Biophys. Acta **8**, 654 (1952); **10**, 540 (1953).

14. NEUBERG, C., and I. MANDL: Beachtliche Wirkung von Salzen organischer Säuren auf unlösliche anorganische Verbindungen. Z. Vitamin-, Hormon- Fermentforsch. **2**, 480 (1948).
15. EVANS, W. D.: Properties of airborne dusts in relation to pneumoconiosis. Bull. Inst. Mining Met. **65**, 13 (1955).
16. –, and D. H. AMOS: An example of the origin of coal-balls. Proc. Geologists' Assoc. (Engl.) **72**, 445 (1961).
17. SZALAY, A.: Cation exchange properties of humic acids and their importance in the geochemical enrichment of UO_2^{++} and other cations. Geochim. Cosmochim. Acta **28**, 1605 (1964).
18. KRANZ, R.: Die geochemische Bedeutung organischer Verbindungen in den Gaseinschlüssen uranhaltiger Mineralien. Naturwissenschaften (in press).
19. MULFINGER, H. O., and H. MEYER: Über die physikalische und chemische Löslichkeit von Stickstoff in Glasschmelzen. Glastech. Ber. **36**, 481 (1963).
20. WINOGRADOV, A. P., K. P. FLORENSKII, and V. F. VOLYNETS: Ammonia in meteorites and igneous rocks. Geochemistry (USSR) 905 (1963).
21. SCHÖNBEIN (1863) and WYROUBOFF (1866): Z. Angew. Chem. **73**, 40 (1961).
22. HEINRICH, F.: Z. Angew. Chem. **33**, 21 (1920); **73**, 40 (1961).

CHAPTER 22

Chemistry of Humic Acids and Related Pigments*

F. J. STEVENSON and J. H. A. BUTLER

Department of Agronomy,
University of Illinois,
Urbana, Illinois and Division of Soils,
C.S.I.R.O., Adelaide, Australia

Contents

I. Introduction

Humic acids and related pigments, collectively referred to as humic substances, are widely distributed in soils, natural waters, marine and lake sediments, peat bogs, carbonaceous shales, lignites, brown coals and other miscellaneous deposits [1–5]. In addition, trace amounts may be present in plants [6] and certain carbonaceous meteorites [7]. SZALAY [8] concluded that the amount of organic carbon in the earth as humic acids (60×10^{11} tons) exceeds that which occurs in living organisms (7×10^{11} tons).

Humic substances are best described as a series of acidic, yellow- to black-colored, moderately high-molecular-weight polymers which have characteristics dissimilar to any of the organic compounds occurring in living organisms. The modern view is that they represent an extremely heterogeneous mixture of molecules, which, in any given soil or sediment, may range in molecular weight from

* Appreciation is expressed to the National Science Foundation (Grants GP 1938 and GA 348) for support of research in organic geochemistry.

as low as 2,000 to perhaps over 300,000 [2]. Humic substances differ from kerogen and brown coal in several important respects, including elemental composition (lower carbon but higher oxygen contents), functional group content (higher COOH and OH contents) and solubility characteristics (soluble in alkali).

From a geological standpoint, humic substances are of importance in the transportation and subsequent concentration of a variety of mineral substances, such as bog ores and nodules of marine strata [5]. In addition, they may be involved in the enrichment and concentration of uranium and other metals, as their cations, in various bioliths, including coal [8]. It is suspected that metallic ions are transported in soils and waters as stable humic complexes. Humic constituents undoubtedly contribute to coals and oil shales.

II. Extraction and Fractionation

Humic substances are normally recovered by extraction with caustic alkali (usually 0.5 *N* NaOH), although in recent years use has been made of mild reagents, such as neutral sodium pyrophosphate [2, 9]. The following fractions, based on solubility characteristics, are subsequently obtained: *humic acid*, soluble in alkali, insoluble in acid; *fulvic acid*, soluble in alkali, soluble in acid; *hymatomelanic acid*, alcohol-soluble part of humic acid; *humin*, insoluble in alkali.

German scientists further divide humic acids into two groups by partial precipitation with electrolyte (salt solution) under alkaline conditions. The first group, the brown humic acids *(Braunhuminsäure)*, are not coagulated by an electrolyte and are characteristic of humic acids in brown coals, peat, and podzolic soils. The second group, the gray humic acids *(Grauhuminsäure)*, are easily coagulated and are characteristic of humic acids in chernozems and rendzinas. In the older German literature, reviewed by KONONOVA [3], considerable attention was given to the so-called "*apocrenic*" and "*crenic*" acids, which were light-yellow fulvic acid-type substances.

A complete fractionation scheme is given in Fig. 1. Instructions for preparing specific fractions from terrestrial soil are given by KONONOVA [3] and STEVENSON [9].

The undesirability of regarding individual components as chemically homogeneous compounds has recently been emphasized [1–5]. According to present-day concepts, the various groups represent part of a system of polymers, and the differences between the fractions are due to systematic variations in elemental composition, acidity, degree of polymerization and molecular weight.

Proposed inter-relationships between the main humic fractions are shown in Fig. 2. Most work is now being confined to the humic and fulvic acids, but no sharp division exists between the two. KONONOVA [3] regards the low-molecular-weight fulvic acids as being simple representatives of humic acids. The humin fraction does not appear to be a separate group—this material probably consists of humic and fulvic acids intimately bound to mineral matter. Hymatomelanic acid may be an artifact produced from humic acids during fractionation.

A major difficulty of all work on humic substances is that of obtaining preparations free of salts, clay and complexed metals (particularly iron and aluminium). Clay-sized particles can be eliminated by high-speed centrifugation. Repeated

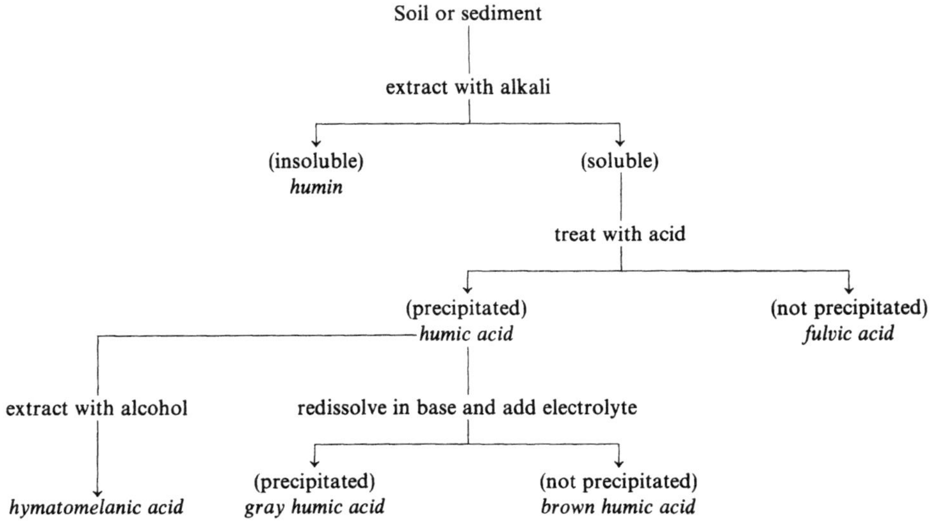

Fig. 1. Fractionation of humic substances

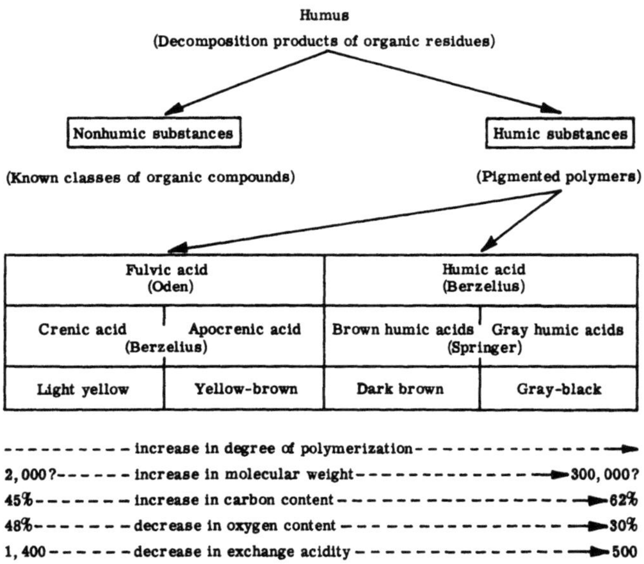

Fig. 2. Classification and chemical properties of humic substances. Adapted from a drawing by SCHEFFER and ULRICH [4]

dissolution and precipitation of humic acids often leads to preparations with low ash contents, but this technique cannot be applied to fulvic acids, which are acid-soluble. Removal of salts and metals from most fractions can be accomplished by appropriate dialysis procedures, although loss of low-molecular-weight compounds invariably occurs. Other desalting procedures include HF treatment, treatment with chelating agents and passage through ion exchange resins. Procedures used for the purification of humic substances have been discussed by DUBACH and MEHTA [2].

In the case of humic acids, the wet precipitate is subjected to freezing and thawing to exude water, and is then freeze-dried. Air-dried preparations are brittle and difficult to redissolve; those that have been freeze-dried occur as soft, pliable powders which dissolve readily in neutral or slightly alkaline aqueous solvents.

III. Biochemistry of the Formation of Humic Substances

The formation of humic acids and related pigments has been the subject of several recent reviews [3–5, 10]. The classical theory, popularized by WAKSMAN, is that they represent modified lignins. Support for this view has been provided by the observation that lignins are relatively resistant to attack by microorganisms, and that humic acids have properties similar to oxidized lignins.

BREGER [11], using RUSSELL lignin as a model, suggested that the initial step involved the opening of heterocyclic rings. His proposed scheme, shown in Fig. 3, produces a basic structure containing double bonds, acidic OH groups and a linear skeletal arrangement of phenylpropane-type units, $(C_6—C_3)_n$.

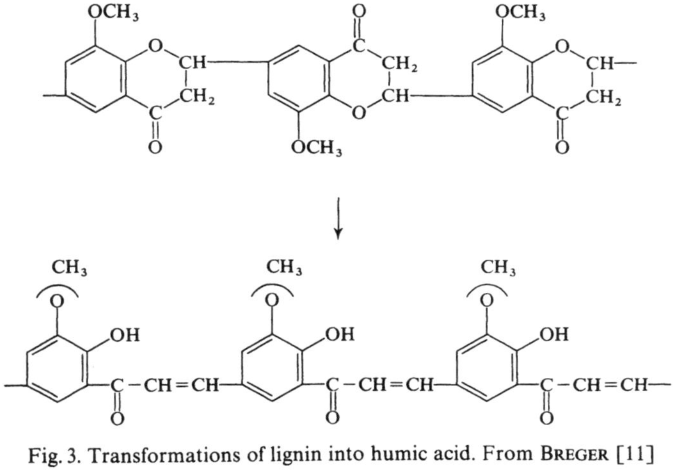

Fig. 3. Transformations of lignin into humic acid. From BREGER [11]

Oxidation of terminal chains would generate COOH groups. In addition, demethylation would result in the formation of o-dihydroxybenzene units which could be oxidized to quinones capable of undergoing condensation reactions, both in the presence and absence of amino compounds. The final product would bear little resemblance to the original lignin molecule.

In recent years, the lignin theory has been severely challenged. According to contemporary views, humus synthesis is a two-stage process which includes: (1) decomposition of all plant components, including lignin, into simple monomers, and (2) polymerization of these simple monomers into high-molecular-weight polymers. Algae, lichens, and mosses, which do not contain lignin, give rise to humic substances.

Most investigators now favor biosynthetic schemes based on the oxidative polymerization of phenolic constituents. Several sources of phenols are postulated, including lignins, tannins, and cellular constituents of microorganisms.

FLAIG [12] proposed the following sequence for humic acid synthesis:

1. Lignin, freed of its linkage with cellulose during the decomposition of plant residues, is subjected to oxidative splitting with the formation of primary structural units.

2. The side chains of these units are oxidized, demethylation occurs, and the resulting polyphenols are converted to quinones by polyphenoloxidases.

3. Quinones arising from lignin, together with those synthesized by microorganisms, react with nitrogenous compounds to form dark-colored polymers.

A schematic diagram showing the formation of humic substances by condensation of phenols and amino acids, as exemplified by the reaction between catechol and glycine, is given in Fig. 4.

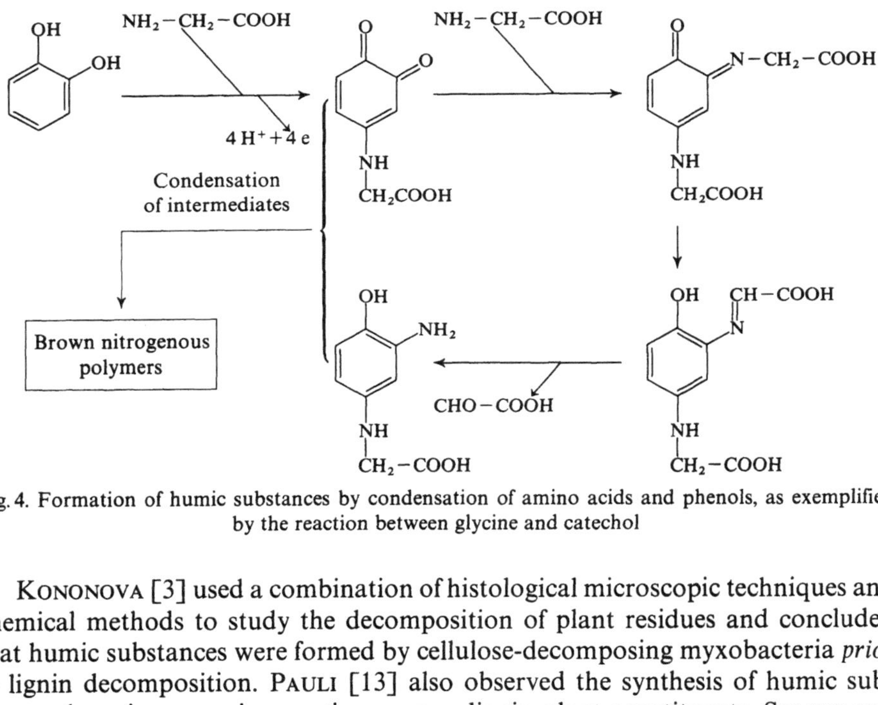

Fig. 4. Formation of humic substances by condensation of amino acids and phenols, as exemplified by the reaction between glycine and catechol

KONONOVA [3] used a combination of histological microscopic techniques and chemical methods to study the decomposition of plant residues and concluded that humic substances were formed by cellulose-decomposing myxobacteria *prior to* lignin decomposition. PAULI [13] also observed the synthesis of humic substances by microorganisms acting on non-lignin plant constituents. SWABY and LADD [14] suggested that humic molecules are formed from free radicals (quinones) produced enzymatically within deceased cells (plant or microbial) while autolytic enzymes are still functioning but before cell walls are ruptured by microbes.

The idea that carbohydrates participate in the formation of humic substances has persisted since the middle of the last century. The reactions involved are believed to be similar to those occurring during the formation of brown polymers by condensation of sugars and amines (the Maillard reaction). The over-all scheme is outlined in Fig. 5.

A major objection to this theory is the slow rate at which sugar-amine condensation reactions occur. However, drastic changes in the environment (freezing and thawing, wetting and drying), together with the intermixing of reactants with mineral material having catalytic properties, may facilitate condensation. In ancient sediments, the time factor becomes of minor importance. STEVENSON and TILO [15] found that the fraction of the organic nitrogen in some deep-sea sediments that occurred as amino acids, decreased with increasing age and concluded that the loss was due to chemical reactions involving sugars and phenols.

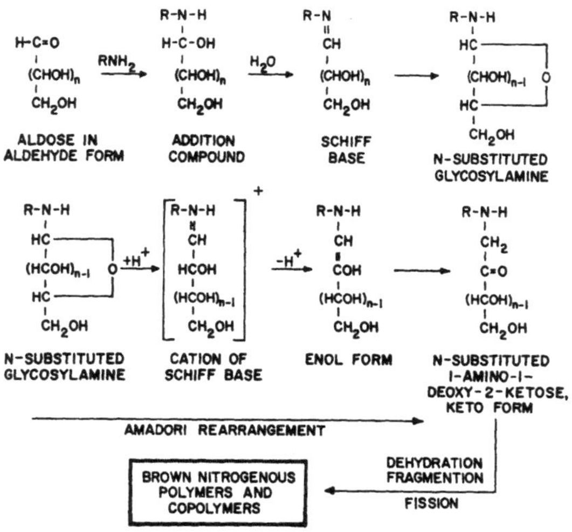

Fig. 5. Sugar-amine condensation reactions

Thus, a completely satisfactory scheme for the synthesis of humic substances in geologic environments has not yet been developed. A lignin pathway may predominate in anaerobic situations (swamps, etc.) whereas biosynthesis from phenols leached from leaf litter may be of considerable importance in certain forest soils. The frequent and sharp fluctuations in temperature, moisture and irradiation in terresterial surface soils under a continental climate may lead to humus synthesis by the Maillard reaction.

IV. Methods of Functional Group Analysis

A variety of functional groups, including COOH, phenolic OH, enolic OH, quinone, hydroxyquinone, lactone, ether and alcoholic OH, have all been reported in humic substances, but there is considerable disagreement as to the amounts present; in some cases even proof of existence is lacking. As DUBACH and MEHTA [2] pointed out, severe problems are encountered in the determination of functional groups because of incomplete reactions, adsorption of reagents, undesirable fractionation during manipulations, sensitivity towards acid and base at high temperatures and the proximity of groups which influence the specificity of reagents. Since the acidities of the various groups overlap, results obtained by methods dependent on ion exchange or pK values must be interpreted with caution. Polycarboxylic acids, for example, exhibit a whole series of dissociation constants which decrease as successive protons dissociate. On the other hand, substituted phenols are often more strongly dissociated than the unsubstituted phenol.

A. Total Acidity

Total acidity of humic substances has been determined by chemical methods and by electrometric titrations in nonaqueous solvents. The most popular chemical method has been by baryta absorption. Briefly, the sample is allowed to react with excess $Ba(OH)_2$, following which the unused base is titrated with standard acid. Reactions carried out in the presence of oxygen lead to high values. The main advantage of the method is its simplicity; unfortunately, arbitrary values are probably secured. Low results have been obtained for OH groups in certain substituted phenols [16, 17], as well as in phenol-formaldehyde resins [16]. However, results for soil humic substances have been in good agreement with data obtained by potentiometric titration [17, 18].

A second method involves an estimate of OCH_3 formed in samples after methylation with diazomethane (CH_2N_2). Methylation is usually carried out in an ether suspension, although sulfolane has recently been used as a solvent [19, 20]. Diazomethane reacts with acidic H of a wide variety of structures, including COOH, phenolic OH, enolic compounds, N—H acidic groups and certain aliphatic alcohols. On the other hand, H-bonded phenolic OH groups may not react [21]. Side reactions include the formation of polymethylene under the catalytic action of heavy metals, such as was observed by FARMER and MORRISON [22].

Following methylation, the samples are analyzed for OCH_3 by the well-known Zeisel method, although the Mather and Pro procedure has also been used [23]. In the latter case, the OCH_3 is split off as methanol, which is recovered by distillation and oxidized to formaldehyde with $KMnO_4$. The formaldehyde is then determined by a colorimetric procedure.

DUBACH *et al.* [24, 25] used still another approach to determine active H in soil humic substances. They employed diborane, B_2H_6, which is believed to react with sterically hindered active H. The method, which involves an estimate of the H_2 produced, is believed to be independent of pK values. For a soil fulvic acid, good agreement was obtained between total acidity determined by this method and the $Ba(OH)_2$ procedure.

The Grignard reagent, methylmagnesium iodide (CH_3MgI), would appear to be of little value for the analysis of humic substances. For coals, more promising results have been obtained using lithium aluminum hydride ($LiAlH_4$) [26].

B. Carboxyl

The Ca-acetate method has been used extensively to determine COOH groups in humic substances [17, 18, 27–30]. Because the procedure involves ion exchange, and since humic substances may contain strongly acidic OH groups, strictly quantitative values may not be obtained. Many substituted phenols, for example, undergo exchange with Ca^{++} ions [16, 17]. DUBACH *et al.* [24] and MARTIN *et al.* [25], in an examination of a series of humic preparations from soil, found that the discrepancy between values obtained for COOH groups by chemisorption and those calculated from active H (diborane) and total OH measurements became increasingly serious as the OH contents of the preparations increased. With humic substances, chelation reactions may lead to inaccurate values.

Another method of estimating COOH groups is methylation and subsequent saponification of the resulting methyl esters. Several methylation procedures have been used, including CH_2N_2, and methanol in dry HCl or H_2SO_4. FARMER and MORRISON [22] found that the latter procedure gave incomplete esterification. BARTON and SCHNITZER [31] used a methyl iodide-silver oxide mixture as their methylating reagent.

Several saponification procedures have been used for the analysis of methyl esters, including distillation of the liberated methanol [23] and determination of the drop in OCH_3 content accompanying saponification [32]. A disadvantage of the latter procedure is that quantitative recovery of material following saponification is difficult. Techniques based on titration of unused alkali with standard acid may give high results due to production of acidic groups during saponification.

WRIGHT and SCHNITZER [18] employed an iodometric method to determine COOH groups in some humic preparations. This technique, which is based on ion exchange, gave higher values than the Ca-acetate method. BLOM *et al.* [21] claimed that results obtained by the method were not reproducible.

Aromatic acids are decarboxylated when they are heated with quinoline in the presence of a suitable catalyst. Application of this method to soil humic substances has given results comparable to the Ca-acetate method [18], indicating that most of the COOH groups are attached to aromatic rings. However, CO_2 can also be released from α-hydroxy aliphatic acids.

Indirect methods for estimating COOH groups involve subtraction of acidic OH groups (to be discussed) from total acidity values. Obviously, results secured by this practice include accumulated errors of both determinations.

C. Total Hydroxyl

The two methods used most frequently to determine total OH content of humic substances are the following: (1) methylation with dimethyl sulfate, $(CH_3)_2SO_4$, and (2) acetylation with acetic anhydride. In the first-mentioned method, the sample is treated repeatedly with dimethyl sulfate in an alkaline solution, after which the resulting precipitate is analyzed for OCH_3 by the Zeisel method [32]. Only phenolic and alcoholic OH groups are believed to be methylated; COOH groups are not esterified. Dimethyl sulfate is capable of reacting with phenolic OH groups that are too weakly acidic to react with CH_2N_2, and this has been used as the basis for determining H-bonded OH groups in humic acids. Unfortunately, results obtained by dimethyl sulfate are difficult to interpret because of possible side reactions in the strongly alkaline solution.

The second method (acetylation) has been used more extensively. The results of BLOM *et al.* [21] and WAGNER and STEVENSON [33] indicate that many early investigators probably used too short a reaction period. For reasons which are not clear, the acetyl content of humic acids methylated with CH_2N_2 and saponified prior to acetylation, is higher than that of unmethylated samples [33].

Since COOH groups are also acetylated with acetic anhydride (formation of mixed anhydrides), a correction has to be made for COOH as determined by an independent method. The possibility that cyclic anhydrides can be formed during acetylation [20, 33] does not seem to have been taken into account in evaluating the method. Primary and secondary amines, as well as sulfhydryl groups, react with acetic anhydride, but WRIGHT and SCHNITZER [18] concluded that interference from these compounds was not serious. Acetylations conducted under reducing conditions (detection of hindered phenols) have given inconclusive results [34].

A quantitative estimate of OH content by acetylation is obtained either by hydrolysis of the excess reagent and titration of the resulting acetic acid [18, 30], or by hydrolysis of the derivative with alkali [25, 32, 33]. MARTIN *et al.* [25] concluded that determination of the unreacted acetic anhydride gave unreliable results, a conclusion previously reached by BLOM *et al.* [21].

FRIEDMAN *et al.* [35] recently proposed a new method for determining OH groups in coal based on the formation of trimethylsilyl ethers, $ROSi(CH_3)_3$. Several highly hindered phenols formed ethers quantitatively under the conditions employed. Adaptation of the method to humic acids and related pigments will require that corrections be made for the formation of trimethysilyl esters from COOH groups. Humic substances may also contain nitrogen-containing functional groups capable of forming derivatives.

D. Phenolic Hydroxyl

In the Ubaldini method, potassium salts of both COOH and phenolic OH groups are formed by heating the material with alcoholic KOH. Carbon dioxide is then bubbled through the solution, and the potassium ion released as K_2CO_3, presumably originating from phenolic OH groups, is estimated by titration. Serious objections include the nonspecificity of the reaction and the danger of hydrolysis during the treatment with hot alcoholic KOH. No information is available as to how results obtained for humic and fulvic acids [28] compare with other methods.

MOSCHOPEDIS [19] pointed out that methods based on coupling of phenolic OH groups with diazonium salts are valid only when the ortho and para positions with respect to these groups are available. Diazo-coupling was believed to account for only part of the total reaction with humic acids.

Indirect methods for estimating phenol OH groups include analysis of samples methylated with CH_2N_2. However, only those phenolic OH groups sufficiently acidic to react with CH_2N_2 will be included with the measurement. Hydroxyquinones of the general types I, II, and III are known to be extremely difficult to methylate with CH_2N_2, because of the high stability of ring structures formed through H-bonding.

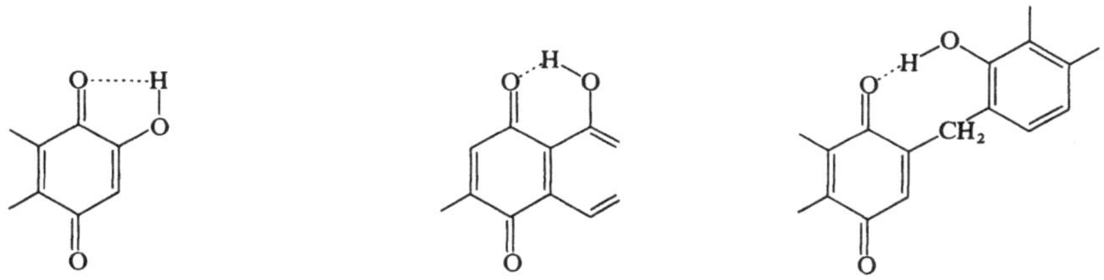

I II III

Until the question of hydroxyquinones in humic substances is resolved, values obtained by analysis of methylated derivatives, or those calculated from total acidity and COOH estimations, should no longer be reported as "phenolic OH".

E. Alcoholic Hydroxyl

The preceeding discussions emphasize that there are no suitable methods for estimating alcoholic OH groups in humic substances. Several indirect methods have been applied, but with variable results. Thus, SCHNITZER and his coworkers [17, 18, 28, 30] reported significant amounts of "alcoholic OH" in humic and fulvic acids by subtracting "phenolic OH", also obtained by difference (total acidity minus COOH by the Ca-acetate method), from total OH as determined by acetylation. On the other hand, DUBACH *et al.* [24] and MARTIN *et al.* [25] failed to detect alcoholic OH groups from the difference between active H (diborane) and total acidity by baryta absorption.

WRIGHT and SCHNITZER [18] observed an increase in "alcoholic OH", but a decrease in C=O, when humic substances were extracted from soil in the presence of oxygen. It was postulated that some form of keto-enol transformation had occurred. Other oxidative changes included a slight decrease in "phenolic OH" and a concomitant increase in COOH.

F. Carbonyl

Most methods that have been used for determining C=O groups in humic substances are based on the formation of derivatives by reaction with such reagents as hydroxylamine, phenylhydrazine, 2,4-dinitrophenylhydrazine, and semicarbazide. Hydroxyquinones of the types shown above (I–III) will not always react with these reagents. The reaction is followed by either measuring the increase in nitrogen content of the derivative [21, 30, 34], or by analysis of the unreacted reagent [18, 36].

Procedures based on the formation of derivatives can be criticized on the grounds that the reagents employed may react without the participation of C=O groups. Humic acids, for example, are known to combine with ammonia to form compounds which are nonhydrolyzable. BLOM *et al.* [21] concluded that hydroxylamine and phenylhydrazine gave high values for C=O in coal humic acids.

A relatively new method for estimating >C=O is based on reduction to —CH_2OH with sodium borohydride ($NaBH_4$). Reduction is carried out in an alkaline solution, and H_2 liberated from the unused $NaBH_4$ is estimated manometrically. The method is reported to be highly specific for C=O groups. MARTIN *et al.* [25] obtained values for some humic and fulvic acids which compared favorably with those reported using the methods mentioned above.

The C=O of quinones is difficult to estimate because of extreme variations in their reactivity. SCHNITZER and SKINNER [34] failed to detect quinones in a soil fulvic acid using reductometric titration and reductive acetylation techniques.

The possible presence of terminal-ring 1,4-quinones (30 to 110 meq/100 g) has been indicated [23] by measuring the increase in nitrogen following methylation with CH_2N_2. This reagent is known to combine with terminal-ring quinones to form pyrazoline rings.

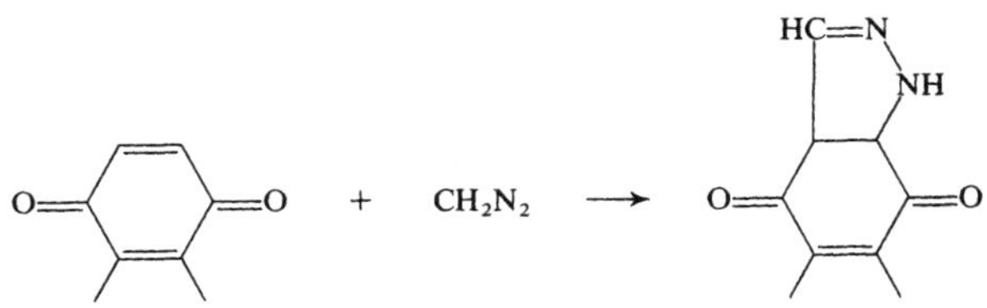

An increase in nitrogen does not constitute unequivocal proof for 1,4-quinones, because CH_2N_2 can also combine with other groups, such as olefinic double bonds.

The treatment of coal humic acids with diazotized sulfanilic acid was found by MOSCHOPEDIS [19] to yield water-soluble addition products. These products were believed to arise by both a free radical mechanism involving quinoid structures and by simple azo coupling with phenolic structures. The former was regarded as the dominant reaction.

G. Ether

Practically all of the humic and fulvic acids examined thus far by the Zeisel method have been found to contain low but variable OCH_3 contents (see next section).

The oxygen not accounted for in COOH, OH, C=O, and OCH_3 groups has often been recorded as existing in unknown ether linkages. This practice has been rejected by DUBACH and MEHTA [2] on the basis that, in some humic acids, four ether linkages would be required per aromatic ring. They suggested that part of the unaccounted-for oxygen exists in extremely stable quinones and lactones.

V. Distribution of Oxygen-containing Functional Groups

As might be suspected from the discussion in the previous section, data reported for functional groups in humic acids and related pigments are difficult to interpret. Not only has there been a lack of standardized methods for their extraction, fractionation and purification, but many of the analytical techniques have lacked specificity. Nevertheless, sufficient information has accumulated to permit tentative conclusions to be drawn regarding the relative abundance of oxygen-containing functions. Since serious gaps in our knowledge are apparent, an organization of existing information will be of value in designing future experiments.

Evidence that humic acids contain significant amounts of COOH groups has come from the finding that values secured by reduction with diborane [24, 25], and by decarboxylation with quinoline [18], agree reasonably well with those obtained by methods based on ion exchange or acidity (chemisorption and methylation-saponification). The occurrence of appreciable quantities of COOH groups has also been established by infrared spectroscopy (see next section).

Because of inadequacies in methods for estimating phenolic OH, alcoholic OH and C=O groups, values reported for these groups should be accepted with reservation. More of the oxygen in humic substances may exist as phenolic OH than has been recorded, because very weak and hindered phenols are difficult to estimate by existing methods. In this paper, the term "acidic OH" will be used rather than "phenolic OH" to refer to the fraction of the total acidity not accounted for as COOH groups. Other OH groups will be referred to as "weakly acidic plus alcoholic OH".

The distribution of oxygen-containing functional groups in some humic and fulvic acids, as recorded in the recent literature, is summarized in Table 1. The values for COOH and acidic OH in the humic acids are in general agreement with data published in the early literature (see review of KONONOVA [3]). For any specific group, a considerable range of values is apparent, even with preparations obtained from the same soil type. Nevertheless, certain trends are evident. The total acidities of the fulvic acids (890 to 1,420 meq/100 g) are unmistakably higher than those of the humic acids (485 to 870 meq/100 g). Both COOH and acidic OH groups (presumed to be phenolic OH) contribute to the acidic nature of these substances, with COOH being the most important. If it is assumed that humic substances consist primarily of C_6 ring units held together by —O—, —CH_2—, —NH—, —N—, and —S— linkages (see section entitled Structural Basis of Humic Acids, this chapter), the conclusion would have to be made that multiple substitution of the ring had occurred. The concentration of exposed acidic functional groups in fulvic acids would appear to be substantially higher than in any other naturally occurring organic polymer.

Since the functional groups listed in Table 1 were determined by a variety of analytical procedures, some of which produce results of questionable accuracy, direct comparisons are difficult. It should be mentioned, however, that not all of the variability between samples is due to differences in experimental techniques. For example, the rather high values shown for acidic OH in the humic preparations examined by SCHNITZER and GUPTA [30], which were from a gray-wooded soil, were obtained using methods identical to those that SCHNITZER and DESJARDINS [27] employed. The latter study was conducted on material recovered from a podzol soil.

Methoxyl and C=O groups seem to be universally present in humic substances. On the other hand, not all samples are reported to contain weakly acidic or alcoholic OH groups. For example, DUBACH *et al.* [24] and MARTIN *et al.* [25] found that active H, as determined by diborane, agreed well with total acidity measurements by reaction with $Ba(OH)_2$, a finding which excluded the presence

Table 1. *Oxygen-containing functional groups in humic and fulvic acids*

Oxygen[a] (percent)	Total acidity (meq/100 g)	COOH (meq/100 g)	Acidic OH (meq/100 g)	Weakly acidic plus alcoholic OH (meq/100 g)	C=O[b] (meq/100 g)	OCH_3[b] (meq/100 g)	Reference[c]
Coal and lignite humic acids							
20.0	613	260	353	111	n.d.	16	LYNCH *et al.* [29]
19.6	485	174	310	225	n.d.	16	LYNCH *et al.* [29]
28.7	730	440	290	0	n.d.	170	MOSCHOPEDIS [19]
Soil humic acids							
37.1	710	280	430	0	190	n.d.	MARTIN *et al.* [25]
35.4	570	150	415	275	90	n.d.	SCHNITZER and DESJARDINS [27]
36.7	870	300	570	350	180	n.d.	SCHNITZER and GUPTA [30]
34.6	570	280	290	300	300	50	WRIGHT and SCHNITZER [18]
Soil fulvic acids							
44.0	1,160	800	360	0	100	n.d.	MARTIN *et al.* [25]
44.7	1,238	908	330	355	310	n.d.	SCHNITZER and DESJARDINS [27]
47.3	1,420	850	570	340	170	n.d.	SCHNITZER and GUPTA [30]
47.0	1,280	610	670	330	300	n.d.	SCHNITZER and GUPTA [30]
44.1	890	610	280	460	310	30	WRIGHT and SCHNITZER [18]
47.7	1,180	910	270	490	110	30	WRIGHT and SCHNITZER [18]

[a] Some oxygen values uncorrected for sulfur.

[b] n.d. = not determined.

[c] Methods used by LYNCH *et al.* and by SCHNITZER and co-workers, are as follows: total acidity, baryta absorption; COOH, Ca-acetate method; acidic OH (reported as phenolic OH), difference between total acidity and COOH; weakly acidic plus alcoholic OH (reported as alcoholic OH), difference between total OH (acetylation) and "phenolic OH"; C=O by oximation; OCH_3, Zeisel procedure. Techniques employed by MARTIN *et al.* are: total acidity (active H), diborane method; COOH, difference between active H (diborane) and OH by acetylation; C=O, reduction with $NaBH_4$. In the latter studies, active H by diborane agreed with total acidity measurements by baryta absorption, a result which was believed to exclude the presence of alcoholic OH.

of alcoholic OH groups. The humic acid from lignite analyzed by MOSCHOPEDIS [19] and WOOD *et al.* [20] also appeared to be free of alcoholic OH groups.

A major difference between the functional group content of humic and fulvic acids is that a smaller fraction of the oxygen in the former can be accounted for in COOH, OH and C=O groups. Table 2 shows that whereas essentially all of the oxygen in some soil fulvic acids was recovered in these groups, a maximum of 74 percent of that in the humic acids was similarly distributed.

The COOH content of humic substances appears to be inversely related to molecular weight. It was pointed out earlier that fulvic acids have lower molecular weights than humic acids. The results given in Tables 1 and 2 show rather conclusively that the proportion of the oxygen which occurs in the form of COOH is highest for the fulvic acids.

BLOM *et al.* [21] prepared a diagram illustrating the distribution of oxygen-containing functional groups in lignites and coals as related to their carbon and oxygen contents. From published data on the carbon, oxygen and functional group

Table 2. *Distribution of oxygen in some humic and fulvic acids from terrestrial soil*

O (percent)	COOH (percent of O)	Acidic OH (percent of O)	Weakly acidic plus alcoholic OH (percent of O)	C=O (percent of O)	Recovered (percent)	Reference[a]
Humic acids						
35.6	13.6	19.0	12.2	4.1	48.9	SCHNITZER and DESJARDINS [27]
36.7	26.2	24.8	15.2	7.8	74.0	SCHNITZER and GUPTA [30]
Fulvic acids						
44.7	65.1	11.8	12.9	11.1	100.9	SCHNITZER and DESJARDINS [27]
47.3	57.5	19.3	11.5	5.8	94.1	SCHNITZER and GUPTA [30]
47.0	41.5	22.8	11.2	10.2	85.7	SCHNITZER and GUPTA [30]

[a] Functional groups were determined as follows: COOH, Ca-acetate method; acidic OH (reported as phenolic OH), difference between total acidity (baryta absorption) and COOH; weakly acidic plus alcoholic OH (reported as alcoholic OH), difference between total OH (acetylation) and "phenolic OH"; C=O by oximation.

contents of humic and fulvic acids it has been possible to extend this diagram to pigments having higher oxygen, but lower carbon, contents. The modification is given in Fig. 6. Admittedly, further refinements are needed, particularly with respect to changes in OH groups. Nevertheless, the diagram provides a comprehensible picture of the diagenetic relationship between humic substances and coal. It is apparent that if humic and fulvic acids are involved in the coalification

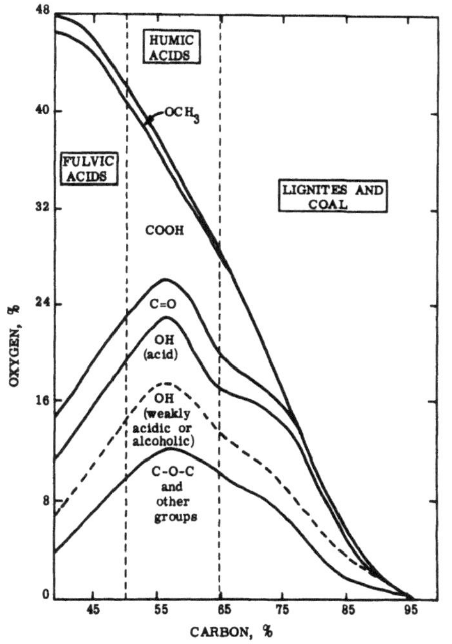

Fig. 6. Oxygen-containing functional groups in fulvic acids, humic acids, coal and lignites as related to carbon and oxygen contents. Adapted from a drawing by BLOM *et al.* [21]

process, the COOH groups disappear first, followed in order by OCH_3 and OH groups.

Unfortunately, the relationship between humic and fulvic acids is not clear. Proposed relationships are illustrated in the following diagram:

Lignin ⟶ Primary structural units

↓ ↓

Humic acids ⇌ Fulvic acids

↓ ↑

Coal Metabolites of microorganisms

As indicated earlier, most investigators now appear to favor synthetic pathways based on condensation of simple organic molecules. However, BREGER [11] maintains that humic acids represent modified lignins, and WRIGHT and SCHNITZER [18] have concluded that fulvic acids are formed from humic acids through an increase in oxygen-containing functional groups at the expense of aliphatic and/or acyclic material.

VI. Structural Arrangement of Functional Groups as Revealed by Infrared Analysis

Main absorption bands of humic substances (Fig. 7a) are in the regions of 3,300 cm^{-1} (H-bonded OH groups), 2,900 cm^{-1} (aliphatic C—H stretching), 1,720 cm^{-1} (C=O stretching of COOH and ketonic C=O), 1,610 cm^{-1} (aromatic C=C and H-bonded C=O), and 1,250 cm^{-1} (C—O stretching and OH deformation of COOH groups). In addition, small bands are often evident at about 1,500 cm^{-1} (aromatic C=C), 1,460 cm^{-1} (C—H deformation of CH_2 or CH_3 groups), and 1,390 cm^{-1} (O—H deformation, CH_3 bending, or C—O stretching).

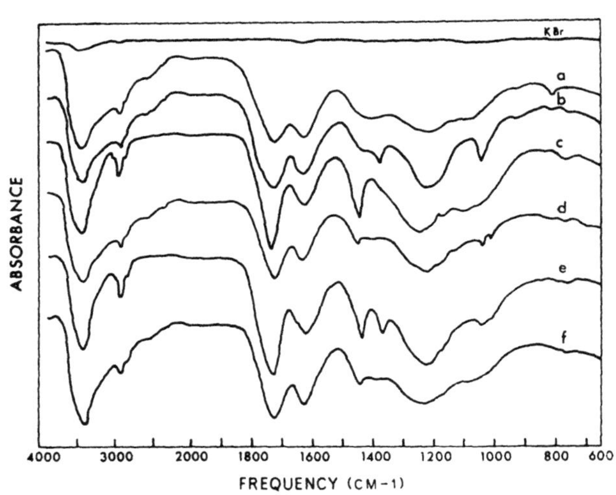

Fig. 7a–f. Infrared spectra of fulvic acid derivatives: a) untreated fulvic acid; b) acetylated; c) methylated; d) methylated and saponified; e) methylated and acetylated; f) esterified. From SCHNITZER [38]

Humic and fulvic acids have similar spectra, the main difference being that the intensity of the 1,720 cm^{-1} band is somewhat stronger in the fulvic acids because of the occurrence of more COOH groups (see previous section). In general, the spectra of humic acids resemble those of brown coals [37, 38].

Methylation of humic substances with diazomethane (Fig. 7c) results in an increase in the intensity of the bands caused by C—H stretching (2,900 cm^{-1}), C=O stretching (1,720 cm^{-1}), and C—H deformation of CH_3 groups (1,460 cm^{-1}). In addition, the 1,250 cm^{-1} band is sharpened considerably because of the formation of C—O linkages by methylation of phenolic OH groups. In the case of acetylation (Fig. 7b), a strong band is introduced at about 1,375 cm^{-1}; this undoubtedly results from CH_3—C deformation of added acetyl groups. The acetylated product also shows enhanced absorption near 1,200 cm^{-1}, due to phenolic and alcoholic acetates.

The significance of these and other changes as they relate to the structure of humic substances is given in the following sections.

A. The 3,300 cm^{-1} Band

Absorption due to O—H stretching nearly always occurs near 3,300 cm^{-1}. There may be some contribution from N—H bands, but in most samples this is likely to be small. In the unassociated state, OH groups absorb near 3,600 cm^{-1}; when associations take place, such as through H-bonding, the frequency is reduced. The band at 3,300 cm^{-1} is accordingly assigned to H-bonded OH groups.

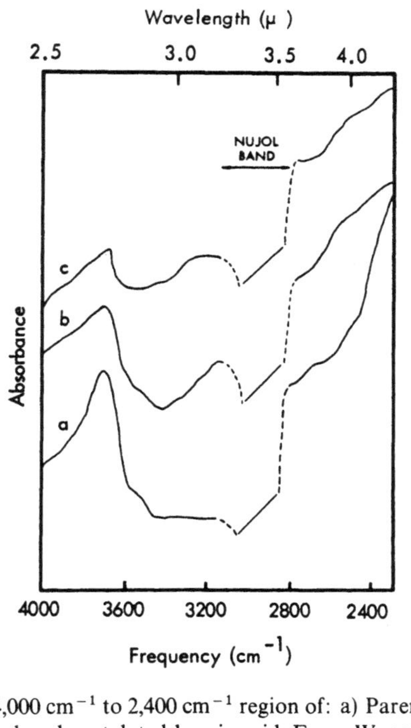

Fig. 8a–c. Absorption in the 4,000 cm^{-1} to 2,400 cm^{-1} region of: a) Parent humic acid; b) methylated humic acid; c) methylated and acetylated humic acid. From WAGNER and STEVENSON [33]

The spectra of methylated or acetylated derivatives (Fig. 7b, c) invariably show prominent residual absorption at 3,300 cm^{-1}. This has also been observed with brown coals, and may be due to OH groups resistant to methylation or acetylation [37–40]. However, interpretations based on the intensity of bands in the OH region must be made with caution because of possible moisture contamination, especially when KBr discs are employed [33, 41].

In experiments using Nujol mulls, WAGNER and STEVENSON [33] found that the spectrum of a carefully dried methylated humic acid still showed a broad, distinct band at about 3,420 cm^{-1} (compare curves a and b of Fig. 8). The intensity of the band was reduced significantly by subsequent acetylation (curve c), and the absorption remaining was centered at a higher frequency (about 3,600 cm^{-1}). These changes could not be discerned in spectra secured using KBr. It is not known whether the residual absorption at 3,600 cm^{-1} was due to unacetylated OH groups or N—H structures.

B. The 1,720 cm^{-1} Band

This pronounced band largely disappears when humic substances are reduced with diborane [25] or are converted to salts [38, 39], thereby providing conclusive evidence that the major part of the absorbance in this region is due to C=O stretching of COOH groups. The residual absorption may be due to saturated open-chain ketones or aldehydes, which also absorb at about 1,720 cm^{-1}. The absence of strong absorption bands between 1,785 cm^{-1} and 1,735 cm^{-1} is reasonable proof that no C=O groups occur as anhydrides or as normal saturated esters.

The 1,720 cm^{-1} band often shifts to a slightly higher frequency (by about 5 cm^{-1}) when humic substances are methylated. This shift has been attributed to the fact that the C=O of esters generally absorb at higher frequencies than the C=O of the corresponding acids [40]. The frequency of both the acid and ester forms are lower than normal, which can be explained by conjugation of the group with double bonds, or to the occurrence of aromatic acids.

WOOD *et al.* [20] found that when a lignite humic acid was treated with acetic anhydride, or heated to above 100° C in sulfolane, bands (1,850 cm^{-1} and 1,785 cm^{-1}) typical of 5-membered (IV) and 7-membered ring (V) anhydrides were formed. It was estimated that nearly 80 percent of the COOH groups (350 meq out of 430 meq/100 g) were either paired on mutually adjacent sites, such as on aromatic rings, or located on attached ring structures of the diphenic acid type.

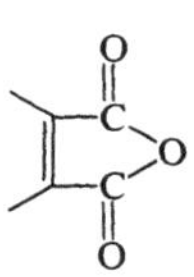

IV

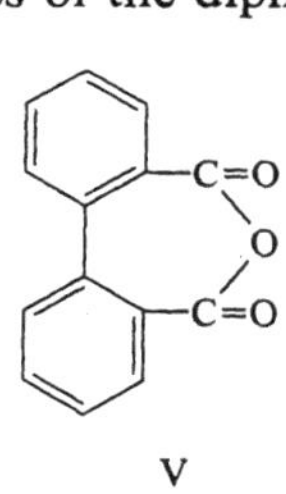

V

In a similar study, WAGNER and STEVENSON [33] found that about one-third of the COOH groups (128 meq out of 389 meq/100 g) in a soil humic acid occurred in sites such that cyclic anhydrides could be formed by heating with acetic anhydride.

SCHWARTZ *et al.* [42, 43] recently reported that some of the COOH groups in dried humic acids existed in ionized form (COO^-). Water and CO_2 were found to be evolved in a constant ratio of 4:1 when humic acid from a lignite was heated at 150° C in a stream of dry nitrogen; the water was believed to exist as the tri-hydrated hydronium ion, $H_3O(H_2O)_3^+$, bound to a COO^- ion (VI).

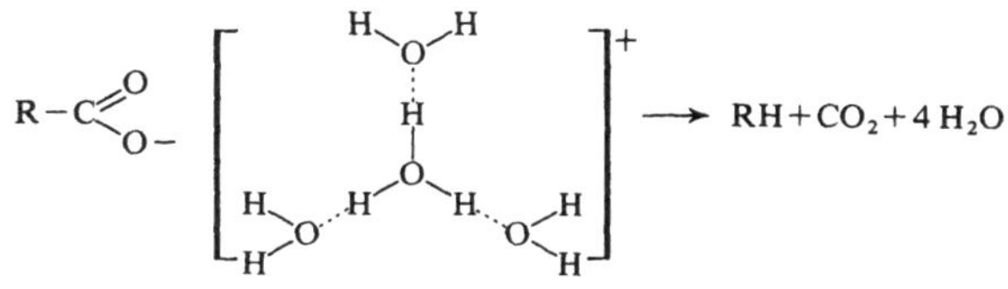

VI

Evidence given for COO^- ions was the presence of characteristic bands (1,550 cm^{-1} and 1,400 cm^{-1}) in the infrared spectra. The work was repeated by FALK and SMITH [44], with negative results. SCHWARTZ *et al.* [42, 43] themselves recognized that humic acids, being weakly acidic, should be unionized, especially in the solid state. A plausible explanation for the divergent results is that the sample analyzed by SCHWARTZ *et al.* contained salts or contaminating metals. On the other hand, SCHNITZER [38] reported that an ash-free fulvic acid preparation contained COO^- ions.

C. The 1,610 cm^{-1} Band

Absorption at about 1,610 cm^{-1} has often been attributed to C=C vibrations of aromatic structures. However, several other reasons can be given for absorption in this region; for example, non-aromatic double bonds and rings formed by H-bonding of OH to C=O of quinones (see structure I–III). Quinones normally absorb between 1,690 cm^{-1} and 1,635 cm^{-1}, but the C=O band shifts to a lower frequency when H-bonding occurs.

If the 1,610 cm^{-1} band is due primarily to C=C vibrations of aromatic rings, one might expect to find a more pronounced band at about 3,030 cm^{-1} for aromatic C—H stretching. However, the absence of a distinct band at this frequency (see Fig. 7a) can be explained by extensive substitution of the ring. Also, the band may be masked by strong absorption due to H-bonded OH groups, which can extend from about 3,660 cm^{-1} to 2,900 cm^{-1}.

Simple aromatic compounds invariably produce a band at 1,500 cm^{-1} which is more intense than the 1,610 cm^{-1} absorption. The reversal of the intensity for humic substances strongly suggests that structures other than aromatic C=C contribute to absorption in this region. Other aromatic absorption bands are too variable in position to be of value.

Particular attention has been given to the possible contribution of C=O stretching to the 1,610 cm^{-1} band of humic acids [19, 40] and coals [37, 45–46]. As indicated earlier, the H-bonded C=O of quinones absorb in this region. The suggestion has been made that the release of H-bonding by methylation with

CH_2N_2 should shift the C=O band to that of the parent quinone. However, attempts to demonstrate a quinone band in methylated humic acids has been unsuccessful. An explanation given for this result is that H-bonded OH groups may not be methylated with CH_2N_2, and, consequently, absorption in this region remains unchanged [19, 37, 40]. MOSCHOPEDIS [19], using humic acids from lignites, attempted to solve this problem by removing H-bonding through acetylation; when this was done, a band at 1,660 cm^{-1}, together with one attributed to the aryl acetoxy group (1,770 cm^{-1}), appeared in the infrared spectrum. This appears to have been the first direct evidence for quinone structures in humic acids. WAGNER and STEVENSON [33] were unable to detect a quinone band in the spectra of acetylated humic acids from soil. However, these investigators used the KBr disc technique and difficulty was encountered in eliminating interference due to moisture.

When humic substances are methylated with CH_2N_2, the intensity of the 1,720 cm^{-1} band is greatly increased, whereas that at 1,610 cm^{-1} appears to be reduced (compare curves a and c in Fig. 7). Several explanations have been given for this result, including: (1) release of H-bonding due to methylation of OH groups, with a shift of C=O absorption from 1,610 cm^{-1} to 1,720 cm^{-1}, (2) increased intensity of the 1,720 cm^{-1} band due to ester formation, and (3) elimination of the stretching mode of COO^- ions [38]. ČEH and HADŽI [39] favor the first explanation, whereas BROOKS *et al.* [37] reject this hypothesis on the basis that a shift of such magnitude (over 100 cm^{-1}) would be extremely rare. FARMER and MORRISON [22] suggested that C=O absorption of COOH groups must extend over a broad range due to varying degrees of H-bonding, and that the esters formed by methylation should absorb within a much narrower range, thereby giving a more intense peak.

It is of interest that the spectrum given in Fig. 7 for an esterified fulvic acid (curve f) does not show the same relationship between the 1,720 cm^{-1} and 1,610 cm^{-1} bands as the sample methylated with diazomethane (curve c), and that saponification of the methylated sample did not return the 1,720 cm^{-1} band to its original position (compare curves a and d). These results are difficult to reconcile on the basis that the 1,610 cm^{-1} band was not affected by methylation with CH_2N_2.

Infrared studies of coals reduced with lithium aluminum hydride [45], or acetylated under reducing conditions [46], indicate that H-bonded C=O groups contribute to the 1,610 cm^{-1} band. On the other hand, SCHNITZER *et al.* [38, 47] found no evidence for quinones in soil fulvic acids by infrared examination of products obtained by reductive acetylation. MOSCHOPEDIS [19] examined some water-soluble derivatives obtained by reacting humic acids with diazotized sulfanilic acid and concluded that the dominant reaction involved an interaction between the diazonium salt and quinone structures. In a subsequent study [48], halogenation of the coupled product was found to result in almost complete elimination of the 1,610 cm^{-1} band and the introduction of new bands at 1,640 cm^{-1} and in the range 2,850 cm^{-1} to 3,000 cm^{-1}. No explanation was given for this unusual result.

Supporting evidence for quinones in humic acids has come from electron paramagnetic resonance (EPR) spectroscopy measurements [49–51]. The spectra obtained strongly indicated the presence of remarkably stable semiquinone ion radicals. The exhaustive survey of ATHERTON *et al.* [49], which was conducted on a wide variety of acid-hydrolyzed humic acids, can be summarized as follows:

1. All humic acids gave spectra of breadth 1.75 to 1.9 gauss, which were susceptible to oxygen effects.

2. The signals were eliminated by reduction with sodium dithionite and recovered by exposure to air, thereby implicating semiquinone ion radicles.

3. The spectra could be divided into two classes, depending upon the pH of the soil or sediment from which the humic acid was derived. Humic acids from acidic environments (pH 2.8 to 4.3), such as mor humus layers, bog peats, and humus-containing horizons of podzols, showed four-lined spectra. On the other hand, humic acids from more basic environments (pH 4.3 to 7.2), such as mull humus layers, fen peats, and most soils, gave ill-defined, structureless spectra without distinct peaks.

4. The spectra were dependent upon the overlying vegetation only in so far as this corresponded to the pH of the soil (see item 3). The nature of the signal was also independent of depth or age of the soil.

The considerations discussed above emphasize the desirability of further research into the origin of the absorption band at 1,610 cm^{-1}. The results obtained thus far appear to favor the view that humic acids contain quinoid structures having an OH group adjacent or peri to each C=O (see structures I to III). A somewhat less positive statement can be made relative to fulvic acids, where a significantly higher proportion of the oxygen can be accounted for in known functional groups.

VII. Structural Basis of Humic Acids

From the data presented, it is apparent that humic substances consist of a heterogeneous mixture of compounds for which no single structural formula will suffice. Each fraction (humic acid, fulvic acid, etc.) must be regarded as being made up of a series of molecules of different sizes, none having precisely the same structural configuration or array of reactive groups. In contrast to humic acids, the low-molecular-weight fulvic acids have higher oxygen but lower carbon contents, and they contain considerably more functional groups of an acidic nature, particularly COOH. Another important difference is that, while the oxygen in fulvic acids can be accounted for largely in known functional groups (COOH, OH, C=O), a high portion of the oxygen in humic acids seems to occur as a structural component of the nucleus (as ether or ester linkages, etc.).

Numerous attempts have been made to devise a structural formula representative of humic acids, but none has proved entirely satisfactory. The widely-quoted model of Fuchs, Fig. 9, while meeting certain requirements relative to the numbers of COOH, OH, and C=O groups, is unsatisfactory for several other reasons, one being that data obtained by X-ray diffraction appear to exclude polycondensed systems containing more than a few rings. A polycondensed system also fails to account for conformational changes implied by viscosity studies [52].

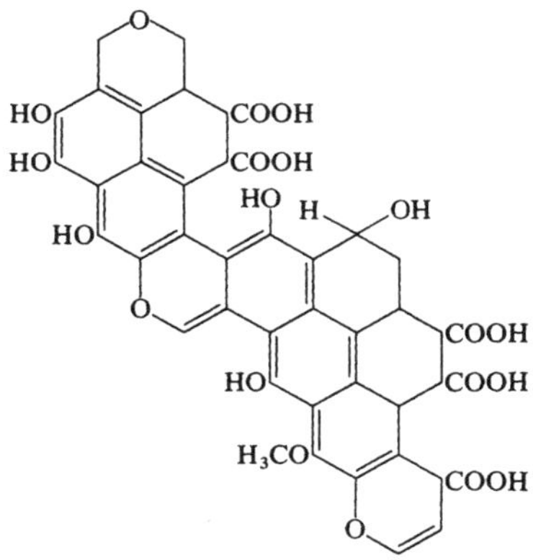

Fig. 9. Structure of humic acid according to FUCHS. See KONONOVA [3] and SWAIN [5]

Contemporary investigators favor a "type" molecule consisting of micelles of polymeric nature, the basic structure of which is an aromatic ring of the di- or trihydroxy-phenol type bridged by —O—, —CH_2—, —NH—, —N—, —S—, and other groups, and containing both free OH groups and the double linkages of quinones. The typical dark color of humic acids, and their ability to be reduced to leucohumic acids with sodium amalgam, is consistent with this concept. In the natural state, the molecule may contain attached proteinaceous and carbohydrate residues.

DRAGUNOV'S formula (Fig. 10) most nearly meets the above requirements. However, a fully acceptable model would also need to account for the occurrence of significant amounts of aromatic COOH groups, some arranged in positions such

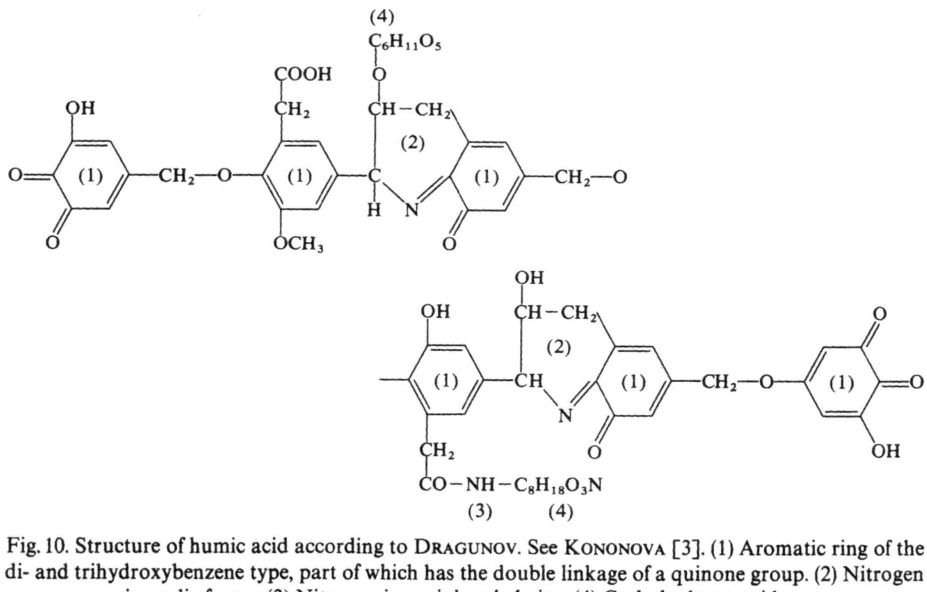

Fig. 10. Structure of humic acid according to DRAGUNOV. See KONONOVA [3]. (1) Aromatic ring of the di- and trihydroxybenzene type, part of which has the double linkage of a quinone group. (2) Nitrogen in cyclic forms. (3) Nitrogen in peripheral chains. (4) Carbohydrate residues

that cyclic anhydrides (see structures IV and V) can be formed by various chemical treatments. The presence of 5-membered ring structures and condensed elements containing two or three rings is also a distinct possibility.

The recent hypothesis of FELBECK [1] that humic acid consists of a central core of pyrone units (VII) held together in a chain by methylene bridges runs into the difficulty that the molecule lacks COOH groups, as well as OH groups which can be methylated readily.

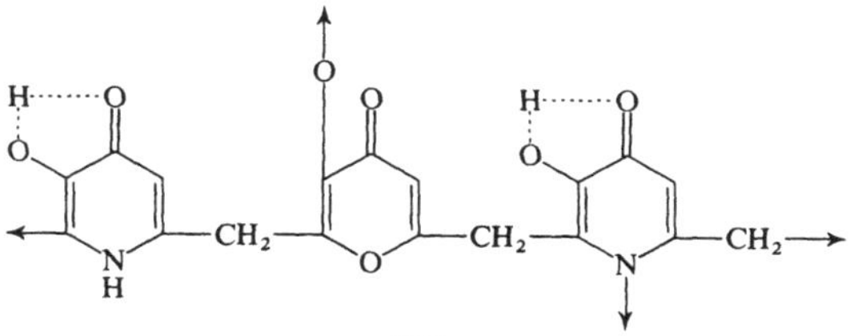

VII

Models based on simple benzenoid systems (VIII to X) bound by oxygen and nitrogen bridges (see discussion of SWAIN [5]) would appear to bear only superficial resemblance to natural humic acids.

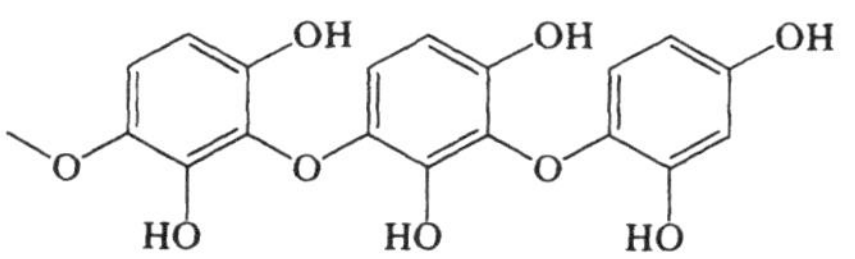

VIII

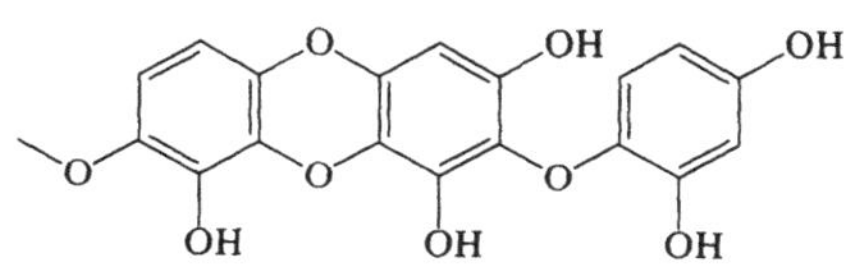

IX

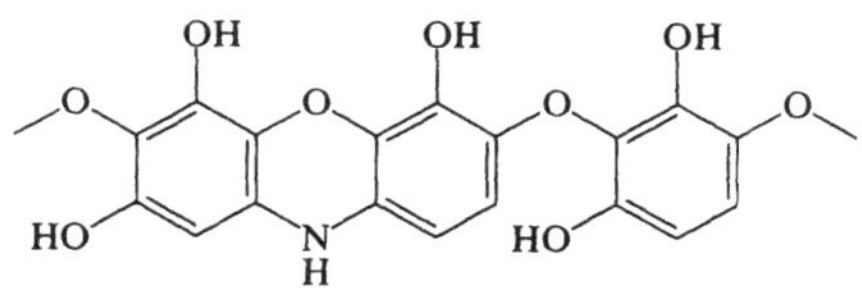

X

VIII. Summary

Although excellent progress has been made in recent years towards the characterization of humic substances, research in this area will have to be expanded if progress is to be maintained. Results obtained thus far indicate that the various fractions (humic acid, fulvic acid, and others) represent part of a system of polymers whose chemical properties (elemental composition, functional group content) change systematically with increasing molecular weight. A solution to the problem of identifying and estimating quinone linkages is vital for a complete understanding of the structures of these constituents. Other major problems remaining to be solved include: (1) nature of bridge units; (2) structural arrangement of reactive groups, and (3) genetic relationships between humic and fulvic acids and their role in the formation of coal and other naturally occurring carbonaceous substances.

References

1. FELBECK, G.T., JR.: Structural chemistry of soil humic substances. Advan. Agron. **17**, 327–368 (1965).
2. DUBACH, P., and N. C. MEHTA: The chemistry of soil humic substances. Soils Fertilizers **26**, 293–300 (1963).
3. KONONOVA, M. M.: Soil organic matter, 2nd ed. New York: Pergamon Press, Inc. 1966.
4. SCHEFFER, F., u. B. ULRICH: Humus und Humusdüngung, Bd. I. Stuttgart, Germany: Ferdinand Enke 1960.
5. SWAIN, F. M.: Geochemistry of humus. In: Organic geochemistry (I. A. BREGER, ed.), p. 81–147. New York: Pergamon Press, Inc. 1963.
6. RAUDNITZ, H.: Occurrence of humic acid in leaves. Chem. Ind. (London) **1957**, 1950–1951 (1957).
7. NAGY, B.: Investigations of the orgueil carbonaceous meteorite. Geol. Foren. Stockholm Forh. **88**, 325–372 (1966).
8. SZALAY, A.: Cation exchange properties of humic acids and their importance in the geochemical enrichment of UO_2^{++} and other cations. Geochim. Cosmochim. Acta **28**, 1605–1614 (1964).
9. STEVENSON, F. J.: Gross chemical fractionation of organic matter. In: Methods of Soil Analysis, part 2 (C. A. BLACK *et al.*, eds.), p. 1409–1421. Madison, Wisconsin: American Society of Agronomy 1965.
10. WHITEHEAD, D. C., and J. TINSLEY: The biochemistry of humus formation. J. Sci. Food Agr. **12**, 849–857 (1963).
11. BREGER, I. A.: Chemical and structural relationship of lignin to humic substances. Fuel **30**, 204–208 (1951).
12. FLAIG, W.: The chemistry of humic substances. In: The Use of Isotopes in Soil Organic Matter Studies, p. 103–127. New York: Pergamon Press, Inc. 1966.
13. PAULI, F. W.: Fluorochrome adsorption studies on decomposing plant residues. I. Decomposition studies. S. African J. Agr. Sci. **4**, 123–134 (1961).
14. SWABY, R. J., and J. N. LADD: Stability and origin of soil humus. In: The Use of Isotopes in Soil Organic Matter Studies, p. 153–159. New York: Pergamon Press, Inc. 1966.
15. STEVENSON, F. J., and S. N. TILO: Nitrogenous constituents of deep-sea sediments. Proc. 3rd. Intern. Meeting on Org. Geochem. (in press).
16. AVGUSHEVICH, I. V., and N. M. KARAVAYEV: Determination of acid groups in humic acids and in some organic compounds. Soviet Soil Sci. (English Transl.) **1965**, 416–422 (1965).
17. SCHNITZER, M., and U. C. GUPTA: Determination of acidity in soil organic matter. Soil Sci. Soc. Am. Proc. **29**, 274–277 (1965).
18. WRIGHT, J. R., and M. SCHNITZER: Oxygen-containing functional groups in the organic matter of the Ao and Bh horizons of a podzol. Trans. 7th Intern. Congr. Soil Sci. **2**, 120–127 (1960).
19. MOSCHOPEDIS, S. E.: Studies in humic acid chemistry. III – The reaction of humic acids with diazonium salts. Fuel **41**, 425–435 (1962).
20. WOOD, J. C., S. E. MOSCHOPEDIS, and W. DEN HERTOG: Studies in humic acid chemistry. II – Humic anhydrides. Fuel **40**, 491—502 (1961).
21. BLOM, L., L. EDELHAUSEN, and D. W. VAN KREVELEN: Chemical structure and properties of coal. XVIII—Oxygen groups in coal and related products. Fuel. **36**, 135–153 (1957).
22. FARMER, V. C., and R. I. MORRISON: Chemical and infrared studies on phragmites peat and its humic acid. Sci. Proc. Roy. Dublin Soc., Ser. A, **1**, 85–104 (1960).
23. BUTLER, J. H. A., and F. J. STEVENSON: Functional groups of soil humic acids. Unpublished.
24. DUBACH, P., N. C. MEHTA, T. JAKAB, F. MARTIN, and N. ROULET: Chemical investigations on soil humic substances. Geochim. Cosmochim. Acta **28**, 1567–1578 (1964).
25. MARTIN, F., P. DUBACH, N. C. MEHTA u. H. DEUEL: Bestimmung der funktionellen Gruppen von Huminstoffen. Z. Pflanzenernaehr. Dueng. Bodenk. **103**, 27–39 (1963).
26. JONES, R., and S. STERNHELL: Chemistry of brown coals. VII—Estimation of active hydrogen. Fuel **41**, 457–469 (1962).
27. SCHNITZER, M., and J. G. DESJARDINS: Molecular and equivalent weights of the organic matter of a podzol. Soil Sci. Soc. Am. Proc. **26**, 362–365 (1962).
28. WRIGHT, J. R., and M. SCHNITZER: Oxygen-containing functional groups in the organic matter of a podzol soil. Nature **184**, 1462–1463 (1959).

29. LYNCH, B. M., J. D. BROOKS, R. A. DURIE, and S. STERNHELL: Chemistry of humic acids formed by alkali treatment of brown coals. Sci. Proc. Roy. Dublin Soc., Ser. A, **1**, 123–131 (1960).
30. SCHNITZER, M., and U. C. GUPTA: Some chemical characteristics of the organic matter extracted from the O and B_2 horizons of a gray wooded soil. Soil Sci. Soc. Am. Proc. **28**, 374–377 (1964).
31. BARTON, D. H. R., and M. SCHNITZER: A new experimental approach to the humic acid problem. Nature **198**, 217–218 (1963).
32. FORSYTH, W. G. C.: The characterization of the humic complexes of soil organic matter. J. Agr. Sci. **37**, 132–138 (1947).
33. WAGNER, G. H., and F. J. STEVENSON: Structural arrangement of functional groups in soil humic acid as revealed by infrared analyses. Soil Sci. Soc. Am. Proc. **29**, 43–48 (1965).
34. SCHNITZER, M., and S. I. M. SKINNER: The carbonyl group in a soil organic matter preparation. Soil Sci. Soc. Am. Proc. **29**, 400–405 (1965).
35. FRIEDMAN, S., M. L. KAUFMAN, W. A. STEINER, and I. WENDER: Determination of hydroxyl content of vitrains by formation of trimethylsilyl ethers. Fuel **40**, 33–36 (1961).
36. SCHNITZER, M., and S. I. M. SKINNER: A polarographic method for the determination of carbonyl groups in soil humic compounds. Soil Sci. **101**, 120–124 (1966).
37. BROOKS, J. D., R. A. DURIE, B. M. LYNCH, and S. STERNHELL: Infra-red spectral changes accompanying methylation of brown coals. Australian J. Chem. **13**, 179–183 (1960).
38. SCHNITZER, M.: The application of infrared spectroscopy to investigations on soil humic compounds. Can. Spect. **10**, 121–127 (1965).
39. DURIE, R. A., and S. STERNHELL: Some quantitative infrared absorption studies of coals, pyrolysed coals, and their acetyl derivatives. Australian J. Chem. **12**, 205–217 (1959).
40. ČEH, M., and D. HADŽI: Infra-red spectra of humic acids and their derivatives. Fuel **35**, 77–83 (1956).
41. THENG, B. K., J. R. H. WAKE, and A. M. POSNER: The infrared spectrum of humic acid. Soil Sci. **102**, 70–72 (1966).
42. SCHWARTZ, D., and L. ASFELD: Solvation of the proton in humic acids. Nature **197**, 177–178 (1963).
43. – –, and R. GREEN: The chemical nature of the carboxyl groups of humic acids and conversion of humic acids to ammonium nitrohumates. Fuel **44**, 417–424 (1965).
44. FALK, M., and D. G. SMITH: Structure of carboxyl groups in humic acids. Nature **200**, 569–570 (1963).
45. FUJII, S.: Infra-red spectra of coal: The absorption band at 1600 cm^{-1}. Fuel **42**, 17–23, 341–343 (1963).
46. BROWN, J. K., and W. F. WYSS: Oxygen groups in bright coals. Chem. Ind. (London) **1955**, 1118 (1955).
47. SCHNITZER, M., D. A. SHEARER, and J. R. WRIGHT: A study in the infrared of high-molecular weight organic matter extracted by various reagents from a podzolic B horizon. Soil Sci. **87**, 252–257 (1959).
48. MOSCHOPEDIS, S. W., S. C. DRIVASTAVA, J. C. WOOD, and N. BERKOWITZ: Homogeneous liquid-phase halogenation of coal and humic acids. Fuel **42**, 338–340 (1963).
49. ATHERTON, N. M., P. A. CRANWELL, A. J. FLOYD, and R. D. HAWORTH: Humic acid—I. ESR spectra of humic acids. Tetrahedron **23**, 1653–1667 (1967).
50. STEELINK, C.: Free radical studies of lignin, lignin degradation products, and soil humic acids. Geochim. Cosmochim. Acta **28**, 1615–1622 (1964).
51. –, and G. TOLLIN: Stable free radicals in soil humic acid. Biochim. Biophys. Acta **59**, 25–34 (1962).
52. MUHKERJEE, P. N., and A. LAHIRI: Polyelectrolytic behavior of humic acids. Fuel **37**, 220–226 (1958).

CHAPTER 23

Soil Lipids

R. I. MORRISON*

The Macaulay Institute for Soil Research,
Aberdeen, Scotland

Contents

I. Introduction

In the present context soils are probably best described as sediments in which plants grow. This definition covers the range from sandy soils containing little organic matter to highly organic soils like peats, but excludes obviously aqueous media, even although these environments are capable of supporting plant growth. The term "lipid" is derived from the Greek word λίπος meaning "fat", but it has no specific chemical connotation. BLOOR [1] defined lipids as substances of biological origin, insoluble in water but soluble in "fat" solvents such as ether, chloroform or benzene; he classified them into *simple lipids*, such as esters of fatty acids with glycerol (glycerides) or with long-chain alcohols or sterols (waxes), *compound lipids*, containing nitrogen, phosphorus or sulphur (e.g., phosphatides), and *derived lipids*, such as free fatty acids, alcohols, hydrocarbons, steroids and isoprenoids. This classification, although not entirely satisfactory, is fairly generally accepted and will be used here with such minor modifications and additions as seem appropriate.

Present trends [2, 3] in the chemical classification of soil organic matter favor its division theoretically into two main groups which in practice may be very difficult to separate: (1) *humic substances*, dark-colored, acidic, high-molecular weight substances, of indefinite structure and produced in soil, and (2) *non-humic substances*, consisting of all classes of organic compounds or their degradation

* The Editors regret to report the death of Dr. MORRISON on Nov. 23rd, 1967. Dr. J. S. D. BACON kindly corrected the proofs of this chapter.

products which also occur in or are produced by living organisms and are of at least fairly well-defined chemical structure. Obviously, lipids fall into this second group which also includes carbohydrates, proteins, lignins, plant pigments and other diverse types of compound too numerous to mention here.

Information on soil lipids is both meager and fragmentary. They have received virtually no systematic attention from soil chemists, no doubt partly because of the almost insuperable difficulty of the separation and identification of components until recently, but also because these lipids seemed of little significance from the point of view of the agronomist. SCHREINER and SHOREY, however, in their now classic work on the isolation of organic substances from soils, extending mainly over the period 1908–1914, obtained several lipid preparations which they were able at least partly to characterize. Since then the only extensive study of the lipids of mineral soils has been that reported by MEINSCHEIN and KENNY [4] in 1957. The lipids, or, more specifically, the waxes, of peats have attracted greater attention because of possible commercial applications, and work in this field has recently been reviewed by HOWARD and HAMER [5]: peat waxes are also discussed as a separate topic in Chapter 24. An extensive review by STEVENSON of the subject of soil lipids was published in 1966 (see Bibliography).

II. Isolation and Characterization of Soil Lipids

A. Total Lipid Content of Soils

It is clear that lipids usually comprise only a small part of the organic matter of soils, but no precise values, either comparative or absolute, can be given since no standard or generally accepted method of determination is available. Much

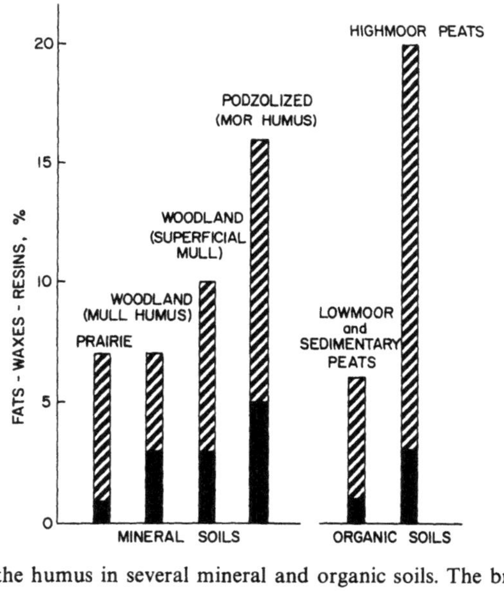

Fig. 1. Lipid content of the humus in several mineral and organic soils. The broken portions of the bars indicate the range of values reported. Figure from STEVENSON (see Bibliography)

information in the literature on the amount of material extracted from soils by lipid solvents cannot be coordinated because of the use of different solvents and conditions. Some indication, however, is given by the amounts of material obtained by the continuous extraction of soils with ether, as in the first stage of the method of proximate analysis of soil organic matter proposed by WAKSMAN and STEVENS [6]. As will be seen later, this method of extraction is neither complete nor selective for lipids but at present it provides the only comparative values available for a fairly wide range of soils. WAKSMAN and HUTCHINGS [7] found that for a variety of soils the amount of ether-soluble material with few exceptions fell within the range 1–5 percent of the total organic matter. Similar results were obtained by SHEWAN [8] for some soils from north-east Scotland. Results obtained by such methods are illustrated in Fig. 1, and they do indicate that quite wide variations in lipid content according to soil type can be expected; high lipid values are typically associated with low pH, and this in turn suggests a correlation with microbial activity. Most of the early work in this field has been reviewed by WAKSMAN [9] and virtually no advance has been made since then. We still lack a reasonably accurate, convenient and generally acceptable method for the determination of total lipid in soils.

B. Extraction

The extraction of lipids from soils is complicated by several factors, but there has been little systematic study to find the most suitable conditions including solvents and pretreatments, for which there are obviously many possible combinations. Although lipids are by definition readily soluble in non-polar solvents such as ether, etc., not all the lipid material in soils can be extracted by such solvents alone. Some lipids may be associated with other organic constituents, e.g., carbohydrates, in such a way as to be insoluble in lipid solvents; some may be complexed with or adsorbed on mineral constituents, especially clays; fatty acids may be present as calcium or other insoluble salts. It is well known that polar solvents such as methanol or ethanol are effective in releasing bound lipids from tissues, and their solvent power may be improved by admixture with selective lipid solvents such as chloroform or ether. The amount of material extracted from soils is greatly increased by the use of such mixtures. Continuous extraction in a Soxhlet apparatus with mixtures of benzene and ethanol in various proportions has often been used to remove lipids from soils and peats. Although this method is convenient and economical in solvent, there is a serious risk that the use of elevated temperatures may have detrimental effects on the extracted lipids. The use of heat in the extraction of lipids from tissues has been found neither necessary nor desirable [10]. MEINSCHEIN and KENNY [4] and others have used a mixture of benzene and methanol (10:1) at room temperature, but it is not certain that all the lipid material is thus extracted. In a study of the extraction of phospholipids from soils, HANCE and ANDERSON [11] compared several methods and found that the use of a succession of solvents extracted more phosphorus than an ether-ethanol mixture. They found also that much less lipid phosphorus was extracted from an air-dry than from a fresh soil, but that the difference could be eliminated by pretreating the dry soil with a mixture of hydrochloric and hydrofluoric acids. In this study only the total phosphorus extracted was determined, but similar studies involving determination of the total lipid extracted and also of the amount extracted by each solvent would be useful.

It seems probable, therefore, that the efficient extraction of lipids from soils will require a preliminary treatment with a mixture of dilute aqueous hydrochloric and hydrofluoric acids, followed by extraction with a mixture of polar and non-polar solvents such as methanol and benzene. The procedure should be kept as simple as possible, compatible with efficient extraction, and, to avoid oxidative degradation or condensation of the extracted material, solvents should be removed at low temperatures ($< 30°$ C) and preferably in an inert atmosphere.

C. Contaminants

The extraction of lipids from soils by such methods inevitably yields dark, waxy products containing much material of a non-lipid nature. The amount of this contaminating material may, especially if continuous extraction has been employed, be greater than the actual lipids. Material extracted from peat by lipid solvents is frequently termed peat bitumen, and it seems useful to use the term "soil bitumen" to describe similar crude lipid extracts from soils in general. In the extraction of lipids from biological tissues [10], the contaminating materials are largely water-soluble and may be removed by extraction with water. With soil bitumen, however, the nature and origin of the contaminating material is obscure; much of it is soluble in methanol, and probably little if any is soluble in water, although information on this point is lacking. At present, however, there is no method of proved efficacy for separating soil lipids from contaminating material.

An obvious approach is to re-extract the bitumen with non-polar solvents such as hexane or light petroleum. Such a method was used by HANCE and ANDERSON [11] to purify crude phospholipid extracts from soils. Peat bitumen has been separated by solvent extraction into three arbitrary fractions [12] (1) "resins", material soluble in methanol, (2) "true wax", material soluble in light petroleum, and (3) "asphalt", material insoluble in both methanol and light petroleum. The relative proportions of these three fractions vary with both the solvent and conditions employed to extract the bitumen. The fraction of soil bitumen soluble in light petroleum is still dark in color, and although it probably contains the bulk of the lipid material, comparison of its infrared spectrum with that of the insoluble fraction [13] suggests that a sharp separation will be difficult to achieve by solvent fractionation alone.

Another approach is the use of adsorption chromatography either alone or to supplement solvent extraction. MEINSCHEIN and KENNY fractionated soil bitumens on silica gel, eluting successively with *n*-heptane, carbon tetrachloride, benzene, and methanol. The first three solvents eluted about 20 percent of the material, and methanol eluted about 70 percent; up to 12 percent was not eluted. The heptane and carbon tetrachloride eluates seemed to be mainly saturated hydrocarbons, and the benzene eluates ester waxes. The more polar lipids, such as long-chain fatty acids or alcohols, were therefore not separated from the contaminating material. Some modification of this method might overcome this objection.

D. Isolation and Identification of Components

Once more no established procedures are available and at this stage only tentative methods can be proffered. A wide range of modern techniques for the isolation and identification of organic substances are considered in Chapter 3 and only those which appear specially applicable to soil lipids will be mentioned here.

Adsorption chromatography in some form is certain to be useful for the separation of classes of substances in soil lipid mixtures. Either column or layer techniques may be employed; columns can deal with relatively large quantities of material, but preparative layer chromatography usually gives better separations. In any event, thin layer chromatography is invaluable for monitoring the effluent from columns. The most generally useful adsorbent for both column and layer chromatography is likely to be silica gel, but other adsorbents such as neutral alumina or magnesium trisilicate may be valuable for certain separations. An automatic fraction collector is a necessity for column chromatography. Jacketed columns which may be heated up to a temperature of about 50° C are useful in the chromatography of materials not readily soluble at room temperature in the solvent of choice.

Crude soil lipid extracts invariably contain much acidic material, and it is probably advantageous to separate acids from non-acids before proceeding with chromatographic or other procedures. Acids may be separated by extraction from a water-immiscible solvent such as benzene with aqueous sodium bicarbonate or suitable buffer solution, but this method frequently results in the production of troublesome emulsions. Alternatively, the acids may be neutralized, in well-cooled benzene-ethanol solution, with ethanolic potassium hydroxide. The addition of excess calcium chloride will then precipitate the acids as calcium salts which, after being dried, can be directly converted to methyl esters by treatment with anhydrous methanol and sulphuric acid. By chromatography of the methyl esters on silica gel or magnesium trisilicate the acids may be separated into such fractions as unsubstituted, hydroxy and aromatic acids. The non-acid material may be chromatographically fractionated without additional pretreatment or it may be converted to saponifiable and unsaponifiable fractions by alkaline hydrolysis. The acids of the saponifiable fraction may be converted to methyl esters and separated as already indicated. The unsaponifiable fraction may be separated into water-solubles, containing moieties like glycerol, inositol, choline, ethanolamine, and "lipid-solvent" solubles, containing derived lipids such as alkanes, alcohols, steroids, etc., which may be separated by adsorption chromatography.

Clathration techniques like urea adduction or the use of molecular sieves may be valuable in separating straight- from branched-chain and/or cyclic compounds prior to the application of adsorption chromatography.

Gas-liquid chromatography (GLC) is a powerful and convenient method for the separation of closely related substances provided they are sufficiently volatile or can be converted to volatile derivatives; complex mixtures of homologous long-chain alkanes, and of methyl esters of long-chain alkanoic acids have, for example, been separated thus. The method is suitable also for quantitative estimation of individual components. When mixtures are insufficiently volatile for GLC, recourse may be made to reversed phase partition chromatography on paper or thin layers.

For the identification of individual compounds after isolation no standard procedure can be advocated. The methods chosen will depend on the nature of the compound and the amount available. More often than not the use of classical methods will be ruled out through scarcity of material. Any of the modern identification methods described in Chapters 3 and 4 may be applicable but GLC, thin-layer chromatography (TLC), infrared and mass spectrometry are likely to be among the most generally useful. IR spectrometry, in addition to its application to the identification of pure substances, is also of value in suggesting probable components of partly fractionated mixtures. The importance of obtaining confirmation of identity of substances by independent methods should not be overlooked.

III. Chemical Nature of Soil Lipids

Information on the nature and abundance of individual lipids in soils is not extensive. Early workers lacked efficient methods to separate and identify closely related components in complex mixtures, and modern techniques have not as yet been extensively applied. The following survey of lipids identified in soils is necessarily brief and should not be considered exhaustive.

A. Glycerides

From the alkaline extract of a silt loam SCHREINER and SHOREY [14] obtained a slightly acid oil which on saponification yielded glycerol and a mixture of unidentified fatty acids, some of which were unsaturated. Similar products, but in smaller amounts, were obtained from other soils. Although since then there have been no reports of the detection or determination of simple glycerides in soils, it seems reasonable to believe that small amounts are usually present.

B. Waxes

Typical waxes are mainly the esters of high aliphatic acids with higher aliphatic alcohols, but cyclic alcohols may also participate. Both the acids and alcohols are usually saturated and unbranched, with carbon numbers ranging from C_{12} to C_{34} in which the even-numbered members predominate. Available evidence indicates that mixtures of waxes comprise a large part of the lipids of soils. Individual components are, however, difficult to separate, and evidence of composition is usually based on identification of degradation products.

MEINSCHEIN and KENNY [4] concluded that the benzene eluates from silica gel column chromatography of benzene-methanol soil extracts consisted largely of waxes. The types of acids and alcohols forming the wax esters were determined by converting the esters to saturated hydrocarbons which were further fractionated by chromatography and analysed by mass spectrometry. Not only aliphatic but tetra-, penta- and hexacyclic hydrocarbons were detected, indicating that the soil wax contained normal fatty acids, normal primary alcohols together with steroids and probably also triterpenoids. The wax fraction examined accounted for about 5 percent of the benzene-methanol extracts.

BUTLER, DOWNING and SWABY [15] obtained from an Australian green soil a wax which was a mixture of hydrocarbons and esters of normal fatty acids with normal primary alcohols. By saponification and column chromatography it was separated into hydrocarbon (38 percent), acid (33 percent) and alcohol (11 percent) fractions. GLC revealed the presence of normal acids and alcohols with carbon numbers from C_{12} to C_{30} with even-numbered members predominating. The principal acid components were C_{22} (13 percent), C_{24} (22 percent) and C_{26} (21 percent); the alcohol mixture was similar.

HIMES and BLOOMFIELD [16] very recently isolated a mixture of waxes from an "organic mat" which had accumulated in a soil as the result of copper-induced inhibition of biological activity. From an examination of the wax by IR, NMR, and mass spectrometry they concluded that the general structure was $CH_3 \cdot (CH_2)_xCOO(CH_2)y \cdot CH_3$. Values obtained for x and y indicated that the acid components were n-C_{14}, n-C_{16}, and n-C_{18}, and the primary alcohols were n-C_{26}, n-C_{28} and n-C_{30}. The carbon numbers of these acids seem surprisingly low and may not be typical of soil waxes in general.

Recently also, MORRISON and BICK [17] saponified the neutral fraction of the lipids from a garden soil and by adsorption chromatography of the methyl esters obtained a fraction consisting of long-chain saturated fatty acids. GLC showed that the acids n-C_{18} to n-C_{34} were present and that about 80 percent were even-numbered. The acids n-C_{24}, n-C_{26}, n-C_{28} and n-C_{30} accounted for about 60 percent. The unsaponifiable fraction contained normal primary alcohols from n-C_{18} to n-C_{30}, about 90 percent even numbered, but mainly n-C_{22}, n-C_{24}, n-C_{26} and n-C_{28}. Similar results were obtained with wax from a peat. The same workers also obtained, from IR spectra, indications of the presence of hydroxy acids in the saponifiable fraction of the neutral lipids from the same sources. These could be present originally as esters, estolides or lactones.

C. Phospholipids

The subject of phospholipids in soils has been reviewed by BLACK and GORING [18] and more recently by ANDERSON [19]. The presence of phospholipids in soils is inferred from (1) phosphorus extracted from soils by lipid solvents and (2) the identification of phospholipid components produced by the acid or alkaline hydrolysis of soil lipid preparations. The amount of phosphorus thus extracted is small, usually 1 to 7 ppm, but 34 ppm has been reported [18]. Lipid phosphorus accounts for less than 1 percent of the organic phosphorus in soils. HANCE and ANDERSON [20] found glycerophosphate, choline and ethanolamine in approximate molar ratios of respectively 1:1:0.2; neither inositol nor serine could be detected. These results suggest that phosphatidyl choline (lecithin; I) is the most abundant phospholipid in soils, followed by phosphatidyl ethanolamine (II).

D. Acids

Crude soil lipid extracts contain much acidic material the precise nature of which is still obscure. A large part consists of what are usually described as resin acids of unknown constitution; but quite a substantial portion seems to consist of a mixture of higher fatty acids both unsubstituted and hydroxylated. From different soils, SCHREINER and SHOREY [21] obtained two acid preparations to which they assigned the same molecular formula ($C_{24}H_{48}O_2$). One which they called paraffinic acid had m.p. 45°–48° C and was obtained from a silt loam; the other, m.p. 80°–81° C, obtained from a peaty soil, they called lignoceric acid from its resemblance to an acid originally obtained from beechwood tar. In the light of present knowledge it seems probable that both of these preparations were mixtures, mainly of unsubstituted long-chain fatty acids.

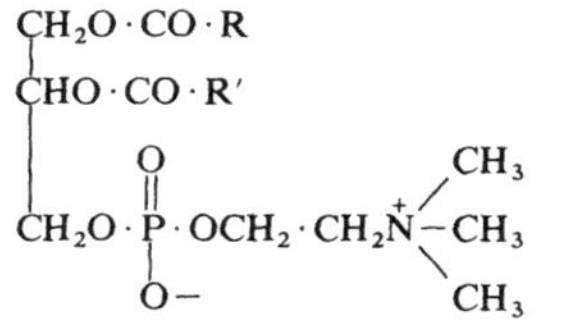

(I) Phosphatidyl choline (lecithin)

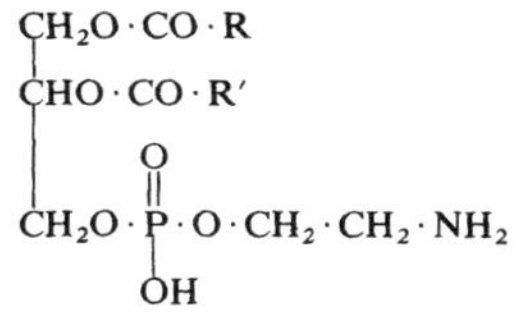

(II) Phosphatidyl ethanolamine

R, R′ = alkyl

Recently MORRISON and BICK [17] obtained a purified lipid preparation from a garden soil which contained about 20 percent of free acids, 55 percent of which consisted of a mixture of fatty acids n-C_{20} to n-C_{34}, with 80 percent of even carbon number. The even-numbered acids, n-C_{24} to n-C_{32}, together made up 75 percent of the fraction.

So far there are no reports of branched-chain acids in soils although they have been found in marine sediments [22].

Free long-chain hydroxy acids are present in soils. SCHREINER and SHOREY [23] identified 9,10-dihydroxystearic acid in 27 out of 84 soils of widely different types. From one soil only, a Maryland silt loam low in organic matter, they obtained also a mono-hydroxystearic acid, m.p. 84°–85° C, which was probably α-hydroxystearic acid, but the position of the hydroxyl group was not rigorously established [24]. Another hydroxy acid preparation, m.p. 109°–110° C, which they called agroceric acid, was obtained from a North Dakota chernozem type soil [21]; it had an elementary composition corresponding to $C_{21}H_{42}O_3$ and may well have been a mixture. MORRISON and BICK [17] have obtained evidence from IR spectra of the presence of a significant free hydroxy-acid fraction in soil lipids, but no individual components have been identified.

It is probable that unsaturated acids are present in soils only in very small amounts. KHAINSKII [25] demonstrated the presence in soil organic matter of both oleic and elaidic acids, and ASCHAN [26] obtained from peat an apparently unsaturated acid, $C_{19}H_{34}O_2$, which he called humoceric acid, but more recent reports are lacking.

Lipid preparations from soils are liable to contain small amounts of aromatic acids such as benzoic, p-hydroxybenzoic or vanillic acids. Larger amounts of phenolic acids are chemically associated with humic substances and may be liberated, at least in part, by acid or alkaline hydrolysis. SWAN [27] has found an Alaskan forest soil containing about 4.6 percent dry weight of dehydroabietin $C_{19}H_{28}$, (III) a cyclic hydrocarbon related to and probably derived from the diterpenoid resin acid, levopimaric acid, commonly present in resins of conifers.

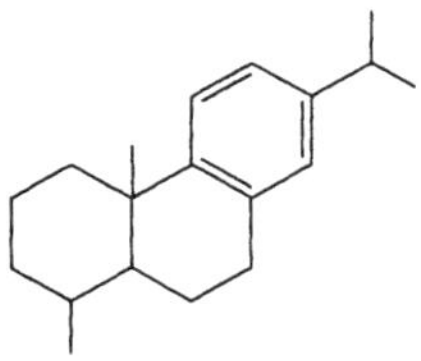

(III) Dehydroabietin

Acids may also be present, especially in mineral soils of relatively high pH, as salts of low solubility such as those of calcium or magnesium; pretreatment of soils with mineral acids usually results in an increase in the amount of material subsequently extracted by lipid solvents, but the precise nature of this additional material has never been studied.

Local concentrations of fatty acids and their insoluble, principally calcium, salts are frequently found at the burial sites of human or other large animal corpses. These local accumulations of lipids have been given the name "adipocire", and have been discussed in some detail by BERGMANN [28].

E. Hydrocarbons

Higher alkanes are probably generally present in soils but, compared with waxes and acids, in relatively small quantities. SCHREINER and SHOREY [29] isolated a paraffinic hydrocarbon from a North Carolina organic soil (27 percent organic carbon). From the method of isolation, its chemical behavior, melting point, and elementary analysis they concluded that it was *n*-hentriacontane ($C_{31}H_{64}$). It is probable, however, that at least small amounts of other alkanes were present. Recently GILLILAND and HOWARD [30] obtained an aliphatic hydrocarbon fraction from a peat bitumen; by mass spectrometry it was shown to consist mainly of the odd-carbon-numbered *n*-alkanes C_{29} (11 percent), C_{31} (40 percent) and C_{33} (34 percent), but it contained also small amounts of all other *n*-alkanes from C_{27} to C_{32}. From an Australian soil BUTLER, DOWNING and SWABY [15] obtained a hydrocarbon fraction which, on GLC study, gave peaks corresponding to *n*-alkanes from C_{17} to C_{24} with odd and even-carbon numbered members in approximately equal amounts. MORRISON and BICK [17] obtained *n*-alkane fractions from both a garden soil and a peat, and separated the components by GLC. Both fractions consisted largely of alkanes from *n*-C_{19} to *n*-C_{33} with about 87 percent odd-numbered. The main components in the fraction from the garden soil were *n*-C_{29} (21 percent), *n*-C_{31} (31 percent) and *n*-C_{33} (15 percent).

Polynuclear aromatic hydrocarbons have been detected in small quantities in soils. KERN [31] isolated chrysene from a garden soil and estimated the concentration as 15 ppm. From UV adsorption studies of certain fractions of the benzene-methanol extracts of two soils MEINSCHEIN and KENNY [4] obtained indications of the presence of alkylbenzenes, alkyl-naphthalenes and alkyl-phenanthrenes, 1,12-benzperylene and coronene. GILLILAND, HOWARD and HAMER [32] have isolated perylene from peat.

BLUMER [33] has obtained evidence that both 3,4- and 1,2-benzpyrene are common and fairly abundant (40–1,300 µg/kg) constituents of soils from rural areas, distant from major highways and industries. He was able also to detect phenanthrene, fluoranthrene, pyrene, chrysene, perylene, anthanthrene, anthracene, triphenylamine, benzanthracene, benzfluorene, 1,12-benzperylene and coronene. It thus seems likely that these hydrocarbons may be indigenous to soils and not exclusively derived from atmospheric fall-out as suggested by COOPER and LINDSEY [34]. It may be significant also that HAWORTH and his collaborators [35] have recently identified many of these aromatic hydrocarbons as products of the distillation of humic acids with zinc dust.

F. Ketones

MORRISON and BICK [36] have isolated long-chain methyl ketone fractions from both a garden soil and a peat. The components were separated by GLC; those from the peat ranged from n-C_{17} to n-C_{33} with 92 percent odd carbon-numbered, and n-C_{25} (26 percent) and n-C_{27} (37 percent) specially abundant; from the soil the range was n-C_{19} to n-C_{35} with 81 percent odd carbon-numbered, mainly n-C_{25} to n-C_{33}. This is the first report of the occurrence of methyl ketones in this range as natural products; the lower members (up to C_{11}) are well known as products of the action of microfungi on milk fatty acids in certain cheeses.

G. Steroids

There is little information on the nature and amount of steroids in soils in spite of their wide distribution in plant and animal tissues. The only steroid positively identified in soils is β-sitosterol (IV), a common sterol of higher plants, which has recently been obtained from a garden soil [28] and from peats [30, 37]. SCHREINER and SHOREY [38, 39] obtained from different soils two preparations which they believed to be sterols since both gave a positive Liebermann reaction. One, which they call *agrosterol*, was a free alcohol with composition corresponding to the formula $C_{26}H_{44}O$; it had, however, a high melting point (237° C) and could

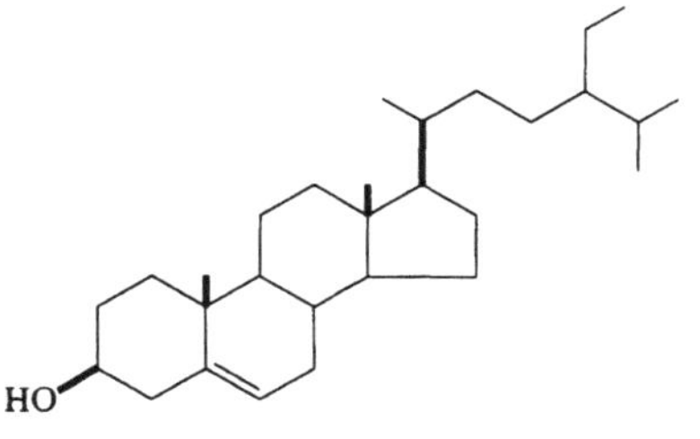

(IV) β-sitosterol
(24α-ethyl, Δ^5 cholesten-3β-ol)

well have been a triterpenoid. The other preparation which they called *phytosterol* had m.p. 135° C and was obtained from a peaty soil by a process involving saponification; it might have been β-sitosterol.

TURFITT [40] in 1943 examined several soils for the presence of free sterols. The sterols were extracted from the soils with acetone, purified by precipitation as digitonides, and estimated colorimetrically by the Liebermann-Burchard reaction. The range of values found was from nil to 12 mg per kg of soil; highest values were found in peats or under other conditions of acidity and lack of aeration, and much lower values in arable soils. On this basis free sterols could account for about 1 percent of the soil lipids. Neither cholesterol, which is common in animals, nor ergosterol, the prevalent sterol in fungi, have been found in soils.

H. Terpenoids

In spite of the abundance and variety of terpenoids in plants, there are few reports of their detection in soils. Friedelan-3β-ol (*epi*friedelanol; V) has been identified in peats by independent workers [30, 37] but it is evidently present in very small quantities. MEINSCHEIN and KENNY [4], in their study of benzene-methanol extracts of mineral soils by mass spectrometry obtained indications of the presence of penta- and hexacyclic compounds which could have been triterpenoids.

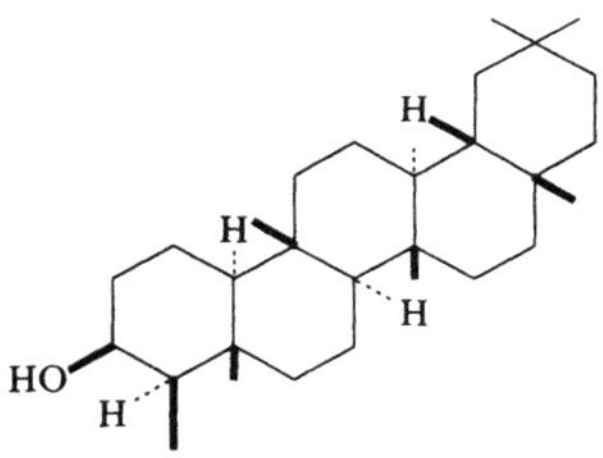

(V) Friedelan-3β-ol
(*epi*friedelanol)

I. Carotenoids

Small amounts of carotenoids have been detected in soils. By the exhaustive extraction of peaty soils with organic solvents, BAUDISCH and EULER [41] obtained carotene (0.6 mg per g of organic matter) and xanthophyll, the latter associated with chalk-containing layers.

J. Miscellaneous

Many other substances of a lipid nature, particularly of plant origin, are likely to be present in soils, but only in very small quantities. Such substances would include tocopherols and porphyrins from higher plants, and polynuclear quinones of fungal origin. Chlorinated insecticides and their degradation products, although not lipids in the strict sense, would be associated with them.

Approximate indications of the amounts and composition of some types of soil lipids are shown in the table. These are based on preliminary results obtained by MORRISON and BICK [17] by methods similar to those outlined in the experimental section at the end of the chapter.

Table. *Some components of soil lipids*

		Phragmites peat (pH 4.0; OM 94 percent)			Garden soil (pH 5.9; OM 8.2 percent)		
Bitumen (ethanol-benzene extract)		5.45 percent of OM			2.69 percent of OM		
Solvent-refined lipid		1.59 percent of OM			0.62 percent of OM		
		mg/100 g OM	a	b	mg/100 g OM	a	b
Acids	*n*-alkanoic acids						
	free	208	24–30	86	70	24–32	80
	esterified	80	18–30	82	51	20–34	81
	hydroxy acids, etc.						
	free	300	unidentified		16	unidentified	
	esterified	95	unidentified		16	unidentified	
Unsaponifiable	*n*-alkanes	8	19–33	14	21	23–33	13
	n-alkan-2-ones	10	17–33	12	40	25–35	19
	n-alkan-1-ols	81	16–30	92	9	20–30	93
Total identified or partly identified		782			233		

OM = organic matter.
a Carbon number range of principal components.
b Mol. percent even carbon-numbered.

IV. Transformations of Soil Lipids

There is little well-founded information on the origins and transformations of soil lipids, and much of what follows is inevitably surmise. Moreover, it is impossible within the available space to consider other than the main implications.

A. Origin of Soil Lipids

Obvious sources of soil lipids are plant and animal residues and the soil microbial population. Vast amounts of plant litter are continually being deposited on the surface of the soil, and are an obvious potential source of a wide variety of lipids. How much of this lipid material is incorporated in the soil will depend upon its resistance to degradation in the particular environmental conditions, and resistance will obviously vary with the type of substance. Recently, CHAHAL, MORTENSEN and HIMES [42] reported results of the incubation in a peat soil for 128 days of C^{14}-labelled rye (*Secale cereale*) tissue. They found that 1.9 percent of the organic matter was extracted with benzene-methanol (10:1) but that this contained 40.3 percent of the residual ^{14}C. This suggests that under these conditions plant lipids are relatively resistant to decomposition; other conditions however might yield different results. Plant cuticle waxes [43] are a likely source of soil waxes, long-chain acids, alcohols and alkanes. Steroids and terpenoids seem likely to be derived mainly from plant sources, but plant glycerides and phosphatides are unlikely to make much contribution.

It is improbable that animal residues contribute much directly to soil lipids. The typical components of animal lipids, glycerides of palmitic, stearic and oleic acids, are rapidly degraded and seem not to be abundant in soils. Adipocire is formed only under special circumstances. Minor contributions by the wax secretions of insects and other terrestrial arthropods can be expected.

Many species of earthworms are usually abundantly present in well-drained mull soils, but they are completely absent from acid soils. The obvious result of their activities is to remove plant residues from the surface and mix them with the soil to a depth of two or three feet. Any direct contribution to soil lipids is likely to be limited.

Little is definitely known of the synthetic contribution of the soil microflora, but obviously it could be significant. Micro-organisms could, for instance, be the main source of the small glyceride and phosphatide components. None of the unusual high molecular weight acids, the mycolic acids, characteristic of the Mycobacteria has yet been detected in soil.

B. Degradation of Lipids in Soils

Micro-organisms must be mainly responsible for the degradation of soil lipids as of soil organic matter in general, but little information is available as to the organisms concerned or the degradative pathways. Degradative acitivity will obviously be at a maximum under eutrophic, aerobic conditions. Glycerides, phosphatides and unsaturated fatty acids are probably quite rapidly attacked by numerous micro-organisms. It seems, however, that the longer chain components (C_{24} and upwards) are relatively resistant to attack as evidenced by their persistence in mineral soils, possibly on account of their very low aqueous solubility. TURFITT [44] demonstrated experimentally the degradation of cholesterol added to soils, and also showed that certain soil organisms, e.g., *Proactinomyces erythropolis*, were able to degrade cholesterol *in vitro* and to grow on it as the sole source of carbon. The first stage in the attack was the conversion of cholesterol to cholest-4-ene-3-one [44].

Although alkanes are regarded as among the most durable of organic compounds, many soil organisms are known to be capable of degrading them [45]. The most active appear to be bacteria or actinomyces belonging to the closely related genera, *Streptomyces*, *Nocardia*, *Mycobacterium*, *Corynebacterium* and *Brevibacterium*, but other bacteria and also microfungi have been found to possess similar ability. It is fairly well established [45] that the most common pathway of oxidation of the alkane chain is by monoterminal oxidation followed by β-oxidation, thus:

$$R\cdot CH_2\cdot CH_2\cdot CH_3 \xrightarrow{[O_2]} R\cdot CH_2\cdot CH_2\cdot CH_2(OH) \xrightarrow{-2H} R\cdot CH_2\cdot CH_2\cdot CH{=}O$$

$$\xrightarrow[-2H]{+HOH} R\cdot CH_2\cdot CH_2\cdot C(=O){-}OH \xrightarrow{\beta\text{-oxidation}} R\cdot COOH + CH_3COOH$$

Other pathways, however, are known; the long-chain methyl ketones found in soils might be formed directly from alkanes, or from acids by β-oxidation and subsequent decarboxylation [46].

It is probable that the original biological discrepancy between odd- and even-carbon-numbered long chain compounds may be gradually reduced as the result of microbial activity [47], but the mechanism of the process is not clear.

In soils where they are abundant, earthworms undoubtedly play an important but indirect part in the transformations of soil organic matter through the effect of their activities on soil structure and aeration. Virtually nothing is known however of the direct effect of earthworms on soil lipids or indeed on soil organic matter in general.

C. Effect of Lipids on Soil Conditions and Plant Growth

It is not known if lipids have any direct effect on soil conditions; it is possible that the resistance to rewetting exhibited by soils which have been dried may be related to the lipid content, but this has never been investigated. There is, however, evidence that the presence in soils of long-chain compounds can result in the increased microbial production by micro-organisms of substances, possibly polysaccharide gums, which increase the stability of soil crumbs [48, 49, 50]. An indirect effect on soil fertility is thus possible.

Certain lipids, on the other hand, may be inimical to plant growth. GREIG-SMITH [51] attempted to explain soil exhaustion as the result of the accumulation of fats and waxes, but this has never been substantiated. SCHREINER and SHOREY [23] found that dihydroxystearic acid was toxic in quite low concentrations to plants growing in nutrient solutions, but FRAPS [52] showed that the injurious effect was much reduced in soil.

V. Conclusion

Present information suggests that the most important soil lipid components are long-chain (C_{22} to C_{34}) alkanoic acids, either unsubstituted or hydroxylated, free or combined as waxes with alcohols or sterols. Smaller amounts of alkanes or methyl ketones of similar chain-length range are also present, together with much smaller amounts of glycerides, phosphatides, and many other substances which may be regarded as lipids in the widest sense. Nevertheless, future studies may well result in some modification of this picture based as it is on inadequate information. Now that most of the technical impediments have been eliminated, progress will largely depend on the topic proving of sufficient interest and importance to attract the attention of investigators, whether in the field of soil science or organic geochemistry.

VI. Experimental Procedures

The following, necessarily brief, description of procedures for the extraction and fractionation of soil lipids is based on methods used by MORRISON and BICK and described elsewhere in greater detail [17, 36]. They are to a large extent tentative, and improvements will undoubtedly result as experience increases. For example, pretreatment of the soil with dilute aqueous HCl/HF mixtures may be found advantageous; extraction with cold solvents may be preferred to Soxhlet extraction; the validity of the arbitrary separation into resin, asphalt, and lipid may be questioned.

A. Extraction and Fractionation

The ground and sieved air-dried soil (1–2 kg) is extracted continuously for 20 hours with benzene-ethanol (2:1 by volume) in a 2 litre Soxhlet extractor. The soluble material (*bitumen*) is in turn continuously extracted with light petroleum (b.p. 60–80° C) in a 100 ml Soxhlet extractor. The soluble material (crude lipid) is heated under reflux with a mixture of methanol and isopropanol (1:1 by volume; 10 ml/g lipid). The hot solution is decanted from the undissolved material (*asphalt*), and the extraction repeated with half the original volume of solvent. The combined extracts are cooled to 5° C for 18 hours and the precipitate (*lipid*) collected, washed with cold methanol and dried. The soluble material (*resin*) may also be collected by evaporation of the solvent.

B. Separation of Free Acids and Conversion to Methyl Esters

The lipid material, thus prepared, is dissolved in benzene (15 ml/g) and neutralized with ethanolic KOH (0.5 *N*) at not more than 5° C. The neutral solution is treated with excess $CaCl_2$ in ethanol, heated, and the precipitated Ca salts collected, washed with benzene, and dried. The methyl esters are prepared by heating the dry, finely-divided Ca salts in anhydrous methanol, benzene, and sulphuric acid.

C. Saponification of the Non-Acid Fraction

The material remaining after removal of the acids as described above is heated under reflux for about 6 hours with 1 *N* ethanolic KOH (5 ml/g) and benzene (10 ml/g). The saponified acids are then precipitated as Ca salts, and converted to methyl esters as in B above. The unsaponifiable material is recovered from the residual solution.

D. Column Chromatography of Methyl Esters

The methyl esters of the acid fractions are further fractionated by chromatography in light petroleum/benzene on a column of magnesium trisilicate (mixed if necessary with sufficient Celite or similar material to give a suitable flow-rate). Fractions are collected automatically and examined by IR spectroscopy, TLC and GLC. Light petroleum-benzene elutes unsubstituted long-chain fatty acids, chloroform elutes hydroxy acids, and chloroform with 5 percent ethanol elutes aromatic acids.

E. Column Chromatography of Unsaponifiable Material

The unsaponifiable fraction is dissolved in hot light petroleum (b.p. 60–80° C) and chromatographed on a column of neutral alumina at 50° C. Light petroleum elutes alkanes closely followed by methyl ketones. Benzene elutes primary alcohols. Unidentified material is eluted by ethanol/benzene. The fractions are examined as in C above.

F. Gas-Liquid Chromatography

Individual components of separated classes of compounds: alkanes, methyl ketones, *n*-alkanoic acids (Fig. 2), primary alcohols, etc., may be separated readily and determined by GLC either as the original compounds or as suitable derivatives. Suitable conditions are: column, 5′ × 1/4″, 2 percent SE 30 + 0.2 percent Carbowax 20 M on Chromosorb W 60/80; col. temp. 225–250° C; sample 0.1 mg.

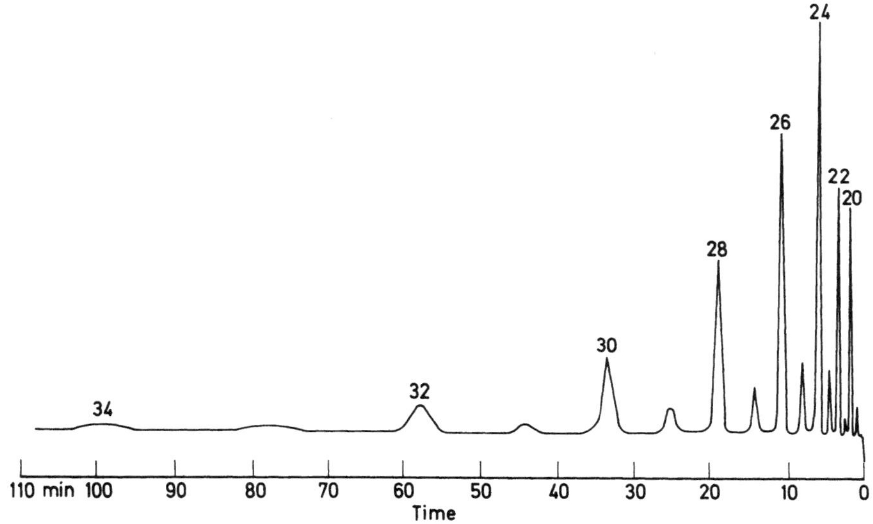

Fig. 2. GLC separation of the methyl esters of bound *n*-alkanoic acids from a garden soil. Pye argon chromatograph: column, 120 cm long × 4 mm internal diameter; 2 percent SE 30 + 0.2 percent Carbowax 20 M on acid-washed silanized Celite 60/80; temp. 250° C; sample 0.1 mg. The figures adjacent to the peaks are the carbon numbers of the respective acids

References

1. Bloor, W. R.: Biochemistry of the fats. Chem. Rev. **2**, 243 – 300 (1925).
2. Kononova, M. M.: Soil organic matter, 2nd English Tr. by T. Z. Novakowski and A. C. D. Newman. Oxford: Pergamon Press, Inc. 1966.
3. Scheffer, F., and P. Schachtschabel: Lehrbuch der Bodenkunde. Stuttgart: Ferdinand Enke 1966.
4. Meinschein, W. G., and G. S. Kenny: Analyses of a chromatographic fraction of organic extracts of soils. Anal. Chem. **29**, 1153 – 1161 (1957).
5. Howard, A. J., and D. Hamer: The extraction and constitution of peat wax. J. Am. Oil Chemists' Soc. **37**, 478 – 481 (1960).
6. Waksman, S. A., and K. R. Stevens: A critical study of the methods for determining the nature and abundance of soil organic matter. Soil Sci. **30**, 97 – 116 (1930).
7. –, and I. J. Hutchings: Chemical nature of organic matter in different soil types. Soil Sci. **40**, 347 – 363 (1935).
8. Shewan, J. M.: The proximate analysis of the organic constituents in north-east Scottish soils. J. Agr. Sci. **28**, 324 – 340 (1938).
9. Waksman, S. A.: Humus. London: Baillière, Tindall & Cox 1936.
10. Sperry, W. M.: Lipide analysis. In: Methods of biochemical analysis, vol. 2 (D. Glick, ed.). New York: Interscience Publishers 1955.

11. HANCE, R. J., and G. ANDERSON: Extraction and estimation of soil phospholipids. Soil Sci. **96**, 94–98 (1963).
12. ACKROYD, G. C.: The extraction, properties and constitution of peat wax. Intern. Peat Symp. Dublin 1954.
13. THOMAS, J. D. R.: Chemistry of peat bitumen: fractionation and infrared studies. J. Appl. Chem. **12**, 289–294 (1962).
14. SCHREINER, O., and E. C. SHOREY: Glycerides of fatty acids in soils. J. Am. Chem. Soc. **33**, 78–80 (1911).
15. BUTLER, J. H. A., D. T. DOWNING, and R. J. SWABY: Isolation of a chlorinated pigment from green soil. Australian. J. Chem. **17**, 817–819 (1964).
16. HIMES, F. L., and C. BLOOMFIELD: Extraction of triacontyl stearate from a soil. Plant Soil **26**, 383–384 (1967).
17. MORRISON, R. I., and W. BICK: The wax fraction of soils: separation and determination of some components. J. Sci. Food Agr. **18**, 351–355 (1967).
18. BLACK, W. A. P., and C. A. I. GORING: Organic phosphorus in soils. In: Agronomy, No. 4 (W. H. PIERRE and A. G. NORMAN, eds.). New York: Academic Press, Inc. 1953.
19. ANDERSON, G.: Nucleic acids, derivatives and organic phosphates. In: Soil biochemistry (A. D. MCLAREN and G. H. PETERSON, eds.). New York: Marcel Dekker 1967.
20. HANCE, R. J., and G. ANDERSON: Identification of hydrolysis products of soil phospholipids. Soil Sci. **96**, 157–161 (1963).
21. SCHREINER, O., and E. C. SHOREY: Some acid constituents of soil humus. J. Am. Chem. Soc. **32**, 1674–1680 (1910).
22. LEO, R. F., and P. L. PARKER: Branched-chain fatty acids in sediments. Science **152**, 649–650 (1966).
23. SCHREINER, O., and E. C. SHOREY: The isolation of dihydroxystearic from soils. J. Am. Chem. Soc. **30**, 1599–1607 (1908).
24. — — The isolation of harmful organic substances from soils. U.S. Dept. Agr. Bur. Soils Bull. **53**, 41 (1909).
25. KHAINSKII, A.: Organic substances in the soil humus. Pochvovedenie **18**, 49–97 (1916) [In Russian].
26. ASCHAN, O.: Humocerinsäure, ein Bestandteil im Torf. Torfwirtschaft **2**, 9 (1921).
27. SWAN, E. P.: Identity of a hydrocarbon found in a forest soil. Forest Prod. J. **15**, 272 (1965).
28. BERGMANN, W.: In: Organic geochemistry, p. 517 (I. A. BREGER, ed.). Oxford: Pergamon Press, Inc. 1963.
29. SCHREINER, O., and E. C. SHOREY: Paraffin hydrocarbons in soils. J. Am. Chem. Soc. **33**, 81–83 (1911).
30. GILLILAND, M. R., and A. J. HOWARD: Some constituents of peat wax separated by column chromatography. Transactions of the 2nd International Peat Congress, Leningrad, 1963. Vol. II, p. 877–886. Edinburgh: H.M.S.O. 1968.
31. KERN, W.: Über das Vorkommen von Chrysen in der Erde. Helv. Chim. Acta **30**, 1595–1599 (1947).
32. GILLILAND, M. R., A. J. HOWARD, and D. HAMER: Polycyclic hydrocarbons in crude peat wax. Chem. Ind. (London) 1357–1358 (1960).
33. BLUMER, M.: Benzpyrenes in soil. Science **134**, 474–475 (1961).
34. COOPER, R. L., and A. J. LINDSEY: Atmospheric pollution by polycyclic hydrocarbons. Chem. Ind. (London) 1177–1178 (1953).
35. CHESHIRE, M. V., P. A. CRANWELL, C. P. FALSHAW, A. J. FLOYD, and R. D. HAWORTH: Humic acid—II. Structure of humic acids. Tetrahedron **23**, 1669–1682 (1967).
36. MORRISON, R. I., and W. BICK: Long-chain methyl ketones in soils. Chem. Ind. (London) 596–597 (1966).
37. MCLEAN, J., G. H. RETTIE, and F. S. SPRING: Triterpenoids from peat. Chem. Ind. (London) 1515 (1958).
38. SCHREINER, O., and E. C. SHOREY: The presence of a cholesterol substance in soil; agrosterol. J. Am. Chem. Soc. **31**, 116–118 (1909).
39. — — Cholesterol bodies in soil; phytosterol. J. Biol. Chem. **9**, 9–12 (1911).
40. TURFITT, G. E.: The microbial degradation of sterols. I. The sterol content of soils. Biochem. J. **37**, 115–117 (1943).
41. BAUDISCH, O., u. H. VON EULER: Über den Gehalt einiger Moor-Erdarten an Carotinoiden. Arkiv Kemi. Miner. Geol. **11**: 21, **A**, 10 (1935).

42. CHAHAL, K. S., J. L. MORTENSEN, and F. L. HIMES: Decomposition products of carbon-14 labelled rye tissue in a peat profile. Soil Sci. Soc. Am. Proc. **30**, 217 – 220 (1966).
43. EGLINTON, G., and R. J. HAMILTON: The distribution of alkanes. In: Chemical plant taxonomy (T. SWAIN, ed.). London: Academic Press, Inc. 1963.
44. TURFITT, G. E.: The microbiological degradation of sterols. IV. Fission of the steroid molecule. Biochem. J. **42**, 376 – 383 (1948).
45. FOSTER, J. W.: Hydrocarbons as substrates for microorganisms. Antonie van Leeuwenhoek J. Microbiol. Serol. **28**, 241 – 274 (1962).
46. CHIBNALL, A. C., and S. H. PIPER: Metabolism of plant and insect waxes. Biochem. J. **28**, 2209 – 2219 (1934).
47. ORÓ, J., D. W. NOONER, and S. A. WIKSTRÖM: Paraffinic hydrocarbons in pasture plants. Science **147**, 870 – 873 (1965).
48. FEHL, A. J., and W. LANGE: Soil stabilization induced by growth of microorganisms on high-calorie mold nutrients. Soil Sci. **100**, 368 – 374 (1965).
49. MARTIN, J. P., J. O. ERVIN, and R. A. SHEPHERD: Decomposition and aggregating effects of fungus cell material in soil. Soil Sci. Soc. Am. Proc. **23**, 217 – 220 (1959).
50. ZOBELL, C. E.: Action of microorganisms on hydrocarbons. Bacteriol. Rev. **10**, 1 – 49 (1946).
51. GREIG-SMITH, R.: The action of wax solvents and the presence of thermolabile bacteriotoxins in the soil. Proc. Linnean Soc. N. S. Wales **35**, 808 – 822 (1910).
52. FRAPS, G. S.: The effect of organic compounds in pot experiments. Texas Agr. Exp. Sta. Bull. **174** (1915).

Bibliography

ALEXANDER, M.: Introduction to soil microbiology. New York: John Wiley & Sons, Inc. 1961.
BERGMANN, W.: Geochemistry of lipids. In: Organic geochemistry (I. A. BREGER, ed.). Oxford: Pergamon Press, Inc. 1963.
DEGENS, E. T.: Geochemistry of sediments; a brief survey. Englewood Cliffs, N.J.: Prentice-Hall, Inc. 1965.
HANAHAN, D. J., F. R. N. GURD, and I. ZABIN: Lipide chemistry. New York: John Wiley & Sons, Inc. 1960.
KONONOVA, M. M.: Soil organic matter, 2nd English Translation by T. Z. NOVAKOWSKI and A. C. D. NEWMAN. Oxford: Pergamon Press, Inc. 1966.
LOVERN, J. A.: The chemistry of lipids of biochemical significance. London: Methuen 1957.
McLAREN, A. D.: Biochemistry in soil science. Science **141**, 1141 – 1147 (1963).
STEVENSON, F. J.: Lipids in soil. J. Am. Oil Chemists' Soc. **43**, 203 – 210 (1966).
WAKSMAN, S. A.: Humus. London: Ballière, Tindall & Cox 1936.

CHAPTER 24

Earth Waxes, Peat, Montan Wax and Other Organic Brown Coal Constituents

V. WOLLRAB and M. STREIBL

Czechoslovak Academy of Sciences,
Institute of Organic Chemistry and Biochemistry,
Prague, Czechoslovakia

Contents

I. Special Isolation and Identification Methods

This section will deal only with special methods bearing directly on the isolation and identification of natural products treated in this chapter, especially methods used in the analysis of waxes.

Extraction of coal or peat is carried out with crushed material using individual organic solvents [1], or mixtures which give better yields than individual solvents. In practice, benzene and mixtures of hydrocarbons or halogen derivatives with polar solvents such as alcohols have proved most effective. The higher extraction yields of these mixtures with polar solvents may be explained by the fact that the vapors of these liquids, such as acetone, alcohols and pyridine, cause swelling and crumbling of coal. The simultaneous increase of hydroxyl groups [2] is now explained by the splitting of hydrogen bonding between OH-groups connecting molecular layers. Extraction yields may be further increased by higher temperatures and pressures during extraction. Excessive preliminary drying of material may result in the reduction of the yield of the substances extracted [3].

Chromatographic Methods

Column adsorption chromatography separates substances according to their relative polarities. The adsorbent used most frequently is silica gel, sometimes partly deactivated with water [4]. Alumina is less suitable for this purpose as it has a greater tendency to hydrolyse [4] waxes (esters).

The following eluotropic series [4] for example, is recommended for the separation of waxy substances: petroleum ether – benzene – chloroform – chloroform/ether – chloroform/ether/methanol. A particularly good separation is achieved by gradient elution [5], i.e., linear increase of the more polar eluent. Sometimes the slight solubility of wax components at normal temperatures causes slow elution and thus an insufficient separation. This problem can partly be obviated by means of chromatography at an elevated temperature, 40–50° C, for example, in a jacketed column [4], but obviously low-boiling solvents cannot be used for this purpose. At higher temperatures, however, the adsorbing capacity of the adsorbent decreases somewhat. It is also advisable in preparative scale chromatography to add solvents previously saturated with water; anhydrous solvents activate the adsorbent by drying it and polar substances may be irreversibly retained. Because of these difficulties, chromatography at elevated temperatures is recommended only in cases where large amounts of wax esters or other poorly soluble substances are present in the mixture. Sometimes functional group tests, e. g., hydroxyl or carboxyl groups [4], are suitable to determine the content of the eluted substances in individual fractions. It is thus possible to detect even relatively small amounts of free acids or possibly even waxes after hydrolysis. Further information on the nature of the fractions may be obtained by various specific chemical reactions and physical methods, especially spectroscopy.

Thin layer chromatography (TLC) is a powerful tool for obtaining a quick survey of substances present in a mixture. On a microscope slide, for example, it is possible to correlate substances from an adsorption chromatographic column with standard substances. Detection of non-volatile components of a waxy nature is preferably carried out with a spray of concentrated sulfuric acid and a subsequent charring [4] of organic substances, for example, under a coil of a *not* wire. Visualization may also be carried out by spraying with a fluorescent dye and observing under UV light. The good separating quality of TLC is often used for preparative separations of waxy substances on larger plates using a thicker layer of adsorbent containing binder and Rhodamine 6 G as the visualizer. When using alumina, it is also possible to work with loose layers [4] according to Mottier [6], since the adsorbent may be poured on the plate in the activated form. For separating saturated compounds from unsaturated substances, both in column chromatography and in TLC, silica gel impregnated with about 20 percent $AgNO_3$ has been used [7]. The silver ion effects a separation of unsaturated and saturated hydrocarbons as well as the partition of various types of unsaturated substances. Thus a good separation occurs with *trans*-olefins which are eluted before *cis*-olefins and olefins with a terminal double bond.

Ion exchange chromatography, suitable for the separation of wax acids from the neutral fractions [4, 8], is carried out on a column with a water jacket (50° C) filled with a strongly basic ion exchanger such as Zerolith FF. The neutral fraction is eluted with benzene, and carboxylic acids, after their displacement with a stronger acid (i.e., mineral acid [4] or acetic acid [8]), are also eluted with benzene. The miscibility between the aqueous phase and benzene is provided by isopropanol which is miscible with the two liquids in an unlimited ratio.

Partition paper chromatography makes possible the separation of homologues of various substances. In separating wax components it is necessary to work on reverse phases at a higher temperature [9, 10].

High-temperature gas chromatography (GLC) is the most effective separation method both for qualitative and quantitative analysis of waxy substances. Because of a high working temperature (up to 350° C), the method requires special apparatus and thermostable phases. Gas chromatography of wax hydrocarbons is the most refined method to date. The most convenient phase appears to be modified [11] Apiezon L which possesses a very powerful separating effect for hydrocarbons; non-polar silicone elastomers of rubbers and phenyl ethers are also frequently used [12, 13]. Sufficient experience has been gathered to ascertain the presence of various types of saturated and unsaturated hydrocarbons in a mixture, especially on the basis of studies with synthetic standard mixtures [14]. It is therefore frequently convenient to convert by a chemical process [11, 15, 16], mixtures of oxygen-containing substances (such as alcohols, esters, acids) to the corresponding hydrocarbons, for analysis. Today even direct analysis of mixtures of methyl esters of wax acids are being carried out, especially on polyester and polar silicone phases [17, 18]; the analysis is in this case confined mostly to the separation of individual homologues since the separation of esters of various types of acids (e.g., n-acids from variously branched ones) has not yet been sufficiently developed. It is possible to expect an increasing application of capillary columns to the analysis of wax components [13].

Finally, it should be noted that GLC generally separates homologues within a class. In a complex mixture, however, the application of column chromatography is necessary to separate the mixture into various classes of compounds.

Clathration and molecular sieving are separation methods utilizing differences in the molecular shape of the substances of partitioned mixtures. Several methods [15, 19] involving clathration with urea are used with long chain paraffins to determine the concentration of branched and cyclic paraffins in a mixture containing n-paraffins. In order to achieve a perceptible concentration of branched hydrocarbons, which are frequently present in the original mixture in very small concentration, the adduction process is repeated several times.

A 5 Å molecular sieve [16, 20] separates n-paraffins from branched or cyclic paraffins. The sieving procedure involves a reflux operation in isooctane, 2,2,4-trimethylpentane, the most suitable solvent. After the refluxing, it is sometimes advisable to immediately reflux the molecular sieves themselves in either pure benzene or ether to remove any branched paraffins from the surface of the sieve.

The adsorbed n-paraffins within the sieve are exchanged for smaller paraffins although not quantitatively. With small samples it is more convenient to dissolve the sieves in hydrofluoric acid. Both clathration and molecular sieving of wax components are difficult because even when a chain is branched at one end, the rest of the free chain is long enough to form a clathrate with urea or to penetrate into a channel of the molecular sieve.

Physical methods for identification are infrared, ultraviolet and nuclear magnetic resonance spectroscopy, mass spectrometry and X-ray analysis. Physical methods are very useful even for the identification of relatively large molecules such as those found in bitumens. IR spectra help establish the presence of a number of functional groups [21–25]. IR [26] and NMR [27] spectroscopy may be useful for ascertaining the nature of a double bond. FUCHS and DIEBERG [28] used IR spectra in the range around 1,300 to 1,150 cm^{-1} for determining the chain length of wax acids. UV spectroscopy offers a convenient way of detecting and analyzing a mixture (e.g., coal [29] and fossils [30]). Similarly, it is possible by means of nuclear magnetic resonance to establish the ratio of aliphatic and aromatic groups and to make conclusions as to the distribution of aliphatic H between methyl, methylene, and methine [31–34] groups. X-ray diffraction studies have also provided useful information on the structure of coal [35–38]. Mass spectroscopy may provide valuable information not only on the molecular weight of an isolated chemical compound but also from its fragmentation pattern, on the structure of the given substance. Although a mass spectrum of a mixture is rather difficult to analyze, a combination of GLC and mass spectra provides a highly useful method of analysis [39, 40].

Discussion

The development in recent years of isolation and identification methods requiring only micro quantities of material has provided a serious approach to many analytical problems. This progress is reflected also in the advance of the analysis of bitumens from coal or peat which represent complex mixtures of relatively large molecules. As compared to "classical" methods of separation (e.g., solubility, fractional crystallization and distillation), recent methods of adsorption chromatography (both column and TLC) or ion exchange are, with their availability and simplicity, far more effective and even can provide a relatively good separation of very complex mixtures. The group separation of complex natural mixtures is one of the fundamental prerequisites for a detailed analysis and successful identification of the greatest possible number of individual substances present in the natural mixture.

II. Specific Geological Organic Materials

A. Earth Waxes

1. Definition and Depositional Changes

Earth waxes (ozocerite, Hatchettite, Scheererite) are natural mixtures of organic substances. They are of dark yellow/brown to black waxy appearance and occur either in veins and nodules or as films on brecciated rocks [41]. The main constituents are hydrocarbons, while the sum of oxygen, nitrogen and sulfur content averages less than 0.5 percent. The occurence of ozocerite near crude oil deposits [41, 42] and the fact that saturated and aromatic hydrocarbons in ozocerite and in the oil fraction of petroleum have similar structures [43–45] suggest that crude oils and ozocerite may originate from the same source and be closely connected generically. Also the common occurence of several derivatives of porphyrin in oil shale, petroleum, and ozocerite support this hypothesis. According to TREIBS [46], mesoporphyrin possibly derived from hemin is present, and this suggests that there is a contribution from animal bodies to the formation of ozocerite. It has been found that certain microorganisms utilize 56–73 percent of the ozocerite-paraffinic and aromatic hydrocarbons, thereby decreasing their content by 12–38 percent with a corresponding increase in resinuous components [47].

2. Chemical Components

The low refractive index, low carbon residue, absence of aromatic peaks and presence of a strong paraffinic peak in IR spectra suggest that ozocerite from Utah [41] is a predominantly paraffinic material. The paraffin mixture usually amounts to 60–90 percent [43] with a high content (40–50 percent) of normal or slightly branched paraffins [48] which form adducts with urea. Cyclic paraffin derivatives are present in the amount of 7 to 8.5 percent [43]. The type of composition of the substances contained varies according to the source; thus, for example, ozocerite from the Carpathian deposits [49] contains up to 18 percent of naphthenes and up to 3.6 percent of aromatic hydrocarbons. Ozocerites also contain small amounts of resinuous compounds. On the basis of spectroscopic measurements of samples of Carpathian ozocerites [44], a number of details, such as the high degree of branching, the number of methyl groups, and substitution positions in monocyclic aromatics have been established for several hydrocarbons in the paraffinic-naphthenic fractions. In another sample of ozocerite [40] high resolution mass spectrometry has indicated the presence of the compound $C_{12}H_{18}S_2$.

B. Peat

1. Definition, Original Chemical Materials and Depositional Changes

Peat is a natural product of decaying plants usually occurring anaerobically. It is a soft organic material with a high water content; frequently it preserves within itself organized plant remains [50]. Its geological age is very little and the transformation of plant material without oxygen in water medium occurs in the initial stage. Presumably, coal in the first stage develops in a similar manner. ALEXENIAN [51] doubts that coal passes through a peat phase since IR spectra show that cellulose participates in the formation of peat and fossil wood but is practically nonexistent in lignite. This conclusion does not seem convincing since the low content of cellulose in lignites proves only that its decay has advanced considerably. Recalling that peat originated only recently, while brown coal was predominant even in the Tertiary era, it appears that peat originated from qualitatively different plant sources and was formed under different climatic conditions. Nevertheless, it is probable that the transformation of organic substances in the initial stage of coal formation was similar to peat.

Raw material for the formation of peat consists of cellulose, lignin, and cork-like tissues which are the main constituents of plants; other substances are resins, waxes, proteins, dyes, etc. The original theories supposed that cellulose, the basic constituent of wood, had the greatest share in the formation of peat and coal [52]. On the other hand, FISCHER and SCHRADER [53] have proposed that the role of cellulose was negligible because cellulose invariably undergoes bacterial decomposition at a rapid rate. A further study [54] states that cellulose is not directly involved in the formation of humic substances but serves only as a nutrient for microorganisms. Using MAILLARD's results [56], ENDERS [55] has shown that sugars in the presence of amino acids can be transformed to humic-like substances, and, further, that in the presence of amino acids or protein decomposition products, both lignin and cellulose can participate in the formation of humic acids when natural conditions become unfavourable for the growth of microorganisms. The fungi and microfauna together with bacteria are undoubtedly active in the humification process [57], which consists of an initial aerobic phase with the disappearance of the surface plant organs, followed by an anaerobic phase in which the decomposition rate is lower and in which mineralization occurs.

The formation of humic acids, which show phenolic properties, can most probably be explained by the transformation of lignin, whose structural units are also phenolic in character. A more difficult problem is to explain the formation of humic acids from cellulose. FLAIG and SCHULZE [58] suppose that the decomposition of cellulose produces reactive ketoacids which show a strong tendency toward aromatization.

Some very stable chemical substances also present in plants (waxes, higher isoprenoids, resins) resist the chemical and microbiological attack and can therefore be found both in peat and brown coals. While cellulose, hemicellulose [59], lignin, and amino acids decrease in the course of humification, humic acids increase. The relatively stable substances listed above accumulate as well, some of them partly subject to changes in the C- and O-content; the C-content is increased and the O-content slightly reduced [54].

Since it has been noted that proteins rapidly disappear in the decomposition of plant material, FUCHS [50] believes that the amino acids identified in peat do not originate from plants but are the products of microbial metabolism which takes place during humification as well.

2. Chemical Components

The chemical components of one sample of peat [60] are as follows: water-soluble material 5 percent; humic acids 41.4 percent; cellulose 5.6 percent; hemicellulose 22.5 percent; lignin 10.5 percent; bitumen 12.9 percent; and other substances 2.1 percent. A typical peat bitumen [61] contained 52.5 percent wax, 20.4 percent paraffins, and 24.6 percent resins.

a) Waxes

In a narrower sense of the term, waxes are esters of long chain acids and alcohols. In a wider sense of the word this group includes all substances accompanying wax esters and having a long paraffinic chain, i.e., hydrocarbons, alcohols, carboxylic acids, ketones, etc. Wax acids as a rule have a greater number of C-atoms ($C_{16}-C_{40}$), while acids from fats have fewer carbon atoms ($C_{10}-C_{24}$).

There is still relatively little information on the detailed composition of peat waxes. On the basis of IR analysis BEL'KEVICH *et al.* [62] believe that apart from paraffins, which predominate in the wax fractions, peat may also contain aromatic hydrocarbons, and it has been found from X-ray studies [63] that the extract even contains cyclic hydrocarbons possessing long side chains. BEL'KEVICH *et al.* [64] separated from peat by fractional crystallization, several alcohols which they suppose to be higher normal alcohols (carnaybyl, arachic, lignoceryl and neoceryl; these terms are now known to refer to mixtures of homologues) on the basis of the determination of iodine-number and melting points. The mixture also contained a sterol $C_{27}H_{40}O$. The results described are not very convincing in view of the methods used. FISCHER *et al.* [65] established the carbonyl-content in peat, using several methods, to be in the range 0.5–1.8 percent. Fractional precipitation and recrystallization [66] helped establish the presence of acids with physical constants very similar to those of arachidic (n-eicosanoic), behenic (n-docosanoic), lignoceric (n-tetracosanoic), montanic (n-octacosanoic), and melissic (n-triacontanoic) acids which form obvious mixtures. KWIATKOWSKI [67], using an alumina column, eluted with a mixture of chloroform-ethanol free acids rich in oxygen which he considers to be hydroxy acids; he eluted allegedly free acids, C_{20}—C_{30}, with the same solvent system to which formic acid was added. It is, however, improbable that more polar hydroxy acids would be eluted from the column before normal acids and, moreover, with less polar solvents. RAUHALA [68] fractionated peat wax by alumina adsorption chromatography. On the basis of the melting points, molecular weights, hydroxyl values and iodine-numbers, he established the presence of monocarboxylic acids, $C_{19}-C_{22}$, and hydroxy acids, C_{20} and C_{21}. RAUHALA observed the occurrence of normal acids and hydroxy acids in peat wax, but adsorption chromatography apparently did not provide a sufficient separation of the various homologues. Thus, identification on the basis of the given constants remains problematic.

However, no description of the fractionation of peat wax appears in the literature in sufficient detail provide definite conclusions.

b) Resins

Very little is known about the substances constituting peat resins. Apart from general data [69] on the occurence of hydrocarbons, hydroxy-compounds and small amounts of acids, the presence of the aromatic hydrocarbon perylene is also described [70].

c) Humic Acids

In this section, both humic acids from peat and from brown coal are discussed. Humic acids are dark amorphous substances, acid in character which are readily soluble in dilute aqueous alkaline solutions and are formed as the primary products of the decomposition of plant materials. They form mixtures of substances with a heterogeneous distribution of [71] molecular weights from 675 to 9,000. By transformation, most probably polycondensation, they give rise to a water-insoluble product – humin. Under the general name of humic substances we also rank fulvo-acids, soluble in water [72], and hymatomelanic acids [71] which in contradistinction to humic acids are very soluble in ethanol. Of all these substances, humic acids deserve primary consideration. They occur both in peat and brown coal to give the characteristic coloring. The structure of humic acids derived from peat and those from brown coals is practically the same; only X-ray diffraction studies [73] showed some differences in the content of side groups and in the more or less ordered arrangement of condensed aromatic rings. In addition to C, H, and O, they also contain N and S. However, since after washing humic acids with hydrochloric acid they no longer contain either nitrogen or sulfur, GANZ [74] assumes that these elements are not proper constituents of humic acids. The presence of a carboxyl group was confirmed by titrimetric analysis and conductivity measurements on the salts formed [71]. By comparing the molecular weights ODEN [75] concluded that humic acids contained 3–4 groups with acid properties, while STADNIKOFF and KOSHEW [76] estimated the number to be 6–7. SEIJI and collaborators [77] found carboxylic and phenolic groups present in regenerated humic acid, employing a non-aqueous potentiometric titration of acid groups. After a complete methylation with dimethyl sulphate, FUCHS [50] established, from the quantity of methoxy groups, that the humic acid he investigated contained three etherifiable OH-groups. IR spectroscopy confirmed [78] that a humic acid extract from peat showed a phenolic OH-group and a C=O vibration band of a carbonyl or carboxyl group. On the basis of chemical reactions, the content of carbonyl groups in the humic acids molecule was estimated by FUCHS and STENGEL [79] at 1–2. A methylene ketonic ($-CH_2CO-$) grouping is probably involved. In humic acids from peat [80], a methoxy group was also found. FUCHS [81] concluded after a bromination experiment that the humic acids contained both double bonds and a hydrogen capable of substitution and dehydrogenation. The oxidative splitting of humic acids has also been studied. From the oxidation of an etherified humic acid with oxygen, 5 percent of vanillin was obtained [71]; after oxidation with air, small amounts of *m*-hydroxybenzoic and 1,3,5-hydroxyisophthalic acids were isolated from the reaction mixture. Treatment with hydrogen peroxide [82], permanganate [83], and ozone [84] as well as with other oxidizing reagents gives rise principally to two types of acid products: low-molecular aliphatic acids, for example, malic, succinic, acetic, tartaric, glycolic, mesoxalic acid, etc., and acids including variously substituted benzene-carboxylic acids, as for example phthalic, trimesic and mellitic acid. The oxidation of subhumic acids with peroxide [85] produces mostly oxalic and melonic acids.

After methylation and esterification, LANDA and EJEM [86] hydrogenated humic acids in the presence of a tungsten disulphide catalyst, and after chromatographic separation of the products, identified the components by means of NMR analysis and IR spectra. They established the presence of higher aliphatic hydrocarbons and of polycyclic hydrocarbons of naphthenic character.

TAKEKAMI and collaborators [87] carried out the hydrolysis of humic acids with Adkins copper-chromite catalyst and a subsequent oxidation of the reaction mixture with nitric acid. They concluded that humic acid contained ether bonds.

On the basis of X-ray studies [88] it is presumed that the humic acid structure is an individual plane consisting of an aromatic network with side chains containing oxygen and other functional groups. Similarly, other authors are of the opinion [89] that molecular layers are arranged in planar configurations 8–12 Å in height and about 25 Å in diameter.

Attempts have also been made to propose speculative structures of humic acids [90, 91] on the basis of the available analytical data.

d) *Other Substances Present in Peat*

Substances also found in peat include cellulose [92], water-soluble fractions [93], HCl-soluble hemicellulose [93], and lignin [94]. SHACKLOCK and DRAKELY [95] established the presence of a number of diamino and monoamino acids in peat extract. Their total content [96] ranges from 2–4,000 ppm with the neutral amino acids being most abundant and alanine having the highest concentration [97]. Substances extracted from peat by water [98] include some aromatic hydroxy- or methoxy-compounds, ketones and acids. In various kinds of peats, the presence of triterpenes and sterols have also been described.

C. Montan Wax

1. Definition

Montan wax [1] is an extract obtained with organic solvents from some kinds of liptobiolithic or saproplitic coals; for example, up to 70 percent may be obtained from pyropyssite. Because of its high content of waxy aliphatic components and its industrial use, montan wax is usually refered to as a special part of bitumens.

2. Chemical Components

The chemical components of montan wax may be separated on the basis of their different solubility in isopropanol into three main fractions: asphalts, insoluble even in boiling solvents; waxes, precipitates from solution after cooling; and resins, soluble even in cold isopropanol. Bohemian montan wax contains, according to WOLLRAB and co-workers [4], 50.5 percent waxy fraction, 30.0 percent resinic fraction, and 19.5 percent asphalts. The wax fraction proper consists of 2.2 percent higher paraffins, 26.5 percent esters and alcohols, 14.2 percent free carboxylic acids; the other substances of the wax fraction have yet to be identified. PRESTING and STEINBACH [99, 100] give the content of German montan wax as 23.5 percent waxy alcohols, 47.7 percent wax-acids, 12–14 percent asphalts and 10–15 percent resinic fraction.

a) *Wax Fraction: Hydrocarbons*

TANNER and collaborators [101] described n-paraffins $C_{23}-C_{28}$ and C_{30} in montan wax. WOLLRAB and co-workers [11] found originally by means of GLC that Bohemian montan wax contained n-paraffins of the homologous series C_{22}–C_{33} with an odd carbon preference; the hydrocarbons C_{29} (39.1 percent), C_{31} (27.5 percent), and C_{27} (13.9 percent) contributed over 80 percent of the mixture. This fraction was examined again [16] and was found to contain an extensive homologous series of n-paraffins, C_{12}–C_{37}, the paraffins at either extreme present only in traces. In addition, two more homologous series of branched paraffins in the range C_{16}–C_{35} were isolated using a 5 Å molecular sieve. By comparing the results of gas chromatography of various synthetic hydrocarbons types [14] with branched alkanes from wax, it was found that one series is composed of iso- (2-methyl) or anteiso (3-methyl) paraffins and the other series is most probably monomethyl alkanes with a methyl group near

Table. *Composition of waxes and free acids in Bohemian montan wax* [11]

Number of carbon atoms[a]	Wax		Free acids (percent)
	alcohols (percent)	carboxylic acids (percent)	
22	14.5	1.8	3.3
23	2.7	0.2	0.7
24	22.8	8.7	5.0
25	2.9	0.9	1.5
26	31.4	23.3	15.7
27	1.2	2.1	5.5
28	17.5	32.7	26.9
29	0.5	2.8	4.8
30	5.4	21.5	25.8
31	0.2	1.6	3.6
32	0.9	4.2	6.2
33	–	–	0.3
34	–	0.2	0.7

[a] All the above mentioned constituents have a straight chain of carbon atoms.

the center of the chain, or possibly dimethyl alkanes with methyl groups near the ends of the chain. The quantity of these branched paraffins is negligible in montan wax. EDWARDS and collaborators [102] established that n-paraffins in the range C_{23}–C_{33} and a further homologous series of alkanes, possibly branched, were present in montan wax. PRESTING and KREUTER [103] from the iodine-number and refractive index on certain hydrocarbon fractions determined that they contained even carbon-numbered olefins. Using Bohemian montan wax STREIBL and SORM [26] proved the presence of *trans*-olefins in the range C_{11} to C_{40} and olefins with a terminal double bond in the range C_{11} to C_{36}, as well as traces of branched olefins; a further type of olefin could have been isoprenic or cyclic in character according to IR spectra.

The distribution pattern of paraffins from Bohemian as well as German montan wax [11, 103] are similar to those of recent high plant waxes [104], from which it probably follows that brown coal arose from plants similar to contemporary higher plants. Entirely different distribution patterns were found in hydrocarbons of montan wax analyzed by EDWARDS and co-workers [102]. In contradistinction to predominating odd carbon number homologues in Bohemian and German montan wax, this sample shows only a smooth distribution with a maximum at n-C_{28}. However, it is necessary to bear in mind that the sample analyzed by EDWARDS and co-workers [102] is an industrially-treated montan wax where the possibility of chemical changes cannot be excluded.

b) Acids

In the past the composition of the wax-acids of montan wax was the subject of some controversy due to poor separation and identification methods which were available at that time. While TROPSCH and KREUTZER [105] thought that after hydrolysis of montan wax they had isolated acids with an odd number of carbons from the acid fraction, HOLDE and BLEYBERG [106] were of the opinion that this fraction contained only n-acids with an even number of carbon atoms. Only FUCHS and DIEBERG [107] proved that acids with an even number of carbons in the range from C_{20} to C_{34} were contained in montan wax. They used adsorption column chromatography for the analysis of commercial montan acid, and based their conclusion on the determination of the acid values, melting points of individual fractions, and comparison of the mean distance of bands in the IR region 1,150–1,300 cm^{-1}. These results were confirmed by means of paper chromatography by KAUFMANN and DAS [108] and by us [9]. We [9], in addition, established that among the components corresponding to even acids there occurred compounds in very small amounts which could be explained by the presence of odd carbon number acids. In contradiction to the other authors, we analyzed separately the free and esterified acids fractions. We obtained both fractions, the esters and free wax-acids, by first making a group separation of the raw material with column chromatography, and followed by chromatography on an ion exchanger [4] (see the diagram). We later established by means of gas chromatography [11]

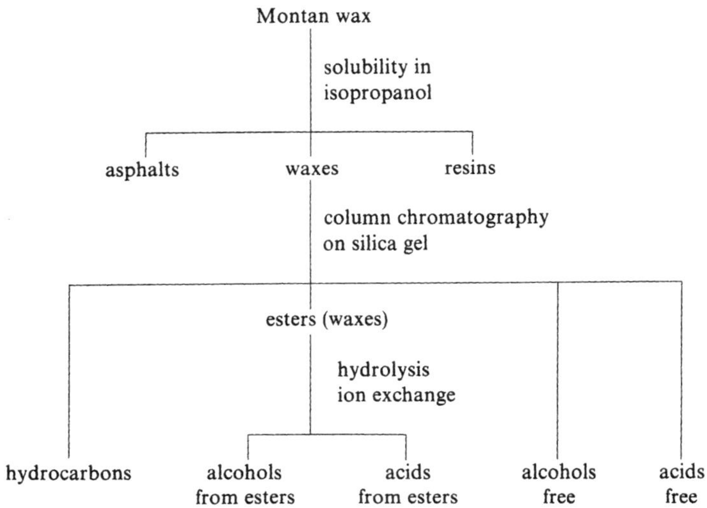

Diagram

that both the free and the bonded acids consisted of a wider homologous series of acids C_{22} to C_{34}, with the odd numbered carbon compounds present in smaller amounts (see the table). Montanic acid (n-C_{28}) is in both fractions the predominant constituent; the qualitative composition of free and bonded acids, therefore, is the same. However, both types show a somewhat different percentage composition (table). In a later investigation [16], we found that the homologous series of n-acids occured in the range $C_{10}-C_{36}$, the low and high carbon number acids occuring only in trace amounts. Moreover, the investigation helped to establish for the first time the presence of two homologous series of branched acids, similar to paraffins; one of them again consists of iso- or anteiso homologues and the other probably of monomethylated acids with the methyl group near the center of the chain, or even carbon number acids twice methylated near the ends of the chain. It is impossible to decide between these two types on the basis of retention values obtained by gas chromatography [14]. Also, HEWETT and collaborators [17] and EDWARDS *et al.* [102] established (by GLC of methyl esters in a saponified bleached montan wax) a homologous series of n-carboxylic acids in the range $C_{16}-C_{35}$. PRESTING and KREUTER [103] also found in German montan wax n-acids in the range $C_{20}-C_{34}$ and believe that acylated hydroxy acids are also present apart from a smaller amount of acylated binary estolides. These bifunctional

$$\mathrm{AcO[\underset{\displaystyle R_1}{\underset{|}{C}H}-CO\cdot O\underset{\displaystyle R_2}{\underset{|}{C}H}]COOH}$$

substances are considered to be the building stones of more complicated substances [109] which then represent the actual and original basis of the wax fraction of montan wax.

c) Ketones

PRESTING [110] described the ketone montanone ($C_{27}H_{55}\cdot CO\cdot C_{27}H_{55}$) in montan wax which he believes, however, was formed secondarily, that is by catalytic effect in the extraction of iron.

d) Alcohols

PSCHORR and PFAFF [111] established in 1920 the presence of the alcohols C_{24}, C_{26} and C_{30} in montan wax on the basis of the melting points of fractionally crystallized acetates. WOLLRAB and co-workers [4] carried out group separation of Bohemian montan wax by column chromatography and then studied the composition of the fraction of ester-bonded alcohols after hydrolysis of the waxes (esters) and removal of combined acids on an ion exchange column (see the diagram). Gas chromatography of the paraffins derived from the alcohols proved again the presence of an extensive homologous series of normal alcohols, $C_{22}-C_{32}$; odd carbon number homologues constitute only a small part (table). The predominate constituents are hexacosanol (31.4 percent) and tetracosanol (22.8 percent). JAROLIMEK *et al.* [16] later carried out a more detailed analysis of this fraction and identified, even further, alcohols present only in trace amount, so that the whole series is constituted by the homologues C_{10}—C_{36}. Similar to the acids, these alcohols also showed the presence of two homologous series of branched isomers. EDWARDS and collaborators [102], who analyzed another montan wax by GLC, describe alcohols as increasing from C_{18} to C_{21}. PRESTING and KREUTER [103] found the alcohols $C_{20}-C_{34}$ in German montan wax by GLC and, moreover, α,ω-diols in the same range, where C_{26}, C_{28} and C_{30} were present in the greatest amount.

e) Waxes

After the group separation of the content substances [4], WOLLRAB *et al.* [11] carried out the hydrolysis of chromatographic fraction of waxes (esters) and identified the wax components (table) by GLC. The complex procedure of Bohemian montan wax [11, 26] is given in the diagram for the sake of clarity and the composition is given in the table.

D. Other Organic Brown Coal Constituents

1. Chemical Changes During the Coalification Process

Coal contains mainly carbon, hydrogen and oxygen. During the coalification process (which from the chemical standpoint seems to be a dehydration process) hydrogen and oxygen decrease and the content of carbon increases [52]. The content of carbon therefore increases [50] from peat (50–60 percent) through brown coal (60–65 percent), black coal (75–90 percent), up to anthracite (92–96 percent). Hydroxyl and carbonyl content decreases [114, 115] during coalification. The cellulose [116] and hemicellulose [117] content diminishes during the initial stage of coalification. The amount of phenolic groups increases with the age of lignites but decreases in brown coal [117]. Both the aromaticity and the size of the aromatic nucleus [118] grow at the expense of the alicyclic part during the progressive metamorphism of coal. Several studies were carried out on the pressure coalification of brown coal in the presence of water [119] as well as on the effects of temperature and ultra-high pressure on the coalification of bituminous coal [120].

2. Bitumen—Definition and Chemical Constituents

The substances which can be extracted with organic solvents from natural fossil material are called bitumens*. FUCHS [50] designated the bitumen obtained by extraction under normal pressure as bitumen A; bitumen B was extracted under increased pressure; and bitumen C is an extract obtained after previous treatment of the original material with hydrochloric acid. The amount of bitumen is highest in peat and brown coal, much smaller in black coal, and in coal containing 90 percent carbon the extractibility ceases altogether [52]. The content of carbon increases similarly in bitumen [50] in the sequence: peat (70–72 percent), brown coal (77–80 percent), black coal (80–85 percent). This order indicates that even the extractable substances are gradually subject to chemical changes, especially dehydrogenation.

OUCHI and IMUTA [121] chromatographed an extract from YUBARI coal on a column of alumina and subsequently identified by GLC hydrocarbons of various types. They found paraffins in the range C_9-C_{31}, predominantly C_{20}. The principle aromatic nuclei detected were benzene, naphthalene, phenanthrene, chrysene and picene, that is, zigzag-cata type condensed rings. The most abundant alkyl derivatives of these aromatic compounds were those having two or three carbon atoms in the alkyl side chain, except for benzene derivatives which had mostly long branched alkyl side chains in the para position. Other compounds obtained were fluoreve, anthracene, pyrene, benzofluorene, fluoranthene, and 1,2-benzanthracene. Inert oxygen was mainly heterocyclic, e.g., diphenyleneoxide and benzo-phenyleneoxide. Extracts from SANGARSK coal [122] contain many more condensed aromatic than nonaromatic hydrocarbons. The IR spectrum of the bitumen from recent lignites [123] shows that alicyclic hydrocarbons, $-COCH_2$ groups or cyclic hydrocarbons, are present.

* Montan wax is a special type of bitumen discussed in the preceding section.

SPENCE and VAHRMAN [124] studied the extract of a bituminous coal and identified by GLC the complete series of straight-chain terminal olefins from $C_{12}-C_{38}$ whose greatest amount is represented by the hydrocarbons $C_{20}-C_{27}$ with C_{23} and C_{24} slightly predominating. The range of carbon numbers and the distribution pattern were identical with the paraffins found in the same aliphatic-alicyclic fraction. Polarograms of coal extracts in dimethylformamide with LiCl as supporting electrolyte show 2–4 poorly defined waves which could possibly be attributed to carbonyl groups and aromatic systems [125]. Solvent extracts of coal were also electrochemically reduced and simultaneously acetylated with the addition of ^{14}C labeled acetic anhydride [126]. It was suggested that the carbonyl contents were in the range 1–2.5 percent and most of the carbonyl groups were ketones.

Extracts of brown coals having high fusain content consisted of 50 percent of substances of the phenolcarboxylic acid type [127]. Resinous acids from lignite bitumen were subjected to hydrogenation in the presence of a copper-chromite catalyst and produced the hydrocarbon $C_{26}H_{42}$ [128]. Lignitic resins appear to be a mixture of amber-like substances (succinoabietic acid and succinic acid) and a self-condensation form of the ester of oxidized abietic acid [129]. Special high sulfur coals of different deposits of the Irkutsk basin vary in their S-content from 15–77 percent. Sulfur in coal with a low content (up to 1.1 percent) is predominantly present in an organically combined form [130].

Regarding studies directed to the occurrence of amino acids in coal, the most detailed investigation seems to be the analysis of amino acids of brown coal from the Trimmersdorf area [131], the percentage there being 10^{-3} percent. Other amino compounds were isolated in a negligible amount (10^{-5} percent).

III. Conclusions and Outlook

Modern isolation and identification methods make possible an ever increasing insight into the chemical composition of fossil raw materials. The heterogeneity of the material and the fact that substances having large and complex molecules are involved make the research very difficult. Older studies dealt with the problems in a speculative manner, and only recently could a serious approach to the solution of these questions be made on the basis of augmenting credible experimental facts. Examples are the analyses of montan waxes, especially those from Bohemian brown coal. These have been carried out on a broader basis and are relatively exhaustive using recent separation and identification methods.

Raw fossil materials still represent an energy base, but they are also used to an ever increasing extent as chemical raw materials. Knowledge of the composition and structure of the substances contained in them makes possible not only a better chemical utilization of these resources but can also provide much information, for example, under what conditions and by what chemical changes these fossil raw materials arose. A further aspect of these rich natural resources may be their nutritional uses which, in view of the present world population explosion, is one of the primary tasks of science. As with petroleum, experiments involving microbial processes have been made on raw materials from coal to evaluate them [132].

IV. Experimental Procedure

For a better demonstration of the methods used for isolation and identification of waxes mentioned in this chapter, an example is presented.

The crude montan wax was previously separated by column chromatography into classes of compounds [14]. Chromatography is performed on a 30 to 100 fold excess of silica gel with a grain size of 30 to 60 microns, which was practically deactivated by means of 15–20 percent water. In this case, chromatography was carried out at an elevated temperature (45° C), while cyclohexane eluted paraffins in the first fractions and esters in the following ones. These were also eluted by

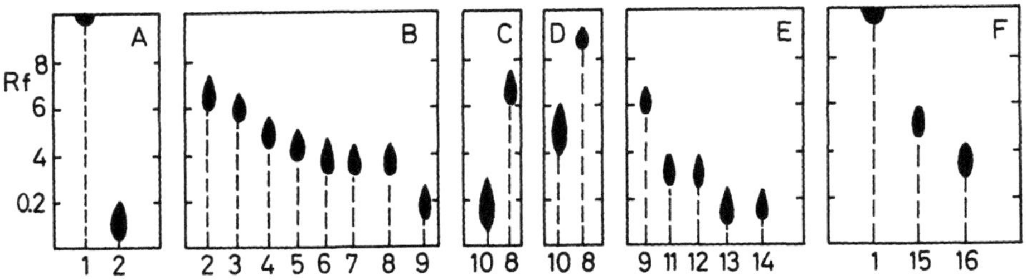

Fig. 1. T.l.c. diagram of some types of substances useful for identification of waxy substances. A Developed with petroleum ether, B with a mixture of petroleum ether and benzene (65:35), C with benzene, D, E with chloroform, F silica gel G impregnated with 20 percent silver nitrate, developed with petroleum ether. 1 non-aromatic hydrocarbons; 2 hexadecyl stearate (hexadecyl octadecanoate); 3 laurone (tricosanone-12); 4 methyl cerotate (methyl hexacosanoate); 5 methyl stearate (methyl octadecanoate); 6 hexadecyl acetate; 7 cholesteryl acetate; 8 stigmasteryl acetate; 9 tricosanol-12; 10 diethyl sebacate; 11 hexacosanol-1; 12 hexadecanol-1; 13 cerotic acid (hexacosanoic acid); 14 palmitic acid (hexadecanoic acid); 15 mixture of *trans*-alkenes (C_{21}—C_{31}); 16 mixture of *cis*-alkenes and alkenes-1 (C_{21}—C_{31})

the subsequent carbon tetrachloride, which eluted also free alcohols and benzene yielded the acids. During chromatography at room temperature, elution is naturally shifted so that, e.g., paraffins are eluted by cyclohexane and esters by carbon tetrachloride or by a mixture of cyclohexane and benzene. The separated fractions were monitored by means of thin layer chromatography. For illustration TLC diagrams of some types of substances are presented in Fig. 1. For an informative TLC analysis, a glass plate with the dimensions of 2.5 × 7.5 cm covered by a 0.4 mm layer of silica gel G (Merck) is convenient. Detection is completed either by carbonization after H_2SO_4 spraying by means of heating a resistance wire 0.5–1 cm above the plate, or by a spray of fluorescent reagent, e.g., Rhodamin 6 G, followed by examination under UV light.

The fractions containing pure waxes (esters) were combined and hydrolyzed by means of a mixture of 0.5 N potassium hydroxide solution in alcohol with benzene (2:1). By boiling the reaction product in diluted hydrochloric acid, the salts of the acids were transformed into free acids and, after cooling, the solid layer of alcohols and acids formed on the surface was removed. The whole mixture was chromatographed on a column [14, 8] of Zerolit FF (on a five-fold weight as compared with that of the sample) which was introduced into the OH-form by means of 5 percent NaOH. The excess of hydroxide was eluted by distilled water and this was removed by propanol-2 and benzene. Then the substance was introduced into the column in benzene solution and alcohols were eluted by means of benzene. The column was eluted successively by means of propanol-2, 5 percent HCl, water, propanol-2; the acids were eluted with benzene.

The acids were investigated informatively by reverse phase paper chromatography [9, 10] (immobile phase, paraffin – paraffin oil, 1:1; mobile phase, 98 percent acetic acid, temperature 75° C). The alcohols [9] were analyzed after their oxidation to acids by means of CrO_3 in acetic acid at 60° C. The course of the separations is presented in Fig. 2.

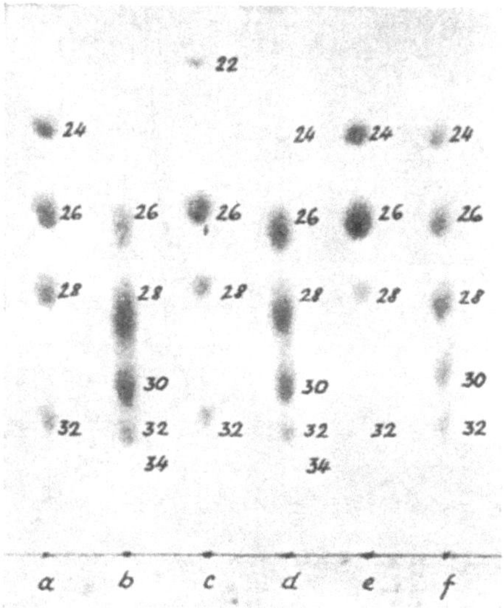

Fig. 2. Paper chromatographic separation of acids and alcohols from waxy esters as well as free acids. a, c, e standards, b free acids, d acids isolated from esters (waxes), f alcohols as products of the hydrolysis of esters (after their oxidation to acids)

The acids were investigated by gas chromatography [11] either as methyl esters (Fig. 3) after methylation with diazomethane, or after their transformation into the structurally equivalent paraffins. Reduction [4] was performed in a series of reactions: methylesters → alcohols → tosylesters → paraffins. Chromatography was carried out on a Pye Argon chromatograph 12,000 at 250° C with an argon flow of 40–60 ml/min, on previously prepared Apeizon L (1 percent on ground tile). Column length 1.2 m, diameter 4 mm.

To be able to establish whether there were acids with branched chains between the normal acids [16], the isolated paraffins were refluxed in isooctane with a 10 to 30 fold excess of 5 Å molecular sieve. After evaporation of the solvent, paraffins were contained in the adduct fraction according to gas chromatography (Fig. 4) and branched chain paraffins were obtained from the filtrate.

The alcohols obtained from the ion exchange resin as the neutral fraction [4] were transformed into their tosylesters. By the reduction of these with $LiAlH_4$, paraffins were obtained which were also investigated by means of gas chromato-

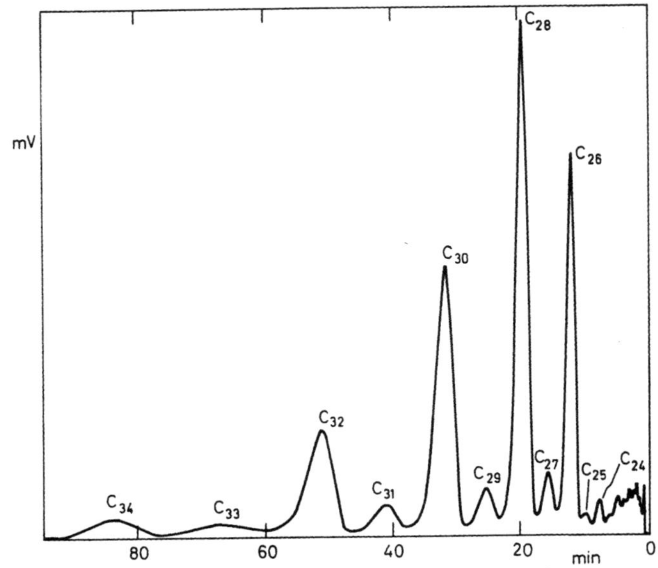

Fig. 3. G.l.c. diagram of the normal acids (as methyl esters) isolated from montan wax esters

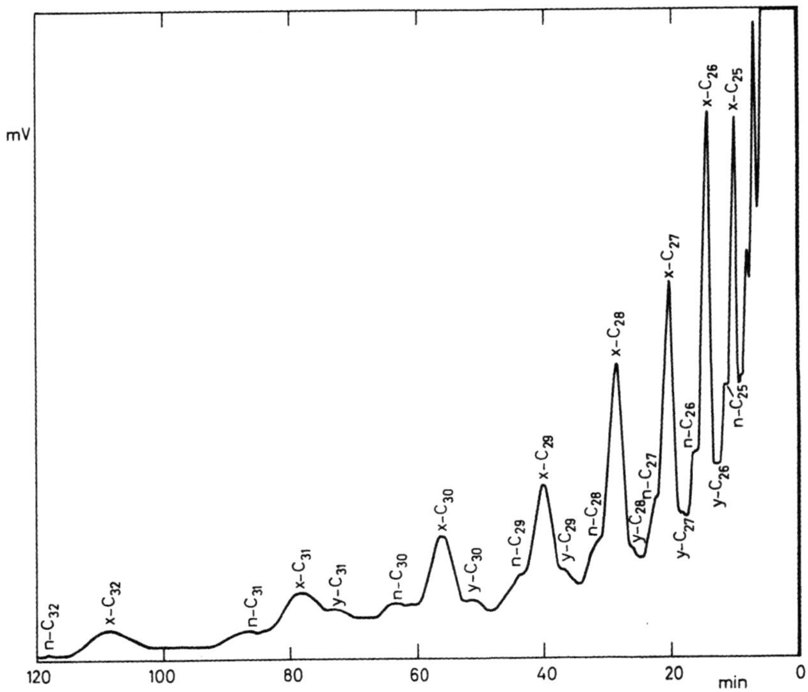

Fig. 4. G.l.c. diagram of acids isolated from montan wax esters reduced to hydrocarbons after treatment with molecular sieve (Linde 5 Å): n-normal alkanes, x-iso(anteiso)alkanes, y-branched alkanes with shorter retention volume

graphy [11]. After sieving with a 5 Å sieve it was found that they also contain small amounts of branched chain alcohols.

References

1. VČELÁK, V.: Chemie und Technologie des Montanwachses, S. 142–161. Praha: Vydavatelství Čsl. akademie věd. 1959.
2. ROGA, B., and M. WECLEVSKA: Hydroxyl group in coal. Brennstoff-Chem. **45**, 334–336 (1964).
3. CIŽINUAITÉ, E., and J. KUDABA: The effect of some factors in peat extraction. Vilniaus Univ. Mokslo Darbai **11**, 67–72 (1956); Chem. Abstr. **54**, 25700 b.
4. WOLLRAB, V., M. STREIBL u. F. ŠORM: Über die Zusammensetzung der Braunkohle. IV. Über die Gruppentrennung des Wachsanteils des Montanwachses mit Hilfe der Chromatographie. Collection Czech. Chem. Commun. **28**, 1316–1325 (1963).
5. ALM, R. S., R. J. P. WILLIAMS, and A. TISELIUS: Gradient elution analysis. Acta Chem. Scand. **6**, 826–836 (1952).
6. MOTTIER, M., et M. POTTERAT: De l'extraction des colorants pour denrées alimentaires avec la quinoleine et de leur identification par chromatographie sur plaque d'alumina. Anal. Chim. Acta **13**, 46–55 (1955).
7. WOLLRAB, V., M. STREIBL u. F. ŠORM: Über Pflanzenstoffe XXI. Über die Zusammensetzung der Kohlenwasserstoffe aus dem Wachs der Rosenblütenblätter. Collection Czech. Chem. Commun. **30**, 1654–1669 (1965).

   MALIUS, D. C.: Recent developments in t.l.c. of lipids. In Progress in the chemistry of fats and other lipids, vol. 8 (3) (R. T. HOLMAN, ed.).
8. PRESTING, W., u. S. JÄNICKE: Zur Anwendung des Ionenaustausches in der Wachsanalyse. Fette, Seifen, Anstrichmittel **62**, 81–87 (1960).
9. WOLLRAB, V., u. M. STREIBL: Über die Zusammensetzung der Braunkohle. V. Beitrag zur Papierchromatographie höherer aliphatischer Carbonsäuren und Alkohole. Collection Czech. Chem. Commun. **28**, 1895–1903 (1963).
10. FIKER, S., u. V. HÁJEK: Papírová chromatografie vyšších nasycených mastných kyselin. Chem. Listy **52**, 549–551 (1958).
11. WOLLRAB, V., M. STREIBL u. F. ŠORM: Über die Zusammensetzung der Braunkohle. VI. Analyse von Wachskomponenten des Montanwachses mittels Hochtemperatur-Gas-Verteilungschromatographie. Collection Czech. Chem. Commun. **28**, 1904–1913 (1963).
12. ORÓ, J., D. W. NOONER, and S. A. WIKSTRÖM: Paraffinic hydrocarbons in pasture plants. Science **147**, 870–873 (1965).
13. – – A. ZLATKIS, S. A. WIKSTRÖM, and E. S. BARGHOORN: Hydrocarbons of biological origin in sediments about two billion years old. Science **148**, 77–79 (1965).
14. JAROLÍMEK, P., V. WOLLRAB u. M. STREIBL: Gas-Verteilungschromatographie einiger höherer gesättigter und ungesättigter Kohlenwasserstoffe. Collection Czech. Chem. Commun. **29**, 2528–2536 (1964).
15. MOLD, J. D., R. E. MEANS, and J. M. RUTH: The higher fatty acids of flue-cured tobacco. Methyl and cyclohexyl branched acids. Phytochemistry **5**, 59–66 (1966).
16. JAROLÍMEK, P., V. WOLLRAB, M. STREIBL u. F. ŠORM: Über die Zusammensetzung des hydrierten Montanwachses. Collection Czech. Chem. Commun. **30**, 880–886 (1965).
17. HEWETT, D. R., P. J. KIPPING, and P. G. JEFFERY: Separation, identification and determination of the fatty acids of montan wax. Nature **192**, 65 (1961).
18. LEO, R. F., and P. L. PARKER: Branched-chain fatty acids in sediments. Science **152**, 649–650 (1966).

    EGLINTON, G., A. G. DOUGLAS, J. R. MAXWELL, and J. N. RAMSAY: Occurence of isoprenoid fatty acids in the Green River shale. Science **153**, 1133–1135 (1966).

    HORNING, E. C., A. KARMAN, and G. C. SWEELEY: Gas chromatography of lipids. In Progress in the chemistry of fats and other lipids, vol. 7 (2) (R. T. HOLMAN, ed.).
19. STREIBL, M., P. JAROLÍMEK u. V. WOLLRAB: Synthesen einiger höherer gesättigter und ungesättigter Kohlenwasserstoffe. Collection Czech. Chem. Commun. **29**, 2522–2527 (1964).
20. EGLINTON, G., P. M. SCOTT, T. BELSKY, A. L. BURLINGAME, and M. CALVIN: Hydrocarbons of biological origin from one-billion-year-old sediment. Science **145**, 263–264 (1964).
21. SCHWARTS, D., L. ASFELD, and R. GREEN: The chemical nature of the carboxyl groups of humic acids and conversion of humic acids to ammonium nitrohumates. Fuel **44**, 417–424 (1965).

22. BEUTELSPACHER, H., and C. NIGRO: Physical properties of humic acids contained in Italian lignites. Agrochimica **6**, 56–73 (1961); Chem. Abstr. **56**, 10466 d.
23. URBANOVSKI, T., W. KUCZYNSKI, W. HOFFMANN, H. URBANIC, and M. WATANOVSKI: The infrared absorption spectra of extracted coals. Bull. Acad. Polon. Sci. Ser. Sci. Chim. Geol. Geograph. **7**, 207–214 (1959); Chem. Abstr. **54**, 16789 d.
24. BROOKS, J. D., R. A. DURIE, B. M. LYNCH, and S. STERNHELL: IR-spectral changes accompanying methylation of brown coals. Australian J. Chem. **13**, 179–183 (1960).
25. ZIECHMANN, W.: Spectroscopic investigations of lignin, humic substances and peat. Geochim. Cosmochim. Acta **28**, 1555–1566 (1964); Chem. Abstr. **62**, 726 g.
26. STREIBL, M., u. F. ŠORM: Über die Zusammensetzung der Braunkohle. XI. Über weitere, namentlich ungesättigte Kohlenwasserstoffe des Montanwachses. Collection Czech. Chem. Commun. **31**, 1585–1595 (1966).
27. BROWNSTEIN, S.: High resolution NMR and molecular structure. Chem. Rev. **59**, 463–496 (1959).
28. FUCHS, W., and R. DIEBERG: Solution of the montanic acid problem by chromatography and IR-spectroscopy. Fette, Seifen, Anstrichmittel **58**, 826–831 (1956); Chem. Abstr. **52**, 5002 g.
29. FRIEDEL, R. A., and J. H. QUEISER: UV-spectrum and aromaticity of coal. Fuel **38**, 369–380 (1959).
30. JAROLÍM, V., K. HEJNO, F. HEMMERT u. F. ŠORM: Über die Zusammensetzung der Braunkohle. IX. Über einige aromatische Kohlenwasserstoffe des Harzanteils des Montanwachses. Collection Czech. Chem. Commun. **30**, 873–879 (1965).
31. TSCHAMLER, H., and E. DE RUITER: Comparative investigation of exinite, vitrinite and micrinite. Brennstoff-Chem. **46**, 106 (1965).
32. TAKEYA GEN, MITSUOMI ITOH, AHIRA SUZUKI, and SUSUMU YOKOYAMA: A study of the structure of pyridine extracts from coals by high resolution nuclear resonance spectroscopy. Bull. Chem. Soc. Japan **36**, 1222–1223 (1963); Chem. Abstr. **59**, 15081 d.
33. TSCHAMLER, H., and E. DE RUITER: Hydrogen in coal and coal extracts. Distribution of aliphatic hydrogen. Brennstoff-Chem. **43**, 212–215 (1962); Chem. Abstr. **58**, 2295 e.
34. YORKE, R. W.: Hydrogen resonance spectra of low temperatures of pure hydrocarbons and of selected coal samples. J. Chem. Soc. **1960**, 2489–2497.
35. CARTZ, L., and P. B. HIRSCH: A contribution to the structure of coals from X-ray diffraction studies. Phil. Trans. Roy. Soc. London, Ser. A **252**, No. 1019, 557–604 (1960).
36. BANERJEE, B. K., P. B. DUTTA, and R. S. DRABEY: X-ray diffraction study of some Indian fusains. J. Sci. Ind. Res. **20 B**, 486–489 (1961).
37. VOLAROVICH, M. P., and K. F. GUSEF: X-ray diffraction analysis of peat. Novye-fiz. Metody Issled. Torfa Sb., 1960, 155–167; Chem. Abstr. **60**, 11806 b.
38. ERGUN, S., and V. H. TIENSUN: Alicyclic structures in coals. Nature **183**, 1668–1670 (1959).
39. HOLDEN, H. W., and J. C. ROBB: A study of coal by mass spectrometry II. Extracts and extractable pyrolysis products. Fuel **39**, 485–494 (1960).
40. REID, W. K.: Use of high-resolution mass spectrometry in the study of petroleum waxes and ozocerite. Anal. Chem. **38**, 445–449 (1966).
RYHAGE, R., and E. STENHAGEN: Mass spectrometry studies in lipid research. J. Lipid Res. **1**, 361–390 (1960).
41. BELL, G. K., and J. M. HUNT: Native bitumens associated with oil shale. In: Organic geochemistry (I. A. BREGER, ed.), p. 363. Oxford: Pergamon Press, Inc., 1963.
42. MSHVELIDZE, N. OZOCERITE: Prirodni resursy. Gruz. SSR. Akad. Nauk Gruz. SSR **2**, 312–315 (1959); Chem. Abstr. **55**, 7190 c.
43. BOIKO, G. YU.: Mineralogy of organic compound deposits in the Carpathians. Dopovodi L'vivsk. Derzh. Univ. **1961**, 134–135; Chem. Abstr. **60**, 1455 e.
44. BOIKO, G. E., L. K. KLIMOVSKIYA, E. V. RYLTSEV, V. V. TURKEVICH, and E. F. YAKSENKO: Infrared absorption spectra of higher liquid hydrocarbons of Carpathian ozocerites. Tr. Ukr. Nauch-Issled. Geol.-Razved. Inst. **1963**, 378–381; Chem. Abstr. **61**, 9334 h.
45. BROOKS, B. T.: Composition of petroleum waxes. In: The chemistry of petroleum hydrocarbons. New York: Reinhold Publ. Division 1954.
46. TREIBS, A.: Organic minerals III. Chlorophyl and heminderivatives in bituminous rocks, petroleums, earth waxes and asphalts. Ann. Chem. **510**, 42–62 (1934).
47. ROZANOVA, E. P., and L. D. SHTURM: Changes in ozocerite composition under the effect of microorganisms. Mikrobiologiya **35**, 138–145 (1966); Chem. Abstr., **64**, 16311 e.

48. SERGIENKO, S. R., and V. KOZYURO: Properties of ozocerites from the Cheleken deposits. Izv. Akad. Nauk Turkm. SSR, Ser. Fiz.-Tekhn. Khim. i Geol. Nauk **1965**, 25–31; Chem. Abstr. **64**, 3224f.
49. BOIKO, G. E.: Mineralogy of ozocerite from the Dzvinyach and Staruniya deposits. Mineralog. Sb. L'vovsk. Geol. Obshchestvo pri L'vovsk. Gos. Univ. **1962**, 449; Chem. Abstr. **59**, 11126b.
50. FUCHS, W.: Die Chemie der Kohle. Berlin: Springer 1931.
51. ALEXANIAN, C.: Mode of formation of coal, results of IR-absorption. Compt. Rend. **246**, 1192–1195 (1958).
52. KREVELEN, D. W. VAN: Geochemistry of coal. In: Organic geochemistry (I. A. BREGER, ed.). Oxford: Pergamon Press 1963.
53. FISCHER, F., u. H. SCHRADER: Entstehung und chemische Struktur der Kohle. Essen: Gerardet 1922.
54. LUKOSHKO, E. S., and V. E. RAKOVSKI: Mechanism of formation of humic substances during turf building. Khim. i Genezis Torfa i Sapropelei, Akad. Nauk Belorussk. SSR, Inst. Torfa 1962, 23–29; Chem. Abstr. **59**, 6938c.
55. ENDERS, C.: Origin of humus nature. Chemie, Die **56**, 281–285 (1943).
56. MAILLARD, L. C.: Formation of humic substances by the action of polypeptides on sugars. Compt. Rend. **156**, 1159–1160 (1913).
57. GROSSE-BRAUCKMANN, G., and D. PUFFE: Microtone-section investigations of peat profile from Teufelsmoor near Bremen. Proc. Intern. Working-Meeting Soil Micromorphol., 2nd., Arnhem Neth. **1964**, p. 83–93; Chem. Abstr. **64**, 15603e.
58. FLAIG, W., and H. SCHULZE: The mechanism of formation of synthetic humic acids. Z. Pflanzenernaehr. Dueng. Bodenk., **58**, 59–67 (1952).
59. BRYDON, J. E.: Chemical composition of the peat bogs of the maritime provinces. Can. J. Soil Sci. **38**, 155–160 (1958).
60. VASIL'EV, S. F., and M. K. D'YAKOVA: Thermal solvent extraction as a method of processing peat into gas, chemical products and motor fuels. Novye Metody Rats. Ispol'z. Mestnykch Topliv, Tr. Soveshch. Riga **1958**, 37–46; Chem. Abstr. **56**, 2666d.
61. KAGANOVICH, F. L., and V. E. RAKUSKII: Selective extraction of peat bitumens of low temp. Vestsi Akad. Navuk Belarusk. SSR, Ser. Fiz. Tekhn. Navuk 1958, 117–122; Chem. Abstr. **57**, 3721c.
62. BEL'KEVICH, P. I., V. M. TSYBULKIN, and E. A. VURKEVICH: A study of the asphaltenes from peat by infrared spectroscopy. Tr. Inst. Torfa, Akad. Nauk Belorusk. SSR **9**, 296–300 (1960); Chem. Abstr. **56**, 1691b.
63. VOLAROVICH, M. P., and K. F. GUSEV: X-ray studies of the structure of peat bitumens and waxes. Kolloidn. Zh. **18**, 643–646 (1956); Chem. Abstr. **51**, 6980a.
64. BEL'KEVICH, P. I., F. L. KAGANOVICH, and E. V. TRUBILKO: A study of the composition of the unsaponifiable part of peat wax by the method of fractional crystallization. Tr. Inst. Torfa, Akad. Nauk Belorusk. SSR **9**, 274–279 (1960); Chem. Abstr. **56**, 2666a.
65. FISCHER, W., G. SCHLUNGBAUM, and P. KADMER: Carbonyl and methoxy groups in lignits and peats. Z. Chem. **4**, 394–395 (1964); Chem. Abstr. **62**, 5102g.
66. KAGANOVICH, F. L., P. I. BEL'KEVICH, and E. V. RAKOVSKII: Composition of peat wax. Tr. Inst. Torfa, Akad. Nauk Belorusk. SSR **7**, 123–130 (1959); Chem. Abstr. **55**, 8815h.
67. KWIATKOVSKI, A.: Chromatographic analysis of peat bitumens. Zeszyty Nauk Politech. Gdansk., Chem. **33** (5), 29–51 (1963); Chem. Abstr. **60**, 5242e.
68. RAUHALA, V. T.: Free hydroxy acids of peat wax. J. Am. Oil Chemist's Soc. **38**, 233–235 (1961).
69. KWIATKOVSKI, A.: Composition of resin components of peat bitumens. Valtion Tek. Tutkimuslaitos, Julkaisu **93**, 37 (1965); Chem. Abstr. **64**, 13973e.
70. GILLILAND, M. R., A. J. HOWARD, and D. HAMMER: Polycyclic hydrocarbons in crude peat wax. Chem. Ind. (London) **1960**, 1357–1358.
71. LISSNER, A., u. A. THAU: Die Chemie der Braunkohle, Bd. I. Halle: VEB Wilhelm Knapp 1956.
72. KUKHARENKO, T. A.: Humic acids of fossil solid fuels. Proc. Symp. Nature Coal, Jealgora, India **1959**, 185–193; Chem. Abstr. **58**, 4337d.
73. KASATOCHKIN, V. I., N. K. LARINA, and O. I. EGOROVA: The structure and properties of humic substances of peat and coal. Zh. Prikl. Khim. **38**, 2059 (1965).
74. GANZ, E.: Preparation of de-ashed humic substances free from N and S. Compl. Rend. **216**, 122–124 (1943).

75. ODÉN, S.: Die Huminsäuren. Leipzig: Steinkopff 1922.
76. STADNIKOFF, G., and P. KORSHEW: Humic acid. Kolloid-Z. **47**, 136–141 (1929); Chem. Abstr. **23**, 3230.
77. SEIJI ARITA, TAKESKITA KENJIRO, and TSUMETARO KATO: Nonaqueous potentiometric titration of acidic groups in regenerated humic acid. Kogyo Kagaku Zasshi **64**, 192–196 (1961); Chem. Abstr. **57**, 4925i.
78. FUIJI SHUYA: Infrared absorption spectra of humic acid. Nenryo Kyokaishi **38**, 267–273 (1959).
79. FUCHS, W., u. W. STENGEL: Zur Kenntnis der Hydroxyl- u. Carboxyl-Gruppen der Huminsäure. Brennstoff-Chem. **10**, 304 (1929).
80. – Über Naturprodukte. Dresden u. Leipzig: Hönig-Festschrift 1923.
81. – Vergleichende Einwirkung von Brom auf Cellulose, Lignin, Holz, Braunkohle und Steinkohle. Brennstoff-Chem. **9**, 348–350 (1928).
82. HARTLEY, R. D., and G. G. LAWSON: Chemical constitution of coal. Ion exchange separation of sub-humic acids with hydrogen peroxide. Fuel **41**, 447–456 (1962).
83. ROY, M. M.: Oxidation of humic acid prepared from Assan coal. Research (London) **10**, 78 (1957).
84. LAWSON, G. J., and J. W. PURDIE: Chemical constitution of coal. Examination of subhumic acids produced by ozonization of humic acid. Fuel **45**, 115 (1966).
85. –, and R. P. HARTLEY: Chemical constitution of coal. Chromatographic examination of sub-humic acids obtained by oxidation of humic acid with H_2O_2. Fuel **41**, 177–183 (1962).
86. LANDA, S., and J. EYEM: Hydrogenation of humic acids from brown coal. Fuel **42**, 265—273 (1963).
87. TAKEGAMI, Y., S. KAJIYAMA, and C. YOKOKAWA: Chemical structure of coal II. Hydrogenolysis of bituminous coal and humic acid in the presence of Adkins copper-chromite catalyst. Fuel **42**, 291–302 (1963).
88. KASATOCHKIN, V. I., and N. K. LARIN: An investigation of humic acid structure in fossil coal. Dokl. Akad. Nauk SSSR **114**, 139–142 (1957); Chem. Abstr. **52**, 1586.
89. AGDE, G., H. SCHURENBERG, and R. JODL: Water binding power form and size of colloidal particles of humic acid. Braunkohle **41**, 545–547 (1942).
90. FUCHS, W.: Untersuchungen über Lignin, Huminsäuren und Humine. Angew. Chem. **44**, 111–120 (1931).
91. KUKHARENKO, T. A.: Humic acids in peat and characteristics of their structure. Tr. Inst. Torfa, Akad. Nauk Belorussk. SSR **3**, 120–132 (1954); Chem. Abstr. **51**, 10032c.
92. HESS, K., u. W. KOMAREVSKY: Über Isolierung und Nachweis von Cellulose in Torf. Angew. Chem. **41**, 541–542 (1928).
93. SCHLUNGBAUM, G., and W. FISCHER: The chemical composition of peats in the GDR. Freiberger Forschungsh. A **254**, 65–86 (1962).
94. STADNIKOW, G., u. A. BARYSCHEWA: Über Lignine einiger Torfbildner und eines Sphagnum-torfes II. Brennstoff-Chem. **11**, 169–171 (1930).
95. SHACKLOCK, C. W., and T. J. DRAKELEY: A preliminary investigation of the nitrogenous matter in coal. J. Soc. Chem. Ind. **46**, 478–481 (1927).
96. SWAIN, F. M.: Limnology and amino acid content of some lake deposits in Minnesota, Montana, Nevada and Louisiana. Bull. Geol. Soc. Am. **72**, 519–546 (1961); Chem. Abstr. **55**, 12195f.
97. –, A. BLUMENTALS, and R. MILLERS: Stratigraphic distribution of amino acids in peat from Cedar Creek bog, Minnesota, and Dismal Swamp, Virginia. Limnol. Oceanog. **4**, 119–127 (1959); Chem. Abstr. **53**, 16860c.
98. WILDENHAIN, W., and G. HENSEKE: Organic compounds of highmoorland peat extracts. Z. Chem. **5**, 457–458 (1965).
99. PRESTING, W., u. K. STEINBACH: Die Zusammensetzung des Rohmontanwachses und seiner Destillate. Bergakademie **5**, 231 (1953).
100. – – Zur Zusammensetzung der Analyse des Rohmontanwachses. Fette, Seifen, Anstrichmittel **57**, 329–335 (1955).
101. TANNER, D. W., A. POLL, J. POTTER, D. POPE, and D. WEST: The promotion of dropwise condensation by montan wax. J. Appl. Chem. **12**, 547–552 (1962).
102. EDWARDS, V. A., P. J. KIPPING, and P. G. JEFFERY: Composition of montan wax. Nature **199**, 171–172 (1963).
103. PRESTING, W., u. T. KREUTZER: Zur Kenntnis der Wachsbausteine des Montanwachses. Fette, Seifen, Anstrichmittel **67**, 334–340 (1965).

104. DOUGLAS, A. G., and G. EGLINTON: The distribution of alkanes. In: Comparative phytochemistry (T. SWAIN, ed.). London: Academic Press 1966.
105. TROPSCH, H., u. A. KREUTZER: Über die Säuren des Montanwachses. Brennstoff-Chem. **3**, 49 (1922).
106. HOLDE, D., u. W. BLEYBERG: Über Säuren des Montanwachses. Brennstoff-Chem. **15**, 311–312 (1934).
107. FUCHS, W., and R. DIEBERG: Solution of the montanic acid problem by chromatography and IR-spectroscopy. Fette, Seifen, Anstrichmittel **58**, 826–831 (1956).
108. KAUFMANN, H. P., u. D. DAS: Die Papierchromatographie auf dem Fettgebiet. Die qualitative und quantitative p.c. Analyse der Wachssäuren. Fette, Seifen, Anstrichmittel **63**, 614–616 (1961).
109. PRESTING, W.: 2. Internationales Kolloquium Paraffine und Wachse. Leipzig 1966.
110. – Zur Kenntnis der Montansäure und des Montanons. Chem. Tech. **4**, 152 (1952).
111. PSCHORR, R., u. J. K. PFAFF: Beitrag zur Kenntnis des Montanwachses mitteldeutscher Schwelkohle. Chem. Ber. **53**, 2147–2150 (1920).
112. STEINBRECHER, H.: Zur Kenntnis des Bitumens, der wasserlöslichen und der pyridinlöslichen Bestandteile einiger Braunkohlen. Brennstoff-Chem. **10**, 198–201 (1926).
113. RUHEMANN, S., u. H. RAUD: Über die Harze der Braunkohle. Brennstoff-Chem. **13**, 341–345 (1932).
114. SARKAR, SAMIR: Oxygen groups in Donets Basin coals. Fuel **41**, 206–208 (1962); Chem. Abstr., **56**, 1450c.
115. KHARITONOV, G. V., M. USUBAKUNOV, and V. P. PURIKOVA: Nature of the chemical composition of petrographic varietes of coals from the Kirgiz SSR. Izv. Akad. Nauk Kirg. SSR **1**, 13–22 (1955). Chem. Abstr. **53**, 7557c.
116. KINNEY, C. R., and E. DOUCETTE: Infrared spectra of coalification series from cellulose and lignin to anthracite. Nature **182**, 785–786 (1956); Chem. Abstr. **53**, 6605e.
117. MURASHKEVICH, T. V.: The changes in properties of certain White Russian SSR lignites in connection with the initial stage of coalification. Sb. Nauch. Rabot, Akad. Nauk Belorussk. SSR, Inst. Fiz.-Organ. Khim, 1959, **7**, 141–149; Chem. Abstr. **54**, 25689h.
118. MARUNDAR, B. K., S. K. CHAKRABARTY, and A. LAHIRI: Some aspects of the constitution of coal. Fuel **41**, 129–139 (1962); Chem. Abstr. **56**, 14540g.
119. LEIBNITZ, E., H. B. KÄNNECKE, and M. SCHRÖTER: Pressure coalification of brown coal in the presence of water IV. J. Prakt. Chem. **6**, 18–24 (1958).
120. PAN, L. S., T. N. ANDERSEN, and H. EYRING: Experimental study of the effects of temperature and ultra high pressure on the coalification of bituminous coal. Ind. Eng. Chem. Process Design Develop. **5**, 242–246 (1966); Chem. Abstr. **65**, 3617h.
121. OUCHI, K., and K. IMUTA: The analysis of benzen extracts of Yubari coal II. Analysis by gas chromatography. Fuel **42**, 445–456 (1963).
122. DANYUSHEVSKAYA, A. J.: The chromatography of hydrocarbon found in the bitumen of the Sangarsk coal. Tr. Nauchn. Issled. Inst. Geol. Arktiki, Mini. Geol. i Okhrany Nedr. SSR **98**, 120–129 (1959); Chem. Abstr. **55**, 9836g.
123. RUSHEV, D., and P. DRAGOSTINOV: Differential thermal analysis and IR-spectral analysis of petrographic ingredients of lignites. Erdoel Kohle **18**, 372–375 (1965).
124. SPENCE, J. A., and H. VAHRMAN: Olefins in coal. Chem. Ind. (London) **1965**, 1522.
125. Brit. Coal Utilization Res. Assocn., Leatherhead, England. Some contributions of polarography in dimethylformamide to the study of the structure of coals. Proc. Intern. Advan. Polarog. Cong. 2nd, Cambridge, England **1959**, 965–973 (Publ. 1960).
126. GIVEN, P. H., and M. E. PEOVA: Investigation of CO-group in solvent extracts of coals. J. Chem. Soc. **1960**, 394–400.
127. KARAVAEV, M. N., Z. A. RUMYANTSEVA, and V. S. POLYANSKAYA: Alcohol-benzene and alcohol extracts of fusinated brown coals. Izv. Akad. Nauk Tadzh. SSR, Otd. Fiz.-Tekhn. i Khim. Nauk **1965** (1), 79–88; Chem. Abstr. **64**, 10984e.
128. KOSO HIGUCHI, and TAKURO ARAI: Resinous acids from lignite bitumen. J. Fuel Soc. Japan **35**, 505–517 (1956); Chem. Abstr. **51**, 2253c.
129. TIETJEN, J. J., and H. L. LOWELL: Chemical constitution of a lignite resin. Am. Chem. Soc. Div. Gas Fuel Chem.: Preprints, **2**, 9–19 (1959); Chem. Abstr. **57**, 4946d.

130. RUMYANTSEVA, O. G.: Chemical-petrographical properties of high sulphur coals of the Irkutsk Basin. Geol. Uglei Sibiri i Dal'nego Vostoka, Akad. Nauk SSSR Sibirsk. Otd., Inst. Geol. i Geofiz. **1965**, 133; Chem. Abstr. **64**, 17299 h.
131. BAJOR, MATHIAS: Amines, amino acids and fats as facies indicators in Lower Rheinisch brown coals and their analytical determination. Braunkohle **12**, 472–478 (1960).
132. SILVERMAN, M. P., J. N. GORDON, and I. WENDER: Food from coal-derived materials by microbial synthesis. Nature **211**, 735 (1966).

Addendum

Recently GLC analysis of wax esters has been performed at 300°–320° C on a carefully conditioned column using 3% OV-1 as the thermostable phase. Thus, the composition of wax esters can now be studied without previous hydrolysis of these esters to the acids and alcohols. (M. STREIBL: unpublished results; c.f. P. J. HOLLOWAY: The composition of breswax alkyl esters. J. Am. Oil Chemist's Soc. **46**, 189–190 (1969).)

CHAPTER 25

Kauri Resins – Modern and Fossil

B. R. THOMAS*

Chemistry Division, DSIR,
Petone, New Zealand

Contents

I. Introduction

The genus *Agathis*, a group of conifers found in the land areas between New Zealand and Malaysia, has several conspicuous resin-forming systems. Thus, the New Zealand species *Agathis australis* (Salisbury) produces resins, often in very large amount, in the heartwood, in resin canals in the bark, in resin canals in the leaves, in the leaf surface layers, and elsewhere. These resins are all different and are mostly quite complex mixtures.

In addition to the straight-forward chemical problems associated with the *Agathis* resins, particular interest is attached to the biosynthesis, evolution and function of these products. Also of interest is the way in which the chemical variations of the products are related to the development of the genus under the influence of major geological events and changes in environment.

Bled resin exudes quite liberally from cut and damaged surfaces of the bark of all species of the genus. It is usually very durable and lumps of "fossil" resin are found in the ground in areas where these trees have grown. The age of this material present in the surface layers probably ranges from tens to thousands of years. Kauri resin from *A. australis* and Manila resin** from *Agathis* species in the Indonesia-Phillipines area have been in the past of considerable commercial significance because of their use in the manufacture of varnishes and linoleum but

* Present address: Organic Chemistry Dept., LTH, Lund, Sweden.

** Manila resin is also called Manila copal or East Indies copal and local varieties may be named after their port of origin, e.g., Pontianak. Bled resin from *A. australis* is known commercially as kauri resin or kauri gum.

only Manila resin is still produced in large amount. Kauri resin from *A. australis* is the main fossil resin. Similar resins are found in Australia, the Phillipines and Indonesia, but Manila resin is mostly recently bled resin. All resin found in the ground, even though quite recent, has usually been termed fossil resin.

Kauri, originally the Maori name for *A. australis*, is now applied to all the southern species and sometimes to all *Agathis* species. This discussion will deal with the resins of the whole group. Many species of *Agathis* give impressive trees and *A. australis* can reach spectacular age and size (trunk diameters of over 7 m).

II. Distribution and Geological History

Agathis and the closely related genus *Araucaria* make up the family *Araucariaceae*. Araucarian remains first appear in the southern hemisphere about 170 million years ago (Jurassic) though *Agathis* only becomes evident as a distinct genus about 40 million years ago [1]. However, when the southern continent subdivided during the Jurassic-Cretaceous periods, both *Agathis* and *Araucaria* must have been present in the area of land which eventually became the long ridge extending from the southern tip of South America through New Zealand to New Guinea. Attenuation of this land bridge finally isolated various populations of the two genera by barriers of ocean or climate, and left them to develop according to local conditions. (For a recent discussion dealing with some aspects of affinity and isolation in the flora of the South Pacific, see [2].)

Kauris, which produce heavy, wind-scattered seed of low viability under adverse conditions, are not well suited to long distance dispersal and, while migration may not be impossible in the Indonesian area, it is less likely across

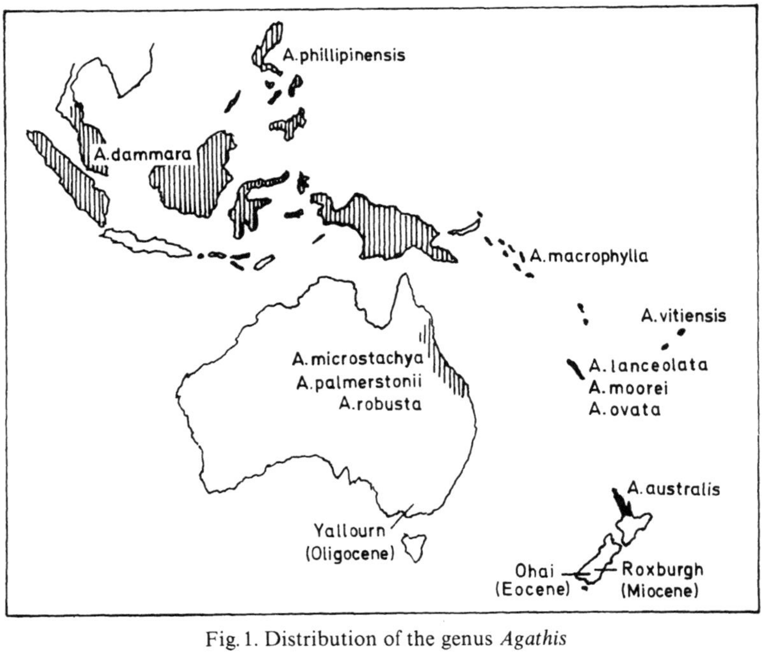

Fig. 1. Distribution of the genus *Agathis*

the ocean barriers further south. The modern distribution of species may accordingly reflect the distribution of forms along the ancient land bridge.

The present range of the genus and the location of the species relevant to this discussion are shown in Fig. 1 [1]. It formerly extended at least to the southern tip of New Zealand and to the south of Australia where a species similar to *A. australis* has been a significant contributor to the enormous brown coal deposits at Yallourn. In New Zealand the glaciations commencing about 1 million years ago, altitude limits and the several thousand km^3 of volcanic debris spread across the center of the North Island relatively recently, have confined *A. australis* to the more stable land surface in the northern part of the North Island.

In the 20,000 km^2 area in which *A. australis* is presently found, surface deposits have yielded more than 500,000 tons of resin, and in some places the presence of timber and layers of resin many meters below the surface indicates a succession of kauri forests over a long period. Resin is also found, sometimes in very large amounts, in association with coal deposits in the same area and in the South Island of New Zealand [3, 4].

Quite apart from their biogenetic and phytochemical interest, kauri resins may thus give information both on the spread and differentiation of the genus and on the changes that can occur in ancient resins.

III. Modern Agathis Resins

A. The Bled Resins of *A. australis*

A. australis gives at least five quite different resins from different tissues. Of these the bled resin has attracted most attention because of the plentiful occurrence and the commercial uses of the fossil material. It is formed in the resin canals of the bark and when these are broken it exudes as a latex to form a protective layer over the damaged surface. This rapidly loses water and some of the monoterpene hydrocarbon present, giving a soft transparent film which gradually hardens further on exposure to air and light. Under some conditions when resin continues to exude from a wound or when repeated damage occurs, very large masses can accumulate, especially in the forks of branches. The resin is either shed as the bark flakes off or is buried when the tree falls and rots away.

The most important constituents of the fresh bled resin are shown in formulae I a–b, II a–e, III a–d, IV a–f and V [5–8]. The resin also contains small amounts of a monoterpene fraction (mainly d-α-pinene [9]), and of a water-soluble fraction [10] that includes some phenolic material [11].

The hardening of the resin layer is largely due to polymerization of the conjugated dienes (II a–e, III a–d). This is probably accompanied by some incorporation of the other components of the resin and by cross-linking by peroxidation processes and polyester formation [10, 12, 13]. The less reactive components will tend to remain and function as plasticiser. In larger masses of resin there can still be 30–50 percent of unpolymerized material after many years [7].

Agathic acid (I a) is present in most *Agathis* bled resins (often as a monomethyl ester, I b or XV c) and is of interest as the first diterpene structure to be elucidated. Abietic acid (V) and its various isomers form the main components of many pine

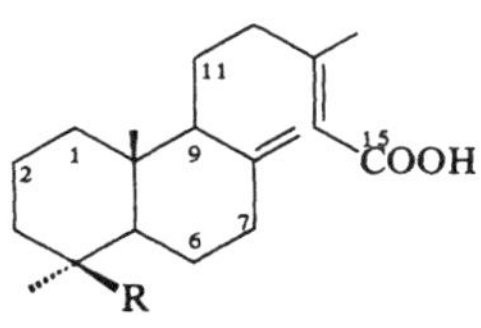

I a R = COOH (agathic acid), b R = $COOCH_3$

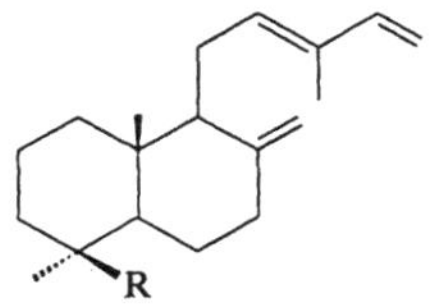

II a R = COOH (*trans*-communic acid), b R = CHO, c R = CH_2OAc, d R = CH_2OH, e R = CH_3

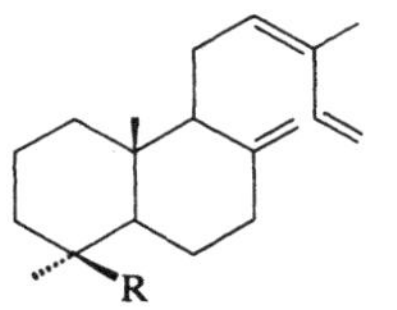

III a R = COOH (*cis*-communic acid), b R = CHO, c R = CH_2OAc, d R = CH_2OH

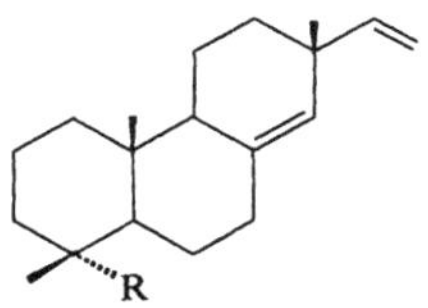

IV a R = COOH (sandaracopimaric acid), b R = COOMe, c R = CHO, d R = CH_2OAc, e R = CH_2OH
f R = CH_3

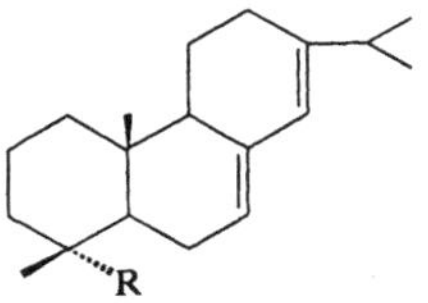

V R = COOH (abietic acid)

resins; communic acid (II a, III a) and related compounds occur in *Pinus*, *Juniperus* and *Abies* and, with the more delicate touch of the modern chemist, will probably be found to be very widespread.

B. The Leaf Resin of *A. australis*

The latex from the resin canals of the leaves of *A. australis* (Fig. 2) contains a single main component to which the acetoxy-communic acid structure IX has been assigned by nmr [8]. Communic acid, which must be the immediate precursor is also present but there are not more than trace amounts of the other compounds that occur in the resin canals of the trunk bark. The leaf surface contains a mixture

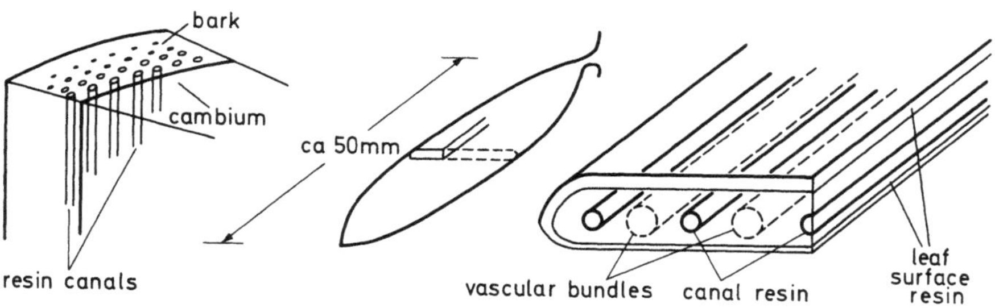

Fig. 2. Location of resin in *A. australis* bark and leaves (diagrammatic)

of (–)-kaurene (VI) [8, 14, 15], the labdatriene VIII [8] and 8β-hydroxy-isopimarene (X) [8] together with smaller amounts of (+)-hibaene (VII) [8, 16], which might be expected to accompany formation of (–)-kaurene. The absence from the products so far isolated of the third isomer which could be formed from the Wenkert type precursor XI [17], tends to suggest that in the present case this does not exist for any significant time and the kaurene is formed by a single smooth sequence of cyclization and rearrangement.

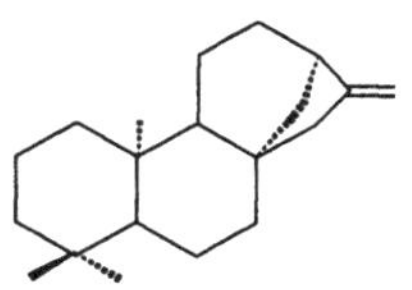

VI (–)-kaurene

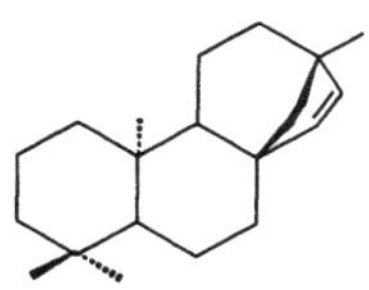

VII (+)-hibaene

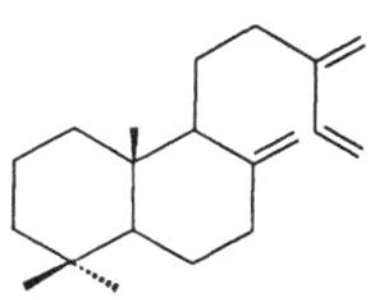

VIII

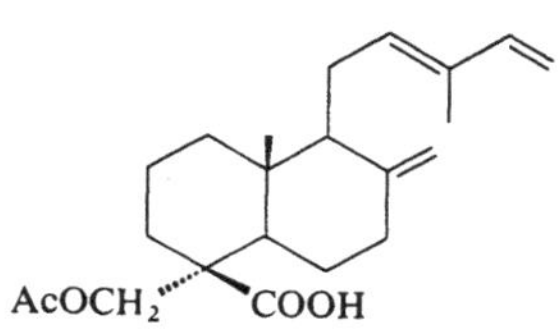

IX

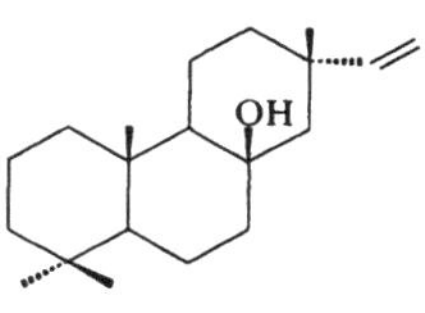

X

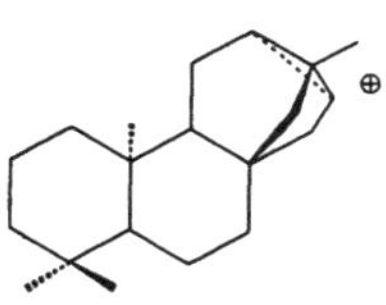

XI

(–)-Kaurene (VI), which was first isolated from *A. australis* leaves, is of particular interest as a precursor of the gibberellins, which are major plant hormones. The function of these leaf surface compounds is again presumably protection. They are probably involved in the formation of the outer surface coating, which protects the leaf against dessication or mechanical damage, but they may act directly against insect or fungal attack. It is also possible that photochemical transformation products of the leaf surface compounds are required for other more specific purposes in the plant.

The bled resin from young twigs is similar to the leaf canal resin but older twigs give resin similar to that from the canals of the trunk. Presumably the difference depends on whether the resin comes from the resin canals in the primary tissue of the stem or from resin canals in the secondary tissue formed by the cambium.

Because they are very exposed to oxidation and microbiological degradation, the leaf surface compounds are unlikely to survive geological conditions, though the more stable kaurene (VI) and hibaene (VII) might be retained for some time in exceptional cases where masses of leaves are rapidly preserved.

C. Bled Resins of Other Species of *Agathis*

Results from the bled resins of other species of *Agathis* are summarized in Table 1. The amounts of communic acid type compounds shown in the table will be low because of their partial conversion to polymer. In general, apart from *A. ovata* and *A. moorei*, all the resins are very similar.

Table 1. *Main types of compound in Agathis bled resins* [a]

Figures where given are approximate percentages (from glc) for typical samples of acetone-soluble resin; compounds of different oxidation level but the same oxygenation pattern are grouped together; + = compound type present, 0 = compound type absent, a space indicates no evidence of presence or absence or (in *A. moorei*) that the compound cannot be classified.

	Agathic type	Abietic type	Communic type	Sandaracopimaric type
A. australis (Salisbury) [7, 8]	35	15	20	10
A. dammara ((Lamb.) Rich.) [8]	75	0	+	5
A. palmerstonii (MUELLER) [21]	2	40	40	0
A. robusta (MUELLER) [21]	0	30	45	0
A. microstachya (BAILEY and WHITE) [21]	40	20	20	0
A. vitiensis (Ben. and Hook.) [8]	+	+	+	+
A. lanceolata (WARBURG) [8]	35	10		35
A. moorei (MASTERS) [8]		+		
A. ovata (WARBURG) [8]	0	0	0	0

[a] *A. dammara* (Lamb.) Rich. (= *A. alba* = *A. loranthifolius*) has sometimes been taken to include a number of forms found in the Malaysia-Indonesia-Phillipines area including *A. phillipinensis*. The botanical problems of the status and nomenclature of these forms will be ignored here. (See reference [40].)

The Queensland species (*A. microstachya*, *A. robusta* and *A. palmerstonii*) examined by CARMAN and co-workers give large amounts of abietic type compounds including the isomers neoabietic acid (XII) and levopimaric acid* (XIII) but no sandaracopimaric acid [18–21]. The dundatholic acid obtained a considerable time ago from *A. robusta* has now been shown to be a mixture of the diterpene acids present in the resin [22], and the dundathic acid obtained at the same time has been shown to be a polycommunic acid [12].

The amount of agathic acid (present mainly as one of the monomethyl esters) varies from zero (in *A. robusta*) to nearly half the resin, and the corresponding alcohols and aldehydes may also be present. *A. dammara* contains other agathic-

* This name is most misleading and would be better changed to levoabietic acid.

type compounds differing with the origin of the resin. A sample from Malaya contained (+)-daniellic acid (XIV) and isocupressic acid (XVa) [8], while commercial samples of *A. dammara* resin (i.e., from Indonesia or the Phillipines) contain instead agatholic acid (XVb) with daniellic acid present in small amount only or entirely absent [8, 23].

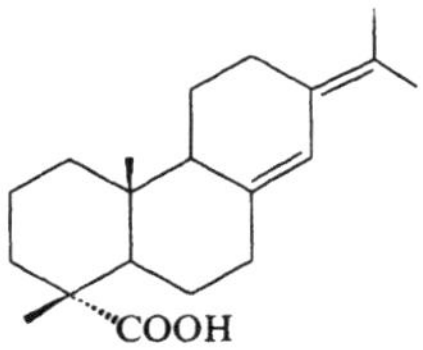

XII Neoabietic acid

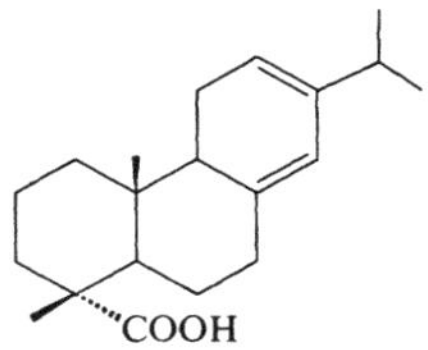

XIII Levopimaric acid

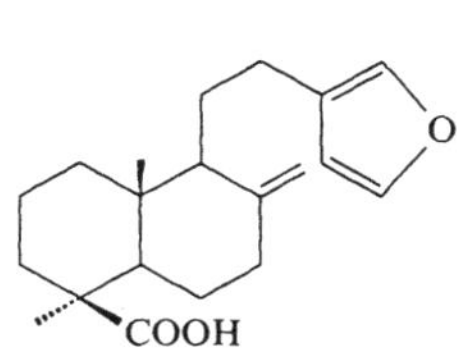

XIV (+)-daniellic acid

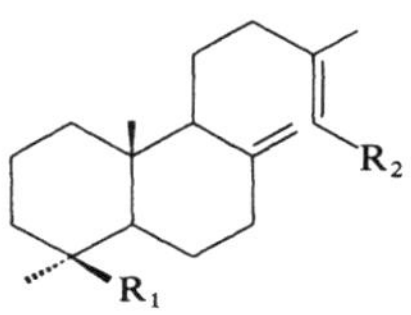

XVa $R_1 = COOH$ $R_2 = CH_2OH$ (isocupressic acid), b $R_1 = CH_2OH$ $R_2 = COOH$ (agatholic acid), c $R_1 = COOH$ $R_2 = COOMe$

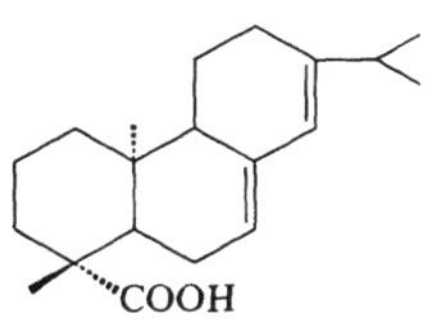

XVI epiabietic acid

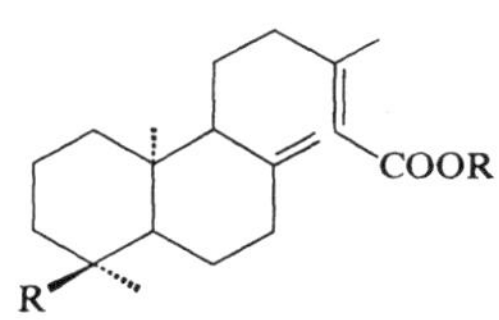

XVIIa R = H R′ = H (copalic acid), b $R = CH_3$ R′ = H, c R = H R′ = COOH

A. moorei gives a very complex mixture which has so far yielded abietic acid (V) and two compounds provisionally assigned, largely on spectroscopic evidence, the structures XVI and XVIIc [8]. The third New Caledonian species, *A. ovata*, which unlike any of the other species described here is a small, scrub-inhabiting form, is still more anomalous in giving a series of compounds with the enantiomeric configuration at C(10) and neither of the C(4) methyls oxidized (e.g., XVIIa–b) [8]. Since it contains no compounds with a communic acid side chain this resin cannot harden in the same way as that from *A. australis.* Possibly it represents a more primitive form of bled resin. Canal resin compounds of enantiomeric configuration would be most unusual in a conifer and these assignments require further confirmation.

D. Biogenesis of *Agathis* Bled Resin and Leaf Resin

The biogenesis of the Agathis bled resin compounds is readily accounted for by the biosynthetic pathway shown in Fig. 3 [24]. The basic precursor, formed by cyclization of geranyl-geranyl pyrophosphate, will be the labdadienyl pyrophosphate XVIIIa. Elimination of the pyrophosphate grouping followed by oxidation will lead to communic acid (IIa, IIIa) or, with cyclization, will give

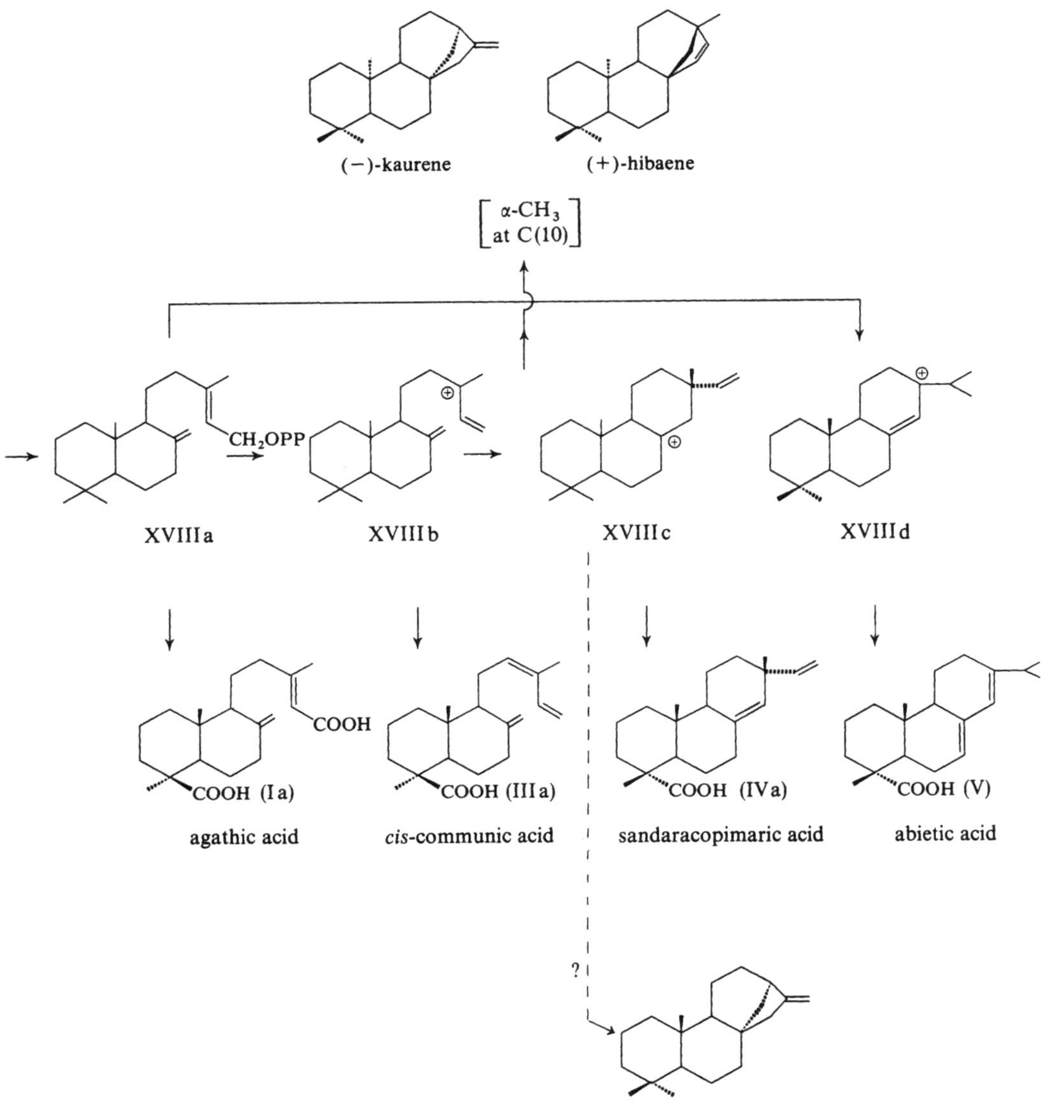

Fig. 3. Biogenesis of *Agathis* bled resin and leaf resin

sandaracopimaric acid (IVa). A more complicated process involving cyclization and rearrangement combined with elimination will lead via the ion XVIIId to abietic acid. The Australian kauris [19, 22] have given three of the four possible abietic acid isomers that could be formed by elimination of a proton from this ion.

The leaf surface compounds fit into the same scheme but in this case the ion XVIIIc must occur in two forms: one with the normal configuration at C(10), which stabilizes largely by addition of a hydroxyl ion, and one with enantiomeric configuration at C(9) and C(10), which stabilizes by cyclization to (−)-kaurene and (+)-hibaene. (+)-Phyllocladene, which would be formed by closure of the fourth ring in the ion XVIIIc with normal configuration at C(10), is not present in significant amount in the leaf surface. It occurs in other conifer leaves [16] but in *A. australis* leaf surface it appears to be displaced by the large amount of 8β-hydroxy-isopimarene formed.

The presence in *A. australis* bled resin of the corresponding hydrocarbons (IIe, IVf) and of some of the intermediates (acetates, alcohols, aldehydes) in the oxidation of the C(4) methyl to a carboxyl group indicates that oxidation takes place after stabilization of the ring C or side-chain structure [8]. In *A. australis* and in most other *Agathis* species examined there appears to be a strict correlation of bicyclic structure with axial C(4) methyl oxidation and of tricyclic structure with equatorial C(4) methyl oxidation [25].

However, the isolation from *A. moorei* of what is apparently the enantiomeric form of epiabietic acid [8] suggests that the specificity is rather for β-oxidation in the bicyclic compounds and α-oxidation in the tricyclic compounds; in *A. moorei* the specificity appears to be much less complete. The oxidation of (−)-kaurene to the gibberellins also involves oxidation of the α-methyl at C(4). The extent to which oxidation of the C(4) methyl goes to completion depends on the type of compound. Thus, appreciable proportions of sandaracopimarinol (IVe) remain but no significant amount of abietinol has been found.

The *Agathis* bled resins can thus be accounted for by two fairly simple processes though the final mixture of compounds can be somewhat complex. There appears to be a close general similarity to the processes occurring in, for instance, pine resin canals. Although pine resin canals are formed towards the inside of the cambium, the resin is similar, differing mainly in the abundance of abietic acid isomers and in the absence of agathic acid derivatives.

The occurrence of diterpenes of normal configuration in the same tissue as (−)-kaurene suggests a more detailed examination of the relationship between the two pathways and of the possibility that the biosynthesis of the normal compounds arose as a variant of a route to (−)-kaurene and the gibberellins.

E. Heartwood Resin of *A. australis*

This resin is composed very largely of a set of diterpenes and a set of phenols [25]. The diterpene set includes 3β-hydroxyisopimaradiene (XIXa) and a series of ketols derived from it by a succession of oxidation steps. The main diterpene component, araucarolone (XIXd) [26] forms 3–5 percent of the heartwood. These heartwood diterpenes must be formed by quite a different route from that leading to the bled resin compounds. Since no significant amount of the corresponding isopimaradiene was found, the oxygen of the 3β-hydroxyl group is probably introduced during or before cyclization; presumably by analogy with steroid processes [27] this will involve a geranyl-geraniol epoxide.

The phenols present include agatharesinol (XXIa) [28, 29], desoxyagatharesinol (XXIb), sugiresinol (XXIc) and hinokiresinol (XXId) [29, 30]. Phenols of this class, although recognized only recently, are apparently widely distributed in conifers. These compounds are of particular interest biogenetically because of

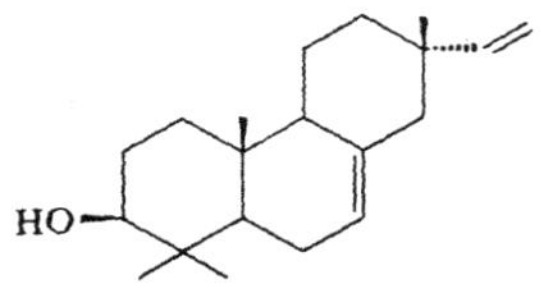

XIXa

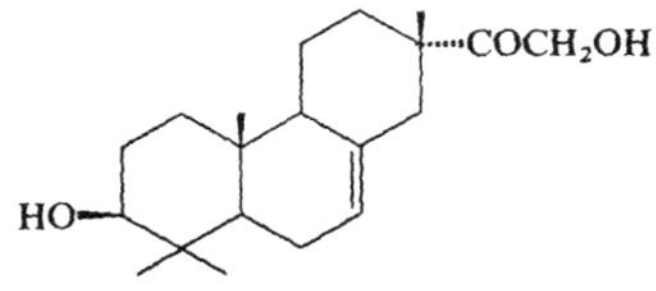

XIXb araucarol

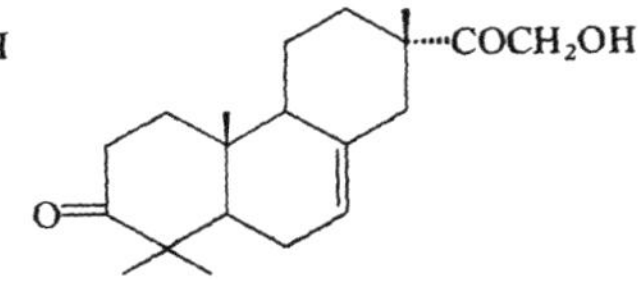

XIXc

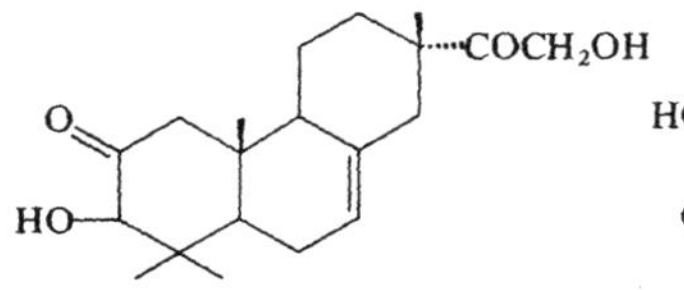

XIXd araucarolone

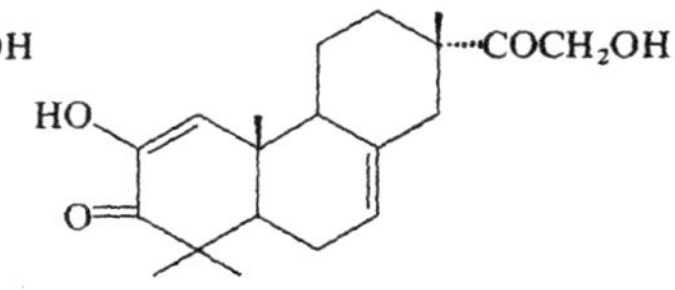

XIXe

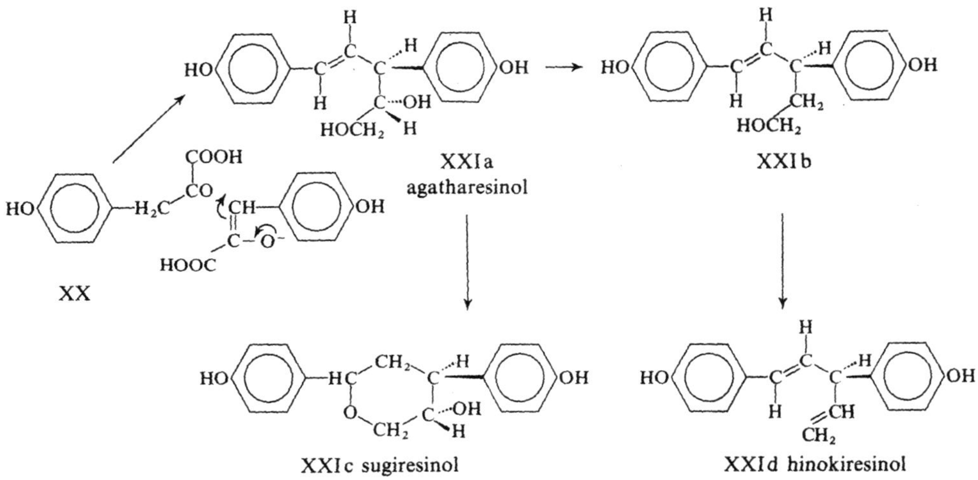

their relationship to lignin and its precursors, though they must be formed by an ionic coupling mechanism (see XX) quite different from that accepted for lignin biogenesis [28].

The heartwood compounds must have a protective function. The sapwood cannot contain more than very small amounts of either the diterpene or the phenolic compounds [8] and is much less durable than heartwood of average resin content.

F. Heartwood Resins of Other *Agathis* Species

The heartwood resins of the other species of *Agathis* that have been examined [8] do not contain any significant amount of araucarolone (XIXd), the main constituent of *A. australis* heartwood. In *A. macrophylla* from the Solomon Islands the main heartwood constituents are araucarol (XIXb) and araucarone (XIXc). Preliminary results [8] (tlc) indicate that *A. vitiensis* heartwood resin is usually similar to that of *A. macrophylla* but that *A. lanceolata*, though it gives a bled resin similar to that of *A. vitiensis* and *A. australis*, gives a wood resin more like that of its companion New Caledonian species, *A. moorei*.

IV. Recent and Ancient Fossil Resins

A. Recent Fossil Resins

Much of the kauri resin that has been produced in New Zealand has been obtained near the surface and has therefore probably ranged in age from less than a few hundred years up to a few thousand years. However, the resin is little different after a very few years from samples that are possibly thousands of years old [7, 8].

Fresh resin from *A. australis* is almost entirely soluble in acetone (90–95 percent), apart from small amounts of impurities and water-soluble gum. As it ages and hardens the amount of acetone-soluble material falls fairly rapidly, and at the end of a year or so the amount of acetone-soluble material can be down to 50 percent. Further change takes place only slowly. The less soluble material appears to be essentially a polycommunic acid containing some communol units [10, 12, 22]. Resin samples with only 30–35 percent unpolymerized material thus no longer contain significant amounts of communic acid type compounds and the acetone-soluble fraction is largely a mixture of agathic, abietic and sandaracopimaric acids together with sandaracopimarinol, some of the corresponding methyl esters and acetates and small amounts of hydrocarbons [8, 13, 31] (Fig. 4).

Essentially nothing is known of the composition of other fossil *Agathis* resins except that hard grades of Manila resin give agathic acid.

A. australis wood resin is present in highest concentration at the heartwood boundary. The amount present decreases fairly rapidly towards the center of the tree and at the center of a 300-year-old tree the amount of extractable material was quite small. The wood is durable under wet conditions and it is possible that information could be obtained from the timber that is found buried deep in swamps. However, this is unlikely to extend very far the period for which information can be obtained.

B. Ancient Resins

Table 2. *Some ancient New Zealand resins*

	Source	Approx. age (M. years)[a]	Geological period	Latitude (°S)
Roxburgh	lignite	20	Miocene	45
Ashers Pit	lignite	30	Oligocene-Miocene	46
Huntly	sub-bit. coal	?40	Eocene	38
Maramarua	lignite	40	Upper Eocene	38
Ohai	sub-bit. coal	60	Eocene	46
Greymouth	bit. coal	?70–100	Cretaceous	42

[a] Ages quoted in this discussion are only approximate estimates. For more accurate ages and for correlations between New Zealand and elsewhere the geological literature should be consulted.

Ancient resins are found at a number of places in New Zealand, nearly all, as would be expected, in association with coal or lignite. Some of these are listed in Table 2 [3, 4]. Resin is also found in Yallourn lignite in Australia. Among the possible sources of the New Zealand resins are *Dacrydium cupressinum* heart-shake resin, mainly podocarpic acid (XXII), *Podocarpus spicatus* heart-shake resin which contains matairesinol (XXIII), or even an extinct *Araucaria*, but the only probable source is some species of *Agathis*. Moreover the resins from Roxburgh and Ashers Pit are found in association with plant remains resembling modern *A. australis*.

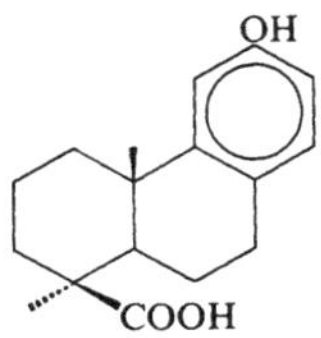

XXII Podocarpic acid

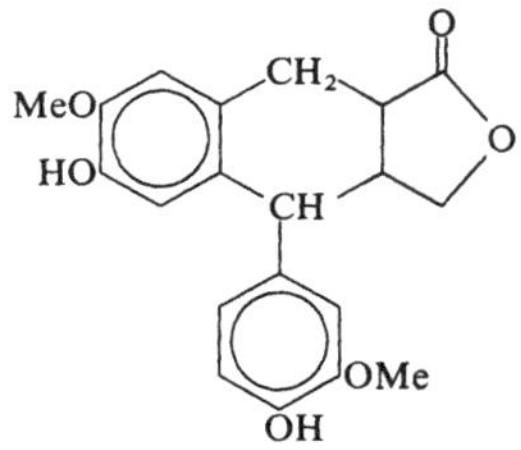

XXIII Matairesinol

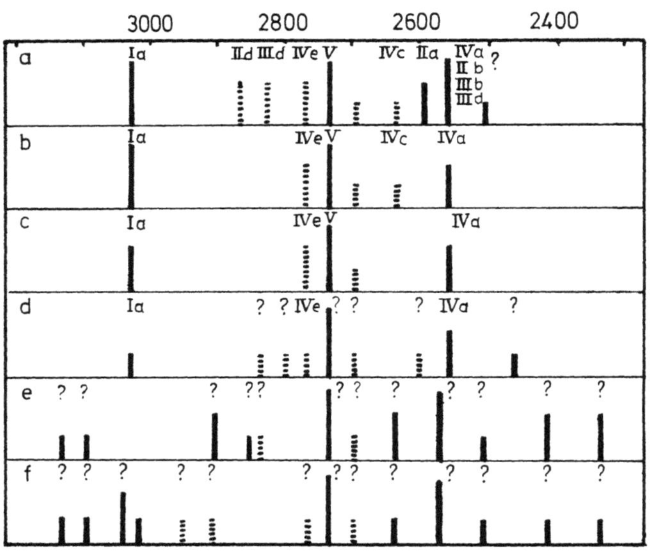

Fig. 4. Typical gas chromatograms of fresh (a, b), recent fossil (c) and ancient kauri bled resins: d, Roxburgh, *ca* 20 M years; e, Maramarua, *ca* 40 M years; f, Ohai, *ca* 60 M years. ⁞ neutral compounds ❙ methylated acids

The interrelationships of these resins are best shown by gas chromatography. Fig. 4 shows gas chromatograms of fresh and fossil kauri resins and of Roxburgh, Maramarua and Ohai resins in all of which the acids present have been converted to the volatile methyl esters by treatment with diazomethane. Roxburgh resin, which still contains about 5 percent of hexane-soluble material (i.e., the fraction which will contain any unchanged material), shows peaks corresponding to agathic acid (or methyl esters), sandaracopimarinol and sandaracopimaric acid, and the presence of these compounds has been confirmed by isolation [8, 32].

As far as peak assignments have been made, the Roxburgh material appears to be in good agreement with what might be expected for a very old kauri resin.

The Maramarua and Ohai resins, although geographically 1,000 km apart and separated in time by possibly up to 20 million years, give very similar gas chromatograms. They differ markedly, however, from the Roxburgh chromatogram, and the sandaracopimarinol and methyl sandaracopimarate peaks are absent.

The results from the Maramarua and Ohai resins cannot easily be harmonized with those from the recent fossil and Roxburgh resins on the provisional time scale given here. However, there is at present no indication of whether this is due to an inaccurate age estimate, to distortion of the time scale by the effects of heat or pressure, to differences in chemical environment or to differences in the resin when first formed. The infrared spectra of all these ancient *Agathis* resins are very similar and are closely related to that of fresh kauri resin (see below).

Further evidence for the origin of these resins has been obtained by selenium dehydrogenation [4] which although far less specific, also gives results consistent with a kauri origin. Dehydrogenation of samples of Roxburgh, Ohai, Huntly and Greymouth resins gives mixtures of pimanthrene (XXIVa) and

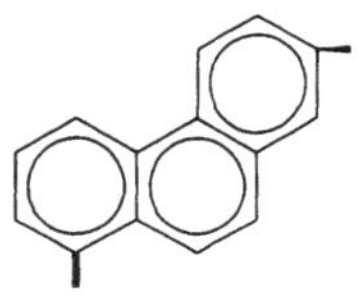

XXIVa Pimanthrene

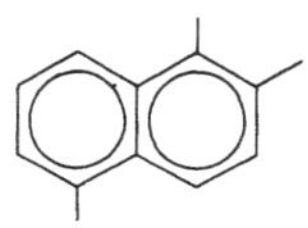

XXIVb Agathalene

agathalene (XXIVb) [4], as would be expected from material derived from agathic and sandaracopimaric acids and similar compounds. Agathic acid on dehydrogenation gives pimanthrene by a cyclization reaction and also gives agathalene by a somewhat unusual partial elimination of the C(9) side-chain [6]. One resin sample (Rotowaro) on dehydrogenation gave a different product [4]. The components were not identified but this sample presumably came from one of the other species that must undoubtedly have coexisted with *Agathis* in the area.

C. Information Available from Ancient Kauri Resins

From the results obtained so far on modern kauri resin the only species that could be clearly distinguished from *A. australis* in an ancient resin are *A. ovata* and possible *A. moorei*. Separation of the other species must depend on much smaller differences, e.g. the absence of abietic type compounds from the Australian *Agathis* species. The separation of *A. australis* from *A. dammara* might also be made on the basis of the ratio of agathic to sandaracopimaric compounds and, for the Malayan *A. dammara*, on the absence of daniellic acid.

In fact, Roxburgh resin appears at present to be consistent only with *A. australis* resin. It would thus be interesting to compare with Yallourn resin of roughly comparable age to see whether there was any reflection of the differences between the modern Australian and New Zealand species. The older Maramarua and Ohai resins are not readily interpreted and therefore might perhaps be best compared with ancient resins of completely different origin to see how far more general geochemical processes account for the compounds present.

If the differences between modern *Agathis* species that are indicated by present results can be confirmed by further work and shown to be significant, then *A. australis* could take up a intermediate position between the Australian species and the Indonesian group. This would not be entirely unexpected since there appears to be a marked botanical boundary between Australia and New Guinea [1, 2, 34] and *Agathis* could have entered both New Zealand and Australia from the south. However, far more evidence is needed for any convincing study of such a relationship.

V. Experimental Techniques

A. Extraction

The unpolymerized material in kauri resins and fossil resins is best extracted by treating the crushed resin with several volumes of acetone at room temperature for 24 hours and then decanting the solution. Older samples and the ancient resins are best extracted with chloroform/acetone (3:1) or chloroform/methanol mixtures, in which the polymeric material is very largely soluble, followed by the addition of an equal volume of hexane to precipitate the polymer. In both cases the extract may need to be purified by further precipitation with hexane from a solution in a very small amount of acetone (this may also remove more polar substances, such as agathic acid).

Resin bleeding from a tree can serve as a pollen trap, but to recover the pollen for analysis it is necessary to dissolve the resin completely. If the sample is not soluble in simple solvent mixtures or in solvents such as methyl ethyl ketone or alcoholic KOH, there are several potentially useful methods for increasing the solubility. The resin can be acetylated (acetic anhydride/pyridine at 100° C for 1 hr. or acetic anhydride/sodium acetate under reflux for 3 hr.) or heat-treated (e.g., 300° C for 5 min) as in varnish technology.

Fossil kauri resin is insufficiently soluble in the oils used for varnish making and is therefore usually heat-treated ("run") to increase its solubility [10]. Agathic acid (Ia) on similar thermal treatment gives noragathic acid (XXV) [10] and on treatment with acid gives isoagathic acid (XXVI) [10, 35], both of which may well be present both in heat-treated resin and in the unpolymerized fraction of the ancient Agathis resins.

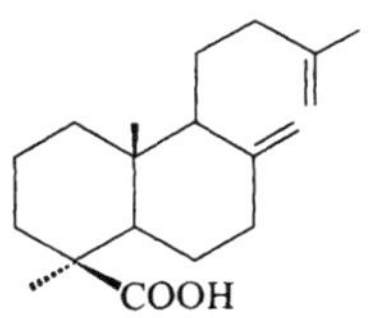

XXV Noragathic acid

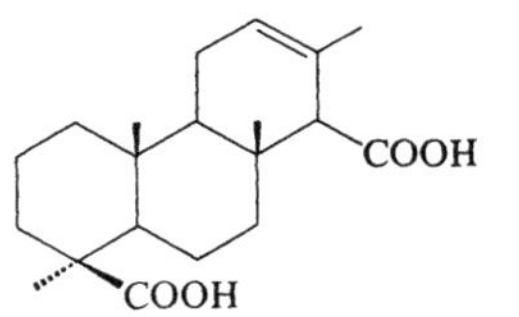

XXVI Isoagathic acid

For wood samples a cold soak extraction (room temp., 24 hr.) of the ground material is most appropriate and convenient and is less harmful than Soxhlet extraction.

B. Separation

The composition of the mixtures of simple diterpenes found in the bled resins is most readily estimated by preparative thin layer chromatography on silica gel (hexane: acetone 70:30, 2–5 mg on a 10 cm wide 250 micron layer or 10–30 mg on a 1 mm layer) followed by gas chromatography. Acids must be converted to methyl esters by diazomethane treatment of fractions containing them before gas chromatography.

Bands on a thin layer chromatogram can be detected by UV light using a fluorescent silica gel layer, supplemented by detection with iodine vapor or a rhodamine spray. For analytical thin layer chromatograms without subsequent recovery a phosphomolybdic acid spray is usually best.

The gas chromatographic results are most conveniently recorded on a retention index scale [36]. Retention times are converted to retention indexes by changing to relative retention times based on a suitable n-alkane and converting this value graphically using a log plot of the relative retention times of a set of n-alkanes. The values given in Fig. 4 were obtained by taking a relative retention time for n-octacosane (n—C_{28}) of 100 and constructing a log plot of the corresponding relative retention times for the C_{24}, C_{28} and C_{31} n-alkanes against 2,400, 2,800 and 3,100 respectively.

Suitable glc conditions for the kauri bled resin diterpenes are: 180° C, 1 percent neopentylglycol succinate on a support treated with hexamethyldisilazane. On a 1 m column this gives peaks with a width at half height of 20–30 retention index units. Greater resolution can of course be obtained with capillary columns. A flame ionization detector or a detector of similar sensitivity is indispensable. (Gas chromatographic data for the investigations centered around *A. australis* were obtained with a 1 m glass column and flow rates of 50–100 cc/min N_2 chosen to give retention times of 1–10 min.)

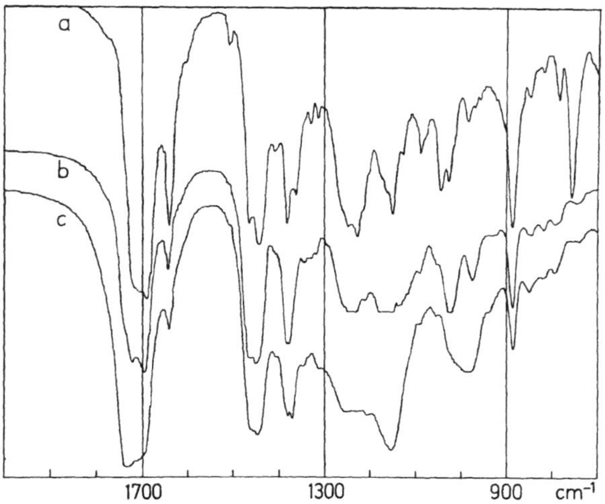

Fig. 5a–c. Infrared spectra of (a) fresh kauri resin, (b) an ancient *Agathis* resin (Maramarua, ca. 40 M. years), and (c) of Baltic amber (? *Pinus* sp. ? 40 – 50 M. years). Resins a and b show a marked relationship. The effects of polymerisation are evident in the loss of olefinic absorption at 760 cm^{-1} and 1,640 cm^{-1}. Resins b and c give infrared spectra that are quite similar but show significant differences in the fingerprint region and in the carbonyl absorption near 1,700 cm^{-1}

Larger amounts of these compounds can be prepared by chromatography on silica gel or silica gel/silver nitrate columns using hexane/acetone or hexane/ether mixtures. The columns may be regenerated with half volumes first of acetone and then of hexane or, for the silver nitrate columns, ether/20 percent hexane first and then pure hexane.

Since many of the compounds described here are sensitive to light and air, they should be kept in a cool, dark place, possibly with antioxidant protection.

C. Identification

Identification of nearly all the kauri resin compounds can be made almost immediately from the nmr spectrum (10–15 mg for a single scan) and the rotation, coupled with consideration of a reasonable biogenetic pathway. Confirmation wherever possible is made by direct comparison. Nmr spectra are particularly valuable because of the directness of the structural information they provide.

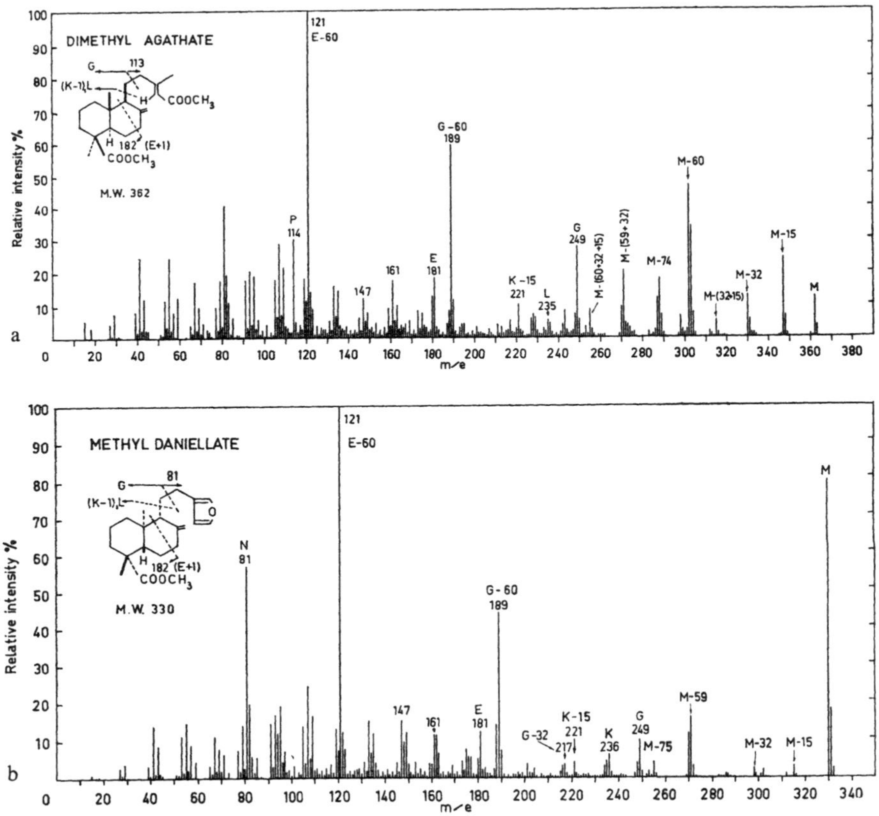

Fig. 6a and b. Use of mass spectra in the identification of resin components. (a) dimethyl agathate, (b) methyl daniellate [38]. Dimethyl agathate and methyl daniellate are closely related but their mass spectra, although similar, are quite characteristic—some compounds, e.g., the C(4) epimers are not distinguished. Given the structure of dimethyl agathate the mass spectrum of the other compound suggests that it also is a dicyclic diterpene and gives a good indication of the nature and location of the modified group

They also give useful information on mixtures and a nmr spectrum of the crude resin gives an excellent check on possible loss and chemical change during work-up.

Identification of the compounds in mixtures of known compounds can usually be obtained from glc and tlc evidence using very small amounts of material. In these cases the identification can be confirmed by comparison of mass [37, 38] or

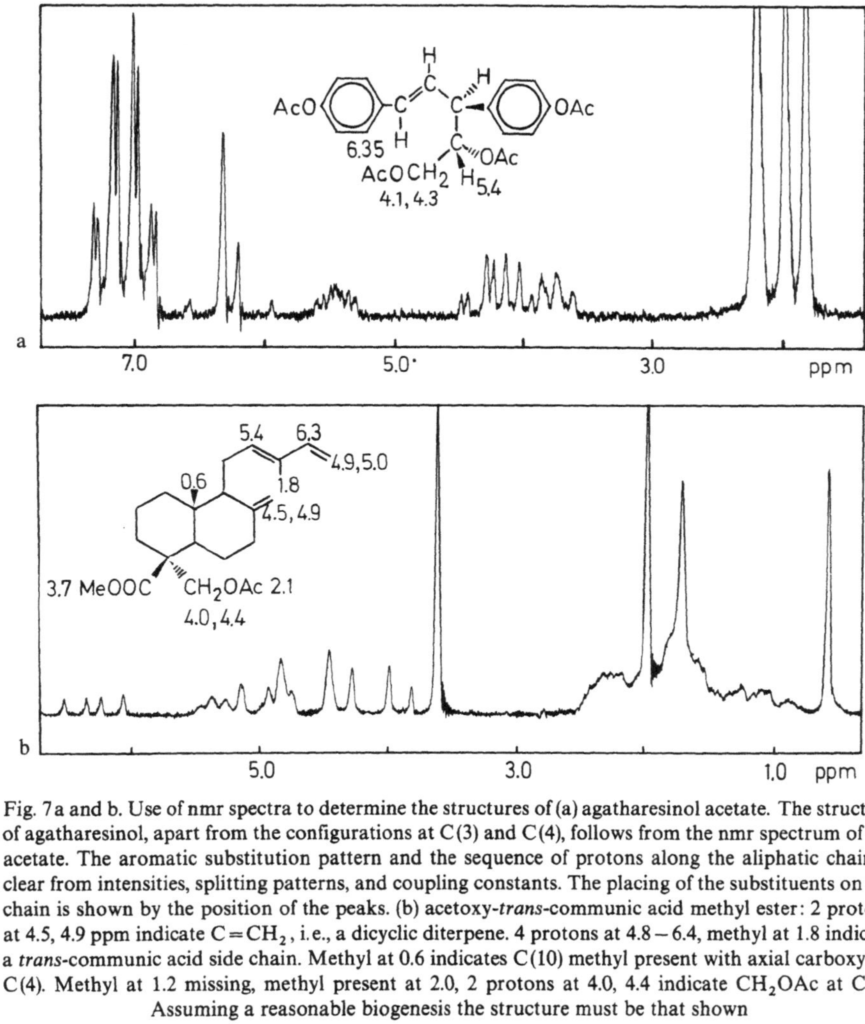

Fig. 7a and b. Use of nmr spectra to determine the structures of (a) agatharesinol acetate. The structure of agatharesinol, apart from the configurations at C(3) and C(4), follows from the nmr spectrum of the acetate. The aromatic substitution pattern and the sequence of protons along the aliphatic chain is clear from intensities, splitting patterns, and coupling constants. The placing of the substituents on the chain is shown by the position of the peaks. (b) acetoxy-*trans*-communic acid methyl ester: 2 protons at 4.5, 4.9 ppm indicate $C=CH_2$, i.e., a dicyclic diterpene. 4 protons at 4.8 – 6.4, methyl at 1.8 indicate a *trans*-communic acid side chain. Methyl at 0.6 indicates C(10) methyl present with axial carboxyl at C(4). Methyl at 1.2 missing, methyl present at 2.0, 2 protons at 4.0, 4.4 indicate CH_2OAc at C(4). Assuming a reasonable biogenesis the structure must be that shown

infrared spectra. The combined gc–ms technique is especially valuable here because of its speed and the very small sample required (10–100 nanograms of each component to be identified).

Some examples of the uses of these techniques are shown in Figs. 5–8. The infrared spectra (Fig. 5) include for comparison a sample of Baltic amber (*Pinus* sp.?) which can be much less soluble than the *Agathis* resins and largely inaccessible to investigation by other techniques. The mass spectra of the methyl esters of agathic acid and daniellic acid (Fig. 6 kindly provided by Dr. C. R. ENZELL) show the excellent fingerprint characteristics of mass spectra combined with useful structural information.

The nmr spectra (Fig. 7) of a heartwood phenol and of the methyl ester of the main leaf-canal compound show how this technique may in a favorable case give

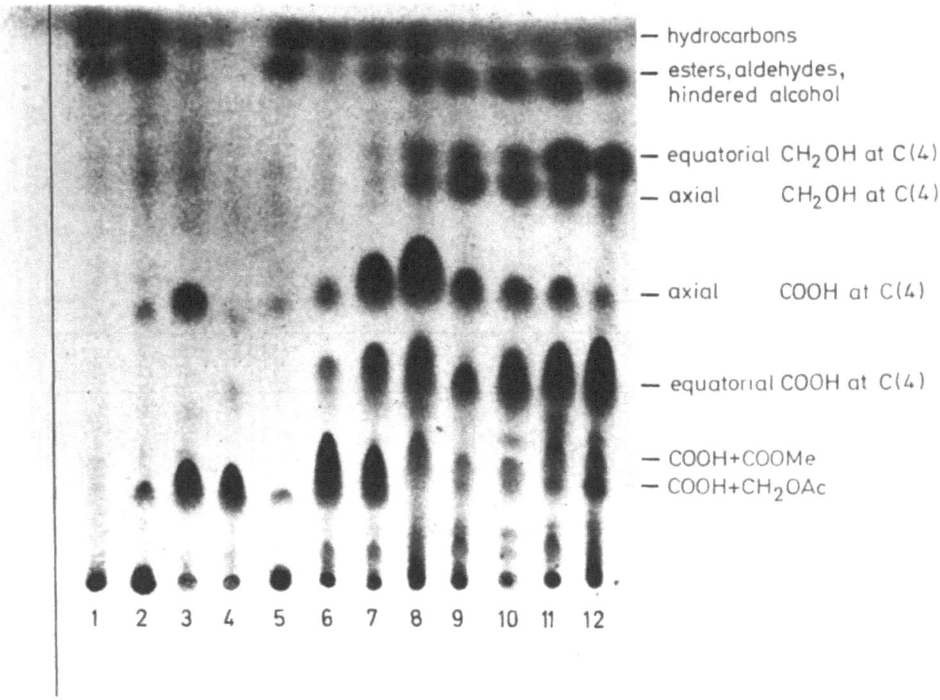

Fig. 8. Thin layer chromatograms (hexane/acetone) showing changes in composition of *A. australis* resin from leaves to bark to fossil resin. 1, leaf surface; 2 and 5, whole leaf; 3 and 4, canal resin from old and young leaves; 6 and 7, canal resin from twigs and stems; 8, 9 and 10, canal resin from branches and trunk; 11, recent bled resin; 12, "fossil" resin

the complete structure (apart from configuration). In the more complex diterpenes biogenetic information is very helpful. The tlc plate (Fig. 8) of a series of samples taken from different parts of a kauri tree illustrates the simplicity and sensitivity of the method for showing the presence of compounds of different types. The interpretation of the polarity in terms of functional groups is simplified in the present case where all the components are diterpenoid.

References

1. FLORIN, R.: The distribution of conifer and taxad genera in time and space. Acta Horti Berg. **20**, 4 (1963).
2. MELVILLE, R.: Continental drift, mesozoic continents and the migration of angiosperms. Nature **211**, 116–120 (1966).

3. EVANS, W. P.: Microstructure of New Zealand lignites. Part III. Lignites apparently not altered by igneous action. New Zealand J. Sci. Technol. **15**, 365–385 (1934).
4. BRANDT, C. W.: The resins from some New Zealand coal-measures. New Zealand J. Sci. Technol. **20**, 306 B – 310 B (1939).
5. TSCHIRCH, A., u. B. NIEDERSTADT: Über den neuseeländischen Kauri buschcopal von *Dammara australis*. Arch. Pharm. **239**, 145–160 (1901).
6. HOSKING, J. R., and L. RUZICKA: Dehydration and isomerisation of agathic dicarboxylic acid. Helv. Chim. Acta **13**, 1402–1423 (1930).
7. THOMAS, B. R.: The chemistry of the order Araucariales. The bled resins of *Agathis australis*. Acta Chem. Scand. **20**, 1074–1081 (1966).
8. – Unpublished data.
9. HOSKING, J. R.: Kauri resins. Rec. Trav. Chim. **48**, 622–636 (1929).
10. – Kauri Resins in Varnishes, Lacquers and Paints. Middlesex: Research Association of British Paint and Varnish Manufacturers 1940.
11. NIERENSTEIN, M.: Gum kauri. The presence of leucomaclurin glycol ether. Pharm. J. **153**, 5 (1944).
12. CARMAN, R. M.: Private communication.
13. GOUGH, L. J.: Conifer resin constituents. Chem. Ind. (London) **1964**, 2059–2060.
14. HOSKING, J. R.: The essential oil from *Agathis australis* (kauri pine). Rec. Trav. Chim. **47**, 578–584 (1928).
15. BRIGGS, L. H., B. F. CAIN, R. C. CAMBIE, B. R. DAVIS, P. S. RUTLEDGE, and J. K. WILMSHURST: Kaurene. J. Chem. Soc. **1963**, 1345–1355.
16. APLIN, R. T., R. C. CAMBIE, and P. S. RUTLEDGE: The taxonomic distribution of some diterpene hydrocarbons. Phytochemistry **2**, 205–214 (1963).
17. WENKERT, E.: Structural and biogenetic relationships in the diterpene series. Chem. Ind. (London) **1955**, 282–284.
18. CARMAN, R. M.: *Agathis microstachya* oleoresin. Australian J. Chem. **17**, 393–394 (1964).
19. CARMAN, R. M., and N. DENNIS: The diterpene acids of *Agathis robusta* oleoresin. Australian J. Chem. **17**, 390–392 (1964).
20. –, and R. A. MARTY: *Agathis microstachya* oleoresin. Australian J. Chem. **19**, 2403–2406 (1966).
21. – – Private communication.
22. –, and D. E. COWLEY: Dundatholic acid. Australian J. Chem. **20**, 193–196 (1967).
23. ENZELL, C. R.: The structures of the diterpenes torulosol, torulosal and agatholic acid. Acta Chem. Scand. **15**, 1303–1312 (1961).
24. RICHARDS, J. H., and J. B. HENDRICKSON: The Biosynthesis of Steroids, Terpenes and Acetogenins. New York: Benjamin 1964).
25. ENZELL, C. R., and B. R. THOMAS: The wood resin of *Agathis australis*. Acta Chem. Scand. **19**, 913–919 (1965).
26. – – Structure and configuration of araucarolone and some related compounds from *Agathis australis*. Acta Chem. Scand. **19**, 1875–1896 (1965).
27. COREY, E. J., W. E. RUSSEY, and P. P. ORTIZ DE MONTELLANO: 2,3-oxido-squalene an intermediate in the biological synthesis of sterols from squalene. J. Am. Chem. Soc. **88**, 4750–4751 (1966); – VAN TAMELEN, E., J. D. WILLETT, R. B. CLAYTON, and K. E. LORD: Enzymic conversion of squalene 2,3-oxide to lanosterol and cholesterol. J. Am. Chem. Soc. **88**, 4752–4753 (1966).
28. ENZELL, C. R., and B. R. THOMAS: Agatharesinol. Tetrahedron Letters **1966**, 2395–2402.
29. – Y. HIROSE, and B. R. THOMAS: Absolute configurations of agatharesinol, hinokiresinol and sugiresinol. Tetrahedron Letters **1967**, 793–798.
30. –, and B. R. THOMAS: Unpublished data.
31. GOUGH, L. J.: Private communication.
32. HOSKING, J. R.: Appendix to reference 3.
33. MCCRINDLE, R., and K. H. OVERTON: The chemistry of the cyclic diterpenoids. Advan. Org. Chem. **5**, 47–113 (1965).

34. FLEMING, C. A.: New Zealand Biogeography. Tuatara **10** (2), 53 – 108 (1962).
35. BORY, S., M. FETIZON et P. LASZLO: Stereochemie dans la serie de l'acide agathique. Bull. Soc. Chim. France **1963**, 2310 – 2322.
36. KOVATS, E.: Gas-chromatographische Charakterisierung organischer Verbindung. Teil I. Retentions indices aliphatischer Halogenide, Alkohole, Aldehyde und Ketone. Helv. Chim. Acta **41**, 1951 – 1932 (1958).
37. BUDZIKIEWICZ, H., C. DJERASSI, and D. H. WILLIAMS: Structure Elucidation of Natural Products by Mass Spectrometry, vol. 2. San Francisco: Holden-Day 1964.
38. ENZELL, C. R., and R. RYHAGE: Mass spectrometric studies of diterpenes, Carbodicyclic diterpenes. Arkiv Kemi. **23**, 367 – 399 (1965); – Mass spectrometric studies of diterpenes. Aromatic diterpenes. Arkiv Kemi. **26**, 425 – 434 (1967).
39. BECK, C. W., E. WILBUR, and S. MERET: Infrared spectra and the origin of amber. Nature **201**, 256 – 257 (1964).
40. DALLIMORE, W., and A. B. JACKSON: A Handbook of Coniferae and Ginkgoaceae (4th ed., revised by S. G. HARRISON). London: Arnold 1966.

CHAPTER 26

Kerogen of the Green River Formation

W. E. ROBINSON

Laramie Petroleum Research Center,
U.S. Bureau of Mines,
Laramie, Wyoming

Contents

I. Introduction

The deposition of plant and animal remains in marine and non-marine environments and the alteration of this debris during subsequent geologic periods by numerous reactions, produced a wide variety of carbonaceous materials. Source material and conditions of deposition were major factors influencing the types of final products formed, which range from anthracite coal to petroleum. Between these two extreme types are numerous other carbonaceous substances of different compositions and properties. Oil shales, kukersite, and torbanite represent types of carbonaceous material that contain various amounts of organic matter, commonly called kerogen, from which high yields of oil can be produced at elevated temperatures, even though the constitutions of the kerogens differ.

Much confusion exists in the use and interpretation of the terms "oil shale" and "kerogen". The term "oil shale", a misnomer, describes organic rich rock that produces significantly large quantities of crude oil by distillation. Based upon a classification of DOWN and HIMUS [1] oil shale is called a kerogen rock. In contrast to kerogen coal, 50 percent or more of the organic material in kerogen rock is converted to oil upon heating to 500° C. The term "kerogen" refers to the organic material present in kerogen rock which is insoluble in common petroleum solvents*.

* The author suggests that the use of oil yield (500° C)/carbon ratios or the use of volatile material (500° C)/carbon ratios would be helpful in classifying the wide range of materials now referred to as kerogen. A high ratio would indicate a pyrobitumen-like material, a low ratio would indicate a coal-like material.

The benzene-soluble organic material present in the kerogen rock is often referred to as bitumen. For convenience when referring to elemental analysis and other tests, the term "kerogen" is associated sometimes with results obtained for the total organic material. Although not precisely correct, the error is small since the soluble material is a minor part of the total organic material.

Table 1. *General description of exposed rocks in the Piceance Creek basin (Cathedral Bluffs) oil-shale area*[a]

Tertiary	Eocene	Green River Formation	Evacuation Creek member. 600 feet thick. Gray and brown fine to medium-grained sandstone with interbedded gray marlstone and a few thin oil-shale beds. Upper part contains lenses of massive sandstone. Member weathers to rounded slopes.	
			Parachute Creek member. 400 to 750 feet thick. Grayish-black, gray, and brown marlstone, including principal oil-shale units and several thin altered tuff and analcite key beds. Includes the Mahogany ledge, which includes the Mahogany bed and Mahogany marker. Weathers light gray and blue; forms cliffs and steep slopes.	
			Garden Gulch member. 200 feet thick. Gray marlstone; some gray and brown shale, and a few thin beds of oil shale. Weathers gray; forms steep slopes and subdued ledges.	Lower shaly member (of southern part of the Cathedral Bluffs oil-shale area). Gray marlstone; a few beds of gray and brown siltstone and limestone. Several thin oolitic and algal beds. Member interfingers with Douglas Creek, Garden Gulch, and lower part of Parachute Creek members. Weathers to slopes and low cliffs.
			Douglas Creek member. Less than 800 feet thick. Buff sandstone and gray shale; a few thin algal limestone beds. Weathers buff; forms slopes and low cliffs.	
	Paleocene?	Wasatch formation. 300 feet thick. Massive buff sandstone; a few interbedded red, brownish-yellow, gray, and maroon shale beds. Weathers to a slope capped by low sandstone cliffs.		
Cretaceous	Upper Cretaceous	Mesaverde formation. Uppermost 500 feet exposed in mapped area. Massive buff sandstone; some shale and coal beds.		

[a] Reported by DONNELL *et al.* [5].

Green River kerogen has been described as having been formed in rather shallow, permanently stratified lakes during the Eocene Epoch [2]. The organic material was derived from the partial degradation of aquatic organisms that grew in these lakes. When the debris fell to the bottom of the lake, it apparently entered a highly reducing environment where it resisted further bacterial degradation. Intimately associated with this organic material are large quantities of mineral matter such as calcite, dolomite, analcite, quartz, pyrite, illite, feldspar, marcasite, and opal, some being washed in while others were formed by precipitation.

The geology of the Green River Formation has been studied extensively by BRADLEY [3], DUNCAN and BELSER [4], DONNELL *et al.* [5], and others of the U.S. Geological Survey. The Green River Formation in the Piceance Creek Basin is divided into four members [3, 5] and short descriptions of the members are given in Table 1. Several distinctive beds occur in the principal oil-shale zone of the Parachute Creek member. The Mahogany marker, an analcitized tuff that averages 0.5 foot in thickness, is a key bed for identification purposes but contains no

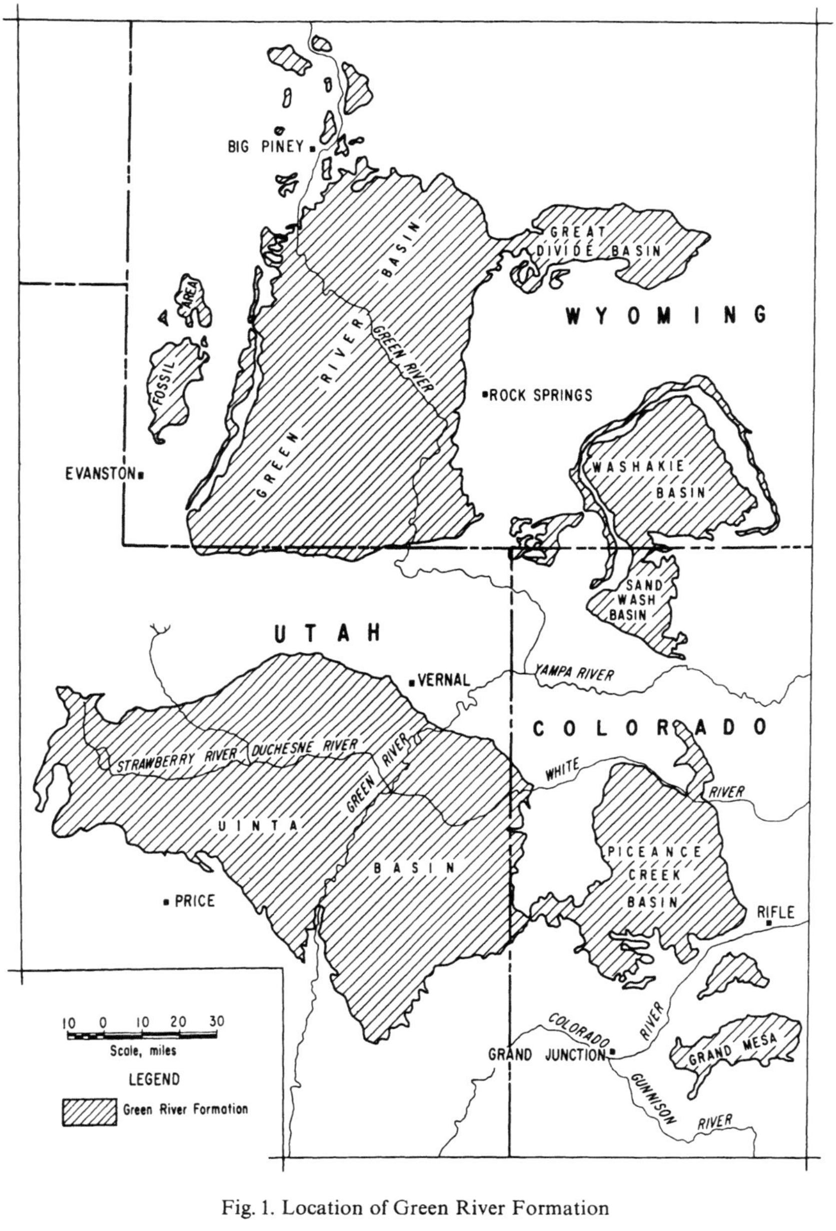

Fig. 1. Location of Green River Formation

oil shale. This marker is 5 to 15 feet above the Mahogany bed, which is the thickest rich oil-shale bed in the Green River Formation. Both the Mahogany marker and the Mahogany bed are in the Mahogany ledge, which contains several beds of distinctive mahogany-colored oil shale and is the richest oil-shale unit in the area. Microfossils present in the Green River Formation have been described by BRADLEY [3].

Green River kerogen as seen under the microscope consists of organic material of two types; the major portion is yellow and the minor portion is brown or brownish black. The yellow structureless organic material occurs in long thin bands or stringers laid parallel to the bedding laminae and is intimately mixed with inorganic mineral, while the brown organic material occurs in thin stringers and as irregular granular masses. The brown organic material constitutes about 5 percent of the total organic material.

Considerable research has been conducted on the Green River Formation kerogen during the past few decades, most of which is reviewed by ROBINSON and STANFIELD [6].

The data presented is discussed in terms of chemical constitutional and physical constitutional analyses, partially summarizing studies of the constitution of the kerogen of the Green River Formation at the U.S. Bureau of Mines laboratory. Unless noted otherwise, the samples studied were from the Piceance Creek basin of Colorado. The other major oil shale basins in the Green River Formation are shown in Fig. 1.

II. Chemical Constitution Analysis

A. Elemental Composition

Based on elemental analyses by SMITH [7], the empirical formula for Green River Formation kerogen at the Mahogany zone level may be represented by $C_{215}H_{330}O_{12}N_5S$ with a formula weight of about 3,200. The atomic H/C of 1.6 shows the average composition of Green River kerogen to be more aliphatic than aromatic. The atomic C/O of 18, the C/N of 43, and the C/S of 215 show the average amount of carbon atoms to hetero atoms.

The assay oil to organic carbon ratio for Green River kerogen is about 0.8. This represents a very high conversion of the organic carbon to a viscous oily product and smaller amounts of gas and carbon residue. This indicates a vast predominance of aliphatic to condensed aromatic structures as the latter type structures are known to produce large amounts of carbon residue and small amounts of oil [8].

B. Oxidation

One method of degrading kerogen to lower-molecular-weight materials for study is by oxidation. BONE *et al.* [9] developed a carbon-balance technique for studying coals. By this method, the organic material is exhaustively degraded in a boiling solution of alkaline potassium permanganate and the amount and distribution of the oxidation products determined. Highly condensed aromatic materials, such as anthracite coals, are oxidized to large yields of nonvolatile, nonoxalic acids which contain benzene carboxylic acids. Fatty materials, on the other hand, are very resistant to alkaline permanganate oxidation. Because of these variations in oxidation behavior, this relatively simple experiment determines differences or similarities between carbonaceous materials. ROBINSON *et al.* [10, 11] found that Green River kerogen was almost completely oxidized to oxalic acid, volatile acids, and CO_2, showing the absence of significant amounts of fatty or condensed aromatic structures. By contrast another kerogen, notably Middle Dunnet (Scottish) kerogen, produced 14 percent nonvolatile-nonoxalic acids from which benzenoid acids were identified. This indicated the presence of aromatic structures, or structures that were oxidized to aromatic acids. Australian torbanite contained about 74 percent material resistant to oxidation, which indicated the presence of considerable fatty remains. The peat to anthracite series [12] was almost completely oxidized to lower-molecular-weight acids, each producing nonvolatile-nonoxalic acids from which benzenoid acids were identified. The production of greater amounts of nonvolatile-nonoxalic acids with increase in the rank of the coal indicated that the amount of condensed aromatic structures increased in the peat to anthracite series. Although this test does not prove the presence of definite structures, it does show differences in structure and indicates that these differences may be due to the aromatic portion of the structure. On this basis, it was concluded that Green River kerogen is essentially nonbenzenoid and contains little or no fatty or algal remains resistant to oxidation.

Alkaline potassium permanganate oxidation is also a convenient laboratory method of producing small quantities of intermediate oxidation products for study. By controlling the extent of oxidation, acids of intermediate molecular weight were produced. The nonprecipitated acids, referred to as filtrate-soluble acids, were converted to n-butyl esters and studied by ROBINSON *et al.* [13]. A total of 20 distillation fractions plus a residue were obtained by vacuum distillation at 7 to 10 Torr. Except for a small amount of residue, the number of acid groups per molecule, based upon saponification equivalents and molecular weights, ranged from 1.9 to 2.4. The residue had 4.8 acid groups per molecule. These results show that the filtrate-soluble acids were predominantly dicarboxylic and suggested a linear rather than a highly condensed structure for Green River kerogen.

Based on mass spectral analyses and other data, the first 10 fractions of n-butyl esters contained 42.4 percent butyl oxalate, 7.1 percent butyl succinate, 6.0 percent butyl glutarate, 1.9 percent butyl adipate, 1.1 percent butyl pimelate, 0.5 percent butyl suberate, and 41.0 percent unidentified product. Physical properties of the unidentified material showed that the normal dicarboxylic acid series did not continue beyond fraction 10, and that the material boiling above this fraction was predominantly alicyclic carboxylic acids.

This unidentified material was reduced to hydrocarbons by ROBINSON and LAWLOR [14] by the following reactions:

$$(1)\quad RCO_2H \xrightarrow[H_2SO_4]{\text{alcohol}} RCO_2C_4H_9$$

$$(2)\quad RCO_2C_4H_9 \xrightarrow[THF]{LiAlH_4} RCH_2OH$$

$$(3)\quad RCH_2OH \xrightarrow[P_2O_5,\,H_3PO_4]{KI} RCH_2I$$

$$(4)\quad RCH_2I \xrightarrow[HCl]{Zn} RCH_3.$$

These reactions gave a product that was amenable to fractionation by solubility in pentane at 0° C, elution chromatography from alumina, wax separation from methylethyl ketone at −5° C, elution chromatography from silica gel and urea adduction of the wax and paraffin oil fractions. The two major fractions were the paraffin oil fraction, which was predominantly naphthenic and represented 29.3 percent of the total fraction, and the polar resins, which represented 23.7 percent of the total fraction. Separation of 4.1 percent of the total fraction by urea adduction indicated that some long-chain fatty acids had been present in the oxidation product. This type of material is somewhat resistant to oxidation and would tend to be concentrated by the oxidation process. The n-alkanes, prepared from the oxidation product without loss of carbon atoms, had a CPI (carbon preference index) ratio of odd to even carbon numbers of 0.7 in the C_{25} to C_{33} range.

Analyses of the alkane fractions of the reduced product showed that the urea adduct wax was predominantly straight-chain alkanes with an average carbon length of 28. By contrast, the nonadduct wax was predominantly naphthenic with 1 to 6 rings per molecule. The adduct paraffin oil was mostly straight-chain material with average carbon chain lengths of 22. The largest single fraction of the product, namely the nonadduct paraffin oil, was analyzed by mass spectrometry and found to be predominantly naphthenic, 1 to 6 rings per molecule. The two nonadduct fractions contained 22 and 25 percent aromatic materials.

Table 2. *Nuclear magnetic resonance analyses of oil fractions*[a]

	Percentage of total CH_2 groups		CH olefinic	CH aromatic	CH_2 groups alpha to O, N, S, or aromatic ring	Terminal CH_3 groups
	straight chain	naphthenic				
Paraffin oil	20	80	None	None	None	Small
Aromatic oil	40	60	None	Small	Large	Small
Polar oil	40	60	None	Small	Small	Small

[a] Obtained with a 40 megacycle oscillator.

Nuclear magnetic resonance analyses of the oil fractions of the reduced product shown in Table 2 indicate the naphthenic character of the oil fractions and show the presence of few side chains with terminal CH_3 groups. The aromatic fraction had a small peak in the region normally assigned to CH groups and a large peak in the region assigned to CH_2 alpha to O, N, S, or aromatic rings, suggesting a high degree of substitution.

Mass spectral analyses of the aromatic oil fraction of the reduced product, which represented 8.7 percent of the oxidation product, indicated that it contained almost exclusively 1-ethylpropyl substitution (see [14] for details of analysis). Based on the results of color tests for aromatics and the method of fractionation used, it is probable that only mononuclear and dinuclear aromatics, including benzenes, tetralins, indanes, and indenes, were present in this fraction.

The infrared spectra of the two resin fractions of the reduced product, which represented 34.6 percent of the total degradation product, are shown in Table 3. Only minor differences were noted in the two materials. Both had strong peaks at 2,900, 1,460, and 1,370 cm^{-1} indicating the presence of methyl-methylene groups. Weak or no absorption at 720 cm^{-1} indicated little or no carbon chains with more than four carbon atoms. The evidence suggests that the resins were predominantly cyclic and had some aromatic character. Both resins had strong absorption at 1,730 cm^{-1} attributed to carbonyl groups. This suggested that a portion of the resin oxygen functional groups remained unreduced.

Table 3. *Infrared spectra of resin fractions*

Band locations (cm^{-1})	Intensity [a]		Possible interpretations
	nonpolar resins	polar resins	
3,400	M–W	M–W	OH (bonded)
2,900	S	S	CH_2, CH_3
1,730	S	S	C=O (ester)
1,680	W	None	C=O (aromatic)
1,580–1,600	M–W	W	C=C (aromatic), carboxylate ion
1,460	S	S	CH_2, CH_3
1,370	S	S–M	CH_3
1,240	W	S–M	ester, ketone
1,120	S	None	alcohol, anhydride
875	W	None	aromatic
870	None	W	aromatic
740	S	None	aromatic
720–730	W	None	alkyl chains

[a] W–weak, M–medium, and S–strong.

Other oxidants, such as ozone [15], nitric acid, periodic acid, hydrogen peroxide, pressure oxidation using gaseous oxygen, and air, have been used to degrade kerogen. Ozone reacts slowly with kerogen suspended in either alcohol or KOH solutions, which suggests the presence of very few olefinic double bonds. Air oxidation of kerogen in open pans at 170° C for 14 weeks produced a total of 49 percent loss in weight and the loss had not leveled off at the end of the heating period. By contrast, coals lose a total of about 10 percent in weight and lignite loses about 25 percent in weight after 300 hours of oxidation.

C. Thermal Degradation

At a temperature of about 325° C, Green River kerogen softens, swells slightly, and becomes darkened. Water is formed at 300 to 405° C together with ammonia, other volatile nitrogen compounds, and hydrogen sulfide. Heavy oil vapors form at about 390° C and continue up to 500 or 600° C.

Green River kerogen was degraded thermally by ROBINSON and CUMMINS [16] at temperatures from 25 to 350° C in the presence of tetralin, a hydrogen-donor solvent. The resulting soluble products were separated by solvent fractionation, chromatographic techniques, and urea adducts. The 25° C extract represented only 4.4 percent of the kerogen, but the 350° C thermal product represented 84.9 percent of the kerogen. Only at 350° C did minor amounts of pyrolytic gas form. The two major fractions of the thermal product were resins and pentane-insoluble material, while the oils and wax represented smaller fractions. The amount of pentane-insoluble material obtained tended to increase with increase in temperature of extraction, while the amount of resins decreased with increase in temperature. Paraffin oil tended to decrease and aromatic oil tended to increase with a rise in temperature.

The elemental compositions of the resins and the pentane-insoluble material prepared at 25 to 350° C in the presence of tetralin were compared. With both materials, it was evident that the ratios of O/C and H/C tended to decrease while the ratio of N/C tended to increase with increase in the temperature of extraction. The ratio of S/C of the resins tended to increase with increase in the temperature of extraction, but the ratio of S/C in the pentane-insoluble material was nearly constant. These results suggest that oxygen structures were quite unstable and were degraded with loss of CO_2, CO, and water. Also, nitrogen structures were quite stable to degradation and required higher temperatures to make them soluble, indicating the presence of considerable amount of double-bonded nitrogen. Hydrogen loss was due to dehydration and dehydrogenation. Nuclear magnetic resonance analyses of the 350° C pentane-insoluble material indicated that it contained approximately 32 percent aromatic and 68 percent alkyl and cycloalkyl structures. A portion of the aromatic content of these extracts was probably formed from the tetralin during the heating process.

At 350° C, in the presence of tetralin, the kerogen was degraded to 10 to 15 percent straight-chain alkanes containing 25 to 30 carbon atoms, 20 to 25 percent cycloalkanes, 10 to 15 percent aromatic structures having an average of 4 rings per molecule, and 45 to 60 percent heterocyclic material. These materials probably represent structures present in the original kerogen and suggest that the kerogen is predominantly a heterocyclic material connected to or associated with smaller amounts of hydrocarbon material consisting of straight-chain alkyl, cycloalkyl, and aryl groups.

When Green River kerogen is pyrolyzed at 500° C, approximately 66 percent oil, 9 percent gas, 5 percent water, and 20 percent carbon residue are formed. Approximately two-thirds of the organic carbon and hydrogen is represented in the oil, two-thirds of the oxygen is converted to gas and water, nitrogen is distributed equally between the oil and the residue, and two-thirds of the sulfur is in the oil and one-third is evolved as gas. The principal gas constituents are H_2S, CO_2, CO, NH_3, H_2, CH_4, C_2H_6, and low-molecular-weight alkenes. The evolution of CO_2, which represents about 1 percent of the total carbon and one-third of the oxygen, may be due to decarboxylation, while CO may be produced from keto groups. The presence of ammonia and hydrogen sulfide in pyrolytic gases may be due to the degradation of terminal nitrogen and sulfur functional groups; however, a large percentage of both bonded nitrogen and bonded sulfur appeared in the oil, which indicates the presence of considerable thermally stable nitrogen and sulfur groups in the kerogen. In other thermal decomposition studies of Green River kerogen, the specific reaction rates of thermal decomposition were reported by HUBBARD and ROBINSON [17].

D. Reduction

Another method of degrading kerogen to soluble products for classification and identification consists of reductive degradation. Green River oil shale was catalytically hydrogenated by HUBBARD and FESTER [18] using hydrogen at 4,200 psig (hot pressure) at 355° C for 4 hours in the presence of stannous chloride catalyst. The hydrogenolysis product was recovered, and fractionated by solvent treatment, chromatography, distillation, formation of adducts and other techniques. The wax recovered, which amounted to 12 percent of the kerogen, was predominantly branched alkanes and cycloalkanes rather than normal alkanes. The paraffin oil fraction was also predominantly cyclic rather than straight chain and represented 29 percent of the kerogen. The major fraction, which represented 40 percent of the Green River kerogen, was the polar oil fraction that contained O, N, and S, and appeared to be essentially nonaromatic in character.

Attempts have been made to reduce Green River kerogen with other reagents. Concentrated hydriodic acid cleaves ethers and other oxygen functional groups. A total of 14.1 percent of soluble degradation product was obtained from the kerogen when treated with HI at 200° C for 24 hours. Comparable tests on Estonian kukersite [19] degraded as much as 90 percent of that kerogen. This suggested that the ether groups present in Green River kerogen are not easily degraded by HI treatment.

Reduction of Green River kerogen with lithium in ethylene-diamine increased the atomic H/C from 1.54 to 1.65 without appreciably changing the solubility of the kerogen.

LAWLOR *et al.* [20] lowered the O/C in Green River kerogen concentrates from 0.08 to 0.05 by reduction using lithium aluminium hydride and increased the H/C from 1.58 to 1.64. The loss of oxygen was reflected in the infrared spectra where the carbonyl absorption was eliminated with accompanying increase in hydroxyl absorption. Pyrite was quantitatively removed from the sample.

A study was made by JENSEN *et al.* [21] of the thermal degradation of Green River kerogen at 370 to 510° C in the presence of petroleum kerosene or a shale-oil gas-oil distillate and at hydrogen pressures from 300 to 4,300 psig. With a high partial pressure of hydrogen, oil that amounted to 126 percent of Fischer assay was obtained.

E. Solvent Treatment

Finely ground Green River oil shale was subjected to thermal degradation at 200° C and autogenous pressure in the presence of associating and nonassociating solvents by SCHNACKENBERG and PRIEN [22]. The total yield of product oil was found to be a function of molecular volume of the solvent. Nitrogen and sulfur contents of the product oil were considerably less than the original kerogen or of assay oil obtained from kerogen and were related to the associating characteristics of the solvent. From this, SCHNACKENBERG and PRIEN postulated that Green River kerogen exists in two forms, one rich in hetero atoms and one predominantly hydrocarbon in nature.

Results of similar extractions at the Bureau of Mines laboratory, using various polar and nonpolar solvents are shown in Table 4. The polar extracts appeared to be enriched in N, S, and O compounds compared to the nonpolar extracts. In a series of other extractions where the same sample of oil shale was extracted successively by various polar and nonpolar solvents, it was found that the total extract increased slightly over single solvent extraction; however, it was apparent that the extracts were of different composition. The methyl alcohol extract contained about 17 percent of the total O, N, and S present in the kerogen when preceded by n-heptane and cyclohexane extraction, compared to 7.3 percent when methyl alcohol was used alone. It appeared that Green River kerogen is only partly soluble because bonding forces present in the kerogen are strong enough to overcome the effect of the few active, functional groups present in the macro-structure.

Table 4. *Solvent extraction of kerogen*

Extractant	Yield, percent of kerogen	O, N, S, extracted, percent of total in kerogen
Acetone	5.3	4.5
n-Heptane	6.1	3.6
Methyl alcohol	8.0	7.3
Benzene	8.5	4.8
Phenol	15.5	15.4
Tetralin	5.4	–
Propylene carbonate	3.0	–
Dipropylene glycol	6.0	–
Polyglycol, Dow P 400	2.2	–
Pyridine	9.4	–
Ethylene diamine	15.6	–

MOORE and DUNNING [23] extracted Green River oil shale successively with solvents of increasing polarity, and determined the interfacial activities, and the porphyrin, nitrogen, and the metal contents of the extracts. Successive extraction with benzene, benzene-methanol mixture, methanol, chloroform-methanol mixture, and pyridine produced 3.7 weight percent of extract. Changing the orders of methanol and chloroform-methanol mixture produced 4.2 weight percent of extract. Infrared analyses of the extracts showed the presence of various amounts of metal-porphyrin complex of iron and nickel with the major part of the porphyrin complexed with iron. The interfacial activities and film-forming tendencies of the extracts generally paralleled their porphyrin and nitrogen contents. The extracts contained from 0.133 to 0.346 weight percent of porphyrin aggregate.

MORANDI and JENSEN [24] identified porphyrins of the phyllo type in Green River oil shale. The porphyrins from at least two homologous series were alkyl-substituted and contained 7 to 13 methylene substituents per molecule and most of the molecules also contained a carboalkoxy group. Their average molecular weights were 508.

The bitumen associated with the kerogen of Green River oil shale was studied by CUMMINS and ROBINSON [25]. The n-alkanes identified represented 1 percent of the bitumen and ranged from C_{13} to C_{33} compounds. Odd-carbon-numbered n-alkanes were present in greater quantity than even-carbon-numbered n-alkanes. The n-alkanes had a CPI (carbon preference index) of 3.6 in the range of C_{25} to C_{33}

compounds. The isoprenoid compounds identified represented 3.4 percent of the bitumen and ranged from C_{15} to C_{20} compounds. The isoprenoid compounds identified were 2,6,10-trimethyldodecane, 2,6,10-trimethyltridecane, 2,6,10-trimethylpentadecane, 2,6,10,14- tetramethylpentadecane, and 2,6,10,14-tetramethylhexadecane.

Table 5. *Assay oil yield and extraction ratios of oil-shale samples at different stratigraphic positions* [26]

Sample number	Stratigraphic position, feet from surface	Assay oil yield, gal per ton	Assay oil/C [a]	Extraction ratio [b]
1	1,036.1–1,056.0	35	0.81	9.2
2	1,056.0–1,080.9	34	–	8.2
3	1,152.0–1,177.6	42	0.87	10.6
4	1,235.5–1,241.1	48	–	6.4
5	1,450.1–1,461.5	39	–	11.1
6	1,597.3–1,628.0	35	0.82	11.5
7	1,668.3–1,695.5	37	–	14.4
8	1,786.0–1,824.8	28	0.73	12.4
9	1,884.1–1,923.5	20	0.79	12.8

[a] Percent assay oil/percent organic carbon.

[b] $\text{Extraction ratio} = \frac{\text{wt. percent benzene extract} \times 100}{\text{wt. percent organic carbon}}.$

Table 6. *Normal alkane and acyclic isoprenoid contents at different stratigraphic positions*

Sample number	Stratigraphic position, feet from surface	Component mg/100 g organic carbon in oil shale				
		normal C_{17}	total n-alkanes	C_{20} phytane	C_{19} pristane	total acyclic isoprenoids
1	1,036.1–1,056.0	58	387	150	37	230
2	1,056.0–1,080.9	21	181	212	55	378
3	1,152.0–1.177.6	10	95	99	20	148
4	1,235.5–1,241.1	18	122	85	20	129
5	1,450.1–1,461.5	157	910	304	181	666
6	1,597.3–1,628.0	264	1,078	326	173	757
7	1,668.3–1,695.5	251	1,378	524	266	1,162
8	1,786.0–1,824.8	77	1,285	332	130	742
9	1,884.1–1,923.5	80	1,739	242	157	742

In bitumens extracted from a 900-foot core of the Green River Formation (Sulfur Creek No. 10), Robinson *et al.* [26] found that the CPI of the n-alkanes decreased from 3.6 to 1.2 with depth. The extraction ratios and pyrolytic assay oil yields were determined for sections of the 900-foot core and the results are shown in Table 5. The assay oil/carbon ratios varied from 0.73 to 0.87. This difference shows some minor variations in the kerogen at different stratigraphic positions. In general, the amount of bitumen extracted from the oil shale increased with increase in stratigraphic position of the oil shale within the formation. There appears to be no relationship between total organic content of the oil shale and the amount of bitumen extracted per gram of oil shale organic carbon. The amount of n-alkanes and chain isoprenoid compounds in milligrams per 100 g of organic carbon at the different stratigraphic positions are shown in Table 6. In general the amount of total n-alkanes in the extracts increased from 387 to 1,739 mg per 100 g of organic carbon in the oil shale with increase in stratigraphic depth from 1,036 to 1,924 feet. A minimum value of 95 mg/100 g of carbon was found at the 1,152-foot–1,177-foot level. The normal C_{17} alkane ranged from 58 to 80 mg/100 g carbon with depth with a minimum of 10 mg at the 1,152-foot–1,177-foot level and a maximum of 264 mg/100 g of carbon at the 1,597-foot–1,628-foot level. The total chain isoprenoid compounds showed similar trends and ranged from 230 to 742 mg/100 g of carbon with a minimum of 129 mg at the 1,235–1,241-foot level and a maximum of 1,162 mg at the 1,668–1,695-foot level of the core. Similarly, the C_{20} phytane and the C_{19} pristane ranged from 150 to 242 mg and 37 to 157 mg respectively with minima of 85 and 20 mg and maxima of 524 and 266 mg at the 1,235–1,241-foot and 1,668–1,695-foot levels of the core. The consistent minima and maxima in bitumen composition at the two positions of the core appear to be geochemically significant.

BURLINGAME *et al.* [27] tentatively identified steranes and pentacyclic triterpanes in a Green River oil-shale bitumen. In a cooperative effort between the Laramie Petroleum Research Center and the British Petroleum Company Ltd., a white crystalline material isolated from the oil-shale bitumen was identified [28] as gammacerane, a pentacyclic triterpane.

F. Functional Group Analysis

A method for determining the carboxyl content of kerogen was developed by FESTER and ROBINSON [29]. The method utilizes an ion exchange reaction whereby the hydrogen ion of the carboxyl function is exchanged with the calcium ion of calcium acetate. The acetic acid liberated by the ion exchange is used as a quantitative measure of the amount of carboxyl function present in the kerogen. Utilization of steam distillation to remove the acetic acid formed forced the reaction to completion, thereby eliminating the reaction period of 2 weeks required to obtain equilibrium, and also facilitated the titration of the acetic acid.

By using conventional methods, FESTER and ROBINSON [30] determined the oxygen functional groups present in Green River kerogen. Carboxyl and ester groups represented 15 and 25 percent of the kerogen oxygen, respectively. An additional 6.5 percent of the kerogen oxygen was accounted for as hydroxyl, carbonyl, and amide oxygen leaving 52.5 percent of the kerogen oxygen accounted for as unreactive oxygen. The unreactive oxygen is probably ether oxygen.

The carboxyl and ester contents were determined for kerogen concentrates from sections of the 900-foot Sulfur Creek No. 10 core and the results are shown in Table 7. The carboxyl oxygen decreased from 18.3 mg carboxyl per gram organic carbon to 4.8 mg per gram organic carbon with increase in stratigraphic position from 1,036 to 1,923 feet. A leveling-off occurred from the 1,668-foot level to the end of the core. Also, a noticeable minimum value of 3.8 mg/g organic carbon was found at the 1,235–1,241-foot level. No consistent trends could be established with the amount of ester oxygen which ranged from a minimum of 17.9 mg to a maximum of 42.9 mg per gram of organic carbon. The total carboxyl and ester oxygen represented 3.4 percent of the kerogen at the top of the core and 1.9 percent of the kerogen at the bottom of the core.

Table 7. *Carboxyl and ester contents of kerogen at different stratigraphic positions* [26]

Sample number	Stratigraphic position, feet from surface	Carboxyl, mg/g carbon	Ester, mg/g carbon[a]
1	1,036.1–1,056.0	18.3	22.2
2	1,056.0–1,080.9	10.1	33.9
3[b]	1,152.0–1.177.6	–	–
4	1,235.5–1,241.1	3.8	26.7
5	1,450.1–1,461.5	6.7	17.9
6	1,597.3–1,628.0	4.0	37.1
7	1,668.3–1,695.5	4.7	26.0
8	1,786.0–1,824.8	4.9	42.9
9	1,884.1–1,923.5	4.8	29.0

[a] Calculated as COOH after saponification.
[b] Insufficient sample.

Brower and Graham [31] showed that Green River kerogen contains one double bond for each 16 to 22 carbon atoms and undergoes, to a very limited extent, substitution reactions typical of aromatic compounds.

G. Hydrolysis

Preliminary tests indicated that approximately 5 to 10 percent of the Green River kerogen is degraded by alkaline hydrolysis, alkaline fusion, or acid hydrolysis. The free fatty acids and the fatty acids obtained by methanolic KOH hydrolysis of kerogen esters were studied by Lawlor and Robinson [32]. Fatty acids ranging from C_{10} to C_{35} compounds were isolated from Mahogany zone oil-shale samples. The distribution of the fatty acids showed a predominance of even over odd carbon numbers of 1.41 in the C_{16} to C_{20} range and 2.05 in the C_{26} to C_{30} range. The data, when compared with n-alkane data, suggested that even-carbon-numbered fatty acids had been converted to odd-carbon-numbered n-alkanes by carbon loss. Also, even-carbon-numbered fatty acids may have been converted to odd-carbon-numbered fatty acids.

Eglinton *et al.* [33] isolated C_{19} and C_{20} chain isoprenoid acids from a core section of Green River oil shale and found a parallelism between the distribution of the isoprenoid acids and isoprenoid alkanes of the same carbon number.

III. Physical Constitution Analysis

The presence of large quantities of inorganic material in oil shale complicates the determination of the physical properties of kerogen. Unlike coal, it is not easy to hand-pick petrographic constituents of kerogen that are nearly free of mineral matter. Attempts have been made to separate the mineral matter from kerogen but only partial success has been achieved. Kerogen concentrates contain about 10 to 15 percent ash unless drastic treatment is used to remove silicates and pyrite. Three types of kerogen concentrates have been available for study: namely, HCl-HF concentrate, attritor concentrate and sink-float concentrate (see Chapter 6).

A. Infrared Analysis

The infrared spectrum of a Green River kerogen concentrate (Fig. 2) shows strong absorption in the region of 2,900 and 1,460 cm^{-1} assigned to methyl-methylene groups and strong absorption in the 1,680 to 1,720 cm^{-1} region assigned to carbonyl groups. The spectrum shows medium absorption in the 1,580 to 1,610 cm^{-1} region assigned to aromatic structures or carboxyl salts and medium absorption in the 1,380 cm^{-1} region assigned to methyl or cyclic methylene groups. Weak absorption in the 720 cm^{-1} region showed the presence of small amounts of chains greater than C_4.

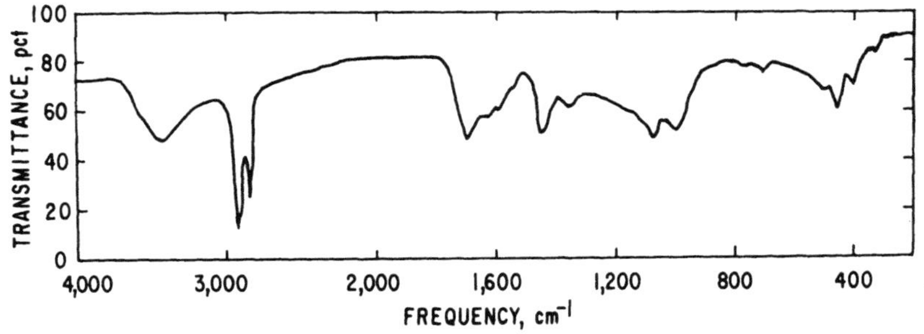

Fig. 2. Infrared spectrum of Green River kerogen

B. Ultraviolet Analysis

Absorption of radiation in the ultraviolet region by Green River kerogen was investigated by McDONALD and COOK [34] by suspending a kerogen concentrate in low melting polyethylene. Using this method, it was shown that the kerogen contains about 5 to 10 percent aromatic structures.

C. X-Ray Diffraction Analysis

X-ray diffraction patterns of a Green River kerogen concentrate (Fig. 3) showed the sample to have maximum diffraction at angle 2 theta of 18.9° which is equivalent to an interlayer spacing of 4.7 Angstrom units. This large spacing compared to graphite (3.4 Å) illustrates the predominantly saturated character of Green River kerogen.

D. Carbon Isotope Ratios

The $^{13}C/^{12}C$ ratios were determined for Green River kerogen concentrates from portions of Sulfur Creek No. 10 core and the results are shown in Table 8. The ratios decreased from about −1.4 to −2.4 permil with increase in depth from 1,036 to 1,241 feet then increased to a maximum of −0.3 permil at the 1,786 to 1,824-foot level, and then decreased to −1.6 at the 1,884 to 1,923-foot level. This minima and this maxima occur at the same position in the core as the minima and maxima previously noted for the alkane and isoprenoid contents of the soluble bitumen.

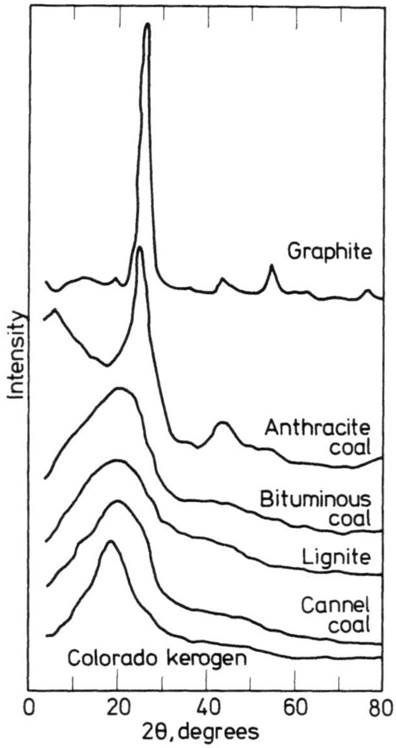

Fig. 3. X-ray diffraction of kerogen and other carbonaceous materials

Table 8. *Carbon isotope ratios of kerogen concentrates at various stratigraphic positions*

Sample number	Stratigraphic position, feet from surface	$^{13}C/^{12}C$ (Δ in 0/00)[a]
1	1,036.1–1,056.0	−1.43
2	1,056.0–1,080.9	−1.81
3	–	–
4	1,235.5–1,241.1	−2.40
5	1,450.1–1,461.5	−1.26
6	1,597.3–1,628.0	−0.70
7	1,668.3–1,695.5	−0.83
8	1,786.0–1,824.8	−0.26
9	1,884.1–1,923.5	−1.59

[a] Determined by Marathon Oil Company using NBS-22 standard.

E. Physical Properties

The density of Green River kerogen measured by the pycnometer method at 20° C in heptane is 1.07 [7].

Free swelling of a Green River kerogen concentrate was less than a No. 1 coking coal [35]. The kerogen concentrate showed plastic properties from 333° to 393° C by the Gieseler Plastometer Test [36].

A Green River concentrate contracted 39 percent when heated from 405° to 445° C (Parry-Potter Dilatometer Test). The kerogen showed partial melting properties between 405 and 445° C.

Calculations by the method of VAN KREVELEN [8] for ring condensation and aromaticity indicated 0.20 ring per carbon atom and 6 to 8 percent aromaticity for Green River kerogen.

IV. Discussion

Based on oxidation studies, Green River kerogen appears to be essentially a nonaromatic material. The nature of the oxidation products indicate that few aromatic structures are present in the kerogen as only nonbenzenoid acids were formed. The filtrate-soluble acids from kerogen oxidation were essentially all dicarboxylic acids. Highly condensed structures would produce polyfunctional acids. This indicates that the kerogen has predominantly a linear-type structure. The kerogen is almost completely degraded to CO_2 and oxalic acid by exhaustive oxidation with alkaline potassium permanganate. As straight-chain and aromatic materials are difficult to oxidize to completion, it appears that these type materials are not the predominant structures. By elimination, it appears that Green River kerogen is predominantly a hetero material or is highly unsaturated. Numerous tests show that the latter is not the case. The nature of the oxidation products suggests that the kerogen contains more cyclic than straight-chain structure and that the heterocyclic structures are oxidatively cleaved at oxygen linkages. There is also evidence that structures present in the kerogen, other than carboxyl groups or acid salts, degrade to carbon dioxide, as heating tests of kerogen acids indicated that additional carbon dioxide was released from the oxidation product after all the carboxyl groups were removed by decarboxylation. Also, a large part of the oxygen is present as "oxygen bridges".

From the results of thermal degradation studies, a number of observations can be made concerning the structure of Green River kerogen. There is some evidence that the kerogen may consist of at least two different structural units. One unit is predominantly hydrocarbon in nature with straight-chains, rings, and some heterocyclic structures that produce most of the oil, wax, and resins by thermal degradation. The other type is essentially a heterocyclic material that is partially converted to oil, but is converted mainly to hydrogen-deficient materials, which in turn yield mostly carbon residue and gas upon being subjected to further degradation. One indication of this is the fact that the total amount of kerogen converted to oil, wax, and resins at 350° and 500° C was approximately the same, but there were large differences in the amount of pentane-insoluble material, gas, and carbon residue produced. About 48 percent of the 350° C extract was insoluble in pentane compared to 6 percent for an extract prepared at 500° C. Only a small amount of kerogen was converted to gas and carbon residue at 350° C compared to 30 percent at 500° C. This indicates that pentane-insoluble compounds are largely converted to gas and carbon residue at 500° C.

The composition of the thermal products indicates that the kerogen is predominantly a heterocyclic material. The kerogen structure is probably linear rather than highly condensed about a given nucleus. Based upon the composition of the low-temperature extracts, Green River kerogen is composed of approximately 5 to 10 percent chain paraffin structures, 20 to 25 percent saturated cyclic structures, 10 to 15 percent aromatic structures, and 45 to 60 percent heterocyclic structures. The aromatic value may be too high because of the formation of aromatic compounds from tetralin during the heating.

Green River kerogen contains functional groups that decompose during pyrolysis to CO_2, CO, NH_3, H_2S, water, and carbon residue. Carbon dioxide in pyrolytic gas suggests the decomposition of carboxyl groups present as acids or acid salts. The amount of CO_2 released during retorting represents approximately 1 percent of the total carbon and 30 percent of the oxygen, while carbon monoxide represents about 0.1 percent of the total carbon and 2 percent of the oxygen. Chemical determination of functional groups indicates the kerogen oxygen to be present mainly in carboxyl, esters, and ether groups.

A study of the products obtained by hydrogenolysis of kerogen indicates that these products contain fewer polar groups than retort oil or low temperature extracts and are in general more paraffinic. However, the same general conclusions concerning the structure of kerogen can be made from this study as from the study of other thermal products.

Although only about 10 percent of the total organic material is soluble in various solvents, some information concerning the nature of Green River kerogen was obtained by solvent studies. The kerogen is only partly soluble because the bonding forces of kerogen are very strong. The results of extraction tests indicate that portions of the organic material exist in at least two different forms, one with a small amount of heteroatoms and one with a large amount of heteroatoms.

The identification of n-alkanes and fatty acids of similar carbon number distribution, acyclic isoprenoid compounds, steranes, and pentacyclic triterpanes shows a biological origin of the organic material present in Green River Formation oil shale. These studies show minor compositional changes in the organic material with stratigraphic position in the formation, suggesting that the kerogen within the Green River Formation has distinct, though minor, differences in structure.

The observed deviations of the carboxyl content, the extraction ratio, the alkane contents, and the $^{13}C/^{12}C$ ratio of the two areas of Sulfur Creek No. 10 core undoubtedly have geochemical significance. The cause of these variations at these locations in the formation are not known but may be indicative of differences in source material or modification differences at the time of deposition or after deposition.

Chemical analyses and the results of X-ray, infrared, and ultraviolet analyses of the kerogen are in agreement and indicate that the structure of Green River kerogen is predominantly naphthenic in nature. The spectral evidence for aromatic, straight-chain, and hetero structures in the kerogen was of similar magnitude to that obtained for chemical or thermal degradation products of the kerogen.

V. Summary

The Green River Formation kerogen is a macromolecular material having predominantly a linearly condensed saturated cyclic structure with heteroatoms of oxygen, nitrogen, and sulfur. Straight-chain and aromatic structures are a minor part of the total kerogen structure.

About one-half of the oxygen is present in the form of acid or ester groups, with the remaining oxygen in the form of ether or "bridge oxygen" linkages. Nitrogen and sulfur heteroatoms probably are present mainly in heterocyclic ring systems.

The identification of moieties obviously related to biological precursors in the soluble organic material associated with kerogen has contributed to the knowledge of the origin of kerogen. The extent to which similar materials contributed to the structure of the insoluble kerogen is still unknown.

References

1. DOWN, A. L., and G. W. HIMUS: The classification of oil shales and cannel coals. J. Inst. Petrol. **26**, 329–348 (1940).
2. BRADLEY, W. H.: The varves and climate of the Green River Epoch. U.S. Geol. Surv. Profess. Papers **158-E**, 87–110 (1929).
3. – Origin and microfossils of the oil shale of the Green River Formation of Colorado and Utah. U.S. Geol. Surv. Profess. Papers **168** (1931).
4. DUNCAN, D. C., and C. BELSER: Geology and oil-shale resources of the eastern side of the Piceance Creek Basin, Rio Blanco and Garfield Counties, Colorado. U.S. Geol. Surv. Map OM, **119**, Oil and Gas Inv. Ser. (1950).
5. DONNELL, J. R., W. B. CASHION, and J. H. BROWN JR.: Geology of the Cathedral Bluffs area, Rio Blanco and Garfield Counties, Colorado. U.S. Geol. Surv. Map OM **134**, Oil and Gas Inv. Ser. (1953).
6. ROBINSON, W. E., and K. E. STANFIELD: Constitution of oil-shale kerogen. U.S. Bur. Mines, Inform. Circ. No. 7968 (1960).
7. SMITH, J. W.: Ultimate composition of organic material in Green River Oil Shale. U.S. Bur. Mines, Rep. Invest., No. 5725 (1961).
8. KREVELEN, D. W. VAN, and J. SCHUYER: Coal Science. Amsterdam, Netherlands: Elsevier Publ. Co. 1957.
9. BONE, W. A., L. HORTON, and S. C. WARD: Proc. Roy. Soc. (London), Ser. A, **127**, 480–510 (1930).
10. ROBINSON, W. E., H. H. HEADY, and A. B. HUBBARD: Alkaline permanganate oxidation of oil-shale kerogen. Ind. Eng. Chem. **45**, 788–791 (1953).
11. – D. L. LAWLOR, J. J. CUMMINS, and J. I. FESTER: Oxidation of Colorado Oil Shale. U.S. Bur. Mines, Rep. Invest., No. 6616 (1963).
12. BONE, W. A., L. G. B. PARSONS, R. H. SAPIRO, and C. M. GROCOCK: Proc. Roy. Soc. (London), Ser. A, **148**, 492–522 (1935).
13. ROBINSON, W. E., J. J. CUMMINS, and K. E. STANFIELD: Constitution of organic acids prepared from Colorado Oil Shale based upon their n-butyl esters. Ind. Eng. Chem. **48**, 1134–1138 (1956).
14. –, and D. L. LAWLOR: Constitution of hydrocarbon-like materials derived from kerogen oxidation products. Fuel **40**, 375–388 (1961).
15. BITZ, M. C., and B. NAGY: Ozonolysis of "polymer-type" material in coal, kerogen, and in the Orgueil Meteorite: A preliminary report. Proc. Natl. Acad. Sci. U.S. **56**, 1383–1390 (1966).
16. ROBINSON, W. E., and J. J. CUMMINS: Composition of low-temperature thermal extracts from Colorado Oil Shale. J. Chem. Eng. Data **5**, 74–80 (1960).
17. HUBBARD, A. B., and W. E. ROBINSON: A thermal decomposition study of Colorado Oil Shale. U.S. Bur. Mines, Rep. Invest., No. 4744 (1950).
18. –, and J. I. FESTER: Hydrogenolysis of Colorado Oil-Shale kerogen. Chem. Eng. Data Ser. **3**, 147–152 (1958).
19. RAUDSEPP, KH. T.: A new method of investigating the chemical structure of combustible minerals and the chemical structure of Estonian Shale-Kukersite. Izv. Akad. Nauk. SSSR, Otd. Tekh. Nauk, **3**, 130–136 (1954).
20. LAWLOR, D. L., J. I. FESTER, and W. E. ROBINSON: Pyrite removal from oil-shale concentrates using lithium aluminium hydride. Fuel **42**, 239–244 (1963).

21. JENSEN, H. B., W. I. BARNET, and W. I. R. MURPHY: The thermal solution and hydrogenation of Green River Oil Shale. U.S. Bur. Mines, Bull. **533** (1953).
22. SCHNACKENBERG, W. D., and C. H. PRIEN: The effect of solvent properties in thermal decomposition of oil-shale kerogen. Ind. Eng. Chem. **45**, 313–322 (1953).
23. MOORE, J. W., and H. N. DUNNING: Interfacial activities and porphyrin content of oil-shale extracts. Ind. Eng. Chem. **47**, 1440—1444 (1955).
24. MORANDI, J. R., and H. B. JENSEN: Comparison of porphyrins from shale oil, oil shale, and petroleum by absorption and mass spectroscopy. J. Chem. Eng. Data **11**, 81–88 (1966).
25. CUMMINS, J. J., and W. E. ROBINSON: Normal and isoprenoid hydrocarbons isolated from oil-shale bitumen. J. Chem. Eng. Data **9**, 304–307 (1964).
26. ROBINSON, W. E., J. J. CUMMINS, and G. U. DINNEEN: Changes in Green River Oil-Shale Paraffins with depth. Geochim. Cosmochim. Acta **29**, 249–258 (1965).
27. BURLINGAME, A. L., P. HAUG, T. BELSKY, and M. CALVIN: Occurrence of biogenic steranes and pentacyclic triterpanes in an Eocene Shale (60 million years) and in an Early Precambrian Shale (2.7 billion years). Proc. Natl. Acad. Sci. U.S. **54**, 5 (1965).
28. HILLS, I. R., E. V. WHITEHEAD, D. E. ANDERS, J. J. CUMMINS, and W. E. ROBINSON: An optically active triterpane, gammacerane, in Green River Colorado Oil-Shale bitumen. Chem. Comm. **20**, 752–754 (1966).
29. FESTER, J. I., and W. E. ROBINSON: Method for determining carboxyl content of insoluble carbonaceous materials. Anal. Chem. **36**, 1392–1394 (1964).
30. – – Oxygen functional groups in Green River Oil-Shale kerogen and trona acids. In: Coal science (R. F. GOULD, ed.). Washington, D.C.: American Chemical Society 1966.
31. BROWER, F. M., and E. L. GRAHAM: Some chemical reactions of Colorado Oil-Shale Kerogen. Ind. Eng. Chem. **50**, 1059–1060 (1958).
32. LAWLOR, D. L., and W. E. ROBINSON: Fatty acids in Green River Formation Oil Shale. Div. of Pet. Chem., ACS, Detroit, Mich., General Papers, No. 1, 5–9 (April 1965).
33. EGLINTON, G., A. G. DOUGLAS, J. R. MAXWELL, J. N. RAMSAY, and S. STALLBERG-STENHAGEN: Occurrence of isoprenoid fatty acids in the Green River Shale. Science **153**, 1133–1135 (1966).
34. McDONALD, F. R., and G. L. COOK: A method of obtaining the ultraviolet and visible spectra of insoluble materials – use of low-molecular-weight polyethylene as a matrix material. U.S. Bur. Mines, Rep. Invest., No. 6439 (1964).
35. ASTM Standard Procedure D 720-46.
36. ASTM Proceedings, 301-5 (1943).

CHAPTER 27

Crude Petroleum *

G. C. SPEERS and E. V. WHITEHEAD

The British Petroleum Company, Limited,
BP Research Centre,
Chertsey Road,
Sunbury-on-Thames,
Middlesex

Contents

I. Introduction

The term "crude petroleum" embraces the naturally-occurring, highly complex, mobile organic mixture, usually predominantly hydrocarbons, which constitutes the commercial crude oil, natural gas and natural asphalt of the petroleum industry. This chapter will be almost entirely concerned with liquid crude oil**. Precise definitions of petroleum or crude oil are impossible, as no two oils are exactly alike and hence a continually different compositional problem exists [1]. It is becoming increasingly evident, however, that most crude oils contain the same series of compounds and that the differences between them are a reflection of the relative amounts of the various compounds which are present [2].

* Permission to publish this paper has been given by The British Petroleum Company Limited.

** That is, the stabilized crude oil after removal of dissolved light hydrocarbons at atmospheric pressure and ambient temperature.

On comparison with dispersed sedimentary organic matter, petroleum appears to be a selective accumulation of the less polar constituents [3, 4]. Petroleum occurrences, in small quantities, are known on all Continents (with the present exception of Antarctica) and indigenous to formations of ages from Precambrian [5, 6] to Pleistocene [7]. Conditions favorable to the genesis and accumulation of commercial amounts of petroleum are, however, very selective [8–10]. General aspects of the geological occurrence of oil and gas accumulations have been reviewed by BROD [11], HEDBERG [7] and WEEKS [10] and their geographic distribution by PRATT and GOOD [12] and ODELL [13]. Crude oils from present day geographic locations, for example Venezuela or the Middle East, are often quite similar in properties and this frequently provides a useful means for their commercial classification.

A petroleum deposit is the result of many complex processes occurring under different geological settings and over the span of geological time. BAKER [14] divides the process into three continuous stages: (1) the origin of petroleum constituents; (2) the primary migration of this organic matter, and (3) its final accumulation in the reservoir. Calculations for some of the compounds in petroleum [15–17] indicate that most crude oils are not in thermodynamic equilibrium, so that a fourth part may conveniently be added to the above process, viz. (4) post accumulation changes, or the "maturation" of the petroleum within the reservoir. The term "petroleum evolution" is sometimes used in considering stages 1–4 in their entirety.

Over the past decade, most investigations have been concerned with the origin of petroleum hydrocarbons (see, for example, Chapter 13) to the exclusion of work on the migration and accumulation of petroleum. The problems involved in these latter processes have recently been reviewed by HODGSON and HITCHON [18], TISSOT [19] and HUNT [20]. Primary migration almost certainly involves the movement of petroleum constituents from low permeability source sediments, where they are present in trace amounts, into porous, permeable reservoir rocks. As sediments become more deeply buried, they lose their interstitial water under the forces of overburden pressure and compaction and, as fluid is expelled, it is reasonable to assume that it carries with it, either in solution or suspension, minute amounts of the various petroleum constituents. Solubility data [21] support this hypothesis, and estimates of the fluid movement during compaction of a sedimentary basin [10, 18] indicate that very large amounts of material can be transported by this mechanism. Additional evidence to support the migration of petroleum constituents in solution or as an extremely dilute colloidal dispersion is provided by the study of "oil" migration in recent sediments of Pedernales, Venezuela [22]. On entering the porous reservoir rock, physical and chemical conditions are postulated to change sufficiently to cause release of the oil. The oil droplets increase in size and are unable to re-enter the water-wet pores of the surrounding impermeable cap-rock.

The mechanism of migration and accumulation of petroleum is thus still mainly speculative and just how petroleum constituents migrate in water, at what stage most of the oil leaves the presumed source beds, and the method of oil release in the reservoir remains to be elucidated.

II. Petroleum Composition – General

The organic geochemist views the origin or "evolution" of petroleum as a particular aspect of the general problem of the fate of organic matter in the Earth. His major approaches to the problem are the following:

(1) A realization that petroleum is almost universally associated with sedimentary rocks and a knowledge of the geological factors that control its occurrence [23–26a].

(2) The study of petroleum composition to see what clues as to likely progenitors are preserved.

(3) The relationship of petroleum to the other organic constituents in sediments, i.e., its organic sedimentological aspects [26b].

This chapter will be almost entirely concerned with aspect (2).

Most petroleums buried under even slight overburden contain dissolved, or natural gas which is evolved under surface conditions [27]. The amount of gas is expressed as a gas/oil ratio (determined under standard conditions) and, for some petroleums, this can be as much as 350:1 measured on a volume:volume basis. Hydrocarbons [28] form a large part of this gas, the amount decreasing rapidly as the carbon number increases from C_1 to C_6, although depending on separation conditions, higher hydrocarbons can be present. Substantial quantities of non-hydrocarbon gases such as H_2S, N_2 and CO_2 are sometimes also present. The components normally determined in natural gases evolved from petroleum are listed in Table 1.

Table 1. *Components of petroleum natural gases*

Non-hydrocarbons	Aliphatic hydrocarbons		Naphthenes and aromatics
	methane	hexanes	cyclopentane
Nitrogen	ethane	(sometimes split into)	methylcyclopentane
Carbon dioxide	propane	2,2-dimethylbutane	cyclohexane
Helium	i-butane	2-methylpentane	benzene
Hydrogen sulfide	*n*-butane	2,3-dimethylbutane	
	i-pentane	3-methylpentane	
	n-pentane	*n*-hexane	
		C_7+	

The individual naphthenes and aromatics are often included in the composite analysis "pentanes" or "hexanes". C_7+ refers to all hydrocarbons of carbon number 7 and above, usually determined by gas chromatography using a column back-flushing technique [29].

The boiling range of the components of petroleum, after removal of dissolved gas, can conveniently be represented as shown in Fig. 1. The full line is the boiling point, at 760 mmHg, of the homologous series of *n*-alkanes [30]. To the left of this line are plotted the best literature values for the most condensed aromatic hydrocarbons with the curve extrapolated beyond coronene. To the right of the *n*-alkane line an approximate boundary for the most highly branched isoalkanes is indicated by the broken line. All known petroleum hydrocarbons are contained

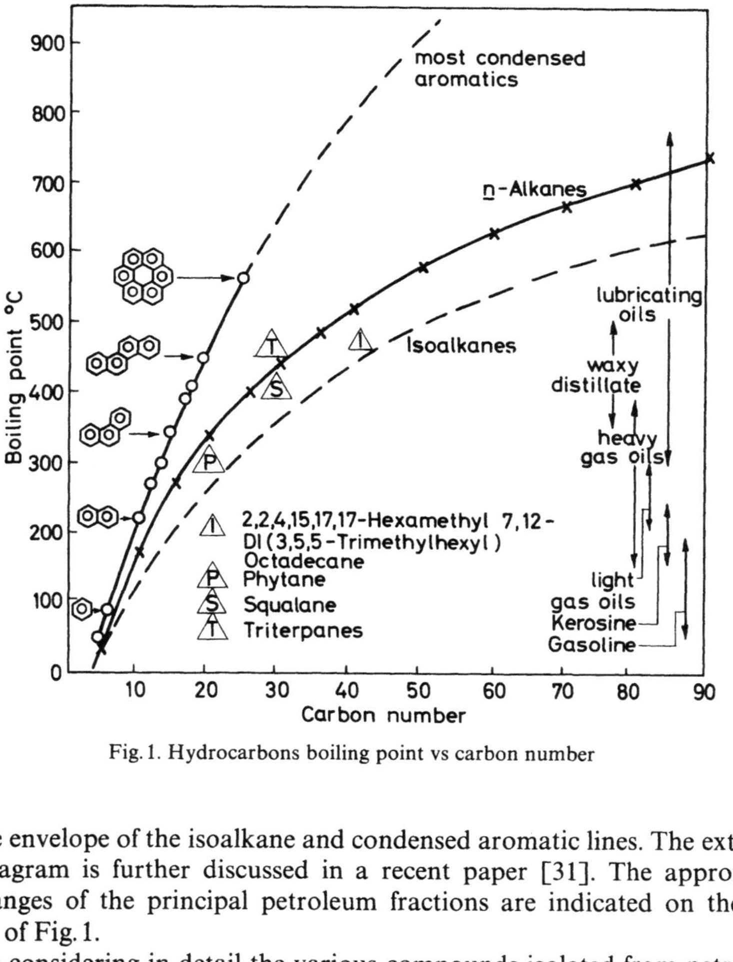

Fig. 1. Hydrocarbons boiling point vs carbon number

within the envelope of the isoalkane and condensed aromatic lines. The extension of this diagram is further discussed in a recent paper [31]. The approximate boiling ranges of the principal petroleum fractions are indicated on the right hand side of Fig. 1.

Before considering in detail the various compounds isolated from petroleum, the methods of separation and analysis used in this work will be briefly reviewed.

III. Separation Techniques

Gas-liquid chromatography (GLC) introduced in 1952, has been extensively developed within the petroleum industry and has had a great impact on the study of petroleum composition. In analytical work, preparative GLC units are now rapidly replacing the distillation techniques used extensively in the past. The great versatility of GLC for separation and particularly analysis, has been reviewed recently [32, 33]. Many of the separating processes that GLC is replacing have been well reviewed by Mair [33]. Fractional distillation has advantages where large quantities of the material to be separated are available and the technique can be used at atmospheric pressure or at pressures as low as a few microns. The problem of constant boiling mixtures can be overcome by adding components which permit separation by extractive or azeotropic distillation [33, 34].

These first two methods principally exploit differences in boiling point to effect the separation of individual compounds. Thermal diffusion [33, 35, 36] of liquids, however, separates by molecular shape, using a thermal gradient established across the narrow annulus formed by two vertical concentric tubes, one of which is heated and the other cooled.

Elution adsorption chromatography, on a wide variety of adsorbents [33, 37, 38] using column and thin layer techniques [39, 40], is used widely for the separation of petroleum samples into compound types.

Segregation of molecular types according to shape and size is effected by a number of organic and inorganic compounds which "sieve" or complex with organic molecules. Molecular sieves [33, 41] are commercially available with effective pore diameters of 3, 4, 5, 9 and 10 Å. Sieves of 5 Å appear to be specific in adsorbing and retaining *n*-alkanes although under certain conditions isoalkanes may be retained [42]. Liberation of the adsorbed *n*-alkane is conveniently accomplished with hydrofluoric acid [43]. Gel permeation chromatography [44] is being applied increasingly to the separation of petroleum components.

Complex formation [45] with organic molecules is widely used for the concentration of individual compounds or compound types. Urea and thiourea [46, 47] adduction is a particularly useful method for separating *n*- and isoalkanes, although certain aromatics [48] and naphthenes will also complex with thiourea. The molecular types which form adducts are listed by McLAUGHLIN [49].

Aromatic hydrocarbons also form complexes with many aromatic nitro compounds [33] and separations exploiting the complex formation of silver ions with centers of unsaturation, i.e., silver ion chromatography [50], are now well known.

IV. Analysis

The petroleum industry has developed a wide range of analyses, designed to reveal product quality, and the average molecular structure of its very complex products and raw material. Determinations of molecular structure rely increasingly upon spectroscopic techniques and the whole spectrum of electromagnetic radiation from X-rays to radio waves is harnessed to this end. X-ray diffraction [51] reveals the absolute molecular geometry of the carbon skeleton of individual molecules, but it is only with the ready access to computers [52] that such studies can be regularly undertaken. Electron and neutron diffraction techniques are applied in the study of catalyst surfaces or to the location of atoms with low scattering power in gases or liquids. The twin techniques of emission [53, 54] and visible absorption spectrometry have been used extensively to study petroporphyrins [55], yielding information on the incorporated metal atom and its ligand respectively.

Infrared spectroscopy [56, 57], especially in the fingerprint region (2–15 μ), often characterizes individual molecules or types [58, 59], it is only now yielding pride of place to mass and nuclear magnetic resonance spectroscopy.

Absorption in the ultraviolet region, depends upon the presence of resonant molecular groups (chromophores) and is particularly suited to the structural determination of petroleum aromatics [60, 61]. Application of the Cotton effect and consideration of rotatory dispersion curves can also provide additional information.

Nuclear magnetic resonance spectroscopy [62] has wide application in the elucidation of petroleum composition [63] and is applied to the study of aromatics [64], porphyrins [55], and olefins [65] as well as saturates. Its chief limitation is that the technique registers the environment of hydrogen atoms within the molecule and the position of the carbon ^{12}C atoms have to be inferred. Carbon ^{13}C is registered directly [66] but analyses are presently restricted by the large sample size required (>0.5 g) and low instrument sensitivity.

The techniques that have been discussed are essentially non-destructive and the sample may be recovered and submitted finally to mass spectrometry [67]. The molecular ions and fragments, produced by bombardment of the molecule with a beam of medium energy electrons, are recorded as mass spectra and rationalized to cracking patterns which provide molecular fingerprints of extraordinary detail on much less than a mg of sample. With high resolution the exact mass of the molecular ion can provide a precise molecular formula, replacing elemental analyses. If the cracking pattern can be exactly matched with that of an authentic sample an immediate identification may be secured. Simplified spectra which frequently show molecular ions alone, are produced at low ionizing voltages [68] or under field ionization conditions [69]. The full potential of mass spectrometry will only be realized with print out accessories [70] and computer [71, 72] correlation of the data. Applications of mass spectrometry within the oil industry have been reviewed recently [73].

Gas-liquid chromatography, essentially a superfractionating device of immense versatility [74] when coupled to the mass spectrometer, has unlimited application in structural analysis, and it is the combination most likely to yield data from which detailed composition may be deduced.

X-ray crystallography is the only known technique which can reveal an exact molecular structure without reference to any authentic sample, although NMR spectroscopy on occasions may yield data sufficient for the certain identification of some individual compounds.

All of the other spectroscopic techniques require calibration and continual reference to authentic samples, which in the high molecular weight ranges are frequently unavailable.

Optical activity of petroleum distillate fractions has been extensively studied [75–77] and the compounds probably responsible for the bulk of the activity have now been identified [78]. The development of new photoelectric polarimeters [79, 80] has enabled smaller samples and dark-colored solutions to be investigated and has made possible the direct measurement of the optical activity of unfractionated crude oils [81].

V. Hydrocarbons

The major hydrocarbon components of natural gas are listed in Table 1. The composition of the other fractions from petroleum can conveniently be considered on the basis of boiling point and carbon number range as illustrated in Fig. 1.

Alkanes account for a large proportion of most petroleum hydrocarbons and by far the most abundant series is the *n*-alkanes. Individual *n*-paraffins with 1 to 33 carbon atoms have been isolated and identified in Ponca City crude [82]. Recent work has shown the presence of *n*-paraffins with carbon numbers from C_{32} to $> C_{60}$ in petroleum wax [83], whilst other reports [84, 85] suggest that *n*-alkanes in the range C_{54} to C_{78} are present in crude petroleums.

Isoalkanes, especially the 2-, 3- and 4-methylalkanes occur in relatively high concentrations in petroleum and individual members up to C_{10} have been isolated from Ponca City petroleum [82]. Isomers of higher carbon numbers have been observed in urea adducts from Agha Jari petroleum fractions [31] and in studies of wax composition [86]. MAIR *et al.* [87] have recently isolated 2,6-dimethyloctane and 2-methyl-3-ethylheptane from the gasoline range of Ponca City crude. These two C_{10} branched hydrocarbons are present in relatively large amounts and could easily be derived from the monoterpenoids. MAIR [88] considers that the distribution of the methylisopropyl benzenes in the same petroleum is also indicative of a derivation from the monoterpenoids.

The abundance of methyl relative to larger alkyl substituents in branched alkanes and cycloparaffin rings seems to be a characteristic of petroleums and is considered by MEINSCHEIN [4] as further evidence in support of the terpenoids being a major source of some petroleum hydrocarbons.

The lower boiling fractions of petroleum in the carbon number range C_4 to C_{11} have been extensively studied, particularly by API Research Project No. 6 [89] whose workers have separated and identified hydrocarbons of the following types: *n*-paraffins, branched paraffins, cyclopentane and alkylcyclopentanes, cyclohexane and alkylcyclohexanes, cycloheptane, bicycloparaffins, adamantane (tricycloparaffin), benzene and alkylbenzenes. Summaries of this work have recently been published by MAIR [82] and ROSSINI [90]. BREGER [91] also lists the various hydrocarbons which have been isolated and BESTOUGEFF [2] discusses the various hydrocarbon types.

The analysis of the lower carbon number range of petroleum, particularly up to C_8, has been made much simpler by the use of gas-liquid chromatography [92] especially capillary column techniques [93, 94]. Analyses of many types of petroleum have been made and geochemical interpretations are discussed by SMITH *et al.* [16], MARTIN *et al.* [95], BIEDERMAN [96] and BOGOMOLOV and SHIMANSKY [97].

The complexity of hydrocarbon types increases with boiling point, and analyses of higher boiling petroleum fractions are frequently quoted on the basis of mass spectrometric hydrocarbon-type analyses [98, 99]. The application of mass spectrometry to petroleum analysis has recently been reviewed by POWELL [100]. As mentioned in the section on Analysis, the combination of gas chromatography and mass spectrometry gives a powerful tool for the study of petroleum composition. A recent example of the use of this technique is the characterization of various mono- and bicycloparaffins from a Californian petroleum naphtha [101]. A detailed account of the many experimental techniques used in petroleum chemistry together with the results obtained on various petroleum fractions is contained in a recent ASTM publication [102].

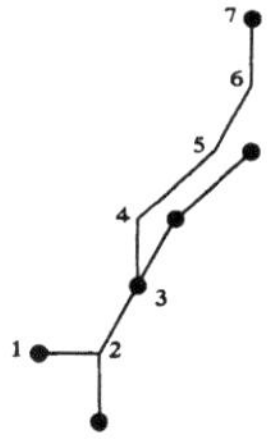
2-Methyl-3-ethyl-heptane

Adamantane

Diamantane

Tetracyclo-(6,3,1,$0^{2,6}0^{5,10}$) dodecane

Farnesane

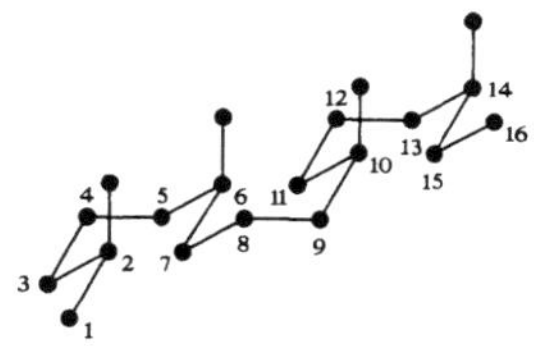
Phytane

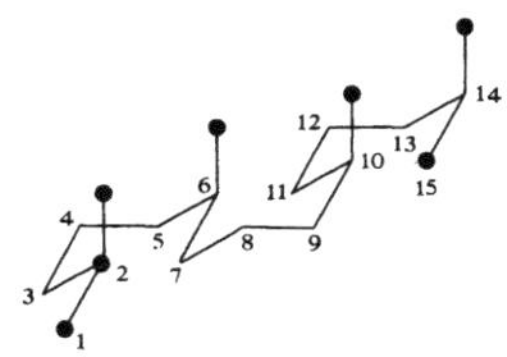
Pristane

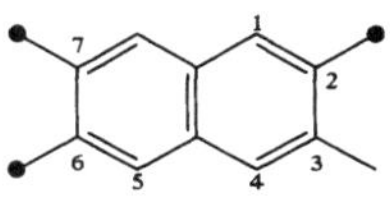
2,3,6,7-Tetramethylnaphthalene

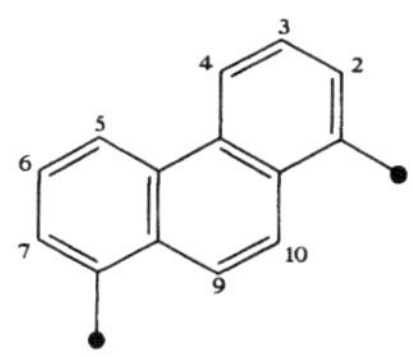
1,8-Dimethylphenanthrene

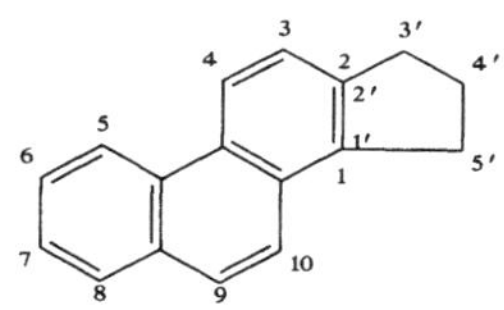
1,2-Cyclopentenophenanthrene

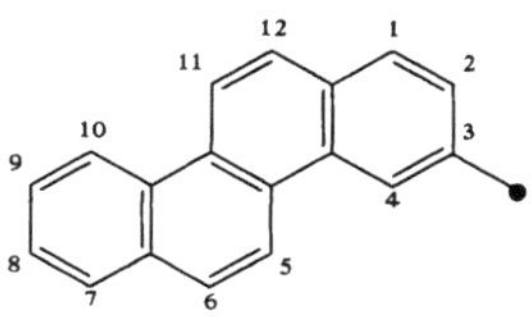
3-Methylchrysene

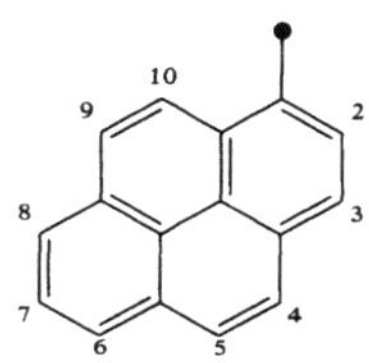
1-Methylpyrene

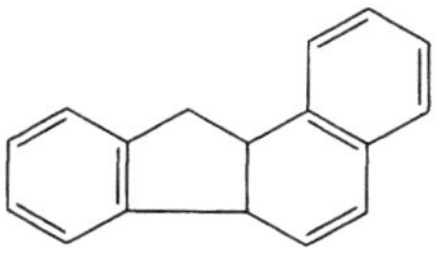
1,2-Benzofluorene

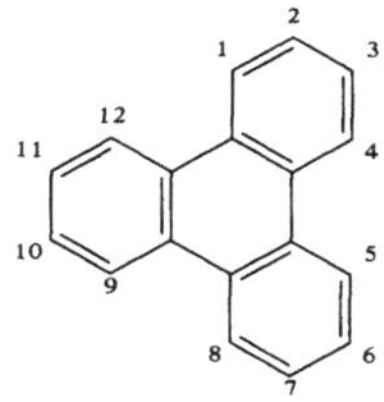
Triphenylene

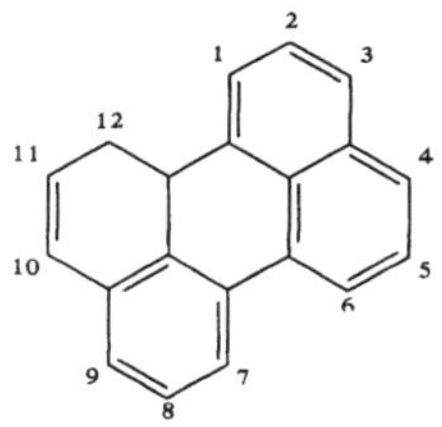
Perylene

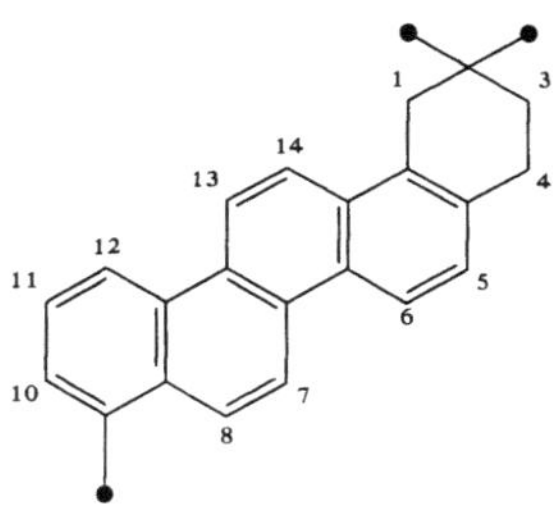
1,2,3,4-Tetrahydro-2,2,9-trimethylpicene

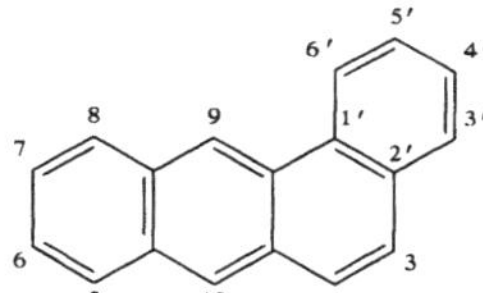
1,2-Benzanthracenes

The alkanes in this figure are presented in a particular conformation that relates them to naturally occurring cyclic terpenes

Fig. 2. Hydrocarbons in petroleum

The isolation of individual compounds becomes increasingly difficult as the boiling point increases and the structures of relatively few compounds above C_{14} are known. One of the most interesting observations in recent years has been the isolation and identification in petroleum distillates of the isoprenoid alkanes, particularly farnesane, pristane and phytane [103–106]. The concentrations of these higher hydrocarbons with a terpenoid structure are again high relative to other types of hydrocarbons in the same carbon number range and support the argument for their biological origin. These hydrocarbons have since been widely used as biological markers in the study of sedimentary organic matter [107, 108].

Several studies have been made of the higher molecular weight aromatic hydrocarbons in petroleum. MAIR and BARNEWALL [109] studied the mononuclear aromatic material from the light gas oil from Ponca City crude and concluded that many of the alkylbenzenes had structures which again suggested a terpenoid origin. MAIR *et al.* [110] also examined the trinuclear aromatic concentrate of a fraction boiling between 305° and 405° C from the same crude oil and isolated 21 individual compounds in the carbon number range C_{14} to C_{18}. These aromatics included 2, 3, 4, 7-tetramethylnaphthalene, phenanthrene and its 1, 2, 3 and 9-methyl and 1,8-dimethyl derivatives, 1,2-cyclopentenophenanthrene and its 3′-methyl derivative (Diels hydrocarbon). Pyrene and its 4-methyl derivative and a series of alkylated fluorenes were also identified. These hydrocarbons are similar to those commonly produced by the dehydrogenation of steroids and other natural products. Several of these compounds were also identified in Agha Jari petroleum [31].

Chrysene and its 3-methyl derivative have been reported in petroleum [111, 112]. Studies on Kuwait crude [112, 113] have also indicated the presence of a series of anthracenes and phenanthrenes, 1-methylpyrene, 1,2-benzofluorene and triphenylene. Perylene has been identified [114] as well as 1,2,3,4-tetrahydro-2,2,9-trimethylpicene [115] and several benzanthracenes [116]. The trimethyl tetrahydropicene is significant as a possible dehydrogenation product of the triterpenoids. The various aromatic types occurring in petroleum distillates boiling between 316° and 538° C have been characterized by LUMPKIN and JOHNSON [117].

In the higher boiling range of petroleum, few cycloalkanes have been identified apart from some alkylcyclopentanes and alkylcyclohexanes [31, 86]. Recent work [78] has shown the presence of a set of optically active triterpanes in petroleum distillates boiling around 475° C and there are good indications that steranes are also present in petroleum [4, 118]. Many of the higher hydrocarbons in petroleum may perhaps be directly related to such natural products as the carotenes and plant prenols.

VI. Sulfur Compounds

The sulfur compounds of known structure that occur in petroleum were listed in 1962 [91] and the highest carbon number included was a C_{14} dibenzothiophene. Since that time the carbon number of known compounds has been increased by two to include a series of alkyldibenzo- and naphtho-benzothiophenes. However, recent reviews of petroleum sulfur compounds [119, 120] reveal that steady

Alkanethiol
1-Pentanethiol

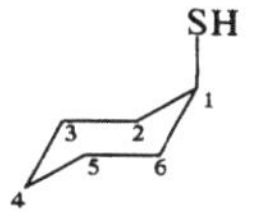

Cycloalkylthiol
Cyclohexanethiol

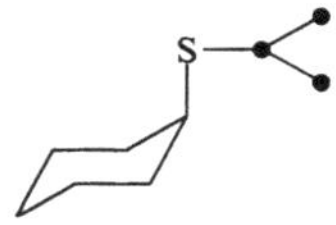

Cycloalkylthiaalkane
(2-methyl-1-thiapropyl)
cyclohexane

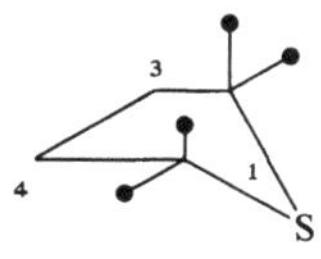

Thiacycloalkane
2,2,5,5-Tetramethyl
thiacyclopentane

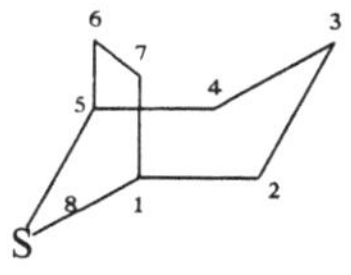

Thiabicycloalkane
8-thiabicyclo [3,2,1] octane

Thiatricycloalkane
Thiaadamantane

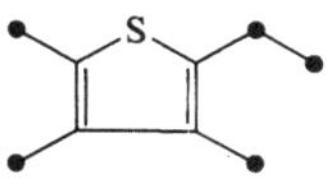

2-Ethyl-3,4,5-trimethyl-
thiophene

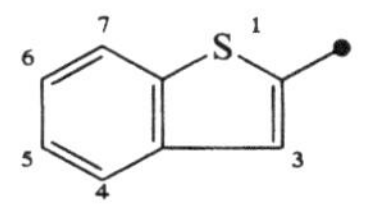

2-Methyl-
benzo [b] thiophene

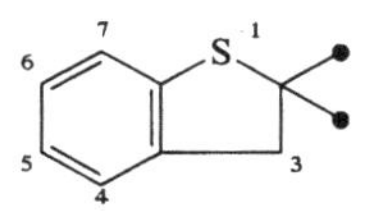

2,2-Dimethyl-1-thiaindan

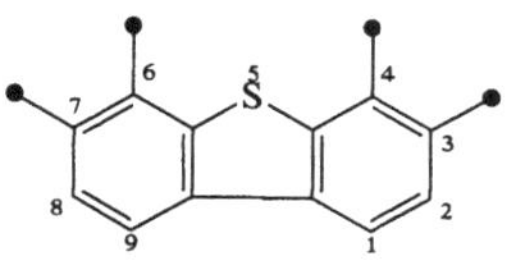

3,4,6,7-Tetramethyl-
dibenzo [bd] thiophene

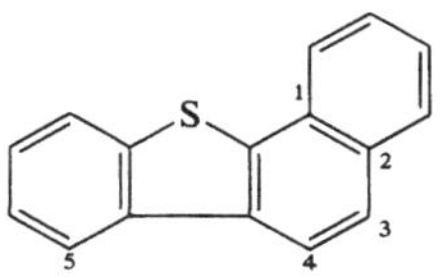

9-Thia-1,2-benzo-
fluorene

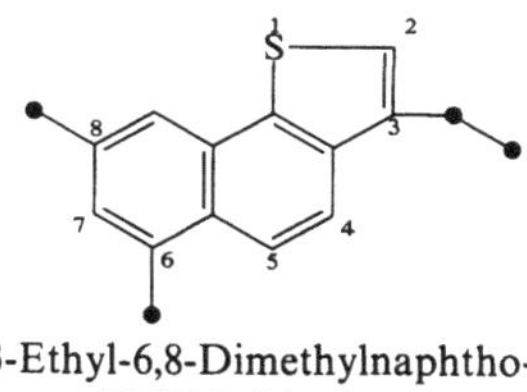

3-Ethyl-6,8-Dimethylnaphtho-
[1,2-b] thiophen

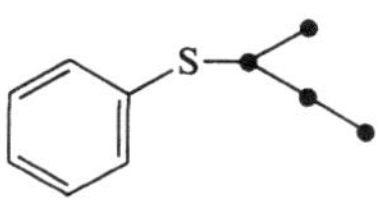

Arylthiaalkane
(2-methyl-thiabutyl)
benzene

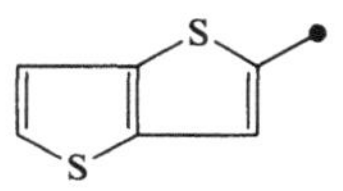

2-Methylthieno-
[3,2-b] thiophene

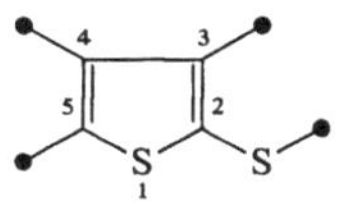

3,4,5-Trimethyl-2 (1-thiaethyl)-
thiophene
(Methyl 3,4,5-trimethyl-2-thienyl
sulfide)

Fig. 3. Sulfur compounds in petroleum

progress has been made in defining more certainly the molecular structure and principal molecular types that are to be found in a wide variety of crude oils.

There are good indications [119] that elemental sulfur is present in some crude petroleums and its presence is thought to indicate that the crude petroleum containing it may not have been subjected to temperatures much above 100° C. The most abundant sulfur compound-types in the lower boiling ranges, below 150° C, appear to be the thiols [121]. Thiaalkanes [122] and cycloalkylthiaalkanes [123] though present, occur in minor amounts. Their place is taken, in distillates boiling up to 250° C, by thiacycloalkanes [120, 124, 125], bi- and tricyclothiaalkanes [120], and thiophenes [126]. These molecular types are supplanted in turn by benzothiophenes [119, 120] and their variants which include the thia-indanes [127], thienothiophenes [128], dibenzo- and naphthobenzothiophenes and naphthothiophenes [120].

Although arylthiaalkanes [129] have been identified, so far very few sulfur compounds – thienothiophenes [128] and 3,4,5-trimethyl-2-(1-thiaethyl)-thiophene [119, 120] with two sulfur atoms per molecule, have been isolated from petroleum, although others are undoubtedly present.

The molecular structure of the sulfur compound-types that are present in the higher boiling ranges of petroleum is not known with certainty, though partial molecular structures have been proposed [130] for the sulfur compounds in a high sulfur, Middle East crude oil. After extensive separation using chromatography, thermal diffusion and chemical treatment, the average molecular types were characterized by chemical and spectroscopic means. The results indicated the presence of 2–3 fused aromatic nuclei and several naphthene rings, with wholly condensed systems predominating. A percentage distribution determined by mass spectrometry [68], has been estimated for sulfur compound types in one petroleum distillate boiling between 347° and 360° C.

The methods used to separate and characterize the wide range of sulfur compound types are described adequately elsewhere [119, 120].

VII. Oxygen Compounds

The principal oxygen-containing compounds which have been isolated from petroleum are the saturated fatty acids and naphthenic acids, although ketones, phenols, ethers, lactones, esters and anhydrides are known to be present. The oxygen content of crude oil increases with boiling point and, like other heteroatoms, the greater part of petroleum oxygen is found in distillates boiling above 400° C. Saturated fatty acids ranging from methanoic to eicosanoic have been isolated from petroleum [91]. In the carbon number range C_1 to C_9, acids with both odd and even carbon numbers are reported as well as isoalkanoic acids [91] with one methyl branch in positions 2 to 5. In the higher boiling ranges alkanoic acids with 14, 16, 18 and 20 carbon atoms are known whereas those with odd carbon numbers do not appear to have been reported. More recently the occurrence of a series of isoprenoid acids C_{11} to C_{20} has been revealed in a Californian petroleum [131, 132].

Naphthenic acids, which can account for some 3 percent weight of some petroleums, have been extensively investigated because of their commercial importance. These acids, occur principally in two ranges of carbon numbers C_6 to C_{12} and C_{14} to C_{19}. The lower range contains chiefly cyclopentyl and cyclohexyl carboxylic acids with several short (C_1 to C_2) substituents, which often include a geminal methyl pair [91, 132, 133]. Less is known about the higher ranges, although these appear to be monobasic acids with an average of 2 to 3 rings per molecule; they may perhaps be related to sesquiterpenes. Much larger molecules, containing sulphur and at least one carboxylic acid group are thought to be present in bitumen [134].

Aliphatic ketones occur in petroleum [135] and, whilst neither cyclopentanone nor cyclohexanone has yet been identified, recently an acetylisopropylmethylcyclopentane was isolated from a Wilmington crude oil [136]. Fluorenones, C_{13} to C_{15}, have been extracted from the higher boiling ranges of the same crude oil by sodium in liquid ammonia [137].

A triterpenoid lactone oxyallobetul-2-ene, isolated from Kuweit crude oil [138], is without doubt related to naturally occurring triterpenoids, and this observation is supported by the identification later on of 2,2,9-trimethylpicene [139]. It may well be that a series of oxygenated triterpanes, like those discovered in brown coal [140], await isolation from petroleum.

Dibenzofurans, long known in coal distillates [141, 142], have been identified in Ponca City crude oil [143] and examples carrying methyl substituents at positions 2 and 3, as well as a 4,6-dimethylisomer, were isolated. Dibenzofuran itself has been reported in petroleum [144].

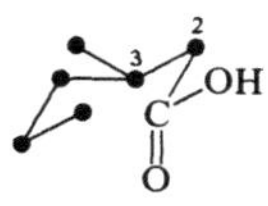

3-Methylhexanoic acid

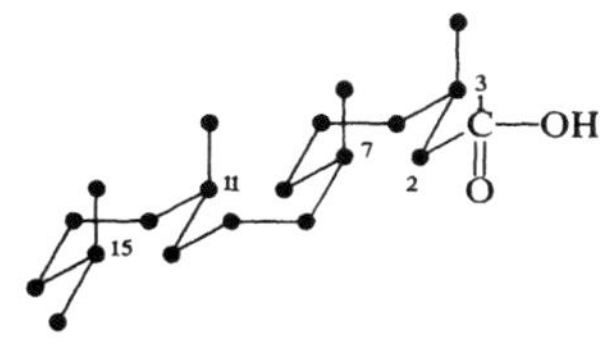

3,7,11,15-Trimethylhexadecanoic acid

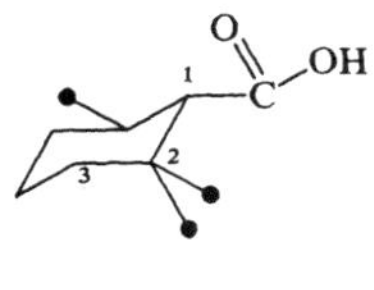

trans-2,2,6-Trimethyl-
cyclohexanecarboxylic acid

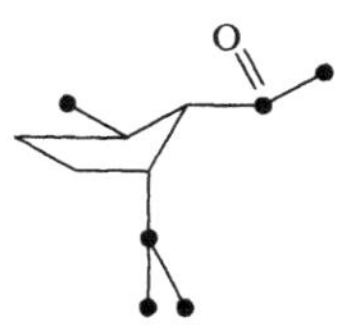

Acetylisopropylmethylcyclopentane

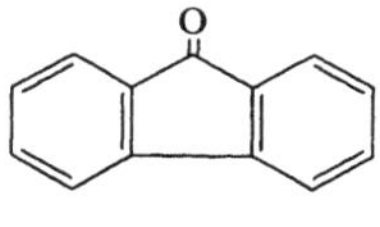

Fluorenone

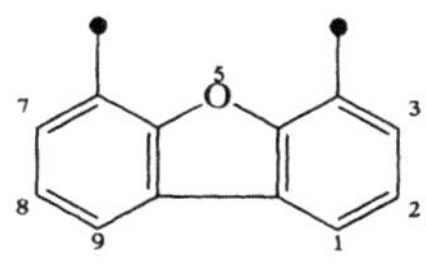

4,6-Dimethyl-
dibenzo [bd] furan

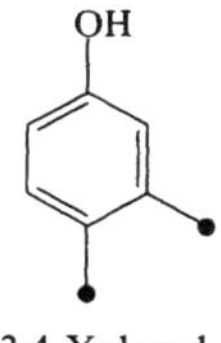

3,4-Xylenol

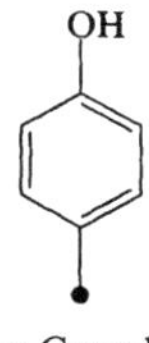

p-Cresol

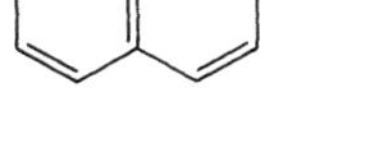

β-Naphthol

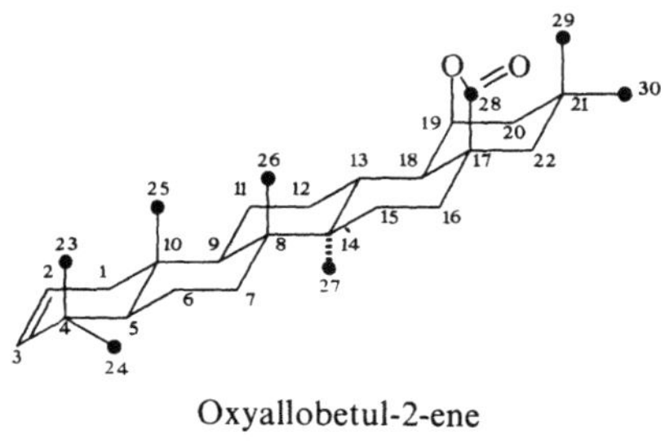

Oxyallobetul-2-ene

Fig. 4. Oxygen compounds in petroleum

The weakly acidic phenols are reported chiefly in lower boiling distillates and phenol, o-, m- and p-cresols, 1,2,3-, 1,2,4-, 1,3,4-, 1,3,5-xylenols, and β-naphthol [91] have all been identified.

If bacteriological modification of plant remains under anerobic conditions is an important process in petroleum formation, as has been suggested by ZOBELL [145], then the presence of oxygen-containing compounds in petroleum may be difficult to explain. On the other hand acidic materials seem to be widespread in sediments and formation waters [22] and these may be a potential source of oxygen compounds, particularly during the migration and accumulation stages.

VIII. Nitrogen Compounds

The nitrogen content of crude petroleums can be as low as 0.01 percent or as high as 0.9 percent by weight, most of this appearing in distillates boiling above 400° C. This nitrogen occurs alone in a variety of molecular environments such as pyridines, pyrroles, quinolines, carbazoles, etc., as well as in combination with other elements: with oxygen as quinolones, with sulfur as thioquinolines and in combination with oxygen and certain metals in the petroporphyrins [55]. These nitrogen compounds are usually considered under two categories defined by the state of the nitrogen atom, which will leave the molecule being considered as basic or non-basic. These terms are defined, and the varied and extensive methods applied to the isolation and identification of these two types are described in two recent papers [146, 147]. The extensive work of LOCHTE and his co-workers has revealed the existence in petroleum of a wide variety of basic nitrogen compounds and this work has been reviewed [148]. The basic nitrogen compounds [148] include pyridines, and the isomers isolated show substituents in all possible positions with a preference for positions 2-, 3- and 5-. The substituents include C_1, C_2, *n* and iso C_3, sec-C_4 and cyclo-C_5, but pyridine itself was not found. A series of quinolines, including quinoline itself, was isolated and positions 2-, 3-, 4- and 8- are favored by the substituents which include C_1, C_2, *n* and iso C_3, and sec-C_4. Systems containing one saturated ring included 5, 6, 7, 8-tetrahydroquinoline and 2- and 5-methyl-6,7-dihydro-1,5-pyrindine. The discovery of 2-(2,2,6-trimethylcyclohexyl)-4,6-dimethylpyridine [148] and very recently of the optically active 2,4,7a,8,8-pentamethylcyclopenta(f)pyrindan [149] lend considerable support to the correlation of petroleum nitrogen compounds with naturally occurring alkaloids [150]. Other basic nitrogen types reported in petroleum are α-hydroxy-benzo- and benzoquinolines, indoloquinolines, phenanthrolines and tetrahydrocarbazolenines [151], although the presence of some of these compounds is in dispute [152], acridines and dihydroacridines are reported in [147] Russian crude oils [153].

Non-basic nitrogen compounds include pyrroles, indoles, carbazoles and benzcarbazoles [154]. The 1,2-benzcarbazole isomer predominates with a minor amount of the 3,4-isomer, whilst the 2,3-isomer is reported to be absent [150]. These observations support an alkaloid source for petroleum nitrogen compounds. Amides with a N-substituted lactam structure are reported in a Russian crude oil [155]. Evidence has been presented for the presence of 2-quinolones and tentatively 2-thiaquinolones [156], whilst analysis of hydrogenated and cracked distillates has revealed phenazines [144] nitriles [157] and possibly pyrroloquinones [158].

A fast sensitive coloumetric method of detecting nitrogen, as ammonia from GLC effluents [159] provides a novel technique for following the distribution of

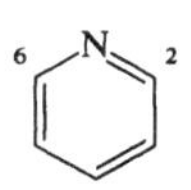
Pyridine

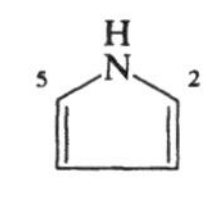
Pyrrole

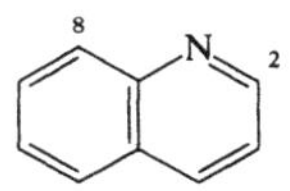
Quinoline

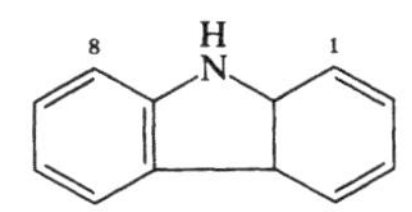
Carbazole

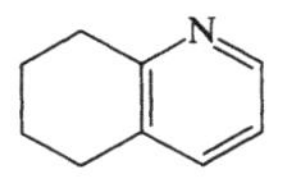
5,6,7,8-Tetrahydroquinoline

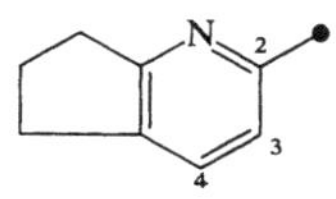
2-Methyl-6,7-dihydropyrindine

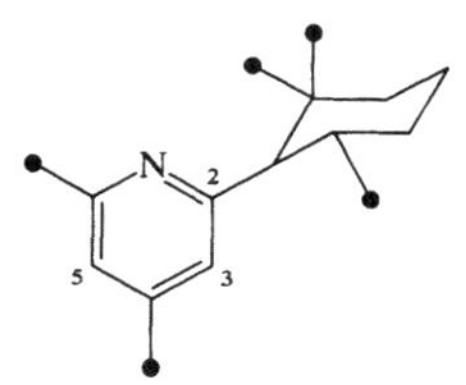
2-(2,2,6-Trimethylcyclohexyl)-
4,6-dimethylpyridine

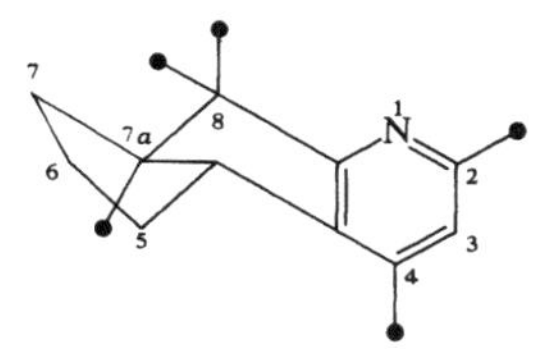
2,4,7a,8,8-Pentamethyl-
cyclopenta [f] pyrindan

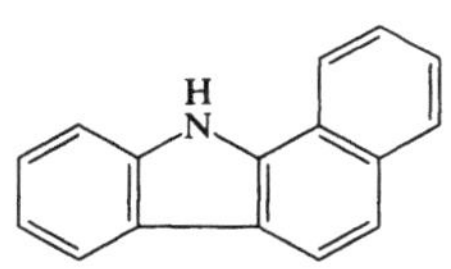
1,2-Benzocarbazole

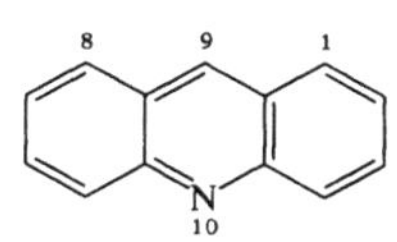
Acridine

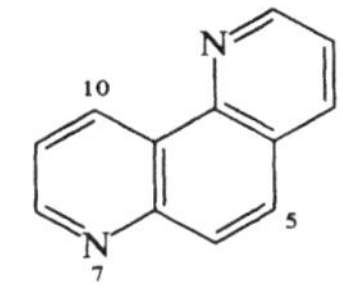
Phenanthroline

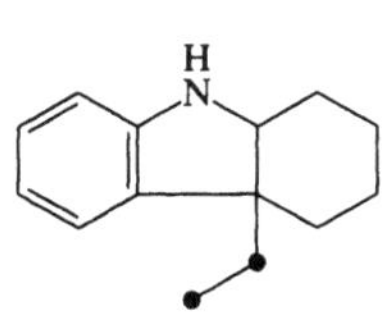
Alkyltetrahydro-
carbazolenine

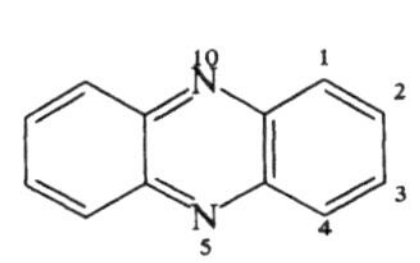
Phenazine

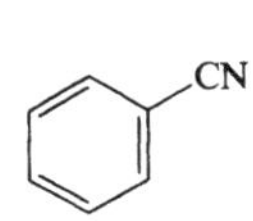
Benzonitrile

Fig. 5. Nitrogen compounds in petroleum

nitrogen compounds in petroleum distillates. Linear elution chromatography [160] and ion exchange resins [154] have been used extensively for the separation of nitrogen compounds and concentration by chemical methods, using perchloric acid [144], sodium aminoethoxide [158], ferric and zinc chloride [157], has been used to advantage.

IX. Porphyrins and Trace Metals

The presence of porphyrins in petroleum and other fossil fuels was discovered over 30 years ago by TREIBS [161], but it is only very recently that the structural type and carbon number distribution of these petroporphyrins has been established by BAKER [162]. He showed that homologs in the carbon number range C_{28} to C_{38} of both deoxophyllo- and etio-series were present. He also found evidence for the presence of a rhodo series. These petroporphyrins (see Chapter 19) may perhaps be related to the plant chlorophylls like those produced by photosynthetic *Chlorobium* bacteria [163]. They are encountered principally as their vanadyl and nickel complexes [16, 164], are quite stable, and are found in petroleum distillates boiling above 500° C. Carboxylated porphyrins have been reported in petroleum [165, 166] but the metal porphyrins of the deoxophyllo- and etio-series appear to be the principal petroporphyrins, whose concentration in petroleum may range from 1 to 1,000 ppm. Several authors [167, 168] have suggested that the transition of naturally occurring pigments in extreme reducing conditions, where metal-exchange, decarboxylation and hydrogenation reactions occur, eventually yield the stable metalloporphyrins found in fossil fuels. The presence of these materials provides strong evidence for the biological origin of part of petroleum.

A variety of methods has been used for the extraction of petroporphyrins. These methods fall broadly into two classes and depend on the recovery on one hand of the free porphyrin by the use of a variety of acids and, recently, of lithium in the presence of ethylene diamine; on the other hand, many methods have been devised to recover the metal porphyrin complexes intact from petroleum. These methods include solvent extraction, elution chromatography from recognized adsorbents and very recently by gel permeation chromatography in non-aqueous systems [44]. High pressure gas chromatography has proved an effective means of separating metalloporphyrins but has never, apparently, been generally applied. These separation methods and other aspects of the role of porphyrins in the geochemistry of petroleum have been fully reviewed by HODGSON and his coworkers [169]. Not all of the nickel and vanadium is combined in petroleum as petroporphyrins [170].

The study of other trace metals in petroleum is complicated by the small amounts present, together with the possibilities of contamination by emulsified formation water, particulate matter from the reservoir, and adventitious metallic constituents from production facilities and sample containers. In a detailed study of uranium and other metals in crude oils from the United States, HYDEN [171] considered that the uranium content of petroleum was probably a direct reflection of the uranium content of sandstone reservoir rocks. He compared other trace metals with vanadium and nickel in an attempt to point indirectly to affinities with hydrocarbons and considered the evidence strongest, in decreasing order, for gallium, molybdenum and cobalt. Some of the vanadium and nickel, and by far the greater part of the remaining metals in petroleum, were considered to be

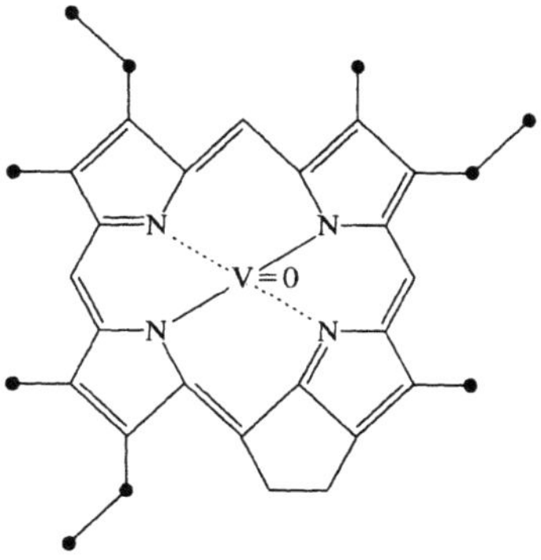

Vanadyldeoxophylloerythroetioporphyrin

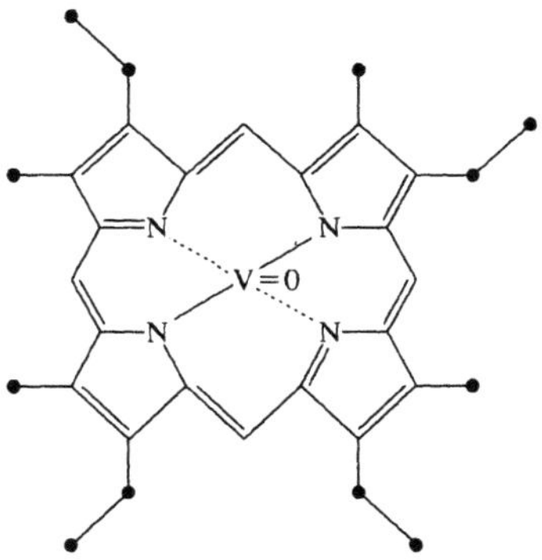

Vanadyletioporphyrin III

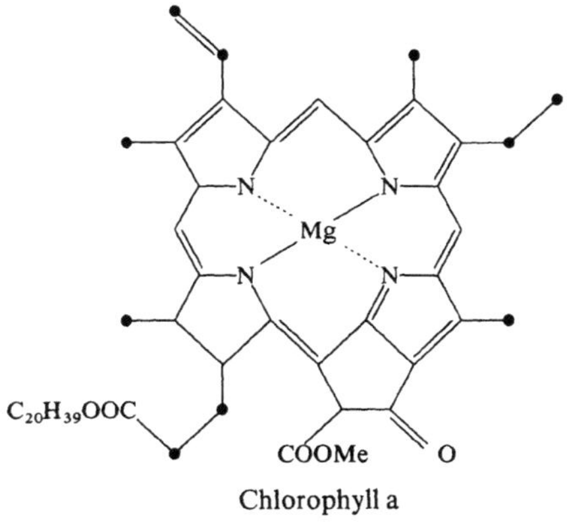

Chlorophyll a

Fig. 6. Porphyrins in petroleum

present in forms other than chelates, possibly as salts of organic acids. The presence of magnesium in West Canadian crudes and its possible significance has been discussed by BAKER and HODGSON [172]. BALL and WENGER [164] found significant amounts of copper in their studies of the ash from 24 petroleums but considered it to be mainly contamination.

The trace metal content of crude oils may serve as an important geological parameter when knowledge is more complete. HYDEN (*loc. cit.*) considers that no association exists between trace metals and reservoir lithology, but that correlations of vanadium, nickel, gallium and molybdenum with oil type and age of reservoir rock are apparent. The relationship of the vanadium/nickel ratio to age has also been noted in the work of KATCHENKOV [173], GULYAEVA [174], SCOTT *et al.* [175], HODGSON [176] and BALL *et al.* [164].

X. Asphalt

The asphaltic components of petroleum are separated by a light hydrocarbon solvent such as *n*-pentane, into a soluble maltene portion and the insoluble asphaltenes. The structure and properties of these two portions will vary with crude oil source and the solvent used for this precipitation. This definition reveals the inadequacy of present knowledge of petroleum asphalt structure. The literature upon which we depend for detailed structural knowledge of petroleum asphaltenes has been reviewed recently by ERDMAN [177]. Spectroscopic techniques, mass [178], infrared, NMR, electron spin resonance (ESR), as well as X-ray diffraction and electron microscopy [179] have all been used to define the indistinct outlines of the asphaltene molecules. The data suggest that the molecular weight of these materials may be as low as $1-5 \times 10^3$ [180, 31]. The asphaltene molecule would appear to carry a core of stacked flat sheets of condensed aromatic rings. Approximately 5 of these sheets, each with some 16 condensed rings, are stacked one above the other with a repeat distance of 3.55 to 3.70 Å giving an overall height for the stack of 16 to 20 Å, the average sheet diameter would appear to be 8.5 to 15 Å. The stacked aromatic sheets show some disorder, probably induced by chains of aliphatic and/or naphthene ring systems, which, linking the edges of the aromatic sheets, tend to be held apart. The repeat distances between these saturated, connecting groups, appears to be 5.5 to 6.0 Å, and the average number of carbon atoms in the rings of both the aromatic and naphthene ring systems would appear to be 6. The condensed aromatic sheets contain oxygen, sulfur and nitrogen atoms and these elements are probably associated with the free radical sites [181] detected by ESR. These sites may well be the centres at which complexed, non-porphyrin, vanadium and nickel is located since SUGIHARA and his co-workers [182] have shown a correlation between sulfur and metal removal from the asphaltenes.

A true understanding of petroleum asphalt will probably not be reached through the study of the asphaltenes in isolation. In the natural colloidal [183] state the asphaltene nucleus is probably associated with adsorbed maltenes in micellar form. The aromatic maltenes, probably through aliphatic side chains permit transition to less polar, aliphatic and naphthenic petroleum components and a stable system in which the inter-molecular adsorption forces are balanced.

XI. Composition and Origin of Petroleum

It is inevitable that any discussion of petroleum composition will turn sooner or later to the problem of the origin of this fossil fuel. Two conflicting theories exist, these presuming either a biogenic or an abiogenic origin and they have been contrasted by KROPTKIN [184]. Sir ROBERT ROBINSON has attempted a reconciliation of these conflicting views with his Duplex [185, 186] theory of origin.

Most geologists and petroleum chemists are persuaded that petroleum is derived from matter of biological origin. In their view the remains of plant and animal life, after burial in argillaceous muds, undergo transformation by aerobic and, in turn, anerobic bacteria. The degraded products, together with bacterial remains, are further transformed under overburden pressure, by chemical and physical processes, at temperatures which probably do not exceed 150° C. The transformation reactions may proceed at the catalytic sites provided by the adjacent rock surfaces, in the presence of water, hydrogen sulfide, sulfur and other inorganic materials. During these processes, the petroleum which is diffusely scattered, accumulates by migration, in reservoirs and finally in oil pools.

Support for a biological origin depends upon a body of circumstantial evidence too large to be discussed here in detail. However some of the more important aspects of this evidence must be defined in passing. Petroleum is always found in, or very closely associated with, sedimentary rocks and none appears to have been found in rocks which predate the arrival of life on earth. TREIBS [161] fully appreciated the importance of the porphyrins which he isolated from petroleum and related them to the living organisms in which the porphyrin nucleus plays such a vital part. Complementary to TREIBS' discovery is the more recent observation of an enrichment in petroleum of the stable ^{12}C isotope of carbon, at the expense of ^{13}C, coupled with the realization that biosynthetic processes, especially those dependent on photosynthesis [187], also produce products enriched in ^{12}C. This ^{12}C enrichment of organic matter relative to inorganic carbonate has persisted throughout geological time. New evidence is being drawn from studies of the precise molecular geometry of the multitude of organic molecules which accumulate in petroleum, and it would seem appropriate to discuss this evidence in greater detail.

Modern analytical techniques can identify essentially all of the components in the low molecular weight ranges of petroleum. They are also revealing in increasing detail a much clearer picture of composition in the higher molecular weight ranges. These studies show that petroleum composition, though most complex, is far simpler than would be expected for a product formed by a random abiogenic process, especially when the astronomical figures for isomers theoretically possible are considered [188]. Furthermore, individual petroleums may be similar in composition but no two are identical, nor do the individual components appear to be in thermodynamic equilibrium [15–17].

A final proof of biogenic origin would probably require the identification and establishment of an unambiguous relationship between each component and its biological precursor. This is no mean task and such rigid proof is never likely, but the current attempts to relate the principal components to likely precursors would appear to offer very significant clues to origin. In addition, a better understanding of the taxonomic distribution of molecular species in biological systems and their relationship to petroleum may well provide a much improved appreciation of the composition of this fossil fuel.

Most petroleums contain more hydrocarbons than any other molecular type and in most, but not all, the principal hydrocarbon type is a homologous series of alkanes. These are chiefly normal, but iso- and anteisoalkanes usually form a very significant proportion of the whole. The presence of these alkanes in both terrestrial [189] and aquatic [190] plants is almost universal. The discovery of isoprenoid alkanes [103, 104, 87] in relatively high concentrations in petroleum has led to their widespread use as biological markers to indicate the presence of past life in a wide variety of geological situations [108]. With the identification of geranylgeraniol in plants [191], it would no longer appear always necessary to relate phytane and pristane directly to chlorophyll, or with the degradation of squalene [105]; indeed, it is probable that in time, free terpenoid precursors for the whole series of isoprenoid alkanes will be revealed in plants.

Prerequisites for biological markers are that (1) they will not be synthesized in significant amounts by abiogenic processes, (2) they should have good chemical stability, and (3) their skeletal features should be obviously and significantly related to biosynthetic sequences. The greater the complexity of such molecules the more information they are likely to convey. Terpenoid hydrocarbons fulfill all of these requirements and the recognition of the precise molecular structure of petroleum steranes [192] and triterpanes [78] would appear to be important since the information contained in their molecular skeletons may well be biologically more specific than that contained in the isoprenoid alkanes. These petroleum steranes and triterpanes appear in abnormal concentration and seem to be chiefly responsible for the high specific rotations of petroleum distillates boiling in the region of 475° C [78]. Optical activity is very closely associated with hydrocarbons and other compounds which are produced biologically [185] and there is a very close similarity between optically active hydrocarbons extracted from marine kelp, algae and bacteria and those that have been isolated from lubricating oil [91].

MEINSCHEIN [4] and MAIR [88] have each proposed plant lipids as source materials for petroleum hydrocarbons, and MAIR has related the skeletal features of a wide variety of petroleum hydrocarbons to naturally occurring compounds which are widely distributed in plants and animals. SNYDER [150] has proposed certain alkaloids as the biogenic precursors of a series of petroleum benzcarbazoles and the skeletal features of many other petroleum nitrogen compounds bear close comparison with a wide variety of plant alkaloids.

Certain oxygen compounds, particularly the isoprenoid alkanoic acids [131] and oxyallobetul-2-ene [138] are either derived unchanged from plants or are so closely related to natural precursors that their biological origin can hardly be disputed.

The relationship that petroleum sulfur compounds have with natural precursors is far less obvious. The simpler thiols, alkyl sulfides and disulfides, certain cyclic sulfides and thiophenes are all known to occur in the plant kingdom [193, 194]. These sulfur compounds are abundant in the lower boiling ranges of some petroleums but virtually absent from others. They do not extend in quantity into the higher boiling ranges of petroleum, but are replaced by a proliferation of aromatic sulfur compounds based on the thiophene ring. Thiophenes [195a] and furans [195b] as well as phenols [196] are abundant in natural products. The thiophene and furan systems appear to be based biosynthetically on natural plant acetylenes [195], however the benzo-, dibenzo- and naphthobenzothiophenes, which occur widely in crude petroleums, do not appear to have a corresponding widespread distribution in the plant kingdom. These thiophenes may perhaps be formed from natural products during petroleum maturation, possibly from phenols as CARRUTHERS [197] has suggested. MAIR [198a] has shown that cholesterol or farnesol when heated at 150° C with sulfur gives rise to hydrocarbons like those found in petroleum. An alternative route might be through the reaction of sulfur [198b], hydrogen sulphide or inorganic sulfides with the naturally occurring oxygen compounds, alkenes or alkynes.

Many organic compounds have been identified in petroleum which have no apparent counterpart among natural products, nor is any relationship immediately obvious. Examples are neopentane and the adamantane family of hydrocarbons. Neopentane would be a logical product from a 1,1-dimethyl-3-, 4- or 5-alkyl-cyclohexa-2,5-diene (the skeleton is common in natural products) by rupture of the two double bonds. MAIR [87] has proposed such a transformation of limonene to explain the abnormally high concentration of 2-methyl-3-ethyl-heptane that he has isolated from petroleum.

Cyclanes, in the presence of Lewis acids, are rapidly converted, under very mild conditions, to adamantanes [199]. A wide variety of cyclanes are present in petroleum from which adamantane, together with some of its homologs, and diamantane [200] have been isolated.

Extrapolating forward from these skeletal associations it would appear reasonable to predict the presence in petroleum and other fossil fuels of a whole range of hydrocarbons in high concentration with skeletal features which will allow them to be related to the carotenoids, phytoquinones, phytoprenols and so on.

Petroleum origin still presents a geological enigma, but it is certain that continuing study of petroleum composition will contribute greatly to its solution.

XII. Experimental Section

It is impossible to record here experimental details of the great variety of analytical techniques that have been used to isolate and identify some of the principal components of crude petroleum. Such details will be found in the many papers that are included in the bibliography. Instead the authors have included, by way of illustration, details of their own work which led to the isolation of a series of triterpanes that are considered to be derived from natural products.

All solvents were redistilled prior to use in clean grease-free equipment and, where appropriate, passed over activated alumina before use.

All glass apparatus was immersed in concentrated chromic acid, rinsed in distilled water and oven-dried.

Teflon sleeves were used occasionally with Quickfit joints, but they were always refluxed overnight with solvent before use to remove surface grease or low molecular weight components.

All melting points were measured on a Koffler hot stage apparatus using standardized thermometers.

A. Distillation

Examination of petroleum distillates* boiling in the range 340–480° C [201] and 480–650° C** indicated that fractions boiling around 475° C contained high concentrations of tetra- and pentacyclic naphthenes, believed to be steranes and triterpanes.

The specific rotation of a Nigerian petroleum distillate increased steadily from $[\alpha]_D^{19} + 1.08$ in the fraction boiling at 410° C (at 760 mm, corrected) and reached a maximum of $[\alpha]_D^{19} + 6.9°$ in the fraction boiling at 478° C. The optical activity then decreased with increase in boiling point until distillates at 550° C (Rota-film) showed a specific rotation of $[\alpha]_D^{19} + 0.3°$.

A detailed examination of a Nigerian petroleum distillate, boiling point 340–480° C, was carried out using the procedures shown in Fig. 7. Details of the procedures follow.

B. Chromatography and Denormalization

Approximately 50 g of distillate (bp 340–480° C) were chromatographed on silica gel (ca. 800 g in a column 3.5 × 175 cm) to yield a naphthene + paraffin ("N + P") fraction. This was denormalized in benzene with 5 Å molecular sieve [43 a] from which the *n*-paraffins were recovered. The denormalized N + P fraction was then freed from traces of oxidation products by passing it over a short column of silica gel, and adducted with urea within a day.

C. Urea Adduction

To a solution of denormalized N + P (17 g) dissolved in redistilled isooctane, was added 20 g of urea, this mixture was initiated with 0.2 ml of methyl alcohol, thoroughly stirred and stood at −25° C for 90 hours. After the mixture had been stirred mechanically for 48 hours at 0–5° C it was allowed to stand for a further 72 hours at −25° C.

Filtration at −25° C gave a filtrate and filter cake, the latter was removed and broken up in 50 ml of fresh (cold, −25° C) isooctane and filtered again. This process was repeated to give two additional filtrates or washes which were combined with the initial filtrate and evaporated at 100° C, 18 mm pressure, until a constant weight of non-adducting oil was obtained. The adduct cake was dissolved in distilled water (50 ml), and the waxy oil which separated was extracted with

* Taken under reduced pressure so that the temperature of the sample being distilled did not exceed 300° C.

** Using a 2 inch single stage ASCO molecular still (Rota-film) (A. F. Smith Company, Rochester, New York, USA) fitted with carbon wipers and operating at 1 micron pressure.

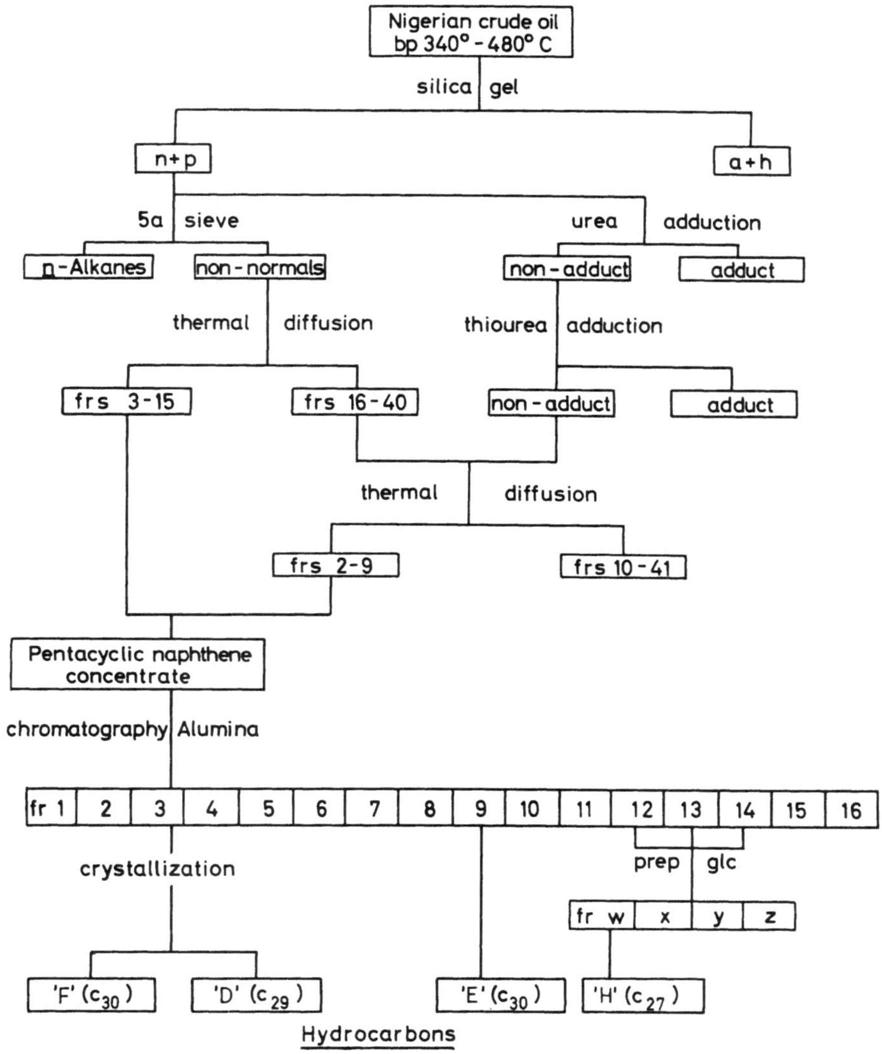

Fig. 7. Separation sequence

4 × 50 ml of redistilled *n*-pentane. The combined extract was washed with water and dried over anhydrous sodium sulphate. This solution, in turn, was evaporated to constant weight at 100° C and 18 mm pressure and yielded a wax. The overall handling loss on adduction was $\not>$ 2 percent.

The urea adducts and non-adducts were analysed by gas-liquid chromatography and mass spectrometry. The course of adduction can be followed if a suitable marker hydrocarbon (e.g., an alkylnaphthalene which can eventually be removed) is incorporated into the denormalized N + P portion under test, aliquots are removed at intervals and analysis by GLC demonstrates the reduction of certain peaks with respect to the unadducted marker peak.

Mass spectra indicated the presence, in the non-adducted (N + P) portion, of a high concentration of the hydrocarbon series C_nH_{2n-6} and C_nH_{2n-8} with carbon numbers in the range C_{27}—C_{31} and a prominent fragment at *m/e* 191.

The adducted portion consisted of normal and "near-normal" alkanes. The adducted "near-normal" alkanes are simple isoalkanes and cyclanes with an unbranched chain of at least twelve to fourteen carbon atoms and from gas-liquid chromatographic data appeared to be mainly 2- and 3-methylalkanes.

D. Thiourea Adduction

The urea non-adduct, freshly recovered from passage over a small column of silica gel, was stirred vigorously at 0° C with a solution of thiourea (5 g; recrystallized from water; m.p. 172–174° C; powdered) in methyl alcohol (50 ml) for 120 hours. The thiourea adduct which precipitated was filtered and washed quickly with cold (− 10° C) isooctane, the filter cake was then broken and stirred with fresh cold (− 10° C) isooctane (50 ml). This suspension was filtered and the cake was washed finally with cold isooctane.

The filtrate and isooctane wash liquors were combined and washed thoroughly with water. After drying over anhydrous sodium sulfate, the isooctane solution was evaporated (water bath, under a stream of nitrogen at 14–20 mm Hg pressure). The residual oil was retained as thiourea non-adduct.

The filter cake was decomposed with hot water (75 ml) and the oil was taken up in isooctane; the solution was thoroughly washed with water, dried over anhydrous sodium sulphate and evaporated to give the thiourea adduct.

Examination of a Libyan distillate (bp 343–460° C) included the preparation of a thiourea adduct from the urea non-adduct of the N + P portion. The gas-liquid chromatograms of the thiourea adduct contained a prominent peak which was exactly reinforced by authentic squalane on SE 30 and SE 52 silicone-gum coated capillary columns, indicating, very tentatively, that this C_{30} isoprenoid is present in the waxy distillate range.

E. Isolation and Examination of Triterpanes

Thermal diffusion [35], in two stages (see Fig. 7) of the non-normals and thiourea non-adducts (51.4 g) yielded two similar concentrates:

Fr 3–15, $(\alpha)_D^{20}$ (in cyclohexane) + 14.5°

Fr 2–9, $(\alpha)_D^{18.5}$ (in cyclohexane) + 17.6°

These were combined and after a preliminary separation, by liquid-adsorption chromatography over alumina, yielded a resin, about 0.7 percent weight on the crude oil. A portion of the gas-liquid chromatogram of this resin, the pentacyclic naphthene concentrate, is reproduced in Fig. 8. The prominent peaks are lettered and these letters are used to identify the hydrocarbons which were eventually isolated from the resin. The resin in turn was separated by liquid-adsorption chromatography over alumina, yielding sixteen fractions. Fractions 1–14, 87 percent by weight of the charge, were eluted by purified *n*-pentane and fractions 15–16, 8 percent by weight of the charge, by purified benzene.

Fraction 3 (0.4072 g; 11 percent by weight of charge, eluted between 14 and 25 percent weight of charge) after crystallization from *n*-pentane gave a crude crystalline sample of hydrocarbon "F" together with a mother liquor. The crude hydrocarbon "F", after crystallization to constant melting point from chloroform, yielded hydrocarbon "F" (5 mg; m.p. 232–234.5° C, corrected, purity by capillary GLC > 98 percent weight). The combined mother liquors were freed from *n*-pentane and chloroform, and after crystallization from ether yielded a crude sample of hydrocarbon "D". This in turn was recrystallized from ether/ethanol, chloroform/ethanol and finally ethyl acetate to yield hydrocarbon "D" (1 mg; m.p. 169–171° C, remelt 178–181° C, corrected, possibly solvent of crystallization present, purity by capillary GLC > 95 percent weight).

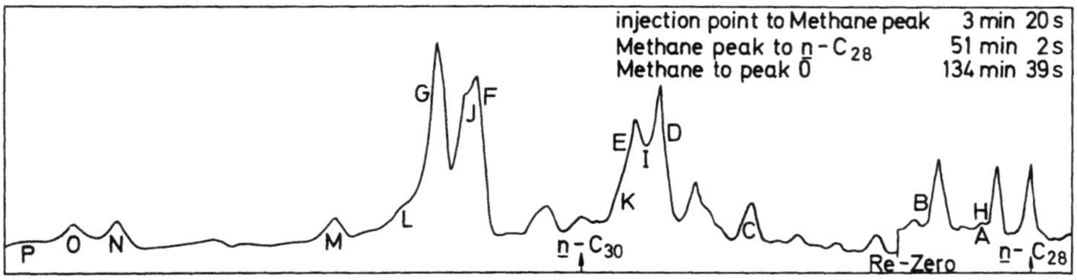

Fig. 8. Portion of gas-liquid chromatogram of pentacyclic naphthene concentrate with added n-C_{28} and n-C_{30}. 148′ glass capillary column 0.01″ ID, coated SE Gum. H_2 carrier gas at 252° C. Flame ionization detector

Fraction 9 (0.1790 g; 5 percent weight on charge at 68 percent weight off) was crystallized from benzene and after several recrystallizations from the same solvent gave hydrocarbon "E" (2–3 mg; m.p. 259.5–260.5° C, corrected, purity by capillary GLC > 98 percent weight).

Fractions 12–14 inclusive (0.4 g; 11 percent weight on charge at 79–87 percent weight off) were combined and separated by preparative scale gas-liquid chromatography (recovery 70 percent by weight from nine runs, Wilkins Autoprep A-700) into four fractions W, X, Y and Z.

Fraction W (0.0374 g) crystallized readily from chloroform to give a crude sample of hydrocarbon "H". After purification over alumina and recrystallization from chloroform/ethyl acetate and finally cyclohexane yielded hydrocarbon "H" (1.5 mg; m.p. 227.5–230.5° C, corrected, purity by capillary GLC > 98 percent weight).

Fraction Y, 0.9030 g, is optically active [$(\alpha)_D^{23} + 17°$ in cyclohexane].

Tests for unsaturation were carried out with tetranitromethane on lupane, shionane, lanostane, friedelane, zeorinane and our hydrocarbons "F", "E" and "H". All gave a negative reaction. Cycloeucalanane, shionene, friedelene, alnusene and euphene, on the other hand, all gave a positive reaction (yellow coloration). These tests indicate that the hydrocarbons "E", "F" and "H", like the authentic steranes and triterpanes are fully saturated and do not contain a cyclopropane ring.

Specific rotations of the various triterpanes were determined using a Bendix Electronics photoelectric polarimeter model 143, modified to reduce noise level and eliminate anomalous color effects. The sensitivity of the instrument was 0.0001° arc, permitting the measurement of specific rotations at concentrations as low as 0.5 mg/ml, using conventional 10 mm low volume cells. Measurements were made on solutions in chloroform at 21° C using the mercury green wavelength at 546 mμ. There was little variation in specific rotation with wavelength in the visible region (450–600 mμ) or with concentration in the range 1–10 mg/ml.

Analytical data determined on these four solid hydrocarbons "F", "E", "D" and "H" are listed in Table 2.

Table 2. *Physical properties of hydrocarbons "F", "E", "D" and "H"*

Hydrocarbon and GLC peak	Accurate mass of molecular ion M	Nuclear magnetic resonance data		GLC data		Optical data $(\alpha)^{21}_{546}$ in chloroform[a]	Empirical formula
		protons in methylene + methine to methyl ± 10 percent	No. of methyl groups per molecule	relative retention distance n-C_{30} = 1.0	purity (percent wt)		
F	412.408	1.1	8	1.10	>98	+22.7 ± 2.4°	$C_{30}H_{52}$
E	412.406	1.1	8	0.95	>98	+21.8 ± 2.0°	$C_{30}H_{52}$
D	398.391	–	–	0.93	>95	–	$C_{29}H_{50}$
H	370.362	1.5	6	0.60	>98	–	$C_{27}H_{46}$
Gammacerane [205]	412.409	1.1	8	1.49	>95	31.9 ± 0.4°	$C_{30}H_{52}$
J	–	–	–	1.11	–	–	–
Lupane	–	1.2	8	1.11	>98	−15 ± 1.6°	$C_{30}H_{52}$

[a] Bendix electronics model 143 photoelectric polarimeter.

High resolution mass spectrometry (AEI-MS-9) confirmed an increase in concentration of the hydrocarbons C_nH_{2n-8} at C_{27}–C_{31} in the resin obtained by thermal diffusion. Prominent fragments in the mass spectrum at m/e 205, 177, 137, 123 and 109 were now dominated by the very intense fragment at m/e 191. DJERASSI [202] notes the frequency with which this fragment at m/e 191 occurs in the mass spectra of triterpenes. In lupane two different fragments contribute to this peak at m/e 191 making it by far the most intense peak in the upper part of the spectrum. One of these fragment species, containing rings A and B of lupane, is classed by this author as the most characteristic fragmentation product of pentacyclic triterpanes.

Mass spectra* of our hydrocarbons "F", "E", "D" and "H" are displayed in Fig. 9 and bear striking resemblance to the three authentic triterpanes shown in Fig. 10. Each spectrum shows a strong molecular ion M, a significant ion at M-15 (methyl) as well as prominent ions at m/e 191, 137, 123 and 109. Hydrocarbon "D" shows, in addition, a prominent fragment at m/e 177, which is the strongest fragment in the spectrum of adiantane. Adiantane, a $C_{29}H_{50}$ pentacyclic nor-triterpane, also exhibits prominent fragments at m/e 205, 191, 137, 123 and 109, but does not reinforce any of the peaks in the gas-liquid chromatogram of our resin.

* Only fragments larger than 5 percent of the base peak are plotted, except in Fig. 12, where, above m/e 235, peaks greater than 2 percent of the base peak are plotted.

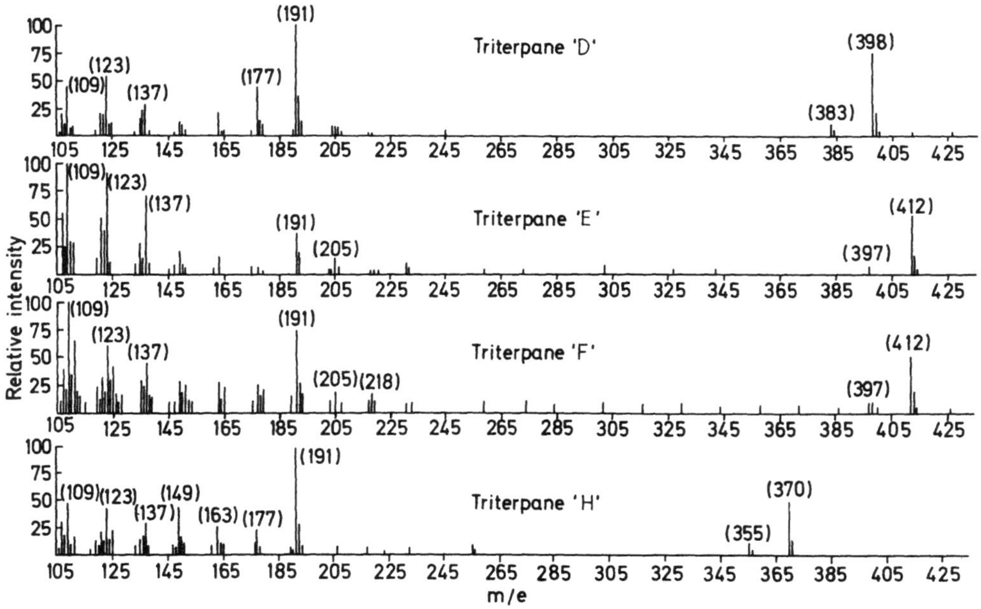

Fig. 9. Mass spectra of triterpanes "D", "E", "F" and "H"

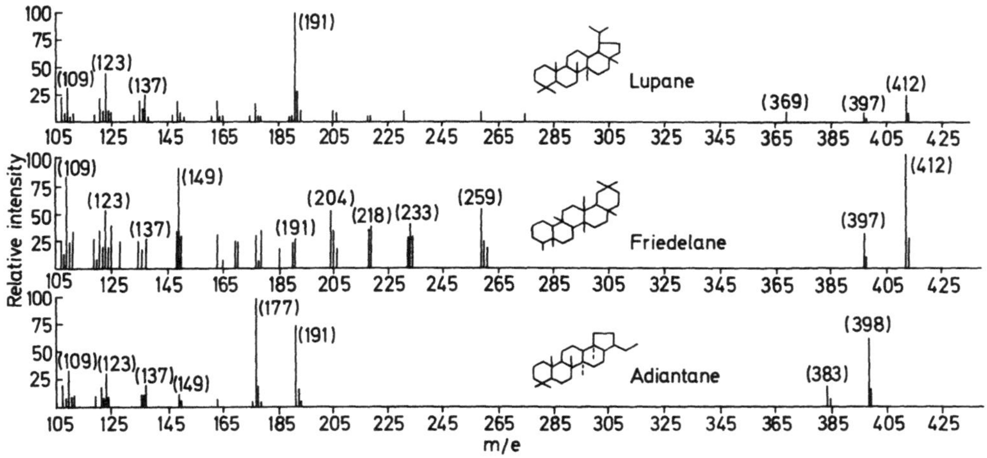

Fig. 10. Mass spectra of authentic triterpanes

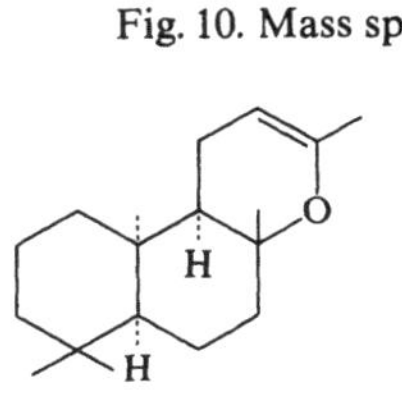

cis-

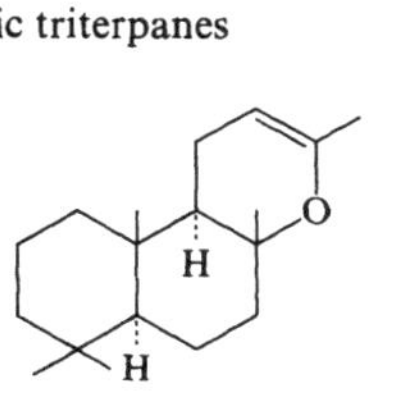

trans-

The mass spectra of *cis*- and *trans*-sclareol oxides [203] also exhibit prominent peaks at m/e 191, 177, 137, 123 and 109. The authors consider that all of these peaks can be related to rings A and B, which, of course, are typical of rings A and B

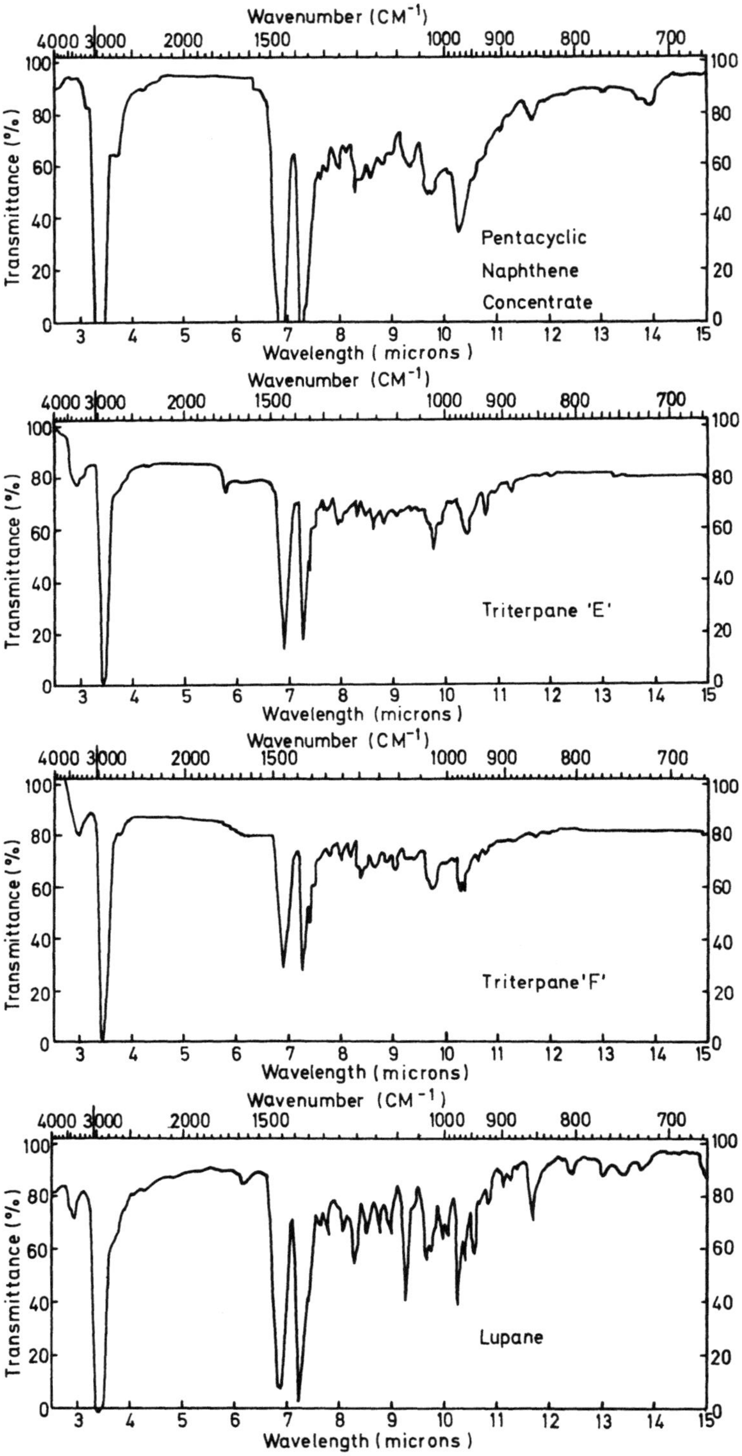

Fig. 11. Infra-red spectra

in many triterpenes. Fragment peaks at m/e 109 and 123 are, however, also prominent in the spectra of steranes.

The accurate masses of the molecular ions listed in Table 2 indicate that the hydrocarbons "F", "E", "D" and "H" all contain five saturated rings (C_nH_{2n-8}) and twenty-seven to thirty carbon atoms.

Nuclear magnetic resonance spectra, were obtained from the very small quantities (1–2 mg) of hydrocarbons "F", "E" and "H" which were available. Although these spectra are not of good quality they are good enough to establish with fair certainty that "E" and "F" each contain eight methyl groups per molecule and that there are six methyl groups per molecule in hydrocarbon "H", Table 2. OURISSON [204] interpreted the principal features of the lupane spectrum, after he had examined the spectra of a large number of lupane derivatives. Our hydrocarbons have NMR spectra which are similar to those of lupane and adiantane.

Infrared spectra of the resin and of the hydrocarbons "E" and "F" isolated from it are shown in Fig. 11. This figure includes for comparison the spectrum of lupane; again the similarities between the unidentified hydrocarbons and the authentic triterpane are quite striking. These spectra also indicate that the samples examined are mainly free from hydroxyl (2.75–3.25 μ), carboxyl (5.75–6.0 μ) and double bonds (6.0–6.3 μ).

Authentic lupane exactly reinforces peak J in the chromatogram of the resin (Fig. 8). However, a component corresponding with peak J has not yet been isolated, and its possible identity with lupane cannot be proved or disproved at this stage.

All of the authentic steranes and triterpanes that we have examined have retention distances (Table 3), which fall between peaks C – P in the gas-liquid chromatogram (Fig. 8), but few of these coincide with any of the labelled peaks in our chromatogram, lupane (peak J) and possibly moretane with peak M, are the exceptions. The presence of a peak at I in Fig. 8 is not immediately obvious, but it was discovered between peaks E and D when the gas-liquid chromatograms of the sixteen fractions separated by liquid-adsorption chromatography over alumina, as already described, were examined.

The indications from all of these data are that a molecular skeleton based on perhydropicene must be considered for the four hydrocarbons that have been isolated.

Hydrocarbon "E" is being examined by X-ray crystallography. The unit cell dimensions and density of the monoclinic crystal gave a molecular weight of 411, in good agreement with the value determined by mass spectrometry (Table 2). Inspection of the Patterson syntheses and the Fourier maps, suggests a nucleus based on perhydropicene. Four of the five rings B, C, D and E appear to be six-membered with *trans*-ring fusions and chair configuration. The configuration of the fifth ring is now established (R value = 14.9 percent) and we hope to publish the detailed structure of hydrocarbon "E" when the present investigation, which is proceeding without the incorporation of a heavy atom, is complete [206].

The cumulative evidence leaves little doubt that hydrocarbons "E" and "F" are C_{30} pentacyclic triterpanes, that "D" is probably a C_{29} nor-triterpane and "H" is a tris-nor-triterpane.

The occurrence of these triterpanes is not limited to Nigerian crude oil. Similar resins have been isolated from Kuwaiti, Iranian and Libyan crude oils. These

Table 3. *Relative retention volumes of unidentified and authentic steranes and triterpanes*

RRV[a] n-C_{30} = 1.0	GLC peak Fig. 3	Authentic hydrocarbon		
0.57		n-C_{28}		
0.59			Cholestane	
0.60	H			
0.61	A			
0.65	B			
0.76			ergostane	
0.84	C			
0.88			onocerane-III	
0.93	D		onocerane-II	
0.94	I			
0.95	E			
0.96	K		fucostane	
0.97				artesane (9:19-9-β-cyclolanostane)
0.99			onocerane-I	
1.00		n-C_{30}		cycloeucalanane
1.03				adiantane (30-nor-hopane)
1.10	F			
1.11	J		lanostane	lupane
1.14	G			
1.17	L			
1.18			shionane (BA-shinolanostane)	
1.24	M			
1.25				moretane
1.38				friedelane (D:A-friedo-oleanane)
1.40				heterolupane (18α-19β-ursane)
1.45	N			
1.49	O			gammacerane
1.53	P			

On a 148′ SE 30 gum capillary column.
[a] Column operating conditions are those noted in Fig. 3.

resins yield capillary gas-liquid chromatograms with the same peaks A—D, G—H and K—P but the proportions in each crude oil are different. The peak identities in each crude were established by blending successively the resin from one crude with that from another and noting the reinforcement of individual peaks in the capillary column gas-liquid chromatograms of the blends. Peaks "E", "F", "I" and "J" prominent in the chromatogram of the Nigerian resin are, however, not seen in the chromatograms of the resins from the other three crude oils. The mass spectrum of each resin shows, however, the prominent fragment ions at m/e 191, 177, 137, 123 and 109 and a high concentration of the hydrocarbon series C_nH_{2n-8} at C_{27}—C_{31} is noticeable in each spectrum.

Our preliminary examination of a tetracyclic naphthene concentrate, obtained by adsorption chromatography over alumina of a urea non-adduct, revealed the presence of a hydrocarbon series C_nH_{2n-6} lying in the carbon number range C_{27}—C_{33}. The mass spectrum showed that the tetracyclic concentrate was incompletely separated from the pentacyclic naphthenes. Such a concentrate segregated from a distillate, bp 340–480° C, from Nigerian crude oil is compared with another segregated from a distillate, bp 350–460° C, from a Libyan crude oil in Fig. 12. The prominent fragments at m/e 217, 218 and 149, which are prominent in the mass spectra of the steranes, are rare in the spectra of pentacyclic triterpanes.

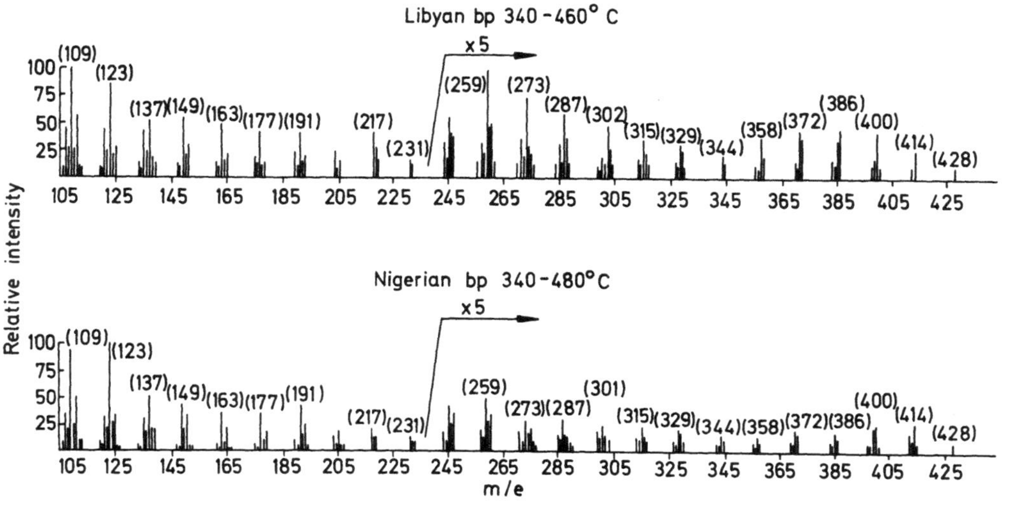

Fig. 12. Mass spectra showing hydrocarbon series C_nH_{2n-6}, steranes, in Nigerian and Libyan crudes

References

1. SMITH, H. M.: Effects of small amounts of extraneous materials on the properties of petroleum, petroleum products and related liquids. Symposium on the major effects of minor constituents on the properties of materials. A.S.T.M. Special Publication No. 304, 62–89 (1962).
2. BESTOUGEFF, M. A.: Petroleum hydrocarbons. In: Fundamentals of petroleum geochemistry, chapt. 3 (NAGY and COLOMBO, eds.). Elsevier 1967.
3. HUNT, J. M.: Distribution of hydrocarbons in sedimentary rocks. Geochim. Cosmochim. Acta **22**, 37–49 (1961).
4. MEINSCHEIN, W. G.: Origin of petroleum. Bull. Am. Assoc. Petrol. Geologists, **43**, 925–943 (1959).
5. MURRAY, G. E.: Indigenous Precambrian petroleum? Bull. Am. Assoc. Petrol. Geologists **49**, 3–21 (1965).
6. BARHOORN, E. S., W. G. MEINSCHEIN, and J. W. SCHOPF: Palaeobiology of a Precambrian shale. Science **148**, 461–472 (1965).
7. HEDBERG, H. D.: Geologic aspects of the origin of petroleum. Bull. Am. Assoc. Petrol. Geologists **48**, 1755–1803 (1964).
8. – World oil prospects – from a geological viewpoint. Bull. Am. Assoc. Petrol. Geologists **38**, 1714–1724 (1954).
9. PERRODON, A.: Sur la notion de province pétrolifère. Rev. Inst. Franc. Petrole Ann. Combust. **16**, 659–677 (1961).
10. WEEKS, L. G.: Origin, migration and occurrence of petroleum. In: Petroleum exploration handbook, chapt. 5 (G. B. MOODY, ed.). New York: McGraw-Hill Book Co. 1961.
11. BROD, I. O.: On principal rules in the occurrence of oil and gas accumulations in the world. Intern. Geol. Rev. **2**, 992–1005 (1960).
12. PRATT, W. E., and D. GOOD (eds.): World geography of petroleum. Princetown: Princetown University Press 1950.
13. ODELL, P. R.: An economic geography of oil. Bell 1964.
14. BAKER, D. R.: Organic geochemistry of the cherokee group in southeastern Kansas and northeastern Oklahoma. Bull. Am. Assoc. Petrol. Geologists **46**, 1621–1642 (1962).

15. SCOTT, D. W., G. B. GUTHRIE, J. P. MCCULLOUGH, and G. WADDINGTON: Isomerization equilibria. The C_2H_6S, C_3H_8S and $C_4H_{10}S$ alkane thiols and sulphides and the methylthiophenes. Preprints, Div. Petrol. Chem., Am. Chem. Soc. **3**, 4, B-5 (1958).
16. SMITH, H. M., H. N. DUNNING, H. T. RALL, and J. S. BALL: Keys to the mystery of crude oil. Proc. Am. Petrol. Inst. **6**, 1–33 (1959).
17. DOBRYANSKY, A., P. F. ANDREYEV, and A. I. BOGOMOLOV: Certain relationships in the composition of crude oil. Intern. Geol. Rev. **3**, 49–59 (1961).
18. HODGSON, G. W., and B. HITCHON: Research trends in petroleum genesis. Paper presented at Eight Commonwealth Mining and Metallurgical Congress. Australia and New Zealand 1965.
19. TISSOT, B.: Problèmes géochimiques de la genèse et de la migration du pétrole. Rev. Inst. Franc. Petrole Ann. Combust. **21**, 1621–1671 (1966).
20. HUNT, J. M.: The origin of petroleum in carbonate rocks. In: Carbonate rocks, chapt. 7, Physical and chemical aspects (G. V. CHILINGAR, H. J. BISSELL and R. W. FAIRBRIDGE, eds.). Elsevier 1967.
21. MCAULIFFE, C.: Solubility in water of paraffin, cycloparaffin, olefin, acetylene, cyclo-olefin and aromatic hydrocarbons. Preprints, Div. of Petrol. Chem. Am. Chem. Soc. Chicago Meeting, 30th August–4th September. 1964, p. 275–287.
22. KIDWELL, A. L., and J. M. HUNT: Migration of oil in recent sediments of Pedernales, Venezuela, 790–817. Habitat of oil (L. G. WEEKS, ed.). Bull. Am. Assoc. Petrol. Geologists (1958).
23. LEVORSEN, A. I.: Geology of petroleum, 2nd ed. San Francisco: W. H. Freeman & Co. 1967.
24. LANDES, K. K.: Petroleum geology, 2nd ed. New York: John Wiley & Sons, Inc. 1959.
25. HOBSON, G. D.: Some fundamentals of petroleum geology. New York: Oxford University Press 1954.
26a. RUSSELL, W. L.: Principles of petroleum geology, 2nd ed. New York: McGraw-Hill Book Co. 1960.
26b. HOBSON, G. D.: The organic geochemistry of petroleum. Earth Sci. Rev. **2**, 257–276 (1966).
27. BURCIK, E. J.: Properties of petroleum reservoir fluids. John Wiley & Sons, Inc. 1957.
28. KATZ, D. K., D. CORNELL, S. A. VARY, R. KOBAYASHI, J. R. ELENBAAS, F. N. POETTMANN, and C. F. WEINAUG: Handbook of natural gas engineering. New York: McGraw-Hill Book Co. 1959.
29. N.G.A.A. Tentative method for natural gas analysis by gas chromatography, Publication No. 2261-61 (1961).
30. KUDCHADKEV, A. P., and B. J. ZWOLINSKI: Vapour pressure and boiling points of normal alkanes C_{21}–C_{100}. J. Chem. Eng. Data **11**, 253–255 (1966).
31. DEAN, R. A., and E. V. WHITEHEAD: The composition of high boiling petroleum distillates and residues. Section V, Paper 9. Proc. 6th World Petrol. Congr., Frankfurt/Main 1963.
32. DIETZ, J. M.: Application of gas chromatography to hydrocarbon analysis. A.S.T.M. Special Technical Publication No. 389. Hydrocarbon Analysis, 153–170 (1965).
33. MAIR, B. J.: Methods for separating petroleum hydrocarbon. Paper for Presentation at The Seventh World Petroleum Congress, Mexico, April 2–8, 1967. Panel Discussion–15, Paper 5.
34. PERRY, E. S., and A. WEISSBERGER: Technique of organic chemistry, vol. 4, Distillation. New York: Interscience Publ. 1965.
35. NATHAN, W. S., and D. M. WHITEHEAD: Thermal diffusion and its application to the analysis of petroleum products, C.R. Thirty-First Congress International Chem. Indust. Liege, September 1958.
36. DICKEL, C.: Separation of gases and liquids by thermal diffusion. In: Physical methods in chemical analysis, vol. IV, p. 261–316. (W. G. BEAL, ed.). New York: Academic Press 1961.
37. HEFTMANN, E. (ed.): Chromatography. New York: Reinhold Publ. Co. 1961.
38. SNYDER, L. R., and B. E. BUELL: Compound type separation and classification of petroleum by titration, ion exchange and adsorption—an index of the acidity, basicity and adsorptivity on alumina of various petroleum compound types. J. Chem. Eng. Data **11**, 545–553 (1966).
39. GAREL, J.-P.: Chromatographie en couche mince B, applications en chemie organique. Bull. Soc. Chim. France 1563–1587 (1965).
40. AKHREM, A. A., and A. I. KUZNETSOVA: Thin layer chromatography. Russ. Chem. Rev. **32**, 366–386 (1963).
41. THOMAS, T. L., and R. L. MAYS: Separations with molecular sieves. Physical methods in chemical analyses, vol. IV, p. 45–97, (W. G. BERL, ed.). New York: Academic Press 1961.

42. WIEL, A. VAN DER: Molekularsiebe und Gas-Flüssigkeits-Chromatographie als Hilfsmittel zur Bestimmung der Konstitution von Paraffin. Erdoel Kohle **18**, 632–636 (1965).

43a. BRUNNOCK, J. V.: Separation and distribution of normal paraffins from petroleum heavy distillates by molecular sieve adsorption and gas-chromatography. Anal. Chem. **38**, 1648–1652 (1966).

43b. MORTIMER, J. V., and L. A. LUKE: The determination of normal paraffins in petroleum products. Anal. Chim. Acta **38**, 119–126 (1967).

44. ALTGELT, K. H.: Gel permeation chromatography of asphalts and asphaltenes. Part I. Fractionation procedure. Makromol. Chem. **88**, 75–89 (1965).

45. HAGAN, Sister M.: Clathrate Inclusion Compounds. London: Reinhold Publ. Corp. 1962.

46. TERRES, E., u. S. N. SUR: Über Harnstoff-Einschlußverbindungen Geradkettiger und Verzweigter Aliphatischer Verbindungen. Brennstoff-Chem. **38**, 330–343 (1957).

47. SCHIESSLER, R. W., and D. FLITTER: Urea and thiourea adduction of C_5–C_{42} hydrocarbons. J. Am. Chem. Soc. **74**, 1720–1723 (1952).

48. MONTGOMERY, D. P.: Thiourea adduction of alkylated polynuclear aromatic hydrocarbons and heterocyclic molecules. J. Chem. Eng. Data **8**, 432–436 (1963).

49. MCLAUGHLIN, R. L.: Separation of paraffins by urea and thiourea. In: The Chemistry of Petroleum Hydrocarbons (B. J. BROOKS, S. S. KURTZ, C. E. BOORD and L. SCHMERLING, eds.), vol. 1, 241–274. New York: Reinhold Publ. Corp. 1954.

50. MORRIS, L. J.: Separations of lipids by silver ion chromatography. J. Lipid Res. **7**, 717–732 (1966).

51. NYBURG, S. C.: X-Ray Analysis of Organic Structures. New York: Academic Press 1963.

52. HOPPE, W.: Automation of x-ray structural analysis, part I. Automation of the measuring process, part II. Automation of the evaluation. Angew. Chem. Intern. Ed. Engl., part I., **4**, 508–516 (1965); part II, **5**, 267–280 (1966).

53. BIRKS, L. S.: X-Ray Spectrochemical Analysis. New York: Interscience Publ. 1959.

54. PATTERSON, G. H.: Petroleum, other elements. Anal. Chem. **37**, 143 R (1965).

55. FALK, J. E.: Porphyrins and Metalloporphyrins, vol. I and II. Holland: Elsevier 1964.

56. RAO, C. N. R.: Chemical Applications of Infra-red Spectroscopy. New York: Academic Press 1963.

57. KENDALL, D. M.: Applied Infra-red Spectroscopy. New York: Reinhold Publ. Corp. 1965.

58. SERGIENKO, S. R.: High molecular compounds in petroleum. Infra-red spectroscopy in the study of high molecular hydrocarbons in crudes 189. Israel Program for Scientific Translations Jerusalem 1965.

59. LE TOURNEAU, R. L.: Hydrocarbon Analysis. Absorption spectroscopy of hydrocarbons. A.S.T.M. Special Technical Publication **389**, 131–152 (1965).

60. SAWIKI, E., W. ELBERT, T. W. STANLEY, T. R. HAUSER, and F. T. FOX: Separation and characterization of polynuclear aromatic hydrocarbons in urban air-borne particles. Anal. Chem. **32**, 810–815 (1960).

61. SNYDER, L. R., and B. E. BUELL: Trace analysis for basic nitrogen in gasoline. Anal. Chem. **64**, 689–691 (1962).

62. EMSLEY, J. W., J. FEENY, and H. SUTCLIFFE: High Resolution Nuclear Magnetic Resonance Spectroscopy, vol. I and II. Oxford: Pergamon Press 1965.

63. ZIMMERMAN, J. R.: Hydrocarbon Analysis. N.M.R. analysis of hydrocarbons and related molecules. A.S.T.M. Special Technical Publication **389**, 103–128 (1965).

64. YEW FU-HSIE, F., R. J. KURLAND, and B. J. MAIR: Chemical shifts of methyl protons in polymethylnaphthalenes. Anal. Chem. **36**, 843–845 (1964).

65. STEHLING, F. C., and K. W. BARTZ: Determination of molecular structure of hydrocarbon olefins by high resolution nuclear magnetic resonance. Anal. Chem. **38**, 1467–1479 (1966).

66. FRIEDEL, R. A., and H. L. RETCOFSKY: Quantitative application of ^{13}C nuclear magnetic resonance: ^{13}C N.M.R. signals in coal derivatives and petroleum. Chem. Ind. (London) 455–456 (1966).

67. BEYNON, J. H.: Mass Spectrometry and its Applications to Organic Chemistry. Amsterdam: Elsevier 1960.

68. LUMKIN, H. E.: Analysis of a trinuclear aromatic petroleum fraction by high resolution mass spectrometry. Anal. Chem. **36**, 2399–2401 (1964).

69. GOMER, R., u. G. WAGNER: Analytische Anwendungsmöglichkeiten des Feldionen-Massenspektrometers. Z. Chem. **197**, 58–80 (1963).

70. MCMURRAY, W. J., B. N. GREEN, and S. R. LIPSKY: Fast scan high resolution mass spectrometry. Operating parameters and its tandem use with gas chromatography. Anal. Chem. **38**, 1194–1204 (1966).

71. TUNNICLIFF, D. D., and P. A. WADSWORTH: A stepwise regression program for quantitative interpretation of mass spectra. Anal. Chem. **37**, 1082–1085 (1965).

72. BIEMANN, K., P. BOMMER, and D. M. DESIDERIO: Element mapping, A new approach to the interpretation of high resolution mass spectra. Tetrahedron Letters **26**, 1725–1731 (1964).

73. BROWN, R. A.: Mass spectrometry of hydrocarbons. Hydrocarbon analysis. A.S.T.M. Special Technical Publication **389**, 68–102 (1965).

74. BEREZKIN, V. G., and O. L. GORSHUNOV: Reaction gas chromatography in analysis. Russ. Chem. Rev. **34**, 470–479 (E), 1108 (R) (1965).

75. FENSKE, M. R., F. L. CARNAHAN, J. N. BRESTON, A. H. CASER, and A. R. RESCORLA: Optical rotation of petroleum fractions. Ind. Eng. Chem. **34**, 638–646 (1942).

76. AMOSOV, G. A.: Optical rotation of petroleums. Trudy VNIGRI New Series No. 5. Contributions to Geochemistry **2–3**, 225–233 (1951).

77. OAKWOOD, T. S., D. S. SHRIVER, H. H. FALL, W. J. MCALEER, and P. R. WUNZ: Optical activity of petroleum. Ind. Eng. Chem. **44**, 2568–2570 (1952).

78. HILLS, I. R., and E. V. WHITEHEAD: Triterpanes in optically active petroleum distillates. Nature **209**, 977–979 (1966).

79. RUDOLPH, H.: Photoelectric polarimeter attachment. J. Opt. Soc. Am. **45**, 50 (1955).

80. GILLHAM, E. J.: A high precision photoelectric polarimeter. J. Sci. Instr. **34**, 435–439 (1957).

81. ROSENFELD, W. D.: Optical rotation of petroleums. J. Am. Oil Chemists' Soc. **44**, 703–707 (1967).

82. MAIR, B. J.: Hydrocarbons isolated from petroleum. Oil Gas J. **62**, 130–134 (1964).

83. LUDWIG, F. J.: Analysis of microcrystalline waxes by gas-liquid chromatography. Anal. Chem. **37**, 1732–1737 (1965).

84. HEADLEE, A. J. W., and R. E. MCCLELLAND: Quantitative separation of West Virginian petroleum into several hundred fractions. Isolation and properties of C_5–C_{27} normal paraffins and other hydrocarbons. Ind. Eng. Chem. **43**, 2547–2552 (1951).

85. DENEKAS, M. O., F. T. COULSON, J. W. MOORE, and C. G. DODD: Materials adsorbed at crude petroleum/water interfaces. Isolation and analysis of normal paraffins of high molecular weight. Ind. Eng. Chem. **43**, 1165–1169 (1951).

86. LEVY, E. J., R. R. DOYLE, R. A. BROWN, and F. W. MELPOLDER: Identification of components in paraffin wax by high temperature gas-chromatography and mass spectrometry. Anal. Chem **33**, 698–704 (1961).

87. MAIR, B. J., Z. RONEN, E. J. EISENBRAUN, and N. G. HORODYSKY: Terpenoid precursors of hydrocarbons from the gasoline range of petroleum. Science **154**, 1339–1341 (1966).

88. – Terpenoids, fatty acids and alcohols as source materials for petroleum hydrocarbons. Geochim. Cosmochim. Acta **28**, 1303–1321 (1964).

89. ROSSINI, F. D., B. J. MAIR, and A. J. STREIFF: Hydrocarbons from Petroleum. New York: Reinhold Publ. Corp. 1953.

90. – Hydrocarbons in petroleum. J. Chem. Educ. **37**, 554–561 (1960).

91. WHITEHEAD, W. L., and I. A. BREGER: Geochemistry of Petroleum. In: Organic Geochemistry (I. A. BREGER, ed.). Oxford: Pergamon Press, Inc. 1963.

92. MARTIN, R. L., and J. C. WINTERS: Composition of crude oil through seven carbons as determined by gas chromatography. Anal. Chem. **31**, 1954–1960 (1959).

93. – – Determination of hydrocarbons in crude oil by capillary column gas chromatography. Anal. Chem. **35**, 1930–1933 (1963).

94. DESTY, D. H., A. GOLDUP, and W. T. SWANTON: Performance of coated capillary columns. Examination of Ponca City crude by gas chromatography. Proc. Instr. Soc. Am. 3rd Nat. Symp. Progr. Trends Chem. Petrol. Instr. 1961, p. 105–134.

95. MARTIN, R. L., J. C. WINTERS, and J. A. WILLIAMS: Composition of crude oils by gas chromatography. Geological significance of hydrocarbon distribution. Proc. 6th World Petrol. Congr., 1963, Section 5, Paper 13.

96. BIEDERMAN, E. W., JR.: Crude oil composition—A clue to migration. World Oil. December, 78–82 (1965).
97. BOGOMOLOV, A. I., and V. K. SHIMANSKY: The origin of the light paraffins in crude oils in the light of composition patterns. In: Advances in Organic Geochemistry (G. D. HOBSON and M. LOUIS, eds.) New York: Pergamon Press, Inc. 1964.
98. BROWN, R. A.: Compound types in gasoline by mass spectrometer analysis. Anal. Chem. **23**, 430–437 (1951).
99. HOOD, A.: The molecular structure of petroleum. In: Mass Spectrometry of Organic Ions, p. 591–635 (F. W. MCLAFFERTY, ed.). New York: Academic Press, Inc. 1963.
100. POWELL, H.: Mass spectrometry in relation to other physical methods of analysis. In: Advances in Mass Spectrometry, vol. 3, p. 621–638 (W. L. MEAD, ed.). Inst. Petrol. Rev. 1966.
101. LINDEMAN, L. P., R. L. LE TOURNEAU: New information on the composition of petroleum. Proc. 6th World Petrol. Congr. 1963, Section V, Paper 14.
102. Hydrocarbon Analysis. A.S.T.M. Special Technical Publications **389** (1965).
103. DEAN, R. A., and E. V. WHITEHEAD: The occurrence of phytane in petroleum. Tetrahedron Letters 768–770 (1961).
104. BENDORAITIS, J. G., B. L. BROWN, and L. S. HEPNER: Isoprenoid hydrocarbons in petroleum. Anal. Chem. **34**, 49–53 (1962).
105. – – – Isolation and identification of isoprenoids in petroleum. Proc. 6th World Petrol. Congr., 1963, Section V, Paper 15.
106. MAIR, B. J., N. C. KROUSKOP, and T. J. MAYER: Compositions of the branched paraffin-cycloparaffin portion of the light gas oil fraction. J. Chem. Eng. Data. **7**, 420–426 (1962).
107. EGLINTON, G., P. M. SCOTT, T. BELSKY, A. L. BURLINGAME, W. RICHTER, and M. CALVIN: Occurrence of isoprenoid alkanes in a Precambrian sediment. In: Advances in Organic Geochemistry, vol. II (1964) (HOBSON and LOUIS, eds.). New York: Pergamon Press, Inc. 1966.
108. JOHNS, R. B., T. BELSKY, A. L. BURLINGAME, P. HAUG, H. K. SCHONES, W. RICHTER, and M. CALVIN: The organic chemistry of ancient sediments. Part II. Geochim. Cosmochim. Acta **30**, 1191–1222 (1966).
109. MAIR, B. J., and J. M. BARNWALL: Composition of the mononuclear aromatic material in the light gas oil range. Low refractive index portion, 230°–305° C. J. Chem. Eng. Data. **9**, 282–292 (1964).
110. –, and J. L. MARTINÉZ-PICÓ: Composition of the trinuclear aromatic portion of the heavy gas oil and light lubricating distillate. Proc. Am. Petrol. Inst. **42**, Section 3, 173 (1962).
111. MOORE, R. J., R. E. THORPE, and C. L. MAHONEY: Isolation of methyl-chrysene from petroleum. J. Am. Chem. Soc. **75**, 2259 (1953).
112. CARRUTHERS, W., and A. G. DOUGLAS: 1,2-benzanthracene derivatives in Kuwait mineral oil. Nature **192**, 256–257 (1961).
113. COOK, J. W., W. CARRUTHERS, and D. L. WOODHOUSE: Carcinogenicity of mineral oil fractions. Brit. Med. Bull. **14**, 132–135 (1958).
114. CARRUTHERS, W., and J. W. COOK: The Constituents of high boiling petroleum distillates. Part I. Preliminary Studies. J. Chem. Soc. 2047–2052 (1954).
115. –, and D. A. M. WATKINS: Identification of 1,2,3,4-tetrahydro-2,2,9-trimethylpicene in an American crude oil. Chem. Ind. (London) **34**, 1433 (1963).
116. – H. N. M. STEWART, and D. A. M. WATKINS: 1,2-benzanthracene derivatives in a Kuwait mineral oil. Nature **213**, 691–692 (1967).
117. LUMPKIN, H. E., and B. H. JOHNSON: Identification of compound types in heavy petroleum gas oil. Anal. Chem. **26**, 1719–1722 (1954).
118. HILLS, I. R., and E. V. WHITEHEAD: Pentacyclic triterpanes from petroleum and their significance. Paper presented at Third Intern. Meeting on Org. Geochem. Sept. 26–28, 1966.
119. THOMPSON, C. J., H. J. COLEMAN, R. L. HOPKINS, and H. T. RALL: Hydrocarbon analysis. Sulfur compounds in petroleum. A.S.T.M. Special Technical Publication **389**, 329–362 (1965).
120. DEAN, R. A., and E. V. WHITEHEAD: Status of work in separation and identification of sulfur compounds in petroleum and shale oil. Paper for presentation at The Seventh World Petrol. Congr. Mexico, April 2–8, 1967. Panel Discussion 23, Paper 7.
121. COLEMAN, H. J., C. J. THOMPSON, R. L. HOPKINS, and H. T. RALL: Identification of thiols in a Wasson, Texas, crude oil distillate boiling from 111°–150° C. J. Chem. Eng. Data. **10**, 80–84 (1965).

122. THOMPSON, C. J., H. J. COLEMAN, R. L. HOPKINS, and H. T. RALL: Identification of some chain sulfides in a Wasson, Texas, crude oil distillate boiling from 111° – 150° C. J. Chem. Eng. Data **9**, 473 – 479 (1964).
123. – – – – Identification of alkyl cycloalkyl sulfides in petroleum. J. Chem. Eng. Data **9**, 293 – 296 (1964).
124. – – – – Identification of some cyclic sulfides in a Wasson, Texas, crude oil distillate boiling from 111° – 150° C. J. Chem. Eng. Data **10**, 279 – 282 (1965).
125. – – – – Identification of naturally occurring cyclic sulfides in a Wilmington, California, crude oil distillate boiling from 111° – 150° C by use of a series of gas-liquid chromatography stationary phases. J. Chromatog. **25**, 34 – 41 (1966).
126. – – – – Identification of some naturally occurring alkyl thiophenes in Wilmington, California, crude oil by use of a series of gas-liquid chromatography stationary phases. J. Chromatog. **20**, 240 – 249 (1965).
127. THOMPSON, C. J., H. J. COLEMAN, R. L. HOPKINS, and H. T. RALL: Identification of thiaindans in crude oil by gas-liquid chromatography, desulphurization and spectral techniques. Anal. Chem. **38**, 1562 – 1566 (1966).
128. HOPKINS, R. L., C. J. THOMPSON, H. J. COLEMAN, and H. T. RALL: Presence of thienothiophenes in Wasson, Texas, crude oil. U.S. Bur. Mines, Rept. Invest. No. 6796 (1966).
129. – R. F. KENDALL, C. J. THOMPSON, H. J. COLEMAN, and H. T. RALL: Identification of alkylaryl sulfides in Wasson, Texas, crude oil. Preprints, Am. Chem. Soc., New York, September 11–16, 1966, vol. 11, p. 7 – 11.
130. BESTOUGEFF, M.: Constitution des composes soufres cycliques du petrole. Proc. 5th World Petrol. Congr., 1959, Section V, Paper 12.
131. CASON, J., and D. W. GRAHAM: Isolation of isoprenoid acids from a Californian petroleum. Tetrahedron **21**, 471 – 483 (1965).
132. –, and A. I. A. KHODAIR: Separation from a Californian petroleum and characterization of geometric isomers of 3-ethyl-4-methyl cyclopentylacetic acid. J. Org. Chem. **31**, 3618 – 3624 (1966).
133. –, and K. L. LIAUW: Characterization and synthesis of a monocyclic eleven-carbon acid isolated from a California petroleum. J. Org. Chem. **30**, 1763 – 1769 (1965).
134. CARO, J. H.: Hochmolekulare Saure Verbindungen aus Erdöl. Erdöl-Z. 3 – 8 (1962).
135. LOCHTE, H. L., and E. R. LITTMAN: The petroleum acids and bases. New York: Chemical Publ. Co. 1955.
136. BRANDENBURG, C. F., D. R. LATHAM, G. L. COOK, and W. E. HAINES: Identification of a cycloalkyl ketone in Wilmington petroleum through use of chromatography and spectroscopy. J. Chem. Eng. Data **9**, 463 – 466 (1964).
137. LATHAM, D. R., C. R. FERRIN, and J. S. BALL: Identification of fluorenones in Wilmington petroleum by gas-liquid chromatography and spectrometry. Anal. Chem. **34**, 311 – 313 (1962).
138. BARTON, D. H. R., W. CARRUTHERS, and K. H. OVERTON: Triterpenoids. Part XXI. A triterpenoid lactone from petroleum. J. Chem. Soc. 788 (1956).
139. CARRUTHERS, W., and D. A. M. WATKINS: The constituents of high-boiling petroleum distillates. Part VIII. Identification of 1,2,3,4-tetrahydro-2,2,9-trimethylpicene in an American crude oil. J. Chem. Soc. 724 – 729 (1964).
140. SŎRM, F., K. HEJNO, M. HORAK, V. JAROLIM, M. STREIBL, and F. HEMMERT: Chem. Ind. (London) 1142 – 1143 (1958). Collection Czech. Chem. Commun. **26**, 459 – 465 (1961); **28**, 2318 – 2326, 2443 – 2453 (1963); **30**, 873 – 879 (1965).
141. ESTEP, P. A., C. KARR, W. C. WARNER, and E. E. CHILDERS: Counter current distribution of high boiling neutral oils from low temperature coal tar. Anal. Chem. **37**, 1715 – 1720 (1965).
142. ANDERSON, H. C., and W. R. K. WU: Properties of compounds in coal carbonization products. U.S. Bur. Mines Bull. No. 606, (1963).
143. YEW, F. F. H., and B. J. MAIR: Isolation and identification of C_{13}–C_{17} alkylnaphthalenes, alkylbiphenyls and alkyldibenzofurans from the 275° – 305° C. Dinuclear aromatic fraction of petroleum. Anal. Chem. **38**, 231 – 237 (1966).
144. HARTUNG, G. K., and D. M. JEWELL: Carbazoles, phenazines and dibenzofuran in petroleum products; methods of isolation, separation and determination. Anal. Chim. Acta **26**, 514 – 527 (1962).

145. ZOBELL, C.: Geochemical aspects of the microbial modification of carbon compounds. In: Advances in Organic Geochemistry, p. 339–356 (U. COLOMBO and G. D. HOBSON, eds.). Pergamon Press, Inc. 1964.
146. JEWELL, D. M., J. P. YEVICH, and R. E. SNYDER: Basic nitrogen compounds in petroleum. Hydrocarbon Analysis. A.S.T.M. Special Technical Publications **389**, 363–384 (1965).
147. LATHAM, D. R., I. OKUNO, and W. E. HAINES: Non-basic nitrogen compounds in petroleum. Hydrocarbon Analysis. A.S.T.M. Special Technical Publications **389**, 385–398 (1965).
148. BALL, J. S., and H. T. RALL: Non-hydrocarbon components of a Californian petroleum. Proc. Am. Petrol. Inst. **41**, 128–145 (1962).
149. HAINES, W. E., G. L. COOK, and G. U. DINNEEN: Techniques for separating and identifying nitrogen compounds in petroleum and shale oil. Paper presented at the 7th Worlds Petrol. Congr. Mexico, April 2–8, 1967. Panel Discussion 23, Paper 1.
150. SNYDER, L. R.: Distribution of benzcarbazole isomers in petroleum as evidence for their biogenic origin. Nature **205**, 277 (1965).
151. JEWELL, D. M., and G. K. HARTUNG: Identification of nitrogen bases in heavy gas oil; chromatographic methods of separation. J. Chem. Eng. Data **9**, 297–304 (1964).
152. SNYDER, L. R.: Qualitative analysis of petroleum and related materials using linear elution adsorption chromatography. Anal. Chem. **38**, 1319–1328 (1966).
153. BEZINGER, N. N., M. A. ABDURAKHMANOV, and G. D. GAL'PERN: The nitrogen compounds of petroleum IV. Group separation of concentrates of nitrogen bases. Petroleum. Chem. U.S.S.R. **1**, 493–505 (1962).
154. SNYDER, L. R., and B. E. BUELL: Characterization and routine determination of certain non-basic nitrogen types in high-boiling petroleum distillates by means of linear elution adsorption chromatography. Anal. Chim. Acta **33**, 285–302 (1965).
155. BEZINGER, N. N., M. A. ABDURAKHMANOV, and G. D. GAL-PERN: The nitrogen compounds of petroleum I. The nature of the neutral nitrogen compounds in petroleum. Petroleum. Chem. U.S.S.R. **1**, 13–19 (1962).
156. COPELIN, E. C.: Identification of 2-quinolones in a Californian crude oil. Anal. Chem. **36**, 2274–2277 (1964).
157. HARTUNG, G. K., and D. M. JEWELL: Identification of nitriles in petroleum products. Complex formation as a method of isolation. Anal. Chim. Acta **27**, 219–232 (1962).
158. DRUSHEL, H. V., and A. L. SOMMERS: Isolation and identification of nitrogen compounds in petroleum. Anal. Chem. **38**, 19–28 (1966).
159. MARTIN, R. L.: Fast and sensitive method for determination of nitrogen. Selective detector for gas chromatography. Anal. Chem. **38**, 1209–1213 (1966).
160. SNYDER, L. R.: Aspects of linear elution chromatography for extraneous materials. Hydrocarbon Analysis. A.S.T.M. Special Technical Publications **389**, 399–416 (1965).
161. TREIBS, A.: Chlorophyll und Haeminderivate in Organischen Mineral-Stoffen. Angew. Chem. **49**, 682–686 (1936).
162. BAKER, E. W.: Mass spectrometric characterization of petroporphyrins. J. Am. Chem. Soc. **88**, 2311–2315 (1966).
163. ARCHIBALD, J. L., D. M. WALKER, K. B. SHAW, A. MARKOVAC, and S. F. MCDONALD: The synthesis of porphyrins derived from chlorobium chlorophylls. Can. J. Chem. **44**, 345–362 (1966).
164. BALL, J. S., and W. J. WENGER: Metal content of twenty-four petroleums. J. Chem. Eng. Data **5**, 553–557 (1960).
165. DUNNING, H. N., J. W. MOORE, H. BIEBER, and R. B. WILLIAMS: Porphyrin, nickel vanadium and nitrogen in petroleum. J. Chem. Eng. Data **5**, 546–548 (1960).
166. MORANDI, J. H., and J. B. JENSEN: Comparison of porphyrins from shale oil, oil shale and petroleum by absorption and mass spectroscopy. J. Chem. Eng. Data **11**, 81–88 (1966).
167. HODGSON, G. W., B. HITCHON, R. M. ELOFSON, B. L. BAKER, and E. PEAKE: Petroleum pigments from Recent fresh-water sediment. Geochim. Cosmochim. Acta **19**, 272–288 (1960).
168. BLUMER, M., and G. S. OMENN: Fossil porphyrins: uncomplexed chlorins in a Triassic sediment. Geochim. Cosmochim. Acta **25**, 81–90 (1961).
169. HODGSON, G. W., and B. L. BAKER: The role of porphyrins in the geochemistry of petroleum. Paper presented at the 7th World Petrol. Congr., Mexico, April 2–8, 1967. Panel Discussion 23, Discussion 23, Paper 4.

170. HOWE, M. W., and A. R. WILLIAMS: Classes of metallic complexes in petroleum. J. Chem. Eng. Data **5**, 106–110 (1960).
171. HYDEN, H. J.: Distribution of uranium and other metals in crude oils. U.S. Geol. Surv. Bull. No. 1100 (1961).
172. BAKER, B. L., and G. W. HODGSON: Magnesium in crude oils of Western Canada. Bull. Am. Assoc. Petrol. Geologists **43**, 472–476 (1959).
173. KATCHENKOV, S. M.: The correlation of petroleum by trace elements. Akad. Nauk. SSSR, Doklady **67**, 503–505 (1949).
174. GULYAEVA, L. A.: Vanadium and nickel in Devonian oils. Akad. Nauk. SSSR Inst. Neft. Trudy. **2**, 73–83 (1952).
175. SCOTT, J.: Trace metals in the McMurray oil sands and other Cretaceous reservoirs of Alberta. Trans. Can. Inst. Min. Met. **57**, 34–40 (1954).
176. HODGSON, G. W.: Vanadium nickel and iron trace metals in crude oils of Western Canada. Bull. Am. Assoc. Petrol. Geologists **38**, 2537–2554 (1954).
177. ERDMAN, J. G.: The molecular complex comprising heavy petroleum fractions. Hydrocarbon Analysis. A. S. T. M. Special Technical Publications **389**, 259–300 (1965).
178. CLERK, R. J., and M. J. O'NEAL: The mass spectrometer analysis of asphalt. A preliminary investigation. Symposium on Fundamental Nature of Asphalt, sponsored by Division of Pet. Chem. of the Am. Chem. Soc. New York, N. Y., September 11–16, 1960, vol. 5, A–5.
179. DICKIE, J. P., and T. F. YEN: Electron microscopic studies on petroleum asphaltenes. Paper presented before the Division of Petroleum Chemistry, Am. Chem. Soc. New York City Meeting, September 11–16, 1966. General Papers, vol. 11, p. 39–47.
180. WINNIFORD, R. S.: Evidence for the association of asphaltenes in dilute solutions. J. Inst. Petrol. **49**, 215–221 (1963).
181. FERRIS, S. W., E. P. BLACK, and J. B. CLELLAND: Aromatic structures in asphalt fractions. Preprints, Symposium on Characterization and Processing of Heavy Ends of Petroleum. Presented before the Division of Petroleum Chemistry. Am. Chem. Soc. Pittsburg, Pa. Meeting, March 23–26, 1966, vol. 11, No. 2, B 130–139.
182. SUGIHARA, J. M., T. OKADA, and J. F. BRANTHAVER: Reductive desulfurization on vanadium and metalloporphyrin contents of fractions from Boscan asphaltenes. J. Chem. Eng. Data **11**, 190–194 (1965).
183. WITHERSPOON, P. A.: Colloidal nature of petroleum. Transaction of the New York Academy of Sciences, Series 2, **24**, 344–361 (1962).
184. KROPOTKIN, P. N.: The geological conditions for the appearance of life on the earth, and the problems of petroleum genesis: Aspects of the origin of life, p. 63–73. Oxford: Pergamon Press, Inc. 1960.
185. ROBINSON, Sir R.: The origins of petroleum. Nature **212**, 1291–1295 (1966).
186. – Duplex origin of petroleum. Nature **199**, 113–114 (1963).
187. ECHLIN, P.: Origins of photosynthesis. Science J. 42–47, April (1966).
188. HENZE, H. R., and C. M. BLAIR: The number of isomeric hydrocarbons of the methane series. J. Am. Chem. Soc. **53**, 3077–3085 (1931).
189. CLARK, R. C., JR.: Occurrence of *n*-paraffinic hydrocarbons in nature. Technical Report No. 66–34. Woods Hole Oceanographic Institution 1966.
190. – Saturated hydrocarbons in marine plants and sediments. Part MSc Thesis. Massachusetts, Institute of Technology 1966.
191. NAGASAMPAGI, B. A., L. YANKOV, and S. DEV: Isolation and characterization of geranylgeraniol. Tetrahedron Letters, 189–192 (1967).
192. BURLINGAME, A. L., P. HAUG, T. BELSKY, and M. CALVIN: Occurrence of biogenic steranes and pentacyclic triterpanes in an Eocene shale (52 million years) and in an early Precambrian shale 2.7 billion years. A preliminary report. Proc. Natl. Acad. Sci. U.S. **54**, 1406–1412 (1965).
193. KJAER, A.: Distribution of sulfur compounds. In: Comparative Phytochemistry (T. SWAIN, ed.). New York: Academic Press, Inc. 1963.
194. – Distribution of sulfur compounds. In: Chemical Plant Taxonomy, p. 453–473. New York: Academic Press, Inc. 1963.
195a. SØRENSEN, N. A.: Chemical taxonomy of acetylenic compounds. In: Chemical Plant Taxonomy, p. 219–251 (T. SWAIN, ed.). New York: Academic Press, Inc. 1963.

195b. DEAN, F. M.: Naturally Occurring Oxygen Ring Compounds. London: Butterworth & Co. 1963.

196. PRIDMAN, J. B. (ed.): Methods in polyphenol chemistry. Proceedings of the plant phenolics. Group Symposium Oxford April 1963. Pergamon Press, Inc. 1964.

197. CARRUTHERS, W., and A. G. DOUGLAS: Constituents of high boiling petroleum distillates. Part IX. 3,4,6,7-tetramethyldibenzthiophene in a Kuwait Oil. J. Chem. Soc. 4077–4078 (1964).

198a. MAIR, B. J., and A. G. DOUGLAS: Sulfur: role in genesis of petroleum. Science **147**, 499–501 (1965).

198b. NELSON, W. L.: Sulfur content of oils throughout world. Oil and Gas **65** (7), 122–124 (1967).

199. FORT, R. C., JR., and P. VON R. SCHLEYER: Adamantane consequences of the diamondoid structure. Chem. Rev. **64**, 277–300 (1964).

SCHLEYER, P. VON R., G. J. GLEICHER, and C. A. CUPAS: Adamantane rearrangements. The isomerization of dihydrocedrene to 1-ethyl-3,4,7-trimethyladamantane. J. Org. Chem. **31**, 2014–2015 (1966).

SCHNEIDER, A., R. W. WARREN, and E. J. JANOSKI: Formation of perhydrophenalenes and polyalkyladamantanes by isomerization of tricyclic perhydroaromatics. J. Am. Chem. Soc. **86**, 5365–5367 (1964).

200. HÁLA, S., and S. LANDA: Isolation of tetracyclo ($6.3.1.0^{2,6}0^{5,10}$) dodecane and pentacyclo ($7.3.1.1^{4,12}0^{2,7}0^{6,11}$) tetradecane (diamantane) from petroleum. Angew. Chem. Intern. Ed. Engl. **5**, 1045–1046 (1966).

201. MURRAY, K. E.: A modified spinning band column for low pressure fractionation. J. Am. Oil Chemists Soc. **28**, 235–239 (1951).

202. BUDZIKIEWICZ, H., J. M. WILSON, and C. DJERASSI: Mass spectrometry in structural and stereochemical problems. XXXII. Pentacyclic triterpanes. J. Am. Chem. Soc. **85**, 3688–3699 (1963).

203. WULFSON, N. S., V. I. ZARETSKII, V. L. SADOVSKAYA, A. V. SEMENOVSKY, W. A. SMIT, and V. S. KUCHEROV: Mass spectrometry of steroid systems. IV. *cis-trans* isomerism of di and tricyclic model compounds. Tetrahedron **22**, 603–614 (1966).

204. LEHN, J. M., and G. OURISSON: Résonance Magnétique Nucleaire de Produits Naturels. 1. Introduction General, Triterpènes de la Séries du Lupane: Les Groupes Méthyles. Bull. Soc. Chim. France 1137–1142 (1962).

205. HILLS, I. R., E. V. WHITEHEAD, D. E. ANDERS, J. J. CUMMINS, and W. E. ROBINSON: An optically active triterpane, gammacerane in Green River, Colorado, Oil Shale bitumen. Chem. Comm. 752–754 (1966).

206. HILLS, I. R., G. W. SMITH, and E. V. WHITEHEAD: Optically active spirotriterpane in petroleum distillates. Nature **219**, 243–246 (1968).

CHAPTER 28

Fundamental Aspects of the Generation of Petroleum

E. EISMA and J. W. JURG

Koninklijke/Shell Exploratie en Produktie Laboratorium, Rijswijk, The Netherlands

Contents

I. Introduction

It is generally accepted that petroleum has been derived from the remains of plants and animals deposited together with fine-grained minerals at the bottom of the sea.

The great similarity between the chemical structures of compounds present in petroleum and those of compounds in living organisms strongly supports the idea that petroleum is of biogenic origin. Development of modern techniques such as gas chromatography, spectrometry and mass spectrometry has permitted the isolation and identification of many of these compounds. From the analysis of the extracts of Recent and ancient sediments and of petroleum it follows that petroleum is a product of the partial conversion of the original organic matter.

The purpose of this chapter is to give a summary of the fundamental aspects of the generation of petroleum. The following aspects will be discussed:

1. Arguments for the biogenic origin of petroleum.

2. The gradual transformation of the original biogenic material.

3. The mechanisms which may have played a role in the generation of petroleum.

4. Model experiments on the decomposition of behenic acid. These experiments suggest that radicals are the intermediates in the formation of petroleum.

5. Some possible implications of the role of radicals in the generation of petroleum.

II. The Biogenic Origin of Petroleum

The chemical evidence supporting the idea that petroleum has been derived from the remains of living organisms may be summarized as follows:

1. As early as 1835 BIOT [1] established that some petroleum fractions show optical activity [2, 3]. The synthesis of optically active organic compounds is considered to occur only in living organisms.

2. The $^{13}C/^{12}C$ ratios in petroleum resemble those of living organic matter more closely than those of atmospheric carbon dioxide or carbonates [4]. Petroleum, however, has a lower ^{13}C content than the organic matter of living organisms. Lipids have been found to be isotopically lighter than the bulk of the organic matter and this is in good agreement with the idea that lipids are very important precursors of petroleum [5].

3. The chemical structures of several compounds present in petroleum show a striking similarity to the structures of compounds present in living organisms, a few examples are discussed below.

In 1934 TREIBS [6] identified the red-colored porphyrin-vanadium complexes in petroleum and rock extracts and suggested that these compounds were derived from chlorophyll or hemin closely related to the prosthetic group of haemoglobin. GLEBOVSKAYA and VOLKENSHTEIN [7] identified a nickel porphyrin in oil shale. Figs. 1, 2, 3 and 4 show the structures of chlorophyll A, hemin and two petroleum porphyrins respectively; the strong structural similarity is obvious. ERDMAN and CORWIN [8], BAKER [9], THOMAS and BLUMER [10] and DUNNING [11] have contributed to a better understanding of the structures of the petroporphyrins. GRANSCH and EISMA [12] have used the presence of porphyrins in West-Venezuelan oils and rock extracts as a means of identifying the oil source rock in West Venezuela.

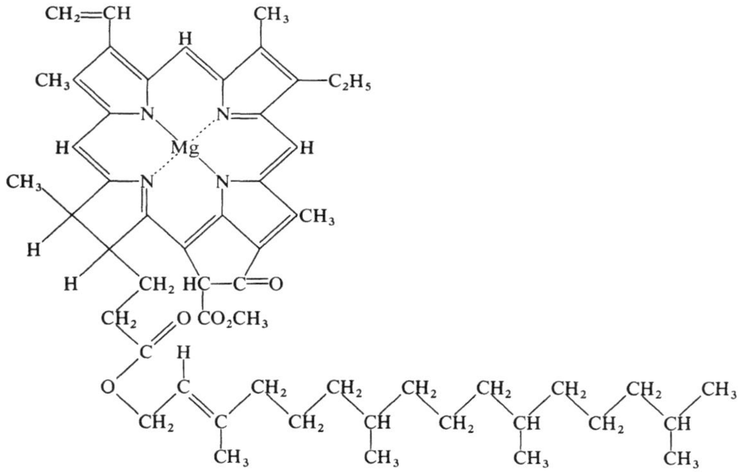

Fig. 1. Chlorophyll-a

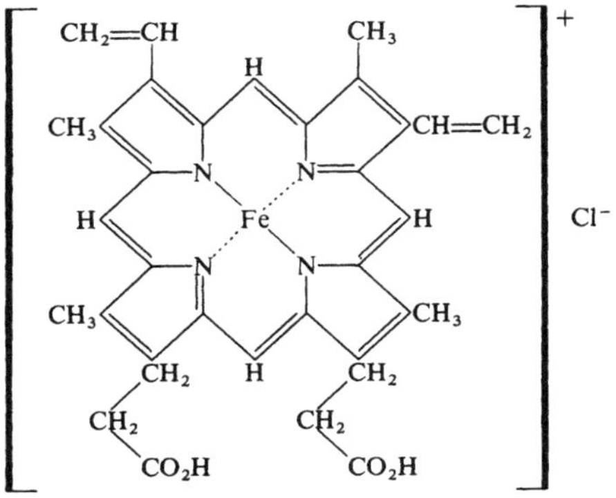

Fig. 2. Hemin

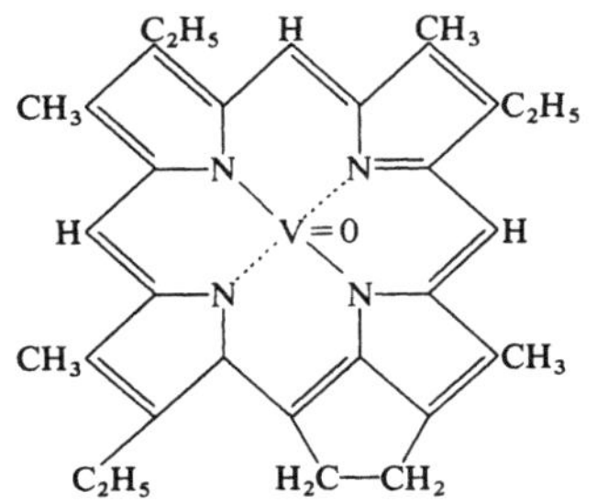

Fig. 3. Desoxo-phylloerythro-etioporphyrin. Vanadyl complex

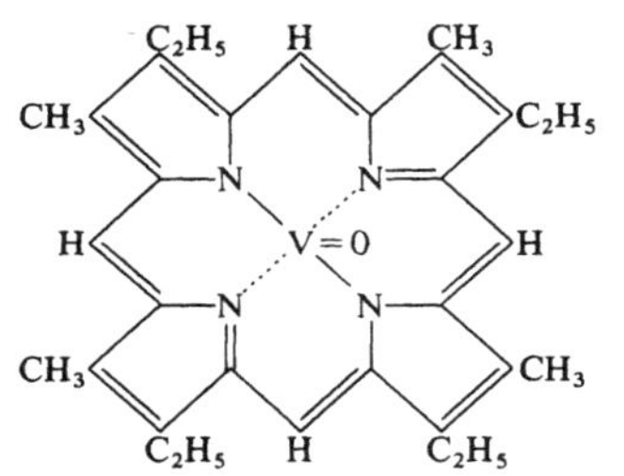

Fig. 4. Meso-etioporphyrin. Vanadyl complex

Phytol, a diterpenic alcohol, which comprises about 30 percent of the chlorophyll molecule is a postulated precursor of pristane and other isoprenoid saturated hydrocarbons present in petroleum and in extracts of sediments [13–15].

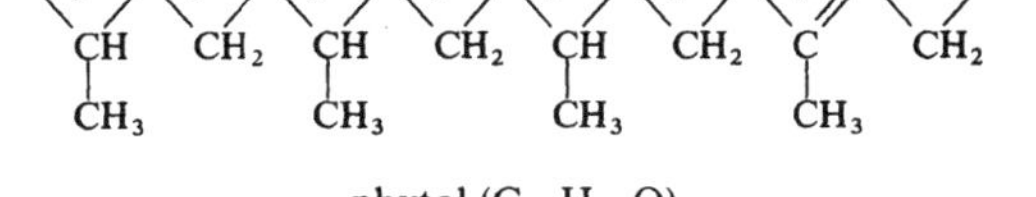

phytol ($C_{20}H_{40}O$)

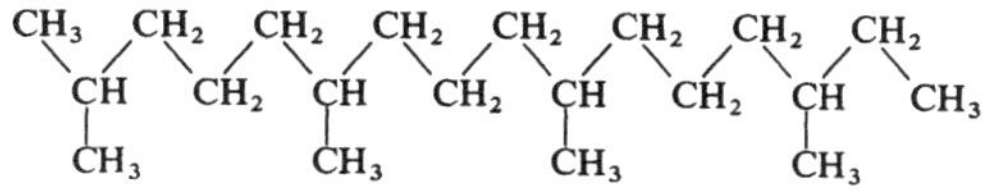

2,6,10,14-tetramethylhexadecane
(phytane, $C_{20}H_{42}$)

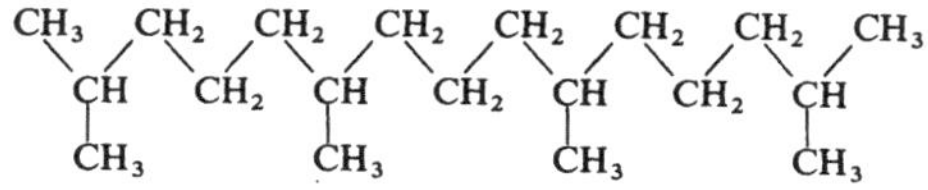

2,6,10,14-tetramethylpentadecane
(pristane, $C_{19}H_{40}$)

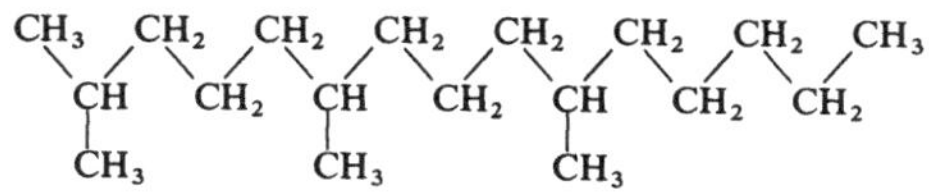

2,6,10-trimethylpentadecane
($C_{18}H_{38}$)

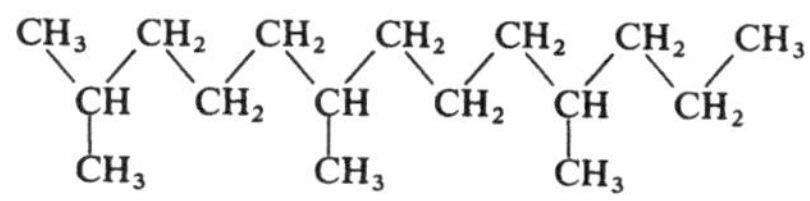

2,6,10-trimethyltridecane
($C_{16}H_{34}$)

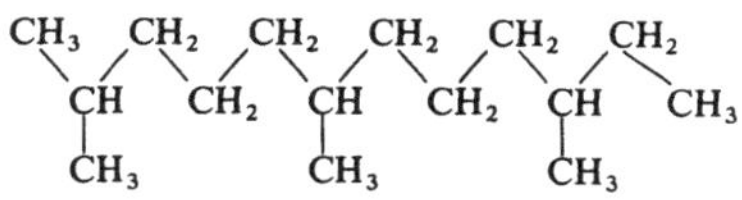

2,6,10-trimethyldodecane
($C_{15}H_{32}$)

Recently, McCarthy and Calvin [16] identified the C_{17} isoprenoid hydrocarbon from the Antrim shale in Michigan. The relative concentration of this compound, however, is low.

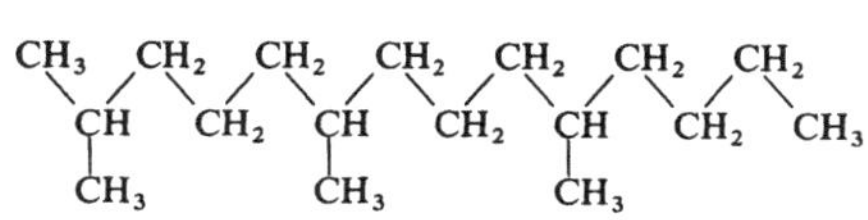

2,6,10-trimethyltetradecane
($C_{17}H_{36}$)

Fatty acids are major compounds in all living matter. In these acids, which occur mainly in fats and waxes, the number of carbon atoms may vary from 4 to 36. In general these acids have an even number of carbon atoms, though a few organisms are known to contain minor amounts of odd-numbered fatty acids [17].

Although a number of living organisms contain some paraffinic hydrocarbons [18, 19], it is generally believed that the bulk of the hydrocarbons, and in particular the n-alkanes present in petroleum, are derived from the fatty acids of living organisms [20–22]. The relation between the fatty acids of the living organism and the n-alkanes present in petroleum will be discussed in detail later.

These examples may suffice to show that important groups of compounds occurring in petroleum show a strong structural resemblance to lipids occurring in living organisms, thus forming the basic evidence for the theory of the organic origin of petroleum.

III. The Gradual Transformation of the Organic Matter

The crucial evidence that part of the organic matter is converted into hydrocarbons has been obtained from investigations of shallow and deep rock extracts.

Discussing the discovery of hydrocarbons in quite Recent sediments, WHITMORE [23] suggested that petroleum represents merely an accumulation of the hydrocarbons synthesised by living organisms. This idea was supported by MEINSCHEIN [24], SMITH [25] and SWAIN [26].

Later work by BRAY and EVANS [27] showed that in Recent sediments high-molecular-weight n-alkanes of odd carbon number predominated over those with an even carbon number, in contrast to corresponding fractions of petroleum, which showed in general no predominance of odd or even carbon number. The inference is that Recent sediments contain a relatively simple mixture of hydrocarbons but that they certainly do not contain the complex mixture that occurs in crude oil and in deep sediments. Similar conclusions result from ERDMAN's [28] work on the low-molecular-weight hydrocarbons. He recently reported that in Recent sediments only methane and n-heptane occur, whereas in ancient sediments and in petroleum all types of low-molecular-weight hydrocarbons are present. PHILIPPI [29] has shown that with increasing depth the distribution of high-molecular-weight naphthenes changes strongly and becomes very similar to that in crude oil. On these grounds a simple accumulation process, such as that proposed by WHITMORE [23], seems unlikely.

The absence of low-molecular-weight hydrocarbons and the relative small concentrations of even-numbered n-alkanes in Recent sediments and the presence of both series of compounds in ancient sediments and in petroleum indicates that a great number of new compounds have been formed from the original biological material. In other words, it is exceedingly probable that the formation of petroleum involves a considerable degree of conversion of the original source material. PHILIPPI [29] has estimated that the temperature for the generation of oil in Miocene sediments should be above 115° C.

IV. The Mechanisms Which May Have Played a Role in the Generation of Petroleum

The next point is how this conversion of organic matter has taken place. Several theories have been suggested which will now be reviewed.

A. The Radioactivity Theory

Various authors have considered the possibility that radiation resulting from the decay of radioactive elements in the sediments could promote the conversion of organic substances into petroleum.

LIND and BARDWELL [30] proved experimentally that, upon bombardment with *alpha*-particles, gaseous hydrocarbons split off hydrogen and give unsaturated oily products. COLOMBO [31] *et al.* obtained similar results with pure hydrocarbons and crude oils. SHEPPARD and BURTON [32] proposed that such a bombardment of fatty acids predominantly resulted in dehydrogenation and decarboxylation.

Since the principal result of the action of radioactivity on hydrocarbons and fatty acids is the formation of hydrogen and unsaturated compounds, a serious objection to the radioactivity theory is the usual absence of hydrogen in natural gas. Moreover, since each *alpha*-particle should have been converted into a helium atom, another objection to the theory is the absence of helium in the vicinity of oil accumulations. Last but not least there is no apparent relation between the presence of radioactive materials and the abundance of petroleum. FAN and KISTER [33] mention that in the sedimentary rocks of the Lower Mississippian Kinderhookian, which is extremely rich in organic material and which is also a very radioactive sedimentary rock, apparently no formation of petroleum has taken place.

B. The Cracking Hypothesis

The hypothesis that low temperature cracking reactions are responsible for the generation of hydrocarbons, as advanced by ENGLER [34], has been supported by BERL [35]. As early as 1867 WARREN and STORER [36] showed that a "hydrocarbon naphtha" could be obtained from the destructive distillation of lime soaps. Similar results have been obtained by KRÄMER [37] and by HÖFER [38].

SEYER [39], studying the results of the kinetic constants for thermal cracking, concluded that within geological times this type of reaction could not have produced petroleum. MONTGOMERY [40] suggested that naturally occurring catalysts, e.g., clay minerals, could accelerate natural hydrocarbon production in such a way that the reaction times become geologically acceptable. The catalysing effect of fine-grained inorganic minerals on the generation of hydrocarbons from the sedimentary organic matter has been extensively discussed by BROOKS [41] and he has suggested that carbonium ions or "catalytic cracking" play an important role in the hydrocarbon production.

BOGOMOLOV [42] and BEDOV [43] studied the influence of silica/alumina catalysts on the decomposition of unsaturated fatty acids. BEDOV reported that in the reaction products no n-alkanes could be identified. The formation of long-chain normal alkanes and low-molecular-weight hydrocarbons as a result of the heating of behenic acid in the presence of clay and water has been reported by JURG and EISMA [44].

From a study of the influence of clay minerals on the transformation of sedimentary organic matter KLUBOVA [45] concluded that the catalytic action of the clay minerals is not due to their structure and chemical composition but to their highly dispersed state.

Thus both "thermal" and "catalytic" cracking have been suggested in the literature as the process by which petroleum is generated from the original source material. Both types of cracking have been investigated extensively in model experiments. The cracking of pure organic compounds has been studied in terms of carbon number distributions and structures of the cracked fragments. On the basis of this work cracking systems are assigned to two fundamental classes, each of which is described by a set of characteristic reactions. Correspondingly, two types of reaction mechanisms are proposed, one a radical mechanism (thermal cracking) based on the work by RICE and KOSSIAKOFF [46] and the other a carbonium-ion mechanism (catalytic cracking) based on the work by GREENSFELDER [47] *et al.* and THOMAS [48].

C. "Thermal Cracking" – A Radical Mechanism

To exemplify this type of mechanism, the cracking of a normal paraffin will be discussed.

The cracking of the normal paraffin is initiated by the loss of a hydrogen atom, due, for example, to a collision. The resulting hydrocarbon radical may immediately crack or it may undergo radical isomerisation. Radical isomerization involves a change in the position of a hydrogen atom, resulting in an energetically more favorable radical. Cracking of either the original or the isomerized radical takes place at the carbon-carbon bond located in the β-position to the carbon atom lacking the hydrogen atom. This cracking produces a primary radical and an α-olefin. The product radical can react in the following ways:

1. It may crack instantaneously, giving ethylene and a new primary radical (β-scission).
2. Radical isomerisation may take place prior to cracking.
3. The radical may abstract hydrogen from a neutral molecule resulting in a n-paraffin and a new radical.

Hence the major products of a n-paraffin subjected to thermal cracking are n-paraffins, α-olefins and ethylene.

The reaction mechanism described is shown in the following scheme:

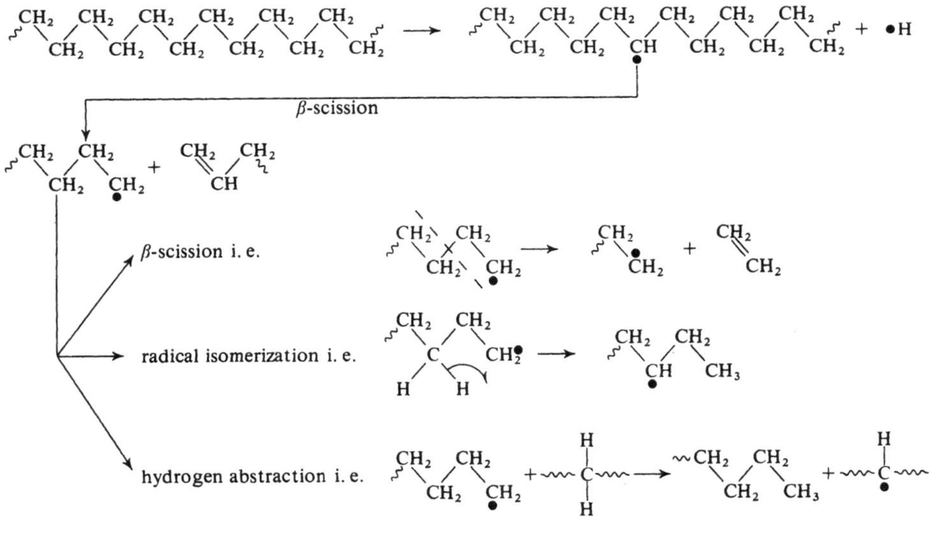

This type of cracking takes place at a relatively high temperature as a homogeneous process. However, it has also been observed that its rate may be increased by the presence of dispersed material such as pure alumina. For a discussion of this phenomenon we refer the reader to GREENSFELDER and TUNG and MCININCH [49].

Where in the following discussion reference is made to reactions of the radical type, it must be understood that these belong to the above class of cracking reactions. It must be stressed that absence of skeletal isomerization is characteristic of this process.

D. "Catalytic Cracking" – A Carbonium Ion Mechanism

In this reaction catalysts are essential participants in the cracking process. The catalysts used to promote the cracking are of the acidic type (e.g., clay and silica/alumina) in contrast to the non-acidic catalysts (e.g., pure alumina) that promote thermal cracking. The acidity of the catalyst refers particularly to its "proton availability". This means that protons are available for reaction with the hydrocarbons undergoing cracking.

The reaction of the protons with the hydrocarbon results in the abstraction of a hydride ion from it; the result is a carbonium ion. Carbonium ions undergo several rearrangements and reactions, the most important of which are:

1. β-splitting, resulting in a new carbonium ion and an olefin.
2. The shift of a methyl group.
3. Hydride ion abstraction from neutral molecules.
4. The shift of a hydrogen atom.

The reactions are illustrated as follows:

Formation of a carbonium ion

1. *β-splitting*

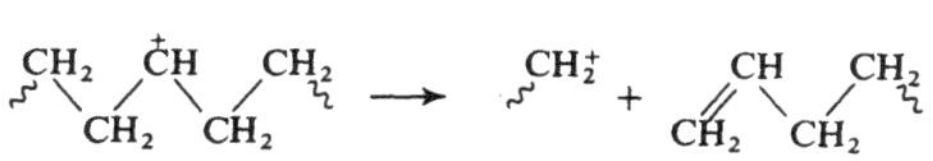

2. *Methyl shift*

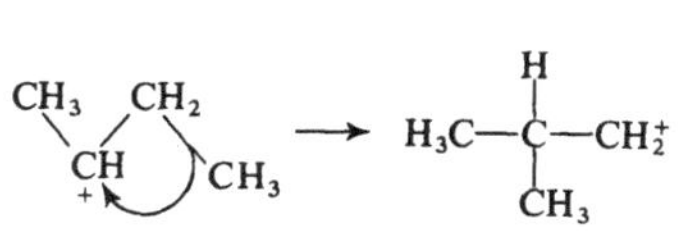

3. *Hydride ion abstraction*

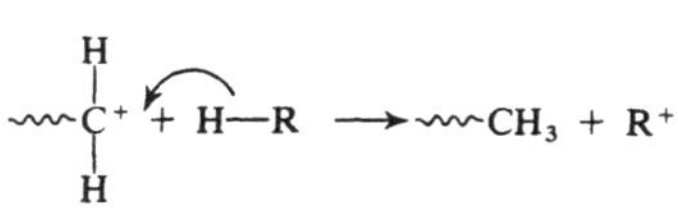

4. *Hydrogen shift*

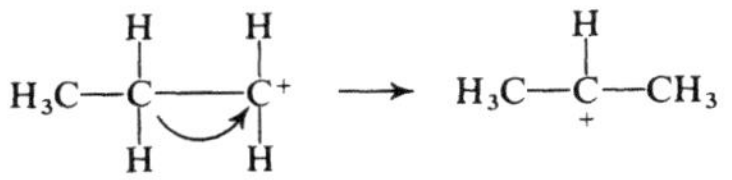

GREENSFELDER *et al.* obtained data concerning the heats of cracking of primary and secondary carbonium ions.

A.

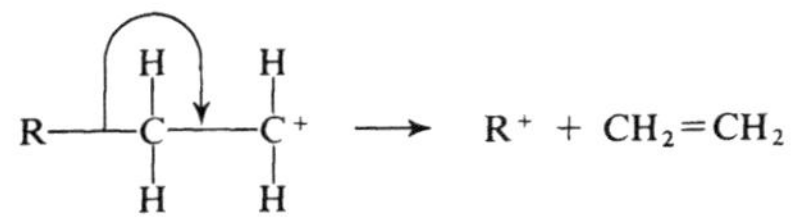

B.

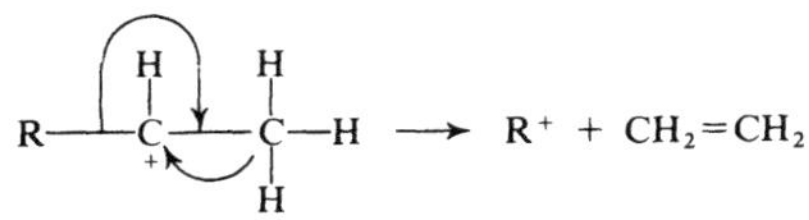

The values of ΔH_{298} in Kcal/mol are given in Table 1.

Table 1. *Heats of cracking of various carbonium ions* [68]

Ion R^+	Value of ΔH_{298}	
	Reaction A	Reaction B
CH_3^+	69.5	85.5
$C_2H_5^+$	35.0	61.0
n-$C_3H_7^+$	22.5	47.5
sec $C_3H_7^+$	8.5	33.5
tert $C_4H_9^+$	−7.5	17.5

The most favorable reaction is thus one in which a primary ion is cracked to yield a tertiary ion with a ΔH_{298} value of −7.5 Kcal/mol. It is clear from these data that methyl or ethyl ions are unlikely fragments in catalytic cracking; reactions producing the secondary propyl and the tertiary butyl ions are energetically more favorable.

In connection with the generation of petroleum both types of cracking reaction have been suggested, the "catalytic cracking" by BROOKS [41] and the "thermal cracking" by KLUBOVA [45] and by JURG and EISMA [44].

V. Model Experiments on the Decomposition of Behenic Acid

The reasons for this investigation and its results are discussed in the following section.

In many young sediments that have never been buried to great depth, there is a predominance of n-alkanes with an odd number of carbon atoms over those with

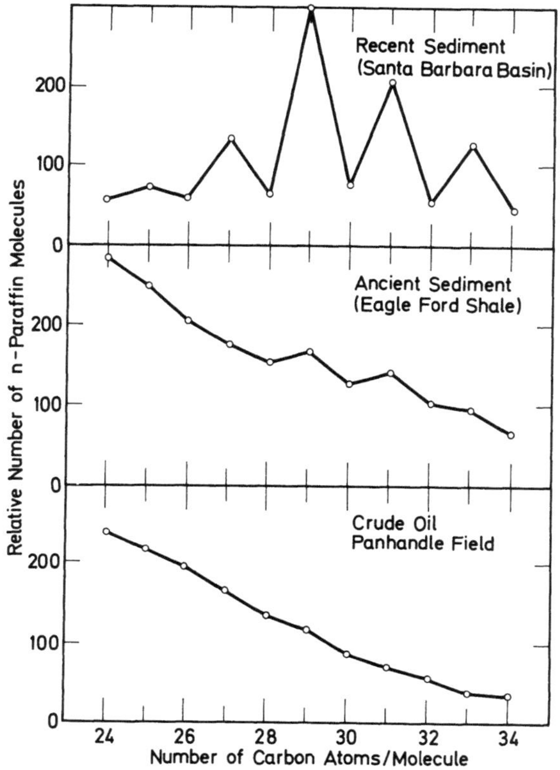

Fig. 5. n-Paraffin distributions for a Recent sediment, an ancient sediment and a crude oil. (After COOPER and BRAY [50].)

an even number of carbon atoms, in the range of 27 to 37 carbon atoms. COOPER and BRAY [50] investigated the distribution of n-alkanes in various Recent and ancient sediments and in petroleum. Their results, some of which are presented in Fig. 5, show that the predominance of odd-numbered n-alkanes decreases with increasing depth and age of the sediments and that in petroleum n-alkanes are smoothly distributed. As already mentioned, it is often suggested that these n-alkanes are derived from the fatty acids initially present in the living organism. These fatty acids, however, have an even number of carbon atoms. Mere decarboxylation of these fatty acids would yield n-alkanes with an odd number of carbon atoms, whereas n-alkanes with an odd and an even number of carbon atoms are found in deep sediments. If therefore the n-alkanes are derived from fatty acids this cannot be as a result of a simple decarboxylation. COOPER and BRAY [50] have also investigated the distribution of the fatty acids present in Recent and in ancient sediments and in petroleum reservoir waters. As shown in Fig. 6 the predominance of fatty acids with an even number of carbon atoms decreases with increasing age and depth of the sediment.

The tendency for the odd predominance of the n-alkanes and the even predominance of the fatty acids to decrease with increasing depth and age of the sediments in which they are found and the smooth distribution of these compounds

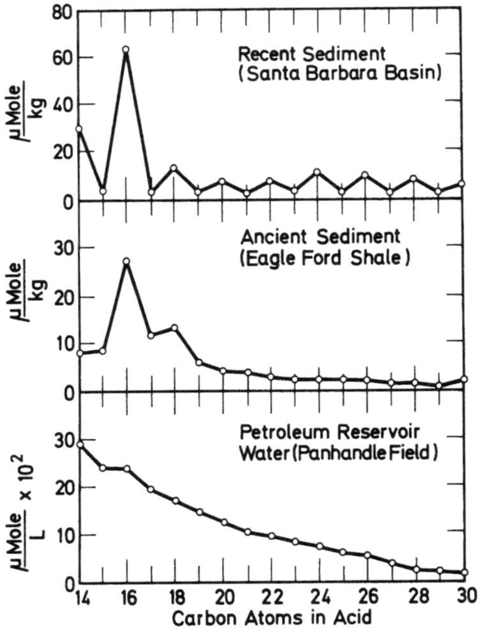

Fig. 6. Comparison of the distributions of fatty acids in a Recent sediment, an ancient sediment, and in water from a petroleum reservoir. (After COOPER and BRAY [50].)

in petroleum and petroleum reservoir waters appear consistent with the generation of a petroleum-like mixture of n-alkanes and fatty acids in the sediment. The parallelism between the disappearance of the odd predominance of the n-alkanes and the even predominance of the fatty acids suggests that there is a relationship between the processes by which these n-alkanes and fatty acids are formed in the sediment.

COOPER and BRAY [50] proposed a mechanism by which odd- and even-numbered n-alkanes and fatty acids can be derived from even-numbered fatty acids. According to them the acids lose CO_2 by decarboxylation, to form an intermediate radical which reacts to give two products, an n-alkane and a fatty acid. Each of these products would have one carbon atom less than the original fatty acid. The acid produced would then undergo the same reactions to form a new acid and a new radical and so on.

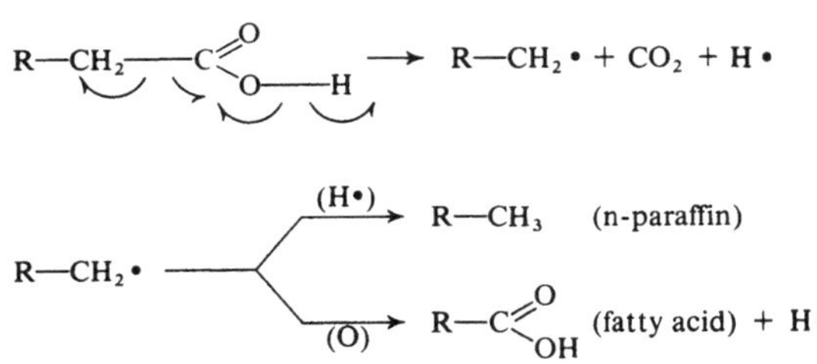

The conversion of the intermediate alkyl radical, $RCH_2^{\bullet}$, to a fatty acid involves an oxidation step. It is not clear how such an oxidation can take place under the conditions prevailing in the sediment.

LAWLER and ROBINSON [51] suggested that even-numbered fatty acids produce odd-numbered fatty acids by two opposing reactions. In the low-molecular-weight range a gain of one carbon atom and in the high-molecular-weight range a loss of one carbon atom would produce odd-numbered fatty acids. These acids can, after decarboxylation, yield n-alkanes.

Experimental evidence is not available to support the suggestions of COOPER and BRAY [50] and of LAWLER and ROBINSON [51].

JURG and EISMA [52] suggested, on the basis of model experiments, a reaction mechanism in which only radicals are the intermediates. They heated behenic acid in the presence of clay and in the presence of clay and water. Tables 2 and 3

Table 2. *Amounts of low-molecular-weight hydrocarbons (μmol) generated by heating 1 g behenic acid in the presence of 2.5 g clay at 200° C for various times* [52]

	94 h	283 h	330 h	976 h	1,848 h
Ethane + ethene	0.03	0.07	–	0.14	0.14
Propane	0.32	0.84	1.00	1.97	2.95
Propene	0.58	0.75	0.64	0.41	0.34
Isobutane	3.04	4.28	3.94	5.94	10.14
n-butane	0.08	0.19	0.23	0.45	0.93
Isobutene + 1-butene	0.13	0.09	0.06	0.04	0.05
2-butene-*trans*	0.08	0.12	0.11	0.09	0.10
2-butene-*cis*	0.05	0.06	0.07	0.05	0.07
Isopentane	1.45	2.62	4.13	4.62	8.27
n-pentane	0.17	0.14	0.28	0.36	0.51
1-pentene	0.01	0.01	0.01	0.01	0.01
2-Me-1-butene	0.03	0.06	0.03	0.03	0.01
2-pentene-*trans*	0.06	0.02	0.09	0.04	0.04
2-pentene-*cis*	0.02	0.11	0.03	0.02	0.01
2-Me-2-butene	0.15	0.10	0.14	0.07	0.05
2-Me-pentane	0.66	1.23	2.41	3.69	4.16
3-Me-pentane	0.21	0.43	0.88	1.31	1.47
n-hexane	0.08	0.12	0.22	0.31	0.39
Total	7.1	11.2	14.2	19.5	29.6

Table 3. *Amounts of low-molecular-weight hydrocarbons (μmol) generated by heating 1 g behenic acid in the presence of 2.5 g clay and 7.5 g water for various times at various temperatures* [52]

Heating-temperature	200° C	250° C	265° C	265° C	±275° C
Heating-time	75 h	275 h	625 h	1300 h	330 h
Ethane + ethene	0.06	0.04	0.21	0.26	0.24
Propane	0.01	0.03	0.08	0.12	0.14
Propene	0.07	0.08	–	0.25	0.16
Isobutane	–	0.02	0.03	0.04	0.03
n-butane	0.01	0.02	0.07	0.09	0.13
Isobutene + 1-butene	0.01	0.03	0.06	0.08	0.06
2-butene-*trans*	–	0.02	0.07	0.10	0.08
2-butene-*cis*	–	0.01	0.05	0.06	0.06
Isopentane	–	0.01	0.01	0.03	0.02
n-pentane	0.02	0.05	0.06	0.09	0.13
1-pentene	–	0.01	0.01	0.09	–

show the amounts of the various low-molecular-weight hydrocarbons generated during heating. Apart from these low-molecular-weight hydrocarbons, n-alkanes with 15 to 34 and fatty acids with 15 to 25 carbon atoms were identified in the reaction mixture.

The conclusion from these experiments is that, because skeletal isomerization is prominent in the experiments without water, carbonium ions act as intermediates in the generation of low-molecular-weight hydrocarbons. Skeletal isomerization is much less pronounced in the experiments where water is present and radicals are likely to be the intermediates in the formation of the low-molecular-weight hydrocarbons.

From two series of experiments in which the behenic acid was heated in the presence of clay, with or without water, at various temperatures (EISMA and JURG [53]), it was concluded that in both series radicals were the intermediates in the formation of the long-chain n-alkanes and fatty acids. Table 4 shows various results of these experiments. The alteration of the whole distribution pattern of the various compounds above 275° C indicates that above that temperature the reaction becomes more complex.

Table 4. *Various data derived from experiments in which behenic acid was heated for 283 hours in the presence of clay at various temperatures* [53]

	200° C	240° C	250° C	275° C	300° C
Total amount of low-molecular-weight hydrocarbons generated (mg) (up to n-C_7)	0.9	2.15	2.84	5.15	15.13
Percentage of n-alkanes	11.0	14.5	15.1	17.9	28.0
Percentage of unsaturates	9.6	4.5	3.9	3.6	7.9
Relative amount of C_2 hydrocarbons	1.6	1.7	2.3	6.8	29
Relative amount of C_3 hydrocarbons	19	26	27	27	41
Relative amount of C_4 hydrocarbons	100	100	100	100	100
Relative amount of C_5 hydrocarbons	62	81	88	94	92
Relative amount of C_6 hydrocarbons	42	57	63	75	74
Relative amount of C_7 hydrocarbons	37	49	51	57	58

From the Arrhenius plot as presented in Fig. 7, it follows that above 275° C a second reaction comes to the fore. The conclusion that this second reaction is a radical reaction is based on the following facts:

1. The relative amount of C_2 hydrocarbons strongly increases.
2. The percentage of n-alkanes strongly increases.
3. The percentage of unsaturated hydrocarbons strongly increases.
4. The predominant reaction has a high activation energy.

In view of these facts the following reaction scheme was developed, which describes the mechanism by which odd- and even-numbered n-alkanes and fatty acids can be formed from an even-numbered fatty acid initially present in the living organism.

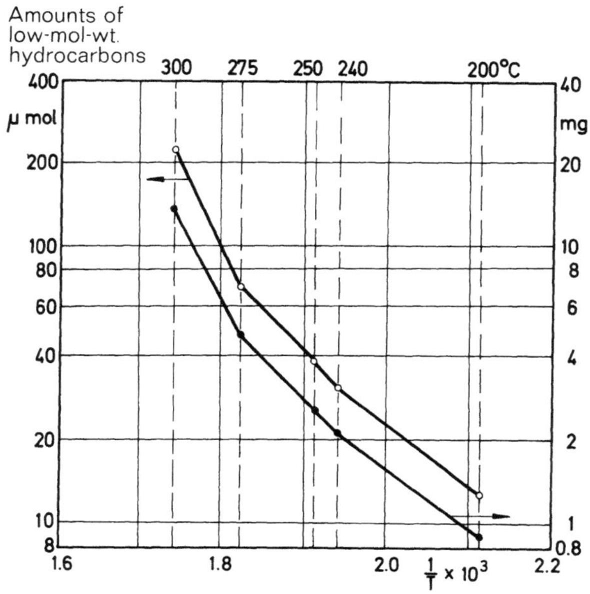

Fig. 7. The Arrhenius plot of the amounts of low-molecular-weight hydrocarbons generated [53] with respect to temperature. Amounts are expressed as micromoles (upper line) and milligrams (lower line)

Reaction scheme

Initiation:	RCOOH	⟶	$\cdot R$
Propagation:	$\cdot R + RCOOH$	⟶	$\cdot (R)COOH + RH$
β-scission:	$\cdot (R)COOH$	↗	R^1 (α-olefin) $+ \cdot R^2COOH$
		↘	$\cdot R^3$ $+ R^4COOH$ (ω-unsaturated fatty acid)
Termination:	$\cdot R + \cdot R^2COOH$	⟶	$(R + R^2)COOH$
	$\cdot R + \cdot R^3$	⟶	$(R + R^3)$

The radicals $\cdot R^3$ and $\cdot R^2COOH$ may also abstract a hydrogen atom from another molecule to form the short-chain n-alkanes and fatty acids. The extended reaction scheme is given in references 52 and 53.

The initiation step of this reaction is given by the decarboxylation of the fatty acid resulting in an alkyl radical. This intermediate will react with the original fatty acid, which is present in a relatively high concentration, to give a n-alkane and a secondary radical of the fatty acid. This secondary radical can split up by β-scission into four products: an α-olefin and a primary radical of a fatty acid or a primary alkyl radical and an ω-unsaturated fatty acid.

The reaction scheme proposed is supported by the following arguments:

1. The C_{21} n-alkane is always the predominating n-alkane formed during the reaction of the behenic acid. Fig. 8 shows the relative amounts of the high-

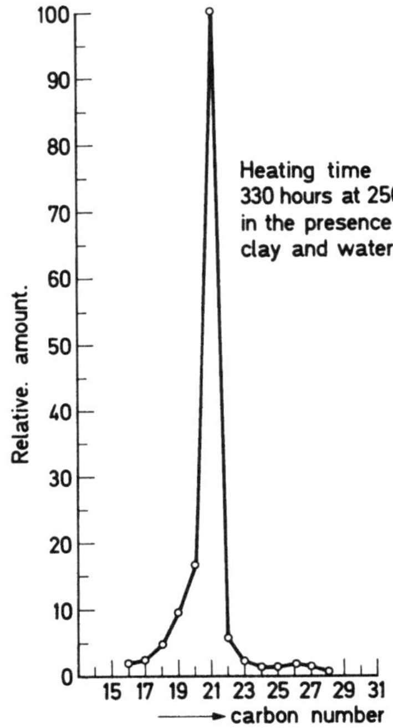

Fig. 8. Relative amounts of n-alkanes generated from behenic acid heated in the presence of clay and water [52]

molecular-weight n-alkanes generated from behenic acid which was heated for 330 hours at 250° C in the presence of clay and water. A similar distribution pattern was obtained when the behenic acid was heated in the presence of clay only.

This distribution indicates that decarboxylation is an important step in this mechanism. That decarboxylation is essential for the generation of n-alkanes with a carbon chain longer than the original acid followed from an experiment in which n-hexadecane was used instead of the behenic acid. After heating n-hexadecane in the presence of clay we were not able to find with our present analytical equipment n-alkanes or other hydrocarbons with a carbon chain longer than n-C_{16}. This implies that the possibility of the formation of hydrocarbons with a carbon chain longer than that of the starting material is less in the case when a n-alkane is used than when a fatty acid is used.

2. In general the absolute amount of unsaturated low-molecular-weight hydrocarbons decreases with increasing heating time, suggesting that unsaturates are intermediates.

3. A radical reaction can be controlled by adding either an initiator or an inhibitor to the reaction mixture. We have chosen to add an initiator, and have used as such 2,2′-azopropane, which generates the $C_3H_7^{\bullet}$ radical.

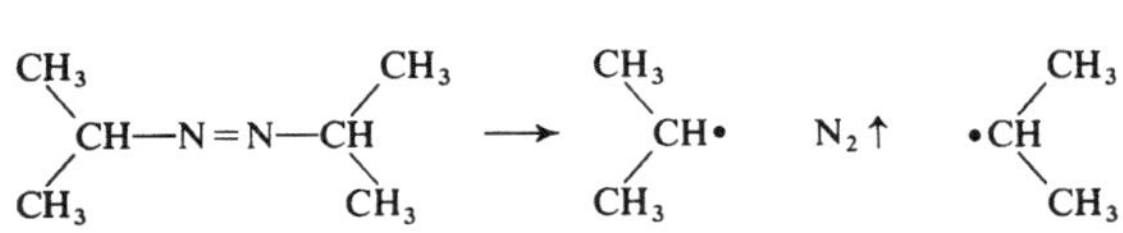

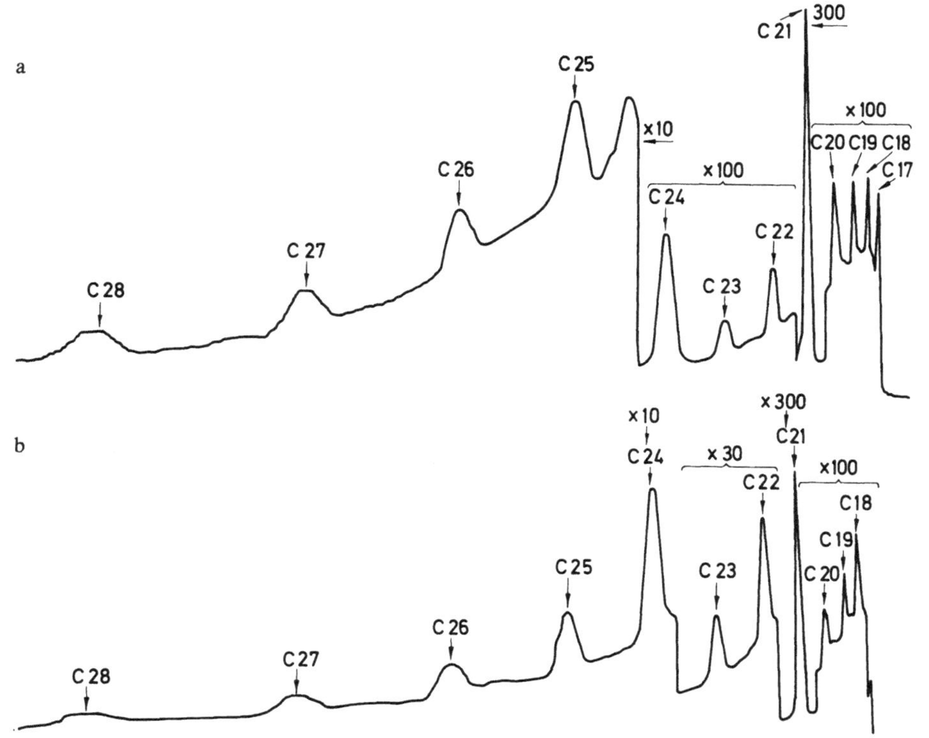

Fig. 9. a Gas chromatogram of the n-alkanes generated during the heating of behenic acid in the presence of 2,2'-azopropane for 300 hours at 200° C. (Peak between n-C_{24} and n-C_{25} not identified.) b Gas chromatogram of the n-alkanes generated during the heating of behenic acid in the presence of clay and water for 300 hours at 200° C

The initiator generates $C_3H_7^{\bullet}$ radicals which initiates the formation of the radicals of the behenic acid.

Fig. 9 gives the gas chromatograms of the n-alkanes generated during the heating of behenic acid in the presence of clay and water or 25 mg 2,2'-azopropane.

In the experiment in which we used the 2,2'-azopropane, the C_{24} n-alkane predominates. This can be attributed to the combination of the C_3 radical derived from the 2,2'-azopropane and the C_{21} radical from the behenic acid.

$$\underset{\text{iso-propyl radical}}{CH_3\text{—}\dot{C}H\text{—}CH_3} \longrightarrow \underset{\text{n-propyl radical}}{CH_3\text{—}CH_2\text{—}CH_2^{\bullet}}$$

$$CH_3\text{—}CH_2\text{—}CH_2^{\bullet} + \text{n-}C_{21}^{\bullet} \longrightarrow \text{n-}C_{24}$$

These experiments indicate that radicals are intermediates in the formation of the long-chain n-alkanes and the fatty acids found as a result of heating behenic acid in the presence of clay and water.

From the previously described experiments it is clear that water largely prevents carbonium-ion reactions. From this it follows that the mechanism suggested by Brooks [41] does not seem very likely and that the clay cannot act as an "acid catalyst". It seems more likely that the highly dispersed state of the clay influences the reaction as proposed by KLUBOVA [45] and that radicals are the intermediates in the formation of petroleum from the original organic matter.

VI. Experimental Procedures

The behenic acid ($C_{21}H_{43}COOH$) was first purified by distillation and several recrystallizations of the methyl ester. The fatty acid recovered after hydrolysis was recrystallized several times. Blank runs with the purified acid did not show the presence of either hydrocarbons or any other fatty acid.

As the clay constituent of the mixture we selected kaolinite; this clay did not contain any detectable amounts of hydrocarbons or fatty acids. The content of organic carbon was 0.04 percent.

Two and a half grams of the clay (air-dry) was thoroughly mixed with one gram of the behenic acid; the mixture was then placed in glass ampoules (Pyrex), sealed off under vacuum and heated. Our method of isolating the reaction products is outlined in the following diagram:

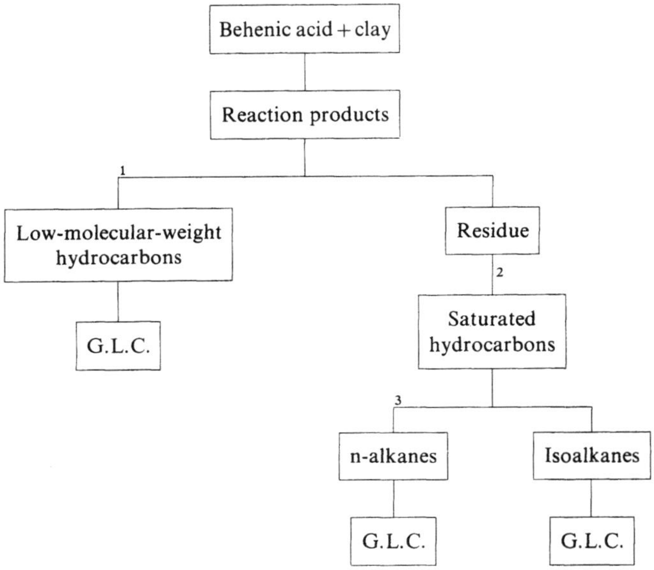

1. The ampoules were opened in an atmosphere of hydrogen. The volatile components were stripped off and frozen out in liquid nitrogen. The low-molecular-weight hydrocarbons of this fraction were analysed by gas chromatography.

2. The residue was extracted with n-pentane. The extract was chromatographed over a column of SiO_2, Al_2O_3, in order to separate the saturated hydrocarbons from the other compounds.

3. The saturated hydrocarbons were separated by urea-adduction into a fraction rich in n-alkanes and a fraction containing the branched-chain and the cyclic hydrocarbons. These fractions were analyzed by gas chromatography. For the urea-adduction at least 4 mg of the saturated hydrocarbon-fraction is required.

When this amount was not available the G.L.C. analysis was done with the saturated hydrocarbon fraction itself.

Our method of isolating the fatty acids from the reaction mixture is outlined in the following diagram:

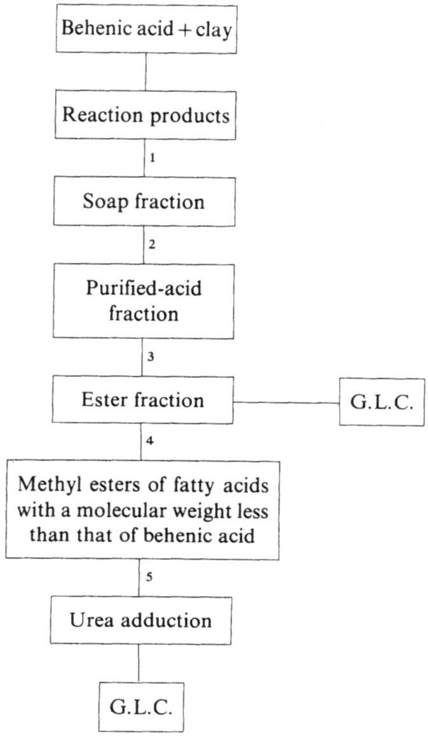

1. The reaction products were treated with alcoholic KOH in order to obtain the soaps.

2. The soaps were acidified with HCl. The fatty acids recovered were purified by chromatography.

3. The purified acids were esterified with BF_3/CH_3OH.

4. Because the methyl ester of behenic acid predominated very strongly we preconcentrated the methyl esters of the fatty acids with carbon chains smaller than the behenic acid by means of preparative-scale G.L.C.

5. These preconcentrated acids were purified by urea-adduction and analyzed by gas chromatography.

VII. Other Possible Implications of the Role of Radicals in the Generation of Petroleum

A. Isoprenoid Hydrocarbons

It has already been mentioned that the saturated isoprenoid hydrocarbons might be related to phytol. BENDORAITIS [13] suggested that during diagenesis the phytol group was split off from chlorophyll by hydrolysis or enzymatically by chlorophyllase. Subsequent reactions, oxidation of the phytol to an acid followed by decarboxylation or reduction of phytol to phytane may have taken place.

BLUMER *et al.* [54] reported that the body fat of copepods of the genus *Calanus* contains from 1 to 3 percent of pristane and suggested that the hydrocarbon might be derived from the chlorophyll of their phytoplankton diet. This author [55] also reported the presence of four isomeric phytadienes in zooplankton of the Gulf of Maine. These hydrocarbons may of course also contribute to the amount of pristane found in petroleum.

It is suggested by MCCARTHY and CALVIN [16] that phytol is the source of the isoprenoid hydrocarbons found in petroleum and in rock extracts. A diagenetic scheme has been postulated by them. The C_{19}, C_{18} and the C_{16} isoprenoid hydrocarbons would require only one carbon-carbon bond cleavage.

The C_{17} isoprenoid hydrocarbon, however, can only be formed by the cleavage of two carbon-carbon bonds located at the same carbon atom:

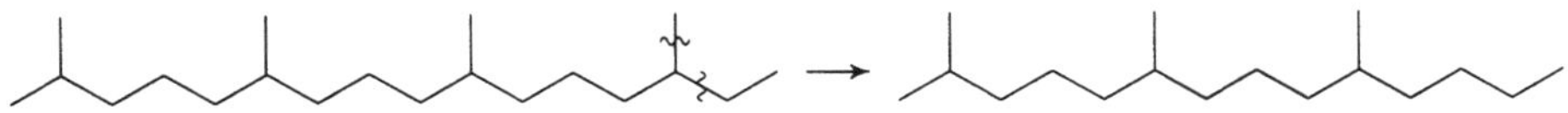

2,6,10-trimethyltetradecane

or, as these authors have also suggested, from squalane. They remark, however, that squalane is possibly a minor source of the isoprenoid hydrocarbons from petroleum.

From the experiments on behenic acid it followed that after heating the acid in the presence of clay and water, n-alkanes with a longer carbon chain than the original acid were formed. If a similar reaction mechanism is applied to the decomposition of an isoprenoid hydrocarbon it is likely that recombination reactions also play a role. The formation of the isoprenoid C_{17} could for instance be represented by the following reactions:

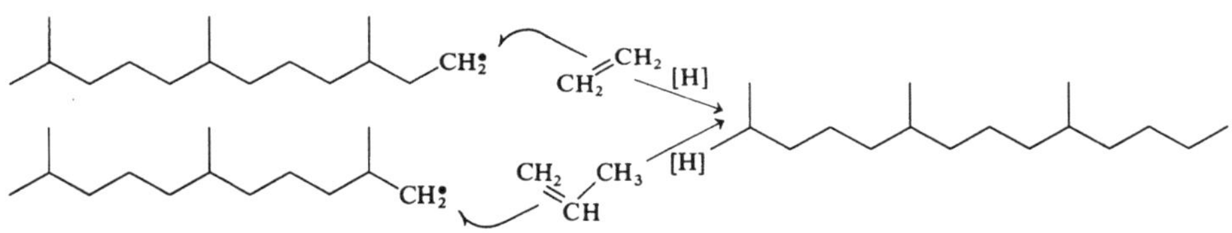

In any case, α-olefins (alk-1-enes) are formed in a relatively high concentration in "thermal cracking" reactions, and from the experiments on behenic acid it follows that the unsaturates decrease with increasing heating time.

MCCARTHY and CALVIN [16] mentioned that the C_{17} isoprenoid hydrocarbon in the Antrim shale was present in a very low concentration, and that this hydrocarbon can only be formed by a cleavage of two carbon-carbon bonds located at the same carbon atom. The reason why the concentration of the C_{17} isoprenoid hydrocarbon is low may be that it can only be formed by a double cleavage and by recombination, while the others, such as C_{19}, C_{18}, C_{16}, etc., can be formed by a single carbon-carbon bond cleavage and by recombinations.

In any case it must be realized that in the formation of isoprenoid hydrocarbons recombination of primary radicals with α-olefins may also be an important process.

B. Porphyrins

Since TREIBS' [6] publications, the structure of porphyrins in petroleum has been the subject of many investigations. He suggested a scheme for the degradation of chlorophyll to desoxo-phylloerythro-etioporphyrin (D.P.E.P.). A degradation according to this scheme yields only one product, viz. D.P.E.P.

Mass-spectrometric studies show that the porphyrins of petroleum are extremely complex mixtures. DEAN and WHITEHEAD [56], THOMAS and BLUMER [10] and BAKER [9] reported that the molecular weight of part of the porphyrins was higher than could be expected from TREIBS' scheme. BAKER [9] suggested that a transalkylation reaction via a radical or an ionic mechanism could help to explain the variations in molecular weight of the different types of porphyrin observed in petroleum. In view of the amount of water present in the sediment, it seems more likely that a radical reaction takes place. If somewhere on the porphyrin nucleus a radical is formed, recombination with another radical or with, say, an α-olefin, may yield a product with a higher molecular weight than could be expected by straight forward degradation as that suggested by TREIBS.

C. Low-Molecular-Weight Hydrocarbons

The low-molecular-weight aromatic and aliphatic hydrocarbons (with the exception of CH_4) which occur in many crude oils are not reported to be substantial constituents of living organisms or metabolic products. Therefore they are from the point of view of petroleum generation of particular interest. According to ERDMAN [57] several groups of compounds stand out as possible precursors for the aromatics, e.g., the terpenes and the conjugated unsaturates such as squalene and the carotenes. Experiments by VAN HASSELT [58], ZECHMEISTER and CHONOKY [59], KUHN and WINTERSTEIN [60], JONES and SHARPE [61] and DAY and ERDMAN [62] have indicated that various aromatics are generated from the above-mentioned precursors under mild thermal conditions. ERDMAN [57] proposed a scheme by which various aromatics can be formed with β-carotene as starting material. It seems probably that this degradation goes via a radical mechanism.

Another group of possible precursors mentioned by ERDMAN [57] for the aromatics are the poly-unsaturated fatty acids, these compounds are common to many forms of aquatic life [63]. Unlike most of the polyene compounds discussed previously, the double bonds in these acids are not conjugated. These bonds do, however, migrate readily into conjugation [64, 65]. These fully conjugated unbranched chains are expected to produce the low-molecular-weight hydrocarbons by condensation reactions just like the carotenes described above. BREGER [22] has pointed out that Diels-Alder reactions may also have played a role and mentions that if this theory is correct it can account for the formation of aliphatic, alicyclic and aromatic hydrocarbons.

During the conversion of all the above-mentioned compounds, aliphatic low-molecular-weight hydrocarbons are certainly also generated. ERDMAN [57] suggested that also the proteins represent an important source of these hydrocarbons. The mechanisms involved are decarboxylation and deamination.

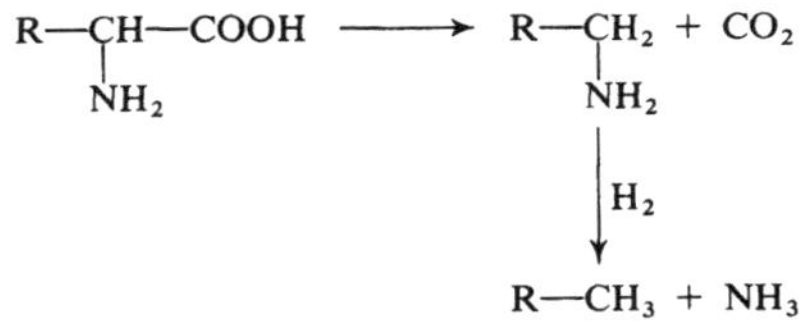

Experiments by VALLENTYNE [66] indicated that the decarboxylation can take place under "Earth" conditions. Any experimental evidence concerning the deamination is unknown, but MARLETT and ERDMAN [67] reported that ammonia, the by-product of the reaction, is abundant in petroleum source rocks, while amines are not present in a substantial amount. ERDMAN's [57] suggestion could account for all the low-molecular-weight hydrocarbons, with the exception of neopentane, which is a relatively rare constituent of petroleum.

The experiments by JURG and EISMA [46] indicated that the saturated straight-chain fatty acids are also a source of the low-molecular-weight aliphatic hydrocarbons and possibly also of the aromatics.

At present we have no means of deciding whether one path is more important than the other. But it seems highly probable that the aliphatic as well as the aromatic low-molecular-weight hydrocarbons are generated from the original material by "thermal cracking" reactions the rate of which is increased by the highly dispersed mineral matter of the sediment.

References

1. BIOT, M.: Memoir sur la polarisation circulaire et sur ses applications à la chimie organique. Mem. Acad. Roy. Sci. Inst. France **13**, 39 (1835).
2. HILLS, I. R., and E. V. WHITEHEAD: Pentacyclic triterpanes from petroleum and their significance. Paper presented at the 3rd Intern. Meeting on Org. Geochemistry, London 1966.
3. LOUIS, M.: L'activité optique des pétroles. Rev. Inst. Franç. Petrole **16**, 263 (1961).
4. DEGENS, E. T.: Geochemistry of sediments. Englewood Cliffs, N.J.: Prentice-Hall, Inc. 1965.
5. SILVERMAN, S. R., and S. E. EPSTEIN: Carbon isotopic compositions of petroleums and other sedimentary organic materials. Bull. Am. Assoc. Petrol. Geologists **42**, 998 (1958).
6. TREIBS, A.: Über das Vorkommen von Chlorophyllderivaten in einem Ölschiefer aus der obern Trias. Ann. Chem. **509**, 103 (1934).
7. GLEBOVSKAYA, E. A., and M. V. VOLKENSHTEIN: Spectra of porphyrins in petroleums and bitumens. J. Gen. Chem. USSR **18**, 1440 (1958).
8. ERDMAN, J. G., and A. H. CORWIN: Nature of the N-H bond in porphyrins. J. Am. Chem. Soc. **68**, 1885 (1946).
9. BAKER, E. W.: Mass-spectrometric characterization of petroporphyrins. J. Am. Chem. Soc. **88**, 2311 (1966).
10. THOMAS, D. W., and M. BLUMER: Porphyrin pigments of a Triassic sediment. Geochim. Cosmochim. Acta **28**, 1147 (1964).
11. DUNNING, H. N.: Geochemistry of organic pigments. In: Organic geochemistry (I. A. BREGER, ed.). New York: Pergamon Press 1963.
12. GRANSCH, J. A., and E. EISMA: Geochemical aspects of the occurrence of porphyrins in West-Venezuelan mineral oils and rocks. Paper presented at the 3rd Int. Meeting on Organic Geochemistry, London 1966.

13. BENDORAITIS, J. G., B. L. BROWN, and L. S. HEPNER: Isoprenoid hydrocarbons in petroleum. Anal. Chem. **34**, 49 (1962).
14. EGLINTON, G., P. M. SCOTT, T. BELSKY, A. L. BURLINGAME, W. RICHTER, and M. CALVIN: Occurrence of isoprenoid alkanes in a Precambrian sediment. In: Advances in organic geochemistry 1964 (G. D. HOBSON and M. LOUIS, eds.). New York: Pergamon Press 1966.
15. ROBINSON, W. E., J. J. CUMMINGS, and G. U. DINNEEN: Changes in Green River oil-shale paraffins with depth. Geochim. Cosmochim. Acta **29**, 249 (1965).
16. MCCARTHY, E. D., and M. CALVIN: The isolation and identification of the C_{17} isoprenoid saturated hydrocarbon 2,6,10-trimethyl tetradecane: Squalane as a possible precursor. Preprints Gen. Papers, Div. Petr. Chem. A.C.S. **11**: 3, Aug. 1966.
17. ACKMAN, R. G.: Occurrence of odd-numbered fatty acids in the Mullet *Mugil cephalus*. Nature **208**, 1213 (1965).
18. CHIBNALL, A. C., S. H. PIPER, A. POLLARD, E. F. WILLIAMS, and P. N. SAHAI: Constitution of primary alcohols, fatty acids and paraffins present in plant and insect waxes. Biochem. J. **28**, 2189 (1934).
19. ORÓ, J., D. W. NOONER, and S. A. WICKSTRÖM: Paraffinic hydrocarbons in pasture plants. Science **147**, 870 (1965).
20. KVENVOLDEN, K. A.: Molecular distributions of normal fatty acids and paraffins in some Lower Cretaceous sediments. Nature **209**, 573 (1966).
21. ABELSON, P. H., T. G. HOERING, and P. L. PARKER: Advances in organic geochemistry (I. A. BREGER, ed.). New York: Pergamon Press 1964.
22. BREGER, I. A.: Diagenesis of metabolites and a discussion of the origin of petroleum hydrocarbons. Geochim. Cosmochim. Acta **19**, 297 (1960).
23. WHITMORE, F. C.: Fundamental research on occurrence and recovery of petroleum. Am. Petrol. Inst., Progr. Rep. 124 (1943).
24. MEINSCHEIN, W. G.: Origin of petroleum. Bull. Am. Assoc. Petrol. Geologists **43**, 925 (1959).
25. SMITH, P. V.: Studies on origin of petroleum: occurrence of hydrocarbons in recent sediments. Bull. Am. Assoc. Petrol. Geologists **38**, 377 (1954).
26. SWAIN, F. M.: Stratigraphy of lake deposits in Central and Northern Minnesota. Bull. Am. Assoc. Petrol. Geologists **40**, 600 (1956).
27. BRAY, E, E., and E. D. EVANS: Distribution of n-paraffins as a clue to the recognition of source beds. Geochim. Cosmochim. Acta **22**, 2 (1961).
28. ERDMAN, J. G.: Personal communication to Dr. J. M. HUNT. Bull. Am. Assoc. Petrol. Geologists **46**, 2246 (1962).
29. PHILIPPI, G. T.: On the depth, time and mechanism of petroleum generation. Geochim. Cosmochim. Acta **29**, 1021 (1965).
30. LIND, S. C., and D. C. BARDWELL: The chemical action of gaseous ions produced by α-particles. IX. Saturated hydrocarbons. J. Am. Chem. Soc. **48**, 2335 (1926).
31. COLOMBO, U., E. DENTI, and G. SIRONI: A geochemical investigation upon the effects of ionizing radiation on hydrocarbons. J. Inst. Petrol. **50**, 228 (1964).
32. SHEPPARD, C. W., and V. L. BURTON: The effects of radioactivity on fatty acids. J. Am. Chem. Soc. **68**, 1636 (1946).
33. FAN, P. H., and T. L. KISTER: Multiple origins of natural gas and oil. Am. Gas Assoc. Monthly **47**, 12 (1965).
34. ENGLER, C.: Die Bildung der Hauptbestandteile des Erdöls. Petrol. Z. **7**, 399 (1911/12).
35. BERL, E.: The origin of petroleum. Am. Inst. Mining Met. Engrs. Inst. Metals Div., Spec. Publ. **920** (1938).
36. WARREN, B. C., and F. H. STORER: Examination of a hydrocarbon-naphta, obtained from the products of the destructive distillation of lime-soap. Mem. Am. Acad. Arts Sci., p. 167 (1867).
37. KRÄMER, G.: Über die Spaltung polymerer Verbindungen; Truxen aus dem Cumaronharz. Chem. Ber. **36**, 645 (1903).
38. HÖFER, H.: Die Entstehung der Erdöle. Petrol. Z. **18**, 1301 (1922).
39. SEYER, W. F.: The conversion of fatty acids and waxy substances into petroleum hydrocarbons. J. Inst. Petrol. **19**, 773 (1933).
40. MONTGOMERY, C. W.: The chemical technology of petroleum (W. A. GRUSE and D. R. STEVENS, eds.). New York: McGraw-Hill Book Co. 1942.
41. BROOKS, B. T.: Active surface catalysts in formation of petroleum. Bull. Am. Assoc. Petrol. Geologists **33**, 1600 (1949).

42. BOGOMOLOV, A. I., and K. I. PANIMA: Low temperature, catalytic transformation of organic compounds on clay. II. Transformation of oleic acid. Transactions of the Soviet Institute for research in geological prospecting, No. 174, 17 (1961).
43. BEDOV, J. A.: Preparation of petroleum-like hydrocarbons by thermal catalytic conversion of fatty acids. Neftekhimya **2**, 313 (1962), Chem. Abstracts 13067[g] (1964).
44. JURG, J. W., and E. EISMA: Petroleum hydrocarbons: generation from a fatty acid. Science **144**, 1451 (1964).
45. KLUBOVA, T. T.: Participation of clay minerals in petroleum formation. Dokly. Geochem. **157**, 146, Chem. Abstracts **61**, 11778[e] (1964).
46. RICE, F. O., and A. KOSSIAKOFF: Thermal decomposition of hydrocarbons. Resonance stabilization and isomerization of free radicals. J. Am. Chem. Soc. **65**, 590 (1943).
47. GREENSFELDER, B. S., H. H. VOGE, and G. M. GOOD: Catalytic and thermal cracking of pure hydrocarbons. Ind. Eng. Chem. **41**, 2573 (1949).
48. THOMAS, C. L.: Chemistry of cracking catalysts. Ind. Eng. Chem. **41**, 2564 (1949).
49. TUNG, S. E., and E. MCININCH: Ion-radical cracking of cumene on alumina. J. Catalysis **4**, 586 (1965).
50. COOPER, J. E., and E. E. BRAY: A postulated role of fatty acids in petroleum formation. Geochim. Cosmochim. Acta **27**, 1113 (1963).
51. LAWLER, D. L., and W. E. ROBINSON: Fatty acids in Green River formation oil shale. A.C.S. Petroleum Chem. Division Preprints, March 1965, vol. 10, No. 1, p. 5.
52. JURG, J. W., and E. EISMA: On the mechanism of the generation of petroleum hydrocarbons from a fatty acid. Paper presented on the 3rd Int. Meeting on Organic Geochemistry, London 1966.
53. EISMA, E., and J. W. JURG: Fundamental aspects of the diagenesis of organic matter and the formation of petroleum. Paper to be presented on the 7th W.P.C. Mexico 1967.
54. BLUMER, M., M. M. MULLIN, and D. W. THOMAS: Pristane in zooplankton. Science **140**, 974 (1963).
55. –, and D. W. THOMAS: Phytadienes in zooplankton. Science **147**, 1148 (1965).
56. DEAN, R. A., and E. V. WHITEHEAD: The composition of high-boiling petroleum distillates and residues. Paper V-9, 6th W.P.C. 1963, p. 261.
57. ERDMAN, J. G.: Some chemical aspects of petroleum genesis as related to the problem of source-bed recognition. Geochim. Cosmochim. Acta **22**, 16 (1961).
58. VAN HASSELT, J. F. B.: Constitution of bisein. Rec. Trav. Chim. **30**, 1 (1911).
59. ZECHMEISTER, L., and L. CHOLNOKY: Paprika colouring matters. IV. Several transformations of capsanthin. Ann. Chem. **478**, 95 (1930).
60. KUHN, R., and A. WINTERSTEIN: Chain contraction and cyclisation in the thermal degradation of natural polyene pigments. Ber. Deut. Chem. Ges. **66**, 1733 (1933).
61. JONES, R. N., and R. W. SHARPE: Pyrolysis of carotene. Can. J. Res., Botany **26**, 728 (1948).
62. DAY, W. C., and J. GORDON ERDMAN: Ionene: a thermal degradation product of β-carotene. Science **141**, 808 (1963).
63. KLENK, E.: The polyenoic acid of fish oils. In: Essential fatty acids, p. 58 (H. M. SINCLAIR, ed.). London: Butterworth 1958.
64. KEPPLER, J. G.: Analysis of unsaturated fatty acids. In: Essential fatty acids, p. 14 (H. M. SINCLAIR, ed.). London: Butterworth 1958.
65. HOLMAN, R. T.: Methods of biochemical analysis, vol. 4, p. 99. New York: Interscience 1957.
66. VALLENTYNE, J. R.: Thermal degradation of amino acids. Carnegie Inst. Wash. Yearbook **56**, 185 (1957).
67. MARLET, E. M., and J. G. ERDMAN: Carbon-nitrogen distribution and nitrogen type relationships in Recent and ancient sediments. Paper presented at the 135th Meeting of the A.C.S. Boston 1959.
68. VOGE, H. H.: Catalytic cracking. In: Catalysis VI, p. 449 (P. H. EMMETT, ed.). New York: Reinhold Publ. Corp. 1958.

CHAPTER 29

Organic Geochemistry of Coal

B. S. COOPER and D. G. MURCHISON

Department of Geology
University of Newcastle upon Tyne
England

Contents

I. Introduction

Organic geochemical studies of coals provide an ideal opportunity for the combination of petrological methods with chemical procedures, since appropriate microscopical techniques will usually establish the presence of many of the petrographic constituents, even in severely altered coals. The considerable progress in the chemical and petrological fields of coal research during the past twenty-five years was stimulated by the urgent need for coal as an energy source in the period immediately following World War II. The solid fuels of greatest economic importance and on which most research effort has been expended are the "hard coals", a term encompassing the bituminous coals, the semianthracites and the anthracites. This chapter is primarily concerned with the organic geochemistry of such coals. Although chemical-petrological correlations are less well established for the more immature fuels, brief reference will also be made here to the geochemistry of brown coals and peats.

In preparing this chapter important earlier reviews of the geochemistry of coal by BREGER [1] and VAN KREVELEN [2] have been consulted. Recourse has also been made to texts by FRANCIS [3], VAN KREVELEN [4] and LOWRY [5], which give a wide coverage of coal science. These sources and the list of key references to individual researches that are mainly of recent origin should be sufficiently comprehensive as a basis for further study.

II. Coalification

A long and widely held view of coals is that they belong to a continuous series of fuels that extends from peats by way of lignites and bituminous coals to anthracites. Passing through the series there is a darkening in color and an increase in lustre of the fuels, which is accompanied by a gradual rise in carbon content and calorific value and a decrease in moisture, volatile matter and oxygen contents, while the proportion of hydrogen remains roughly constant until in semianthracites and anthracites it falls at an increasing rate. Proponents of continuous coalification divide the process into two stages. The first is of a relatively short-lived biochemical nature that is primarily operative during peat formation, when chemical changes to the accumulating plant debris are dominated by the influence of bacterial activity. The second phase of the process is of longer duration and is variously known as the "metamorphic", the "dynamo-mechanical" or the "geochemical" stage in which both chemical and physical changes are governed purely by the interplay of time, temperature and pressure. Within the concept of a geochemical stage, the relative importance of these three factors is disputed. HUCK and KARWEIL [6], M. and R. TEICHMÜLLER [7, 8] and KUYL and PATIJN [9], for example, believe that only time and temperature are of consequence with pressure generally inhibiting advance in chemical coalification. WHITE [10] and STADNICHENKO [11] maintained that tectonic stress and thrust pressures were all-important in promoting the coalification of Appalachian coal seams, but ROBERTS [12] has suggested that the various classes of coals could have formed as a result of metamorphic distillation processes caused either by magmatic heat from igneous intrusions or frictional heat due to orogenic forces.

The concept of a biochemical stage of coalification followed by a geochemical stage is attractive and useful and will be accepted in this chapter. DRYDEN [13], however, in a relatively recent review of theories of coalification has pointed out that there is still no incontrovertible evidence to support any particular theory of coalification. Thus, instead of a continuous series it would be possible at the other extreme for each class of coal to have a quite separate origin from plant material. And again, FUCHS [14] has argued an intermediate hypothesis, but one that is purely biochemical. He regards brown coals, formed under aerobic conditions, as the conclusion of one coalification process, while bituminous coals and anthracites follow another coalification track in response to an anaerobic biochemical environment. Which of these various hypotheses is correct has still to be demonstrated. Attempts to reproduce the natural coalification process under laboratory conditions have only been partly successful. The majority of these experiments has involved heating plant constituents in the presence of water in autoclaves at temperatures up to 400° C. Hydrothermal coalification will not, however, alter the starting materials beyond the stage of the low-rank bituminous coals. To progress to higher levels of coalification requires high mechanical pressures with confinement, as far as possible, of any gaseous products within the system. Such artificial coalification experiments must suffer from the important limitation that reaction rates are highly accelerated and, furthermore, the bypassing of the early decompositional stages, when the activity of bacterial and other biochemical agencies would be dominant, may also be an important experimental omission.

III. Distribution of Coals

Coals are known in rocks of all ages from the Precambrian to the Recent, but only after the establishment of a land flora towards the end of the Silurian, approximately 400 million years ago, did large-scale accumulations of coals become possible. Since then there have been two great coal-forming periods in the earth's history: the first extending through Carboniferous and Permian times and lasting for approximately 120 million years, and the second of shorter duration, mainly in the Tertiary period, with the most widespread coal accumulation occurring some 20 million years ago. Coals also formed in many parts of the world during the Mesozoic era, but quantitatively these deposits are of much less significance. Tertiary coals are dominantly lignites and brown coals, whereas those of Permo-Carboniferous times are mainly bituminous coals, semianthracites and anthracites.

IV. Deposition of Coal Seams

The vast majority of coal seams originated in plant debris that accumulated *in situ* as swamp peats, many of which must have been of considerable thickness and laterally very extensive. Many of the great coalfields of the world are of paralic type, that is, they are characterized by the presence of marine or near-marine sediments in parts of the succession. Often they formed in depressed coastal belts that were bound up with geosynclinal development. In the Carboniferous, because of the rhythmic tectonic movements that affected these belts, producing an alternation of periods of slow subsidence with times of more rapid sinking, the coal seams of the paralic basins are generally thin and never more than a few meters in thickness. Each seam is a member of a characteristic succession of sediments that is repeated many times, with the more marine or estuarine sediments above the coals and the terrestial fluviatile deposits immediately underneath the seams. The literature on cyclic sedimentation is extensive and for a detailed treatment a symposium edited by Merriam [15] should be consulted. Westoll [16] has recently reviewed rhythmic sedimentation in relation to coal-bearing strata. Modern paralic swamps do exist, for example, those off the southern coast of New Guinea and the Everglades-Mangrove complex of southwestern Florida. The diachronous and transgressive nature of sediments in such modern swamps should be noted [17–20].

Peat swamps may also develop in regions away from coastal belts, and in some isolated tectonically-controlled basins, coal-seam thicknesses much in excess of those for any single seam in paralic deposits can develop. The limnic or lacustrine coal basins are those which have no connection with the open sea. Such basins can form due to a variety of geological causes such as tectonic movements, subsurface solution or glacial activity. Usually the basins are of limited extent, but they may in some cases show a gradual transition to paralic basins. Coal-forming peats may also be deposited at high altitudes. Plumstead [21] has contrasted the Palaeozoic coals of the northern and southern hemispheres making particular reference to the Highveld Coalfield in the Transvaal. This coalfield developed in Permo-Carboniferous times at an altitude of between 5,000 and 6,000 feet and many hundreds of miles from the sea; the oldest coals in the field were closely associated with an extensive glaciation.

V. Factors Influencing the Coal-Forming Facies

A number of factors, which are now briefly considered, will affect the coal-forming facies and will have a varying influence on the geochemistry of the resulting organic deposit. A more extended treatment of the effects of depositional, geographical and botanical influences on coal genesis has been given by Teichmüller [22].

A. Stage of Evolutionary Development of the Flora

The flora that gave rise to the Permo-Carboniferous coal seams, while it must have been luxuriant and widespread, was restricted and less varied in its plant types than in later geological periods. By Tertiary times the peat-forming plant communities had become richer and more diversified and were also adapted to a wider range of ecological conditions. Thus, Tertiary coals contain many less megaspores and microspores than do Carboniferous coals, which are, however, much richer in petrographic constituents of the inertinite group (see Table 1) [22].

In considering the organic geochemistry of the Permo-Carboniferous coals, it has to be assumed that the chemical nature of the contributing primitive flora was similar to that of present-day plants, because few representatives of the late Palaeozoic and early Mesozoic floras are extant. There is a similarity in the physical structure of the cells of Recent and ancient plants.

B. Climate

The warmer and wetter the climate, the more luxuriant is the flora and the more the forest swamp replaces reed and grass swamps. Increasing warmth also causes an increase in the rate of decomposition processes, but, even so, swamp peats still continue to accumulate in semitropical and tropical climates. Jacob [23] has demonstrated that Eocene brown coals, which were laid down in a tropical climate, are more strongly decomposed than Miocene brown coals that were deposited under subtropical conditions.

The distance of the open ocean from the coal swamp will also influence the humidity and consequently the type and luxuriance of the vegetation. A rather different flora might be expected to contribute to the peats of inland limnic basins than to paralic deposits. The Permo-Carboniferous coal swamps of the southern hemisphere (Gondwana coals) were influenced by post-glacial conditions which produced lower average temperatures than those occurring in the Palaeozoic coal basins of the northern hemisphere. PLUMSTEAD [21] has suggested that the petrography of South African coals reflects a dry climate and extensive oxidation of the coal-forming peat. The Gondwana coals are clay-rich and finely detrital, indicating absence of a forest vegetation or deposition in lakes [24–26].

C. Plant Associations and Environments

Any coal seam represents a period of prolonged peat accumulation during which time both environmental conditions and the plant communities contributing to the peat may alter radically with marked effect on the geochemistry of the deposit. In the lower Rhine brown coals of Miocene age, TEICHMÜLLER [27] has distinguished between peat types originating in a) reed swamps with many areas of open water and herbaceous plants, b) very wet swamps with subaquatic mud that contain a tree association of *Nyssa*, *Taxodium* and *Glyptostrobus*, c) drier Myricacean-Cyrillacean bush swamps, and d) the driest environment of all formed by forests of *Sequoia*. Fig. 1 shows the different coal types that arise from these swamp peats. The gyttja consists of small resistant plant fragments that have floated to their depositional site. Reed coals are lignin-poor and heavily decomposed with frequent evidence of fungal attack. Of the coals formed from forest vegetation, the resin- and tannin-impregnated tissues from *Sequoia* are best preserved.

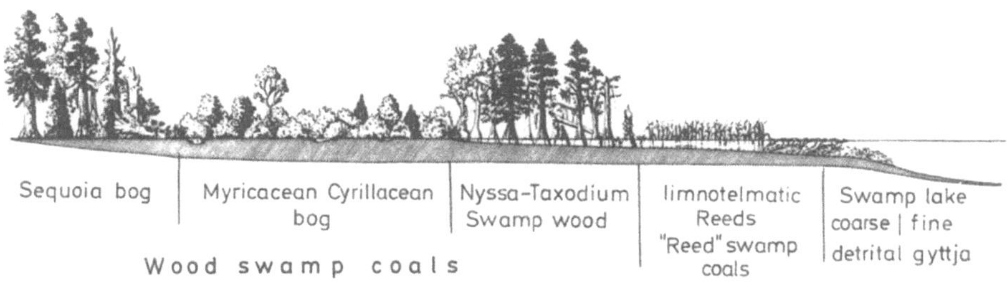

Fig. 1. Swamp and coal types of the Miocene brown coals in the lower Rhine district of Germany. (After TEICHMÜLLER [27])

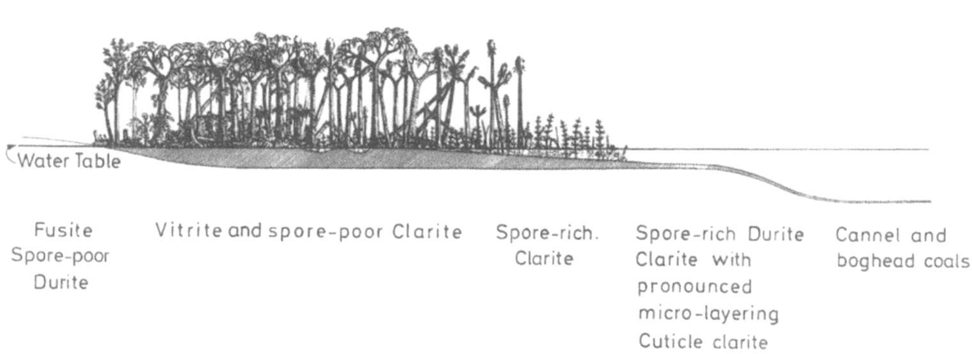

Fig. 2. Facies of different Carboniferous coal types in the northern hemisphere. (After TEICHMÜLLER [22])

The palaeoecology of Carboniferous peats has been extensively considered by SMITH, particularly in relation to the petrography and palynology of the coals [28–30]. TEICHMÜLLER [22] has also illustrated the different peat facies of Carboniferous coals in the northern hemisphere (Fig. 2). Names of the coal types forming in the different parts of the depositional area are given and will be referred to again later. Vitrite and spore-poor clarite develop primarily from forest vegetation, while the spore-rich clarites and durites, as well as cannel and boghead coals, contain an increasing proportion of allochthonous resistant elements, such as spores, resins and cuticles, which have drifted or been blown into the peat.

The observations above indicate that lateral variations are common in accumulating peat-swamp types and may be particularly noticeable within paralic basins. Provided that there is no shift in the relative positions of land and sea, the different swamp environments would presumably remain fairly constant in position over long periods of time. With, however, the inevitable marine transgressions and regressions in coastal belts, the positions of the swamp types shift, which accounts for the vertical variations in fossil peat and coal types that are found within coal seams.

D. Nature of Swamp Waters and Exclusion of Oxygen

The importance of the environmental waters in the formation of coals was discussed many years ago by WHITE [31]. In the drier parts of the swamps, in areas covered by oxygenated waters, or in regions where the water table fluctuates strongly, such as has been suggested in the high-altitude Gondwana swamps of South Africa [21], there is the strong possibility of oxidation of the peat components. Thus, doppleritic-type humic gels occur along with highly carbonized constituents in the *Sequoia* peats of the lower Rhine brown coals [32]. Dessication cracks may also be observed in the coal [22]. Aerobic conditions at the time of peat formation are also indicated by evidence of fungal activity.

While much of the plant material must accumulate in an environment that is essentially stagnant and in which little decomposition of the tissues occurs, influx of marine or calcium-rich waters may cause rapid degradation of the humic materials. SPACKMAN [33] states that, despite the ample supply of lignified tissues, these are rapidly broken down in the saline mangrove swamps of southeast Florida to give amorphous and fibrous peats and not the wood and bark peats that might be anticipated. Calcium-rich water raises alkalinity and increases bacterial activity, as well as encouraging dopplerite formation, and if oxygenated conditions also prevail, then decomposition will be very severe.

VI. Petrographic Constituents of Hard Coals

A. Classification

Table 1 gives a summary of the European Stopes-Heerlen system of coal petrographic nomenclature as it existed at the time of publication of the second edition of the "International Handbook of Coal Petrography" [34]. The nomenclature has suffered some modifications at meetings of the International Commission for Coal Petrology since 1963 and will probably undergo further evolution as a result of petrographic research before the next edition of the Handbook is published. No reference is made here, however, to any new term that is not yet included in the published literature. Other systems of nomenclature do exist and are currently in use, but the Stopes-Heerlen system provides a perfectly adequate basis for subsequent discussion in this chapter.

Table 1. *Summary of coal petrographic nomenclature*

Macerals		Microlithotypes	
Maceral	Maceral group and symbol	Microlithotype	Principal groups of constituent macerals in the microlithotypes
Collinite Telinite	Vitrinite (Vt)	Vitrite	Vt
		Vitrinertite	Vt+I
Micrinite (fine-grained) Micrinite (massive)		Microite	I (micrinite dominant)
Semifusinite Fusinite Sclerotinite	Inertinite (I)	Fusite	I (except micrinite)
Cutinite Resinite Sporinite Alginite	Exinite (E)	Liptite	E
		Clarite Durite Duroclarite Clarodurite	Vt+E I+E Vt+E+I I+E+Vt

The Stopes-Heerlen system is primarily based upon examination of polished surfaces of bituminous coals, semianthracites and anthracites using reflected light with oil immersion objectives. "Macerals" are the elementary microscopical constituents of coals and are analogous to the minerals of other rocks. They can be grouped according to certain similarities in their petrographic properties, although it is not implied that the macerals within each of the groups vitrinite, exinite and inertinite are identical; this should be clear from the accompanying photomicrographs (Figs. 3a–k). "Microlithotypes" are typical associations of macerals. They are delimited on the basis of a minimum band width (50 microns) in the coal which agrees satisfactorily with their technological behavior. The microlithotypes are also of considerable use in environmental studies.

B. Progenitors and Genesis

Vitrinite is the most abundant maceral of humic coals and is the dominant constituent of the bright, black and lustrous bands that characterize "hard coals". The maceral occurs in two forms: telinite (Fig. 3a) which shows the cell structure

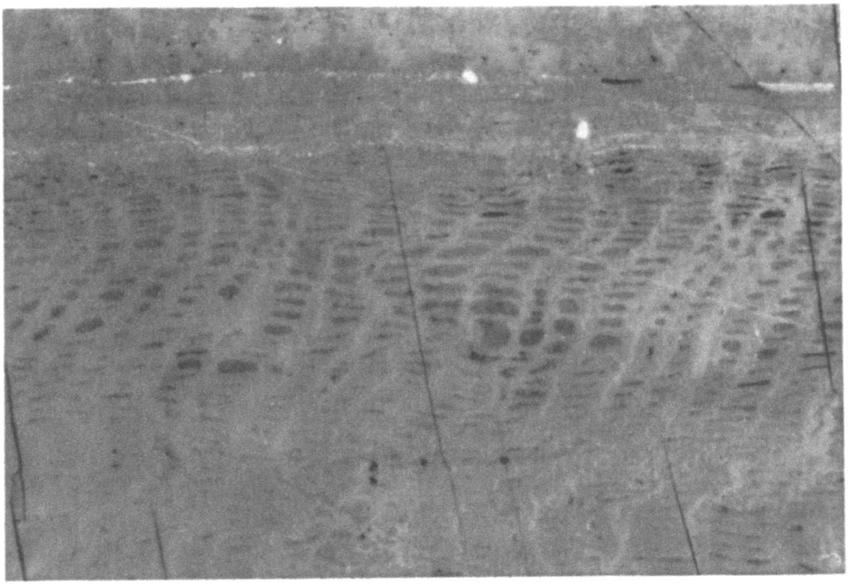

Fig. 3a (× 760)

Fig. 3. a) Vitrinite: telinite in which the cell cavities are filled with resinite. b) Vitrinite: collinite which appears as a structureless, medium gray substance. The higher-reflecting constituents all belong to the inertinite group. c) Inertinite: semifusinite of high reflectivity displaying cellular structure and fragmentation that illustrates its brittle nature. d) Inertinite: thin bands and lenticles of semifusinite of varying reflectivity which alternate with darker gray, structureless collinite. e) Inertinite: granular micrinite which is associated with vitrinite that shows faint traces of cell structure. f) Inertinite: high-reflecting massive micrinite which is structureless and associated here with vitrinite. g) Inertinite: sclerotinite whose structure in this case has probably originated from a carbonised resin globule. h) Exinite: sporinite composed of dark gray, low-reflecting fragments of megaspores, which show a granular appearance, and much smaller microspores with similar optical properties that are associated with vitrinite and inertinite. i) Exinite: sporinite – a microspore associated with vitrinite and inertinite, much of the finely granular material of high reflectivity being micrinite. j) Exinite: resinite in the form of an irregular globule of extremely low reflectivity. k) Exinite: cutinite with the cuticular structure still surrounding the degraded leaf tissue which has been transformed to vitrinite. All photomicrographs of relief-polished surfaces immersed under oil of refractive index 1.520

of wood or bark, and collinite (Fig. 3b), which is structureless in reflected light. The cell cavities in telinite may be filled with a variety of other macerals, including collinite.

Besides occurring in discrete bands and as a cell filling, the structureless collinite is found as a matrix, a cement or impregnation and may also fill fissures in the coal. While the origin of telinite is clear, that of collinite is not, but there is evidence that collinite arises from more than one source. STACH [35] suggests that collinite is similar to peat dopplerite, originating by precipitation from humic solutions, but he also states that the maceral can form by the gelification of humified wood and periderm. Whatever may be the precise origins of collinite, it is widely held that the group maceral vitrinite primarily originates in the massive cellular tissues of plants and chemically is the altered product of a substantial

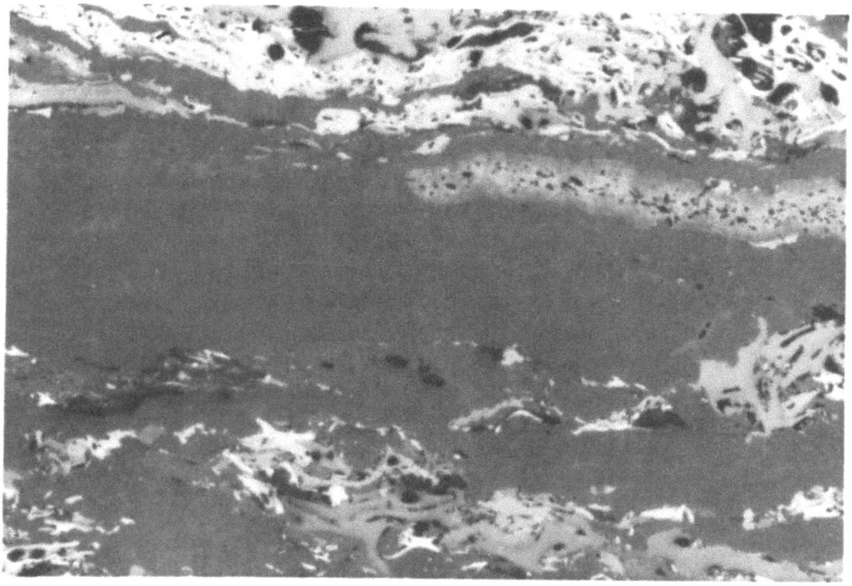

Fig. 3 b (× 300) (see p. 706 for legend)

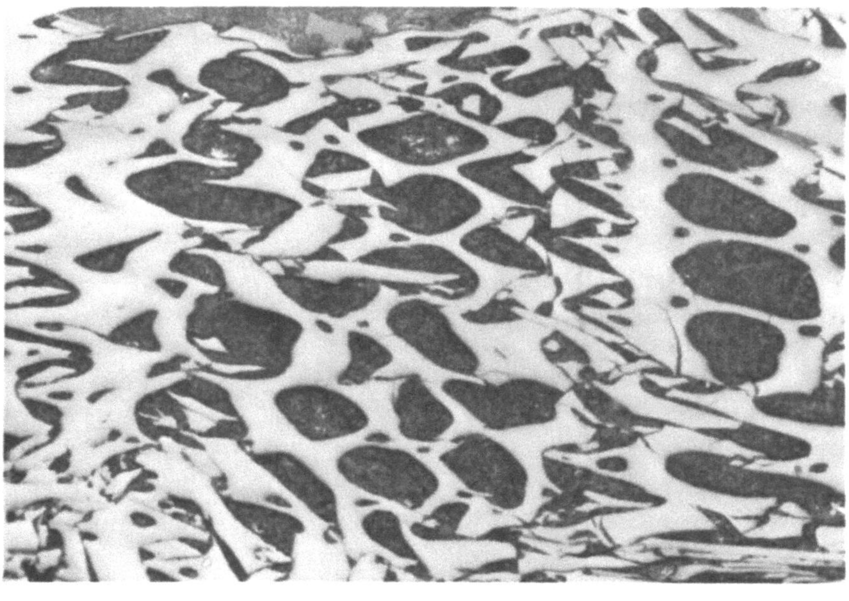

Fig. 3 c (× 300) (see p. 706 for legend)

contribution of lignin, cellulose and nitrogen-bearing compounds, such as proteins, to the coal peat.

Within the inertinite group, whose members are all characterized by a higher reflectivity than vitrinite of the same coal, semifusinite (Figs. 3 c and d) and fusinite also show the distinct cellular structure of wood and sclerenchyma, but differ

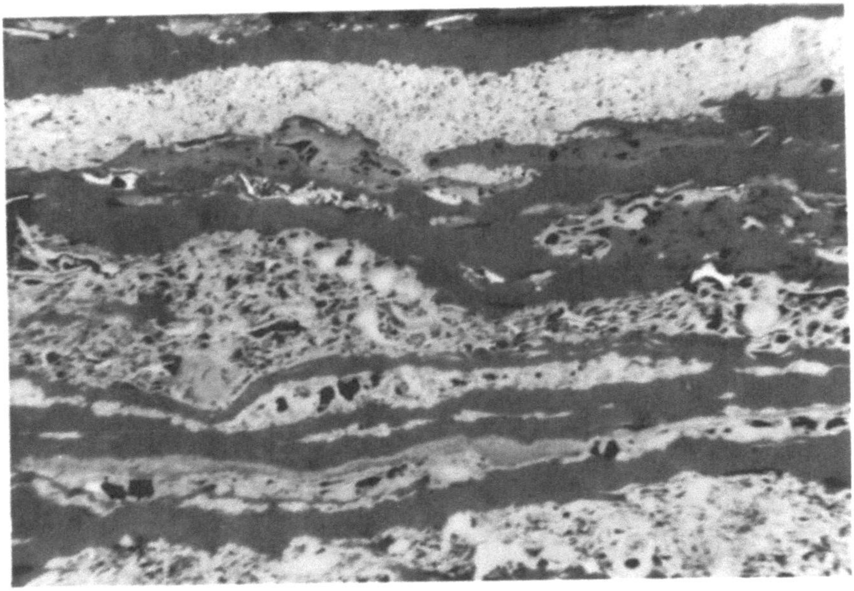

Fig. 3d (× 290) (see p. 706 for legend)

Fig. 3e (× 750) (see p. 706 for legend)

from one another in their reflectivity and color. The similarity of these macerals to fossil charcoal has been widely emphasized and forest fires have been invoked as a possible cause of formation. Independent researches on the specific heats [36] and stable free radical contents [37, 38] of these macerals from coals of the northern hemisphere indicate their exposure to moderately high temperatures

Fig. 3f (× 315) (see p. 706 for legend)

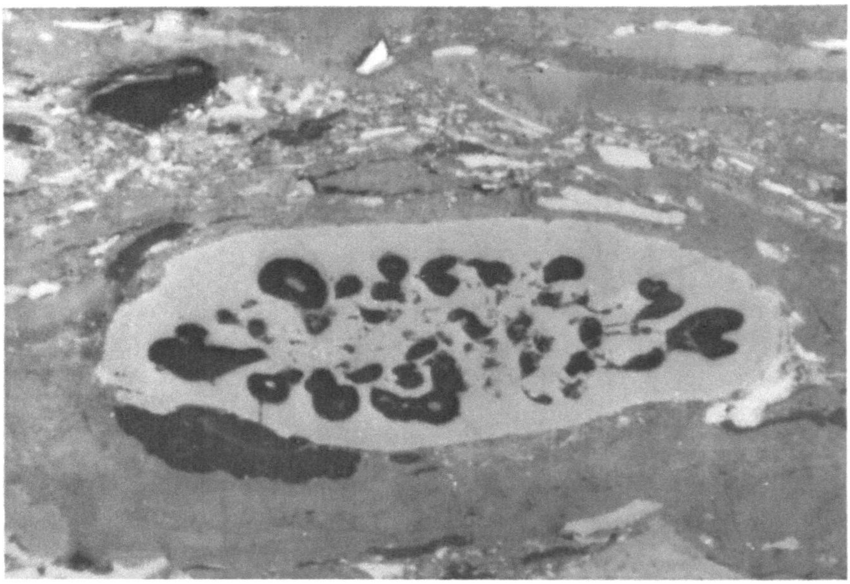

Fig. 3g (× 375) (see p. 706 for legend)

before their incorporation in the coal-forming peat. TERRES *et al.* [36] suggest that the same macerals in the southern hemisphere have developed as a result of coalification processes that took place at much lower temperatures in the aqueous phase under varying conditions of pH. It does not appear that there is one mode of formation for semifusinite and fusinite. Thus, either a varying contribution of

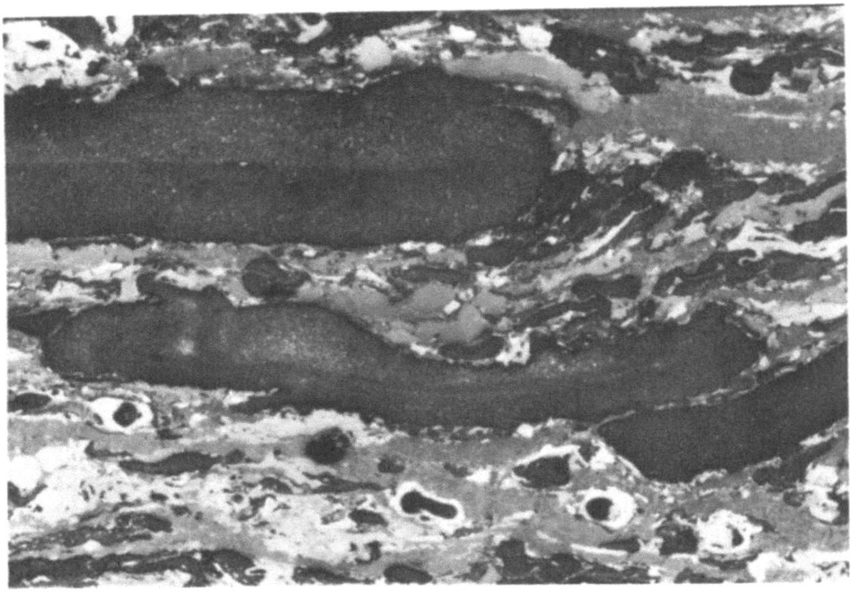

Fig. 3h (× 285) (see p. 706 for legend)

Fig. 3i (× 760) (see p. 706 for legend)

highly carbonized material was blown or drifted into the coal swamp, or material of similar character developed from wood or bark tissues very soon after their incorporation in the peat due to biochemical factors.

Micrinite (Figs. 3e and f) is another maceral forming very early in the history of the coal. SMITH [29] has summarized the views of various workers, many of

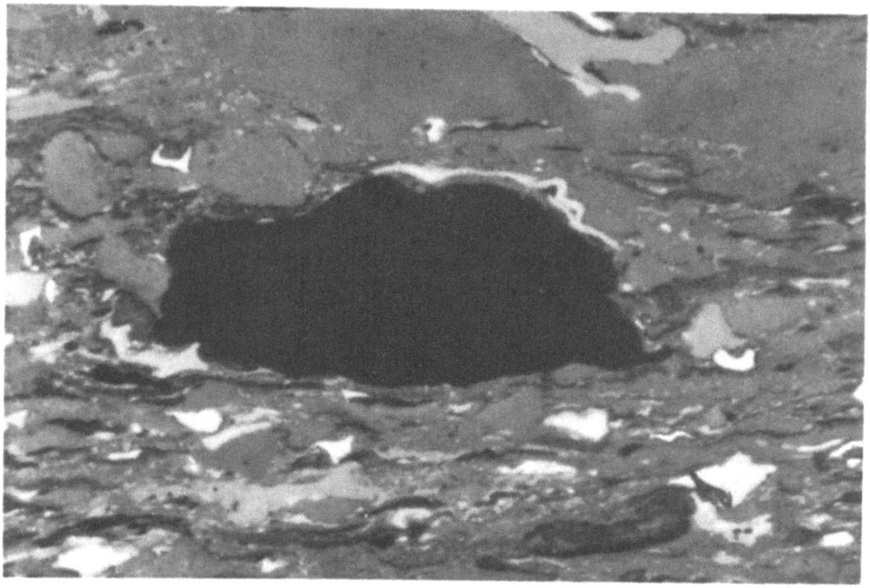

Fig. 3j (× 340) (see p. 706 for legend)

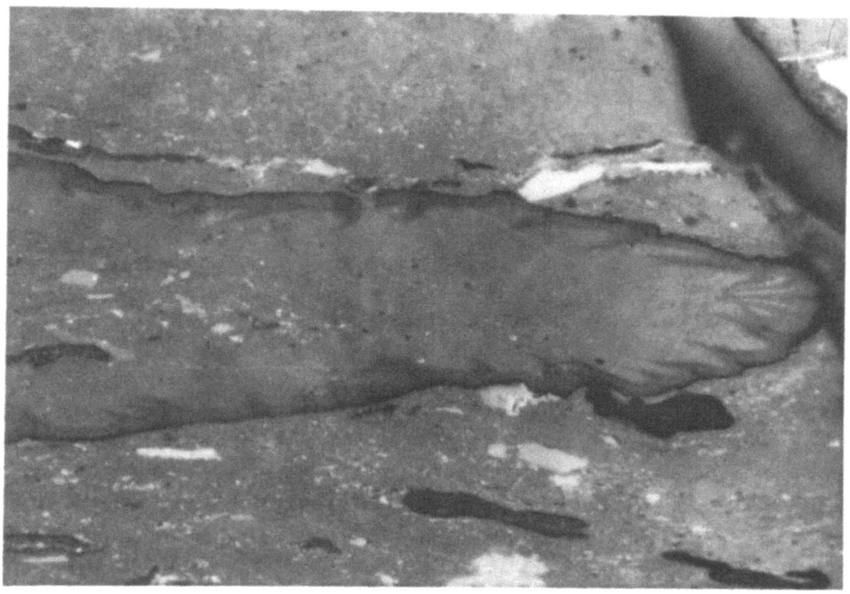

Fig. 3k (× 750) (see p. 706 for legend)

whom regard micrinite as being of aerobic origin. In hydrothermal, artificial-coalification experiments, lignin in an acid medium yielded low-hydrogen products similar in composition to micrinite [39].

Sclerotinite (Fig. 3g) is of less significance quantitatively, but is still widespread in coals, and is formed from plectenchyma, sclerotia or fungal spores. A number of workers [40–43] have commented on the similarity in appearance of sclerotia to resins that have been carbonized and which have vacuolated due to loss of volatiles.

The origins of macerals of the exinite group are more obvious from their morphology. All members show lower reflectivity than that of vitrinite in the same coal. Sporinite (Figs. 3h and i) is the dominant member of the group, occurring ubiquitously in Carboniferous coals, and is comprised of megaspore and microspore exines that are flattened parallel to the coal banding. Cutinite (Fig. 3j) is formed from cuticles, recognizable as narrow strips with serrated edges, and resinite (Fig. 3k) represents the remains of plant resins and waxes which are present as discrete bodies with oval, circular and rectangular cross-sections, as impregnations, as cleat fillings or as cell infillings. Alginite is composed of algal bodies and is only encountered in allochthonous coals such as algal cannels or torbanites. The spores and cuticles bring to the coal-forming peat, lignin and tannins as well as polymerization products of wax alcohols and waxy acids. Terpenes and the resin acids (Fig. 4d), which are genetically related to the sterols, are contributed by the resins and sterols. These compounds and fats are important chemical constituents in algal deposits.

Moving from the bituminous coals to the semianthracites and then the anthracites, the macerals become increasingly similar to one another in properties observable under the microscope. Many of the macerals can be recognized in less mature fuels, notably brown coals, but these coals are likely to prove much more complex petrographically and also chemically. Their petrographic nomenclature is not yet so well systematized as that of the bituminous coals.

VII. The Biochemical Stage

The different environmental factors discussed earlier in this chapter have their most important influence during the biochemical stage of coal formation. For the ultimate development of thick and extensive coal seams, it is essential that the original chemical substances in the peat have not been heavily oxidized, a situation recognized over thirty years ago by WHITE [31]. The actual chemical substances which accumulate during degradation of plant tissue are still controversial. Because cellulose is the major component of woody plants, it has been favored as the progenitor of many of the coal constituents. The suggestion, however, of FISCHER and SCHRADER [44], that lignin is the main contributor, has been widely accepted. Studies of modern peats are equivocal, because both cellulose and lignin may be preferentially degraded by microorganisms depending upon environmental conditions. Peats of temperate climates and soil humus both lose cellulose rapidly, but according to GIVEN [45], in red mangrove swamps of the Everglades, lignin is degraded before cellulose. DAVIS [46], however, has shown preferential enrichment of lignin with depth in reed peats of the same area.

Cellulose (Fig. 4a) is a polymer built of glucose units, whereas lignin is a polyphenol built up from units of phenylpropane derivatives such as coniferyl, sinapyl and coumaryl alcohols (Fig. 4b) by condensation and dehydrogenation

(a)

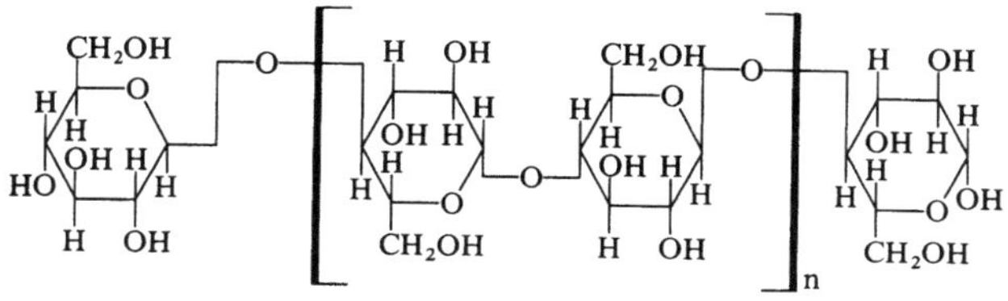

(b)

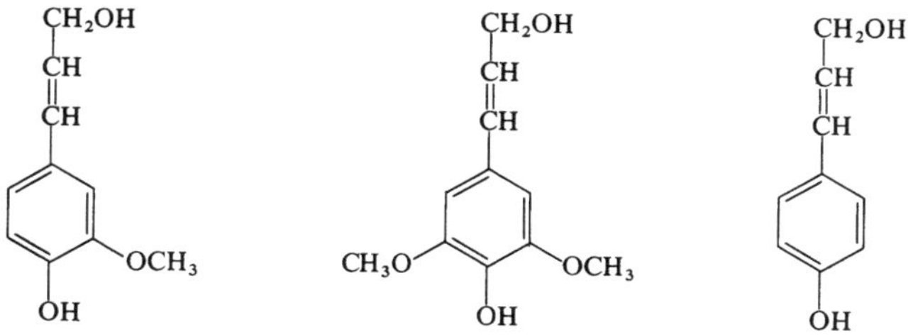

(c)

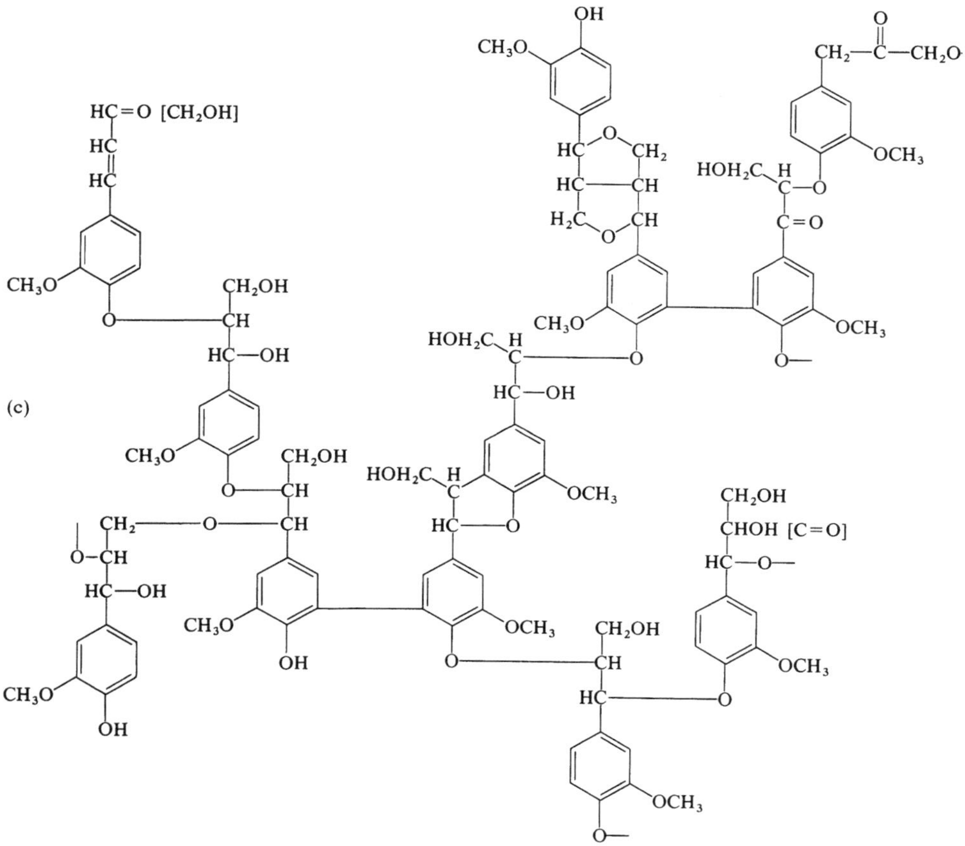

Fig. 4a–d. Structural formulae of: a) Cellulose; b) Coniferyl, sinapyl and *p*-coumaryl alcohols (left to right); c) Lignin (after ADLER [48]); d) Abietic, dextropimaric and agathic acids (left to right)

(d)

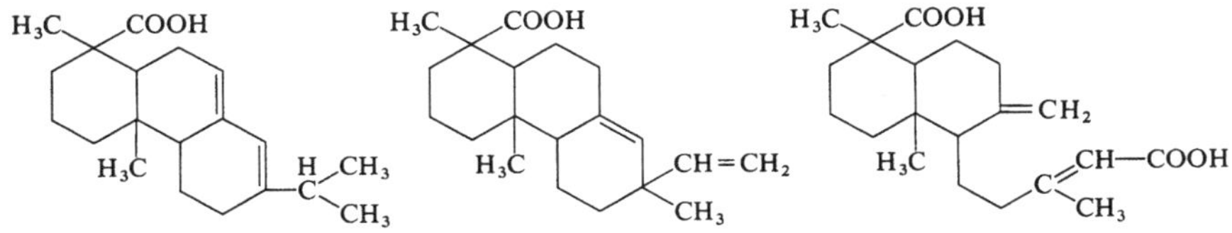

Fig. 4d

within the plant [47, 48]. The structure of lignin seems to require benzene nuclei joined through ether and carbon-carbon bonds with aliphatic, hydroxyl and methoxyl side chains. A suggested structure for pine lignin, based on coniferyl alcohol, is illustrated in Fig. 4c; approximately half the linkages are through carbon-carbon bonds. VAN KREVELEN has suggested that controversy about the actual progenitor of vitrinite and certain other macerals is of little relevance in the light of modern research. Carbohydrate decomposition products, for example, the keto-acids, can easily aromatize, while the phenols derived by fungi from lignin may be further oxidized by microorganisms to quinones, which themselves undergo ring cleavage to aliphatic compounds [49]. Thus, lignin and cellulose can both be degraded to mixtures of aromatic and aliphatic compounds. The net result is decomposing plant material permeated with these oxygenated and partially unsaturated compounds which are undergoing condensation and polymerization.

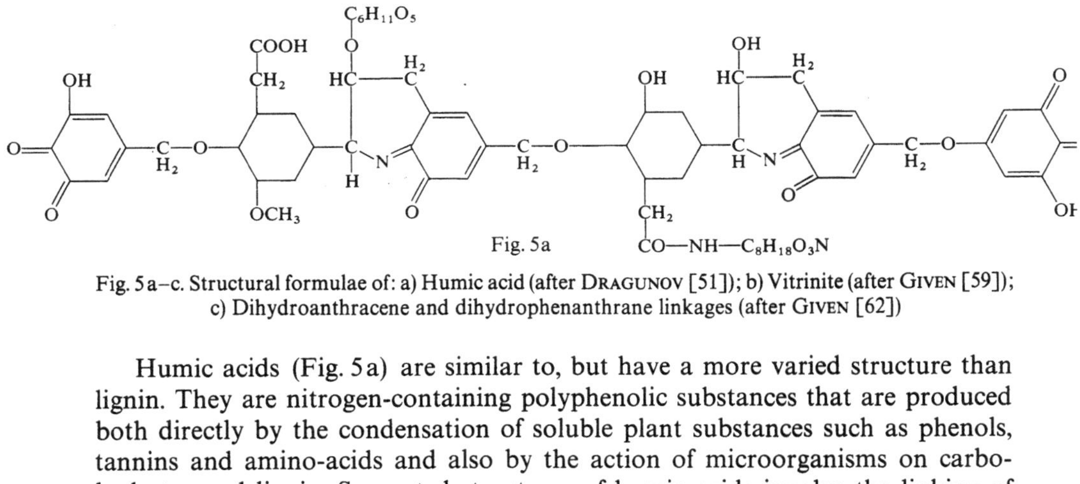

Fig. 5a

Fig. 5a–c. Structural formulae of: a) Humic acid (after DRAGUNOV [51]); b) Vitrinite (after GIVEN [59]); c) Dihydroanthracene and dihydrophenanthrane linkages (after GIVEN [62])

Humic acids (Fig. 5a) are similar to, but have a more varied structure than lignin. They are nitrogen-containing polyphenolic substances that are produced both directly by the condensation of soluble plant substances such as phenols, tannins and amino-acids and also by the action of microorganisms on carbohydrates and lignin. Suggested structures of humic acids involve the linking of benzene nuclei by oxygen and carbon-carbon bonds and also by nitrogen of amino-acids [50, 51]. Humic acids have a lower aliphatic content than lignins, but contain more phenolic hydroxyl, carboxyl and quinone groups; their side chains are linked to saccharides, fatty acids, amino-acids and nitrogen bases. They exhibit a wide range of solubility, but during ageing condense to insoluble products or form insoluble products with cellulose and lignin. Humic substances become an increasingly larger fraction of modern peat as the peat matures and the content of lignin and cellulose decreases.

It is during the biochemical stage that many minor elements of the transition series are brought into the coal matrix by formation of organo-metallic com-

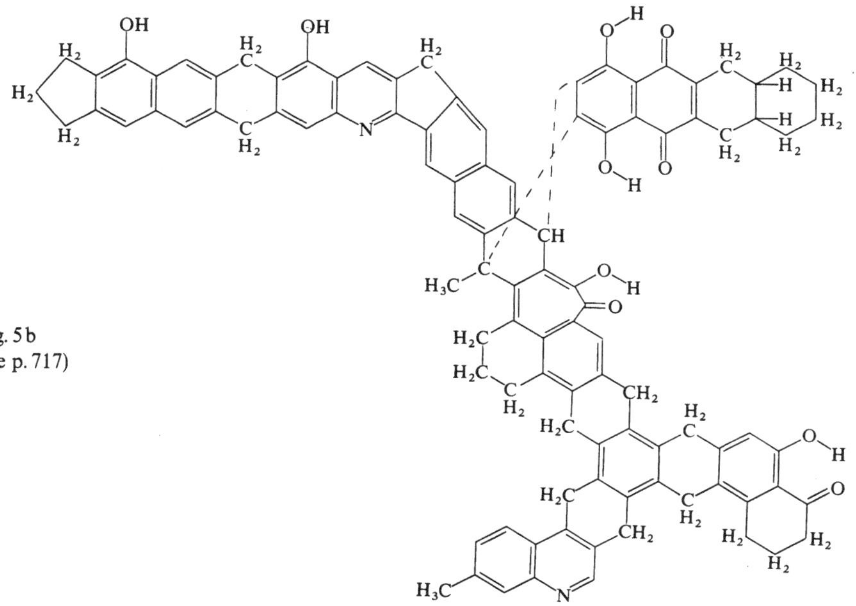

Fig. 5b
(see p. 717)

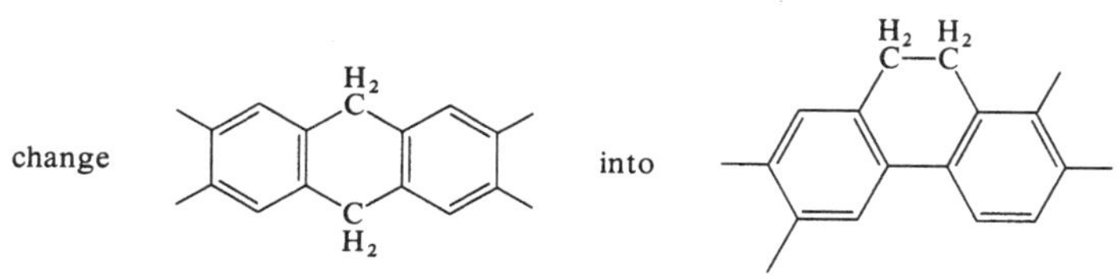

Fig. 5c
(see p. 717)

plexes [52, 53]. Although the plants have already concentrated certain elements, the presence of chelators, such as 1, 2, dihydroxybenzenes and amino-acids, in the developing peat, causes a large fixation of trace elements from the formation waters. Of these chelates those of vanadium and germanium are the most stable, and as more water passes through the peat mass, these two elements tend to displace other metals in the complexes.

VIII. The Geochemical Stage and the Molecular Structure of Coal Constituents

As coalification proceeds, changes in the elementary compositions of whole coals see an increase in carbon content and a concomitant decrease in the level of oxygen; hydrogen remains between 5 and 6 percent up to the rank level of the semi-anthracites, after which this element is lost at an increasing rate. Nitrogen fluctuates between 1 and 2 percent throughout the whole rank series, and sulfur, while variable, is generally less than 1 percent in normal coals. Methane, carbon dioxide and water are regarded as the main reaction products of the coalification process. Large amounts of water are lost in the early stages of coalification, but as rank rises, water becomes of decreasing importance as a product. The ratio of methane to carbon dioxide increases as coalification progresses and in the later stages the conversions to semianthracite and anthracite can be accounted for primarily by loss of the methane and a small quantity of water.

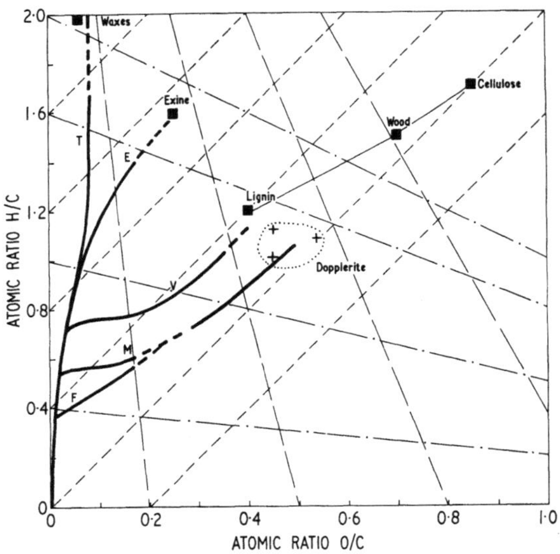

Fig. 6. Development lines for the coal macerals (after VAN KREVELEN [2, 4, 55]. — — — demethanation; —·—·— decarboxylation; - - - - - - - dehydration. T – Torbanite; E – Exinite; V – Vitrinite; M – Micrinite; F – Fusinite

The changes to whole coals described above do not uniformly affect the three maceral groups. Although the bulk of the plant material is broken down in the biochemical stage, leaving only remnants or shadows of botanical structure, constituents that ultimately go to form the exinite group survive this early period in the history of the coal with little change [54].

As stated previously, other tissues suffer considerable alteration at this time, or even earlier in their history, and so progress to become members of the inertinite group. The further course of the coalification of the different materials available after the biochemical stage is best followed on the useful diagram proposed by VAN KREVELEN [55] in which the elementary compositions of constituents are plotted in terms of their atomic H/C and O/C ratios. Fig. 6 shows the coalification tracks followed by the exinites, vitrinites and micrinites, as well as the relationship of these bands to different progenitors. The extensions of the vitrinite band towards lignin and the micrinite band towards dopplerite are based on analyses of alkaline extracts and their residues from humic acids [56]. VAN KREVELEN [4] suggests that micrinite may have formed by humic acid flocculations. Lines of demethanation, decarboxylation and dehydration are shown on the diagram and the successive processes undergone by the various petrographic constituents can easily be traced. Vitrinite, for example, undergoes dehydration which is followed in rapid succession by decarboxylation and then demethanation.

DRYDEN [57] has given one of the most recent and extensive critical reviews of the chemical constitution of "hard coals" in which he discusses the evidence for particular molecular structures that have been suggested for the petrographic constituents. He has indicated that an increasing number of numerical parameters have now to be satisfied by any model proposed for coal or its organic constituents and he has also summarized the experimental data to which any such molecular

Table 2.
Estimated average parameters for the maceral groups at two levels of coalification (after DRYDEN [57, 58]*)*

Parameter	Vitrinite		Exinite	Micrinite
percent C, pure coal (Parr basis)	82.5	90	$\sim$v	$>$v
percent volatile matter	39	24.5	$>$v	$<$v
H/C (atomic)	0.76	0.65	$>$v	$<$v
H_{ar}/H_{al}	0.23	0.54	$<$v	$>$v
H_{CH_3}/H_{al} (IR, oxidation)	0.21 (0.1–0.3)	0.23 (0.1–0.3)	–	–
$(H/C)_{al}$ (atomic)	$\sim$1.65	$\sim$1.6	–	–
CH_2 bridges: $C_{ar}-CH_2-C_{ar}$	absent	absent	–	–
Percent substitution on aromatic periphery	50+	30	–	–
R_a (aromatic rings per cluster)	$\ngtr 3$	$\sim$4	$\lesssim$v	$\gtrsim$v
C'_{ar} (aromatic C atoms per cluster)	$\ngtr 13$	16	$\sim$v	$\sim$v
Average stacking number of parallel aromatic clusters	1.3	1.8	–	–
O as hydroxyl, percent of coal	7.0	2.0	$<$v	$>$v
O as carbonyl, percent of coal	2.3	0.9	–	–
Carbon atoms per free radical	$\sim$8,000	$\sim$3,000	$<$v	$>$v
f_a (fraction of carbon as aromatic) derived from above	0.69–0.70	0.79–0.80	$<$v	$>$v

H_{ar}—Aromatic hydrogen.
H_{al}—Aliphatic hydrogen.
v—Vitrinite.

model must conform (Table 2) [58]. The results of Table 2 were accumulated by extensive recent research which has included dehydrogenation and oxidation studies, X-ray diffraction, infrared and nuclear magnetic resonance investigations, all of which have been most valuable in elucidating more precisely the carbon-hydrogen skeletal structure of the coal molecule. Polarography and functional group analytical methods have also helped to determine the character of the oxygen-containing groups in coals.

One of the most recent molecular structures proposed is that of GIVEN [59] for a vitrinite of 82 percent carbon, shown in Fig. 5b as a non-planar molecule based on dihydroanthracene. BROWN, LADNER and SHEPPARD [60] and BROWN and LADNER [61], however, soon after the publication of GIVEN's model, demonstrated by nuclear magnetic resonance studies that it is most unlikely that methylene bridges exist in coals. GIVEN [62], in a later paper, agreed and proposed modification to an isomeric type of structure based on dihydrophenanthrene, in which no methylene bridges are involved (Fig. 5c). GIVEN also cites two further points of evidence in favor of dihydrophenanthrene rather than dihydroanthracene structures in vitrinites. Firstly, some conjugation exists between the two benzene rings in dihydrophenanthrene and this would help to explain the long wavelength electronic absorption and high refractive indices of vitrinites. Secondly, the phenanthrene nucleus and several nuclei of the diaryl type, but few anthracene nuclei, were detected by MONTGOMERY and HOLLY in polycarboxylic aromatic acids produced by oxidation of high-rank coals [63].

The size distribution of aromatic clusters has not yet been finally established. Polarographic studies of coal extracts indicate the presence of many biphenyl, naphthalene, phenanthrene and triphenylene structures, while X-ray investigations suggest an almost equal distribution of about half the carbon present in the clusters between one, two and three ring structures. There is, however, considerable evidence that points also to the presence of appreciable quantities of larger ring systems.

The two main oxygen-containing groups in bituminous coals are the hydroxyl (predominantly phenolic) and the carbonyl, their proportions varying with the rank of the coal, but generally in total accounting for between 70 and 90 percent of oxygen in the coal. The remainder of the oxygen is present as ether, heterocyclic oxygen or ester groupings. Alcoholic hydroxyl, carboxyl and methoxyl groups are low in amount in coals of bituminous rank or higher, but alcoholic hydroxyl groups have been detected in brown coals [64], which also contain groups that can undergo the keto-enol transition [65].

Compared with the data available on the character and distribution of carbon, hydrogen and oxygen in coals, there is relatively little information on nitrogen and sulfur functions. Nitrogen is assumed to occur in heterocyclic ring systems. Various forms of organic sulfur such as thiophenol, sulphide, disulphide and heterocyclic, have been distinguished with 70 percent of the organic sulfur probably occurring in heterocyclic aromatic ring structures [66, 67]. IYENGAR, GUYA and BERI [68] in studies on high-sulfur Tertiary coals from Assam suggested that sulfur was present in three forms, in SH or disulphide forms, in all probability in C=S groupings and more tentatively as the thiophene or RSR form.

There has, of course, always been a concentration of experimental work on vitrinite, the predominant and most easily separable petrographic constituent of coals. Considerable chemical data do exist, however, for the two other maceral groups exinite and micrinite (inertinite), which were reviewed by BROWN [69]. Table 2 gives a recent concise and semiquantitative summary of the variation of some fundamental chemical parameters of vitrinite, exinite and micrinite (inertinite) in two coals. All data produced for the maceral groups show that they become similar to one another with rising coalification, until at 94 percent carbon there is little or no difference apparent between them. It is important to remember that in the lower-rank coals there may also be significant differences between the individual macerals of a single group. Little is known, for example, of the detailed chemistry of resinite, cutinite and alginite at different rank levels. In "float and sink" concentrates of exinites these three minor petrographic constituents have been "swamped" by the spores that were invariably present in the coals in much higher proportions than other members of the exinite group. Most exinite chemical data for bituminous coals thus refer primarily to sporinite. The final constitution of micrinite or inertinite "float and sink" concentrates will also vary depending upon the petrographic constitution of the coal used. (See Appendix).

Mechanical separations of resinite on a semimicro-scale are possible under a stereomicroscope [70]. MURCHISON and JONES [71] have shown that there is general agreement between the coalification tracks of sporinite and resinite, but that these diverge from the track suggested by VAN KREVELEN for exinite. MURCHISON [72] has also demonstrated the similarity of the infrared spectra of resinites and sporinites in bituminous coals. Such resinites have a highly aliphatic or alicyclic structure, which is also characteristic of resinites in fuels of lower rank. The absorption pattern of resinites from brown coals and lignites is, however, markedly different to resinites from higher-rank coals in the presence of an intense carbonyl band and the absence of a band at 1,600 cm^{-1} that is a prominent feature of the spectra of bituminous-coal macerals, coal extracts and of lignin. If coalification is a continuous process, then marked and rapid structural changes to resinites at the hard brown-coal stage of rank would have to take place to bring about the development of the spectral pattern of resinites found in high-volatile bituminous coals. This is possible. There is resistance by the exinites to change in the earlier stages of the coalification process, but pyrolysis experiments and the rapidly changing appearance of spores observed in microscopical preparations of bituminous coals of different rank suggest that when chemical decomposition begins in members of the exinite group, it takes place very rapidly. Alternatively, there is also the possibility that the precursors of Tertiary brown-coal resinites had a different chemical constitution to the progenitors of the resinites in Carboniferous coals.

Few data are available on cutinites in coals. LEGG and WHEELER [73] analysed cutinite from the Russian "paper coal" of the Moscow Basin and from an English durain, and concluded that the maceral contained neither water-soluble constituents, nor the cellulosic component normally found in cuticles. More recently NEAVEL and MILLER [74] have shown a divergence of plots for cutinites of Indiana "paper coal" from the exinite coalification track as defined by VAN KREVELEN. No details for molecular structure are available.

The fatty nature of the algal coals, and particularly alginite, which results in substantially higher volatile and hydrogen contents than for normal coals of equivalent rank, has long been recognized. Until very recently there was little information available on the detailed chemical structure of these allochthonous coals, although their petrology was well understood. MAXWELL [75] has now produced an extensive chemical study of the Scottish torbanite (alginite) from Linlithgow, and DOUGLAS, EGLINTON and HENDERSON [76] refer to the presence of pristane and phytane in the branched and cyclic alkane fraction in the solvent extract of this coal at room temperature; no alkenes were observed. A series of normal fatty acids in the range C_{10}—C_{28} has been observed in untreated torbanite and normal fatty acids from C_8—C_{21} accompanied by a second series of α, ω-dicarboxylic acids have been identified in demineralized torbanite [77].

IX. Concluding Remarks

The molecular structures described above for a particular petrographic coal constituent at different levels of coalification can only be regarded as "average" structures for that maceral. Microscopical evidence abounds of substantial changes in properties of petrographic constituents over quite small distances within coals. Direct chemical proof of structural modification is generally not available, but variations in reflectivity and morphology can be taken as evidence of chemical alteration. Examples of a few such property changes are the well-known vitrinite to semifusinite transition, the sometimes marked variations affecting even very small entities of the exinite group [40–43, 78], and the pronounced reflectivity rise in vitrinites caused by inclusions of radioactive minerals [79, 80] (Fig. 7). The last effect has been attributed to radiochemical dehydrogenation producing cross-linking of adjacent alicyclic units rather than aromatization of the maceral [80].

Data are still lacking on the chemical nature of many of the coal constituents and particularly as to how their structure varies in relation to different depositional and post-diagenetic environments. Much may be learned from extensive investigations of recent swamp environments similar to those undertaken by SPACKMAN and his colleagues [20]. Increasingly sophisticated equipment for chemical studies is also becoming available, and for the chemical petrologist, instruments like the laser-micropyrolysis system described by VASTOLA, GIVEN, DUTCHER and PIRONE [81] may open completely new fields in the study of organic sediments. Despite the recent contractions in the coal industries of a number of major coal-producing countries, geological, chemical and geochemical interest in coals has never been so widespread and intense as it is at present. Substantial progress in these areas of coal science seems almost inevitable.

a

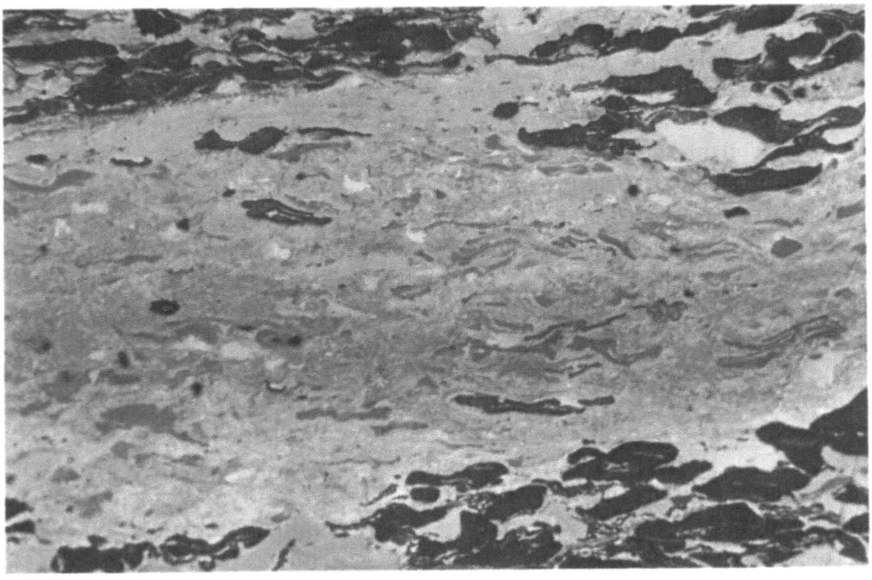

(× 360)

Fig. 7. a) Low-reflecting, thick-walled microspores closely associated with thin-walled, high-reflecting microspores from low-rank bituminous coal (after BELL [82]). b) Megaspore showing a transition in optical properties from exinite to vitrinite of the same coal (after BELL and MURCHISON [78]). c) Exinite in a low-rank coal showing transition in optical properties and morphology to semifusinite of varying character (after BELL and MURCHISON [78]). d) Vitrinite from the Old Red Sandstone showing increased reflectivity around inclusions that are probably radioactive phosphates. The unaffected vitrinite is of lower reflectivity and granular in appearance. All photomicrographs of relief-polished surfaces immersed under oil of refractive index 1.520

Fig. 7b
(× 42)

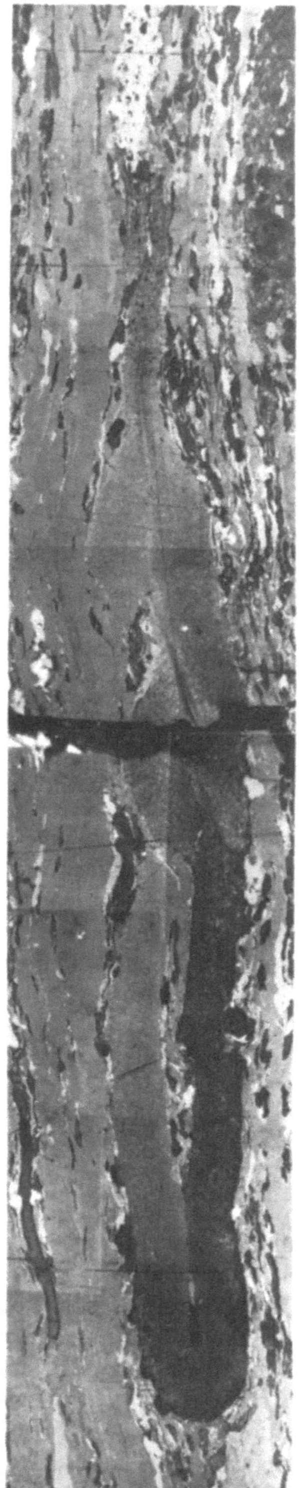

Fig. 7c
(× 123)

Fig. 7d ($\times$ 720)

X. Appendix—Maceral Separation

The extensive physical and chemical investigations of the three maceral groups have been possible because of their successful separation from one another by essentially non-destructive techniques. The methods employed have almost invariably involved very fine grinding of the coal followed by concentration of the different maceral groups in liquids of varying specific gravity. For example, DORMANS, HUNTJENS and VAN KREVELEN [83] used differing concentrations of aqueous solutions of zinc chloride to achieve their separations, while KRÖGER and his collaborators [84] employed carbon tetrachloride-toluene mixtures of varying density.

These methods of separation allowed the collection of valuable data on the properties and chemical constitution of the macerals. It was recognized that there was the possibility of oxidation of certain of the coal constituents during the necessary prolonged, preliminary fine-grinding of the coal, or of some chemical alteration when organic liquids were used to separate fractions of differing density. There is little evidence, however, of such effects in the mass of data that has been produced. Nowadays many large-scale separations of the maceral groups are carried out by grinding the coals to controlled particle size in inert atmospheres and using ultra-high speed centrifuging in water for separation.

Successful "sink and float" separations depend upon a high initial concentration in the coal of the required maceral group. The method cannot therefore be used to separate for chemical study minor petrographic constituents whose densities are similar to those of the dominant maceral within a group. Thus, MURCHISON and JONES [70], when investigating resinite in bituminous coals, employed a simple mechanical method on a semimicro-scale to separate this maceral from coal types which carried high proportions of sporinite of similar density.

Chemical methods of extraction of petrographic constituents from coals have not been widely employed, presumably because of the fear of chemically altering the macerals. NEAVEL and MILLER [74] were forced to use a 10 percent solution of potassium hydroxide to free cutinite from weathered vitrinite in Indiana "paper coal", but stated that their chemical data still coincided with the exinite regressions published by VAN KREVELEN [4]. A method that will have to be increasingly applied if certain petrographic constituents are to be investigated chemically is the use of hydrochloric and hydrofluoric acids to remove syngenetic carbonates and silicates that are intimately associated with the organic matter. The method is standard and accepted in the study of kerogens [85], but has not been widely used in coal geochemistry. DOUGLAS *et al.* [77] have already employed the method in an investigation of fatty acids in the rich algal coals called torbanites. Similar treatment had little if any effect on the band intensities of infrared spectra of mineral-free resinites that were used as controls in studies of the absorption patterns of demineralized alginites [86]. Wider application of the method would thus seem feasible and safe.

References

1. BREGER, I. A.: Geochemistry of coal. Econ. Geol. **53**, 823–841 (1958).
2. KREVELEN, D. W. VAN: Geochemistry of coal. In: Organic geochemistry, ed. I. A. BREGER, p. 183–247. London: Pergamon Press 1963.
3. FRANCIS, W.: Coal, 2nd edn. London: Edward Arnold Ltd. 1961.
4. KREVELEN, D. W. VAN: Coal. Amsterdam: Elsevier Publishing Company 1961.
5. LOWRY, H. H.: Chemistry of coal utilization. New York: John Wiley & Sons, Inc. 1963.
6. HUCK, G., and J. KARWEIL: Physikalisch-chemische Probleme der Inkohlung. Brennstoff-Chem. **36**, 1–11 (1955).
7. TEICHMÜLLER, M., and R.: Geological causes of coalification. In: Coal science, ed. R. F. GOULD. Advances in chemistry series, No. 55, p. 133–155. Washington: American Chemical Society 1966.
8. – Geological aspects of coal metamorphism. In: Coal and coal-bearing strata, ed D. G. MURCHISON and T. S. WESTOLL. Edinburgh: Oliver and Boyd 1968.
9. KUYL, O. S., and R. J. H. PATIJN: Coalification in relation to depth of burial and geothermic gradient. 4ᵉ Congr. de Strat. et de géol. du Carbonifère. Paris, 1958, 357–365 (1961).
10. WHITE, D.: Progressive regional carbonisation of coal. Trans. Am. Inst. Mining Met. Engrs. **71**, 253–281 (1925).
11. STADNICHENKO, T.: Progressive regional metamorphism of the Lower Kittanning coal bed, Western Pennsylvania. Econ. Geol. **29**, 511–543 (1934).
12. ROBERTS, J.: Transition stages in the formation of coals. 1 and 2. Coke Gas **18**, 25–28, 68–72 (1956).
13. DRYDEN, I. G. C.: How was coal formed? Coke Gas **18**, 123–127, 138, 181–184 (1956).
14. FUCHS, W.: Neuere Untersuchungen über die Entstehung der Kohle. Chemiker-Ztg. **76**, 61–66 (1952).
15. MERRIAM, D. F.: Symposium on cyclic sedimentation. Kansas geol. Surv. Bull. 169 (1964).
16. WESTOLL, T. S.: Sedimentary rhythms in coal-bearing strata. In: Coal and coal-bearing strata, ed. D. G. MURCHISON and T. S. WESTOLL. Edinburgh: Oliver and Boyd 1968.
17. SCHOLL, D. W.: Sedimentation in modern coastal swamps, Southwestern Florida. Bull. Am. Assoc. Petrol. Geologists **47**, 1581–1603 (1963).
18. – Recent sedimentary record in mangrove swamps and rise in sea level over the southwestern coast of Florida: Pt. 1. Marine Geol. **1**, 344–366 (1964).

19. SCHOLL, D. W.: Recent sedimentary record in mangrove swamps and rise in sea level over the southwestern coast of Florida: Pt. 2. Marine Geol. **2**, 343–346 (1964).
20. SPACKMAN, W., C. P. DOLSEN, and W. RIEGEL: Phytogenic organic sediments and sedimentary environments in the Everglades-mangrove complex. Palaeontographica **117** B, 135–152 (1966).
21. PLUMSTEAD, E. P.: The Permo-Carboniferous Coal Measures of the Transvaal, South Africa – an example of the contrasting stratigraphy in the southern and northern hemispheres. 4e Congr. de strat. et de géol. du Carbonifère, Heerlen, 1958, 545–550, Tome II (1961).
22. TEICHMÜLLER, M.: Die Genese der Kohle. 4e Congr. de strat. et de géol. du Carbonifère, Heerlen, 1958, 699–722, Tome III (1962).
23. JACOB, H.: Untersuchungen über die Beziehung zwischen dem petrographischen Aufbau von Weichbraunkohlen und der Brikettierbarkeit. Freiberger Forschungsh. A **45**, Berlin (1956).
24. MARSHALL, C. E.: Coal petrology. Econ. Geol., 50th Anniv. vol., 757–834 (1955).
25. PUTZER, H.: Die Steinkohlenvorkommen Brasiliens. Glückauf **91**, 227–237 (1955).
26. TAYLOR, G. H., and S. WARNE: Some Australian coal petrological studies and their geological implications. Proc. Intern. Comm. Coal Petrol. 1958, 75–83 (1960).
27. TEICHMÜLLER, M.: Rekonstruktionen verschiedener Moortypen des Hauptflözes der niederrheinischen Braunkohle. In: Die Niederrheinische Braunkohlenformation, ed. W. AHRENS. Fortschr. Geol. Rheinland-Westfalen 1 and 2, 599–612 (1958).
28. SMITH, A. H. V.: The palaeoecology of Carboniferous peats based on the miospores and petrography of bituminous coals. Proc. Yorkshire geol. Soc. **33**, 423–474 (1962).
29. – Palaeoecology of Carboniferous peats. In: Problems in Palaeoclimatology, ed. A. E. M. NAIRN, p. 57–75. London: Interscience Publishers 1965.
30. – Zur Petrologie und Palynologie der Kohlenflöze des Karbons und ihrer Begleitschichten. In: Paläobotanische, kohlenpetrographische und geochemische Beiträge zur Stratigraphie und Kohlengenese. A Symposium. Fortschr. Geol. Rheinland-Westfalen **12**, 285–302 (1964).
31. WHITE, D.: Role of water conditions in the formation and differentiation of common (banded) coals. Econ. Geol. **28**, 556–570 (1933).
32. TEICHMÜLLER, M.: Zum petrographischen Aufbau und Werdegang der Weichbraunkohle (mit Berücksichtigung genetischer Fragen der Steinkohlenpetrographie). Geol. Jahrb. **64**, 429–488 (1950).
33. SPACKMAN, W.: The maceral concept and the study of modern environments as a means of understanding the nature of coal. Trans. N. Y. Acad. Sci., Ser. II, **20**, 411–423 (1958).
34. International Committee for Coal Petrology. International handbook of coal petrography, 2nd edn. Paris: Centre National de la Recherche Scientifique 1963.
35. STACH, E.: Der Collinit der Steinkohlen. Geol. Jahrb. **74**, 39–62 (1951).
36. TERRES, E., H. DÄHNE, B. NANDI, C. SCHEIDEL, and K. TRAPPE: Die Entscheidung der Frage der Entstehung von Faserkohle auf Grund ihrer spezifischen Wärmen. Brennstoff-Chem. **37**, 269–277, 342–347, 366–370 (1956).
37. AUSTEN, D. E. G., D. J. E. INGRAM, P. H. GIVEN, and L. W. HILL: Electron spin resonance study of pure macerals. In: Coal science, ed. R. F. GOULD. Advances in chemistry series, No. 55, p. 344–362. Washington: Amer. Chem. Soc. 1966.
38. GIVEN, P. H., and C. R. BINDER: The use of electron spin resonance for studying the history of certain organic sediments. In: Advances in organic geochemistry, 1964, ed. G. D. HOBSON and M. C. LOUIS, p. 147–164. London: Pergamon Press 1966.
39. SCHUHMACHER, J. P., F. J. HUNTJENS, and D. W. VAN KREVELEN: Chemical structure and properties of coal. XXVI. studies on artificial coalification. Fuel **39**, 223–234 (1960).
40. KOSANKE, R. M., and J. A. HARRISON: Microscopy of resin rodlets of Illinois coal. Circ. Ill. Geol. Surv. No. 234, 1–14 (1957).
41. TAYLOR, G. H., and A. C. COOK: Sclerotinite in coal – its petrology and classification. Geol. Mag. **99**, 41–52 (1962).
42. STACH, E.: Die biochemische Inkohlung des Exinits im Mikrobild. Brennstoff-Chem. **43**, 72–78 (1962).
43. – Resinit and Sklerotinit als Kennzeichen der Kohlenflözfazies. 5e Congr. de strat. et de géol. du Carbonifère, Paris, 1963, 1003–1013 (1964).
44. FISCHER, F., and H. SCHRADER: Entstehung und chemische Struktur der Kohle. Essen: Girardet 1922.
45. GIVEN, P. H., W. SPACKMAN, A. COHEN, J. IMBALZANE, E. CASIDA, and T. HISCOTT: Botanical, chemical and microbiological studies of a accumulation processes in the Everglades of

Florida. In: Advances in organic geochemistry, 1966, ed. G. D. HOBSON. London: Pergamon Press (in Press).

46. DAVIS, J. H.: The ecology and geologic role of mangroves in Florida. Carnegie Inst. Wash. Publ. **517**, 303–412 (1940).
47. FLAIG, W.: Effect of microorganisms in the transformation of lignin to humic substances. Geochim. Cosmochim. Acta **28**, 1523–1537 (1964).
48. ADLER, E.: The status of research on lignin. Z. Papier **15**, 604–609 (1961).
49. FLAIG, W.: Humic substances and coalification. In: Coal science, ed. R. F. GOULD, p. 58–68. Washington: Amer. Chem. Soc. 1966.
50. STEELINK, C.: Free radical studies of lignin, lignin degradation products and soil humic acid. Geochim. Cosmochim. Acta **28**, 1615–1622 (1964).
51. DRAGUNOV, S. S.: In: Soil Organic Matter (M. M. KONOVA), p. 65. London: Pergamon Press 1961.
52. ZUBOVIC, P.: Physico-chemical properties of certain minor elements as controlling factors in their distribution in coal. In: Coal science, ed. R. F. GOULD, p. 221–231. Washington: Amer. Chem. Soc. 1966.
53. MANSKAYA, S. M., and T. V. DROZDOVA: Geochemistry of organic substances. P. 155–281. London: Pergamon Press 1968.
54. SCHOPF, J. M.: Variable coalification: the processes involved in coal formation. Econ. Geol. **43**, 207–225 (1948).
55. KREVELEN, D. W. VAN: Graphical-statistical method for the study of structure and reaction processes of coal. Fuel **29**, 269–284 (1950).
56. KREULEN, D. J. W., and F. J. KREULEN-VAN SELMS: Humic acids and their role in the formation of coal. Brennstoff-Chem. **37**, 14–19 (1956).
57. DRYDEN, I. G. C.: Chemical constitution and reactions of coal. In: Chemistry of coal utilization, suppl. vol., ed. H. H. LOWRY, p. 232–295. New York and London: John Wiley & Sons, Inc. 1963.
58. – Carbon-hydrogen groupings in the coal molecule. Fuel **41**, 55–61, 301–304 (1962).
59. GIVEN, P. H.: The distribution of hydrogen in coals and its relation to coal structure. Fuel **39**, 147–153 (1960).
60. BROWN, J. K., W. R. LADNER, and N. SHEPPARD: A study of the hydrogen distribution in coal-like materials by high-resolution nuclear magnetic resonance spectroscopy. I. The measurement and interpretation of the spectra. Fuel **39**, 79–86 (1960).
61. – – A study of the hydrogen distribution in coal-like materials by high resolution nuclear magnetic resonance spectroscopy. II. A comparison with infra-red measurement and the conversion to carbon structure. Fuel **39**, 87–96 (1960).
62. GIVEN, P.: Dehydrogenation of coals and its relation to coal structure. Fuel **40**, 427–431 (1961).
63. MONTGOMERY, R. S., and E. D. HOLLY: Decarboxylation studies of the structures of the acids obtained by the oxidation of bituminous coal. Fuel **36**, 63–75 (1957).
64. STERNHELL, S.: Further aspects of the chemistry of hydroxyl groups in Victorian brown coals. Australian J. Appl. Sci. **9**, 375–379 (1958).
65. BROOKS, J. D., R. A. DURIE, and S. STERNHELL: Chemistry of brown coals. II. Infra-red spectroscopic studies. Australian J. Appl. Sci. **9**, 63–80 (1958).
66. – Organic sulfur in coal. J. Inst. Fuel. **29**, 82–85 (1956).
67. ROY, M. M.: Organic sulfur in groups in high sulphur Assam coal. Naturwissenschaften **43**, No. 21, 497 (1956).
68. IYENGAR, M. S., S. GUHA, and M. L. BERI: The nature of sulfur groupings in abnormal coals. Fuel **39**, 235–243 (1960).
69. BROWN, J. K.: Macerals. B.C.U.R.A. Bull. **23**, 1–17 (1959).
70. MURCHISON, D. G., and J. M. JONES: Properties of the coal macerals: Elementary composition of resinite. Fuel **42**, 141–158 (1963).
71. – – Resinite in bituminous coals. In: Advances in organic geochemistry. Proceedings of the International Meeting in Milan, 1962, ed. U. COLOMBO and G. D. HOBSON, p. 49–69. London: Pergamon Press 1964.
72. – Properties of the coal macerals: Infrared spectra of resinites and their carbonized and oxidized products. In: Coal science, ed. R. F. GOULD. Advances in chemistry series, No. 55, p. 307–331. Washington: Amer. Chem. Soc. 1966.
73. LEGG, V. H., and R. V. WHEELER: Plant cuticles. Part II. Fossil plant cuticles. J. Chem. Soc. **127**, 2449–2458 (1929).

74. NEAVEL, R. C., and L. V. MILLER: Properties of cutinite. Fuel **39**, 217–222 (1960).
75. MAXWELL, J. R.: Studies in organic geochemistry. Ph. D. Thesis, University of Glasgow (1967).
76. DOUGLAS, A. G., G. EGLINTON, and W. HENDERSON: Thermal alteration of the organic matter in sediments. In: Advances in organic geochemistry, 1966, ed. G. D. HOBSON. London: Pergamon Press (in Press).
77. –, K. DOURAGHI-ZADEH, G. EGLINTON, J. R. MAXWELL, and J. N. RAMSAY: Fatty acids in sediments including the Green River shale (Eocene) and Scottish Torbanite (Carboniferous). In: Advances in organic geochemistry, 1966, ed. G. D. HOBSON. London: Pergamon Press (in Press).
78. BELL, J. A., and D. G. MURCHISON: Biochemical alteration of exinites and the origin of some semifusinites. Fuel **45**, 407–415 (1966).
79. STACH, E.: Radioaktive Inkohlung. Brennstoff-Chem. **39**, 329–331 (1958).
80. ERGUN, S., W. F. DONALDSON, and I. A. BREGER: Some physical and chemical properties of vitrains associated with uranium. Fuel **39**, 71–77 (1960).
81. VASTOLA, F. J., P. H. GIVEN, R. R. DUTCHER, and A. J. PIRONE: A laser-micropyrolysis system for the study of organic sediments and inclusions. In: Advances in organic geochemistry, 1966, ed. G. D. HOBSON. London: Pergamon Press (in Press).
82. BELL, J. A.: The petrographic character and sedimentary environment of the Harvey-Beaumont seam of Northumberland and North Durham. Ph. D. Thesis. University of Newcastle upon Tyne (1966).
83. DORMANS, H. N. M., F. J. HUNTJENS, and D. W. VAN KREVELEN: Chemical structure and properties of coal. XX. Composition of the individual macerals (vitrinites, fusinites, micrinites and exinites). Fuel **36**, 321–333 (1957).
84. KRÖGER, C., and F. KUTHE: Über die Isolierung der Steinkohlengefügebestandteile aus Glanz- und Mattkohlen von Ruhrflözen. Glückauf **93**, 122–135 (1957).
85. FORSMAN, J. P., and J. M. HUNT: Insoluble organic matter (kerogen) in sedimentary rocks of marine origin. In: Habitat of oil, ed. L. G. WEEKS, p. 747–778. Tulsa: Amer. Assoc. Petrol. Geol. 1958.
86. MILLAIS, R., and D. G. MURCHISON: Properties of coal macerals: Infrared spectra of alginites. Fuel **48**, 247–258 (1969).

CHAPTER 30

Pre-Paleozoic Sediments and Their Significance for Organic Geochemistry

PRESTON E. CLOUD, Jr.

Department of Geology,
University of California,
Santa Barbara, California

Contents

I. Introduction

All valid inferences about the history of the earth derive from or take into account information obtained by various means from the different rocks or rock-successions that make up the geologic record (including meteorites). The materials of organic geochemistry also are found within these rocks. Such materials are significant only insofar as they are interpreted in the context of the local succession and total geologic record. Substances of interest to the organic geochemist, however, are mostly confined to sedimentary rocks and the following discussion will be limited to pre-Paleozoic (or "Precambrian") sedimentary rocks—referred to as "sediments".

From radiometric dating we know that the pre-Paleozoic comprises roughly the first seven-eighths of the earth's history—from the origin of the earth about 4.5 to 4.8 billion* years (B.Y.) ago [1], to the beginning of the Paleozoic about 0.6 to possibly as much as 0.7 B.Y. ago. It is possible to deal with this record in a short space primarily because the first billion years or so of it are missing and the remainder is as yet not well understood. Nevertheless, we do know many rocks and areas of investigation that should be of interest to organic geochemists, and it is the purpose of this brief report to suggest what some of these are.

* In this chapter, billion = 10^9.

II. Distribution of Relatively Little-Altered Sediments

As an original sedimentary deposit is deformed and metamorphosed, materials of interest for organic geochemistry become altered, by heat and pressure, and migrate. It thus seems advisable at this stage in the investigation of pre-Paleozoic organic components to focus on the least-altered rocks, as well as on those most likely to contain materials suitable for investigation.

From a casual perusal of textbooks and general writings, one might infer that little-altered pre-Paleozoic sediments are rare. Instead, they are fairly widespread on all continents, especially in the younger pre-Paleozoic sedimentary basins at the margins of and exterior to the crystalline shields, and also include very old rocks, reaching ages of more than 3 B.Y. old in eastern South Africa. Relatively unaltered ancient sediments are particularly widespread in Australia, where the pre-Paleozoic consists mainly of a series of great sedimentary basins and only a small part in the southwest is a crystalline shield in the sense that the term "Precambrian shield" is used on other continents. The pre-Paleozoic of Australia is being studied energetically by geologists of the federal and provincial surveys, and a wealth of information is now beginning to appear in the form of explanatory notes and reports accompanying the 1/250,000 geological map series of the Bureau of Mineral Resources, Geology, and Geophysics. Meanwhile the most intensive investigations of little-altered pre-Paleozoic sediments that have been published to date are probably those of the Soviet geologists dealing with Siberia and the Russian Platform. Here large groups of investigators at the Geological Institute in Moscow and the Institute for Precambrian Studies in Leningrad (both of the Akademia Nauk, U.S.S.R.) are investigating the stratigraphy and paleontology of the rocks [2–11], although little organic geochemistry seems to have been done as yet. The best known and most complete sequence of pre-Paleozoic sediments from the viewpoint of radiometric ages and chronologic succession, however, is that of South Africa [12].

There would be little point here in attempting to outline the major features of the pre-Paleozoic sediments for the different continents. For that the reader should refer to summary volumes currently being prepared (or already issued) under the editorship of LOTZE and SCHMIDT [13] and of RANKAMA [14, 15].

What is of interest here is to consider briefly the kinds of sediments found and the broad conclusions to which this leads, and what may be some of the potentially fruitful opportunities for research in the organic geochemistry of these ancient rocks.

III. Nature of the Sedimentary Record

A. Nomenclatural Problems

It is well to begin this section with a few observations on the nomenclature of the pre-Paleozoic rocks (Table 1). From the viewpoint of vital processes (as well as other features) the earth's history can be divided into two unequal but meaningful major divisions. The first seven-eighths is alternatively called Precambrian or Cryptozoic. The last eighth of earth history (younger than 600 to 700 million years ago), with abundant visible records of multicellular animal life, is called, collectively, the Phanerozoic. Neither term is entirely satisfactory. Although

Table 1. *Major divisions of earth history*

Eons			Eras	Approximate age in years × 10^9
Phanerozoic			Cenozoic	
				0.07
			Mesozoic	
				0.2
			Paleozoic	
				0.6–0.7
Pre-Paleozoic or "Precambrian"	Cryptozoic	Proterozoic	Late Proterozoic	
				1.8–2.0
	"Cryptozoic"		Early Proterozoic	
				2.5–2.7
		Archean		
				3.3–3.6
Records unknown				
				4.6 ±

Accumulation of the planets in process.

Phanerozoic-Cryptozoic makes a nice contrast, Cryptozoic applies logically only to the younger Precambrian. Precambrian itself gets us into even worse semantic difficulties. In the formal sense in which currently employed, it signifies to many geologists (including this writer) not merely the record antecedent to the Cambrian, but the record antecedent to the Paleozoic and to the Phanerozoic. It is bad enough when we speak of "post-Precambrian" (as is commonly done), but we get into real difficulties when we try to discuss rocks or events that we might consider including in the Paleozoic Era but antecedent to the Cambrian Period (currently the basal Paleozoic period). Such rocks would then literally be pre-Cambrian rocks of post-Precambrian age!

Perhaps the best way out of this dilemma, without proliferating jargon, is to reinstate the terms Archean and Proterozoic for main divisions of the pre-Paleozoic, as is now being widely done and as shown on Table 1. The major terms then become Archean, Proterozoic, and Phanerozoic. The facts that they are not coordinate in etymology or time included and that we use them for both time and rocks may seem deplorable to some; but the objective here is unambiguous communication without undue violence to existing nomenclatural ecology, and these terms will accordingly be used in the sense implied above and in Table 1.

Not all semantic problems are eliminated by this device, to be sure. For instance, is it proper to designate as Archean the unrecorded history of the earth between completion of accumulation at about 4.5 to 4.8 B.Y. and the first dated event thereafter at 3.5 to 3.6 B.Y. [16]. That problem, fortunately, does not concern us here, because we will discuss only rocks that exist and events that are known to have happened. Of course, when referring to Archean and Proterozoic together, the term in most common use is Precambrian, but I find it less ambiguous to choose between Cryptozoic and pre-Paleozoic, and I prefer the latter because it is the least ambiguous of all. Eventually a comprehensive new terminology may be needed, but until more and better information is available movements in this direction are likely to be short-lived.

B. Kinds of Sediments Found and Their Implications for the Evolution of the Biosphere

As observed by PETTIJOHN in 1943 [17] and by others before and since 1960 (e.g., JAMES [18]), the kinds of sedimentary rocks found in the pre-Paleozoic are much the same as those found in Paleozoic and younger rocks. What is of interest in the present connection is that they occur in different proportions at different times and that these proportions show a rough evolutionary sequence [19, 20] that can be related to the concommitant evolution of vital processes. Silica precipitates decrease and carbonate sediments increase upward from older to younger rocks, with transition from abundant chert to abundant carbonate at 2 B.Y. or more [20]. Rhythmic open-water precipitates of alternating Fe-poor and Fe-rich siliceous rocks called banded-iron-formation (BIF) are essentially restricted to rocks between 3 to 3.2 and 1.8 to 2 B.Y. old [20–22], in contrast to younger and down-ranging iron formations of other kinds and different genesis. Red beds – detrital deposits of terrestrial or near-shore origin in which ferric oxides coat the individual grains and fill part of the space between grains – occur mainly or entirely in rocks younger than the youngest thick and extensive BIF [20]. Uraninite grains that may be detrital and should have undergone alteration in the persisting presence of free O_2 are found in rocks as young as 2 B.Y. but no younger [23]. Sedimentary sulfate deposits (gypsum and anhydrite), with trivial and perhaps with relatively late exceptions, appear not to occur in pre-Paleozoic rocks [20]. There are secular increases upward through known Archean and Proterozoic sediments in the ratios Fe_2O_3/FeO, K^+/Na^+, Ca^{2+}/Mg^{2+}, and in the ratio of residual sediments such as relatively pure quartz-sandstones to immature sediments such as graywackes and arkoses [19]. A striking feature of the Archean sediments is the "singular scarcity of true quartzite and almost total lack of limestone" [17], both of which are abundant in rocks younger than about 2 B.Y.

How does this translate to terms relevant to organic geochemistry? To begin with, the prevalence of siliceous precipitates and rarity of carbonate rocks among the older sediments discriminates between alternate models of the atmosphere under which the abiogenic evolution of organic compounds, and subsequent biosynthesis, presumably took place. Moreover, inasmuch as the oldest dated rocks so far known are all crystalline, there is little hope of finding an antecedent record of sediments and some reason to infer a cataclysmic event (such as lunar capture?) at perhaps 3.5 to 3.6 B.Y. which obliterated older records and reset radiometric clocks (by heat transfer from energy conversion?). Life as we know it presumably originated (or reoriginated) subsequent to that event. Alternate models for the atmosphere of biosynthesis, leaving out an extension backward of present conditions, are (1) an atmosphere rich in methane and ammonia, and (2) an atmosphere of juvenile volcanic gases with little or no methane and ammonia. For a variety of reasons neither could have contained more than trace quantities of free oxygen, since this reactive element is almost certainly inimical to pre-biotic chemical evolution of critical organic compounds, to biosynthesis, and to persistence of life in the absence of oxygen-controlling enzymes. If NH_3 were in the atmosphere, it would also be in the hydrosphere. The resultant high pH would favor precipitation of $CaCO_3$ and $CaMg(CO_3)_2$ and oppose the precipitation of SiO_2, whereas the record of the oldest sediments shows the reverse. If NH_3 were negligible there is little reason to invoke CH_4. Nor is there evidence of it in the form of the extensive carbon deposits that one would expect to arise from the recombination of CH_4 in an oxygen-poor atmosphere. Thus a primary atmosphere of juvenile volcanic gases such as CO, CO_2, H_2O, N_2, H_2, and probably HCl is likely – at least for the time for which we have a sedimentary record and during which life presumably originated.

The chronologic distributions of BIF and red-beds (as well as possibly detrital uraninite) are consistent with a dependent relation between BIF and oxygen-producing vital processes. The fact that the oldest red-beds appear to follow or overlap slightly with the youngest BIF at about 1.8 to 2 B.Y. before the present, suggests that this may represent the time of introduction in green-plant photosynthesizers of oxygen-absorbing and peroxide-reducing enzymes (the advanced cytochromes and catalases) and the first large scale evasion of O_2 from hydrosphere to atmosphere [20, 22]. The appearance of the first thick and extensive sedimentary sulfate deposits near the base of the Paleozoic is consistent with the concept of a relatively large increase in the amount of atmospheric oxygen at about that time [24]. The increased ratios upward of Fe_2O_3/FeO, K^+/Na^+, Ca^{2+}/Mg^{2+}, and residual to immature sediments, is consistent with increasing importance upward of vital processes in both the hydrosphere and interstitial to soils.

The foregoing implies a wealth of opportunities to seek a relevant body of information in organic geochemistry and to construct rational models of biochemical evolution. Let us next consider some of these potentialities.

C. Deposits of Special Interest for Organic Geochemistry

In choosing an ideal rock for research in the organic geochemistry of the pre-Paleozoic, one would normally seek a rock of known age in which remains of life or other evidences of vital processes were to be found and which had minimum likelihood of addition, subtraction, or gross alteration of materials.

Inasmuch as there are as yet no records of unequivocal metazoan fossils in rocks of undoubted pre-Paleozoic age, the demonstrable record of pre-Paleozoic life is confined to structurally preserved microbiological remains and sedimentary structures similar to those accreted under the influence of known organisms in younger rocks – the so-called stromatolites or layered domal, conical, sphenoidal, digitate, or multiformed deposits of $CaCO_3$, $CaMg(CO_3)_2$, or SiO_2.

Published records of structurally preserved microbiological (algal and bacterial) remains of pre-Paleozoic age are as yet not numerous. The best known is from approximately 2 B.Y. old biogenic chert of the Gunflint Iron Formation in southern Ontario [22, 25–29]. The oldest is from the carbonaceous shales and cherts of the more 3 B.Y. old Figtree Series of eastern South Africa [30, 31]; the biologically most pristine is from chert of the younger Proterozoic Bitter Spring's Formation of central Australia [32]. These occurrences, plus a few others which I regard as including demonstrable or highly probable nanofossils, are shown in chronologic sequence in Table 2.

Other published occurrences known to me [41, 42] involve varying degrees of doubt greater than those listed here. This is not to reject them out of hand, but

Table 2. *Main published occurrences of structurally preserved microbiological remains in rocks of known pre-Paleozoic age**

Approximate age in billions of years	Lithology and rock-unit	Region	Representative source of information
Younger Proterozoic	shale, mudstone, and siltstone of various units	Eastern Baltic, Russian Platform, Siberia	TIMOFEEV (1959, 1960 a and b, 1966) [7–9]
Younger Proterozoic	chert of Bitter Springs Formation	Central Australia	BARGHOORN and SCHOPF (1965) [32]
Proterozoic	Uchusk Series Gonamsk beds	Ayan-Maysk	VOLOGDIN and DROZDOVA (1964) [40]
1	Nonesuch Shale	N. Michigan	BARGHOORN *et al.* (1965) [27] EGLINTON *et al.* (1964) [33]
1.2	limestone of Belt Series	W. central Montana	PFLUG (1964, 1965) [34–35]
1.3	carbonaceous siltstone of Muhos Formation (Jotnian)	W. central Finland	TYNNI and SIIVOLA (1966) [36]
1.9	biogenic chert of Gunflint Iron Formation	S. Ontario	TYLER and BARGHOORN (1954) [25] BARGHOORN and TYLER (1965) [26] CLOUD (1965) [22] CLOUD and HAGEN (1965) [29] BARGHOORN *et al.* (1965) [27]
2.1	chert in Brockman Iron Formation	W. Australia	LA BERGE (1966) [43]
2.7	pyrite nodules in carbonaceous shale lens of Soudan Iron Formation	N.E. Minnesota	CLOUD *et al.* (1965) [37] MEINSCHEIN (1965) [38] BELSKY *et al.* (1965) [39]
3.2	carbonaceous shale and chert in Figtree Series	E. South Africa	PFLUG (1966) [30] BARGHOORN and SCHOPF (1966) [31]

* This list, compiled in June 1967, needs a number of new occurrences and references to bring it up to date.

simply to indicate that I do not find the evidence so far presented sufficiently convincing to list them here. Nor do I necessarily consider that *all* objects illustrated and described from all the occurrences listed are of unequivocally vital origin. The supposedly extensive occurrence of nano-fossils in banded iron formations as reported by LA BERGE [43], for instance, is entirely expectable and probably true, but it needs further study and documentation. Although a number of the structures figured by him are very persuasive fossils, others resemble structures of non-biogenic colloidal origin (common in cherts of various ages) or mineral aggregates. Nevertheless, structurally preserved microbiological remains clearly are to be found far back in the records of the primitive earth. These include morphologically well-preserved procaryotes (non-mitosing cells without nuclear membrane), such as blue-green algae, bacteria, and perhaps procaryotic fungi, in rocks 1.8 to 2 B.Y. old and older. Eucaryotes (having mitosing cells and nuclear membrane) have been found so far only in younger Proterozoic rocks [9, 10, 32].

Apart from specific fossiliferous rocks such as those listed in Table 2, it would seem profitable, from the viewpoint of either organic geochemistry or the search for additional microbiological remains, to pay special attention to very-fine-grained sediments or probable chemical precipitates that contain non-graphitic carbonaceous matter which may represent the degradation products of organisms living at the time. Where such organisms have been sealed within a crystallizing silica gel, as seems to have been the case with cherts in the Gunflint and Bitter Springs beds, they may preserve their original morphology and retain organic molecules resulting from the lysis of original vital products. Fine-grained cherts that are black from contained organic matter are thus the most promising type of sediment in which to search, provided such cherts are primary or very early replacement products. Unfortunately, later silification of stromatolites and the post-depositional or even post-lithification formation of chert nodules or other masses from silica in solution in migrating interstitial waters is very difficult to distinguish from penecontemporaneous chert, leading inevitably to many disappointments in the search for early records of life and its degradation products.

Sediments other than chert that have yielded pre-Paleozoic microbiological remains or products, include cryptocrystalline and oolitic limestone, as well as similarly well-preserved shale, mudstone, and siltstone. In narrowing down the target area within so vast a searching ground as the pre-Paleozoic, it seems only sensible to concentrate on samples of similar nature – especially those of a dark color attributable to dispersed organic matter, or containing hydrocarbons (both solid and liquid hydrocarbons occur in pre-Paleozoic rocks) or dispersed small pyrite spherules or grains. It would, for instance, involve a much higher element of risk in an already risky area to search for organic matter in white, red, or coarse-grained detrital sediments or in dolomite or other recrystallized sediments. Among the fine-grained rocks, evidence of biological sedimentary processes, as in the construction of stromatolites, is an additional factor indicating opportunity for the preservation of biological materials, although sediments called stromatolitic by geologists are not invariably of biological origin.

D. Opportunities for Research

Apart from simply undertaking a systematic survey of what may be found in sediments known to contain microbiological remains or otherwise believed to be hopeful, some prospects for particular investigation deserve mention.

It has been stated in different places recently that evidence of life has been found in the oldest sediments known on earth. That observation presumably applies to carbonaceous shales and cherts in the Figtree Series of South Africa. Beneath the Figtree, however, there is another 35,000 feet or more of sediments at the base of the Swaziland System – the Onverwacht Series. The detailed stratigraphy and regional relations of this vast pile of sediments and basic volcanics are only now being worked out by M. J. and R. P. Viljoen of the University of Witwatersrand. We know already, however, that it locally contains carbonaceous shale and chert which, although by no means unaltered, offer at least some slim hope of adding another chapter to the record of pre-Paleozoic life*.

Sedimentary sulfate deposits reported to be associated with pre-Paleozoic rock-sequences (as well as little-altered pyritic sediments) should be investigated more extensively as to their ratios of sulfur isotopes, in particular for the implications of such ratios for microbiology, photosynthetic mechanisms, and atmospheric oxygen. Deposits perhaps suitable for such investigations are reported from the younger Proterozoic of Arctic Canada [44] and central Australia [45], as well as from 1.2 B.Y. old rocks in New York [46].

The many elegant studies of the branched isoprenoids, steranes, and other alkanes such as have been carried out by CALVIN and co-workers, ORÓ and associates, and MEINSCHEIN and collaborators, should be extended to more of the promising samples available – the coal in the Michigamme Shale, for instance [47], as well as the numerous little-altered carbonaceous shales of the Northern Territory, Western Australia, and northwest Queensland in Australia. They should also be checked and extended by studies of the stable isotopes of carbon, especially utilizing HOERING's method of cross-checking the carbon isotope ratios of extractable substances against non-extractables [48] as a means of enhancing or reducing confidence regarding the possible endemism of the volatiles.

Other classes of lipids should be isolated and studied more extensively. A new chapter in the organic geochemistry of the pre-Paleozoic, and possibly in the story of biogenesis, could result from such studies.

And finally, a special opportunity is perhaps to be found in the geochemical investigation of pre-Paleozoic sedimentary phosphate deposits [49]. Trace quantities of phosphate certainly occur in a number of pre-Paleozoic sediments, and even small phosphatic nodules are to be found locally in some younger Proterozoic deposits such as the Torridonian of Scotland. It is not generally known outside the U.S.S.R. and China, however, that sedimentary phosphorite deposits of economic and near-economic significance are reported to be widespread in rocks of late Proterozoic age (RIPHEAN and SINIAN) in the eastern U.S.S.R. and China. Such deposits, at least, would comprise an interesting target for research in organic geochemistry. Less interesting, perhaps, but not completely devoid of interest, are the now altered, but presumably original, sedimentary phosphate deposits mentioned by DAVIDSON [49] as being associated with marble and other crystalline rocks of Archean age and containing up to 4 to 5 percent P_2O_5.

Indeed the range of uninvestigated and insufficiently investigated materials and problems, and the diversity of applicable techniques and interests suggests that we are on the verge of a new era in the investigation of pre-Paleozoic rocks; organic geochemistry and paleomicrobiology will continue to play inter-related and significant roles and will increase the scope and number of investigators. Studies in these fields, together with related investigations of the geochronology, physical geochemistry, stratigraphy, sedimentology, petrology, and

* Since this was written in June 1967 structures of probable biologic origin have been described from the Onverwacht Series by A. E. J. ENGEL and others: Algae-like forms in the Onverwacht Series, South Africa – Oldest recognized life-like forms on Earth. Science **161**, 1005 – 1008 (1968).

regional geology of the Archean and Proterozoic successions, are now beginning to unscramble the first seven-eighths of earth history. When a comprehensive summary of decipherable pre-Paleozoic history is written, we may confidently expect it to reveal and in large part to be based on a rich succession of microbiotas and biogeochemical episodes.

References

1. TILTON, G. R., and R. H. STEIGER: Lead isotopes and the age of the earth. Science **150**, 1805–1808 (1965).
2. SEMIKHATOV, M. A.: The suggested stratigraphic scheme for the Precambrian. Izv. Akad. Nauk USSR, Geol. Ser. **4**, 70–84 (1966).
3. KOMAR, V. A.: Upper Precambrian stromatolites in the north of Siberian Platform and their stratigraphic significance [in Russian]. Akad. Nauk USSR, Geol. Inst. 1966.
4. KELLER, B. M.: Problems of the later Precambrian [in Russian]. Priaoda **9**, 30–38 (1959).
5. MENNER, V. V.: On the nomenclature problem of the upper Precambrian group. Intern. Geol. Congr., 21st Copenhagen, 18–23 (1960).
6. REITLINGER, E. A.: Microscopic organic remains and problematica of the ancient beds of the South Siberian Platform [in Russian]. Intern. Geol. Congr., 21st Copenhagen, 140–146 (1960).
7. TIMOFEEV, B. V.: Ancient flora of the Baltic area and its stratigraphic significance [in Russian]. Tr. Inst. for the All-Union Sci. Invest. and Prospecting of Petroleum **129**, 320 p., (1959).
8. – Spore and phytoplankton from Proterozoic and early Paleozoic of Eurasia [in Russian]. Intern. Geol. Congr., 21st Copenhagen, 177–188 (1960).
9. – Precambrian spores [in Russian]. Intern. Geol. Congr., 21st, Copenhagen, 138–149 (1960).
10. – Microphytological investigations of the ancient strata [in Russian]. Akad. Nauk USSR, Laboratory of Precambrian Geology 1966.
11. TOMKEIEFF, S. I.: The Rhiphaean system and the structure of the Russian platform. Proc. Geol. Soc. London **1501**, cviii–cxii (1953).
12. NICOLAYSEN, L. O.: Stratigraphic interpretation of age measurements in Southern Africa. In: Petrologic studies – A volume to Honor A. F. Buddington. Geol. Soc. Am. 569–598 (1962).
13. LOTZE, F., and K. SCHMIDT (eds.): Prakambrium, erster Teil, Nördliche Halbkugel. Stuttgart: Ferdinand Enke 1966.
14. RANKAMA, KALERVO: The Precambrian, vol. 1. New York: Interscience Publ. 1963.
15. – The Precambrian, vol. 2. New York: Interscience Publ. 1965.
16. CATANZARO, E. J.: Zircon ages in Southwestern Minnesota. J. Geophys. Res. **68** (7), 2045–2048 (1963).
17. PETTIJOHN, F. J.: Archean sedimentation. Bull. Geol. Soc. Am. **54**, 925–972 (1943).
18. JAMES, H. L.: Problems of stratigraphy and correlation of Precambrian rocks with particular reference to the Lake Superior Region. Am. J. Sci. **258-A**, 104–114 (1960).
19. ENGEL, A. E. J., and C. G. ENGEL: Continental accretion and the evolution of North America. In: Krishnan volume, Advancing frontiers in geology and geophysics (A. P. SUBRAMANIAM and S. BALAKRISHA, eds.). Indian Geophysical Union 1964.
20. CLOUD, P. E., JR.: Pre-metazoan evolution and the origins of the Metazoa, p. 1–72. In: Evolution and environment (E. T. DRAKE, ed.). New Haven: Yale University Press 1968.
21. LEPP, HENRY, and S. GOLDICH: Origin of Precambrian iron formations. Econ. Geol. **59**, 1025–1060 (1964).
22. CLOUD, P. E., JR.: Significance of the Gunflint (Precambrian) microflora. Science **148**, 27–35 (1965).
23. HOLLAND, H. D.: Model for the evolution of the earth's atmosphere. In: Petrologic studies – a volume to honor A. F. Buddington. Geol. Soc. Am. 447–477 (1962).
24. BERKNER, L. V., and L. C. MARSHALL: History of major atmospheric components. Proc. Natl. Acad. Sci. **53** (6), 1215–1225 (1965).
25. TYLER, S. A., and E. S. BARGHOORN: Occurrence of structurally preserved plants in Precambrian rocks of the Canadian shield. Science **119**, 606–608 (1954).
26. BARGHOORN, E. S., and S. A. TYLER: Microorganisms from the Gunflint Chert. Science **147**, 563–577 (1965).

27. Barghoorn, E. S., W. G. Meinshein, and J. W. Schopf: Paleobiology of a Precambrian shale. Science **148**, 461 – 472 (1965).
28. Oro, John, D. W. Nooner, A. Zlatkis, S. A. Wikstrom, and E. S. Barghoorn: Hydrocarbons in a sediment of biological origin about two billion years old. Science **148**, 77 – 79 (1965).
29. Cloud, P. E., Jr., and H. Hagen: Electron microscopy of the Gunfline microflora – preliminary results. Proc. Natl. Acad. Sci. **54** (1), 1 – 8 (1965).
30. Pflug, H. D.: Structured organic remains from the Figtree series of the Barberton Mountain Land. Univ. Witwatersrand, Econ. Geol. Res. Unit, Inf. Cir. **28** (1966).
31. Barghoorn, E. S., and J. W. Schopf: Microorganisms three billion years old from the Precambrian of South Africa. Science **152**, 758 – 763 (1966).
32. – – Microorganism from the late Precambrian of Central Australia. Science **150**, 337 – 339 (1965).
33. Eglinton, Geoffrey, P. M. Scott, Ted Belsky, A. L. Burlingame, and Melvin Calvin: Hydrocarbons of biological origin from a one-billion-year-old sediment. Science **145**, 263 – 264 (1964).
34. Pflug, H. D.: Niedere Algen und ähnliche Kleinformen aus dem Algonkium der Belt-Serie. Ber. Oberhess. Ges. Natur-Heilk. Gießen, Naturw. Abt. **33**, 403 – 411 (1964).
35. – Organische Reste aus der Belt-Serie (Algonkium) von Nordamerika. Z. Palaont. **39**, 10 – 25 (1965).
36. Tynni, Risto, and Jaakko Siivola: On the Precambrian microfossil flora in the siltstone of Muhos, Finland. Comp. Rend. Soc. Geol. Finlande **38**, 127 – 133 (1966).
37. Cloud, P. E., Jr., J. W. Gruner, and H. Hagen: Carbonaceous rocks of the Soudan Iron Formation (early Precambrian). Science **148**, 1713 – 1716 (1965).
38. Meinshein, W. G.: Soudan Formation: organic extracts of early Precambrian rocks. Science **150**, 601 – 605 (1965).
39. Belsky, Ted, R. B. Johns, E. D. McCarthy, A. L. Burlingame, W. Richter, and Melvin Calvin: Evidence of life processes in a sediment two and a half billion years old. Nature **206**, No. 4983, 446 – 447 (1965).
40. Vologdin, A. G., and N. A. Drozdova: Some species of algae from the Gonamsk suite, Uchusk Series of the Proterozoic of the Ayan-Maysk Region of the far east [in Russian]. Dokl. Akad. Nauk SSSR **159** (1), 114 – 116 (1964).
41. Madison, K. M.: Fossil protozoans from the Keewatin sediments. Trans. Illinois State Acad. Sci. **50**, 287 – 290 (1957).
42. Marshall, C. G. A., J. W. May, and C. J. Perret: Fossil microorganisms – possible presence in Precambrian shield of Western Australia. Science **144**, 290 – 292 (1964).
43. LaBerge, G. L.: Microfossils and Precambrian Iron-Formations. Bull. Geol. Soc. Am. **78**, 331 – 342 (1966).
44. Thorsteinsson, R., and E. L. Tozer: Banks, Victoria, and Stefansson Islands, Artic Archipelageo. Can. Dep. Mines Tech. Surv., Geol. Surv. Can. Mem. **330** (1962).
45. Stewart, Alistair: Personal communications.
46. Brown, J. S., and A. E. J. Engel: Revision of Grenville stratigraphy and structure in the Balmat-Edwards district, Northwest Adirondacks, New York. Bull. Geol. Soc. Am. **67**, 1599 – 1622 (1956).
47. Tyler, S. A., E. S. Barghoorn, and L. P. Barrett: Anthracitic coal from Precambrian upper Huronian black shale of the Iron River district, Northern Michigan. Bull. Geol. Soc. Am. **68**, 1293 – 1304 (1957).
48. Hoering, T. C.: The stable isotopes of carbon in the carbonate and reduced carbon of Precambrian sediments. Carnegie Inst. Wash., Yearbook **61**, 190 – 191 (1961).
49. Davidson, C. F.: Phosphate deposits of Precambrian age. Mining Mag. (London) **109**, 205 – 208 (1963).

General Reference

Burlingame, A. L., Pat Haug, Ted Belsky, and Melvin Calvin: Occurrence of biogenic steranes and pentacyclic triterpanes in an Eocene Shale (52 million years) and in an early Precambrian shale (2.7 Billion years): a preliminary report. Proc. Natl. Acad. Sci. **54**, 1406 – 1412 (1965).

CHAPTER 31

Organic Derivatives of Clay Minerals, Zeolites, and Related Minerals

ARMIN WEISS

Institute of Inorganic Chemistry
The University of Munich, West Germany

Contents

I. Introduction

In the field of Biopoesis which is concerned with the abiotic events leading to the origin of life, organic derivatives of minerals are of increasing interest. They allow, *inter alia*, the accumulation of organic matter, the stabilization of metastable compounds by means of complex formation, or the alteration of organic material by catalysis. Of special interest are those minerals which either due to small particle size exhibit large surface areas or for structural reasons have boundary surfaces inside the crystal which are easily accessible from the outside. For this reason, the present work will principally be concerned with minerals that exhibit the aforementioned characteristics. They include within the silicates, minerals such as montmorillonites, vermiculites, micas, swelling chlorites, kaolinite, mixed layer minerals, and zeolites with a wide open channel network. In addition, organic derivatives of uranium mica, swelling vanadates and phosphates, and hydroxyaluminates will briefly be discussed. In view of the fact that little is known on naturally occurring allophanes, these amorphous compounds are just mentioned.

II. Types of Compounds and General Properties

Among minerals that exhibit accessible surfaces within the crystal and which contain exchangeable ions, several types of organic derivatives can be distinguished.

Type 1: *Organic derivatives of the outer surface.* Along outer surfaces of minerals, atoms or ions cannot utilize the same linkage elements as they can within the crystal because co-ordination polyhedra are either only partially developed or atoms or groups of atoms different from those present within a crystal may substitute. For instance, in case of the SiO_4 tetrahedra positioned along a crystal edge, an OH-group will frequently proxy for an oxygen; alternatively, in case of the Al(O, OH) octahedra, a water molecule may substitute for an oxygen or hydroxyl ion, respectively.

These arrangements characteristic for outer surfaces are reactive.

The silanol groups Si-OH may react as follows:

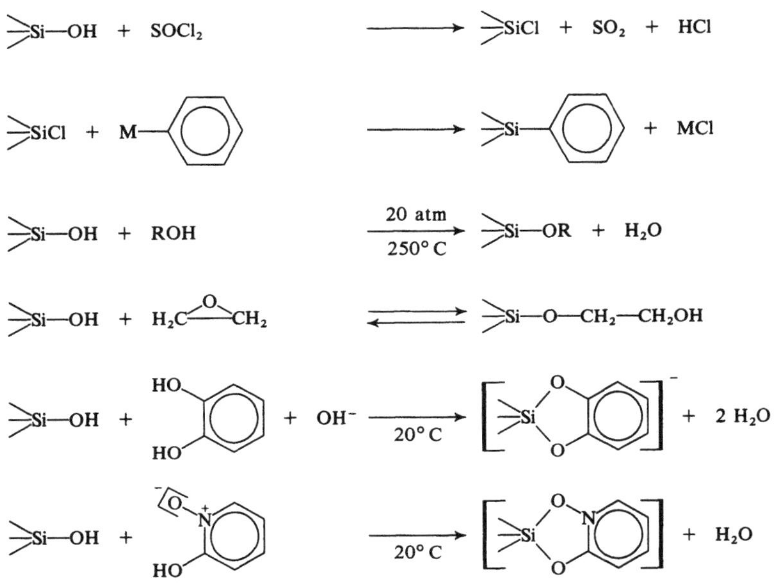

and yield esters or derivatives with silicon-carbon bonds. However, these derivatives are rather unstable in aqueous systems [1–6]. Esters become rapidly hydrolyzed except for derivatives containing pyrocatechol or other o-diphenols which under alkaline conditions form stable anionic silicon complexes [7].

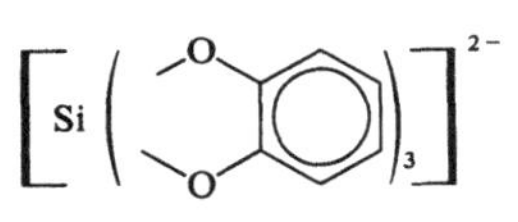

Tropolones, e.g. tropolone and thujaplicine [8, 9], yield stable cationic complexes

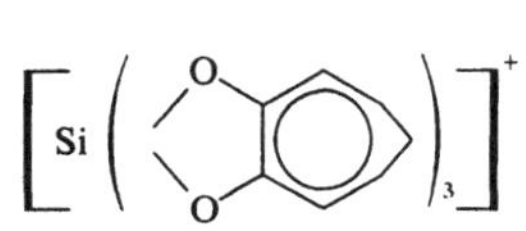

in the pH range between 4 and 6. In both instances, the stability is due to the formation of five- and six-fold oxygen co-ordinated silicon. Although the Si-C linkage is relatively stable, such linkage elements present in surface compounds are only moderately stable in aqueous systems. This is due to a gradual hydrolysis and the resulting elimination of R-$Si(OH)_3$ groups which are soluble and thus can be carried away or may become re-condensed again.

Organic derivatives produced in the vicinity of co-ordination polyhedra surrounding the aluminium ions located along crystal edges will preferentially form with acids, for instance, the long chain sulfonic and fatty acids. In a first step these are fixed as exchangeable anions, then in a second step additional neutral molecules are taken up [10]. Most crucial in this reaction is the hydrophobic bonding, that is, the destruction of the water structure as a result of the presence of long chain hydrophobic residues. These bonding forces are weak. Consequently, the anionic exchange equilibrium is strongly shifted towards the OH-compounds making the adsorption conditions rather unfavorable for the formation of organic derivatives [11]. Stronger linkages may come into existence by employing suitable complex-forming agents. At higher concentrations, however, they will rapidly lead to a complete decomposition of the silicates.

Surface compounds of this type also occur on allophanes; yet, the formation of Si–O–C or Si–C linkages is only favored at extremely low water vapor partial pressures. Adsorption from the gaseous state becomes noteworthy only at relatively high pressures. Frequently, there is a direct relationship between the total amount of fixed organic matter and the specific outer surface area. Inasmuch as this area rarely exceeds 30 to $50 m^2/g$ even for fine crystalline minerals, such organic derivatives are less important as the ones which will be discussed below.

Type 2: *Minerals having exchangeable organic cations*. Cations in exchange position within a crystal may become substituted by certain organic cations [12–18]. In case the inside exchange positions are only accessible through rigid channels as is true for the zeolites, the size of the substituting organic cations is limited by the width of the channel opening. In contrast, if the cations are positioned between layers having variable spacings, essentially all sizes of organic cations may penetrate into the crystals. The larger the size of a cation, the wider is the resulting spacing between neighboring layers. In case of very large isometric cations, the exchange is no longer quantitative because the area available in a monolayer within the lattice plane is insufficient. The linkage is electrostatic. It is enhanced if hydrogen bonds can operate or additional hydrophobic interactions come into play as is the case for long chain *n*-alkylammonium ions [19].

Organic derivatives of this type are commonly rather stable. In nature they may lead to a significant enrichment of the organic matter [20].

Type 3: *Minerals having polar organic molecules as solvate*. A third type of organic compounds can be generated in the presence of certain polar organic molecules. Exchangeable inorganic cations are frequently solvated by water molecules which in turn can be replaced by polar organic molecules such as alcohols or amines [21–25]. Since the hydration energy of cations is larger than their solvation energy with organic molecules, derivatives will only form if the organic reaction partners are available in high concentrations. Once formed, they are kinetically stable if the polar molecules exhibit distinct hydrophobic properties.

Type 4: *Derivatives by complex formation with exchangeable cations.* Closely related to Type 3 are compounds in which the exchangeable cations inside the crystal react with suitable complex-forming compounds. Inasmuch as the effective charge of the cations is frequently lowered hereby, protons have to compensate the charge deficiency. Such complexes are easily formed in particular with exchangeable transition metal ions. Complex formation is decisively influenced by steric relations. Namely, planar complex compounds are preferentially generated in reactive minerals with layer-structures, whereas bulky three-dimensional complexes will not form. In case of octahedral complexes, layer oxygen atoms may occupy the 5th and 6th position of the complex. As a consequence, complexes which in the free state would be unstable, become stabilized inside the crystal. In all instances the diffusion of the complex-forming compounds into the lattice and the back diffusion of the reactive cations out of the lattice compete with each other. In turn, under extreme conditions the formation of the complex may take place either inside or along the outer surface of the crystal [26–34].

Type 5: *Derivatives having exchangeable organic cations and polar organic molecules in the form of swelling agents.* Since almost any type 2 representative is capable of swelling with any polar organic molecule, this group is particularly broad. For montmorillonites alone more than 9,000 different derivatives have been studied [35]. They are formed relatively simply and permit an extraordinary enrichment of organic matter. The catalytic properties of the minerals, however, are less pronounced for this type of derivative in comparison with the four other types of derivatives discussed before [36–38].

All the types of derivatives mentioned so far may liberally combine among each other. Research is complicated by the fact that this may occur even within a single crystal. Thus uniformly appearing minerals may give rise to organic derivatives of the mixed-layer-type.

III. Organic Compounds of Various Minerals

A. Montmorillonites, Vermiculites, and Micas

The same basic structural details are involved in montmorillonites, vermiculites, and micas (Fig. 1) [39–43]. For this reason we will treat them simultaneously.

Within a layer, silicon by surrounding itself with 4 oxygens forms a tetrahedron of which three oxygens are also a part of neighboring tetrahedra. In this way two-dimensional infinite layers of tetrahedra are formed having the general composition of Si_2O_5. Two of these tetrahedral layers, opposite and parallel to each other, are joined by cations co-ordinated to oxygens at the tetrahedra tips. The individual cations, intending to be octahedrally co-ordinated, will complete the octahedron by the introduction of OH groups.

For each Si_4O_{10} unit, there are three octahedral vacancies. To balance the negative charge, six positive ones are required which can be supplied by either three 2-valent or two 3-valent cations. The first choice will result in the formation of a so-called trioctahedral type and the second one a dioctahedral silicate.

The individual silicate layers generated in this way are electrically neutral. Due to isomorphous substitution, however, negative charges may develop. For

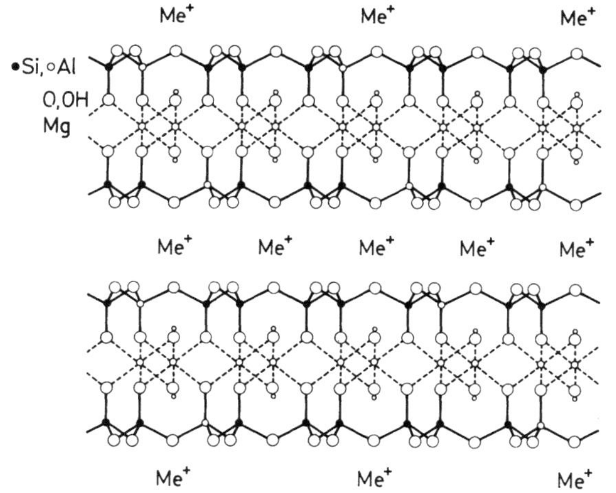

Fig. 1. Structure of Mica-type Layer-Silicates (Projection along a-axis)

instance, 3-valent octahedrally co-ordinated cations may be exchanged for 2-valent metals, and aluminium or other 3-valent cations of comparable ionic size may proxy for silicon in tetrahedral positions. To compensate the charge deficiency, the equivalent amount of cations must be introduced into interlayer positions. On account of the solvation tendency of these interlayer cations and according to the special requirements of the solvate molecules, the basal spacing will more or less increase, a phenomenon commonly referred to as one-dimensional or intracrystalline swelling of the minerals under discussion [40].

The standardized crystallochemical formula of montmorillonites, vermiculites, and micas can be expressed as follows:

$$M^{z+}_{(x+y)/z} \cdot (Y)_n \cdot \underbrace{[(Me_I^{2+}, Me_{II}^{3+})^{(6-y)+}_{2-3}(OH)_2Si_{4-x}Al_xO_{10}]^{(x+y)-}}_{\text{Silicate layer}}$$

Cations in interlayer position	Swelling liquid	Silicate layer

The physical-chemical properties are principally determined by

(a) the net sum $(x+y)$ of the charge of cations substituted in tetrahedral (x) and octahedral (y) positions,

(b) the valence (z) of the exchangeable cations,

(c) the kind of octahedral cations Me_I and Me_{II} and

(d) the ratio $x:y$.

Factors c and d, however, are less crucial than factors a and b [44, 45]. In case the negative charge is introduced mainly by the substitution of Si^{4+} by three-valent metals $(x \gg y)$, the charge type is referred to as *beidellitic;* in the reverse situation $(y \gg x)$ we have the *montmorillonitic* type.

For montmorillonite minerals the charge $(x+y)$ per structural unit $(Si, Al)_4O_{10}$ is in the order of about 0.25 to 0.55; the corresponding range for vermiculites is

Table 1. *System of the mica silicates* $M^{z+}_{(x+y)/z}(Y)_n\{Me^{2+}_{I}, Me^{3+}_{II})^{(6-y)+}_{2-3}(OH)_2Si_{4-x}Al_xO_{10}\}^{(x+y)-}$

layer charge $(x+y)$ per $(Si, Al)_4O_{10}$ unit $x+y=$	group	subgroup	dominant type of cations within the octahedral layer Me_I and/or Me_{II}	silicates of the dioctahedral series	silicates of the trioctahedral series
0	–	–	Al^{3+}	Pyrophyllite	
			Mg^{2+}		talcum
			Fe^{2+}		minnesotaite
0.25–0.55	montmorillo-		Al^{3+}	montmorillo-	–
$x \ll y$	nite or			nite	
	montmorine	montmorillo-	Cr^{3+}	wolchonskoite	–
	or smectite	nite	Mg^{2+}	–	hectorite
			Al^{3+}	beidellite	–
$y \ll x$	–	beidellite	Fe^{3+}	nontronite	–
			Mg^{2+}		saponite
			Zn^{2+}		sauconite
			Ni^{2+}		pimelite
			Cu^{2+}		medmontite
0.56–0.69	vermiculite	vermiculite	Al^{3+}	dioctahedral	–
$y \ll x$			Mg^{2+}	vermiculite	vermiculite
			Fe^{2+}		jefferisite
0.66–0.88	mica	illite	Al^{3+}	illite	–
$y \ll x$			Mg^{2+}		trioctahedral
					illite
			Fe^{2+}		glauconite
0.89–1.1	–	mica	Al^{3+}	muscovite	–
$x \approx 0$				and paragonite	
			Mg^{2+}		phlogopite
			Fe^{2+}		biotite
1.8–2.0	–	brittle mica	Al^{3+}	margarite	–
$x \approx 0$				and ephesite	
			Mg^{2+}		xanthophyllite

0.55 to 0.70; above the 0.70 level the micas start. The lower charged silicates have a predominantly montmorillonitic charge distribution. An increase in the amount of $(x+y)$ will favor the beidellitic type which is actually caused by its more uniform charge distribution relative to the montmorillonitic type. The most common minerals are summarized in Table 1.

In the charged layer silicates $(x+y>0)$ electrostatic forces add to the van der Waals attraction between adjacent layers. The electrostatic interactions increase with the net amount of $(x+y)$, so that finally the hydration energy of the cations in interlayer position is no longer sufficient to enforce the uptake of water into the interlayer space against the electrostatic forces simply by increasing the lattice dimensions. The solvation energy per interlayer cation is lowered at higher charges of the layer where for reasons of space limitations, the solvation halo cannot fully develop and thus the water dipoles cannot be oriented favorably for all the

Table 2. *Micaceous silica compounds capable of cation exchange and swelling*

layer charge $(x+y)$ per $(Al, Si)_4O_{10}$-unit	subgroup	cation exchange (mval/100 g)			swelling inside the crystal
		natural mineral	Na^+, Mg^+, Ca^{2+} as interlayer cations in the mineral	K^+ as interlayer cation in the mineral	
0	talcum pyrophyllite	0–3	–	–	no swelling
0.25–0.55 $y \gg x$ $x \gg y$	montmorillonite beidellite	60–120	60–120	60–120	swelling occurs with all inorganic cations and most of the organic cations. alkaloids inhibit or retard the swelling
0.56–0.69 $x \gg y$	vermiculite	100–200	100–200	without complex forming unit: 0–25[a]; with complex forming unit for K^+: 100–200; with $R-NH_3^+$: 100–200	K^+, Rb^+, Cs^+, NH_4^+ ions inhibit swelling
0.66–0.88 $x \gg y$	illite	0–35[a] depending on the size of the particles	-200	without complex forming unit for K^+: 0–35[a]; with complex forming unit for K^+: up to 200; with $R-NH_3^+$: up to 200 (very slowly)	swelling with $[H_3N \cdot (CH_2)_xNH_3]^{2+}$ and $(C_xH_{2x+1} \cdot NH_3)^+$ $x>8$. At temperatures below appr. 40° C, swelling may occur too with inorganic cations except for K^+, NH_4^+, Rb^+, Cs^+, Tl^+
0.89–1.1 $x \gg y$	mica	0–20[a] depending on the size of the particles	-250	with complex forming unit up to 250; with $R-NH_3^+$: up to 250	like illite
1.8–2.0 $x \gg y$	brittle mica	0–20[a] depending on the size of the particles	0–20[a]	0–20[a] in the presence of complex forming units; with $[H_3N \cdot (CH_2)_xNH_3]^{2+}$ up to 380	swelling only with $[H_3N \cdot (CH_2)_xNH_3]^{2+}$ $x=10$ within a narrow range

[a] only the outer surface is involved, the ion exchange thus depending on the size of the particles.

interlayer cations. For this reason the interlayer cations are less mobile and more difficult to exchange at higher charges of the layer. Geometry and polarizability cause a preference of potassium, rubidium, cesium, and ammonium ions which leads to a contraction of the basal spacing to approximately 9.8 to 10.8 Å. These cations are considerably less mobile than either sodium, lithium, magnesium, or calcium ions. As a result, the highly charged layer silicates occur in nature most commonly with potassium. It is noteworthy that in those minerals even potassium may become replaced by using suitable complex-forming compounds for potassium ions. Exchange can also be accomplished for primary organic alkylammonium ions [45]. A survey on the general ion exchange and swelling characteristics of micaceous minerals is presented in Table 2.

B. Compounds with Organic Swelling Liquids

By treating water-containing micaceous minerals having inorganic interlayer cations with methanol or ethanol, the water molecules gradually become displaced by the alcohols if the charge of the silicate layers does not exceed 0.60 [25]. One should mention, however, that long chain alcohols cannot be easily introduced because they are only moderately or are not miscible with water. In contrast, methanol and ethanol can be replaced by long chain alcohols. This replacement is particularly pronounced if ethanol is first substituted by butanol which subsequently may become replaced by longer chain alkanols [23, 24, 46]. This phenomenon can simply be attributed to a decrease in the solubility of alkanols in water as a function of increased chain length, a fact which complicates or prevents the penetration of molecules with hydrophobic alkyl functions into the water-containing interlayer space. In addition, the lowering of interactions between neighboring layers as a function of lattice expansion has a decisive influence.

Table 3. *Layer distance of Ca-montmorillonites in the presence of n-alkanols*

swelling alcohol $n-C_xH_{2x+1}OH$ $x=$	layer distance (Å) at 20° C			
	montmorillonite from Geisenheim	montmorillonite from Cyprus	nontronite from Kropfmühl	beidellite from Unterrupsroth
6	27.5	27.4	28.1	28.0
8	31.0	31.4	32.7	30.9
10	36.3	37.3	36.9	39.0
12	45.3	46.3	46.1	45.4
14	48.8	49.1	51.5	48.3
16	51.6	51.9	55.3	53.3
18	59.8	58.2	61.8	59.0

The arrangement of the alcohols in interlayer position is only little influenced by the charge of the layers (Table 4), whereas the kind of cations has a noticeable effect. This suggests that both the solvation of cations and silicate layers is important. The experimentally determined increase in spacing with increase in carbon number of the alcohols indicates that the alkyl chains are oriented perpendicular to the silicate layers but not very regularly (Table 3). This is caused by the abnormal temperature dependence of the lattice spacings. Within a certain temperature range the distance between the layers is lowered abruptly (Tables 4 and 5). The temperatures where these sudden contractions take place rise with increase in melting

Table 4. *Influence of charge density and exchangeably linked cations on the basal spacing of montmorillonites*

Temperature	Layer distance (Å) in the presence of *n*-tetradecanol													
	montmorillonite from Geisenheim with exchangeably linked		montmorillonite from Cyprus with exchangeably linked		nontronite from Kropfmühl with exchangeably linked						beidellite from Unterrupsroth with exchangeably linked			
°C	Mg^{2+}	Ca^{2+}	Mg^{2+}	Ca^{2+}	Li^{+}	K^{+}	Mg^{2+}	Ca^{2+}	Sr^{2+}	Ba^{2+}	Mg^{2+}	Ca^{2+}	Sr^{2+}	Ba^{2+}
10					51.3									
20	51.7	48.8	48.2	46.6	51.3	48.8	51.4	51.5	49.6	49.5	51.7	48.3	48.1	48.2
30	51.6	48.7	48.1	46.5	50.9	48.0	51.4	50.7	48.3	49.6	51.3	47.1	48.0	48.1
40	51.0	48.4	47.8	45.7	50.8	47.3	50.3	49.6	48.2	50.1	51.0	46.9	47.6	47.6
50	49.1	40.6	46.8	44.4	51.3	46.7	49.7	48.6	45.1	49.7	50.3	46.3	46.6	46.3
60	44.2	39.3	41.6	40.0	51.3	46.1	46.1	43.8	39.6	48.5	49.5	44.7	43.6	45.8
70	42.9	38.4	40.7	38.1	48.5	37.4	41.7	40.5	38.3	39.7	41.9	40.5	40.9	44.3
80	42.4	37.6	39.8	38.0	44.0	35.8	40.8	39.0	37.8	38.3	40.0	39.4	40.6	41.0[2]
90	42.4	39.5			43.7	33.8	40.2	38.5	37.5	37.2	38.3	39.0	40.5	40.6
100			39.4			34.1	39.3		37.1	36.2	36.7	38.8	40.4	
110						34.1	39.1		36.9		35.0	38.6		

Table 5. *Temperature at which a sudden lattice change occurs (° C)*

swelling alcohol $n\text{-}C_xH_{2x+1}OH$	montmorillonite from Geisenheim with		montmorillonite from Cyprus with		nontronite from Kropfmühl with						beidellite from Unterrupsroth with			
$x=$	Mg^{2+}	Ca^{2+}	Mg^{2+}	Ca^{2+}	Li^{+}	K^{+}	Mg^{2+}	Ca^{2+}	Sr^{2+}	Ba^{2+}	Mg^{2+}	Ca^{2+}	Sr^{2+}	Ba^{2+}
6	−6.7	0.0	−10.0	−4.7	?	?	?	?	?	?	?	−15.8	?	–
8	−9.0	+8.5	+1.5	−8.0	−9.3	+5.3	−15.0	?	+20.0	+0.5	?	0.0	+13	–
10	−7.0	−1.0	+19.5	−5.5	+12.5	+10.7	+5.0	−3.5	−0.5	+8.5	−5.0	−1.5	+23.5	–
12	+25.7	+40.0	+44.0	+22.5	+33.6	+61.0	+25.4	+46.0	+24.0	+52.5	+56.0	+61.0	+65.8	+65.0
14	+54.5	+45.0	+56.5	+54.0	+70.4	+70.0	+61.0	+61.0	+54.0	+62.5	+65.3	+64.7	+57.0	+73.8
16	+67.3	+66.0	+72.0	+80.0	+60.7	+76.0	+66.5	+65.3	+64.5	+67.7	+87.3	+76.0	+64.0	+84.7
18	+80.0	+73.3	+80.0	+78.7	+70.0	+82.0	+90.7	+54.0	+76.0	+95.0	+74.3	+92.0	+84.7	+103.0

point, which increases with the length of the alkyl chain. The jump-temperature for a given alcohol is especially influenced by the kind of cations in interlayer position. Since the strong reduction in lattice dimensions is principally caused by the reversible loss of alcohol, the strong influence of solvating cations becomes obvious. The number (n) of the adsorbed alcohol molecules per unit formula is in the order of 2.0 for the low temperature form and 1.25 for the high temperature form. The shorter the chain of the alcohols (*n*-hexanol and shorter), the larger are the deviations from these values [47].

Glycol, glycerol or polyalcohols become much better incorporated into the interlayer space than the mono-functional alcohols. Apparently the choice for multiple cross-linkages involving hydrogen bridges can be attributed to this preference [21, 22, 48–52].

Analogous to the alcohols, the aldehydes, ketones, carbonic acids, acid amides, amino acids, sugars and polymers can become incorporated as neutral molecules into the interlayer space. Work in this area, although not too extensive, already shows interesting details. According to these studies, the organic compounds which can form hydrogen bridges to the oxygen atoms of the silicate layers are preferentially adsorbed. For this reason, amino acids are of special interest; they will be treated in a separate section. In contrast, the sugars have little tendency to become adsorbed unless concentrated sugar solutions are employed [53, 54, 20, 55, 56].

Polar basic compounds such as indole, quinol, pyridine and imidazole and their respective derivatives are also easily taken up. Apart from the adsorption as solvent, the adsorption of protons and the corresponding cation exchange become important [57–61, 35]. This can be inferred by the vivid interaction which is partly manifested by the additional water adsorption, and by the changes in pH of the equilibrium solution [62]. Complex formation, too, is involved which particularly favors the incorporation of these bases by cations of the transition metals present in interlayer position.

C. Micaceous Layer Silicates Having Organic Cations

The number of compounds of this group is unusually large. Virtually any organic cation having less than 6 Å as its smallest dimension can be introduced by ion exchange. Whether or not the exchange is quantitative simply depends on the charge of the silicate layer and the size of the cation to be introduced. For instance, primary alkylammonium ions $n\text{-}C_xH_{2x+1}NH_3^+$ for any given carbon number can be quantitatively exchanged whereas the alkaloid codein is only quantitatively accepted as cation by the lower charged silicates (Table 6). The relationship between charge of the layer, size and type of cation, and the essential exchange capacity is based on the fact that for energetic reasons cations in interlayer positions can only become arranged in single layers except for strongly asymmetric structures with the charge on one end; here the cations may also form double layers [63].

Research is particularly plentiful for the *n*-alkylammonium compounds which have industrial applications [64–71]. The ions of phosphonium, arsonium,

Table 6. *The influence of asymmetric molecules on the exchangeability on micaceous silicates*

	micaceous silicate					
	hectorite from Hector/Calif. $x+y=0.25$	montmorillo-nite from Geisenheim $x+y=0.33$	montmorillo-nite from Cyprus $x+y=0.42$	beidellite from Unter-rupsroth $x+y=0.44$	saponite from Gro-schlattengrün $x+y=0.55$	batavite from Kropfmühl $x+y=0.66$
versible cation change with $^+$, Na^+, Mg^{2+}, Ca^{2+} equiv./100 g[a]	69	91	108	112	142	166
tion exchange with $-NH_3^+$ (exchange mequiv./100 g)[a, b]	69	90	109	111	140	163
tion exchange with dein (exchange in equiv./100 g)[b]	68	89	108	110	70	82

[a] for better comparison, the exchange capacities have been related to the sodium compounds.
[b] furthermore, due to hydrophobic interaction together with the anions long chain alkylammonium ions may come adsorbed. Since this does not represent a real ion exchange, the amount has not been considered.

stibonium, sulfonium and oxonium are less well studied. From the existing data, however, it appears that their properties are similar to the ammonium compounds of comparable molecular structure. The difference in stability of the free compounds towards higher temperatures and hydrolysis is also maintained in interlayer position.

The bond stability of the alkyl compounds steeply decreases in the direction:

$$R \cdot NH_3^+ > R_2 \cdot NH_2^+ > R_3 \cdot NH^+$$

Quarternary alkylammonium ions R_4N^+ on the other hand behave quite differently; strongly asymmetric ions such as trimethylcetyl- or dimethyl-di-*n*-octadecylammonium have to be grouped with the primary ammonium ions [35, 72]; in contrast, symmetric ions follow the secondary ammonium ions. Due to space restrictions, the bonding strength of $R_2NH_2^+$, R_3NH^+, and symmetric R_4N^+ ions decreases rapidly with an increase in the charge of the silicate layer [35, 73].

Equilibrium for the lower charged silicates becomes established in a matter of minutes or hours; but for the higher charged silicates it may take up to 14 months. Surprisingly, long-chain compounds require less time for equilibration with higher charged silicates than is needed for their short-chain counterparts. This feature can be attributed to the rearrangement of structural positions inside the crystals. In all instances the chance to form hydrogen bridges is the most crucial factor for the fixation of organic cations in interlayer position. In view of the fact that oxygen atoms will form ditrigonal six-membered rings, cations having a trigonal symmetry are favored with respect to hydrogen bridge formation, for example:

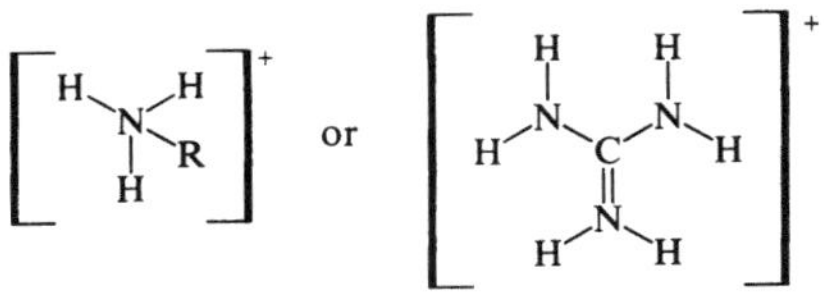

Once formed such compounds only reluctantly allow the exchange of the organic cations for an inorganic cation [19, 74]. On the other hand, the organic cation can be initially picked up by the silicate from extremely dilute solutions.

The high accumulation capacity for these organic cations even in the presence of brines is a function of the extreme shift of the exchange equilibrium to one side of the reaction. In addition, the enrichment and storage processes become favored if hydrophobic interactions take place.

The structural arrangement of the organic cations is only sufficiently known for the *n*-alkylammonium ions. The charge of the layers and the chain length of the ions are hereby of decisive influence.

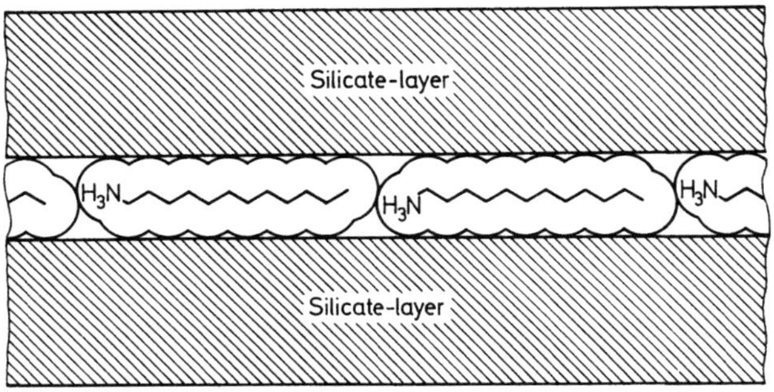

Fig. 2a. Monolayers of *n*-alkylammonium ions in the interspace of *n*-alkylammonium layer silicates

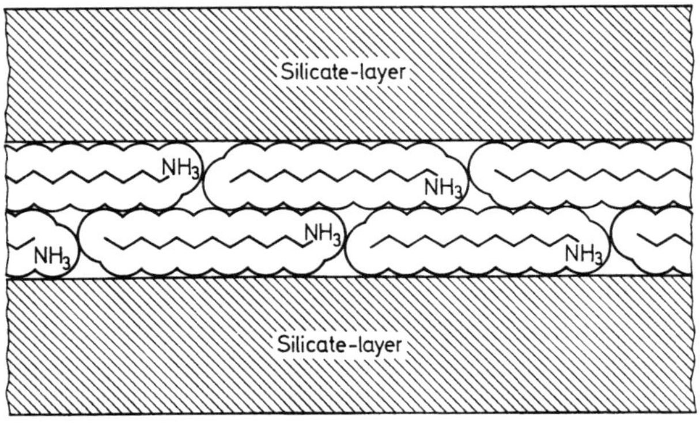

Fig. 2b. Double layers of *n*-alkylammonium ions in the interlayer space of *n*-alkylammonium layer silicates

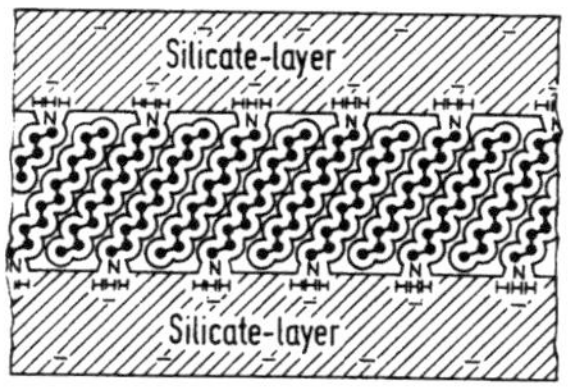

Fig. 2c. Arrangements of *n*-alkylammonium ions in the interlayer space of *n*-alkylammonium layer silicates with tilted alkyl-chains (56°)

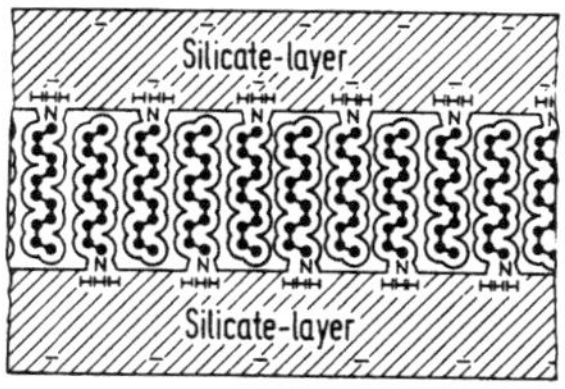

Fig. 2d. *Cis-trans*-conformation of alkyl-chains in the interlayer space

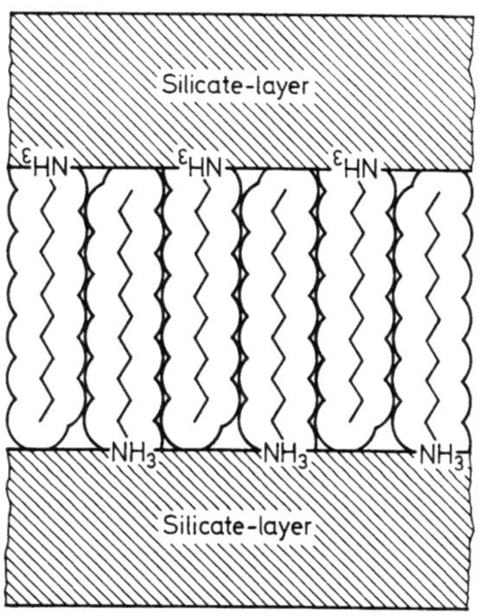

Fig. 2e. "Paraffin type" structures of *n*-alkylammonium ions in the interlayer space of *n*-alkylammonium layer silicates

With layer charge density below $(x+y)=0.67$ the cations form monolayers. With increasing length of the alkyl chain the spacing remains constant until the closest packing of cations within the monolayer is achieved (Fig. 2a) [64, 65]. After that a second layer will form, whereby the spacing abruptly rises by the amount of the van-der-Waals-radius of the alkyl chain and remains constant until the second layer is completely filled up (Fig. 2b). Further increase of the chain length will only result in a super-imposition and tilting of the alkyl chains because the formation of a third cationic layer is energetically not feasible. The higher the layer charge, the wider the angle of inclination (Fig. 2c and d). Therefore, the layer charge can be estimated independently both by means of the angle of inclination and the carbon number of the alkyl chains at just completed mono- or double-layer respectively [75, 76, 150].

Beyond $(x+y)=0.67$, even short chain alkylammonium ions are tilted. The mean lattice expansion per carbon atom (Δd) increases, the higher the layer charge. With $(x+y)$ close to 1.0, Δd is about 1.26 – 1.27 Å per carbon atom. This value corresponds to monolayers of stretched alkyl chains oriented perpendicular to the silicate layers (Fig. 2e) [35].

Although the mean lattice expansion per carbon atom is rather constant with given layer charge, the individual basal spacings rise in a regular but alternating fashion. For instance in batavite the spacing rises by 2.0–2.1 Å per carbon atom in going from an ammonium ion with an odd numbered *n*-alkyl chain to an even-numbered ion, whereas in the opposite direction the increase only amounts to 0.0–0.1 Å (mean lattice expansion $\Delta d = {}^1/_2\ (2.1+0.0) \approx 1.05$ Å $\approx 1.27 \cdot \sin 56°$) (Fig. 2c and d) [17, 63]. The odd-even pattern is governed by the possibility of the methyl end-group to immerse into the cavity of the six-membered oxygen ring.

Therefore this pattern strongly depends upon the charge density and minor changes in the dimensions of the layers [148].

In the past the significance of the layer charge has frequently been neglected and data in disagreement with the "expected" pattern were unwarrantly attributed to experimental errors of other authors. It is true, however, that the use of improper, i.e. less crystalline, starting material will indeed introduce considerable errors. Even high purity grade amines should be analyzed by gas chromatography. Research is further complicated by the fact that even within the same crystal micaceous silicates may contain layers of different charge; in the case of the *n*-alkylammonium derivatives this will yield regular basal reflections for alkyl chains of certain length, whereas a non-integral series having all qualifications of mixed-layer-structures may be produced by others.

In comparison to the mono-alkylammonium ions, the alkylene di-ammonium ions are more tightly linked to the silicates although kinetically their exchange proceeds less rapidly, particularly for the α,ω-diammonium ions [45]. Polyammonium ions and the later discussed peptides and proteins share this characteristic.

Alkaloid cations too are very tightly fixed particularly in the lower charged montmorillonites; they can become adsorbed from rather dilute solutions. With excess montmorillonite, the equilibrium concentration of many alkaloids in solution can be lowered below the limit of detection. The relationship between effective exchange capacity and layer charge has already been presented in Table 6. Additional solvation is blocked by alkaloids whereas a zeolitic uptake of liquids is still possible except in those cases where closest cation packing is observed [63].

Arylamines are less tightly linked than alkylamines. Once incorporated, however, they are more difficult to exchange for reasons of their oxidative alteration. The catalytic oxidation phenomenon will be treated separately. In contrast N-pyridinium compounds and cations having longer hydrophobic residues become preferentially adsorbed [77].

D. Derivatives Having Organic Cations and Organic Solvent Molecules

Due to the introduction of organic interlayer cations, the interlayer space will become organophilic in character. Consequently, solvation involving many polar compounds may proceed. For this group again, a host of data is available involving the *n*-alkylammonium compounds [35, 78–84].

Conditions are particularly simple and clear in case of the number of carbons in the cation and the swelling fluid being equal. Under such circumstances no major difference between variably charged silicates can further be recognized (Table 7), and it becomes unimportant whether the swelling fluid is an *n*-alkylamine, an alkanol, an aldehyde, *n*-alkylnitrile or an *n*-alkylcarbonic acid respectively.

The pronounced van der Waals interactions that are established between the long-chain *n*-alkyl residues will bring about, in the expanded state, the uptake of an equal number of *n*-alkyl molecules per mineral area unit irrespective of the layer charge. Assuming the closest packing, the van der Waals forces hereby amount to approximately 0.6–0.8 kcal/M $-CH_2-$ [76]. The chains become so tightly packed that each one has only about 21 Å^2 for its disposal, which corresponds to 2.05–2.08 alkyl chains per $(Si, Al)_4O_{10}$ unit (Table 8).

In comparison to the lower charged compounds, the higher charged ones contain more alkyl cations and less polar neutral molecules per $(Si, Al)_4O_{10}$ unit. Hectorite, for example, has 0.25 cations and 1.82 alcohols per unit; the corresponding values for vermiculite are 0.67 vs. 1.40, and for muscovite 0.92 vs. 1.10 per unit [63, 35].

Swelling liquids and cations are arranged in two layers; the alkyl chains are stretched, and the spacing expands for each additional carbon by 2 × 1.26 Å which corresponds actually to the ideal *trans-trans* chains (Fig. 3).

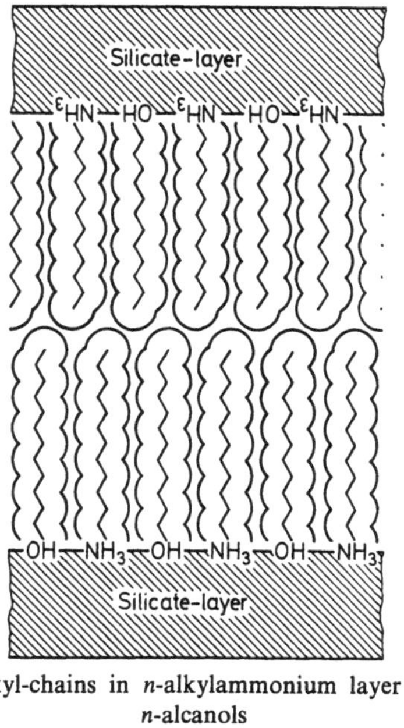

Fig. 3. Arrangement of the alkyl-chains in *n*-alkylammonium layer silicates after swelling under *n*-alcanols

Table 7. *Layer distances after swelling in n-alkyl*

micaceous silicate	layer distance (Å)			
	n-$C_6H_{13}NH_3^+$		n-$C_8H_{17}NH_3^+$	
	$+n$-$C_6H_{13}OH$	$+n$-$C_6H_{13}NH_2$	$+n$-$C_8H_{17}OH$	$+n$-$C_8H_{17}NH_2$
hectorite ($x+y=0.25$)	25.5_3	25.6_9	30.1_1	30.3_2
montmorillonite ($x+y=0.33$)	26.0_1	26.0_6	29.9_1	29.9_8
vermiculite ($x+y=0.67$)	25.49	25.8_4	30.3_2	30.5_1
illite ($x+y=0.78$)	25.6_8	25.9_4	29.8_6	29.9_6
muscovite ($x+y=0.92$)	25.4_6	25.6_2	29.9_1	29.9_3
biotite ($x+y=0.87$)	25.3_8	25.6_3	30.0_4	30.2_4

Table 8. *Proportion of* $\frac{R\cdot NH_3^+}{X}$ *per* (Si, Al)$_4$O$_{10}$ *in swollen n-alkylammonium derivatives of micaceous silicates*

swelling solvent X	hectorite from Hector/Calif.	mont-morillonite from Geisenheim	mont-morillonite from Cyprus	beidellite from Unter-rupsroth	saponite from Groschlat-tengrün	batavite from Kropf-mühl	muscovite from Norway
n-C_4H_9OH upto n-$C_{18}H_{37}OH$	$\frac{0.25}{1.82}$	$\frac{0.33}{1.74}$	$\frac{0.40}{1.66}$	$\frac{0.58}{1.50}$	$\frac{0.57}{1.50}$	$\frac{0.67}{1.40}$	$\frac{0.92}{1.10}$
n-$C_4H_9NH_2$ upto n-$C_{18}H_{37}NH_2$	$\frac{0.25}{1.83}$	$\frac{0.33}{1.75}$	$\frac{0.40}{1.68}$	$\frac{0.58}{1.48}$	$\frac{0.57}{1.47}$	$\frac{0.67}{1.39}$	$\frac{0.92}{1.13}$
n-C_3H_7CHO upto n-$C_{17}H_{35}CHO$	$\frac{0.25}{1.77}$	$\frac{0.33}{1.69}$	$\frac{0.40}{1.63}$	$\frac{0.58}{1.45}$	$\frac{0.57}{1.46}$	$\frac{0.67}{1.35}$	$\frac{0.92}{1.10}$
n-C_3H_7COOH upto n-$C_{17}H_{35}COOH$	$\frac{0.25}{1.75}$	$\frac{0.33}{1.66}$	$\frac{0.40}{1.60}$	$\frac{0.58}{1.42}$	$\frac{0.57}{1.43}$	$\frac{0.67}{1.33}$	$\frac{0.92}{1.09}$

Under reduced pressure, particularly if the reacting compounds are ground up together, a metastable state is often achieved whereby the expansion averages 1.05–1.06 Å per carbon atom. This suggests folded *trans-trans* chains or twisted *cis-trans* chains.

At higher temperatures this type of arrangement will take place at normal pressure and even in presence of excess swelling fluid. The sum of the numbers of alkylammonium cations and polar neutral molecules only amounts to 1.60–1.75

npounds of n-alkylammonium derivatives of micaceous silicates

$C_{10}H_{21}NH_3^+$		n-$C_{12}H_{25}NH_3^+$		n-$C_{14}H_{29}NH_3^+$	
n-$C_{10}H_{21}OH$	$+n$-$C_{10}H_{21}NH_2$	$+n$-$C_{12}H_{25}OH$	$+n$-$C_{12}H_{25}NH_2$	$+n$-$C_{14}H_{29}OH$	$+n$-$C_{14}H_{29}NH_2$
.4	37.8	43.4	43.6	49.3	49.3
9	37.1	43.2	43.5	49.3	49.4
1	37.6	44.8	43.7	48.7	49.3
9	37.2	43.4	43.5	48.7	49.2
7	36.9	43.2	43.5	49.8	49.3
1	37.3	43.4	43.6	49.0	49.3

per $(Si, Al)_4O_{10}$ unit. This agrees with the larger requirement in area for the folded *trans-trans* chain and the twisted *cis-trans* chain, respectively.

The situation is more complicated in case the alkyl chain of the cation is longer or shorter than the chain of the polar swelling fluid. Frequently, short-chained cations do not swell at all [85, 35, 86]. The critical length of the alkyl chain is hereby a function of the layer charge. At $(x+y)$ equal to 0.25, at least eleven carbon atoms are required for swelling; at $(x+y)$ equal to 0.33, at least nine carbons will be needed; whereas at $(x+y)$ equal to 0.40, a minimum of seven carbons is essential. For $(x+y)$ equal to 0.43 and 0.67, n-hexylammonium and n-propyl ammonium ions respectively are still satisfactory.

In case the carbon number of the alkyl chain in the swelling fluid exceeds that of the cation, the layer charge will strongly affect the spacing which can be expressed:

$$d = 1.4\,\{(x+y)\cdot(n_K - n_A) + 2n_A\} + 9.6\ [\text{Å}]$$

whereby: d = basal spacing; $(x+y)$ = layer charge; n_K = carbon number for cation; and n_A = carbon number in swelling fluid. This equation is strictly valid for the swelling with n-alkanols and thus permits the determination of the layer charge of the silicate on the basis of one single measurement of d [76].

Polar compounds with branched alkyl chains, and aromatic and cyclic compounds exhibit a less clear-cut picture [87, 88]. The orientation of these molecules depends on the nature of the polar group as well as on the size and the form of the individual molecule itself. For this reason, no straightforward stoichiometrical relationships are discernible. Published information on derivatives of ethers, ketones, aldehydes, nitriles, halogen compounds, esters, and hydrocarbons is rather plentiful. Aromatic hydrocarbons can become easily incorporated whereas alkenes do only so in the presence of traces of polar compounds. In contrast, aliphatic hydrocarbons have no swelling properties; they are only picked up in the case when interlayer channels are created by organic cations.

Generally, non-linear organic molecules do penetrate less effectively into the crystal lattice the higher the layer charge. For the lower charged silicates, the sorption capacity is frequently determined by the increase in volume which in essence is a result of the readjustment of the tilted alkyl chains of the cations into a vertical position. Because of this, molecules of variable size and form will frequently yield the same spacing. Dodecylammonium-beidellite (locality: Unterrupsroth/West Germany) becomes expanded from 18.06 Å to 27.5–28.5 Å in the presence of the following compounds: 2-nitrophenol, 4-nitrophenol, N-dimethylformamide, aniline, 3-picoline, ethyl acetate, dimethyl ketone, methyl ethyl ketone, methyl phenyl ketone, di-i-butyl ketone, cyclohexanone, cycloheptanone, methyl cyclohexanone, and *n*-propyliodide. The calculated spacing for *n*-dodecylammonium-beidellite having stretched *n*-dodecylammonium ions oriented perpendicular to the silicate layers amounts to 27.4–28.0 Å [35].

The *n*-alkylammonium derivatives of micaceous silicates react perfectly with the lipids [89]. Simple correlations are established between the length of the fatty acid residues, the length of the alkyl chains in the cations, the layer charge, and the spacings. For lipids having short fatty acid residues, e.g. triacetin, the spacing is identical to that with glycerol and it remains constant for a wide range of $(x+y)$. If the length of the fatty acid residue however is comparable in size to that of the alkyl residue in the cation, the spacing increases linearly with rise in layer charge. This implies, that the fatty acid residues, analogous to the alkyl chains of the cations, will become progressively stretched. For fatty acid residues which exceed the alkyl chains of the cations far in length, e.g. tristearin in derivatives of dodecylammonium compounds, the spacing is largely independent of the layer charge (Table 9).

Table 9. *Swelling of some n-alkylammonium silicates in the presence of symmetric triglycerides*

interlayer cation $n\text{-}C_xH_{2x+1}NH_3^+$	layer distance (Å)					
$x=$	hectorite from Hector/ Calif.	mont-morillonite from Wyoming	mont-morillonite from Geisenheim	mont-morillonite from Cyprus	beidellite from Unter-rupsroth	vermiculite from South Africa
	a) *swelling in the presence of triacetin*					
10	26.0	–	26.0	26.2	26.4	27.0
12	28.1	28.2	28.2	28.3	28.6	29.1
18	31.2	–	31.4	31.6	32.1	33.3
	b) *swelling in the presence of tricaprylin*					
10	31.6	–	32.2	33.8	34.4	35.5
12	35.1	35.7	36.2	37.0	37.2	40.5
18	39.6	–	40.1	40.5	44.1	44.5
	c) *swelling in the presence of tristearin*					
10	50.3	–	50.6	50.6	50.8	52.6
12	52.5	52.6	52.6	52.8	52.8	54.7
18	58.0	–	58.2	58.2	58.8	59.4

With given layer charge and length of the alkyl residue of the *n*-alkylammonium ion, the layers expand drastically, if the length of the alkyl chain of the fatty acid residue exceeds a minimum length. The increase in basal spacing, however, is not constant. This phenomenon is related to the temperature controlled orientation of lipids in interlayer position, which is different below and above the respective melting point of the lipid compound. In all instances, the oxygens of the glycerol component tend to become closely linked to the silicate layer.

From basal spacings and density measurements, the content of adsorbed lipids can be estimated. For instance, a ton of montmorillonite having calcium in exchangeable position can incorporate 620 kg of triolein, and a ton of *n*-octadecylammonium-montmorillonite can accommodate up to 1,400 kg of tristearin in interlayer position. By means of this mechanism lipid compounds may escape microbiological degradation [90].

E. Organic Derivatives of Allevardite, Stevensite, and other Mixed Layer Silicates

As previously shown, micaceous silicate layers of different charge can alternate with each other in a random fashion. As long as the charge differences are small, the incorporation of inorganic cations in interlayer position will find its reflection only by way of a continuous swelling process. In contrast, if the charge differences are large, the different properties become obvious, particularly if the silicate layers follow in a regular manner as is the case for stevensite and allevardite.

Most likely, in stevensite charged trioctahedral layers ($x+y=0.33$–0.40) alternate with uncharged trioctahedral layers of talc. This may, with respect to organic compounds, explain the similar behaviour of stevensite and low-charged montmorillonite.

Allevardite is composed of micaceous silicate layers. Concerning the charge distribution, three structural propositions can be made:

(1) alternating, the interlayer spaces are pyrophyllite-like (no cations) and mica-like (high cation-density)

(2) each second interlayer space conforms with that in vermiculite; and

(3) the interlayer cation density alternates between that observed in montmorillonite and that in mica.

All three suggestions are in agreement with the fact that glycol and glycerol can be easily accomodated structurally. Yet, cation exchange involving long chain *n*-alkylammonium ions leads to various reaction products as a function of the time of reaction. For instance, an exposure for six weeks will only affect the interlayers resembling montmorillonite and the basal spacing will be exactly 10 Å above that of similarly treated montmorillonite, i.e. a montmorillonite and a mica basal spacing ($x+y=0.42$). However, a nine month treatment will also exchange the potassium ions in the high-charge-density interlayer positions for the long chain alkyl ammonium ions. The spacing is then an intermediate of two basal spacings, corresponding to charge densities of $x+y=0.42$ and $x+y=0.94$ respectively.

F. Aquacreptite

Aquacreptite is a rare trioctahedral mica which contains some clusters of magnesium hydroxide in interlayer position. This will put this mineral into the group of the swelling chlorites. Surprisingly, however, little swelling can be observed, except where the $Mg(OH)_2$ clusters are leached from the system by complexing agents. Yet, the observed lattice expansion remains far behind the values commonly observed for layers of that charge. The explanation for this behavior is the presence of tetramethylenediamine (putrescine) and pentamethylenediamine (cadaverine) in positions of interlayer cations. Microbial degradation can produce these amines. Since these compounds are strongly fixed, analogous to small di- and poly-ammonium ions such as trimethylene-diammonium, sperminium, or spermidinium, they can prevent swelling of a micaceous mineral. In turn, putrescine and cadaverine will be protected against further decay.

G. Organic Derivatives of Swelling Chlorites

Chlorites are structurally distinguished from micas by the presence of a complete octahedral hydroxide layer, *i.e.*, $M^{2+}(OH)_2$ or $M^{3+}(OH)_3$, in place of the interlayer cations, and by their neutral charge. Since the cations of the hydroxide layer are already octahedrally coordinated to OH^- ions, no further hydration and consequently no swelling is possible, except in the case of corrensite [91–93]. One has thus to assume that corrensite has either an incomplete hydroxide layer with some cations of normal solvation properties, or that a negative charge of the mica structure has only been partly compensated by positively charged hydroxide layers. The latter requires that additional cations be incorporated between the silicate and hydroxide layers as a consequence of charge compensation.

Both cases appear to be present in nature. The first type of corrensite can pick up only neutral (uncharged) polar molecules, due to the absence of exchangeable cations within the crystal. Particularly glycerol is readily incorporated with the basal spacing rising by about 4 to 5 Å. The second type of swelling chlorite contains cations in exchange position within the mineral allowing their substitution by certain cations such as primary *n*-alkylammonium ions. Since smaller cations only form a monolayer in this case, additional swelling with polar organic molecules is restricted. The increase in spacing up to 0.8 Å is simply a consequence of a zeolite-type uptake of solvent medium.

Due to the fact that the charge of the layers is relatively small, ion exchange reactions proceed slowly because the activation energies for change of lattice sites must become large in return for the wide distances observed between the fixed negative charges. Swelling chlorites have occasionally been reported which actually represent simply micaceous silicate layers having organic compounds in interlayer position. These organic compounds have a high thermal stability so that even a short-time exposure to temperatures up to 540° C will not destroy them. A longer exposure in presence of oxygen, however, will eliminate the organic matter and a 10 Å spacing, corresponding to mica, will be obtained. The organic component may be easier detected if the silicate is destroyed with potassium dihydrogenphosphate at a pH of 2 to 3, which will result in the formation of taranakite.

Thermal behaviour and X-ray characteristics of artificial tropocollagen-montmorillonite are similar to the so-called swelling chlorite. Using montmorillonite from Moosburg/Germany as starting material, the tropocollagen-montmorillonite resembles the corrensite described by LIPPMANN [91]. Using montmorillonite from Wyoming/USA, the reaction product corresponds more closely to the mineral reported by STEPHEN and MACEWAN [92, 93] (Table 10). Although both kinds of montmorillonite have the same mean charge distribution, the montmorillonite from Moosburg, in contrast to that from Wyoming, contains differently charged layers that follow each other. Thus, little can be said at this time regarding the swelling chlorites and their organic derivatives. Their wide abundance in nature and their structural and chemical diversity make them excellent research objects in the area of organic geochemistry.

Table 10. *Comparison of properties of swelling chlorites and of tropocollagen-montmorillonites*

treatment of the sample	layer distance in Å			
	corrensite from LIPPMANN	tropocollagen-montmorillonite from montmorillonite from Geisenheim	Wyoming	swelling chlorite after STEPHEN and MACEWAN
natural, air-dry	28.3	27.9	28.1	28
with glycerol	32.5	32.3	32.4	32
$\frac{1}{2}$ hour 540–550° C	~28 flattened	~27,5 flattened	14 flattened	13.8

H. Organic Derivatives of Kaolinites

The kaolinite group comprises the following minerals: kaolinite, nacrite, dickite, anauxite, halloysite (endellite), chrysotile, serpentine (lizardite), antigorite, amesite, and garnierite. The incorporation of cronstedtite, chamosite, and greenalite into the kaolinite group is still uncertain.

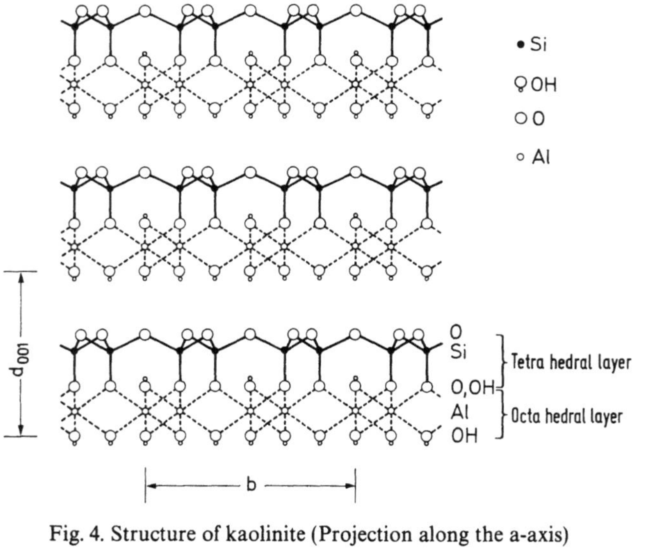

Fig. 4. Structure of kaolinite (Projection along the a-axis)

Apart from surface adsorptions, only the true kaolinite minerals are capable of the formation of organic derivatives. The structural organization is straightforward. Silicon is tetrahedrally coordinated with oxygen, three of which are also part of neighboring tetrahedra, sharing corners so that two-dimensional layers of six-membered tetrahedra rings are formed. The fourth oxygen of each tetrahedron is oriented (in the same manner) to one side of the layer and replacing one OH per octahedron of a gibbsite layer (Fig. 4). Thus, in contrast to micaceous silicates, the kaolinite layer is composed only of one tetrahedral and one octahedral layer; hence, a polar character is achieved.

The layers are electrically neutral and cross-linked via hydrogen bonds and dipole-dipole interactions. Accordingly organic compounds, which themselves form strong hydrogen bonds, can intrude between the kaolinite layers. With regard to the mode of preparation these derivatives fall into two categories:

(1) type I-derivatives, and

(2) type II-derivatives.

Type I means a direct reaction between kaolinite and the compound to be intercalated, whereas type II-derivatives can only form when the spacing has already increased (two-step reaction) [94]. To distinguish between these two kinds of derivatives is frequently difficult, because the formation of a derivative according to a type I-reaction is sometimes only kinetically delayed [95].

Concerning the binding forces, in type I-reaction, three groups of reactive molecules can be distinguished:

(1) Compounds such as urea, formamide, acetamide [96], hydrazine [97, 94, 98] or imidazole [95] that exhibit strong hydrogen bonding. Donor and acceptor groupings for hydrogen bonds have to be at least 1.5 Å apart. For example, among the hydrogen bond breakers studied, hydrazine which most strongly reacts with kaolinite clearly shows that the dipole character of the total molecule is not crucial for this kind of reaction. One should perhaps mention that the temperature rises to about 300° C by reacting 5 kg of kaolinite with 1 kg of hydrazine hydrate. The significance of hydrogen bridges in these derivatives is furthermore supported by infrared spectroscopy. Interpretation, however, is not completely unbiassed, since the bonding of the OH^- groups in the kaolinite lattice is affected by different kinds of incorporations, and because the N–H bonds of the incorporated molecules can interact not only with the silicate layer but also among each other.

(2) Alkali salts of lower fatty acids, in particular potassium, rubidium, cesium, and ammonium acetate [99, 100], as well as potassium propionate and potassium cyanoacetate [94]. Large monovalent cations are hereby favored; their ease of polarization and their low hydration energy strongly suggest the participation of London-energy. The difficulty of incorporating the analogous salts of silver and thallium into the kaolinite lattice, and the type I-reaction with sodium propionate but not with sodium acetate, still require explanations. The fixation of acetate ions in substitution for OH^- groups which is similar to that observed in basic aluminium acetates, has not been established so far; upon washings, a small amount of acetate, however, will be strongly retained in the lattice [99, 97].

(3) Compounds having betaine or betaine-like mesomeric structure, for example:

dimethylsulfoxide

pyridine-N-oxide

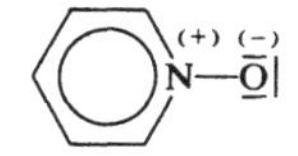

The perfect reaction of kaolinite with these kinds of compounds suggests that silicate layers are not only held together *via* hydrogen bridges but primarily by means of dipole-dipole interactions [95].

Between the aforementioned three groups there are intermediates. For instance, ammonium acetate can best be structurally accommodated between the first and the second group. In accordance, ammonium acetate-kaolinite exhibits characteristics which are unknown for the acetates of potassium, rubidium, and cesium, such as the formation of a hydrogen-acetate kaolinite having a spacing of 17.2 Å [95]. The potassium salt of the picolinic acid-N-oxide can be looked at as an intermediate between the second and the third group. N-methylformamide and N-methylacetamide are positioned between the first and the third group. These compounds are still capable of hydrogen bonding but simultaneously fit also the betaine pattern:

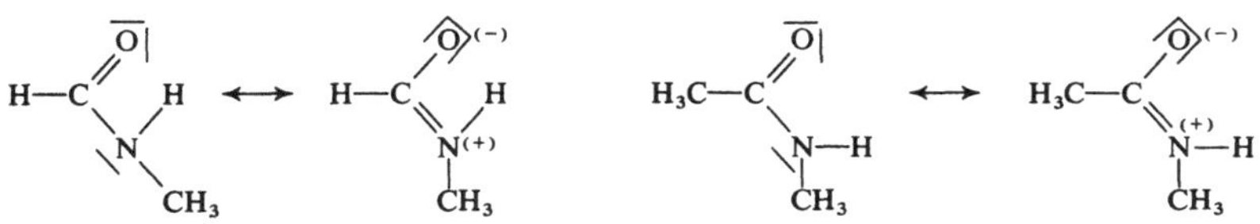

Aside of these derivatives, there are many more that do come into existence by a type II-reaction only. In order for the substitution to proceed, high concentrations are required.

So far, more than 280 different kaolinite derivatives have been reported of which 220 are of organic descent. Most prominent are those having substituted acid amides, alkyl amines, alkylene diamines, pyridine, imidazole and its derivatives, morpholines and its derivatives as well as other heterocycles, purines and pyrimidines, amino acids, peptides, and amino acid amides [101, 102].

The quantity of organic compounds which can be accommodated between the kaolinite layers can be estimated by the following equation:

$$m\,[\text{g/g kaolinite}] = \Delta d \cdot \frac{1}{\rho_K \cdot d_K} \cdot \rho_x$$

whereby Δd = increase in spacing; d_K = spacing of kaolinite starting material: ρ_K = density of kaolinite; and ρ_X density of intercalated compound in its crystalline state.

Organic molecules incorporated into the kaolinite lattice can easily be leached out. Heating will result in condensation phenomena. For instance, ammonium acetate-kaolinite will yield acetamide at 65° C, and the urea compound yields biuret-kaolinite at slightly higher temperatures. Ammonium salts of amino acids are transformed to peptides on prolonged heating at 130° C leading to molecular weights in the order of 2,000 to 3,000. Their release from the kaolinite by means of water washings is rather slow.

Kaolinite, nacrite, dickite, and halloysite resemble each other strongly with regard to the formation of organic derivatives. Only a few exceptions are known. For instance, halloysite reacts faster, and remains in a hydrated form after washing, whereas the kaolinite hydrates are unstable and can be isolated only with difficulty. Furthermore, the typical fire-clay varieties of kaolinite (e.g. kaolinite from Il Provins, France and from Pugu, Tanganyika) do not react at all [102, 103].

In nature only one urea kaolinite has been found so far [98].

I. Derivatives of Hydroamesite

Amesite is a trioctahedral analogon of kaolinite and occurs in nature in a partly hydrated form. This mineral is distinguished from the rest of the trioctahedral kaolinites in being able to form derivatives identical to those observed for kaolinite.

J. Organic Derivatives of Uranium Micas

The uranium micas resemble morphologically the micas. Chemically, however, they represent uranyl phosphates, uranyl arsenates, and uranyl vanadates. The structural formula is written as follows:

$(M^{z+})_{1/z}$	$\cdot\ (Y)_n$	$\cdot\ \{UO_2 \cdot XO_4\}$	$X = P, As, V$
exchange-able cations	swelling liquid	anionic, rigid layer	

Uranyl ions and XO_4 tetrahedra form two-dimensional anionic layers having a tetragonal or rhombic symmetry. Cations (M^{z+}) are placed in interlayer positions to compensate the negative charges; they may become solvated resulting in lattice expansion. Being solvated, these cations can become exchanged for cations present in a solvent medium [27]. This phenomenon was formerly not recognized, and different names were assigned to the same mineral having different exchangeable cations (Table 11). In all fairness, however, the various naturally occurring

Table 11. *Minerals of the uranium mica group*

interlayer cation M^{z+}	names of uranium micas with the layer units		
	$[UO_2PO_4]^-$	$[UO_2AsO_4]^-$	$[UO_2VO_4]^-$
H^+, Al^{3+}	sabugalite		
Mg^{2+}	saleeite	novacekite	
Te^{2+}	bassetite	kahlerite	
Cu^{2+}	torbernite	zeunerite	
Ca^{2+}	autunite	uranospinite	{thujamunite (sincosite)
Ba^{2+}	uranocircite	heinrichite	francevillite
UO_2^{2+}, H^+		trögerite	
K^+		abernathyite	carnotite
Co^{2+}		kirchheimerite	

cations are rather tightly fixed, thus giving the appearance of distinct mineral species on first sight. With regard to structural principles and swelling properties, uranium micas resemble micas. The amount of their negative charge corresponds to that in muscovite and biotite. As a result uranium micas form similar organic derivatives.

Due to the high layer charge, derivatives having inorganic cations in interlayer position and neutral organic molecules are rare. Only small and polar organic molecules such as methanol, glycol, glycerol, or formamide can easily enter the mineral [27, 28]. By means of cation exchange, however, many organic cations can become incorporated. Particularly well-studied derivatives are those which involve:

$$RNH_3^+;\quad R_2NH_2^+;\quad R_3NH^+ \text{ and } R_4N^+.$$

Amino acids and peptides can be bound as cations, preferentially lysine and arginine.

Rapid ion exchange is accomplished with *n*-alkylammonium ions and the exchange equilibrium is progressively shifted to the side of the alkylammonium compound as a function of increase in chain length. The spacing increases by about 1.26 to 1.27 Å per C atom, indicating that the alkyl chains are positioned

Table 12. *The influence of the number of C-atoms in straight-chain alkylammonium ions and straight-chain swelling liquids on the layer distance of different synthetic uranium micas*

total number of C-atoms of ammonium ion and of the swelling liquid	number of C-atoms of the intercalated *n*-alkyl-ammonium ions	number of C-atoms of the primary *n*-alcohols, intercalated as swelling liquid	number of C-atoms of the primary *n*-alkylamine, intercalated as swelling liquid	layer distances (Å)		
				uranyl-phosphate	uranyl-arsenate	uranyl-vanadate
6	4	2		15.4	15.4	
	6			16.6	16.8	
8	4	4		18.6	18.7	
	4		4	18.9	18.9	
	6	2		18.4	18.1	
	8			18.6	18.8	
12	4	8		22.7	22.6	
	6	6		22.6	23.5	
	6		6	22.8	23.3	
	8	4		22.7		
	12			22.5	23.1	23.4
14	4	10		24.6	25.0	
	6	8		25.2	25.7	26.3
	8	6		24.8	25.0	
	12	2		24.5	24.3	
16	4	12				
	6	10		27.8	28.0	31.7
	8	8		29.2		
	8		8	29.8	30.1	
	12	4		27.0	27.4	
18	6	12			29.4	
	8	10		32.5		
	12	6		29.4	27.8	
	18			28.0	28.8	27.7

perpendicular to the layers of the uranium mica and that they are arranged in the form of monolayers. The spacing is further increased in the presence of polar *n*-alkyl compounds, and the total expansion of the lattice depends only on the sum of the number of carbon atoms in the cationic alkyl chains and the swelling medium. This phenomenon is due to the fact that the number of molecules picked up from the swelling liquid equals the number of exchangeable alkylammonium ions (Table 12).

Swelling is also accomplished with nitriles, ketones, pyridine, benzene, toluene and various other compounds. A list of a selected number of compounds is presented in Table 13 [104–106].

Dialkyl- and trialkyl-ammonium ions will become only partially exchanged for reasons of too large space requirements. Structural openings will develop between the cations allowing a zeolitic uptake of solvents with only minimal change in volume.

Analogous to the micas, aromatic amines are less tightly fixed to the uranium micas than are the primary alkyl amines. Within the mineral they will become easily oxidized and firmly bound [104, 107].

The behavior of tetragonal and orthorhombic uranium micas, with respect to small inorganic interlayer cations is clearly distinguished, for instance by potassium fixation or swelling. In contrast these differences become negligible where large organic cations are involved [104]. Under these circumstances, the differently termed uranium micas can only be grouped according to phosphate or arsenate predominance, with the exception of the vanadates which produce slightly larger spacings.

Table 13. *Layer distance in synthetic uranium micas after adsorption of*

state of swelling	layer distance (Å) of				
	uranylphosphate after ion exchange with				
	n-butyl-ammonium	*n*-hexyl-ammonium	*n*-octyl-ammonium	*n*-dodecyl-ammonium	*n*-stearyl-ammonium
dried at 65° C	14.2	16.6	18.6	23.5	28.0
with water	23.6	19.8	21.2	31.8	
benzene	18.6	21.1	24.5	32.8	
toluene	17.9		24.6	33.5	44.0
nitrobenzene	17.1	21.5	24.2		34.6
pyridine	16.0	21.0		30.0	
acetonitrile	17.0		22.2		32.2
methyl ethylketone		22.9	24.0		31.0
n-butylamine	18.9				
n-hexylamine		22.8			
n-octylamine			29.8		
ethyl alcohol	15.4	18.4		24.5	33.4
n-butyl alcohol	18.6	20.3	22.7	27.0	32.6
n-hexyl alcohol		22.6	24.8	29.4	35.6
n-heptyl alcohol	21.5	23.6	27.0	32.1	39.2
n-octyl alcohol	22.7	25.2	29.2	34.4	40.6
n-decyl alcohol	24.6	27.8	32.5	37.6	45.3
n-dodecyl alcohol					

In studies concerned with biopoesis, it has been suggested that in a number of rock formations a quantitative correlation exists between the total amount of organic substances and the uranium content. It was frequently implied, then, that these organic compounds here originated from abiogenic starting material by means of radioactive exposure [108]. This is not the place to review these claims critically. However, these inferences should be challenged in those instances where uranium micas or similar swelling uranium compounds such as uvanite or rauvite (of next section) are involved. These compounds are rather effective in concentrating organic ammonium ions from rather dilute solutions.

On the other hand, there is no doubt that radiation can alter included organic compounds up to a certain extent. It has been shown for instance, that the polymerization of acryl compounds or styrolamines in interlayer position can be started by radiation. The mineral rauvite appears to be of particular interest in this respect.

K. Organic Derivatives of Uvanite and Rauvite

The minerals uvanite and rauvite are related to the uranium micas. Their structural formula is as follows:

$(M^{z+})_{1/z}$	·	$(H_2O)_n$	·	$[VO_2 \cdot V_3O_9]^-$
exchangeable cations		swelling medium (H_2O)		anionic layer

n-alkylammonium ions in the presence of various swelling solvents

uranylarsenate after ion exchange with				uranylvanadate after ion exchange with		
n-hexyl-ammonium	*n*-octyl-ammonium	*n*-dodecyl-ammonium	*n*-stearyl-ammonium	*n*-hexyl-ammonium	*n*-dodecyl-ammonium	*n*-stearyl-ammonium
16.8	18.8	23.1	28.8		23.4	27.7
			27.0	19.6		32.0
21.6			45.0			
22.0		33.5	44.0			42.8
22.2			37.6			
			26.2			
			30.4			
23.3						
	30.1					
18.1		24.3	28.5	21.1		32.0
20.6		27.4	34.0	21.4		
23.5	25.0	27.8	38.6			39.0
25.0		32.0	39.6	27.2		40.2
25.7				26.3		
28.0				31.7		42.8
29.4		39.9	46.8			

They are distinguished from the vanadium uranium micas by substitution of one $[VO_4]^{3-}$ for either one $[V_3O_9]^{3-}$ or one $[V_3O_6(OH)_6]^{3-}$ ion. The symmetry of the anionic complex is strictly trigonal. It is conceivable, however, that dihydrogen-orthovanadate anions are present which in the X-ray diffraction pattern may have been masked by the high scattering power of the uranium ions [109].

Rauvite is distinguished from uvanite by its higher radium content which in turn results in a change of color from yellow to black and in a highly distorted crystal lattice. The amount of uranyl is also lower.

According to the structural formula, the charge density on the anionic uvanite layers is smaller than that observed for the uranium micas; it corresponds more closely to that of the vermiculites. As a result, vermiculite and uvanite exhibit similar properties in the formation of organic derivatives.

In the presence of inorganic interlayer cations, only a few polar organic compounds are readily taken up, *e.g.*, glycol, glycerol, or formamide. Subsequent to the exchange of inorganic cations *versus* organic ones (which is rather efficient due to the steep inclination of the exchange isotherms), an additional swelling is observed in the various polar organic compounds (Table 14). No values are listed for the dry samples since complete drying will result in the distortion of the layers and the (001) reflections become diffuse. The swelling behavior does not change even subsequent to the distortion of the layer [109].

Basic proteins such as salmin or clupein are preferentially bound to uvanite. Apparently the flexibility of the uvanite layers has a stimulating effect on the strong interaction with the proteins.

Synthesis of uvanite is easy to accomplish. This suggests that uvanite readily forms in nature but in small concentrations. Since this mineral has a small particle size of the order of 0.05 μm, its presence is difficult to ascertain.

Table 14. *Swelling of alkyl-ammonium uvanites*

exchangeable cations	layer distance (Å) in the presence of							
	ethylene glycol	nitroethane	$C_nH_{2n+1}NH_2$ with equal number of C atoms as the cations	*n*-butanol	*n*-octanol	*n*-decanol	*n*-dodecanol	*n*-hexadecanol
$n\text{-}C_8H_{17}NH_3^+$	22.2	23.8	28.6	21.2	26.8	32.3	–	–
$n\text{-}C_{10}H_{21}NH_3^+$	24.1	25.0	33.1	23.2	28.4	33.8	38.2	46.1
$n\text{-}C_{12}H_{25}NH_3^+$	26.0	26.6	40.0	26.9	30.2	35.6	40.1	47.8
$n\text{-}C_{14}H_{29}NH_3^+$	27.3	28.1	45.7	27.8	32.0	37.4	41.1	48.8
$n\text{-}C_{18}H_{37}NH_3^+$	32.2	30.3	54.0	–	34.1	40.0	–	–
$n\text{-}C_{16}H_{33}N(CH_3)_3^+$	30.3	31.0	–	30.7	35.1	41.8	45.3	–

L. Organic Derivatives of Swelling Vanadates

The group of the swelling vanadates comprises the minerals hewettite, metahewettite [110] and melanovanadite. The structural formula of hewettite is written as follows:

$(M^{z+})_{(2+x)/z}$	·	$(H_2O)_n$	·	$(V^{5+}_{6-x}\,V^{4+}_x O_{16})^{(2+x)-}$
exchangeable cations		swelling medium (H_2O)		anionic polyvanadate layer

The naturally occurring mineral has M^{z+} calcium ions incorporated in the form of interlayer cations. Occasionally, 5-valent vanadium may be substituted by 6-valent molybdenum; the complete formula for the anionic layer thus should be written:

$$(V^{5+}_{6-x-y}\,V^{4+}_x Mo^{6+}_y O_{16})^{(2+x-y)-}.$$

In synthetic minerals, y can go up to 0.25 without structural alterations [111, 112].

Melanovanadite is a deep-black mineral; its chemical composition is exceedingly variable in terms of V^{5+}/V^{4+} ratio. Air-dried material has a higher ratio than the moist outcrop material; this feature can simply be attributed to air oxidation. The most probable structural formula is listed below:

$(M^{z+})_{3/z}$	·	$(H_2O)_n$	·	$[V^{4+}_x V^{5+}_{12-x} O_{27}(OH)_{9-x}]^{3-}$
exchangeable cations		swelling medium (H_2O)		anionic polyvanadate layer

In both hewettite and melanovanadite, cations in exchange position can be readily substituted by organic cations. Furthermore, organic interlayer cations permit an additional swelling in the presence of many polar solvents. The swelling properties of *n*-dodecylammonium hewettite and *n*-dodecylammonium melanovanadite are compared to each other in Table 15. Both compounds exhibit highly charged layers comparable to those formed in muscovite (*ca.* 21.6 Å^2/e in hewettite); for melanovanadite the correct charge density is not so well established but it falls in about the same range [109].

Table 15. *Examples of organic derivatives of hewettite and melanovanadite with n-$C_{12}H_{25}NH_3^+$ ions adsorbed between the layers*

swelling liquid	layer distance (Å)	
	n-$C_{12}H_{25}NH_3^+$-hewettite	n-$C_{12}H_{25}NH_3^+$-melanovanadite
mineral, dried	24.0	23.2
H_2O	27.0	diffuse
n-$C_6H_{13}OH$	35.3	36.0
n-$C_8H_{17}OH$	36.2	36.1
n-$C_{10}H_{21}OH$	39.6	39.3
n-$C_{12}H_{25}OH$	41.6	41.7
$CH_3 \cdot CH_2-NO_2$	26.3	27.7
n-$C_4H_9OOC \cdot CH_3$	33.2	34.6

M. Organic Derivatives of Taranakite

Taranakite, named after its occurrence in Taranaki, New Zealand, has been relatively little investigated. This is surprising because taranakite is rather abundant in cultivated acid soils where it develops from clay minerals and phosphorus fertilizers. It is capable of incorporating potassium and ammonium ions which are only gradually released and substituted by other inorganic cations. The structural formula for this mineral is only known in some approximation:

$K_3^+(H_3O^+)_x$	·	$(H_2O)_n$	·	$Al_5(PO_4)_{2+x}(HPO_4)_{6-x}^{(3+x)-}$
exchangeable cations		swelling medium (H_2O)		anionic aluminium phosphate layer

The formulation of the structure is actually based on cation exchange experiments involving organic cations. Thermal degradation suggests that part of the water is bound within the anionic layer.

The charge density of the aluminium phosphate layers is apparently very high because the exchange reaction is equally as slow as the one for muscovite and biotite. Long chain amines react at a faster rate in comparison to the shorter chain variety. The fact that alkyl ammonium compounds cannot swell further can be attributed to the high layer charge. Namely, the cations have already developed two steeply inclined double layers between the aluminium phosphate layers. However, if only the H_3O^+ ions are converted to *n*-alkylammonium ions by means of free amine, a monolayer is built up. These samples are capable of swelling, for instance with alcohols [109].

N. Organic Derivatives of Zeolites

The previously discussed minerals have a two-dimensional continuous layer as structural element in common. This may explain the similar properties of their organic derivatives in spite of their chemical diversities. Most crucial are the layer charges, the symmetry of the charge distributions, the possibility of charge delocalization within a layer, and the mechanical flexibility of the layers themselves.

In contrast, zeolites exhibit different structural principles with regard to their organic derivatives. The classical zeolites are constructed by means of SiO_4 tetrahedra which are interconnected in such a way as to give rise to an open, but rigid channeltype three-dimensional skeleton. Substitution of Si^{4+} for Al^{3+} will result in an excess of negative charge, which in turn has to be compensated for by cations. These are placed in the open channel network and may either become hydrated or solvated, or may become replaced by organic cations. Different from the layer type-minerals which can structurally expand, a zeolite can only accommodate ions or molecules that are smaller than the cross-section of its smallest channel opening because of the structural rigidity. Consequently, zeolites may function as well-defined molecular sieves [113, 114].

This property for natural zeolites to act as molecular sieves has first been described by MACBAIN [113] and by SZIGETTI [115] for synthetic ones. Particularly the detailed work of BARRER has stimulated their industrial application and extensive research in this field within the last 20 years [114].

Zeolites, varying in the size of their channel openings, have been synthesized according to required specification. In nature, they are highly abundant in sediments, and particularly well crystallized in basaltic rocks. The intracrystalline

Table 16. *Zeolite silicates*

zeolite	chemical formula	cavity volume for H_2O cm^3/cm^3 zeolite	channel openings and number of (Al, Si)O_4-tetrahedrons per ring (in parenthesis)
analcite type			
analcite	$Na_{16}[Al_{16}Si_{32}O_{96}] \cdot 16\ H_2O$	0.18	–
wairakite	$Ca_{18}[Al_{16}Si_{32}O_{96}] \cdot 16\ H_2O$	0.18	–
viseite	$Na_2Ca_{10}[Al_{20}Si_6P_{10}H_{36}O_{96}] \cdot 16\ H_2O$	0.18	–
mordenite type			
mordenite (ptilolite)	$Na_8[Al_8Si_{40}O_{96}] \cdot 24\ H_2O$	0.29	6.7 × 7.0 (12) 2.9 × 5.7 (8)
dachiardite	$Na_5[Al_5Si_{19}O_{48}] \cdot 12\ H_2O$	0.31	3.7 × 6.7 (10); 3.6 × 4.8 (8)
epistilbite	$Ca_3[Al_6Si_{18}O_{48}] \cdot 15\ H_2O$	0.33	3.2 × 5.3 (10); 3.7 × 4.4 (8)
ferrierite	$(Na_2, Mg)_3[Al_6Si_{30}O_{72}] \cdot 18\ H_2O$	0.27	4.3 × 5.5 (10); 3.4 × 4.8 (8)
bikitaite	$Li_2[Al_2Si_4O_{12}] \cdot 2\ H_2O$	0.20	3.2 × 4.9 (8)
phillipsite type			
phillipsite	$(K, Na)_{10}[Al_{10}Si_{22}O_{64}] \cdot 2\ H_2O$	0.36	4.2 × 4.4 (8); 2.4 × 4.8 (8); 3.3 (8)
harmotome	$(K_2, Ba)_2[Al_4Si_{12}O_{32}] \cdot 12\ H_2O$	0.34	–
garronite	$(Na_2Ca)_3[Al_6Si_{10}O_{32}] \cdot 14\ H_2O$	0.41	–
gismondite	$Ca_4[Al_8Si_8O_{32}] \cdot 16\ H_2O$	0.49	2.8 × 4.9 (8); 3.1 × 4.4 (8)
chabasite type			
chabasite	$(Ca, Na_2)_2[Al_4Si_8O_{24}] \cdot 12\ H_2O$	0.46	3.7 × 4.2 (8)
gmelinite	$(Na_2Ca)_4[Al_8Si_{16}O_{48}] \cdot 24\ H_2O$	0.45	6.9 (12)
offretite	$(Ca, Na_2)_4[Al_8Si_{14}O_{36}] \cdot 13\ H_2O$	0.36	6.3 (12)
erionite	$(Ca, Na_2)_{4.5}[Al_9Si_{27}O_{72}] \cdot 27\ H_2O$	0.36	3.6 × 4.8 (8)
levynite	$Ca_3[Al_6Si_{12}O_{36}] \cdot 18\ H_2O$	0.39	3.3 × 5.1 (8)
cancrinitehydrate	$Na_6[Al_6Si_6O_{24}] \cdot 7\ H_2O$	0.30	6.2 (12)
sodalitehydrate	$Na_6[Al_6Si_6O_{24}] \cdot 7\ H_2O$	0.30	2.6 (6)
faujasite type			
faujasite	$(Na_2, Ca)_{32}[Al_{64}Si_{128}O_{384}] \cdot 256\ H_2O$	0.54	7.4 (12)
paulingite	$(K_2, Ca, Na_2)_{152}[Al_{152}Si_{520}O_{1344}] \sim 700\ H_2O$	0.49	3.9 (8)
heulandite type			
heulandite	$Ca_4[Al_8Si_{28}O_{72}] \cdot 24\ H_2O$	0.33	2.4 × 6.1 (8); 3.2 × 7.8 (10); 3.8 × 4.5 (8)
clinoptilite	$(Ca, Na_2, K_2)_3[Al_6Si_{30}O_{72}] \cdot 24\ H_2O$	0.34	–
stilbite	$Ca_4[Al_8Si_{28}O_{72}] \cdot 28\ H_2O$	0.37	4.1 × 6.2 (10)
brewsterite	$(Si, Ba, Ca)_2[Al_4Si_{12}O_{32}] \cdot 10\ H_2O$	0.33	2.3 × 5.0 (8); 2.7 × 4.1 (8)
natrolite type			
natrolite	$Na_{16}[Al_{16}Si_{24}O_{80}] \cdot 16\ H_2O$	0.21	2.6 × 3.9 (8)
scolecite	$Ca_8[Al_{16}Si_{24}O_{80}] \cdot 24\ H_2O$	0.29	–
mesolite	$Na_{16}Ca_{16}[Al_{48}Si_{72}O_{240}] \cdot 64\ H_2O$	0.27	–
gonnadite	$Na_4Ca_2[Al_8Si_{12}O_{40}] \cdot 12\ H_2O$	0.31	–
thomsonite	$Na_4Ca_8[Al_{20}Si_{20}O_{80}] \cdot 24\ H_2O$	0.33	2.6 × 3.9 (8)
edingtonite	$Ba_2[Al_4Si_6O_{20}] \cdot 8\ H_2O$	0.30	3.5 × 3.9 (8)
other zeolites			
ashcroftine	$(Na_2, K_2, Ca)_{80}[Al_{160}Si_{200}O_{720}] \cdot 320\ H_2O$	0.47	–
laumontite	$Ca_4[Al_8Si_{16}O_{48}] \cdot 16\ H_2O$	0.35	–
yugawaralite	$Ca_2[Al_4Si_{10}O_{28}] \cdot 6\ H_2O$	0.21	–
welsite	$(Ca, Ba, Na_2, K_2)[Al_2Si_3O_{10}] \cdot 3\ H_2O$	0.31	–

Table 17. *Possibilities for the separation of molecules according to size in zeolitic molecular sieves*

growing molecular size →							
H_2O, He	Ne, Ar CO, O_2, N_2, NH_3	Kr, Xe CH_4, C_2H_6 CH_3OH, CH_3CN CH_3NH_2 CH_3Br CO_2 C_2H_2	C_3H_8 n-C_4H_{10} n-C_xH_{2x+2} C_2H_5Br C_2H_5OH $C_2H_5NH_2$ CH_2Br_2 CHF_3 $(CH_3)_2NH$ CH_3I	CF_4 C_2F_6 CF_2Cl_2	*iso*-C_4H_{10} *iso*-C_xH_{2x+2} $CHCl_3$ $CHBr_3$ CHI_3 $(CH_3)_2CHOH$ n-C_xF_{2x+2} $(C_2H_5)_3N$ $(CH_3)_4C$ C_6H_6 cyclohexane thiophene naphthalene quinoline pyridine	1, 3, 5-tri- ethylbenzene, decahydro- chrysene	$(n$-$C_4F_9)_4N$
←→ analcite							
← Ca- und Ba-mordenite, levinyte	→						
← Na-mordenite		→					
← chabasite rich in Ca, gmelinite				→			
← faujasite					→		
← synthetic zeolite X						→	

volume for water can be used as a criterion to estimate their capacity for the uptake of solvent media. The diameters of the pores (last column Table 16) are crucial, regarding size and shape of the molecules to be adsorbed. They can be altered according to the size of the exchangeable cations. The adsorption properties of zeolites with respect to limits in the molecular size of the guest molecules are shown in Table 17 [114].

Within the structural framework, Al-tetrahedra are always separated by SiO_4 tetrahedra. The Al:Si ratio will determine the charge distribution and will affect the separation properties for compounds that have polar character.

O. Complex Formation with Inorganic Cations in the Interlayer Space

Transition metals in interlayer position may interact with complexing agents. For instance, nickel ions in the presence of methylglyoxime will yield the well-known red planar complex. Since the complex is neutral, hydronium ions have to bring about the charge balance. Phenylmethylglyoxime also reacts with nickel ions but only along the crystal edges [28].

Similar to the nickel ions, magnesium ions in interlayer position will form a faint violet complex with chinalizarin. Basal spacing measurements permit insight into the structure of the complexes [28].

Complex formation and ion exchange compete with each other, if the exchangeable cations belong to the transition metals and the reaction is with alkylamines, alkylene diamines or polyamines, respectively [29–33]. For weakly basic compounds such as the bases of the purines and pyrimidines, complex formation significantly enhances the amount which can be adsorbed [24].

P. Derivatives with Amino Acids, Peptides, Proteins, Purines, and Pyrimidines

Proteins:

The fact that proteins can become incorporated in montmorillonites and other micaceous layer silicates [116–121, 149], has for many years been used to eliminate clouding proteins from beer and wine [26]. The fixation characteristics of proteins to minerals not only varies with the type of mineral but also with the kind of proteins involved (Table 18). Aside of their basic properties, the proportions of hydrophobic residues in proteins are determining factors. It is noteworthy that pre-expanded sodium montmorillonites have a larger exchange capacity for proteins than the non-expanded variety. X-ray diagrams, however, suggest a lower crystallinity in such instances. The reaction products can simply be described in terms of cationic and anionic polyelectrolyte precipitates. For the strongly basic protamines (Table 19), the uptake increases strongly with rising pH of the equilibrium solution. This suggests that the protein is picked up as a cation. Increase in pH raises the ratio of free amino groups to ammonium groups. Consequently, more protein becomes incorporated in order to substitute all interlayer cations by ammonium ions [35, 122].

Table 18. *Intercalation of proteins in micaceous silicates,* (1) *Ca-montmorillonite from Geisenheim,* (2) *nontronite from the Ficht mine,* (3) *montmorillonite from Unterrupsroth,* (4) *Na-montmorillonite from Geisenheim*

protein	basic concentration g protein/1 l solution	pH value	linked protein mg protein/ 1 g silicate	layer distance of the dehydrated sample (110° C)
salmin[a]	6.0	8.0	266(1)	16.9
		8.3	289(3)	17.3
		11.4	285(2)	17.5
serum albumin[a]	3.1	5.1	215(1)	12.5
		5.2	607(4)	16.0
egg albumin[a]	20.0	4.0	215(1)	14.8
		4.0	412(4)	15.5
		4.0	548(3)	15.8
zein[a]	15.3	3.9	95(3)	13.5
		4.0	107(1)	12.6
		6.1	83(2)	12.4
gliadin[a]	3.8	6.1	82(2)	12.4
		6.8	138(3)	15.0
		6.8	168(1)	13.2
gelatin[b]	10.0	2.2	62	14.3
human albumin[b]	10.0	2.2	49	14.2
pepsin[b]	8.0	2.2	26	13.6
hemoglobin[b]		2.6	50	

[a] G. Koch, dissertation, Technische Hochschule Darmstadt 1960.
[b] A. Weiss, dissertation, Technische Hochschule Darmstadt 1953.

Table 19. *Cation exchange of micaceous silicates with protamines (salmin from herring sperma)*

micaceous silicate	equivalent area Å²/elementary charge	pH-value during the exchange		linked protein mg protein/g silicate		layer distance (Å)	
		initial value	final value			dehydr. at (110° C)	with H_2O
montmorillonite from Geisenheim	75	2.5	2.7	218	222	14.8	17.1
		4.6	5.9	218	248	15.7	17.2
		8.7	8.0	218	262	15.9	18.0
		11.2	8.5	252	267	16.9	18.4
		12.1	11.4	488	266	16.9	18.3
nontronite from the Ficht mine	60	2.5	2.7	263	258	15.3	18.0
		4.6	3.9	263	250	15.9	19.0
		8.7	7.2	263	250	16.1	18.9
		11.2	7.7	303	272	17.5	19.3
		12.1	11.4	588	285	17.5	20.3
beidellite from Unterrupsroth	57	2.5	2.7	316	263	15.5	18.8
		4.6	5.4	316	258	17.0	19.8
		8.7	8.3	316	289	17.3	20.1
		11.2	10.0	363	283	17.3	20.1
		12.1	11.9	706	386	17.3	20.2

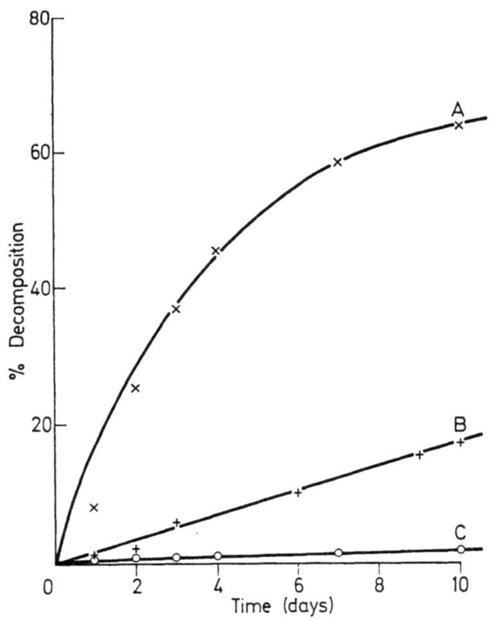

Fig. 5. Decomposition of gelatine by soil-bacteria (A = gelatine, B = mixture of gelatine and montmorillonite, C = gelatine-montmorillonite-complex)

The stability of proteins in the interlayer space of montmorillonite has already been discussed by ENSMINGER and GIESEKING [118]. Results for gelatin-montmorillonites and soil microorganisms have been given by PINCH and ALLISON [123] (Fig. 5).

Peptides and Amino Acids:

High molecular weight peptides exhibit characteristics similar to proteins; low molecular weight peptides resemble amino acids in their affinity to clay minerals.

Basic amino acids are preferentially picked up as cations. The linkage is considerably lower than for the proteins because the van der Waals interaction decreases with a decrease in molecular size [26].

Neutral amino acids can also become incorporated as cations. In order to substitute all inorganic cations for amino acids, a low pH and high amino acid

Table 20. *Intercalation of amino acids into micaceous silicates*

intercalated amino acids	concentration of the exchange solution (g amino acid/l solution)	pH-value of the exchange solution	intercalated amount of amino acid (mval/100 g silicate)	layer distance after dehydration at 110° C (Å)
glycine	1.9	2.0	5.0	–
	10.0	2.5	75.0	12.6
	30.0	2.2	4.0	11.7
	37.5	5.6–6.4	120–180	12.5
	50.0	6.5	51.2	–
	375.0	2.0	60.0	13.5
α-alanine	10.0	2.5	–	13.16
	445.0	2.0	15.0	14.5
β-alanine	2.2	2.0	30.0	–
	44.5	5.6–6.4	120–180	
α-amino butyric acid	206.0	2.0	20.0	15.2
β-amino butyric acid	206	2.0	90.0	15.2
γ-amino butyric acid	2.5	2.0	45.0	–
	50.0	5.6–6.4	120–180	–
	206	2.0	90.0	15.1
norvaline	587	2.0	25.0	14.6
δ-amino valeric acid	2.9	2.0	60.0	–
ε-amino capronic acid	3.3	2.0	73.0	–
	61.0	5.6–6.4	120–180	
	263.0	2.0	90.0	
hydroxyproline	10.0	2.5		13.6
lysine	30.0	2.2	13.0	
	73.0	2.5	120–180	
arginine	10.0	2.5	87.0	13.16
phenylalanine	10.0	2.5		13.67

concentration is required. Above pH 6, amino acids will still be picked up; the inorganic cations, however, can no longer be completely substituted. Complex formation begins and is favored if transition metals are present (Table 20) [124–130].

Acidic amino acids will only be fixed in the form of neutral molecules. Yet in the presence of 3-valent interlayer cations, particularly stable cationic complexes will result. In most instances amino acids will only yield monomolecular layers. The same relationships hold true for glycyl-glycine, whereas diglycyl-glycine, and tri-glycyl-glycine will form monolayers at low concentration and double layers at a higher one [131].

Published reports should be cautiously looked upon, because in many experiments microbial degradation has not been ruled out. Micro-organisms may decarboxylate amino acids and form amines which are far more tightly fixed in interlayer position relative to their amino acid precursor, and hence become enriched in the mineral.

Q. The Bases of the Purines and Pyrimidines

Thymine, uracil and their nucleosides (concentration: 0.8–1.3 molar) will not be taken up by sodium and calcium montmorillonites. Adenine, hypoxanthine, cytosine, and uracil will be taken up to a greater extent, the stronger basic the compounds are. In any event, the free bases will be more tightly fixed than their nucleosides.

Analogous to Na- and Ca-montmorillonites, the incorporation of these organics into the Co^{2+}, Ni^{2+}, Cu^{2+}, and Fe^{3+} montmorillonites will proceed as a cation exchange reaction in the acid pH-range. Under neutral and moderately alkaline conditions, however, complex formation takes place. The trend of complex formation is as follows:

$$Cu^{2+} \gg Ni^{2+} > Co^{2+} \quad \text{and}$$

$$\text{adenine} \gtrapprox \text{hypoxanthine} > \text{purine} \gg \text{6-chloropurine}$$ [132, 133].

R. Organic Derivatives of Calcium-Aluminium-Hydroxoaluminates

The bulk of the organic derivatives described so far are only stable in the acid to neutral pH range. In some instances, where complex formation may proceed, the stability range extends into the alkaline pH range.

The pH conditions on the prebiotic earth were most probably alkaline. This might suggest that the inferences drawn in this work are of no direct interest as far as the origin of life is concerned. It should be emphasized, however, that in the proximity of volcanoes, due to the discharge of for instance boric acid and free CO_2, acid conditions may be anticipated. Furthermore, there are certain organic derivatives that are preferentially generated under alkaline conditions, *e.g.*, in the presence of calciumhydroxyaluminates. These minerals are frequently associated with basaltic rocks in the form of ettringite and hydrocalumite. They are layer compounds capable of intracrystalline swelling with organic molecules. Acid compounds such as carboxylic acids and amino acids become preferentially incorporated, but they will be released under acidic conditions via decomposition of the layer structure [134].

Little is known regarding the organic derivatives of this group of minerals. Aside of compounds with neutral layers there are most likely also some minerals having negatively charged layers. This would demand balancing by cations, e.g. organic ones. Concerning the origin of life, these derivatives should be more fully investigated.

IV. Degradation of Organic Derivatives of Clay Minerals

Organic compounds in the interlayer space of layer silicates are to a large extent protected against microbial degradation. Only oxygen may introduce oxidative alterations within the silicates. For this reason, these minerals may have certain significance regarding the origin of petroleum. Organic compounds may become adsorbed and enriched from rather dilute solutions in a rather specific manner. Experiments have shown that a ton of montmorillonite may incorporate up to 550 kg of protein, 260 kg of polypeptides, 150 kg amino acids, 100 kg amines, 620 kg lipids (or 320 kg lipids and 240 kg proteins), 150 kg fatty acids, 300 kg glycerol, or 200 kg carbohydrates. In contrast, a non-specific adsorption will only accommodate 5 to 10 kg of organics per ton of montmorillonite [89, 90]. Besides layer minerals having organic cations in the interlayer position, zeolites can accommodate finely dispersed oil droplets [135].

Finally, the silicate layers may participate in the catalytic alteration of the incorporated organic compounds. Indeed, in the absence of oxygen a thermal degradation will yield a mixture of compounds that resembles crude oil. At temperatures between 220° and 280° C, the reaction proceeds so rapidly that within a few hours preparative quantities of these reaction products can be obtained. Most common among these products are hydrogen, methane, carbon monoxide, carbon dioxide and a mixture of paraffins, cycloparaffins, olefins and aromatics. Increase in the temperature of degradation will lower the yield in olefins, while the amount in aromatic hydrocarbons increases. The overall analyses of the liquid and gaseous decomposition products of protein-montmorillonites, Ca-fats-montmorillonites, or protein-fats-montmorillonites correspond to those of crude oil [20, 136]. At temperatures below 300° C, montmorillonites still retain the capacity to expand and polar organic compounds may still enter the lattice. This will result in random (001) interference series of the type reported from montmorillonites of crude oil deposits. Extensive exposure to heat (250° to 300° C) will result in the release of ammonia, and a protonated montmorillonite will remain. Protons move from interlayer position towards octahedral vacancies, resulting in formal electrically neutral layers and the failure to expand. Analytically, this silicate would have to be grouped with pyrophyllite, but the basal spacing is considerally larger and resembles that of the illites [90] (9.9 to 10.1 Å). Such illites having an extremely low potassium content have frequently been observed in the proximity of petroleum deposits [137].

For complexes, whose steric configuration fits the silicate pattern, the thermal stability in interlayer position can be enhanced. An example is the porphyrin montmorillonite. Porphyrin actually blocks the swelling capability and is kept stable up to about 280° C, while porphyrin in a free state will already decompose at 220° C. For this reason the presence of porphyrin can no longer be used as a criterion for the formation of petroleum below 220° C, a statement which has frequently been made in the past [138].

V. Catalytic Effects

The aforementioned minerals are of great interest with respect to the catalytic properties which they exhibit in the interlayer space of micaceous minerals or the channel openings of zeolites, respectively. For instance, aromatic amines which have been introduced as exchangeable cations will become easily oxidized [139–145]. The kind of reaction product will depend on the charge of the silicate anions. Aniline will become oxidized to a red ion at low charge density ($x+y=0.33$), a deep blue ion at medium ($x+y=0.55$) and a black ion at high charge density ($x+y=0.70$). In all instances, oxygen in the air, which diffuses through the openings between the arylammonium ions, is the means for oxidation. The high oxidation potential is most likely related to a change in the bonding state of the oxygen molecule. Magnetic data indicate that in the interlayers as well as in structural channels the molecular oxygen looses its paramagnetism. In interlayer position oxygen in higher concentrations is most likely present in the form of O_4, and in the small openings of zeolites partly as O_n [63, 35].

Inasmuch as cationic constituents of the anionic layer can participate reversibly in redox-reactions, the catalytic properties become modified, since with the oxidation state the charge density is also altered. Examples are the organic derivatives of layer-vanadates ($V^{5+}+e \rightleftharpoons V^{4+}$).

Beside the oxidase effect, a protease effect has been noticed which also is a function of the charge density. Proteins can become perfectly exchanged in lower charged layer compounds, whereas in highly charged layer compounds having H_3O^+ ions in the form of interlayer cations, proteins become hydrolyzed into peptides and eventually into amino acids. Since the smaller fragments are less tightly bonded relative to the high molecular proteins, they may become easily replaced for the high molecular original starting material present in the equilibrium solution and the reaction may continue [63].

The ε-amino and guanido groups of the proteins interact with the H_3O^+ ions present in interlayer position and result in the formation of ammonium ions. The effective interaction between these ammonium ions and the fixed negative charges of the layers forces peptide bonds ($-CO-NH-$) to approach closely the free H_3O^+ ions, and a simple acid hydrolysis will be accomplished. In those instances where the number of H_3O^+ ions is small or the number of ε-amino and guanido groups in the proteins is large, all H_3O^+ ions will be consumed for the ammonium formation and acid hydrolysis does not take place. Thus, the size of the individual fragments released during this kind of reaction is a function of the amount of basic amino acids in the protein and the charge density of the anionic layers.

Under suitable conditions amino acids may condense in the interlayer spaces [130, 125]. For montmorillonites this reaction will take place only above 180° C, whereas for kaolinite (in the presence of low vapor pressure) condensation can already be achieved at about 90° C [95]. The catalytic effectiveness of free SiO_2 is well known from numerous papers and is not discussed here. However, it seems noteworthy, that on its surface sugars will be altered into aminosugars, if ammonia is present. Those in turn may become incorporated in the form of cations in interlayer positions of clay minerals and thus protected [146, 147].

VI. Outlook

Organic derivatives of clay minerals, zeolites and related minerals are of considerable academic and technical interest. They are structurally and chemically quite variable and thousands of compounds are already known. The steric and stoichiometric relationships which determine their formation are rather unique. Over the last 15 years their production for technical purposes has been well developed.

Of particular significance are these compounds with respect to the origin of crude oil and questions concerning the origin of life. In this context it must be considered that these minerals may concentrate and store metastable organic compounds which may be transported in this preserved form and suddenly released in high concentration upon change in the pH condition of the environment or due to the partial or complete destruction and recrystallization of the mineral matter.

Since within individual crystals layers with different charge density stacked randomly, for instance in mixed layer minerals, a high catalytic activity of redox reactions, formation and cleavage of peptide bonds and selective permeation may occur within the same molecular frame work. The formation of pyrimidine and aminosugars seems to be realized in systems similar to these.

References

1. Deuel, H., G. Huber u. R. Ibey: Organische Derivate von Tonmineralien. Helv. Chim. Acta **33**, 1229–1232 (1950).
2. — Organic derivatives of clay minerals. Clay Minerals Bull. **1**, 205–211 (1952).
3. Weiss, A.: Der Kationenaustausch bei den Mineralen der Glimmer-, Vermikulit- und Montmorillonitgruppe. Z. Anorg. Allgem. Chem. **297**, 257–286 (1958).
4. — Der Kationenaustausch bei Kaolinit. Z. Anorg. Allgem. Chem. **299**, 92–120 (1959).
5. Aragon de la Cruz, F.: Interlamellar sorption in a methylated montmorillonite. Nature **205**, 381–382 (1965).
6. —, and R. Espinoza: Interlamellar sorption in a "phenyl-montmorillonite". Proc. Intern. Clay Conf. Jerusalem (1), 247–250 (1966).
7. Weiss, A., G. Reiff u. Al. Weiss: Zur Kenntnis wasserbeständiger Kieselsäureester. Z. Anorg. Allgem. Chem. **311**, 151 (1961).
8. Muetterties, E. L., and C. M. Wright: Tropolone and aminotroponium derivatives of the main group elements. J. Am. Chem. Soc. **86**, 5132–5137 (1964).
9. Weiss, A., and A. Herzog: Thujaplicin-Komplexe der Kieselsäure. In: Diplomarbeit, A. Herzog, Heidelberg 1967.
10. Flegman, A. W.: Adsorption of sodium dodecyl sulfate by kaolinite. Intern. Clay Conf. Proc. Conf. Stockholm (1963) 333–336 (publ. 1965).
11. Weiss, A., A. Mehler, G. Koch u. U. Hofmann: Über das Anionenaustauschvermögen der Tonmineralien. Z. Anorg. allgem. Chem. **284**, 247–271 (1956).
12. Smith, C. R.: Base-exchange reactions of bentonites and salts of organic bases. J. Am. Chem. Soc. **56**, 1561–1563 (1934).

13. HENDRICKS, S. B.: Base-exchange of the clay mineral montmorillonite for organic cations and its dependence upon adsorptions due to van der Waals' forces. J. Phys. Chem. **45**, 65–81 (1941).
14. ERBRING, H., u. H. LEHMANN: Austauschreaktionen an Na-Bentoniten mit großvolumigen organischen Kolloidionen. Kolloid-Z. **107**, 201–205 (1944).
15. GRIM, R. E., W. H. ALLAWAY, and F. L. CUTHBERT: Reaction of different clay minerals with some organic cations. J. Am. Ceram. Soc. **30**, 137–142 (1947).
16. ALLAWAY, W. H.: Differential thermal analysis of clays treated with organic cations as an acid in the study of soil colloids. Soil Sci. Soc. Am. Proc. **13**, 183–188 (1949).
17. WEISS, A., A. MEHLER u. U. HOFMANN: Kationenaustausch und innerkristallines Quellungsvermögen bei den Mineralen der Glimmergruppe. Z. Naturforsch. **11**b, 435–438 (1956).
18. COWAN, C. T., and D. WHITE: The mechanism of exchange reactions occurring between sodium montmorillonite and various N-primary amine salts. Trans. Faraday Soc. **54**, 691–697 (1958).
19. WEISS, A., E. MICHEL u. AL. WEISS: Über den Einfluß von Wasserstoffbrückenbindungen auf ein- und zweidimensionale innerkristalline Quellungsvorgänge. In: Hydrogen bonding, p. 495–508. Pergamon Press 1958.
20. —, u. G. ROLOFF: Die Rolle organischer Derivate von glimmerartigen Schichtsilikaten bei der Bildung von Erdöl. Internat. Clay Conf. Stockholm, vol. 2, 1963, p. 373–378. Pergamon Press (pub. 1964).
21. BRADLEY, W. F.: Molecular associations between montmorillonite and some polyfunctional organic liquids. J. Am. Chem. Soc. **67**, 975–981 (1945).
22. — R. A. ROWLAND, E. J. WEISS, and C. E. WEAVER: Temperature stabilities of montmorillonite- and vermiculite-glycol complexes. Natl. Acad. Sci. Natl. Res. Council, Publ. No 566, 348–355 (1958).
23. BRINDLEY, G. W., and M. RUSTON: Adsorption and retention of organic material by montmorillonite in the presence of water. Am. Mineralogist **43**, 627–640 (1958).
24. —, and S. RAY: Clay-organic studies. VIII. Complexes of Ca-montmorillonite with primary monohydric alcohols. Am. Mineralogist **49**, 106–115 (1964).
25. MCEWAN, D. M. C.: Complexes of clays with organic compounds. I. Complex formation between montmorillonite and halloysite and certain organic liquids. Trans. Faraday Soc. **44**, 349–367 (1948).
26. WEISS, A., u. U. HOFMANN: Batavit. Z. Naturforsch. **6**b, 405–409 (1951).
27. — — Reaktionen im Innern des Schichtgitters von Uranglimmern. Z. Naturforsch. **7**b, 362–363 (1952).
28. — Reaktionen im Innern von Schichtkristallen. Diss. Darmstadt 1953.
29. BODENHEIMER, W., L. HELLER, B. KIRSON, and SH. YARIV: Organometallic clay complexes II. Clay Minerals Bull. **5**, 145—154 (1962).
30. — B. KIRSON u. SH. YARIV: Metallorganische Tonkomplexe. 1. Mitt. Israel J. Chem. **1**, 69–77 (1963).
31. — L. HELLER, B. KIRSON, and SH. YARIV: Organometallic clay complexes. IV. Nickel and mercury aliphatic polyamines. Israel J. Chem. **1**, 391–403 (1963).
32. — — — — Organometallic clay complexes. III. Copper-polyamine-clay complexes. Proc. Intern. Clay Conf. Stockholm **2**, 351–363 (pub. 1965).
33. — —, and SH. YARIV: Organo-metallic clay complexes. VI. Copper-montmorillonite-alkylamines. Proc. Intern. Clay Conf. Jerusalem **1**, 251–262 (1966); **2**, 171–173 (1967).
34. YARIV, SH., W. BODENHEIMER, and L. HELLER: Organometallic clay complexes. V. Fe(III)-pyrocatechol. Israel J. Chem. **2**, 201–208 (1964).
35. WEISS, A.: Mica-type layer silicates with alkylammonium ions. Clays Clay Minerals **10**, 191–224 (1963).
36. FRIEDLÄNDER, H. Z.: Spontaneous polymerization in and on clays. Am. Chem. Soc. Div. Polymer Chem. Reprints **4**, 300–306 (1963).
37. SUITO, E., M. ARAKAWA, and S. KONDO: Adsorbed state of organic compounds in organobentonite. Chem. Res., Kyoto Univ. **44**, 316–324, 325–334 (1966).
38. WHITE, D.: The structure of organic montmorillonite and their adsorptive properties in the gas phase. Clays Clay Minerals **12**, 257–266 (1963).

39. Pauling, L.: Structures of micas and related minerals. Proc. Natl. Acad. Sci. U.S. **16**, 123–129 (1930).
40. Hofmann, U., K. Endell u. D. Wilm: Struktur und Quellung von Montmorillonit. Z. Krist. Mineral. **86**, 340–348 (1933).
41. — — — Über die Natur der keramischen Tone. Ber. Deut. Keram. Ges. **14**, 407–438 (1933).
42. — — — Röntgenographische und kolloidchemische Untersuchungen über Ton. Angew. Chem. **47**, 539–558 (1934).
43. Hendricks, St. B., and M. E. Jefferson: Crystal structure of vermiculites and mixed vermiculite-chlorites. Am. Mineralogist **23**, 851–862 (1938).
44. Weiss, A., G. Koch u. U. Hofmann: Zur Kenntnis von Saponit. Ber. Deut. Keram. Ges. **32**, 12–17 (1955).
45. — Die innerkristalline Quellung als allgemeines Modell für Quellungsvorgänge. Chem. Ber. **91**, 487–502 (1958).
46. Hoffmann, R. W., and G. W. Brindley: Clay-organic studies. II. Adsorption of nonionic aliphatic molecules from aqueous solution on montmorillonite. Geochim. Cosmochim. Acta **20**, 15–29 (1960).
47. Weiss, A., u. G. Pfirrmann: In G. Pfirrmann, Kolloidchemische und röntgenographische Untersuchungen an Huminsäuren. Diss. Heidelberg 1968.
48. Foster, W. R., and J. M. Waite: Adsorption of polyoxyethylated phenols on some clay minerals. Am. Chem. Soc., Div. Petro. Chem. Reprints-Symposia, Dallas, Texas **1**, 39–44 (1956).
49. Tettenhorst, R., C. W. Beck, and G. Brunton: Montmorillonite-polyalcohol-complexes. Proc. Natl. Conf. Clays Clay Minerals **9**, 500–519 (1962).
50. Emerson, W. W.: Complex formation between montmorillonites and high polymers. Nature **176**, 461 (1955).
51. — Organo-clay complexes. Nature **180**, 48–49 (1957).
52. —, and M. Ranchach: The reaction of polyvinylalcohol with montmorillonite. Austral. J. Soil Res. **2**, 46–55 (1964).
53. Greenland, D. J.: The adsorption of sugars by montmorillonite. I. X-ray studies. J. Soil Sci. **7**, 319–328 (1956). II. Chemical studies. J. Soil. Sci. **7**, 329–334 (1956).
54. Lynch, D. L., L. M. Wright, and L. J. Cotnoir: The adsorption of carbohydrates and related compounds on clay minerals. Soil Sci. Soc. Am. Proc. **20**, 6–9 (1956).
55. Emerson, W. W.: Complexes of calcium-montmorillonite with polymers. Nature **186**, 573–574 (1960).
56. Schott, H.: Interaction of nonionic detergents and swelling clays. Kolloid-Z. **199**, 158–169 (1964).
57. Greene-Kelly, R.: An unusual montmorillonite complex. Clay Minerals Bull. **2**, 226–232 (1955).
58, 59. — Sorption of aromatic organic compounds by montmorillonite. I. Orientation studies. Trans. Faraday Soc. **51**, 412–424 (1955). II. Packing studies with pyridine. Trans. Faraday Soc. 425–430 (1955).
60. — Birefringence of Montmorillonite complexes. Nature **184**, 181 (1959).
61. Carthew, A. R.: Use of piperidine saturation in the identification of clay minerals by differential thermal analysis. Soil Sci. **80**, 337–347 (1955).
62. Olphen, H. van, and C. T. Deeds: Stepwise hydration of clay-organic complexes. Nature **194**, 176–177 (1962).
63. Weiss, A.: Organische Derivate der glimmerartigen Schichtsilicate. Angew. Chem. **75**, 113–148 (1963).
64. Jordan, J. W.: Organophilic bentonites. I. Swelling in organic liquids. J. Phys. Colloid. Chem. **53**, 294–306 (1949).
65. — Alteration of the properties of bentonites by reaction with amines. Mineral. Mag. **28**, 598–605 (1949).
66. — B. J. Hork, and C. M. Finlayson: Organophilic bentonites. II. Organic liquid gels. J. Phys. Chem. **54**, 1196–1208 (1950).
67. Weiss, A., A. Mehler u. U. Hofmann: Zur Kenntnis von organophilem Vermiculit. Z. Naturforsch. **11** b, 431–434 (1956).

68. SCHNEIDER, H., u. H. ZIESMER: Zur Sorption von n-Alkylammoniumionen an Vermiculit. Naturwissenschaften **50**, 592–593 (1963).

69. ZIESMER, H.: Appearance of definite expansion steps of the vermiculite lattice during the sorption of the n-dodecyl ammonium ion. Wiss. Z. Tech. Hochsch. Chem. Leuna-Merseburg **6**, 421–426 (1964).

70. CHAUSSIDOU, J., and R. CALVET: Evolution of amine cations adsorbed on montmorillonite with dehydration of the mineral. J. Phys. Chem. **69**, 2265–2268 (1965).

71. POKHODNYA, G. A., and N. V. VDOVENKO: Kinetics of sorption of octadecylammonium acetate on minerals. Kolloidn. Zh. **27** (1), 90–94 (1965) [Russ.].

72. FRANZEN, P.: X-ray analysis of an adsorption complex of montmorillonite with cetyltrimethyl ammonium bromide (lissolamine). Clay Minerals Bull. **2**, 223–225 (1955).

73. TAHOUN, S. A., and M. M. MORTLAND: Complexes of montmorillonite with primary, secondary and tertiary amides. Soil Sci. **102**, 314–321 (1966).

74. BECK, C. W., and G. BRUNTON: X-ray and infrared data on hectorite-guanidines and montmorillonite-guanidines. Clays Clay Minerals **8**, 22–28 (1960).

75. WEISS, A., u. I. KANTNER: Über eine einfache Möglichkeit zur Abschätzung der Schichtladung glimmerartiger Schichtsilicate. Z. Naturforsch. **16**b, 804–807 (1960).

76. —, u. G. LAGALY: Ein einfaches Verfahren zur Abschätzung der Schichtladung quellungsfähiger, glimmerartiger Schichtsilicate. Kolloid-Z. u. Z. Polymere **216/17**, 356–361 (1967).

77. MEHLER, A.: Über die Einlagerung von Alkylammoniumionen in Minerale der Montmorillonitgruppe und in Vermikulit. Diss. Darmstadt 1956.

78. GUPTA, P. K. S.: The orientation of aliphatic amine cations on vermiculite. Washington Univ., Univ. Microfilms, Order No 65–6823, 135 pp.

79. JONES, W. D., and P. K. SEN GUPTA: Vermiculite-alkylammonium-complexes. Am. Mineralogist **52**, 1706–1724 (1967).

80. BARRER, R. M., and J. S. S. REAY: Sorption and intercalation by methylammonium bentonites. Trans. Faraday Soc. **53**, 1253–1261 (1957).

81. —, and L. W. R. DICKS: Synthetic alkylammonium montmorillonites and hectorites. J. Chem. Soc. 1523–1529 (1967).

82. —, and D. M. MACLEOD: The activation of montmorillonite by ions exchange and sorption complexes of tetraalkylammonium montmorillonites. Trans. Faraday Soc. **51**, 1290–1300 (1955).

83. GARRETT, W. G., and G. F. WALKER: Swelling of some vermiculite-organic complexes in water. Clays Clay Minerals **9**, 557–567 (1962).

84. OVCHARENKO, J. F. D., N. S. DYACHENKO, N. V. VDOVENKO, A. B. OSTROWSKAYA, and YU. J. TARASEVICH: Adsorption properties of organosubstituted vermiculite. Ukr. Khim. Zh. **33** (1), 49–56 (1967) [Russ.].

85. KINTER, E., and S. DIAMOND: Characterization of Montmorillonite saturated with short-chain amine cations. I. Interpretation of basal spacing measurements. Clays Clay Minerals **10**, 163–173 (1963). — II. Interlayer surface coverage by the amine cations. Clays Clay Minerals **10**, 174–190 (1963).

86. ROWLAND, R. A., and J. E. WEISS: Bentonite-methylamine complexes. Clays Clay Minerals **10**, 460–468 (1961) (pub. 1963).

87. MOLL, F.: The orientation of some cyclic amine cations on vermiculite. Univ. Microfilm Order Nr 64–2444, 174 pp.

88. STREET, G. B., and D. WHITE: The adsorption of phenol by organoclay derivatives. J. Appl. Chem. **13**, 203–206 (1963).

89. WEISS, A., u. G. ROLOFF: Über die Einlagerung symmetrischer Triglyceride in quellungsfähige Schichtsilicate. Proc. Intern. Clay Conf. **1**, Jerusalem 1966, 263–275.

90. — — In: G. ROLOFF, Über die Rolle glimmerartiger Schichtsilicate bei der Entstehung von Erdöl und Erdöllagerstätten. Diss. Heidelberg 1965.

91. LIPPMANN, F.: Keuper clay from Zaisersweiher. Heidelberger Beitr. Mineral. Petrog. **4**, 130–134 (1954).

92. STEPHEN, I., and D. M. C. MACEWAN: Swelling chlorites. Geotechnique (London) **2**, 82—83 (1950).

93. — — Chloritic minerals of unusual type. Clay Minerals Bull. **1**, 157—162 (1951).

94. WEISS, A., W. THIELEPAPE, W. RITTER, H. SCHÄFER und G. GÖRING: Zur Kenntnis von Hydrazin-Kaolinit. Z. Anorg. Allgem. Chem. **320**, 183–204 (1963).
95. — — u. H. ORTH: Neue Kaolinit-Einlagerungsverbindungen. Proc. Intern. Clay Conf. 1966 Jerusalem, **1**, 277–293.
96. — Eine Schichteinschlußverbindung von Kaolinit mit Harnstoff. Angew. Chem. **73**, 736 (1961).
97. — W. THIELEPAPE, G. GÖRING, W. RITTER u. H. SCHÄFER: Kaolinit-Einlagerungs-Verbindungen. Intern. Clay Conf. Stockholm 1963 (**1**) 287–305 (1963).
98. — Ein Geheimnis des chinesischen Porzellans. Angew. Chem. **73**, 755–762 (1963).
99. WADA, K.: Lattice expansion of kaolinminerals by treatment with potassium acetate. Am. Mineralogist **46**, 78–91 (1961).
100. ANDREW, R. W., M. L. JACKSON, and K. WADA: Intercalation as a technique for differentiation of kaolinite from chlorite minerals by x-ray diffraction. Soil Sci. Soc. Am. Proc. **24**, 422–424 (1960).
101. WEISS, A., u. W. THIELEPAPE: In: W. THIELEPAPE, Kaolinit-Einlagerungsverbindungen. Diss. Heidelberg 1966.
102. RANGE, K. J., A. RANGE u. A. WEISS: Zur Existenz von Kaolinit-Hydraten. Z. Naturforsch. **23**b, 1144–1147 (1968).
103. WEISS, A., K. J. RANGE u. H. LECHNER: Klassifizierung von Halloysiten und Kaoliniten mit unterschiedlicher Fehlordnung durch ihr Reaktionsvermögen und Verhalten bei sehr hohen Drucken. Proc. Intern. Clay Conf. 1966, Jerusalem **2**, 8 (1967).
104. —, u. K. HARTL: Über organophile Uranglimmer. Z. Naturforsch. **12**b, 351–355 (1957).
105. — — u. U. HOFMANN: Zur Kenntnis von Monohydrogen-uranylphosphat $HUO_2PO_4 \cdot 4H_2O$ und Monohydrogenuranylarsenat $HUO_2AsO_4 \cdot 4H_2O$. Z. Naturforsch. **12**b, 669–671 (1957).
106. — F. TABORSZKY, K. HARTL u. E. TRÖGER: Zur Kenntnis des Uranminerals Trögerit. Z. Naturforsch. **12**b, 356–358 (1957).
107. HARTL, K.: Zur Chemie der Uranglimmer. Diss. Darmstadt 1958.
108. KRANZ, R. D.: Die geochmische Bedeutung organischer Verbindungen in den Einschlüssen uranhaltiger Mineralien. Naturwissenschaften **54**, 469 (1967).
109. WEISS, A., u. K. J. HILKE: Uvanit, ein Uranylvanadat mit Schichtstruktur und innerkristallinem Quellungsvermögen. Angew. Chem. **77**, 347 (1965).
— — In: K. J. HILKE, Innerkristallin quellungsfähige Vanadate, Wolframate und Phosphate. Diss. Heidelberg 1966.
110. — K. HARTL u. E. MICHEL: Zur Konstitution der Vanadinminerale Hewettit und Meta-Hewettit. Z. Naturforsch. **166**, 842–843 (1961).
111. WEISS, AL., E. MICHEL u. A. WEISS: Kationenaustausch und eindimensionales innerkristallines Quellungsvermögen von Polyvanadaten mit Schichtstruktur. Angew. Chem. **73**, 707 (1961).
112. — — u. T. FODNES: Kationenaustausch und innerkristallines Quellungsvermögen von kettenförmigen Polyvanadaten $[Me^+(VO_3)^-]_n$. Naturwissenschaften **49**, 11–12 (1962).
113. MCBAIN, J. W.: Main principles in colloid chemistry. Kolloid-Z. **40**, 4 (1926).
114. BARRER, R. M.: Inorganic inclusion complexes. In: L. MANDELORN, Non-stoichiometric compounds, p. 310–437. New York: Academic Press 1964.
115. SZIGETI, P.: Über sogenannte negative Adsorption und Dampfdruckisothermen an Permutiten und Tonen. Kolloid-Beih. **38**, 99–176 (1933).
116. ENSMINGER, L. E., and J. E. GIESEKING: The adsorption of proteins by montmorillonitic clays. Soil Sci. **48**, 467–473 (1939).
117. — — The adsorption of proteins by montmorillonitic clays and its effect on base-exchange capacity. Soil Sci. **51**, 125–132 (1941).
118. — — Resistance of clay-adsorbed proteins to proteolytic hydrolysis. Soil Sci. **53**, 205–209 (1942).
119. TALIBUDEEN, O.: Interlamellar adsorption of protein monolayers on pure montmorillonoid clays. Nature **166**, 236 (1950).
120. — The technique of differential thermal analysis. I. The quantitative aspect. J. Soil Sci. **3**, 251–256 (1952). II. Organic complexes of clays. J. Soil Sci. **3**, 256–260 (1952).

121. TALIBUDEEN, O.: Complex formation between montmorillonoid clays and amino-acids and proteins. Trans. Faraday Soc. **51**, 582–590 (1955).
122. WEISS, A., u. G. KOCH: In: G. KOCH, Diss. Darmstadt 1960.
123. PINCH, L. A., and F. E. ALLISON: Resistance of a protein-montmorillonite complex to decomposition by soil microorganisms. Science **114**, 130–131 (1951).
124. WALKER, G. F., and W. G. GARRETT: Complexes of vermiculite with amino acids. Nature **191**, 1389 (1961).
125. FRIPIAT, J. J., P. CLOOS, B. CALICIS, and K. MAKAY: Adsorption of amino acids and peptides by montmorillonite. II. Identification of adsorbed species and decay products by infrared spectroscopy. Proc. Intern. Clay Conf. Jerusalem **1**, 233–246 (1966).
126. SIESKIND, O.: Adsorption complexes formed by montmorillonite and certain aminoacids. Adsorption isotherms at pH 2 and 20°. Compt. Rend. **250**, 2228–2230 (1960).
127. — The adsorption complexes formed in acid media between montmorillonite-H and certain amino acids. Compt. Rend. **250**, 2392–2393 (1960).
128. — and R. WEY: Adsorption of amino acids by montmorillonite-H. Effect of relative positions of the two functional groups $-NH_2$ and $-COOH$. Compt. Rend. **248**, 1652–1655 (1959).
129. HAXAIRE, A.: Diss. Univ. Nancy 1956.
130. CLOOS, P., B. CALICIS, J. J. PRIPIAT, and K. MAKAY: Adsorption of aminoacids and peptides by montmorillonite. I. Chemical and X-ray diffraction studies. Proc. Intern. Clay Conf. Jerusalem **1**, 223—232 (1966).
131. GREENLAND, D. J., R. H. LABY, and J. P. QUIRK: Adsorption of glycine and its di-, tri-, and tetrapeptides by montmorillonite. Trans. Faraday Soc. **58**, 829–841 (1962).
132. LAILACH, G. E., T. D. THOMPSON, and G. W. BRINDLEY: Absorption of pyrimidines, purines and nucleosides by Li-, Na-, Mg- and Ca-montmorillonites (to be published).
133. — — — Absorption of pyrimidines, purines and nucleosides by Co-, Ni-, Cu- and Fe(III)-montmorillonites (to be published).
134. DOSCH, W., and H. ZUR STRASSEN: Investigation of tetracalcium aluminate hydrates. I. The various hydrate stages and the effect of carbonic acid. Zement-Kalk-Gips **18**, 233–237 (1965).
135. OLPHEN, H. VAN: Clay organic complexes and the retention of hydrocarbons by source rocks. Am. Chem. Soc., Div. Petrol. Chem. Reprints **8**, A 41–49 (1963).
136. CHAUSSIDOU, J.: Diskussion, Intern. Clay Conf. Stockholm 1963, **2**, 338 (publ. 1964).
137. FRIEDRICH, H., u. U. HOFMANN: Die Tonminerale in Oelschiefer und in dem Trinidad-Asphalt. Z. Naturforsch. **21** b, 912–920 (1966).
138. WEISS, A., u. G. ROLOFF: Hämin-Montmorillonit und seine Bedeutung für die Festlegung der oberen Temperaturgrenze bei der Bildung des Erdöls. Z. Naturforsch. **19** b, 533–534 (1964).
139. HAUSER, E. A., and M. B. LEGGETT: Color reactions between clays and amines. J. Am. Chem. Soc. **62**, 1811–1814 (1940).
140. ENDELL, J., R. ZORN u. U. HOFMANN: Über die Prüfung auf Montmorillonit mit Benzidin. Angew. Chem. **54**, 376–377 (1941).
141. KRÜGER, D., u. F. OBERLIES: Struktur und Farbreaktionen von Montmorillonit. Naturwissenschaften **31**, 92 (1943).
142. MIELENZ, R. C., and M. E. KING: Identification of clay minerals by staining tests. Proc. Am. Soc. Test. Mater. **51**, 1213–1233 (1951).
143. HAMBLETON, W. W., and C. G. DODD: A qualitative color test for the rapid identification of the clay mineral groups. Econ. Geol. **48**, 139–146 (1953).
144. DODD, G.: Dye adsorption as a method of identifying clay. Proc. 1st Natl. Conf. on Clays and Clay Minerals 1952, Calif. Dept. Nat. Resources, Div. Mines Bull. **169**, 105–111 (1955).
145. RAMACHANDRAN, V. S., and K. P. KACHER: Thermal decomposition of dye clay mineral components. J. Appl. Chem. **14**, 455–460 (1964).
146. WEICKER, H., and R. BROMMER: Ammination of sugars during chromatography on Kieselgel-H (silica-gel) by use of ammonia-containing solvent media. Klin. Wschr. **39**, 1265–1266 (1961).

147. WEICKER, H., and J. GRÄSSLIN: Galactosamine synthesis from tagalose on Kieselgel-H in aqueous ammoniacal medium. Hoppe-Seylers Z. physiol. Chem. **337**, 133–136 (1964).
148. WEISS, A., u. G. LAGALY: In: G. LAGALY, Untersuchungen von Quellungsvorgängen in n-Alkylammonium-Schichtsilicaten. Diss. Heidelberg 1966.
149. MCLAREN, A. D.: The adsorption and reactions of enzymes and proteins on kaolinite I. J. Phys. Chem. **58**, 129–137 (1954).
—, and G. H. PETERSEN: Montmorillonite as a caliper for the size of protein molecules. Nature **192**, 960–961 (1961).
150. LAGALY, J., and A. WEISS: Determination of the layer charge in mica-type layer silicates. Proc. Intern. Clay Conf. 1969, Tokio, p. 61. Israel Progr. Scient. Transl. Ltd., Jerusalem (1969).

Subject Index

Typesetting and printing: Universitätsdruckerei H. Stürtz AG Würzburg